COMPARATIVE ANATOMY
of the Vertebrates

Ninth Edition

COMPARATIVE ANATOMY
of the Vertebrates

GEORGE C. KENT
Louisiana State University

ROBERT K. CARR
Ohio University

Boston   Burr Ridge, IL   Dubuque, IA   Madison, WI
New York   San Francisco   St. Louis
Bangkok   Bogotá   Caracas   Lisbon   London   Madrid   Mexico City
Milan   New Delhi   Seoul   Singapore   Sydney   Taipei   Toronto

McGraw-Hill Higher Education

*A Division of The **McGraw-Hill** Companies*

COMPARATIVE ANATOMY OF THE VERTEBRATES
NINTH EDITION

Published by McGraw-Hill, an imprint of The McGraw-Hill Companies, Inc., 1221 Avenue of the Americas, New York, NY 10020. Copyright © 2001, 1997 by The McGraw-Hill Companies, Inc. All rights reserved. No part of this publication may be reproduced or distributed in any form or by any means, or stored in a database or retrieval system, without the prior written consent of The McGraw-Hill Companies, Inc., including, but not limited to, in any network or other electronic storage or transmission, or broadcast for distance learning.

Some ancillaries, including electronic and print components, may not be available to customers outside the United States.

 This book is printed on recycled, acid-free paper containing 10% postconsumer waste.

1 2 3 4 5 6 7 8 9 0 QPH/QPH 0 9 8 7 6 5 4 3 2 1 0

ISBN 0–07–303869–5

Vice president and editor-in-chief: *Kevin T. Kane*
Publisher: *Michael D. Lange*
Senior sponsoring editor: *Margaret J. Kemp*
Developmental editor: *Donna Nemmers*
Marketing managers: *Michelle Watnick/Heather K. Wagner*
Project manager: *Susan J. Brusch*
Production supervisor: *Enboge Chong*
Coordinator of freelance design: *David W. Hash*
Cover designer: *Annis Leung*
Cover image: *© PhotoDisc, Vol. 6, zebra*
Senior photo research coordinator: *Carrie K. Burger*
Photo research: *Shirley Lanners*
Compositor: *Shepherd, Inc.*
Typeface: *10/12 Berkeley*
Printer: *Quebecor Printing Book Group/Hawkins, TN*

Cover image: A pair of Burchell's zebras that range today from southern Ethiopia to eastern South Africa and Angola. Although it is the most common zebra, its range has diminished due to hunting and reduction of habitat. The southern-most subspecies is extinct and the Mozambique subspecies is endangered.

Library of Congress Cataloging-in-Publication Data

Kent, George C. (George Cantine), 1914–
    Comparative anatomy of the vertebrates / George C. Kent, Robert K. Carr. — 9th ed.
        p.   cm.
    ISBN 0–07–303869–5
        1. Vertebrates—Anatomy.   2. Anatomy, Comparative.   I. Carr, Robert K.   II. Title.

QL805 .K43   2001
571.3′16—dc21   00–038680
                                                                      CIP

BRIEF CONTENTS

CONTENTS

5   EARLY CRANIATE MORPHOGENESIS, 86

6   INTEGUMENT, 105

16 NERVOUS SYSTEM, 387

17 SENSE ORGANS, 429

PREFACE

Students in the life sciences have the opportunity to study a variety of organisms, both large and small, as individuals or in populations, with one defining commonality—anatomy. Anatomy is a composite of molecules, cells, tissues, organs, systems, and their functions. Comparative anatomy is the study of structural patterns and the influence of evolutionary design. This text presents functional and comparative morphology within a developmental and evolutionary framework.

The scientific method advocates that the scientist observe, question, hypothesize, predict, test, and summarize the results. Thus, it follows that the learning process should start with the first step—observation. Although we cannot repeat the history of all anatomical observations, we do provide a clear and concise synopsis of comparative anatomy. In addition, we introduce comparative anatomy as an integrative science—a holistic approach that fosters an understanding of organisms, their parts and functional systems, development, evolution, and ecology.

The advantage of an integrated presentation is that we can demonstrate the reciprocal importance of other biological disciplines in understanding comparative anatomy and evolution. This approach includes the combination of data from classical comparative anatomy with current developmental, ecological, and theoretical research. It exposes students to both underlying principles and to current theory. They will be challenged to take their observations and question the world around them. Hypotheses are presented along with examples of alternatives, which encourage the student to explore on their own or allow the instructor to guide the course in a specific direction.

Comparative Anatomy of the Vertebrates is designed for biology majors, preprofessional students in medicine, dentistry, and allied health fields, and for liberal arts students interested in the phylogeny, embryonic development, and adult anatomy of vertebrates, including humans. Although function is covered in a fairly comprehensive manner, the physics and chemistry that drive these are not.

TEXT FEATURES

- Each chapter begins with an outline that provides a quick overview of the key topics to be covered.
- Summaries conclude each chapter with a reiteration of key principles that aid in reviewing and repetition.
- Selected Readings have been updated to reflect current research.
- NEW "Links to the Internet" have been added to each chapter to further explore specific topics. Hyperlinks for each chapter are found at: www.mhhe.com/zoology (click on "WebLinks" under the cover of *Comparative Anatomy*, 9th edition).
- NEW Questions have been added at the end of each chapter. Students will be challenged to answer in detail or to apply concepts and the use of analytical skills.
- NEW Glossary provides definition of boldface terms and includes prefixes, suffixes, roots, and stems as a comprehensive resource of vocabulary.

CHANGES IN THIS EDITION

NEW co-author, Robert Carr, Department of Biological Sciences, Ohio University, accepted the challenge and responsibility of writing and researching the many updates and improvements in illustrations and chapter content for the ninth edition. Dr. Carr describes the revision:

A goal of any course is to provide the students with the tools to access the scientific literature. The changes in this edition represent an effort to integrate current research by updating systematic relationships and taxonomy. Among these are phylogenetic relationships for the majority of taxa covered in the text, character transformations presented phylogenetically, the position of developmental genetics in understanding evolution, new fossil discoveries, revised cleavage and early development in mammals, muscle development in the craniate head, a summary of phylogenetic changes in the respiratory and circulatory systems, segmentation patterns of the craniate of the head, and homeobox gene evolution. New tables are added to summarize complex or confusing details. These include neural crest derivatives, dermal scale terminology, properties of muscle fibers, nitrogenous wastes, evolutionary solutions to osmotic challenges, human and craniate cranial nerves, functional components of nerves, and components of hypothetical segments of the craniate head.

ORGANIZATION

The first five chapters provide the foundation for the remaining chapters. In chapter 1, the student is provided with a look from the large-scale picture to details of the major organ systems within vertebrates. It also introduces students to those features that are retained by inheritance from ancestors ("The Craniate Body") and those that are unique to each group ("Craniate Characteristics" and "Vertebrate Characteristics").

Chapter 2 examines evolutionary concepts and distinguishes between pattern and process. Because we will be evaluating groups of organisms, it is important to clearly define the basis on which organisms are united. Equally important are the rules for the naming of groups once they are determined. "Systematics and Taxonomy" provides both an overview of traditional and phylogenetic systematics and reviews the current debates on taxonomy. Finally, a history of anatomy is provided, and a number of words to ponder are presented. These terms represent either uncommon terms, new to the student, or commonly misused terminology.

Chapter 3 is a brief introduction to protochordates and introduces craniates in "The Origin of Craniates."

Chapter 4 provides the cast of characters—a who's who among the craniates. This chapter provides both the taxonomy and phylogenetic relationships used throughout the text.

It should be remembered that the pattern of adult anatomical features we present in this book are the result of development and may still be influenced by change even into adulthood. Additionally, embryonic structures and their developmental transformations (ontogenetic trajectories) represent additional patterns for comparison among taxa and understanding relationships. A key evolutionary innovation for the craniates was the origin of neural crest, an embryonic tissue that has major consequences for adult craniate anatomy. This and other unique developmental features cannot be appreciated by the student without an overview of development in general—the focus of chapter 5.

The remaining chapters discuss the craniate body and, when appropriate, with references to outgroups (closely related protochordates). The universality of basic patterns and the adaptive nature of adult morphology are highlighted. Graphic summaries of character evolution are presented to provide the student with both a sense of the details and a hypothesis of evolutionary character transformations. This pattern of representing isolated details of anatomy as character phylogenies is continued in other chapters. In chapter 9, the section "A Question of Homology" brings to the student's attention the debate concerning the naming of dermal bones among fishes and tetrapods.

In chapter 11, the nature and origin of branchiomeric somatic muscles are clarified with special attention given to the pattern of muscle development in the craniate head (somitomeric and somitic origins). Twitch and tonic muscle fibers are introduced to provide a classification of fiber type that is consistent across all craniates.

The presence of filter feeding in relatives to craniates and in the lamprey larva has been used as a model for procuring food in the fossil agnathans. In chapter 12, an alternative hypothesis is presented uniting the evolution of craniate sensory structures with the origin of an active predator.

Principles of diffusion are introduced in chapter 13 to integrate this and the following chapter, which discusses the circulatory system.

Chapter 14 outlines the new understanding of development in the circulatory system. A systematic summary of respiration and circulation is introduced to unite and clarify the major changes discussed in chapters 13 and 14. Respiratory strategies from amphioxus to amniotes are presented and are related to the shared and unique structures used in these behaviors.

A complete revision of the cranial nerves is presented in chapter 16. The traditional presentation is an incomplete picture that fails to recognize the complex and composite nature of cranial nerves in Craniata. The functional organization of these nerves is updated. A new section on the segmentation of the craniate head is presented to summarize our current understanding. Included are discussions on genes, skull, pharynx, cranial nerves and brain, and a special section is presented ("Registry of Segments: A Hypothetical Segment?").

Chapters 5 and 18 are frequently unassigned due to the lack of classroom time. Evolutionary developmental biology is providing key insights into the evolutionary process and is at the cutting edge of current research. Walter Garstang, in 1922, foresaw these advances in his statement that "Ontogeny does not recapitulate phylogeny: it creates it." If time does not permit the inclusion of chapter 5, hints of development are included in each chapter when pertinent. As an alternative, assignment of one or more *selected* discussions (such as gastrulation in an amphioxus as an introduction to the digestive system) will provide valuable insight into the development of the digestive tract with minimal expenditure of additional study time. Students in premedicine and allied health fields may opt to read about insulin, thyroxine, or gonads as endocrine organs (for example, in chapter 18).

As in the past, some reviewers have expressed opinions that there is "too much" or "too little" detail. We have tried to keep foremost in mind the diversity of the audience, being aware that instructors will omit some discussions and provide enrichment in others in accordance with the needs of their students and their own professional expertise. We have endeavored to keep the subject matter current to facilitate the integration of comparative anatomy with other subjects.

A feature of the text is the frequency with which illustrations may be repeatedly cited, including in other chapters. Viewing an illustration more than once, and in different contexts, reinforces what we try to teach—the interdependence of body systems. It also enhances visual memory and evokes spontaneous recall of earlier subject matter.

REVIEWERS

We acknowledge the helpful suggestions and comments from these reviewers of the ninth edition of this text:

Deborah K. Anderson—*St. Norbert College*
Robert E. Bleiweiss—*University of Wisconsin, Madison*
William R. Eckberg—*Howard University*
Raj V. Kilambi—*University of Arkansas*
Timothy A. Lyerla—*Clark University*
Kenneth E. Nuss—*University of Northern Iowa*
David Quadagno—*Florida State University*
Sue Simon Westendorf—*Ohio University*

COMPARATIVE ANATOMY
of the Vertebrates

Introduction

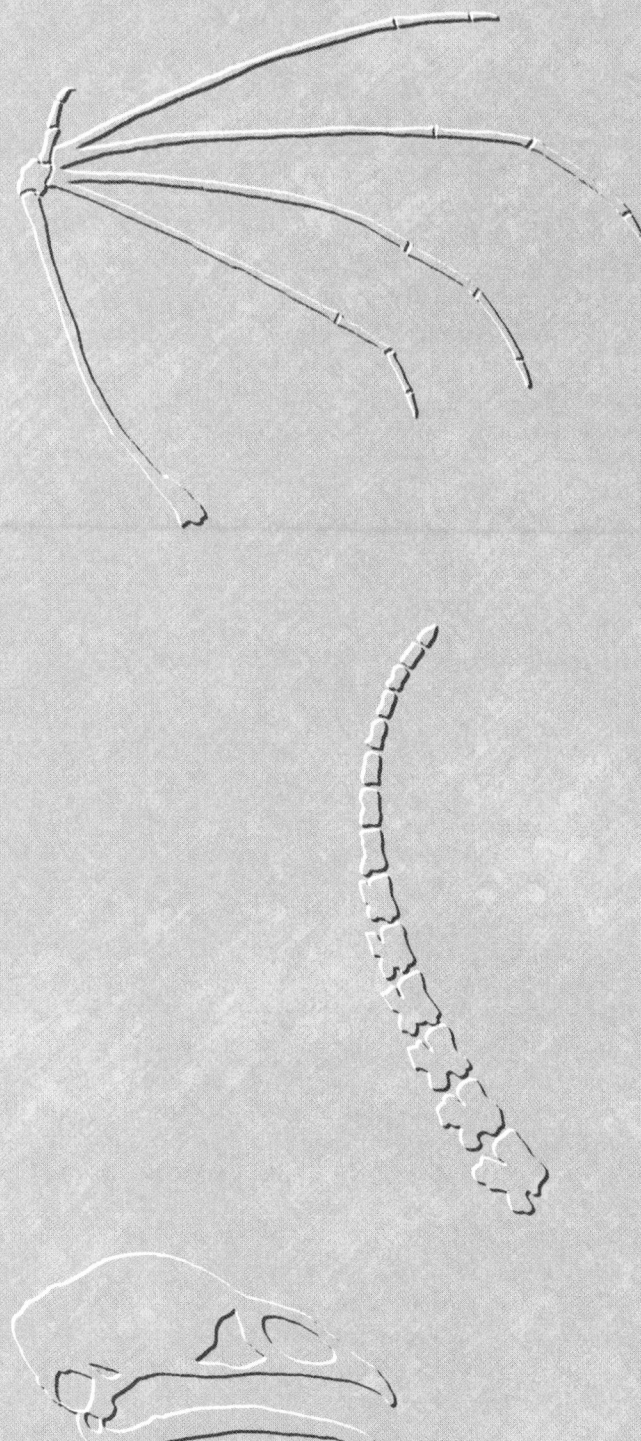

This chapter is to the rest of the book as an overture is to an extended musical score. It is a taste of things to come. It commences as an orientation to the taxonomic status of vertebrates in the animal kingdom and outlines their unique and not-so-unique characteristics. Thereafter, it provides thumbnail sketches of the organ systems that will be studied in detail in later chapters. Along with the next chapter, it is a preparation of the mind designed to minimize surprise in the journey ahead. Bon voyage!

OUTLINE

omparative vertebrate anatomy is the study of the structure of vertebrates (descriptive morphology) and of the functional significance of structure (functional morphology). Because structure entails development of the individual (ontogenesis), and individuals have an ancestral history (phylogenesis), the discipline embraces these areas of inquiry as well. Every area of biological inquiry provides relevant data to the discipline. Ecology, embryology, genetics, molecular biology, serology, biochemistry, and paleobiology are all sources of valuable data. Geology is an indispensable source. Because of the physical limitations of a single text, the contribution of some of these areas will be mentioned only in passing. The thrust of the chapters will be the organs and systems, their roles in survival, their embryogenesis, and their historical background in geological time. The latter entails consideration of the phylogenesis of vertebrates in general. In studying vertebrates, it is important to consider related organisms. So the craniates (hagfish and vertebrates) will be compared to protochordates.

To the extent that comparative anatomy is concerned with phylogenesis, it is a study of history and of animals that no longer inhabit the earth and are known to us only by a fossil record. It is interested in the survival value of structure (an adaptation), of the struggle for compatibility with an ever-changing environment, of the invasion of new territory by those most effectively equipped for survival, and of the extinction of species. The history of vertebrates, including humans, is a fascinating story from which is developing a genealogy based on the data just described. Comparative anatomy addresses curiosity about the origin of species, including our own. The generalizations and conclusions arrived at in the discipline add to the enlightenment of the human mind.

THE PHYLUM CHORDATA: THE BIG FOUR

It is conventional to think of animals as falling into two categories—those lacking vertebral columns, or invertebrates, and animals with vertebral columns, or vertebrates. Such a dichotomy, although valid, does not recognize a group of small marine animals that are transitional between invertebrates and vertebrates—the protochordates—which we will be studying shortly. Protochordates have no vertebral column, but they share with vertebrates and with no other animals a combination of four other morphological features—a **notochord**, a **dorsal hollow central nervous system**, a **postanal tail**, and an **endostyle** (a glandular groove in the floor of the pharynx). These characteristics are so fundamental in the architecture of vertebrates that they are among the first to appear in vertebrate embryos. Indeed, without most of them, no vertebrate could proceed beyond the

earliest stages of embryonic development (the postanal tail is relatively less critical to development). Because of the primacy of these structures in protochordates and vertebrates alike, these two groups have been incorporated into a single taxon, or classification category, the **phylum Chordata**. The taxonomic relationship of protochordates and vertebrates is as follows:

Kingdom Animalia
  Phylum Chordata
    Subphylum Urochordata
    Subphylum Cephalochordata
    Subphylum Craniata
      Hagfish (craniates without vertebrae)
      Vertebrata (craniates with vertebrae)

Note that vertebrates are a subgroup of a larger taxon that includes hagfish (Craniata). Many traditional Linnean classifications equate craniates with vertebrates (they include hagfish despite the absence of vertebrae). A second phylum of protochordates, sister group to Chordata, is discussed in chapter 3.

Chordates are animals that have a notochord in the embryo stage at least. Craniates are chordates with a neurocranium (braincase). Vertebrates are chordates with vertebrae. Vertebrae appear during embryonic development after the notochord has formed. Subsequently, they reinforce the notochord or replace it functionally.

THE CRANIATE BODY: GENERAL PLAN

All craniates conform to a generalized pattern of anatomic structure representing a collection of primitive (plesiomorphic) and unique (derived) anatomical features. This is revealed by dissection and is the result of the expression of similar DNA molecules inherited during the course of evolution. Craniates also exhibit similar, but not identical, patterns of embryonic development. This, too, is a result of common ancestry. Both morphology and developmental processes have been altered during the passage of time, which, as it lengthens, provides increasing opportunities for genetic changes that result in anatomic diversity. Yet, despite these changes, innumerable primitive structural and developmental similarities still exist. These similarities and diversities will be examined in detail in later chapters. In this chapter, we will discuss only the structural highlights that characterize craniates in general.

Regional Differentiation

The typical craniate body consists of three regional components—head, trunk, and postanal tail. Concentrated on or within the **head** are special sense organs for monitoring the external environment; a brain that is at least large enough to receive and process essential incoming information and to provide appropriate stimuli to the body musculature; jaws in some species for

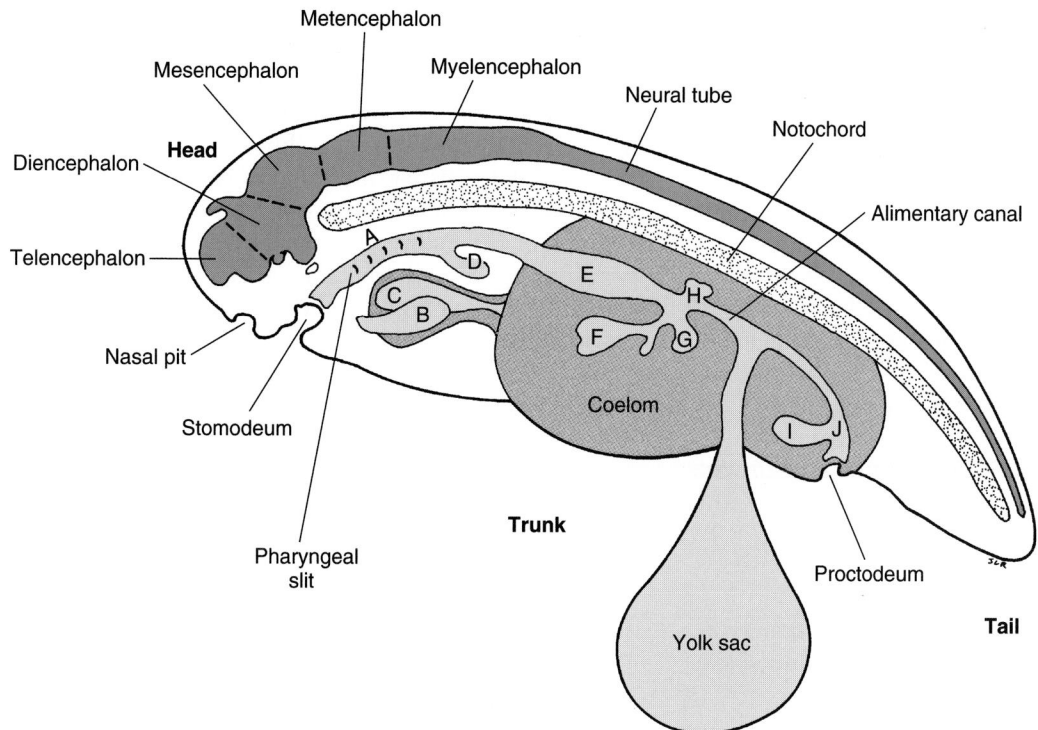

FIGURE 1.1

Sagittal section of a generalized craniate embryo. **A,** pharynx (*light red*) with pharyngeal slits; **B** and **C,** ventricle and atrium of heart; **D,** diverticulum that gives rise to the lung in tetrapods and swim bladder in fishes; **E,** stomach; **F,** liver bud and associated gallbladder; **G,** ventral pancreatic bud; **H,** dorsal pancreatic bud; **I,** urinary bladder of tetrapods; **J,** cloaca separated from the proctodeum by a cloacal membrane. The stomodeum is separated from the pharynx by a thin oral plate. The differentiated brain has five major subdivisions; telencephalon and diencephalon (forebrain), mesencephalon (midbrain), and metencephalon and myelencephalon (hindbrain).

acquiring, retaining, and macerating food; and, in fishes, gills for respiration. Expansion of the brain over hundreds of millions of years has resulted in larger braincases that have become increasingly movable independent of the trunk during that time. **Cephalization** has developed to a greater degree in craniates than in any other group of animals.

The **trunk** contains a cavity, the **coelom,** which houses most of the viscera (figs. 1.1 and 1.2). Surrounding the coelom is the **body wall** consisting chiefly of muscle, vertebral column, and ribs. The body wall must be opened to expose the viscera. In many, but not all, craniates paired pectoral and pelvic appendages (fins or limbs) are associated with the trunk. The **neck** is a narrow extension of the trunk of amphibians, reptiles (including birds), and mammals, and lacks a coelom. It consists primarily of vertebrae, muscles, spinal cord, nerves, and elongated tubes—esophagus, blood vessels, lymphatics, trachea—that connect structures of the head with those of the trunk.

The **tail** commences at the anus or vent; hence, it is postanal. It consists almost exclusively of a caudal continuation of body wall muscles, axial skeleton, nerves, and blood vessels. Some adult craniates lack a postanal

tail, although it is present in all embryos. The swimming larvae of frogs, toads, and wormlike amphibians (caecilians) have a tail, but it is resorbed at metamorphosis. Modern birds have tails reduced to a nubbin, but the first birds had long tails (see fig. 4.28). Human beings have a vestigial postanal tail early in embryonic life (see fig. 1.10). Remnants remain in adults as the "tailbone" (coccyx).

Two pairs of **appendages**—pectoral and pelvic, supported by an internal skeleton and operated by contributions from the trunk musculature—are characteristic of most vertebrates, but again, they are sometimes vestigial or have been completely lost. The earliest known craniates lacked one or both pairs of these appendages, with the latter condition seen in living jawless craniates (agnathans). We do not see pelvic appendages until the origin of gnathostomes (craniates with jaws).

Bilateral Symmetry and Anatomic Planes

Craniates have three principal body axes: a longitudinal (anteroposterior) axis, a dorsoventral axis, and a left-right axis (shared among members of Bilateria). With reference to the first two, structures at one end of the axis are different from those at the other end. The left-right axis terminates in identical structures at each end. Thus the head differs from the tail, and the dorsum differs from the venter, but right and left sides are mirror images of each other. An animal with this arrangement of body parts exhibits **bilateral symmetry**.

It is sometimes convenient to discuss parts of the craniate body with reference to three **principal anatomic planes**. Two axes define a plane. The left-right and the dorsoventral axes establish the transverse plane. A cut in this plane is a **cross section** (fig. 1.3, fish). The left-right and longitudinal axes establish the frontal plane. A cut in this plane is a **frontal section**. The longitudinal and dorsoventral axes establish the sagittal plane. A cut in this plane is a **sagittal section**. Sections parallel to the sagittal plane are **parasagittal**. Acquainting oneself with these concepts is a simple exercise in anatomy and logic.

Metamerism

Craniates exhibit the primitive feature of **metamerism**, the serial repetition of structures in the longitudinal axis of the body. It is clearly manifested in craniate embryos (see fig. 16.6, somite) and is retained in many adult systems. No external evidence is seen in most adult reptiles (including birds) and mammals because the skin is not metameric. If, however, the skin is stripped from the trunk and tail of fishes, most amphibians, and some reptiles, one sees a series of muscle segments that are reflections of the embryonic metamerism (see fig. 11.5). In addition, the serial arrangement of vertebrae, ribs, spinal nerves, embryonic kidney tubules, and segmental arteries and veins, which will be studied later, are further expressions of the fundamental metameric nature of craniates.

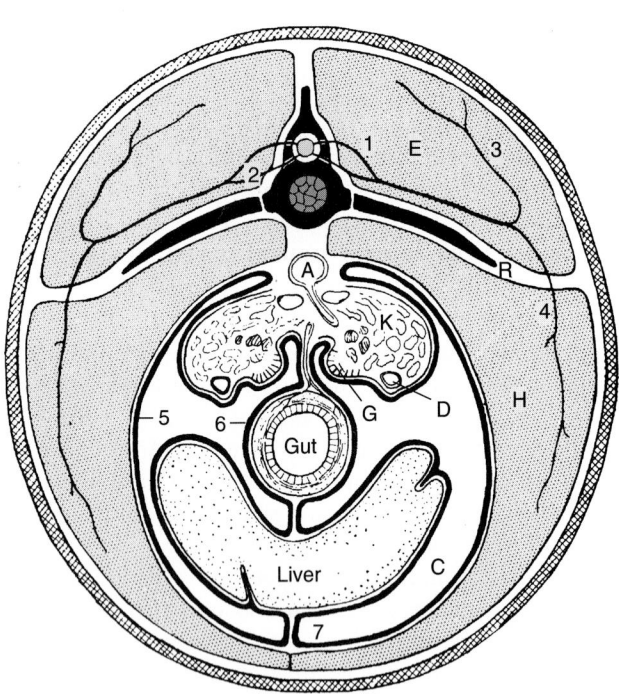

FIGURE 1.2

Cross section of a generalized craniate trunk. **A,** dorsal aorta, giving off renal artery to kidney; **C,** coelom; **D,** kidney duct; **E,** epaxial muscle; **G,** future gonad (genital ridge); **H,** hypaxial muscle in body wall; **K,** kidney; **R,** rib projecting into a horizontal skeletogenous septum from the transverse process of a vertebra (*black*). **1,** dorsal root of spinal nerve; **2,** ventral root; **3,** dorsal ramus of spinal nerve; **4,** ventral ramus; **5,** parietal peritoneum; **6,** visceral peritoneum; **7,** ventral mesentery. A remnant of the notochord (*dark red*) lies within the centrum of a vertebra. The spinal cord (*light red*) lies above the centrum surrounded by a neural arch (*black*). Body wall muscle is shown in *light red*.

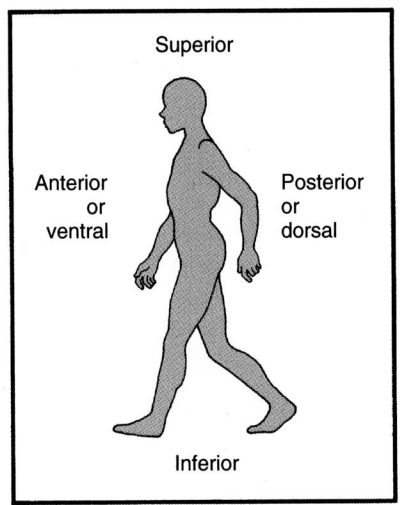

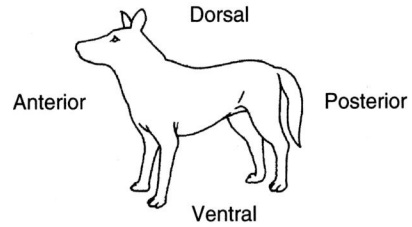

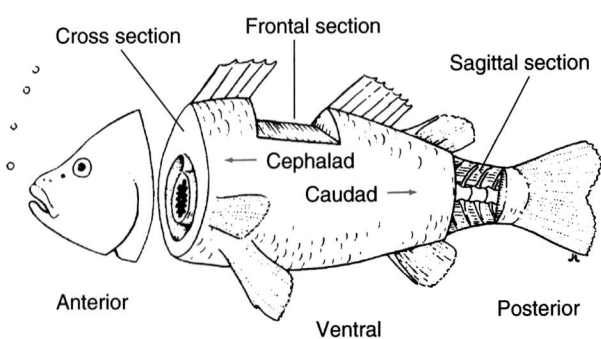

FIGURE 1.3

Terms of direction and position and planes of sectioning of the craniate body. Terms employed in human anatomy are shown in the box.

CRANIATE CHARACTERISTICS

In addition to retaining a number of primitive characters (e.g., bilateral symmetry, deuterostome development, and the chordate "big four"), craniates exhibit a unique combination of morphological features: (1) cranium; (2) a three-part brain; (3) neural crest and its derivatives; (4) paired external sense organs (e.g., olfactory, optic, otic with a single semicircular canal, and lateral line system with unicellular sense organs); and (5) cartilage.

VERTEBRATE CHARACTERISTICS

Vertebrates are characterized by a number of morphologic features: (1) a vertebral column (primitively seen as isolated elements associated with an unrestricted notochord); (2) two semicircular canals; (3) electroreception; (4) lateral line system with multicellular neuromasts; and (5) a number of additional soft tissue specializations. If a bizarre organism were to be discovered in the abyssal depths of the oceans, these features, in combination, would admit this creature to the vertebrate hierarchy.

STRUCTURES COMMON TO CRANIATES

We will now examine some of the primitive and unique features of craniates and vertebrates. The notochord and vertebral column will be discussed together because they occupy the same site. We will then consider briefly a few other craniate features.

Notochord and Vertebral Column

The **notochord**, a chordate feature, is the first skeletal structure to appear in craniate embryos. At its peak of embryonic development, it is a rod of living cells located immediately ventral to the central nervous system and dorsal to the alimentary canal extending from the midbrain to the tip of the tail (see fig. 1.1). The part of the notochord in the head becomes incorporated in the floor of the skull, and, except in hagfish, the part in the trunk and tail becomes surrounded by cartilaginous or bony **vertebrae**. These provide more rigid support for the body than does a notochord alone. A typical vertebra (fig. 1.4) consists of a **centrum** that is deposited around the notochord, a **neural arch** that forms over the spinal cord, and various **processes.** In the tail, a **hemal arch** may surround the caudal artery and vein (see fig. 8.1*b, c,*).

The fate of the notochord in adult craniates is variable. In almost all fishes, it persists the length of the trunk and tail, although usually constricted within each centrum (see fig. 8.2). The same is true in many urodeles and as a primitive condition in some lizards (see fig. 8.3). However, in modern reptiles (including birds) and mammals, the notochord is almost obliterated during development. A vestige remains in mammals within the intervertebral disks that separate successive centra. The vestige consists of a soft spherical mass of connective tissue called the **pulpy nucleus** (see fig. 8.11*d*). Modern reptiles lack even this vestige.

In lampreys, the notochord grows along with the animal, and paired **lateral neural cartilages** become perched on the notochord lateral to the spinal cord (fig. 1.5). These cartilages are reminiscent of neural arches, but whether they are primitive vertebrae, vestigial vertebrae from an ancestor that had a typical vertebral column, or an entirely different structure is not

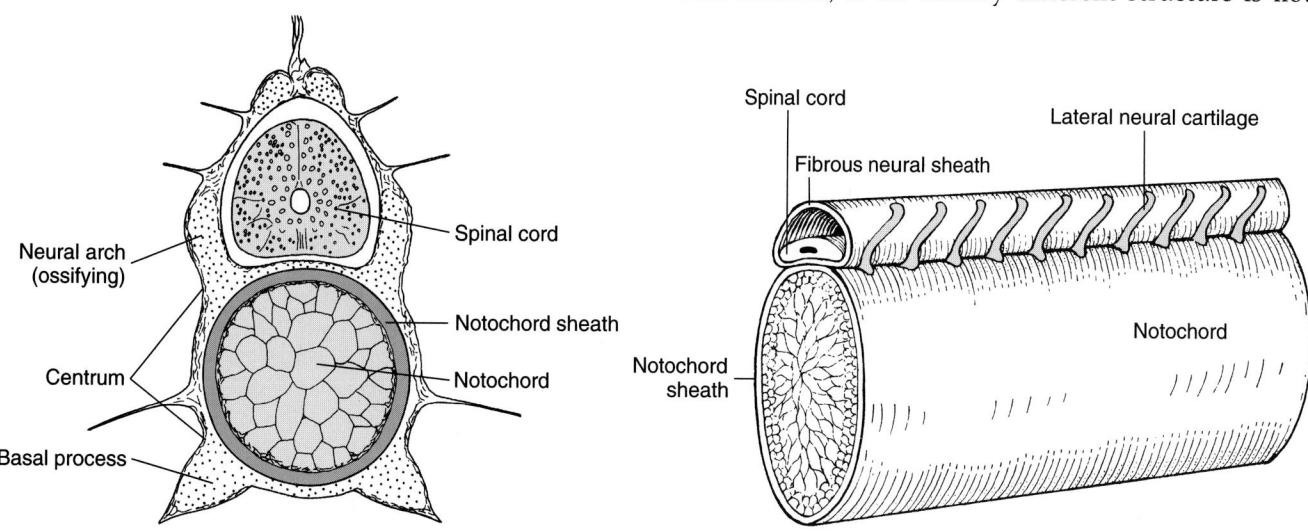

FIGURE 1.4
Transverse section of the vertebral column of a very young trout.

FIGURE 1.5
Lateral neural cartilages (*red*) and notochord of a lamprey.

FIGURE 1.6

Basic pharyngeal architecture as exemplified by a composite craniate embryo. The dorsal aorta is paired in the head. The ventral aorta is unpaired. A series of pharyngeal pouches has evaginated from the lateral walls of the digestive tract. Six aortic arches (*red*) connect the heart and ventral aorta with the dorsal aorta. (Typically, the first aortic arch disappears before the sixth one has formed in gnathostomes.) Although a lung does not form in all craniates, it is an ancient structure and is represented by a swim bladder, in many fishes. The anterior end of the dorsal, hollow nervous system is enlarging to form the brain. The notochord commences at the level of the midbrain. All reptile (including bird) and mammal embryos exhibit a cephalic flexure at the level of the midbrain.

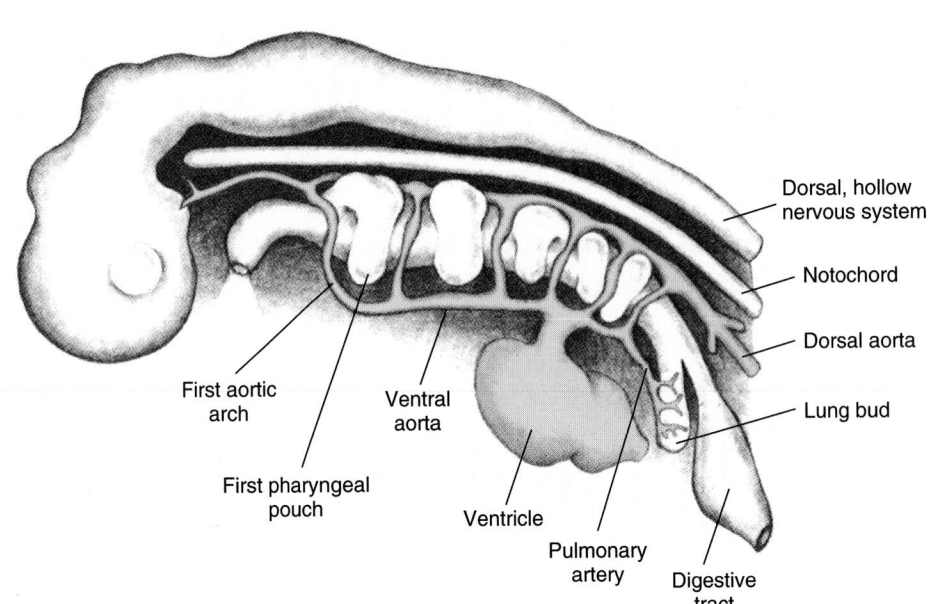

clear. When a notochord persists as an important part of the adult axial skeleton, it develops a strong outer elastic and inner fibrous **notochord sheath.**

The notochord has been disappearing as an adult structure, but development of a notochord in every chordate embryo is a reminder that the craniates and some protochordates have a common ancestry.

Pharynx

A pharynx that is perforated by openings to either the exterior or an atrium is common to the hemichordates (Chapter 3) and chordates, suggesting their common ancestry. The pharynx (pronounced *fair-inks*) is a vital part of the craniate embryo. It produces the gills of fishes, the lungs of tetrapods, the skeleton and musculature of the jaws, the endocrine glands that regulate metabolic rates in every cell of the body and that maintain appropriate calcium levels in the bones, other tissues, and the circulating blood. It gives rise to the middle ear cavity of tetrapods. Last, but of utmost importance, it provides, in humans, during fetal life and for a short time thereafter, the initial cells of the immune system, which, if destroyed, leads to death. A basic blueprint for the architecture of the craniate pharynx is expressed in all craniate embryos from fishes to human beings. We will now examine that blueprint.

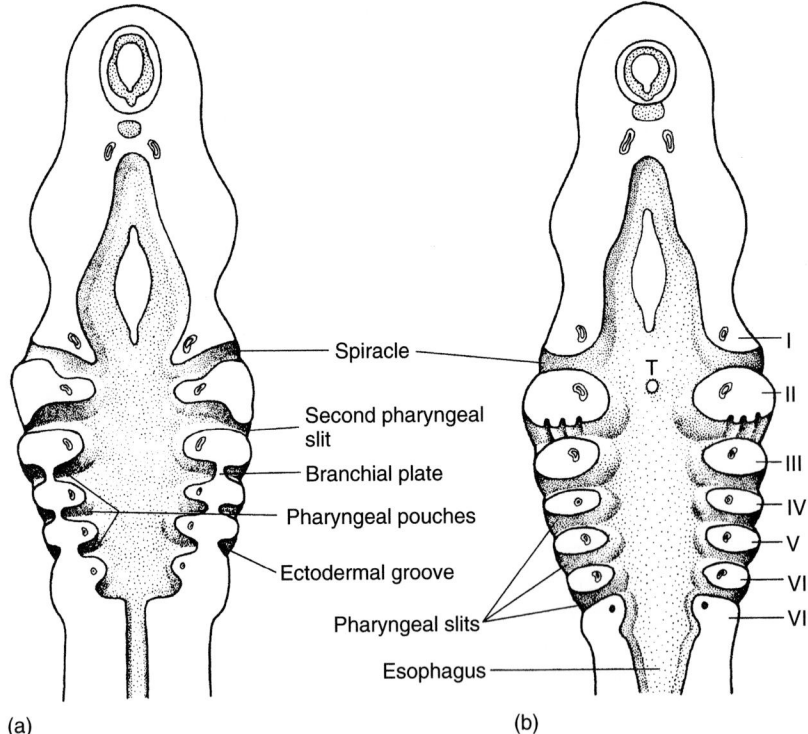

(a)   (b)

FIGURE 1.7

Early pharyngeal development in embryonic shark, looking down onto pharyngeal floor. (*a*) Early stage. Formation of pharyngeal pouches and slits. (*b*) Later stage. Pharyngeal arches **(I–VII). T,** Site of thyroid evagination.

Pharyngeal Pouches and Slits

A series of **pharyngeal pouches** arises as diverticula of the endoderm of the foregut, and the pouches grow toward the surface of the animal (fig. 1.6). These pouches establish the limits of the pharynx. Simultaneously, an **ectodermal groove** grows toward each pouch (figs. 1.7

and 1.8). Soon, only a thin membrane, the **branchial plate**, separates the groove from the pouch. If, and when, the branchial plate ruptures, a passageway is formed between the pharyngeal lumen and the exterior of the animal. This passageway is a pharyngeal slit. The slits may be permanent, as in fishes in which they are exits for respiratory water from the gills, or they are temporary, as in most tetrapods. Eight is the largest number of pouches that develop in any jawed craniate, and that number is found only in some basal sharks. Living agnathans have as many as 15 slits (fig. 1.9).

Pharyngeal slits are temporary if the animal is going to live on land. Of the six pharyngeal pouches that form in frog embryos, four give rise to gill slits in tadpoles. These slits close when the tadpole metamorphoses into a frog. In reptiles (including birds) and mammals, no gills develop in the pouches, and the slits are transitory. Of the five pouches that develop in chicks, the first three rupture to the exterior and close again. Only one or two of the more anterior pouches of mammals may rupture. Cervical fistulas occasionally seen in human beings are usually the result of the failure of the cervical sinus, housing the third and fourth slits, to close.

Although the pharyngeal pouches of tetrapods rarely give rise to permanent slits, the first one becomes the auditory tube and middle ear cavity of tetrapods, and the second persists as the pouch of the palatine tonsil of mammals. The walls of several pouches give rise to endocrine tissue in all craniates.

Pharyngeal Arches

Each embryonic pharyngeal pouch or slit is separated from the next by a column of tissue known as a **pharyngeal arch** (figs. 1.7*b* and 1.10). Each arch typically contains four basic components, or the blastemas from which these components develop. They are (1) *supportive skeletal elements*, illustrated in an adult shark in figure 9.1; (2) *striated muscles that operate the arch* (see fig. 11. 24*a*); (3) *branches of fifth, seventh, ninth, and tenth cranial nerves*, which innervate the muscles and provide sensory input to the brain (see fig. 16.30); and (4) an *aortic arch* that connects the ventral and dorsal aortas (see fig. 1.6). These components are found also immediately anterior to the first pouch or slit and, with

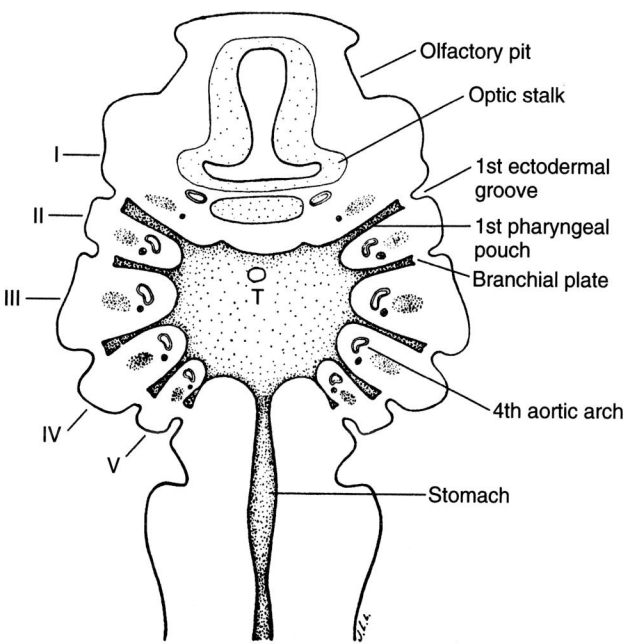

FIGURE 1.8

Frontal section of embryonic frog pharynx. I to **V**, Pharyngeal arches; **T**, Site of thyroid evagination.

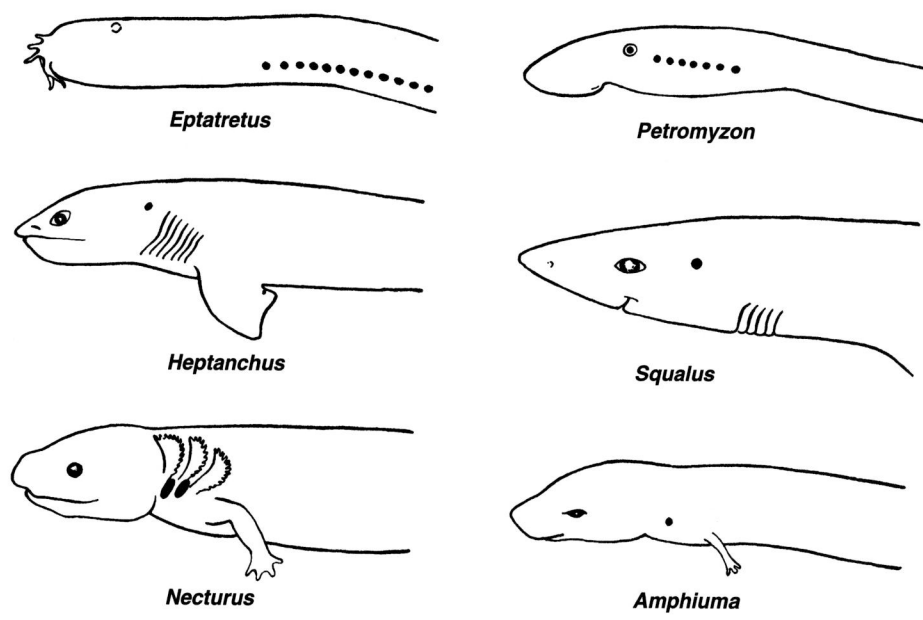

FIGURE 1.9

Persistent pharyngeal slits in selected agnathans (*top*), sharks (*center*), and tailed amphibians (*bottom*).

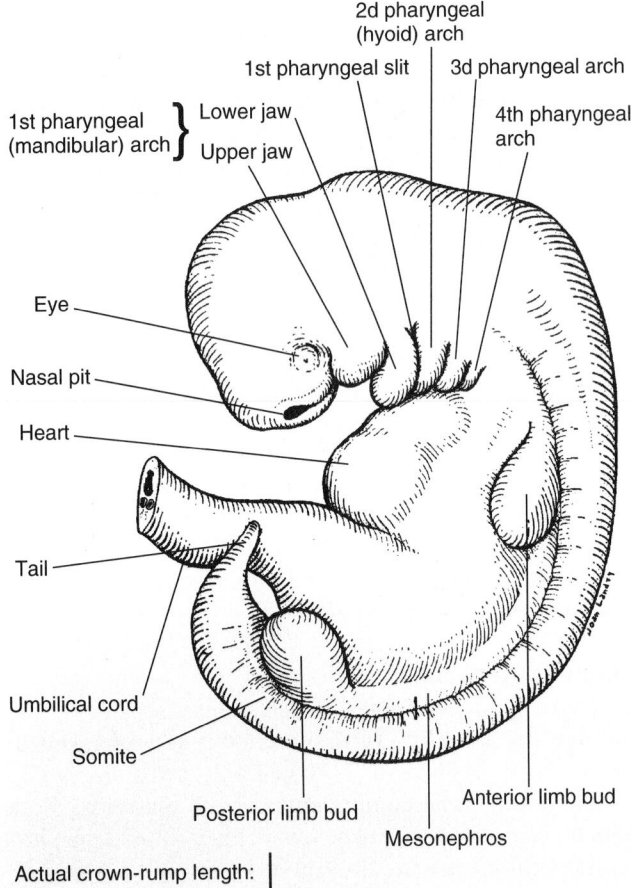

2d pharyngeal (hyoid) arch

1st pharyngeal slit

3d pharyngeal arch

1st pharyngeal (mandibular) arch } Lower jaw / Upper jaw

4th pharyngeal arch

Eye

Nasal pit

Heart

Tail

Umbilical cord

Somite

Posterior limb bud

Anterior limb bud

Mesonephros

Actual crown-rump length: |

FIGURE 1.10
Human embryo approximately 4½ weeks after fertilization (5 mm stage).

some omissions, immediately caudal to the last. The boundaries of a pharyngeal arch can be identified from the exterior when there are ectodermal grooves or pharyngeal slits to serve as landmarks, and only then. If the grooves disappear or the slits close, as they do in tetrapods, the boundaries of the arches are lost and the components become reoriented.

The skeleton of the pharyngeal arches, whether in fishes or adult tetrapods, is referred to collectively, for identification purposes, as the **pharyngeal** (or **visceral**) **skeleton**. The muscles of the arches are referred to as **branchiomeric** in recognition of their primitive relationship to gills.

The upper and lower jaws and associated muscles, nerves, and vessels constitute the first, or **mandibular arch**. It develops anterior to the first pouch. Behind the first pouch is the second, or **hyoid arch**. The remaining arches are referred to by number, starting with the third (3rd–7th visceral arches; alternatively these latter arches may be referred to as the 1st–5th branchial arches representing unmodified branchial arches in fishes). The modifications that developed in the pharynx as craniates shifted from life in water to life on land constitute a fascinating chapter in craniate history. Many of the chapters that follow will describe this evolutionary metamorphosis.

Dorsal Hollow Central Nervous System

The central nervous system consists of brain and spinal cord and contains a central lumen, the **neurocoel**. The dorsal location and the lumen result from the observation that the central nervous system typically arises as a longitudinal **neural groove** in what will become the dorsal surface of the embryo (figs. 1.11 and 5.9c). The groove closes over or rolls up and sinks beneath the surface to become a hollow **neural tube** located dorsal to the notochord (see figs. 1.1 and 1.6). The tube is wider anteriorly, and this part becomes the brain with its ventricles. The process of forming the neural tube is **neurulation**.

In living agnathans and neopterygian fishes (bowfins, gars, and teleosts), the basic pattern of neurulation has become slightly modified. Instead of forming a groove, the surface ectoderm dorsal to the notochord proliferates a wedge-shaped **neural keel** (see fig. 5.9e). Eventually the keel separates from the surface and a cavity forms within it by rearrangement of the cells in its interior. The result is a typical dorsal hollow central nervous system.

Neural groove

Neural fold

Nephrogenic mesoderm

Ectoderm

Endoderm

Notochord

Mesodermal somite

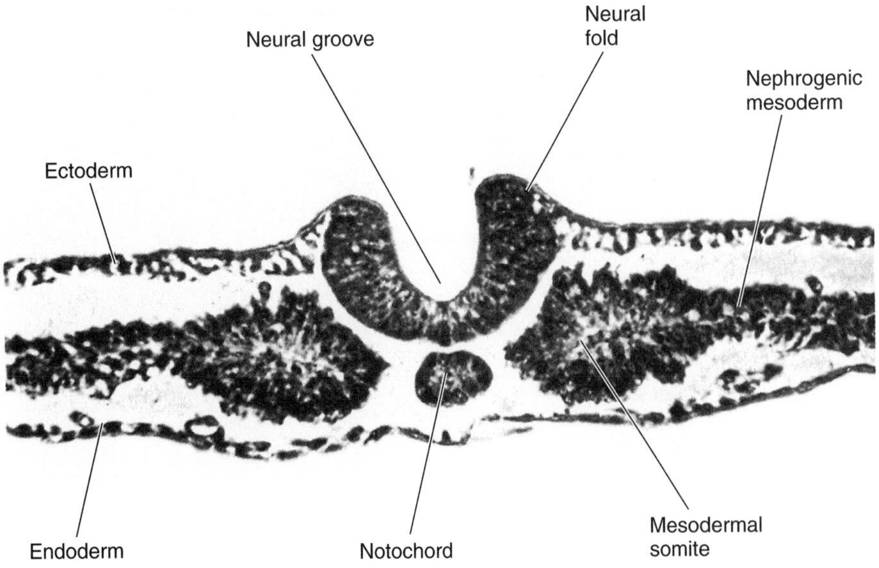

FIGURE 1.11
Cross section of 24-hour chick embryo. The neural groove and mesodermal somites of a whole chick embryo of a similar age are seen in figure 5.9c.

Cranial and spinal nerves connect the central nervous system with the various organs of the body. The nerves, along with associated ganglia and plexuses, constitute the **peripheral nervous system.** The spinal nerves of most craniates are metameric (see fig. 16.9), arising at the level of each embryonic body segment and passing to the skin and muscles derived from that segment, and to the viscera. Up to 16 cranial nerves arise from the brain of fishes and amphibians and 13–18 in reptiles and mammals (changes are the result of gains, losses, or fusion of cranial nerves or their components). Two gains (or reorganizations) in tetrapods are spinal nerves that became "trapped" within the skull.

OTHER CRANIATE CHARACTERISTICS

In the preceding section, we noted features that collectively characterize craniates and vertebrates alone. Other features, not unique to craniates or even to chordates, include bilateral symmetry and metamerism, which were also discussed earlier. The paragraphs that follow provide thumbnail sketches of some of the other characteristics of craniate animals.

Integument

The skin of craniates consists of epidermis and dermis, and the epidermis is multilayered (see fig. 6.2). Many varieties of defensive, lubricatory, nutritive, pheromonal, and homeostasis-maintaining glands develop from the epidermis and empty on the surface in one taxon or another. The epidermis of terrestrial craniates forms a variety of cornified (horny) appendages such as spines, reptilian scales, feathers, hair, claws, and hooves. Bone in the dermis provides a heavy coat of armor for catfishes and other relics of the Mesozoic Era and contributes to the dense, sharp, pointed scales of sharks and to the thin, flexible scales of modern bony fishes. Dense fibrous (collagen-bearing) tissue is common in the dermis. In bovines, the dermis is made into leather

through a training process. The epidermis of terrestrial craniates has a surface layer of dead (cornified) cells that protect these species from becoming dehydrated as a result of exposure to air.

Respiratory Mechanisms

Most craniates carry on external respiration (exchange of respiratory gases between the animal and the environment) by means of highly vascularized membranes (gills) located on the pharyngeal arches or derived from the pharyngeal floor (lungs, see lung bud fig. 1.6). In some species, respiration takes place through the skin or the lining of the oral and pharyngeal cavities. In embryos that develop within a porous eggshell or within the body of a parent, gaseous exchange usually occurs via special extraembryonic membranes that lie just inside the shell or that are in contact with the lining of the mother's uterus. (Chapter 5 includes discussions of oviparity, viviparity, and extraembryonic membranes.)

Coelom

Like the body of many invertebrates, the craniate **trunk** is built like a tube within a tube, the outer tube (body wall) and inner tube (gut) being separated by a cavity, the coelom (see fig. 1.2). In fishes, amphibians, and some nonavian reptiles, the coelom is partitioned into a **pericardial cavity** housing the heart, and a **pleuroperitoneal cavity** housing most of the other viscera, including the lungs in tetrapods (fig. 1.12, fish and amphibian). The pericardial and pleuroperitoneal cavities in these craniates are separated by a fibrous **transverse septum.** In other reptiles and mammals, the heart occupies a pericardial cavity, each lung occupies a separate **pleural cavity,** and the digestive tract caudal to the esophagus occupies an **abdominal (peritoneal) cavity** (see fig. 1.12, mammal). In most male mammals, a pair of **scrotal cavities** houses the testes. The scrotal cavities arise as evaginations from the abdominal cavity.

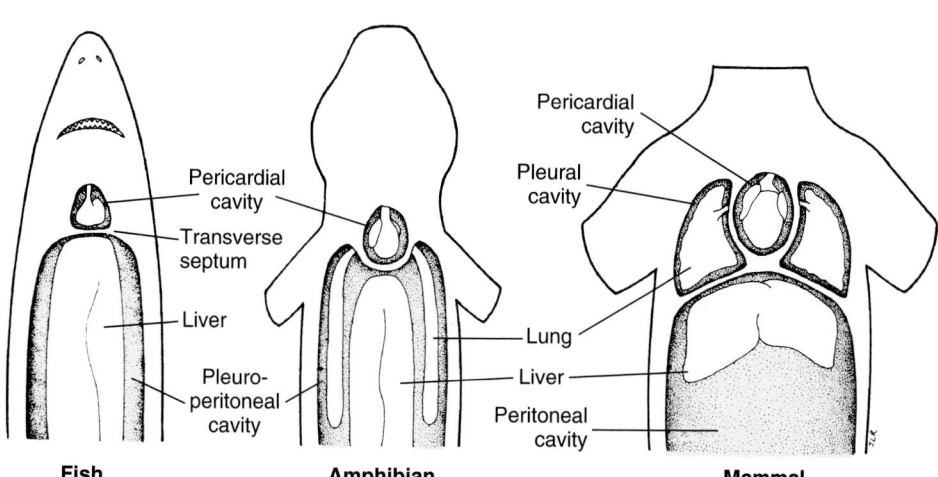

FIGURE 1.12
Chief subdivisions of the coelom of craniates.

The coelom is totally enclosed by a **peritoneal membrane** that lines the body wall and invests the coelomic viscera (**parietal** and **visceral peritoneum**, 5 and 6 in fig. 1.2). These membranes are continuous via **dorsal mesenteries** and, when present, via **ventral mesenteries.** The few visceral organs that do not develop mesenteries—chiefly the kidneys—lie against the dorsal body wall just external to the peritoneum. Therefore, they are **retroperitoneal** (in some species, the kidneys may be suspended in the coelom [e.g., in *Necturus*]).

Digestive System

The digestive system consists of a digestive tract and several auxiliary (accessory) organs, chiefly a liver, gallbladder, and pancreas. The digestive tracts have specialized regions for the acquisition, processing, temporary storage, digestion, and absorption of food and for elimination of the unabsorbed residue. These include the oral cavity, pharynx, esophagus, stomach, and intestine. The latter is often coiled or has within it a spiral valve, either of which increases the absorptive area without increasing the body length (see fig. 12.1, shark). The tract may also exhibit one or more diverticula, or ceca, that play various roles in the digestive process.

The digestive tract in most craniates terminates in a **cloaca**, a common chamber that also receives the urinary and reproductive ducts. It opens to the exterior via a **vent.** In modern fishes and a few tetrapods, the cloaca becomes very shallow or nonexistent. In mammals other than monotremes, the embryonic cloaca becomes subdivided into two or three passageways, each with its own exit to the exterior (see fig. 15.48*b* and *c*). The exit from the intestine is then called the **anus.**

Urogenital Organs

Kidneys and gonads arise close together in the roof of the coelom (see fig. 1.2G and K), and the two systems share certain passageways. Kidneys (nephroi) are the chief organs for elimination of water in those species in which this is necessary. (The ease of diffusion in an aquatic environment is seen as a reduction of water elimination in many marine teleosts. In contrast, the reduced elimination of water in desert animals represents a highly derived system for resorption.) The kidneys also assist in maintaining an appropriate electrolyte balance in the blood. In the most basal fishes, fluid wastes accumulate in the coelom and are removed by simple kidney tubules reminiscent of the nephridia of earthworms. In most craniates, however, more complex kidney tubules collect fluids directly from blood capillaries. In both instances, the tubules lead to a pair of longitudinal ducts that empty into the cloaca or a urinary bladder or, less commonly, lead directly to the exterior.

Reproductive organs include gonads, ducts, accessory glands, storage chambers, and copulatory mechanisms. Early in development all craniate embryos are bisexual,

having gonadal and duct primordia for both sexes. If the animal is genetically constituted to become a female, the gonad primordia develop into ovaries, and only the female ducts differentiate. If the animal is to become a male, the gonad primordia become testes, and the male ducts differentiate. The duct system associated with the opposite sex largely disappears. Jawless craniates lack reproductive ducts. Their sperm and eggs are shed into the coelom and exit via a urogenital papilla located immediately behind the anus.

Circulatory System

The circulatory system pervades almost every cubic millimeter of the body of the smallest and largest craniate, a condition necessitated by logistics. Whole blood, consisting of plasma and formed elements—chiefly red and white blood cells and platelets—is confined to arteries, veins, capillaries, and sinusoids. Sinusoids are broad channels rather than tubes. They are a prominent feature of many invertebrate circulatory systems and are more common in fishes than in tetrapods. Craniates also have a system of lymph vessels that collect some of the interstitial tissue fluids and conduct them to large veins (see figs. 14.41 and 14.42).

The heart, which forms immediately ventral to the embryonic pharynx, remains in that location, close to the gills, in fishes. In tetrapods it is displaced somewhat caudad during subsequent development. The heart pumps blood forward into a **ventral aorta**, which is foreshortened in tetrapods, through **aortic arches**, and into a **dorsal aorta**, in which blood courses caudad. Branches from the anteriormost aortic arches carry blood to the head (see fig. 1.6).

Fishes have a **single-circuit heart.** Blood passes from the heart to the gills, where oxygen is acquired, then to the body tissues, where oxygen is given up, and then back to the heart to be recirculated. The loss of gills and reliance on lungs eventually resulted in **two-circuit hearts.** In these craniates, oxygenated blood in the heart is pumped to all parts of the body where oxygen is released, after which the blood, low in oxygen, returns to the heart. This blood is then pumped to the lungs where it regains oxygen, after which it returns to the heart. Oxygen-rich blood is kept separate from oxygen-depleted blood within the heart by partitions in the chambers. Other adaptations for separation of oxygenated and deoxygenated blood in one craniate species or another are described in chapter 14.

Skeleton

Cartilage, bones, and ligaments constitute a jointed framework of rigid or semirigid components that gives the body its shape, protects vital organs, and provides a site for attachment of locomotor and other muscles (see figs. 8.24 and 10.31). The framework consists of a longitudinal **axial skeleton**, principally skull and vertebral column; a **pharyngeal skeleton**, best-developed in fishes, in which it supports the gills; and an **appendicular skeleton.** As in invertebrates, mineralization of ordinary connective tissues gives skeletal components their rigidity.

Muscles

The axial skeleton is moved principally by trunk and tail muscles that are metameric (see fig. 11.5). The appendicular skeleton is operated primitively by bud-like extensions of the body wall muscles onto the fins or limbs (see fig. 11.17). Muscles of the pharyngeal arches (branchiomeric muscles) operate the pharyngeal skeleton. All of these muscles appear **striated** when viewed by light microscopy (see fig. 11.2). **Cardiac** muscle—that of the heart—is a special variety of striated muscle tissue. **Nonstriated (smooth) muscles** are found principally in the walls of hollow viscera other than the heart and in the walls of tubes and vessels. Striated muscles other than cardiac generally can be operated independent of reflex control to the extent that higher brain centers are available for providing stimuli. They are therefore often referred to as voluntary muscles. Cardiac and smooth muscles respond only to reflex control.

Sense Organs

Craniates have a wider variety of sense organs (receptors) than any other animals, which in craniates are concentrated anteriorly on the head. These monitor the constantly changing external and internal environments. **Exteroceptors** monitor the external environment; they include **mechanoreceptors, chemoreceptors, electroreceptors,** and **thermoreceptors,** and **receptors for radiation** in the visible and infrared spectra. Mechanoreceptors are stimulated chiefly by vibrations of varying amplitudes in the water or air, by barometric pressure, or by direct contact of another object with the skin (touch). With the exception of receptors for radiation, exteroceptors are distributed widely on the head, trunk, and tail of fishes. In tetrapods, exteroceptors are generally confined to the head, except those for touch and temperature. **Proprioceptors** monitor the activity of muscles, joints, and tendons. **Visceral receptors** monitor the rest of the internal environment. All sense organs report to the central nervous system. By "keeping in touch," the central nervous system is able to regulate skeletal muscle, visceral muscle, and glandular activity in a manner conducive to survival.

Summary

1. Chordates are animals with a notochord, a dorsal hollow central nervous system, an endostyle (a pharyngeal feature that is glandular, in part), and a postanal tail. Uniting hemichordates with chordates is the possession of pharyngeal pouches in embryos at least.

2. Craniates are chordates with a cranium, paired external sense organs, neural crest, a three-part brain, and cartilage. They further exhibit metamerism, marked cephalization, and a closed circulatory system with a single- or double-circuit heart. They respire by gills, lungs, skin, or the lining of the oral cavity or pharynx; as reptile or mammal embryos, they respire by means of extraembryonic membranes in contact with a porous eggshell or the lining of the maternal uterus. Members include the hagfish and vertebrates.

3. Vertebrates are those craniates with two semicircular canals in the ear, electroreception, a lateral line system with multicellular neuromasts, and a vertebral column—a skeletal series of bones or cartilages.

4. A notochord is the first skeletal structure to appear in craniate embryos. In lampreys, it is surmounted by lateral neural cartilages. In other vertebrates, it becomes surrounded by cartilaginous or bony vertebrae after which the notochord may be reduced to remnants or disappear.

5. Pharyngeal pouches tend to rupture to the exterior to form temporary or permanent pharyngeal slits. These persist as gill slits in fishes and larval amphibians. They close permanently in other vertebrates.

6. A pharyngeal arch lies between each pharyngeal slit, anterior to the first and immediately caudal to the last. The arches support gills when the latter are present. The first two are mandibular and hyoid arches.

7. Pharyngeal arches contain skeletal components, branchiomeric muscles, branches of the fifth, seventh, ninth, or tenth cranial nerves, and an aortic arch.

8. The dorsal hollow central nervous system generally arises as a neural groove that sinks into the dorsal body wall to become a neural tube and, subsequently, a brain and spinal cord. The peripheral nervous system consists of nerves, ganglia, and plexuses.

9. A wide variety of sense organs monitor the external and internal environment. The information they acquire is employed by the central nervous system to regulate skeletal muscle and visceral muscle and gland activity in a manner conducive to survival.

Critical Thinking Questions

In these first few questions, we will employ the comparative method using the human condition as a point of reference. Most students have been exposed to human anatomy in introductory courses or possess a wealth of personal observations. As we introduce new structures or new organisms throughout the text, we will build on your existing knowledge. How is the organism similar or how is it different? Of the similarities that are unique to a subset of known organisms, how do the structures compare to what we know, how are they built, how do they function, and how are they organized? In what initially appears as a disorganized mass of details, you will develop an integrated knowledge of comparative anatomy.

1. Given your knowledge of human anatomy, what characters unique to chordates are present in humans? How have they been modified?
2. The tissues and organs in the human body possess a recognizable size, shape, and position relative to other structures. How would these features compare to other mammals, to other amniotes, and to other vertebrates?
3. What are the superior to inferior regions in the human body, and how would they compare to the anterior to posterior regions in a fish?
4. In reviewing the various figures in this chapter, what developmental features in humans are similar to those of other craniates even if they are not preserved in the adult?
5. In the section on "Other Craniate Characteristics," what comparisons can be made between the human condition and the patterns seen in other craniates for each of the discussed systems? How general are human features?
6. In comparing humans to other terrestrial craniates, we see the presence of a neck in contrast to the condition in fishes. What explanation can you give for the origin of a neck in tetrapods?

Links to the Internet

Visit the zoology website at http://www.mhhe.com/zoology to find live Internet links for each of the references listed below.

1. Introduction to the Scientific Method. A good explanation of the steps of the scientific method.
2. The Scientific Method.
3. Visible Human Slice and Surface Server. A home page with links to views inside the human body.
4. The National Library of Medicine's Visible Human Project. Links to information, pictures, and more relating to the project.
5. Phylum Chordata, from the University of Minnesota.
6. Animal Diversity Web, University of Michigan. Phylum Chordta. General characteristics of chordates, with the following links to urochordates and vertebrates.
7. Chordata. Arizona's Tree of Life Web Page. An introduction, pictures, characteristics, phylogenetic relationships and references on chordates.

2 Concepts, Premises, and Pioneers

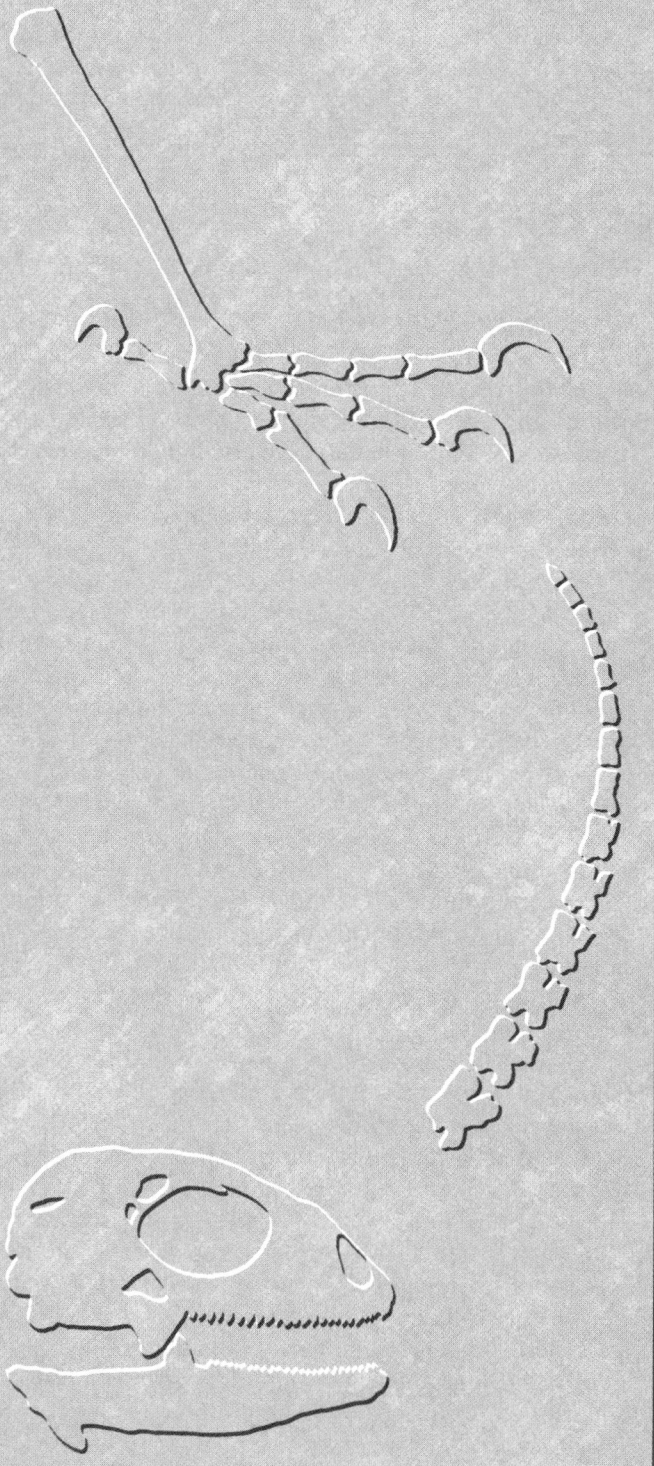

This chapter provides background for understanding the major concepts and premises upon which the discipline of modern comparative vertebrate anatomy is based. It includes mention of a few of the pioneers in the study of vertebrate anatomy following the Renaissance and concludes by calling attention to some abstract terms that may be misunderstood. The selected readings at the end provide additional input from easy-to-read articles to in-depth discussions about some of the concepts.

OUTLINE

Pattern and Process
   Homology and Homoplasy
   Serial Homology
   Analogy
Adaptation
Speciation
Evolutionary Convergence
Development
   Ontogeny and Phylogeny
   von Baer's Law
Metamorphosis and Heterochrony
Systematics and Taxonomy
Organic Evolution and Evolutionary Selection
Anatomy from Galen to Darwin
Words to Ponder

As stated earlier, one aspect of the discipline of comparative vertebrate anatomy is the study of the structure of living and extinct species. The former reveals similarities and differences among animals of today. The latter reveals what vertebrates were like in the past. Assembling the data in a geological time frame reveals the panorama of changes that have taken place from the distant past to the immediate present. That panorama has led to formulation of certain concepts that we will now examine. Their full import may not be apparent this early in the course; insight will grow as the terms are used in relevant contexts. Some, if not all, may be recalled from previous studies in the biological sciences.

PATTERN AND PROCESS

As we observe nature and the organisms around us, it is obvious that many taxa are quite distinct and are easily recognized as belonging to discrete groups. What are of special interest are those features that appear to be similar between disparate taxa. A number of questions surface: What are these similarities, and how do we account for their presence? The patterns of similarity may be in terms of common structure (form, materials, components, construction), function (how it works, biological role), and development (developmental trajectories, regulatory processes, contributory tissues, etc.). A separate question is the process or processes that gave rise to this pattern of similarity. One explanation for similarity is that the structure has been inherited from a common ancestor (homology). All other mechanisms (e.g., convergence or accident) result in nonhomologous similarity (homoplasy).

Homology and Homoplasy

In the middle of the sixteenth century, Pierre Belon, a French physician and naturalist, published a woodcut of the skeletons of a bird and a human side by side (fig. 2.1). The caption explains that it represents the bones of a human "put in comparison" with those of a bird, drawn to the same size in order to show "how great is the relationship [kinship] of one to the other." At that time, it was believed in Western civilization that similarities in appearance (for example, of the more than 20,000 species of teleost fishes, or the more than 60 families of perching birds, or the superficial resemblance of salamanders and lizards, horses and deer, monkeys and apes) were a manifestation of a divine architectural plan in the mind of the Creator while creating animals from fishes to mammals.

As the concept of organic evolution spread among scholars and enlightened laymen, they realized that *common ancestry* accounts for most similarities, and *adaptive modifications* account for most differences. Consequently a growing corps of anatomists in universities in western Europe devoted its research to the comparative study of the systems of the craniate body with the intent of adding to the knowledge of craniate phylogeny. These studies gave rise to the concept of **homology**—homologous structures (*homologues*) in different species.

Initially, structures in the same location with the same function and appearance were said to be homologous. Thus, most of the bones in the bird and mammal illustrated by Belon would be homologues. But we now know, as a result of studies in comparative embryology, that not all look-alikes are homologous, and many homologues bear no resemblance to one another. As phrased by the embryologist Boyden (1943), a homologue is "the same organ in different animals under every variety of form and function." The stapes in the middle ear of mammals and the hyomandibular cartilage that suspends the jaws and hyoid skeleton from the braincase of sharks are homologous, although one would not realize this from their appearance, location, or function. The precaval vein of cats and the right common cardinal vein of basal craniates are homologous, as are the intermaxillary bones of human embryos (which become fused with the maxillary bones before birth) and the premaxillary bones of adult apes.

If not all look-alikes are homologous, and homologues may bear no resemblance to one another in location, appearance, or function, how can we establish that two organs in two different taxa are homologous? Comparative embryology can often provide the answer, because structures in two different animals are homologous if they come from the same embryonic precursor (i.e., anlage, blastema, fundament). When this criterion is unproductive, other sources of data may be useful. For example, if the two structures are muscles, do they have the same innervation? If confronted with a fossil, useful clues would be the relative location of foramina in bones or impressions left in a bone by an overlying organ such as a sensory canal. In the instance of the mammalian stapes and the hyomandibular cartilage of sharks, that they are homologous was not difficult to establish once embryos of a critical developmental stage were available. The evidence in that instance is indisputable. A final test of homology is the congruence between the structure's distribution and the hypothesis of relationships for the taxa possessing the structure. There are structures that do not appear to be homologues, but their individual history can be traced back to a single structure in a common ancestor (this final test requires an *a priori* knowledge of relationships).

All remaining similarities that cannot be explained by homology are the result of **homoplasy** (homoplastic structures).

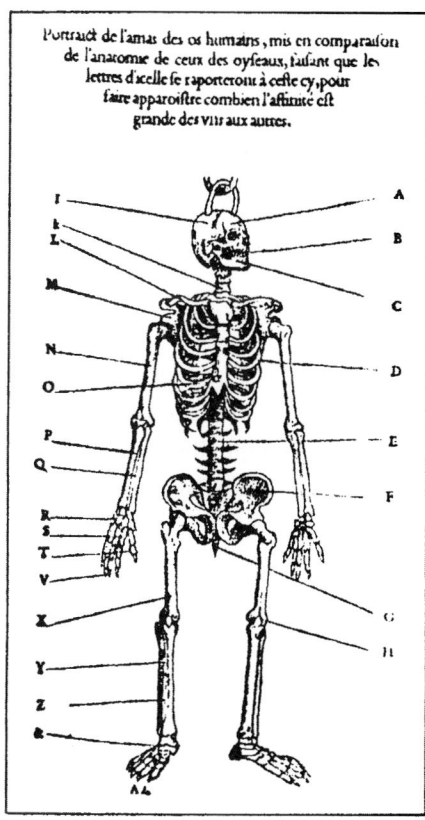

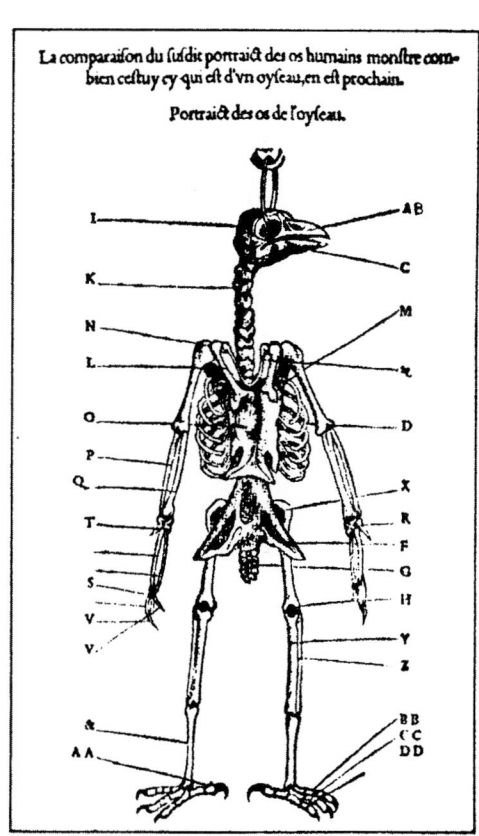

FIGURE 2.1
Comparative anatomical woodcuts
by Belon, 1555.

Serial Homology

Because metamerism is characterized by repetition of paired structures in the long axis of the body, it should be pointed out that serial structures—spinal nerves, vertebrae, muscle segments, for instance—are homologous *as a* series with the same series in another species, but the *individual components of the series are not homologues*. For example, a spinal nerve that supplies a muscle of the anterior fin is not homologous with a spinal nerve that supplies a muscle of the posterior fin—not even in the same animal. They do not come from the same embryonic metamere. The term **serial homology** is applied to metameric structures in a single organism and represents a pattern of segmental equivalence. Serial homologues have little relevance in phylogenetic studies; however, they are important when we try to understand the pattern of segmentation (e.g., segmentation of the craniate head) and when we consider phylogenetic changes in segmental patterns—both are areas of intensive ongoing research.

A related problem arises in comparing the digits of two different tetrapods. *Collectively*, the digits of the manus (hand) of a frog, bird, and horse are homologous, but the first, second, or last digit of one species, if there are more than one digit, may not be homologous with the first, second, or last digit of another. The single digit of the anterior limb of modern horses, for example, appears to be the middle digit of generalized tetrapods and of ancestral horses. The question "Which digit is missing?" is posed with reference to the tailed amphibian *Necturus* in figure 10.23.

Analogy

The term **analogy**, which refers to coincidental resemblance, is probably familiar because it has application in many disciplines. In zoological usage, *two structures that have the same function are analogous.* Thus, the jaw teeth of bony fishes, human teeth, and the horny ectodermal "teeth" of the jawless lamprey are analogous, as are the horns of cattle and those of rhinoceros; so, too, are the thymus gland of vertebrates and the bursa of Fabricius of birds (chapter 18). The definition of analogy has not been standardized in the current literature. Richard Owen, a renowned comparative anatomist of the nineteenth century, was one of the first to clearly delineate the usage of these terms. He did not consider analogy and homology to be mutually exclusive (i.e., a homologous structure that shares a similar function is an analogous homologue [nonhomologous structures that share a similar function represent analogous homoplasies]).

ADAPTATION

The infinitive *to adapt,* like the word *analogy,* should be familiar because it is a common concept. Adaptations, whether a modification of a man-made instrument that enables its use in a new, specific context, or a modification having the same effect on a living organism, are analogous phenomena. A *biological* **adaptation** is *a hereditary modification of a* **phenotype** (the assortment of morphological features of an organism) *that increases the probability of survival.* Because the phenotype is an expression of the **genotype** (the array of genes), adaptations are believed to be the result of environmental pressures that, by natural selection, propagate genetic mutations that have survival value. Natural selection is discussed later in the chapter. The streamlined fishlike body of an ichthyosaur (an extinct marine reptile, fig. 10.28) participated in adapting the reptile to a lifestyle in the sea. It was, therefore, an adaptation. The thick, scaly, water-impervious skin of reptiles (descendants of semiaquatic amphibians) is an adaptation that inhibits water loss (dehydration) in animals living in the air. The craniate body from fishes to humans is a complex of adaptations—old and sometimes expendable, or recent (in geologic time) and useful, if not vital. In evaluating or comparing adaptations, it is important to remember that any structure need only be sufficiently effective to ensure survival to pass on genes to the next generation. Due to limited resources, the evolutionary modification of any one adaptation in an organism may inflict a cost on other aspects of the organism. So, modifications are continually balanced by natural selection.

There is no evidence that adaptive changes occur because of need, present or future. Such a concept is an example of **teleological reasoning.** It attributes natural phenomena to a *preconceived design or purpose.* An example of teleological reasoning is, "Birds have wings in *order that* they may fly." The alternative is, "Birds have wings and *therefore they can* fly." Teleology implies a supernatural intelligence inherent in nature that guides natural phenomena. In contrast, what we see in nature is that populations that are locked into a changing environment and need a genetic mutation to remain successful and do not acquire one become extinct. Science attributes these adaptations, when they occur, to fortuitous genetic changes that are induced by natural laws that govern the behavior of matter.

The term **preadaptation** (protoadaptation) has been applied to traits that have enabled a phenotype to meet a new environmental challenge before it materializes. There are an abundance of examples of such traits, but they were not adaptations to a future need. They increased the chances of survival in an existing environment. Lungs were present in many fishes long before they were called on to function on land. They enabled fishes to obtain oxygen from the atmosphere when the dissolved oxygen content of the water became too low to sustain life. Subsequently, they enabled the initial short excursions of fishes onto land. Lungs became adapted for a sustained terrestrial existence over millions of years of subsequent evolution. Unlike adaptation, preadaptation is not a biological phenomenon. The term might be misconstrued as having a teleological implication.

SPECIATION

The formation of a new species is preceded in almost every instance, if not invariably, by geographical isolation of a population from other populations of the same species. This could occur, for instance, in a migrant species in which some of the migrants become isolated from the others for one reason or another. This appears to be what happened to the landlocked freshwater lamprey with a marine name, *Petromyzon marinus dorsatus,* which lives in the Great Lakes of North America. Its very close relatives, *Petromyzon marinus arinus,* are marine lampreys that migrate into freshwater streams to spawn, then return to the sea until sexual maturity. There are several possible explanations—all conjectural—for the isolation of these two specific groups, but they are irrelevant in the present discussion. The important point is that further exchange of genes between the two groups has become impossible in the present circumstances. Consequently, chance mutations, recombination of genes, and random **genetic drift** (spontaneous irregular change in gene frequencies) that characterize all genetic complexes will result ultimately in genetic incompatibility between the two. Thereafter, if brought together again they would no longer be able to produce viable offspring. *Geographical isolation coupled with genetic change is the matrix out of which new species evolve.* Geographical isolation provides the time necessary for the species to drift apart genetically. The length of time required for speciation depends on many factors, but the eventual emergence of new species, given the foregoing conditions, seems inevitable. The result of continuous speciation is phylogenesis, the formation of new taxa.

EVOLUTIONARY CONVERGENCE

When two or more unrelated organisms, such as ichthyosaurs (see fig. 10.28) and dolphins (see fig. 4.50) occupy similar environments, whether concurrently or millions of years apart, and acquire similar morphological features as adaptations to that environment, the phenomenon is termed **evolutionary convergence** (a form of homoplasy). The streamlined body, dorsal fins, and paddlelike structure of the forelimbs of ichthyosaurs and the same features in dolphins are examples of evolutionary convergence; so, too, are webbed feet in the frog, duck, and platypus. Evolutionary convergence produces look-alike features that are not the result of inheritance from a common ancestor. Convergent features are often the focus of studies on how organisms have evolutionarily adapted to a specific environment. In the aquatic example above, there are physical limits on how a bilateral organism can be streamlined to limit the effects of drag. In other cases, convergence is less pronounced because the parameter that may provide a selective advantage can be achieved in one of several ways.

DEVELOPMENT

Why emphasize developmental biology in a textbook on comparative anatomy? The structures described in this book are the end product of development and may still be influenced by continued growth. Patterns of development provide a further point of comparison among organisms (patterns of similarity). The unique development of the dorsal hollow nerve cord in teleosts provides evidence that unites them in a common group. The presence of extraembryonic membranes, specifically the amnion, provides the evidence that is used to unite the synapsids and reptiles in the group Amniota.

Changes in the timing of developmental events (heterochrony) occur within species and between ancestors and descendents. Delay of metamorphosis in some frogs and salamanders in response to environmental stress is a well-recognized phenomenon. Heterochrony has been proposed as a mechanism of evolutionary change in a number of lineages. An isolated population of ammocoetes (larval lampreys) has acquired sexual maturity eliminating metamorphosis into a parasitic "adult" (see fig. 3.17). Work is progressing to determine if this heterochronic process is reversible or represents a speciation event. A number of heterochrony hypotheses have been proposed to explain the origin of craniates (see fig. 3.17).

It is important to remember that the evolutionary changes in adult structures (e.g., the origin of the tetrapod limb from a finlike precursor) require modification of the ontogeny of these structures. The missing part of the story has been the developmental mechanisms that might mediate these changes. Small-scale changes were believed to accumulate, resulting in the step-by-step origin of new structures. Associated with these transitions were scenarios (a sequence of interrelated hypotheses) to associate functions with intermediate stages. Large-scale changes were believed to be most likely lethal. That is, major changes early in development would have a cascading effect resulting in what was known as "hopeful monsters." Our current understanding of the developmental process leads us to an understanding of how some large-scale changes may have occurred. Northcutt (1996) has suggested that the origin of neural

crest in craniates may represent one of these unique changes that is not preceded by intermediate precursors. Maybe there is now hope for those "hopeful monsters." Remember, however, any change—whether small or large—is subject to the selective pressures facing the organism as it develops and makes a living.

Arthur (1997), in his review of new advances in evolutionary developmental biology, suggested a circular relationship among the various disciplines that contribute to our understanding of evolutionary patterns (fig. 2.2). Early studies centered on "classical evidence" and were followed by an emphasis on population and quantitative genetics to explain the gradual step-by-step changes seen in much of evolution. Controversies over gradualistic versus punctuated changes often centered on differing interpretations of the rate of microevolution. No clear mechanism could explain large-scale changes (the "hopeful monsters" of some authors), so much effort was spent looking for evolutionary intermediate stages in both living and fossil organisms. The incorporation of developmental genetics and related mutational and embryological studies into the analysis of evolutionary patterns has bridged the gap between "classical" and "modern Darwinian" approaches.

Ontogeny and Phylogeny

Ontogeny (ontogenesis) is the developmental history of an organism. It begins with embryogenesis, includes postembryonic changes attributable to aging, and ends at death. The primary operants are the genes. In our discussions, we will be concerned chiefly with embryogenesis.

Phylogeny (phylogenesis) is the evolutionary history of a taxon. A **taxon** is a related group of organisms that constitute a taxonomic (classification) unit such as a family, order, or class. Phylogeny relates a taxon to ancestral taxa in an evolutionary line. The operant in the establishment of an evolutionary lineage is **speciation,** the emergence of new species from preexisting ones. Ontogeny occupies a single lifetime. Phylogeny requires hundreds of thousands to hundreds of millions of years. *Change* is the universal phenomenon that characterizes both processes.

In the chapters that follow, the morphology—shape, form, and structure—of organisms will be correlated with ontogeny and, to the extent current insight warrants, with phylogeny.

von Baer's Law

All early craniate embryos exhibit a similar architectural pattern both of structure and development. This observation was made by the embryologist Karl Ernst von Baer, who said that *features common to all members of a major taxonomic group of animals*—the craniates, for example—*develop earlier in ontogeny than do features that distinguish subdivisions of the group* (classes, orders,

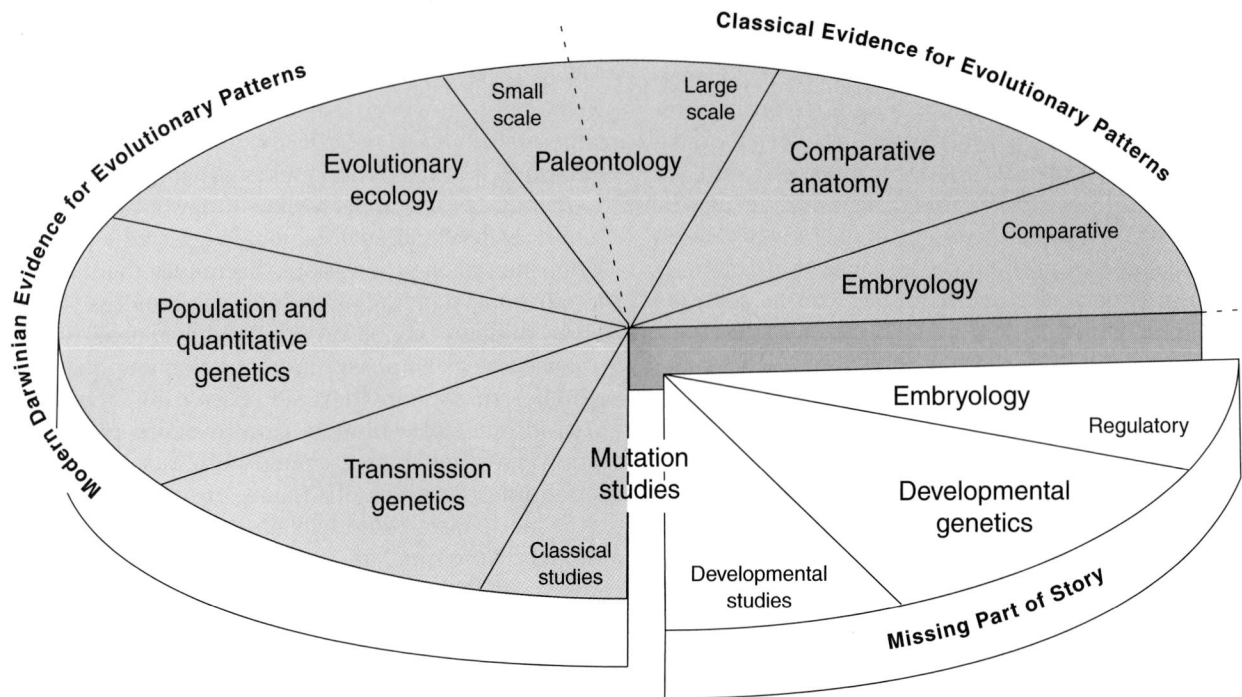

FIGURE 2.2
Relationships among the various disciplines in evolutionary biology. Modified from Arthur, 1997.

genera, and species). This is known as von Baer's law. An example of this is the early development of a notochord, dorsal hollow central nervous system, pharyngeal pouches, and aortic arches in all members of the subphylum. These early features become gradually modified in the direction of the species as development progresses.

A corollary of this law evolved following the increasing acceptance of the concept of organic evolution. As it stands today, the corollary is that *features that develop earliest in ontogeny are the oldest phylogenetically, having been inherited from early common ancestors, and features that develop later in ontogeny are of more recent phylogenetic origin.* (An exception to this premise is seen in the extraembryonic membranes, which form early in development, but represent a derived feature characteristic of Amniota.)

Historically, as indicated by the work of von Baer, there have been efforts to determine if there are common patterns among the ontogenetic trajectories of diverse taxa. A refuted hypothesis suggested that "ontogeny recapitulates phylogeny" in that adult features representing the phylogeny of an organism are repeated in its ontogeny. This idea was refined and restated to imply that early development is conserved and that related groups of organisms share this conserved pattern (von Baer's first observation). The pattern is best demonstrated by the work of Ernst Haeckel who in the nineteenth century compared and figured the ontoge-

netic stages in eight different craniates showing a common pattern of early development for forms as diverse as humans and fish. From this common point, each trajectory diverges. Recent studies have indicated that the earliest stages (early cleavage of the zygote) are also variable resulting in a bottleneck in developing craniates where they pass through a conserved stage called the pharyngula (a stage where the segmented pattern is evident, and the pharyngeal pouches are apparent; a tailbud stage has also been proposed as the conserved stage).

Richardson et al. (1997) have called this interpretation into question and rejected the data collected by Haeckel. It now appears that individual structures in development may be integrated by genetic and inductive events, but the structures of any developmental compartment are capable of independent variation. Therefore, the suite of characters that are used to define the pharyngula or tailbud stages do not exist at the same time of development in a comparison among craniates. Thus, the size, number of somites or pharyngeal arches, and the presence of limbs or rudiments of other organs are not consistent between craniate species and are each subject to heterochronic processes. A striking example is that seen in the zebrafish (a teleost) where the development of pharyngeal arches occurs after the tailbud stage. What appear to be conserved are the patterns of developmental regulatory genes and the inductive events that occur among primary tissues prior to the differentiation of developmental compartments. The developmental relationships among the segmented components of the craniate head will be the focus of a discussion in chapter 16.

METAMORPHOSIS AND HETEROCHRONY

Naturalists have been interested for centuries in tailed amphibians that become sexually mature and reproduce while retaining one or more larval features, including external gills and gill slits (see fig. 4.20, *Necturus* and *Siren*). These species have been characterized as exhibiting "incomplete metamorphosis." Of the eight families of urodeles, every one has at least a few individuals or populations that demonstrate a retention of larval features. However, this pattern is not confined to amphibians, or even to vertebrates. It is seen in invertebrate taxa, including larvaceans (see fig. 3.5), which reproduce at a morphological stage similar to the larval stage of other related taxa (compare a larvacean with an ascidian larva).

Incomplete metamorphosis is an advantage conferred on urodeles that might perish if forced, by loss of gills, to abandon an aquatic environment while the terrestrial environment is unfriendly. Neuroendocrine responses to environmental conditions can temporarily slow down, or speed up, the hormonal component of the control of metamorphosis in any species (see fig. 18.5), thereby adapting an organism to the exigencies of the environment. Inheritance plays a significant role in species such as *Necturus,* in which a genetically altered pattern of metamorphosis has been imposed in the course of evolution.

These changes are the result of the dissociation of the timing of some of the morphological changes of metamorphosis from the timing of maturation of the germ cells, so that the latter occurs earlier, whether induced by environmental factors, heredity, or both. Among the most visible morphological changes in amphibians are resorption of the gills and closure of the slits. Less visible are the remodeling of the skeleton and muscles of the pharyngeal wall and tongue, alterations of body shape and the integument, and other changes that adapt the organism for survival in a terrestrial environment. Individuals stop short of completing these morphological changes, temporarily or permanently. The dissociation of sexual maturation from the morphological changes of metamorphosis may become a permanent feature, fixed in the genes, in populations in which geologic or physiologic isolation provides an environment in which natural selection can operate. In those instances, new species may evolve. These new species exhibit a concept known as paedomorphosis.

In the above examples, two scenarios of change are proposed. First is the change in the patterning of development within a species (a temporary alteration of development). Second, the latter case, is the difference between species (alterations between ancestor and descendant). The terminology characterizing interspecific and intraspecific changes is in a state of flux (see Reilly et al., 1997 for a review of these changes). All of the changes represent differences in the rate or timing of developmental events or **heterochrony**. We will limit our discussion to those changes between ancestors and descendants (table 2.1).

Paedomorphosis is defined as ontogenetic changes whereby certain larval or immature features of the immediate ancestor become the end product of metamorphosis in the descendant species. The descendant species possesses paedomorphic characters. Knowledge of ancestral morphology is inferred from the morphology of fully metamorphosed species in the same taxon. Most populations *of Cryptobranchus allegheniensis* come close to completing metamorphosis, ancestral larval features in some populations being limited to retention of a single gill slit. A closely related genus in the same family, *Andrias japonicus,* which is postulated to reflect the adult state of ancestral cryptobranchids, exhibits complete metamorphosis. *C. allegheniensis* is therefore considered to possess paedomorphic characters.

Some populations of axolotls (Family Ambystomatidae) that live in montane (mountainous) environments (populations of *A. mexicanum* and *A. gracilis,* for example) have complete retention of larval morphology throughout life. Other Ambystomatidae (*A. tigrinum,* for example) exhibit complete metamorphosis. Because the

TABLE 2.1 Heterochronic Classification and Processes

	Heterochronic Classes*	Heterochronic Processes
Truncated development (descendant mimics ancestral juvenile)	Paedomorphosis	Decrease rate. Truncate development time. Postpone start of development.
Extended development (descendant surpasses ancestral development)	Peramorphosis	Increase rate. Elongate development time. Start development early.

*Only heterochronic classes relating to ancestor-descendant relationships are discussed. For the terminology of intraspecific changes, see Reilly et al., 1997.

ancestral species exhibited full metamorphosis, as inferred from the morphology of *A. tigrinum* and its closest relatives, the montane populations of *A. mexicanum* and *A. gracilis* demonstrate paedomorphic characters. The paedomorphism in *A. mexicanum* is due to insufficient levels of pituitary thyroid stimulating hormone (TSH) and thyroid hormone itself. However, complete metamorphosis can be induced by injection of TSH or thyroid hormone. We may infer from these observations that modern *A. mexicanum* has inherited mutations from ancestors that now limit the output of hormones necessary for metamorphosis.

The red spotted salamander of the woods of the eastern United States, a phase in the life cycle of most populations of *Notophthalmus viridescens*, exhibits a very different pattern of metamorphosis from those described above. In the course of ontogeny, the larvae undergo morphological changes that enable the population to emerge from the ponds and assume a juvenile (sexually immature) existence on land far in advance of maturation of the gonads. These changes include remodeling of the skull, loss of gills and gill slits, development of lungs, and cornification of the skin. Cornification protects the **eft** (see also chapter 4), the terrestrial juvenile, from water loss to the air. Their adoption of a moist location under rocks or rotting logs further promotes water retention. Efts remain on land for 1 to 3 years before the onset of sexual maturity, which is signaled by a "water drive" initiated by prolactin. At that time, the efts undergo a final metamorphosis. Their anatomy is altered in the direction of an air-breathing aquatic organism, they migrate back to the pond, the gonads mature, and reproduction takes place. Metamorphosis in these populations does not include a heterochronic stage.

In some localities, the larvae of *Notophthalmus* remain in the pond, the gonads mature, reproduction occurs, but gill resorption is delayed indefinitely (see fig. 4.21*e*). Members of the population remain in the pond throughout life. The population exhibits at least a temporary stage of intraspecific heterochrony, but the species is not paedomorphic unless at some time in the future these changes become genetically fixed resulting in a new species.

Heterochrony has been proposed in a number of hypotheses for the phylogenetic origin of craniates including paedomorphic modification of tunicates or cephalochordates to evolve an ammocoetelike craniate common ancestor (see figs. 3.1*a,b,c* and 3.17).

SYSTEMATICS AND TAXONOMY

Systematics is the process by which organisms are grouped, whereas taxonomy involves the conventions for applying names to those groups. The criteria for grouping organisms have changed over time. In pre-Darwinian times, organisms were placed in groups based on both an overall similarity and the possession of unique features. As a result, birds with their unique possession of feathers and their overall similarity in design were grouped together distinct from other reptiles (remember that birds are a subgroup within the larger group called reptiles). These early groupings were established prior to Darwin and our current understanding of phylogenetic relationships. Phylogenetic systematics is a grouping strategy, which attempts to arrange organisms in historical entities (that is, a group based on a common ancestor and all of its descendants [a monophyletic group]).

Most students are familiar with Linnean taxonomy, which uses a system of hierarchical groupings (phylum, class, order, family, genus, and species), the latter two forming the binomial system of taxonomy. Each level can be further divided by the addition of an appropriate prefix (e.g., subphylum, infraclass, superorder, etc). An international code of nomenclature provides detailed guidelines for the binomial system and only general guidance for higher level groupings.

Phyla represent divisions based on major body plans that are easily recognized. Most students can distinguish such phyla as arthropods, cephalopods, and chordates. Classes typically represent recognizable morphologies within phyla. In subphylum Craniata, we easily recognize Mammalia, Amphibia, Reptilia, and Aves. In the latter case of Aves and Reptilia, their distinctness obscures their true relationships. The morphological boundaries between these classes become unclear when we consider fossils. In the case of the fossil *Archaeopteryx*, initial interpretations ranged from an ancient bird to a (nonavian) dinosaur with feathers (suggesting an independent origin of feathers). Similar blurring occurs in the so-called mammal-like reptiles in the lineage leading to modern mammals. It is obvious that initial descendants from a speciation event will be nearly indistinguishable. With the breakdown of morphological boundaries between groups, our understanding of Linnean groupings is less clear. Why distinguish birds from other reptiles when the fossil record demonstrates one of the best known transitions?

So, how can we organize groups of animals? Current systematists group organisms based on their historical (phylogenetic) relationships. The advantages of this methodology are fivefold: (1) All groupings are equivalent in that they all represent monophyletic groupings, unlike Linnean classes based on traditional systematics, which may be paraphyletic, monophyletic, or represent a subgroup subjectively removed from other members of a larger group. As in traditional (Linnean) classification, the groups are arranged hierarchically. (2) Groups are based on the evidence of shared derived features (synapomorphies) rather than overall similarity. We recognize mammals as those taxa that share the possession of hair, a single bone in the lower jaw, and the presence of three ear ossicles. The presence of eyes or other plesiomorphic (primitive) features provide no useful information. (3) Phylogenies represent hypotheses of relationships that are subject to testing within the scientific method versus subjective groupings. The argument that birds and mammals are more closely related by reason of their shared endothermy can be tested. (4) The reliance on shared derived features as the evidence for relationships necessitates a clear and concise understanding of the anatomical features being evaluated. (5) Finally, the methodology facilitates our understanding of the polarity of character change over time (e.g., it helps us to understand the evolutionary changes that occurred in the origin of the mammalian heart).

Within phylogenetic systematics, cladistics is the methodology used by most systematists. Underlying cladistics is a number of assumptions: (1) the hierarchy of relationships is knowable and can be represented as a branching diagram. (2) Only those characters that diagnose a group are useful, (i.e.), primitive characters or those that are shared by all members of a group are not useful in determining relationships within the group under study. (3) It is the distribution of derived characters on a branching diagram (congruence) that determines whether they have independently evolved (homoplasy) or are similar due to common ancestry (homology). (4) Finally, the distribution of characters follows the principle of parsimony. In choosing between alternative hypotheses, the most economical solution is preferred (i.e., the tree with the minimum number of evolutionary changes to account for the available data [derived features]). Although debatable whether evolution is parsimonious or not, cladistics does provide a reproducible methodology (a requirement of the scientific method) whose assumptions can be tested.

The branching diagram of cladistics is the **cladogram.** Even if rejecting the underlying principles of cladistics, the cladogram provides a useful graphic tool to represent the relative degree of relationships among organisms. Accompanying cladograms are a number of terms that are common in research literature and necessary for discussion (fig. 2.3). A cladogram consists of a branching diagram with **terminal taxa** indicated at the ends of branches (A–E, fig. 2.3*a*). Adjacent taxa represent sister taxa (e.g., taxon D is the sister group to taxon E, and taxon C is the sister group to taxon (D,E)). The point of bifurcation between sister taxa represents a **hypothetical ancestor.** The branching pattern indicates relationships with taxa D and E being more closely related to each other than either is to the outgroup C or B. Carrying this one step further, taxa C and (D,E) are more closely related than either is to B.

A **monophyletic group** (or **clade**) is one that includes a common ancestor and all descendants (represented by circles 1 and 2—solid lines, fig. 2.3*a*). A **paraphyletic**

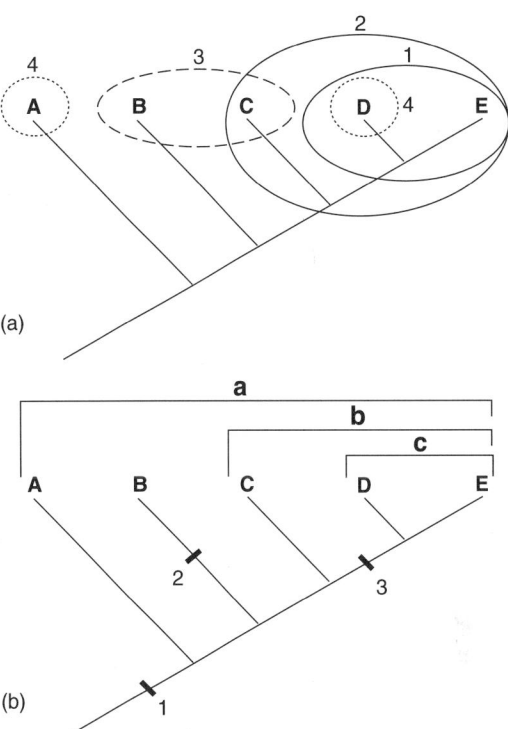

FIGURE 2.3

Terminology applied to cladograms. (*a*) A–E represent terminal taxa. Circles 1 and 2 represent monophyletic groups, 3 is a paraphyletic group, and 4 is a polyphyletic group. (*b*) a–c represent higher taxonomic groups. 1–3 represent synapomorphies. For example, character 1 provides evidence for the monophyly of taxon (A–E), character 2 diagnoses taxon B, and character 3 diagnoses taxon (D,E).

group is one that includes a common ancestor, but one or more descendants is omitted (represented by circle 3—dashed line, where B and C share a common ancestor, but descendants D and E are excluded [fig. 2.3*a*;] if birds are excluded from reptiles, as in the Linnean class Reptilia, then the class represents a paraphyletic group. Note that paraphyletic groups are often based on subjective criteria). A **polyphyletic group** is one that does not share an immediate common ancestor (represented by circle 4—dotted line; uniting birds and mammals based on endothermy would represent a polyphyletic group). A convention used in this text is demonstrated in figure 2.3*b*. Synapomorphies (shared derived characters) are indicated as tick marks on the branches (1–3). Brackets are used to indicate higher taxonomic units (a–c). It is important to distinguish the evidence for relationships (synapomorphies) from the names applied to the taxon diagnosed by a synapomorphy. Amniota is a taxonomic group diagnosed by the presence of an amnion. The group name is arbitrary, whereas the characters provide the evidence for relationships.

ORGANIC EVOLUTION
AND EVOLUTIONARY SELECTION

The concept of organic evolution is a single, simple, un-equivocal, yet widely misconstrued premise, which states: The plants and animals on earth have been changing, and the ones around us today are descendants of those that were here earlier.

The conclusion that the animals and plants have been changing is based in part on geological evidence. The evidence includes the fossilized remains of plants and animals preserved in sedimentary rock formations. It indicates that, if a human from today could be transported back several hundred million years, the plants and animals would be exotic. Five hundred million years ago, one would see mostly water, and on land there would be no plants. No fish could be caught with a fishing pole because the fish would be ostracoderms, which most likely were filter feeders. Three hundred and fifty million years ago, fishing would have improved, but there would be no trout, perch, or salmon. The land would be higher and drier, and there would be mosses and other simple land plants. Labyrinthodonts would be lumbering in and out of swampy waters, but there would be no frogs or toads. Three hundred million years ago, the land would be still higher, and the swamps would be luxuriantly forested with seed ferns and conifers but no flowering plants. Basal reptiles would be basking in the sun, but there would be no lizards or snakes. One hundred fifty million years ago, there would be birds with teeth; the earth would be heavily populated with reptiles, dinosaurs would be abundant and specialized, and a small minority population of ancestors to modern mammals, known chiefly from their teeth, jaws, and skulls, would have been struggling to survive. But there would be no cats, rats, or monkeys. It is in these geological findings that the theory of organic evolution has its roots, although not all of them.

The second part of the premise, which says that *the animals and plants around us today are offspring of the animals of yesterday,* is axiomatic. Life comes from preexisting life.

This is the theory of organic evolution, stripped of all satellite theories. It does not say that multicellular animals came from protozoa; it does not say that humans came from a monkey. It does not say where humans came from—that is another theory, or several other theories, but they are not the theory of organic evolution. Neither does the theory say what caused, or what causes, species to change. That, too, is another theory. The theory does not state how life began or how the universe began. There are theories about beginnings, but they are not the theory of organic evolution. These theories belong to scientific disciplines beyond the scope of this book.

The theory of organic evolution might also be called the "theory of the mutability of species." There is only one alternative, the "theory of the immutability of species." Proponents of the latter theory must insist that every species on earth today is precisely like it was when it first appeared, and that the first member of the species appeared *de novo,* coming from no previous living organism. To consider the possibility of even a single change ("fishes appeared and from these came all kinds of fishes") is to abandon the theory of the immutability of species in favor of organic evolution. If it is accepted that one change can occur, it must be accepted that two changes can occur. And once this step in logic has been taken, there is no limit to the number of changes that can occur.

The concept of evolution appeared in early Greek writings, including those of Aristotle. However, under the restraints of scholasticism, the philosophy and theology of the Middle Ages of Western civilization, the concept was not nurtured, and it remained dormant until the Renaissance when inquiry into natural phenomenon revived. **Jean Baptiste de Lamarck** (1744–1829), a French naturalist, was among the first of the Renaissance naturalists to publicly espouse the idea that species are not immutable and that complex species evolved from simpler ones. Lamarck propounded the **Doctrine of Acquired Characteristics** to explain how evolution worked. It postulated that characteristics acquired through the use or disuse of morphological structures are inheritable. Thus, when a part is employed in successive generations, it becomes stronger and better adapted for a specific role. Conversely, when a part is neglected in successive generations, it tends to become vestigial. This is how Lamarckism would account for the observation that the olfactory nerves of whales commence development in embryonic life, then regresse and become vestigial. A lung-breathing animal could not inhale under water without drowning. Therefore, according to the theory, the lack of use of the olfactory apparatus resulted in its disappearance. Vampire bats have an esophagus so small that they cannot swallow anything other than liquids. According to the Doctrine of Acquired Characteristics, the esophagus would have become too small to swallow solids because the only nourishment that passed through it for generations was mammalian blood. The doctrine as proposed has not been acceptable because there is no known explanation as to how use or disuse of a part in individuals can be translated into alterations of the hereditary code stored in the sperm and eggs. The advent of modern genetics made this aspect of Lamarckism, at least for the present, untenable.

Charles Darwin (1809–1882), born 65 years after Lamarck, had a different explanation to account for evolutionary change. Darwin observed that variation in individual organisms is universal; no two individuals are identical. This is the starting point for Darwin's **Theory of Natural Selection** propounded in 1859. Individual variations, Darwin believed, result in varying degrees of success in competition between individuals and with the physical conditions of life. Darwin summarized by stating that "preservation of favorable variations and the rejection of injurious variations I call Natural Selection" and paraphrased the concept as the "struggle for existence" in which the fittest survive. Darwin preceded the advent of genetic theory and thus could not explain natural selection in terms of the molecular units of inheritance. In the age of genetics, natural selection means that random genetic mutations and recombinations that render an organism better able than others to cope with the environment are more likely than other gene complexes (genotypes) to be passed on to another generation. *The concept of natural selection (survival of the fittest) is Darwin's most important contribution to evolutionary theory.* Natural selection is a major route to speciation. Whether or not it is the only route is not known. Current evolutionary research is examining the concept that selection analogous to Darwinian natural selection operates at several levels in addition to the genes (Lieberman and Vrba, 1995). Called a hierarchical theory, it embodies the essence of Darwin's concept, expanded to work at several hierarchical levels (Gould, 1982).

The first paragraph of this discussion cited the *premise* of evolutionary theory. With the advent of genetics, the *mechanics* of the process of evolution have gradually become clearer. Our current insight is a synthesis of the basic premise set forth in the first paragraph of this discussion, the concept of natural selection, and the insight gained through subsequent research in genetics, comparative morphology of plants and animals, paleontology, systematics, biochemistry, molecular biology, and other disciplines. Natural selection is a natural, rather than a supernatural, explanation for the variety and number of species on the planet (conservatively estimated at 5 million [Myers, 1985]), just as the law of gravity is a natural explanation for the attraction between the earth and the moon. A natural explanation may gain acceptance from an accumulation of supporting data gathered by independent observers. Conversely, a particular natural explanation will always be subject to rejection pending future observations. By contrast, supernatural explanations are generally not falsifiable and thus fall outside the bounds of scientific inquiry.

ANATOMY FROM GALEN TO DARWIN

Cave dwellers had some knowledge of the internal organs of mammals, as did the Babylonians and ancient Egyptians, who practiced surgery and embalming. Embalming, which involved removal of most of the internal organs, was an advanced art. Aside from a small number of Egyptian medical papyri dated to about 3000 B.C., the oldest anatomical works in Western civilization were written during the last 400 years B.C. by Greek philosophers and physicians. These works were incomplete, mostly superficial, and often imaginative. Anatomy of the classical era culminated in the works of **Galen**, a Greek philosopher-physician who practiced in Rome between A.D. 165 and 200. He assembled all available Greek anatomical writings, supplemented them with his own dissections of apes from the Barbary coast (human dissection was at that time prevented by public opinion and superstition), and, in addition, wrote more than 100 treatises on medicine and human anatomy. Shortly thereafter, scholasticism took over, and during the next 1,300 years Galen's descriptions were considered infallible, and dissent was punishable. Subservience to authority became so accepted that (as has been said, probably partly in jest) if a scholar wanted to know how many teeth horses have, he saddled up and rode 100 miles, if necessary, to the nearest library to see what Galen said. (No doubt the peasants looked into the horse's mouth in an application of the scientific method.)

It was not until the fifteenth century that **Leonardo da Vinci** (1452–1519) and other Italian artists began to make anatomical observations of their own (at the peril of excommunication from their church), and a renaissance in human and animal anatomy began. In 1533 a young Flemish medical student named **Vesalius** attended classes at the University of Paris where Galen's works on human anatomy were read (by a "reader") while the professor tried, often with embarrassing lack of success, to harmonize Galen's descriptions with the dissection. (Recall that many of Galen's descriptions of humans were written from dissections of apes.) After 3 years, Vesalius quit Paris, earned a degree at Padua, stayed on to teach, and recorded his own anatomical observations. In 1543, he published *De humani corporis fabrica* (*On the Structure of the Human Body*), ushering in a renewed interest in anatomy of animals in general. Galen, however, influenced his *Fabrica,* to a large degree, and many of the errors of Galen were perpetuated. Vesalius's illustration of the hyoid and kidney were actually drawn from a dog, which was Vesalius's favorite specimen for dissection. However, he was not interested in mammals other than humans beyond the extent to which they might contribute to his knowledge of major organs of the human body. His interest was in descriptive human anatomy, not in species differences.

Eight years after publication of Vesalius's anatomy, **Pierre Belon** (1517–1564), a French naturalist and physician, published an important work on the dissection of cetaceans and other marine animals, invertebrate and vertebrate, including fishes. (Because cetaceans lived in the sea and exhibited adaptations for the marine environment, they were, to Belon, fishes, despite his detailed study of their anatomy that is otherwise mammalian.)

Belon was also interested in birds, saying he never missed an opportunity to dissect one and had examined the internal parts of 200 different species. In 1555 he published his classic illustration of a human and bird skeleton side by side, showing that the skeletons correspond almost bone for bone (see fig. 2.1). He published the illustration because he had made the observation and thought the general public would be enlightened. Like other public figures of his day, he was a teleologist. He attributed the similarity of the skeletons, like the convergent evolutionary similarities that he observed between fishes and cetaceans, to the manifestation of a basic architectural plan, or *archetype,* in the mind of the Creator. Today, the term *basic architectural plan* refers to the generalized pattern of development and morphology that has been inherited from the earliest craniate ancestors.

The distinction of being considered the founder of comparative anatomy (and also of paleontology) should probably be awarded to Georges Leopold Chretien Frederic Dagobert **Cuvier**, Baron de Montbeliard (1769–1832), a French naturalist who studied at Stuttgart. A prolific author in the field of natural history, his works include, among many others, nine volumes entitled *Lecons d'anatomie comparee* (Paris, 1801–1805), four volumes of paleontological observations on fossil skeletons of quadrupeds, a geological description of the environs of Paris, and a work that is now a classic, a four-volume 1816 abridgement of his course in comparative anatomy at the Ecole Centrale du Pantheon in Paris, which he titled *Le regne animal distribue d'apres son organisation pour servir de base a l'histoire naturelle des animaux et d'introduction l'anatomie comparee* (*The animal kingdom arranged in accordance with its structure for the purpose of serving as a basis for the natural history of animals and as an introduction to comparative anatomy*). In this volume, he summed up his lifetime of research on the morphology of fossil and living animals, which he divided, in the role of taxonomist, into four taxa, Radiata, Mollusca, Articulata, and Vertebrata, based on differences in structure of the skeleton and organs. (Note that Cuvier's classification was based solely on phenotypes. Species were considered immutable.) His inspiration throughout life came from a copy of the first edition of Linnaeus's *Systema naturae* (1735), which in Cuvier's early years was his entire scientific library. His final contribution was a 22-volume masterpiece, *Historie naturelle des poissons* (*Natural history of fishes*), published over a 4-year span ending in 1832. He died in 1832 in Paris at the age of 56 after an energetic life dedicated to natural history and teaching, including the chancellorship at the University of Paris. Throughout life, he steadfastly maintained the absolute immutability of species, objecting vehemently to the mutability ideas of Lamarck (a contemporary), of **Geoffroy Saint-Hillaire** (1772–1844), with whom he had collaborated in several published works, and of others who began to envision the possibilities of changes in species with time.

Cuvier would have been a leading Darwin adversary in the great uproar in scientific circles, in academia, among theologians, and, consequently, in the press, that was evoked by the publication of Charles Darwin's masterpiece in 1859, *On the Origin of Species by Means of Natural Selection, or the Preservation of Favoured Races in the Struggle for Life.* The 1,250 copies sold out on the day of issue. The first four chapters explain the *artificial selection* of domesticated animals and plants by humans, comparing this with *natural selection* as a consequence of the struggle of undomesticated species for existence. The fifth chapter discusses variations and causes of modification other than natural selection. The next five chapters examine the difficulties that stand in the way of a belief in hereditary changes (evolution) in species generally, as well as in natural selection as an operating factor. The last three chapters present the evidence for evolution from the field of paleontology, geographical distribution of species, comparative anatomy, embryology, and vestigial organs. The uproar and almost universal condemnation of Darwin was comparable to that resulting from Copernicus's treatise (1543) hypothesizing that the earth revolves around the sun. (Copernicus's treatise was probably completed by 1530, but was circumspectly withheld by him until he was on his deathbed.) Neither Copernicus's nor Darwin's theory conformed to the accounts in the Judeo-Christian Book of Genesis, and both were therefore considered a denial of the authority of God. Darwin's theory was also seen as a denial of the story of Adam and Eve and of creation of the earth in seven days.

With the advent of the study of genetics the mechanism whereby change in phenotypes takes place was elucidated, and humankind was given a natural explanation for the diversity of species on earth.

WORDS TO PONDER

The following words are sometimes employed in discussions of phylogeny. You will read them, hear them, and use some of them. There will be differences of opinion with reference to the connotations of most of them. However, if calling attention to these abstract terms stimulates thoughtful discussion, inclusion of them will have been justified.

Primitive is a relative term. It refers to a beginning or origin. A primitive trait is one that appears in a stem ancestor from which arose an array of subsequent species, some of which may retain the trait. The notochord is primitive because it occurred in the first chordates. However, one cannot always be certain that a given *structure* is primitive. For example, the lateral neural cartilages of lampreys (see fig. 1.5) are primitive only if they reveal an original condition from which typical vertebrae later evolved. They could represent a secondary reduction from some unknown vertebral condition.

The term should not be applied to organisms. Although lampreys may retain a number of primitive characters, they are highly specialized parasites. Their invasion of the Great Lakes in North America devastated the local commercial fishing industry. (How did a "primitive" form outcompete so many "advanced" taxa?)

Generalized refers to structural complexes that, at least in some of the descendants, have undergone subsequent adaptation to a variety of conditions. The hand of an insectivore was, and remains, a generalized mammalian hand. It was competent to evolve into the wing of a bat, the hoof of a horse, the flipper of a seal, and the hand of a primate. A generalized group of animals has demonstrated that it was genetically suitable for divergent evolution (that is, evolution in many directions). Labyrinthodonts were generalized tetrapods. The term *generalized* connotes a state of potential adaptability.

A **specialized** condition is one that represents an adaptive modification. Craniate wings are specializations of anterior limbs, and beaks are specialized upper and lower jaws. Beaks (fig. 2.4) may be needlelike and tubular for extracting nectar from flowers (hummingbirds); chisel-like for drilling (woodpeckers); hooked for piercing and tearing prey (predatory birds, such as hawks and eagles); sturdy and spearlike, for capturing small fish and other aquatic animals in marshes (herons); long, spearlike, but less sturdy, used for nocturnal foraging for grubs and worms (kiwi); or long and upcurved for sweeping through shallow waters and mud in search for small crustaceans, molluscs, and other marsh inhabitants (avocets). Increased specialization connotes increased adaptation. The greater the specialization, the less may be the potential for further adaptive changes.

Derived, or **modified,** connotes any state of change from an ancestral condition, a mutated state. If the presence of bone is a primitive trait, wholly cartilaginous skeletons are modifications of the condition. The modification (loss of the potential to form bone) was a specialization if it adaptively modified the animal. Perhaps not all modifications are adaptive (students of speciation disagree strongly about this), but if they are not, they may portend the demise of the species because any change is statistically more likely to make the animal less competitive.

Higher and **lower** are used to express the relative position of major taxa on a conventional phylogenetic scale. They are less informative when applied below the class level. Birds and mammals evolved from a common ancestor and hence are said to be higher than the ancestral species. In this context, the words have some meaning. Sometimes the terms are used to express relative mutational distance of a given taxon from a common ancestor when compared with some other taxon—but are mammals to be considered higher than birds? In this context, the term may be misleading. The terms may also be meaningless when used to compare a genus within one taxon with a genus within another taxon, as when comparing a modern frog with a perch, or a human being with a hummingbird. Each would be incapable of competing in the other's environment. With the introduction of cladograms, the relative position of taxa can change when nodes are rotated (this rotation has no effect on relationships [(e.g., A(BC) has the same grouping pattern as (BC)A)]). Although still used in the literature, higher and lower have no context in the latter case.

Sister group, outgroup, and **ingroup** are used to describe subunits of a cladogram. The group of interest ((B,C) in fig. 2.5a) is also known as an ingroup. An outgroup is any group outside the clade of interest, whereas a sister group is the immediate outgroup that shares the common ancestor with the ingroup.

Basal, stem, and **crown groups** are terms that further describe a cladogram. Basal taxa represent those lineages that diverge at a point close to the common ancestor within a clade. A stem group or lineage consists of those basal taxa not included in the crown group (fig. 2.5b). A crown group is a group that includes all living members of a clade and their closest common ancestor (fig. 2.5b). Crocodiles and birds would form a crown group for Archosauria (chapter 4).

Simple is a relative term connoting a lack of complexity of component parts. A simple state is not necessarily

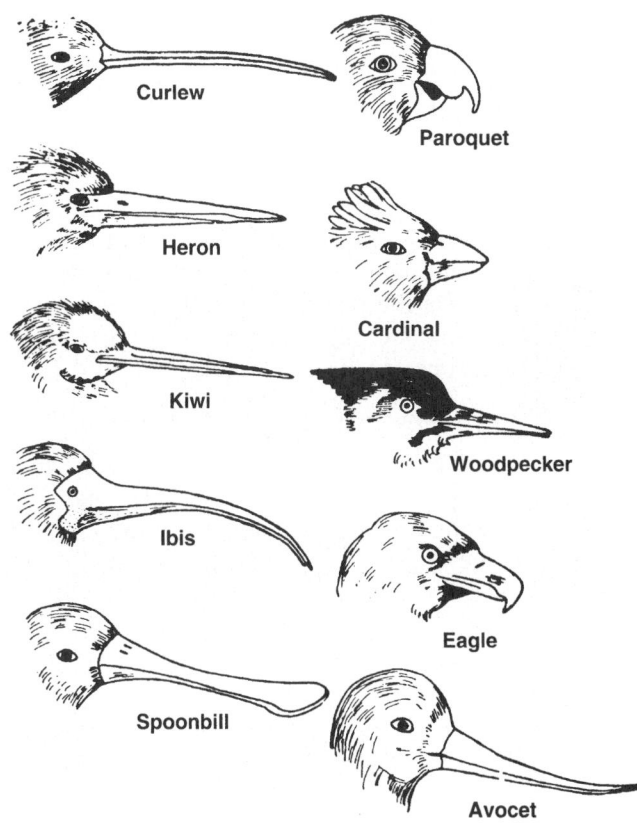

FIGURE 2.4

Adaptations of the beak for feeding in selected birds. Not to scale.

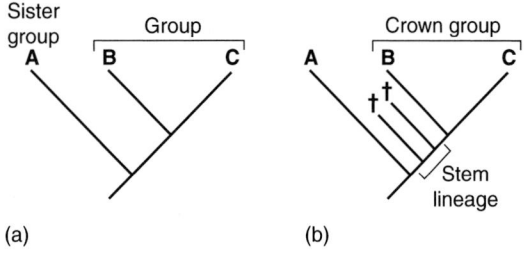

FIGURE 2.5

Parts of a cladogram. (*a*) The group of interest (B,C) is compared to its sister group (A). (*b*) A crown group is the group formed by all the descendants of the closest common ancestor for the living members of a clade. The stem group or lineage includes those basal members of a clade excluded from the crown group. (†) indicates an extinct taxon.

a primitive one. The skull of a human being is simple compared to that of a teleost, but it is not primitive. The primitive may also be far from simple.

The term **advanced** should connote a modification in the direction of further adaptation. Unfortunately the word has overtones connoting progress and hence is subjective and misleading. It is a matter of opinion of one species—humans—whether or not a modification in another species represented or represents progress. The phrase *more recent* or *more specialized* is more informative than the term *advanced*.

Degenerate is another value-judgment word. For example, it is sometimes applied to lampreys by those who think lampreys have lost jaw skeletons, paired appendages, bone in the skin, and other characteristics of typical vertebrates. However, the condition of the living agnathans represents an adaptation to a semiparasitic state and as such may better be characterized as "specialized." Unless lampreys never had jaws, these agnathans may have specialized themselves right into a state of neosimplicity! To call them "degenerate" would seem to discount the value of adaptive modification. Degenerate seems to be a term that should be avoided.

The words **vestigial** and **rudimentary** require explanation. A phylogenetic remnant that was better developed in an ancestor is vestigial. The pelvic girdle of whales is said to be vestigial because ancestors of whales were tetrapods with functional tetrapod appendages. The yolk sac of the mammalian embryo is vestigial. The term *rudimentary* is used in two different senses, phylogenetic and ontogenetic. In the phylogenetic sense, structures that became more fully exploited in descendants are said to be rudimentary in the phylogenetic precursor. For example, the lagena of the inner ear of fish is sometimes referred to as a rudimentary cochlea because it evolved into a cochlea in later forms. In the ontogenetic sense, a structure that is undeveloped or not fully developed is said to be rudimentary. The muellerian duct may be considered rudimentary in most male craniates. It is not always possible to be certain whether a structure should be called rudimentary or vestigial. The pseudobranch of the shark *Squalus acanthias* is vestigial if it represents a gill that, in ancestral sharks, was a full-fledged functional gill surface.

If the foregoing thoughts trigger discussion, no matter how dissonant, the space employed in presenting them will have been well utilized. It must not be forgotten that words are sounds made by humans to connote a concept. Otherwise, they may be nothing but noise. Or, do you wish to object, on a semantic basis, to the word *noise?*

Summary

Definitions may briefly paraphrase those in the text.

1. Ontogeny is the developmental history of an individual.
2. Phylogeny is the evolutionary history of a taxon.
3. Homology is the manifestation of homologous structures in different species.
4. Homologous structures are derived from the same ancestral anlage. They may be similar or broadly dissimilar morphologically and functionally.
5. Analogous structures have similar functions and may be either homologous or homoplastic.
6. An adaptation is a hereditary modification that increases the chances of survival.
7. Natural selection is the nonrandom differential reproduction of genotypes as a result of the selective effect of the environment.
8. Speciation, the formation of new species, is, in most cases, the result of genetic changes over time typically in isolated populations of preexisting species.
9. Evolutionary convergence is the evolution of similar structures in unrelated taxa as a result of mutations that are adaptive to similar environments.
10. Homoplasy is the presence in different species of structures that look alike but are not similar due to common ancestry. The structures are said to be homoplastic.
11. Heterochrony is the temporary or permanent retention of new or larval traits in sexually mature individuals, populations, or species. These traits are due to changes in the rate and/or timing of developmental events.
12. Paedomorphosis is an evolutionary process wherein certain larval or immature features of the immediate ancestor become the end product of metamorphosis in the descendant species.
13. Peramorphosis, in contrast to paedomorphosis, results in a descendant that exceeds ancestral development.
14. Systematics is the discipline and practice of ordering organisms into hierarchies that reflect their morphological similarities and phylogenetic history.

15. Taxonomy is the process and rules by which we apply names to the groups determined in a systematic analysis. An international code provides the rules for binomial nomenclature.
16. Organic evolution is the changes that have taken place in animals and plants through historical time.
17. Teleology ascribes natural phenomena to purposeful design rather than to natural laws.
18. Darwin's major contribution was the Theory of Natural Selection, an alternative to Lamarck's Doctrine of Acquired Characteristics.
19. Georges Cuvier was the founder of the discipline of comparative anatomy.

Critical Thinking Questions

1. What is the relationship of pattern and process in comparative anatomy?
2. Distinguish the comparison of structures between ancestors and descendants from comparison of structures within an individual. What term implies similarity due to common ancestry? What term implies developmental, structural, or positional similarities between segmental components in a metameric organism?
3. Distinguish homology and homoplasy. What is the relationship of these terms to analogy?
4. What is heterochrony? In comparing ancestor-descendant relationships, what heterochronic processes are possible? Can you think of an example for each? (Recognize that these hypotheses are based on a limited number of observations at this point, but it is useful to recognize the potential importance of developmental changes in evolution.)
5. If the function of taxonomy is to provide a universal language of science to communicate ideas, then does it matter that Linnean classes are not equivalent if their content is understood? Why or why not?
6. In comparing two cladograms, indicated here as nested sets, what are the similarities or differences in the hypotheses of relationships? Draw these as branching diagrams: (A (B (C,D))) and (A ((D, C) B)).

Selected Readings

Arthur, W.: The origin of animal body plans: A study in evolutionary developmental biology. Cambridge, 1997, Cambridge University Press.

Ayala, F. J.: Teleological explanations in evolutionary biology, *Philosophical Society* (London) **37**:1, 1970.

Boyden, A.: Homology and analogy: A century after the definitions of "homologue" and "analogue" of Richard Owen, *The Quarterly Review of Biology* **18**:228, 1943.

Browne, J.: Charles Darwin. Voyaging. Vol. 1 of a biography. New York, 1995, Knopf.

Cole, J. E.: A history of comparative anatomy. London, 1944, Macmillan.

Davis, J. I.: Phylogenetics, molecular variation, and species concepts, *Bioscience* **46**(7):502, 1996.

de Beer, G. R.: Homology, an unsolved problem. London, 1971, Oxford University Press.

Gaffney, E. S., Dingus, L., and Smith, M. K.: Why cladistics? *Natural History,* June 1995.

Gould, S. J.: Darwinism and the expansion of evolutionary theory, *Science* **216**:380, 1982.

Gould, S. J.: Ontogeny and phylogeny. Cambridge, MA, 1985, The Belknap Press of Harvard University.

Hall, B.K., editor: Homology: The hierarchical basis of comparative anatomy. San Diego, 1994, Academic Press.

Hickman, C. P., and Roberts, L. S.: Animal diversity. Dubuque, IA, 1995, Wm. C. Brown. Discussions on traditional and cladistic taxonomies.

Lewis, R. W.: Teaching the theories of evolution, *American Biology Teacher* **48**:344, 1986. A discussion of Darwin's two major theories—the descriptive theory of descent and the mechanistic theory of natural selection.

Lieberman, B. S., and Vrba, E. S.: Hierarchy theory, selection, and sorting: A phylogenetic perspective, *Bioscience* **45**:394, 1995.

May, E.: One long argument: Charles Darwin and the genesis of evolutionary thought. Cambridge, MA, 1991, Harvard University Press.

Moore, J. A.: Science as a way of knowing: Evolutionary biology, *American Zoologist* **24**:419, 1984.

Myers, N.: The end of the lines, *Natural History* **94**(2):2, 1985. A discussion of duration of species and extinction rates.

Nelson, J. S.: Fishes of the world, ed. 3. New York, 1994, John Wiley and Sons.

Northcutt, R. G.: The origin of craniates: Neural crest, neurogenic placodes, and homeobox genes. In Gans, C., Kemp, N., and Poss, S., editors: The lancelet (Cephalochordata): A new look at some old beasts, *Israel Journal of Zoology* **42**:S-273, 1996.

Reilly, S. M., Wiley, E.O., and Meinhardt, D.J.: An integrative approach to heterochrony: The distinction between interspecific and intraspecific phenomena, *Biological Journal of the Linnean Society* **60**:119, 1997.

Reilly, S. M.: Ontogeny of cranial ossification in the eastern newt, *Notophthalmus viridescens* (Caudata, Salamandridae), and its relationship to metamorphosis and neoteny, *Journal of Morphology* **188**:215, 1986.

Richardson, M. K., Hanken, J., Gooneratne, M. L., Pieau, C., Raynaud, A., Selwood, L., and Wright, G. M.: There is no highly conserved embryonic stage in vertebrates: Implications for current theories of evolution and development, *Anatomy and Embryology* **196**:91, 1997.

Ridley, M.: Evolution. Boston, 1993, Blackwall Scientific Publications.

Sanderson, M. J., and Hufford, L., editors: Homoplasy: The recurrence of similarity in evolution. San Diego, 1996, Academic Press.

Shaffer, H. B., and Voss, S. R.: Phylogenetic and mechanistic analysis of a developmentally integrated character complex: Alternate life history modes in Ambystomid salamanders, *American Zoologist* **36**:24, 1996.

Steele, E. J.: Somatic selection and adaptive evolution. On the inheritance of acquired characteristics, ed. 2. Chicago, 1981, University of Chicago Press.

Stock, D. W., and Witt, G. S.: Evidence from 18S ribosomal RNA sequences that lampreys and hagfishes form a natural group, *Science* **257**:787, 1992.

Thorington, R. W., Jr.: Books, *BioScience* **44**(10):705, 1994. Cites problems of validation and confirmation in phyletic reconstruction of mammalian orders.

Whiteman, H. H.: Evolution of facultative paedomorphosis in salamanders, *Quarterly Review of Biology* **69**(2):205, 1994.

Links to the Internet

Visit the zoology website at http://www.mhhe.com/zoology to find live Internet links for each of the references listed below.

1. Systematics. A series of slides explaining various concepts related to systematics.

2. Glossary of Phylogenetic Systematics. An extensive glossary of the terminology necessary to understand systematics.

3. Genetics and Evolution. This is from a MCAT prep course on-line. The link to information on natural selection is particularly useful pertaining to this chapter.

4. Evolution Entrance. This site, from the U.C. Berkeley Museum of Paleontology is an introduction to Darwin's theories, and has many links to other useful sites.

5. Journey into the World of Cladistics. Information and links to more on cladistics. Hot linked to a well-done glossary of terms.

6. PHYLIP Home Page. This is a free computer-based phylogenetic systematics program from the Department of Genetics at the University of Washington.

7. Haeckel's Drawings. A reevaluation of Haeckel's "phylotypic" stages.

8. Growth. A description and diagrams illustrating allometric growth, isometric growth, and related topics.

3 Protochordates and the Origin of Craniates

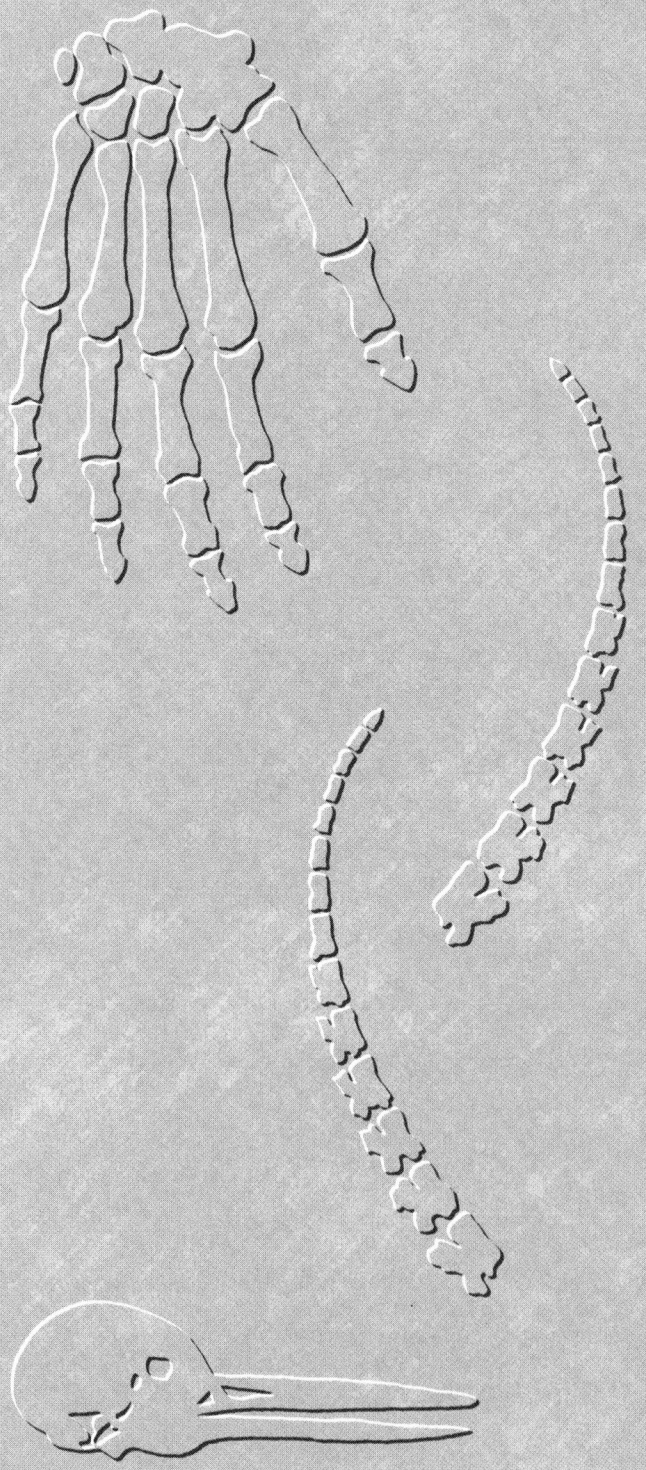

In this chapter, we will speculate concerning the invertebrate origins of craniate animals. For background, we will gain some knowledge concerning protochordates—small aquatic animals with a notochord but lacking craniate features. We will then hypothesize concerning their affinities with craniates. At the end of the chapter, we will examine a larval lamprey, for the reason that it bears a remarkable resemblance to protochordates, has been mistaken for a protochordate, yet grows to become an eel-like lamprey.

OUTLINE

No study of vertebrate anatomy would be complete without providing some knowledge about the closest relatives of vertebrates, the protochordates and hagfish. A sessile sea squirt squirting water out of its excurrent siphon seems a far cry from even a simple amphioxus, let alone the most basal jawed fishes. Yet, sea squirts, amphioxuses, fishes, and tetrapods exhibit a number of fundamental structural features in common, bilateral symmetry, deuterostomous development, pharyngeal slits, a notochord, a dorsal hollow central nervous system, a postanal tail, and an endostyle (fig. 3.1). The latter four features are unique to the chordates. Notwithstanding the fact that the notochord becomes encased within vertebrae in vertebrates and the pharyngeal slits close sooner or later in tetrapods, the manifestation of the chordate fundamental structures in all of them during ontogeny indicates that vertebrates are more closely related genetically to protochordates and hagfish than to any other known animals. Therefore protochordates may provide some clues to the invertebrate origins of animals with backbones.

PROTOCHORDATES

In recognition of the postulate that protochordates and vertebrates are closely related, Ernst Haeckel in 1874 proposed the establishment of the **phylum Chordata** incorporating three subphyla. **Urochordata** (sea squirts and certain other invertebrates), **Cephalochordata** (the amphioxuses), and **Vertebrata** (Haeckel included hagfish within the vertebrates). The first two subphyla are protochordates, a convenient term, although without taxonomic status. All are marine organisms. Our interest in them stems from the realization that they and the vertebrates probably had common

ancestors. Therefore we will examine them briefly, emphasizing the morphologic features that prompt this hypothesis. The place of protochordates in the chordate hierarchy as proposed by Haeckel and generally accepted today is as follows (typical Linnean hierarchies are included in parentheses where appropriate):

> Deuterostomata
> > Echinodermata (phylum)
> > Hemichordata (phylum)
> > Chordata (phylum)
> > > Urochordata (subphylum)

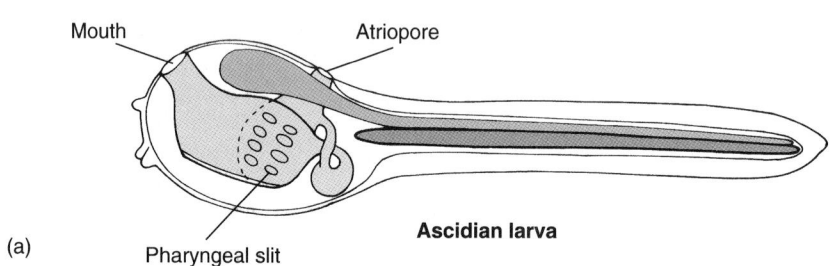

(a)   **Ascidian larva**

Mouth · Atriopore · Pharyngeal slit

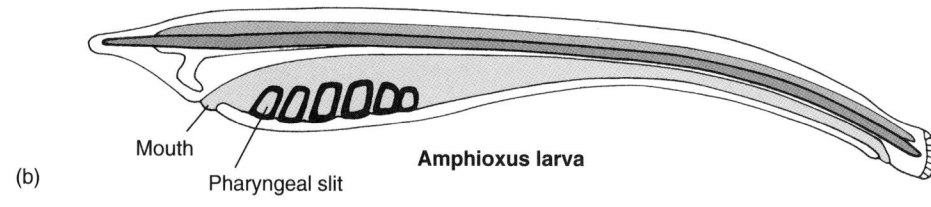

(b)   **Amphioxus larva**

Mouth · Pharyngeal slit

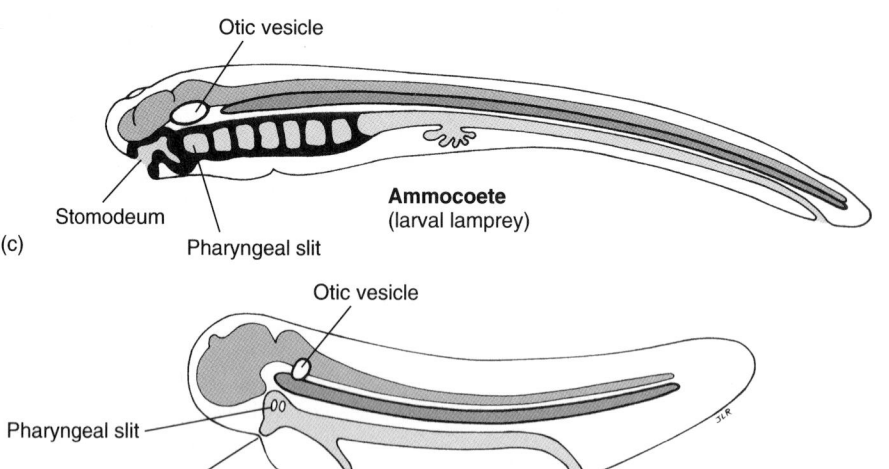

(c)   **Ammocoete** (larval lamprey)

Otic vesicle · Stomodeum · Pharyngeal slit

(d)   **Larval amphibian**

Otic vesicle · Pharyngeal slit · Stomodeum

FIGURE 3.1

Chordate larvae showing basic architectural pattern. (*a*) and (*b*) are protochordates; (*c*) and (*d*) are vertebrates. *Dark red,* notochord; *medium red,* dorsal nervous system; *light red,* alimentary canal. In (*a*) a larval sea squirt, part of the left wall of the pharynx has been removed to show the slits in the right wall opening to the atrium. A comparison of these larval forms with adult craniate structures will help to form the basis of a hypothesis of craniate origins at the end of this chapter (figs. 3.14 and 3.17).

Ascidiacea. Sea squirts (class)
Larvacea. Larvaceans (class)
Thaliacea. Thaliaceans (class)
Cephalochordata. Lancelet or amphioxus
    (subphylum)
Craniata (subphylum = subphylum Vertebrata
    of Haeckel)
Myxini. Hagfish
Vertebrata. Craniates with vertebrae

Basal Deuterostomes (Echinoderms and Hemichordates)

To better understand evolutionary changes within the protochordates and craniates, it is useful to step back and look at the larger picture—to see those features that have been inherited by the groups of interest. Deuterostomes are a subset of the larger group Animalia and share common features in development. These primitive patterns of development are retained by the chordates.

Echinoderms

Echinoderms, known from the Cambrian to the present and represented by as many as 20 classes, include among their living representatives the starfish (sea, brittle, and feather stars), sea urchins, sea cucumbers, sea daisies, and crinoids. Uniting this diverse group is the presence of a unique calcium carbonate skeleton, and in all but the most basal members, the presence of a secondary radial symmetry (bilateral symmetry being primitive).

Hemichordates

The hypothesis for the relationships of hemichordates is debated, ranging from a sister group relationship with echinoderms + chordates, to the sister group of chordates, to *incertae sedis*. The phrase *incertae sedis* means "of uncertain status." It denotes that the taxonomic status of a group of organisms is widely disputed. Among such groups are acorn worms—marine organisms that live in mud under shallow waters and, although fragile when handled, have been known to reach 1.5 meters (5 ft) in length (fig. 3.2a). They have invertebrate and chordate features and have been classified at one time or another in each category. They are here grouped with the chordates (see fig. 4.3).

In 1870 Karl Gegenbaur established a new taxon, **Enteropneusta**, to accommodate acorn worms. Four years later, Ernst Haeckel established the phylum Chordata to include all organisms with notochords. After 10 more years, William Bateson (1884) added enteropneusts to Haeckel's phylum Chordata, believing them to have a notochord. He named the new group **Hemichordata**. His decision was based on the following considerations.

1. Acorn worms have a strand of nerve cells and fibers in a longitudinal groove of ectoderm on the dorsal surface of the collar and extending to the end of the body. In the collar, the groove sinks beneath the surface to become a **collar nerve cord** with a continuous or discontinuous lumen. This dorsal nerve strand and the tendency to have a lumen in the collar region was believed by Bateson to be an expression of the same phenomenon that induces the formation of a dorsal neural groove and tube in chordates (see fig. 1.11).

2. Acorn worms have slits that open to the exterior in the lateral walls of the foregut. Pharyngeal slits are basic features of chordates.

3. Acorn worms have a short diverticulum of the foregut, the **stomochord**, which extends forward into the proboscis (fig. 3.2b). Bateson hypothesized that the diverticulum may be homologous with the notochord of chordates. Evidence for this is lacking.

Bateson discounted certain observations that indicate a close kinship with echinoderms. Enteropneusts have bilaterally symmetrical larvae (**tornaria**) so similar to

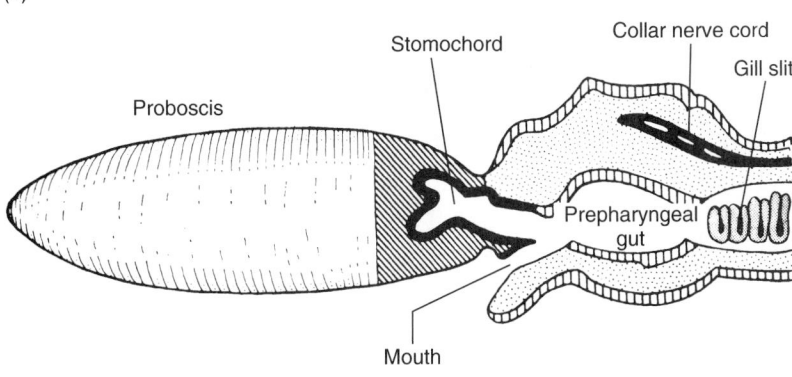

FIGURE 3.2

An acorn worm, *Saccoglossus*. (*a*) Entire worm. (*b*) Head, sagittal section (except proboscis).

some echinoderm larvae that nineteenth-century biologists frequently confused the two. This suggests a possible affinity of echinoderms and enteropneusts. But even if this is an erroneous inference, enteropneust larvae do not resemble ascidian or cephalochordate larvae and could not be mistaken for them. Second, similarities between enteropneusts and echinoderms in certain developmental processes, in muscle proteins, and in other traits seem to link enteropneusts closer to echinoderms than to chordates. Third, a detailed examination of the nervous systems of enteropneusts reveals a system quite unlike the hollow central nervous system of larval sea squirts and cephalochordates. In addition to the dorsal nerve strand, a second strand lies in the coelomic floor, as in invertebrates. A circumintestinal nerve ring at the caudal boundary of the collar unites dorsal and ventral strands. But nowhere, collar or trunk, is there a region that can be regarded as a *central* nervous system (that is, a center that has been shown to integrate stimuli and responses). The collar nerve cord is not a brain in either the invertebrate or chordate sense, having no neurohistologic specialization, and neither receiving nor giving off nerves (*bundles* of nerve fibers).

The similarities between enteropneusts and echinoderms cited above coupled with the highly improbable homology of stomochord and notochord, and the enigmatic nature of the enteropneust nervous system prompt a view that enteropneusts should be classified close to echinoderms, close to protochordates, but in a taxon independent of both. The taxon Hemichordata includes, in addition to enteropneusts, one or more classes that have little resemblance to one another or to enteropneusts. Suggesting the grouping of hemichordates with chordates is the presence of pharyngeal slits and a blood vascular system (see fig. 4.3).

Urochordates

Urochordata is a group of marine chordates in which the notochord is confined to the locomotor tail of the free-living larval stage (see fig. 3.1*a*). Of the three classes, only one, the ascidians, undergoes complete metamorphosis, resorbing the tail and the notochord before becoming sexually mature. Larvaceans retain the tail and notochord throughout life and reproduce in a larval-like state. Most thaliaceans have no larval stage, no notochord, and no tail. Urochordates are enclosed within a delicate, nonliving, and sometimes beautifully colored transparent tunic that accounts for the alternate name, tunicates. They are **filter feeders**, their food consisting of particulate organic matter that is suspended in the seawater and filtered out before the water passes over the gills.

Ascidians

Sea squirts are the best known urochordates (fig. 3.3). There are solitary and colonial populations. The larvae are tiny (0.5 mm to 11.0 mm), they have a fleeting exis-

tence (as little as a few minutes and not longer than a few days), and they do not feed (they subsist on nutrients stored in the tissues). The body consists of a trunk containing immature viscera and a muscular notochord-supported tail, the function of which is to provide locomotion during the few hours of its free-swimming existence. The notochord is a stiff rod composed of a small number of epithelial cells arranged in a single layer around a central core of matrix and surrounded by a filamentous sheath. Alongside of the notochord in solitary species are about 36 uninucleate striated muscle fibers. (Striated muscle fibers of craniates are multinucleate.) Other species have as many as 1,600. There are no tendons. The nervous system consists of a dorsal hollow nerve cord, several ganglia, and nerves. A **sensory vesicle** associated with the brain houses an **otolith** that stimulates nerve endings for statoreception. A light-sensitive **ocellus** (a pigment-protected receptor cell) is usually present in its wall. In some species, neither definitive blood cells nor a functional heart differentiates until metamorphosis.

Respiratory water enters the larval pharynx via the mouth, passes over gills lining the pharyngeal slits, and enters a chamber, the **atrium**, surrounding the pharynx. Oxygen-depleted water is ejected via an **atriopore**. At metamorphosis, three **adhesive papillae** with sticky secretions attach the larva to a permanent substrate.

At metamorphosis the notochord is resorbed and becomes a source of nutrients during metamorphosis. The nervous system is altered in location and structure, and a rearrangement of the viscera takes place (fig. 3.4). The

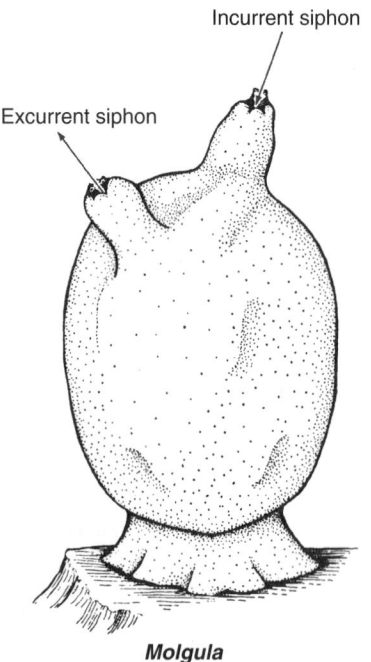

Molgula

FIGURE 3.3
Adult sea squirt.

larval mouth becomes an **incurrent siphon**, and the atriopore becomes an **excurrent siphon**. The forceful discharge of jets of water from the latter whenever the polyp is irritated inspired the name sea squirt. At other times, water flows steadily out of the siphon. Food particles are filtered out of the incoming water stream, becoming trapped in mucus in the **endostyle**, a ciliated glandular groove in the pharyngeal floor directed toward the entrance to the short esophagus. Cilia and esophageal papillae sweep the mucus and entrapped

food particles into the stomach, whereas respiratory water passes over the gills and into the atrium, a fluid-filled collecting chamber surrounding the pharynx. Atrial muscles discharge the deoxygenated water to the exterior via the excurrent siphon. The polyp has now become a filter feeder. Arising from each end of the now functioning heart is a blood vessel. Blood is pumped alternately forward for a few beats, then backward, the heart pausing before reversing the flow.

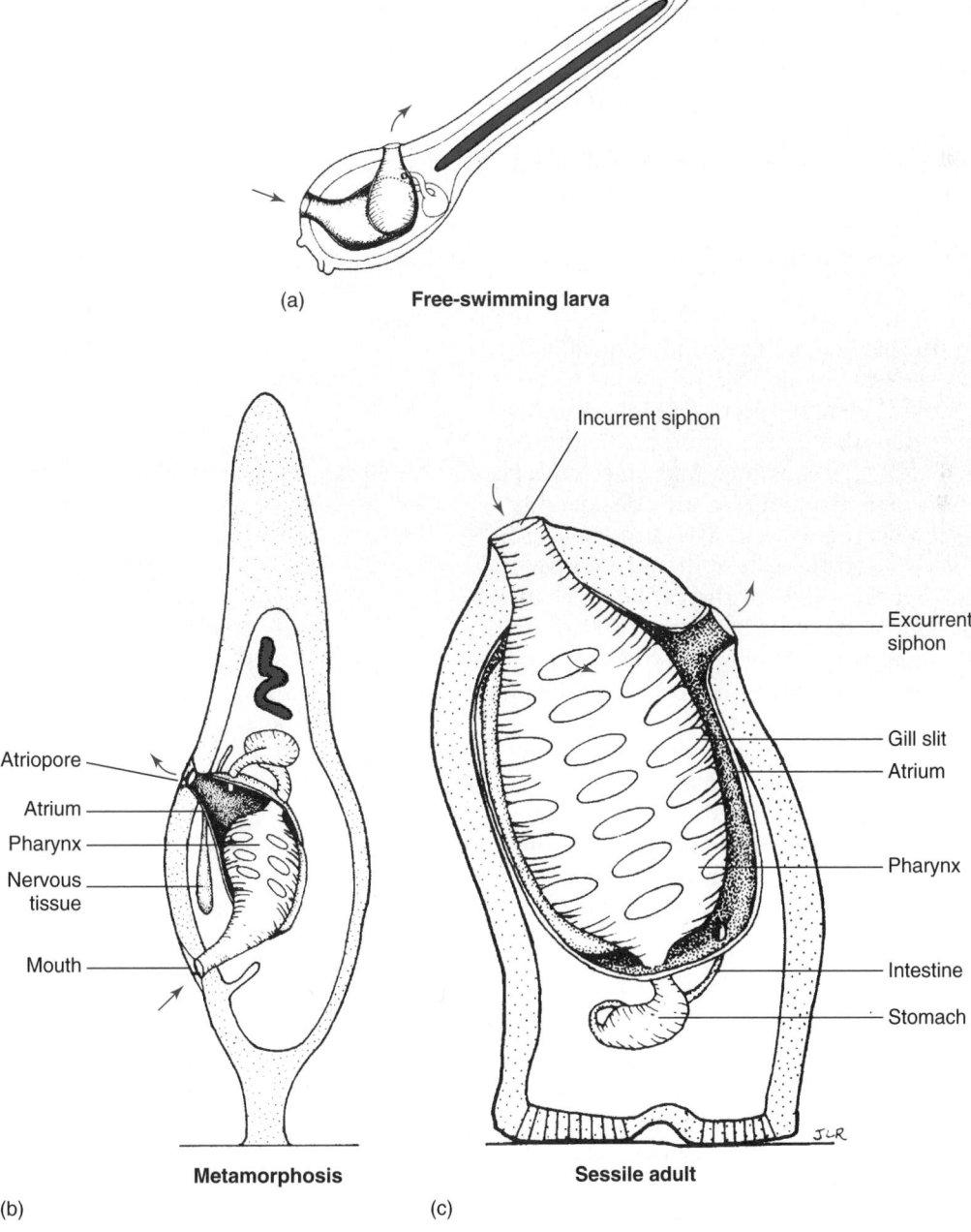

(a)   **Free-swimming larva**

Metamorphosis   **Sessile adult**

(b)   (c)

FIGURE 3.4

Metamorphosis in a sea squirt. In (*b*) and (*c*), the atrial wall has been removed to show pharyngeal slits opening into atrium. Arrows indicate direction of water flow. The notochord (*red*) disappears during metamorphosis.

It was not until the middle of the nineteenth century that these polyplike, sometimes colonial, marine animals were discovered to have a free-living larval stage with a notochord and a dorsal nervous system. From the viewpoint of craniate kinships, there is more to be learned from the larvae than from adult sea squirts.

Larvaceans

Larvaceans are found most often as plankton in surface waters (100 meters or less), but may live so deep in the sea that sunlight penetrates the area only dimly. They have a body and a long, flat locomotor tail supported by a notochord that together measure no more than 8 mm (1/4 in) in length (fig. 3.5). The organism is surrounded by a giant mass of fragile transparent mucus that is secreted by the larva. The mucus serves as a filter that excludes coarse particulate matter that is suspended in the seawater and admits only the finest organic particles to the immediate vicinity of the incurrent aperture. A pulsating membrane propels these into the branchial basket (pharynx) along with dissolved oxygen in the water. The tuniclike mass is abandoned at frequent intervals as the filter becomes clogged with debris.

Thaliaceans

Thaliaceans are free-living individuals and colonial individuals in alternate generations. The organisms resemble adult ascidians in many respects. The pharynx, or branchial basket, like that of sea squirts, constitutes most of the body; the digestive tract and other organs are miniscule in comparison. There are considerably fewer slits in the pharyngeal wall. The organisms are cylindrical in shape, and the incurrent and excurrent apertures are at opposite ends of the cylinder. Thaliaceans have no tail, locomotion being by a jet stream from the excurrent aperture. A colony may be linear, or it may consist of a floating water-filled sac the walls of which are composed of organisms that arise as a string of buds proliferated by a sexual individual. The incurrent aperture of a colony member projects into the sea; the excurrent aperture empties into the saccular chamber. Colonies move only with water currents. Only one order of thaliaceans has a free-living larval stage, and this is of very short duration. No notochord has been described.

Cephalochordates

Amphioxus means "sharp at both ends." Any member of the subphylum Cephalochordata may be called an amphioxus, or **lancelet** (little spear), but the generic name for the lancelet commonly studied in the laboratory is *Branchiostoma* (fig. 3.6). *Asymmetron* is the only other genus in the subphylum.

Lancelets are marine organisms found a short distance out from sandy beaches throughout most of the globe. They quickly burrow into sand with eel-like movements, make a U turn, and then emerge until only

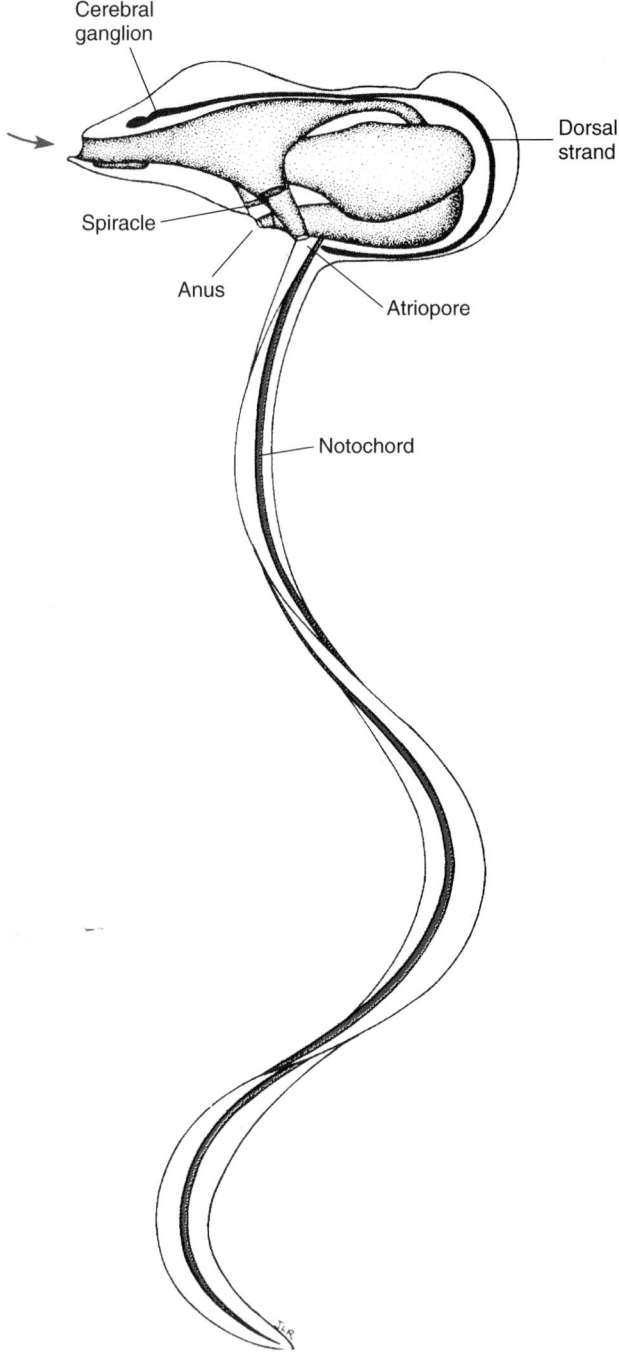

FIGURE 3.5

A larvacean. *Red* arrow indicates entrance to pharynx. The immense tunic has been removed.

the oral hood area is protruding for filter feeding. Adults vary from less than 2 cm to more than 8 cm in length, the largest being *Branchiostoma californiense*. Off the coast of southern China, lancelets are collected in quantity and sold as a table delicacy. A living amphioxus is semitransparent but becomes opaque when immersed in preserving fluids.

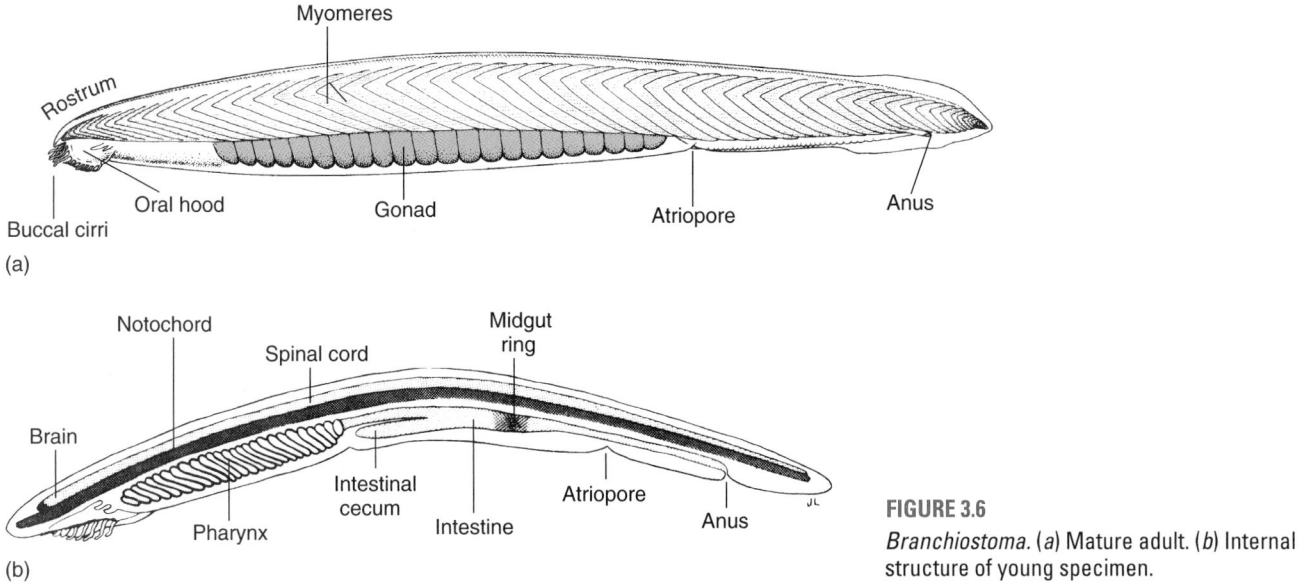

FIGURE 3.6
Branchiostoma. (*a*) Mature adult. (*b*) Internal structure of young specimen.

Locomotor Musculature and Skin

The body is practically all trunk. It exhibits noticeable metamerism because of the muscle segments, or **myomeres**, which lie just under the thin skin, extend the entire length of the organism, and constitute the bulk of the body wall. These, collectively, are the locomotor muscles of an amphioxus. Individual muscle fibers are uninucleate as in ascidians (again contrast this with craniate multinucleate fibers). Each myomere is separated from the next by a **myoseptum (myocomma)**, a connective tissue partition that is the origin and insertion of the longitudinal muscle bundles that constitute each myomere. Because myomeres are <-shaped, a cross section of the body intersects several myomeres (fig. 3.7). The integument consists of a single layer of epidermal cells and a thin dermis (see fig. 6.1), which facilitates the diffusion of oxygen (amphioxus utilizes cutaneous respiration). A layer of subcutaneous connective tissue separates the muscles from the skin.

Pharyngeal Slits

The **pharyngeal slits** of an amphioxus do not open directly to the exterior but into a fluid-filled cavity, the atrium, which surrounds the pharynx laterally and ventrally (fig. 3.7*c*). Pharyngeal water, relieved of food particles, streams through the slits into the atrium, which expands due to the action of the transverse musculature in the atrial floor. The same muscles expel the water to the exterior via a midventral atriopore (see fig. 3.6). The number of slits varies but exceeds 60 in adults. Respiration takes place through the skin. Unlike craniates, the pharynx does not serve a respiratory function. Oxygen-rich water entering the pharynx is rapidly depleted by diffusion to the point where there is a local oxygen debt. This debt is due to the demands of the ciliated cells lining the pharynx, the slow movement of water, and the

location of blood vessels within the collagenous pharyngeal skeleton minimizing the effectiveness of diffusion across the pharyngeal surface. The term **gill** implies a respiratory function and is best not used when referring to amphioxus.

Notochord

The **notochord** extends from the tip of the rostrum to the tip of the tail (see fig. 3.6*b*). In contrast to all other chordates, it consists of muscular discs arranged like a long roll of coins and separated by fluid-filled spaces. The muscle fibers in each disc are disposed transversely and have cytoplasmic extensions that pass through the tough notochord sheath and end on the surface of the spinal cord. There the extensions receive motor impulses from neurons within the cord. Contraction of the notochordal muscles increases the stiffness of the chord. Continuation of the notochord to the very tip of the rostrum—unlike in any other chordate—may be an adaptation that facilitates burrowing in the sand. The only other skeletal structures are the fibrous rods that support the pharyngeal bars, buccal cirri, and the middorsal finlike ridge.

Nervous System and Sense Organs

An amphioxus, like a craniate, has a central nervous system consisting of a hollow brain and a dorsal nerve cord containing a central canal, both lined by a nonnervous supportive membrane, the **ependyma**, and, as in craniates, the caudal end of the cord consists of ependyma alone. A single connective tissue membrane, the **leptomeninx** (see the glossary) surrounds the brain and cord.

Only two brain subdivisions can be identified in contrast to three in craniates (see fig. 16.13*a*). In an amphioxus, the notochord extends anterior to the brain; in

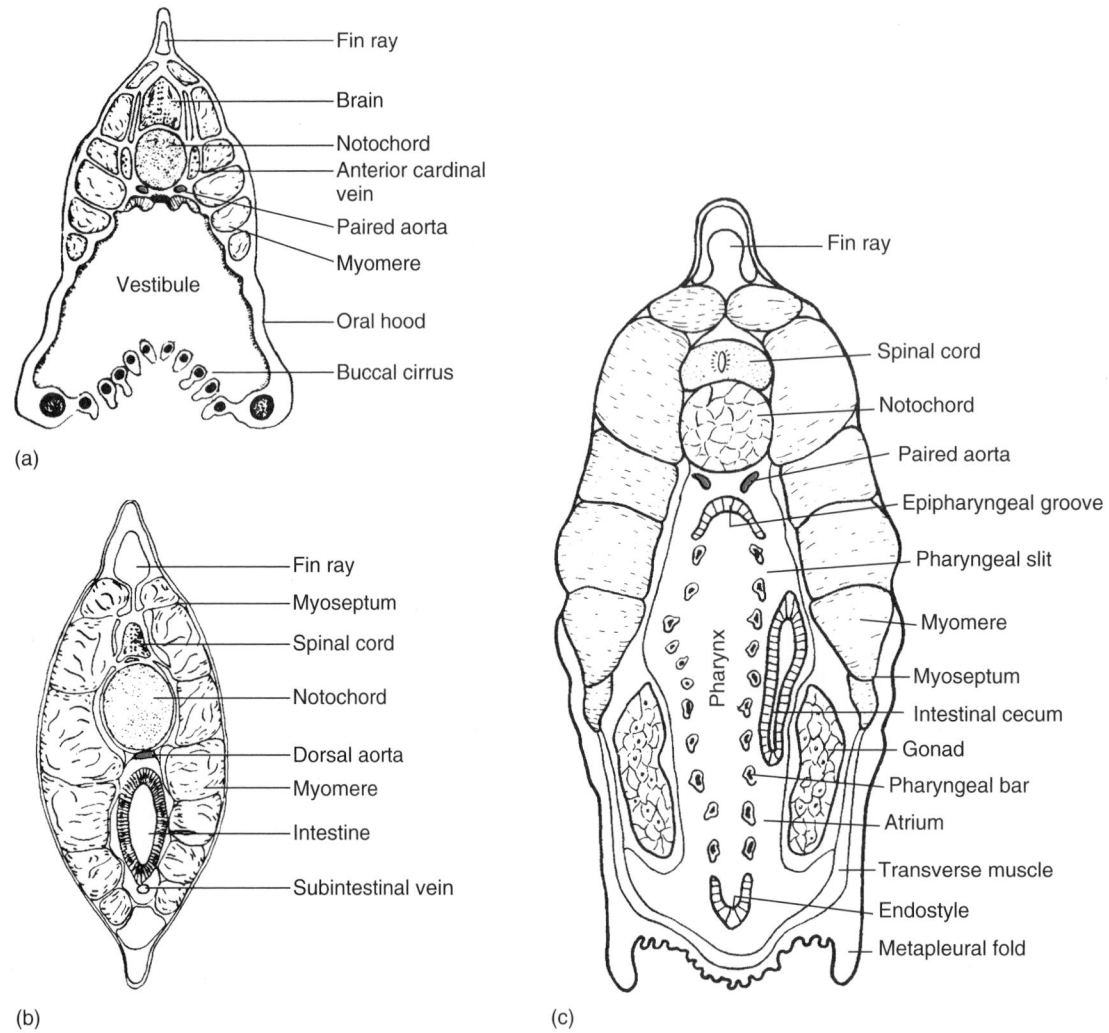

FIGURE 3.7
Cross sections of an amphioxus. (*a*) Anterior to mouth. (*b*) Posterior to atriopore. (*c*) Level of pharynx. The significance of the metapleural folds is unknown.

craniates, it stops at the midbrain. Does this indicate the absence of a forebrain? Attempts at homologizing the parts of the brain with those of craniates have not been entirely successful at the level of gross morphology; however, mapping the expression of homeobox genes is proving to be a useful tool in recognizing homologous features in the head of amphioxus and craniates. Whether or not the nerves that supply the pharynx should be considered cranial nerves complicates the problem. If the branchial nerves are omitted, there are seven pairs of **cranial nerves.** If the branchial and oral nerves are included, there are 39 or more pairs. Craniates have 10 to 18 cranial nerves (or more if functional components are considered as discrete nerves). The absence of semicircular canals, eyes, lateral-line system, and foramen magnum (there being no skull) deprives us of landmarks that would be helpful. Because of these difficulties, it is not clear where the brain ends and the cord begins.

Spinal nerves in cephalochordates, unlike those of craniates, consist solely of dorsal roots. These contain sensory fibers from the integument and sensory fibers, from and motor fibers to the visceral organs. Similar to the innervation of pharyngeal (branchiomeric) muscles in craniates, the dorsal root contains fibers that innervate the transverse (pterygial) muscle in the atrial floor. Instead of ventral roots conducting spinal nerve fibers to the muscles of the myomeres, as in craniates, tubes in amphioxus conduct cytoplasmic extensions of the muscle fibers to the spinal cord. There they receive their innervation from motor neurons whose nerve fibers remain within the cord! Dorsal roots and ventral tubes are metameric but not directly opposite one another because left and right myomeres are not exactly opposite one another.

The relatively small size of the brain is correlated with the paucity of special **sense organs.** There are no

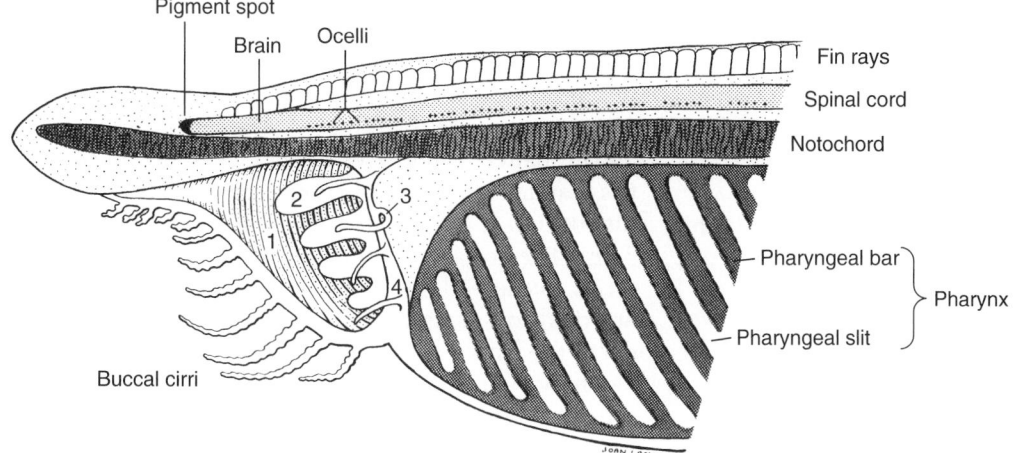

FIGURE 3.8
Cephalic end of an amphioxus shown in sagittal section. **1,** vestibule bounded by oral hood; **2,** part of the wheel organ projecting into vestibule; **3,** velar tentacle; **4,** velum.

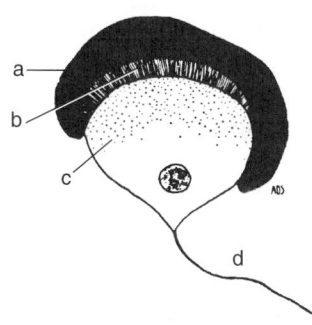

FIGURE 3.9
Ocellus (light receptor) from spinal cord of an amphioxus.
a, melanocyte; **b,** apical border of receptor cell; **c,** receptor cell; **d,** process for conduction of impulse.

retinas, semicircular canals, or lateral-line organs. However, chemoreceptors are abundant on the buccal cirri and velar tentacles, where they monitor the incurrent water stream. Chemoreceptors are also scattered on other surfaces of the body, the tail being more sensitive than the trunk. Tactile receptors, which elicit withdrawal, are present in the skin of the entire body.

The most prominent sense organs are the light-sensitive, pigmented ocelli embedded within the ventrolateral walls of the spinal cord throughout its length (figs. 3.8 and 3.9). Each ocellus consists of a receptor cell and a caplike melanocyte. The melanocyte lies between the receptor cell and the incoming light rays·and is packed with large melanin granules. A conducting process extends from the base of the receptor cell to the central nervous system.

Food Processing

Cephalochordates, like sea squirts, are filter feeders. A collecting chamber for seawater, the **vestibule,** is bounded laterally by an **oral hood** and caudally by a perpendicular membranous **velum** perforated by the mouth, which opens into the pharynx (see fig. 3.8). The vestibule is broadly open to the sea ventrally. Cilia on the pharyngeal surface of the pharyngeal bars create a steady flow of water through the mouth and into the pharynx. A set of stubby projections in the vestibule, the **wheel organ,** is covered with sticky mucus that retrieves some of the heavier food particles that miss the mouth, and it directs these through the mouth and into the pharynx along with the water stream. **Buccal cirri** partially strain the water as it enters the vestibule, and chemoreceptors on their surface monitor it chemically.

Once food is in the pharynx, it is processed as follows: In the pharyngeal floor, there is an endostyle or hypobranchial groove, (see fig. 3.7c), and in the roof there is an **epipharyngeal groove.** On the pharyngeal bars, ciliated **peripharyngeal bands** connect the two grooves. The cells of these bands and grooves secrete mucus. Food particles trapped in the mucus are incorporated into a stringy food cord that is propelled by cilia dorsally into the epibranchial groove and then caudad into the midgut behind the pharynx. There it is temporarily arrested by the **midgut ring** (see fig. 3.6b) and mixed with digestive enzymes. Some of the digesting foodstuffs then pass beyond the ring into the hindgut, and some are propelled forward by cilia into the **intestinal cecum.** This evagination arises in the same manner as the craniate liver, but the two are not homologous and they also differ in function. The cecum secretes enzymes for digestion within the lumen of the cecum and intestine, and its lining cells phagocytose the smallest food particles and digest them by *intracellular* digestion. The intestine terminates at the anus.

Coelom

The coelom is not a prominent cavity in adults. It becomes laterally compressed as additional pharyngeal bars form in the elongating larvae. Consequently, only traces remain in a mature amphioxus, and they are difficult to find in stained cross sections. Small pockets of

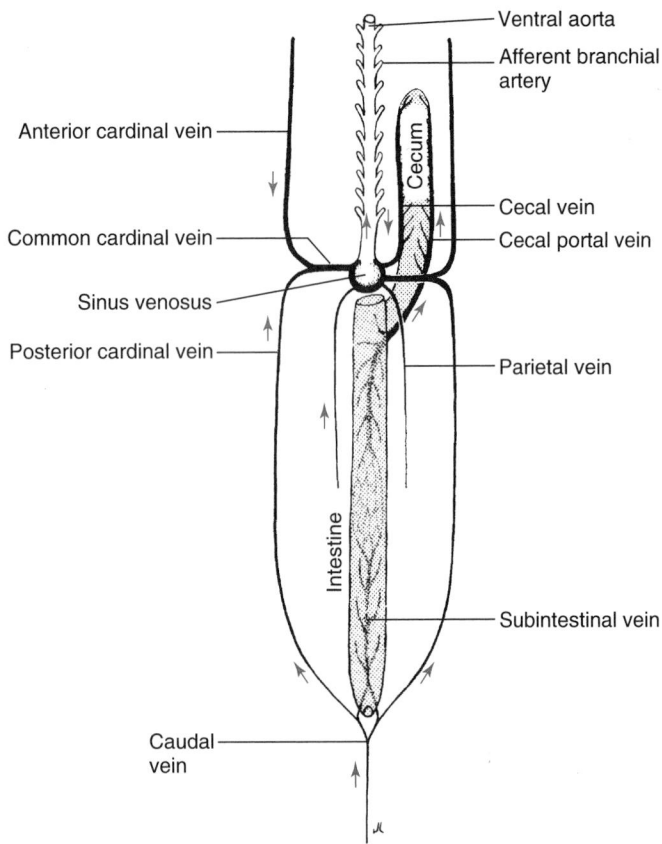

Ventral aorta
Afferent branchial artery
Cecum
Anterior cardinal vein
Cecal vein
Common cardinal vein
Cecal portal vein
Sinus venosus
Posterior cardinal vein
Parietal vein
Intestine
Subintestinal vein
Caudal vein

FIGURE 3.10
Basic venous channels and ventral aorta of an amphioxus, dorsal view. The cecum has been rotated 90 degrees around the long axis. The cecal portal vein in life is ventral to the cecum, and the cecal vein is dorsal.

coelom persist lateral to the epibranchial groove, in the pharyngeal bars, and just internal to the body wall musculature. However, the early larval coelom is prominent and metameric because it forms from metameric outpocketings of the dorsolateral roof of the embryonic archenteron (see fig. 5.3).

Circulatory System

Unlike a craniate, an amphioxus has no heart, and the blood is a colorless serum that lacks blood cells, platelets, and other formed bodies. A bulbous venous sinus (**sinus venosus**, fig. 3.10), analogous to the first chamber of the craniate heart, is the termination of the venous system and the beginning of arteries. Deoxygenated blood from all parts of the body empties into this sinus then continues forward beneath the pharynx in an unpaired vessel, the **ventral aorta**. The ventral aorta gives off arteries that ascend in the pharyngeal bars and then become tributaries of the paired **dorsal aortae**, located above the pharyngeal bars. This is basically the pattern we find in gnathostome fishes (see fig. 14.16). The details differ. The cecal vein, ventral

aorta, and bulbous swellings at the base of the afferent branchial arteries participate in pumping blood through the circulatory system. The sinus venosus is not a pump. Muscular contractions during body movements may contribute to the flow.

The paired aortae give off branches to the rostrum, pass caudad above the pharynx (see fig. 3.7*c*), and unite just behind it to form an unpaired dorsal aorta (see fig. 3.7*b*). The aortae distribute blood by paired **parietal arteries** to the body wall and by median vessels (**visceral arteries**) to the viscera. The dorsal aorta continues caudal to the anus as the **caudal artery**.

The venous channels are similar to those of embryonic craniates (fig. 3.10). From the capillaries of the tail, a **caudal vein** courses forward and divides into right and left posterior cardinal veins. These pass forward in the lateral body wall to a point just behind the pharynx. Here the posterior cardinal veins meet the anterior cardinal veins from the rostrum and pharyngeal wall. The blood then enters a common cardinal vein leading to the sinus venosus. Parietal veins drain the body wall of the trunk. These also terminate in the sinus venosus.

Drainage from the visceral organs is via a median subintestinal vein that receives blood from the caudal vein. The subintestinal vein passes cephalad along the ventral surface of the intestine. There it breaks up into smaller channels, receives tributaries from the intestine, and reconvenes to continue forward as a **portal vein** ending in the capillaries of the cecum. From the cecum, the contractile cecal vein pumps blood to the sinus venosus.

Elimination of Metabolic Wastes

Amphioxus has no compact organ known as a kidney. Metabolic wastes are collected by cells known as **cyrtopodocytes**, intermediate in design between the solenocytes (protonephridia) of invertebrates and the podocytes of craniates. The cyrtopodocytes are clustered below the paired aortae lateral to the epibranchial groove (fig. 3.11). Each cyrtopodocyte has (1) footlike projections (**pedicels**) that make contact with an adjacent **glomerulus** (an arterial rete), and (2) a long strand of microvilli that extends across a remnant of the coelom and terminates in the lumen of a **nephridial tubule**. The tubules are tiny ductules that collect fluids from a cluster of cyrtopodocytes and empty into the atrium via tiny pores. A single long flagellum from the cyrtopodocyte lies at the core of the microvillar strand and, like microvilli, projects into the nephridial tubule. Collectively, flagella presumably propel tubule fluids to the atrium. The atrium opens to the exterior via the atriopore.

Solenocytes are common among invertebrates, whereas podocytes are found in craniates. The details of how the intermediate cyrtopodocytes work in an am-

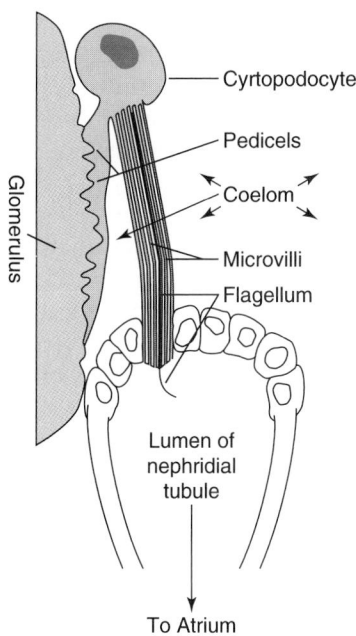

FIGURE 3.11
One of a cluster of amphioxus cyrtopodocytes with pedicels in intimate contact with the thin wall of a glomerulus. Cyrtopodocytes lie lateral to the epibranchial groove and ventral to the paired aortae.

phioxus are unknown. Contact of the pedicels with a glomerulus suggests that glomeruli are the source of one or more components of what is called urine in craniates. The fact that microvillar strands project across a coelomic space—especially one that has been spared when many parts of the larval coelom are greatly reduced during metamorphosis—may or may not be significant; coelomic fluid is the collection site for similar wastes in many multicellular invertebrates and in larval hagfishes.

Gonads

Mature gonads are visible through the body wall (see fig. 3.6a) and bulge into the water-filled atrium, into which sperm or eggs are shed. The gametes are then flushed into the sea via the atriopore along with respiratory water and metabolic wastes. The amphioxus is **dioecious**; that is, ovaries and testes do not develop in the same individual.

Amphioxus and the Craniates Contrasted

Although an amphioxus resembles a craniate in many respects, differences are evident. An amphioxus has almost no cephalization and no paired sense organs; it has a notochord but no vertebral column (even hagfish may have some vertebral features in the tail); it has pharyngeal slits but in large numbers, emptying into an atrium; it has a dorsal hollow central nervous system, but the brain lacks the major craniate subdivisions; it has a seg-

mented musculature, but the segments extend to the anterior tip of the head; it has a two-layered skin, but the outer layer is only one cell thick; it has arterial and venous channels similar to the basic channels of craniates but no muscular heart; and it is coelomate, but the coelom is greatly restricted in adults. A hypothetical relationship of amphioxus to other protochordates and to craniates is diagrammed in figure 3.14. In evaluating this figure, remember that it is not the differences that are important but the shared derived features that can be used to hypothesize relationships.

THE ORIGIN OF CRANIATES

The oldest craniates include recently discovered vertebrate fossils from the Lower Cambrian of China (Shu et al., 1999). The vast majority of early vertebrate fossils, which have provided the greatest amount of anatomical detail, consist of those taxa possessing bony hard parts. These are the ostracoderms (figs. 3.12 and 4.5). These strange fishes, 2 cm to 2 m long and of diverse appearances, had no jaws, most were without paired fins, and many of the earliest ones were assumed to be filter feeders. Broad bony plates in the skin formed a protective shield over the head and much of the trunk, inspiring the nickname "armored fishes." Where there were no bony shields, smaller bony scales fitted tightly together over the body like tiles. Fossils that can be clearly identified as ostracoderms date back to the beginning of the Ordovician, after which ostracoderms became an abundant and very diversified group. One order is described in chapter 4.

The broad outlines of vertebrate evolution after the Ordovician have been remarkably well ascertained (see fig. 4.2). The jawless ostracoderms were succeeded in the seas by jawed fishes, and amphibians eventually became established on land. The perplexing problem is, who were the ancestors of ostracoderms? Will the Lower Cambrian Chinese fossils fill this role or do we need to look deeper?

Because ostracoderms were chordates, we can look for clues in the protochordates that are with us today and in the fossil record that immediately preceded ostracoderms. What clues to vertebrate ancestry might we find in living protochordates? Earlier in this discussion, an amphioxus was contrasted with vertebrates with respect to differences in their morphologic traits. Let us now look at these cephalochordates, emphasizing similarities rather than differences.

Cephalochordates have a notochord, pharyngeal slits, a dorsal hollow central nervous system with brain and cord, a metameric body wall musculature, a two-layered skin, and arterial and venous channels similar to those of fishes and to the early embryonic vessels of tetrapods. Cephalochordates are deuterostomous and coelomate. Finally, cephalochordates are filter feeders, as were

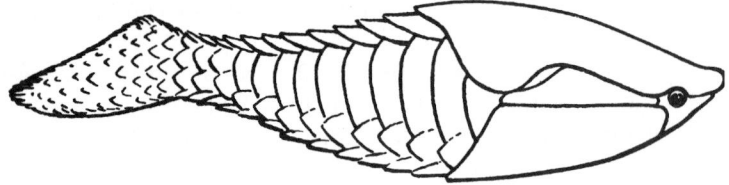

FIGURE 3.12
One of many varieties of heterostracan ostracoderms.

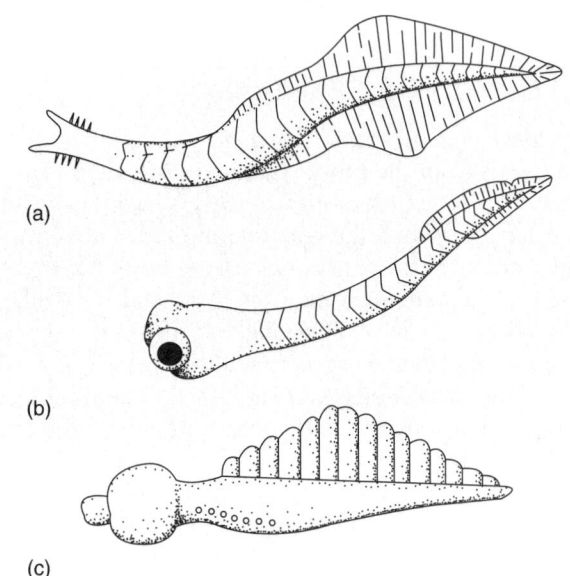

(a)

(b)

(c)

FIGURE 3.13
Reconstructions of three fossil protochordates. (*a*) *Pikaia,* a Cambrian cephalochordate. (*b*) A conodont, a mid-Cambrian to Triassic animal of uncertain affinities. Their phosphatic "tooth" plates have been interpreted as bone, which may indicate they are protovertebrates. (*c*) *Yunnanozoon,* a Cambrian hemichordate. Note the presence of a proboscis, collar, and gill pores similar to that found in acorn worms. Not to scale.

many early ostracoderms. These similarities bespeak close genetic ties between the ancestors of cephalochordates and those of vertebrates. But did cephalochordates precede ostracoderms in the earth's waters?

The answer is, "Yes and no!" Early twentieth-century paleontologist C. D. Walcott unearthed, in a limestone quarry 8,000 feet up in the Canadian Rockies, several dozen specimens of soft-bodied, laterally compressed, ribbon-shaped fossils about 5 cm (2 in) long that resembled an amphioxus. The stratum from which they were recovered was the Middle Cambrian. They had zigzag myomeres and a notochord that extended the length of the body. This recovery places this specific chordate, *Pikaia gracilens* (fig. 3.13), in the earth's waters considerably earlier than the ostracoderms. However, the Lower Cambrian Chinese fossils, if correctly identified as vertebrates, predate the Middle Cambrian rocks from which *Pikaia* was recovered by millions of years. This

does not preclude our hypothesis of relationships but simply moves the timing of origin to the Lower Cambiran or possibly to the Precambrian. Gould (1989) has related the broad lessons in natural history that are implicit in Walcott's discoveries.

Although we know, or believe we know on the basis of current data, that protochordates preceded craniates in the course of natural history, we have to speculate concerning the lineages that might have led from prechordate invertebrates to protochordates on the one hand, and to craniates on the other. Surely, cephalochordates *as we know them today* were not genetic ancestors of the first craniates. We must consider the observation that echinoderms, like vertebrates (hagfish lack bone), have mineralized tissue in their mesoderm (not in their ectoderm, as in molluscs, for example); that echinoderms, like cephalochordates, form their mesoderm and coelom as outpocketings of their archenteron; that echinoderms, enteropneusts, protochordates, cephalochordates, and craniates are all deuterostomes, a trait found in only one other invertebrate taxon (Chaetognatha); and that all have larvae in their history. On the basis of these and other morphologic and ontogenetic observations, it is rational to hypothesize (1) that genetic ties unite one or more of these in unknown lineages, (2) that the characters shared by protochordates and craniates are derived from a common ancestor, and (3) that the bony dermal armor of ostracoderms must have developed late in the Cambrian in one or more soft-bodied invertebrates or unossified early craniates that already had a notochord, dorsal hollow nervous system, and pharyngeal slits. Figure 3.14 depicts one postulated phylogeny. A cautionary note should be interjected: Morphologic similarities may, in one or several instances, be attributable to evolutionary convergence (homoplastic traits).

THE AMMOCOETE: A VERTEBRATE LARVA

The ammocoete larvae of lampreys (eel-like jawless fishes) provide compelling support for the concept that vertebrates share a common ancestor with protochordates. These larvae metamorphose very slowly, spending from 2 to 6 years before metamorphosis is completed and the gonads mature. They were at one time classified as protochordates with the generic name *Ammocoetes*. Only later was it discovered that they are vertebrate larvae.

Ammocoetes have a notochord that commences at the midbrain and continues to the end of the body (see fig. 3.1*c*). The dorsal hollow nervous system develops in the same manner as that of an amphioxus, but the ammocoete brain has three vesicles instead of two. Seven pairs of gill slits open to the exterior. The body wall musculature is disposed as overlapping myomeres that provide locomotion.

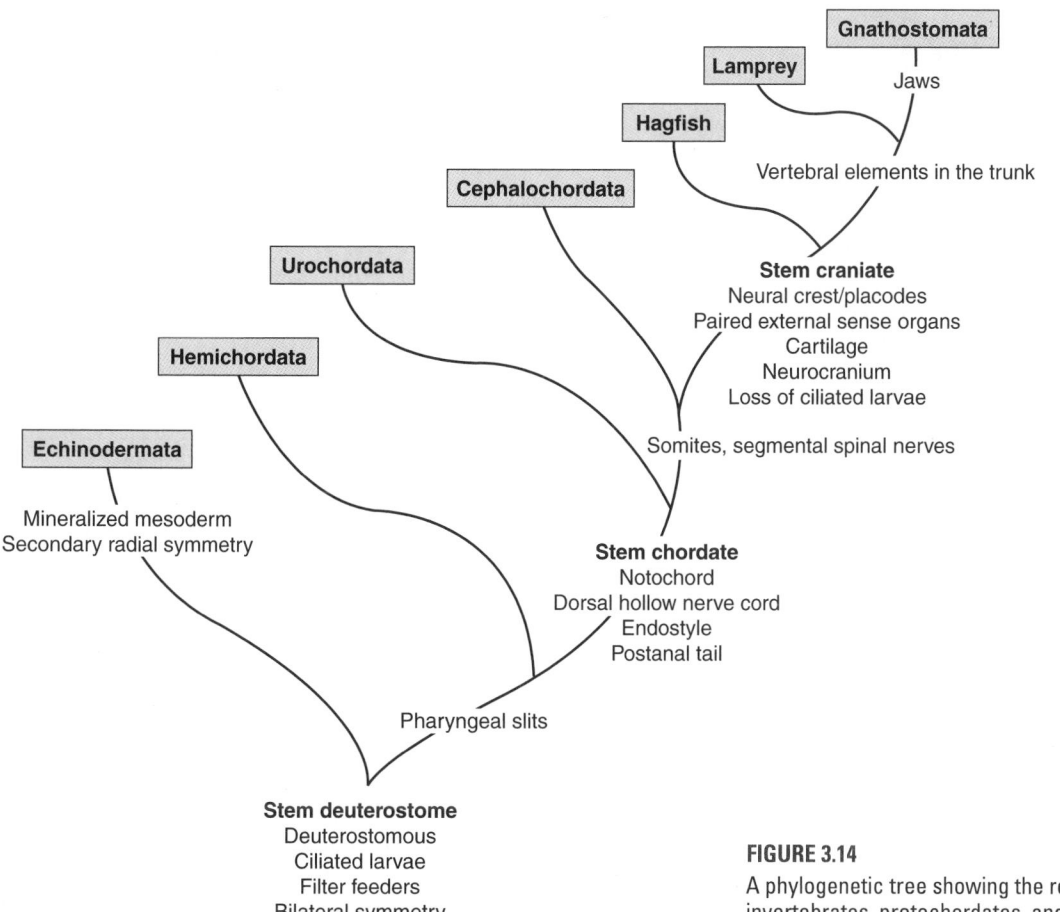

FIGURE 3.14

A phylogenetic tree showing the relationships among invertebrates, protochordates, and craniates.

Ammocoete larvae, like protochordates, are filter feeders. Food particles in the respiratory water stream, chiefly diatoms, are ensnared in mucus produced by the endostyle, which is homologous with the hypobranchial groove (endostyle) of an amphioxus. In ammocoetes, however, the groove eventually sinks into the floor of the pharynx to become a **subpharyngeal gland** (see fig. 18.15). Beyond the ammocoete pharynx is a *solid* liver diverticulum. An amphioxus has a *saccular* diverticulum at the same location, but the two do not have the same functions.

The circulatory system of an ammocoete larva is remarkably similar to that of an amphioxus, except that an amphioxus has no muscular heart. The ammocoete heart consists of a sinus venosus, atrium, and ventricle. Blood in both organisms is pumped forward in the ventral aorta, dorsad into the pharyngeal bars, then to the dorsal aorta. Venous blood returns from the head via anterior cardinal veins and from most of the trunk and tail via posterior cardinals. All converge on paired common cardinals that empty into the sinus venosus (figs. 3.10 and 3.15). Blood from the intestine returns via a ventral intestinal (subintestinal, portal) vein that ends

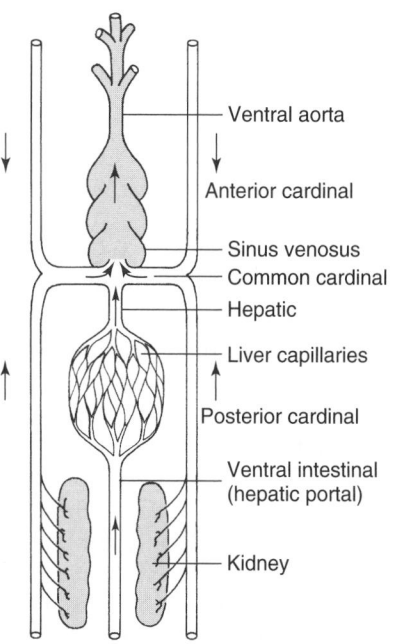

FIGURE 3.15

Venous channels of an ammocoete. See also figure 3.10.

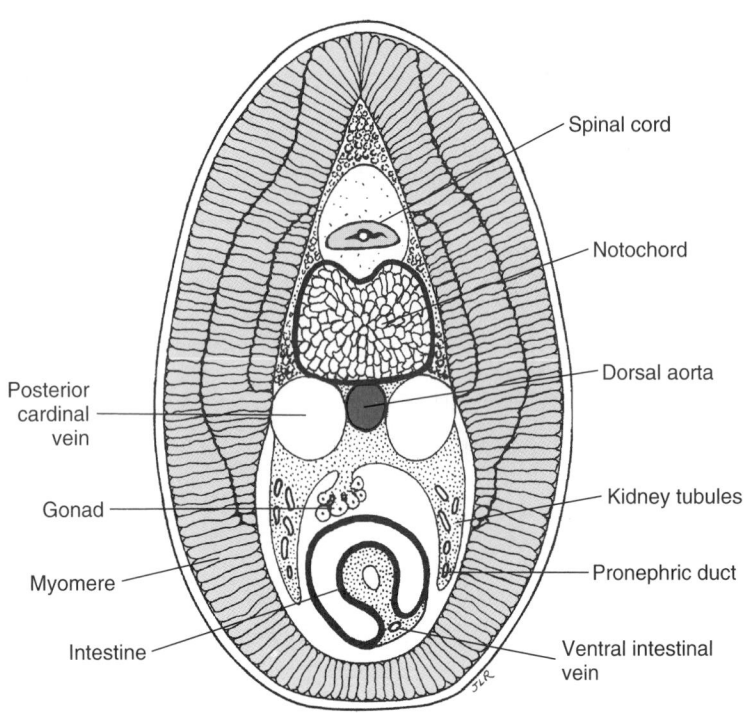

Spinal cord

Notochord

Dorsal aorta

Posterior
cardinal
vein

Kidney tubules

Gonad

Pronephric duct

Myomere

Intestine

Ventral intestinal
vein

FIGURE 3.16

Relatively old ammocoete in cross section just behind the liver. Several
myomeres are cut in a single section because they curve caudad in passing
inward from the skin.

in the capillaries of the liver of an ammocoete and in
the cecal diverticulum of an amphioxus. This blood
then continues via a hepatic or cecal vein to the sinus
venosus.

Ammocoete larvae exhibit craniate features that are
not present in an amphioxus, among which are special
sense organs and a craniate means of eliminating excess
tissue fluids and metabolic wastes. Among early developing sense organs are **otic vesicles**, which are precursors of a primitive membranous labyrinth (inner ear,
but not for hearing), and an **olfactory sac** for chemoreception (associated with feeding). Two middorsal light
receptors (**pineal** and **parapineal organs**, also known as
median eyes, see fig. 17.20) lie in tandem beneath the
skin on top of the head. They monitor light rays, but
there are no retinas, and the lenses cast no image. The
function of these light receptors is somewhat analogous
to that of protochordate ocelli, but they are complex organs, not mere cells. Paired lateral eyes eventually form,
but they remain rudimentary. (Eventually, they lie under
the thick skin of adults, and are useless.) Whether or
not the paired eyes are vestiges (chapter 2) depends on
the phylogenetic history of lampreys. Special neuromast
organs, discussed in chapter 17, lie in the skin of the
head and open onto the surface via pores.

Ammocoete larvae have a different mechanism for
elimination of fluid wastes than do protochordates. The
larvae have a primitive craniate kidney that, in an early

larva, consists of three to six tubules with
funnel-like **nephrostomes** that collect body
fluids for processing and excretion (figs.
3.16 and 15.1). These fluids are *expressed
into the coelom* from glomeruli. The tubules,
which increase in number as the larvae
elongate, lead to longitudinal ducts that
empty to the exterior near the anus. Between the kidneys is an elongated rudimentary gonad that is paired in early larvae and
is unpaired later. The epidermis is multilayered as in other craniates (see fig. 6.5).

HETEROCHRONY AND THE
RELATIONSHIP BETWEEN
AMPHIOXUS AND VERTEBRATES

When amphioxus and an ammocoete are
compared, it is important to remember that
we are comparing a larval vertebrate with an
adult cephalochordate (reexamine fig. 3.1).
Although amphioxus retains a number of
plesiomorphic (primitive) features, as a living species it is far removed from its common
ancestor with vertebrates even though the
discovery of *Pikaia gracilens* may suggest that
the differences are not that great. What distinguishes amphioxus from an ammocoete?
Primarily, amphioxus possesses an atrium and an increased number of pharyngeal slits but lacks the numerous craniate and vertebrate synapomorphies. Additionally,
an ammocoete is a premetamorphic larva. What evolutionary processes might account for the protochordate-
vertebrate transition? In the development of amphioxus,
the atrium forms from an outpocketing of the ectoderm to
enclose the developing pharynx. The pharyngeal slits
form in three stages: (1) the formation of primary slits,
(2) division of the primary slits doubling their number,
and (3) terminal (posterior) addition of slits with
growth resulting in 100 to 200 pharyngeal slit pairs depending on the species. Deletion of atrial development
and truncation of pharyngeal slit development would
result in a craniatelike descendant with a reduced number of slits directly exposed to the external environment. If craniate and vertebrate derived features are
added, then it is easy to envision the transition to an
ammocoetelike vertebrate ancestor. Finally, how do we
reconcile an ammocoetelike vertebrate ancestor with a
larval ammocoete (fig. 3.17)? Lampreys are not "primitive" organisms but represent highly specialized parasites, so successful that introduction of lampreys into
the Great Lakes destroyed the commercial fishing industry. Sport fishing has made a comeback only through
continued eradication programs for lampreys. When
comparing lampreys with their fossil relatives, it is clear
that their parasitic lifestyle is a derived character, and

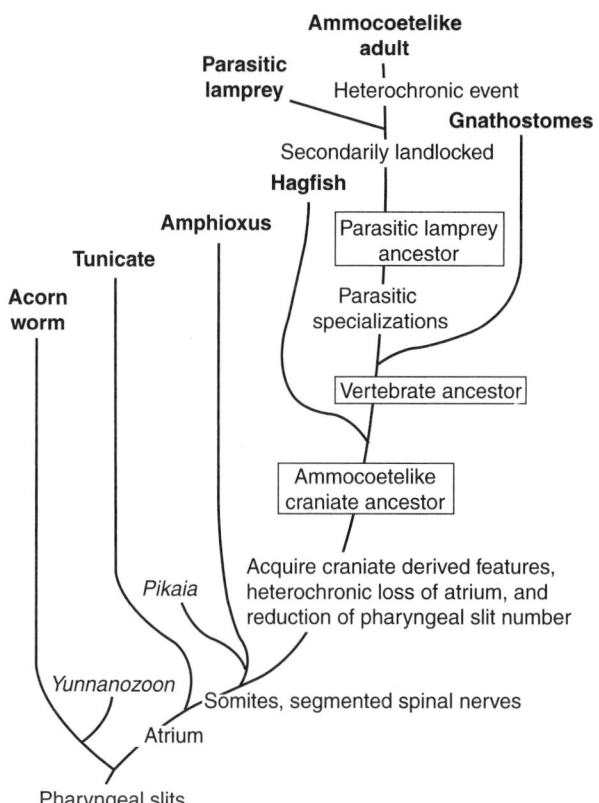

FIGURE 3.17

A hypothetical scenario for the evolutionary transition from an amphioxuslike ancestor to modern lampreys. Extant organisms in bold. *Pikaia* (a Cambrian cephalochordate), *Yunnanozoon* (a possible Cambrian hemichordate), synapomorphies, and hypothetical ancestors are shown. Heterochronic events account for the loss of the amphioxus atrium, reduced pharyngeal slit number, and sexual maturity in an ammocoetelike lamprey.

the ammocoete lifestyle represents a primitive behavior. In areas of the eastern United States, there are isolated populations of landlocked lampreys that do not metamorphose into a parasitic adult. These ammocoetes achieve sexual maturity in the larval state and may represent the formation of a descendant species from their marine parasitic ancestors.

Summary

1. Protochordates are bilaterally symmetrical, filter-feeding marine chordates with a notochord, pharyngeal slits, a dorsal hollow central nervous system, and no vertebral column. There are two subphyla, Urochordata and Cephalochordata.
2. Urochordates include (1) sea squirts (ascidians), in which the notochord is confined to the tail of larvae, (2) larvaceans, which do not become sessile possess a notochord confined to the tail, and (3) thaliaceans. The latter have no tail and no notochord.
3. Cephalochordates are the lancelets (amphioxuses). The notochord extends the length of the body to the tip of the postanal tail.
4. Hemichordates are invertebrates with pharyngeal slits, a longitudinal dorsal nerve strand that does not constitute a central nervous system, a ventral nerve strand characteristic of major invertebrate phyla, and no notochord. Best known are the acorn worms (Enteropneusta), which are sometimes included in the phylum Chordata.
5. The basic chordate features common to protochordates and craniates and the similarity of the details of the embryonic development of these features bespeak close genetic ties between the two groups. It is rational to postulate that craniates are derived from, or share common ancestors with, protochordates.
6. Ammocoetes are the larvae of lampreys. They have many features of an amphioxuslike protochordate.
7. One hypothesis derives ammocoetelike craniate and vertebrate ancestors through a number of heterochronic processes that resulted in the loss of primitive characters (atrium, large numbers of pharyngeal slits). Contradicting this hypothesis, current geologic data place protochordates in the Middle Cambrian and the earliest known vertebrates in the Lower Cambrian.

Critical Thinking Questions

1. What are the living protochordates? Why is Hemichordata excluded from chordates?
2. What are the "big four" or the synapomorphies for Chordata?
3. What structures found in amphioxus are homologous to human features? If they have been modified, then how?
4. If cephalochordates are the sister group to craniates, based on fossil evidence what is your estimate for the time of origin for craniates (at least minimal or youngest estimate)? How would you explain a lack of fossils if they were not found in the time frame of your estimate?
5. Why use a larval lamprey as a model for the hypothetical craniate ancestor? Does this imply that the parasitic adult lamprey is a good model for the lifestyle of early craniates? If not, how has the adult been modified from its common ancestor with other craniates?
6. Can you develop a heterochronic hypothesis for the evolution of craniates from a sessile tunicate? What evidence might support your hypothesis?

Selected Readings

Barrington, E. J. W., and Jefferies, R. P. S., editors: Protochordates, *Symposium of the Zoological Society of London* **36**, 1975.

Cloney, R. A.: Ascidian larvae and the events of metamorphosis, *American Zoologist* **22**:817, 1982.

Flood, P. R.: Fine structure of the notochord of amphioxus, *Symposium of the Zoological Society of London* **36**:81, 1975.

Gans, C., Kemp, N., and Poss, S., editors: The lancelets (Cephalochordata): A new look at some old beasts (IVth International Congress of Vertebrate Morphology), *Israel Journal of Zoology* **42**, supplement, 1996.

Gee, H: Before the backbone, views on the origin of the vertebrates. London, 1996, Chapman & Hall.

Gould, S. J.: Wonderful life. New York, 1989, Norton.

Harrison, F. W., and Ruppert E. E., editors: Microscopic anatomy of invertebrates, vol. 15, hemichordata, chaetognatha, and the invertebrate chordates. New York, 1997, Wiley-Liss Inc.

Jeffries, R. P. S.: The ancestry of the vertebrates. London, 1986, British Museum (Natural History).

Mallatt, J.: The suspension feeding mechanisms of the larval lamprey, *Petromyzon marinus, Journal of Zoology* (London) **194**:103, 1981.

Northcutt, R. G., and Gans, C: The genesis of neural crest and epidermal placodes: A reinterpretation of vertebrate origins, *Quarterly Review of Biology* **58**:1–28, 1983.

Ruppert, E. E: Evolutionary origin of the vertebrate nephron, *American Zoologist* **34**: 542–53, 1994.

Shu, D–G., Luo, H–L., Morris, S. C., Zhang, X–L., Hu, S–X., Chen, L., Han, J., Zhu, M., Li, Y., and Chen, L–Z: Lower Cambrian vertebrates from south China, *Nature* **402**: 42–46, 1999.

Willmer, P: Invertebrate relationships patterns in animal evolution. Cambridge, 1994, Cambridge University Press.

Links to the Internet

Visit the zoology website at http://www.mhhe.com/zoology to find live Internet links for each of the references listed below.

1. Metazoa. This site hooks you up to all of the sites at the Arizona's Tree of Life site. Very easy to navigate, and more information here than you'll be able to absorb!

2. Animal Diversity Web, University of Michigan. Phylum Chordata. General characteristics of chordates, with the following links to urochordates and vertebrates.
   Urochordates.
   Vertebrates.

3. *Phylum Chordata, from the University of Minnesota.*

4. Introduction to the Urochordata. This University of California at Berkeley Museum of Paleontology site provides photographs and information on the biology and classification of the urochordates.

5. Ascidian News. This site is an online newsletter focusing on the biology of the urochordates. It provides links to other ascidian sites.

4

Parade of the Craniates in Time and Taxa

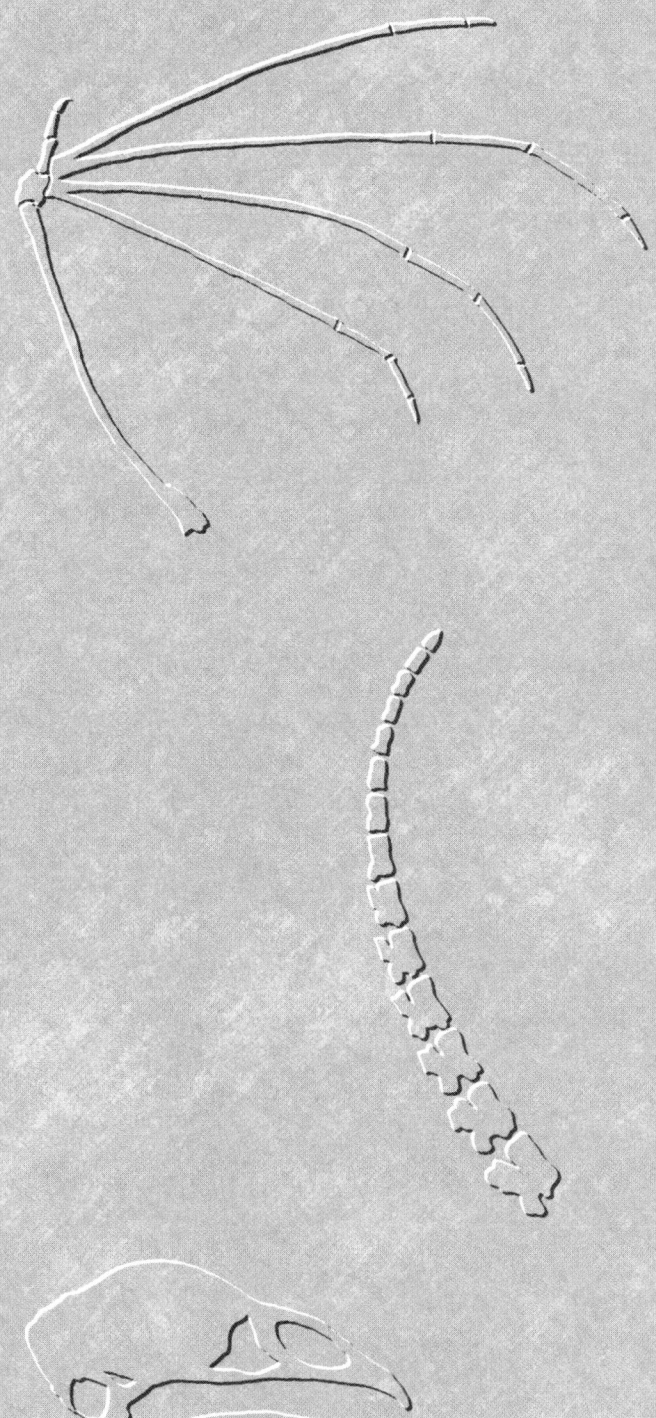

This chapter is designed to introduce craniates that are referred to in later chapters, relating briefly their natural history and their postulated ancestry so they will not be total strangers when you meet them again. The chapter also provides a glimpse of the diversity of craniate life on our planet today. It may prove rewarding to revisit the chapter should you meet an organism with which you are not acquainted.

OUTLINE

There are approximately 50,000 known species of animals with vertebral columns. Fortunately for Noah, more than half of these are fishes. However, many amphibians and some reptiles and mammals share with fishes the freshwater ponds and streams as their preferred or permanent abode, and a much smaller number share the seas. Amphibians that can tolerate saltwater are rare. There are some. They don't live in the sea, but they forage there. Among reptiles a few snakes, turtles, and iguanid lizards are permanently marine, but oviparous females must emerge to deposit their eggs. Among mammals, cetaceans and sirenians never leave the sea, and marine carnivores (pinnipeds) leave the sea only to breed. Although no birds live in the sea, many are dependent on marine organisms for food. All the foregoing animals other than fishes, however, are descendants of terrestrial ancestors and have returned to water. In the chapters that follow, we will examine the adaptations that make these different lifestyles possible.

CRANIATE TAXA

The classification scheme in this chapter is a phylogenetic taxonomy in that we will discuss organisms in historical groupings (i.e., those that share a common ancestor and all descendants [monophyletic groups]). Figure 4.1 shows the phylogenetic relationships of a number of better-known living taxa. Some informal groupings are so deeply imbedded in the literature that we will continue to use those terms as a convenience (e.g., agnathans for agnathous craniates, fish for nontetrapod craniates, and amphibians for anamniote tetrapods). Our choice of a phylogenetic taxonomy reflects its usage in most current research literature and the need for students to be familiar with these changes. Not all taxonomies are identical. A conventional evolutionary taxonomy is listed in the synoptic classification of craniates in the appendix. Not all taxa in a conventional evolutionary classification are monophyletic. You are urged to read the short discussion of systematics in chapter 2 before going farther into this chapter. It will prepare you to understand why all classifications

are not identical. It will also provide insight into the problems faced by scholars who are trying to reconstruct phylogenetic lines that began over 400 million years ago.

Taxa consist of species that are designated by a latinized *binomial* name, an invention of the Swedish naturalist Carl von Linné (Latin, Linnaeus). The Latin name enables zoologists of all nationalities to understand one another without translation when a specific animal is mentioned. For instance, not every zoologist knows what is *un chat, die Katze,* or *el gato*. All these words refer to the domestic cat, but *Felis catus* is the binomial name for the species. The generic name *Felis* separates certain cats—mountain lions, European wildcat, domestic cat, and others—from lions, tigers, leopards, and so forth, which bear the generic name *Panthera*. *Felis catus* separates domestic cats from mountain lions (*F. concolor*) and from European wildcats (*F. sylvestris*).

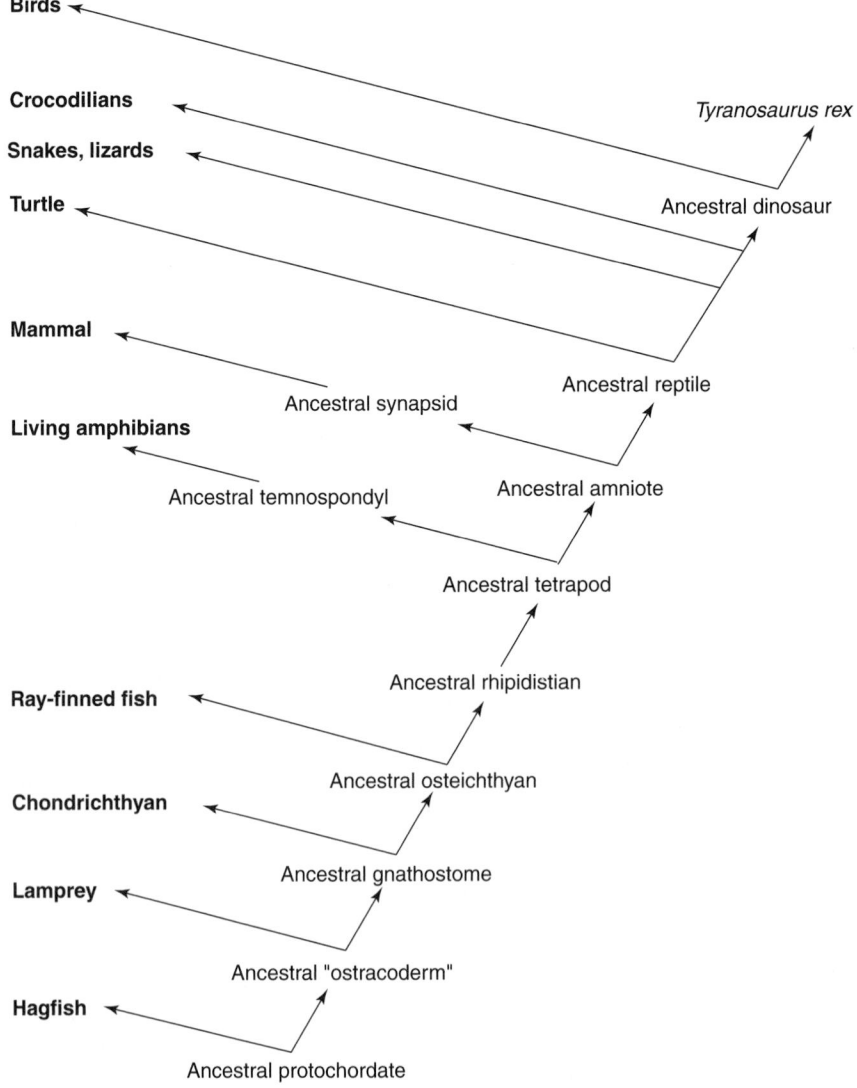

FIGURE 4.1

A phylogenetic tree for the mainstreams of craniate evolution leading to modern taxa (bold left column). Unlike a cladogram a hypothetical common ancestor is indicated at the nodes.

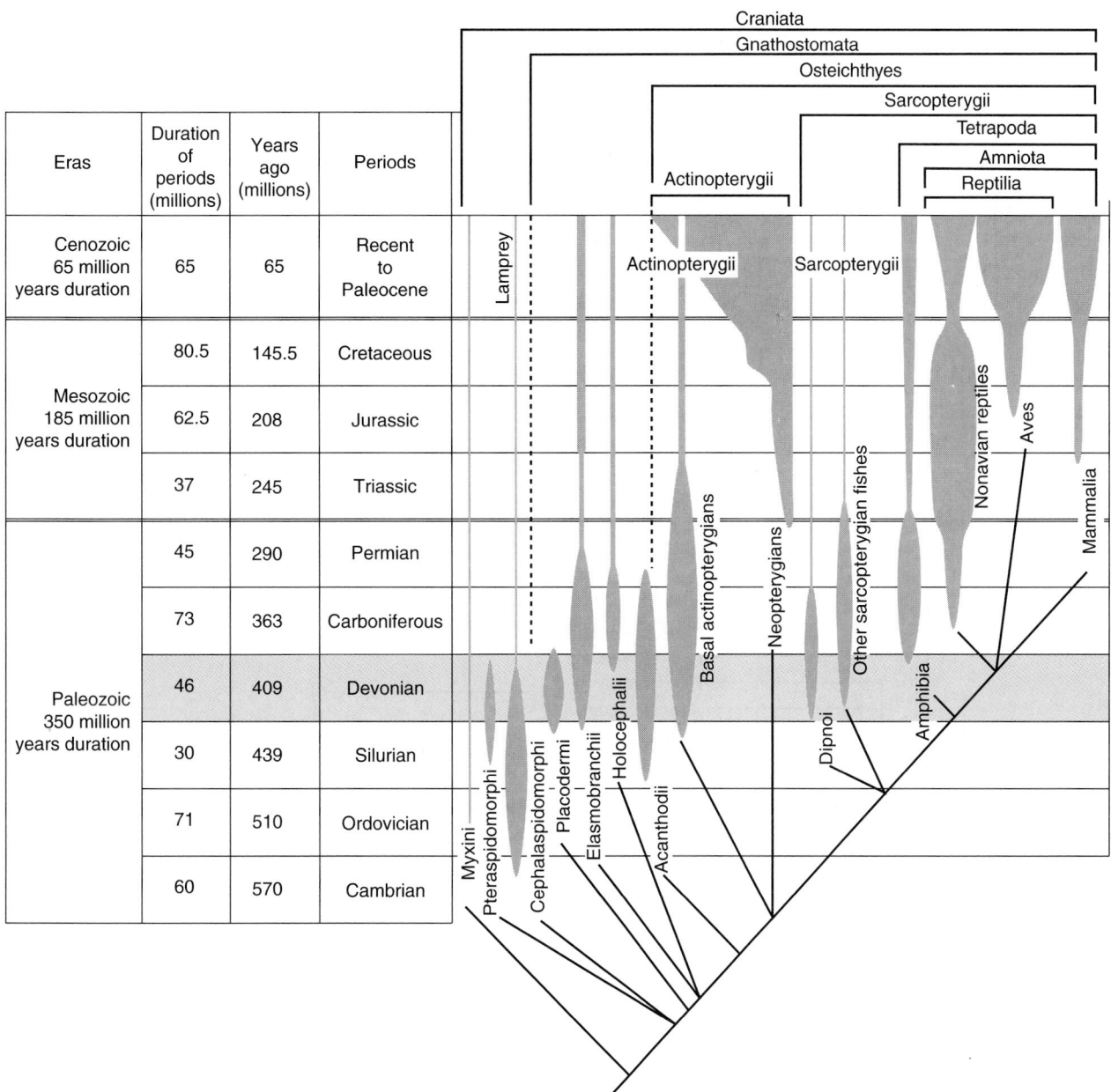

FIGURE 4.2

Range and relative abundance of craniates through geologic time. Four billion years of Earth's existence preceded the Cambrian period. The two pairs of double lines denote times of significant extinctions. The Devonian (*red*) is known as the age of fishes. A hypothesis of relationships (cladogram) is superimposed at the base of the figure. The times of bifurcation on the cladogram are relative, not absolute estimates. Note that the fossil evidence for sister groups does not always extend equally into the past. All figures are approximations. Finally, note the alternative hypothesis for agnathan relationships in figure 4.4.

Figure 4.2 shows the range and relative abundance of some of the major categories of craniates in geologic time. Figure 4.3 summarizes graphically (using a clado-gram) the postulated phylogeny of many of the crani-ates that are discussed in this text.

AGNATHA

Agnathans include two contrasting groups: the extinct, heavily armored ostracoderms and the living hagfishes

and lampreys. Two newly described Chinese fossils from the Lower Cambrian may represent unossified basal ver-tebrates (Shu et al., 1999). All other craniates are gnathostomes (fig. 4.4).

Ostracoderms

Ostracoderm is a general term that has been applied to a diverse assembly of ancient armored craniates. The term has no taxonomic status and prior to the recent discovery of the Chinese fossils, ostracoderms were the

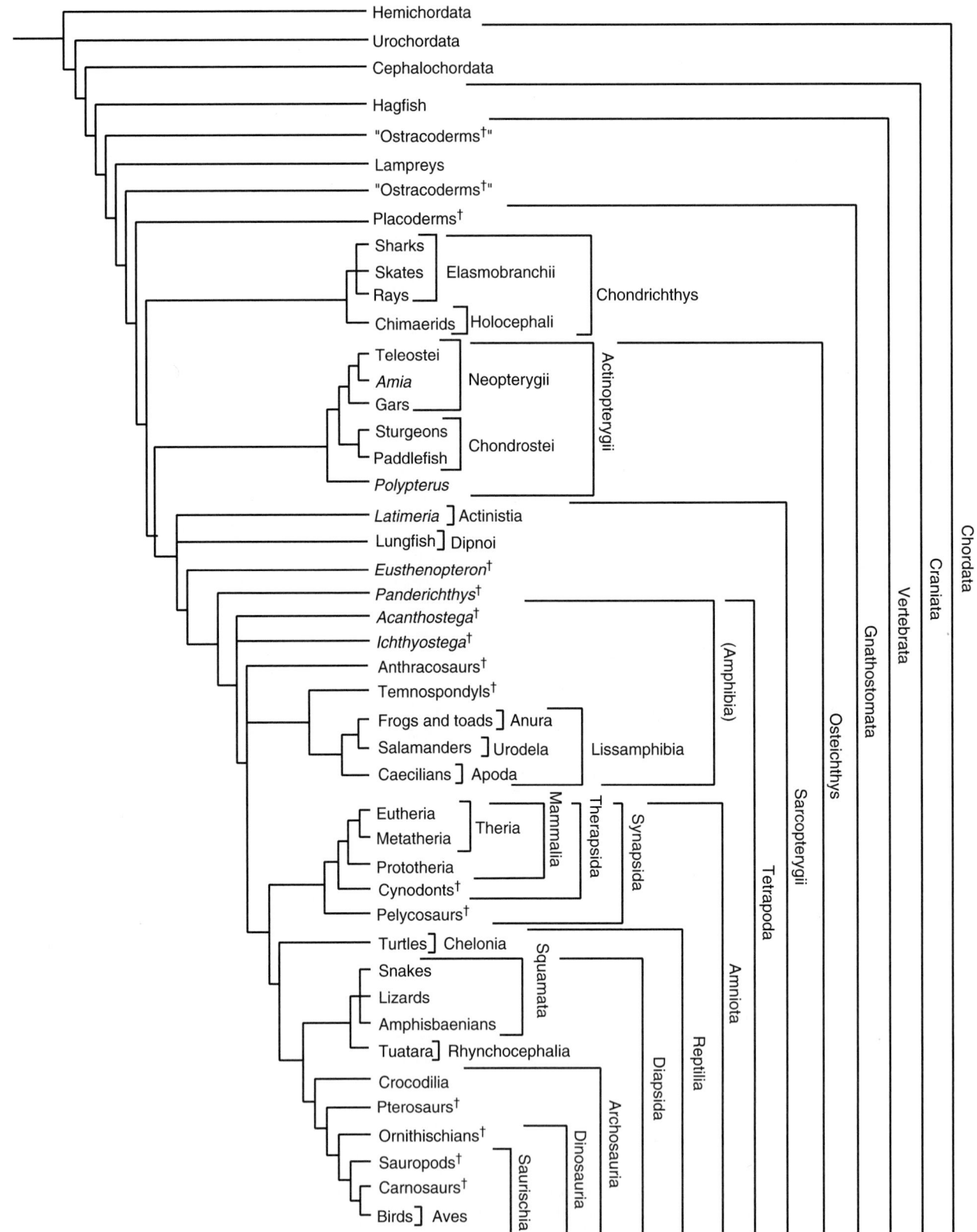

FIGURE 4.3

A hypothesis for the relationships of many of the taxa discussed in this text. Finer-scale cladograms are presented in the discussion of specific taxa. (†) indicates an extinct taxon, and (taxon name) indicates a paraphyletic group.

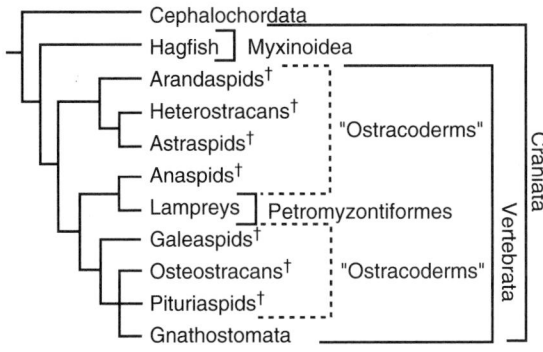

FIGURE 4.4

A hypothesis for the relationships of representative basal agnathous craniates. Cephalochordata represents the outgroup, and Gnathostomata is added for comparison.

(a)

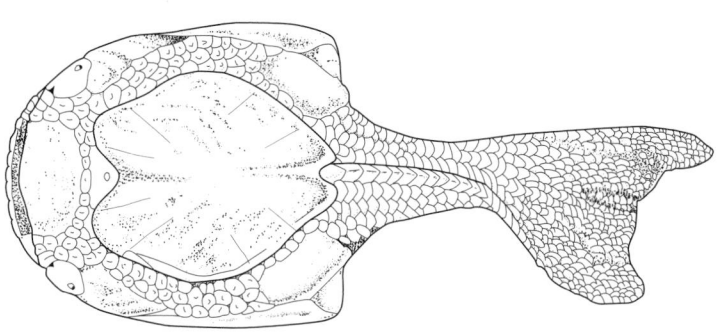

(b)

FIGURE 4.5

Ostracoderms of the order Pteraspidiformes. (*a*) A fossil from a Silurian-Devonian formation in Canada. (*b*) Dorsal view of a reconstruction of the flat-bodied *Drepanaspis*
(x 1/3).

(*a* and *b*) From *Vertebrate Paleontology and Evolution*, by Carroll. Copyright © 1988 by W. H. Freeman and Company. Used with permission.

oldest known craniates. Ostracoderms date back at least to the middle of the Ordovician, if not to the Cambrian. The entire body was covered with bony dermal armor consisting of broad plates and small tilelike scales (fig. 4.5). The plates were largest on the head where they formed a bony shield. Ostracoderms lacked jaws, and most lacked paired fins. Many were only 2 or 3 cm long (1 cm = 0.3937 in); a few attained a length of up to 2 m. What appear to have been the earliest ostracoderms were marine species and are clearly associated with a marine invertebrate fauna.

Knowledge of the anatomy beneath the fossilized head shield of an osteostracan (a cephalaspidomorph ostracoderm [Cephalaspidomorphi], the appendix) is due largely to the painstaking research of E. A. Stensiö, an early twentieth-century paleontologist at the University of Stockholm. His studies revealed that the head skeleton was a more or less flattened denticle-covered bony shield with four dorsal apertures (fig. 4.6). Two of these accommodated a pair of upward-staring eyes, a third accommodated the median, or pineal eye, and a small, anterior opening was a single naris from which a nasohypophyseal duct led to an olfactory sac, and beyond. The shield turned under along its lateral edges, where it was then replaced beneath the gills by tilelike scales. Between the anterior edge of the shield and the scales, a very small mouth opened into the oropharyngeal chamber, which was lined by gills. A curved row of external gill slits extended from the corners of the mouth to the caudal margins of the head shield. In addition to the shield, the head contained an endoskeleton of endochondral bone and considerable cartilage. Little is known of the postcranial skeleton except that the tail was heterocercal like that of most Paleozoic fishes (see fig. 10.17*a*).

The origin of ostracoderms is unknown. Possible phylogenetic ties with protochordates and basal craniates were discussed in the preceding chapter. Their numbers dwindled, and they disappeared at the end of the Devonian, having been replaced by jawed fishes.

Living Agnathans

Hagfishes and lampreys, the living agnathans (fig. 4.7), may not be as closely related to one another as their superficial resemblance suggests. Hagfishes, of which there are 42 living species, have been given class status (Class **Myxini**) in a superclass Agnatha in one cladistic taxonomy of fishes (Nelson, 1994), and lampreys have been tentatively included with the ostracoderms in the Class **Cephalaspidomorphi**. (The glossary may be consulted for derivation of the term.) This taxonomy has been employed in the appendix. It is a tentative postulate. Figure 4.4 provides a current hypothesis of relationships suggesting that the hagfishes are the sister group to Vertebrates. Among living taxa, lampreys would be the sister group to gnathostomes. When considering fossils, it becomes clear that the cephalaspidomorphs represent a paraphyletic grouping. Other methods of studying phylogenetic relationships among extant taxa (comparison of the sequence of nucleotide bases from ribosomal DNA, for example) have yielded other postulates (Forey and Janvier, 1993). Hagfishes and lampreys are generally referred to by the informal descriptive term **cyclostomes.**

Living agnathans have a prominent notochord that serves as the sole axial skeleton throughout life. They have no paired fins, no skeleton comparable to that of jaws, no vertebral column comparable to the typical vertebrate spine, and no bone anywhere in the body—no bony skeleton, no integumentary armor or scales, and no bony teeth. They have only one (hagfishes) or two (lampreys) semicircular ducts instead of the three found in jawed vertebrates, and a single nostril that is connected with a single olfactory sac. They exhibit adaptations for parasitism, including a buccal funnel and a rasping "tongue."

Hagfishes

Hagfishes are living marine agnathans with a shallow buccal funnel that lacks rasping denticles. The funnel is surrounded by a ring of stubby fingerlike papillae. Hagfishes are chiefly bottom-feeding scavengers whose diet includes a variety of small invertebrates and the viscera of dead or weakened fishes. A canal leads from the median nostril to the olfactory sac and then continues to the pharyngeal cavity, carrying a respiratory stream of water (see fig. 13.1a). The eyes, unlike those of lampreys, are vestigial and covered by opaque skin.

Myxine glutinosa, the Atlantic hagfish, has six pairs of porelike gill pouches (occasionally five or seven) that open into a common efferent duct (see fig. 13.1a). *Eptatretus stouti,* common off the coast of California, has 10 to 15 pairs of gill pouches that open directly to the exterior. Hagfishes

do not enter freshwater, and the larvae stay within the egg membranes until metamorphosis. This may be an adaptation to the saltwater environment in which the eggs are laid.

Lampreys

Lampreys have not changed appreciably since the Carboniferous (compare lampreys in figs. 4.7 and 4.8). A large buccal funnel lined with horny denticles helps keep the parasitic adult lamprey attached to the host while a tonguelike cartilaginous rod covered with horny teeth rasps the flesh of the victim, leaving only skin and skeleton. A nasohypophyseal duct leads from the median nostril to the olfactory sac and then terminates blindly in a nasohypophyseal sac (see fig. 13.1b). Seven pairs of gill pouches open separately to the exterior via

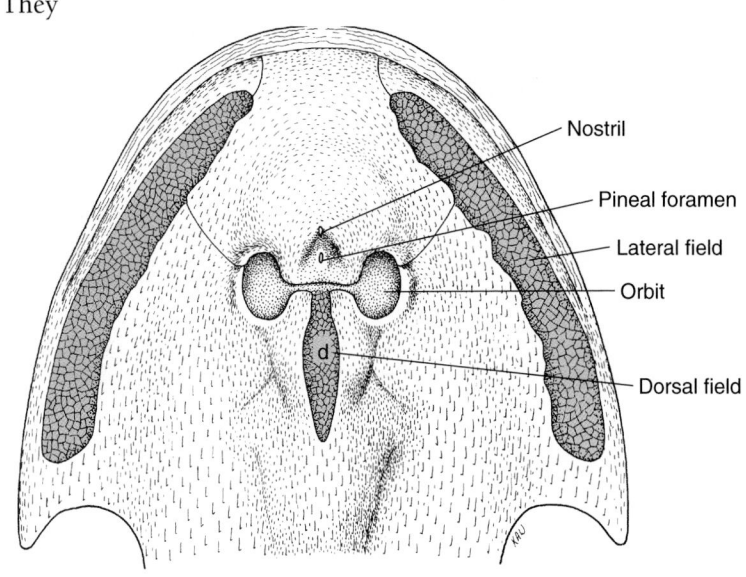

FIGURE 4.6

The bony denticulated head shield of the ostracoderm *Cephalaspis.* The dorsal and lateral fields may have been sites of exteroceptors and electric organs, respectively.

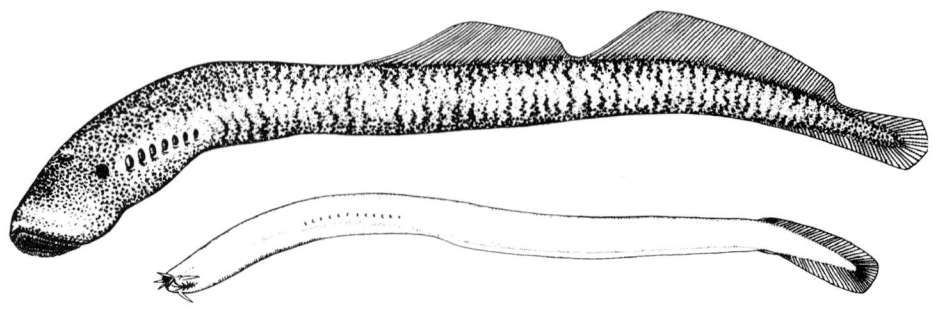

FIGURE 4.7

Lamprey *Petromyzon* above, hagfish *Eptatretus* below.

porelike gill slits that conduct water *into* as well as *out of* the pouches, thus freeing the buccal funnel for feeding on the host.

Petromyzon marinus is a species of **anadromous** lampreys. Adults live in the sea but migrate upstream to lay their eggs. In about three weeks, the small, free-living ammocoete larvae emerge. After several years in freshwater, the larvae have fully metamorphosed and, as juveniles, they migrate to the sea. There they attain sexual maturity and become physiologically adapted for the freshwater journey back to the spawning place. A land-locked population of this species inhabits the freshwater Great Lakes of North America. The freshwater brook lampreys are not parasitic. They cease feeding upon reaching sexual maturity, spawn, and like all other lampreys, they die shortly thereafter.

Suggesting a relationship between lampreys and gnathostomes (= Vertebrata) is the presence of vertebral elements in the trunk, two semicircular canals, electroreception, and a number of additional characters.

FIGURE 4.8
Fossil lamprey *Mayomyzon* from the Carboniferous.

GNATHOSTOMES: PLACODERMS

Three groups of gnathostomes evolved in the Paleozoic: placoderms (figs. 4.2, 4.9), chondrichthyans (figs. 4.2, 4.10, 4.11), and teleostomes (acanthodians, figs. 4.2, 4.12, and osteichthyans, figs. 4.2, 4.13–4.15). Two abundant groups of unusual jawed fishes, placoderms and acanthodians (see figs. 4.9 and 4.12), are present in the Paleozoic. The placoderms represent the sister

Dunkleosteus

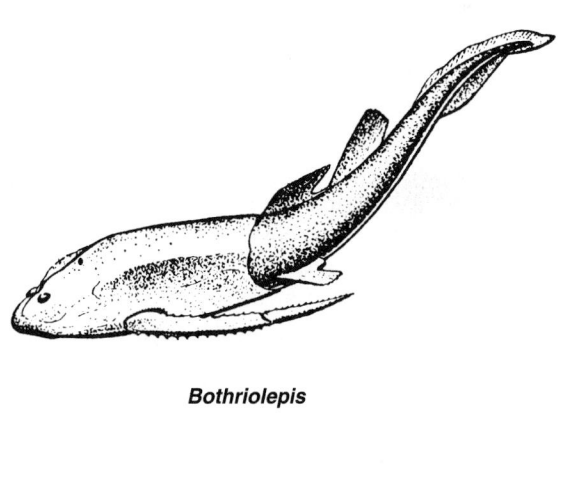

Bothriolepis

FIGURE 4.9
Two armored Devonian placoderms. *Dunkleosteus* is an arthrodire; *Bothriolepis* is an antiarch. *Dunkleosteus* attained an adult length of 15–20 feet. *Bothriolepis* is shown about one-third natural size.

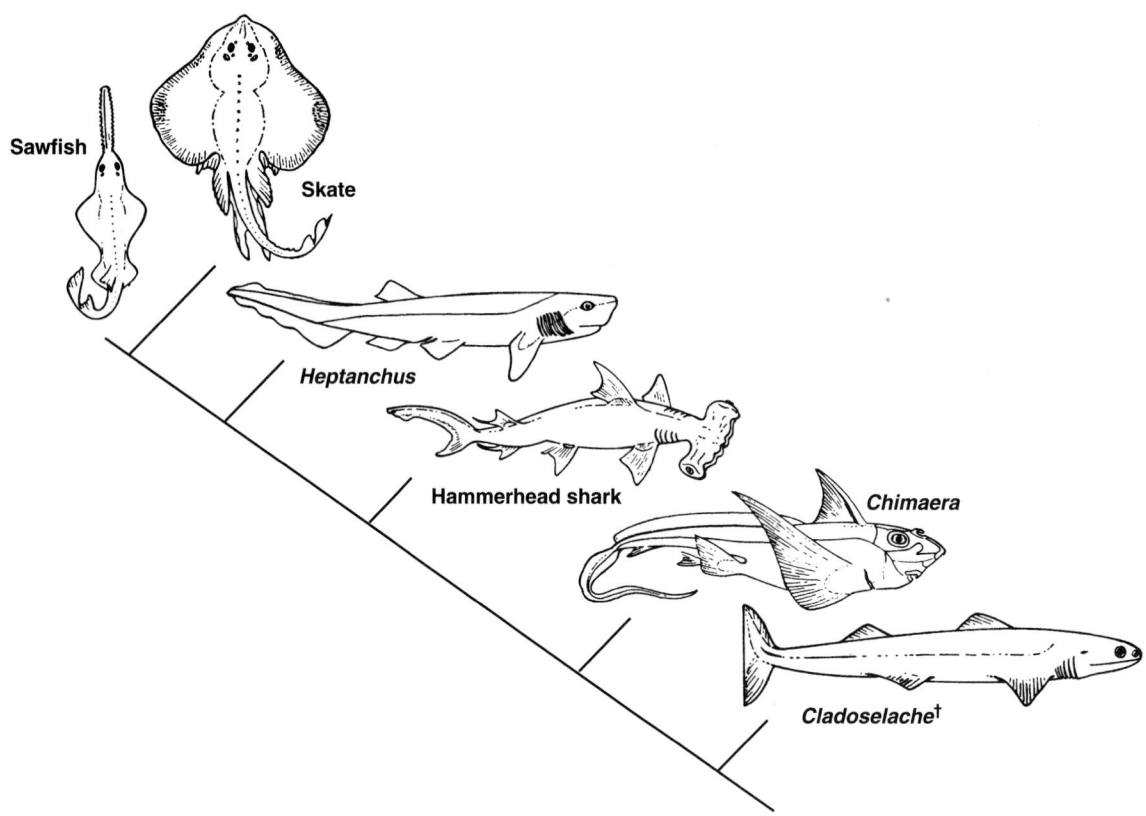

FIGURE 4.10
One hypothesis for the relationships of representative Chondrichthyes. This hypothesis represents one of two alternatives in figure 4.11 (left cladogram). (†) indicates an extinct taxon.

group of chondrichthyans + teleostomes and evolved from ostracoderm ancestors. Among the teleostomes, the acanthodians are the sister group to the osteichthyans (lobe-finned and ray-finned fishes).

Placoderms, named for their bony dermal plates, had paired pectoral and pelvic fins (see fig. 4.9). Among the better known are the **arthrodires** (see fig. 4.9, *Dunkleosteus*). Heavy bony shields covered the head and gill region, and another covered part of the trunk. The shields met in movable joints, thus the group name (*arthro* [G], jointed; and *dir* [G], neck). The remainder of the body had small bony scales or, in later species, was naked. Arthrodires appear to have been active predators. **Antiarchs** (see fig. 4.9, *Bothriolepis*) were small armored placoderms with eyes on top of the head and a flattened belly, features that suggest they may have been bottom feeders. Some placoderms reached a length of 15–20 feet, but the majority was much shorter. Placoderms flourished throughout the Devonian along with the acanthodians and disappeared rapidly thereafter.

The phylogenetic relationships of placoderms to other jawed fishes have been a matter of conjecture. Evidence at present suggests that placoderms are a basal gnathostome group—sister group to other gnathostomes (see fig. 4.3). Placoderms disappeared, leaving no descendants.

CHONDRICHTHYES (CARTILAGINOUS FISHES)

Living Chondrichthyes, the cartilaginous fishes, are usually placed in two subclasses, **Elasmobranchii,** which includes sharks, rays, skates, and sawfishes, and **Holocephali,** the chimaeras, or ratfishes (see figs. 4.10 and 17.2*b*). They have no bone in their body other than in their scales and teeth. The mouth is on the ventral surface rather than terminal, except in Paleozoic sharks such as *Cladoselache* (see fig. 4.10). Their unique dermal scales, called placoid, are composed of a basal plate and a bony spine of dentin that protrudes through the epidermis, giving the skin of sharks the texture of sandpaper. The skeleton of the pelvic fins of males is modified to form claspers used in transferring sperm to the female reproductive tract, fertilization being internal. The eggs are macrolecithal, and in oviparous species they are often encased in a horny or leathery shell with tendrils that entwine around vegetation as holdfasts.

Elasmobranchs

The taxon Elasmobranchii consists of many orders of sharks, Paleozoic to modern, and of skates, rays, and sawfishes. The gill slits are exposed (naked) rather than covered by a fleshy or bony operculum, and there are usually five pairs, although the sharks *Hexanchus* and *Heptanchus* have six and seven pairs, respectively. Seven slits may have been the primitive number in sharks because that was the number in *Cladoselache*. It is the largest number in any jawed fish. Anterior to the first gill slit is usually a spiracle that bears in its walls a miniature gill-like surface (pseudobranch). The caudal fin of Paleozoic sharks was heterocercal, the predominant type in the Paleozoic, and that of today's sharks, skates, and rays has not changed (see fig. 10.17*a*).

Dissection of a shark (usually in the order **Squaliformes**) is a valuable point of departure for study of comparative craniate morphology because of its generalized anatomy. If one were seeking a living blueprint of a generalized craniate—an architectural pattern that could provide preliminary insight into the systems of the bodies of all craniates—it would be difficult to find a better example than a shark. Most frequently studied is the viviparous spiny dogfish of the Atlantic, *Squalus acanthias*, named for the prominent spine associated with each dorsal fin. *Squalus suckleyi* is the Pacific spiny dogfish.

Rays, skates, and sawfishes—**Rajiformes**—are compressed dorsoventrally, and the anterior fins are attached all along the sides of the head and trunk, forming broad, undulating, winglike locomotor organs (see fig. 4.10). The five gill slits are ventral (see fig. 13.7*a*), but the spiracle, which is the chief incurrent route for respiratory water, remains on the dorsum. Most rays and skates subsist on molluscs scooped up from muddy bottoms, and the location of the spiracle removes it from the mud and debris that is stirred up by bottom feeding. Manta rays, on the other hand, are filter feeders. The tail of Rajiformes has changed from a muscular locomotor organ to a lean, often whiplike, organ of defense and, in some species, of offense. In electric rays (torpedos), it houses an electric organ capable of a high-voltage discharge that immobilizes prey. Stingrays have rows of spines along their tails that inflict wounds and often contain a poison. Sawfishes, not to be confused with sawfish sharks, have the rajiform shape to a lesser degree. The sword is used for impaling small fishes and disturbing the sea bottom in a search for burrowing animals.

Relationships among living elasmobranchs are presented in figure 4.11. Alternative hypotheses are presented in the literature, one suggesting that sharks (Squaliformes) are not a natural grouping (left cladogram in fig. 4.11).

Holocephalans

Chimaeras lack scales on most of the surface, they have a fleshy operculum that hides their gill slits, and the spiracle is closed. The upper jaw, unlike that of elasmobranchs, is solidly fused with the cartilaginous braincase, and instead of teeth, hard flat bony plates on the jaws crush molluscan shells, the diet of chimaeras being similar to that of rays.

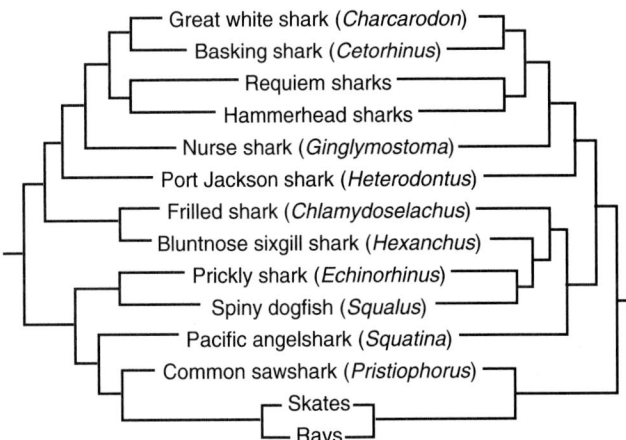

FIGURE 4.11
Alternative hypotheses for the relationships of living elasmobranchs. The right cladogram retains modern sharks as a monophyletic group. Alternatively, the left cladogram suggests that sharks are paraphyletic.

ACANTHODIANS AND OSTEICHTHYANS (SPINY FISHES AND BONY FISHES)

Acanthodians and **osteichthyans** are united in the group Teleostomi that represents the sister group to the chondrichthyans (see fig. 4.3).

Acanthodians

Acanthodians (see fig. 4.12) are extinct spiny fishes, as the name implies. Stout hollow spines were associated with the median and paired fins, and in some species an additional series of paired spines, probably associated with fin-like membranes, extended along the lateral body wall. As in ostracoderms and placoderms, the body was covered by bony armor consisting of small scales and, on the head, dermal plates. Most acanthodians were only a few inches long, but some reached 5 feet in length. The skeleton consisted of bone and cartilage. Unlike sharks, they had a large operculum overlying the gill slits (primitively consisting of elongate scales with ancillary gill covers followed by a derived single opercular cover). Acanthodian spines have been found in the early Silurian; fossils are abundant from the Devonian and Carboniferous, after which acanthodians disappear from the geologic record.

Osteichthyans

The osteichthyans include all the bony fishes that we are familiar with and their descendants the tetrapods. Osteichthyans are characterized by an air sac (lung or swim bladder) that may be secondarily lost and the presence of large units of dermal bone on the head and shoulder girdle. The group is subdivided based on the structure of the paired appendages (fins)—**Actinopterygii** (the ray-finned fishes) and **Sarcopterygii** (the lobe-finned fishes and their tetrapod descendants).

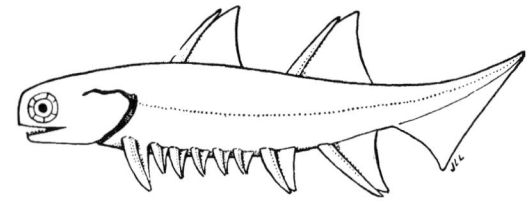

FIGURE 4.12
An acanthodian.

ACTINOPTERYGII (RAY-FINNED FISHES)

Actinopterygians are ancient and modern bony fishes whose membranous fins are supported by slender fin rays radiating from basal skeletal elements within the body wall (see fig. 10.13). Therefore, the fins are not lobed. Their gill slits are covered by a bony operculum, an air sac (swim bladder or lung) is usually present, and internal nares are lacking. During the Paleozoic, their bony dermal armor and scales were overlain with a form of enamel called ganoin (see fig. 6.29c), and their caudal fins, like those of Paleozoic sharks, were heterocercal (see fig. 4.13, sturgeon). Both traits have all but disappeared in more recent species. Living actinopterygians can be evaluated by considering two groups, the basal members and **Neopterygii**.

Basal Actinopterygians

Among the basal actinopterygians are the oldest known ray-finned fishes. They are chiefly **paleoniscoids**—Paleozoic fishes that flourished from the Devonian until the beginning of the Mesozoic, after which only relicts (remnants) remained (see fig. 4.2). These are the African freshwater fishes *Polypterus* and *Calamoichthys*, and the sturgeons and paddlefishes. Representative species of these relicts are seen in figure 4.13.

Polypterus and *Calamoichthys* have features of the paleoniscoids—large ganoid scales, a well-ossified endoskeleton, and air sacs (lungs) that are connected to the pharynx by an air duct that enables aerial respiration. They are sometimes called African lungfishes, but they shouldn't be confused with dipnoans.

Sturgeons and **paddlefishes** have an endoskeleton that is largely cartilaginous because most of the embryonic cartilage is not replaced by bone as in other bony fishes. The scales lack ganoin, and the skin of paddlefishes is naked except for small bony scales on the tail. The taxon **Chondrostei** is a mostly extinct monophyletic assemblage to which sturgeons and paddlefish are assigned (see fig. 4.13).

Neopterygians

The clade Neopterygii includes the surviving Mesozoic ray-finned fishes—gars (two genera, including *Lepisosteus*) and bowfins (one species, *Amia calva*)—and the vast array of more recent ray-finned fishes, the **teleosts** (see fig. 4.14). Gars and bowfins, classified as non-monophyletic holosteans in conventional evolutionary taxonomies, are placed in distinct clades (Ginglymodi and Halecomorphi, respectively). Teleosts are placed in the taxon Teleostei (see fig. 4.3 and the appendix).

Gars and **bowfins** are freshwater fishes. The trunk and tail of gars are covered with ganoid scales that are modifications of the ganoid scales of paleoniscoid fishes (see figs. 6.29c and 6.30). The trunk and tail of *Amia* have modern fish scales. Much of the endoskeleton is ossified in both genera, but the braincase (neurocranium) remains largely cartilaginous throughout life.

Teleosts are the most recently evolved ray-finned fishes and hence are referred to as "modern." From the perspective of geologic time, they are modern indeed, although some have been around for 65 million years, more or less. Their scales, which are in the dermis, have become very thin and flexible; the dermal bones of the skull are generally thinner and more numerous than in other bony fishes; the jaws and palate have become more independently maneuverable; the pelvic fins in many species are far forward; and the body has become altered in innumerable other ways, resulting in organisms that occupy all aquatic niches on the planet. The number of species has been exploding since the Mesozoic (see fig. 4.2), and the more than 23,000 extant species now constitute 96 percent of all living fishes.

There are long, slim teleosts that have lost paired appendages; short, fat ones with sails; transparent ones; fishes that stand on their tails; fishes with both eyes on the same side of the head; fishes that carry lanterns, that climb trees, that carry their unhatched eggs in their mouth, that appear to be smoking pipes, that possess periscopic eyes; and hundreds of other bizarre genera. They inhabit the abyssal depths far out from the continental shelves, cavort in modest brooks, and make nightly sorties onto land. They display a range of coloration, although a relatively small number of pigments assisted by a myriad of light-dispersing crystals are responsible for all hues.

SARCOPTERYGII (LOBE-FINNED FISHES)

Sarcopterygians are bony fishes that have a prominent fleshy lobe at the base of their paired fins and include their tetrapod descendants whose appendages have been modified into limbs. The lobefin contains part of the fin skeleton (see figs. 4.15 and 10.44a). Most lobe-finned fishes also have internal nares that open into the oropharyngeal cavity, and they retain a gas-filled air sac. The gill slits are covered by a bony operculum that grows caudad from the second pharyngeal arch. There are two major clades: (1) **Actinistia**, the coelacanths and (2) **Rhipidistia**, which includes **Dipnoi**, the lungfishes, and the ancestors of amphibians. The two were

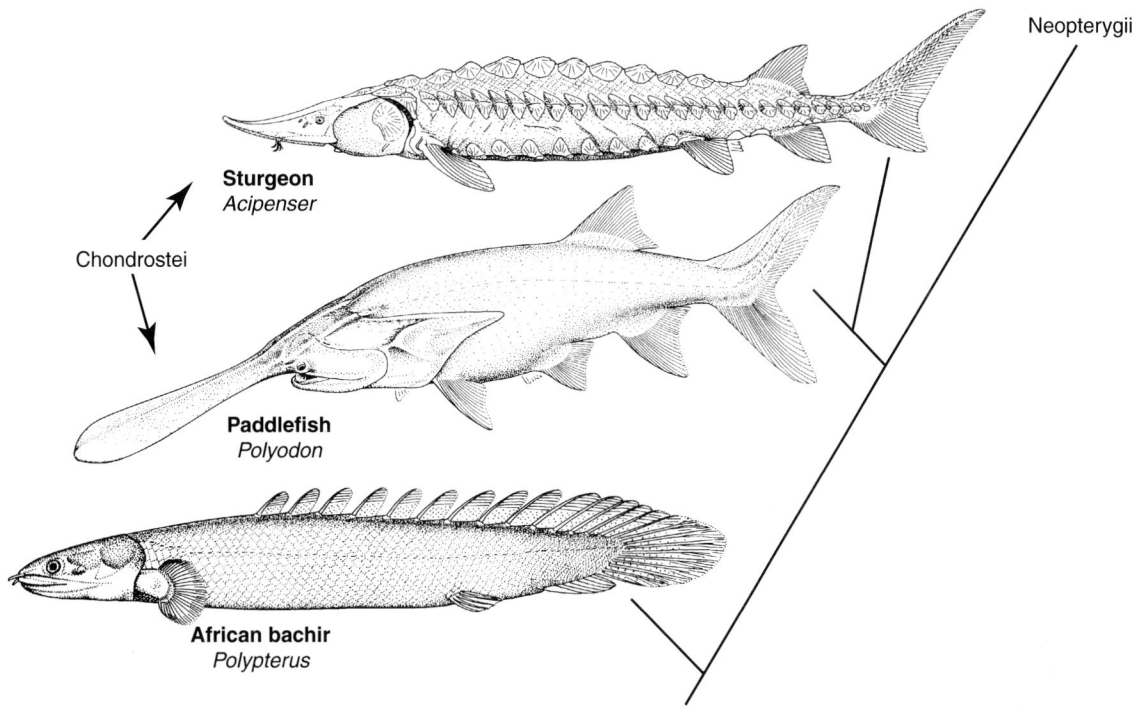

FIGURE 4.13
A hypothesis for the relationships of actinopterygians. Three basal taxa are figured.

distinct in the early Devonian. Their pre-Devonian ancestors are unknown.

Actinistians

All actinistians are extinct with one exception, the relict coelacanth, *Latimeria*. The coelacanths were thought to be extinct for more than 65 million years until the amazing discovery of *Latimeria* in the deep waters off the coast of Madagascar. Once thought a key to the understanding of the origin of terrestrial craniates, *Latimeria* has been the focus of many studies.

Rhipidistians

The skeletal elements within the fin lobes of rhipidistians correspond closely to the proximal skeletal elements of early tetrapod limbs (see fig. 10.44); the skull was similar to that of the first amphibians (see fig. 9.8); they had air sacs that were probably used at least occasionally as lungs; and most had internal nares, although these may not have been used for breathing. A current hypothesis of relationships proposed by Cloutier and Ahlberg (1996) places the lungfishes within the rhipidistians. Alternative hypotheses differ in that lungfishes are excluded and placed either as a sister group to actinistians + rhipidistians or as the sister group to rhipidistians (fig. 4.3 leaves the relationships of lungfish as unresolved). Conventional interpretations for the environment in which these organisms lived propose a freshwater habitat. However, new analyses (both sedimentological and geochemical)

suggest a marginal marine habitat for many of the fossil rhipidistians including fossil lungfish.

Dipnoans

These lobe-finned fishes are sometimes referred to as "true" lungfishes to distinguish them from those few ray-finned fishes that use air sacs for respiration in times of drought only. There are three living genera, *Protopterus* from Africa, *Lepidosiren* from Brazil, and *Neoceratodus* from Australia. The gills are inefficient, and *Protopterus* and *Lepidosiren* would suffocate if held under water because they are dependent on air and their air sacs for oxygen. *Neoceratodus*, on the other hand, utilizes gills except when the oxygen content of the water is low. During the wet season, dipnoans inhabit streams and swamps, but when the rains cease and the tropical sun dries up their environment, the African and Brazilian species dig deep burrows in the mud and spend the dry hot season in a state of aestivation. Their lowered metabolism minimizes water loss and reduces the need for nutrients and oxygen.

Dipnoans resemble amphibians in many respects. In both dipnoans and amphibians (also seen in *Amia* and *Polypterus*), the swim bladder or lungs have ducts leading to the pharynx with blood supplied by a branch from the sixth aortic arch instead of from the dorsal aorta as in teleosts. The atrium of the heart is partially divided into two chambers. Both usually have a larval stage with external gills, and both have internal nares.

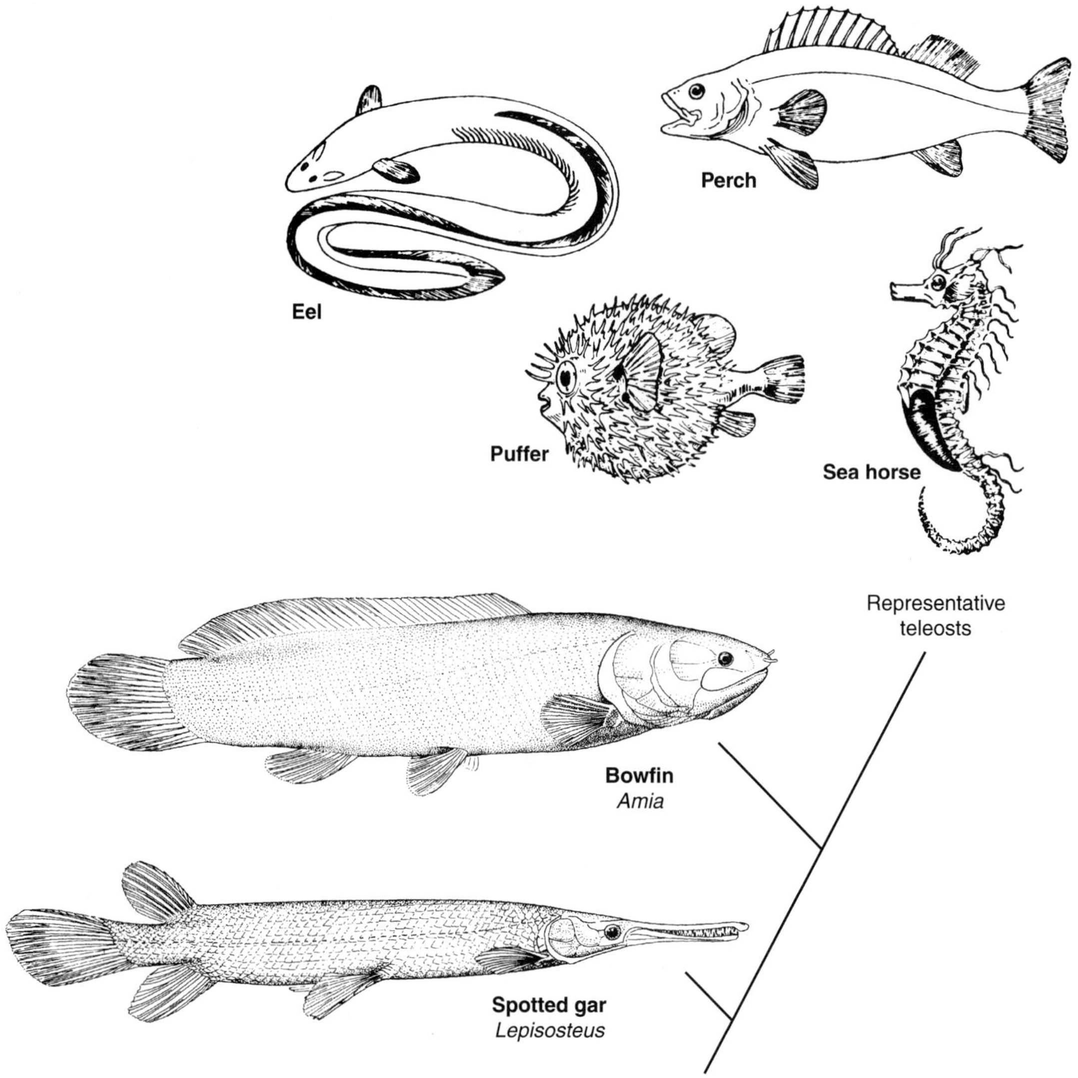

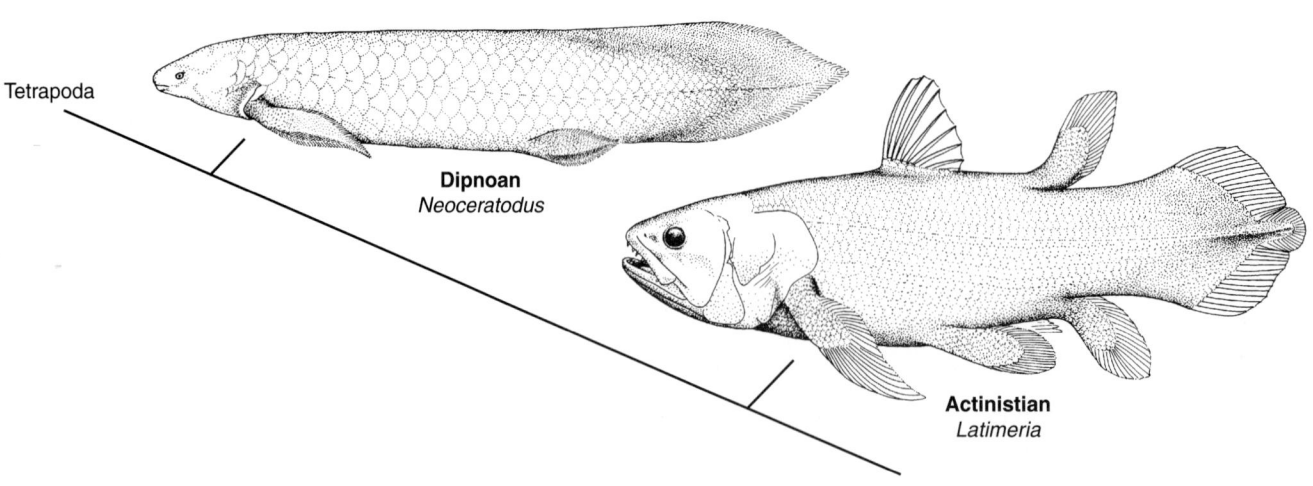

FIGURE 4.14
A hypothesis for the relationships of neopterygians.

FIGURE 4.15
A hypothesis for the relationships of extant sarcopterygians. A modern dipnoan and a relict actinistian are figured.

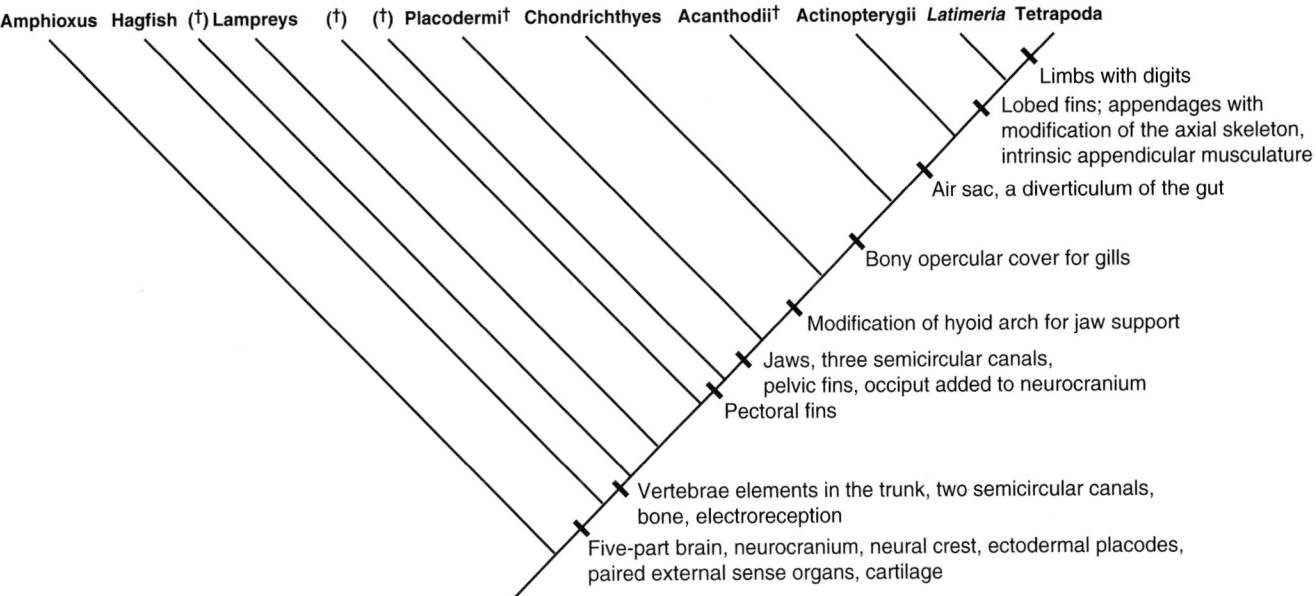

Amphioxus Hagfish (†)Lampreys   (†)   (†) Placodermi† Chondrichthyes Acanthodii† Actinopterygii *Latimeria* Tetrapoda

Limbs with digits

Lobed fins; appendages with modification of the axial skeleton, intrinsic appendicular musculature

Air sac, a diverticulum of the gut

Bony opercular cover for gills

Modification of hyoid arch for jaw support

Jaws, three semicircular canals, pelvic fins, occiput added to neurocranium

Pectoral fins

Vertebrae elements in the trunk, two semicircular canals, bone, electroreception

Five-part brain, neurocranium, neural crest, ectodermal placodes, paired external sense organs, cartilage

FIGURE 4.16

Summary hypothesis for the relationships of basal craniates with amphioxus used as an outgroup. Three unnamed extinct basal agnathans are indicated to give some sense of the relationships of numerous fossil forms to living agnathans. Some of the shared derived characters (synapomorphies) that characterize individual clades are indicated to the right of the cladogram. (†) indicates an extinct taxon.

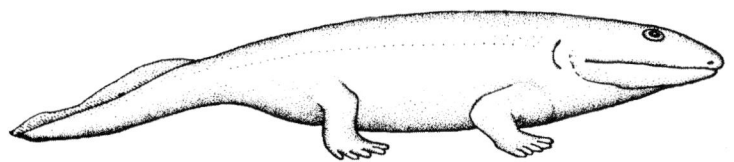

FIGURE 4.17

Reconstruction of the early labyrinthodont *Ichthyostega* from the Devonian of Greenland. The hind limbs had seven digits and the forelimbs had at least six, rather than the five that had been found at the time the skeleton was reconstructed.

Remember that if we accept lungfishes as rhipidistians, they are, among living taxa, the sister group to tetrapods. It is not until we consider the numerous fossil taxa that we appreciate the true relationship between lungfishes and tetrapods. Figures 4.3 and 4.16 are postulated lineages of fishes.

AMPHIBIA

Basal tetrapods have been variably subdivided although the relationships among these groups remain unclear. A number of basal taxa share a unique labyrinthine folding of the dentine seen in the cross section of their teeth. This shared feature was the basis for placing this diverse group in Labyrinthodontia (recognized now as a paraphyletic group). By the Carboniferous, many groups are recognized including **Temnospondyli**, **Anthracosauria**, and **Microsauria**. What is the relationship of living amphibians (**Lissamphibia**) to these groups? One hypothesis suggests that lissamphibians are the sister group to Temnospondyli (see fig. 4.3). Alternatively, lissamphians may have evolved from a temnospondyl ancestor (see fig. 4.18). What we do know from our knowledge of amphibian relationships is that the class

Amphibia, as traditionally defined, is a paraphyletic group that omits its amniote descendants. A correlation between the origin of amphibians and a terrestrial life is unclear; however, successive mutations and natural selection increasingly adapted basal amphibian descendants for terrestrial life culminating with the origin of the amniotes.

Labyrinthodonts

The oldest amphibians were the swamp-dwelling labyrinthodonts (fig. 4.17). Among labyrinthodonts, *Ichthyostega* was the oldest and appeared in the Devonian.

Labyrinthodonts were a large, widely dispersed, and diverse assemblage, and their kinships are unclear because of a lack of fossil evidence. The structure of labyrinthodont vertebrae has played a prominent role in attempts to reconstruct phylogenetic lineages and, especially, to identify the labyrinthodont ancestors of amniotes. On the basis of the morphology of labyrinthodont vertebrae, paleontologists have been of the opinion that fossil amphibians with stereospondylous and embolomerous vertebrae were not in the amniote line (see fig. 8.10).

Labyrinthodonts had many features seldom seen in modern amphibians. Among them were minute bony scales in the dermis of the skin, a fishlike tail supported by dermal fin rays, and skulls similar to those of rhipidistian fishes. Grooves in skull bones that lay just under the skin show that labyrinthodonts, like their aquatic ancestors, had a sensory canal system of neuromast organs that monitored the aquatic environment. Today's aquatic amphibians have this system; terrestrial species lose it at metamorphosis. Labyrinthodonts were as small as today's newts and as large as today's largest crocodiles. One or another of them was ancestral to the first amniote.

Among the early amphibians there was a diverse assortment of small forms that became extinct before the last of the labyrinthodont amphibians. Among them were species that lacked limbs, species resembling salamanders, and species with large, bizarre, triangular skulls unlike those of any extant amphibian. One subgroup, the microsaurs, is important to our understanding of the origin of living amphibians.

Temnospondyls

Temnospondyli was a group that was common in the Permian with its fossil record extending back to the Mississippian (Lower Carboniferous). Members of the temnospondyls have achieved a number of skeletal similarities to modern frogs and salamanders suggesting their close relationship (fig. 4.18). A collection of remarkably preserved growth series suggests that Lissamphibia evolved as the result of heterochronic processes. A number of lissamphibian skeletal features and their relatively smaller size can be explained as the retention of juvenile ancestral temnospondyl features (paedomorphosis). The condition in caecilians does not fit easily into this scenario, possibly suggesting an independent origin from microsaurs.

Microsaurs

Microsaurs represent a diverse group of fossil forms known from the Pennsylvanian (Upper Carboniferous) to the Lower Permian. They share a number of skeletal features with caecilians, which may suggest either a close relationship (fig. 4.18) or convergence on an elongate body form specialized for burrowing.

Lissamphibians

Lissamphibians include the three groups of extant amphibians: **Apoda (Gymnophiona)**, the limbless, burrowing caecilians; **Urodela (Caudata)**, the tailed amphibians; and **Anura (Salientia)**, the frogs, toads, and tree toads. The taxon also includes Triassic and Jurassic anurans whose skeletons resemble those of today's frogs and toads. Given a monophyletic origin of amphibians, alternative hypotheses (fig. 4.18) suggest that lissamphibians are either monophyletic (a common tem-

nospondyl ancestor) or diphyletic (apodans descended from a microsaurian ancestor). The fossil record of apodans, for the most part, consists solely of vertebrae from the Cenozoic and Late Cretaceous; however, the best-preserved material is from the Jurassic of Arizona. This material consists of both cranial and postcranial skeleton showing an animal with an apodan-style skull but with limbs, although reduced. The weight of evidence supports the monophyly of Lissamphibia.

Apodans

Apodans (caecilians) are circumtropical, limbless amphibians that, except for aquatic species, live in burrows in swampy locations (fig. 4.19). Their eyes are small and, in some of the 50 or more species, buried beneath the bones of the skull. (Was this the cause, or is it the effect, of a subterranean habitat? The explanation is surely less simplistic.) Some species have minute scales in their dermis. Most caecilians are about 30 cm long, but some attain a length of more than a meter and have as many as 250 vertebrae. The vent is almost at the end of the body; hence, the tail is very short. Terrestrial species lay large yolky eggs, and the larval stage is passed in the egg envelopes. Several aquatic genera are viviparous.

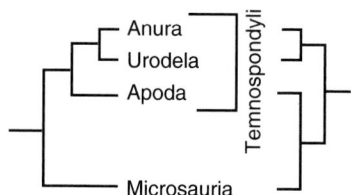

FIGURE 4.18

Alternative hypotheses for the relationships of Lissamphibia and extinct amphibians. The left cladogram suggests that a monophyletic group of modern amphibians is most closely related to members of the extinct temnospondyls. In contrast, the right cladogram suggests that lissamphibians are diphyletic with Apoda more closely related to the extinct Microsauria.

FIGURE 4.19

An apodan. The annular structure probably aids in burrowing.

Urodeles

There are eight families of urodeles (table 4.1), six of which are represented in figure 4.20. Two families, **Proteidae** and **Sirenidae** (represented by *Necturus* and *Siren*) are perennibranchiate; that is, they retain larval gills and one or more gill slits throughout life. The other families may have a few perennibranchiate populations in one locality or another, and sometimes in the same pond with fully metamorphosed individuals. In some localities, perennibranchiates undergo complete metamorphosis when administered iodinated compounds or thyroid hormone.

Necturus is the only genus of Proteidae in North America. *Necturus* attains sexual maturity in about 5 years. *Proteus,* its European relative, is blind and lives in dark caves. *Siren* (family Sirenidae) lacks hind limbs and lives in muddy ditches and weed-choked lakes or ponds.

Plethodon, like most other members of its family, loses its gills at metamorphosis but fails to develop lungs. It lives in moist sites and, except in the tropics, lays its eggs under damp logs or in caves. Respiration is via the skin. Larvae hatch with legs already formed and may never enter water.

Amphiuma is an eel-like urodele up to 1 m long, but with tiny appendages. *Amphiuma pholeter* has one digit. *A. means* has subspecies with two or three digits.

Cryptobranchus looks ferocious because of its broad flat head and wrinkled skin, which often conceals a single gill slit, if one is present.

Hynobius, a basal urodele, and *Ambystoma* are terrestrial genera. However, *A. mexicanum,* the Mexican axolotl, is an aquatic perennibranchiate that can be caused to discard its gills experimentally. (Xolotl was the Aztec god of twins and monsters.)

Notophthalmus and *Salamandra* are true salamanders (family **Salamandridae**). *Salamandra atria* is viviparous. *Notophthalmus viridescens* (fig. 4.21) has an interesting life history. After several months of an aquatic larval existence, the red-spotted newt loses its gills and gill slits, sprouts legs, and, now an **eft**, leaves the pond for life on land. The skin develops a thick cornified layer that blocks the openings of the sensory canal system and skin glands, the body assumes a bright orange-red color, and a series of dorsolateral black-bordered red spots appear. The red eft stage lasts 1 to 3 years, depending on the locality.

At the approach of sexual maturity, the eft, stimulated by the hormone prolactin, joins a mass migration to freshwater ponds where mating occurs and eggs are deposited. The tail commences to change from round to laterally compressed with dorsal and ventral keels, the thick cornified layer of epidermis is shed, exposing the openings to mucous glands and the sensory canal system, and the body gradually becomes olive green above and light yellow below, a protective coloration appropriate for life in a freshwater pond. The animal is now a **newt.** In some localities, the larvae remain in the pond, mature, and retain vestigial gills throughout life.

Anurans

Anurans are tailless amphibians in which several caudal vertebrae are fused into one elongated urostyle (see fig. 8.13). No single morphological trait distinguishes frogs from toads. The most representative frogs are the long-legged, slender-bodied members of the family **Ranidae.** The most representative toads are in the family **Bufonidae.** Toads are more terrestrial than frogs. Tree toads, also called tree frogs, are in the family **Hylidae.** Anurans breathe with lungs and skin, and most are amphibious. A few, such as

TABLE 4.1 Distribution of Gills, Pharyngeal Slits, and Lungs among Adult Urodeles

| Family | Representative Genera | Number of Pairs | | |
		Gills	Slits	Lungs
Proteidae	*Necturus*	3	2	Yes
Amphiumidae	*Amphiuma*	0*	1	Yes
Hynobiidae	*Hynobius*	0*	0	Occasionally
Cryptobranchidae	*Cryptobranchus*	0*	1	Yes
Salamandridae	*Notophthalmus*	0*	0	Yes
Ambystomatidae	*Ambystoma*	0*	0	Yes
Plethodontidae	*Plethodon*	0*	0	No
Sirenidae	*Siren*	3	3 to 1	Yes

*Some individuals are occasionally perennibranchiate

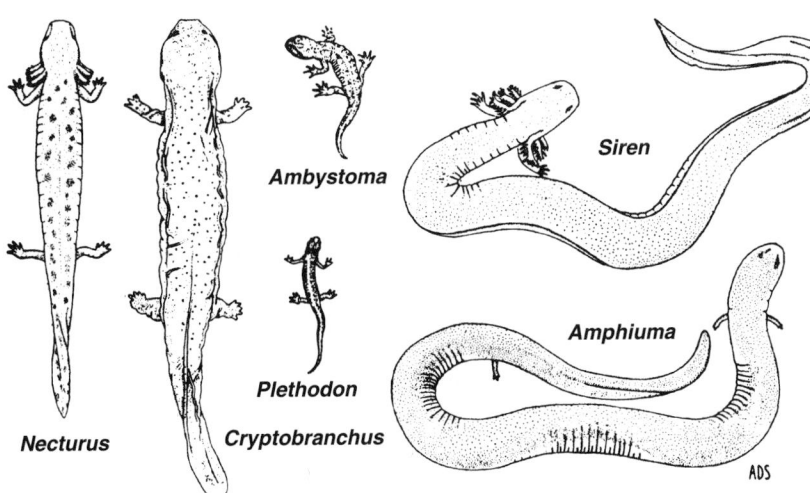

FIGURE 4.20
Representative urodeles.

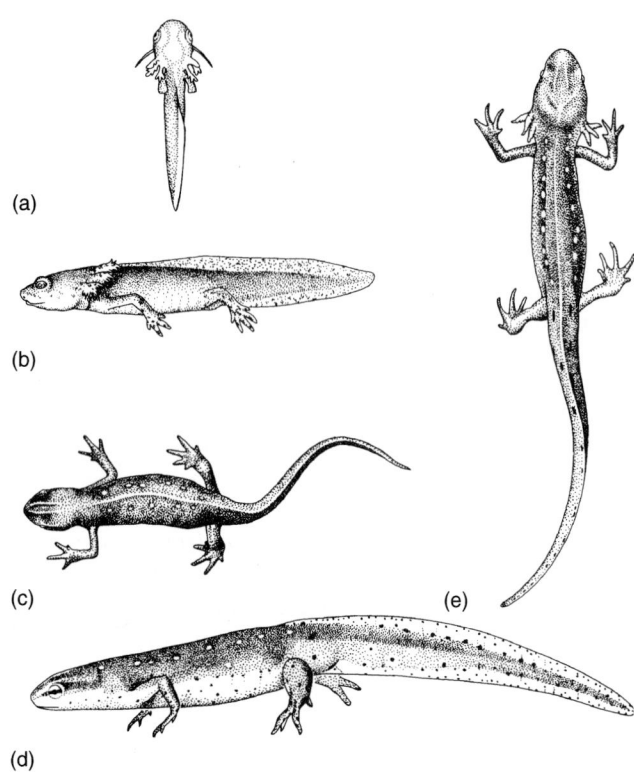

(a)

(b)

(c)

(e)

(d)

FIGURE 4.21
Life history of *Notophthalmus*. (*a*) Newly hatched larva (7 mm), balancers behind eyes, gills only slightly branched, forelimbs are blunt buds, hind limbs not formed, keeled back and tail. (*b*) Late larva. (*c*) Terrestrial eft. (*d*) Adult male (95 mm). (*e*) Heterochronic adult. All to same scale, except (*a*).

Rana cancrivora, the crab-eating frog of Thailand, can tolerate saltwater.

Most anurans breed only during or shortly after a rain and deposit their jelly-encapsulated eggs in water-filled ditches or in ponds. Tadpoles emerge from the jelly envelopes and further develop in the water. Robber frogs, however, lay their eggs in rain-filled crevices, in a pile of leaf mold, or at the base of a grass hummock away from the water's edge and above flood level. The larval stage therefore lacks an aquatic environment. As an adaptation, the tadpoles remain within the jelly envelopes until metamorphosis. The interval between egg laying and completion of metamorphosis is much shorter than in other anurans. Some species of tree toads also do not lay their eggs in bodies of water. Instead, they carry developing eggs in a brood pouch under the skin of their back, and fully metamorphosed tree frogs emerge through an opening in the skin. A similar condition is found in aquatic frogs of the genus *Pipa.* The East African toad *Nectophrynoides vivipara* is, as the name implies, viviparous. As many as 100 young develop in the uterus.

The earliest known anuran, *Triadobatrachus* (= *Proto-batrachus*) is from the Triassic. The skull was quite similar to that of today's anurans, and the body was similarly

shortened as a result of a reduction that had already taken place in the number of trunk vertebrae. The ribs were not as foreshortened, the tibia and fibula were not fused, and bony scales covered the abdomen.

Anthracosaurs

Anthracosauria, a relatively small Paleozoic group, is thought to be in a direct line to amniotes. Their fossil record extends from the Mississippian to the Triassic.

AMNIOTES (REPTILES AND SYNAPSIDS)

Amniota, characterized by their possession of extraembryonic membranes, consists of two sister clades—Reptilia and Synapsida (figs. 4.3 and 4.22). The definition of the taxonomic group Reptilia has changed over time and is not standardized among current textbooks, often leading to confusion for the student. Initially, Reptilia referred to living turtles, snakes, lizards, and tuatara (Rhynchocephalia). With the inclusion of fossils, the term expanded to incorporate all nonmammalian and nonavian amniotes, thus the term *mammal-like reptiles* for synapsids. With the advent of phylogenetic systematics, amniotes were subdivided into Synapsida and Reptilia (including birds). Due to historical confusion, some current researchers have recommended the use of Sauropsida in place of the name Reptilia.

REPTILIA (SAUROPSIDA)

From some group of early labyrinthodonts, perhaps anthracosaurs, there arose during the early Carboniferous a group of tetrapods destined to be named Amniota more than 300 million years later. From their common ancestor two groups evolved, the synapsids and reptiles. Within reptiles two subgroups of fossil and living reptiles are recognized—**Anapsida** and **Diapsida**. From them evolved a varied and abundant assemblage of descendants (figs. 4.22 and 4.23) who established themselves securely on land, conquered the air, and reinvaded the seas, flourishing as a tetrapod majority for 260 million years until the mass extinction of most reptiles that ended the Cretaceous. Remaining are a relatively few descendants, the **turtles**, an ancient group that thus far has doggedly persevered in the struggle for existence; *Sphenodon,* a lizardlike relict; **modern lizards**, recent additions to reptilian society; **snakes** and **amphisbaenians**, which are lizards deprived of their appendages; **crocodilians**; and **aves**. These survivors mirror the accumulative effects of mutations and of a changing environment on basal reptiles and their descendants.

Reptiles underwent changes that made them better adapted for a permanent terrestrial existence than were their amphibious ancestors. This trend toward a terrestrial lifestyle was paralleled among the synapsids.

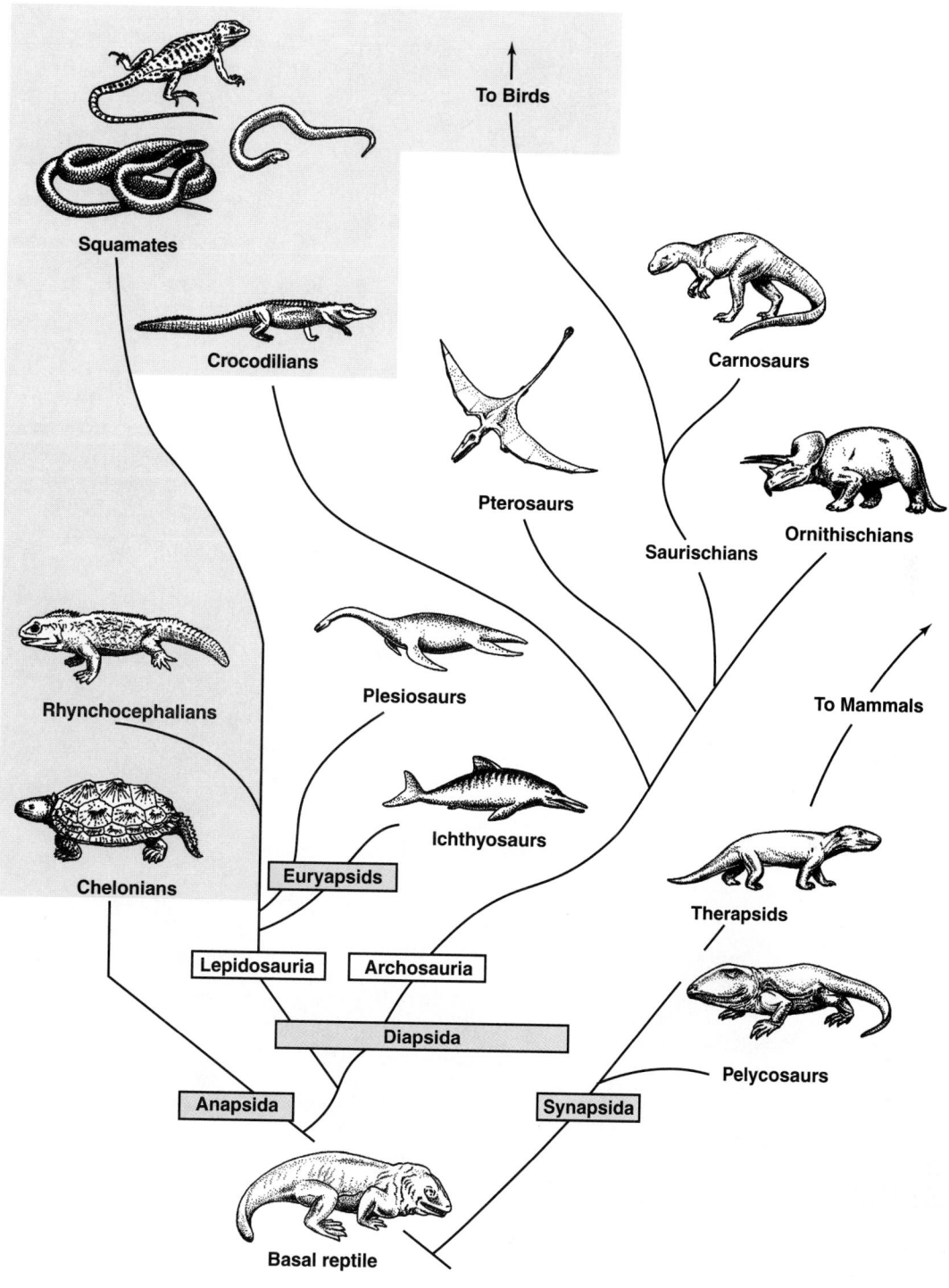

FIGURE 4.22

Postulated relationships of selected amniotes. *Red* groupings reflect taxa based on the architecture of the temporal region of the skull. Note that the euryapsids represent a paraphyletic group within the diapsids that have secondarily acquired a single temporal fenestra. *Gray* indicates surviving reptiles (sauropsids).

Perhaps the major change was acquisition of three extraembryonic membranes, the **amnion**, **chorion**, and **allantois**, which emancipated them from the necessity of returning to water to lay their eggs (refer to chapter 5, "Extraembryonic Membranes"). The amnion is a fluid-filled membranous sac in which the embryo develops

(see fig. 5.13*b*). The fluid is secreted by the lining of the sac. To state the situation figuratively, instead of the mother having to go to water to deposit her eggs, the embryo is provided with its own private pond, and the egg can be laid on land. Thus, these amphibian descendants became the first amniotes. The chorion and allantois

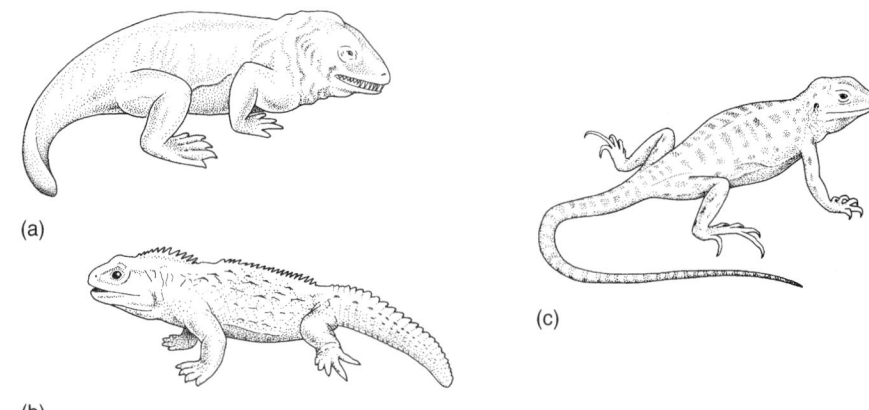

FIGURE 4.23
(*a*) Basal extinct reptile. (*b*) *Sphenodon*— a "living fossil." (*c*) Modern iguanid lizard.

FIGURE 4.24
Siamese crocodile emerging from egg.
Courtesy of Metrozoo, Dade County, FL.

constitute a vascularized membrane that lies against the porous eggshell, another amniote innovation, taking the place of larval gills for respiration.

Because of these three extraembryonic membranes, oviparous amniotes were able and, indeed, forced, to lay their eggs on land. The young hatch fully formed, without a larval stage, ready to seek their own food (fig. 4.24). Not only were amniotes liberated from returning to water to lay their eggs, aquatic oviparous species must go onto land to do so because porous eggs would become waterlogged in an aqueous environment.

Reptiles exhibit other adaptations to terrestrial life. The body surface is covered with a thick layer of cornified epidermal cells that is organized into plaques, shields, or surface scales, unlike the bony dermal scales of fishes and early amphibians. The scales are impervious to water, which results in water conservation, a necessity for animals living in air. Reptiles elaborated the development of a neck by further specialization of several postcranial vertebrae. This modification, combined with their single occipital condyle, enables reptiles to scan the horizon. The pelvic girdle now articulates with two sacral vertebrae, providing a stouter brace for more powerful hind limbs. The digits are supplied with claws, and a new kidney, the metanephros, has come into existence as a modification of the amphibian kidney. The heart is partially or completely divided into right and left chambers, thereby separating the systemic and pulmonary circulations to an extent not achieved in anamniotes.

Nonavian living reptiles, like fishes and amphibians, are **ectotherms**; that is, they cannot maintain a more or less constant body temperature in the face of variations in the ambient temperature. *Sphenodon* and some lizards have retained a parapineal organ that is exposed to the environment and aids in thermoregulation by monitoring the duration of exposure of the animal to the sun's rays (see fig. 17.19). In contrast, birds and most likely some of their dinosaur ancestors were **endotherms**.

These then are the reptiles: scaly, clawed, mostly terrestrial tetrapods and the feathered birds that, excepting viviparous species, lay macrolecithal, shell-covered (**cleidoic**) eggs on land, the embryos of which develop within an amnion, and the young of which are hatched fully formed. We will look briefly at two subgroups, Anapsida and Diapsida (including Aves, a subgroup of diapsids), which have been erected on the basis of the number of fossae in the temporal region of the skull. Anapsids have none and diapsids have two, with modifications in some recent species and in some extinct marine reptiles. (You may wish to consult the appendix for the derivation of the terms.)

Anapsids

Anapsids have no fossae in the temporal region of the skull (see figs. 9.15 and 9.16). This was the primitive condition exhibited in basal reptiles including Chelonia, all of whom are included in the taxon. The anapsids represent a paraphyletic grouping with the extinct members of this group placed in the Captorhinida (also a paraphyletic grouping). Basal reptiles are known from the Middle Pennsylvanian (see fig. 4.2). The diapsid condition was an evolutionary descendant from an anapsid ancestor and is recognized in the fossil record from the Upper Pennsylvanian. Turtles are the sole living members of the anapsids and diverged from other anapsids in the Triassic.

Turtle skulls are enigmatic. There is no evidence of a temporal fossa, but unlike in any other reptile, there is a deep cavitation on each side of the midline at the rear of the skull (see fig. 9.14*b*). The phylogenetic history of this condition is unclear. Consequently, in some taxonomies, turtles (**Testudinata**) are assigned a separate subclass status. Their relationship to the diapsids is also debated. Most classifications (including the one in this text) place turtles as the sister group to diapsids; how-

ever, recent molecular studies propose a diapsid origin suggesting that their anapsid condition is secondary.

Diapsids

Diapsida includes all living reptiles except turtles. It also includes a vast array of extinct reptiles that ruled the land, swam the seas, and flew in the air during the Mesozoic. The subtaxa **Archosauria** and **Lepidosauria** include extant reptiles. To consider additionally the many fossil taxa, diapsids can be divided into two clades—Lepiodosauromorpha (includes Lepidosauria and the extinct marine **plesiosaurs** [Sauropterygia] and **ichthyosaurs**, [see fig. 4.22]) and Archosauromorpha (includes Archosauria and a number of fossil taxa).

Lepidosaurs

There are two extant groups of Lepidosaurs: rhynchocephalians and squamates (see figs. 4.3 and 4.22). **Rhynchocephalians** are primitive lizardlike reptiles but with quite different scales, teeth, and internal morphology than modern lizards. **Squamates** are the modern lizards, snakes, and amphisbaenians. Rhynchocephalians retain the diapsid skull of their ancestors. Squamates have a diapsid skull that has been adaptively modified.

Rhynchocephalians The only surviving rhynchocephalian is *Sphenodon*, commonly known as a tuatara (see fig. 4.23*b*). *Sphenodon* feeds on small vertebrates and insects and attains a length of 0.75 m (2.5 ft). They become sexually mature at 20 years of age, and their normal life span may exceed 60 years. They may face extinction in their only surviving habitats in New Zealand and a few adjacent coastal islands.

Squamates Of all living reptiles, lizards are the most versatile. Because of well-developed appendicular muscles and a suitably constructed skeleton, some run agilely on their hind limbs; some are amazing broad jumpers; some, the nocturnal geckos, have suction discs on their toes that adhere to smooth vertical surfaces; and some like *Draco*, the flying dragon, glide through the air on rib-supported extensions of the lateral body wall. A few lizards are either limbless or have only vestiges of limbs. Some lizards are blind, and some have transparent eyelids (**spectacles**). There is a third eyelid, the **nictitating membrane**, and the teeth are in sockets. Iguanid lizards are more generalized. *Iguana* is an arboreal, strictly vegetarian species that enters seawater to forage on seaweed. The largest lizard, the Komodo dragon of Indonesia, reaches 2.75 m (9 ft) in length and 115 kg (250 lbs) in weight. The horned toad of the North American desert, *Phrynosoma*, is a lizard.

Snakes evolved from lizards. Having lost limbs, they have acquired other modes of locomotion, including a unique set of muscles that connects the ribs with the

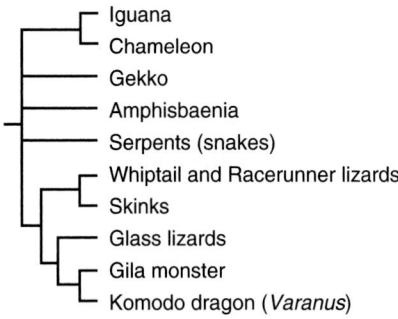

FIGURE 4.25
A hypothesis for the relationships of representative squamates. Although unresolved, this hypothesis suggests that living "lizards" represent a paraphyletic grouping.

large ventral scales, or scutes. Despite the absence of limbs, snakes inhabit mountains, deserts, trees, freshwater, and the sea. Most sea snakes are viviparous, thereby eliminating the necessity to return to land to lay their eggs.

Amphisbaenians are subterranean lizards 30 to 70 cm long (1 to 2 ft or more), mostly limbless, with annulated bodies like those of the burrowing apodans. The eardrum and eyes are covered with opaque skin. There are about 120 species.

The relationships of living squamates are presented in figure 4.25 as an unresolved hypothesis. Although snakes and lizards are often considered as distinct groups, the hypothesis presented in figure 4.25 is not so unusual when it is recognized that snakes are derived lizards.

Plesiosaurs and Ichthyosaurs
Plesiosaurs (**Sauropterygia**) and ichthyosaurs are marine reptiles. They were contemporaries of the dinosaurs but, unlike the latter, they had a dorsal temporal fossa on each side (a "euryapsid" skull). It has been postulated that the single dorsal fossa is a result of the loss of a lower fossa that was present in a diapsid ancestor. For that reason, these reptiles are classified with the diapsids (see fig. 4.22). It is also hypothesized that the euryapsid condition was a result of evolutionary convergence and that the two groups were not closely related.

Archosaurs
Archosauria is defined as the group of all descendants from the common ancestor for crocodiles and birds (a crown-group definition), thus containing crocodilians, pterosaurs, and dinosaurs. They were the dominant land craniates during the Mesozoic. Today's descendants are the modern crocodilians and birds (see fig. 4.22).

Crocodilians
Known from the Middle Triassic, modern crocodilians include the subtropical and tropical alligators, crocodiles, caimans, and a single species of gavials. Crocodiles can be distinguished from alligators by the shape of the snout, which is more slender and triangular in crocodiles, broad and rounded in alligators, and by the enlarged fourth tooth of the lower jaw, which in crocodilians fits into a notch on the upper jaw and can be seen when the mouth is closed, whereas in alligators it fits into a pit medial to the upper jaw tooth line and cannot be seen. Caimans and alligators look very much alike. The gavial snout is long and very slender because the left and right halves of the mandible are united by a long symphysis that extends all the way back to the fifteenth tooth.

Pterosaurs
This extinct group of flying archosaurs, like birds, had pneumatic bones, but their wings were more like those of bats, being supported by an elongated fourth finger (see fig. 10.27b). The largest had a wingspread of 10 m (35 ft).

Saurischian and Ornithischian Dinosaurs
Dinosaurs can be subdivided into two major clades that are differentiated by the structure of their pelvic girdles. Those of ornithischians are structurally similar to those of birds, their ischial and pubic bones being parallel to each other and directed caudad as opposed to those of most saurischians (fig. 4.26). Dinosaurs came in many sizes and body builds, and many were bipedal. Massive ones did not represent the majority. Many saurischians were swift predatory carnivores, whereas ornithischians were herbivores lacking front teeth, although some had horny beaks. A small unidentified bipedal saurischian is thought to have given rise to birds. This hypothesis assumes that the birdlike pelvis of ornithischians and that of birds is a product of convergence.

Aves
Birds are endothermic saurischian dinosaurs with feathers (unless feathers are more widely distributed among dinosaurs, thus making them a plesiomorphic character for birds—see following discussion). **Endothermy** is the ability to maintain a relatively stable body temperature despite fluctuations in the ambient temperature. A less precise term is *warm blooded*. Birds are descendants of a bipedal archosaur, perhaps a saurischian dinosaur. Bipedal saurischians could stand on their hind limbs, which freed the forelimbs for other functions including, ultimately, flight in birds (independently achieved in pterosaurs). Birds have retained reptilian scales on their legs, feet, and beak, a single occipital condyle, and a diapsid skull that has been modified by partial or complete loss of the zygomatic arch.

Feathers are exquisitely structured keratinized integumentary appendages that replace reptilian scales wherever feathers are present. Feathers make flight possible in birds. Whereas pterosaurs and bats have a wing membrane stretched between the forelimbs and trunk, or even the tail, that serves as an airfoil, feathers attached to the bones of the forearm and hand are the only airfoil birds have. Feathers also insulate against seasonal heat and

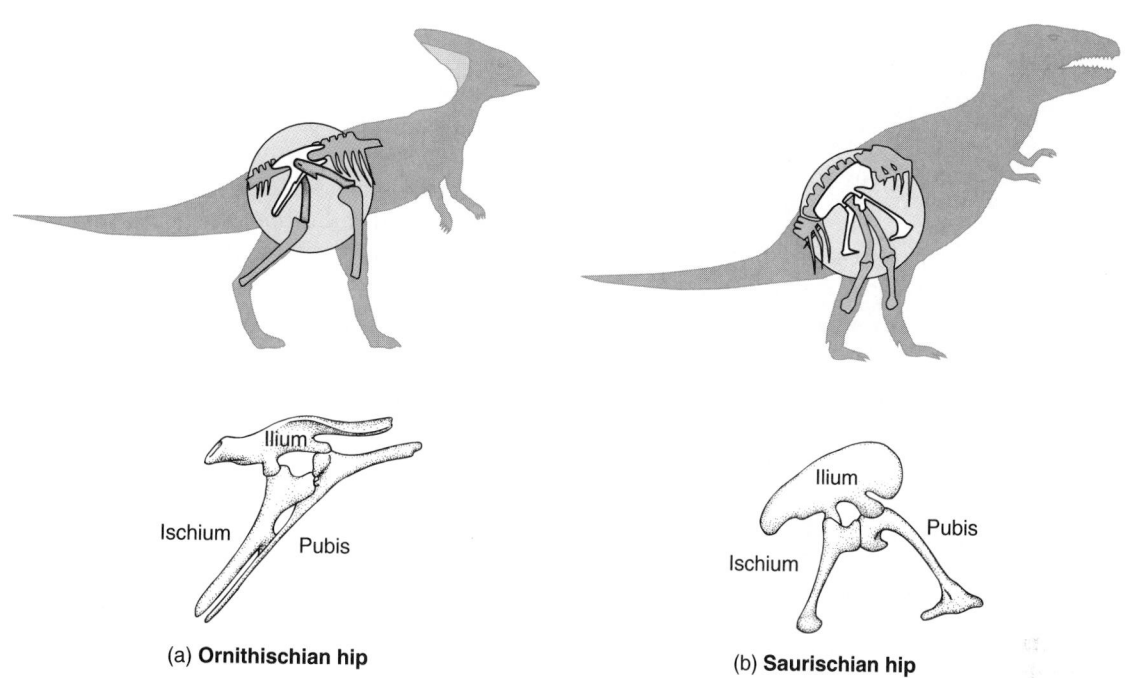

(a) **Ornithischian hip**   (b) **Saurischian hip**

FIGURE 4.26

Lateral view of the pelvis distinguishing two dinosaur groups. (*a*) A birdlike pelvis in which the pubis is parallel to the ischium and points to the rear. (*b*) A lizard pelvis—the pubis points ventrad. Note that birds (a subgroup of the saurischians) have independently acquired a birdlike pelvis.

high-altitude cold. (When birds produce excessive heat during flight, it is eliminated in expired air.) Feather pigments facilitate recognition by other members of the species and often provide protective coloration. Plumage coloration can vary by age and sex, thereby signaling the reproductive state of the individual.

Feathers have been equated with birds since they were first grouped together in Aves (considered a synapomorphy of birds in most phylogenetic hypotheses). However, recent fossil finds in China and elsewhere are shedding new light on the origin of feathers (Ackerman, 1998). *Sinosauropteryx, Caudipteryx,* and *Protarchaeopteryx* from China possessed feathers. The protofeathers of the former consisted simply of downlike filaments. *Caudipteryx* and *Protarchaeopteryx* had feathers, although it is assumed that they were not capable of powered flight. Our knowledge of these forms and their relationships is unclear. Whether they are birds (extending the possession of feathers to a larger but monophyletic group) or not (indicating feathers had a wider distribution among dinosaurs) must await further research.

In addition to feathers and wings, birds have other adaptations for flight. Body weight has been reduced in several ways. Long bones have become slender, and most bones, including vertebrae, lack the central marrow, which leaves cavities that contain air-filled extensions of air sacs from the lungs. The skull has been lightened by a thinning of the compact layers of the membrane bones, but it remains durable because su-

tures have been eliminated in adults, and the spongy bone that remains is architecturally strong. The bones of the wrist, palm, and digits have been reduced in number. Teeth do not develop, no urinary bladder develops, and the large intestine has been shortened. All these modifications, and others, have reduced the cost of flight in terms of energy expended. There are two generally recognized subclasses of birds, **Archaeornithes** and **Neornithes.**

Archaeornithes include a number of basal birds representing a paraphyletic assemblage. Two fossil birds about the size of a crow were found in Late Jurassic limestone deposits in Bavaria, Germany, in the nineteenth century. These were given the appropriate name *Archaeopteryx,* the first birds (fig. 4.27). The first specimen was complete except for a few cervical vertebrae, the right foot, and lower jaw. The second specimen was even more complete, and five additional specimens, not exactly alike, have been recovered elsewhere. *Archaeopteryx* had a long tail (supported by the axial skeleton as in nonavian reptiles), thecodont teeth on both jaws, and feathers on the wings and tail that were no different from today's. The skull was more similar to other reptiles than to modern birds, the nostrils were far forward, there was no beak, and the braincase had not expanded to accommodate an enlarged brain. The cervical vertebrae were not saddle shaped at the ends as in today's birds, trunk vertebrae were not rigidly fused, the synsacrum was not well developed, and the sternum was small, except in the last specimen recovered. The small sternum, which

FIGURE 4.27
Fossil *Archaeopteryx* embedded in limestone. Wings and their skeleton are near bottom of photograph; the long feathered tail is at top.
Courtesy of Dr. John Ostrom, Yale Peabody Museum. Portrait O.C. Marsh, photograph *Archaeopteryx*.

could not have accommodated strong pectoral muscles for sustained flight, suggests that these birds may have soared more than they flew. Figure 4.28 contrasts some of the more obvious features of the skeleton with that of a pigeon.

In 1986, two crowlike fossil skeletons 75 million years older than *Archaeopteryx* were discovered in a mudstone quarry in Texas. These specimens, for which the genus *Protoavis* was proposed, were more dinosaur-like than *Archaeopteryx*, and they had smaller wings. Whether or not they should be regarded as birds is in dispute. A *Protoavis* feather would be helpful!

Finally, the recent discovery of Early Cretaceous Chinese fossils with feathers (*Sinosauropteryx, Caudipteryx,* and *Protarcheopteryx*) raises again a number of questions concerning both the origins of feathers and of birds. Despite their more recent Early Cretaceous age, these taxa may possess more primitive features of flight than seen in the Late Jurassic *Archaeopteryx*. These new finds represent an exciting new period in the discussion of feathers, birds, and flight.

Neornithes includes three major subgroups, **Odontognathae, Palaeognathae,** and **Neognathae.** Odontognaths were toothed marine birds that used the land as a base for marine operations. The only known odontognaths are *Hesperornis* and *Ichthyornis* species. The first *Hesperornis* fossils were recovered from rocks formed 75 million or more years ago at the bottom of what was then a shallow North American sea extending inland northward as far as Kansas. Similar fossils have since been recovered in Europe. *Hesperornis* was covered with small, hairlike feathers. It had vestigial wings, so it could not fly, but it had stout legs for wading, and it was a good diver. Its diet consisted of fishes that it caught with its sharp, pointed teeth. *Ichthyornis* was an active flier that was able to go far offshore to feed.

Palaeognaths, known generally as ratites, have small incompetent wings, but they have powerful leg muscles that enable them to run well. They are descendants of active fliers. Many are known only as fossils, having been depleted by human society. Among current survivors are the rheas, ostriches, emus, and cassowaries. The extinct moas of New Zealand were more than 3 m (8 ft) tall, and they laid eggs more than 30 cm (1 ft) in diameter. Ostrich eggs weigh about 3 pounds.

Neognaths are, for the most part, birds that have a large carina to which relatively massive flight muscles attach. They are therefore generally known as carinates. There are about 10,000 species, and they include all living birds except the palaeognaths. The largest living carinate is the Andean condor, with a wingspread of 3 m (10 ft) and weight of up to 14 kg (30 lbs). The Giant Teratorn, a vulture that lived in Argentina 5 million years ago, had a wingspread of more than 7 m (25 ft) and an estimated weight of 23 kg (50 lbs). Penguins have a large carina, but their forelimbs have become flippers, so they cannot fly. However, they are powerful swimmers.

Many carinates are annual migrants, a behavioral adaptation to predictable annual climatic changes. The Arctic tern spends several months above the Arctic Circle and the remaining months in the Antarctic, traveling 22,000 miles round trip each year! During migration, birds move in mass flights, often at night and at an elevation of approximately 600 m (2,000 ft). Migration is physiologically demanding, and birds have become able to store and rapidly metabolize large energy reserves. An internally regulated annual cycle, entrained by seasonal environmental cues, coordinates the physiological and behavioral events associated with migration. How birds navigate has long challenged ornithologists. The sun, stars, barometric pressure, polarized light, naturally fluctuating magnetic fields, and low-frequency sound all seem to provide input, probably along with other cues at any time or location.

The number of bird species increased greatly in the Late Mesozoic. The spread of human habitation with the resulting destruction of the foraging and breeding sites is the major threat to birds in the twenty-first century.

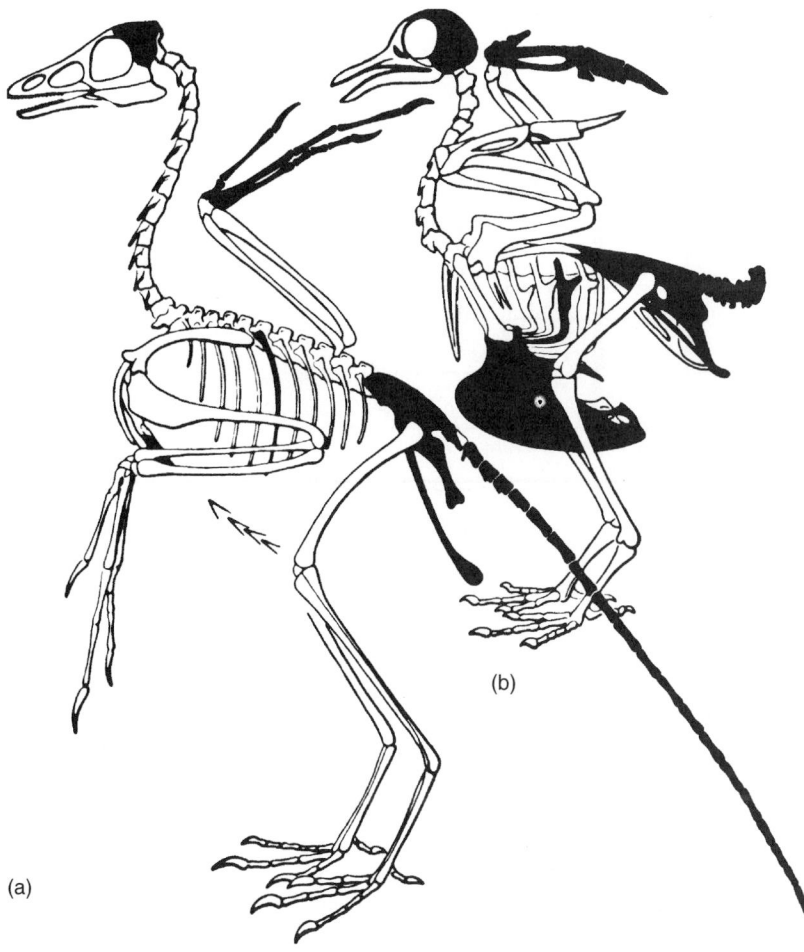

(b)

(a)

FIGURE 4.28

(*a*) *Archaeopteryx,* from the Jurassic.
(*b*) A carinate bird (pigeon) for comparison.

From E.H. Colbert & M. Morales, *Evolution of the Vertebrates: A History of Backboned Animals Through Time,* 4th edition. Copyright © 1991 John Wiley & Sons, New York, NY. Reprinted by permission of John Wiley & Sons, Inc.

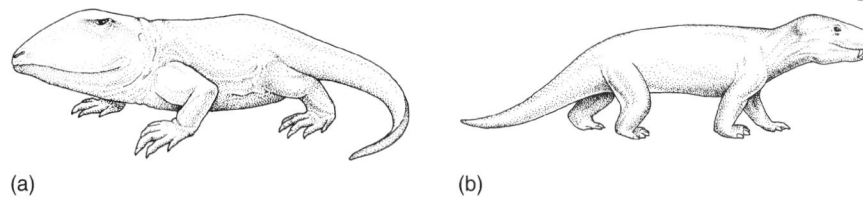

(a)   (b)

FIGURE 4.29

(*a*) Pelycosaur, a Carboniferous synapsid. (*b*) A Triassic therapsid about the size of a large dog. Nothing is known of the skin.

SYNAPSIDS

Synapsids were diverging from anapsid ancestors during the Carboniferous. They are the group from which mammals emerged during the Triassic. The synapsid lineage began with **pelycosaurs**, which were transitional between amphibians and later members of the group.

The pelycosaurs were succeeded by the **therapsids** (fig. 4.29), from within which mammals evolved. Mammals retain primitive therapsid features such as two occipital condyles, a secondary palate, and heterodont dentition with incisors, canines, and grinding molars (cheek teeth). The dentary was the largest bone in the lower jaw, presaging the condition in modern mammals (see fig. 9.39 stem therapsid). The distribution of selected amniotes in geologic time is shown in figure 4.30.

Mammalia

Mammals (a subgroup of therapsids) appeared at the end of the Triassic. They are amniotes with a synapsid skull, hair, and, except monotremes, mammary glands and nipples. Further distinguishing modern mammals from other craniates are the single dentary bone on each side of the lower jaw articulating with the squamosal bone; three bones in the middle ear cavity; a muscular diaphragm separating thoracic and abdominal cavities; sweat glands (in most mammals); absence of an adult cloaca in all but the monotremes; heterodont dentition (except in toothed whales); only two sets of teeth (milk teeth and a permanent set) instead of continual replacement as in other craniates; biconcave, enucleate red blood cells that are circular (except in camels and llamas); loss of the right fourth aortic arch; a sound-collecting lobe (pinna, or auricle), accessory to the outer ear; a more specialized larynx; and extensive development of the cerebral cortex.

Because of many variations in limb structure, mammals have been able to achieve a greater diversity of habitats than any other tetrapod. They burrow in the ground, hop, lumber, or gambol over the plains, scramble over mountain crags, swing through trees, propel themselves through the air in true flight, and inhabit the oceans, each lifestyle made possible by modifications of body structure.

Some taxonomies divide mammals into two major groups—**Prototheria**, which lays eggs and has a cloaca throughout life, and **Theria**, which gives birth to its young (fig. 4.31). Living prototherians are placed in the **Monotremata**. Therians are subdivided into two groups—those that have a yolk sac placenta (**Metatheria**) and those that have a chorioallantoic placenta (**Eutheria**).

Monotremata

Monotremes and therians are thought to have diverged early in mammalian evolution. Their name mirrors the observation that the cloaca has a single opening to the exterior (see fig. 15.48*a*). The sole surviving monotremes are the *platypus,* or duckbill, *Ornithorhynchus,* and two genera of spiny anteaters, *echidnas,* all from Australia or nearby Tasmania and New Guinea (fig. 4.32). Like reptiles, they lay heavily yolked eggs, they have a ventral mesentery that extends the length of the abdominal cavity, the testes are confined to the abdominal cavity, the outer ear has no pinna, and the brain lacks the great transverse fiber tract, the corpus callosum, that connects the two cerebral hemispheres in other mammals, except marsupials. The malleus and incus are larger than in other mammals, resembling the articular and quadrate bones of extinct therapsids. They have no nipples; a milky fluid exudes from modified sweat glands onto tufts of hairs in shallow pits on the abdomen, from which the young lick it up. Monotremes are endothermic, but their body temperature is less stable than that of most therian mammals.

Duckbills live in pairs in burrows up to 18 m (60 ft) long, in the banks of creeks and rivers, and spend most of their life in the water. They have webbed feet that enable them to walk on muddy river bottoms and have a soft, rubbery, sensitive, beaklike muzzle with which they detect invertebrates, especially molluscs, in the mud and dredge them. The food is then stored in a cheek pouch until it is crushed later with horny teeth that replace a prenatal set of regular ones. During breeding seasons, the female moves out of the burrow and constructs a new burrow just above the waterline, where she lays one to three eggs in a nest she has constructed. The eggs are nearly round, about 2 cm in diameter, and covered with a pliable white shell. She incubates the eggs for 10 to 14 days, after which they hatch.

Echidnas are terrestrial and have long, sticky tongues and stout claws that are used to gather ants and termites. Except on the abdomen, they are armed with sharp quills interspersed among coarse hairs, and they roll into a ball for protection. Usually, a single egg, about 4 mm in diameter, is incubated in a temporary pouch that develops as a thin flap of abdominal skin in females. The milk-secreting glands are on the abdomi-

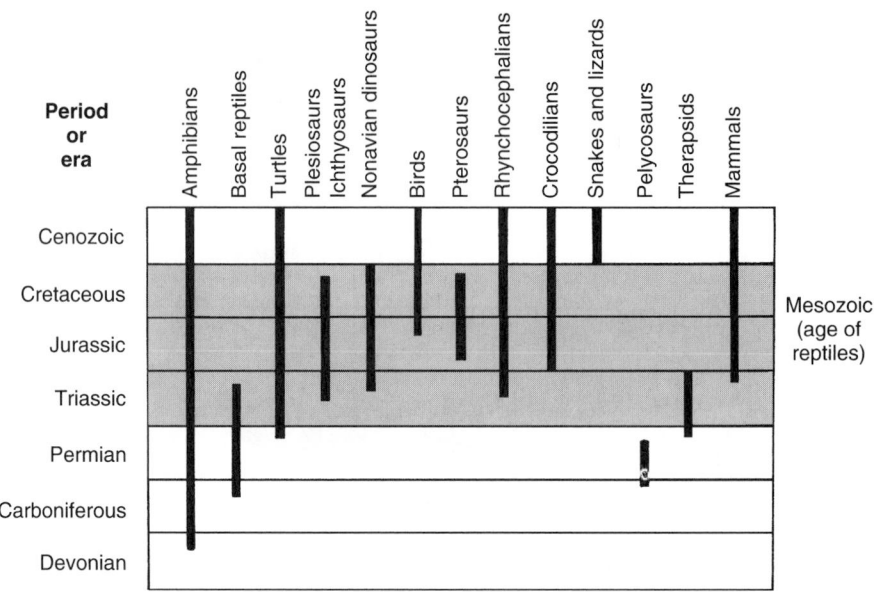

FIGURE 4.30
Range of selected amniotes through time.

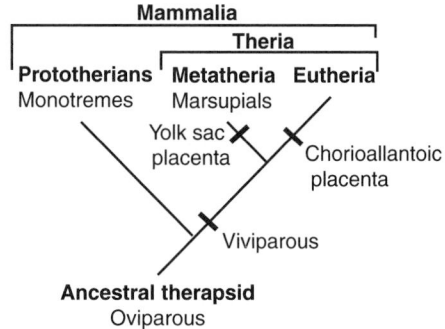

FIGURE 4.31
Major categories of living mammals.

nal wall in the pouch, and the young hatch and are carried in the pouch for several weeks until they can forage for food.

Marsupialia

Marsupials (fig. 4.33) are mammals in which the fetal yolk sac (in contact with the chorion) serves as a placenta. The young are born in almost a larval state and are incubated and nursed after birth in a maternal abdominal pouch (**marsupium**) of muscle and skin until they are old enough to be independent. The walls of the pouch are supported by two slender marsupial (epipubic) bones that project forward from the pelvic girdle. In several South American genera, the pouch is incomplete or absent. Newborn young make their way to the pouch by squirming and wriggling and by using the claws on their forelimbs, which are considerably larger than the hind limbs at birth. The lips of neonates are sealed at the angles of the mouth, which therefore is

Platypus

Echidna (spiny anteater)

FIGURE 4.32
The two surviving monotremes, the platypus *Ornithorhynchus* and a spiny anteater.

only a small, circular opening. Once the young has taken a nipple into its mouth, the tip of the nipple swells, and the young cannot easily drop off.

Marsupials and monotremes were the only mammals in Australia, with the possible exception of bats, until the arrival of the first human inhabitants. The people brought with them rats and mice, which inhabited the wooden sailing vessels. Australia at one time was part of a huge supercontinent that included the present Antarctica, Africa, and South America. The breakup of this supercontinent into several land masses eventually separated Australia from Antarctica, and these two continents drifted apart. The absence of eutherian fossils in Australia indicates that eutherians were not present when that land mass separated from Antarctica.

Marsupials diverged from other therians in the Late Cretaceous. They appear to have arisen in what is now the northwestern part of North America, and they arrived in Australia before it was isolated. Their southbound trail crossed a land bridge to South America and eventually reached Australia via Antarctica (fig. 4.34). (The present Central American land bridge was formed later.) Metatherian fossils of increasingly later ages have been found in North America along the southbound route.

While Australian marsupials were undergoing an extensive adaptive radiation, those on other continents were facing extinction in competition with evolving eutherians. With the exception of opossums and a few other genera in North and South America, marsupials

are now found only in Australia, Tasmania, and New Guinea, and on a few offshore islands. Marsupial fossils found in Europe attest to their presence there until about 15 million years ago. Continental breakup provided the geographical isolation that fostered marsupial radiation in Australia and eutherian radiations elsewhere.

Among marsupials are kangaroos, wallabies, Tasmanian wolves, bandicoots, wombats, anteaters, and phalangers. These and other marsupials resemble eutherian mammals—wolves, foxes, bears, rabbits, mice, cats—in surprising detail. Some phalangers resemble "flying" squirrels, and there are marsupial moles. The term *evolutionary convergence* does not apply to this phenomenon because metatherians and eutherians shared ancestral therian genes. The term *parallel evolution* has been applied to this phenomenon. It connotes that similar traits have evolved in isolated taxa whose common ancestor lacked such traits.

Insectivora

The members of this basal order are generalized mammals (fig. 4.35). Although at one time they were abundant, they are represented today by relatively few survivors, including moles, shrews, and hedgehogs. There are numerous examples of insectivory among mammals in other orders, but the latter are thought to represent examples of convergent feeding strategies. Most insectivores subsist on a diet of insects, worms, and other small invertebrates.

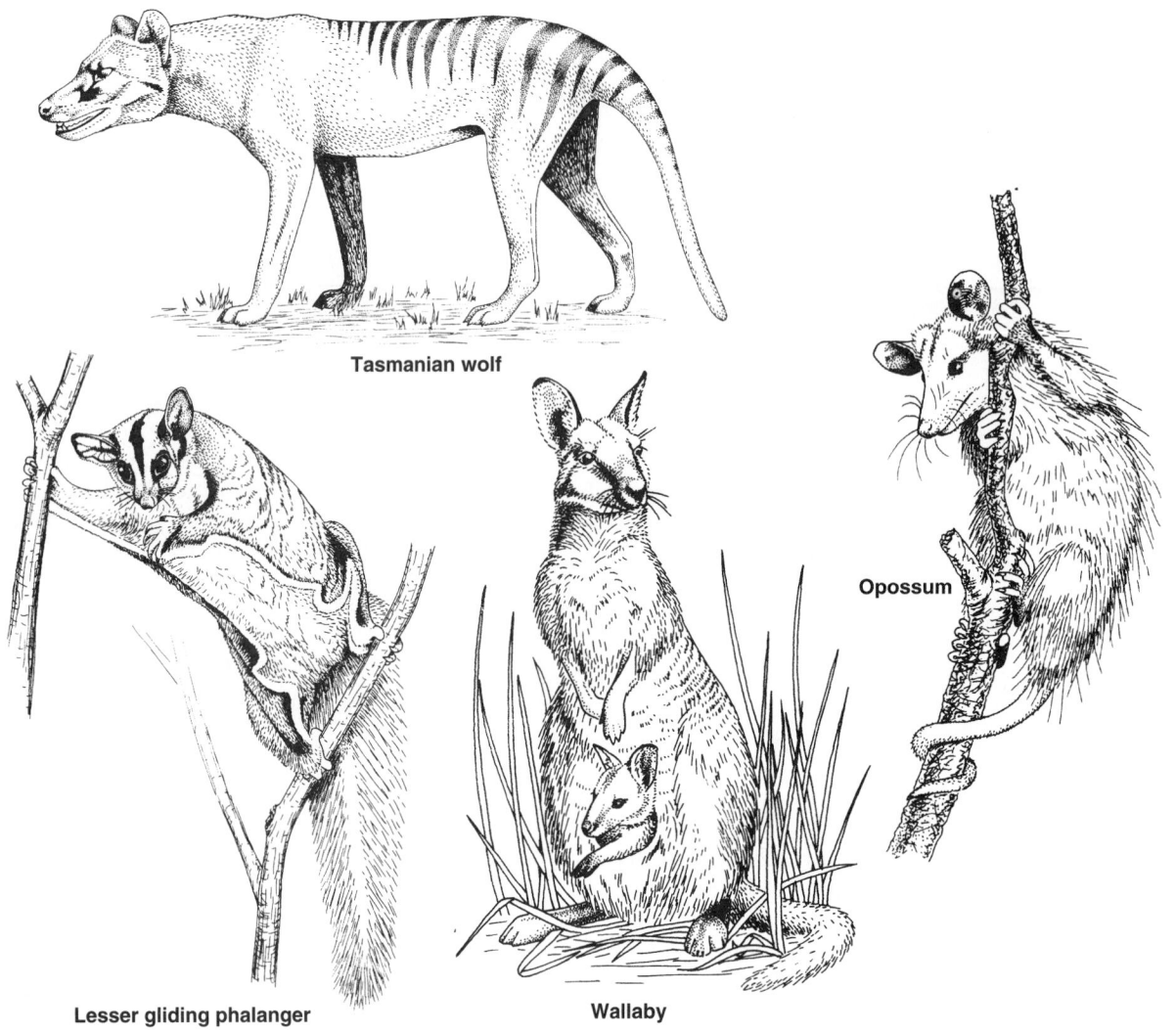

FIGURE 4.33
Four marsupials.

Among primitive traits of the members of this order are a flat-footed (plantigrade) stance; five toes; smooth cerebral hemispheres; small, sharp, pointed teeth with incisors, canines, and premolars poorly differentiated; a large embryonic allantois and yolk sac; and, in some species, a shallow cloaca. The testes are retained in the abdominal cavity in some genera (a primitive trait), and they never descend fully into scrotal sacs in any genera.

Moles have short, stout anterior limbs, with forefeet that are broad and more than twice as large as the hind feet, which is an adaptation for digging (see fig. 10.21). The neck is short, and the shoulder muscles are so powerful that the head and trunk seem to merge. Their tiny eyes are practically useless, but an acute sense of smell locates distant food, and the sensitive tip of the elongated snout tells them when they have encountered it. Shrews superficially resemble mice. They are shy, busy little fighters with a keen sense of hearing. They have an elongated, bewhiskered, sensitive snout, and their incisor teeth are long and curved.

Xenarthra

Xenarthrans are New World insectivorous mammals that are considerably more specialized than members of the order Insectivora. They are **armadillos** (fig. 4.36a), **sloths**, and **South American anteaters**. None have incisor or canine teeth, and cheek teeth, when present, are peglike and lack enamel, although histological evidence of enamel organs has been reported. Anteaters are completely toothless. Enlarged front claws are used for digging into ant nests or mounds and, conveniently for sloths, for hanging from the limbs of trees, which are their major habitats. Armadillos are notable for always giving birth to identical quadruplets from a single fertilized egg. They are also the only mammals that develop a true bony dermal armor.

Tubulidentata

A single species of Central and South African insectivorous **aardvarks** comprises the order Tubulidentata

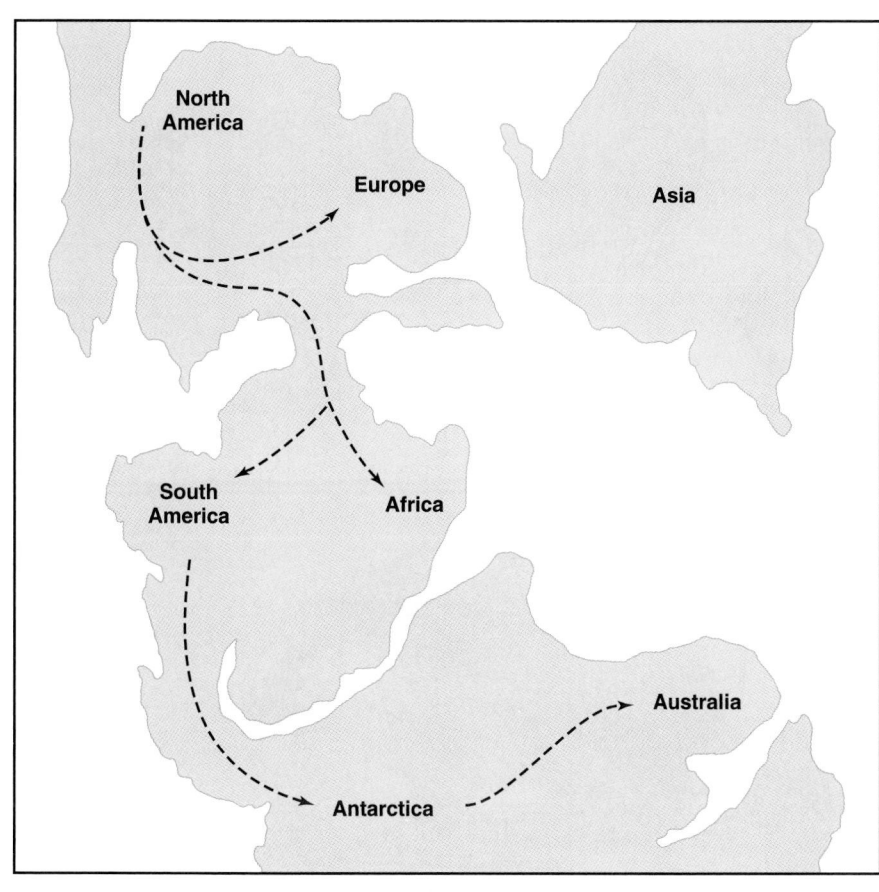

FIGURE 4.34
Postulated dispersal of evolving marsupials during the Mesozoic.

(fig. 4.36*b*). Its elongated snout, long sticky tongue, and strong claws on the front feet facilitate routing out and capturing insects. Aardvark teeth, like those of many other insectivores, are peglike, lack enamel, and have shallow roots.

Pholidota

Another anteater, the **pangolin** of Africa (fig. 4.37), is toothless and peculiarly scaly. It belongs to the order Pholidota. Its scales are made of keratin and appear to be agglutinated hairs. Typical hairs grow between the scales. There is only one genus, *Manis.*

Chiroptera

Bats (fig. 4.38) comprise a large mammalian order that is probably derived from a basal insectivore. They, pterosaurs, and birds are the only known craniates to have achieved true (powered) flight. The wing, or **patagium,** is a double membrane of skin stretched along the length of the body between the trunk, forelimbs, and hind limbs, and extending from there to the tail. It incorporates four greatly elongated, clawless fingers. The first digits, or thumbs, project from the anterior margins of the wing membranes and bear claws that aid in clambering about on rough vertical surfaces. All five digits of the hind limbs bear claws, and these are used for hanging upside down, wings folded, from crevices in caves, from

small branches, or within hollow trees. The pectoral muscles are strong, and the sternum is keeled, although not as much as in carinate birds. All the bones are slender, but not pneumatic. Nipples, usually a single pair, are limited to the thoracic wall. Bats have large external ear auricles (pinnas), and their facial glands are prominent, giving the head a bizarre appearance. The development of a wing in pterosaurs, birds, and bats is indicative of convergent evolution. The bumblebee bat, weighing less than 2 gm, is the smallest mammal on earth.

Bats are insectivorous, frugivorous (fruit eaters), or sanguinivorous (subsisting on the blood of other mammals). Vampire bats have received attention because of sanguinivorous habits. Incisor teeth occur only on the upper jaw, and there is only one pair. They are razor sharp and point toward one another so they slit the skin of prey. As blood oozes from the wound, the bat licks it up without awakening the sleeping victim, which is usually a domestic animal. Associated with the vampire bat's practice of taking only fluid nourishment is the very small lumen of the esophagus, through which no solid food could possibly pass.

Primates

Primates (prī-mā′tēz) are primarily arboreal mammals that arose as an offshoot of Cretaceous insectivore stock. One of many different classification schemes divides them into two suborders, **Prosimii** and **Anthropoidea.**

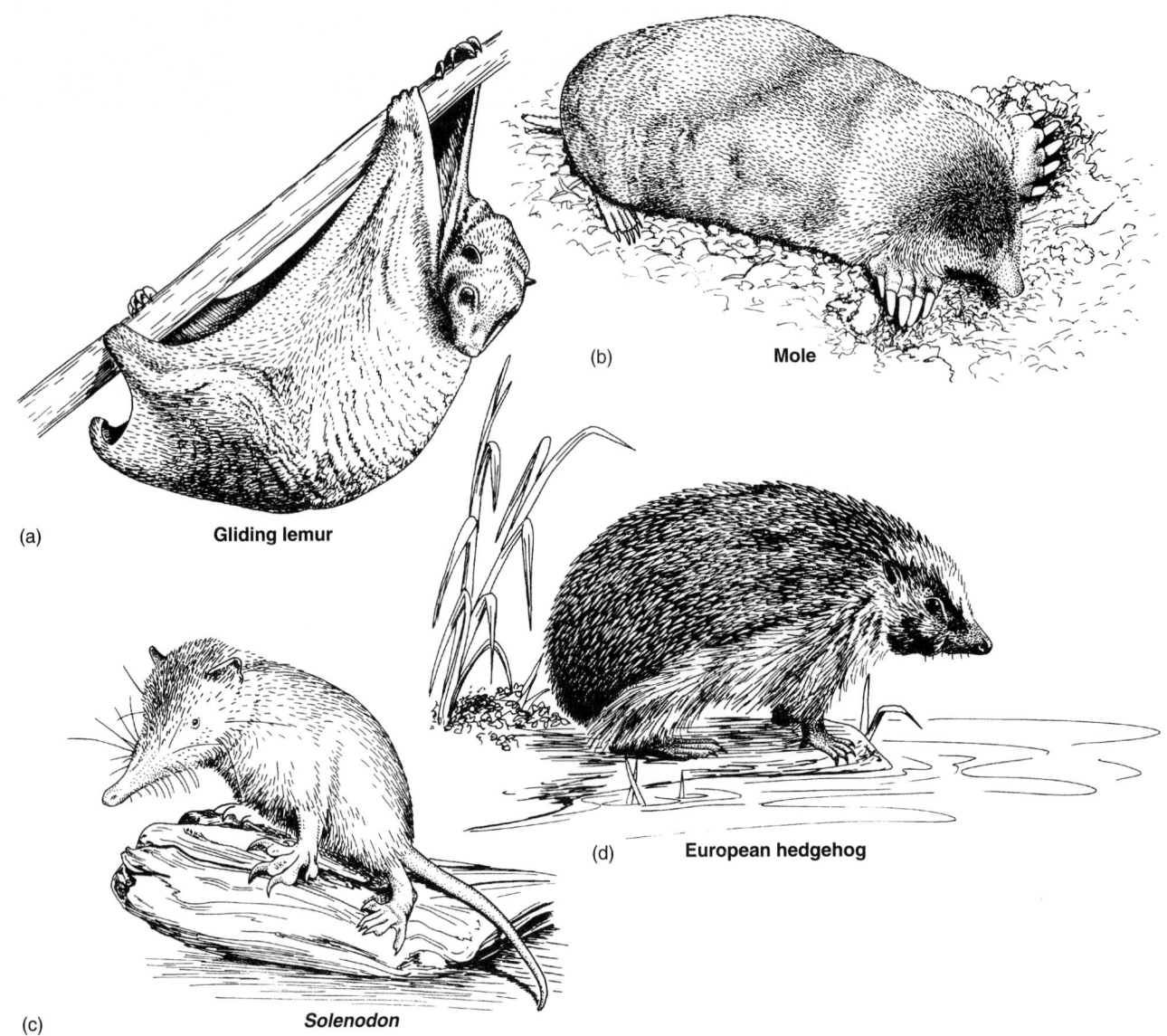

(a)   **Gliding lemur**

(b)   **Mole**

(c)   *Solenodon*

(d)   **European hedgehog**

FIGURE 4.35
Four members of the order Insectivora.

Lemurs, lorises, and tarsiers are prosimians. Monkeys, apes, and humans are anthropoids.

Among primate specializations is the grasping hand, built so that the thumb can touch the tips of the other four fingers of the same hand. The big toe is also opposable in most primates, and at least some digits have nails instead of claws. The cerebral hemispheres are larger than in any other mammal. There is frequently only one pair of nipples. Among primitive features are a flat-footed stance, five digits, a large clavicle, a central carpal bone in the wrist of many primates, and a generalized mammalian dentition.

Prosimians

Prosimians are arboreal and mostly nocturnal primates found in the tropics of the Old World. **Lemurs** (from the Latin word for "ghost") receive their name from the habit of swinging silently through the forest at night. Unlike other primates, they have the long axis of their head in line with the vertebral column (fig. 4.39*b*). The largest lemurs are the size of a domestic cat. **Lorises,** from India, Sri Lanka, and Southeast Asia, have no tail, and the index finger is vestigial. In the same family are the bush babies and pottos.

Tarsiers (fig. 4.39*a*) resemble anthropoids to a greater degree than do the other prosimians. Their head is more nearly balanced at right angles to the vertebral column, the snout is truncated, and their eyes are close together and directed forward so there is an overlap in the left and right visual fields. All fingers have nails, as do all toes except the second and third. Their placenta is deciduate; that is, the fetal membranes become rooted into the wall of the maternal uterus, as in anthropoids.

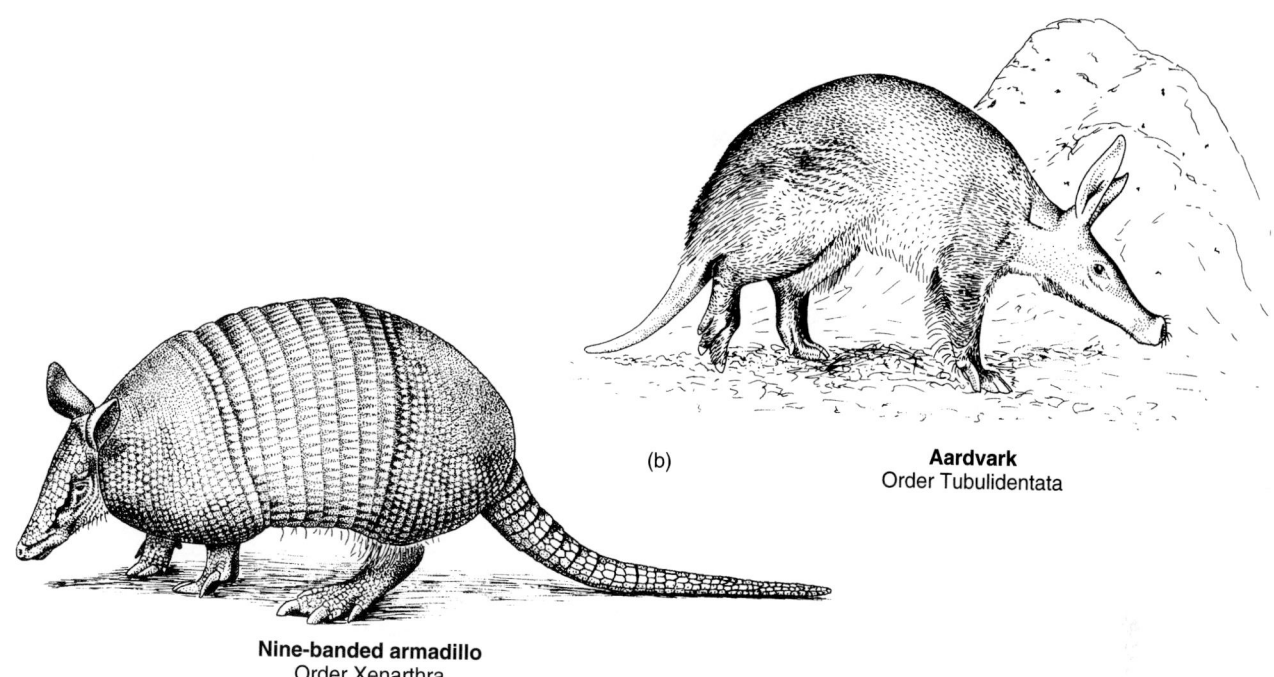

(b)

Aardvark
Order Tubulidentata

Nine-banded armadillo
Order Xenarthra

(a)

FIGURE 4.36
Two anteaters.

FIGURE 4.37
A pangolin, the scaly anteater.

Anthropoids

There are two groups of anthropoids, platyrrhines and catarrhines. They are differentiated on the basis of the direction that the nostrils open. Those of platyrrhines are separated by a broad internarial septum and open to the side (fig. 4.40). The nostrils of catarrhines are close together and open downward.

Platyrrhines are the New World monkeys and marmosets. They include capuchins (*Cebus*), spider monkeys (*Ateles*), and howler monkeys (*Alouatta*). Howler monkeys are named for the screeching cries they produce from an enlarged larynx and hyoid bone (see fig. 13.14).

Catarrhines are Old World monkeys, apes, and humans. The Old World monkeys are in a separate superfamily, **Cercopithecoidea**. Among Old World monkeys are baboons and macaques, or rhesus monkeys. The abbreviation "Rh," used in designating human blood types, was derived from the latter, in which it was first detected. Apes (chimps, gorillas, orangutans) are in the pongid family, and humans are hominids (family **Hominidae**, see the appendix). Catarrhines have no tail. A well-developed tail is present in fetuses. In humans, the tail disappears except for remnants during the third month of fetal life.

The skull of all anthropoids is set at right angles to the vertebral column, the eyes are directed forward and are close together, the cerebral hemispheres are highly developed, there are 32 teeth in the permanent set, the placenta is deciduate, and, usually, a pregnancy produces only one infant. Despite any specializations, anthropoids, including humans, are much less specialized than many other mammals—whales and perissodactyls, for example.

Since the discovery of Neanderthal man (*Homo neanderthalensis*), who lived in Europe, the Middle East, and parts of central Asia until about 30,000 years ago, fossil remains of much earlier hominids have been found in considerable numbers, particularly in Africa. With each new discovery, the problem arises of whether to include the new member in an existing species or to erect a new one, or even a new genus. The criteria for classifying other craniates are applied to the classification of hominids and are subject to the same contingencies. To date, the oldest fossil considered to be a hominid is *Ardipithicus*, about 4.5 million years old. The oldest *Australopithicus* is about 4 million years old. Sufficient *Australopithicus* bones have been recovered to estimate that these individuals were about 5 ft tall and had a humanlike skeleton, facial bones of an ape, and an ape-sized cranial cavity. They made crude tools of stone or bone. *Homo erectus*, 1.5 million years old, is thought to be the immediate ancestor of both *Homo neanderthalensis* and *Homo sapiens* (modern humans). *Homo sapiens*, about 100,000 years old, lived in Europe at the same time as the Neanderthals and eventually replaced them on that continent. The history of humans includes continual intraspecific warfare. It is possible this behavior brought about the extinction of the Neanderthals.

With the emergence of humans from an earlier anthropoid, many changes occurred. An **S**-shaped curve in the vertebral column permitted an erect human posture; the facial angle became less acute (fig. 4.41); the teeth, especially the canines, became smaller; the frontal lobes of the cerebral hemispheres enlarged, resulting in a larger braincase and a more prominent forehead; the eyebrow ridges became reduced; the nose became more prominent; the tail became confined to embryonic stages; the arms became shorter; a metatarsal arch developed in the otherwise flat feet; the big toe moved in line with the other toes and ceased to be opposable; articulate speech became structurally possible; and many other changes occurred.

It is to the development of the massive frontal lobes of the cerebral hemispheres, to the opposable thumb, and to articulate speech that humans owe their present dominance in the animal kingdom. With fingers they can construct instruments of offense and defense and equipment to lighten burdens, and with them they write symbols that convey experiences and techniques to the corners of the earth and to generations unborn. With voices they communicate with contemporary fellow creatures and exchange ideas with delicate shades of meaning. With larger brains, they associate sensory stimuli that are currently received with those recalled from earlier experiences, and after meditation elect a

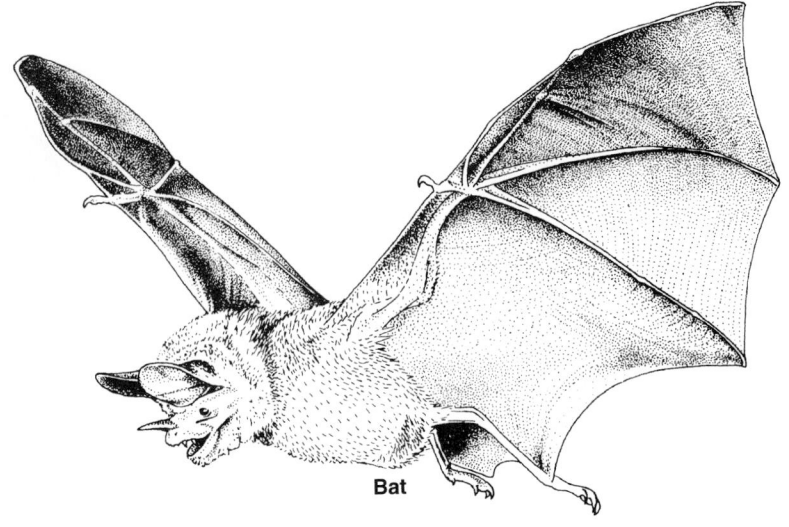

Bat

FIGURE 4.38
A chiropteran.

mode of action that they call "intelligent." With their brains, too, they can enjoy the esthetic beauty of the universe, search for ultimate truth, and dream of Utopia, which, but for their residual animal nature, might be their heritage.

Lagomorpha

There are only two families of lagomorphs (fig. 4.42), the pikas and the hares and rabbits, all of which are herbivores. Lagomorphs differ from rodents in having *two* pairs of incisors on the upper jaw, a small pair lying immediately behind, not alongside of, a much larger pair. The large front pair is rodentlike and grows throughout life. The smaller pair lacks cutting edges.

Rabbits differ from hares in being born in a fur-lined nest, blind, virtually hairless, and helpless. Hares (fig. 4.42*a*) are born without benefit of a nest, fully formed, and with eyes that open within an hour. They are able to hop around before they are a day old. Some hares of the genus *Lepus* acquire white fur in winter; no rabbits do. Jackrabbits of western North America and snowshoe rabbits of the Arctic are hares, but the so-called Belgian hare is a rabbit! Rabbits and hares have a split upper lip, which inspired the term *harelip* for the human anomaly.

Pikas (fig. 4.42*b*) live above the timberline in western North America and northern Asia. They differ from rabbits and hares in having a smaller body, shorter ears, and fore and hind limbs of about equal length.

Rodentia

Rodents comprise the largest mammalian order and are distributed worldwide. They have a *single* pair of long, curved incisor teeth on each jaw that is used for gnawing. These teeth have enamel on their outer surface only, which provides a chisel-like edge as the softer dentin

FIGURE 4.39
Two prosimians: (*a*) *Tarsius,* (*b*) *Lemur.*

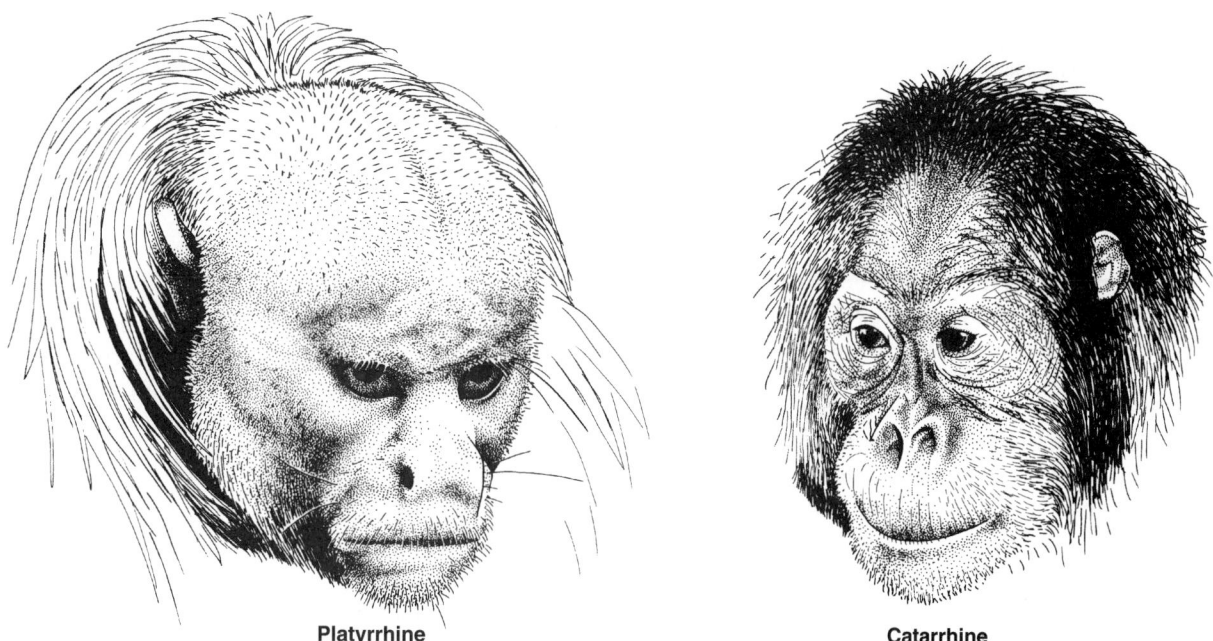

Platyrrhine   **Catarrhine**

FIGURE 4.40
Two anthropoids. The platyrrhine is a scarlet-faced red ouakari, a small South American monkey with a nonprehensile tail. The catarrhine is an orangutan.

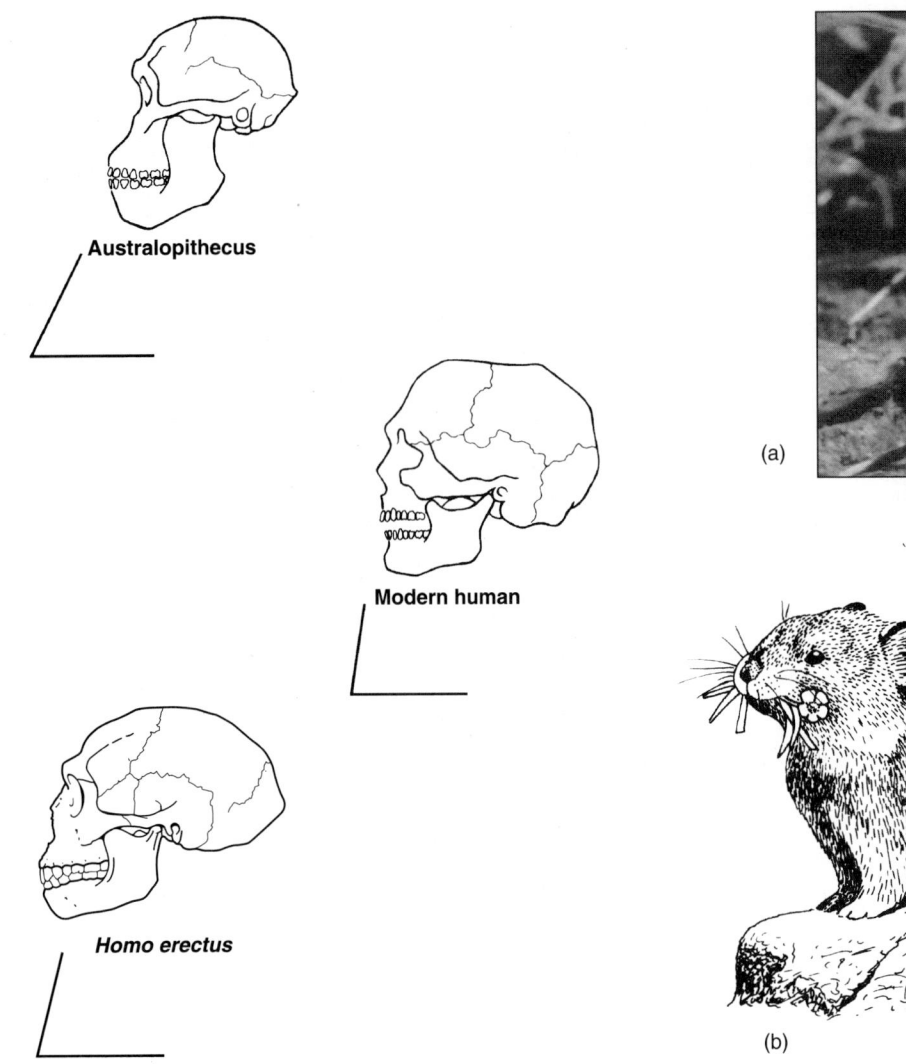

FIGURE 4.41
Facial angles of three hominids. The angles are shown beneath the skulls.

FIGURE 4.42
Two lagomorphs. (*a*) The hare, *Lepus,* is shown in its brown summer coat. (*b*) A pika, *Ochotona.*
(*a*) © Leonard Lee Rue III.

wears away. The teeth grow throughout life. Canine teeth are absent, so there is a diastema, or a stretch of toothless jaw, between the incisors and the first grinding teeth. Rodents are cellulose eaters, and at the beginning of the large intestine is a long, coiled cecum that houses commensal cellulose-digesting microorganisms.

There are two suborders. **Sciurognathi** is subdivided into three infraorders (Sciuromorpha, Castorimorpha, and Myomorpha). **Sciuromorpha** includes squirrels, marmots (woodchucks), and prairie dogs. **Myomorpha** includes pocket gophers, micelike rodents such as rats, voles, hamsters, and lemmings. Suborder **Hystricognathi** consists of a single infraorder **Caviomorpha**. This group includes cavys, nutria (coypus), chinchillas, and porcupines (fig. 4.43).

Carnivora (Fissipedia)

Carnivora is a large, diverse terrestrial group, some species of which have powerful jaws and sharp upper canine teeth capable of spearing and tearing flesh. The saber-toothed tiger, now extinct, was an extreme example. There are typically 42 permanent teeth. (Human beings have 32.) Premolars tend to be tricuspid; molars have four cusps. The clavicles are reduced, vestigial, or missing altogether. The cerebral cortex is unusually convoluted compared to most mammals. A combination of these and other morphologic traits—not a meat-eating, or carnivorous, diet—define the order. Most terrestrial carnivores have five toes (a few have four) with sharp and sometimes retractable claws (e.g., seen in most species of cats).

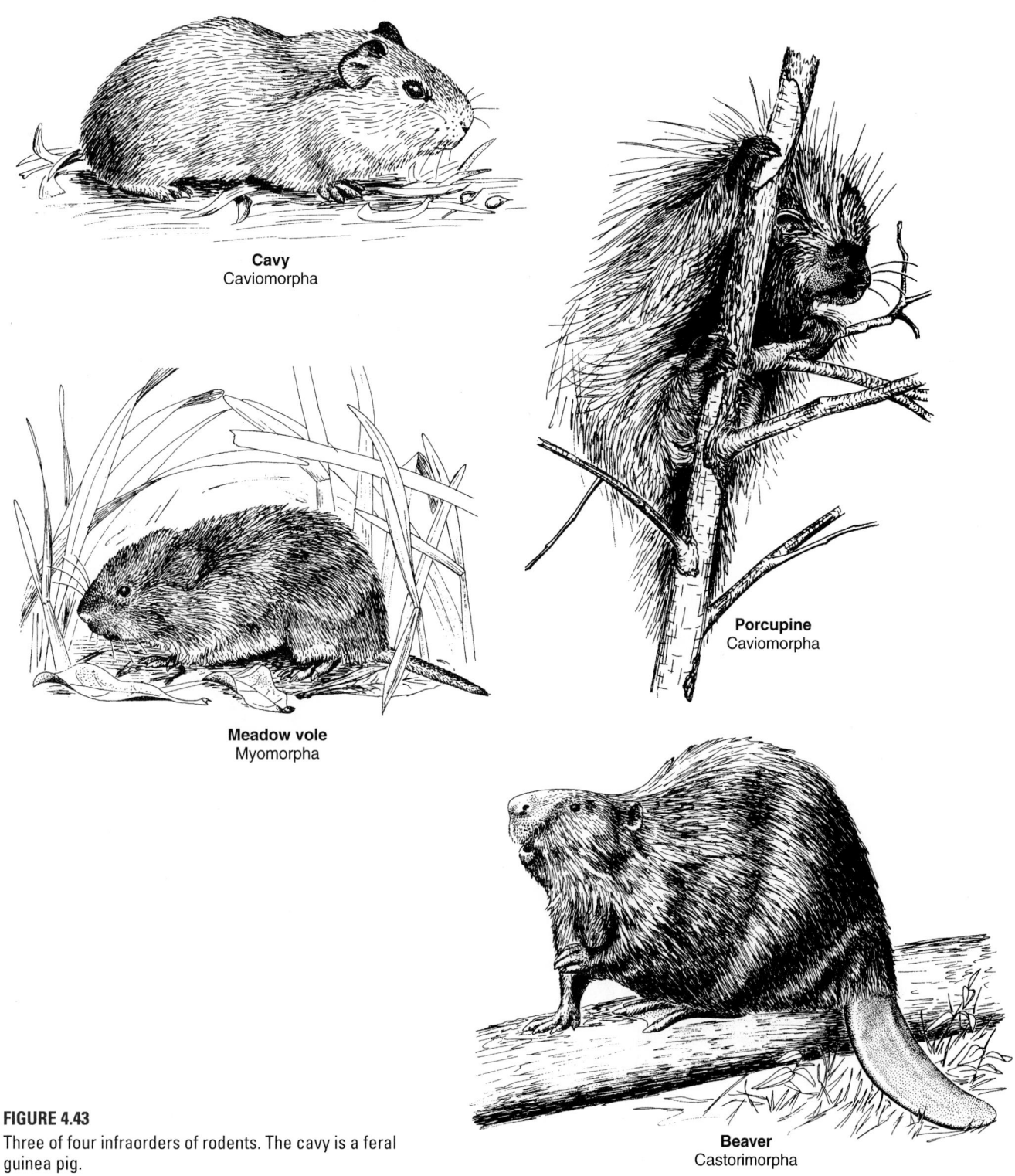

FIGURE 4.43
Three of four infraorders of rodents. The cavy is a feral guinea pig.

There are currently eight families of carnivores: **cats** (domestic cats, panthers, tigers, lynx, others), **civets, hyenas, canines, bears** (including **pandas,** although some authors consider pandas to be more closely related to raccoons), **raccoons, mongooses** and **mustelids** (mink, otter, skunks, ferrets, badgers, oth- ers), all with large scent glands near the anus (fig. 4.44). Not all members of the order are carnivorous, or even predominantly so. For example, bears other than polar bears (fish eaters) are partly herbivorous, and pandas eat only nuts, roots, bamboo shoots, and other succulent vegetation.

Raccoon

Panda

African civet

River otter
Mustelid

Spotted hyena

FIGURE 4.44
Representatives of five families of terrestrial carnivores.

Pinnipedia

These are carnivores that are morphologically and physiologically adapted for an aquatic life. They are closely related to the terrestrial carnivores but have been separated for a considerable time. They feed on fish, squid, molluscs, and crustaceans. Pinnipeds are the earless (wriggling) seals, eared (fur) seals, sea lions, and walruses (fig. 4.45). Earless seals lack the pinna of the outer ear characteristic of most mammals, but they have excellent hearing, nonetheless. All pinnipeds leave the water at least once a year, at breeding time. Females give

FIGURE 4.45
Harbor seals (marine carnivores).
© Art Wolfe/Tony Stone Images.

birth a year after mating, and the young are born on land, unable to swim. Among pinniped adaptations for life in the water are webbed paddlelike limbs known as flippers. The digits are mostly enclosed within the distal end of the flipper and, unlike those of land-dwelling carnivores, they usually lack claws.

Seals and sea lions seek their food deep in the oceans, including beneath the dense Antarctic ice pack. Northern elephant seals have been recorded at depths of over 900 m (3,000 ft), where they may remain for as long as an hour. The time spent at the surface between dives for replenishing their oxygen supply is only a few minutes, during which time the oxygen in the inhaled atmospheric air is transferred from the lungs to the bloodstream, to be released as needed. The animals then exhale before going back down, and the lungs collapse at 40 m (130 ft). Walruses, on the other hand, are shallow divers.

Ungulates and Subungulates

Ungulates are mammals that walk on the tips of their toes, protected by hoofs. There are two orders: Perissodactyla and Artiodactyla. They have no more than four toes on each foot, and some, such as horses, have only one. Ancestral ungulates had five, the generalized tetrapod number. Reduction in the number of toes is documented by extensive studies of the evolution of horses, whose small Eocene ancestors had four toes in front and three behind.

Ungulates are herbivores. Their molar teeth are high crowned and deeply grooved for grinding vegetation, and there is little morphologic difference between premolars and molars. Ungulates have no clavicle. Its absence facilitates grazing. Ungulates are the only mammals with horns.

Subungulate is a term that has been applied to elephants (Proboscidea), hyraxes (Hyracoidea), and manatees (Sirenia), which are discussed shortly. The taxonomic status of these groups, like that of many mammals, is debatable.

Perissodactyla

There are three families of perissodactyls: **horses and horselike mammals, tapirs,** and **rhinoceros** (fig. 4.46). They walk on the hoofed tips of one, three, or occasionally four toes and are distinguished by the observation that most of the body weight is borne on a single digit (see figs. 10.33, rhinoceros and horse, and 10.34). This is a **mesaxonic foot**. Perissodactyls are usually called odd-toed ungulates, but tapirs and some rhinos have four toes on the forefeet.

Artiodactyla

Artiodactyls are ungulates in which the weight of the body is borne on two toes (fig. 4.47). This is a **paraxonic foot** (see fig. 10.34). Living artiodactyls have an even number of toes, but at least one extinct artiodactyl had five toes on the forelimb. Artiodactyls include pigs, hippopotamuses, peccaries, cattle, camels, deer, antelope, and giraffes. With the exception of pigs, hippopotamuses, peccaries, and camels, they have stomachs divided into not fewer than three chambers, sometimes four. They bolt their food, which then passes to the rumen, the first segment of the stomach. Such animals are **ruminants**. At their leisure, they force undigested food balls back up the esophagus and masticate this cud more thoroughly.

Hyracoidea

The order consists of three genera of **hyraxes** (fig. 4.48), with four digits on the forefeet and three on the hind feet. Although hyraxes are flat-footed (plantigrade), all digits except one end in small flat hoofs. The teeth of grazing members are high crowned like those of ungulates. Because of these and other features, some systematists include hyraxes with the perissodactyls. The upper lip is split (harelip), and the incisor teeth grow continuously, as in rodents.

Proboscidea

The order Proboscidea includes elephants, mastodons, and their relatives. They have a proboscis, scanty hair, and thick, wrinkled skin. The incisor teeth of one or both jaws are elongated to form tusks, canine teeth are absent, and the molars are large grinders, as in ungulates. Proboscideans are bulky animals, and the limbs are almost vertical pillars of bone and muscle. They have five toes that end in thick, hooflike nails. An elastic pad on the back of each toe bears much of the body weight. The pad is not present in ungulates.

Because of the structure of the teeth and feet, proboscideans have been considered subungulates. But applying gene sequence data of nuclear and mitochondrial genes leads to the postulate that elephants are no more closely related to ungulates than they are to other orders of mammals.

Sirenia

Manatees and dugongs, known also as sea cows, constitute the order Sirenia (fig. 4.49). Like ungulates, they are strictly vegetarian and are thought to be descendants of early ungulate stock. The forelimbs are paddles, but within them the tetrapod complement of bones is intact. The hind limbs are lost, but there are internal skeletal vestiges of them and of the pelvic girdle.

Indian rhinoceros

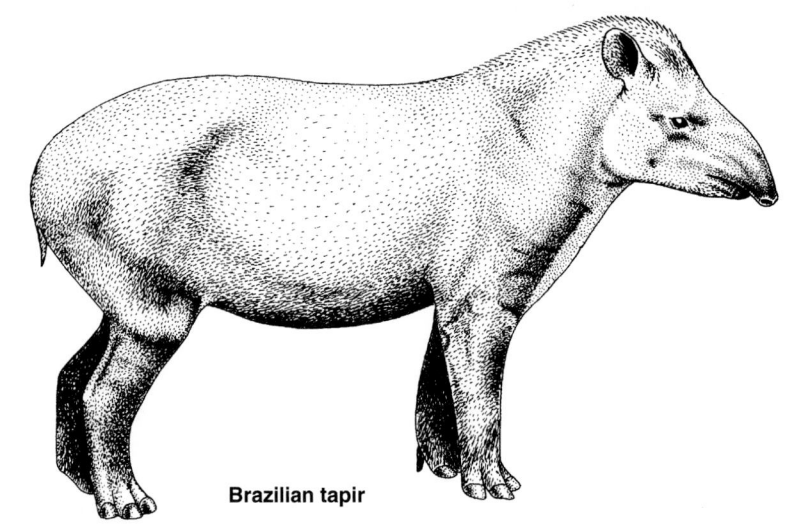

Brazilian tapir

FIGURE 4.46
Two perissodactyls.

Peccary

FIGURE 4.47
An artiodactyl.

FIGURE 4.48
A rock hyrax, order Hyracoidea.

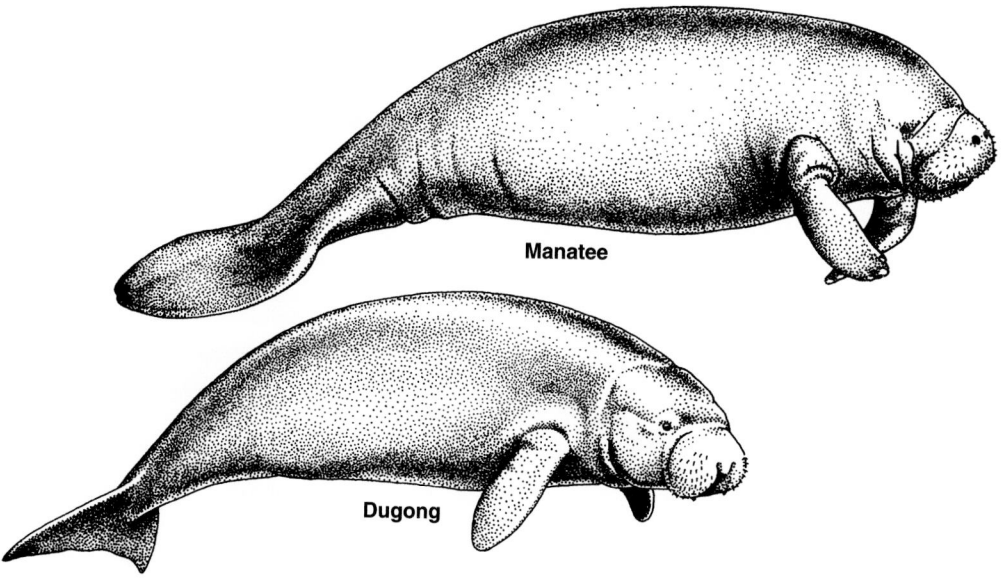

FIGURE 4.49
Representatives of the order Sirenia. Notice the notch in the fluke of the dugong, and vestigial nails on the flippers of the manatee.

half of the trunk provides most of the thrust for locomotion. The forelimbs are paddle-like, with the phalanges embedded in the distal end. Hind limbs and girdles are either mere vestiges embedded in the body wall or, as in some species, they are entirely missing. Early Cenozoic ancestors had small, but complete, hind limbs. A dorsal fin, also of fibrous tissue, is present in some species.

The frontal and nasal bones are short, with the result that the nostrils are on top of the head, sometimes united to form a single large blowhole. The nostrils have valves that close underwater. The thoracic diaphragm is exceptionally muscular, and the whale's waterspout, composed mostly of water vapor, may last from three to five minutes. The vapor forms in cold air when warm respiratory air is expelled at the water's surface. A heavy layer of subcutaneous fat (blubber) conserves body heat in the cold ocean.

Most cetaceans have teeth, but whalebone whales have, instead, frayed horny sheets of whalebone (baleen) hanging from the palate (see fig. 6.27). These strain several tons of small fish, crustaceans, or plankton, depending on the species, out of the sea each day. A few hairs occur on the muzzle of cetaceans, but elsewhere the skin is naked. Babies are born in the sea. The mother or an assistant bites the umbilical cord to separate the neonate from the afterbirth, and the mother nudges the baby to the surface for the first inhalation of air. Thereafter, the baby hangs onto one of the four inguinal teats as the mother swims about.

Cetaceans have good eyesight, but they also scan their environment by echolocation, which can tell them even the texture of another object. They have an excellent sense of taste but no sense of smell, because their olfactory nerves do not develop. They would be useless, anyway; if the animal were to inhale under water, it would drown. Cetaceans communicate with one another by distinct whistles (at least 16 in bottle-nosed dolphins), snapping the jaws, and slapping the water with the flukes.

The snout is covered with coarse bristles, and the rest of the body is naked except for a few scattered hairs. The flukelike tail is a result of evolutionary convergence. Sirenians and cetaceans are the only completely adapted aquatic mammals. All others must return to rookeries to breed.

Cetacea

Today's cetaceans—whales, dolphins, and porpoises—are permanently marine mammals (fig. 4.50). Their tail consists of two horizontally oriented fleshy lobes, or flukes, composed of connective tissue. The posterior

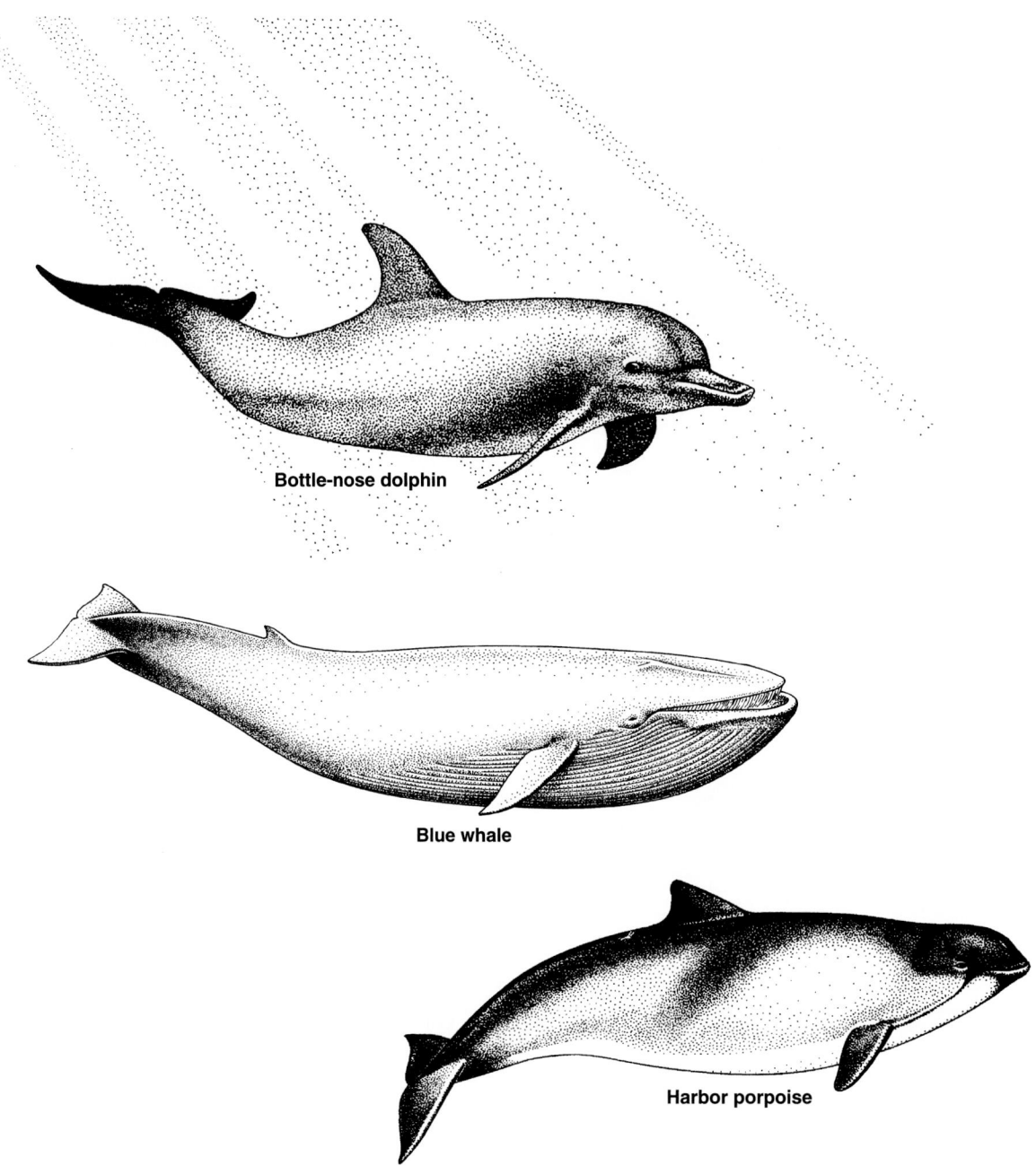

FIGURE 4.50
Representative cetaceans.

VARIATION AMONG INDIVIDUALS

In this chapter, we have had a glimpse of the diversity of craniate life. However, variation is not restricted to differences between species. Variation occurs within populations and even among littermates. Indeed, no two products of sexual reproduction are identical. The number of kinds of gametes or of homozygous genotypes possible within n pairs of genes is 2^n. The number of different kinds of genotypes possible is 3^n. Even if there were only 100 pairs of genes, 48 digits would be needed for writing the number of possible genotypes. And there are thousands and thousands of pairs of genes. In addition, there may be as many as 400 mutational sites on a single gene. Therefore, the number of possible combinations of hereditary traits staggers the imagination. For example, the gastrohepatic artery in a population of domestic cats is sometimes 20 mm long, sometimes 1 mm long, or the gastric and hepatic arteries may arise independently, hence no gastrohepatic artery. The number of variations that actually occur in arterial channels alone in any population is such that one could identify a single shark, or a single human being, from the arterial channels alone. In practice, each of us does better than that! We identify a single human being simply by looking at a face or at a print made by a fingertip. We could, of course, recognize an individual organism by looking at a map of the sequence of nucleotides in its DNA. Ongoing, random genetic mutations and recombinations of genes resulting from sexual reproduction maintain variation in populations. These processes then provide the raw material for evolution. Genetic variation, coupled with natural selection, is the matrix out of which new species evolve.

Summary

1. Agnathans (jawless craniates) are the extinct ostracoderms and the living cyclostomes (hagfishes and lampreys) thus representing a paraphyletic group of basal craniates. Newly described Lower Cambrian fossils from China represent the oldest known vertebrates. Ostracoderms are the oldest known vertebrates with bone. They were covered with dermal armor, and many lacked paired appendages (paired pectoral appendages are found in some fossil agnathans).

2. Acanthodians, named for their spiny fins, and placoderms, named for the bony plaques in the skin, were early jawed fishes (first known in the fossil record in the Lower Silurian and the Upper Silurian, respectively). They flourished in the Devonian.

3. Chondrichthyes (cartilaginous fishes), another early gnathostome group, include elasmobranchs and holocephalans (first known from the Lower Silurian). The former has a spiracle, naked gill slits, and placoid scales that become teeth on the jaws. Holocephalans have nearly naked skin, a fleshy operculum, and bony plates on the jaws instead of teeth.

4. Bony fishes (Osteichthyes) are in two clades, lobe-finned fishes and their tetrapod descendants (Sarcopterygii) and ray-finned fishes (Actinopterygii). They have ganoid, ctenoid, or cycloid scales, and a bony operculum. Most lobe-fins have internal nares.

5. Sarcopterygians include actinistians (e.g., *Latimeria*) and rhipidistians (which may include dipnoans depending upon the hypothesis of relationships and tetrapods). Amphibians evolved from a rhipidistian ancestor.

6. Actinopterygians include *Polypterus,* chondrosteans (sturgeons, paddlefishes), and neopterygians (bowfins and gars, which are surviving Mesozoic fishes, and modern teleosts). In modern fishes, thin flexible scales have replaced the bony armor of early fishes.

7. Amphibians include: labyrinthodonts (the earliest tetrapods), a number of Carboniferous groups (e.g., temnospondyls, anthracosaurs, and microsaurs), and lissamphibians (Triassic and modern amphibians).

8. Modern amphibians are anurans (frogs, toads), urodeles (tailed amphibians), and apodans (limbless species). Anurans and urodeles are scaleless, have a glandular skin, typically deposit their eggs in the water, and have an aquatic larval stage. A few are viviparous. Apodans have minute dermal scales.

9. Reptiles (including birds) and mammals are amniotes. Embryos develop within an amniotic sac in association with extraembryonic membranes.

10. Living reptiles are ectotherms (with the exception of birds) with cornified epidermal scales, plaques, or scutes, and with claws. They typically lay large yolky eggs surrounded by a porous shell. Some are viviparous. Birds are characterized by the presence of feathers.

11. There are three subgroups of amniotes: anapsid reptiles, diapsids, and synapsids. Anapsids have no temporal fossa and represent those basal reptiles that retain this primitive condition. Anapsids include stem reptiles and turtles. Diapsids have two temporal fossae and include all living reptiles except turtles. Synapsids have one temporal fossa and include mammals.

12. Archosaurs are a subgroup of diapsids. They include winged reptiles (pterodactyls), crocodilians, and dinosaurs (remember that birds are an evolutionary subgroup of Dinosauria).

13. Lepidosaurs are another subgroup of diapsids. It includes sphenodons and squamates (lizards, snakes, and amphisbaenians). Other subtaxa include large extinct aquatic reptiles (plesiosaurs and ichthyosaurs).

14. Birds (Aves) are endothermal archosaurs with feathers. The oldest known birds (Archaeornithes) appeared in the mid-Mesozoic. All others are Neornithes. Living birds are in two subtaxa, ratites (Palaeognathae) and carinates (Neognathae). The former is flightless. The latter have a large sternal carina.

15. Mammals are craniates with hair and mammary glands. They include one group of oviparous mammals (Monotremata: platypuses and echidnas), one group that employs the yolk sac as the fetal portion of a placenta (Marsupialia: opossums, kangaroos, others), and the remaining mammals, which have a chorioallantoic placenta ("true placentals"). The order Primates includes Old and New World monkeys, apes, and extinct and modern humans.

Critical Thinking Questions

1. Using either nested brackets as on a cladogram or an indented outline, what are the hierarchical relationships among the following taxa: craniates, amniotes, mammals, actinopterygians, chondrichthyes, Aves, vertebrates, sharks, dinosaurs, and sarcopterygians? (Use monophyletic definitions where possible.)

2. What is the evidence that supports the statement that lampreys are more closely related to gnathostomes than to hagfishes?

3. What are the three living sarcopterygian subgroups?

4. Name a neopterygian that you might have eaten for dinner lately.

5. Why is Amphibia a paraphyletic group? Are Lissamphibia monophyletic?

6. What is the phylogenetic systematics definition of Reptilia? What is the sister group to Reptilia?

7. What plesiomorphic features are present in fossil birds that suggest their origin among dinosaurs?

8. What are the implications of a saurischian origin of birds on our interpretation of hip structure in dinosaurs?

9. If euryapsids are correctly placed among diapsids, then how do we account for the presence of only a single temporal fenestra in plesiosaurs and ichthyosaurs?

10. Is the statement, monotremes evolved an oviparous reproductive behavior, an accurate statement?

11. As a mammal yourself, can you name one example for each order of mammals? (Taxonomic terms without a visual image to accompany them are much more difficult to learn. You should find examples for all the taxa in this chapter that you are not familiar with.)

Selected Readings

Ackerman, J.: Dinosaurs take wing, *National Geographic* **194**(1):76, 1998.

Ahlberg, P. E.: Origin and early diversification of tetrapods, *Nature* **368**:507, 1994.

Brodal, A., and Fange, R., editors: The biology of Myxine. Oslo, Norway, 1963, Universitetsforlaget.

Carr, R. K.: Placoderm diversity and evolution, Bulletin du Muséum National d'Histoire Naturelle, Paris, 4e sér., **17** (1–4):85–125, 1995.

Carroll, R. L.: Vertebrate paleontology and evolution. New York, 1988, W. H. Freeman and Company.

Chen, J.-Y., Dzik, J., Edgecombe, G. D., Ramsköld, L., and Zhou, G.-Q.: A possible early Cambrian chordate, *Nature* 377:720.

Chiappe, L. M.: A diversity of early birds, *Natural History*, p. 52, June 1995.

Cloutier, R., and Ahlberg, P. E.: Morphology, characters, and the interrelationships of basal sarcopterygians, chapter 17 *in* Stiassny, M. L. J., Parenti, L. R., and Johnson, G.D. (editors), Interrelationships of fishes, San Diego, 1996, Academic Press.

Dineley, D. I., and Loeffler, E. J.: Ostracoderm faunas of the Delorme and associated Devonian-Silurian formations, North West Territories, Canada, Palaeontological Society. Special Papers in Palaeontology, Number 18 published by The Palaeontological Association, London, 1976.

Feldhamer, G. A., Drickhamer, L. C., Vessey, S. H., and Merritt, J. F.: Mammalogy: Adaptation, diversity, and ecology. Boston, 1999, McGraw-Hill.

Forey, P., and Janvier, P.: Agnathans and the origin of jawed vertebrates, *Nature* **361**:129, 1993.

Forey, P., and Janvier, P.: Evolution of the early vertebrates, *American Scientist* **82**:554, 1994.

Gauthier, J., Kluge, A.G., and Rowe, T.: Amniote phylogeny and the importance of fossils, *Cladistics* 4:105–209, 1988.

Gingrich, P. D.: The whales of Tethys, *Natural History*, p. 86, April 1994.

Hickman, C., and Roberts, L.: Animal diversity, Dubuque, IA, Wm. C. Brown Publishers, 1995.

Hou, L., Zhou, Z., Martin, L. D., and Feduccia, A.: A beaked bird from the Jurassic of China, *Nature* 377:616, 1995.

Janvier, P.: Early vertebrates, Oxford, 1996, Clarendon Press.

Jørgensen, J. M., Lomholt, J. P., Weber, R. E., and Malte, H., editors: The biology of hagfishes. London, 1998, Chapman & Hall.

Lee, M.: The turtle's long lost relatives, *Natural History*, p. 63, June 1994.

Lombard, R. E., and Sumida, S. S.: Recent progress in understanding early tetrapods, *American Zoologist* **32**:609, 1992.

Long, J. A.: The rise of fishes, 500 million years of evolution. Baltimore, 1995, The Johns Hopkins University Press.

Maisey, J. G., Discovering fossil fishes. New York, 1996, Henry Holt and Company.

Nelson, J. S.: Fishes of the world, ed. 3. New York, 1994, John Wiley and Sons.

Norell, M., Chiappe, L., and Clark, J.: New limb on the avian family, *Natural History*, p. 39, September 1993.

Nowak, R. M., Walker's mammals of the world, ed. 6. Baltimore, 1999, The Johns Hopkins University Press.

Pough, F. H., Andrews, R. M., Cadle, J. E., Crump, M. L., Savitzky, A. H., and Wells, K. D.: Herpetology. Upper Saddle River, NJ, 1998, Prentice Hall.

Proctor, N. S., and Lynch, P. J.: Manual of ornithology, avian structure and function. New Haven, CT, 1993, Yale University Press.

Shu, D-G., Luo, H-L., Morris, S. C., Zhang, X-L., Hu, S-X., Chen, L., Han, J., Zhu, M., Li, Y., and Chen, L-Z: Lower Cambrian vertebrates from south China, *Nature* **402**: 42–46, 1999.

Stiassny, M. L. J., Parenti, L. R., and Johnson, G. D., editors: Interrelationships of fishes. San Diego, 1996, Academic Press.

Stock, D. W., and Witt, G. S.: Evidence from 185 ribosomal RNA sequences that lampreys and hagfishes form a natural group, *Science* **257**:787, 1992.

Szalay, F. S.: Evolutionary history of the marsupials and an analysis of osteological characters. Cambridge, 1994, Cambridge University Press.

Szalay, F. S., et al., editors: Mammal phylogeny: Placentals, vol. 2. New York, 1993, Springer-Verlag.

Taylor, G.: Winging it, *New Scientist* **163**(2201), 1999.

Thomson, K. S.: Living fossil: The story of the Coelacanth. New York, 1991, W. W. Norton.

Weishampel, D. B., Dodson, P., and Osmólska, H., editors: The Dinosauria. Berkeley, 1990, University of California Press.

Wilson, E. O.: The diversity of life. Cambridge, MA, 1992, The Belknap Press/Harvard University Press.

Links to the Internet

Visit the zoology website at http://www.mhhe.com/zoology to find live internet links for each of the references listed below.

1. Craniata. This is a starting off point at the Arizona's Tree of Life site that leads you to information on all craniates.

2. Phylum/Major Group Index to Zoological Record, Taxonomic Hierarchy. A starting point from Biosis to go to nearly all taxonomic gorups.

3. Animal Diversity Web, University of Michigan. Phylum Chordata. General characteristics of chordates, with the following links to urochordates and vertebrates.

4. Introduction to the Chordates. University of California at Berkeley, Museum of Paleontology. Images, photos, systematics. Links to a vast array of specialized links on various vertebrate groups.

5. Phylum Chordata. A great starting place to find a large number of links, from the University of Minnesota.

6. SeaWorld/Busch Gardens Animal Information Database. Much information on aquatic animals, and some terrestrial organisms as well.

5

Early Craniate Morphogenesis

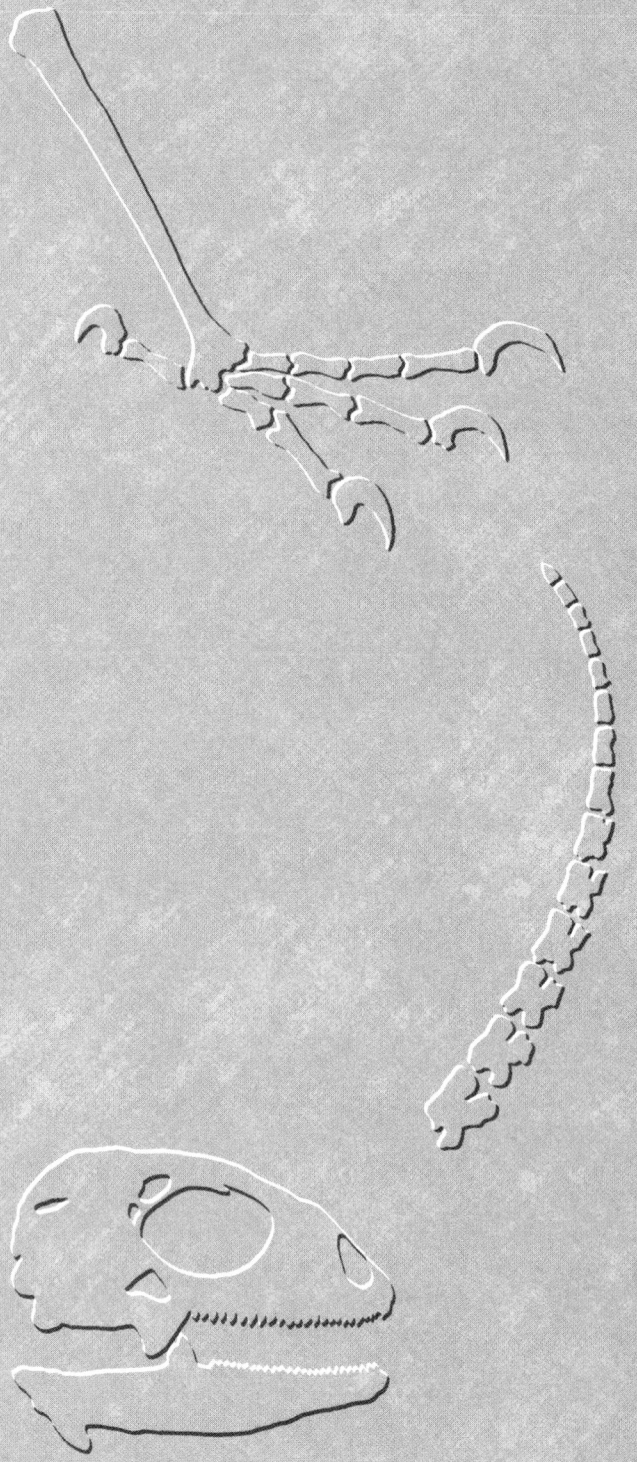

In this chapter, we will learn that all craniates pass through similar, though not identical, early formative stages. We will discover the steps whereby a hollow ball of totipotent embryonic cells is converted into an early embryo with three indispensable germ layers, and the contribution of each layer to the adult body. At the end, we will examine the extraembryonic membranes and their roles before hatching or birth.

OUTLINE

 pecimens studied in craniate morphology laboratories are viewed in a single time frame, the immediate present. A bit of knowledge from two additional time frames will enrich our understanding of morphology: the specimen's remote past, which requires knowledge of its ancestors; and its recent past, which includes its early ontogeny. Chapter 4 introduced the ancestors. The present chapter recounts the highlights of early embryogenesis. The total picture—phylogeny, ontogeny, and morphology—is more meaningful than is morphology alone, if only because the first two account for the last.

CRANIATE EGGS

Egg Types

Eggs of craniates vary in the amount of yolk they contain and in the distribution of yolk within the egg (table 5.1). Eggs with very little yolk, such as those of the amphioxus and placental mammals, are **microlecithal**. Eggs with moderate amounts of yolk, such as those of freshwater lampreys, basal actinopterygian and neopterygian fishes, lungfishes, and amphibians, are **mesolecithal**. Eggs with massive amounts of yolk, such as those of marine lampreys, elasmobranchs, teleosts, reptiles, and monotremes, are **macrolecithal**.

The yolk in microlecithal eggs is evenly distributed throughout the cytoplasm as fat droplets and small yolk globules. Eggs with an even distribution of yolk are said to be **isolecithal**. In mesolecithal and macrolecithal eggs, the large yolk mass tends to be concentrated at one end, the **vegetal pole** (fig. 5.1, *red*). The opposite pole, containing the nucleus and relatively yolk-free cytoplasm, is the **animal pole**. Eggs in which cytoplasm and yolk tend to accumulate at opposite poles are **telolecithal**.

Oviparity and Viviparity

Animals that spawn (lay their eggs) are said to be **oviparous**. Eggs of oviparous species contain sufficient nourishment in the form of yolk, and sometimes albumen, to support development into a free-living organism that is able to take food orally. If the yolk is massive, the young may hatch essentially fully formed, as in

birds. If there is less yolk, the young hatch in a larval state, as in frogs. When there is very little yolk, as in the egg of the amphioxus, the free-living, self-nourishing state must be achieved very quickly after the egg is deposited. Accordingly, the amphioxus hatches into an externally ciliated, free-swimming embryo 8 to 15 hours after fertilization, at which time the notochord is a mere ridge in the roof of the primitive gut, and there are no pharyngeal slits.

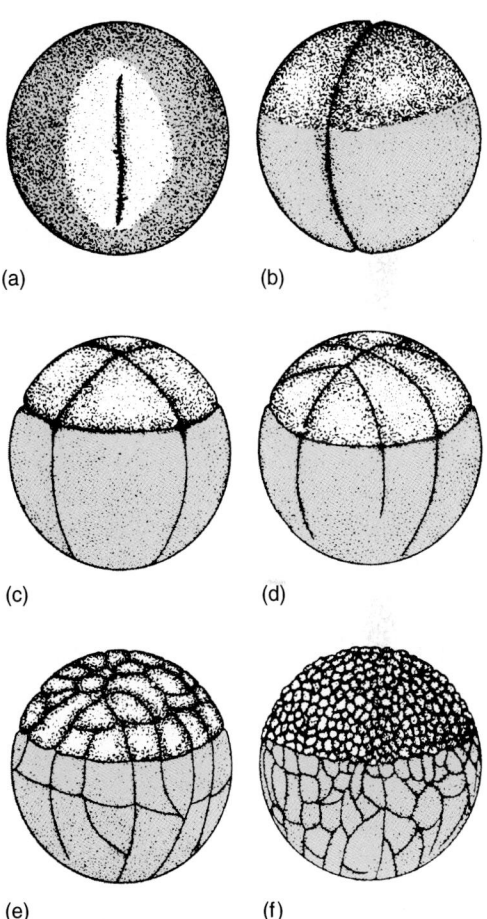

FIGURE 5.1

Segmentation and blastula formation in a mesolecithal egg (amphibian). (*a*) Initial cleavage of egg, seen from above. (*b–f*) Successive cleavages resulting in a blastula. *Gray* indicates animal pole, the cells of which contain little yolk; *red* is vegetal pole. In (*f*), gastrulation by epiboly and involution is about to commence.

TABLE 5.1 Classification of Eggs Based on the Characteristics of the Yolk

		Terminology	Example
Amount of Yolk	Little	Microlecithal	Amphioxus, therian mammals
	Moderate	Mesolecithal	Many fishes, amphibians
	Large	Macrolecithal	Sharks, monotremes
Distribution of Yolk	Even	Isolecithal	Therian mammals
	Concentrated at one pole	Telolecithal	Amphioxus and most nontherian craniates

Animals that give birth to offspring are said to be **viviparous**. The relationship between mother and embryos varies from one in which the mother provides only protection and oxygen, nourishment having been stored in the egg, as in viviparous sharks, to one in which the embryos are dependent on maternal tissues for all nourishment, for oxygen, and for carrying away waste products of metabolism, as in placental mammals. The term **ovoviviparity** has been coined to designate the former viviparous condition. **Euviviparity** designates the viviparous condition in which the embryo cannot develop without nourishment being constantly supplied by the mother from maternal tissues. There are many intermediate conditions. Reptilian embryos that are initially ovoviviparous may require nourishment from the mother before pregnancy has terminated. A reptilian egg in a viviparous species has a shell, but it is an uncalcified semipermeable membrane.

Viviparity in one degree or another has evolved in every class of craniates except agnathans and birds. It has developed independently more than 12 times in teleost fishes, at least 100 times in lizards, and at least six times in snakes. In fact, 20 percent of all squamate species are viviparous. There are no viviparous crocodilians or turtles. Viviparity occurs in 40 families of sharks and rays, and there is fossil evidence for viviparity in extinct holocephalans and basal actinopterygians. Unborn fossilized young have been found in the uteri of the mesozoic marine reptile *Ichthyosaurus*. There are viviparous urodeles and anurans that inhabit a terrestrial environment rather than a pond and give birth to fully metamorphosed young, and there are viviparous aquatic apodans.

The dogfish shark, *Squalus acanthias*, is an example of an ovoviviparous organism. The unborn pups are nourished entirely by the yolk from their own yolk sac (see fig. 5.12), but they receive oxygen from greatly enlarged and highly tortuous blood vessels in the uterine lining of the mother. The pups may be removed from the mother two to three months before birth (the gestation period is 20 to 22 months) and will complete their development, utilizing the nourishment from their yolk sac, provided they are supplied with oxygenated seawater.

In viviparous teleosts, the egg may be fertilized and the young may develop in the ovarian follicle, there having been no ovulation. The young of *Gambusia* develop in this way, or the embryos may develop in the ovarian cavity (see fig. 15.33). Among adaptations of the embryo for development within the follicle is temporary enlargement of the *embryo's* pericardial sac, gut, or some other organ, which absorbs oxygen and nutrient-rich juices from ovarian tissues. Some species that develop in the ovarian cavity sprout long villuslike projections (**trophotaeniae**) from the lining of their embryonic gut; these protrude through the vent into the surrounding nutrient-rich medium within the cavity. In still other teleosts, the young develop for a time in the follicle and exhibit an enlarged pericardial sac, then pass into the ovarian cavity and develop absorptive villi. In some species, developing eggs or larvae, which are present in enormous numbers in teleosts, are ingested by other larvae, thus guaranteeing a source of nourishment for a portion of the larval population.

Maternal tissues, under the influence of hormones, exhibit diverse adaptations for viviparity. The wall of the ovarian follicle occupied by teleost embryos may develop vascular folds or villi that secrete nutrients, and these may even project into the mouth or opercular cavity of the embryos. In *Dasyatis americana*, a euviviparous sting ray, the gravid uterus is lined with villi 2 to 3 cm long, which produce a copious secretion that nourishes the embryo. Secretions of uterine glands provide nourishment for temporarily unimplanted blastocysts of mammals and perhaps for implanted blastocysts throughout pregnancy in perissodactyls and artiodactyls. **Histotrophic (embryotrophic) nutrition** is the term applied to nutrition by *glandular secretions* from maternal tissues, as contrasted with nutrition supplied via a placenta.

Internal and External Fertilization

In viviparous craniates, fertilization takes place within the body of the female. Fertilization is also internal whenever eggs are covered by an impenetrable shell before being extruded, as in reptiles.

In oviparous fishes, frogs, and toads, external fertilization is the rule despite an aquatic environment that dilutes the sperm. This is possible only because millions of sperm are shed over the eggs as the latter are being extruded. In apodans and urodeles, however, fertilization is usually internal even in oviparous species. Some male urodeles deposit a gelatinous packet of sperm (**spermatophore**) in the immediate vicinity of congregated females during the mating ritual. The spermatophore is generally picked up by the lips of a female's cloaca, although in some species it is placed in the cloaca by the male. The sperm eventually escape from the jelly and fertilize eggs, which are passing down the oviduct. In some species, the sperm are stored in crypts (**spermathecae**) in the female tract and fertilize eggs that are ovulated at a later date when the environment is more suitable for embryonic development. The adaptation has a genetic basis. Packaging sperm in spermatophores makes it possible to convey sperm to females despite a lack of copulatory organs.

EARLY DEVELOPMENT OF REPRESENTATIVE CHORDATES

Cleavage and the Blastula

Early cell divisions of the zygote are initiated by fertilization of the egg and are referred to as **segmentation**, or **cleavage**. As a result, the zygote becomes subdivided into smaller and smaller cells that form a usually hollow

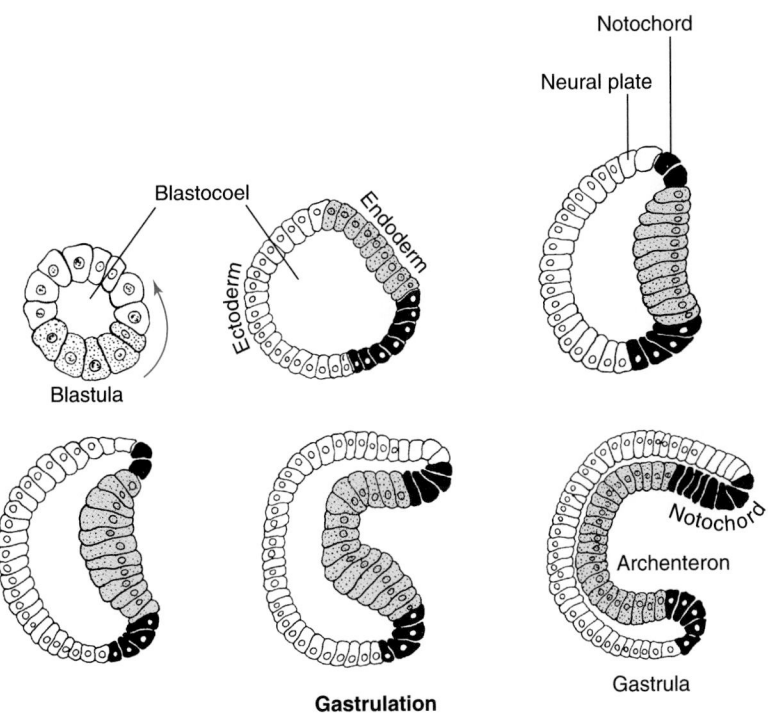

FIGURE 5.2

Blastula and gastrulation by involution in amphioxus, sagittal section. Cells in *black* surround the blastopore (entrance to archenteron) and are the most active, mitotically. *Red* indicates presumptive endoderm in the blastula, definitive endoderm in the gastrula. The notochord establishes the dorsum and the long axis of the body.

sphere, or **blastula.** Each cell of the blastula is a **blastomere,** and the cavity is the **blastocoel** (fig. 5.2).

When there is very little yolk in an egg, the blastomeres are approximately of equal size (fig. 5.2, blastula). When cell division is impeded by a moderate amount of yolk, as in amphibian eggs, the yolk-laden cells at the vegetal pole divide more slowly and are larger (see fig. 5.1). When there is a massive yolk, as in the eggs of reptiles, segmentation is confined to the animal pole and results in a **blastoderm** perched like a skullcap on the massive yolk (see fig. 5.6). The embryo develops from this blastoderm.

The eggs of therian mammals have almost no yolk and do not show the typical pattern of animal and vegetal poles seen in taxa with yolk. The pattern of cleavage in mammals is unique (see fig. 5.10*a–e*). The first cleavage is meridional, as in other craniates. At this point, cleavage begins to differ, with each daughter cell from the first division cleaving at right angles to each other, one meridionally and the other equatorially. Cell division after this time is not always synchronous, producing at times odd numbers of cells (rather than the typical pattern of 2-4-8-16-32-64). By the 16-cell stage, an **inner cell mass** is present that will form the body of the embryo. At the 32-cell stage, the outer layer is expanded around the blastocoel and forms the **trophoblast.** Be-

cause of the cystlike nature of the mammalian blastula, it is called a **blastocyst.** The trophoblast is capable of absorbing nutrients from the uterine fluids and will be the site of contact with the uterus when implantation occurs. Eventually, the trophoblast will contribute to the formation of the placenta and chorion. Meanwhile, the inner cell mass forms an embryonic disc and progresses through development in a manner similar to that seen in birds—the retention of a primitive pattern inherited from ancestors that had substantial yolk.

As a result of cleavage, craniates typically pass through a hollow, spherical stage of development known as a blastula. During this stage, areas of distinctive developmental potential—the rudiments of future ectoderm, endoderm, mesoderm, and notochord—are established.

Gastrulation: Formation of the Germ Layers and Coelom

Gastrulation in chordates is a dynamic process of cellular movements whereby presumptive endoderm, mesoderm, and notochord cells of the blastula flow to the interior, thereby generating the three germ layers from which all future tissues and organs form. In the process, bilateral symmetry is established. These cellular migrations are referred to as **formative,** or **morphogenetic, movements.** Rapid cell proliferation maintains a continuing supply of additional cells for these formative movements. The developmental processes that constitute gastrulation are induced and regulated by inductor substances (morphogens) that will be discussed shortly. Because gastrulation in an amphioxus is unencumbered by yolk, we will examine the process in this chordate first.

Amphioxus

In an amphioxus, presumptive endoderm from the surface of the blastula invaginates into the blastocoel (fig. 5.2, *red*). The process, **involution,** produces the earliest gut, or **archenteron,** and results in a gastrula (fig. 5.2). The entrance to the archenteron is the site of involution, the **blastopore.** Soon, presumptive notochord cells from the rim of the blastopore flow inward to become the beginning of the notochord, which is located temporarily in the roof of the archenteron (fig. 5.2, gastrula). Then, presumptive mesoderm cells form bands of undifferentiated mesoderm lateral to the notochord (fig. 5.3*a*). At this stage, the amphioxus embryo consists chiefly of an outer layer of ectoderm and an inner layer of endoderm that surrounds the archenteron. The roof

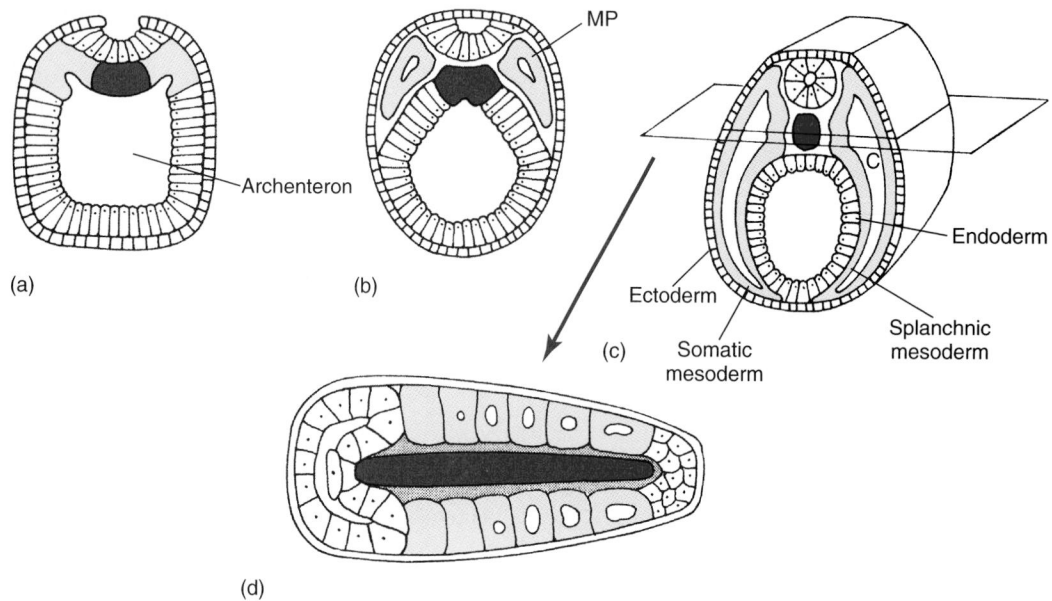

FIGURE 5.3

Mesoderm, *light red,* and coelom formation in an amphioxus. (*a*) Longitudinal bands of mesoderm lie in the dorsolateral wall of the archenteron lateral to the notochord (*dark red*). (*b*) mesodermal pouches (**MP**) have formed. (*c*) Pouches have grown ventrad between ectoderm and endoderm to form a coelom, **C**. (*d*) Early larva in frontal section showing segmented coelom.

of the archenteron is the elongating notochord and the parallel bands of mesoderm (fig. 5.3*a*). (The notochord is generally regarded as mesoderm.)

Once mesoderm has been established in the dorsolateral roof of the archenteron, the actively mitotic bands fold upward, separate from the endoderm, and form a series of hollow **coelomic pouches (mesodermal pouches)** commencing at the *anteriormost* end of the band (fig. 5.3*a* and *b*). About the time two pouches have formed, the embryo hatches into a free-swimming larva. The larva continues to elongate partly as a result of proliferation of additional cells from the rim (**dorsal and ventral lips**) of the blastopore. The mesodermal bands elongate, and additional coelomic pouches form.

After a coelomic pouch has formed, it grows ventrad, pushing between the ectoderm and endoderm (fig. 5.3*c*). The mesoderm of the two pouches, left and right, eventually meet underneath the gut. The outer wall of each pouch lies against the ectoderm and is **somatic mesoderm**. Together with the ectoderm, it constitutes the **somatopleure**, or **body wall**. The inner wall of each pouch lies against the endoderm and is **splanchnic mesoderm**. Together with the endoderm, it constitutes the **splanchnopleure**, which gives rise to the digestive tract. The cavity between somatic and splanchnic mesoderm is the **coelom** (fig. 5.3*c*).

The coelom of a larval amphioxus is segmented for a while because of its origin from a series of pouches (fig. 5.3*d*). Later, the cavities within successive pouches become confluent, *starting at the anterior end,* establishing a single coelomic cavity on each side. Still later, the

right and left coelomic cavities become confluent beneath the gut. The germ layers are now positioned to participate in the formation of organs, a process called **organogenesis**.

Frog

In craniates other than therian mammals, gastrulation is complicated by the presence of moderate to large quantities of yolk in the cells of the vegetal pole. This unwieldy yolk impedes simple involution, such as seen in an amphioxus. Still, a gastrula forms by alternative cell movement. In frogs, which have mesolecithal eggs, a process known as **epiboly** occurs in which the small cells of the animal pole grow downward over the larger cells of the vegetal pole, thus effectively tucking the yolk inside and forming a blastopore (fig. 5.4*a–c*). Yolk cells then constitute the archenteric floor (figs. 5.4*d–f* and 5.5).

Notochord and Dorsal Mesoderm   As involution gets under way, a stream of undifferentiated cells flows over the dorsal lip of the blastopore to establish in the roof of the archenteron a narrow band of **chordomesoderm** (see fig. 5.4*f*). A notochord organizes from the midline portion of the chordomesoderm and remains, temporarily, in the roof of the archenteron. The mesoderm immediately lateral to the notochord gives rise to a pair of longitudinal bands of mesoderm that parallel the notochord. These bands become the **dorsal mesoderm**. They subsequently segment and reorganize to form hollow **mesodermal somites** (see fig. 5.9*b*, *c*, and *d*). These

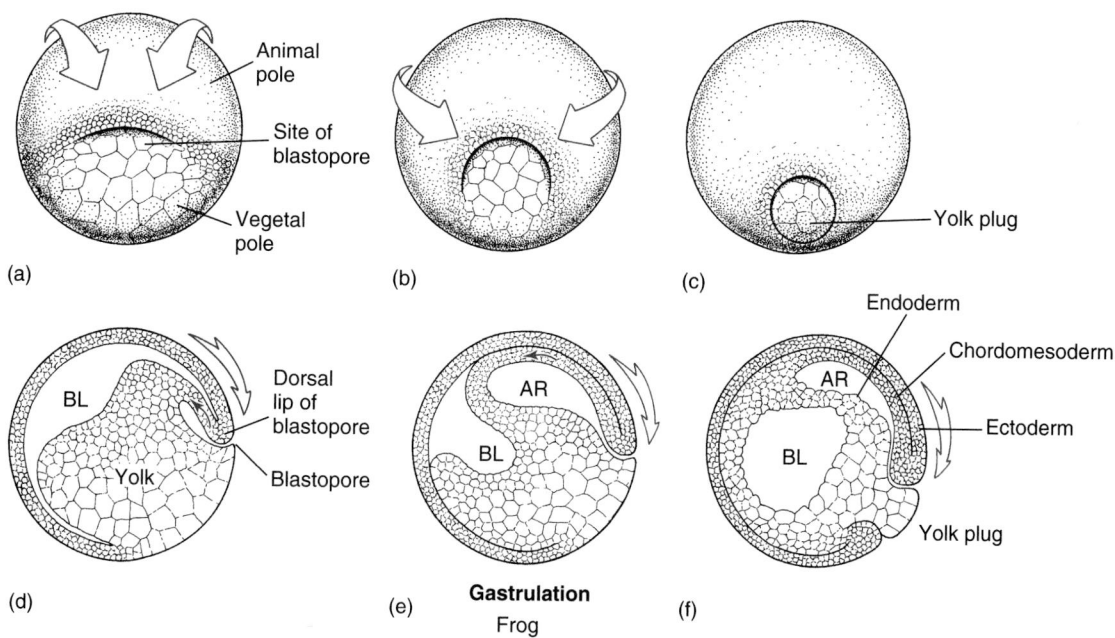

FIGURE 5.4

Epiboly, involution, and the blastopore of a frog. (*a–c*) Epiboly of ectoderm to yolk plug stage. (*a*) is an advanced stage of figure 5.1 (*d–f*) Hemisections of (*a–c*) showing involution at blastopore. (*f*) Compare figure 5.5. **AR**, archenteron; **BL**, blastocoel. Arrows indicate growth of cells.

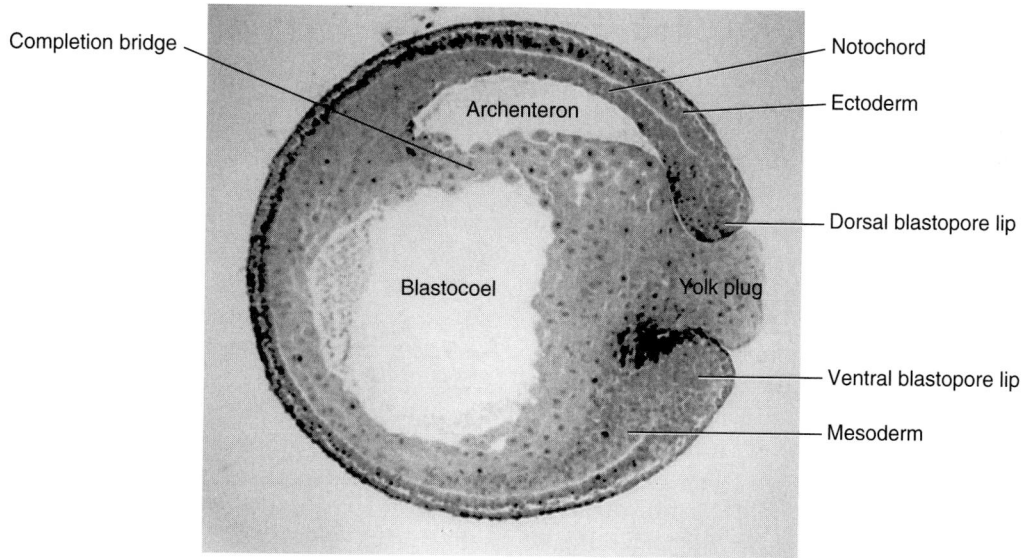

FIGURE 5.5

Early frog gastrula. Notochord is differentiating from chordomesoderm. Redistribution of weight has rotated embryo 90 degrees counterclockwise from orientation in figure 5.4*c*.

From Phillips, JB: *Development of Vertebrate Anatomy* 1975 (Fig. 7.5c page 137).

are the source of the body wall musculature. The formation of the notochord and dorsal mesoderm in frogs is essentially the same as in an amphioxus. Both arise from streams of undifferentiated cells that flow over the dorsal lip of the blastopore, and both become established in the roof of the archenteron. The rest of the mesoderm

forms in a slightly modified manner possibly dictated by the presence of the yolk.

Endoderm and Ectoderm   At the same time that the notochord and dorsal mesoderm are differentiating, presumptive endoderm streams to the interior, fans out, and provides, with the yolk, a lining for the archenteron. The cells on the surface of the gastrula, along with additional ones that are being rapidly proliferated, become the ectoderm.

Intermediate and Lateral-Plate Mesoderm

The cells that give rise to mesoderm other than that of the somites reach the interior by flowing over the lateral and ventral rims of the blastopore. Once inside, they stream forward as an advancing sheet of rapidly expanding **lateral-plate mesoderm**, pushing its way cephalad between ectoderm and endoderm. Unlike dorsal mesoderm, the lateral-plate mesoderm does not segment and, except for a narrow strand paralleling the dorsal mesoderm, it soon splits into two lamina—an outer sheet, the **somatic mesoderm**, and an inner sheet, the **splanchnic mesoderm** (see fig. 5.9b). The cavity between the two sheets is the **coelom**. As in an amphioxus, somatic mesoderm plus ectoderm constitute the **somatopleure (body wall)**; splanchnic mesoderm plus endoderm constitute the **splanchnopleure**. The strand of lateral-plate mesoderm immediately lateral to the mesodermal somites (that which does not participate in the splitting process) is **intermediate (nephrogenic) mesoderm** (see fig. 5.9b). It gives rise to the kidney tubules and the ducts of the urogenital system.

Chick

We will use the chick to illustrate gastrulation in amniotes with macrolecithal eggs. But first, we must review the term *blastoderm*.

Blastoderm: Epiblast and Hypoblast

A macrolecithal egg has a massive yolk. Epiboly would take a long time to tuck it into the interior. As an adaptation, nature has created an alternate means of achieving a **triploblastic (three germ-layered) organism**. The adaptation is a blastoderm, a result of segmentation of the cytoplasm at the animal pole only. At first, the blastoderm is simply a thin multilayered mound of cells perched on top of the massive yolk like a skullcap (fig. 5.6). This blastoderm then organizes into an upper sheet of cells, the **epiblast**, and a lower sheet, the **hypoblast**, as seen in figure 5.6. Cells from the periphery of the hypoblast slowly grow outward over the surface of the yolk, then downward around it to become part of the **endodermal lining** of a yolk sac. The hypoblast does not contribute to the developing embryo (it is limited to the yolk sac and its stalk). The epiblast is then the presumptive tissues of the developing embryo. Initial cells that migrate inward through the primitive streak (discussed in the next section) include the future endoderm

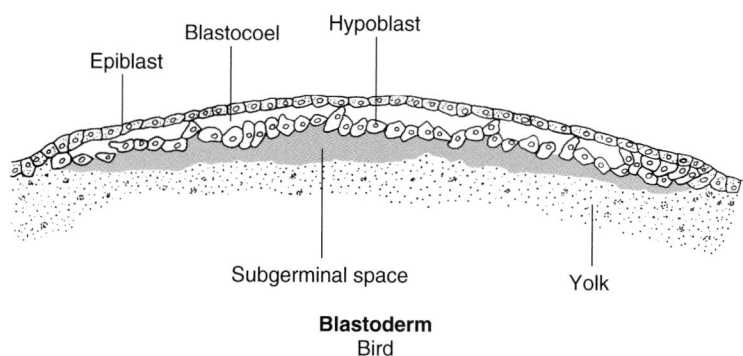

Blastoderm
Bird

FIGURE 5.6

Blastoderm at animal pole of avian egg, sagittal section. Future head end is to left. Only a fraction of the massive uncleaved yolk is shown.

that displaces the hypoblast. This is followed by the migration of future mesoderm between the epiblast and endoderm. All three primary tissues also migrate laterally within their respective layers to eventually form extraembryonic membranes.

Notochord and Mesoderm Formation   Before looking at the details of initial notochord and mesoderm formation, it is necessary to examine two additional structures that develop within the blastoderm: (1) a multilayered longitudinal **primitive streak** (fig. 5.7) and (2) **Hensen's node**, a thickened nodule of closely packed blastoderm cells that defines the caudal end of the future embryo (figs. 5.7 and 5.8). The primitive streak and Hensen's node are the functional equivalent of the blastopore of eggs with less yolk.

From Hensen's node, a **notochordal process** pushes forward beneath the epiblast (fig. 5.8), while other cells stream forward alongside the notochord to establish the dorsal mesoderm (fig. 5.8, *dark red*). Concurrently, lateral-plate mesoderm streams from both sides of the primitive streak, between the ectoderm (epiblast) and the endoderm, and onto the surface of the yolk in all directions except the region of the future head (fig. 5.8, *arrows*). While the dorsal mesoderm is segmenting to form somites (fig. 5.9c), the lateral-plate mesoderm is splitting into somatic and splanchnic mesoderm (fig. 5.9d). The resulting cavity between the two is the coelom.

By the end of the second day of incubation, the splanchnopleure, which has spread outward onto the surface of the yolk to form a yolk sac, has produced the first blood cells and a network of delicate **vitelline (omphalomesenteric) vessels** in the **area opaca** (fig. 5.9c). These vessels collect yolk globules and transport them to the simple, twitching, **S**-shaped heart (see figs. 14.14d–i and 16.6). The heart then propels this early stream of cells and nutrients to the rapidly growing tissues within the embryo.

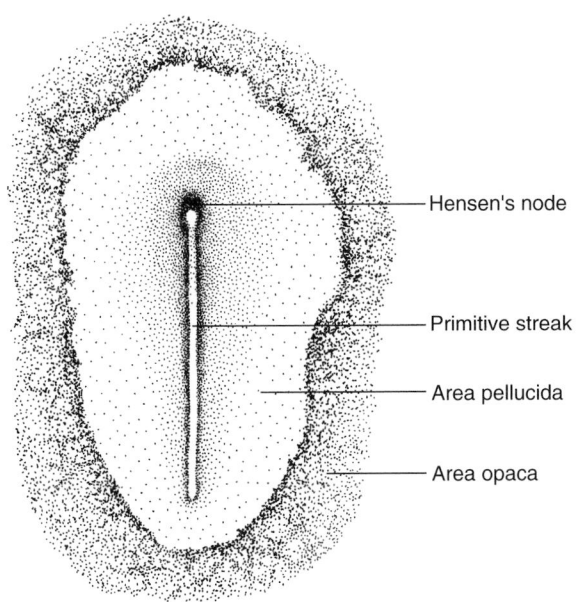

FIGURE 5.7

Primitive streak stage of chick embryo at 16 hours of incubation. The tissue seen is the animal pole of the egg. It was removed from the yolk with scissors, fixed, stained, and rendered translucent. The area opaca is the site in the yolk sac where blood islands are forming. It is therefore opaque when viewed with transmitted light. The area pellucida was not attached to the yolk; hence, it transmits light readily.

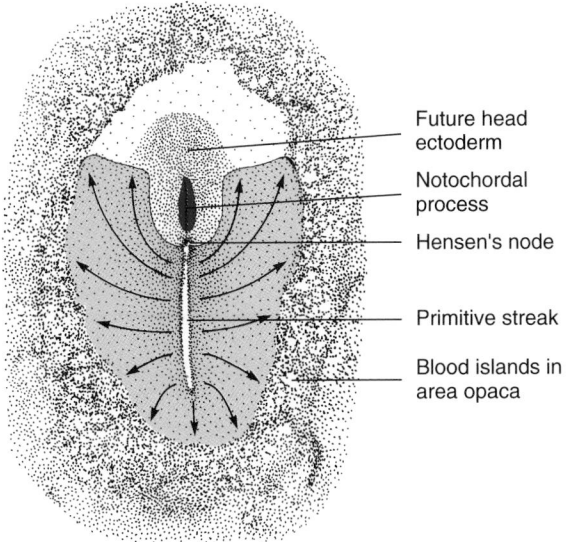

FIGURE 5.8

Chick embryo at 18 hours of incubation. Arrows indicate morphogenic streaming of lateral-plate mesoderm (*light red*) from the primitive streak. *Dark red* indicates dorsal mesoderm from Hensen's node. Figure 5.9*c* shows primitive streak in an embryo about 8 hours older.

Overview   As a result of complex morphogenetic movements and accompanying cell proliferation, the chick establishes ectoderm and endoderm; dorsal, intermediate, and lateral-plate mesoderm; a coelom; and the beginning of an extraembryonic yolk sac. For a look at a later stage showing the relationship between the splanchnopleure, yolk sac, and trunk, you may wish to look at figure 5.13*a*.

The processes of gastrulation described above for craniates with mesolecithal and macrolecithal eggs differ in details among the craniate groups that exhibit them. Variations are most numerous among bony fishes because there are so many divergent species.

Placental Mammals

At the time gastrulation commences in mammals, the blastocyst exhibits an inner cell mass (fig. 5.10*e–f*). The wall of the blastocyst of mammals does not contribute to formation of the embryo; the embryo develops from the inner cell mass. The blastocyst wall, called a trophoblast, establishes contact with, or implants itself into, the lining of the maternal uterus (in humans, at about 14 days). It is the first membrane through which the **conceptus** (embryo plus extraembryonic membranes) receives nourishment and oxygen directly from maternal tissues. Any earlier nutrition is histotrophic (see earlier discussion in "Oviparity and Viviparity").

The method of formation of the first endoderm is not precisely the same in all placentals. It is, however, a product of the inner cell mass, whether by cell migration or delamination. In either case, the inner cell mass is eventually converted into epiblast and primitive endoderm. It is then called an **embryonic disk** or **blastodisk**. Thereafter, there is much streaming of ectoderm, presumptive mesoderm, and endoderm from the embryonic disk, resulting in establishment of the germ layers and, simultaneously, of the extraembryonic membranes. Among the latter is a yolk sac (fig. 5.10*g*) for yolk that does not exist. The trait bespeaks the common ancestry of mammals and reptiles.

Before concluding the discussion on craniate gastrulation, a final word must be added. Because of the diversity of the specific processes whereby various craniates achieve a bilaterally symmetrical, three germ-layered stage with notochord and neural tube rudiments, it is not possible to specify a common end point for the process of gastrulation that would be applicable to all species. Gastrulation places the germ layers in their respective positions. When this has been accomplished, gastrulation is complete, and organogenesis can proceed.

Neurulation

Presumptive neural ectoderm occupies a wide band on the dorsal surface of the elongating gastrula, overlying the notochord. As elongation of the gastrula continues,

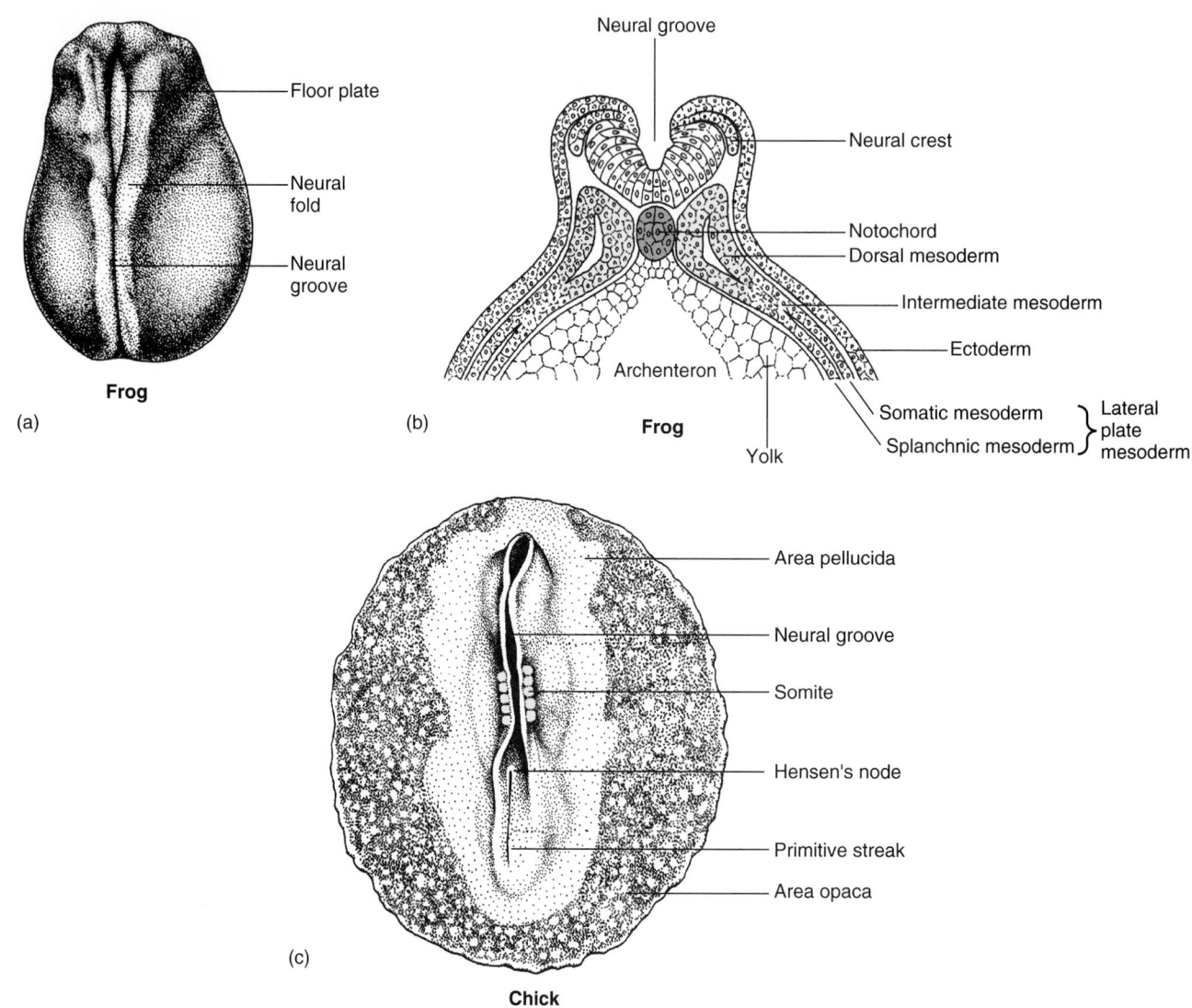

FIGURE 5.9
Neurulation in a frog, chick, and teleost. (*a*) Frog neurula near end of gastrulation, removed from its jelly envelope. The nearly closed blastopore is at the caudal end of the neural grove. (*b*) Cross section of (*a*) at level of future midgut. The coelom is the space between the somatic and splanchnic mesoderm. (*c*) Chick embryo at about 26 hours of incubation. (*d*) Cross section of (*c*) at level of labeled somite. (*e*) Neural keel in a teleost. Continued

the band thickens to become a **neural plate** (**floor plate**, in the terminology of the developmental biologist). The lateral borders of the neural plate become elevated to form a pair of neural folds bounding a neural groove. The embryo is now a **neurula** (see fig. 5.9*a*). The anterior end of the neural groove is widest, being the future brain. The rest is future spinal cord. The floor plate sinks into the dorsum of the embryo, and the neural folds grow toward one another above the groove, meet in the middorsal line, then fuse, converting the neural groove into a **neural tube.** The process of closure commences near the caudal end of the brain and sweeps cephalad and caudad. Finally, only a small opening remains at each end of the tube (**anterior** and **posterior neuropores**). These later close. As a result of formation of a neural tube by closure

of the neural groove, the central nervous system is ectodermal, dorsally located, and hollow. The cavity is the **neurocoel.** It remains as the ventricles of the brain and central canal of the spinal cord. The process that establishes the neural tube is **neurulation.**

Neurulation differs among the craniate classes in minor details. The major variation is seen in living agnathans and ray-finned fishes, in which neural folds do not form, and the floor plate does not roll into a tube. Instead, it becomes a wedge-shaped **neural keel** that invades the dorsal body wall above the notochord (see fig. 5.9*e*). Subsequently, the wedge becomes detached from the overlying ectoderm and organizes a neurocoel by rearrangement of the interior cells. The result is a typical dorsal hollow central nervous system.

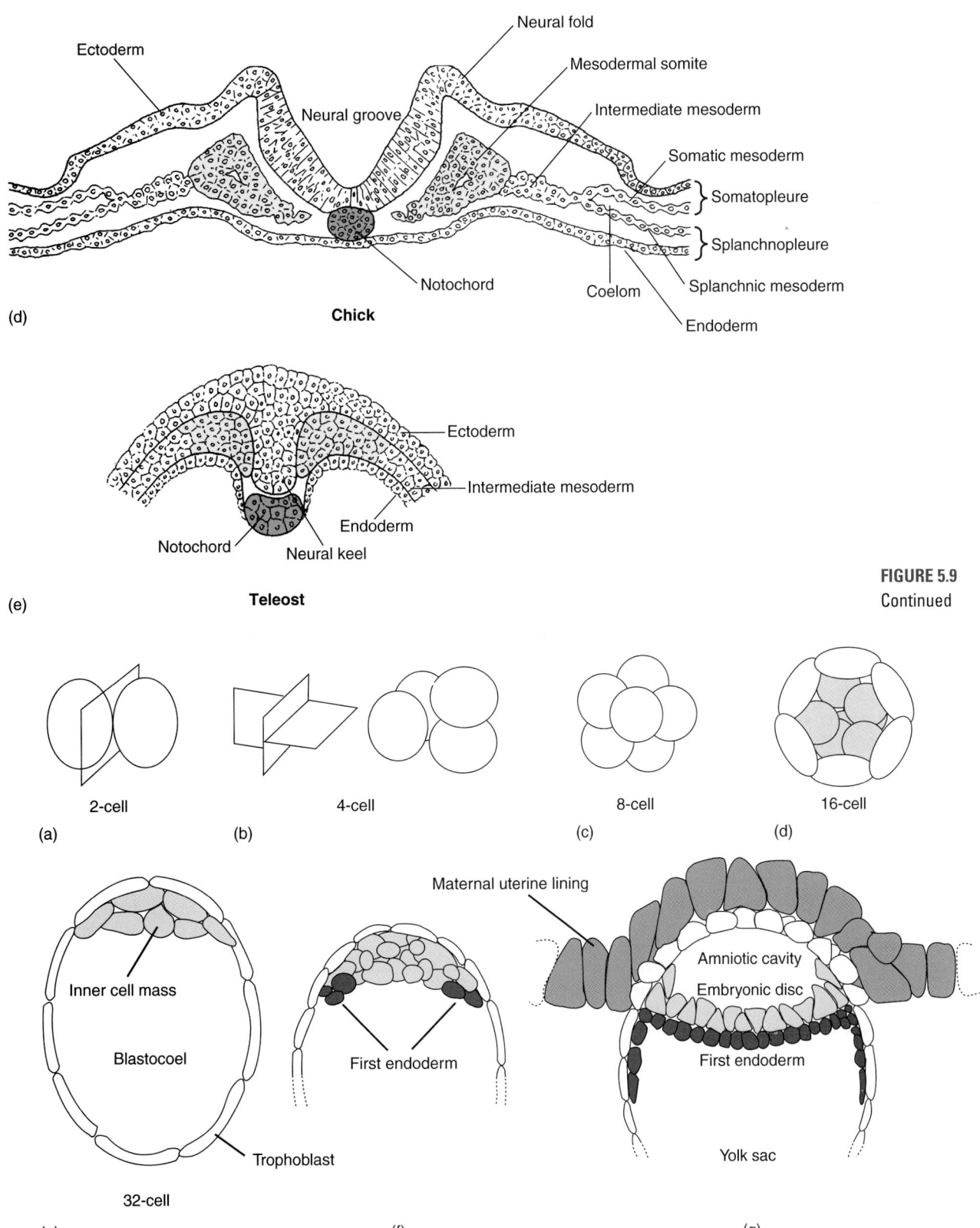

FIGURE 5.9
Continued

FIGURE 5.10

Cleavage, blastula, and gastrulalike stages of a mammal. (*a*) First cleavage is meridional; note the indicated plane of division. (*b*) Rotational cleavage. In the next round of cleavage, one cell is divided meridionally (left cell of *a*) and the other equatorially (right cell of *a*). Planes of cleavage are indicated. (*c*) 8-cell stage. (*d*) 16-cell stage with formation of an inner cell mass (*light red*) that forms the embryo. (*e*) 32-cell stage blastocyst showing the developed trophoblast (the placenta and chorion are derivatives of the trophoblast). (*f*) Later stage showing differentiation of the first endoderm (*dark red*). (*g*) Implantation within the maternal uterine lining (*medium red*) showing the gastrula with continued development of the endoderm and beginning formation of the amniotic cavity and yolk sac. Not to scale.

Induction of Differentiation: Morphogens

Early in the twentieth century, Hans Spemann, using transplantation of tissues in amphibian gastrulas to induce differentiation of an organ in an ectopic (abnormal) location, demonstrated that the dorsal lip of the blastopore is an **organizer area** that establishes the long axis of the craniate body. This was the first demonstration that preexisting cells govern the fate of contiguous ones. Experimental biologists using similar techniques later demonstrated the organizer effects of one embryonic tissue after another in species from fruit flies to fishes to chicks to mice. Chapter 17 cites induction of the lens of the eye from embryonic ectoderm in the presence of a retina. What could not be answered with techniques then available was how induction brought about its effect. Obviously, it had a genetic basis, but how, specifically, were genes expressed? Answers are now coming from molecular biology.

Genes similar to those that are important in fruit fly development are the key molecular choreographers of the major events of gastrulation, neurulation, and certain later embryonic developments in craniates. A number of cascading interactive genes, through a process of signal and response, establish segmental domains within the developing craniate (as also seen in the fruit fly). One step in this process involves a number of genes that determine the fates of individual segments once they are determined by earlier processes. These segmental genes in fruit flies and craniates are called **homeotic genes**. An important complex of homeotic genes includes the **Hox gene clusters** of craniates (refer to "Segmentation of the Craniate Head" in chapter 16). Other regulatory genes are also present and are often named after their homologues in the fruit fly. In the fruit fly embryo, a gene causes the ventral cuticle region to take on a bristly appearance like that of a hedgehog, thus the name "hedgehog gene." The homologous gene in craniates is sonic hedgehog (although the function of this gene in craniates is quite different from that in the fruit fly). These homeotic genes have been shown, thus far, to participate in differentiation of the notochord, the early neural tube, and fins or limbs in craniates. The molecular basis for induction of differentiation is a protein encoded by the genes. The protein is called a **morphogen** (the signal for the cascading genes). Without morphogen, there is no induction. There is evidence that in at least some inductive processes, more than one morphogen is involved. Whether or not these participate together in producing a common final inductor, or whether they regulate parallel processes, is not yet known.

Chordin is a secreted morphogen encoded by genes in the dorsal lip of the blastopore during early proliferation of presumptive notochord cells. It therefore has potent body axis-forming activity in the early stages of gastrulation. The transcription of chordin is activated by growth factors that also induce formation of dorsal mesoderm in the *neurula stage*. At the *tailbud stage* (two days of development in *Xenopus*, the experimental amphibian), chordin is expressed only in the tailbud region, which is the last population of cells with chordin-organizing activity. The observation that chordin is a protein consisting of 941 amino acids, and that it is only one of several morphogens regulating differentiation in the dorsal axial region of the early gastrula, attests to the complexity of the problem of exploring induction processes.

Formation of the first rudiment of the central nervous system, the floor plate (see fig. 5.9), requires inductive signals from the notochord. The morphogen in this instance is not a diffusible secretion; induction requires direct contact of the notochord with the overlying ectoderm. Differentiation of the components of the neural tube, which occurs later, is accomplished by diffusible morphogens. The floor plate of the gastrula participates with the notochord in inducing differentiation of motor neurons in the ventral half of the neural tube. (This half, illustrated in fig. 16.7, becomes the **basal plate of the spinal cord**, and is so labeled. It is not to be confused with the floor plate of the gastrula.) It is this area of the neural tube that is in contact with the notochord. Ventral portions of the developing brain are also responsive. Induction of components of the dorsal half of the tube appears not to be the province of the notochord.

The same genes that act early in gastrulation initiate, later, the development of pectoral fins, chick wings, and the anterior limbs of mice. (Limb and fin buds are illustrated in figs. 11.17 and 11.18.) Participating with the hedgehog genes is a growth factor produced in the thickened ectodermal ridge overlying the mesoderm of the limb bud. Removal of the ridge halts limb bud development, but the (fibroblast) growth factor, when implanted, restores the function of the epidermal ridge, inducing the continued proliferation of mesenchyme that keeps the limb bud growing outward from the body. Substituting anterior limb bud mesoderm for the blastema of the posterior bud in chick embryos has resulted in reversal of the position of wings and hind limbs. Research on development of the hand of a chick indicates the probability that successive morphogens are responsible for the development of successive digits.

If a homeotic gene is defective, the result may be a severe developmental anomaly. For instance, failure of dorsal closure of a section of the neural tube, a condition known as **spina bifida** in human infants, is a result of failure of a normal inductive process early in ontogeny. Anomalies of the arm, hands, or digits have a similar etiology.

Mesenchyme

Mesenchyme is an unorganized accumulation of dendritic (branching), undifferentiated, embryonic cells that compose much of an early embryo. During organogenesis, under the influence of morphogens, mesenchyme becomes organized in aggregates known as **blastemas.** These gradually assume the general outlines of a future organ, and the cells therein commence to differentiate to form appropriate tissues. As organogenesis proceeds, mitosis of the differentiating cells participates in furthering the growth of the organ.

Although mesenchyme is primarily embryonic, some mesenchyme formation continues into adulthood, albeit at an increasingly slower rate as aging progresses. These are reserve cells that can replace some varieties of worn or damaged tissues, especially connective tissues. Many specialized tissues—epithelial, for instance—are replaced solely by mitosis of existing cells in that tissue. Epidermal cells are replaced only by mitosis of the already differentiated cells of the basal layer of the epidermis. Most mesenchyme arises from mesoderm. Mesenchyme of ectodermal origin, a minor source, is known as **mesectoderm.** If the source is neural ectoderm (neural crest, for example), it is referred to as **neurectoderm.** A few of the many varieties of connective tissue that arise from mesenchyme are listed in figure 7.1.

Blastema cells are totipotent—they can be induced to become any specialized cell that the inductor substance specifies. This accounts for ectopic organs induced in transplantation experiments on embryos.

FATE OF THE ECTODERM

Embryonic ectoderm gives rise to (1) the entire nervous system and some of its membranes, (2) the lens and retina of the lateral eyes and the sensory epithelium of the other special sense organs that monitor the external environment, (3) the epidermis and its derivatives, including glands that lie within the dermis, (4) the lining of the stomodeum and proctodeum and their derivatives, and (5) a broad spectrum of other tissues, particularly in the head and pharyngeal regions, that result from migration of mesenchyme from neural crest and ectodermal placodes. These other tissues include cartilage, bone, and other connective tissues, which elsewhere arise from mesoderm.

Stomodeum and Proctodeum

Formation of the oral cavity commences with a **stomodeum,** a midventral invagination of the ectoderm of the embryonic head that develops at the anterior end of the foregut (see figs. 1.1 and 17.6a). The two cavities are temporarily separated by a thin partition, the **oral plate,** the rupture of which provides an opening to the digestive tract from the exterior. Stomodeal ectoderm lines the anterior portion of the oral cavity and covers at least the anterior part of the tongue. There is, however, some overriding of stomodeal ectoderm by endoderm,

so we cannot say with certainty to what extent either ectoderm or endoderm contributes to some of the oral glands that evaginate near the junction of the two.

Stomodeal ectoderm provides the cells that secrete the enamel of mammalian teeth; however, the enameloids of other vertebrates are secreted by cells that are of neural crest origin. An evagination (sometimes a club-shaped growth) known as **Rathke's pouch** evaginates from the roof of the stomodeum before the oral plate ruptures, and the pouch gives rise to the adenohypophysis (see fig. 18.6a).

An ectodermal invagination, the **proctodeum,** similar to the stomodeum, develops in relation to the embryonic hindgut. The proctodeum and hindgut are separated temporarily by a **cloacal plate** (see fig. 1.1).

Neural Crest

As the neural groove sinks into the dorsal body wall and the neural folds meet above the groove to form a tube, some of the ectoderm at the site of union of the folds separates from the surface ectoderm and from the neural tube to form a pair of mesenchymal ribbons paralleling the neural tube on each side. Shortly thereafter, the ribbons become segmented, forming a cluster of cells in each metamere and several clusters lateral to the midbrain and hindbrain. These neurectodermal clusters are **neural crest** (see fig. 5.9b). They are induced by protein growth factors evidently synthesized in the dorsal walls of the neural groove.

Neural crest cells proliferate rapidly, giving rise to a large number of descendants, many of which migrate to areas remote from the crests (table 5.2). Some of the cells in the head stream ventrad into developing pharyngeal arches where they form blastemas for the skeleton of the arches, including that of the jaws. Others contribute to the anteroventral portion of the neurocranium and to some of the membrane bones that ensheathe the neurocranium. Still others invade the anlagen of the thyroid, parathyroid, thymus, and ultimobranchial bodies that are developing from the epithelium of the pharyngeal pouches or pharyngeal floor. The lingual cartilages and branchial basket of agnathans are products of neural crest mesenchyme. Induction of many of these cranial derivatives appears to be initiated by morphogens from head ectoderm or endoderm. Some of the derivatives of cranial neural crest, such as cartilage and bone, are of mesodermal origin elsewhere in the body.

Neural crest mesenchyme in the head or trunk streams ventrad around the notochord to form autonomic ganglia and the specific cells of the adrenal glands that produce adrenaline (see fig. 16.35). Other emigrants aggregate around the developing neural tube to invest it with meninges other than the dura mater. Still others become pigment cells not only in the skin, but in muscles and in many internal organs other than the retina, brain, and spinal cord.

TABLE 5.2   Primary Derivatives of Neural Crest

	Trunk Neural Crest	Cranial Neural Crest
Nervous System		
Sensory	Spinal ganglia	Ganglia of V, VII, IX, X, LL (lateral-line ganglia and neuromasts), Rohon-Beard cells
Autonomic	Postganglionic neurons	Postganglionic neurons
Neuroglial	Schwann cells	Schwann cells
Pigment Cells	Melanocytes	Melanocytes
Glandular	Adrenal medulla	Carotid body
Skeleton	Distal unpaired fins (teleosts and chondrichthyans)	Most of preotic skull, pharyngeal skeleton
Connective Tissue	None	Cornea of eye, dermis, tooth dentine, walls of aortic arches
Muscle	None	Ciliary muscles, dermal smooth muscles, aortic arch smooth muscles

Refer to Hall, 1999, for a detailed discussion of our current knowledge on neural crest.

One might suppose, with all this emigration of cells, that there would be few remaining in the neural crest. This is not so. Mitosis continually replenishes the supply until organogenesis is well underway. A very large number in the crests remain there to become the cell bodies of sensory neurons in discrete groupings (ganglia) on the pathway of spinal and cranial nerves (see fig. 16.5a and b). Others migrate distally and centrally along the course of the sensory nerve fibers, contributing to the neurilemma, a living membrane that ensheathes all nerve fibers outside the central nervous system.

Ectodermal Placodes

Ectodermal placodes are paired localized thickenings of the embryonic ectoderm that sink beneath the skin to one degree or another and give rise to neuroblasts and to the sensory epithelia of certain sense organs. Any other ectodermal thickenings that are characterized as placodes—those, for example, that become the lens of the lateral eyes (lens placode, see fig. 17.6b)—do not give rise to neurectoderm. Presumptive neurectodermal placodes can be grouped as follows:

1. A pair of **nasal (olfactory) placodes** forms just above the stomodeum at the anterior end of the brain. These placodes sink into the head and become part of the lining of a pair of nasal pits that open to the exterior via nostrils (see figs. 1.1 and 13.7). Some of the cells of these placodes become neurosensory cells in the olfactory epithelium. Their processes grow into the olfactory bulb of the brain and constitute, collectively, the olfactory nerves.
2. A pair of **otic placodes** in the ectoderm lateral to the hindbrain sink deep into the head to become otocysts (see fig. 17.6a and b). These are precursors of the sensory epithelia of the membranous labyrinth (inner ears).
3. A pair of **optic placodes**, which interact with an outgrowth of the brain, form the lens of the eye.
4. A series of **epibranchial placodes** (P_{VL}, chapter 16) on the side of the head (named for their location in fishes) sink into the head to a location alongside the midbrain and hindbrain, where they might be mistaken for neural crest. These placodes, like neural crest, contribute neuroblasts that become sensory cell bodies in one pair or more of the ganglia on cranial nerves VII, IX, and X. Their fibers innervate the taste buds. The number varies with the species, augmenting the cell bodies coming from neural crest cells.
5. A **linear series of placodes** (P_{DL}, chapter 16) extending the length of the trunk and tail of fishes and amphibians, and continuing onto the head in a variable pattern, sink into the dermis to become the neuromast organs of the cephalic and lateral-line canal systems. Other placodes on the head of fishes, not part of the linear series, become electroreceptive epithelia.

From the discussions on neurulation, neural crest, and ectodermal placodes, it may be evident that the components of the entire nervous system, central and peripheral, many of their supportive and nutritive cells, and the sensory epithelia of all the special sense organs other than taste buds are of either neural plate, neural crest, or ectodermal placode origin. It is also evident that the embryonic ectoderm exerts a much greater influence on differentiation and organogenesis than might be supposed from its superficial position in early embryos.

FATE OF THE ENDODERM

Endoderm gives rise to the epithelium of the entire alimentary canal and its evaginations between the stomodeum and proctodeum. The epithelioid components of the parathyroid glands, ultimobranchial glands, thymus, and the lining of the auditory tube and middle ear cavity are endodermal because of their origin from pharyngeal pouches. Midventral evaginations of the pharynx, including thyroid, lungs, and swim bladders and their ducts, if any, are lined with endoderm or have endodermal components. Caudal to the pharynx, the endoderm evaginates to form crop sacs, liver, gallbladder, pancreas, and various gastric and intestinal ceca. Most urinary bladders and the urogenital sinus of mammals are lined with endoderm, because these are derived from the cloaca. Any other structure that arises as an evagination of the embryonic foregut, midgut, or hindgut has endodermal components.

FATE OF THE MESODERM

Dorsal Mesoderm (Epimere)

The somites collectively constitute most of the dorsal mesoderm. They are aligned beside the notochord and neural tube the entire length of the embryonic trunk and tail, and develop in the head in varying numbers (see figs. 5.9c and d and 16.6). A somite exhibits three regions, myotome, sclerotome, and dermatome (fig. 5.11). **Myotomes** contribute mesenchyme that migrates into the body wall of the trunk and tail to form skeletal muscles. **Sclerotomes** give rise to the vertebral column and ribs; those of head somites give rise to parts of the neurocranium and to one or more sense capsules, depending on the species. **Dermatomes** contribute mesenchyme that forms the dermis of the skin of the dorsum. (Most of the dermis in the body arises from lateral-plate mesoderm). The dorsal mesoderm (paraxial mesoderm) continues anterior to the somites as incompletely segmented **somitomeres.** Unlike the somites of the body, the somitomeres lack dermatome and sclerotome. The products of these latter tissues are derived in the anterior head from neural crest (table 5.2). The remaining myotomal component of the somitomeres gives rise to the branchiomeric and extrinsic eye muscles (see chapter 11).

Lateral-Plate Mesoderm (Hypomere)

Lateral-plate mesoderm is confined to the trunk and consists of somatic and splanchnic mesoderm (see figs. 5.9b and d and 5.11). Somatic mesoderm gives rise chiefly to the connective tissues and blood vessels of the body wall and to the skeleton of the body wall, girdles, and limbs. The dermis of the body wall is its outermost product, and the parietal peritoneum (see fig. 1.2, 5) is the innermost. Splanchnic mesoderm gives rise to smooth muscles and connective tissue of the digestive tract and of its outpocketings, to the heart, and to the blood vessels of the viscera. The visceral peritoneum (see fig. 1.2, 6) is its outermost derivative.

Intermediate Mesoderm (Mesomere)

Intermediate (nephrogenic) mesoderm consists of a pair of longitudinal ribbons of unsegmented mesoderm extending the length of the trunk just lateral to the somites (see fig. 5.9b and d). It gives rise to kidney tubules and the longitudinal ducts of the urogenital system.

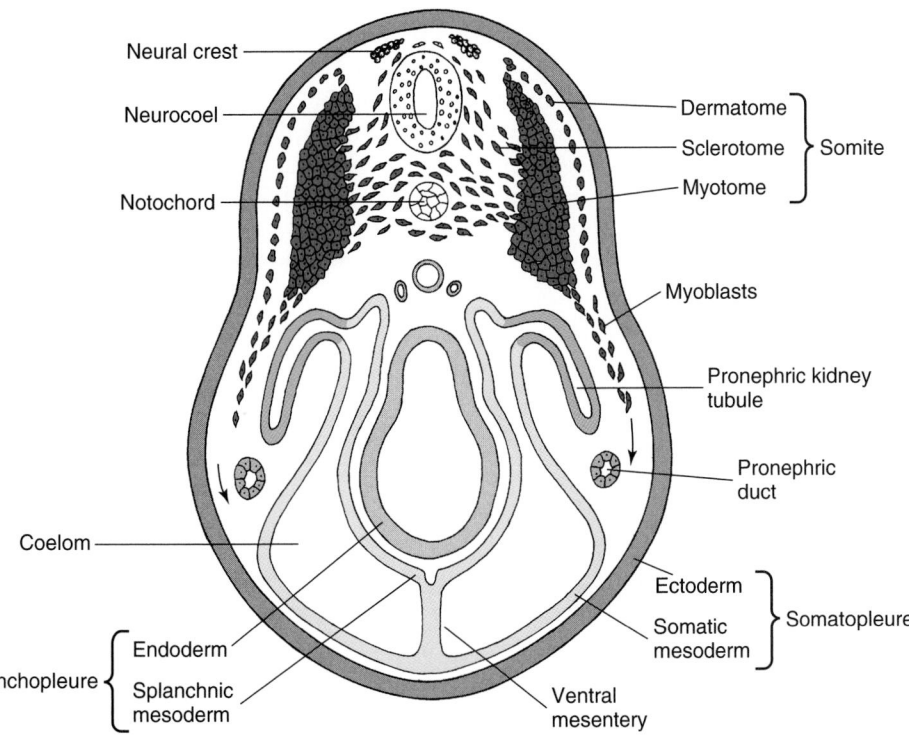

FIGURE 5.11

Vertebrate embryo in cross section showing fate of the mesoderm. *Dark red,* mesodermal somite (dorsal mesoderm); *medium red,* derivatives of intermediate mesoderm; *light red,* lateral-plate mesoderm. Sclerotome cells are streaming toward notochord and neural tube to form a vertebra; myotomal cells (myoblasts) are streaming into lateral body wall (*arrows*) to form myotomal muscle; dermatome will form dermis of skin of dorsum.

SIGNIFICANCE OF THE GERM LAYERS

There appears to be nothing intrinsic to the earliest cells of the germ layers that determines the direction of differentiation of their descendants. The early generations are totipotent. The factors that drive differentiation are the morphogens, but we have no idea what induces Spemann's organizer area to start putting them out. That puzzle may not be solved for a long time. The germ layers, positioned by an inherent genetic code, constitute the initial cellular organization that enables organogenesis to proceed.

EXTRAEMBRYONIC MEMBRANES

Most embryonic craniates are provided with special membranes that extend beyond the body. These extraembryonic membranes arise early in ontogeny and perform important services for the embryo until hatching or birth. The chief extraembryonic membranes are the **yolk sac, amnion, chorion,** and **allantois.** The last three are found only in reptiles (including birds) and mammals. The yolk sac is the most primitive.

Yolk Sac

The yolk sac surrounds the yolk (figs. 5.12 and 5.13). It empties into the midgut and is usually lined by endoderm, although not in bony fishes. The sac is highly vascular, and its vessels (**vitelline arteries** and **veins**) are confluent with circulatory channels within the embryo proper. Yolk particles in the sac are usually digested by enzymes secreted by the lining of the sac and then transported to the embryo by vitelline veins. (In sharks, yolk also enters the intestine directly from the sac, propelled by rapidly beating cilia that line the sac and stalk.) The yolk sac becomes smaller as the embryo grows larger. As it shrivels, it slowly disappears into the ventral body wall. The intracoelomic remnant of the sac is finally incorporated into the wall of the midgut, or it may remain as a small diverticulum.

In embryonic sharks, a large diverticulum of the yolk sac develops within the coelom close to where the sac empties into the duodenum. In pups ready to be born and with the yolk sac almost completely retracted, the diverticulum is easily demonstrated. It is distended with yolk, which has also spilled over into the spiral intestine. This remaining yolk nourishes the newborn pup for several days until it is able to obtain food from the environment.

Despite the fact that no yolk is present in therian mammals, the embryos develop a yolk sac—a reminder of their genetic relationship with egg-laying reptiles. In humans, a vestige of the yolk sac (Meckel's diverticulum) remains in about 2 percent of the adult population. Its average position is on the small intestine, 30 cm above the ileocolic valve. Its average length is about 5 cm.

Because the yolk sac in viviparous fishes and amphibians is highly vascularized and lies close to maternal tissues, it often serves as a membrane for absorbing oxygen from the parent. After the yolk in the sac is depleted, or if there is none, the sac may absorb nourishment from maternal tissues. When functioning in either capacity, it constitutes a **simple yolk sac placenta.**

Amnion and Chorion

The embryos of reptiles and mammals develop within two membranous sacs, an amniotic sac composed of an amnion, and a chorionic sac composed of a chorion. These two membranes are formed when upfoldings of the embryonic somatopleure meet above the embryo and fuse, forming the amniotic sac around the embryo. The chorion forms with continued growth of somatopleure and surrounds both the amniotic sac and the yolk sac (see fig. 5.13a). Figure 15.45 shows fetuses within these sacs in the uterus of a pregnant dog. The amnion and chorion, along with the semiporous eggshell, enabled ancestral amniotes and their descendants to lay their eggs on land, which had become their primary environment.

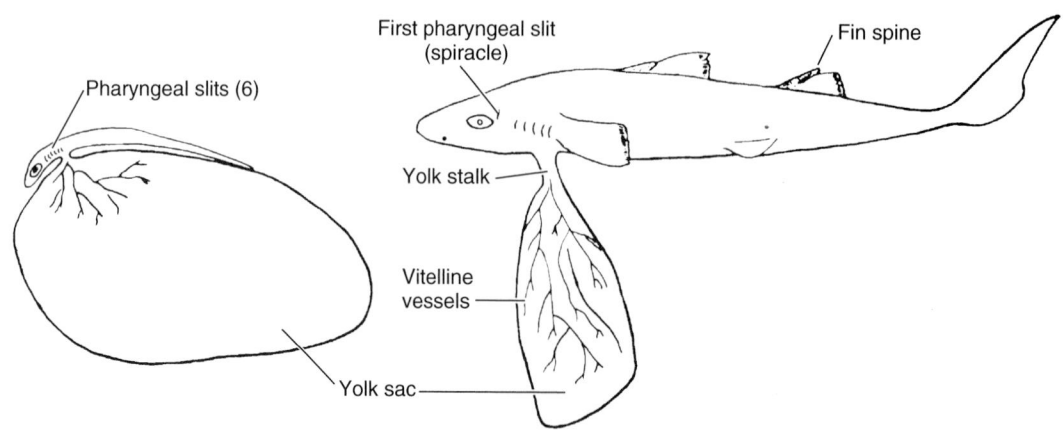

FIGURE 5.12
Dogfish embryos with yolk sacs, removed from the uterus.

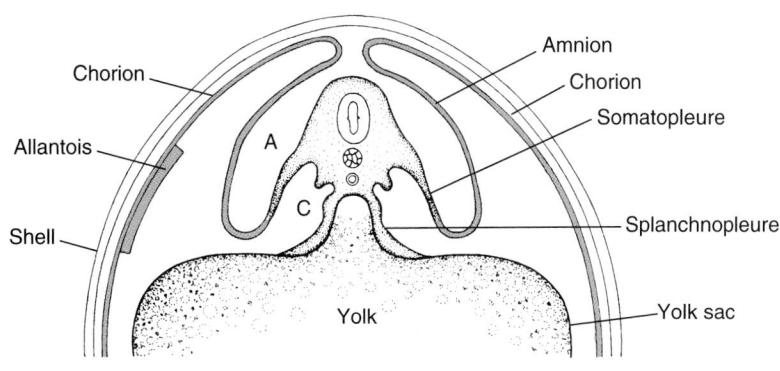

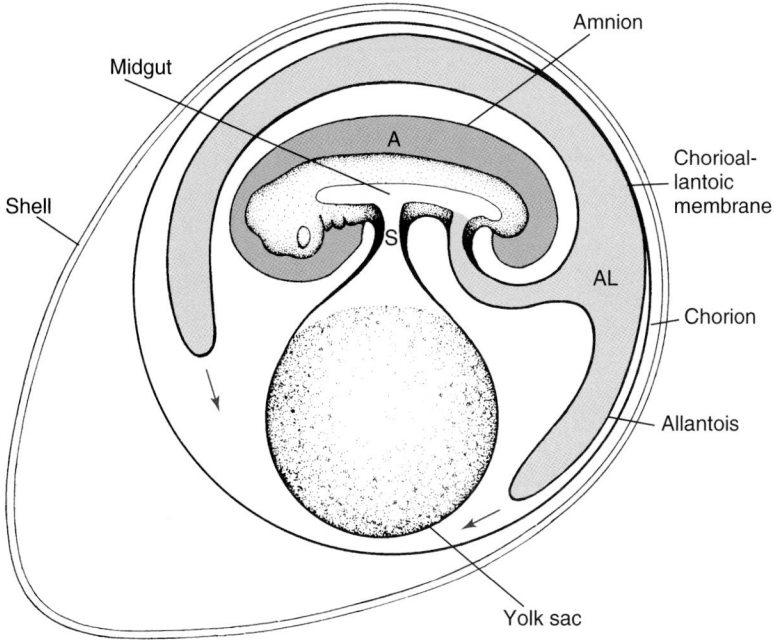

FIGURE 5.13
Extraembryonic membranes of an egg-laying amniote. (*a*) Cross section of embryo showing amnion and chorion arising from amniotic folds of somatopleure. Position of the allantois is indicated on left of drawing. (*b*) Relationships of the allantoic sac (**AL**), which is growing in direction of arrows. **A,** amniotic cavity; **AL,** allantoic cavity; **C,** intraembryonic coelom; **S,** yolk stalk, an evagination of the midgut.

Surrounding the embryo within the amniotic sac is a slightly salty **amniotic fluid** consisting chiefly of metabolic water from embryonic tissues. Turtle eggs that are laid in moist locations absorb considerable water from the environment through the shell, but this is not a common source of amniotic fluid. When the embryonic kidneys begin to function, nitrogenous wastes are added to the fluid. Amniotic fluid buffers the fetus against mechanical injury and, in oviparous species, helps prevent dessication of the embryo.

The chorion lies in intimate relationship with either the eggshell or the lining of the mother's uterus, depending on the species. Thus, the amnion with its amniotic fluid protects the fetus, and the chorion keeps it in communication with its source of oxygen and, in viviparity, its source of nutrients.

Allantois

The allantois is an extraembryonic membranous sac that arises typically as a midventral evagination of the embryonic cloaca (see fig. 5.13*b*). It ordinarily grows until it comes in contact with the inner surface of the chorion to form a **chorioallantoic membrane** (see fig. 5.13*b*). The chorioallantoic membrane in reptiles and monotremes is in contact with the eggshell where, except in those viviparous squamates that depend on the mother late in pregnancy for at least some nourishment, it serves chiefly as a *respiratory organ*. In eutherian mammals, the chorioallantoic membrane is in direct contact with the lining of the maternal uterus, where it constitutes the fetal part of a **chorioallantoic placenta.** Here it performs two additional functions, being also the site of *transfer of nutrients* from maternal tissues to the embryo and of *transfer of metabolic wastes* from embryo to mother.

The base of the allantois—the part closest to the cloaca—becomes the **urinary bladder** of amniotes (see fig. 14.38). In mammals, the part extending between the tip of the bladder and the umbilicus may remain after birth as a **middle umbilical ligament,** or **urachus** (see fig. 15.49*f*); the portion within the umbilical cord is discarded at birth. It is likely that the allantois is the old amphibian urinary bladder with added functions.

Placentas

The term *placenta* in its broadest sense refers to any region in a viviparous organism where maternal and embryonic tissues of any kind are closely apposed and that serves as a site for exchanges between parent and embryo. In this sense, viviparous fishes that develop in an ovarian follicle sometimes exhibit a placenta. In a more restricted sense, a placenta is an organ composed of (1) the highly vascular region of an extraembryonic membrane (yolk sac, choriovitelline membrane, chorioallantoic membrane, or chorion alone) and (2) the associated highly vascular lining of the maternal uterus.

The yolk sac frequently serves as part of a placenta in viviparous amphibians and fishes such as the dogfish

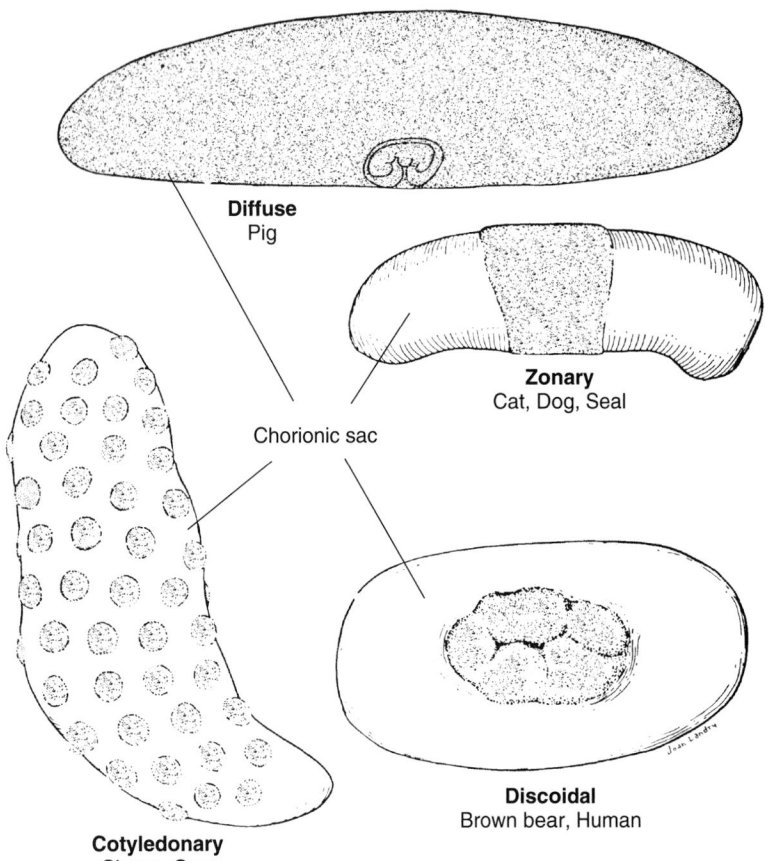

Diffuse
Pig

Zonary
Cat, Dog, Seal

Chorionic sac

Discoidal
Brown bear, Human

Cotyledonary
Sheep, Cow

FIGURE 5.14
Chorionic sacs of mammals showing placental areas. Pig embryo surrounded by the amnion can be seen through the exceptionally thin chorionic sac.

shark. These craniates have no amnion, chorion, or allantois, and a simple yolk sac placenta lies in immediate contact with the maternal uterus. In reptiles that depend on viviparity near the end of fetal life, a soft, thin eggshell lies between the fetal membranes and the uterine lining, but it does not impede the essential functions of a placenta. In most marsupials, the yolk sac, rather than the allantois, lies against the chorion as part of a **choriovitelline placenta**. As stated earlier, eutherian mammals have a chorioallantoic placenta. These mammals have an umbilical cord connecting the fetus with the placenta (see fig. 14.38). In some of them (cats, rabbits, humans, others), the allantois extends only part way out the umbilical cord before dwindling and terminating as a blind sac, but its vessels, the allantoic (umbilical) arteries and veins, continue toward, and vascularize, the chorion.

The intimacy of the relationship between fetal and maternal tissues in mammals varies greatly. In marsupials and most ungulates, the extraembryonic membranes lie in simple contact with the uterine lining (**endometrium**), and at birth the fetal membranes simply

peel away without any shedding of the lining. This is a **contact** or **nondeciduous placenta**. In more intimate relationships, **chorionic villi**, which are fingerlike outgrowths of the chorionic sac, become rooted to one degree or another into the endometrium, or even dangle into uterine blood sinuses, as in humans. When the fetal part of such a placenta disengages at parturition, the *invaded portion* of the uterine lining, the **decidua**, is shed, and some bleeding occurs. This is a **deciduous placenta**. In any case, the extraembryonic membranes and any sloughed uterine tissues are delivered as the afterbirth.

Chorionic villi are distributed on the surface of the mammalian chorionic sac (fig. 5.14) in isolated patches (**cotyledonary placenta**), in a band encircling the sac (**zonary placenta**), in a single large discoidal area (**discoidal placenta**), or diffusely over the entire surface of the chorion (**diffuse placenta**). The mammalian placenta is one source of hormones essential for the maintenance of pregnancy.

Summary

Definitions may paraphrase those in the text.

1. Eggs are micro-, meso-, or macrolecithal, depending on the quantity of yolk. They are isolecithal or telolecithal, depending on the distribution of cytoplasm and yolk within the egg.
2. Oviparity is a condition in which unfertilized or fertilized eggs are shed to the exterior for development and hatching.
3. Viviparity is a condition in which offspring are born alive. The developing organisms may or may not be dependent on the parent for nourishment before birth.
4. Viviparity is found in some members of every craniate class except agnathans and birds.
5. Euviviparity is a viviparous condition in which offspring are dependent on the parent for nourishment throughout pregnancy.
6. Ovoviviparity is a viviparous condition in which offspring are not dependent on the parent for nourishment throughout pregnancy.
7. Fertilization is internal in viviparous species, in urodeles and apodans, and in species that cover the egg with a shell.
8. Cleavage follows fertilization. It produces a blastula with a blastocoel.

9. Gastrulation is characterized by notogenesis, neurulation, establishment of bilateral symmetry, and formation of the germ layers. The basis of the process is morphogenetic streaming of undifferentiated cells.

10. Gastrulation is initiated by involution in cephalochordates, epiboly in species with mesolecithal eggs, and delamination of a blastoderm in species with macrolecithal eggs.

11. In therian mammals, an inner cell mass gives rise to the embryo. Gastrulation is by delamination or a modification thereof.

12. A trophoblast membrane develops from the blastula wall, providing initial embryo and uterine contact.

13. Neurulation results in formation of a brain and spinal cord.

14. Morphogens are inductor (signaling) proteins that induce organization of the germ layers and participate in differentiation of specialized tissues from embryonic mesenchyme. Morphogens are the expression of specific homeotic genes.

15. Mesenchyme is an embryonic tissue consisting of undifferentiated, totipotent, embryonic cells that require inductor factors for differentiation.

16. Ectoderm contributes epidermis and its derivatives, the entire nervous system, epithelia of special sense organs such as lateral eyes and inner ears, the lens of paired eyes, the lining of the stomodeum and proctodeum, and a broad spectrum of miscellaneous tissues that differentiate from neural crest.

17. Neural crest give rise to the sensory ganglia on the roots of cranial and spinal nerves, to autonomic ganglia, to many of the skeletal components of the skull and pharyngeal arches, to most pigment cells, and to numerous miscellaneous tissues.

18. Ectodermal placodes give rise to neurosensory cells of the olfactory epithelium, to sensory epithelia of the inner ear, to a variety of mechano- and electroreceptive organs of fishes, and to some neuroblasts in the ganglia of cranial nerves VII, IX, and X that innervate the taste buds.

19. Endoderm gives rise to the epithelium of the digestive tract and to the epithelioid components of the organs that evaginate from it.

20. Dorsal mesoderm forms mesodermal somites consisting of sclerotome, dermatome, and myotome. These give rise, respectively, to vertebrae and the ribs, the dermis of the dorsum, and skeletal muscles other than those of the pharyngeal arches.

21. Lateral-plate mesoderm splits into somatic and splanchnic layers.

22. The somatopleure consists of somatic mesoderm plus ectoderm. The splanchnopleure consists of splanchnic mesoderm plus endoderm.

23. The coelom is the cavity between the somatic and splanchnic mesoderm. It is initially segmented in cephalochordates.

24. Intermediate mesoderm gives rise to kidney tubules and to the ducts of the urogenital system.

25. In viviparity, highly vascularized embryonic tissues (pericardial sac, gills, villi, opercular lining, extraembryonic membranes) are sites for absorption of essential substances.

26. The chief extraembryonic membranes are yolk sac, amnion, chorion, and allantois.

27. The yolk sac serves as a simple yolk sac placenta in viviparous anamniotes and as part of a choriovitelline placenta in marsupials.

28. Amniotic fluid bathes the amniote embryo, preventing desiccation and helping to prevent mechanical trauma.

29. The chorion is applied to the eggshell or to the lining of the maternal uterus. In that latter case, it is part of the placenta.

30. The allantois is a midventral evagination of the cloaca of amniotes. It participates in formation of chorioallantoic placentas. The part proximal to the cloaca becomes the urinary bladder. The distal part becomes the urachus in some mammals.

31. Chorioallantoic placentas are found in eutherian mammals. They may be deciduous or nondeciduous. Villi may be arranged in zonary, cotyledonary, discoidal, or diffuse patterns.

Critical Thinking Questions

1. Why study development as part of a course on comparative anatomy? What does it contribute to this study?

2. What is the relationship between ontogeny and phylogeny?

3. Craniate eggs can be classified based on the yolk they contain. Describe the distribution and quantities of yolk seen in craniate eggs and include the appropriate terms.

4. What are the primary processes associated with cleavage and blastulation? What structures are formed?

5. What are the primary processes associated with gastrulation? What structures are formed?

6. What are the primary processes associated with neurulation? What structures are formed?

7. Describe the effects of yolk on the patterns of development in a chicken.

8. How is chicken and mammal development similar? Why?

9. What are the unique features of development seen in mammals?

10. What are the fates of the individual derivatives of ectoderm (general ectoderm, ectodermal placodes, neural ectoderm, and neural crest)?
11. What are the three primary divisions of mesoderm (other than chordamesoderm), and what structures develop from each division?
12. How does the paraxial mesoderm (adjacent to the axial skeleton) of the head and trunk differ?
13. Which muscles are derivatives of the head paraxial mesoderm?
14. What are the extraembryonic membranes? What function do they serve?

Selected Readings

Arthur, W.: The origin of animal body plans, Cambridge, 1997, Cambridge University Press.

Bolker, J. A.: Comparison of gastrulation in frogs and fish, *American Zoologist* 34:313, 1994.

Carlson, B. M.: Human embryology and developmental biology, St. Louis, 1994, Mosby.

Coen, E.: The Art of genes, how organisms make themselves, London 1999, Oxford University Press.

Gilbert, S. F.: Developmental biology, Sunderland, 1997, Sinauer Associates.

Graveson, A. C.: Neural crest: Contributions to the development of the vertebrate head, *American Zoologist* 33:424, 1993.

Hall, B. K.: The neural crest in development and evolution, New York, 1999, Springer-Verlag.

Horner, J. R., and Weishampel, D. B.: Dinosaur eggs: The inside story, *Natural History,* p. 61, December 1989.

Maderson, P. F., editor: Developmental and evolutionary aspects of the neural crest. New York, 1987, John Wiley & Sons.

Meier, S.: Development of the chick embryo mesoblast: Morphogenesis of the prechordal plate and cranial segments, *Developmental Biology* 83:49, 1981.

Moore, J. A.: Science as a way of knowing—*Developmental Biology, American Zoologist* 27:415, 1987.

Müller, G. B., and Wagner, G. P.: Homology, *Hox* genes, and developmental integration, *American Zoologist* 36:4, 1996.

Nelsen, O. E.: Comparative embryology of the vertebrates. New York, 1953, The Blakiston Co., Inc. This comprehensive work has not been equaled.

Northcutt, G.: The origin of craniates: neural crest, neurogenic placodes, and homeobox genes. In Gans, C., Kemp, N., and Poss, S., editors: The lancelets: A new look at some old beasts, *Israel Journal of Zoology* 42:S-273–S-313, 1996.

Patten, B. M.: Early embryology of the chick, ed. 5. New York, 1971, McGraw-Hill Book Co.

Patten, B. M.: Early embryology of the pig, ed. 3. New York, 1948, McGraw-Hill Book Co.

Patten, B. M., and Carlson, B. M.: Foundations of embryology, ed. 5. New York, 1988, McGraw-Hill.

Roelink, H., et al.: Floor plate and motor neuron induction by vhh-1, a vertebrate homolog of hedgehog expressed by the notochord, *Cell* 76:761, 1994.

Sasai, Y., et al.: *Xenopus* chordin: A novel dorsalizing factor activated by organizer-specific homeobox genes, *Cell* 79:779, 1994.

Shine, R.: Young lizards can be bearable, *Natural History,* p. 34, January 1994. Patterns of viviparity in lizards.

Wourms, J. P.: Viviparity: The maternal-fetal relationship in fishes, *American Zoologist* 21:473, 1981.

Links to the Internet

Visit the zoology website at http://www.mhhe.com/zoology to find live Internet links for each of the following references.

1. Initial Development. From Sperm and Egg to Embryo. Includes modules such as "Close Encounters of the Zygotic Kind" and "Developmental Biology in the Bedrooms of the Nation."
2. Zygote: Info Link. Information on embryogenesis.
3. Embryo Development Overview. This site lets you view human development from conception to week 38.
4. Society for Developmental Biology. This site includes many valuable websites of the members of the society.
5. Bill Wasserman's Developmental Biology Page. Many links are found in "web resources."
6. The Foundations of Developmental Biology. An on-line segment of a course dealing with development. A plethora of links.

Integument

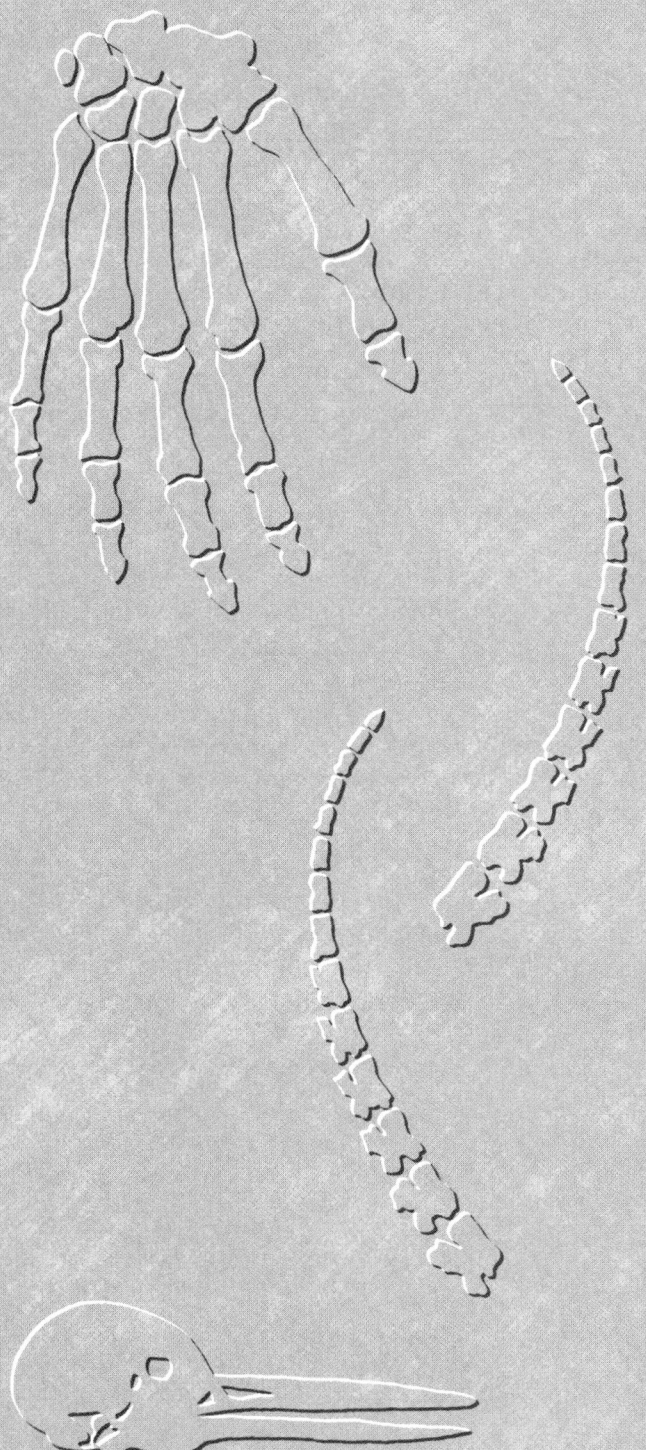

In this chapter, we will study the basic structure of the integument. We will examine the epidermis of aquatic craniates then see how it has been modified for life in air. We will also note some of the remarkable structures that form from epidermis. After that, we will examine the dermis and learn that bone is present in much modern skin and that its absence is a specialization. As a recapitulation, we will review the integument group by group. Finally, we will be reminded of some of the roles the integument plays in survival.

t would be difficult to imagine an organism the surface of which is not structured specifically for its role as an interface between organism and environment. Even amoebae have plasma membranes that wall off the protoplasm from the surrounding water. The integument of multicellular animals subserves many functions that contribute to the survival of the organism. It covers all external surfaces, including the exposed portion of the eyeball, where it is usually transparent and is called the **conjunctiva**, it covers the eardrum, and it is directly continuous with the mucous membranes that line all passageways opening onto the surface.

The integument of craniates including aquatic and terrestrial forms conforms to a basic morphologic pattern, consisting of a multilayered **epidermis** derived from ectoderm and a **dermis** derived chiefly from mesoderm. Variations in the morphologic features of the epidermis and dermis of the several craniate groups involve (1) the relative number and complexity of skin glands, (2) the extent of differentiation and specialization of the most superficial layer of the epidermis, and (3) the extent to which bone develops in the dermis. The skin of an amphioxus exhibits epidermis and dermis, but the epidermis is only one cell thick (fig. 6.1).

PREVIEW: SKIN OF THE EFT

As an introduction to the integument, we will take a quick look at a skin that is unencumbered by scales, feathers, or hair—the skin of the red eft, the juvenile land stage of the aquatic urodele *Notophthalmus* (fig. 6.2). It does not reflect an ancient feature of craniate skin, the presence of bony plates or scales in the dermis, but it illustrates the contrasting adaptations of craniate skin for life on land as opposed to life in water.

The epidermis of the eft is a stratified (multilayered) epithelium. The columnar cells in the basal, or **germinal**, layer are constantly undergoing mitosis, replacing those lost from the surface. Proliferation from the basal layer causes older cells to be pushed outward. As they approach the surface, they synthesize **keratin**, a scleroprotein that is insoluble in water, and they become flattened (**squamous**). They are then said to be **keratinized**, or **cornified**. Keratinization causes the cells to die. In efts, the cornified layer is thin in contrast to the thicker cornified stratum of terrestrial craniates.

The glands of the skin develop from the epidermis and, in efts, are simple multicellular sacs that bulge into the dermis, where they are in the immediate vicinity of capillaries that supply nutrients and oxygen, and carry away metabolic wastes. They are able to synthesize mucus, but in the eft stage they are quiescent, although not totally inactive.

The dermis of efts consists chiefly of connective tissue that supports the bases of the glands, blood vessels, lymphatics, small nerves, and pigment cells. It adheres closely to the underlying body wall muscle.

The life cycle of *Notophthalmus,* described in chapter 4, dramatically illustrates the direct correlation between a terrestrial habitat, lack of surface mucus, and the phenomenon of keratinization. The larvae of this species live in water, have many active mucous glands, and have no cornified epidermis. When they metamorphose and assume life on land, most of the skin glands become quiescent, keratinization of the epidermis commences, and it persists as long as the eft lives on land. At

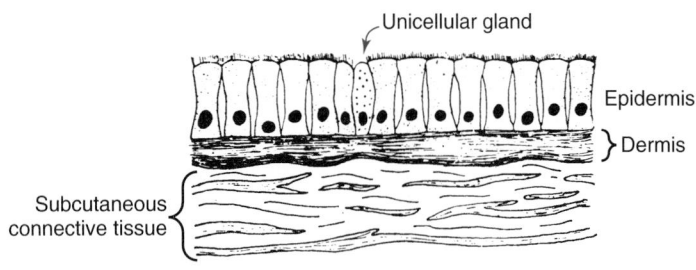

FIGURE 6.1
Skin of a young amphioxus.

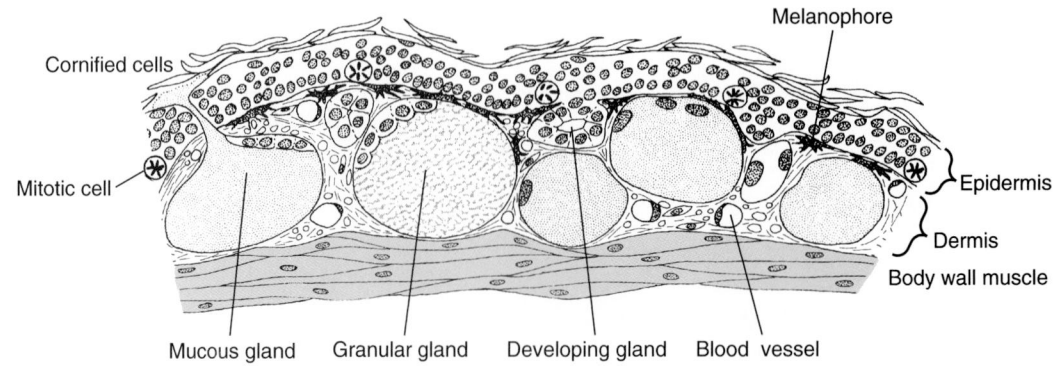

FIGURE 6.2
Skin of the red eft
Notophthalmus.

the onset of sexual maturity, the eft migrates back to water, the mucous glands again become active, and the cornified cells are shed and do not reappear. The skin of larvae and aquatic adults resembles that of fishes, whereas the skin of efts is like that of terrestrial amphibians.

THE EPIDERMIS

Having completed our preview, we are better able to study the integument in detail. We will study the epidermis and the dermis separately because the two have different roles. The avascular epidermis is the interface between organism and environment, which is reflected in its variable structure from fishes to amniotes. The highly vascular dermis provides physiologic support for the interfacing epidermis.

Two kinds of nonliving coverings overlie the living epidermis of craniates. In fishes, especially teleosts, and in aquatic amphibians, it is a thin coat of **mucus** that is continually being replenished. The role of this mucus cuticle is speculative. In terrestrial craniates the covering is a layer of dead, water-impervious cornified cells, the stratum corneum (fig. 6.3). Its survival value is clear. It minimizes water loss through skin that is exposed to air.

The epidermis is glandular to some degree in all craniates. The epidermis of most fishes and amphibians is exceptionally so. The most common epidermal glands of fishes and larval amphibians are unicellular, whereas those of metamorphosed amphibians and amniotes are predominantly multicellular. Although of epidermal origin, multicellular skin glands from fishes to amniotes invade the dermis, where they are in the immediate vicinity of blood capillaries that provide for their metabolic needs.

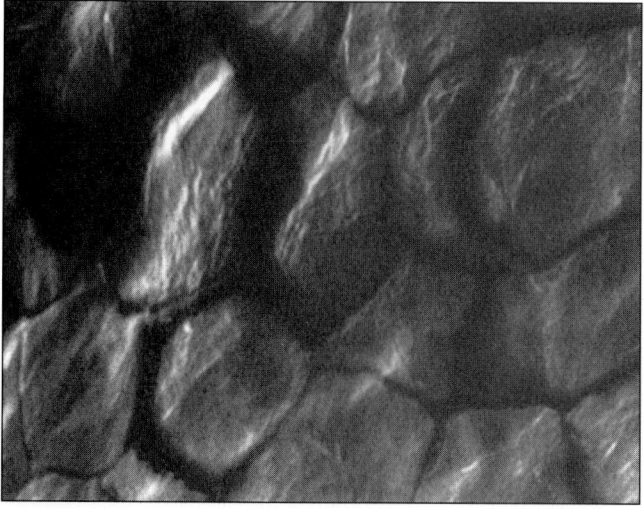

FIGURE 6.3
Keratinized (cornified) cells of the stratum corneum of human skin.
Courtesy Johnson & Johnson Research, New Brunswick, NJ.

Epidermis of Fishes and Aquatic Amphibians

The predominant feature of the epidermis of most fishes and aquatic or semiaquatic amphibians is the abundance of epidermal glands. Scales are not in the epidermis; they are beneath it.

Epidermal Glands

Fishes   Most of the integumentary glands in fishes are single cells, such as **goblet cells** (named for their shape) that secrete only mucus, and a variety of more specialized single cells, some with a granular cytoplasm (figs. 6.4 and 6.5). The **granular cells** of fishes secrete mucus and additional ingredients of mostly unknown functions. Multicellular glands (fig. 6.6) are not abundant in fishes, but those that are present contribute mucus and other ingredients including, in one species or another, thick slime or alkaloids that repel predators. Alkaloids, which are products of some of the granular cells, are not common in aquatic craniates. They lose effectiveness when diluted with environmental water. Slimy mucus is secreted in quantity in response to stressful external stimuli, particularly when life is endangered by a predator. This is especially applicable to hagfishes, aptly called slime eels, whose slime glands are surrounded by striated muscle fibers. The lungfish *Protopterus* surrounds itself with a slimy dessication-retarding cocoon before aestivating in a burrow during the dry season. Small amounts of nonslimy mucus are secreted more or less constantly, especially in teleosts, forming the mucus cuticle. Embedded in the cuticle may be some of the alkaloid or other ingredients from granular cells. Some female teleosts secrete a nutritious mucus that is eaten by the hatchlings. The mucus secreted by cells at the base of the stinger in sting rays contains an irritating toxin. The role of the mucus secreted by the lining of the ampullary neuromast organs of fishes (see fig. 17.2a) is unknown.

Aquatic Amphibians   The epidermal glands of aquatic amphibians are mostly multicellular mucous or granular glands (figs. 6.7 and 6.8). Tailed amphibians that are semiaquatic (that is, those whose primary habitat is a pond or stream but who may emerge on occasion) have the largest number of integumentary glands. Their mucous secretions keep the skin moist in air, enabling it to function as a respiratory membrane, as it does in water. Mucous glands on the digits of some frogs and tree toads serve as holdfasts, and the swollen mucous glands on the thumb pads of male anurans during the reproductive season assist in restraining the female during amplexus.

Photophores

Some of the multicellular glands of deep-sea teleosts have become light-emitting organs, or **photophores.**

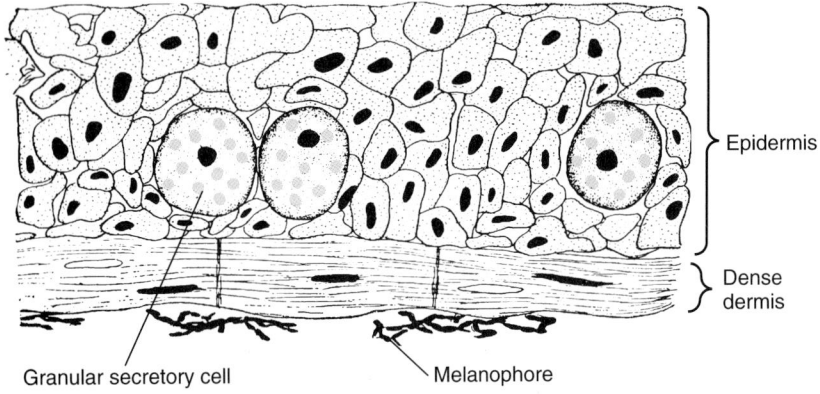

Epidermis

Dense dermis

Granular secretory cell   Melanophore

FIGURE 6.4
Skin of a larval lamprey.

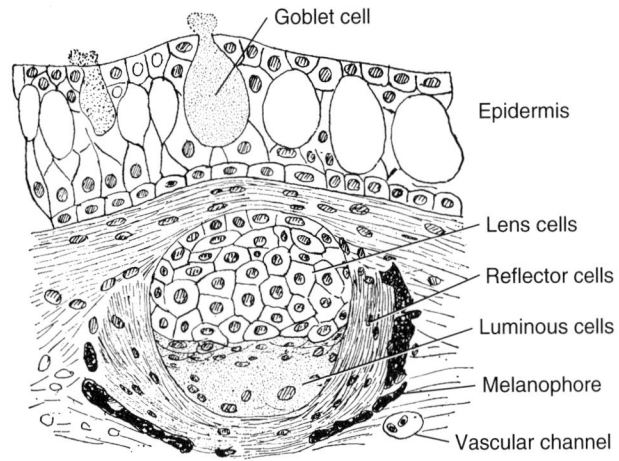

Goblet cell

Epidermis

Lens cells

Reflector cells

Luminous cells

Melanophore

Vascular channel

FIGURE 6.5
Light organ (photophore) of a luminous fish.

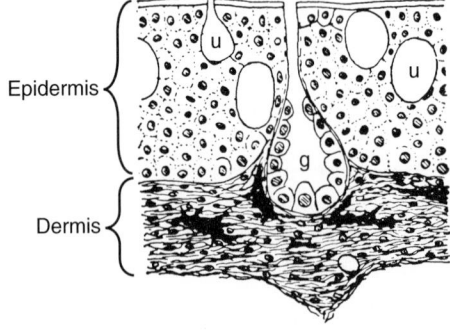

Epidermis

Dermis

FIGURE 6.6
Glandular skin of the dipnoan *Protopterus.* Multicellular, **g,** and unicellular, **u,** glands in the epidermis. Melanophores are seen in the dermis.

Like other multicellular skin glands, photophores arise in the epidermis and invade the dermis. In one variety (see fig. 6.5), the upper part of the gland consists of modified mucous cells that serve as a magnifying lens. Beneath the lens, other cells are the source of the light. Surrounding the base of the photophore, in the dermis, are a blood sinus and a concentration of melanophores (melanin pigment cells). The blood sinus is the source of raw materials needed for bioluminescence. The emitted light most often is produced in the manner of fireflies: the enzymatic action of luciferase on the phosphorus-containing compound luciferin. The "cold" light emitted is not intense and is of many hues. It serves for species and sex recognition, sometimes as a lure, a warning, or an aid to concealment by countershading. Photophores are not limited to teleosts, but they are rare in other fishes. The beautiful color patterns of the skin of fishes seen in aquaria are the result of pigment cells in the dermis.

Keratin

Most fishes synthesize little or no keratin in their epidermis, but some aquatic urodeles, especially those that venture onto land, have a thin desiccation-impeding stratum of cornified cells. In lampreys, conical cornified epidermal spines and "teeth" develop in the buccal funnel and on the rasping lingual cartilage (fig. 6.9). Hagfishes also possess cornified "teeth." Anuran tadpoles have rows of horny toothlike structures surrounding the entrance to the mouth that enable them to feed by scraping vegetation during their herbivorous larval stage. Tadpoles shed these at metamorphosis. Calluslike caps develop on the toes of aquatic urodeles subjected to buffering in mountain streams. In general, however, keratin is a feature of the skin of terrestrial craniates, not of fishes and aquatic amphibians.

Epidermis of Tetrapods

The stratified epithelium that constitutes the epidermis of terrestrial tetrapods reflects a long history of gradual adaptation from life in water to life in air. According to Peterson's (Unnatural) Law, "Anything that can happen will happen." And it practically did! Could one have predicted feathers? hair? mammary glands? sweat? It was epidermal adaptations such as these, and less complicated ones, that participated in fostering the diversity of craniate life on land. Let us look, therefore, at the two primary epidermal features of terrestrial tetrapods—integumentary glands and the stratum corneum.

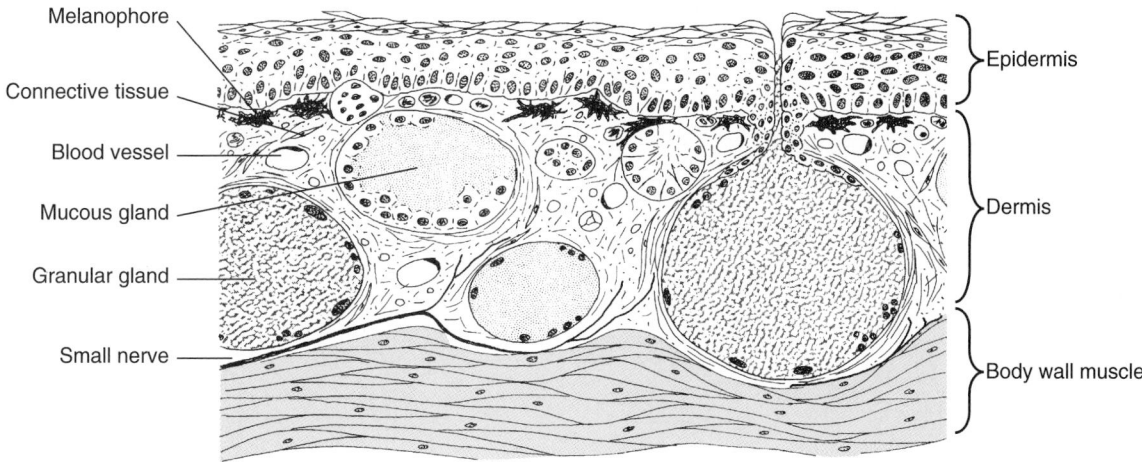

FIGURE 6.7
Frog integument. The multicellular glands are in various stages of development. A thin stratum corneum covers the living epidermis.

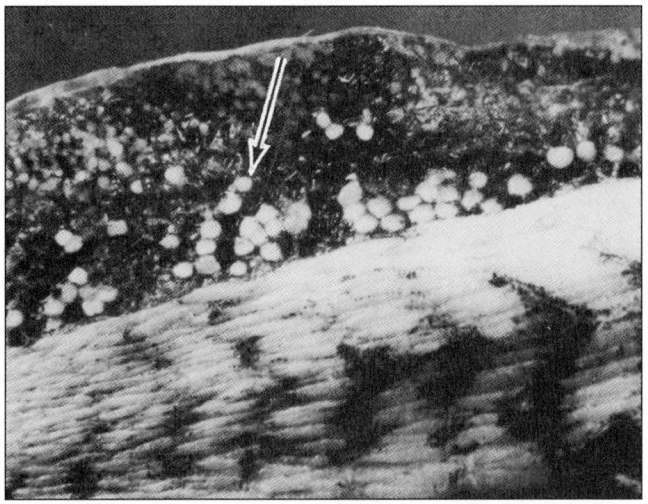

FIGURE 6.8
Undersurface of skin reflected from the underlying muscle in tail of a *Necturus*. Mucous glands (*arrow*) are seen projecting into subcutaneous tissue just external to the tail musculature.

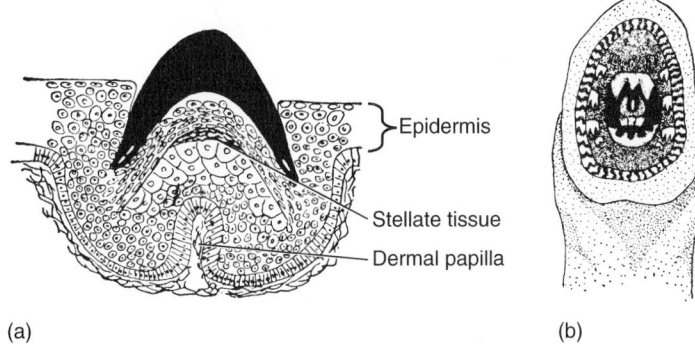

FIGURE 6.9
Lamprey teeth and buccal funnel. (*a*) Cornified tooth (*black*). The stellate tissue may be the beginning of a replacement tooth. (*b*) Buccal funnel. The horny teeth are shown as white against a dark background. The tip of the retracted tooth-bearing lingual cartilage (tongue) occupies the rear of the cavity.

Epidermal Glands

The epidermal glands of well-adapted tetrapods are either **saccular** or **tubular**. The sacs are known as **alveoli**. A variety of simple and compound saccular (**alveolar**) and tubular glands are illustrated in figure 6.10. Saccular glands are variously complex expansions of the simple saccular glands of dipnoans and amphibians. Tubular glands are uncommon in the skin of nonmammals but are abundant in mammals. However, they are ubiquitous in the digestive tract from fishes to human beings.

Glands vary in the way their cells liberate the substances they synthesize. There are three basic categories. The cells of **merocrine glands,** the most common variety, secrete their products via the cell membrane, and the cell remains intact. Most sweat glands of humans are merocrine. **Holocrine glands** are those in which the cells themselves constitute the secretion. Oil glands of birds and the sebaceous glands of mammals are holocrine. **Apocrine glands** exhibit an intermediate condition. The secretion accumulates in the apical portion of the cell, which is then pinched off along with some cytoplasm. The cell then repairs itself. Mammary glands are apocrine. Goblet cells do not fit precisely in any of these three categories. They rupture at the apex, and the mucus oozes out. The cell eventually repairs itself numerous times before being discarded.

Mucous Glands   Mucus-secreting epidermal glands have practically disappeared among terrestrial tetrapods except in mammals, in which they are confined to a few sites where lubrication of a surface is essential. Unicellular glands probably lost survival value when, as in efts, they became covered by a continuous sheet of cornified cells that rendered them functionless. Also, the synthesis of mucus other than in limited quantities would dehydrate a craniate lacking constant access to drinking water or to moisture that could

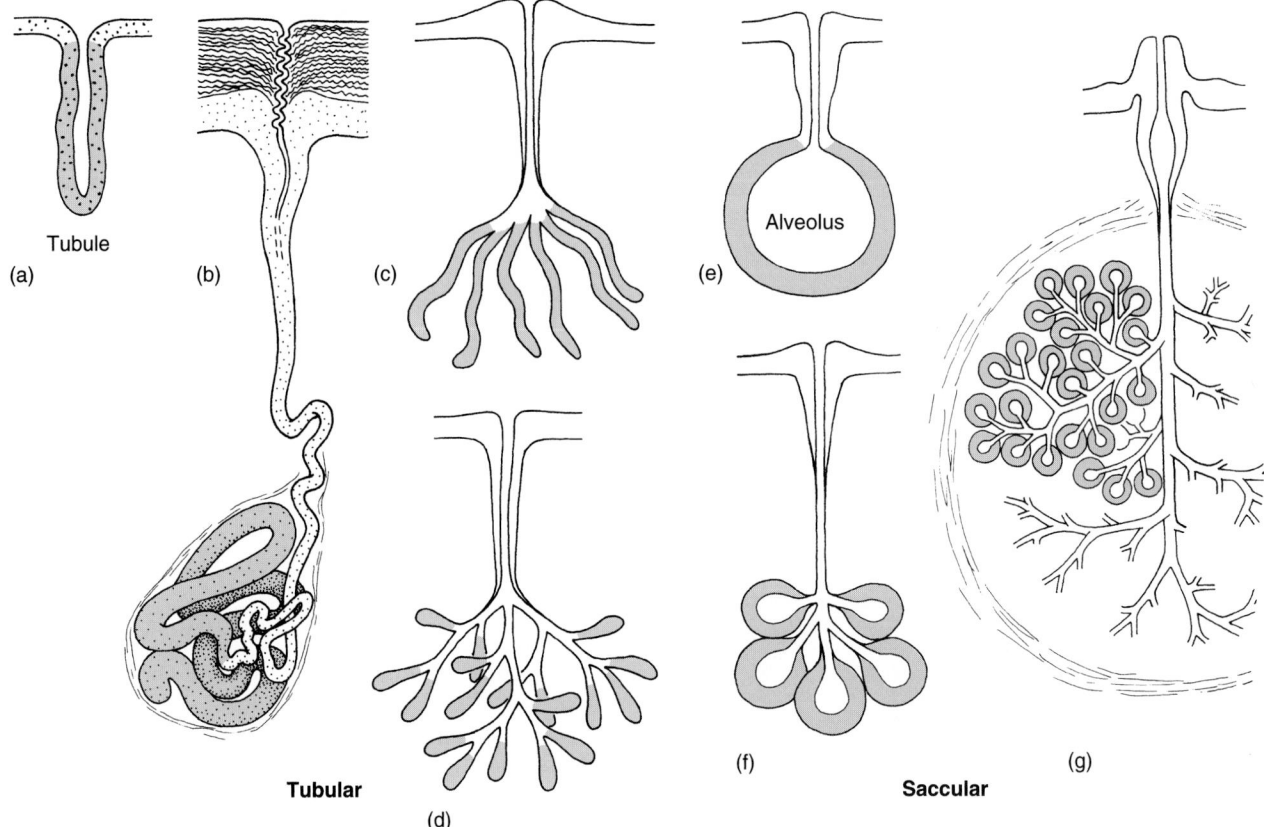

FIGURE 6.10

Major morphologic varieties of multicellular glands. (*a*) Simple tubular. (*b*) Coiled tubular. (*c*) Simple branched tubular. (*d*) Compound tubular. (*e*) Alveolus of a simple saccular gland. (*f*) Simple branched saccular. (*g*) Compound saccular (compound alveolar). *Red* shows secretory epithelia. Not to scale.

be absorbed through the skin. Among early craniates that ventured onto the land, those that had fewer mucous glands and a thicker stratum corneum surely had an advantage. Natural selection would have favored such a suite of characters.

Granular Glands   Granular glands are present with little variety in toads, which are terrestrial amphibians, and in a larger variety but not abundantly in reptiles. They are absent in birds and mammals. **Granular glands** secrete irritating or toxic alkaloids that are defensive, and they ward off predators. Granular glands are the source also of many, but not all, pheromones. **Pheromones** are substances that, when released into the environment, affect the behavior or physiology of other organisms of the same or different species. They may signal the sex of an organism, identify members of the same population, or leave trails on a substrate for the information of other members of the species. Some reptilian pheromones attract insects on which the reptile then feeds, and some sound alarms.

Granular glands are generally restricted to a localized area of the body. In toads, they are associated with warty skin, chiefly on the legs and back. The secretion can be irritating to human skin and highly distasteful to preda-

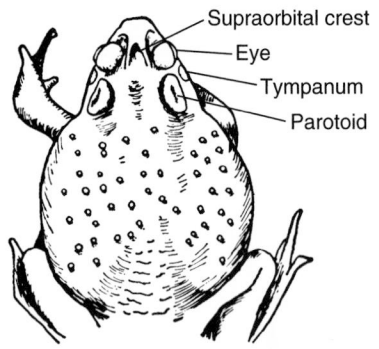

FIGURE 6.11

Warty skin of toad (*Bufo*). A granular gland, the parotoid, is seen behind the eye.

tors. Many toads also have a prominent granular **parotoid gland** behind the eye (fig. 6.11). Granular glands surround the vent of lizards and snakes, and most of these secrete pheromones. The only other integumentary glands in lizards are **femoral glands** on the medial aspect of the hind limbs of males. These secrete a substance that hardens to form temporary spines that restrain the female during copulation. Musk turtles exude a yellowish fluid from two glands on each side of the trunk at the lower border of the carapace. In crocodil-

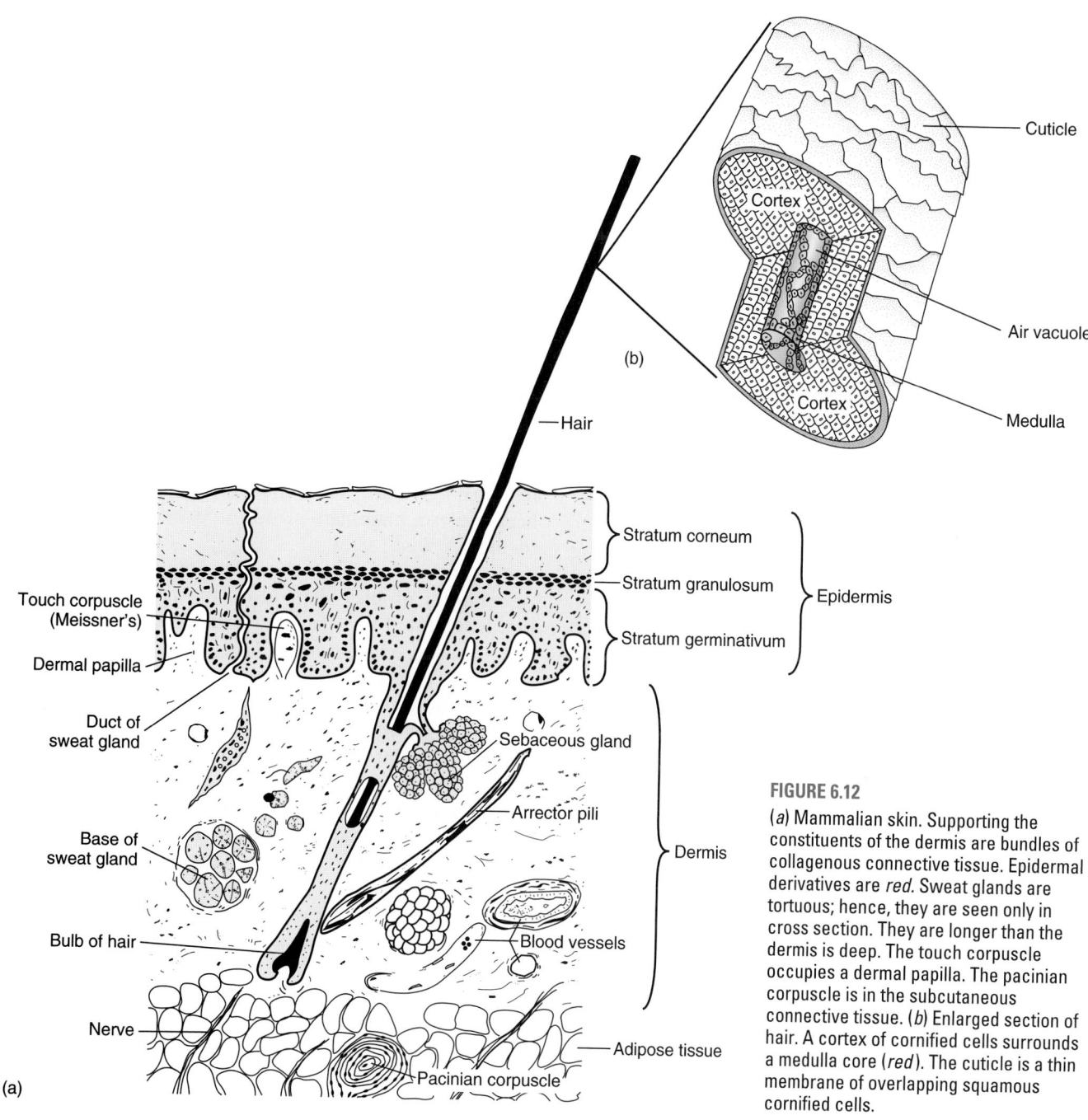

(a)

FIGURE 6.12

(*a*) Mammalian skin. Supporting the constituents of the dermis are bundles of collagenous connective tissue. Epidermal derivatives are *red*. Sweat glands are tortuous; hence, they are seen only in cross section. They are longer than the dermis is deep. The touch corpuscle occupies a dermal papilla. The pacinian corpuscle is in the subcutaneous connective tissue. (*b*) Enlarged section of hair. A cortex of cornified cells surrounds a medulla core (*red*). The cuticle is a thin membrane of overlapping squamous cornified cells.

ians, a row of glands of unknown function, probably pheromonal, extends along the back.

Avian Oil Glands   Very few integumentary glands are found in birds, and the secretions of all have an oily component. The **uropygial gland** is a prominent swelling at the rump immediately behind the pygostyle (see fig. 8.20). It is largest in aquatic birds and prominent also in domestic fowl. The secretion, which is water-repellent, is transferred to feathers during preening. Smaller oil glands line the outer ear canal and encircle the vent in some species.

Sebaceous Glands   Integumentary glands reach a peak in complexity and specialization in mammals. All appear to be variants of two functional types, sebaceous and sudoriferous. **Sebaceous glands** are alveolar glands with an oily exudate. They are present wherever there are hairs, and the secretion, **sebum**, is usually exuded into hair follicles (fig. 6.12). Fur, including human hair, glistens after brushing because of the oil. In the outer ear canal, **ceruminous glands** secrete **cerumen**, which in association with hairs traps insects that otherwise might go deeper into the canal and touch the painfully sensitive eardrum. **Meibomian glands** assist in moistening the conjunctiva of the eye. They are embedded in a plate of dense connective tissue (the palpebral tarsus) in each eyelid, and their long ducts open via a row of minute foramina along the border of the lid just internal to the eyelashes. If a duct (there are up to 30) occasionally becomes occluded, it causes a **chalazion**, an inflamed swelling on the conjunctival surface of the lid. Sebaceous glands open independent of hairs on the lips, glans penis, labia minora, and skin surrounding the nipples.

Sudoriferous Glands   Sudoriferous (sweat) glands are coiled tubular glands that extend deep into the dermis of mammals (see figs. 6.10*b* and 6.12). Their secretions ooze onto the surface via tortuous canals that open as pores. The cooling effect of the evaporation of sweat, their predominant secretion, is thermoregulatory. Sweat glands in furry mammals are confined to the least furry regions (for example, the feet of cats and mice, lips of rabbits, and the side of the head of bats). In hippopotami, they are only on the ears, which are above the waterline when a hippo is in its preferred habitat. Some mammals lack sweat glands. Among these are pangolins, whose epidermis is scaly (see fig. 4.37); sirenians and cetaceans, which are marine mammals; and echidnas, which are monotremes. Ciliary glands that open into the follicles of eyelashes are sudoriferous glands. Humans, with less hair than most mammals, have the largest number of sweat glands per centimeter of surface area.

Scent Glands   Sebaceous and sudoriferous glands both produce a variety of scents, all of which may be pheromones. Glands on the feet of goats leave scent trails that other members of the species are able to recognize, anal glands of skunks drive away enemies, and anal glands of male musk deer signal the sex. Kangaroo rats have sebaceous glands on the back, where they are exposed when the back is arched in defense. Male elephants have a temporal gland that swells during breeding seasons. Natives say that elephants are dangerous to humans at that time. A gland above the eye of a peccary looks like a navel. Some male lemurs have a hardened patch of skin on the forearm, under which is a gland the size of an almond. Most of the odors at a well-maintained mammalian zoo are caused by scent glands, not by unhygienic conditions in the pens. One species, *Homo sapiens*, takes the pheromone from the anal gland of a musk deer, adds other odorants, and dabs it behind the ears. It is called perfume, but it performs the role of a pheromone. Not all pheromones are products of integumentary glands.

Mammalian urine, for example, contains substances that also serve as pheromones.

Mammary Glands   **Mammary glands** are compound alveolar glands that develop in both sexes from **milk lines**, a pair of elevated ribbons of ectoderm that extend along the ventrolateral body wall of the fetus from axilla to groin (fig. 6.13*a*). Patches of future mammary tissue develop at one or more sites along each milk line, depending on the animal, and invade the dermis (fig. 6.14). They later spread *beneath* the dermis, and a nipple forms above each patch. As females approach sexual maturity, rising titers of female steroid hormones cause the juvenile duct system to spread and branch. Later, during pregnancy, a battery of hormones induces the formation of enormous numbers of alveoli at the ends of the branching duct system (figs. 6.10*g* and 6.15*b* and *c*). Like sebaceous glands, from which mammary glands of viviparous mammals appear to have been derived, mammary glands produce a secretion that includes lipids.

The number and location of mammary glands, and therefore of nipples, depend on the number of young that is typical of the species and on the survival value of one location over another. Cats, dogs, pigs, rodents, edentates, and many other mammals have axillary, thoracic, abdominal, and inguinal nipples that develop along most of the milk line. During pregnancy in these species, adjacent glands may expand toward each other until they have formed two long continuous masses of considerable weight. Species with smaller litters have fewer nipples. Insectivores and some lemurs have one thoracic and one inguinal pair. Flying lemurs and marmosets have a single pair, in the armpits. Monkeys, apes, and humans have one pair located where the nursing baby can be protected in the mother's arms while she monitors the environment for enemies. Cetaceans have a single pair near the groin, where the baby por-

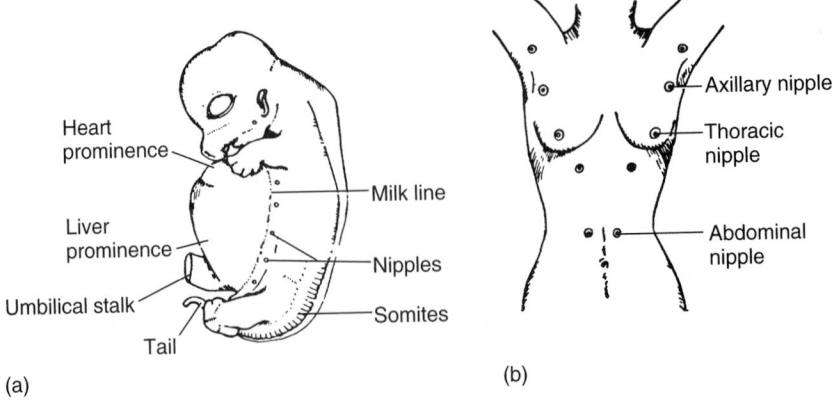

(a)

(b)

FIGURE 6.13
(*a*) Milk line and nipples in a 20 mm pig embryo. (*b*) Sites of supernumerary nipples in a human.

poise or whale can hold on and nurse while the mother feeds, surfaces, and dives. Nutrias, which are at home in water, have four nipples on the back, and the babies ride along above the waterline while nursing. The nipples originate from a ventrolateral milk line but are displaced dorsad by differential growth rates on the two sides of the milk lines during ontogenesis. Supernumerary nipples may develop in any mammal (see fig. 6.13b). Although only three ducts are shown in the nipple in figure 6.15b, there are actually as many terminal ducts in a human nipple as there are lobes in the gland, which number from 15 to 25. Each lobe consists of many compound lobules. In ungulates, ducts from all lobes empty into a single terminal duct. At the base of the nipple, terminal ducts have a **cystern** in which milk accumu-

lates after having been "let down" from the lobes. Suckling drains the cysterns. The hormone **oxytocin** from the pituitary (chapter 18) is responsible for the smooth muscle contractions that cause milk letdown.

Monotremes do not develop mammary glands like those of viviparous mammals, and they do not have nipples. Instead, in both sexes, what appear to be modified sweat glands produce a nutritious secretion that is lapped off an overlying tuft of hairs by the young (fig. 6.15a).

Stratum Corneum

Keratinization provides protection against desiccation on land. With the acquisition of an amnion, which completed the liberation of craniates from the water, the stratum corneum became increasingly specialized in various regions of the body for protection against abrasion, for offense and defense, and finally, as an adjunct to thermoregulation. Early specializations—scales, claws, and horny protuberances—were followed by hair and feathers, the latter being the most remarkable of all. A mammalian stratum corneum is seen in figure 6.12.

Epidermal Scales Epidermal scales are repetitious thickenings of the stratum corneum that are found only in amniotes. In squamates—lizards and snakes—the stratum corneum is disposed on overlapping folds of the epidermis (figs. 6.16 and 6.17). The cells on the exposed surface of each fold are especially heavily cornified, forming overlapping scales. Thinning of the stratum corneum at scale joints permits mobility of the skin. The scales of lizards sometimes assume bizarre shapes (figs. 6.18 and 17.14).

The continuity of the stratum corneum is readily demonstrable in crocodilians, which have small, heavily cornified, nonoverlapping scales on

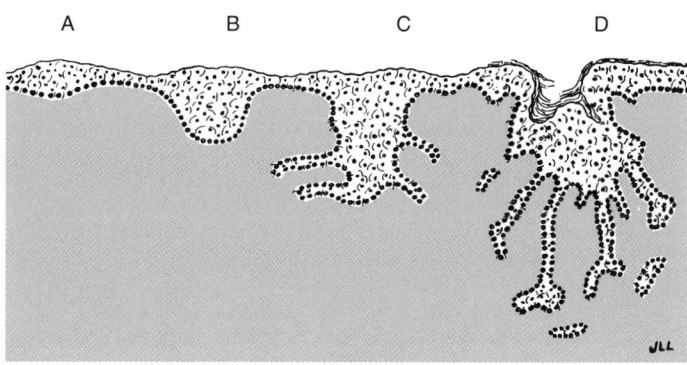

Mammary gland
Morphogenesis

FIGURE 6.14
Successive stages of mammary gland development. **A,** Equivalent to a human embryo at 6 weeks; **B,** equivalent to a human embryo at 9 weeks and to a mouse embryo at 16 days; **C,** intermediate stage; **D,** at birth. *Gray* area represents dermis.

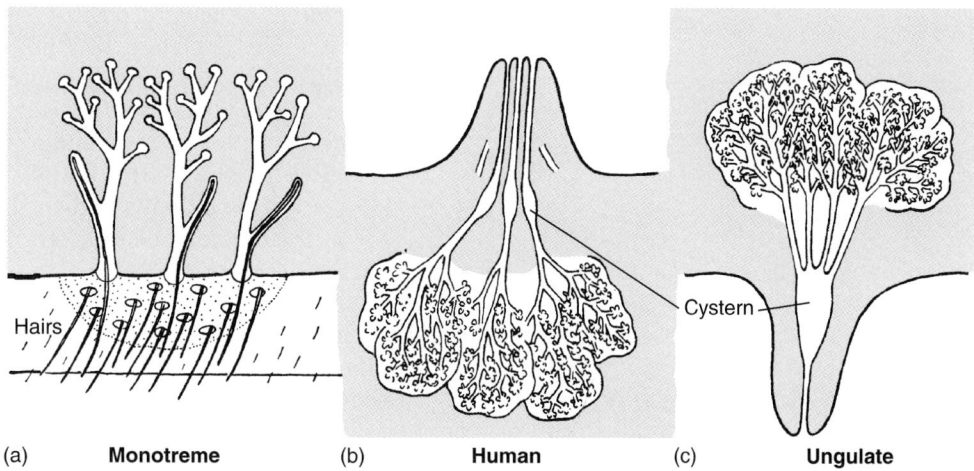

(a) **Monotreme** (b) **Human** (c) **Ungulate**

FIGURE 6.15
Mammary glands, ducts, and nipples. The monotreme lacks nipples, and the glands resemble modified sweat glands.

(a)

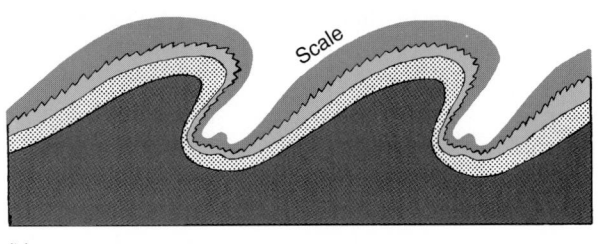

(b)

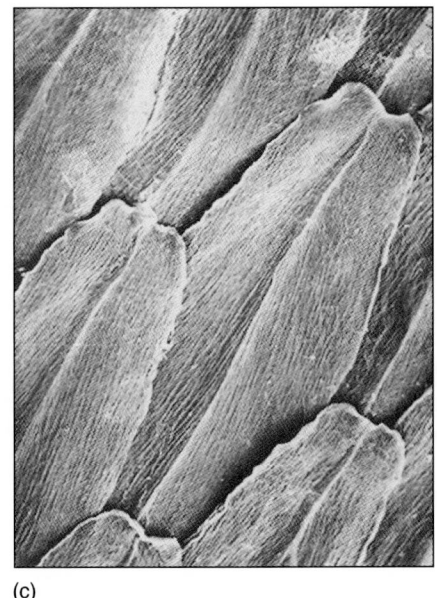

(c)

FIGURE 6.16

Squamate scales. (*a*) Collared lizard. The entrance to the outer ear canal is seen as a dark crescent behind the corner of the mouth. (*b*) Diagrammatic section of skin of a snake or lizard. **1,** Dermis; **2,** actively mitotic layer of epidermis; **3,** newly cornified layer of epidermis; **4,** older cornified layer to be shed at next molt. (*c*) Scales from the banded water snake, *Nerodia fasciata.*

(*a*) © John Cancalosi/Peter Arnold, Inc.

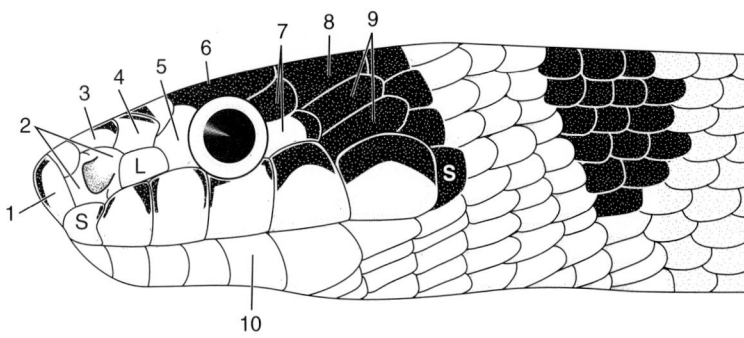

FIGURE 6.17

Epidermal scales on head of milk snake *Lampropeltis triangulum.* **1,** rostral; **2,** nasal; **3,** internasal; **4,** prefrontal; **5,** preocular; **6,** supraocular; **7,** postoculars; **8,** parietal; **9,** anterior temporals; **10,** fourth of seven infralabials; **L,** loreal; **S,** first and last supralabials.

the limbs, and larger, more or less regular thickenings on the rest of the body (see fig. 6.35*a*).

Large, thin, quadrilateral or polygonal scales are called **scutes.** Snakes have them on the belly, where they are used for locomotion. Turtles have thick scutes on the plastron, which slides along the ground, thinner ones on the carapace (fig. 6.19), and small scales elsewhere. Turtle scutes and scales do not overlap.

Epidermal scales in birds develop where there are no feathers: the facial area, legs, and feet. Armadillos have hair and scales interspersed over the entire body (figs. 4.36 and 6.20), but the scales of most mammals are confined to the legs and tail, as in rats and beavers. The scales of pangolins appear to be agglutinated hairs, and are of recent origin rather than inherited from a scaled ancestor (see fig. 4.37).

Lizards and snakes have two distinct layers of stratum corneum, an inner layer in the process of being deposited and an outer layer that will be shed at the next molt (see fig. 6.16*b*). In lizards, the outer layer flakes off in large patches, but in snakes the outer layer of the entire body, including the spectacle, a thick lenslike conjunctiva, is shed in one piece. Crocodilians and turtles do not molt; their stratum corneum wears off gradually, like that of humans.

Claws, Hoofs, and Nails   Claws, hoofs, and nails are modifications of the stratum corneum at the ends of digits. Claws first appeared in basal amniotes and have persisted in birds and most mammals. (The "claws" of the African clawed toad, an amphibian, and reptilian claws are not homologous.) Claws evolved into nails in primates, and into hoofs in ungulates. Claws, hoofs, and nails have the same basic structure. They consist of two

FIGURE 6.18
Warty and spiny skin of the Coast horned lizard, *Phrynosoma*. Skin color and pattern resemble its substrate habitat.
© H. Armstrong Roberts.

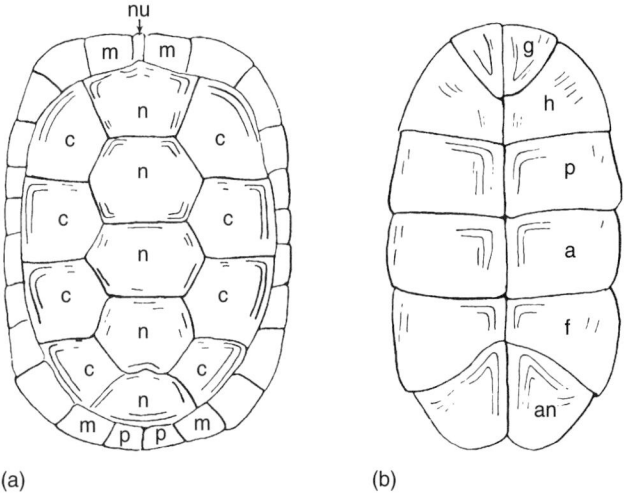

(a)   (b)

FIGURE 6.19
Epidermal scales of the turtle *Chrysemys*. (*a*) Carapace. (*b*) Plastron. On the carapace, the scutes shown are **c**, costals; **m**, marginals all around the periphery, including the nuchal (**nu**), and pygals (**p**); **n**, neurals. On the plastron, the scutes shown are **g**, gulars; **h**, humerals; **p**, pectorals; **a**, abdominals; **f**, femorals; **an**, anals.

curved parts, a horny dorsal plate, the **unguis**, and a softer ventral plate, the **subunguis** (fig. 6.21). The two plates wrap partially around the terminal phalanx, which is usually pointed when associated with a claw, blunt when associated with a hoof or nail. A still softer calluslike, cornified pad, the **cuneus** (called the "frog" by equestrians), is frequently present in ungulates, partially surrounded by the subunguis. The thick, horny unguis of a hoof is U- or V-shaped, and, because it consists of dead cells, a shoe can be nailed into it. The unguis of a nail has become flattened, and the subunguis is

much reduced. As a result, nails cover only the dorsal surfaces of digits, but if permitted to grow, nails become clawlike.

Although claws in birds are often thought of as associated only with the feet, sharp claws are frequently borne on one or more digits of the wings (ostriches, geese, some swifts, and others). Young hoatzins use claws on the wings for climbing about on the bark of trees, but the claws cease growing and disappear at maturity. *Archaeopteryx* had three claws on each wing. Only squamate claws are shed. Claws, hoofs, and nails of other animals are worn down by abrasion.

Feathers   Feathers are remarkably complicated cornified epidermal appendages of the integument. There are three morphologic varieties: contour feathers, down feathers (plumules), and hairlike feathers (filoplumes). The roles of feathers are described in chapter 4.

Morphologic Varieties of Feathers   Contour feathers are the conspicuous feathers that give a bird its contour, or general shape (fig. 6.22). A typical contour feather consists of a horny **shaft** and two flattened **vanes**. The base of the shaft is the **calamus** (quill). The vane-bearing segment is the **rachis**. Each vane consists of parallel rows of **barbs** that have **barbules** and **flanges** (fig. 6.22*b*). The barbules have **hooklets** that interlock with the flanges on the next barb, stiffening the vane (fig. 6.22*c*). When a contour feather is ruffled, the barbules have become unhooked. Preening rehooks the hooklets. On the smaller contour feathers of the wing (covert feathers), hooklets are missing from the lower barbules, so this region of the feather is fluffy. In ostriches and a few other birds, all feathers are fluffy.

Arising from the **superior umbilicus**, a notch at the base of the rachis of a contour feather, is an **afterfeather** (fig. 6.22*a*). It is usually shorter than the main feather, but in emus and cassowaries the two are the same length, resulting in "double feathers."

Although contour feathers cover most of the body, the follicles from which they grow are usually disposed in feather tracts, or **pterylae** (fig. 6.23). A few birds, including ostriches and penguins, lack these.

Inserting on the walls of the feather follicles in the dermis are smooth erector muscles (**arrectores plumarum**), which, along with extrinsic integumentary muscles, enable a bird to fluff its feathers.

Down feathers are small, fluffy feathers lying underneath and between contour feathers. Very young birds

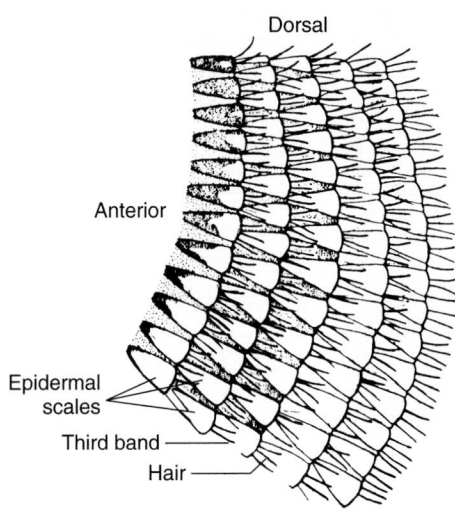

FIGURE 6.20
Epidermal scales and interspersed hairs of armadillo skin. See also figure 4.36.

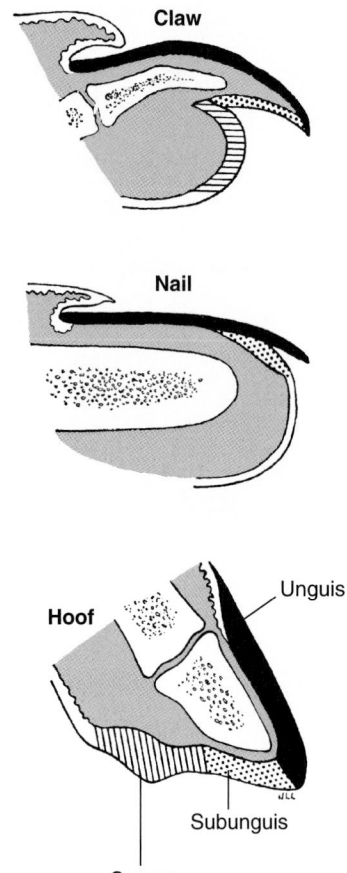

FIGURE 6.21
Claw, nail, and hoof with terminal phalanx, sagittal section.

lack contour feathers and are covered with down. Down feathers have a short calamus (fig. 6.22*d*) with a crown of barbs that lack hooklets. Eiderdown, used in pillows, is the down feathers of the eider duck.

Filoplumes are hairlike feathers consisting of a shaft and a few barbs and associated barbules at the tip (fig. 6.22*e*). They are scattered across the skin among contour feathers. Their follicles are richly supplied with tactile nerve endings. The long colorful feathers of a peacock are conspicuous examples of filoplumes. **Bristles** resemble filoplumes but lack any terminal barbs. A few barbs do occur at the base of the shaft. Bristles occur on the head and neck where they screen eyes and ear and nasal openings of foreign matter. Those around the mouth are also tactile receptors, conveying information such as a possible insect meal for a flycatcher.

Development of a Feather

Feather development is initiated by formation of a **dermal papilla**, which is a mound of mesodermal cells in the dermis that indents the undersurface of the epidermis and induces mitotic activity in its basal layer (fig. 6.24*a*). Growth of the papilla, and its inductor effect on the overlying epidermis, results in a pimplelike elevation, the **feather primordium**, on the surface of the skin (fig. 6.24*b*). This is the first indication that a feather is organizing. The dermal papilla becomes vascularized, and thereafter it is the source of nutrients and oxygen for the developing feather. It is also a collecting site for metabolic wastes that are then carried away in the blood stream. As the feather primordium elongates, a pit lined with epidermis, the **feather follicle**, develops around its base (fig. 6.24*c*).

At the base of the feather follicle, a mitotically active growth zone proliferates tall columns of epidermal cells that push toward the distal tip of the growing feather

between the dermal papilla and the epidermis, now a **feather sheath**. These epidermal columns separate from one another, cornify, and develop into barbs. A growing feather still surrounded by its sheath is a **pinfeather**. When the feather sheath splits open, the fluffy barbs stretch out of their cramped quarters, and the shaft elongates.

When the feather is full grown, the dermal papilla in the shaft dies and becomes pulp. The living basal portion of the papilla withdraws from the base of the shaft, leaving an opening, the **inferior umbilicus** (see fig. 6.22). New feathers develop from reactivated dermal papillae that have already given rise to feathers. In many birds—perhaps in all—old feathers are passively pushed out of the follicles by incoming feathers.

Origin of Feathers

One hypothesis about how feathers originated is that they arose from reptilian scales because early developmental stages are similar. Development of both is initiated by formation of a vascularized dermal papilla, after which the overlying epidermis thickens and then forms a scale or feather. However, subsequent development of the two is dissimilar, and the hypothesis is viewed by some with skepticism.

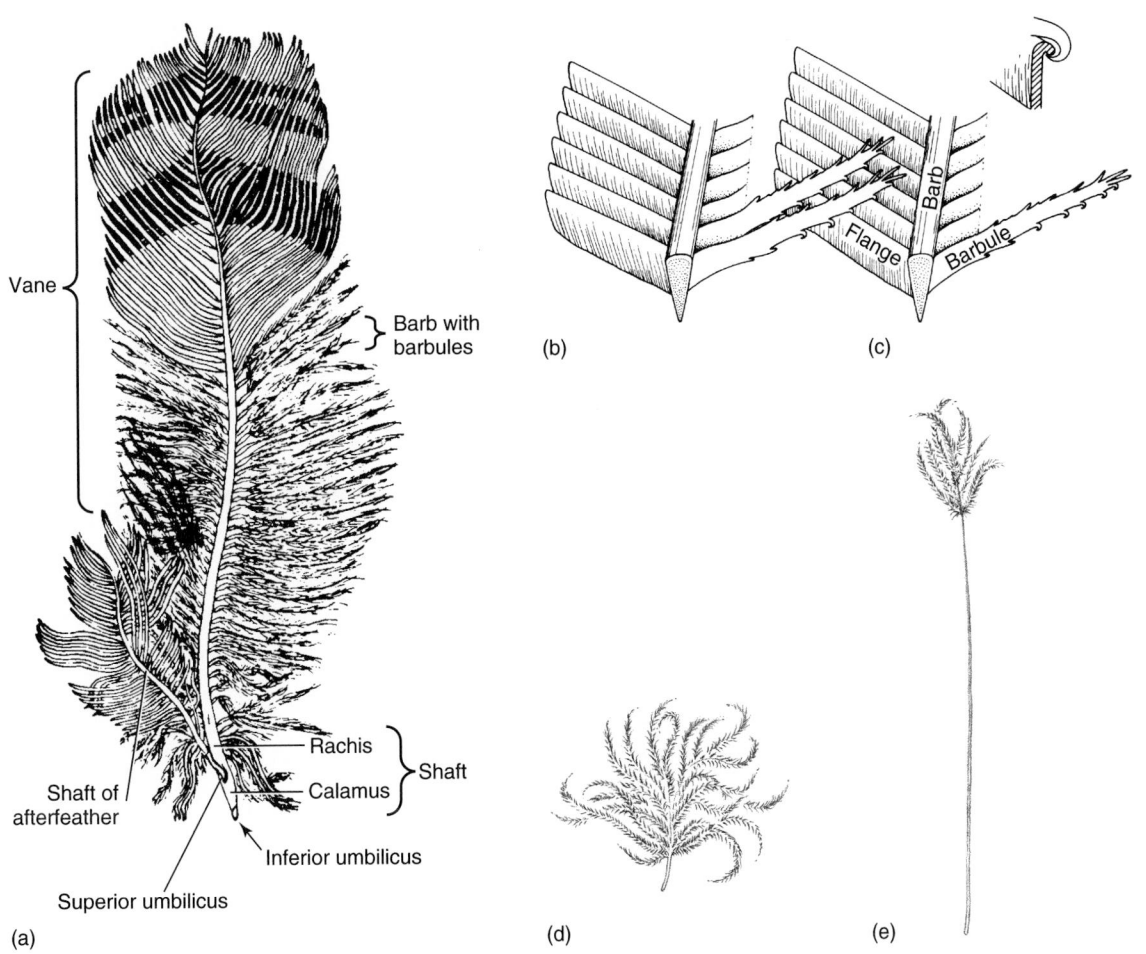

FIGURE 6.22

(*a*) Contour feather from a grouse. (*b*) Two successive barbs showing two barbules interlocked by hooklets with flanges. (*c*) Cross section of a flange showing interlock. (*d*) A down feather. (*e*) A filoplume.

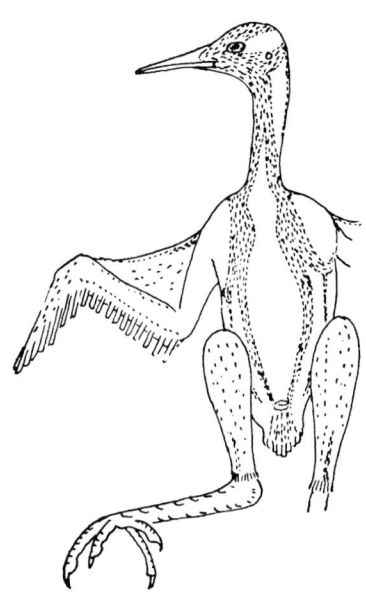

FIGURE 6.23

Feather tracts.

However, the discovery of possible **protofeathers** in the Chinese *Sinosauropteryx* that appear as a filamentous cover may add to the validity of a scale-to-feather scenario (refer to the Aves discussion in chapter 4). The only other hypothesis that is justified at present is that, like hairs, they may be new evolutionary structures lacking an ancestral precursor.

Hair   Hairs, like feathers, are keratinized appendages of the skin. They may form a dense, furry covering over the entire body, or there may be only one or two bristles on the upper lip, as in some whales. Where fur is dense, there are usually short, fine hairs (underfur) as well as long, coarse ones.

Hairs have an insulating effect when dense enough, and they are also sensitive tactile organs. The root of each is surrounded by a basketlike network of sensory nerve endings, and displacement of the root initiates a train of sensory impulses to the brain. Disturb a single hair on the back of your hand and note the sensation evoked. Vibrissae (stiff long whiskers on the face of many mammals) perform this role exclusively.

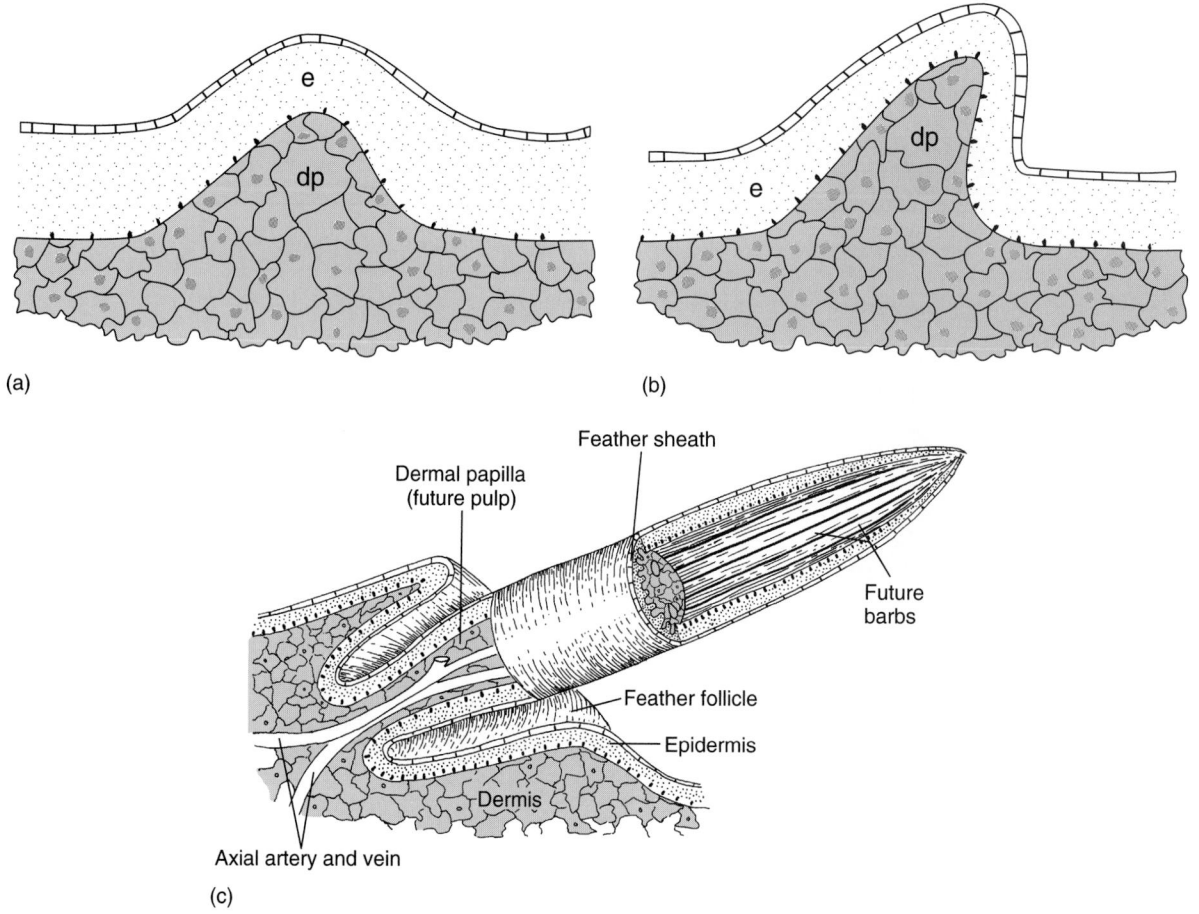

(a)

(b)

(c)

FIGURE 6.24
Growth of a contour feather. (*a*) Germinal dermal papilla **dp.** (*b*) Feather primordium. (*c*) Pin feather. **e** is epidermis.

Morphology of a Hair Hairs grow from hair follicles and elongate as a result of continual mitosis in the **bulb** (see fig. 6.12), within which a highly vascular dermal papilla provides nutrients and oxygen required for growth. The **root** is the part within the follicle where the hair cells are cornifying and dying but where the hair has not yet separated from the follicular wall. The remainder of the hair is the **shaft**, which commences just below the openings from the sebaceous glands. Within the follicle, the shaft is surrounded by sebum from the sebaceous gland.

A single hair consists of dense keratin from disintegrated or partially disintegrated cornified cells, trapped air vacuoles, and varying quantities of melanin granules that were released when the hair cells disintegrated. (The source of the granules and an explanation of hair color are discussed later under "Dermal Pigments.") Each hair is covered by a membranous **cuticle** composed of thin, transparent, cornified squamous cells (scales) arranged like shingles with free edges directed away from the root. Coarse or spiny hairs contain a **medulla** composed of irregular, shrunken cornified cells separated by large amounts of air and connected by intercellular bridges of keratin (see fig. 6.12*b*).

Inserted on the wall of each hair follicle is a tiny smooth muscle, the **arrector pili** (see fig. 6.12*a*). When the arrectores pilorum contract, the hairs are elevated and the skin is pulled into tiny mounds resembling "gooseflesh." Elevation of the hairs gives carnivores a ferocious appearance. It is also a device for thermoregulation, increasing the insulating effect of the fur.

The scales of pangolins, and perhaps the horns of rhinoceros, may be evolutionary products of agglutinated hairs. Other modifications of hair include bristles, the spines of spiny anteaters, and porcupine quills.

Development of a Hair The first manifestation of a developing hair is a cylindrical ingrowth of epidermis into the dermis (fig. 6.25*a*). Soon thereafter a dermal papilla, histologically identical with that of a developing feather, organizes at the base of the ingrowth, becomes vascularized, and the two become intimately associated. Thereafter, the papilla plays the same role as in feather development. Utilizing oxygen and nourishment provided by the vessels of the dermal papilla, the hair primordium grows deeper and deeper into the dermis. The base of the primordium eventually becomes bulbous, cornified cells begin to form at that site, and a hair shaft

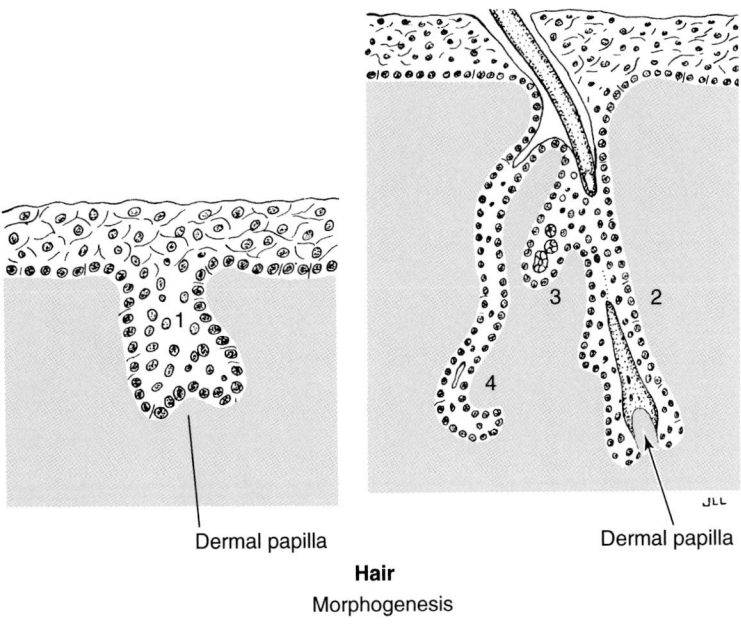

Hair
Morphogenesis

FIGURE 6.25
Successive stages in the development of hair and associated glands. **1,** Initial epidermal ingrowth into dermis, dermal papilla not yet organized; **2,** hair follicle surrounding growing hair shaft; **3,** developing sebaceous gland; **4,** developing sweet gland. The discontinuous appearance of the hair shaft is an artifact of sectioning. *Gray area,* dermis.

begins to rise out of a hair follicle (see figs. 6.12 and 6.25*b*).

Hairs are not permanent. After an interval of time—one or many months, and often seasonally, depending on the species—proliferation of hair cells in the follicle ceases temporarily, and the hair becomes loose in the follicle. Subsequently, when formation of hair in the follicle resumes, the old hair is shed if it has not already fallen out. When all follicles in an area permanently cease activity, the area becomes bald.

Origin of Hair   The phylogenetic origin of hair is conjectural. Hairs grow in isolated groups of two dozen or more where there are no scales, and grow between scales in areas such as the tail and legs of rodents. The interscalar distribution of hair in armadillos is seen in figure 6.20. Among anthropoids, some monkeys have hair in groups of three, apes have groups of five, and humans have groups of three to five, arranged linearly. Examining the back of a human hand, near the base of the thumb, will reveal the linear arrangement of the groups in that location. Where scales are present, the interscale distribution of hairs makes possible the receipt of tactile stimuli despite a scaly skin. This distribution of tactile hairs is reminiscent of the interscale distribution of sensory pits in lizards, where each pit houses a tactile bristle (**protothrix**). The interscale distribution of hairs may reflect the amniote origin of mammals, but tactile hairs and tactile protothrixes seem to be products of convergent evolution. Feathers and hairs may be **neo-**

morphs (structures that have no ancestral precursor).

Feathers, Hair, and Dermal Papillae
We have seen that dermal papillae are essential to the early formation of feathers and hair. In the morphogenesis of a feather, a dermal papilla appears first, and this is followed by activity in the ectoderm. In the morphogenesis of a hair, an ectodermal invagination appears first, and this is followed by formation of a dermal papilla. Feathers have been induced on the feet of chicks where scales ordinarily develop if dermis from a potential feathery area is transplanted to a site under the epidermis of an embryonic foot at a critical time during organogenesis. Even the extraembryonic chorionic epithelium of a chick embryo has been induced to form feather papillae or embryonic scales by carefully timed transplants of appropriate dermis. Whether or not the dermis induces the initial ectodermal ingrowth that signals development of a hair is not known.

The development of placoid scales and vertebrate teeth is also dependent on dermal papillae (see fig. 12.12). The dermal papillae at the interface of the epidermis and dermis of fully differentiated mammalian skin (see fig. 6.12) should not be confused with the embryonic dermal papillae we have been discussing here.

Horns and Antlers   Many ungulates are endowed with organs of offense, defense, and display, the horns and antlers. The term *horn* means that the surface is composed of keratin. Three varieties of mammalian horns meet this criterion: **bovine horns, hair horns** (fig. 6.26), and the **horns of pronghorn antelopes.** Antlers and giraffe horns are not "true" horns.

Bovine Horns and Pronghorns   Artiodactyls of the family Bovidae (oxen, cows, sheep, goats, true antelopes, others) and pronghorn antelopes have true horns. They consist of a *core of dermal bone covered by a sheath of horn.* Bovine horns are usually curved or recurved and are never shed. When the sheath is removed from the underlying bone by surgery, the sheath is hollow (fig. 6.26). Bovine horns are usually found in both sexes, but there are exceptions. Polled cattle have lost their horns by selective breeding. The chief difference between bovine horns and pronghorns is that pronghorns are branched, and the horny covering, but not the bony core, is shed annually.

Hair Horns   Rhinoceroses have hair horns (fig. 6.26). These differ from other horns in being composed of agglutinated keratinized hairlike epidermal fibers that form a solid horn perched on a roughened area of the nasal bone. Both sexes have them, and they are not shed. Some African rhinos have two horns, one behind the other.

Antlers and Giraffe Horns

Antlers are characteristic of the deer family. They are not cornified structures but dermal bone attached to the frontal bone. New growing antlers are said to be "in velvet" because they are covered with a soft vascular skin and velvety hair. Antlers develop only in males, with the exception of caribou and reindeer. Typically at the approach of autumn, when antlers are full-grown, the blood supply to the velvet is occluded at the base of the antlers, and the skin, unable to be maintained, rubs off, leaving bare bone. Later, at the end of the rutting season, when male territory no longer must be defended and testosterone concentration declines in the circulation, the antlers are shed with some help from the deer, who evidently find them annoying. They are replaced annually.

The "horns" of giraffes resemble stunted antlers. They are short bony projections of the frontal bones that remain in velvet throughout life.

Baleen and Other Cornified Structures
Toothless whales have from 100 to 400 broad, thin, horny sheets of oral epithelium called baleen, or whalebone (fig. 6.27), that hang into the oral cavity from the palate along its length. Each sheet is fringed along the edge, and the fringes act like combs or sieves that strain food out of the water passing between them. The sheets of the huge right whale exceed 3 m in height. Differences in the arrangement and configuration of the sheets are correlated with feeding habits. A blue whale swims up to a dense swarm of small fish or crustaceans, opens its mouth, and takes in water and food until a huge pouch under the tongue, pharynx, and chest wall is filled with about 70 tons of food and water. The pouch is then emptied by forcing the water through the sieve and back into the sea. The collected food is swallowed leisurely through a very small throat. Other baleen whales skim the surface for plankton with the mouth open or feed near the bottom. Baleen is continually being worn away and replaced, like hair or fingernails.

Rattlesnake **rattles** are rings of horny stratum corneum that remain attached to the tail after each molt (fig. 6.28a). **Beaks** are covered with a horny sheath, and roosters' **combs** are covered with a thick, warty stratum

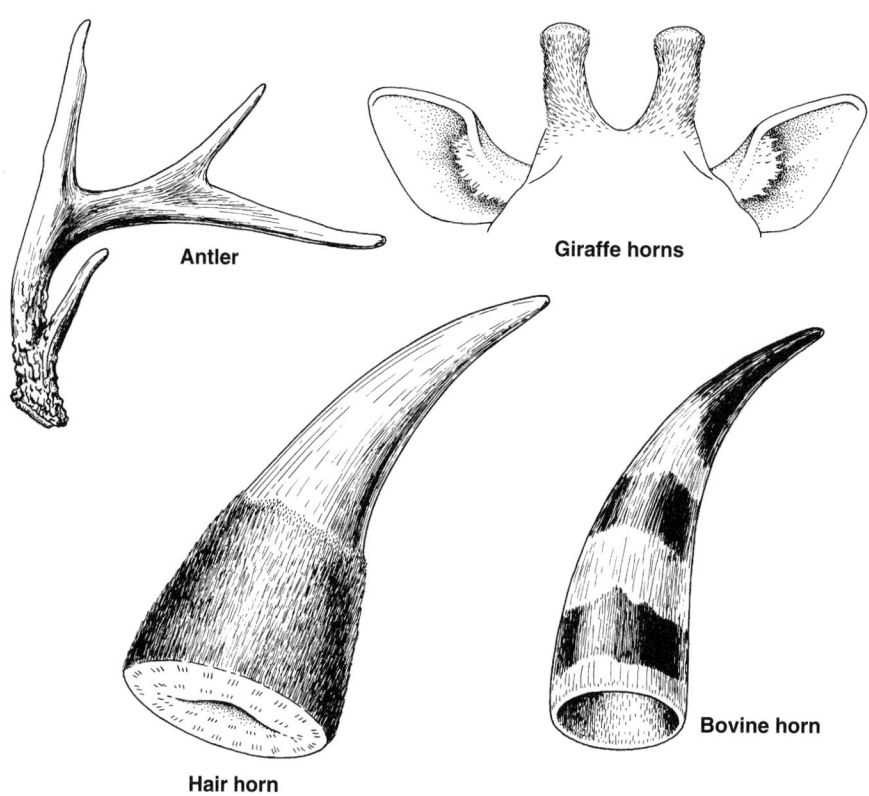

FIGURE 6.26
Mammalian horns and antlers. Note that the "horns" of giraffes are not true horns, but instead are antlers.

corneum. Monkeys and apes sit on thick **ischial callosities,** and camels kneel on **knee pads. Tori** are epidermal pads that most mammals other than ungulates walk on (fig. 6.28b and c). Cats "pussyfoot" by retracting their claws and walking stealthily on tori. At the ends of digits, tori are called **apical pads. Corns** and **calluses** are temporary thickenings of the stratum corneum that develop where the skin has been subjected to unusual friction.

The structure of the mammalian epidermis is described later in the chapter under "The Integument from Fishes to Mammals."

THE DERMIS

The basic component of dermis, whether of fishes or human beings, is collagenous connective tissue that holds the other components in place and provides tensile strength (see fig. 7.2). Other components include, universally, blood vessels, small nerves, and pigment cells; and in one species or another, lymphatics, naked and encapsulated exteroceptors, the bases of multicellular glands; and in endotherms, the bases of hairs or feathers and their erector muscles. In addition, *the dermis has an ancient and persistent potential to form bone,* and bone is a constituent of the dermis in some members of every vertebrate class except birds. Early fishes had so much bone in the skin that they are called

(a)

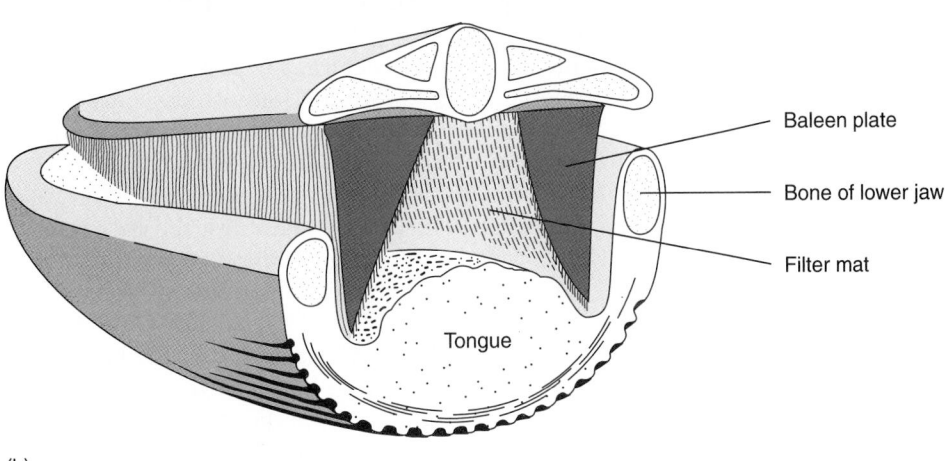

(b)

FIGURE 6.27

(*a*) Humpback whale snout showing outside edges of approximately 125 baleen plates. (*b*) Diagrammatic section through the head of a rorqual whale. The filter mat lines the oral cavity with the frayed edges of numerous baleen plates.

(*a*) © Thomas Kitchin/Tom Stack & Associates. (*b*) From Nigel Bonner, *Whales of the World,* 1989. Copyright © 1989 Cassell PLC, London. Reprinted by permission.

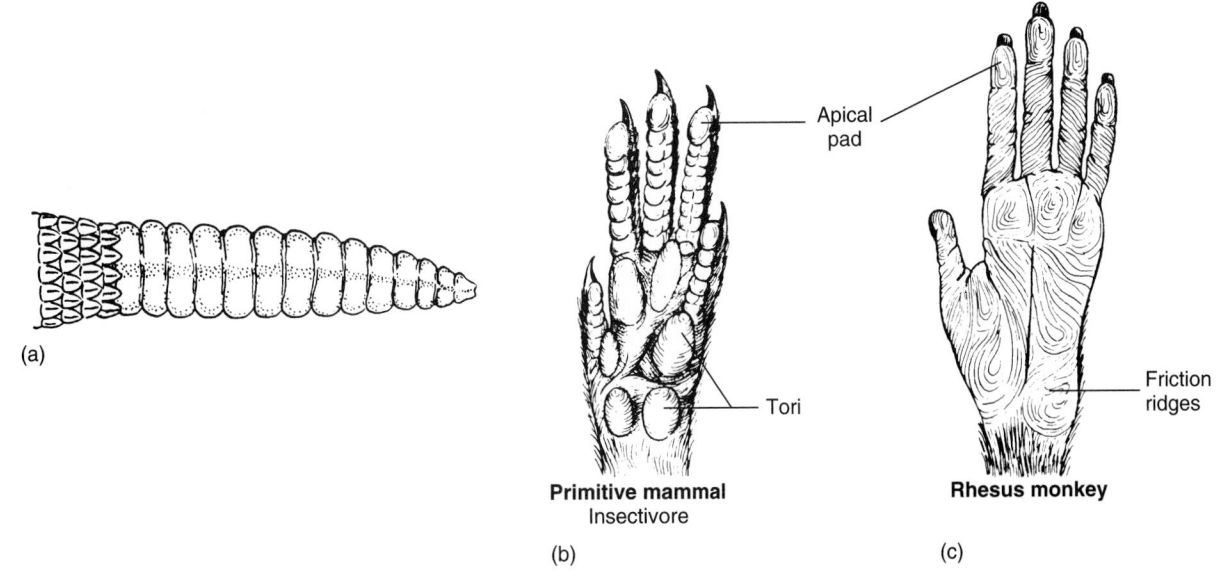

FIGURE 6.28

Products of the stratum corneum. (*a*) Rattlesnake rattle. (*b*) Tori. (*c*) Friction ridges.

armored fishes. When bone is lacking in the dermis, it is because bone salts were not deposited on the collagenous matrix. Cowhide and pigskin, from which leather is made, the dermis of leatherback turtles, the skin of lampreys—all are exceptionally tough because the collagen bundles are densely packed as though prepared for a shower of bone salts that, for lack of an enzyme or some other reason, never materialized.

The Bony Dermis of Fishes

The bony armor of early fishes was disposed as broad bony plates or smaller bony scales over most of the body, immediately beneath the epidermis (see figs. 4.5 and 4.9). The structure of this **dermal bone** varied in histologic details, but a generalized pattern consisted of lamellar bone, spongy bone, dentin, and a surface layer of an enamellike substance of one physical nature or another (fig. 6.29a). Frequently, the surface of the armor was sculptured by knobby or spiny elevations of the dentinal layer known as **denticles** (3 plus 4 in fig. 6.29a). Lamellar bone is compactly structured bone that has been deposited in successive layers, or lamellae, a common method of bone deposition. Spongy bone, on the other hand, is penetrated by blood channels of macroscopic size, which give the bone a spongy appearance. Dentin is yet another variety of bone: These varieties of bone are discussed in more detail in the next chapter.

Dermal armor was protective, but it may also have served as a storage site for calcium and phosphates that could be retrieved by the bloodstream when needed. In time, the large plates on the trunk and tail gave way to the smaller and thinner scales of later fishes, whereas those on the head became part of the skull, and those immediately behind the last pharyngeal arches became a part of the skeleton of the shoulder girdle.

Dermal plates and scales (table 6.1) are classified as **placoid, rhomboid,** and **elasmoid.** Rhomboid scales, named after their shape, retain the primitive four layers of dermal bone (fig. 6.29a). Within osteichthyans, two subtypes of rhomboid scales are discerned, ganoid and cosmoid scales. Elasmoid scales are independently seen in most teleosts, *Amia, Latimeria,* and dipnoans.

Cosmoid plates and scales were present on Lower and Middle Paleozoic fishes, including early lobe-finned fishes, and resembled the dermal armor seen in figure 6.29a. Cosmine is not a distinct tissue, but it represents a complex canal system associated with the dentine and enamel layers. No living fishes have cosmoid scales.

Ganoid plates and scales covered Paleozoic actinopterygian fishes when they were at their zenith. Ganoine is a form of enamel. The earliest ganoid scales (paleonoscoid, fig. 6.29c) are found today only on the basal actinopterygians *Polypterus* and *Calamoichthyes*. The Mesozoic brought modifications of these scales, which are mirrored in the two genera of living basal neopterygians. *Amia* and gars have lost the spongy bone and dentin (fig. 6.29). In gars, these very hard ganoin-covered scales still constitute a continuous body armor over the entire trunk and tail (fig. 6.30). On the head, they form bony plates and smaller scalelike bones (see fig. 9.11a). The scalelike bony plates of *Amia* lack ganoin and are confined to the head. Elasmoid flexible scales cover the rest of the body.

Elasmoid scales are formed from thin laminar bone that may be associated with a fibrous plate (see fig. 6.29). Two subtypes are typically recognized, ctenoid and cycloid scales. Phylogenetically derived elasmoid

TABLE 6.1 Terminology for the Various Forms of Scales

Type*	Distribution	Definition
Ectodermal		
Keratinized Scales	Reptiles, mammals	Unmineralized epidermal elements (fig. 6.16)
Dermal		Mineralized dermal elements
PLACOID	Elasmobranchs	Layers 4, 3, 1 of figure 6.29; a dentine crown with a pulp cavity and covered by enameloid/enamel
RHOMBOID	Osteichthyans	Layers 4, 3, 2, 1 of figure 6.29
Cosmoid	Fossil sarcopterygians	A specialized canal system associated with dentine and enamel; not a tissue, but a structural complex
Ganoid	Actinopterygians	Layers 4, (3), 1 of figure 6.29; layer 4 composed of ganoine, a form of enamel
ELASMOID	Most teleosts, *Amia, Latimeria,* and Dipnoi	Layer 1 of figure 6.29; thin lamellar bone that may be associated with a fibrous plate
Ctenoid	Teleosts (perciforms)	Possesses small spines; a derived scale
Cycloid	Teleosts, *Amia, Latimeria,* and Dipnoi	A composite group of ctenoidlike scales lacking spines

*Terminology after Francillon-Vieillot et al., 1990.

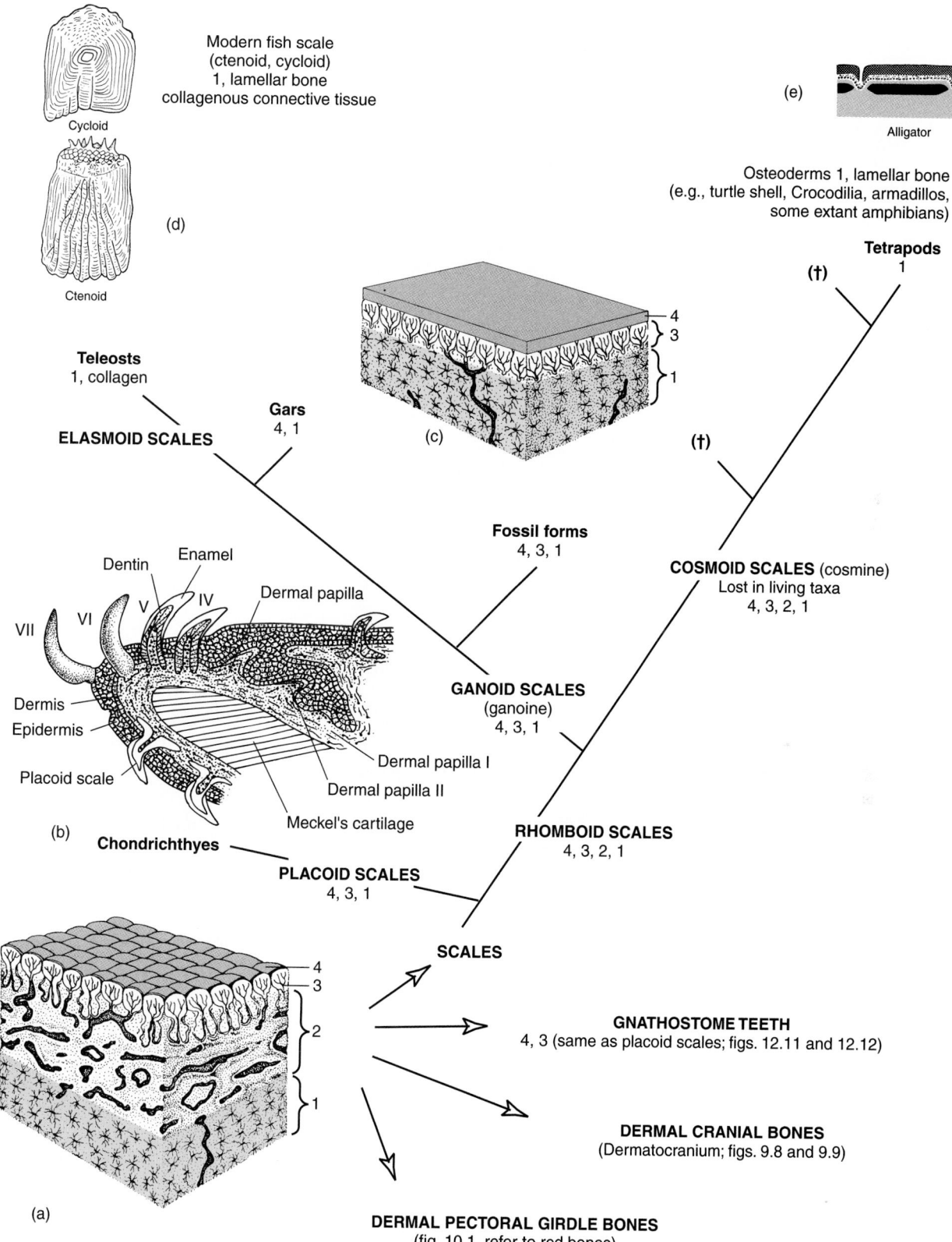

Modern fish scale
(ctenoid, cycloid)
1, lamellar bone
collagenous connective tissue

Cycloid

(e)

(d)

Ctenoid

Alligator

Osteoderms 1, lamellar bone
(e.g., turtle shell, Crocodilia, armadillos,
some extant amphibians)

Tetrapods
1
(†)

Teleosts
1, collagen

Gars
4, 1

ELASMOID SCALES

(c)

(†)

Dentin   Enamel

Dermal papilla

V   IV

VII   VI

Dermis

Epidermis

Placoid scale

Dermal papilla I

Dermal papilla II

Meckel's cartilage

(b)   **Chondrichthyes**

PLACOID SCALES
4, 3, 1

COSMOID SCALES (cosmine)
Lost in living taxa
4, 3, 2, 1

Fossil forms
4, 3, 1

GANOID SCALES
(ganoine)
4, 3, 1

RHOMBOID SCALES
4, 3, 2, 1

SCALES

GNATHOSTOME TEETH
4, 3 (same as placoid scales; figs. 12.11 and 12.12)

DERMAL CRANIAL BONES
(Dermatocranium; figs. 9.8 and 9.9)

(a)

DERMAL PECTORAL GIRDLE BONES
(fig. 10.1, refer to red bones)

FIGURE 6.29

Derivatives of primitive dermal bone. The primitive state (*a*) includes: **1**, lamellar bone; **2**, spongy bone; **3**, dentin; **4**, enameloid/enamel. **3** plus **4** constitute a denticle. (*b*) Placoid scales and teeth of a shark. The scales consist of a denticle (**4, 3**) and a basal plate (**1**). Cosmine represents a specialized canal system associated with dentin and enamel. (*c*) Ganoine is a form of enamel in ganoid scales. (*d*) Elasmoid scales. (*e*) A tetrapod osteoderm (alligator). (†) Indicates an extinct taxon.

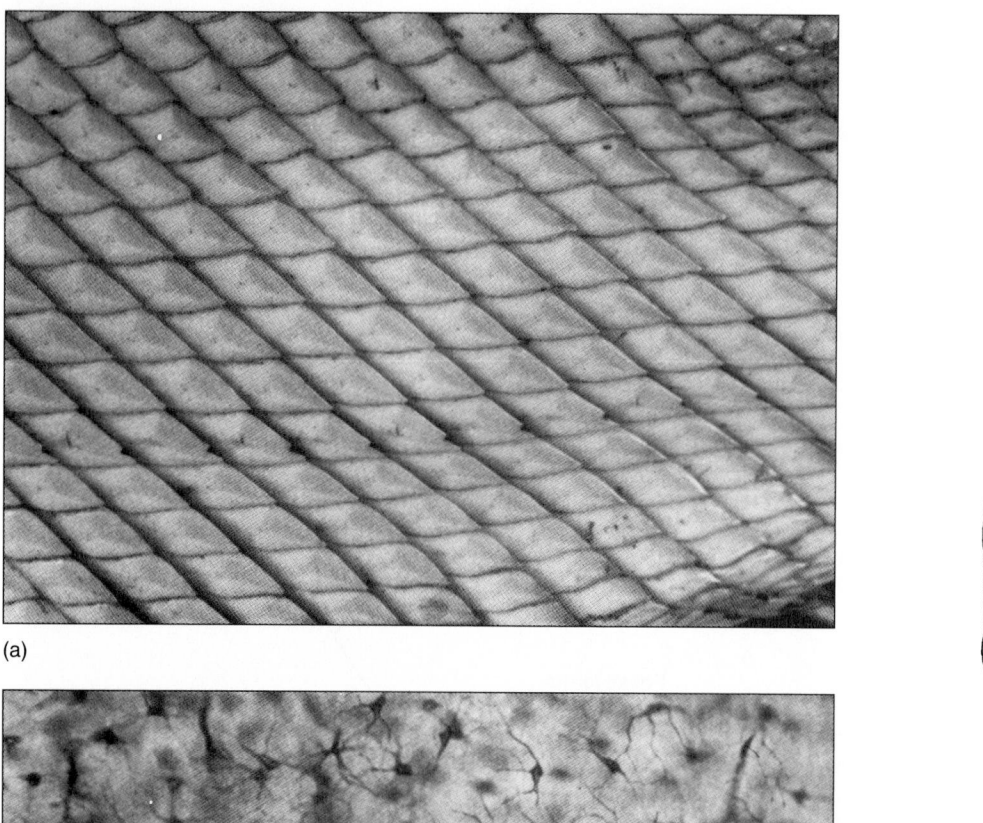

(a)

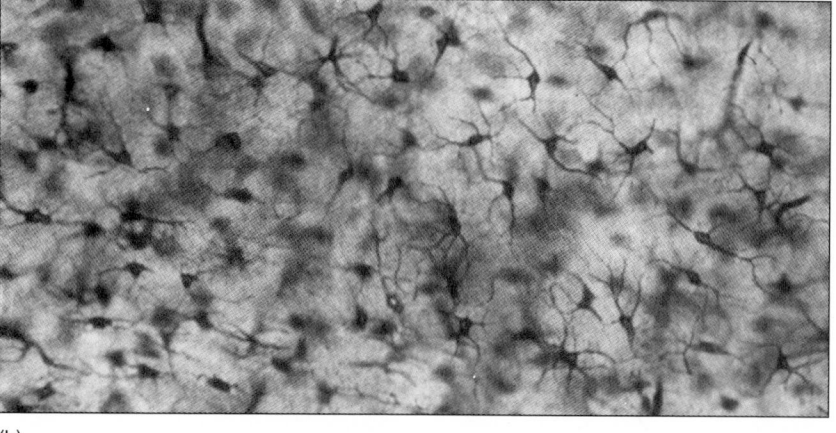

(b)

FIGURE 6.30
Ganoid scales. (*a*) Scales on trunk of a gar. The tail is to the right. (*b*) Canaliculi and lacunae within a single scale (microscopic).

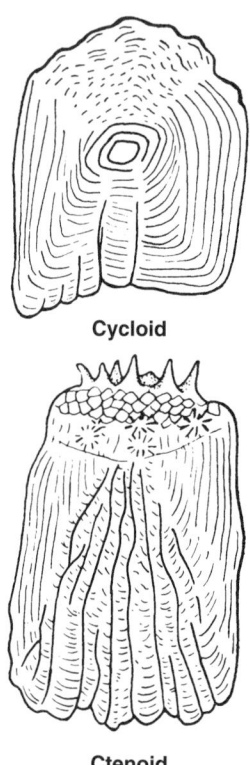

Cycloid

Ctenoid

FIGURE 6.31
Modern flexible fish scales. The upper edges are caudal free borders.

scales may include osteoderms and spines. **Cycloid and ctenoid scales** are found on teleosts and modern lobe-finned fishes (dipnoans and *Latimeria*). There is little difference between them except that the ctenoid scales have a comblike (ctenoid) free border (fig. 6.31). These modern scales consist of a very thin layer of acellular lamellar bone underlain by a plate of dense collagen (see fig. 6.29*d*). As a result, they are flexible and translucent. The overlying epidermis is very thin. Their arrangement in the dermis is seen in figure 6.34.

The ability to form scales has been lost by lampreys, catfish, bony eels, and some other fishes; nevertheless, scale anlagen develop transitorily in the embryos. In contrast to scaleless teleosts, sea horses and pipefish have dense bony plates in their dermis over most of the body surface.

Placoid scales were present on Paleozoic sharks, and are still present on today's sharks, skates, and rays. They have the same structure as paleoniscoid scales, consisting of lamellar bone, dentin, and an enameloid covering (see fig. 6.29*b*). But arising from the flat bony **basal plate (root)** in the dermis is a **spine** of enameloid-covered dentinoid that erupts through the epidermis (fig. 6.32). The spine is a specialized denticle. It is hypothesized that placoidlike scales of some unknown agnathan ancestor became teeth at the edges of the jaws in gnathostomes.

Dermal Ossification in Tetrapods

The outstanding difference between the dermis of one vertebrate class and another is the extent to which each contains bone. When the first tetrapods lumbered onto land, they brought with them versions of the cosmoid scales of their lobe-finned ancestors. Some labyrinthodonts, and even some basal amniotes, had large bony plates in the dermis; others had minute bony scales. The latter are called **osteoderms** in tetrapods.

Among amphibians, caecilians and some tropical toads still have osteoderms. Those between furrows in caecilian skin are microscopic, are perpendicular to the surface, and lie in circumferential bands separated by glandular skin. They are just large enough to be seen with the naked eye when loosened with a scalpel (fig. 6.33).

Crocodilians have oval osteoderms, especially along the back, where they are often associated with cornified crests (figs. 6.34, alligator, and 6.35). The young of some species of lizards have osteoderms under the epidermal scales of the head until some time after hatching or birth, after which they usually fuse with the underlying membrane bones of the skull. Turtles other than leatherbacks are truly armored, being boxed in by large bony plates that meet in immovable sutures (fig. 6.36). The carapace and plastron are united by bony lateral bridges that must be sawed through to expose the viscera.

Among extant mammals, armadillos alone have dermal armor. It consists of identical small polygonal bones immovably united and extending almost to the midventral line. The bone is covered by epidermal scales (see fig. 6.20). Bone forms in the skin of other mammals, including human beings, as a pathologic condition. An earlier observation is worth repeating: *The dermis of vertebrates has an ancient and persistent potential to form bone.*

Dermal Pigments

Chromatophores are cells that contain pigment granules. They are found in many locations of the body, but those in the skin are in the dermis and are responsible for skin coloration. All have processes that are permanent extensions from the cell body, and in some the pigment granules can be dispersed into the processes or aggregated in the vicinity of the nucleus by neutrotransmitters and by hormones such as intermedin and melatonin. Chromatophores with dispersible granules are

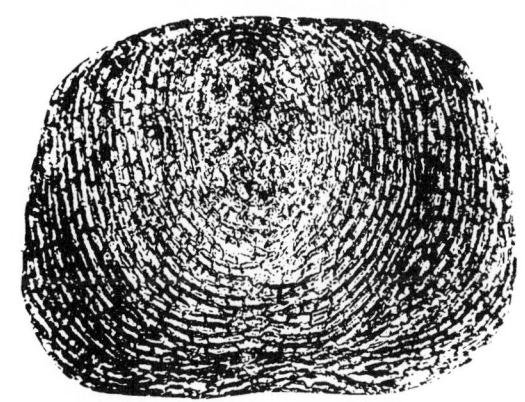

FIGURE 6.33
A single osteoderm from caecilian skin.

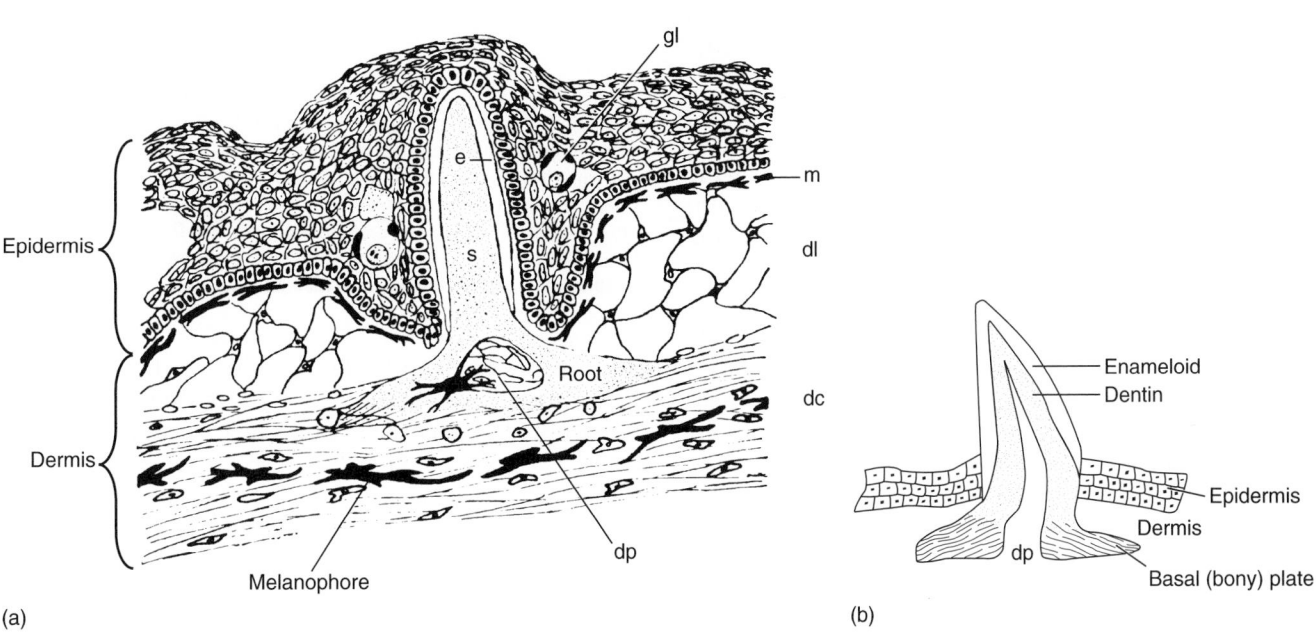

(a)   (b)

FIGURE 6.32
(*a*) Unerupted placoid scale in skin of embryonic shark. Dermis consists of compact layer, **dc,** and loose layer, **dl; dp,** dermal papilla (pulp) within root and spine; **e,** enameloid; **gl,** unicellular epidermal gland; **m,** melanophore; **s,** spine, composed of dentin, enameloid, and a core of pulp. (*b*) Erupted placoid scale, a spiny denticle.

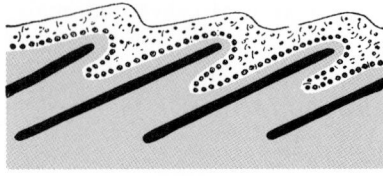

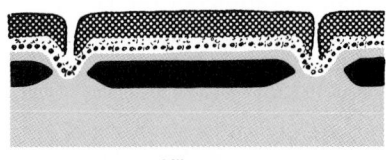

FIGURE 6.34

Site of dermal scales (*black*) in a teleost and of osteoderms in a crocodilian.

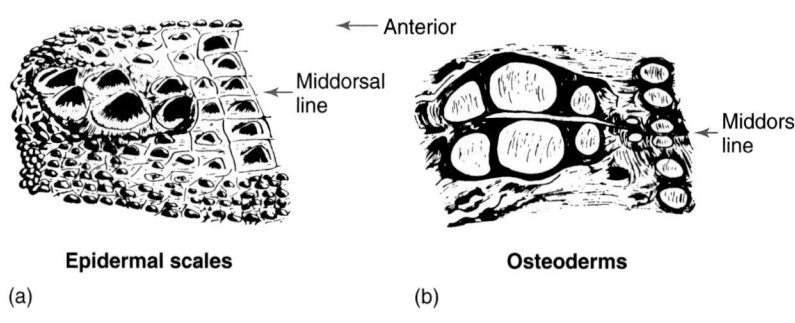

FIGURE 6.35

Alligator skin. (*a*) From dorsum of neck, showing epidermal scales. (*b*) Same section turned upside down to show osteoderms (dermal scales) embedded in skin beneath the tall crests. The middorsal line may be used as a reference point.

responsible for **physiologic color changes**, which are rapid changes such as are seen in chameleons.

Chromatophores are named for the color of their granules. **Melanophores** contain melanin granules, which are varying shades of brown. These are located in the cytoplasm in organelles known as **melanosomes**. **Xanthophores** contain yellow granules. **Erythrophores** contain red granules. Xanthophores and erythrophores are sometimes called lipophores because the granules are soluble in lipid solvents. **Iridophores** contain a prismatic substance, guanine, which reflects and disperses light, producing silvery or iridescent skin.

The pigment granules in hair and feathers, and any in the epidermis, are not in the cells that produced them for the following reason: Deep within the dermis, the branched processes of dermal melanophores ramify among the epidermal cells of the growth zones of feather and hair follicles and actively *inject* pigment granules into cells that are being added to the growing hair or feather. Similarly, melanin granules in the mammalian epidermis have been injected into cells of the germinal layer.

Hairs receive only melanin granules, and hair color (black, brown, red, blonde) is attributable to both the specific distribution and density of the granules and to the presence of two structural forms of melanin. Adding to the variation in color is the number of air vacuoles in the medulla of the hairs. Gray or white hair is the result of large numbers of air vacuoles and few melanin granules.

Feathers receive brown, yellow, and red pigments. Despite what appear to be blue feathers, craniates have no blue granules. When viewed under a light microscope by *transmitted* light, the "blue" feather is brown, the color of the melanin granules beneath the prismatic layer. The blue color observed in *reflected* light is a dis-

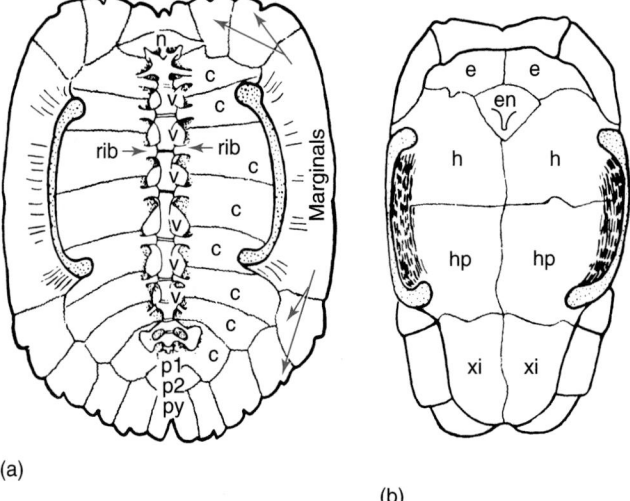

FIGURE 6.36

Dermal plates of the carapace (*a*) and plastron (*b*) in the turtle *Chrysemys,* from an internal view. On the carapace, the plates shown are **c**, costals, united with ribs; **n**, nuchal; **p1** and **p2**, precaudals; **v**, vertebrals (six of the eight are labeled). Marginals encircle the periphery and include the pygal, **py.** On the plastron, the plates shown are **e**, epiplastrons; **en**, entroplastrons; **h**, hyoplastrons; **hp**, hypoplastrons; **xi**, xiphiplastrons.

persion phenomenon, like the blue of the sky. A red feather is red even when viewed in transmitted light because of red pigment granules. The iridescence of feathers is also a dispersion phenomenon.

Physiologic color changes occur only in ectotherms, chiefly in teleosts, anurans, and some lizards. Dispersal of granules creates a pigmentary blanket that masks any underlying pigments; aggregation exposes the latter. Inasmuch as not all varieties of chromatophores respond similarly to the same stimuli, various color combinations result. A functional association of dermal chromatophores is illustrated in figure 6.37.

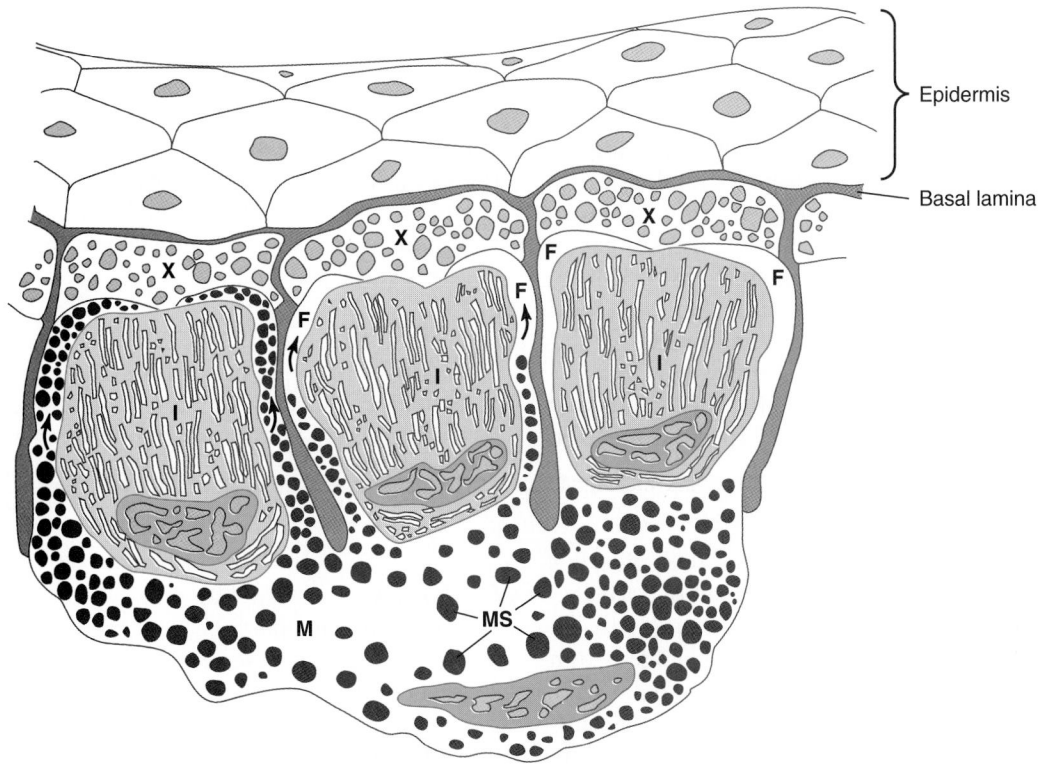

FIGURE 6.37

Dermal chromatophore unit of the frog *Hyla cinerea* in transverse section. There are three types of pigment cells: xanthophores, **X**; iridophores, **I**; melanophore, **M**. *Arrows* show melanosomes (**MS**), dark granules, moving into four of the six-finger (**F**) processes alongside the iridophores. Dispersal of melanosomes into fingers will temporarily darken the skin in a darkened environmental background.

Most craniates cannot change color reflexly because the granules in their chromatophores cannot aggregate and disperse. These animals change color only as pigment granules are synthesized in response to long-term environmental or hormonal stimuli such as exposure to sunlight ("getting a tan," for example) or when hairs or feathers are shed and replaced with others. These are **morphologic color changes.** They are generated most commonly by seasonal biological rhythms that regulate hormone synthesis, for example, the summer coat color in *Lepus* (see fig. 4.42*a*) and its winter white coat.

Pigment cells are not confined to the skin. They are found in many deep locations, including the meninges and striated muscles. Pigment cells in these locations contain only melanin granules and are usually referred to as **melanocytes.** The granules are permanently dispersed. All pigment cells arise from neural crest.

Not all skin color is due to pigment. The deep scarlet color of the face, forehead, and scalp of red ouakari monkeys from Amazon River valley (see fig. 4.40*a*) is due to an exceptionally rich network of blood capillaries lying immediately under the epidermis. When circulation ceases, the color disappears.

THE INTEGUMENT FROM FISHES TO MAMMALS: A GROUP-BY-GROUP OVERVIEW

Agnathans

Epidermis

Living agnathans lack scales, but their multilayered epidermis is characterized by an abundance of unicellular mucous glands, which earned hagfishes the descriptive name "slime eels" (see fig. 6.4). All layers are mitotic, including the surface cells. The only cornified structures are horny denticles in the lamprey buccal funnel and the biting-scraping cornified teeth of living agnathans (see fig. 6.9). These are periodically shed and replaced.

Dermis

The dermis is thinner than the epidermis, but it is exceptionally tough because of very compactly interwoven bundles of collagenous connective tissue. It contains many melanophores and adheres tightly to the underlying body wall musculature, especially at the myosepta.

Cartilaginous Fishes

Epidermis

The epidermis of cartilaginous fishes (see fig. 6.31) consists of more layers of cells than does that of living agnathans, the cells are compactly disposed, and unicellular glands are less abundant, except in chimaeras. An aggregate of goblet cells at the base of the stinger in sting rays have been modified to secrete toxins. Multicellular glands are few in number and in restricted locations, such as at the base of the claspers of males. Light-emitting organs, or photophores, in the dermis in a few species are modified multicellular epidermal glands that have invaded the dermis, then lost their connection with the epidermis.

Dermis

The dermis is considerably thicker than the epidermis and is composed of two more or less well-defined layers. The integument of elasmobranchs is unique in the presence of placoid scales, inherited unchanged from paleozoic sharks. Placoid scales have a bony basal plate in the dermis and a spine that penetrates the epidermis from below and projects above the surface of the skin, giving it a sandpaper texture. Just under the epidermis is a continuous sheet of melanophores, which is more dense on the back than on the belly, except in deep-sea species. This distribution makes a shark in partially illuminated waters less visible from both above and below. Other types of pigment cells may also be present.

Chimaeras have lost the scales over most of the surface of the body, and because they have many more mucous glands than elasmobranchs, their skin is slippery, rather than having a sandpaper texture.

Bony Fishes

The skin of most extant bony fishes, especially dipnoans and teleosts, have a relatively large number of mucous glands, and they have dermal scales of one kind or another, except placoid.

Epidermis

Epidermal glands are chiefly unicellular mucous glands that maintain a constant thin mucous coating on the surface of the skin (see fig. 6.5). Multicellular glands (see fig. 6.6) are relatively sparse. Some of them secrete the mucoid cocoon that envelopes aestivating lungfishes in dry seasons, impeding desiccation. A few are granular glands that secrete an irritating or, to some species, poisonous alkaloid. Many teleosts that inhabit the deep, dark recesses of the seas have variously disposed and sometimes bizarre **photophores** that facilitate species recognition or serve as lures or warnings (see fig. 6.5).

Dermis

The dermis is characterized by the presence of ancient ganoid scales, or modern cycloid or ctenoid scales (see figs. 6.29, 6.30, and 6.31). Ganoid or modified ganoid scales are found on several relict basal ray-finned fishes such as *Polypterus* and on gars (basal neopterygians), often in the form of plates on the head and scales on the body and tail. *Amia*, however, has modern fish scales on the body. Teleosts, except for a few naked ones, have ctenoid or cycloid scales (sometimes both) on the same fish; some scaleless teleosts develop abortive scales during embryonic life.

Amphibians

The skin of amphibians differs from that of fishes in three major respects: (1) Scales are absent except in a few species; (2) epidermal glands are generally abundant and predominantly multicellular; and (3) except in toads, the epidermis exhibits only an incipient stratum corneum.

Epidermis

The epidermis is glandular, and highly so in aquatic or semiaquatic species (see figs. 6.2 and 6.7). Unicellular cells secrete mucigens, but most mucous glands are multicellular. Urodeles and anurans differ in the proportion of time they spend on land and in the water. In aquatic amphibians that are infrequently exposed to air, many of which breathe partly with gills, integumentary mucus keeps the skin moist while the animal is out of the water. Produced in abundance when the animal is being attacked, mucus makes the body slimy, facilitating escape from prey. Granular glands are common but not abundant in toads (see fig. 6.11, paratoid). Their irritating or toxic secretions aid in the survival of these land-adapted anurans.

The epidermis of species that spend a considerable time on land exhibits a layer of cornified cells of varying thickness that impede desiccation. Toads have the thickest layer; permanently aquatic species have little or none. Efts have a thin covering until they return to water as mature adults, at which time the cornified cells are shed, epidermal glands hypertrophy, and the skin becomes smooth and slippery.

Cornified appendages are rare. Calluslike caps form on the ends of the digits of urodeles subjected to buffeting in mountain brooks. Tadpoles have horny teeth that rasp aquatic mosses and algae, which is their larval diet. At metamorphosis, the horny teeth are shed and the adult becomes insectivorous.

Dermis

The dermis is firmly attached to the underlying musculature over much of the body in apodans and urodeles, but the relationship of skin and muscle in anurans is unique. Separating the skin from the underlying muscles in anurans are voluminous broad subcutaneous lymph sinuses, the survival value of which is conjectural (see fig. 14.43). The skin of anurans is therefore less intimately attached to the underlying tissues than in other craniates. Dermal chromatophores in some species are capable of eliciting physiologic color changes (see fig. 6.37).

Tiny bony scales are embedded in the dermis of some caecilians (see fig. 6.33), often in bands alternating with bands of glandular epidermis. A few tropical toads have bony scales in the dermis of the back.

Nonavian Reptiles

The skin of reptiles has a thick stratum corneum with a variety of specialized cornified features. Integumentary glands are relatively sparse, and the dermis of many reptiles has bony deposits varying from small dermal scales (osteoderms) to large bony plates.

Epidermis

The epidermis is unique in having a thick, water-impervious stratum corneum with many localized modifications, including epidermal scales, scutes, beaks, rattles, claws, plaques, and spiny crests (see figs. 6.16 through 6.18, 6.19, and 6.28*a*). These features represent the ultimate adaptation of craniate skin to life in an arid and often hostile environment. All these are modifications of a continuous sheet of stratum corneum.

The scales of squamates overlap because they form on the surface of overlapping folds of epidermis (see fig. 6.16). Periodically, snakes shed the entire outer layer of stratum corneum in one piece, or "cast," in a process called **ecdysis**. Lizards slough it off in patches. Other reptiles wear it off by abrasion. The scales of amphisbaenians are disposed in concentric rings that encircle the body, imparting an annulate appearance to the body similar to that of caecilians. Scutes on the ventral surface of snakes and on the plastron of turtles are smooth, quadrilateral, or polyhedral sheets of stratum corneum that protect the ventral surface from abrasion. (The scutes on the plastron and carapace of turtles are identified by name in fig. 6.19.) The stratum corneum of crocodilians is warty, crested, or composed of nonoverlapping scales or thick plaques, depending on the region of the body. In all species, small nonoverlapping epidermal scales cover the limbs, tail, neck, and other surfaces that are not otherwise protected.

Epidermal glands are almost entirely granular and are restricted to localized regions of the body where their protective and pheromonal secretions will be most effective. Inasmuch as granular glands secrete only in response to specific discontinuous stimuli, the skin of reptiles is exceptionally dry.

Dermis

The dermis is notable for the presence of dermal bone. Most conspicuous is the shell of turtles, consisting of carapace, plastron, and lateral bridges that connect the two. Carapace and plastron are composed of large dermal plates arranged in a pattern, but one that does not coincide with that of the overlying scutes (compare

figs. 6.19 and 6.36). Soft-shelled and leatherback turtles have a flexible leathery skin because of the absence of ossification in the dermis. Crocodilians have osteoderms in localized regions of the body (see fig. 6.35). Along the dorsum, they are overlaid by prominent cornified crests that give the back an awesome prehistoric appearance. A few lizards have small osteoderms underlying the epidermal scales. Osteoderms are absent in snakes.

Birds

The term *thin-skinned,* sometimes applied figuratively to humans, is literal when applied to birds. The epidermis and dermis constitute a delicate membrane that is loosely applied to the underlying muscle and is therefore quite mobile. Only on the feet and head is the skin relatively thick and intimately attached. In these regions are the horny beak, epidermal scales, spurs, and claws. Elsewhere the body is clothed in feathers, the most complex of all the derivatives of the craniate stratum corneum (see figs. 6.22 to 6.24).

Epidermis

Epidermal scales, similar to the small scales of turtles, are limited to the feet and the base of the beak. The beak is covered with a horny sheath similar to that of turtles. It may consist of several segments, and it may extend dorsally over the nasal region to form a frontal shield. The beak sometimes bears horny toothlike protuberances that are correlated with specific diets, and which take the place of teeth, which are missing in today's birds. Sharp claws are present on the toes, and one or two are found on digits of the wings of species whose habits include climbing over bark, as do young hoatzins, or hanging from crags, as do chimney swifts.

Integumental glands are lacking except for two varieties, a prominent uropygial gland at the base of the foreshortened tail, best developed in water birds, and small oil glands that line the outer ear canal of some domestic fowl and sometimes encircle the vent. The oily secretion of the uropygial gland is applied by the beak to the feathers during preening.

Dermis

The dermis supports the feather follicles and erector muscles. Its exceptional thinness, and the mobility of the skin as a whole, may be correlated with the thermoregulatory role of feathers, which can be fluffed to provide insulation for the body. Birds have no osteoderms. Some, however, such as male gamecocks, develop a spur of dermal bone on each ankle, fused to the tarsometatarsus. In some other species, both sexes have a spur on the carpometacarpus of the wrist.

Mammals

The notable features of mammalian skin are (1) hair, (2) a greater functional variety of epidermal glands than in any other craniate class, (3) a relatively thick cornified epidermis, and (4) a dermis that, unlike earlier craniates, is many times thicker than the epidermis.

Epidermis

The mammalian epidermis in general consists of three strata, an actively mitotic **stratum germinativum** (the basal layer), a relatively thin **stratum granulosum**, and a **stratum corneum** (see fig. 6.12). The granulosum, when present, contains granules of keratohyaline, a precursor of the keratin found in the stratum corneum. The stratum corneum is thickest on the parts of the forelimbs and hind limbs that come in contact with the substrate (hoofs, apical pads, knee pads, and tori, for example). In humans, it is thickest on the palms and on the soles. Although corns and calluses are a result of friction due to usage, the thick stratum corneum of the palms and soles is hereditary. It develops in the fetus. A fourth cellular layer, the **stratum lucidum**, is present in the epidermis of the palms and soles. It is a layer of cells rendered translucent by the dissolution of the keratohyaline granules of the granular layer in those locations.

Hairs are cornified epidermal appendages that arise from hair follicles that extend deep into the dermis (see fig. 6.12). Hair color results from combinations of varying concentrations of melanin granules and air vacuoles within the shaft.

Epidermal glands are of two major types, sebaceous, which synthesize oily secretions, and sudoriferous, which are predominantly sweat glands. Mammary glands of viviparous mammals appear to be modified sebaceous glands. The milk glands of monotremes appear to be modified sudoriferous glands. The variety of glands found at junctions between cornified and noncornified integumentary epithelia—glands that lubricate the eyeball, for example—are mostly sebaceous.

The stratum corneum gives rise in various species to scales, claws, and horns. Epidermal scales are found on the tail and feet of opossums and many rodents; they also develop in fetal bears and European hedgehogs but are shed into the amniotic fluid before birth. Claws have become hoofs in ungulates and flat nails in primates and a few other orders (see fig. 6.21). Horns are either composed entirely of keratin, as are the unique hair horns of rhinoceros, or they have a horny sheath overlying a bony core, as in bovines and pronghorn antelopes (see fig. 6.26).

Scalelike skin unlike any other is seen in armadillos and pangolins. Armadillos have a continuous sheet of horny scales overlying a similarly sculptured sheet of dermal bone, making armadillos truly armored mammals (see figs. 4.36 and 6.20). Pangolins have epidermal scales that give the impression of having originated as clumps of agglutinated hairs (see fig. 4.37). The horny baleen of toothless whales is an adaptation for straining its food of small organisms out of seawater (see fig. 6.27).

Dermis

The exceptional thickness of mammalian dermis is a result of the presence of many hair follicles, erector muscles of the hairs, sweat and sebaceous glands, the requisite supportive connective tissue, and its exceptional vascularity. The rich blood supply is essential to satisfy the metabolic demands of these structures. It also regulates heat loss from the body surface through vasoconstriction and vasodilation. Encapsulated sensory nerve endings are abundant in dermal papillae that indent the underside of the epidermis (see fig. 6.12). The dermal bone of armadillos was referred to earlier. Antlers and giraffe horns are dermal derivatives.

The skin of most mammals is hidden by fur; dermal pigments other than those in hairs play little role in skin coloration. The vividly colored ischial callosities of baboons are among the exceptions. In human beings, melanophores protect against the rays of the sun in proportion to their density.

Mammalian skin is separated from the underlying muscle by a cushion of loose connective tissue, the **superficial fascia**. Contour-shaping adipose tissue (fat), of which whale blubber is an extreme example, accumulates in this subcutaneous tissue.

SOME ROLES OF THE INTEGUMENT

The craniate integument is an organ, just as are the liver and gill, and it plays a variety of roles in the survival of a species. Some of the roles are primary; others are supplementary. Among the roles in one species or another are protection, exteroception (sensitivity to environmental conditions), respiration, excretion, thermoregulation, participation in locomotion, maintenance of homeostasis (regulation of the ratio of water to salts in body fluids), nourishment of the young, sexual and species signaling, and additional miscellaneous functions. The paragraphs that follow bring together ideas that have been advanced in other contexts throughout the chapter.

The **protective role** of the integument is primary. Dermal armor protects the internal organs from mechanical injury, glands secrete slimy or noxious substances, pigments provide protective coloration and a barrier to the rays of the sun, the bristling fur of a threatened mammal and the ruffled plumage of a bird give these animals an ominous appearance, and claws, horns, spiny protuberances, needles—all confer advantages in the struggle for existence.

The **exteroceptive role** of the integument is protective in its most primitive state. Naked nerve endings are stimulated when a foreign object comes in contact with the skin. This specific information is a danger alert and, in living agnathans, is the only role of cutaneous receptors. In other fishes and tetrapods, more complex nerve endings provide additional information, but even in humans, some of this is essential for survival. Cutaneous receptors are discussed in chapter 17.

Respiration via the skin supplements that via gills or lungs in many amphibians, particularly aquatic urodeles, which may acquire as much as three-fourths of their oxygen from the water in this manner. Plethodontid salamanders rely entirely on skin, having neither gills nor lungs. Scales and a thick stratum corneum are not conducive to cutaneous respiration.

Excretion of carbon dioxide in some aquatic amphibians is entirely via the skin, even in species in which oxygen is being acquired via gills. Sweat glands are a supplementary route for excretion of nitrogenous wastes in mammals that have them. In many fishes, ammonia (one form of nitrogenous waste) is easily diffused across the gill epithelium and other tissues exposed to the aquatic environment.

Thermoregulation is partly a function of the skin of endotherms. Fur and feathers insulate against cold, sweat cools by evaporation, and dilation of small blood vessels within the dermis increases heat loss by radiation. When heat conservation is necessary, these same vessels constrict.

Locomotion is subserved by adhesive pads, by claws that assist in climbing, scutes that assist in slithering, and feathers that provide an airfoil for flight. Integumentary webs stretched between the digits of many species that frequent ponds or the seashore facilitate locomotion in sand, muck, or water. Other integumentary webs permit flight in bats.

The **maintenance of homeostasis** is assisted in some fishes by dermal scales, which serve as reservoirs for calcium and phosphate molecules that can be drawn on as needed. Chloride-secreting glands dispose of excess chloride ions acquired from a marine environment, and the stratum corneum of terrestrial species conserves water. Under the influence of posterior pituitary hormones, the skin of aestivating lungfishes, and of toads and some other craniates holed up in moist locations during dry weather, absorbs water from moist surroundings, thereby preventing desiccation.

Nourishment is provided to some teleost hatchlings that feed on mucus secreted by the mother's skin, and mammary glands nourish newborn mammals.

Pheromones and sex-linked **skin coloration** signal the sex or serve for species identification by other members of a population. Some pheromones serve as alarms. **Vitamin D** is synthesized from ergosterol in the skin as a derived feature in some animals, including primates. One can no doubt list other roles performed by the integument. No wonder that an earlier biologist characterized the integument as "a jack of all trades."

Summary

The section entitled "The Integument from Fishes to Mammals" provides an additional summation.

1. The craniate integument consists of a nonvascular epidermis and a vascular dermis.
2. The epidermis of most fishes has a thin mucous cuticle. That of terrestrial craniates is covered by keratinized cells that retard desiccation.
3. Unicellular mucous glands abound in the skin of most fishes; multicellular ones appear in aquatic amphibians. Granular glands produce irritating alkaloids; they are characteristic but sparse in toads and reptiles (absent in birds). Oil glands are few in birds, abundant in mammals, where they are chiefly sebaceous. Mammary glands of viviparous mammals are probably modified sebaceous glands. Those of monotremes resemble sudoriferous glands.
4. Photophores are multicellular light-emitting epidermal glands.
5. The stratum corneum of amniotes produces a variety of cornified appurtenances, including epidermal scales, scutes, claws, hoofs, nails, horns, baleen, rattles, horny teeth and beaks, feathers, and hair.
6. Contour feathers typically consist of an afterfeather and a main feather, each with an axial shaft supporting two vanes. Vanes consist of barbs, barbules, flanges, and hooklets. Down feathers lack hooklets and are therefore fluffy. Filoplumes are little more than a threadlike shaft. Feathers are erected by arrectores plumarum.
7. Hairs arise out of hair follicles and consist of a bulb, root, and shaft. Arrectores pilorum insert on the walls of the follicles.
8. Hairs have become modified as spines, quills, bristles, and vibrissae, and components of the scales of scaly anteaters. Rhinoceros horns may be agglutinated hairs.
9. The basic constituent of dermis is collagenous connective tissue. It supports blood vessels, lymphatics, nerves, chromatophores, and, in one species or another, the bases of multicellular glands, encapsulated receptors, feather or hair follicles, and their erector muscles.
10. The dermis has an ancient potential to form bony plates or scales, which persists in one or more living members of every vertebrate class except birds.
11. The dermal scales of extant fishes are placoid (cartilaginous fishes), ganoid (some basal actinopterygians and neopterygians), and flexible ctenoid or cycloid scales (most teleosts, *Amia*, *Latimeria*, and dipnoans). Cosmoid scales were present in extinct lobe-finned fishes.

12. Chromatophores are permanently branched dermal pigment cells. Physiologic color changes are reflexive and result from movement of pigment granules into or out of the processes. Morphologic color changes are slow and depend chiefly on seasonal pigment synthesis.

Critical Thinking Questions

1. What changes occurred in the integument during the transition from protochordates to craniates?
2. Name a gland found in aquatic organisms that is retained in terrestrial tetrapods. Has its function been modified? If so, how?
3. Mammary glands are believed to have been derived from sebaceous glands. What evidence supports this hypothesis? What additional evidence might you look for?
4. Many structures have a function that serves a biological role (they are adaptive). Can you hypothesize an adaptation that might be associated with the origin of feathers? With the origin of mammalian hair?
5. Why aren't giraffe "horns" true horns?
6. What vertebrate structures have incorporated dermal bone?
7. Trace the history of dermal bone in the evolution of scales.
8. Why are teeth considered composite structures both histologically and developmentally?
9. What functions are associated with the integument?

Selected Readings

Appleby, L. G.: Snakes shedding skin, *Natural History* 89(2): 64, 1980.

Birch, M. C., editor: Pheromones. New York, 1974, Elsevier/ Excerpta Medica/North Holland.

Francillon-Vieillot, H., de Buffrénil, V., Castanet, J., Géraudie, J., Meunier, F. J., Sire, J. Y., Zylberberg, L., and de Ricqlés, A.: Microstructure and mineralization of vertebrate skeletal tissues. In Carter, J. G., editor: Skeletal biomineralization: Patterns, processes and evolutionary Trends, vol. 1, chapter 20. New York, 1990, Van Nostrand Reinhold.

Goss, R. J., illustrated by Wendy Andrews: Deer antlers, regeneration, function, and evolution. New York, 1983, Academic Press.

Nelson, D. O., Heath, J. E., and Prosser, C. L.: Evolution of temperature regulatory mechanisms, *American Zoologist* 24:791, 1984.

Rao, K. R., Fingerman, M., and others: Chromatophores and color changes, *American Zoologist* 23:461, 1983.

Regal, P. J.: The evolutionary origin of feathers, *The Quarterly Review of Biology* 50:35, 1975.

Sengel, P.: Morphogenesis of skin. Cambridge, England, 1976, Cambridge University Press.

Links to the Internet

Visit the zoology website at http://www.mhhe.com/zoology to find live Internet links for each of the references listed below.

1. Skin conditions from Yahoo. Links to more information on topics ranging from skin cancer to fish-hook removal to cutaneous anthrax.
2. Skin Conditions. Interesting links to information ranging from benign dandruff, a new medication for poison ivy, to genital warts.
3. Department of Dermatology, Univ. of Iowa. More images than you would probably want to see with regard to dermatological problems.
4. Integument and Its Accessory Organs. Photomicrographs of the skin. Also links to histological units from North Harris College.
5. Histology—The Web Laboratory. Links to units on many systems of the body, including the skin. Requires a free plug in to access the units. From Ohio State.
6. Comparative Vertebrate Anatomy—The Integument. Good information on specialized structures of the skin, such as scales, feathers, and hoofs.

7

Mineralized Tissues: An Introduction to the Skeleton

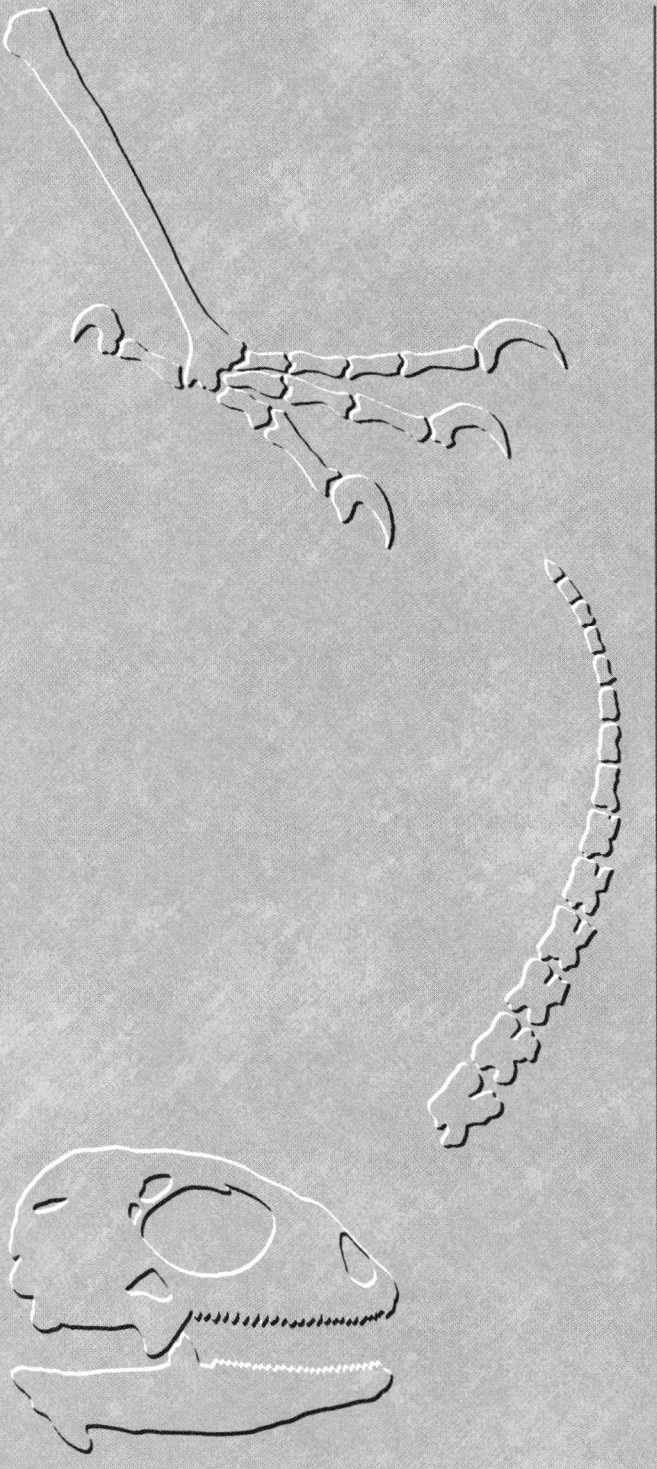

In this chapter, we will be introduced to the various types of skeletal tissues, see how they are formed, and look briefly at their invertebrate origins. We will discover that mineralized tissues have a role in homeostasis, and we will learn why growing skeletons must undergo continual remodeling.

he skeleton is composed of mineralized connective tissue and of ligaments, tendons, and bursae. The mineralized tissue for the most part is bone, but there is also dentin (often considered a variety of bone), cartilage, and enamel or enameloid substances. Osteoblasts produce bone, odontoblasts produce dentin, chondroblasts produce cartilage, and ameloblasts produce most enamels. These specialized cells arise from less-differentiated scleroblasts that arise from mesenchyme (fig. 7.1).

A preliminary step in formation of skeletal tissues is the synthesis of collagen by fibroblasts. Collagen is a proteinaceous **fibril** demonstrable only by high-power electron microscopy. Fibrils aggregate to form **collagen fibers** that are visible with light microscopy. The fibers form dense **collagen bundles** that are woven into a compact network of dense connective tissue like that found in dermis, tendons, and ligaments (fig. 7.2). It is on such a network that minerals are deposited to form cartilage and bone.

BONE

The gross structure of a long bone is shown in figure 7.3. Bone tissue is composed of a matrix of collagenous fibers, the spaces between which have been impregnated with **hydroxyapatite crystals** that are composed of calcium, phosphate, and hydroxyl ions [$3Ca_3 (PO_4)_2 \cdot Ca(OH)_2$]. The crystals are deposited under the influence of osteoblasts. A **cementing substance** composed of water and mucopolysaccharides binds the crystals to the collagenous matrix. In most bone, the osteoblasts ultimately become trapped by the bone they have laid down around them, and, as osteocytes, occupy tiny fluid-filled pools, or **lacunae**, of interstitial fluid (figs. 7.3 and 7.4). Interconnecting the lacunae are fluid-filled canals, **canaliculi**, that house protoplasmic processes extending from the osteocytes. The fluid in the lacunae and canaliculi contains calcium and phosphate ions, which

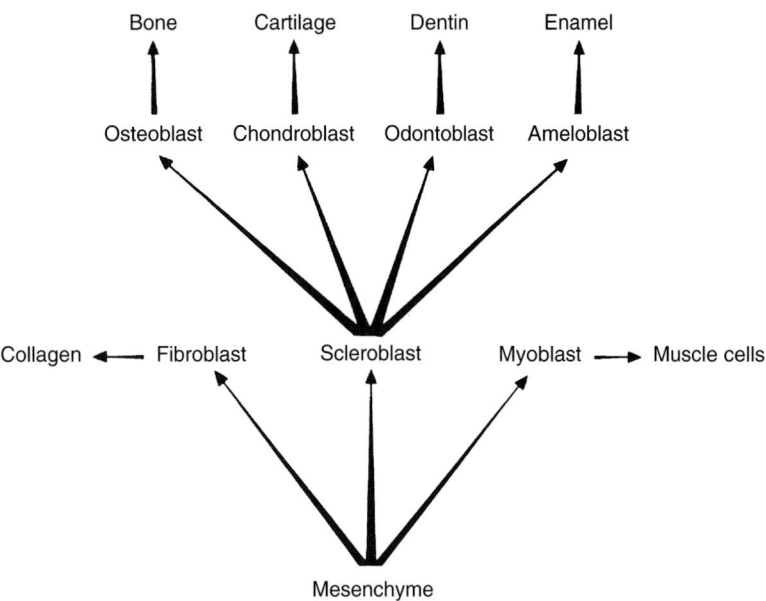

FIGURE 7.1
Some products of mesenchyme.

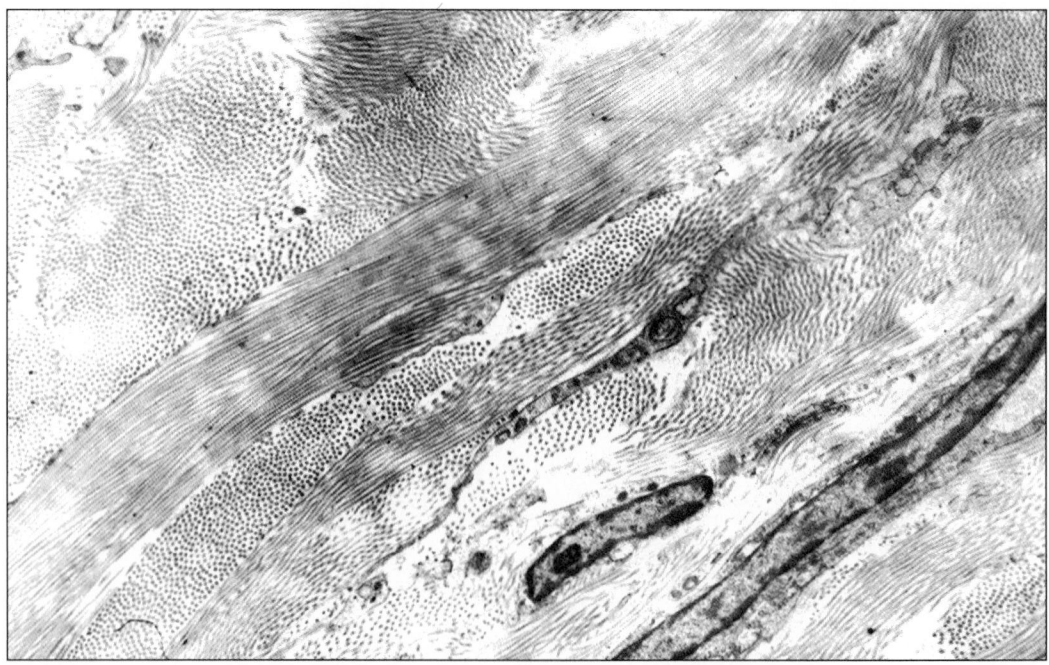

FIGURE 7.2
Bundles of collagen fibers in a tendon. The individual fibers in each bundle appear as dots in cross section. The long nuclei are in fibrocytes (mature fibroblasts).

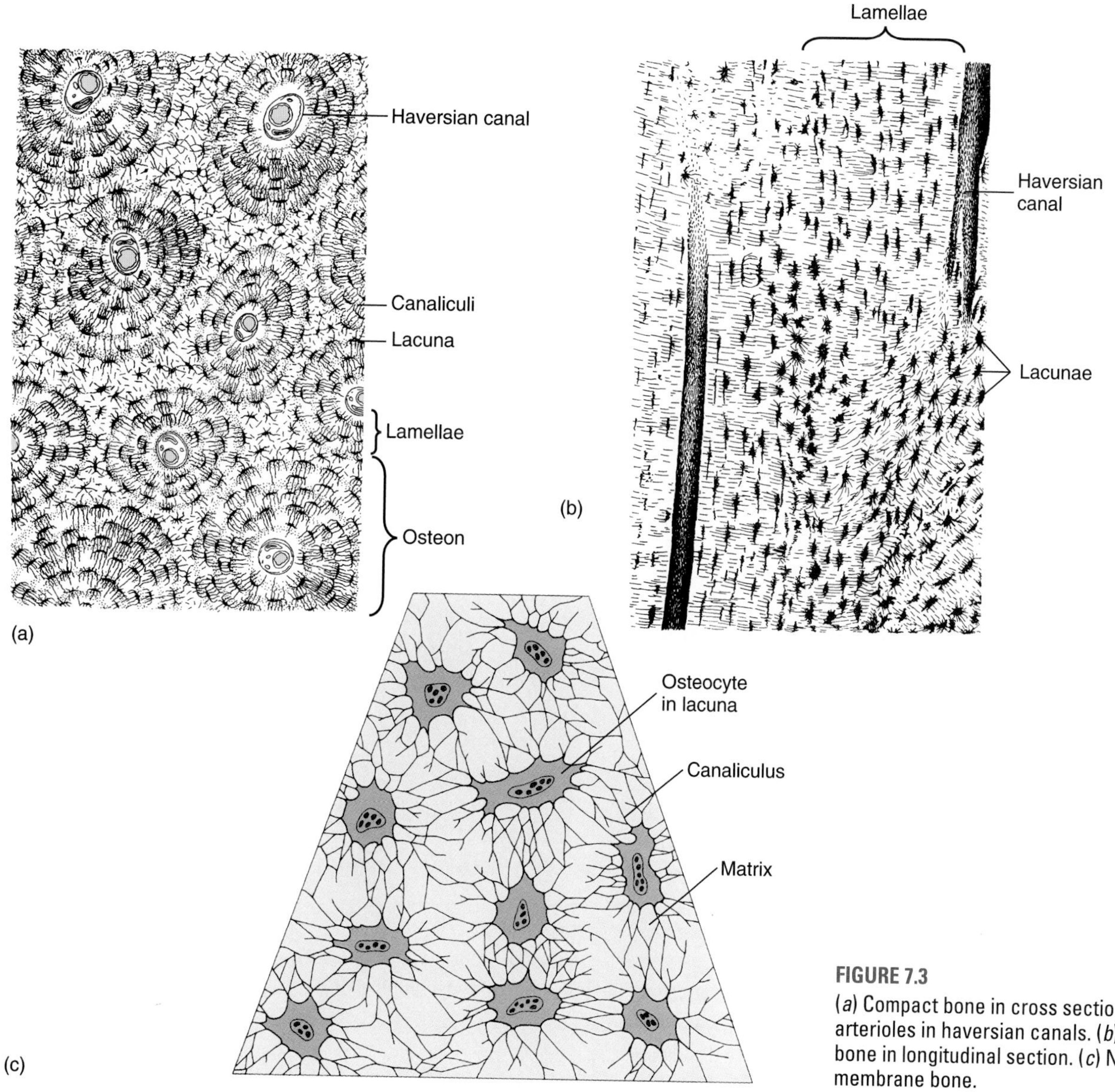

Haversian canal

Canaliculi

Lacuna

Lamellae

Osteon

(a)

Lamellae

Haversian canal

Lacunae

(b)

Osteocyte in lacuna

Canaliculus

Matrix

(c)

FIGURE 7.3

(*a*) Compact bone in cross section. *Red,* arterioles in haversian canals. (*b*) Osteon bone in longitudinal section. (*c*) Nonlamellar membrane bone.

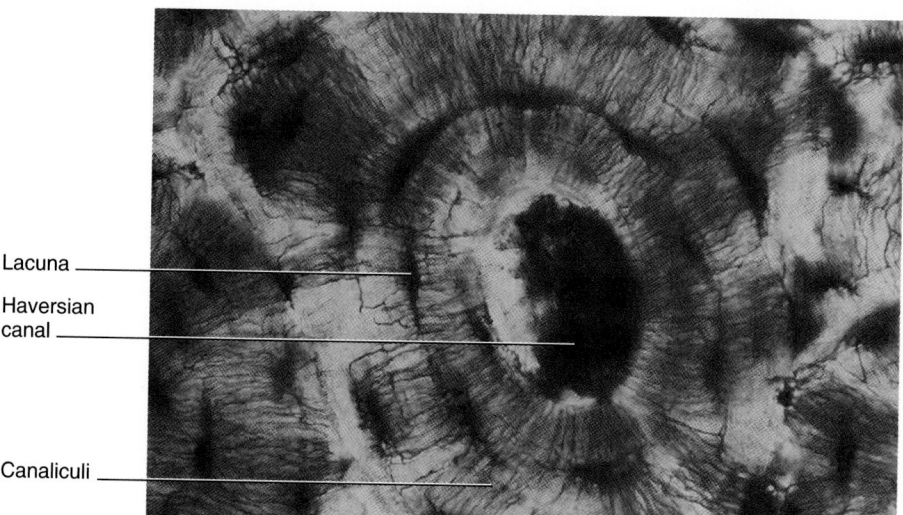

Lacuna

Haversian canal

Canaliculi

FIGURE 7.4

Cross section of one osteon (haversian system).

© R. Calentine/Visuals Unlimited.

are constantly being deposited on
the collagenous matrix or withdrawn from it, depending on the
level of serum calcium.

Compact Bone

Compact bone consists of layers
(**lamellae**) of mineralized collagenous bundles arranged concentri-
cally around a **haversian canal** (see
fig. 7.3*a*). Haversian canals contain
an arteriole, a venule, a lymph vessel, and nerve fibers. The canal and
its surrounding lamellae constitute
an **osteon**, or **haversian system**
(fig. 7.4). The blood vessels are responsible for the configuration of
the haversian systems, and because
the vessels branch and rebranch,
haversian systems do likewise. The vessels in the canals
are continuous with those in the bone marrow. In the
shaft of long bones (fig. 7.5), the haversian systems are
more or less parallel to the long axis of the bone (see fig.
7.3*b*), an architectural feature that minimizes the likelihood of fracture under normal external stress. The
lamellae of **periosteal bone** are formed by osteoblasts on
the inner surface of the **periosteum**, the dense fibrous
membrane that covers all bones except at their articular
surfaces. Haversian systems are characteristic of amniotes. Most amphibians, a few reptiles, and some small
insectivores and rodents lack them.

Spongy Bone

Spongy, or **cancellous, bone** consists of bony trabeculae
and bone marrow (fig. 7.5). **Trabeculae** are an assemblage of beams, bars, and rods that, like architectural
trusses, form a rigid framework that provides maximum
strength at areas of stress. They are composed of irregu-
larly arranged lamellae without haversian systems. **Marrow** occupies the cavities between trabeculae. It consists
of a reticulum of connective tissue fibers that support
blood vessels, nerve fibers, and adipose tissue (yellow
marrow). The hemopoietic tissue (red marrow) in some
bones produces red blood cells and some types of white
blood cells. The marrow cavities are lined with a thin
connective tissue membrane, the **endosteum**. The endosteum has many features of a periosteum and has the
capacity to deposit bone, or remodel it. A core of spongy
bone and marrow sandwiched between two layers of
compact surface bone characterizes **flat bones**, such as
ribs, scapulae, and membrane bones of the skull.

Dentin

Dentin (dentine) has the same constituents as compact
and spongy bone, but the odontoblasts are not trapped
in lacunae during osteogenesis. They retreat as they

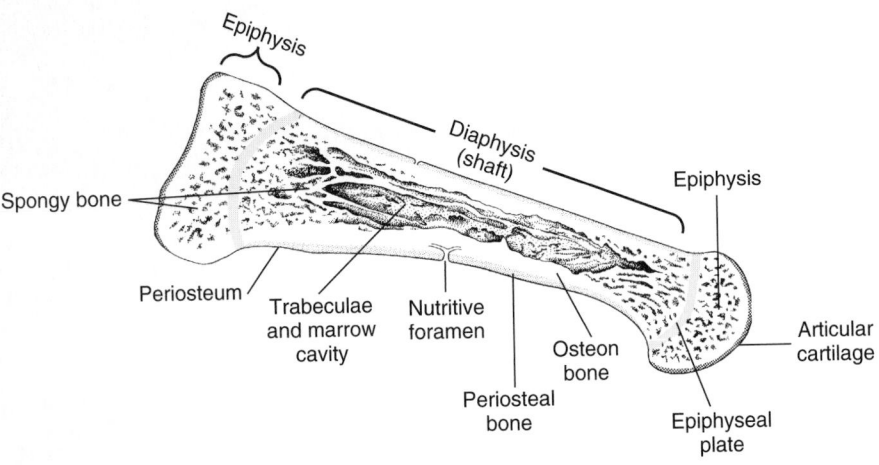

FIGURE 7.5

A metatarsal bone in longitudinal section. Epiphyseal plates in *light red.*

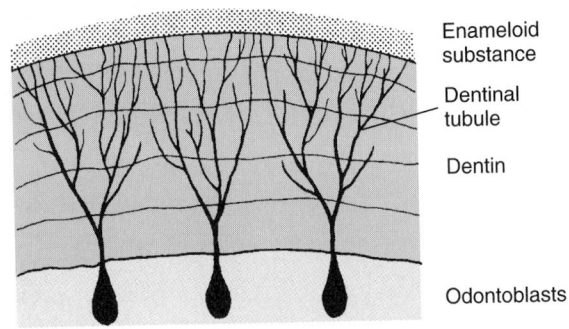

FIGURE 7.6

Dentin covered with enameloid. The dentinal tubules contain
processes of the odontoblasts.

deposit dentin; hence, they are at its inner border
(fig. 7.6; contrast the condition seen in some fossil taxa
where the odontoblasts may be trapped in lacunae
[e.g., placoderms]). However, they leave behind protoplasmic processes in canaliculi. In dentin, the canaliculi
are called **dentinal tubules**. The tubules extend all the
way to the surface of the dentin. Dentin forms only in
the outer layer of the dermis, just beneath the epidermis, and is frequently coated on its surface by enamel
or enameloid (see fig. 6.29*a–c*). The enameloid of placoid scales is produced by odontoblasts; hence, it is a
very hard dentin. Although a constant feature of the
dermis of the earliest vertebrates, dentin is found today
only in the scales of basal ray-finned and elasmobranch
fishes and in teeth.

Acellular Bone

There is one type of bone in which the osteoblasts not
only retreat as they deposit bone but in addition leave
behind no processes or canaliculi. It is the thin layer of

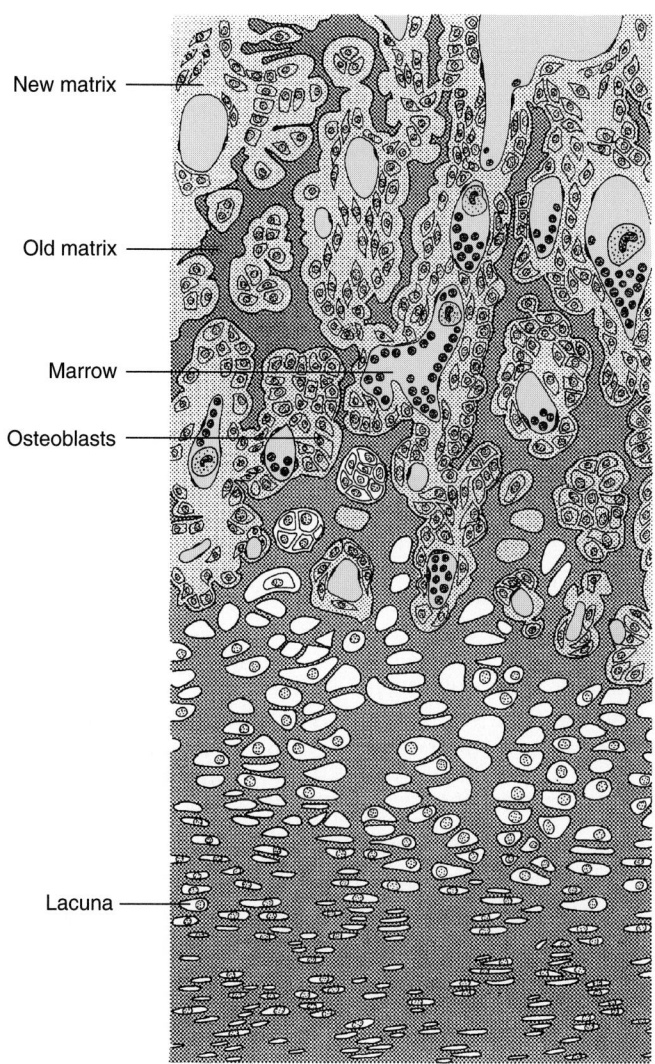

New matrix

Old matrix

Marrow

Osteoblasts

Lacuna

Spongy bone

Dechondrification and
bone deposit

Hyaline cartilage

FIGURE 7.7

Site of endochondral ossification. *Red,* blood vessels of bone marrow invading hyaline
cartilage (old matrix) and digesting it in preparation for deposit of new matrix (*light gray*) by
osteoblasts.

Membrane Bone

Bone deposited directly within a membranous blastema without having been preceded by a cartilaginous model is **membrane bone.** Intramembranous ossification gives rise to certain bones of the lower jaw, skull, and pectoral girdles; to dentin and other bone that forms in the dermis of the skin; to vertebrae in teleosts, urodeles, apodans only; and to a few miscellaneous bones. Membrane bone may be compact or spongy, and lamellar or non-lamellar. Because of the arrangement of the blood vessels that participate in its deposition, it lacks haversian canals (see fig. 7.3*c*). Periosteal bone is membrane bone.

Any bone derived ontogenetically or phylogenetically from the dermis of the skin may be called a **dermal bone.** The term specifies its history, not its histological features. It is membrane bone. Many, though not all, membrane bones are of dermal origin, as for example, the bones that constitute the dermatocranium, the bones that invest Meckel's cartilage of the lower jaw, and the membrane bones of the pectoral girdles. These bones are phylogenetic derivatives of the integument. Not much bone remains in the dermis of recent vertebrates, but that which does—the bony plates and scales of early vertebrates—arises in situ.

acellular bone (**aspidin**) that constitutes the fibrous plates of the flexible scales of modern fishes and the cementum of vertebrate teeth. The term *aspidin* was first applied to the acellular dermal bone found in fossil heterostracens (see fig. 4.4) and its presence in teleosts is due to convergence.

Membrane Bone and Replacement Bone

Before bone or cartilage can be deposited, a preskeletal blastema must develop. A **blastema** is any aggregation of mesenchyme that, given the appropriate stimulus, differentiates into some tissue, such as muscle, cartilage, or bone. Once a preskeletal blastema has formed, some of the mesenchyme cells become fibroblasts and secrete collagen. Others become either osteoblasts or chondroblasts and secrete the enzymes essential for formation of bone or cartilage.

Replacement Bone

Replacement bone is deposited where hyaline cartilage already exists (fig. 7.7). In the process, the existing cartilage undergoes degenerative changes and disappears. The processes of endochondral and intramembranous ossification are the same in that both consist of impregnation of a collagenous matrix with hydroxyapatite crystals. The difference is that in endochondral ossification, cartilage must be removed before bone can be deposited. In both cases, the immediate result is formation of temporary spongy bone. The latter, in turn, is eroded and replaced by compact bone, spongy bone, or a marrow cavity, depending on its location.

The process of endochondral ossification of a typical long bone of a tetrapod limb commences midway in the shaft (**diaphysis**) of a miniature cartilaginous model of the bone-to-be, and proceeds toward the two ends, or **epiphyses.** Soon after the diaphyseal center becomes active, one or more

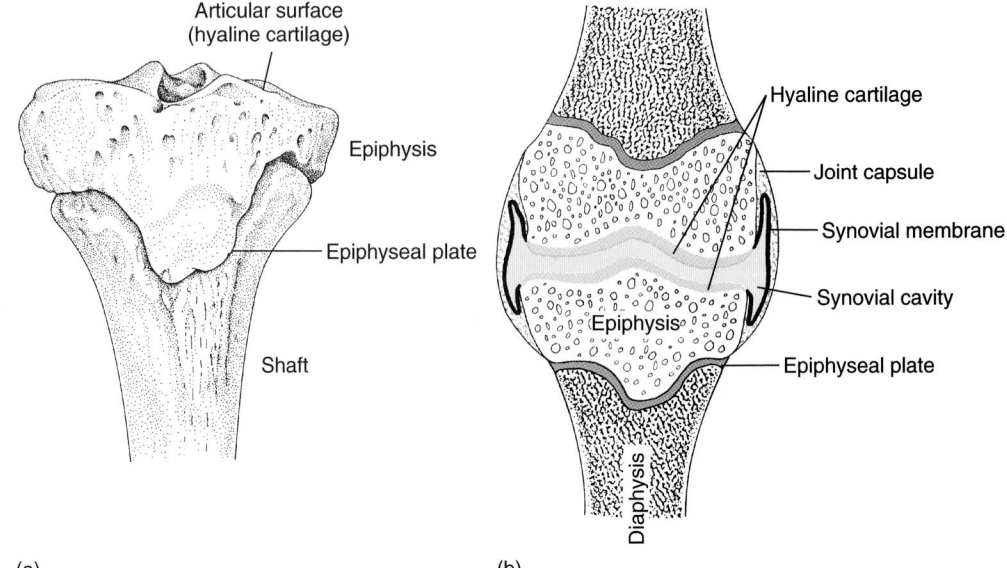

FIGURE 7.8

(*a*) Proximal end of tibia of youth aged 16 years. The epiphyseal plate is unossified; hence, the bone is still elongating.
(*b*) Diagrammatic section of a simple diarthrodial joint, ligaments removed. *Red* is synovial fluid.

(a)

(b)

endochondral ossification centers appear in each epiphysis. Additional cartilage continues to be deposited in advance of all zones of ossification, and the rate of deposit of cartilage equals the rate of ossification. As a result, the shaft continues to grow in length while simultaneously becoming ossified. As the individual grows older, however, deposition of cartilage gradually slows, eventually leaving only a narrow cartilaginous **epiphyseal plate** between the diaphysis and each epiphysis (figs. 7.5 and 7.8). Chondrogenesis continues within this plate for a time, and the bone continues to elongate so long as the epiphyseal plates are elaborating cartilage. In ectotherms, this is throughout life. In birds and mammals, the epiphyseal centers cease elaborating cartilage shortly after the individual attains sexual maturity, the epiphyseal plates ossify, the epiphyses become sutured to the shaft, and the bone stops growing. Bones of other shapes exhibit other combinations of ossification centers (see fig. 9.5). Growth in diameter of the bones takes place when bone is deposited from the inner surface of the periosteum.

While endochondral ossification is going on within a bone, periosteal bone is being deposited at the surface. As a result of this activity combined with continual resorption and remodeling (to be discussed shortly), a bone reaches adult shape, size, and proportions. Details of the histogenesis of bone will be found in textbooks on histology.

Cartilaginous fishes either no longer have the genetic codes necessary for endochondral ossification or there is some factor that prevents gene expression for ossification.

CARTILAGE

Cartilage resembles bone in that it is formed within a matrix of collagenous connective tissue, and the cells lie in lacunae. The intercellular matrix, however, contains a sulfated mucopolysaccharide instead of hydroxyapatite crystals. Unlike bone, cartilage has no canaliculi demonstrable by light microscopy and no blood vessels of its own. The lacunar cells (**chondrocytes**) are supplied with oxygen and nutrients by diffusion from the nearest capillaries adjacent to the cartilage.

Cartilage formation is initiated when chondroblasts deposit their mucopolysaccharide on a preexisting collagenous matrix. The mesenchyme surrounding the area of chondrification organizes a bounding membrane, the **perichondrium**. Thereafter, additional cartilage is laid down within the blastema by mesenchyme from the perichondrium (appositional growth), and by fibroblasts and chondroblasts that have arisen by mitosis from parent cells within the developing mass (interstitial growth). Chondroblasts become chondrocytes after being trapped in lacunae. The perichondrium is constantly restructured as the mass enlarges and changes shape.

Hyaline cartilage is the least differentiated variety and is the precursor of replacement bone. It is translucent because all components have the same refractive index to light. Little hyaline cartilage remains after the growth of replacement bones has ceased. Thereafter, it is found chiefly on the articular surfaces of bones within the joints of tetrapods (fig. 7.8*b*). Most any other hyaline cartilage that forms, whether in fishes or tetrapods, is likely to be transformed eventually into fibrocartilage, elastic cartilage, or calcified cartilage.

Cartilage with exceptionally thick, dense collagenous bundles in the interstitial matrix is **fibrocartilage**. The intervertebral discs of mammals are fibrocartilage. **Elastic cartilage** contains, in addition to collagenous fibers, a network of elastic fibers. In mammals, it is found in the pinna of the ear, in the walls of the outer ear canal, in the epiglottis, and elsewhere. **Calcified cartilage** is formed when calcium salts are deposited within the interstitial substance of hyaline cartilage or fibrocartilage. It is often mistaken for bone. The jaws of sharks, no matter how large, are calcified cartilage.

SKELETAL REMODELING

Bones not only provide mechanical support for the vertebrate body but, with scales and teeth, constitute important storage places for calcium and other mineral salts. They therefore participate in maintaining homeostasis. Calcium is constantly being deposited and withdrawn in response to dietary intake and cellular demands. When serum calcium levels are rising, excess calcium that is not excreted is deposited along with inorganic phosphate as hydroxyapatite crystals. When serum calcium is falling, calcium is withdrawn from the skeleton and other depots, thereby maintaining a normal serum calcium level. Withdrawal is regulated by parathyroid hormone and calcitonin; deposit may not be under endocrine control.

In addition to bone resorption in response to lowered serum calcium levels, cartilage and bone are constantly being resorbed and replaced in another process, skeletal remodeling. For example, a skull the size of that of a newborn child could not accommodate a 21-year-old brain without growing (fig. 7.9). But growth by accretion would not provide a larger brain cavity; it would only provide a thicker skull. The entire skull must be continuously remodeled by constant resorption of existing cartilage and bone and replacement of these in a new alignment to provide a wider, higher skull with a cavity large enough to accommodate the growing brain. This entails not only remodeling of the cranium but also of the facial bones and lower jaw. Cartilage and bones elsewhere in the body are also subject to remodeling. Although remodeling is a more active process in growing skeletons, it continues to some degree throughout life, accelerating during periods of skeletal repair (fig. 7.10).

Remodeling of bones also occurs in response to mechanical stress resulting from continuing use of inserting muscles or from weight bearing. Within long bones possessing haversian systems, invasion of the diaphysis by blood vessels results in resorption with deposition of secondary haversian systems. The orientation of these secondary systems is related to fields of stress within the bone. A bone becomes thicker at sites where stress is high, thin when stress is withheld. Roughened surface areas, bony ridges, and other prominences to which muscles attach enlarge with sustained muscle use. The arrangement of the latticelike components of spongy bone (see fig. 7.5) bears a striking resemblance to lines of force generated by the bone's natural load.

Just as mechanical stress results in bone remodeling, affecting the shape and architecture of existing bones, a correlation exists between mechanical stress and ossification in certain tendons, ligaments, and connective tissue

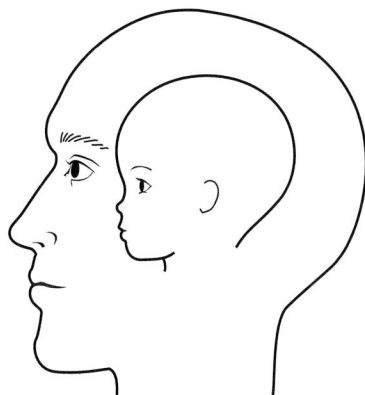

FIGURE 7.9
Comparative size of skull of newborn and 21-year-old human being. Continued remodeling enlarges the brain cavity (hence, the size and shape of the growing skull).

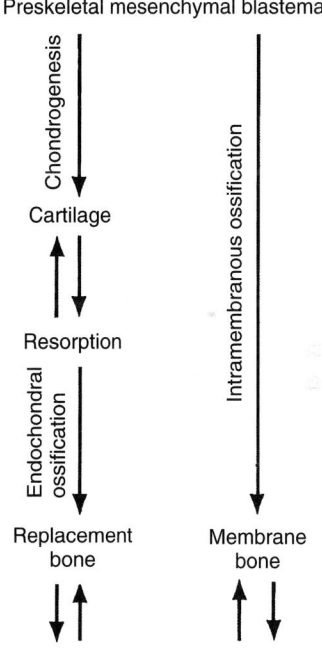

FIGURE 7.10
Steps in osteogenesis and remodeling.

seams. Heterotopic bones, to be discussed shortly, are instances of stress-correlated ossifications.

TENDONS, LIGAMENTS, AND JOINTS

Tendons and ligaments consist chiefly of closely packed bundles of collagen (see fig. 7.2). **Tendons** connect muscles with bones and have a shiny white appearance. The collagen bundles are so arranged as to offer maximal resistance to the tension created when a muscle contracts. Where the tendon is attached to muscle, the collagenous bundles of the tendon are directly continuous with those of the fibrous muscle sheath (**epimysium**), and where it attaches to cartilage or bone, they are continuous with the collagenous bundles of the perichondrium or periosteum. **Ligaments** connect bone to bone. The arrangement of the collagen bundles is less regular, but they are directly continuous with those of the periosteum. Tendons and ligaments that are flat and wide are **aponeuroses**. The mammalian scalp consists largely of an aponeurosis, the gala aponeurotica. The longest ligament in mammals is the nuchal ligament in the back of the neck of grazing herbivores. Extending between the occiput of the skull and the neural spines of some of the thoracic vertebrae, it provides a resilient mechanical support for the head. It is least developed in mammals with short necks.

The term *ligament* is sometimes applied to fibrous mesenteries or cords that support visceral organs or hold them in place, as for instance, the falciform ligament of the liver and the round ligament of the ovary. However, these are not skeletal structures.

In some species, certain tendons and ligaments become mineralized. Turkeys, for example, have ossified tendons in their legs, and ornithischian dinosaurs had them millions of years ago. **Sesamoid cartilages** or **bones** are mineralized nodules in tendons or ligaments. Best known is the patella, or kneecap. It is endochondral bone in some species, membrane bone in others.

A joint, or **arthrosis**, is the site where two bones or cartilages meet. The joint is a **diarthrosis** (see fig. 7.8*b*) if it is freely movable in one or more planes and the articular surfaces are covered with hyaline cartilage, and if the joint is enclosed in a fibrous capsule lined by a synovial membrane that secretes a lubricatory fluid. Ligaments hold the bones in their proper alignment in the joint. The articulations between the upper and lower jaws of early bony fishes were diarthroses. These same joints and the hinge joint at the elbow and knee of mammals are diarthroses.

An **amphiarthrosis** permits limited movement. Resilient fibrocartilage unites the components of the joint, and fibrous joint capsules keep the bones properly aligned. There may or may not be a cavity within the capsule, but there is no synovial membrane. The joints between the centra of mammalian vertebrae are amphiarthroses.

A sutured joint is a **synarthrosis**. A suture is an irregular jagged seam at the junction of two bones that renders the joint immovable. The joints in the roof of the mammalian skull are synarthroses (see fig. 9.27). If the suture between two bones becomes obliterated during development, as in the skull of birds, the condition is an **ankylosis**. The premaxilla and maxilla of the human embryo ankylose and as a result cannot be distinguished as two bones in an adult. In cats, they are sutured but not ankylosed. A **symphysis** is a joint in the midline of the body, in which bilateral bones are separated by a pad of fibrocartilage, and movement is severely restricted, if not impossible. The pubic symphysis of pregnant female mammals becomes movable by hormone-initiated dissolution of the fibrocartilage shortly before the onset of labor.

The terms *diarthrosis, amphiarthrosis,* and *synarthrosis* were coined by human anatomists to classify mammalian joints. Subsequent studies of other vertebrates revealed considerable diversity in the anatomy of joints that permit a degree of movement in one plane or another but lack synovial cavities. Examples include the hyostylic joints between the hyomandibular cartilages and the skull of modern sharks and the joints between segments of the entoglossal and paraglossal bones in the tongue of many birds. Special terms have been coined to designate these other varieties. The range of articulations in birds is much broader than in mammals.

MINERALIZED TISSUES AND THE INVERTEBRATES

Mineralized tissues are not limited to vertebrates. In fact, two-thirds of the living species of animals that contain mineralized tissues are invertebrates. The matrix is collagen, and it is as old as the sponges, but the inorganic crystals are more often calcium carbonate than calcium phosphate.

Cartilage is found among invertebrates, including squids and some gastropods, although it appears to be lacking in the outgroups to craniates (echinoderms and protochordates). Cartilage is present in hagfishes, but bone, dentin, and enameloid are restricted to vertebrates (first seen in Ordovician ostracoderms, but secondarily lost in lampreys). This distribution would suggest that cartilage might have preceded bone as a structural tissue in craniates; however, the fossil record of hagfish is unknown prior to the Carboniferous so that the antiquity of bone might be equal to that of cartilage (for a summary of alternative hypotheses, see Hall, 1999:41).

REGIONAL COMPONENTS OF THE SKELETON

The skeleton may be divided regionally into the following components, which include associated ligaments and tendons. These are the subjects of the next three chapters.

> Axial skeleton
>   Notochord and vertebral column
>   Ribs and sternum
>   Skull and visceral skeleton
> Appendicular skeleton
>   Pectoral and pelvic girdles
>   Skeleton of paired fins and limbs
>   Skeleton of median fins of fishes

HETEROTOPIC BONES

In addition to the skeletal components outlined earlier, miscellaneous **heterotopic bones** develop by endochondral or intramembranous ossification in areas subject to continual stress in amniotes. They are usually missing from routinely prepared skeletons. Among heterotopic bones are an **os cordis** in the interventricular septum of the heart of deer and bovines, a **baculum** (os penis) in the septum between the spongy bodies of the penis of dogs, basal primates, and many other mammals (fig. 7.11), and an **os clitoridis** in many female mammals. In walruses, the baculum is nearly 60 cm long.

A heterotopic bone forms in the gizzard of some doves, in the tongue of at least one species of bats, in the gular pouch of a South American lizard, in the muscular diaphragm of camels, in the syrinx of some birds (pessulus, see fig. 13.16), and in the upper eyelid of crocodilians (adlacrimal, or palpebral, bone). A similar plate of fibrous tissue, the tarsal plate of the eyelid, develops in human beings. A rostral bone develops in the snout of swine and is used in routing in soil, and a cloacal bone develops in the ventral wall of the cloaca of some lizards. All these and others, including sesamoid bones, are incidental bones located at foci of stress and lacking phylogenetic significance.

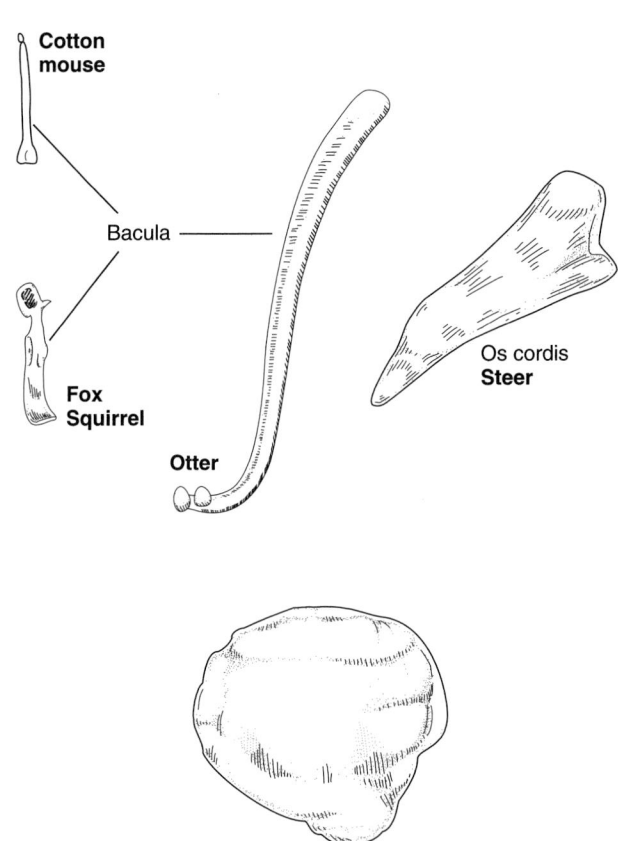

FIGURE 7.11
Some heterotopic bones.

Summary

1. The chief mineralized tissues are bone (from osteoblasts), cartilage (from chondroblasts), and enamel-like tissues (from ameloblasts or odontoblasts).
2. Collagen is a proteinaceous fibril disposed in dense bundles that form a compact matrix upon which the mineralizing constituents of cartilage or bone are deposited.
3. Bone is a tissue consisting of collagen, hydroxyapatite crystals, cementing substance, and osteocytes in lacunae that are usually connected by canaliculi.
4. Bone tissue is compact or spongy (cancellous). The lamellae of deep compact bone (osteon bone) are concentric around haversian canals. The lamellae of periosteal bone are not part of a haversian system.
5. Spongy bone is characterized by the presence of trabeculae and marrow cavities lined by an endosteum. It constitutes the core of most bones.
6. Ossification may be intramembranous, producing membrane bone, or endochondral, replacing preexisting cartilage and yielding replacement bone.
7. Endochondral ossification of the long bones of tetrapod limbs proceeds from a single diaphyseal center in the shaft and from epiphyseal centers in each epiphysis. The final site of ossification, if any, is in the epiphyseal plates.
8. Bones are surrounded by a membranous periosteum.
9. Dentin is a form of bone with peripheral odontoblasts, dentinal tubules, and typically no lacunae (present in placoderms). It is found in the skin and teeth.
10. Mature acellular bone lacks cells. It is found in modern fish scales and extinct heterostracans.
11. Dermal bones are membrane bones originating ontogenetically or phylogenetically from the dermis of the skin.
12. Cartilage consists of a collagenous matrix, a mucopolysaccharide in the interstices of the matrix, and chondrocytes in lacunae. It may be hyaline, fibrocartilage, or elastic cartilage and is sometimes calcified. Cartilage is surrounded by a membranous perichondrium.
13. Bone and cartilage are constantly being deposited and resorbed, thereby facilitating orderly growth, homeostasis, and tissue repair.
14. Tendons attach muscles to bone. Ligaments connect bone to bone. Both may be calcified or contain sesamoid bones.
15. Joints are diarthroses, amphiarthroses, synarthroses, or symphyses. Ankyloses are junctions

between bones in which the sutures have been obliterated.

16. Cartilage, bone, dentin, and enameloid were present in ostracoderms, although the relative antiquity of cartilage and ossified tissues is debated.

17. Heterotopic bones form in miscellaneous locations by endochondral or intramembranous ossification.

Critical Thinking Questions

1. In the development of the femur in mammals, how does this bone grow from its embryonic start to its adult size?

2. In relation to long bone growth, why is tackle football more dangerous for a prepubescent versus postpubescent child?

3. Large organisms are subject to major ontogenetic shifts in stress on their developing bones. How can a bone adapt to these changes?

4. What is a heterotopic bone? What function does it serve?

5. Can you develop a hypothesis to explain the relative time of origin for cartilage and bone? What evidence do you have to support your hypothesis?

6. What functions are associated with bone? Can any of these functions be logically associated with the initial origin of bone? If not, can you hypothesize a biological role?

Selected Readings

Burt, W. H.: Bacula of North American mammals, *University of Michigan Museum of Zoology Miscellaneous Publications* **113**:1–76, 1970.

Chinsamy, A., and Dodson, P.: Inside a dinosaur bone, *American Scientist* **83**:174, 1995.

Currey, J.: Comparative mechanical properties and histology of bone, *American Zoologist* **24**:5, 1984.

Francillon-Vieillot, H., de Buffrénil, V., Castanet, J., Géraudie, J., Meunier, F. J., Sire, J. Y., Zylberberg, L., and de Ricqlés, A.: Microstructure and mineralization of vertebrate skeletal tissues. In Carter, J. G., editor: Skeletal biomineralization: Patterns, processes and evolutionary trends, vol. 1, Chapter 20. New York, 1990, Van Nostrand Reinhold.

Hall, B. K.: The neural crest in development and evolution. New York, 1999, Springer-Verlag.

Kemp, N. E.: Organic matrices and mineral crystallites in vertebrate scales, teeth, and skeleton, *American Zoologist* **24**:965, 1984.

Moss, M. L.: Skeletal tissues in sharks, *American Zoologist* **17**:335, 1977.

Northcutt, R. G., and Gans, C.: The genesis of neural crest and epidermal placodes: A reinterpretation of vertebrate origins, *The Quarterly Review of Biology* **58**(1): 1, 1983.

Patterson, C.: Cartilage bones, dermal bones and membrane bones, or the exoskeleton versus the endoskeleton. In Andrews, S. M., et al., editors: Problems in vertebrate evolution. New York, 1977, Academic Press.

Links to the Internet

Visit the zoology website at http://www.mhhe.com/zoology to find live Internet links for each of the references listed below.

1. Histology—The Web Laboratory. Links to units on many systems of the body, including cartilage and bone. Extensive coverage of microstructure, from Ohio State.

2. Cartilage and Bone. Clickable index of histological sections of connective tissues, and accompanying informative text. The text includes an innovative clickable quiz for interactive learning.

3. Jay Doc Histo Web. The University of Kansas (the Blue Jays) Histology site. You can click on cartilage and bone to view photomicrographs and electron micrographs of histological sections. Expanded views show much detail.

4. Cartilage and Bone. A large number of photomicrographs with descriptive text from Loyola University Medical Education Network (LUMEN).

5. Endochondral Ossification. A large number of photomicrographs with descriptive text from Loyola University Medical Education Network (LUMEN).

Vertebrae, Ribs, and Sterna

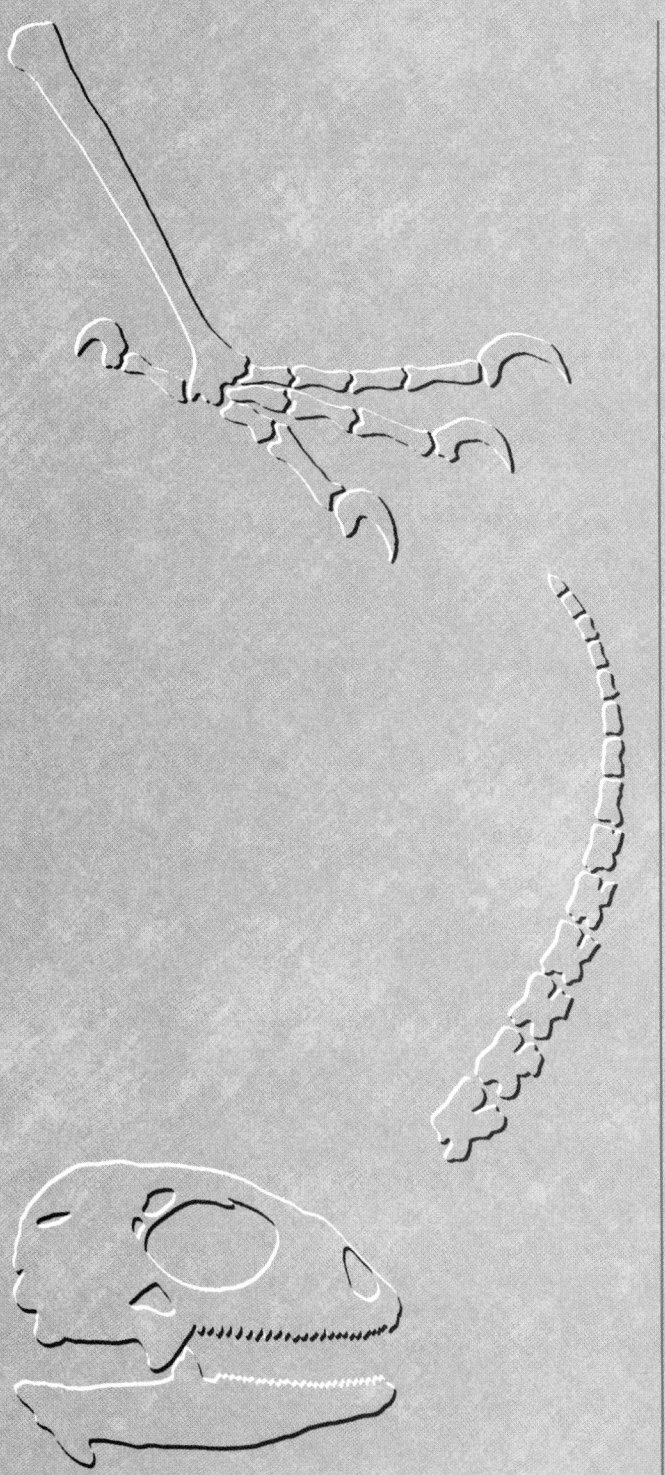

In this chapter, we will look at some typical and some not-so-typical vertebral columns of fishes and tetrapods, discuss their ontogeny and evolutionary history and note the advantages conferred by regional specialization of the column in tetrapods. Next we will study the ossified myosepta commonly known as ribs. Last, we will examine the diversity and roles of amphibian and amniote sterna and speculate concerning their phylogenetic origins.

he vertebral column, skull, ribs, and sterna, along with their ligaments, constitute the major components of the axial skeleton. The vertebral column forms around, and sometimes invades, the embryonic notochord during ontogenesis. Ribs and the tetrapod sternum form in the lateral and ventral body walls, respectively.

VERTEBRAL COLUMN

The vertebral column is the keystone of the vertebrate skeleton. It is a segmented, more or less flexible, arched rod flanked by axial musculature, to which the head is attached, from which the rest of the body is suspended in fishes, and from which the trunk is suspended between anterior and posterior limbs in most tetrapods. It provides a protective bony tunnel for the spinal cord, and it is an essential architectural component of the axial locomotor apparatus in vertebrates other than agnathans (the unusual condition in hagfish and lampreys is discussed later). In no vertebrate is its locomotor role more readily demonstrable than in fishes and limbless tetrapods.

Fishes are surrounded by an environment that offers considerable resistance to forward locomotion and a buoying effect on the body. During locomotion, resistance is almost universally overcome in fishes by pushing laterally against the water with rhythmical, undulating, side-to-side movements of the trunk and tail (see fig. 11.8). These movements are brought about by segmental muscles attached to the vertebrae, or to the ribs, and myosepta. The articulations between vertebrae in most fishes permit only side-to-side flexibility of the column. When vertebrates ventured onto land, they brought with them the fishlike method of locomotion, but it was an awkward way to move about. Unlike water, land is beneath the animal rather than surrounding it, it is not level, and it is littered with obstacles that must be avoided or clambered over. Eventually, selective forces altered the articulations in the column in such fashion as to provide dorsoventral flexibility, which is more suited to locomotion on land, while also providing a strong skeletal bowlike arch for suspending the trunk above the ground between forelimbs and hind limbs. Also associated with a shift to dorsoventral flexion is the functional separation of the muscles used in locomotion from those used in ventilation of the tetrapod lung. These changes were achieved at the expense of some side-to-side flexibility, and they were correlated with regional specialization of the vertebral column. These and other adaptations are discussed in the sections that follow.

Centra, Arches, and Processes

Modern vertebrae typically consist of a **centrum, neural arch,** and one or more processes, or **apophyses,** that project from the arch or centrum (fig. 8.1). Centra occupy the position occupied early in ontogeny by the notochord. Neural arches are perched on the centra, and the successive arches and their interconnecting ligaments enclose a long **vertebral canal (neural canal)** occupied by the spinal cord. **Hemal arches,** known also as chevron bones in amniotes, are inverted beneath the centra of the tail and house the caudal artery and vein (fig. 8.1*b*, *c*, and *e* chevron bone).

Transverse processes (diapophyses) are the most common. They articulate with ribs that project laterad into the horizontal skeletogenous septum separating the epaxial and hypaxial muscles of the body wall (see figs. 1.2 and 8.5). Transverse processes serve as attachments for some of the muscles that extend or flex the vertebral column. **Zygapophyses** are paired processes at the cephalic end of trunk vertebrae (**prezygapophyses**) and at their caudal end (**postzygapophyses**), chiefly in tetrapods (fig. 8.1*d–f*). The articular facets of prezygapophyses face dorsad and articulate with the ventrad-facing facets of the postzygapophyses just ahead of them. This interlocking arrangement limits dorsoventral flexion of the column in the region of the trunk where they are present. Tetrapod tails are highly flexible because zygopophyses are absent. **Parapophyses** are lateral projections from the centra of a few tetrapods. When present, they serve as the articulation site for the capitulum of a bicipital rib, as illustrated in figure 8.23*b*. (This is not the usual site, however, as will be seen when we study ribs.) **Hypapophyses** are prominent midventral projections from the centra of snakes and a few other amniotes (fig. 8.1*d*). They are sites of attachment for certain muscles or tendons. Still other processes develop in occasional species.

Morphogenesis of Vertebrae

A typical vertebra arises from mesenchyme cells that stream out of the sclerotomes of mesodermal somites, surround the notochord and neural tube, and produce the blastema for a future vertebra (see fig. 5.11). Chondroblasts that differentiate within the blastema subsequently deposit a cartilaginous centrum and neural arch, and, in the tail, a hemal arch. The result is a cartilaginous vertebra with the notochord constricted within each centrum. Later, except in cartilaginous fishes, the cartilage is generally removed, and bone is deposited. Remnants of the notochord may remain in the centrum to varying degrees, depending on the species. In Chondrichthyes, the cartilaginous vertebrae are never replaced by bone. In teleosts, urodeles, and apodans, membrane bone rather than cartilage is deposited in the perichordal blastema (see fig. 8.3), but the arches arise from cartilage.

In many fishes, and in amphibians except anurans, not only is cartilage or membrane bone deposited *around* the notochord (**perichordal cartilage** or **bone**), but chondroblasts invade the notochord sheath and

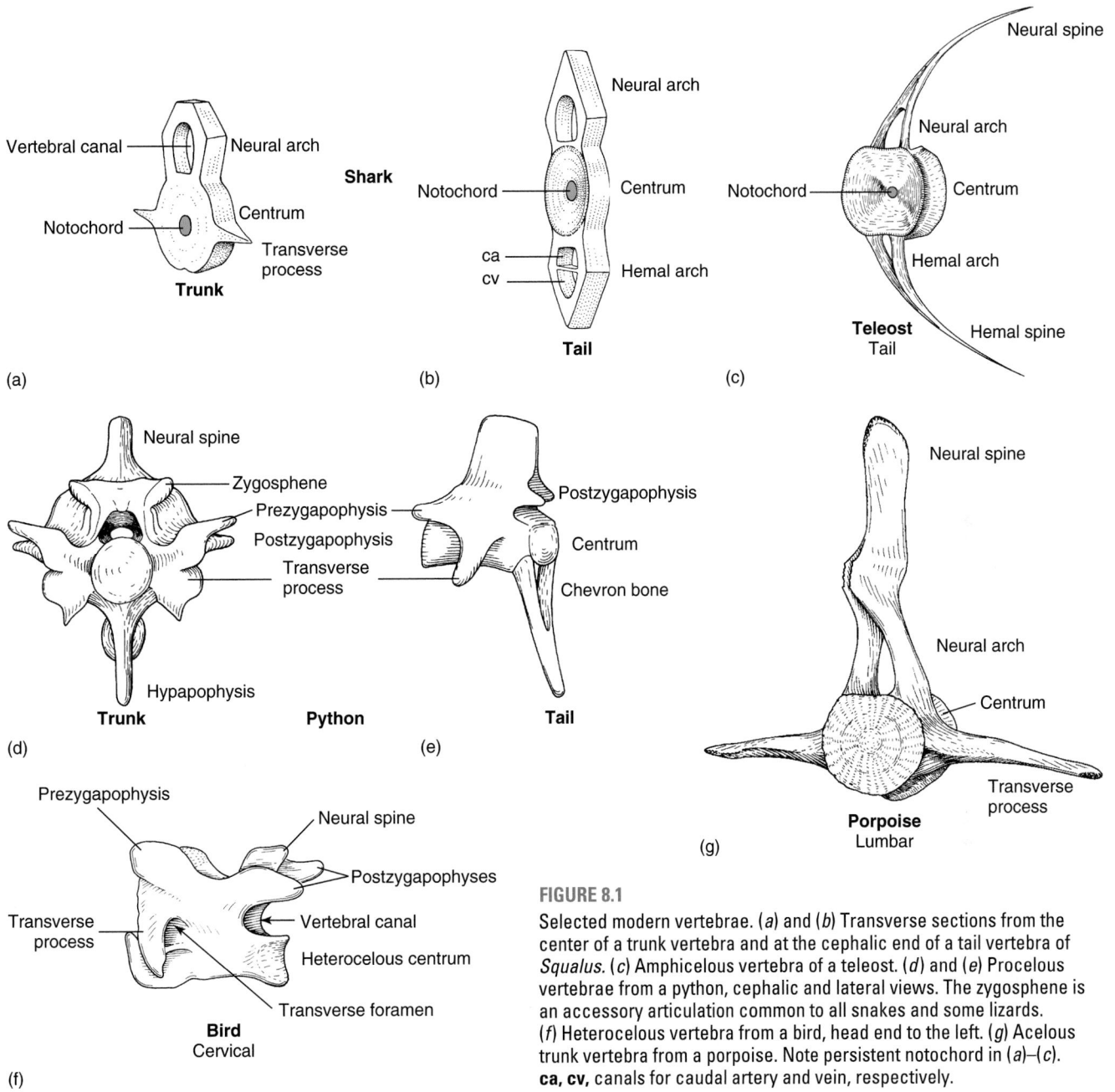

FIGURE 8.1

Selected modern vertebrae. (*a*) and (*b*) Transverse sections from the center of a trunk vertebra and at the cephalic end of a tail vertebra of *Squalus*. (*c*) Amphicelous vertebra of a teleost. (*d*) and (*e*) Procelous vertebrae from a python, cephalic and lateral views. The zygosphene is an accessory articulation common to all snakes and some lizards. (*f*) Heterocelous vertebra from a bird, head end to the left. (*g*) Acelous trunk vertebra from a porpoise. Note persistent notochord in (*a*)–(*c*). **ca, cv,** canals for caudal artery and vein, respectively.

deposit cartilage in the sheath and sometimes in the notochord itself. Consequently, the centra of these vertebrates contain **chordal cartilage** that may calcify or ossify (figs. 8.2, 8.3, and 8.8). The result of these variations in the process of centrum formation can be illustrated by contrasting centra of urodeles with those of anurans. The centra of urodeles (and also of apodans) consist of perichordal membrane bone, chordal cartilage, and, within the core, unaltered notochordal tissue (fig. 8.3). The centra of anurans, on the other hand, consist entirely of replacement bone deposited in a perichordal blastema. The notochord disappears. The difference has been interpreted by some paleontologists as supporting evidence for a diphyletic origin of amphibians.

In forming centra, scleroblasts from the caudal half of one somite and the cephalic half of the next somite stream to an intersegmental location around the notochord to establish the perichordal blastema. Consequently, centra are typically intersegmental, and myomeres are segmental (fig. 8.4). However, chiefly in the tail of some fishes and in the tail of a few primitive tetrapods, there are two centra per segment, a condition known as **diplospondyly.** The condition may provide the tail with increased flexibility for locomotion. Generally, in fishes, flexibility of the vertebral column is facilitated by elasticity of the intervertebral ligaments rather than by intervertebral joints, as in tetrapods.

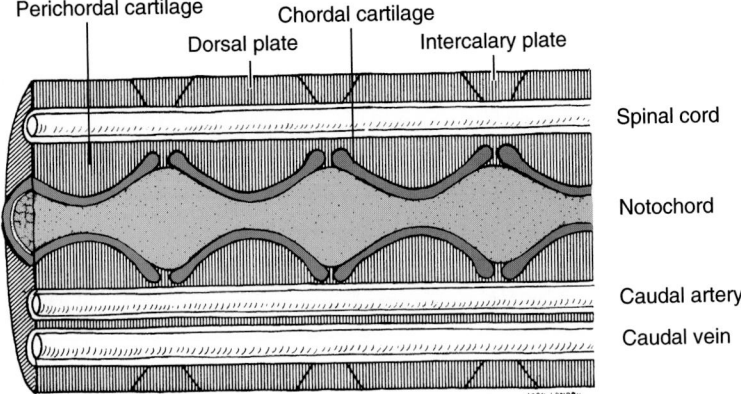

FIGURE 8.2
Caudal vertebral column of *Squalus,* sagittal section. Calcified chordal cartilage has been deposited in the notochord sheath (*dark red*). Intercalary plates are seen from lateral view in figure 8.5.

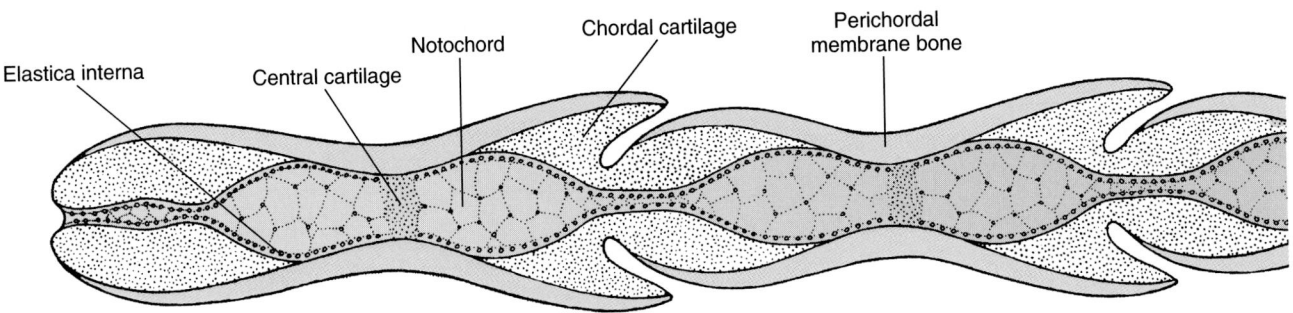

FIGURE 8.3
Two trunk vertebrae from the plethodontid salamander *Gyrinophilus,* frontal section, head to the left. The chordal cartilage is within the notochord sheath. The elastica interna is the innermost derivative of the sheath.
Source: Data from J. S. Kingsley, *Outline of Comparative Anatomy of Vertebrates,* 1920, The Blakiston Company, Philadelphia, after Widersheim.

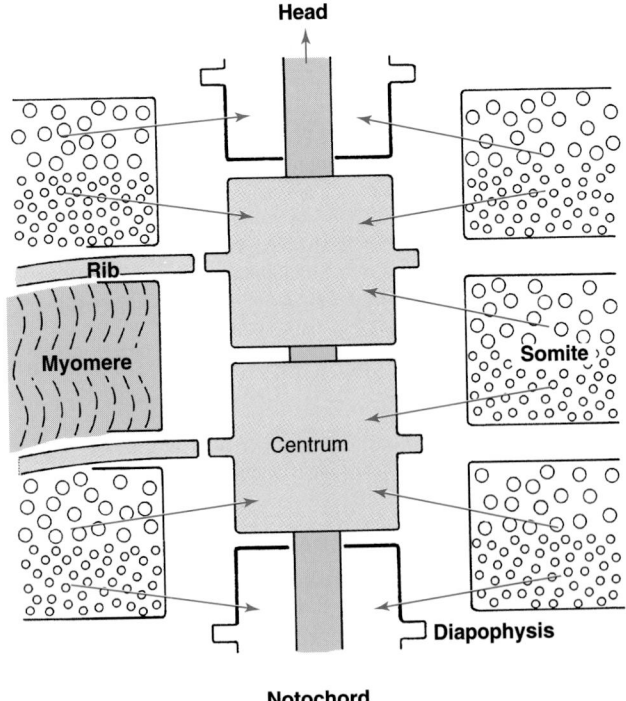

FIGURE 8.4
Intersegmental (intersomitic) position of centra and myosepta or ribs. Each centrum is a contribution from two successive somites (*arrows*).

The Vertebral Columns of Fishes

Fish vertebrae are highly diverse, as might be expected in view of the enormous number of species that evolved—greater than in any terrestrial group of craniates—and the opportunity for subsequent genetic mutations during the lengthy Paleozoic Era. We will look at just a few interesting morphological variants. They are attributable to varying developmental factors, among which are the extent to which scleroblasts invade and thicken the notochord sheath and the notochord proper, the extent to which the notochord persists within the adult column, the number of chondrification and ossification centers that develop in the perichordal blastemas, and the extent to which separate centers remain independent or coalesce during ontogeny. As for a single species, however, the vertebral column exhibits only two major morphologic regions of specialization, trunk (**dorsals**) and tail (**caudals**). The vertebrae in each region generally show gradual modifications between the head and vent, where there may be a fairly abrupt change, and between the vent and the end of the tail.

Living agnathans have a strange vertebral column if, indeed, they may be said to have one. The only skeletal elements are **lateral neural cartilages** (see fig. 1.5), one or two pairs per body segment, depending on the species. Caudally, the lateral neural cartilages fuse to

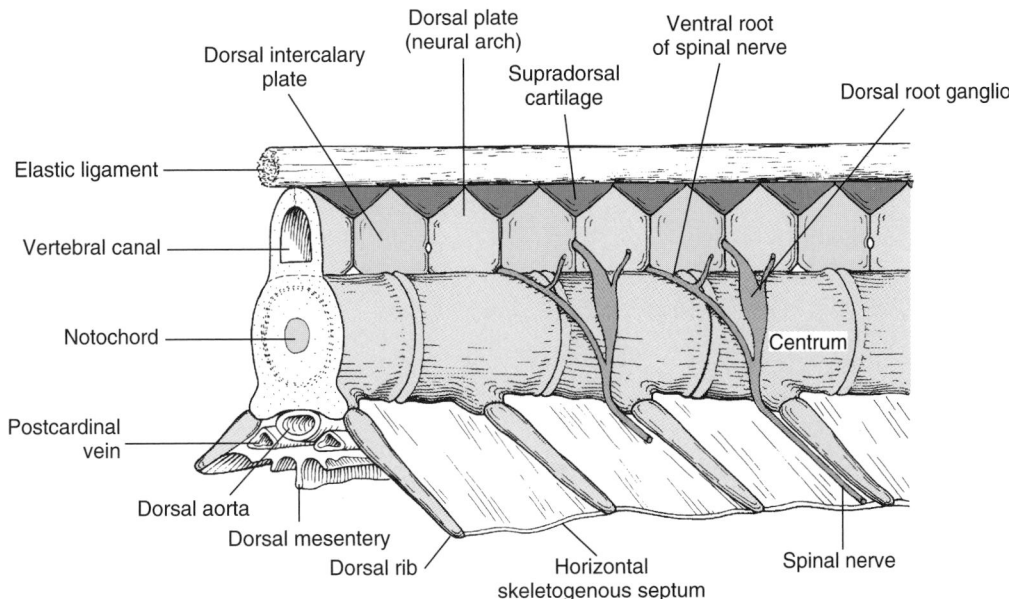

FIGURE 8.5
Vertebral column and associated structures of trunk of *Scyllium,* a shark. Each rib articulates with a transverse process of a centrum.

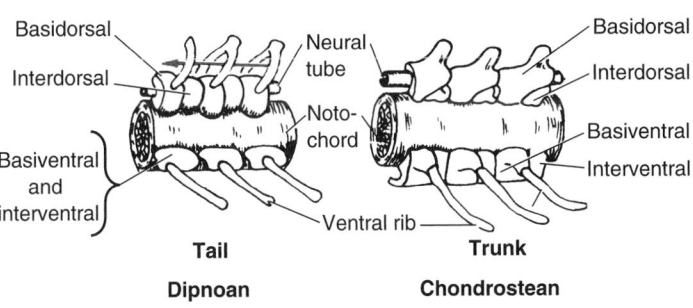

FIGURE 8.6
Vertebral components in adult dipnoan (*Neoceratodus*) and chondrostean (sturgeon). *Arrow* in the dipnoan indicates canal occupied by a supportive longitudinal ligament.

form a single longitudinal dorsolateral cartilaginous plate perforated by foramina for spinal nerves. In hagfishes, lateral neural cartilages are limited to the tail, although some researchers consider these elements to represent fused fin supports and not vertebral elements. Lateral neural cartilages may be vestigial vertebrae or primitive vertebrae, or they may bear no phylogenetic relationship to vertebrae. Whether you call them vertebrae depends on how you wish to define *vertebra.* In the classification used here, we define *vertebrae* as vertebral structures of the trunk and tail, thus uniting lampreys and gnathostomes in the group Vertebrata. How would *you* define the term?

In sharks, the notochord is present throughout the length of the adult vertebral column and is constricted within each centrum (see fig. 8.2). The centra, composed of chordal and perichordal cartilage, are concave at each end. The centrum that is concave at each end is said to be **amphicelous** (see fig. 8.1*a,b*). The vertebral

canal (fig. 8.5) consists of paired **dorsal plates** that constitute a neural arch, paired **dorsal intercalary plates** between the arches, and, in some species, paired wedge-shaped **supradorsal cartilages** (fig. 8.5). *Squalus acanthias* lacks supradorsals. The neural arches and intercalary plates of *Squalus* are perforated by foramina for the dorsal and ventral roots of spinal nerves and for blood vessels. In *Scyllium,* they emerge between plates. Hemal arches consist of paired **ventral plates. Ventral intercalary plates** are common between hernal arches. A fibroelastic ligament, present in all fishes, overlies and connects the neural spines along the entire length of the vertebral column.

A few fishes have exceptional columns. Dipnoans, Chondrostei, and the coelacanth *Latimeria* develop no centra. The notochord is present and unconstricted, and its thick fibrous sheath contains little cartilage or bone (fig. 8.6). Associated with the notochord in each body segment are paired **basidorsal, basiventral, interdorsal,** and **interventral cartilages** that are products of separate chondrification centers in the embryonic perichordal blastema. An extinct holocephalan's "vertebral column" consisted of thick rings of calcified cartilage that had been deposited in the notochord sheath (fig. 8.7). The rings were more numerous than the body segments.

Teleosts have well-ossified amphicelous vertebrae (see fig. 8.1*c*). The core of each centrum is a dumbbell-shaped vacuole where the notochord previously existed, and the space between successive vertebrae is occupied by a porous cartilagelike material that probably includes notochordal remnants. The centra and neural arches of successive vertebrae are interconnected by a complex system of collagenous and elastic ligaments that facilitate lateral undulation of the body for locomotion. In the posterior trunk and tail, these ligaments are frequently ossified to form long, delicate, bony rods. The neural spines are often very tall, and they are sometimes surmounted by **supraneural bones** (see fig. 8.24, *Perca*). A variety of processes, mostly unique to fishes, protrude from the arches and centra.

Elasmobranchs and numerous bony fishes have two centra and two sets of neural and hemal arches in each metamere of the tail. A few have duplication of these in the trunk also. Consequently, the number of centra in those regions is double the number of myomeres and spinal nerves, a diplospondylous condition. *Amia* has what appear to be two centra per body segment in the postanal portion of the vertebral column (fig. 8.8), but only one bears arches. The other is termed an intercentrum, although it may not be homologous with hypocentra (intercentra) of other fishes or tetrapods. The embryonic basidorsal and basiventral cartilages are incorporated in the centra, and the interdorsal and interventrals are incorporated in the intercentra.

No movement is possible in fishes between the first vertebra and the skull. They are united by cartilage or nonelastic connective tissue. The caudal-most vertebrae are part of the skeleton of the caudal fin, as described in chapter 10.

Evolution of Tetrapod Vertebrae

Unlike the vertebral column of modern tetrapods, which in the trunk consists of a centrum and a neural arch in each body segment, the columns of some rhipidistian fishes and early labyrinthodonts consisted of several bones per segment. Generally, these were a **hypocentrum** (intercentrum of some authors) a large, median, U-shaped anterior bone that cradled the notochord (fig. 8.9); a pair of **pleurocentra**, small wedges of bone overlying the notochord dorsolaterally; and independent left and right laminae of bone lateral to the spinal cord that, collectively, provided a neural arch. Each lamina rested in a notch between the pleurocentrum and the hypocentrum. The notochord was continuous throughout the length of the body but constricted at the level of each hypocentrum. A "vertebra" of this nature is said to be **rachitomous** (consists of several pieces).

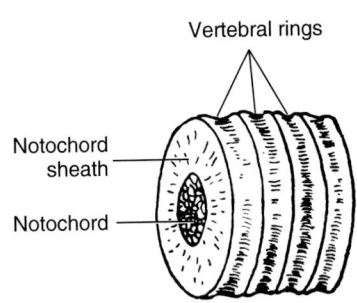

FIGURE 8.7

Vertebral rings from an extinct holocephalan. Neural arches have been removed.

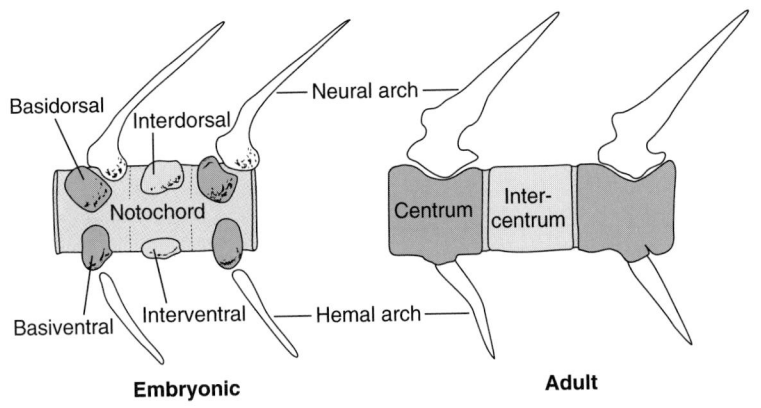

FIGURE 8.8

Tail vertebrae of *Amia*. A centrum and intercentrum develop in each body segment. The intercentrum is apparently not homologous to similarly named structures (or hypocentra) of other fishes and tetrapods.

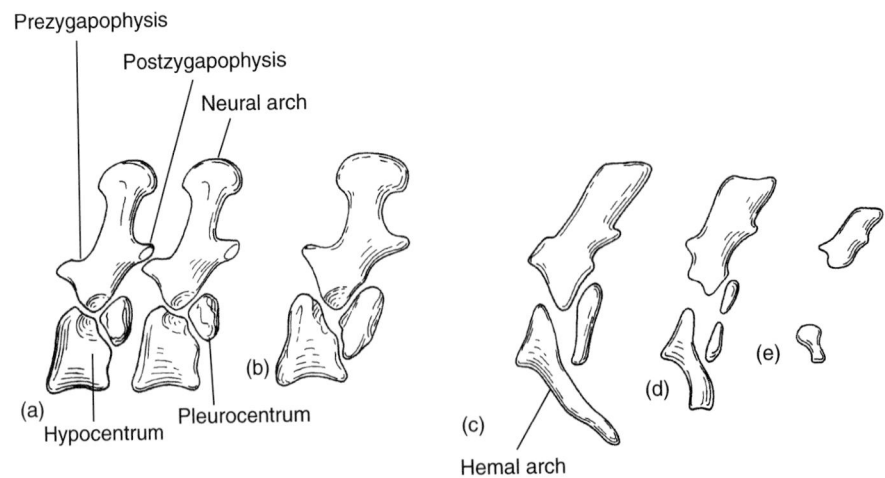

FIGURE 8.9

Rachitomous vertebrae from a primitive labyrinthodont, left lateral view. (*a*) and (*b*) Trunk. (*c*)–(*e*) Caudal. Hypocentra (intercentrum of some authors).

Source: Data from E.S. Goodrich, *Studies on the Structure and Development of Vertebrates*, 1930, Macmillan and Company, Ltd., London.

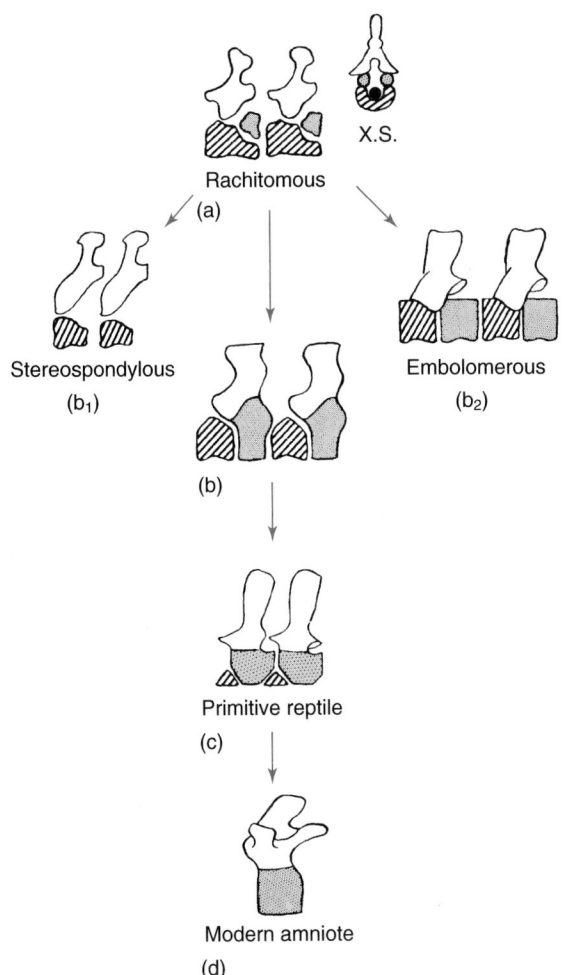

FIGURE 8.10

Postulated modifications leading from labyrinthodont to modern amniote vertebrae. (*a*), (*b*), (*b₁*), and (*b₂*) Four varieties of labyrinthodont vertebrae. **X.S.,** Cross section of (*a*), notochord in *black.* The rachitomous type was present in rhipidistian fishes and in the earliest tetrapods. (*b*) is from an extinct amphibian thought to be in the reptile line. *Diagonal lines,* hypocentrum (intercentrum); *red,* pleurocentrum.

Later tetrapod vertebrae with the exception of those of urodeles and apodans appear to be modifications of the rachitomous condition. The change appears to have been characterized by increased prominence of pleurocentra and concomitant reduction of hypocentra (fig. 8.10).

Today, as in the distant past, each centrum begins ossification at several paired and median loci in the perichordal blastema. As separate ossification centers expand, they coalesce, unlike in Devonian times. The findings of paleontology, comparative anatomy, and developmental biology all point to the same conclusion: A modern adult tetrapod vertebra is a composite structure.

The vertebrae of urodeles and apodans lack evidence of pleurocentra or hypocentra, and for this and numerous other reasons, the phylogenetic history of these am-

phibians is debated. Whether or not anuran centra are pleurocentra or hypocentra has not been ascertained.

The vertebrae of generalized urodeles, apodans, *Sphenodon,* and basal lizards such as geckos are amphicelous, as are the vertebrae of most fishes. The centra are concave at each end, and vestiges of the notochord persist in the intervertebral joints. In most other tetrapods, the concavity has disappeared anteriorly, posteriorly, or at both ends of the centra, resulting in **opisthocelous, procelous,** or **acelous** vertebrae, respectively (fig. 8.11).

The vertebrae of anurans and modern nonavian reptiles are procelous. This condition is thought to have evolved from a buildup of a hypocentrum of chordal cartilage between two centra and the eventual coalescence of the intercentrum with the centrum ahead of it. This would have resulted in procelous vertebrae. Coalescence of the hypocentrum with the centrum immediately behind it would have resulted in the opisthocelous centra seen in salamanders (see fig. 8.3). The vertebrae of mammals are acelous, having a concavity at neither end. Instead, between successive vertebrae is a fibrocartilaginous **intervertebral disc,** and in its core is a remnant of the embryonic notochord known as a **pulpy nucleus.** Humans are aware of the disc only when it becomes displaced and presses on a spinal nerve. These modified intervertebral joints of tetrapods permit more flexibility to the column than do the amphicelous vertebrae of fishes. An intervertebral joint permitting still more flexibility will be described when we study the highly specialized heterocelous neck vertebrae of birds.

Regional Specialization in Tetrapod Columns

With the advent of life on land, the vertebral column underwent regional specialization. Tetrapod limbs push against the earth. Initially, they did so purely for locomotion; later they were also supporting the body above the ground. The opposing physical force exerted by the earth in response to the push of a tetrapod posterior appendage is transmitted via the pelvic girdle to one or more of the hindmost trunk vertebrae (fig. 8.12). These vertebrae became appropriately modified and are now called **sacral** (figs. 8.13, 8.14, 8.17, 8.21, and 10.10). The pelvic fins of fishes have no such relationship to the vertebral column.

Survival on land with its irregular contours and scattered obstructions to vision, such as rocks or vegetation, was facilitated in tetrapods by increased mobility of the head with its special sense organs, including eyes that can scan the horizon. This was achieved to a small degree in amphibians by development of a somewhat more mobile joint between the first vertebra and the skull. The innovation was improved in many amniotes by shortening or eliminating ribs on some of the more anterior vertebrae and by increasing the mobility of the intervertebral joints in that region, which became the neck. These are **cervical vertebrae** (fig. 8.14). In amniotes other than turtles and snakes, long ribs have become restricted to vertebrae in the anterior portion of the trunk immediately behind the neck, thereby preserving a bony cage that houses the thoracic viscera and participates in external respiration. These are **thoracic vertebrae.** The remaining vertebrae anterior to the sacral are **lumbar** (fig. 8.14). Vertebral columns that have become modified by life in water are shown in figure 10.31.

Among tetrapods, snakes have the longest columns with as many as 400 or more vertebrae and almost no regional specialization. Such a column provides snakes with an elongated, multijointed body that can be coiled or drawn into sinuous curves for locomotion without limbs. Apodans, also limbless, have as many as 250 or more vertebrae, and some urodeles have 100. The awkward leaping anurans have the shortest columns and limited flexibility. In turtles and birds, only the cervical and caudal segments of the column are flexible; most of the trunk vertebrae are rigidly fused to the synsacrum in birds (see figs. 8.18, 8.19b, and 8.20) and to the cara-

pace in turtles (see fig. 6.36). The role of the vertebral column and its associated musculature in locomotion is discussed in chapters 10 and 11.

The Craniovertebral Junction and Neck Vertebrae

The first and only cervical vertebra of modern amphibians lacks processes, and its cranial end bears two smooth concave facets that articulate with two knoblike occipital condyles. The condyles of a urodele are illustrated in figure 9.12b. The absence of processes and the nature of the amphibian craniovertebral joint permits limited dorsoventral rocking of the skull, a function that is not possible in fishes. The two-facet articulation is a change from the earliest amphibians, who had only one occipital condyle.

Amniotes have a larger number of cervical vertebrae than amphibians, which provides them with a long, flexible neck. The first two vertebrae are modified in such a fashion as to permit considerably more independent movement of the head. The first vertebra, or **atlas**, is ringlike because its centrum is missing (see fig. 8.27). On its cephalic end are one or two deep con-

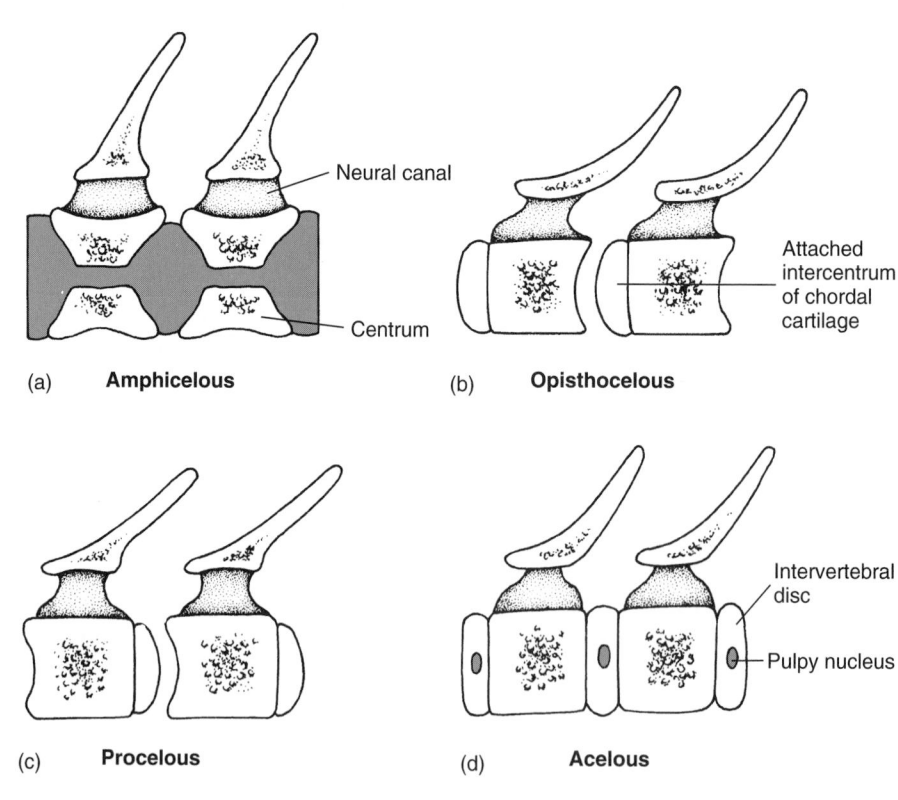

FIGURE 8.11
Vertebral types based on articular surfaces of centra. Midsagittal section, head to the left. The complex heterocelous vertebra of birds is seen in figure 8.1. The *red* areas are occupied by the notochord in (*a*) and by notochordal vestiges in (*d*).

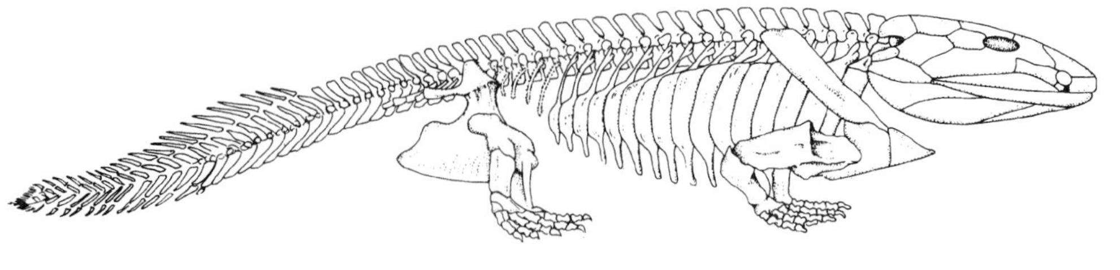

(a) **Amphicelous**
(b) **Opisthocelous**
(c) **Procelous**
(d) **Acelous**

Neural canal
Centrum
Attached intercentrum of chordal cartilage
Intervertebral disc
Pulpy nucleus

FIGURE 8.12
Ichthyostega, an early labyrinthodont 3 feet long, reconstructed from recovered fossil skeletal parts.
Source: Data from Jarvik, *Scientific Monthly,* Vol. 80, p. 152, March 1955.

cavities (**superior articular facets**) for articulation with the single occipital condyle of reptiles or with the two of mammals. These are condyloid joints in which the skull rocks as in nodding "yes." Except in snakes, the centrum of the atlas has become the **odontoid process** of the second vertebra, or **axis**. The process projects forward to rest on the floor of the atlas where in mammals, it is held in place by a **transverse atlantal ligament** (fig. 8.15*a*). The skull and atlas rock as a unit on the odontoid process, which is the axis of rotation. Rocking is facilitated by the reduction or absence of zygapophyses on the atlas. The odontoid process of *Sphenodon* chon-

drifies as an independent element and unites with the axis only after hatching.

A bony **proatlas**, resembling a neural arch and derived from a pair of bilateral cartilaginous blastemas, is interposed between the skull and atlas in crocodilians, *Sphenodon,* and the insectivorous spiny hedgehogs (figs. 8.15*b* and 8.16). This element appeared first in basal amniotes and was present in the synapsid ancestors of mammals. Some modern lizards have a fibrous membrane instead of a bone at that location.

A flexible neck is a valuable asset in gathering food and avoiding enemies on land. It not only extends the horizon but also increases the number of degrees in its arc. A human, for example, moving nothing but the head, can sweep the horizon over an arc of more than 180 degrees.

Flexibility of the neck is exceptional in birds because of the nature of the articulations of their vertebrae. The caudal ends of the centra are saddle-shaped, having a convexity in the right-left axis and a concavity in the dorsoventral axis (see fig. 8.1*f*). The cephalic end of the next more posterior centrum is shaped to accommodate this configuration. Such a vertebra is said to be **heterocelous**. These vertebrae enable both lateral and dorsoventral flexion of the neck. Heterocelous centra, combined with the atlas-axis complex and the largest number of cervical vertebrae in any tetrapod—12 commonly, 25 in swans—enable some birds to turn their heads 180 degrees to the rear, enlarging the arc of their horizon to that of a complete circle.

Turtles also have a uniquely flexible neck. Ball-and-socket joints between successive procelous centra enable the entire head and neck to be completely retracted into the shell by flexing the neck into several folds dorsoventrally or, in "side-necked" turtles, sidewise, during retraction. However, like other living nonavian reptiles, turtles have only eight cervical vertebrae. In contrast to the high flexibility of the neck, the remainder of the column of birds and turtles, except a short section of caudal vertebrae, is totally rigid.

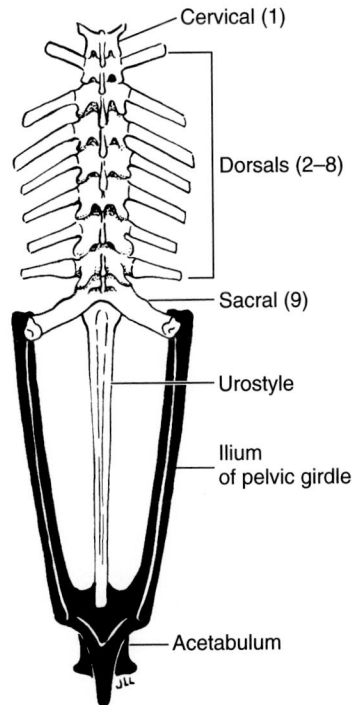

FIGURE 8.13

Vertebral column and pelvic girdle of an anuran. The long ilium of the girdle is braced against the sacral vertebra.

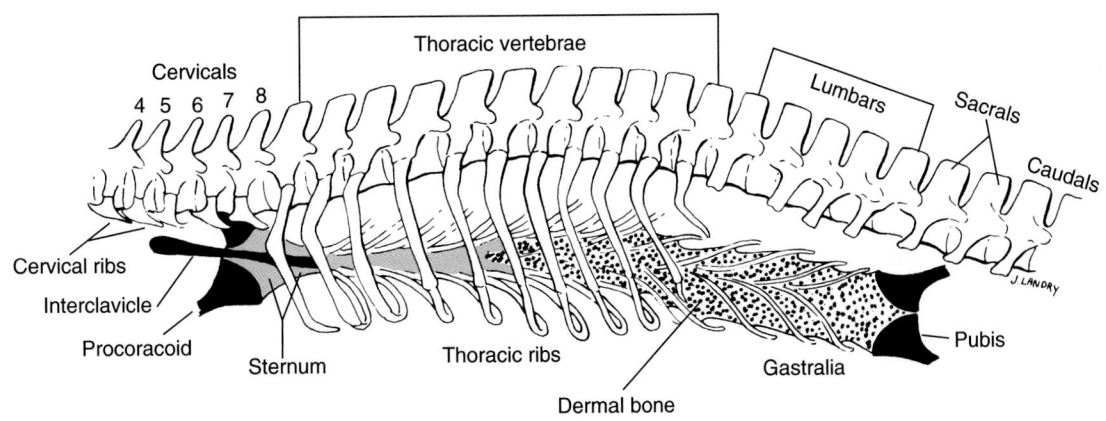

FIGURE 8.14

Vertebrae and ribs of an alligator.

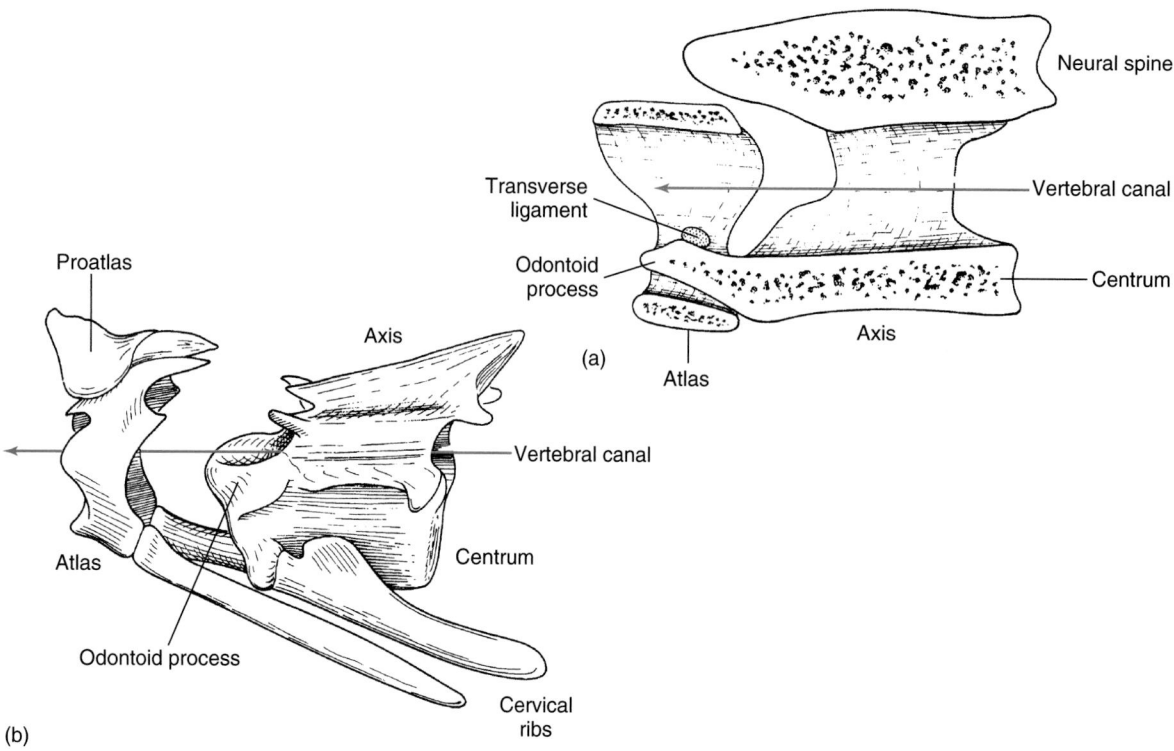

FIGURE 8.15

Atlas, axis, and odontoid process of a cat, (*a*) and a crocodilian, (*b*) head to the left. (*a*) is a sagittal section. The transverse atlantal ligament has been severed where it overlies the odontoid process. In (*b*), the vertebrae have been slightly separated for clarity. Note that the second cervical rib is bicipital.

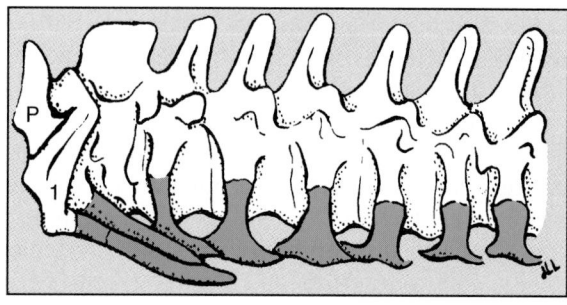

FIGURE 8.16

The eight cervical vertebrae, eight cervical ribs (*red*), and proatlas, **P,** of an alligator, left lateral view. **1,** atlas and attached first rib. Immediately behind the first rib is the rib of the axis. The ribs are fused to transverse processes.

Mammals almost always have seven cervical vertebrae. This is as true of the stubby, rigid neck of whales as it is of the neck of the tallest giraffe. The sole exceptions are three edentates (two sloths and the Great Anteater) with six, eight, or nine, and manatees with six. The cervical vertebrae of marsupial moles are shortened and more or less fused together (an adaptation to life in a burrow?). The same is true of armadillos and most cetaceans. It is the length of the centra, not their number, that determines neck length in mammals.

Stabilizing the Hind Limbs: Sacrum and Synsacrum

Sacral vertebrae bear stout transverse processes that are strong enough to bear the thrust of the pelvic girdle as the hind limbs push against the earth in locomotion (see fig. 8.13). Ankylosed to the distal ends of the transverse processes may be equally stout ribs (see fig. 10.10). Amphibians have one sacral vertebra, living reptiles including most birds and opossums have two, and most mammals have three to five. However, there are perissodactyls with up to eight and edentates with 13. The sacral vertebrae of mammals usually ankylose to form a single bony complex, the **sacrum** (fig. 8.17). Sacral vertebrae do not differentiate in tetrapods that lack hind limbs.

In modern birds, the last thoracic vertebra, all lumbars, the sacrals, the first few caudals, and their ribs ankylose to form one adult complex, the **synsacrum.** The latter fuses with the pelvic girdle. Thus, a compact pelvis (figs. 8.18, 8.19, and 8.20) is formed that provides a rigid brace for the teeter-totterlike stance of birds. The thoracic vertebrae anterior to the synsacrum also unite more or less completely, so there is little flexibility in the avian backbone caudal to the neck. This axial rigidity minimizes the number of muscles needed to keep the body streamlined during flight, thereby reducing the cost in terms of energy expended. However, the historical selective forces responsible for it may have

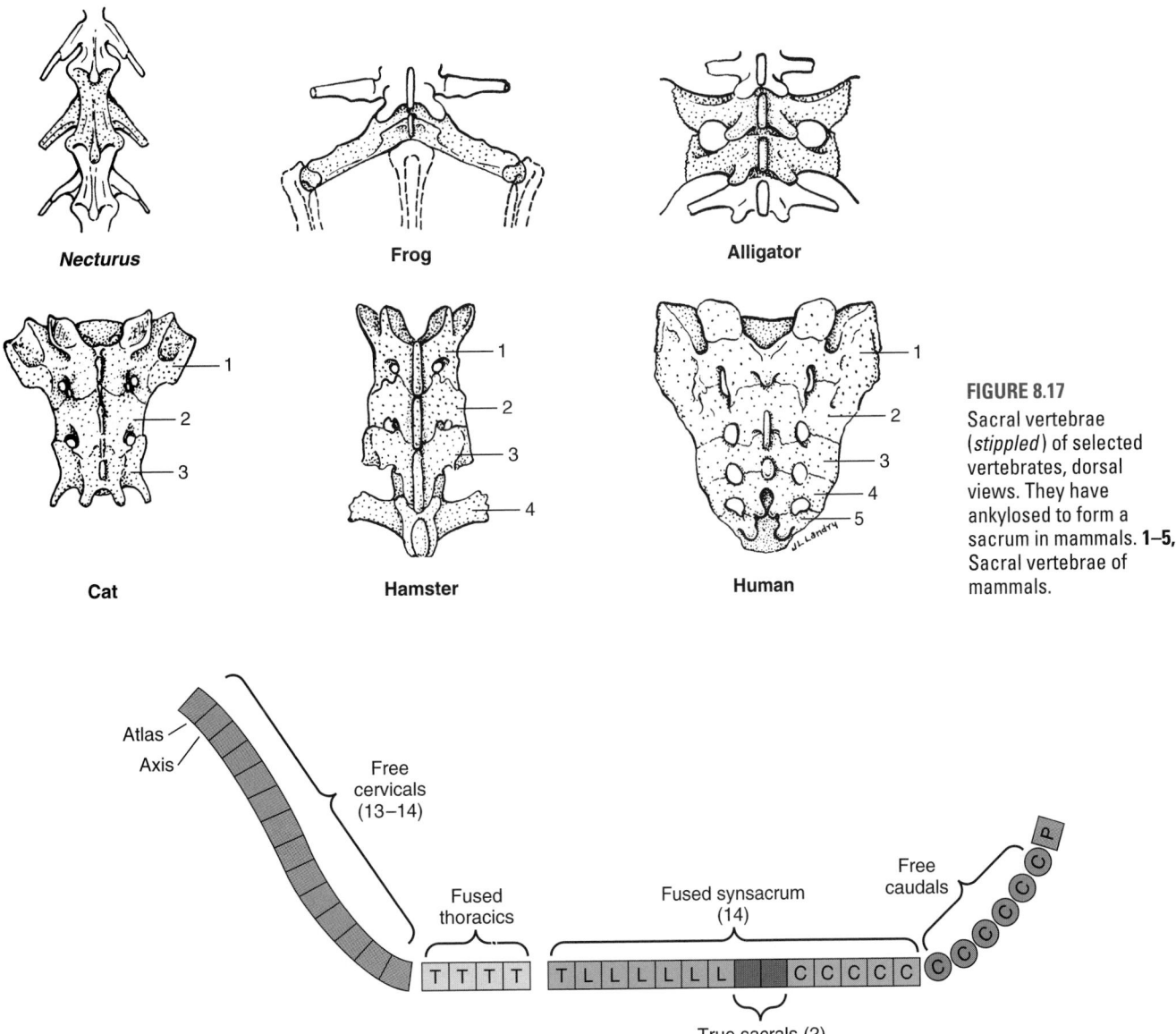

Necturus

Frog

Alligator

Cat

Hamster

Human

FIGURE 8.17
Sacral vertebrae (*stippled*) of selected vertebrates, dorsal views. They have ankylosed to form a sacrum in mammals. **1–5,** Sacral vertebrae of mammals.

FIGURE 8.18
Diagram of vertebral column of pigeon. **T,** Thoracic; **L,** lumbar; **C,** caudal; **P,** pygostyle, composed of four fused vertebrae.

been quite different from what appear today to be its advantages. *Archaeopteryx* provides us a view of a rudimentary synsacrum (fig. 8.21). The trunk vertebrae were not fused, and the pelvic girdle (ilium, ischium, and pubis) was proportionately smaller than in today's birds. The weight of the body when standing was counterbalanced to a considerable degree by the long tail shown in figure 4.28.

Among mammals, armadillos have a synsacrum. It consists of up to 13 fused sacral and caudal vertebrae.

Tail Vertebrae: Urostyle, Pygostyle, and Coccyx

Caudal vertebrae in early tetrapods probably numbered 50 or more. In modern ones, the number is much reduced and highly variable, dictating how effective the tail may be for one use or another. Toward the end of

the tail in all tetrapods, the arches and processes become progressively shorter and rudimentary until finally the vertebrae consist solely of small cylindrical centra (fig. 8.22).

Anurans have a unique **urostyle** at the end of the vertebral column (see fig. 8.13). It develops from a continuous elongated perichordal cartilage at the base of the larval tail, and it grows and ossifies after the tail is lost at metamorphosis. That it is composed of postsacral vertebrae is evident from the one or more coalesced centra, vestigial arches, transverse processes, and nerve foramina that are part of the cranial end of the urostyle in some species.

Among reptiles, the caudal vertebrae of lizards deserve mention. Many lizards, when captured by the tail, break off to the end distal to the site of capture and

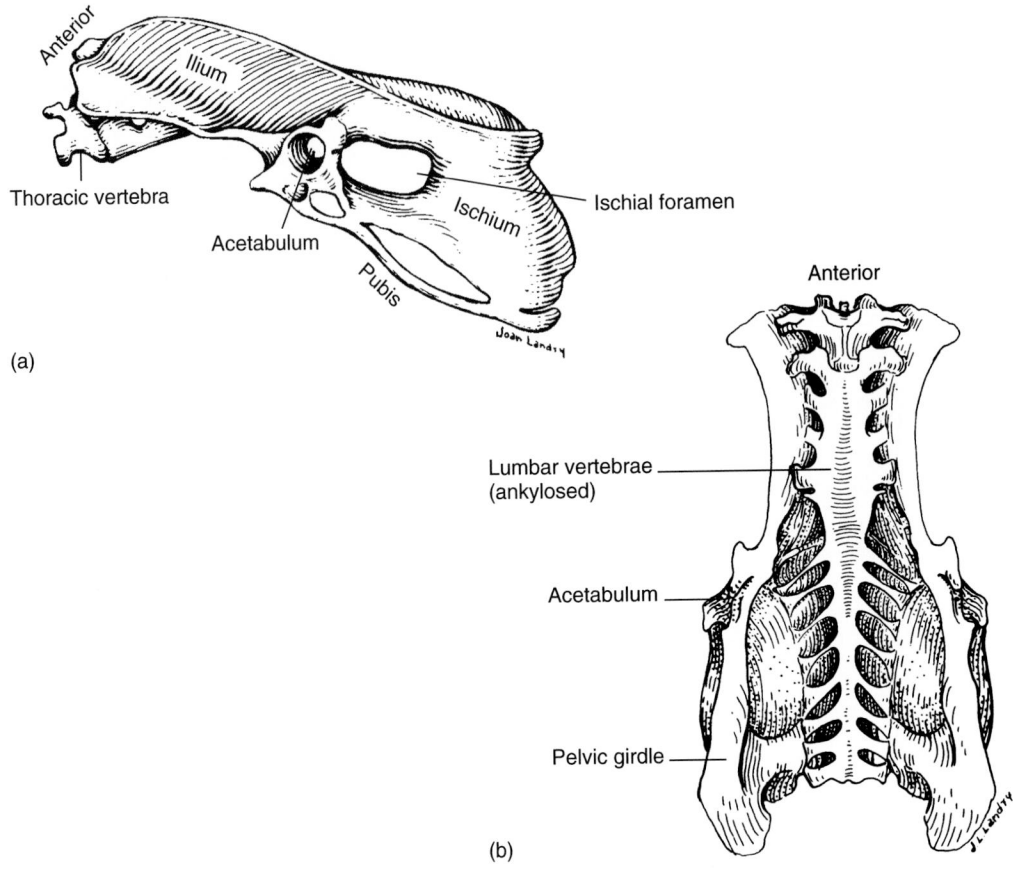

(a)

(b)

FIGURE 8.19
Synsacrum and pelvic girdle of the guinea hen. (*a*) Left lateral view. (*b*) Ventral view. The acetabulum accommodates the head of the femur.

scurry away. The tail then regenerates. This **autotomy** is implemented by a zone of soft tissue that divides each tail vertebra into cephalic and caudal sections, the location being at the level of a myoseptum. Sudden opposing reflex muscle contractions on each side of the septum cause the break.

Archaeopteryx had a long tail, and modern birds still have a tail, although it is not prominent. Pigeons have 15 caudal vertebrae. Five are ankylosed with the synsacrum, six are independent, and the last four are fused to form a **pygostyle**, which is the skeleton of the visible part of the tail (see fig. 8.20). The pygostyle develops as four separate cartilaginous centra.

Among mammals, there are as few as three caudal vertebrae and as many as 50. The tail of sperm whales has 24. Apes and humans have four or five vestigial caudal vertebrae comparable to the pygostyle of birds. These caudal vertebrae lack arches, but most of them have rudimentary transverse processes. The last three or four diminish progressively in size and, in humans, they fuse with one another to form a rigid **coccyx** at about 25 years of age. The centra of the coccyx are still identifiable, but the last one is a mere nod-

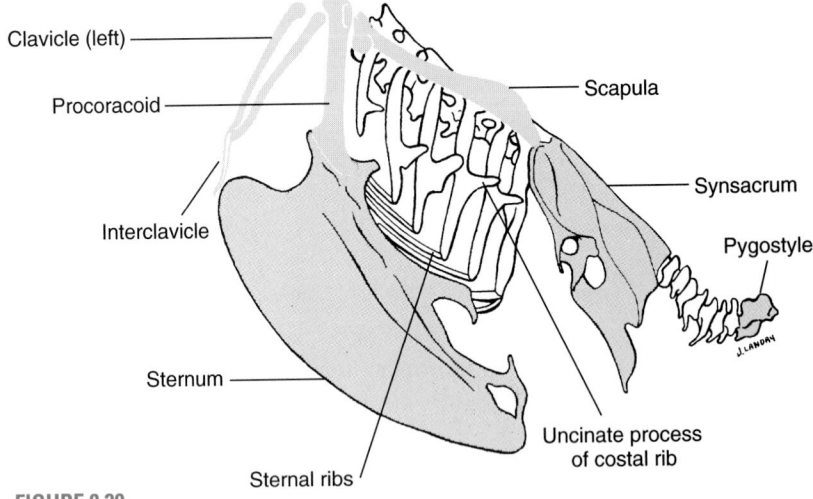

FIGURE 8.20
Skeleton of trunk, tail, and pectoral girdle of pigeon, head to the left. Bones of the pectoral girdle are *red*. The ventral keel of the sternum is the carina. The clavicles and interclavicle comprise the furcula, or wishbone. The first rib illustrated is the second and last of the two cervical ribs.

ule of bone. People whose "tailbone" is recurrently sore have fractured the coccyx at some earlier date in a fall that caused them to land on the end of their spine. In contrast to hominoids, adult rhesus monkeys have a prehensile tail the length of which averages nearly 50 percent of the monkey's sitting height.

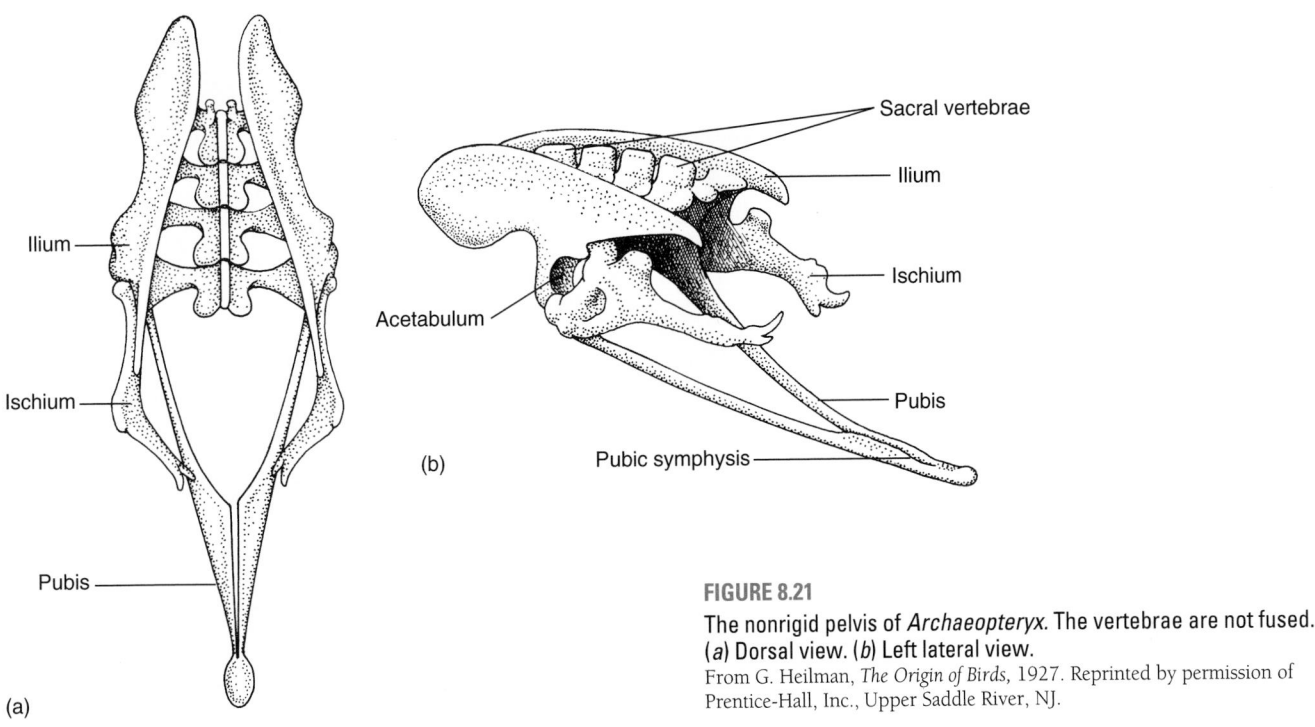

FIGURE 8.21

The nonrigid pelvis of *Archaeopteryx*. The vertebrae are not fused. (*a*) Dorsal view. (*b*) Left lateral view.

From G. Heilman, *The Origin of Birds*, 1927. Reprinted by permission of Prentice-Hall, Inc., Upper Saddle River, NJ.

FIGURE 8.22

Complete set of tail vertebrae from a hamster, left lateral view.

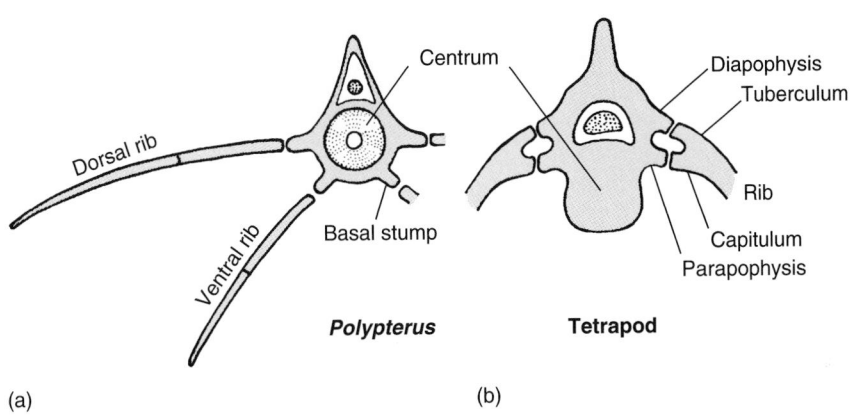

FIGURE 8.23

Dorsal and ventral ribs of *Polypterus* and bicipital rib of a tetrapod. A more usual articulation site of the capitulum of tetrapods is seen in figure 8.25. The basal stump in (*a*) and the parapophysis in (*b*) are not homologues.

RIBS

Ribs articulate with vertebrae and extend into the body wall. The rib is formed intersegmentally by scleroblasts from two successive mesodermal somites in the same manner as centra (see fig. 8.4). In contrast to that portion adjacent to a vertebra, the origin for the distal rib has been debated. Ruijin et al. (2000) have confirmed a sclerotomal origin for the entire rib, but have noted a necessary interaction between sclerotome and myotome for the successful development of a rib. Most bony ribs are of endochondral origin. The so-called abdominal ribs, or **gastralia**, of some reptiles (see fig. 8.14) are not

ribs, but riblike membrane bones that may be remnants of an ancestral bony dermal exoskeleton.

Fishes

The basal actinopterygian *Polypterus* and some teleosts (perch, for example) have two sets of ribs, dorsal and ventral, associated with each trunk vertebra (fig. 8.23*a*). **Dorsal ribs** pass laterad into the horizontal skeletogenous septum that separates the epaxial and hypaxial muscles (see figs. 1.2 and 8.5). **Ventral ribs** develop in myosepta and arch ventrad in the lateral body wall just external to the parietal peritoneum as in the mammalian thorax. In fishes, they stop short of the midventral line.

Commencing at the vent, beyond which there is no coelom, each pair of ventral ribs, or their basal stumps, approach one another beneath the centrum and unite to form the hemal arches of the tail (see fig. 8.1*c*).

Most fishes have only ventral ribs (fig. 8.24, except perch). Sharks and a few others have only dorsal ribs. Skates, chimaeras, and a few unusual teleosts such as sea horses have no ribs. Neither have living agnathans, but this is probably correlated with the absence of centra. Fish ribs strengthen the myosepta to which the longitudinally disposed locomotor muscles of the myomeres (muscle segments) are attached (see fig. 11.5*c*).

Tetrapods

In early tetrapods, ribs were associated with vertebrae from the atlas to nearly the base of the tail (see fig. 8.12). They later became restricted in length and number, the short ones often ankylosing with transverse processes (see fig. 8.16). The long ribs of the anterior thoracic region acquired a ventral anchorage, the sternum. Typical long tetrapod ribs occupy the same position as the ventral ribs of fishes, in that they encircle the coelomic cavity rather than extending laterally within the horizontal skeletogenous septum. Although the tetrapod rib and ventral fish rib share a similar position they are not homologous structures (the tetrapod rib appears to be homologous with the dorsal rib of fishes).

Most tetrapod ribs are **bicipital**, having a dorsal head, or **tuberculum**, and a ventral head, the **capitulum** (fig. 8.25). This is solely a tetrapod feature. The tuberculum in early tetrapods articulated with the extremity of a transverse process and the capitulum articulated with the hypocentrum. With reduction in size of the hypocentrum that took place in amniotes over time, the articulation site of the capitulum became altered. Thereafter, the capitulum articulated in one of several fashions: (1) on two adjacent demifacets (half-facets), one on the caudal end of a centrum and the other on the cephalic end of the next more posterior centrum; (2) with a parapophysis, when the latter was present (see fig. 8.23*b*); or (3) in a facet on a single centrum. The site often changes along the length of the trunk. In some mammals, the tuberculum becomes reduced to a vestige in one region of the column or another. In some other tetrapods, crocodilians, for example, there was a trend toward fusion of the two heads until only the tuberculum remained. In figure 8.16, the tuberculums of a crocodilian are fused to the tips of transverse processes.

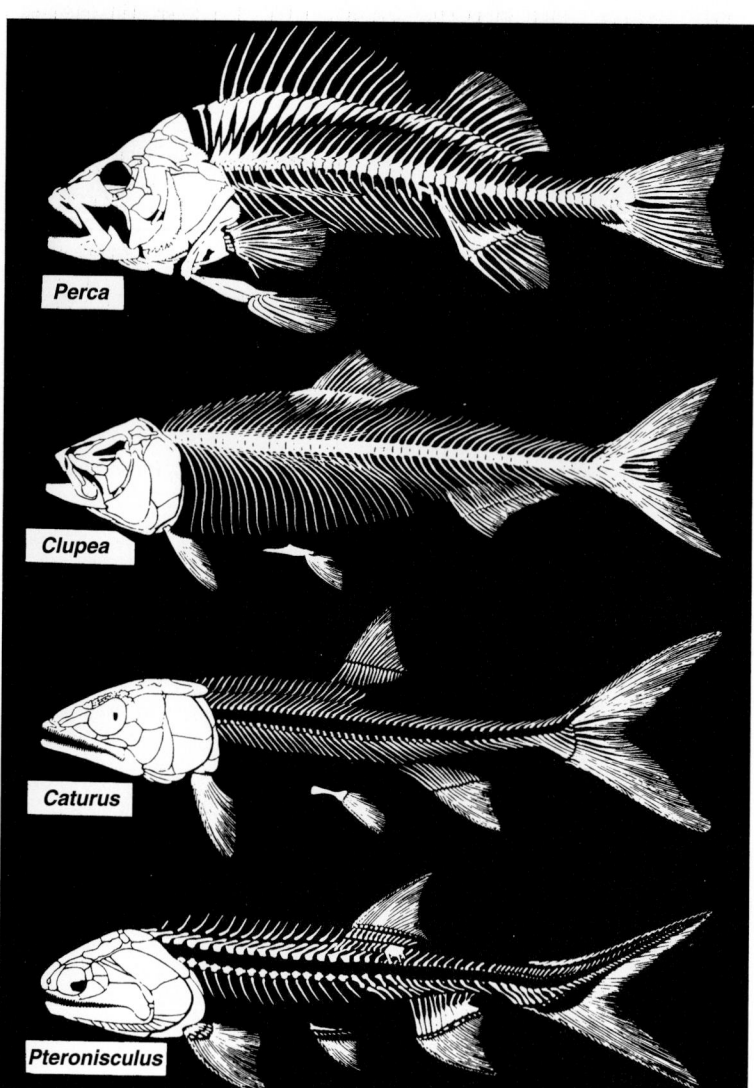

FIGURE 8.24

The skeleton in actinopterygian fishes. *Pteronisculus* was a Triassic basal neopterygian; *Caturus* was a Mesozoic relative of *Amia; Clupea* is a generalized teleost, and *Perca* (perch) is a specialized teleost.

From *Evolution of the Vertebrates*, by Colbert, 3rd edition. Copyright©1980 John Wiley & Sons. Reprinted by permission of John Wiley & Sons, Inc.

Typical thoracic ribs of amniotes consist of two parts, a **costal rib** adjacent to the vertebrae, and a distal, more ventral, **sternal rib** (see fig. 8.20). At least some of the ventral ribs generally articulate with the sternum. Sternal ribs may remain cartilaginous, as in humans, in which case they are also referred to as costal cartilages. A skeletal enclosure for the thoracic viscera of amniotes is provided by the vertebral column, costal ribs, sternal ribs, and the sternum. No such skeletal cage exists in amphibians.

Amphibians

All the ribs of anurans and urodeles have become very short and, in anurans, ankylosed to transverse

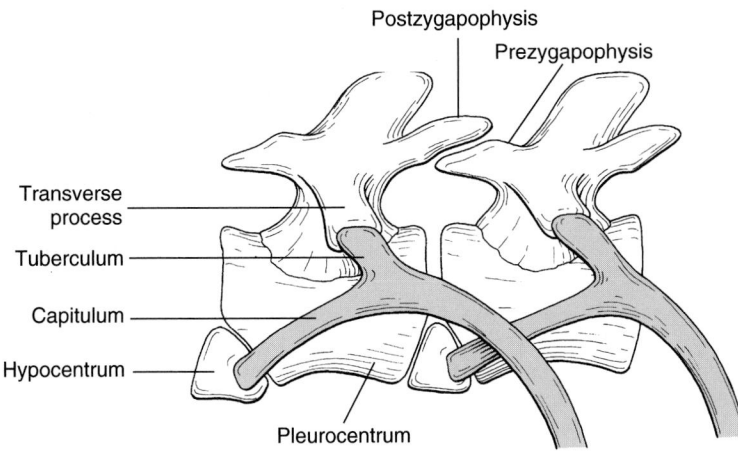

Postzygapophysis
Prezygapophysis
Transverse process
Tuberculum
Capitulum
Hypocentrum
Pleurocentrum

FIGURE 8.25

Two thoracic ribs (*red*) from a basal tetrapod showing articulation with vertebral column.

Source: Data from E.S. Goodrich, *Studies on the Structure and Development of Vertebrates,* 1930, Macmillan and Company, Ltd., London.

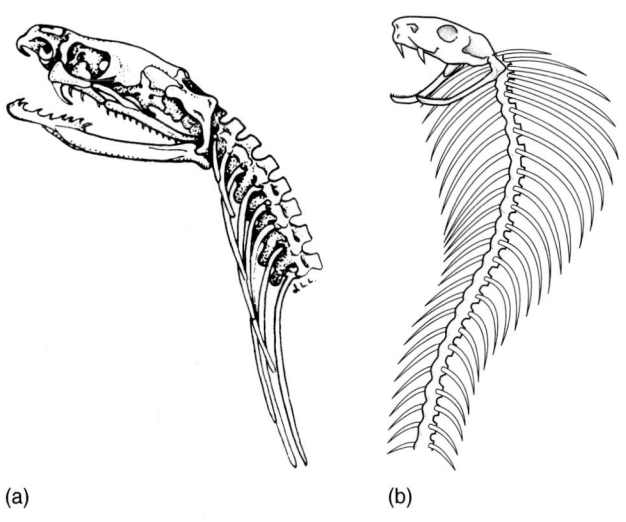

(a)   (b)

FIGURE 8.26

Cervical ribs of a cobra, a "spreading adder." (*a*) Lateral view. (*b*) Frontal view with ribs spread.

processes. By contrast, the ribs of the limbless, burrow-dwelling apodans are quite long. They are present in association with all except the first vertebra and a few of the more posterior ones and, as in snakes, they play a vital role in locomotion. (Recall that apodans have no postanal tail.) Among amphibians, only the ribs of anurans (see fig. 8.13) are not bicipital.

Nonavian Reptiles

Lizards and crocodilians have long ribs on many of the trunk vertebrae and short ribs in most of the neck. Crocodilian ribs are illustrated in figure 8.14. *Sphenodon*, a basal lizardlike reptile, has long ribs in most of the trunk and ribs continue into the tail. Geckos, which

are relatively unspecialized modern lizards, have ribs associated with every cervical vertebra; more specialized lizards usually lack them on the atlas and axis. A half dozen or more of the posterior ribs of the trunk of the gliding lizard *Draco* are greatly elongated. They can be rotated outward to elevate a broad fold of body wall skin, the **patagium**, which then becomes a winglike membrane on which the lizard soars. When not in use, the patagium is folded against the side of the body.

Turtles have no cervical ribs, and the ribs of the trunk are fused with the costal plates of the carapace (see fig. 6.36). The two sacral ribs are not fused with the carapace. They are short, and their expanded distal ends are ankylosed to the ilia of the pelvic girdle (fig. 10.10). They brace the girdle against the vertebral column. Slender, riblike processes may be associated with some of the caudal vertebrae.

Snakes have long, curved ribs beginning at the second vertebra and continuing far into the tail. There is no sternum for ribs to attach to; however, the ventral ends of the ribs have ligamentous connections with integumentary scutes. These ribs participate in locomotion. Long, curved ribs in the neck of cobras can be rotated outward like those of the trunk of *Draco*, causing a fold of loose skin in the neck to "spread" (fig. 8.26). The fold is then inflated with air from the lungs.

Birds

The first two pairs of ribs in many carinates articulate with the last two cervical vertebrae at the base of the neck. They are generally movable and lack a sternal segment. The next five pairs are thoracic ribs with bony sternal segments. Thoracic ribs constitute the major skeleton of the thoracic basket (chest cavity). These ribs are thin, flat, and broad and most of them bear broad **uncinate processes** (see fig. 8.20). Thin broad ribs and uncinate processes provide birds with a lightweight but sturdy thoracic body wall skeleton for the attachment of the powerful muscles required for flight. Ribs caudal to the thoracic basket are ankylosed to the underside of the synsacrum, as are all trunk ribs in turtles. (There is an old saying that birds are turtles with feathers and wings! Sounds good, but it is phylogenetically untenable. Birds evolved within archosaurs.)

Although ribs are confined to the trunk in adults, developing embryos exhibit transitory rudimentary ribs deep into the tail. Uncinate processes are found also in some lizards and were present in some early labyrinthodonts (see fig. 8.12).

Mammals

Recognizable ribs in mammals are generally confined to the thorax. The number ranges from nine pairs in some

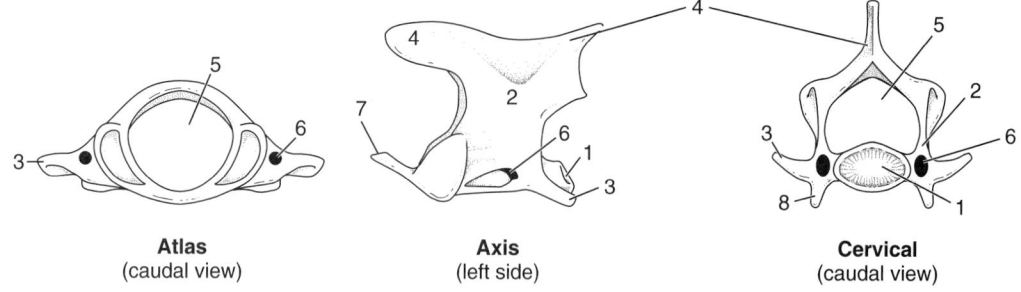

Atlas
(caudal view)

Axis
(left side)

Cervical
(caudal view)

FIGURE 8.27
Mammalian cervical vertebrae. **1**, Centrum; **2**, pedicle; **3**, transverse process; **4**, neural spine; **5**, vertebral canal; **6**, transverse foramen; **7**, odontoid process; **8**, vestige of a cervical rib with two heads enclosing a transverse process.

whales to 24 in some sloths; 12 pairs are common. Where the number is larger than 10, the rest are "floating"; that is, the **costal cartilages** fail to reach the sternum. Humans and apes other than orangutans have 12 pairs of thoracic ribs, and apes have an additional two pairs of lumbars. Humans frequently have an additional rib, cervical or lumbar, as an anomaly, but it usually remains undiscovered unless revealed by an X-ray examination.

Studies of mammalian embryos reveal that two heads of a vestigial bicipital rib develop in association with each cervical vertebra. (A bicipital rib is illustrated in fig. 8.25.) One of the heads is attached to a transverse process; the other to a centrum. The two heads create between them a **transverse foramen** on each cervical vertebra (fig. 8.27). The consecutive foramina provide a bony tunnel, the **arteriovertebral canal**, which transmits a vertebral artery from its source at the base of the neck to the brain. Transverse foramina develop in birds, but the number of vertebrae involved depends on the species. The vertebral artery exists in other reptiles but is not enclosed in a canal.

TETRAPOD STERNUM

The sternum is a tetrapod structure and predominantly amniote (fig. 8.28). It is of endochondral origin. Its presence, size, and anatomical relationships are correlated with the extent to which the forelimbs are employed in locomotion. It serves chiefly as a base against which the pectoral girdles and, in amniotes, the ribs are braced, and as the anatomical origin of the ventral muscles of the forelimbs. The sternum is largest in pterosaurs, carinate birds, and bats, which require strong muscles for flight. It is much reduced or absent in tetrapods lacking forelimbs. It has acquired additional roles in some tetrapods, such as assisting in breathing movements in birds.

The sternum of amphibians is well differentiated only in anurans (fig. 8.28c). Whether or not *Necturus* has a sternum depends on one's definition. It consists of more

or less isolated nodules of cartilage associated with the linea alba immediately caudal to the coracoid cartilages of the pectoral girdle and extending into the proximal ends of adjacent myosepta. A small but discrete sternal element is present in other salamanders at the same location. Apodans have no trace of a sternum.

Lizards are agile climbers and runners. As a correlate, their strong pectoral girdles are firmly braced against a substantial shield-shaped sternum of cartilage or replacement bone (fig. 8.28d). The sternum of the lumbering crocodilians is a simple cartilaginous plate attached to the procoracoid of the pectoral girdle (see fig. 8.14). Caudally, it is continuous with an expanse of a mineralized fibrous membrane incorporating gastralia. Turtles have no sternum. The two halves of the pectoral girdle are united ventrally by fibrous ligaments or by a mere wedge of cartilage.

The sterna of modern birds that are capable of flight have developed an enormous keel, or **carina**, to which the massive pectoral flight muscles are attached (see fig. 8.20). The development of a carina in pterosaurs and birds is an instance of convergent evolution. The sternum of *Archaeopteryx* was small and lacked a keel, an indication that these first birds were not strong fliers.

The sternum of terrestrial mammals, including humans, is composed of a series of bony segments, or **sternebrae** (fig. 8.28f) and is comparatively narrow in comparison to that of reptiles. The last sternebra, **xiphisternum**, bears a cartilaginous or bony **xiphoid process**. Marine mammals tend to have a relatively short sternum, only a small number of ribs reach the sternum, and individual sternebrae tend to be fused. This condition in cetaceans may be correlated with the galloping mode of swimming (chapter 10) necessitated by the lack of functionally competent anterior flippers.

Efforts to determine the nature of the earliest sterna have been frustrated in that there is little evidence of a sternum in labyrinthodont amphibians. Either it was not preserved (perhaps because it was cartilage?), or it had not yet come into being. The tendency of collagenous tissues—which includes myosepta and the linea alba—to undergo ossification when subjected to contin-

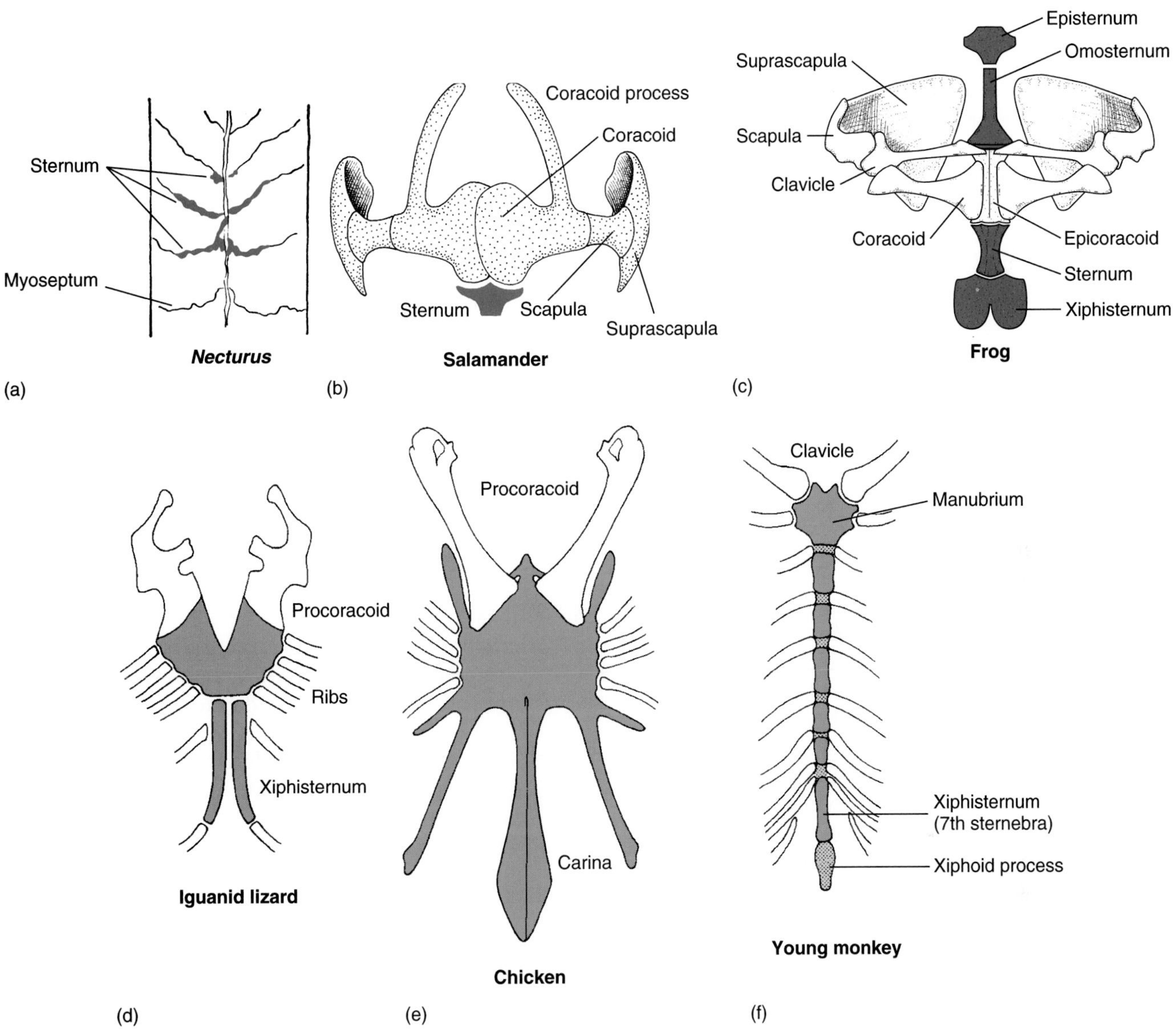

FIGURE 8.28

Tetrapod sterna (*red*). *Stipple* indicates cartilage.

ued mechanical stress was pointed out in chapter 7. It may be that ossification in the cephalic portion of the linea alba that ultimately produced sterna was a response to sustained mechanical stresses resulting from locomotion on land. A sternum may have arisen several times in the evolution of tetrapods, perhaps independently in urodeles, in anurans, and in an amphibian line leading to amniotes.

The amniote sternum arises as paired mesenchymal bars that later unite and undergo chondrogenesis. In many mammals, **presternal** and **suprasternal** blastemas also develop (fig. 8.29). The presternal blastema contributes to the manubrium, and the suprasternal centers sometimes do. The latter may give rise to independent **suprasternal ossicles.** Some humans have suprasternal ossicles, but they are not found unless there is an occasion to x-ray the sternoclavicular joint.

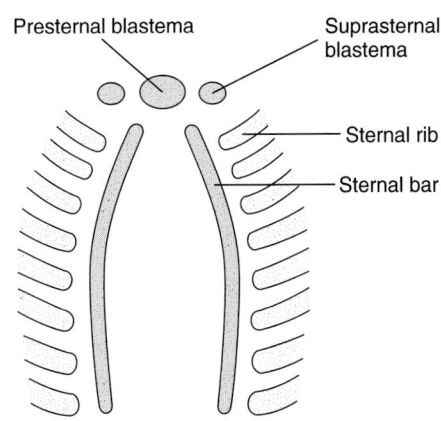

FIGURE 8.29

Mesenchymal blastemas that contribute to the amniote sternum (*red*). Presternal and suprasternal blastemas develop in mammals only. The ventral ends of developing ribs are also shown.

Summary

1. Typical vertebrae consist of centra, neural arches, one or more processes, and, in many tails, hemal arches.

2. Transverse processes (diapophyses) are present in most vertebrates. Tetrapods of one species or another have basapophyses, hypapophyses, parapophyses, and zygapophyses.

3. Fishes have only trunk and tail vertebrae. The trunk vertebrae of tetrapods are subdivided into cervical, dorsal, and sacral vertebrae. Dorsals are further subdivided into thoracic and lumbar when long curved ribs are limited to the anterior trunk.

4. In fishes, the notochord persists within the adult vertebral column from skull to tip of tail. It is usually constricted within centra, expanded between centra, producing an hourglass-shaped amphicelous vertebra.

5. Vertebrae are constructed around, and sometimes within, the notochord by mesenchyme cells from the sclerotomes of mesodermal somites. The notochord sheath may be infiltrated with chordal cartilage and surrounded by perichordal cartilage that usually ossifies to become replacement bone. Membrane bone is deposited in teleosts, urodeles, and apodans. No replacement of cartilage occurs in Chondrichthyes.

6. Notochordal remnants often remain within the vertebral column in all vertebrate classes. In mammals, a vestige persists in the intervertebral discs as a pulpy nucleus.

7. A few fishes have aberrant columns, lacking unified centra (dipnoans, some chondrosteans, *Latimeria*) or have calcified cartilaginous rings instead of typical centra (holocephalans).

8. Agnathans have no centra or arches. Lateral neural cartilages are perched above the notochord. If truly present in hagfish, they are limited to the tail.

9. Diplospondyly is common in many fishes.

10. Early tetrapod vertebrae were rachitomous, consisting of independent arches, a hypocentrum, and two pleurocentra. The notochord was present and uninterrupted throughout the column.

11. Amphicelous centra are concave at both ends because the persistent notochord is constricted within centra and expanded between them. They are less specialized than other types and are found in most fishes and in urodeles, apodans, and some basal lizards.

12. The centra of the remaining tetrapods are opisthocelous (salamanders), procelous (anurans and modern nonavian reptiles), heterocelous (in the neck of birds), and acelous (mammals). These increase the flexibility of the spine.

13. Amphibians have one cervical vertebra. The number is higher in reptiles and still higher in birds. Mammals typically have seven. The first two vertebrae in amniotes are atlas and axis. A proatlas is found in a few reptiles and mammals.

14. Sacral vertebrae bear stout transverse processes that brace the pelvic girdles against the vertebral column. Amphibians have one, reptiles two, and mammals three to five sacral vertebrae. The number generally reflects the stress transmitted to the pelvic girdle from the hind limbs.

15. Sacral vertebrae in amniotes usually unite to form a sacrum. The sacrum of birds unites with adjacent lumbar and caudal vertebrae to form a synsacrum.

16. Caudal vertebrae often bear hemal arches or chevron bones. Neural and hemal arches and chevron bones become vestigial or disappear near the end of the tail.

17. Caudal vertebrae form a urostyle in anurans, a pygostyle in birds, and a coccyx in apes and humans.

18. Ribs, like myosepta, develop intersegmentally from cartilaginous blastemas. Most fishes have only ventral ribs; sharks and a few others have only dorsal ribs. A few fishes, such as *Polypterus* and perch, have both. Agnathans have none.

19. Tetrapod ribs are generally bicipital with a tuberculum and capitulum. Most thoracic ribs have dorsal (costal) and ventral (sternal) segments.

20. Snakes have long movable ribs beginning at the axis and continuing far into the tail. They are essential for locomotion.

21. Short ribs are sometimes attached to the distal ends of transverse processes, as in anurans.

22. A vestigial head of a missing rib in the neck of mammals becomes the roof of a transverse foramen that, with others, creates a vertebroarterial canal housing a vertebral artery.

23. Sterna are tetrapod structures that facilitate life on land. They are reduced or missing in species lacking forelimbs, and in turtles. The sternum of birds capable of flight has a carina. Mammalian sterna exhibit sternebrae.

24. Amniote sterna arise as paired sternal bars that generally unite. An unpaired embryonic presternal blastema may contribute the manubrium sterni. Paired blastemas contribute independent suprasternal ossicles in some mammals. These occur as anomalies in humans.

Critical Thinking Questions

1. Differentiate the cervical, thoracic, lumbar, sacral, and caudal vertebrae of a typical tetrapod.

2. In the shift from lateral undulation to dorsoventral flexion, what vertebral structures help to stabilize the vertebral column?

3. What is the relationship between an individual mammalian vertebra and the segmented somites from which it arises (remember that the axial skeleton is derived from somitic sclerotome)? What functional advantage might this relationship provide?

4. What is a rhachitomous vertebra? How are your vertebrae related to it?

5. As you nod off while reading this or turn your head in response to a noise, what modifications of your cervical vertebrae allow these motions?

6. Give at least two examples of how the vertebral column has been further stabilized (beyond that discussed in question 2 above).

7. What is the relationship between the two ribs found in some teleosts and the ribs of tetrapods?

8. Why is the origin of a sternum associated with tetrapods? In other words, why don't tetrapod ancestors appear to possess a sternum?

Selected Readings

Readings relating to the role of the vertebral column in locomotion are at the end of chapter 10.

Carrier, D. R.: The evolution of locomotor stamina in tetrapods: Circumventing a mechanical constraint, *Paleobiology* 13(3):326, 1987.

Goodrich, E. S.: Studies on the structure and development of vertebrates. London, 1930, The Macmillan Co., Ltd. (Reprinted by The University of Chicago Press, Chicago, 1986.) Although the theoretical discussions are out of date, the work is rich in skeletal morphology and abundantly illustrated.

Heilmann, G.: The origin of birds. New York, 1927, Appleton-Century-Crofts, Inc.

Hoffstetter, R., and Gasc, J. P.: Vertebrae and ribs of modern reptiles. In Gans, C., Bellairs, A. d'A., and Parsons, T. S., editors: Biology of the reptilia, vol. 1. New York, 1969, Academic Press.

Jarvik, E.: Basic structure and evolution of vertebrates, vol. 1. New York, 1980, Academic Press. An in-depth study of *Amia* and the early fossil vertebrate *Eusthenopteron*.

Panchen, A. L.: The origin and evolution of early tetrapod vertebrae. In Andrews, S. M., Miles, R. S., and Walker, A. D., editors: Problems in vertebrate evolution. New York, 1977, Academic Press.

Ruijin, H., Zhi, Q., Schmidt, C., Wilting, J., Brand-Saberi, B., and Christ, B.: Scleotomal origin of the ribs, Development 127: 527–532, 2000.

Links to the Internet

Visit the zoology website at http://www.mhhe.com/zoology to find live Internet links for each of the references listed below.

1. Bones of the Body. A clickable list of all of the bones of the human body, then a picture labels the parts, and displays answers when clicked upon. Very useful for all of these skeletal chapters, although the detail for some bones is limited, particularly the skull.

2. Postcranial Skeleton. Click on the portion of the body you wish to examine, then click again on each bone for great detail of the human skeleton.

3. Axial Skeleton. A series of informative slides on the vertebral column.

4. Abnormal Spinal Anatomy. Information on human "back" problems, with a link to information on normal anatomy.

5. Bones of the Vertebral Column. Nice labeled photos of vertebrae. Can click to see lateral, posterior, anterior views.

6. Rib phylogeny. Information on ribs of vertebrates.

7. Rib. Just two pictures, but an excellent photograph of the articulation of the vertebra and the rib.

8. Morphology. A great picture of the skeletal system of a fish, which includes the multitude of vertebrae and ribs, as well as the skull and fin supports.

9

Skull and Visceral Skeleton

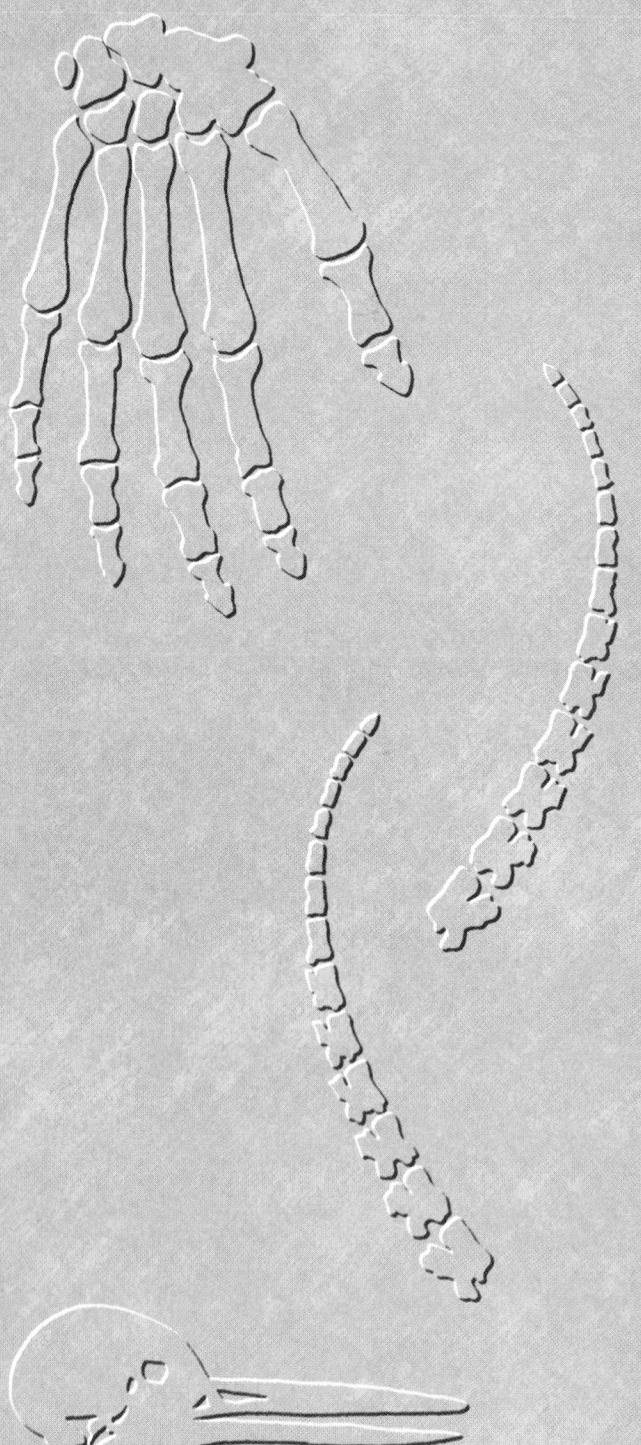

In this chapter, we will find that craniates from fishes to human beings form their head skeleton out of three components: an embryonic cartilaginous braincase, ancient dermal armor in vertebrates, and contributions from the branchial skeleton. We will learn how these components are assembled and note the new functions achieved by the branchial skeleton when vertebrates took up life on land.

OUTLINE

 he word *skull* is seldom misunderstood by the layperson. To him or her, it is the bony structure that Hamlet held in his hand and gazed at dolefully as he spoke his now famous words, "Alas, poor Yorick." To the morphologist, however, the term poses a problem because of the intimate relationship in fishes between the skeleton that protects the brain and special sense organs of the head and the skeleton of the jaws and branchial arches. The latter has been inherited in modified form by descendants including, alas, poor Yorick. For this reason, morphologists may avoid the term *skull* and refer instead to (1) the **neurocranium**, or **primary braincase**; (2) the **dermatocranium**; and (3) the **visceral skeleton**, or **splanchnocranium**. The neurocranium and splanchnocranium of early craniates were formed in cartilage. This pattern is still seen in the development of all craniates. With the origin of bone in vertebrates, the neurocranium and splanchnocranium ossified to various degrees, and membrane bones were added to the integument surrounding the skull and jaws. In this chapter, for practical purposes, *skull* will mean what the layperson probably thinks it means, but without the lower jaw. We can then classify the parts of the cranial skeleton of craniates as follows.

Skull (cranial skeleton)
  Neurocranium
  Dermatocranium
Visceral skeleton
  Embryonic upper jaw cartilage (palatoquadrate)
    and its replacement bones
  Embryonic lower jaw cartilage
    (Meckel's) and its replacement
    and investing bones
  Skeleton of the branchial arches

The upper jaw is visceral skeleton being part of or derived from the first visceral arch (fig. 9.1). In bony vertebrates, it becomes incorporated into the skull as development progresses.

NEUROCRANIUM

The neurocranium (sometimes called endocranium, chondrocranium, or primary braincase), is the part of the skull that (1) protects the brain and certain special sense organs, (2) arises as cartilage, and (3) is subsequently partly or wholly replaced by bone except in cartilaginous fishes. The neurocrania of all gnathostomes develop in a similar manner, as will be described here. It commences as several independent cartilages that later expand and unite to form a cartilaginous braincase.

Cartilaginous Stage

Parachordal and Prechordal Cartilages and Notochord

The neurocranium commences as a pair of **parachordal** and **prechordal** cartilages (fig. 9.2a) underneath the brain. Parachordal cartilages parallel the anterior end of the notochord beneath the midbrain and hindbrain. Prechordal cartilages (**trabeculae cranii**) develop anterior to the notochord underneath the forebrain. The parachordal cartilages expand across the midline toward each other and unite. In the process, the notochord and parachordal cartilages are incorporated into a single, broad, cartilaginous **basal plate**. The prechordal cartilages likewise expand and unite across the midline at their anterior ends to form an **ethmoid plate**.

Sense Capsules

While parachordal and prechordal cartilages are forming, cartilage also appears in two other locations: (1) as an **olfactory (nasal) capsule** partially surrounding the olfactory epithelium and (2) as an **otic capsule** completely surrounding the otocyst, which is the developing inner ear (fig. 9.2a and b). The olfactory capsules are incomplete anteriorly, because water (in fishes) or air (in tetrapods) must have access to the olfactory epithelium. The walls of the olfactory and otic capsules are perforated by foramina that transmit nerves and vascular channels.

An **optic capsule** forms around the retina, but it is not the orbit, or skeletal socket, in which the eyeball lies. It is the **sclerotic coat** of the eyeball. Although this

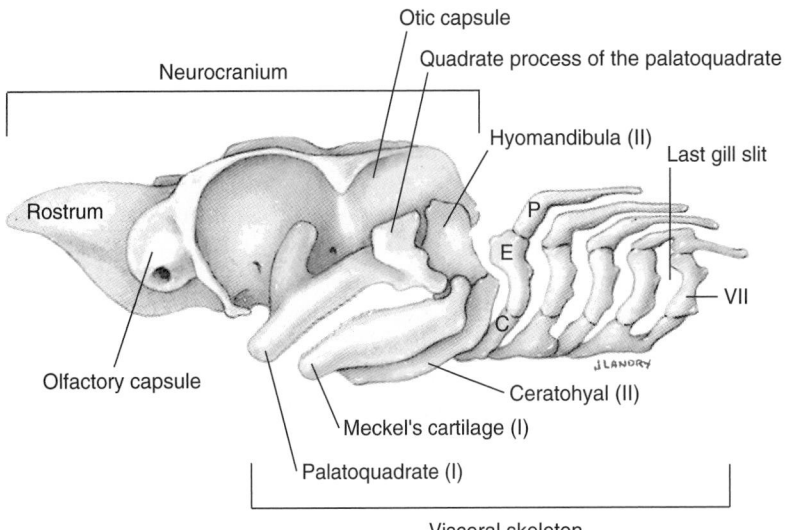

FIGURE 9.1

Skull and visceral skeleton of the shark, *Squalus acanthias*. **I, II,** and **VII,** Skeleton of first, second, and seventh pharyngeal arches. **C,** ceratobranchial; **E,** epibranchial; **P,** pharyngobranchial cartilages of the second gill arch. Labial cartilages, gill rakers, and gill rays are omitted. The hyomandibula suspends the entire visceral skeleton from the skull at the otic capsule.

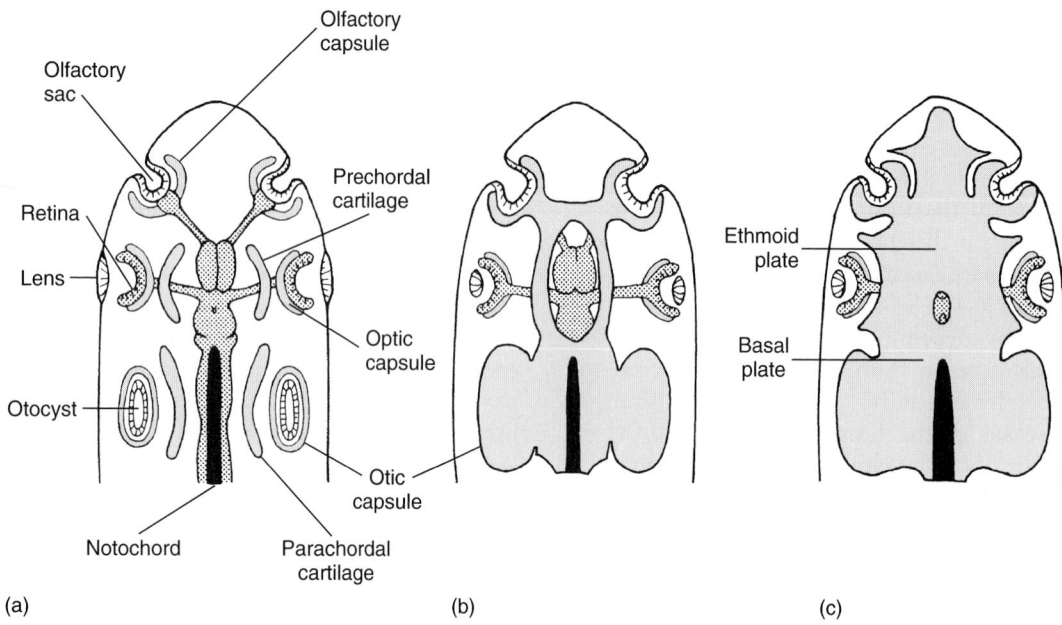

FIGURE 9.2

Early stages in development of a cartilaginous neurocranium, as seen from a ventral view. In (*a*), the notochord is seen underlying the midbrain and hindbrain. In (*b*), the notochord has been incorporated into the caudal floor of the neurocranium (basal plate). In (*c*), a cartilaginous floor has been completed beneath the entire brain, and a hypophyseal fenestra remains between the ethmoid and basal plates. The optic capsule becomes the sclerotic coat of the eyeball.

capsule is fibrous in mammals, cartilaginous or bony plates often form within the sclerotic coat (see fig. 17.17). This is an ancient condition, having been present in some ostracoderms, placoderms, and basal osteichthyans. Because the optic capsule does not fuse with the rest of the neurocranium, the eyeball is free to move independent of the skull. Therefore the sclerotic coat is not conventionally considered part of the neurocranium.

Completion of the Floor, Walls, and Roof

The expanding ethmoid plate unites anteriorly with the olfactory capsules, and the expanding basal plate unites with the otic capsules, which are lateral to the hindbrain. The ethmoid and basal plates also expand toward one another until they meet to form a floor upon which the brain thereafter rests (fig. 9.2*c*). In the latter process, there remains in the midline between the two plates a **hypophyseal fenestra** accommodating the hypophysis and internal carotid arteries, the latter being en route to the brain. Later the fenestra is usually reduced to a pair of foramina transmitting the arteries. Further development of the neurocranium primarily involves construction of cartilaginous walls around the brain and, as a primitive condition in some craniates, a cartilaginous roof (**tectum**) above the brain with one or two prominent fenestrae (fig. 9.3*c*). The cranial nerves and blood vessels are already present by this time, and cartilage is deposited in such a manner as to leave foramina for

these structures. The largest is the foramen magnum in the rear wall of the neurocranium. As the brain, blood vessels, nerves, and sense organs grow, remodeling of the neurocranial cartilage takes place.

The preceding basic pattern of development recurs in all craniate classes, producing a cartilaginous neurocranium that houses the brain, olfactory epithelia, and inner ear (fig. 9.3). The mesenchyme that gives rise to the neurocranium comes from at least two sources. The prechordal cartilages are formed from neural crest ectoderm that streams ventrad in front of the developing eyestalks (optic stalks) that connect the brain with the developing eyeball. The mesenchyme that forms the parachordal cartilages, occiput, and encapsulates the inner ear, in part, comes from sclerotome (epimeric mesoderm). The source of the mesenchyme that forms the remaining bulk of the neurocranium is derived from neural crest.

Cartilaginous Neurocrania of Adult Craniates

Living Agnathans

The several cartilaginous components of the embryonic neurocranium remain more or less independent throughout life (see fig. 9.35). An olfactory capsule, otic capsules, a basal plate, and a notochord (not fused with the basal plate) are identifiable, and several other cartilages of unknown homology are also present. The roof above the brain remains unchondrified and therefore fibrous.

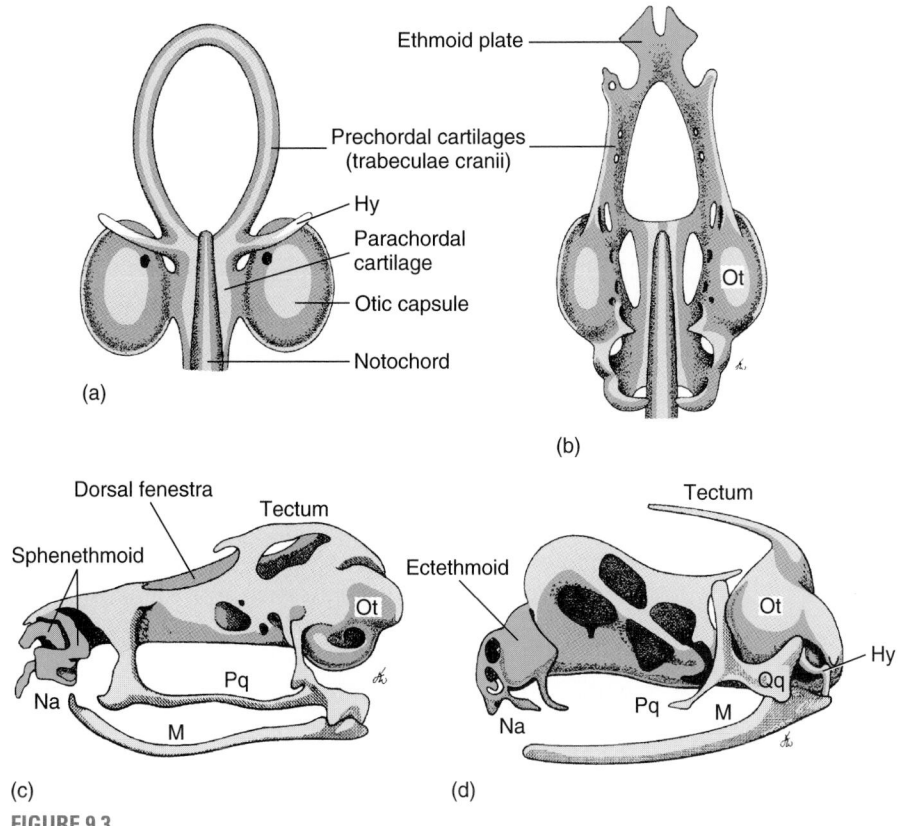

FIGURE 9.3

Cartilaginous neurocrania of early ammocoete larva (*a*), young urodele *Amphiuma* (*b*), recently metamorphosed frog (*c*), and young *Sphenodon* (*d*). (*a*) and (*b*) dorsal views; (*c*) and (*d*), left lateral views with associated visceral arch components. *Dark red,* ethmoid plate and its derivatives; *light red,* palatoquadrate and Meckel's cartilages; **Hy,** hyoid arch component; **M,** Meckel's cartilage; **Na,** nasal capsule; **Ot,** otic capsule; **Pq,** palatoquadrate cartilage; **Qu,** quadrate region of palatoquadrate cartilage.

On each side of the foramen magnum, an occipital condyle is the site of the immovable articulation between the occipital region of the neurocranium and the first vertebra. On the posterodorsal aspect of the neurocranium, a depression, the **endolymphatic fossa,** exhibits two pairs of foramina that house **endolymphatic** and **perilymphatic ducts,** respectively. The endolymphatic ducts open onto the surface of the head. A well-developed, totally cartilaginous adult skull, devoid of dermal bones, is found only in Chondrichthyes.

Bony Fishes

Although a cartilaginous neurocranium is an embryonic structure in teleosts and tetrapods, it persists throughout life as cartilage in most chondrosteans, basal neopterygians other than gars, and dipnoans. In order to see it one must strip away the dermal bones that overlie it (see fig. 9.7). In teleosts and tetrapods, however, the embryonic cartilaginous neurocranium is replaced partially or wholly by endochondral bone as development progresses. The ossification centers where this takes place will be discussed next.

Neurocranial Ossification Centers

The process of endochondral ossification within the neurocranium occurs more or less simultaneously at numerous separate ossification centers. The specific number of centers varies among species, but four regional groups are universal. These groups—occipital, sphenoid, ethmoid, and otic—will be discussed next. They are illustrated in a fetal pig in figure 9.4. The composite bones that develop from these centers in humans are shown in figure 9.6.

Occipital Centers

The cartilage surrounding the foramen magnum may be replaced by as many as four bones. One or more endochondral ossification centers ventral to the foramen magnum produce a **basioccipital bone** underlying the hindbrain (figs. 9.4 and 9.5*a*). Centers in the lateral walls of the foramen magnum produce two **exoccipital bones.** Above the foramen, a **supraoccipital bone** may

Cartilaginous Fishes

The neurocranium of *Squalus acanthias* is typical of cartilaginous fishes (see fig. 9.1). It constitutes a high water mark in the development of neurocrania because the embryonic components unite to form a boxlike adult cartilaginous braincase or **chondrocranium** in which the basic components have almost lost their identity. Walls are fully developed and in gnasthostomes we see a posterior occipital wall for the first time. The brain is completely roofed by cartilage. The last portion of the roof to chondrify is an area just behind the rostrum, and in young skulls this may still be an unossified fenestra. (Some elasmobranch species have a fenestra at this location throughout life.) The otic capsules are inextricably fused into the posterolateral walls of the braincase, and the olfactory capsules are firmly united to it anteriorly. The notochord is visible on the ventral aspect as a ridge extending cephalad from the base of the foramen magnum. The hypophysis is cradled in a cartilaginous pocket, the sella turcica, beneath the brain. The neurocranium projects forward beyond the olfactory capsules as the rostrum.

develop. In mammals, all four occipital elements usually fuse to form a single **occipital bone.** In modern amphibians, one or more of these may remain cartilaginous, although they were bony in stem amphibians.

The neurocranium of tetrapods articulates with the first vertebra via one or two **occipital condyles.** Stem amphibians had a single condyle borne chiefly on the basioccipital bone. Living reptiles, including birds still have a single condyle. Modern amphibians and mammals diverged from the early tetrapod condition, gradually shifting the condyles from the median basioccipital to the two exoccipitals. What the survival value may have been is purely speculative.

Sphenoid Centers

Ossification of the sphenoid region occurs independently in the synapsid and reptilian lineages. A number of terms have been applied to these ossifications, and some confusion remains. The embryonic cartilaginous neurocranium underlying the midbrain and pituitary gland ossifies to form a **basisphenoid bone** (anterior to the basioccipital). In mammals, a **presphenoid bone** (see fig. 9.4) ossifies anterior to the basisphenoid. Thus, a bony platform consisting of the basioccipital and the sphenoid bones underlies the brain. The sidewall in the sphenoid region above the basisphenoid is formed by an additional presphenoid ossification in mammals. In archosaurs (crocodiles and dinosaurs), an independent **laterosphenoid bone** forms the lateral ossification of the sphenoid region. A separate interorbital septum forms in archosaurs as the **orbitosphenoid bone** (see fig. 9.23a). The **alisphenoid bone,** of at least some mammals, helps to form the lateral wall, but it is derived from the palatoquadrate cartilage rather than the neurocranium. The sphenoid elements in mammals (basisphenoid, presphenoid, and alisphenoid) may remain separate or unite to form a single **sphenoid bone** with "wings" (fig. 9.6). The pituitary rests in the sella turcica of the basisphenoid region. No replacement bones develop above the brain.

Ethmoid Centers

The ethmoid region lies immediately anterior to the sphenoid and includes the ethmoid plate and olfactory (nasal) capsules (see fig. 9.3). Of the four major potential ossification centers in the cartilaginous neurocranium (occipital, sphenoid, ethmoid, and otic), the ethmoid tends to remain cartilaginous in tetrapods from amphibians to mammals. Most basal tetrapods develop

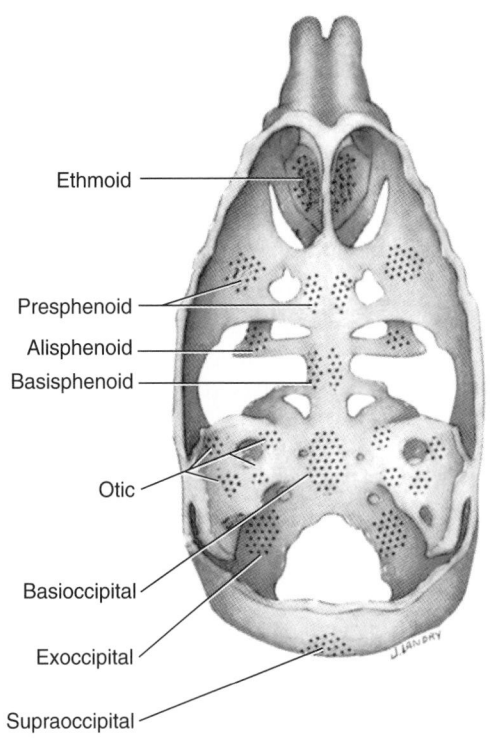

FIGURE 9.4

Cartilaginous neurocranium of a fetal pig with chief endochondral ossification centers of mammals superimposed (*dots*). The neurocranium is complete as shown, there being no cartilage above the brain. The ethmoid center becomes the cribriform plate perforated by olfactory foramina. The otic centers are in the otic capsule. The alisphenoid center is derived from palatoquadrate cartilage in mammals.

FIGURE 9.5

Typical mammalian endochondral ossification centers (*dots*) and intramembranous centers (black reticulum) in two cat skull bones. (*a*) Occipital bone (caudal view). (*b*) Right temporal (medial view). The petrous portion (*red*) houses the inner ear.

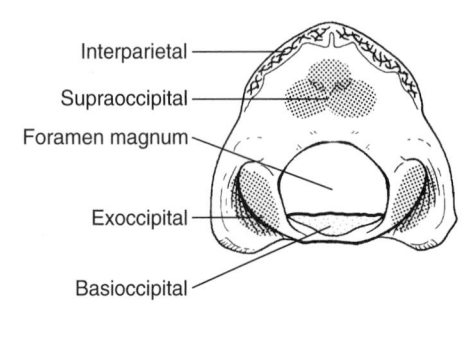

(a)

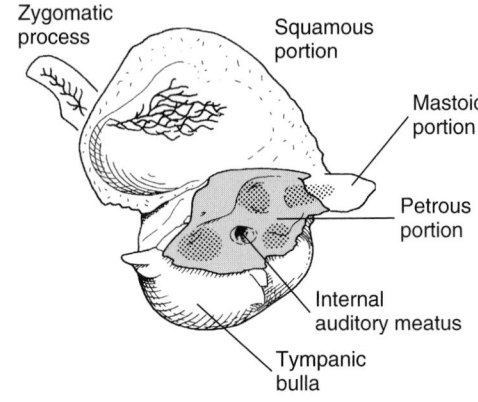

(b)

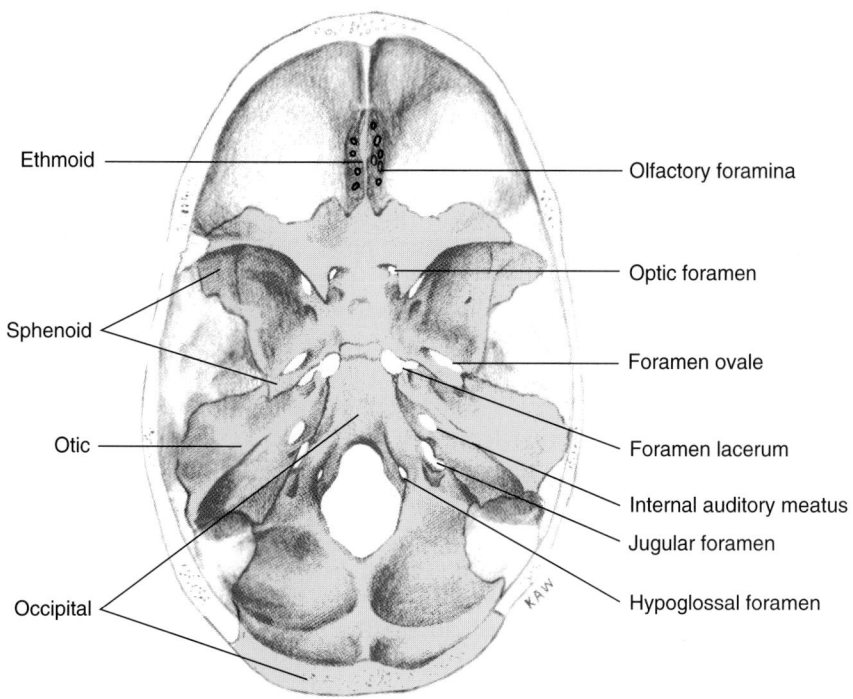

Ethmoid

Olfactory foramina

Optic foramen

Sphenoid

Foramen ovale

Foramen lacerum

Otic

Internal auditory meatus

Jugular foramen

Hypoglossal foramen

Occipital

FIGURE 9.6

Bony neurocranium (*red*) of human skull. The calvarium (roof) of the skull has been sawed off, and the view is looking down into skull from above. Major endochondral ossification centers are labeled at left. The sphenoid bone exhibits an anterior (lesser) wing enclosing the optic foramina and a posterior (greater) wing between the two label lines for the sphenoid bone. The olfactory foramina are in the cribriform plate of the ethmoid.

no ethmoidal ossification centers whatsoever. Ossification centers in amniotes are chiefly **mesethmoid** bones, which contribute to the otherwise cartilaginous nasal septum of birds and mammals, and in birds, to the anterior portion of the interorbital septum; one or more scroll-like **turbinal** bones (**conchae**) in the walls of the nasal passageways of most reptiles including birds, and mammals; and in mammals, a **cribriform plate** perforated by **olfactory foramina** that transmit bundles of olfactory nerve fibers from the olfactory epithelium to the brain (fig. 9.6). In anurans, a **sphenethmoid** is the sole bone arising in the ethmoid and sphenoid regions. An **ectethmoid** develops in the lateral walls of the nasal passageway of *Sphenodon* (see fig. 9.3*d*). It should be noted here, however, that not all cartilages in the nasal passages of amniotes are derived from ethmoid ossification centers. Among these that are not are the winglike alar and small sesamoid cartilages that stiffen the lateral walls of the human nose, thereby keeping the walls from collapsing against the nasal septum during inhalation.

Otic Centers

The cartilaginous otic capsule surrounding the membranous labyrinth is replaced by several bones with such names as **prootic, opisthotic,** and **epiotic.** One or more of these may unite with adjacent replacement or membrane bones. For example, in frogs and most nonavian reptiles, the opisthotics fuse with the exoccipitals, and in birds and mammals, the prootic, opisthotic, and epiotics all unite to form a single **periotic** or **petrosal bone.** The petrosal, in turn, may unite with the squamosal, a membrane bone, to form a **temporal bone** (see fig. 9.5*b*). Six ossification centers have been described in the otic capsule of a human fetus.

GENERALIZED DERMATOCRANIUM

The membrane bones of the skull collectively constitute the dermatocranium. After considering how the dermatocranium probably originated, we will examine its basic architecture in generalized vertebrates. The insight thus gained will enable us to relate the dermatocranium to the rest of the skull in modern tetrapods. Because the skulls of modern fishes are highly specialized and vary widely, we will mention them only occasionally.

How It May Have Begun

Much of the body of the earliest vertebrates was encased in a bony dermal armor. As far back as ostracoderms, this armor varied widely in the extent to which it covered the body—head and anterior trunk only, head and entire trunk, or head, trunk, and tail; in the relative size of the bones or scales making up the armor—large shields, smaller plates, minute scales; and in the relative size of the plates or scales on the head contrasted with those on the rest of the body. There is also evidence that some ostracoderms—cephalaspids, at least—passed through one or more cycles of expansion and reduction of their dermal armor, going from small scales to increasingly larger plates, then more or less reversing the trend; and that the reduction was less complete on the head. The large bony plates in the skin of early *jawed* fishes also gave way to smaller scales, first on the trunk and tail and eventually on the head. However, bony plates are still present in the skin of the head. These plates are an integral part of the skull (fig. 9.7). Note also the cheek plates in the skull of a gar (see fig. 9.11*a*). They are either skull bones or integumentary scales, whichever you prefer to call them. But whatever one calls them does not alter the fact that these bones in the skin are also part of the skull. The neurocranium is the endoskeleton; the dermal bones constitute an

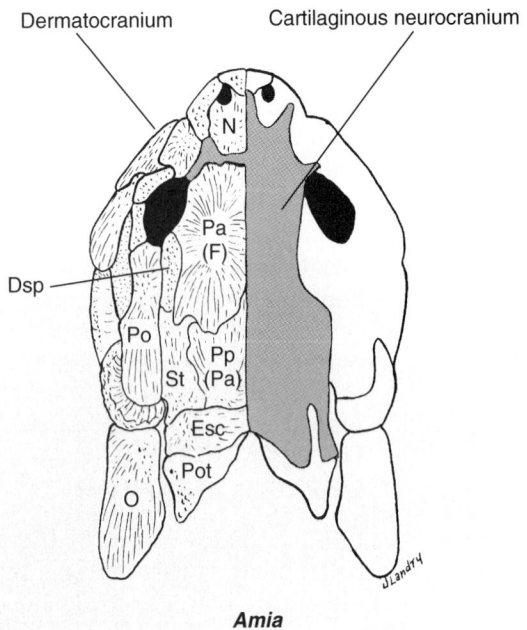

Dermatocranium   Cartilaginous neurocranium

Dsp

Amia

FIGURE 9.7
Skull of *Amia calva,* dorsal view. Dermal bones have been removed on the right side to reveal underlying cartilaginous neurocranium. The abbreviations in parentheses represent the classical nomenclature applied to these bones (refer to "A Question of Homology" section in the text). **Dsp,** dermosphenotic (intertemporal of some authors but represents an infraorbital bone that fuses to the roof in ontogeny); **Esc,** extrascapular; **F,** frontal; **N,** nasal; **O,** opercular; **Pa,** parietal; **Po,** postorbital; **Pp,** postparietal; **Pot,** posttemporal; and **St,** supratemporal. The bones anterior to the nasals are ethmoids. Premaxillae are not visible in this view.

exoskeleton. It is the latter that constitutes the dermatocranium in these "living fossils."

In modern vertebrates from fish to humans, these membrane bones of the head no longer ossify from mesenchyme in the *dermis,* but from *subdermal* mesenchyme. The inductors that convert this mesenchyme to fibroblasts and scleroblasts may be the same ones that induced ossification in the dermis of earlier generations. *It is these membrane bones that constitute the modern dermatocranium from fish to humans.* Ongoing research on mesenchymal cell migration and inductor substances reinforces the concept that the dermatocranium is derived from the dermal armor of the earliest vertebrates.

Its Basic Structure

For convenience of discussion, the generalized dermatocranium can be divided into (1) bones that form above and alongside the brain and neurocranium (roofing bones), (2) dermal bones of the upper jaw (marginal bones), (3) dermal bones of the primary palate, and (4) opercular bones.

A Question of Homology

If we look at the pattern of dermal bones in gnathostomes, it is apparent that the dermal bones of placo-

derms are not homologous to those in bony fishes. Even among bony fishes, there has been some debate among researchers over the naming of elements within the dermatocranium. This is in part a historical artifact. The bones of modern ray-finned fishes were named, based on a comparison with mammals, prior to an understanding of the phylogeny of basal tetrapods and basal osteichthyans. A main point of contention concerns the central roofing bones (described in detail later) that consist of frontals, parietals, and postparietals. These roofing bones in mammals can be traced through basal tetrapods (fig. 9.8c) to their lobe-finned outgroups (fig. 9.8b). Through this comparison, we recognize that the frontal bone is a new structure among lobe-finned fishes. We are now faced with the problem of naming roofing bones in ray-finned fishes. There are two possible hypotheses: (1) ray-finned and lobe-finned fishes share via their common ancestor parietal and postparietal bones with the frontal being unique to later sarcopterygians, and (2) the roofing bones evolved independently and are not homologous (and should be given distinct names). Fossil evidence supports the former hypothesis, and the naming convention used in this text reflects current hypotheses of homology. The student should be aware that the application of names is not consistent within the research literature and in textbooks. It has been argued that changing names to reflect homologies would only lead to confusion. Which is more confusing to you—a frontal that is sometimes a parietal, or the use of the same name for homologous structures?

Roofing Bones

An early pattern of roofing bones is seen in rhipidistian fishes, and this pattern was inherited almost intact by labyrinthodonts (fig. 9.8). Roofing bones in these vertebrates provided a protective shield over the brain and special sense organs with openings only for the nares, paired eyes, and a median (parietal) eye. In rhipidistians, a series of paired and unpaired scalelike bones extended along the middorsal line from the nares to the occiput, overlying the brain, olfactory capsules, and any other components of the neurocranium that developed in that area. In labyrinthodonts, the unpaired bones were missing, and a series of paired **nasals, frontals, parietals,** and **postparietals** (dermoccipitals) took their place. The postparietals have disappeared as independent bones in modern amphibians, but they may have coalesced with endochondral supraoccipitals that appeared first in early reptiles. (The bone caudal to the parietal was named dermoccipital in recognition of this possibility.) A parietal foramen housing the median eye is still present in many fishes, amphibians, and lizards.

Forming a ring around the orbit in the generalized skull were a **lacrimal, prefrontal, postfrontal,** and **jugal (infraorbital).** The lacrimal bone derives its name from

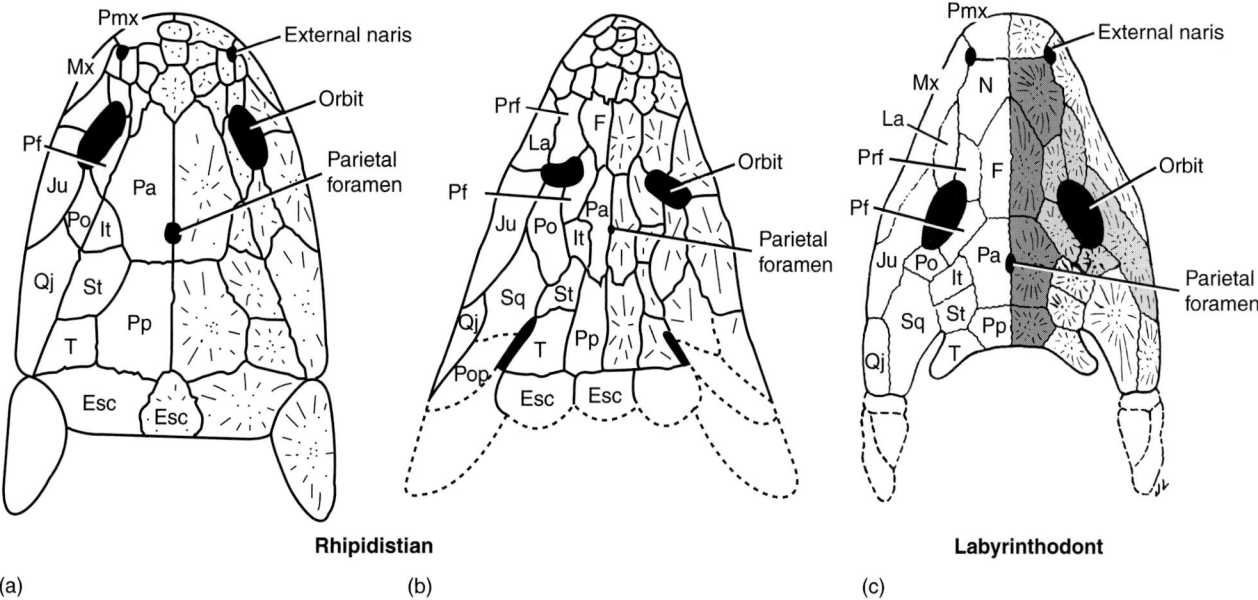

FIGURE 9.8

Early dermal bone patterns from which tetrapod dermatocrania probably evolved. (*a*) Skull of the rhipidistian fish *Eusthenopteron.* Note the midline bones and small, scalelike bones in the rostral region. The parietal foramen houses a median eye. (*b*) Skull of the rhipidistian fish *Panderichthys.* Note the origin of frontal bones that are missing from actinopterygians and basal sarcopterygians (fig. 9.7 and *a* above). *Broken lines* indicate missing bones. (*c*) Skull of a Carboniferous labyrinthodont. *Broken lines* indicate missing opercular bones.
Esc, Extrascapular; **F,** frontal; **It,** Intertemporal; **Ju,** jugal; **La,** lacrimal; **Mx,** maxilla; **N,** nasal; **Pa,** parietal; **Pf,** prefrontal; **Pmx,** premaxilla; **Po,** postorbital; **Pp,** postparietal; **Prf,** prefontal; **Qj,** quadratojugal; **Sq,** squamosal; **St,** supratemporal; and **T,** temporal. In (*c*), midline roofing bones are *dark red;* periorbital bones are *light red.*
(*b*) Redrawn from Schultze, 1996.

its relationship to the nasolacrimal duct of amniotes, which drains excess fluid, including tears (lachrymae), from the nasal corner of the eye to the nasal canal. At the posterior angle of the skull were **intertemporal, supratemporal, tabular,** and lower down, **squamosal** and **quadratojugal** bones. Labyrinthodonts developed a longer facial area, or snout, than is seen in rhipidistians by elongating the bones between the external nares and orbits. This could have been correlated with altered methods of gathering or capturing food or manipulating it in the mouth.

Dermal Bones of the Upper Jaw

The palatoquadrate cartilages, the only upper jaws that cartilaginous fishes develop (see fig. 9.1), are the embryonic precursors of the upper jaw of bony vertebrates. In the latter, these cartilages become overlaid, or ensheathed, by tooth-bearing dermal bones, the **premaxillae** and **maxillae,** which then become part of the margins of the dermatocranium (figs. 9.8 and 9.10). Thus, the completed upper jaw complex of bony vertebrates becomes part of the skull. The lower jaw will be discussed with the other components of the visceral skeleton.

Primary Palatal Bones

The primary palate is the roof of the oropharyngeal cavity of fishes and of the oral cavity of basal tetrapods. In

sharks, it is cartilaginous, being the floor of the neurocranium on which the brain rests. In bony vertebrates, membrane bones are applied to the underside of the neurocranium and to any part of the palatoquadrate cartilages (upper jaw cartilages) occupying the area. These membrane bones then become the major components of the primary palate.

In rhipidistian fishes and early tetrapods, these membrane bones of the palate were an unpaired **parasphenoid** beneath the sphenoid region of the neurocranium, paired **vomers** beneath the ethmoid region, and paired **palatines, ectopterygoids,** and **pterygoids** laterally (fig. 9.9). The last three invested the palatoquadrate cartilage to one degree or another. Primitively, teeth formed on all these bones, and several of them still bear teeth in extant basal vertebrates. Internal nares pierced the palate anterolaterally. A primary palate is still present with modifications in all tetrapods (see fig. 9.13), but in those that also develop a secondary palate the primary palate remains in the roof of the nasal passageway (see fig. 9.18).

Opercular Bones

The operculum is a flap of tissue that arises as an outgrowth of the hyoid arch and extends caudad over the gill slits. It is membranous in holocephalans and absent in elasmobranchs. In bony fishes, it is stiffened by squamous

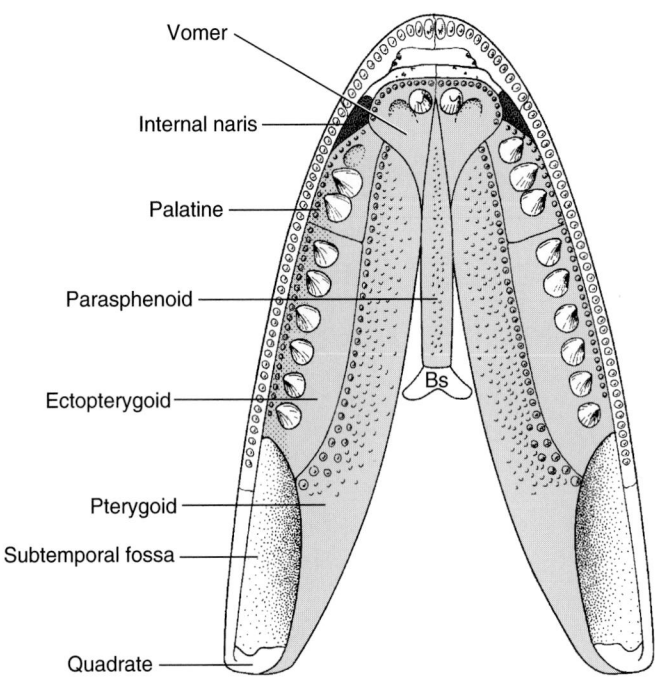

(a)

Rhipidistian

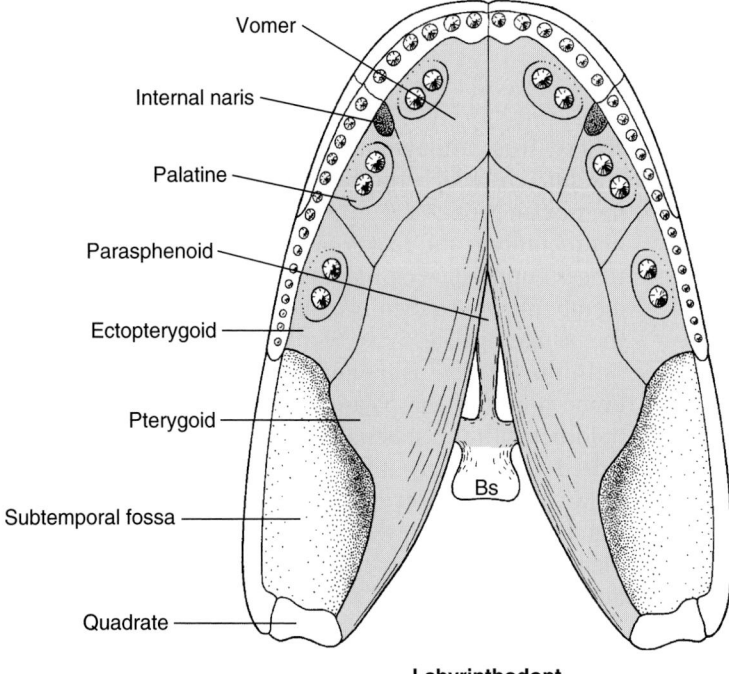

(b)

Labyrinthodont

FIGURE 9.9

Primary palate (*red*) of a rhipidistian and a late Paleozoic labyrinthodont. Small denticles cover the parasphenoid and pterygoid bones in (*a*). Bs, basisphenoid.

plates of dermal bone. Most constant are a large **opercular**, smaller **preoperculars** overlying the site of articulation of the upper and lower jaws, **suboperculars**, and **interoperculars** (fig. 9.10). In some of the more basal bony fishes, one or more **gular bones** (fig. 9.10) lie in the opercular membrane in the floor of the opercular chamber. They are represented in more specialized ray-finned fishes and dipnoans by **branchiostegal rays** located in caudally directed flaps, the **branchiostegal membranes** of the operculum. No opercular or gular plates or vestiges of them remain in tetrapods.

NEUROCRANIAL-DERMATOCRANIAL COMPLEX OF BONY FISHES

We will now examine the neurocranium and dermatocranium as they combine architecturally to contribute to the skulls of bony fishes. Later we will study the skeleton of the visceral arches to complete the picture.

Basal Actinopterygians

Sturgeons and spoonbills (paddlefishes) have a skull that justifies the name Chondrostei. The neurocranium remains almost completely cartilaginous throughout life. The only traces of endochondral ossification in the neurocranium of sturgeons are isolated sites in the otic capsules and in the sphenoid bone where it contributes to the walls of the orbit. Elsewhere, the neurocranium remains cartilaginous. The spoonbill is not much different. Its unusual bill is an extension of the cartilaginous rostrum. Numerous overlying dermal bones hide the cartilaginous neurocranium and the cartilaginous components of the upper jaws and hyoid arch.

Polypterus retains more of the primitive features of Paleozoic fishes than do sturgeons and spoonbills. The neurocranium is well ossified and is overlaid in the middorsal line by paired nasals, parietals, and postparietals of dermal origin (fig. 9.10). Lateral to these is a linear series of small, mostly unnamed bony dermal plates. Premaxillae and maxillae overlie the endochondral palatoquadrate bones of the first visceral arch, and a preopercular bone overlies the site of articulation of the upper and lower jaws. Other opercular bones complete the dermatocranium laterally, and a pair of primitive gular bones of dermal origin lie in the opercular membranes.

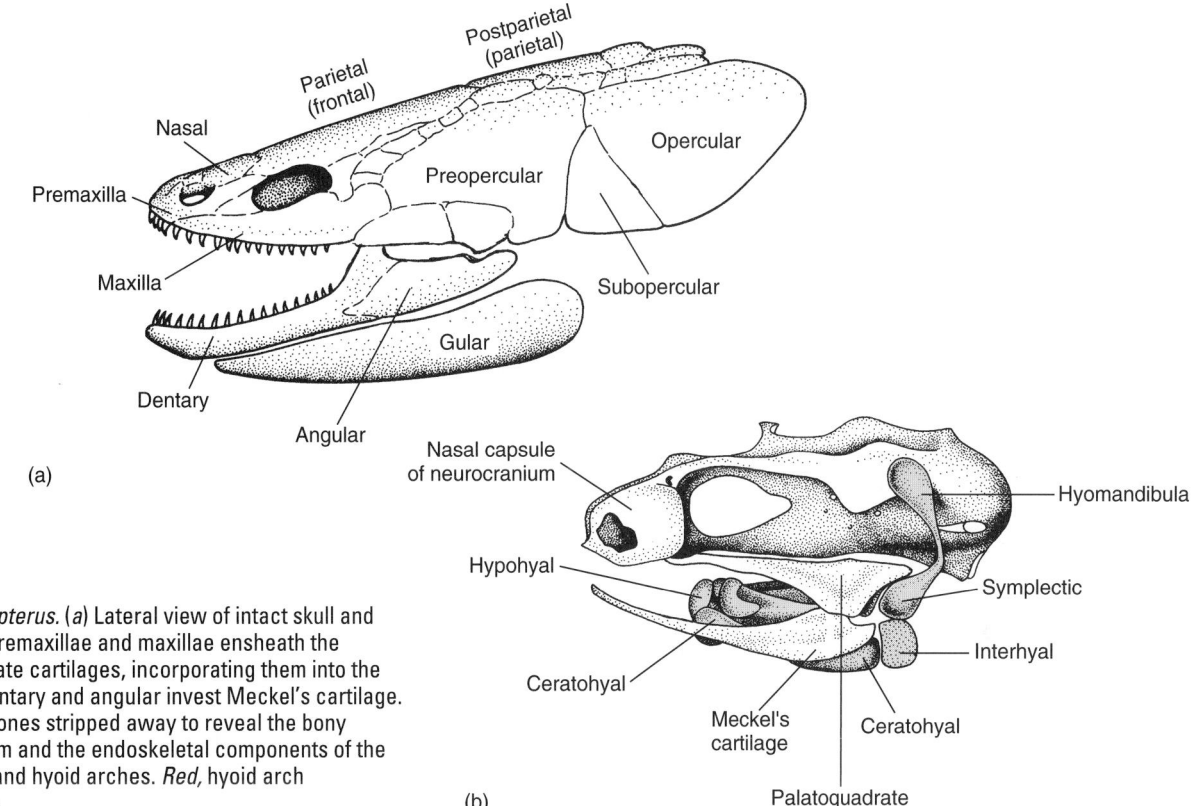

FIGURE 9.10

Skull of *Polypterus*. (*a*) Lateral view of intact skull and lower jaw. Premaxillae and maxillae ensheath the palatoquadrate cartilages, incorporating them into the skull. The dentary and angular invest Meckel's cartilage. (*b*) Dermal bones stripped away to reveal the bony neurocranium and the endoskeletal components of the mandibular and hyoid arches. *Red*, hyoid arch components.

Basal Neopterygians

The skulls of *Amia* and gars, like those of chondrosteans, bear little resemblance to the skulls of most modern fishes (figs. 9.7 and 9.11*a*). The dermal bones are grooved and pitted, exhibiting the sculpturing effects of the overlying skin. They can be characterized equally well as skull bones or dermal scales. The cheek plates of gars are especially compelling evidence that the dermatocranium may have been derived from earlier dermal scales. The neurocranium of *Amia* remains highly cartilaginous; in gars it is bony. A detailed account of the anatomy of the skull of *Amia* has been provided by E. Jarvik (1980).

Teleosts

The skulls of most teleosts are highly specialized and architecturally diverse. This is correlated with the diverse feeding habits of the group, which includes species that harvest plant tissues or capture animal life in every conceivable niche of the streams, lakes, and oceans of the world. A combination of highly maneuverable jaws and palates, coupled with the largest number of bones in any modern vertebrate skull, accounts for the anatomic diversity.

The skull of a carp is illustrated in figure 9.11*b*. Like many teleost skulls, it is compressed laterally and vaulted dorsally. The names of most of the bones will be recognized from earlier discussions. The maxillae, premaxillae, dentary, articular, quadrate, and symplectic are associated with the jaws and hyoid arch and will be discussed with the visceral skeleton. The posttemporal is the dorsalmost segment of the pectoral girdle. It has been secondarily incorporated into the skull. The mesethmoid, epiotic, and pterotic often incorporate some membrane bone along with the usual replacement bone. Branchiostegal rays in paired branchiostegal membranes beneath the gill chamber take the place of the more primitive gular bones of early ray-finned fishes.

The neurocranium in most teleosts is fully ossified except for the olfactory capsules. However, in cyprinids, which include the carp, neurocranial ossification is less than complete, and the islands of cartilage seen on the surface of the carp skull are unossified regions of the neurocranium or, at some sites, new cartilage.

Despite the kinetic nature of many teleost palates and the architectural disruptions that this has entailed, the dermal bone complement of the palate reflects that of ancestral bony fishes. The vomer, parasphenoid, and pterygoid bones are common components.

As mentioned earlier, names given to bones of the skull many years ago have been assigned on the basis of location (palatine), shape (pterygoid), or some other characteristic (squamosal) rather than on the basis of whether they are endochondral or intramembranous in origin. A bone of the same name in two species may be

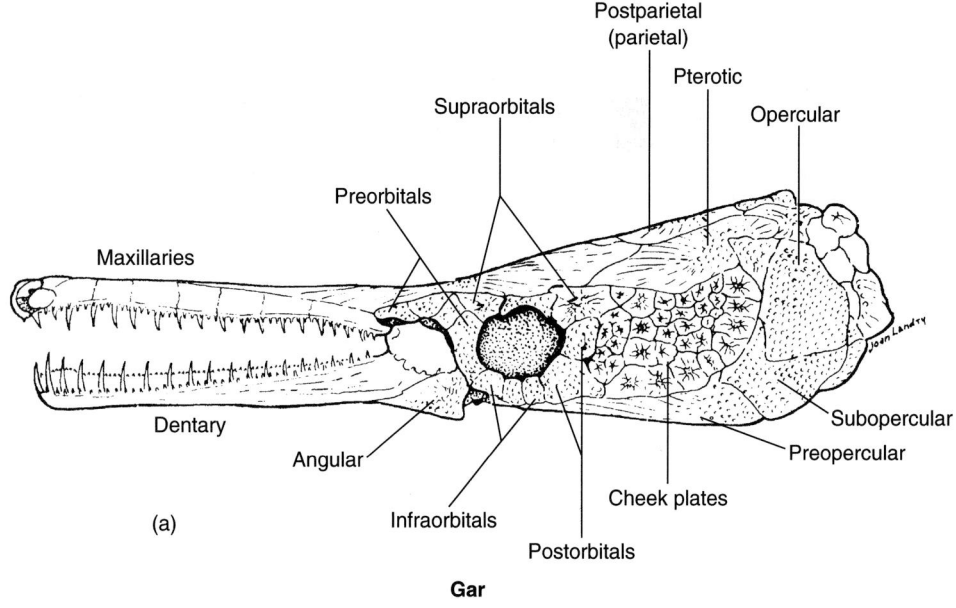

Gar

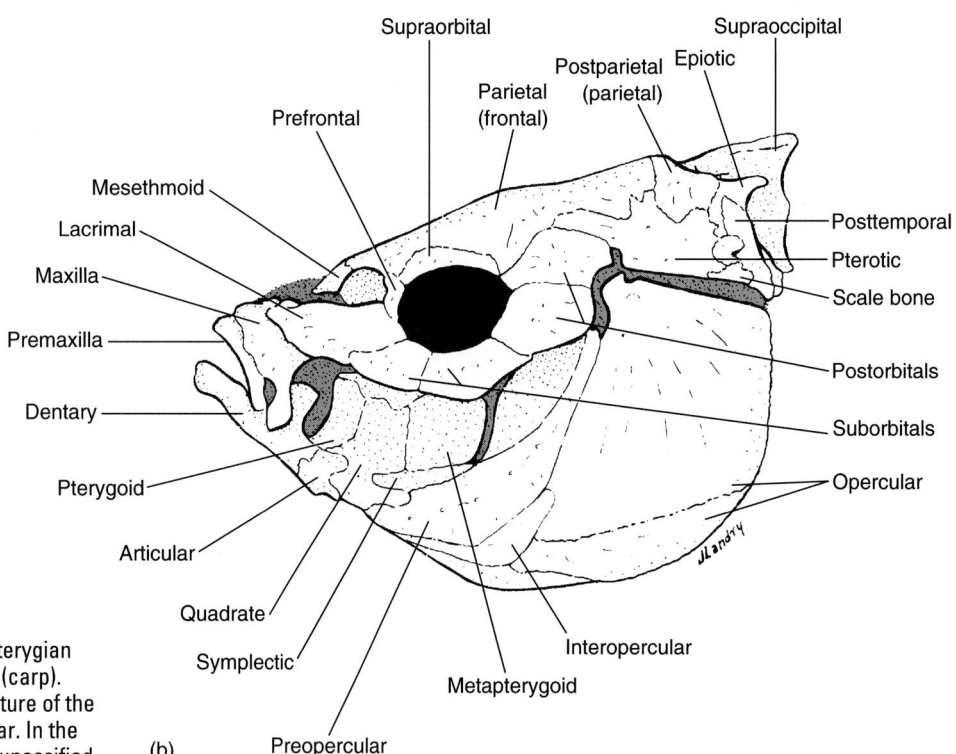

FIGURE 9.11
Skull of a basal neopterygian (gar) and of a teleost (carp). Note the scalelike nature of the cheek plates in the gar. In the carp, *red* represents unossified cartilage.

(b)

Carp

of endochondral origin in one species and of dermal origin in another. In such instances, the bones are not homologues. On the other hand, homologues sometimes have different names in different species, which poses problems of synonymy. Some of the bones labeled on the carp skull in figure 9.11b are of doubtful homology.

Dipnoans

Following their first appearance in the early Devonian, dipnoans did not experience the burst of speciation sub-

sequently undergone by teleosts. Consequently, changes in the architecture of dipnoan skulls have been more conservative. With reference to the dermatocranium, time has taken lungfishes in a direction quite different from that of teleosts, whose dermal bones increased in number and decreased in size. The dermatocranium of lungfishes has evolved from a large number of scalelike dermal bones in fossils to a relatively few broad bony plates in today's species. The neurocranium is largely cartilaginous like that of some chondrosteans and basal

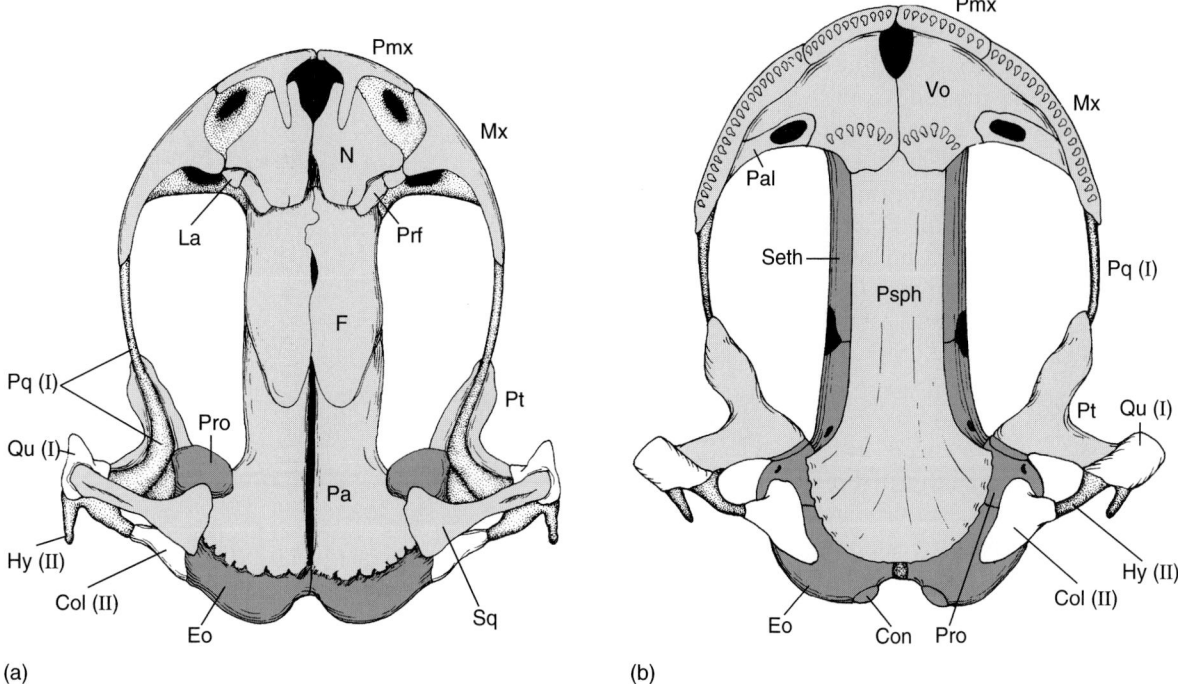

FIGURE 9.12

Skull of *Ranodon,* a urodele in the family Hynobiidae. (*a*) Dorsal view. (*b*) Palatal view. *Light red,* dermal bones; *dark red,* neurocranial bones; *stipple,* cartilage. **Col,** columella; **Con,** occipital condyles; **Eo,** exoccipital; **F,** frontal; **Hy,** a dorsal segment of the hyoid skeleton; **La,** lacrimal; **Mx,** maxilla; **N,** nasal; **Pa,** parietal; **Pmx,** premaxilla; **Pq,** palatoquadrate cartilage; **Prf,** prefrontal; **Pro,** prootic; **Psph,** parasphenoid; **Pt,** pterygoid; **Qu,** quadrate; **Seth,** sphenethmoid; **Sq,** squamosal; **Vo,** vomer; **I** and **II** indicate origin from the first or second visceral arch.

From I.I. Schmalhausen, *The Origin of Terrestrial Vertebrates.* Copyright © 1968. Academic Press, Orlando, FL. Reprinted by permission.

neopterygians, rather than bony, as in teleosts. The palate accommodates openings of the nasal canal into the oral cavity just behind the mouth, as in other rhipidistians.

NEUROCRANIAL-DERMATOCRANIAL COMPLEX OF MODERN TETRAPODS

We can now look at the neurocranial-dermatocranial complex as it has evolved in modern tetrapods. A disproportionate amount of time will be spent on amniotes because they have diversified more than other tetrapods and because during the Mesozoic the skulls of some of them underwent architectural changes that were incorporated into the skulls of birds and mammals.

Amphibians

The skulls of modern amphibians are considerably modified from those of labyrinthodonts, although they are still flattened (**platybasic**) compared with the vaulted skulls of amniotes. The neurocranium is incomplete dorsally (see fig. 9.3*c*), and much of it remains cartilaginous except in apodans. The rigidity of the apodan skull may be correlated with burrowing. The only replacement bones in anurans and urodeles (excepting the columella, when ossified) are a sphenethmoid, two

prootics, and two exoccipitals, each of the latter bearing a condyle. In some perennibranchiate urodeles, even a sphenethmoid fails to ossify.

Abutting against the otic capsule in amphibians (and other tetrapods) is a skeletal rod, the **columella** (stapes), a middle ear ossicle that conducts sound waves from an eardrum to the capsule. The phylogenetic origin of the columella will be discussed under "Visceral Skeleton."

The dermatocranium is quite incomplete. The bones that surrounded the orbit of early amphibians have been lost except for the lacrimals and prefrontals of basal urodeles (fig. 9.12*a*). Also missing are the primitive bones of the temporal region from the orbits caudad (intertemporals, supratemporals, tabulars, and postparietal). This loss of ensheathing bones leaves the otic capsule (prootic) exposed dorsally and laterally. Only the squamosal and sometimes a quadratojugal remain in this region. Premaxillae and maxillae are usually present to ensheath the upper jaw cartilages at the margin of the dermatocranium, but in perennibranchiates even maxillae may fail to develop.

The primary palate has been altered. In anurans and a few urodeles, large palatal vacuities have evolved beneath the orbit, reducing the palatines to transverse splinters that brace the upper jaw against the palate anteriorly and reducing the pterygoids to a pair of bipartite bones that brace the upper jaw against the braincase posteriorly (figs. 9.12*b*, and 9.13, *Rana*). Because of the palatal vacuities, the large eyeballs of anurans, which protrude just above the waterline to monitor the environment in an otherwise submerged frog, can be retracted until the membranous floor of the orbit bulges into the oral cavity. In urodeles, the parasphenoid has become exceptionally broad (figs. 9.12*b*, and 9.13, *Necturus*), and palatal vacuities are lacking in some orders.

Nonavian Reptiles

The skulls of stem reptiles were little changed from those of labyrinthodonts, and some primitive features are still present in living reptiles. Among these are a well-ossified neurocranium, a single occipital condyle, a large complement of membrane bones, and, in rhynchocephalians and many lizards, a parietal foramen housing a median eye (fig. 9.14*c*). Figure 17.19 is a photograph of the median eye of an iguana lizard. Major changes from stem reptile skulls included the appearance of temporal fossae and development of a partial or complete secondary palate. Temporal fossae and secondary palates were inherited by birds.

The fully ossified neurocranium of an adult alligator consists of two exoccipital bones, a supraoccipital, a basioccipital that bears a single condyle for articulation with the atlas, laterosphenoids, ethmoids, and several otics that arise from separate ossification centers in the otic capsules. None of the otic bones are on the surface, being overlaid by the squamosal, a dermatocranial bone. Some of the otic bones are coalesced with an adjacent occipital. The chiefly cartilaginous neurocranium of a juvenile *Sphenodon* is seen in situ in figure 9.14*c*. It will ossify further with age, but more cartilage remains in adult sphenodons and lizards than in other reptiles.

The roofing and marginal bones of the dermatocranium of three reptiles are seen in figure 9.14. Of living reptiles, crocodilians retain the largest number of membrane bones. Turtles have the most enigmatic skull, as will be explained shortly.

Temporal Fossae

A temporal fossa (temporal fenestra) is a cavernous opening in the temporal region of some amniote skulls bounded by one or more bony arches (fig. 9.15). Stem

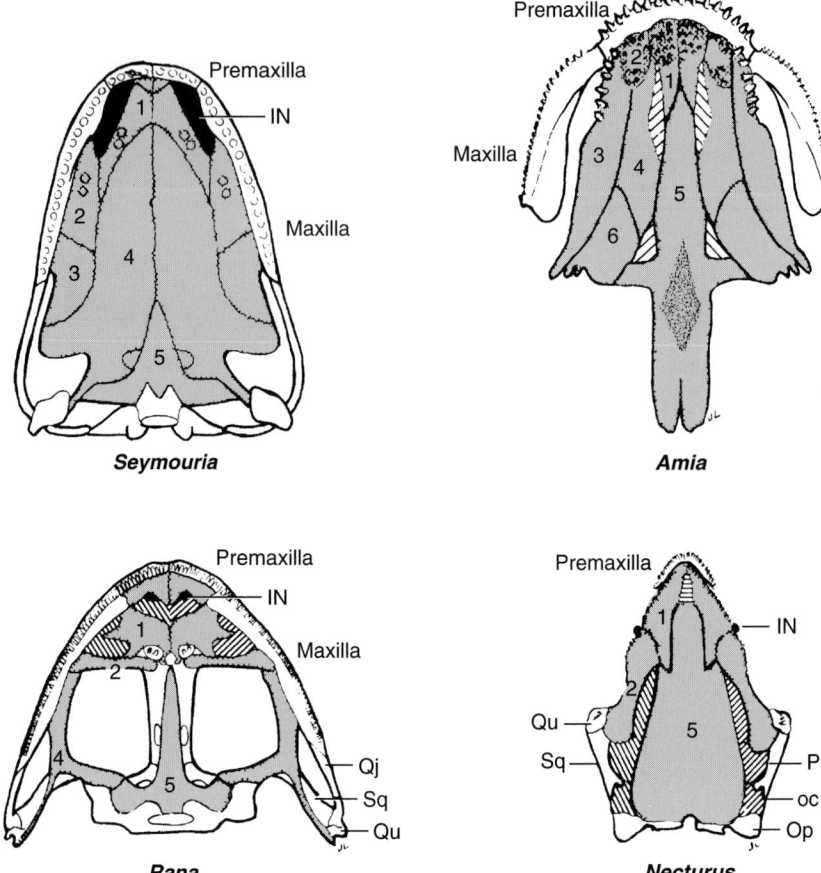

Primary Palates

FIGURE 9.13

Primary palates (in color) of a reptilelike amphibian (*Seymouria*), a basal neopterygian (*Amia*), and two amphibians. Cartilage is indicated by *diagonal lines;* internal nares are *black.*
1, vomer; **2,** palatine (in *Necturus,* palatopterygoid); **3,** ectopterygoid; **4,** pterygoid; **5,** parasphenoid; **6,** epipterygoid. **oc,** cartilaginous portion of otic capsule; **Op,** opisthotic; **Pro,** prootic; **Qu,** quadrate; **Qj,** quadratojugal; **Sq,** squamosal; **IN,** internal naris.

reptiles had none (a retention of the primitive condition), so their skulls are **anapsid**; that is, they lack temporal arches. Today, among living amniotes only turtles are anapsid (fig. 9.16). Synapsida developed a single **lateral temporal fossa** surrounded by the postorbital, squamosal, and jugal bones, the last two forming an underlying **infratemporal (zygomatic)** arch. This synapsid skull was transmitted to mammals. In humans, the cheek-bones are the zygomatic arches (see fig. 9.28).

Archosaurs and the predecessors of *Sphenodon*, lizards, and snakes had superior and inferior temporal fossae. When two fossae are present, the skull is said to be **diapsid**. (For derivation of the term, see the glossary.) The lower arch corresponds to the zygomatic arch of mammals. The upper arch, or **supratemporal**—the arch between the two fossae—consists of parts of the postorbital and squamosal bones. Crocodilians and *Sphenodon* still have diapsid skulls (figs. 9.14*a* and

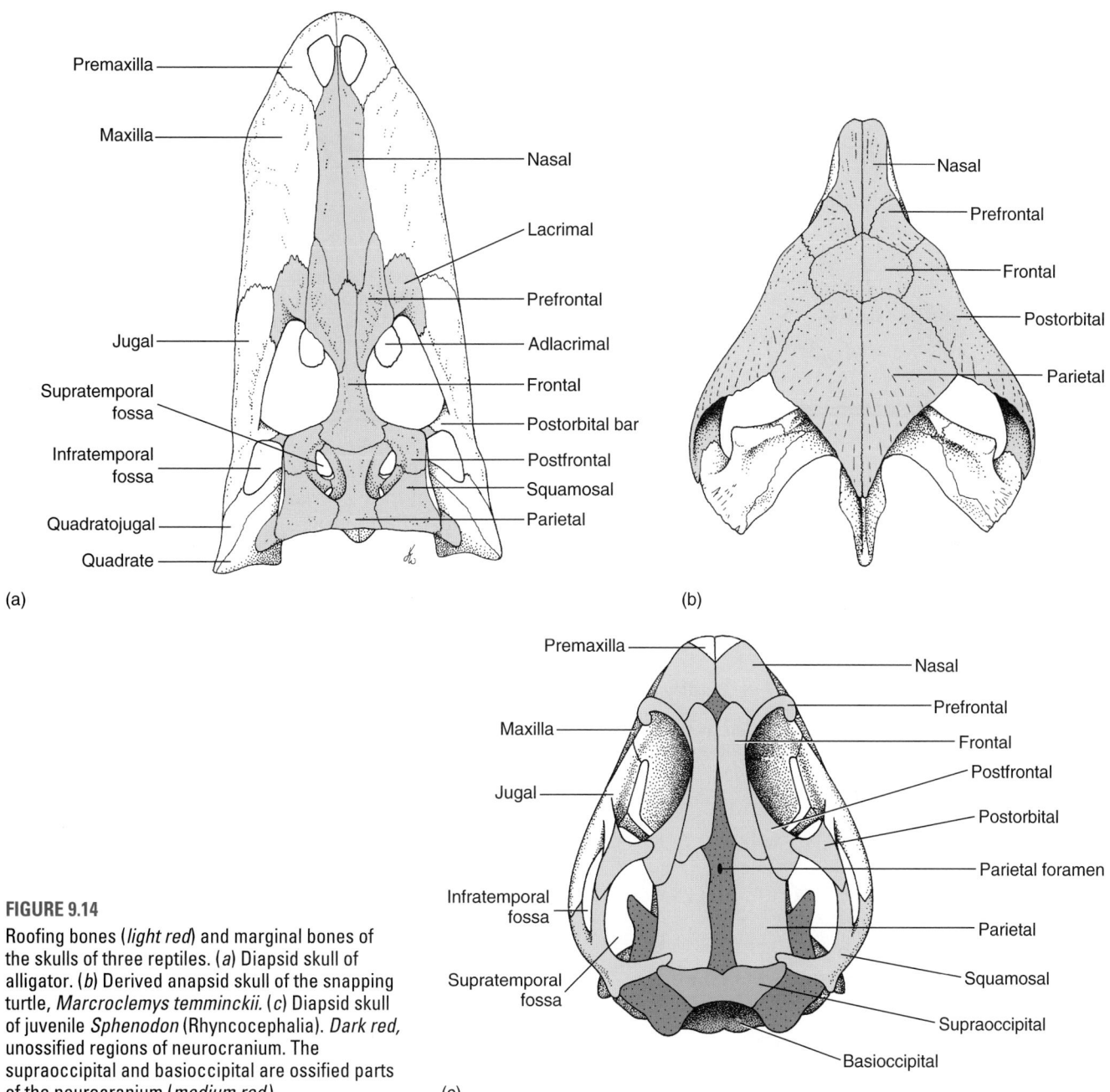

Premaxilla

Maxilla

Nasal

Lacrimal

Prefrontal

Jugal

Adlacrimal

Supratemporal fossa

Frontal

Infratemporal fossa

Postorbital bar

Quadratojugal

Postfrontal

Quadrate

Squamosal

Parietal

(a)

Nasal

Prefrontal

Frontal

Postorbital

Parietal

(b)

Premaxilla

Nasal

Maxilla

Prefrontal

Frontal

Jugal

Postfrontal

Postorbital

Parietal foramen

Infratemporal fossa

Parietal

Supratemporal fossa

Squamosal

Supraoccipital

Basioccipital

(c)

FIGURE 9.14

Roofing bones (*light red*) and marginal bones of the skulls of three reptiles. (*a*) Diapsid skull of alligator. (*b*) Derived anapsid skull of the snapping turtle, *Marcroclemys temminckii.* (*c*) Diapsid skull of juvenile *Sphenodon* (Rhyncocephalia). *Dark red,* unossified regions of neurocranium. The supraoccipital and basioccipital are ossified parts of the neurocranium (*medium red*).

9.17*a*), but modern lizards have lost part or all of the lower arch, and snakes have lost both arches (fig. 9.17*b* and *c*). This loss has left a cavernous void in the posterolateral walls of squamate skulls. These are **modified diapsid skulls.** Reduction or loss of the arches along with the acquisition of intracranial joints facilitated **cranial kinesis,** the movement of one section of a skull independent of others.

Ichthyosaurs and plesiosaurs (fig. 4.22) had one temporal fossa, dorsally located, that resembled the dorsal fossa of diapsids in its anatomic relationships (compare figs. 9.15*c* and *d*). However, it is likely that this **euryapsid** condition is an instance of evolutionary convergence in lineages distinct from that leading to other reptiles. All euryapsid reptiles are extinct (see fig. 4.22).

The temporal region of a turtle skull is an enigma. Absence of temporal fossae suggests a primitive condition. However, there has been extensive loss of dermal bones and excavation of the temporal region dorsal to the otic capsule (see fig. 9.14*b*), a condition more pronounced in some families than in others. Supratemporal, tabular, and postparietal bones are missing, postorbitals and postfrontals have united, and parietals give the impression of having receded from the rear, leaving a wide vacuity that is best seen from a caudal or ventral view.

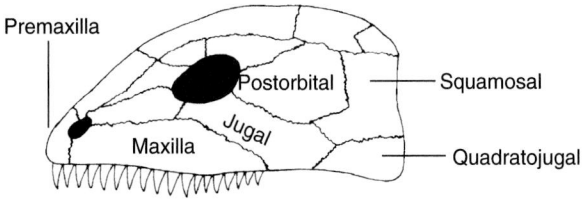

Anapsid (stem reptile)

(a)

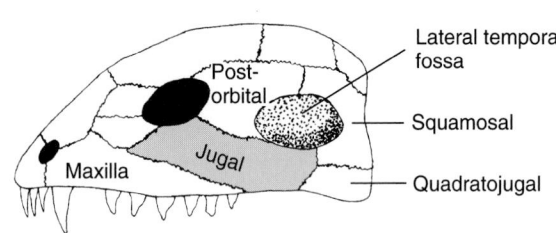

Synapsid

(b)

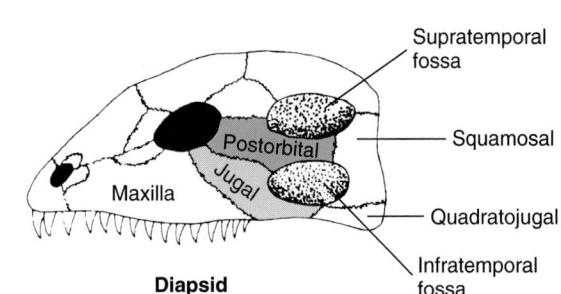

Diapsid

(c)

Euryapsid (ichthyosaurs and plesiosaurs)

(d)

FIGURE 9.15
Temporal fossae. *Light red,* infratemporal arch component; *dark red,* supratemporal arch component. Euryapsids are extinct lepidosaurs (see fig. 4.22).

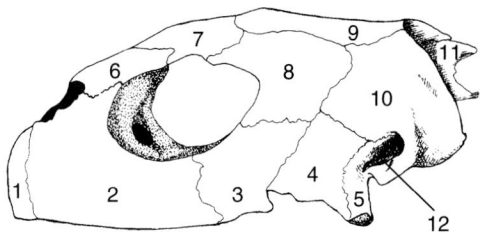

FIGURE 9.16
Anapsid skull of small sea turtle, lateral view, lower jaw removed.
1, premaxilla; **2,** maxilla; **3,** jugal; **4,** quadratojugal; **5,** quadrate; **6,** prefrontal; **7,** frontal; **8,** postorbital; **9,** parietal; **10,** squamosal; **11,** supraoccipital; **12,** middle ear cavity.

and early tetrapods to seize food, bite off pieces, and close the mouth to prevent return of the food to the environment. The food was swallowed whole. Temporal fossae provided space for an expanding amniote adductor muscle to shorten and thicken during contraction; they provided access to the surface of the dermatocranium so that one portion of the adductor, the temporalis muscle, could acquire more leverage by spreading upward onto the temporal region of the skull; and they provided a zygomatic arch to which another portion of the adductor, the masseter, could acquire an anchorage. Figure 11.24c shows the temporalis and masseter in relation to the zygomatic arch and temporal fossa of a primate. These powerful new adductors, assisted by other muscles of the mandibular and hyoid arches, make possible the complex side-to-side, forward-backward, and rotary chewing movements seen in herbivorous mammals that grind grasses or chew their cud and in carnivores that macerate flesh and crush bones.

Secondary Palates

A secondary palate is a horizontal partition that partially or completely divides the primitive oral cavity into separate oral and nasal passageways, thereby displacing the **internal nares (posterior choanae)** caudad. Embryonic development of a secondary palate in a mammal is illustrated in figure 9.18. In vertebrates with a secondary palate, the primary palatal components remain in the roof of the nasal passageway, usually with a reduced complement in membrane bones. For example, no parasphenoid bone is present in mammals.

In crocodilians, medially directed, shelflike **palatal processes** of the premaxillae, maxillae, palatine, and pterygoid bones meet in the midline to form a long bony secondary palate with internal nares far to the rear (fig. 9.19). In other reptiles, not all the palatal processes reach the midline, so the secondary palate is incomplete. Varying degrees of completion can be seen by examining several species of turtles (fig. 9.20). When the secondary palate is incomplete, the respiratory airstream is channeled in a fairly deep longitudinal groove, the

Temporal fossae provide space and surfaces in functionally advantageous positions for accommodating the powerful adductor muscles needed to operate the lower jaws of amniotes. In labyrinthodonts and extinct basal reptiles, the adductor mandibulae (chief levator muscle of the lower jaw) was confined to cramped quarters internal to the temporal region of the dermatocranium. (In cartilaginous fishes, as seen in figure 11.24a, it lies just under the skin of the head because there is no dermatocranium.) At best, this muscle enabled bony fishes

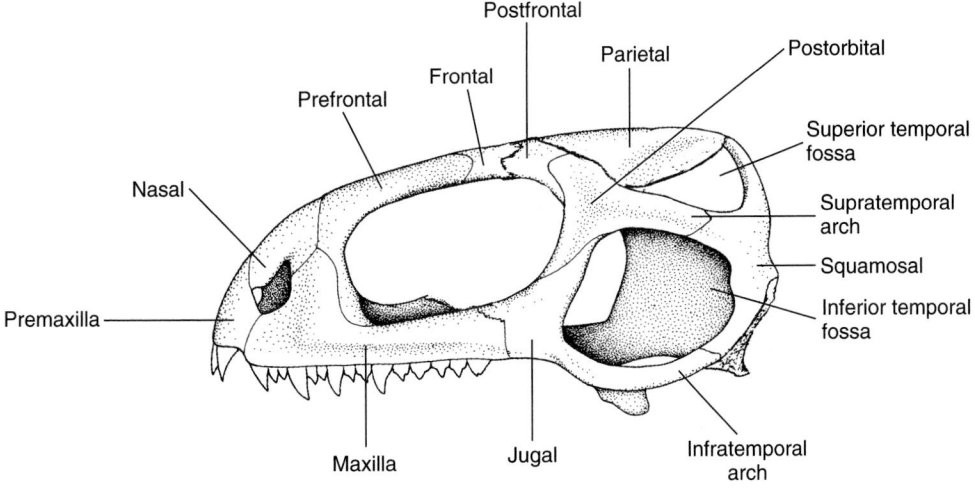

(a) *Sphenodon*

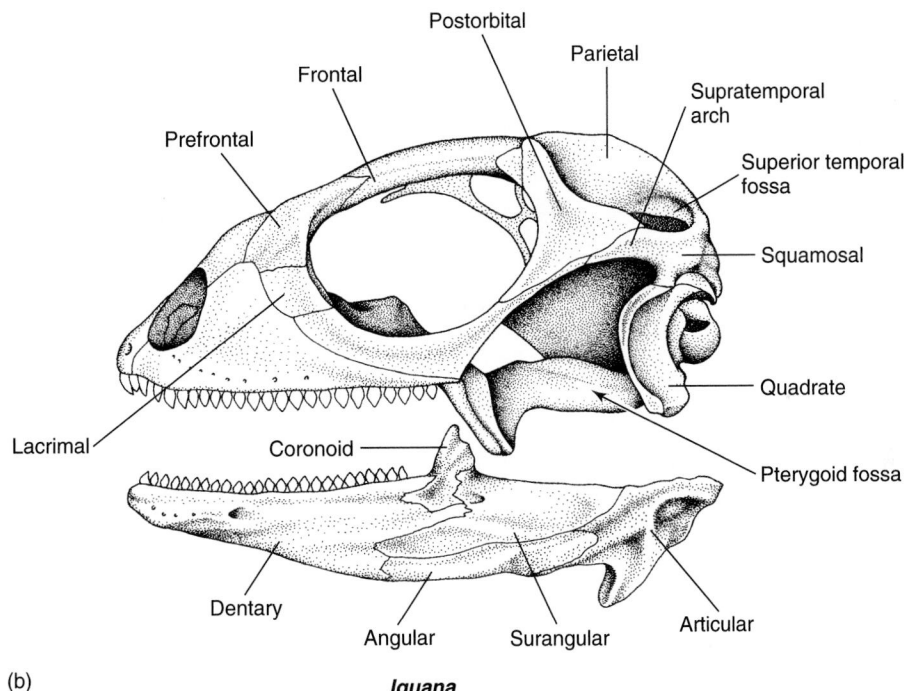

(b) *Iguana*

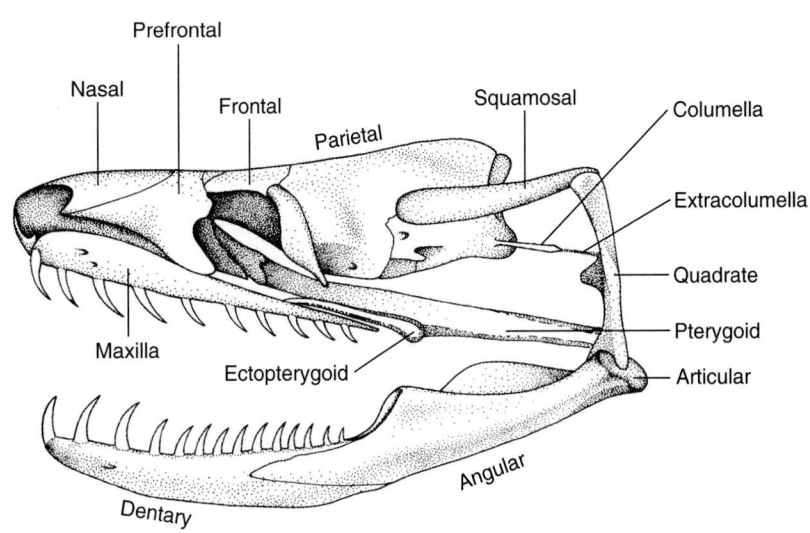

(c) *Boa*

FIGURE 9.17

Diapsid (*a*) and modified diapsid (*b, c*) skulls of three reptiles. In (*b*), the lower arch is missing. In (*c*), both arches are missing.

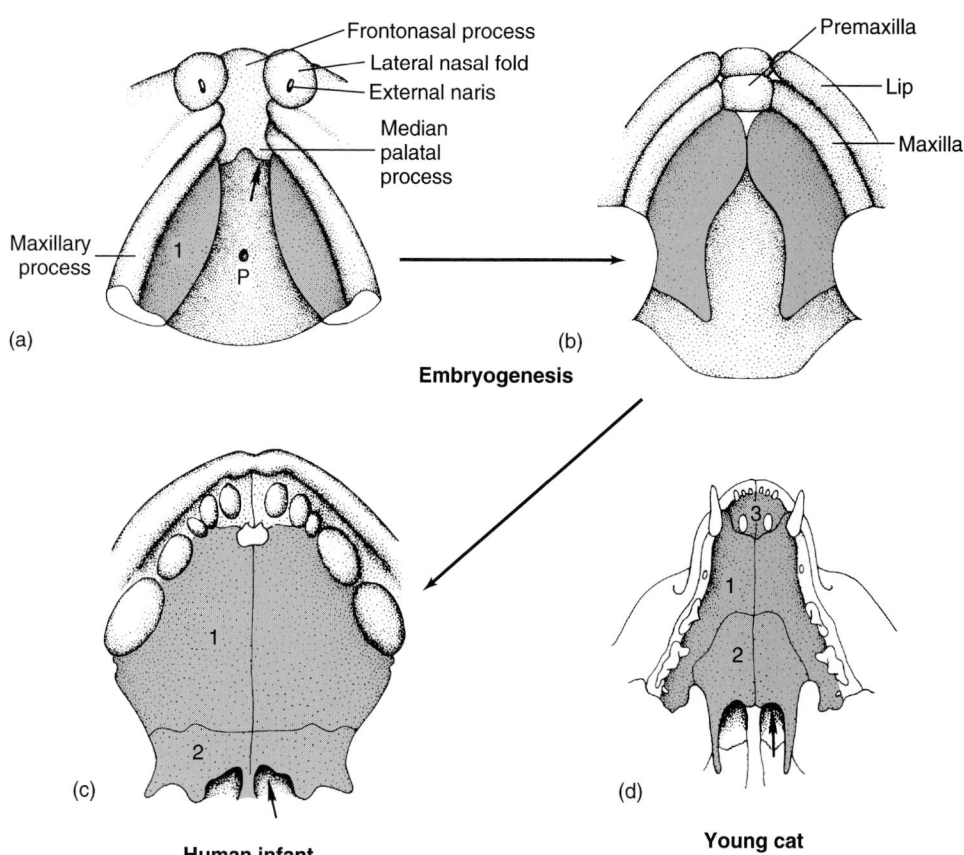

(a)

(b)

Embryogenesis

FIGURE 9.18

(*a*) to (*c*) Formation of secondary palate (*red*) in humans. (*d*) Secondary palate of young cat for comparison. *Arrows* point to posterior choanae (internal nares). **1,** palatal process of maxilla; **2,** palatal process of palatine bone; **3,** palatal process of premaxilla. In (*a*) (fetus approximately 18 weeks old), the palatal process of the maxillae are growing toward the midline, forming a secondary roof in the oral cavity. In (*b*), the processes have met anteriorly. In (*c*), the palate is complete. **P,** evagination site of anterior pituitary.

(c)

Human infant

(d)

Young cat

Secondary palates

palatal fissure, in the roof of the oral cavity (fig. 9.20*b* and *c*). The borders of the fissure are fleshy **palatal folds,** seen also in birds (see fig. 12.4). In mammals, the secondary palate extends all the way to the pharynx, but the caudal portion—the soft palate—lacks bone (see fig. 12.3*b*).

Cranial Kinesis

Cranial kinesis, or kinetism, is the movement of a functional component of a skull independent of another component. It is made possible by the presence of movable intracranial joints between the two components. Teleosts, lizards, snakes, and birds are especially well endowed with such joints. For instance, most of them raise the upper jaw and palate as a unit independent of the neurocranium when they open their mouth. In some teleosts, even the two sides of the upper jaw can be moved independently. These movements cannot be accomplished by crocodilians, for example, or by frogs, salamanders, or mammals. However, Paleozoic fishes, many labyrinthodonts, early reptiles, and the ancestors of mammals had kinetic skulls, and large numbers of living vertebrates do. As a result of chance mutations, kinetism has probably evolved independently many times during vertebrate evolution.

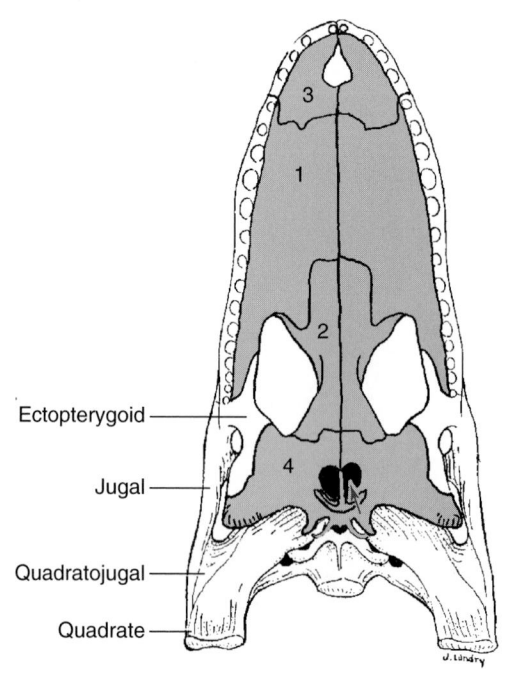

FIGURE 9.19

The long rigid secondary palate of an alligator (*red*). **1, 2,** and **3,** palatal processes of maxilla, palatine and premaxilla bones, respectively; **4,** pterygoid bone. *Arrow* indicates an internal naris.

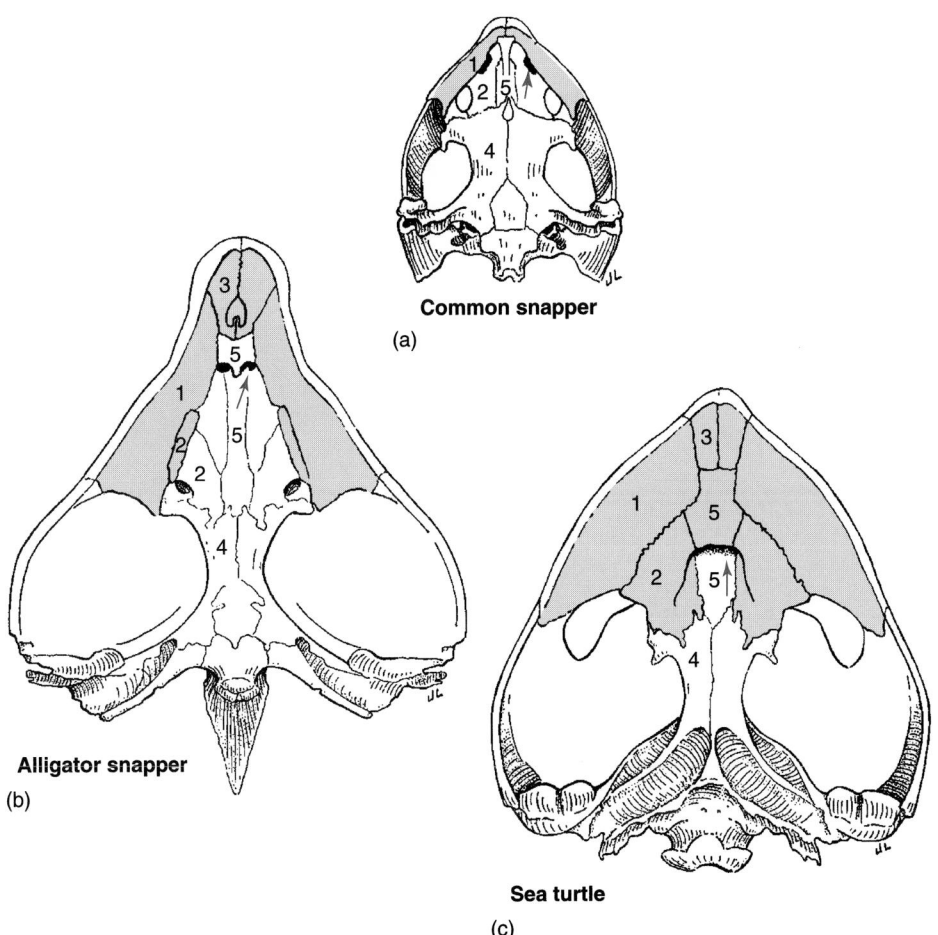

Common snapper
(a)

Alligator snapper
(b)

Sea turtle
(c)

FIGURE 9.20

Species differences in the secondary palates (*red*) of turtles. (*a*) *Chelydra serpentina.* (*b*) *Macroclemys temminckii.* (*c*) *Lepidochelys olivacea*, temporal and supraoccipital regions omitted. **1**, maxilla; **2**, palatine; **3**, premaxilla; **4**, pterygoid; **5**, vomer. **1, 2, 3, 5** in *red* are secondary palatal processes only. Note that in (*a*), only the maxilla has palatal processes, and these are rudimentary. *Arrows* point to an internal naris.

Cranial kinesis (fig. 9.21) is correlated primarily with food-getting and manipulation of food within the oral cavity. It enables small vegetarian teleosts to scrape algae from underwater objects, and it enables plankton-feeding and other small fish to obtain food by suction. (Details are described in chapter 12.) Cranial kinetism enables some squamates, especially vipers, to open their mouth wide enough to swallow prey larger than their own head. A movable quadrate, palate, upper jaw, and some of the bones in front of or above the orbit contribute to the process, along with a hinge joint between the kinetic unit and the rest of the braincase. The palate and temporal region of squamates and some species of birds have become so modified for feeding that the primitive dermatocranial-splanchnocranial complex has become greatly disrupted. The palatal complex illustrated in figure 9.22 exemplifies this disruption. The names applied to the bones in red are those kinetic elements assigned to a primary palate, the parasphenoid of the primary palate is not kinetic, and there are no other palatal bones.

Table 9.1 contrasts the skulls of early tetrapods with those of modern amphibians and modern reptiles with respect to the palates and other traits.

Birds

The avian skull is fundamentally reptilian but exhibits modifications correlated with flight, altered feeding habits, and a much larger brain. Some of the roofing bones have been lost, and the remaining dermal bones are much thinner; in carinate skulls, most of the sutures have been obliterated. Therefore, to identify individual bones, one must use the skull of a ratite or a nestling.

One might think of the avian skull as comprising two functional regions (fig. 9.23): at the rear, a solid bony box (neurocranium and dermatocranium) that houses the brain, olfactory organs, eyeball, and equilibratory and hearing complex, and therefore protects all structures needed for input and processing of information; and at the front, a food procuring and handling area, the elongated beak and palate.

The posterior component is highly vaulted, bulging outward and arching upward alongside the greatly enlarged brain. The orbits are situated where they achieve maximal arcs of visibility, and they reflect the size of the massive eyeballs. The parietal foramen has closed. The neurocranium is incomplete dorsally and, like that of other reptiles, well ossified, the only cartilage being in the olfactory capsule and in the mesethmoid component

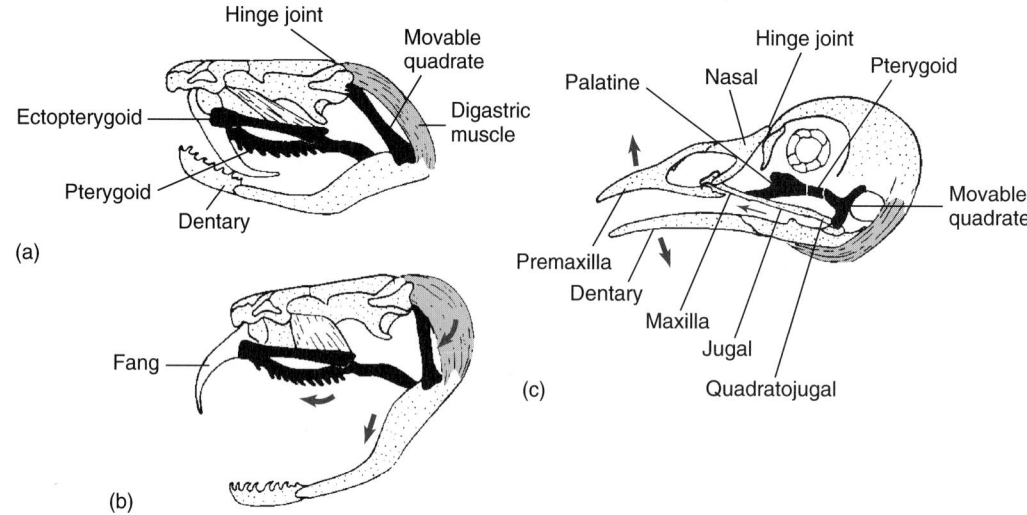

FIGURE 9.21

Cranial kinetism in the palate of snakes (*a* and *b*) and birds (*c*). *Arrows* indicate direction of movement of components. Digastric muscle (*red*) contraction lowers the mandible.

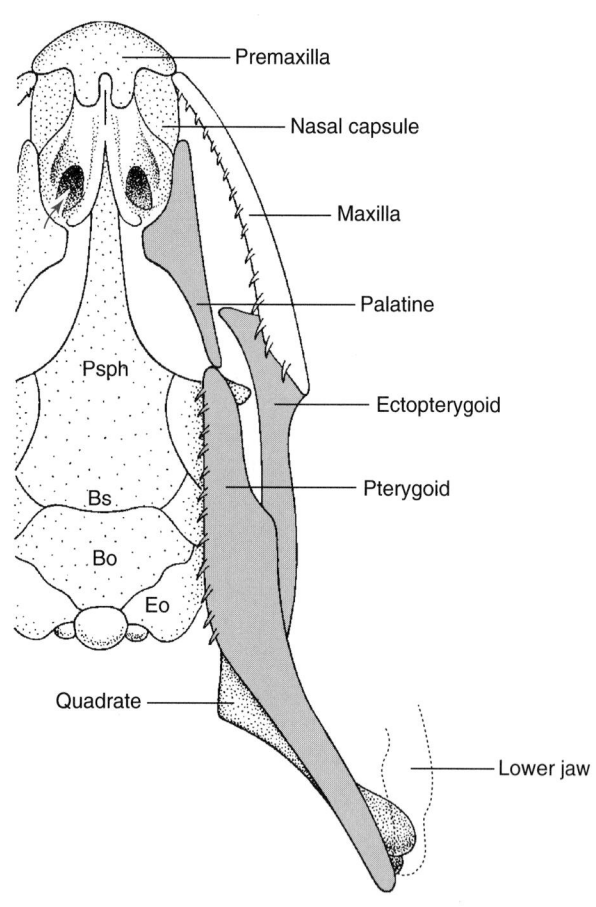

FIGURE 9.22

Kinetic palatal complex of the snake *Tropidonotus*. **Bo,** basioccipital; **Bs** and **Psph,** ankylosed basisphenoid and parasphenoids of primary palate; **Eo,** exoccipital. Vomer bones, not visible, lie anterior to the parasphenoid. The left and right palatine and pterygoid bound a median palatal fissure. *Arrow* indicates an internal naris. The posterior part of the lower jaw is drawn as being transparent to show the site of articulation with the quadrate.

of the interorbital septum. As in other reptiles, the neurocranium bears a single occipital condyle.

The skull is modified diapsid, the arch between the superior and inferior temporal fossae having been lost. The resulting fossa is broadly open to the rear and is confluent with the orbit in front. A preorbital fossa, separated from the orbit by the lacrimal bone, is present, as it was in other dinosaurs. The infratemporal arch, consisting of a primitive complement of jugal and quadratojugal bones, is intact but very slender (fig. 9.23).

The kinetic palate resembles that of squamates except that the ectopterygoids have been lost (fig. 9.24). When a bird with a kinetic palate opens its mouth by lowering the lower jaw, the quadrate bone is pushed forward and the motion is transmitted to the upper beak via either a movable palate, a movable zygomatic arch, or a combination of these. The parasphenoid is immobile, being fused to the basisphenoid. There are a multitude of variations in the structure of beaks and palates and in their anatomic relationships to one another and to the braincase. For this reason, palatal structure has been used as one basis for establishing major avian taxa.

Not all bird skulls are equally kinetic. Birds such as woodpeckers that subject the beak to rough usage usually have minimal kinetism. The ethmoid region of the neurocranium and a narrow arch of roofing bones between the orbits bear the shocks generated when woodpeckers drill in solid wood. Some adaptations of the beak for procuring food are illustrated in figure 2.4.

Mammals

Major features that differentiate mammalian skulls from those of other amniotes are emergence of the dentary as the sole bone of the lower jaw, an altered site of articulation of the lower jaw with the braincase, alterations in

TABLE 9.1 Skulls of Early Tetrapods Contrasted with Those of Modern Amphibians and Reptiles with Reference to a Few Selected Characteristics

	Early Tetrapods	**Modern Reptiles**	**Modern Amphibians**
Neurocranium	Well ossified	Well ossified	Mostly cartilage
	One condyle	One condyle	Two condyles
Primary palate	Complete complement of dermal bones	Relatively complete	Fewer
	Parasphenoid small	Small	Large in urodeles
	Vacuity small	Small	Large in anurans
Secondary palate	None	Partial or complete	None
Temporal fossae	None	Present except turtles	None
Dermal roofing bones	Complete complement	Some reduction	Extensive reduction
Parietal foramen	Present	Present in some	Confined to larvae
Marginal bones	Complete complement	Usually complete	Fewer
Bones ensheathing Meckel's cartilage	Numerous	Numerous	Fewer

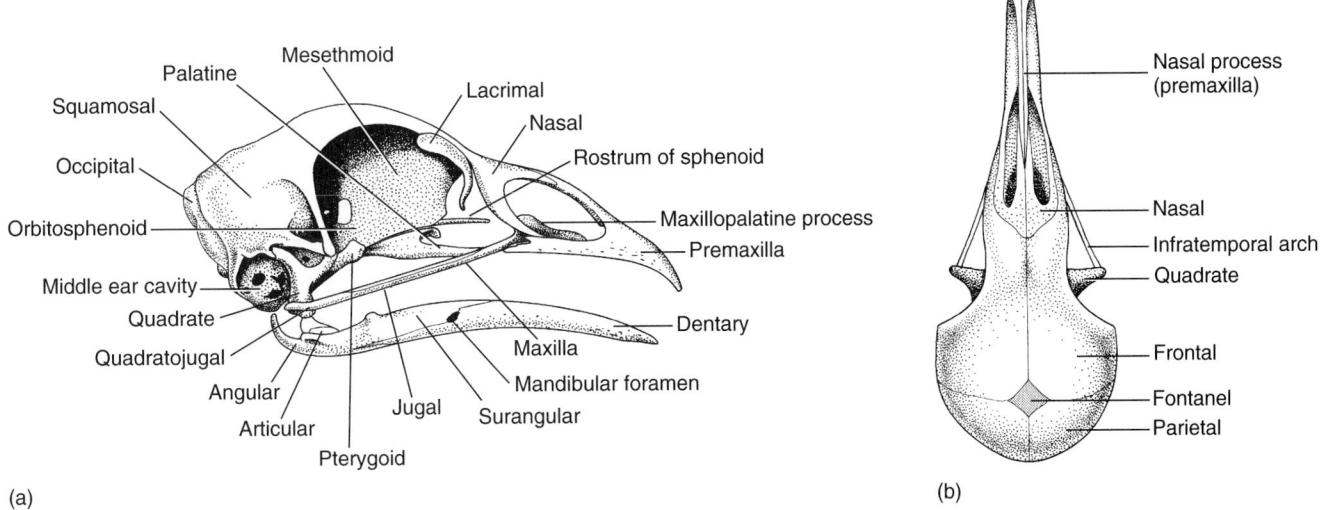

(a)   (b)

FIGURE 9.23

(*a*) Skull of adult chicken, lateral view. (*b*) Skull of young pigeon, dorsal view. The fontanel in (*b*) (*gray*) undergoes intramembranous ossification, and sutures are obliterated as pigeon matures.

the secondary palate, and the presence of three bones (ear ossicles) in the middle ear cavity. The mammals, as members of the Synapsida, differ from reptiles in their modifications in the temporal region of the dermatocranium. The skull has become increasingly domed as the cerebral hemispheres have ballooned dorsally, laterally, and caudad.

The neurocranium is incomplete dorsally (see fig. 9.4), with the result that membranous soft spots, or **fontanels**, can be felt in the heads of newborn human infants until ossification of the roofing bones is complete (fig. 9.25). Fontanels enable the fetal skull to be molded as necessary during delivery through the narrow birth canal. One or more small **bregmatic bones** may ossify in the frontal fontanel in some species, and a single bregmatic bone may develop as an anomaly in human skulls. Paracelsus called this bone the "antiepileptic bone" because he believed it served as a

pop-up valve for relieving pressure within the cranial cavity.

Ossification centers in the neurocranium of a fetal pig are shown in figure 9. 4. The basioccipital, basisphenoid, and presphenoid bones form a floor on which the brain rests; exoccipitals, alisphenoids, and lateral extensions of the presphenoids form partial side walls; and a supraoccipital completes, along with the basioccipital and exoccipitals, a bony ring that resembles a vertebra. Ethmoid cartilages and bones house the olfactory epithelium, underlie the olfactory bulbs of the brain, constitute much of the nasal septum (mesethmoid), and are perforated by foramina for olfactory nerve bundles (see figs. 9.4, 9.6, and 9.30). Ossification centers in the otic capsules unite solidly to form petrosal (periotic) bones beneath the overgrown temporal lobes of the mammalian cerebral hemispheres. They are often incorporated into a temporal bone complex. Each exoccipital

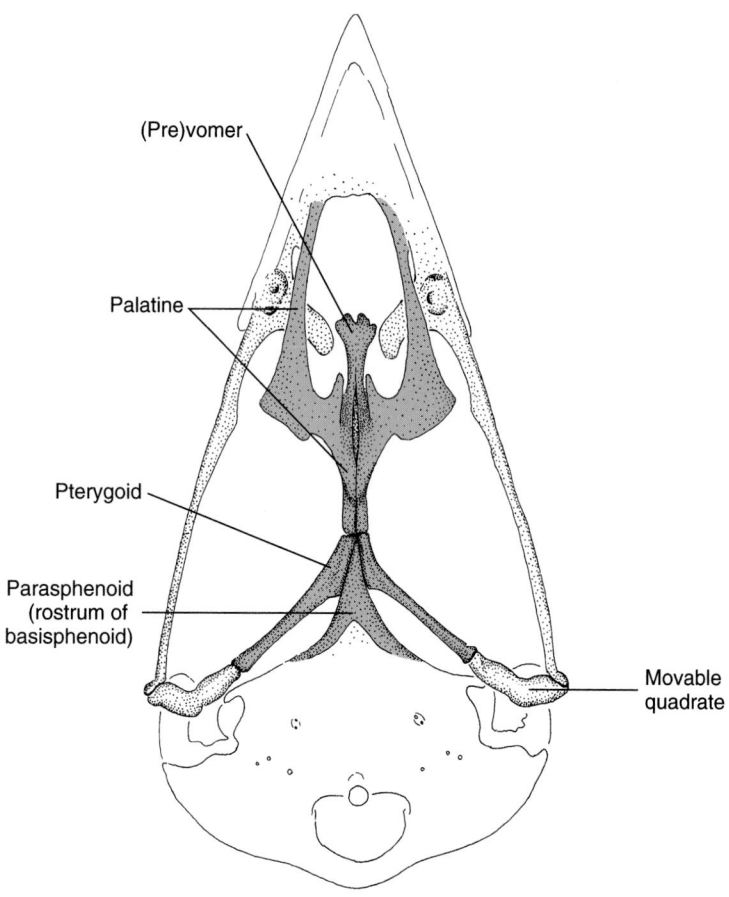

FIGURE 9.24

Palatal complex of a bird, the cotinga, ventral view. The parasphenoid is immovable. The homology of the prevomer is conjectural.

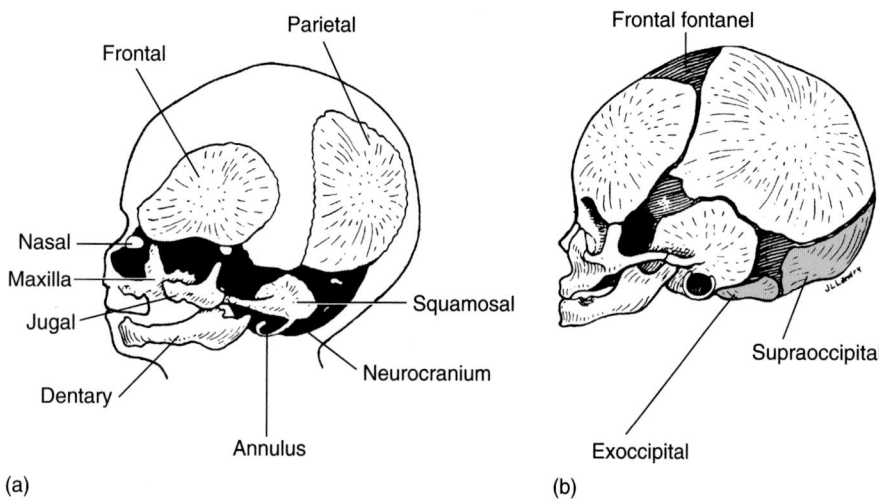

(a)   (b)

FIGURE 9.25

Two stages in the development of a human skull. (*a*) Intramembranous ossification is under way. The cartilaginous neurocranium (*black*) is incomplete lateral to and above the brain. (*b*) Intramembranous ossification has progressed, but "soft spots" (fontanels) remain where there is no membrane bone. Neurocranial bones are *red*.

bone bears a condyle inherited from therapsid ancestors. The neurocranial derivatives of a generalized mammalian skull are diagrammed in figure 9.26.

The sequence in mammals of centrumlike basioccipital, basisphenoid, and presphenoid bones and their dorsal wings (the exoccipitals, alisphenoids, and lateral aspect of the presphenoids) resembles a series of three successive vertebrae that are incomplete dorsally. It inspired Goethe's Vertebral Theory of Origin of the Skull. Goethe's observation was partially correct, but his conclusion lacked supporting data (refer to "Segmentation of the Craniate Head" in chapter 16).

The dermatocranium is represented in mammals by paired premaxillae, maxillae, jugals (malars), nasals, lacrimals, and squamosals; by paired or unpaired frontals and parietals; and by an unpaired interparietal, all of which are paired in embryos and, in some species, in neonates. A postparietal (dermoccipital) was present in *Homo erectus* (a likely immediate ancestor of humans) and is still present in some human populations, chiefly Mongolians. It is sometimes called an **Inca bone** because it was common among Inca Indians (fig. 9.27). Premaxillae are not identifiable in adult human skulls because they unite with the maxillae early in embryonic life, a discovery made by the poet-biologist Goethe, much to his "unspeakable joy." The zygomatic arch varies from massive to slender, depending on the magnitude of the force exerted on it by the masseter muscle. It has become extremely delicate, or even incomplete, in some insectivores.

The temporal complex of mammals consists of numerous components of intramembranous and endochondral origin (fig. 9.28). The **squamous portion** is the squamosal of lower tetrapods. A **tympanic bulla** (fig. 9.29) consists of two parts, a **tympanic** and **entotympanic**. The tympanic (annulus tympanicus) surrounds the tympanic membrane (eardrum) as a bony ring. It is derived, according to evidence from embryonic opossums, from the angular bone of nonmammalian amniotes. (The annulus of anurans is not a homologue. It is thought to be derived from the palatoquadrate cartilage.) A large portion of the bulla, when present, is represented by

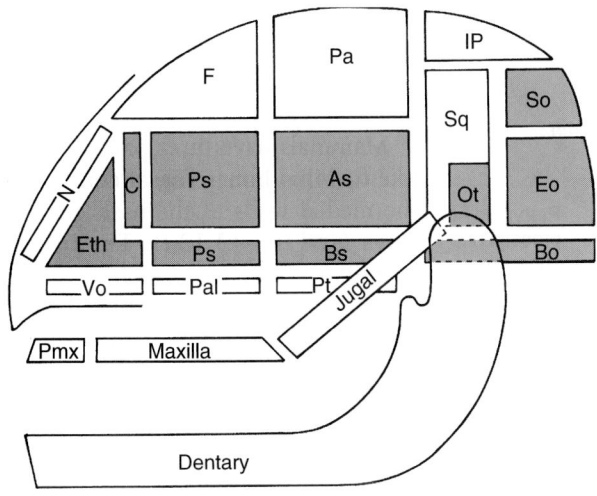

FIGURE 9.26

Chief endochondral bones (*red*) of a generalized mammalian skull. The vomer, palatine, and pterygoid are parts of the primary palate. The premaxilla, maxilla, and palatine processes to the secondary palate. **As,** Alisphenoid (pleurosphenoid); **Bo,** basioccipital; **Bs,** basisphenoid; **C,** cribriform plate of ethmoid; **Eth,** ethmoid, perpendicular plate; **Eo,** exoccipital; **F,** frontal; **IP,** interparietal; **N,** nasal; **Ot,** otic (petrous); **Pa,** parietal; **Pal,** palatine; **Pmx,** premaxilla; **Ps,** presphenoid; **Pt,** pterygoid; **So,** supraoccipital; **Sq,** squamosal; **Vo,** vomer.

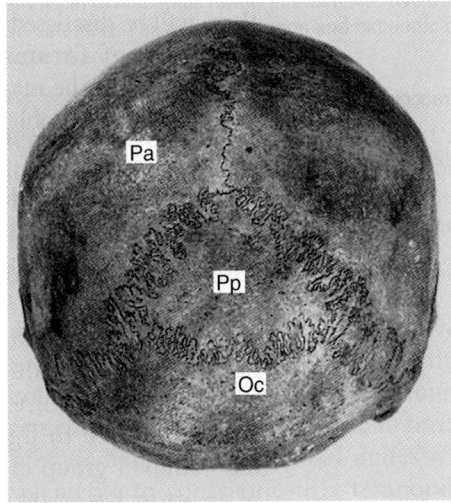

FIGURE 9.27

Postparietal (Inca) bone in a human skull from the Aleutian Islands. **Oc,** Occipital; **Pa,** parietal; **Pp,** postparietal.
Courtesy William S. Laughlin.

the entotympanic, a cartilage replacement ossification, which is a new structure in some mammals. The **petrous portion** (fused prootic, ophisthotic, and epiotic) is the ossified otic capsule (see fig. 9.30). As a primarily internal structure, it may still be seen externally as the **mastoid process** of the mastoid portion. The bulk

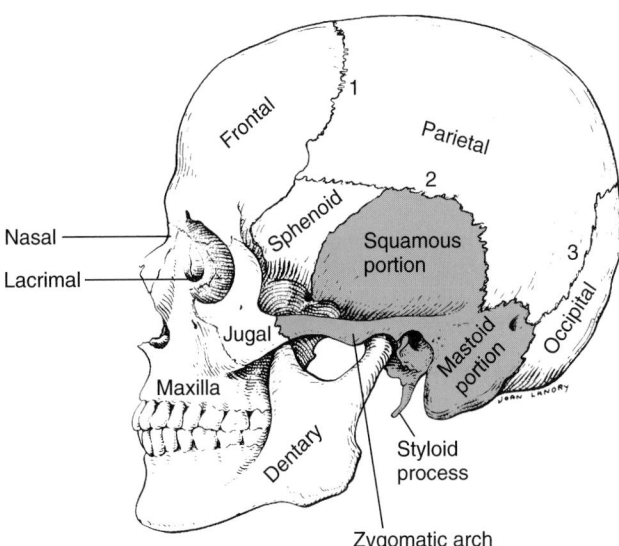

FIGURE 9.28

Temporal bone of human skull. **1, 2, 3:** coronal, squamosal, lambdoidal sutures, respectively.

of the **mastoid portion** of endochondral origin is new in mammals (see fig. 9.28). Although the tympanic and petrous portions are separate bones in some mammals, they frequently unite to form a **petrotympanic bone,** as in rabbits, and the petrotympanic may unite with the squamosal to form a temporal bone. A dorsal segment of the hyoid arch skeleton coalesces in some species with the temporal bone to become a **styloid process** (see fig. 9.28). The composite nature of the temporal bone is diagrammed in figure 9.31.

The squamosal bone has become a new site of articulation of the lower jaw with the skull in mammals. The shift from the quadrate-articular joint of other vertebrates is correlated with expansion of the dentary bone during mammalian evolution (see fig. 9.39).

Air-filled **cranial sinuses** are often found within the maxilla, frontal, sphenoid, and ethmoid bones (see fig. 9.30f and **s**). The frontal sinuses of sheep and goats extend into the horns. When male goats butt heads at speeds of up to 35 miles (60 km) per hour as part of the mating ritual, the arched walls of the sinus act as a bony brace that shunts shock waves to the vertebral column and away from the brain via the bones of the skull. Inflammation of the sinuses (sinusitis) is a common ailment in human beings.

Of the inherited primary palate, an unpaired vomer lies at the base of the nasal septum (see fig. 9.30). The septum consists of a mesethmoid bone and more or less cartilage. A **nasal process of the palatine** develops in the lateral wall of the nasopharynx where it contributes to the wall of the orbit, and a palatal process of the palatine bone contributes to the secondary palate. The pterygoids are reduced to small, winglike, **pterygoid**

TABLE 9.3 Reduction in Number of Dermal Bones Investing Meckel's Cartilage in Early Vertebrates Contrasted with Later Ones

Fishes			Tetrapods			
			Basal	Modern		
Basal	Rhipidistians	Teleosts	Labyrinthodonts	Reptiles	Amphibians	Mammals
Dentary	Dentary	Dentary*	Dentary	Dentary	Dentary	Dentary
Angular	Angular	Angular†	Angular	Angular	Angular‡	
Surangular	Surangular		Surangular	Surangular		
Infradentary§	Splenial		Splenial	Splenial	Splenial‡	
Infradentary	Coronoid		Coronoid	Coronoid		
Infradentary	Prearticular	Derm-articular‖	Prearticular			
Infradentary			Intercoronoid			
Infradentary			Precoronoid			
Infradentary			Postsplenial			

Primitive conditions had a larger number of bones than derived ones. Reptiles have retained more of the primitive elements than other modern tetrapods.

*Dentary incorporates mentomeckelian of endochondral origin in some teleosts.
†May be absent.
‡Sometimes incorported in an angulosplenial.
§Variable number.
‖May include articular of cartilage origin.

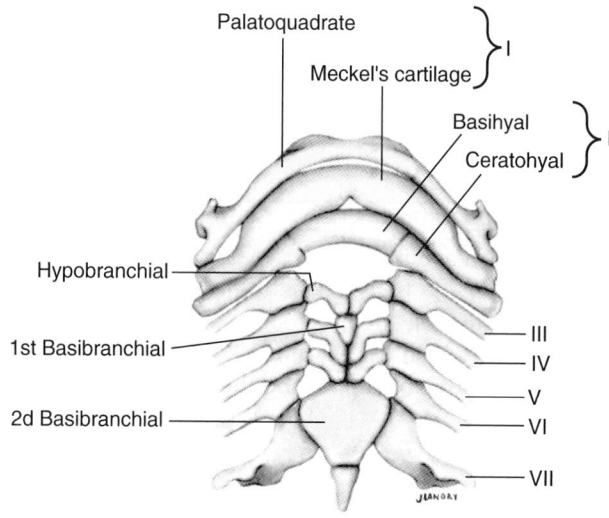

FIGURE 9.32

Visceral skeleton of *Squalus acanthias,* ventral view.
III to VII, Ceratobranchial cartilages of the third to seventh pharyngeal arches. See also figure 9.1.

dorsally and gill-bearing **ceratohyals** laterally (figs. 9.1 and 9.33*c*).

At the angle of the mouth, Meckel's cartilage and the palatoquadrate cartilage articulate with one another and with the hyomandibular cartilage in a movable joint united by ligaments (see fig. 9.1). The dorsal end of the hyomandibula is bound by ligaments to the otic capsule and suspends the jaws and the entire branchial skeleton from the neurocranium. This is hyostylic jaw suspension.

Jaw Suspension in Fishes

The jaw-hyoid complex of fishes must necessarily be braced against some support. The nearest is the braincase. In most elasmobranchs (see fig. 9.1) and most bony fishes, the hyomandibular cartilage is braced against the otic capsule, and the posterior end of the palatoquadrate cartilage is braced against the hyomandibula. This condition is known as **hyostylic jaw suspension,** or **hyostyly.** A more primitive condition is seen in some older sharks in which the hyomandibula and one or more processes of the palatoquadrate are braced independently against the braincase. This is known as **amphistyly.** A third variant, and perhaps the oldest, possibly found in a very early shark and, independently evolved with modifications, in chimaeras and lungfishes, is known as **autostyly.** Here the palatoquadrate is attached to the neurocranium, and the hyomandibula plays no role in jaw suspension. The structural details of jaw suspension among fishes, ancient and modern, vary greatly, and an expanded terminology has been devised for these variants. The anatomic relationship of the palatoquadrate cartilage to the hyomandibula and to the neurocranial-dermatocranial complex is correlated with the diet of the species.

Bony Fishes

The visceral skeleton of bony fishes resembles that of sharks in basic morphology (fig. 9.34). The major differences are that the embryonic palatoquadrate and Meckel's cartilages in bony fishes become invested by membrane bones during development, the hyoid skeleton may consist of a greater number of segments, and the embryonic cartilages of the gill arches are ultimately

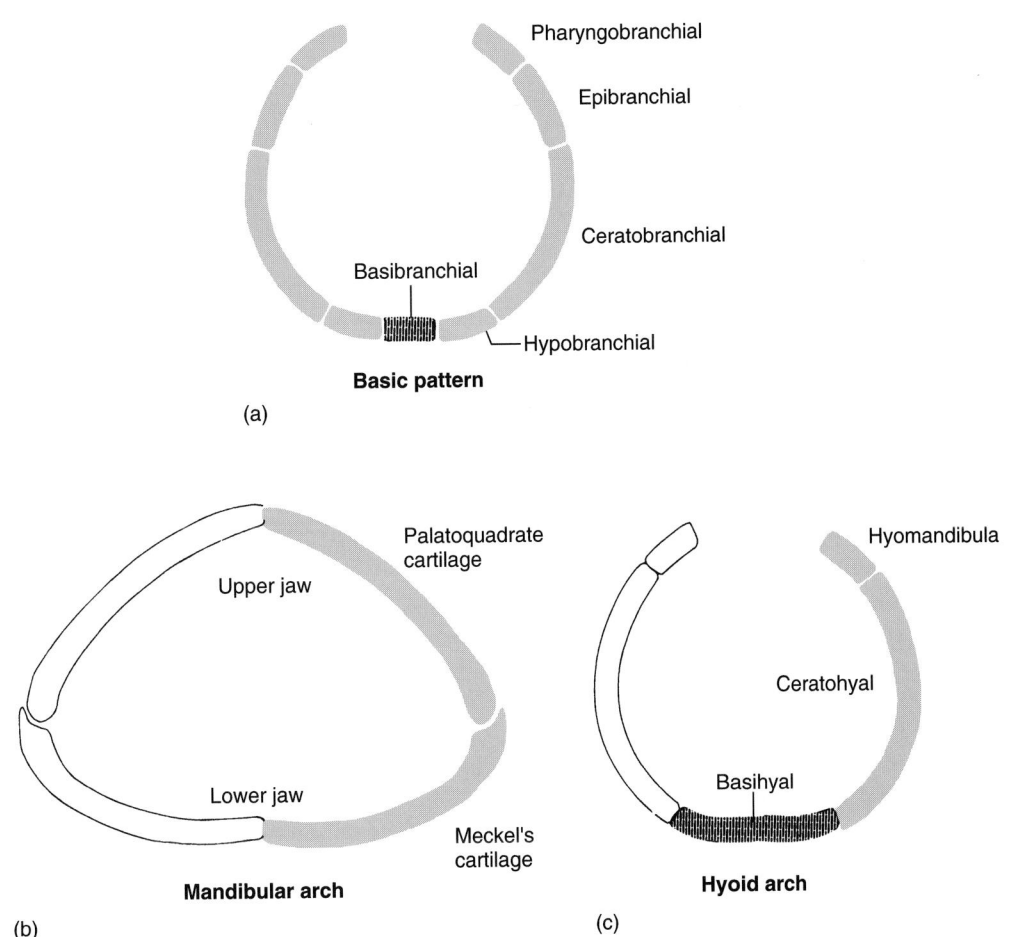

FIGURE 9.33
Skeletal components of a generalized branchial arch, (*a*), and of the mandibular and hyoid arches of *Squalus acanthias*, (*b*) and (*c*). The basihyal of sharks is paired in embryos.

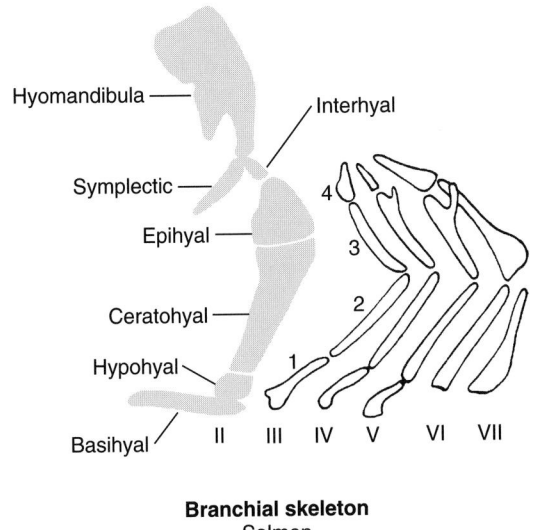

FIGURE 9.34
Visceral skeleton of a salmon, jaws removed. Hyoid cartilages are in *red.* The basihyal is unpaired. **1** to **4**, hypobranchial, ceratobranchial, epibranchial, and pharyngobranchial elements of the third arch. The hyomandibula articulates with the otic capsule.

replaced by bone. The site of articulation between the upper and lower jaws of bony fishes and of all other vertebrates except mammals is the same as in cartilaginous fishes; the posterior ends of the palatoquadrate cartilages, whether or not ultimately replaced by bone, articulate throughout life with the posterior ends of Meckel's cartilages, whether or not these ends are replaced by bone. To this day, there has been no change in this Paleozoic relationship in gnathostomes other than in mammals.

To gain a general impression of the fate of the embryonic palatoquadrate cartilage in bony fishes, the reader is urged to look again at figure 9.10. In most bony fishes, (1) this cartilage becomes ensheathed by two membrane bones, a premaxilla and maxilla; (2) the portion in the roof of the oropharyngeal cavity, the palatal portion, develops two or three sites of dermal ossification including palatine and ectopterygoid bones (see fig. 9.9*a*); and (3) the posterior tip of the palatoquadrate cartilage ossifies to become a **quadrate** bone. The entire complex of ensheathing and replacement bones unites with the dermatocranium to one degree or another, depending on the species.

Meckel's cartilages ossify at their caudal ends to become **articular bones**. The remainder of the cartilage becomes invested by several membrane bones including a **dentary** and **angular**. The mandible of the more basal *Amia* consists of dentary, angulars, surangulars, **prearticulars**, and four pairs of **coronoids**. All but the angular and surangular bear teeth.

The replacement bones of the **hyoid arch** of a relatively primitive condition seen in a teleost are shown in figure 9.34. The **symplectic** and **interhyal** are ossification centers in the hyomandibular cartilage, and the **epihyal** is an ossification center in the ceratohyal. The symplectic usually articulates with the quadrate bone or with the quadrate and lower jaw, where it participates in cranial kinetism. The hyomandibula has been lost in lungfishes.

The skeleton of a typical bony **gill arch**, like the gill arches of a shark, consists of four segments, **hypobranchial**, **ceratobranchial**, **epibranchial**, and **pharyngobranchial**, the latter in the pharyngeal roof. Only the middle two segments bear gills. The pharyngeal borders of the branchial arches are covered with bony dermal plates having flattened or pointed denticles. The size and arrangement of the plates and denticles are correlated with the nature of the diet and the method of food handling in the pharynx.

Feeding Mechanisms in Bony Fishes

Early voracious jawed fishes had wide mouths, the hinge of the jaws was far back under the skull, and the upper jaw was fused with the braincase and therefore incapable of independent movement. As a result, in feeding, the mandible was simply lowered and then snapped shut on prey in the manner of some sharks. The portion of the prey that was inside the orobranchial chamber was swallowed.

With the appearance of the type of kinetism seen in teleosts, more complex mechanical means of procuring food became possible. As an accompaniment of these changes, the large adductor mandibulae muscle inserted farther and farther forward on the mandible, an additional array of jaw muscles developed, and the mouth became narrower and more oval. The effect of these changes was that the jaws of the most highly specialized teleosts can be thrust forward and employed in feeding by inertial suction. The mechanism works equally well in herbivorous and carnivorous fishes, although not in precisely the same manner.

Feeding by inertial suction in the herbivorous freshwater teleost *Petrotilapia* has been described by Liem. In its natural habitat, this fish feeds on algae attached to submerged objects; in an aquarium, it feeds equally well on commercial fish food. Abduction of the hyomandibula expands the orobranchial chamber, the lower jaw falls open, and the opercular cavity is then expanded, in that sequence, creating forward suction. As a

result, zooplankton in the water immediately ahead are slowly drawn into the mouth. The lower jaw is then raised, the upper jaw remains briefly protruded, and the hyomandibula is adducted, which compresses the chamber. The two phases of the pharyngeal pump, expansion and compression, occupy together about 600 milliseconds. Activity of the muscles was recorded by electromyography. By manipulating the hyoid skeleton and jaws appropriately, the protruded jaws can be directed upward for collecting food on the surface or downward for food on the substrate.

A similar mechanism operates in predatory fishes. When the hyomandibula is drawn forward, the symplectic forces the kinetic elements of the upper jaw and palate to slide forward, and the premaxillae and dentary bones close on the prey. A process of the maxilla usually becomes attached by a ligament to the dentary bone, so that movement of one participates in displacing the other. The details of the anatomic relationships and the mechanics of their operation are as numerous as the taxa that exhibit them. It should be pointed out, however, that snapping and biting has not gone out of style. One-half or more of the teleosts have nonprotrusible jaws.

The theory has been advanced that vertebrate jaws are former gill arches that became modified, along with the hyoid arch, for predatorial feeding as an alternative to filter feeding, which was the ancestral method of obtaining nourishment. A student examining a shark and its visceral arches will see the basis for this idea. In fact, (1) the hyoid arch does contain a demibranch in *Squalus*, (2) the spiracle is a pharyngeal slit as are all gill slits, and (3) the nerves and arteries of the mandibular and hyoid arches are the most anterior of a series that is repeated in each gill arch, as will be seen in later chapters. The theory is an interesting speculation. It is unlikely that it will ever be validated or falsified.

Living Agnathans

The visceral skeleton of living agnathans is totally unlike that of jawed fishes (fig. 9.35). There are no identifiable palatoquadrate cartilages, Meckel's cartilages, hyoid cartilages, or branchial arch cartilages. A V-shaped **lingual cartilage** (dental plate) bearing horny teeth and located in the floor of the buccal cavity moves in and out of the buccal funnel, serving as a rasping tonguelike organ. It is operated by protractor and retractor muscles that are attached to a narrow, segmented, immobile basal plate cartilage lying immediately beneath it (not illustrated). There is no evidence that any of the cartilages associated with the feeding apparatus are derived from ancestral visceral arches. The lingual cartilage, or "tongue," and associated cartilages are correlated with the unique method of feeding characteristic of living agnathans. The rest of the pharyngeal skeleton consists of a branchial basket, less well developed in

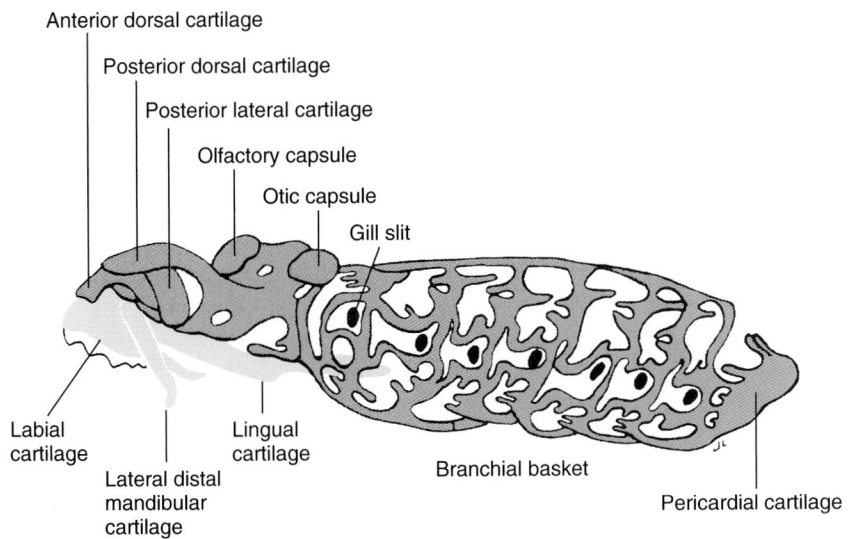

Anterior dorsal cartilage
Posterior dorsal cartilage
Posterior lateral cartilage
Olfactory capsule
Otic capsule
Gill slit
Labial cartilage
Lateral distal mandibular cartilage
Lingual cartilage
Branchial basket
Pericardial cartilage

Neurocranium and visceral skeleton
Lamprey

FIGURE 9.35

Neurocranium, branchial basket, and feeding apparatus (*red*) of a lamprey. The olfactory capsule is a midline structure; otic capsules are paired. The lingual cartilage is also referred to as a tongue.

hagfishes, which is simply a fenestrated framework of cartilage immediately under the skin. The caudal end of it in lampreys is located at the level of the heart.

Tetrapods

With life on land, the visceral skeleton underwent profound changes from that of jawed fishes. Some previously functional parts were deleted, and those that persisted performed new and sometimes unexpected functions. It is these modifications, which took place over the course of millions of years, that we will examine in the pages that follow. But first, we will look at some of these changes in a frog tadpole, where they occur during the course of two or three days!

Larval frogs have six pairs of visceral cartilages, the last four of which bear gills (fig. 9.36a). The cartilages of the gill arches meet ventrally in a **hypobranchial plate.** During metamorphosis (fig. 9.36b), the third, fifth, and sixth visceral cartilages regress. The hypobranchial plate enlarges and, along with the first basibranchial, becomes incorporated into the body of the hyoid. The latter thereafter is a broad cartilaginous and bony plate in the buccopharyngeal floor. The ceratohyal cartilage of the second arch becomes a slender anterior horn, or cornu, of the hyoid apparatus, and the cartilage of the fourth arch becomes a posterior horn. As a result of these and other changes, a visceral skeleton initially adapted for branchial respiration becomes converted, in the span of a few days, to one adapted for life on land. Among its new roles is serving as an anchorage for the muscular tetrapod tongue.

The components of the branchial skeleton of a larval salamander are illustrated in figure 9.37. The number in urodeles is usually reduced at metamorphosis, as in frogs. The branchial skeleton of *Necturus,* a perennibranchiate urodele, is illustrated in figure 9.41a. In the pages that follow, we will trace the adaptations that have taken place in the jaws and branchial arches of tetrapods and especially of amniotes, which abandoned a free-living, larval stage along with the gills that helped sustain it.

Fate of the Palatoquadrate and Meckel's Cartilages

We have seen in bony fishes that the embryonic palatoquadrate cartilages become ensheathed laterally by premaxillae and maxillae and ventrally by dermal bones of the primary palate; that the palatal portions may contribute in one fish species or another to the otherwise dermal primary palate; and that the posterior ends of the palatoquadrate cartilages usually ossify to become quadrate bones at the hinge of the jaws. Except for contributing to the primary palate, the story is the same for amphibians and amniotes. The quadrate remains as the site of articulation of the cranium with the lower jaw in all nonmammalian tetrapods. In that location in squamates and birds, it is part of the kinetic mechanism of the skull (see fig. 9.21). The quadrate bone in mammals becomes the **incus,** a middle ear ossicle.

The embryonic Meckel's cartilages of amniotes may continue to grow and become a prominent core of cartilage within the adult mandible, as in turtles and crocodilians (fig. 9.38). More often, little remains in adults. However, the cartilages become ensheathed by dermal bones that included, primitively, on each side, a dentary, angular, **surangular, splenial,** one or more coronoids, and a prearticular (dermarticular). Modern amniotes have a smaller number, and mammals have only the dentary. As in bony fishes, the posterior end of each Meckel's cartilage ossifies to become an articular bone at the hinge of the jaws, except in mammals. In mammals it becomes the **malleus,** another middle ear ossicle.

Expansion of the Dentary and a New Jaw Joint in Mammals

As explained earlier, temporal fossae enabled the adductor mandibulae muscles of synapsids to enlarge, subdivide, and acquire anatomic origins on the temporal region of the skull and on the zygomatic arch. The increased mass of these new muscles was accompanied

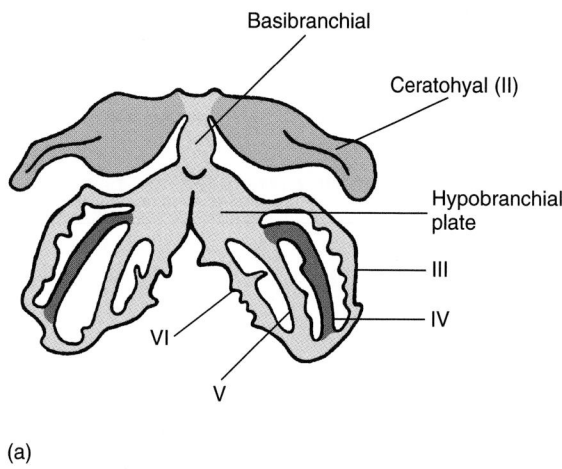

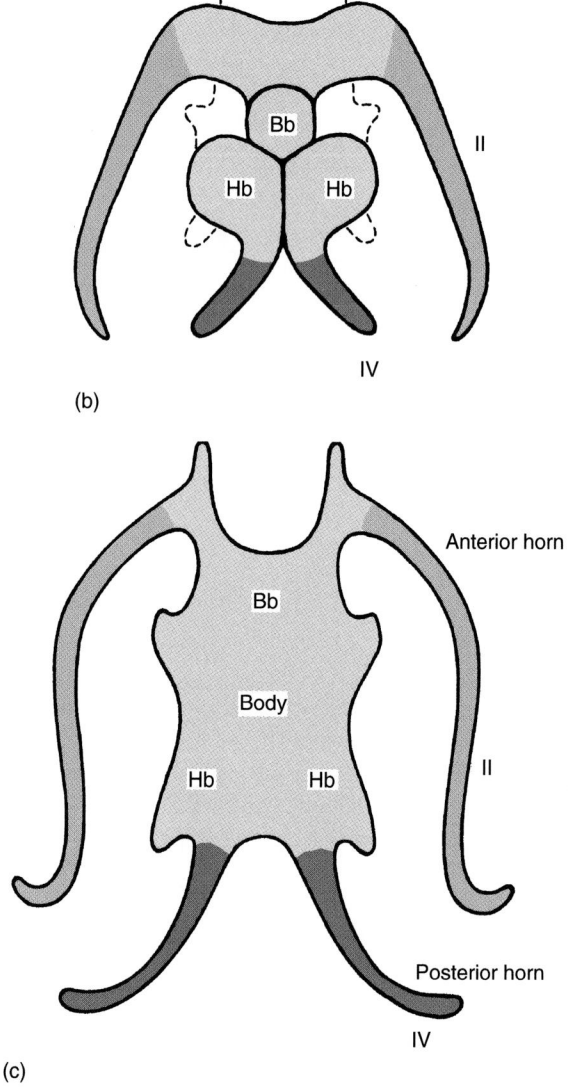

FIGURE 9.36

Changes in hyoid and branchial skeleton during metamorphosis in a frog. (*a*) Larval skeleton. **II–VI**, skeleton of second through sixth visceral arches. **III–VI**, bear gills. (*b*) Late metamorphosis. (*c*) Hyoid of a young frog. **Bb**, basibranchial contribution. **Hb**, hypobranchial contribution. *Dashed line* in (*b*) outlines new cartilage. *Red* shades designate homologous horns.

by expansion of the dentary bone and formation of a ramus on which the temporalis muscle inserted (figs. 9.39 and 11.24c). Eventually the other dermal bones of the mandible disappeared, the articular bone became an ear ossicle, and mammals were left with a mandible consisting solely of two dentary bones.

Expansion of the dentary brought it close to the squamosal, and there it eventually established a new site for articulating with the skull (see fig. 9.28). For a while, the lower jaw articulated with the skull at two sites, the primitive one (articular against the quadrate) and the new one (condyloid process of the dentary against the squamosal (see fig. 9.39). The two existed side by side in *Eozostrodon*, an immediate outgroup to Mammalia from the Triassic. Eventually the articular and quadrate were "captured" by the middle ear cavity, and only the new articulation remained in mammals. The shape, slant, and relationships of the condyloid process of the mandible differ with the demands made on it by the feeding habits of the various mammalian orders.

Ear Ossicles from the Hyomandibula and Jaws

It will be recalled that the hyomandibular cartilage of sharks is interposed between the quadrate cartilage and the otic capsule and that the latter houses the inner ear (see fig. 9.1). In an autostylic jaw, the hyomandibular bone can be dispensed with, and that is what happened to it in dipnoans. It persisted in tetrapods, the jaws of which are also autostylic, but became dissociated from the quadrate and attached, instead, to what was destined to become a tympanic membrane. It continued to abut against the otic capsule, which houses the inner ear, and it became surrounded by a portion of the first pharyngeal pouch, now known as the middle ear cavity. The hyomandibula thereby became the first ear ossicle, or columella (stapes), of tetrapods. Thereafter it con-

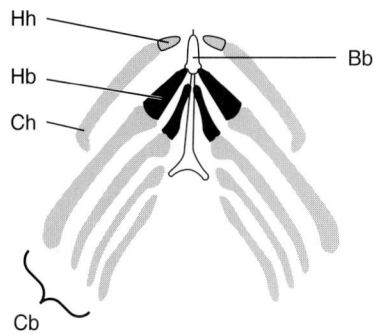

FIGURE 9.37
Hyoid and branchial skeleton of *Ambystoma* larva.
Cb, ceratobranchials; **Ch,** ceratohyal; **Hb,** hypobranchials, *black;*
Hh, hypohyal; basibranchials, **Bb,** *white.*

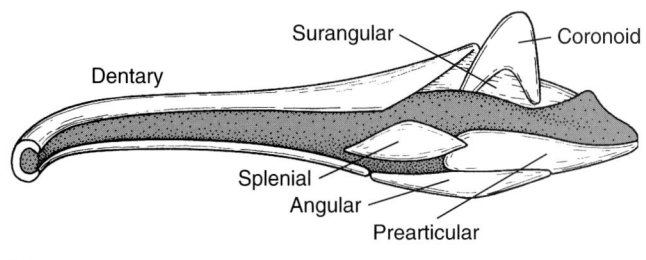

(a)

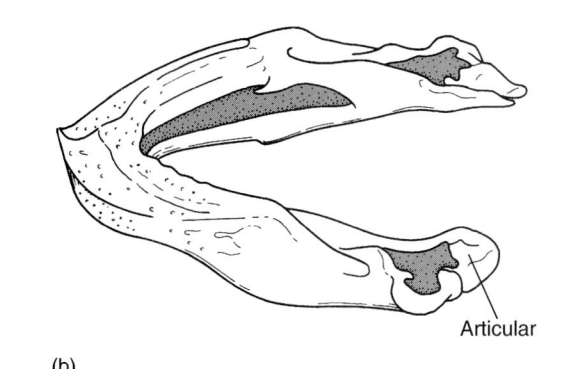

(b)

FIGURE 9.38
Meckel's cartilages (*red*) in lower jaw of embryonic lizard.
(*a*) Medial view and within mandible of adult sea turtle, (*b*). The
cartilage of the lizard is becoming ensheathed by dermal bones and
will disappear with further development of the jaw. Its posterior end
will become the articular bone. The prearticular is partly
endochondral in (*a*).

ducted sound waves from the eardrum, across the middle
ear cavity, to the inner ear. Figure 9.12*b* shows a segment
of the hyoid cartilage (Hy II) and columella (Col II) still
interposed between the quadrate (Qu I) and the otic
capsule (Pro) in a urodele whose middle ear cavity is
vestigial and who lacks a tympanic membrane.

When the dentary bone of therapsids acquired an articulation with the squamosal, the articular and quadrate
bones were freed to perform a new function or to disappear. There is no doubt that the caudal end of Meckel's

cartilage becomes the malleus of mammals. The embryonic Meckel's cartilage in one mammalian species after
another can be seen projecting into the area where middle ear cavitation is proceeding (fig. 9.40), and its cartilaginous posterior tip can be seen to separate, ossify, and
become the malleus.

The evidence that the quadrate bone of synapsids became the incus of the middle ear of mammals is less direct, but the facts are these: (1) The articular and
quadrate bones have been articulating in a diarthrosis
since the time of the earliest jawed fishes and were
doing so in therapsid reptiles; (2) the articular has separated from the lower jaw and is now the malleus; (3) the
quadrate has disappeared from the upper jaw; and
(4) the articular still articulates in a diarthrosis with a
bone (the incus), that, if it is not the quadrate, is of unknown homology (see fig. 17.12). In the nineteenth
century, C. Reichert proposed the theory that mammalian ear ossicles are derived from the jaws. Early in
the twentieth century, the theory was modified by
E. Gaupp to include the origin of the stapes from the
dorsal tip of the hyoid arch. Subsequent knowledge
from paleontology and embryology has strengthened
the Reichert-Gaupp theory, and it is now generally accepted (see E. Jarvik, 1980: 161).

Amniote Hyoid

The hyoid of amniotes consists of a **body** in the pharyngeal floor just anterior to the larynx and of two or three
horns (**cornua**) in the pharyngeal walls (fig. 9.41*c–h*).
We have seen that the hyoid of anurans is derived at
metamorphosis from basibranchial and hypobranchial
cartilages, from the skeleton of the hyoid arches, and
from one of the larval gill-bearing arches (see fig. 9.36).
In amniotes, it is derived from homologous anlagen.
The cartilages of the second arch become anterior
horns, and those of the third arches and sometimes the
fourth become additional horns.

In lizards and birds (fig. 9.41*c* and *e*), an elongated
bony process, the **entoglossus,** extends from the body
forward into the long darting tongue. In some male
lizards, a similar process extends caudad into the
dewlap, or gular pouch. Snakes have no hyoid. In fact,
the entire branchial skeleton is vestigial.

The entoglossus and caudal horns of woodpeckers, in
association with the tongue, is a remarkable tool for impaling grubs (figs. 9.42 and 12.6*a*). From the base of the
entoglossus, two long, flexible caudal horns loop caudad then dorsad around the occipital region of the skull
and forward under the scalp all the way to the **lore,** the
space between the eyes and the bill, where both horns
dip into the right nasal passageway. They then continue
forward in the nasal canal a short distance farther and
terminate. The termination of both horns in a single
nasal passageway is surprising and lacks a functional
explanation. When impaling grubs, the horns are

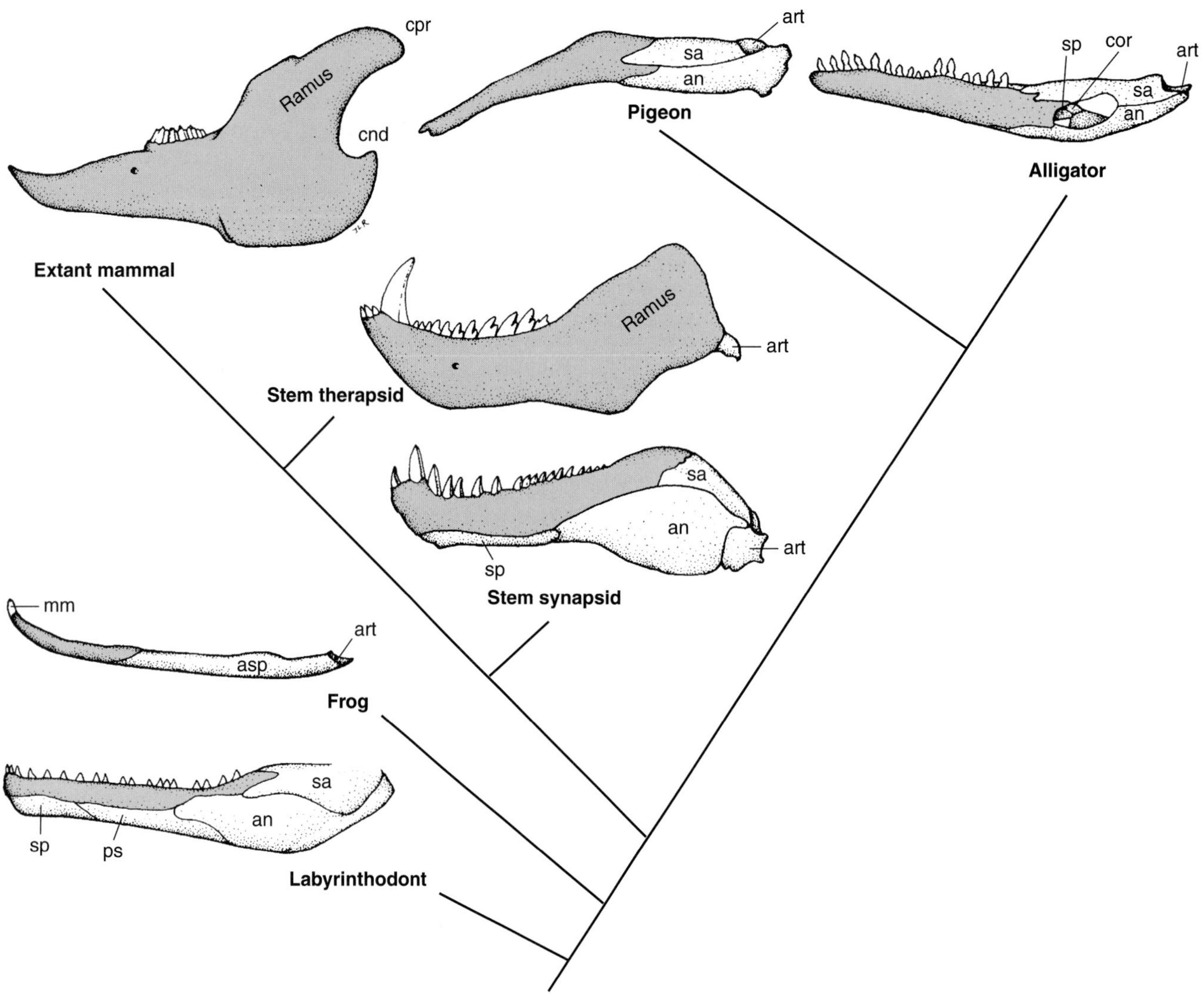

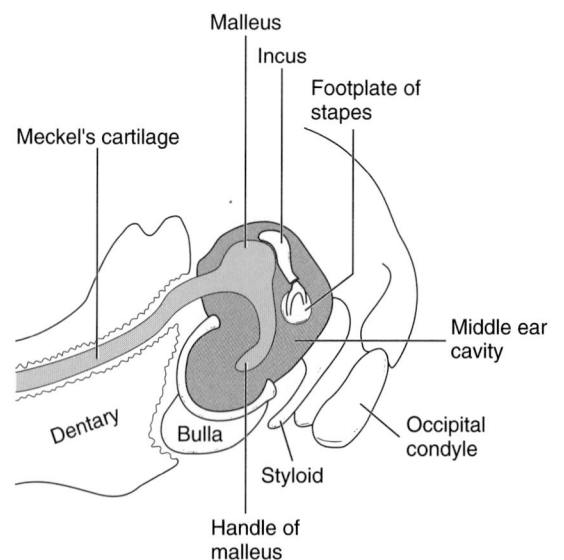

FIGURE 9.39

The mandibles of tetrapods, left lateral views, on a cladogram showing the hypothesized relationships among represented taxa. The dentary (*red*) became larger among synapsids; other bones were reduced and finally lost. All are dermal bones except mentomeckelian and articular. **an,** angular; **art,** articular (cartilage in frog); **asp,** angulosplenial; **cnd,** condyloid process for articulation with squamosal; **cor,** coronoid; **cpr,** coronoid process; **mm,** mentomeckelian; **ps,** postsplenial; **sa,** surangular; **sp,** splenial.

FIGURE 9.40

Caudal end of Meckel's cartilage surrounded by the developing middle ear cavity in a mammalian embryo.

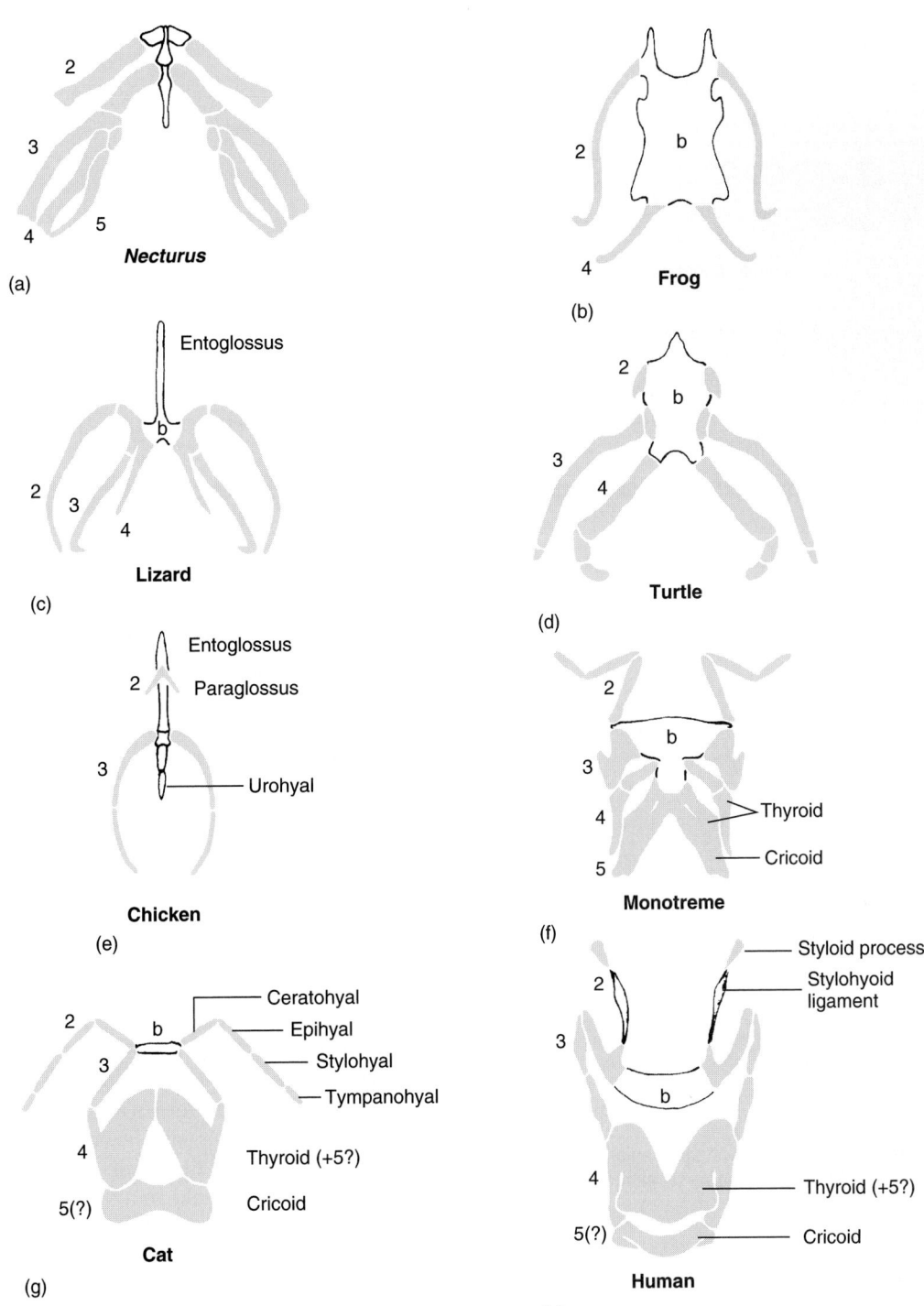

FIGURE 9.41
Skeletal derivatives of the second through fifth pharyngeal arches in selected tetrapods. **b,** Body of hyoid. **2** to **5,** Derivatives of arches **2** through **5** (*red*). The projections from the body in (*b*) through (*h*) are horns of the hyoid. In (*e*), the body of the hyoid extends forward into the tongue as an entoglossus to which are attached two paraglossals (**2**).

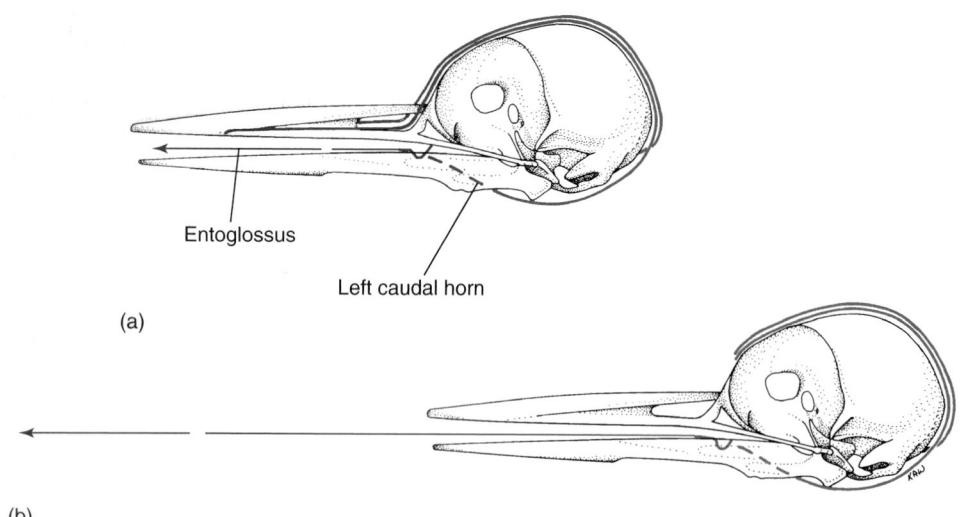

(a)

Entoglossus

Left caudal horn

(b)

FIGURE 9.42
Caudal horns and entoglossus of the hyoid of a hairy woodpecker. (*a*) Tongue retracted. (*b*) Tongue extended. The paired caudal horns lie close together and appear as one when extended.

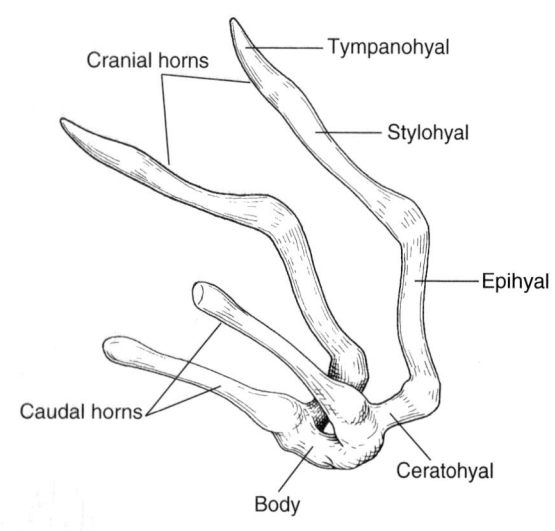

FIGURE 9.43
Hyoid of a dog. Caudal horns articulate with the cranial horns of the thyroid cartilage (see fig. 13.12). Cranial horns terminate at the tympanic bullae.

straightened by an accelerator muscle that shoots the tongue into the prey. Elastic recoil of the horns immediately returns the tongue to the mouth with impaled food.

The hyoid of mammals (figs. 9.43 and 9.44) has cranial horns from the second arch and caudal horns from the third (see fig. 9.41*g,h*). In dogs and cats, the cranial horns are the longer ones (**greater horns**) and are composed of four segments. The dorsalmost, or **tympanohyal** (fig. 9.43), ends in a notch in the tympanic bulla. (The stapes, which became detached from the tympanohyal during phylogeny, is inside the bulla.) In rabbits, the anterior horns are the short ones (**lesser**

horns), and the tympanohyal is represented by a slender stylohyal bone embedded in the tendon of insertion of the posterior belly of the stylohyoid muscle. In humans, too, the anterior horns, equivalent to ceratohyals, are the lesser horns; an unossified stylohyoid ligament represents epihyal and stylohyal segments; and the tympanohyal is attached to the temporal bone where it is known to anatomists as the styloid process (figs. 9.28 and 9.45).

The hyoid anchors the highly mobile tongue of tetrapods, is the skeleton for the buccopharyngeal pressure pump used in respiration in anurans, provides attachment for some of the extrinsic muscles of the larynx, has subtle effects on lower jaw movements, and is the site of attachment of muscles that participate in swallowing. The associated hypobranchial and branchiomeric muscles of amniotes approach the hyoid from many directions—lower jaw, larynx, sternum, clavicle, the temporal region of the skull, and elsewhere. These muscles stabilize the hyoid in a single position or move it forward, backward, up, or down.

Laryngeal Skeleton

Nearly all tetrapods have **cricoid** and **arytenoid cartilages** or replacing bones (see fig. 13.11), and mammals have **thyroid cartilages** or **bones** in addition (figs. 9.44, 9.45, and 13.13*b*). Thyroid cartilages arise from mesenchyme of the fourth pharyngeal arch and, perhaps, the fifth. Cricoid and arytenoid cartilages are probably products of the fifth arch. Because the caudal end of the pharyngeal arch series has been subject to reduction during evolution, it is not surprising that problems are encountered in relating the caudalmost laryngeal cartilages to specific arches. The laryngeal skeleton is discussed in more detail in chapter 13.

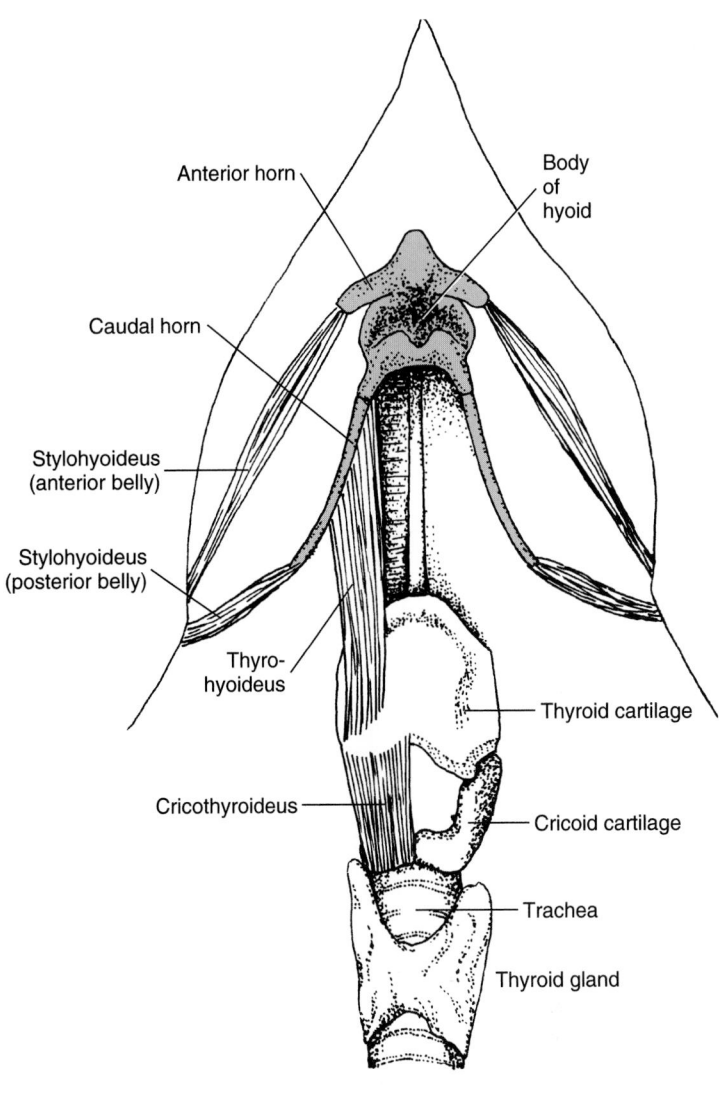

FIGURE 9.44

Hyoid (*red*), larynx, and associated structures of a rabbit, ventral view.

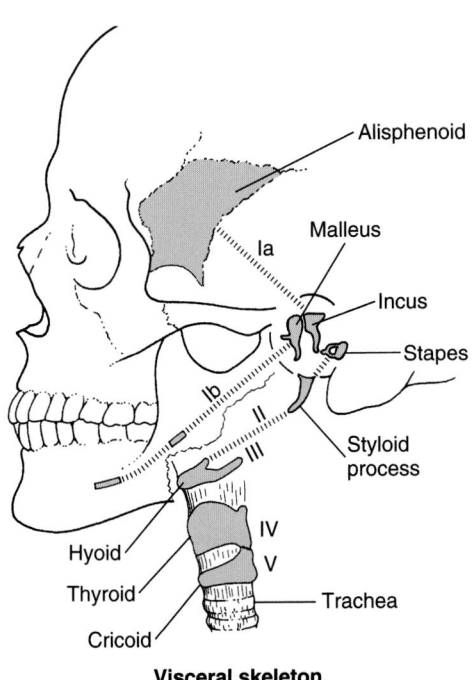

Visceral skeleton

FIGURE 9.45

Derivatives of the human visceral skeleton (*red*). **Ia,** *Broken line* connects derivatives of the palatoquadrate cartilage. **Ib,** *Broken line* connects vestiges and derivatives of Meckel's cartilage. **II,** *Broken line* from stapes to styloid process to lesser horn of hyoid connects derivatives of hyoid arch. Section between lesser horn and styloid process is site of a stylohyoid ligament. **III–V,** Derivatives of the third, fourth, and fifth arches. **III** is at tip of the posterior (greater) horn.

PERSPECTIVE

It is evident that the visceral skeleton is a functional complex that was associated primitively with feeding and branchial respiration. In tetrapods it has been modified in part for transmission of airborne sounds, for attachment of tongue muscles, and for support of vocal cords. These adaptations illustrate how the combined effects of mutations and selection alter ancient structures and perhaps produce new functions. Some of the skeletal derivatives of the visceral arches in representative living vertebrates are listed in table 9.4.

TABLE 9.4 Skeletal Derivatives of Pharyngeal Arches in Sharks and *Approximate* Homologues in Bony Vertebrates

Arch	Shark	Teleost	*Necturus*	Frog	Reptile	Mammal
I	Meckel's cartilage Palatoquadrate	Articular* Quadrate Epipterygoid Metapterygoid	Articular Quadrate Palatal cartilage	Articular Mentomeckelian† Quadrate Annulus tympanicus (?)	Articular Quadrate Epipterygoid	Malleus Incus Alisphenoid
II	Hyomandibula	Hyomandibula	Rudimentary	Columella (stapes)		
	Ceratohyal	Symplectic Interhyal Epihyal Ceratohyal Hypohyal	Ceratohyal	Styloid process in mammals Anterior horn of hyoid		
	Basihyal	Basihyal	Hypohyals	Body of hyoid Entoglossus in reptiles		
III	Pharyngobranchial Epibranchial Ceratobranchial Hypobranchial	Pharyngobranchial Epibranchial Ceratobranchial Hypobranchial	Epibranchial Cereatobranchial	Body of hyoid	2nd horn of hyoid	
IV	Branchial skeleton	Branchial skeleton		Second horn of hyoid	Last horn of reptiles	Thyroid cartilages
V	Branchial skeleton			Cricoids and arytenoids (?) (precise homologies unknown)		
VI	Branchial skeleton			Not present		
VII	Branchial skeleton					

*Sometimes part of derm-articular.
†Of intramembranous origin in some species.

Summary

1. During ontogeny (from a dual neural crest and mesodermal origin), the neurocranium is constructed of prechordal and parachordal cartilages, notochord, and cartilaginous olfactory and otic capsules knit together and completed by cartilaginous walls and, in basal craniates, a fenestrated roof above the brain. An optic capsule remains independent of the rest of the neurocranium and sometimes forms scleral rings.

2. In living agnathans and cartilaginous fishes, the neurocranium remains cartilaginous throughout life. It ossifies at one or more sites in bony vertebrates.

3. The chief centers of ossification in the neurocranium are occipital, sphenoid, ethmoid, and otic. Basic replacement bones are exoccipital, basioccipital, supraoccipital; basisphenoid, presphenoid, orbitosphenoid, laterosphenoid; mesethmoid, cribriform plate, ethmoturbinals; and prootic, opisthotic, epiotic, or petrosal. These may be reduced in number by fusion.

4. The neurocrania of teleosts, apodans, and amniotes are well ossified. Those of dipnoans, basal ray fins, and most amphibians possess considerable cartilage.

5. A single occipital condyle is found in ancient tetrapods and in modern reptiles. Modern amphibians and synapsids have two.

6. The membrane bones of the skull constitute the dermatocranium. They are vestiges of ancient dermal armor.

7. The chief dermal bones of primitive tetrapod skulls were (1) roofing bones—nasal, frontal, parietal, postparietal; intertemporal, supratemporal, tabular,

squamosal, quadratojugal; lacrimal, prefrontal, postfrontal, postorbital, infraorbital (jugal); (2) dermal bones of the upper jaw—premaxilla and maxilla; (3) primary palatal bones—parasphenoid, vomer, palatine, endopterygoid, ectopterygoid; and (4) opercular bones. A partietal foramen was present.

8. Teleosts have the largest number of dermatocranial bones; dipnoans have large dermal bony plates. The dermatocrania of other bony fishes resemble those of their Devonian ancestors.

9. Modern amphibian skulls have lost many membrane bones. Anurans have large palatal vacuities. Apodan skulls have been modified the least.

10. Modern reptiles retain well-ossified neurocrania, a single occipital condyle, many membrane bones, and, in lizards, a parietal foramen. Among specializations are temporal fossae, a partial or complete secondary palate (turtles, some lizards, and crocodilians), and cranial kinesis.

11. Temporal fossae result in diapsid (crocodilians and *Sphenodon*), synapsid (mammals), or euryapsid skulls (ichthyosaurs and plesiosaurs). Turtle skulls are anapsid; those of lizards, snakes, and birds are modified diapsid.

12. Secondary palates arise as horizontal processes of premaxillae, maxillae, and palatine bones. They are complete to the midline and caudad only in crocodilians and mammals.

13. Cranial kinesis is found in skulls having intracranial joints. It is especially characteristic of teleosts, snakes, lizards, and birds. It is correlated with modes of procuring food.

14. Bird skulls are modified diapsid. Dermal bones are numerous; sutures have been obliterated except in ratites. The skulls are thin, domed, and have large orbits. Jaws are elongated to form a beak.

15. Mammalian skulls have a single temporal fossa. The dentary is the sole bone of the lower jaw, and it articulates with the squamosal. The quadrate and articular have become middle ear ossicles. The braincase is expanded, and the dermatocranial bones are reduced in number. There is a full complement of neurocranial bones with some fusions. A temporal complex combines numerous components that are separate in other vertebrates.

16. The living agnathan visceral skeleton is unique. There are no palatoquadrate or Meckel's cartilages and no branchial arches. A cartilaginous branchial basket is located under the skin of the pharyngeal region.

17. The visceral skeleton of elasmobranchs consists of palatoquadrate and Meckel's cartilages, hyoid cartilages, branchial cartilages, and median ventral basihyal and basibranchial cartilages.

18. Jaw suspension in fishes is mostly hyostylic. Some older sharks were amphistylic. Chimaeras, lungfishes, and tetrapods are autostylic.

19. The hyoid skeleton of bony fishes consists of a larger number of components than in sharks, and the skeleton of the caudalmost gill arches may be reduced.

20. The embryonic palatoquadrate cartilages in bony vertebrates are invested by membrane bones. Their caudal ends become quadrate bones, except in mammals. Meckel's cartilages are invested, and their caudal ends become articular bones, except in mammals.

21. The articular and quadrate bones constitute the jaw joints in all nonmammalian gnathostomes (that ossify this articulation). In mammals, they become middle ear ossicles (malleus, incus).

22. In nonmammalian tetrapods, the hyomandibula becomes a columella (stapes). The remainder of the second arch skeleton, that of the third, and sometimes part of the fourth gives rise to horns of the hyoid. Part of the fourth arch and perhaps the fifth contribute to the skeleton of the larynx.

Critical Thinking Questions

1. What components make up the skeletal structures of the head? Give one example for each component.

2. Craniates receive their name from the presence of a cranium. Do all craniates possess a fully developed cranium, or if not, what major steps occurred in the phylogenetic development of a "complete cranium"?

3. What criteria would you use to name similar bones among different taxa? Defend your criteria.

4. What is the organization of "skull" bones in a basal tetrapod? In general, how do teleosts and mammals compare to basal tetrapods?

5. What are the ossifications of the neurocranium? Name the regions that become ossified and give at least one example for each region.

6. What are the phylogenetic patterns for temporal fossae? Include the primitive condition.

7. Crocodilia and Mammalia independently evolve a secondary palate. What functional role may be related to the origin of these structures?

8. What is the phylogenetic history of the temporal bone that we find in mammals?

9. Describe a typical branchial arch skeleton. What are the serial homologues of the components of a branchial arch in the mandibular and hyoid arches?

10. Compare a shark lower jaw with your jaw.

11. What is the evolutionary trend for the lower jaw in synapsids? What lower jaw characters are unique to mammals within the synapsid lineage?

12. What is the phylogenetic history of your three ear ossicles?

Selected Readings

Readings on feeding mechanisms will be found also in Selected Readings, chapter 12.

Allin, E. F.: Evolution of the mammalian ear, *Journal of Morphology* 147:403, 1975.

Carroll, R. L.: The hyomandibular as a supporting element in the skull of primitive tetrapods. In Panchen, A. L., editor: The terrestrial environment and the origin of land vertebrates. New York, 1980, Academic Press.

de Beer, G. R.: The development of the vertebrate skull. Oxford, 1937, The Clarendon Press. Reprinted with a foreword by Brian K. Hall and James Hanken by The University of Chicago Press, Chicago, 1985.

Drysdale, T. A., Elinson, R. P., Graveson, A. C., Hanken, J., Herring, S. W., Langille, R. M., Noden, D. M., Ramirez, F., Solursh, M., and Webb, J. F.: Development and evolution of the vertebrate head, a symposium, *American Zoologist* 33 (4):417, 1993.

Gans, C., and Parsons, T. S., editors: Biology of the reptilia, vol. 4. New York, 1973, Academic Press.

Goodrich, E. S.: Studies on the structure and development of vertebrates. London, 1930, The Macmillan Co., Ltd. Reprinted by The University of Chicago Press, Chicago, 1986.

Gorniak, G. C., Gans, C., Radinsky, L., Hylander, W. L., Herring, S. W., English, A. W., and Byrd, K. E.: Mammalian mastication: An overview, *American Zoologist* 25:289, 1985.

Greaves, W. S.: The mammalian jaw mechanism—the high glenoid cavity, *American Naturalist* 116:432, 1980.

Hanken, J., and Hall, B. K., editors: The skull, vols. 1–3. Chicago, 1993, The University of Chicago Press.

Herring, S. W.: Formation of the vertebrate face: Epigenetic and functional influences, *American Zoologist* 33:472, 1993.

Jarvik, E.: Basic structure and evolution of vertebrates, 2 vols. New York, 1980, Academic Press.

Kuhn, H.-J., and Zeller, U., editors: Morphogenesis of the mammalian skull. New York, 1987, Paul Parey Scientific Publishers. A monograph in the series Mammalia depicta, a supplement to the International Journal of Mammalian Biology. Included are special problems of the skulls of monotremes, marsupials, and whales.

Langille, R. M.: Formation of vertebrate face: Differentiation and development, *American Zoologist* 33:462, 1993.

Liem, K. F.: Adaptive significance of intra- and interspecific differences in the feeding repertoires of cichlid fishes, *American Zoologist* 20:295, 1980.

Lombard, R. E., and Bolt, J. R.: Evolution of the tetrapod ear: An analysis and reinterpretation, *Biological Journal of the Linnaean Society* 11:19, 1979.

Radinsky, L. B.: Patterns in the evolution of ungulate jaw shape, *American Zoologist* 25:303, 1985.

Reilly, S. M.: Ontogeny of cranial ossification in the eastern newt, *Notophthalamus viridescens* (Caudata: Salamandridae), and its relationship to metamorphosis and neoteny, *Journal of Morphology* 188:315, 1986.

Reilly, S. M., and Lauder, G. V.: Atavisms and the homology of hyobranchial elements in lower vertebrates, *Journal of Morphology* 195:237, 1988.

Rowe, T.: Co-evolution of the mammalian middle ear and neocortex, *Science* 273:651, 1996.

Schmalhausen, I. I.: The origin of terrestrial vertebrates (Translated from the Russian by Leon Kelso). New York, 1968, Academic Press.

Smith, K. K.: Integration of craniofacial structures during development in mammals, *American Zoologist* 36:70 (1996).

Stahl, B. J.: Vertebrate history: Problems in evolution. New York, 1974, McGraw-Hill Book Company.

Trueb, L: A summary of osteocranial development in anurans with notes on the sequence of cranial ossification in *Rhinophrynus dorsalis* (Anura: Pipoidea: Rhinophrynidae), *South African Journal of Science* 81:181, 1985.

Links to the Internet

Visit the zoology website at http://www.mhhe.com/zoology to find live Internet links for each of the references listed below.

1. The Skull Practical Exam. This is designed to teach you the bones and landmarks of the skull. You can toggle back and forth between question and answer mode.
2. Bones of the Skull. Clickable images of the bones, and then close up, labeled photographs. Foramina are well done.
3. Skull Collection. A myriad of vertebrate skulls, albeit small in size. These are not labeled.
4. Skull Module. A thorough treatment of the bones of the skull, combining labeled bones and descriptive text.
5. Axial Skeleton System Review. Good photos of bones of the skull, with nice enlargements for more detail, although unlabeled.
6. SkullSun Company. You can order skulls from this site, but better yet, see photographs of a wide variety of skulls, with dental formulas included. Ever wonder what the skull of a red breasted toucan looks like? Check it out!
7. Special Topics. Great images of skeletons, primarily skulls of mammals and dentition types.

10

Girdles, Fins, Limbs, and Locomotion

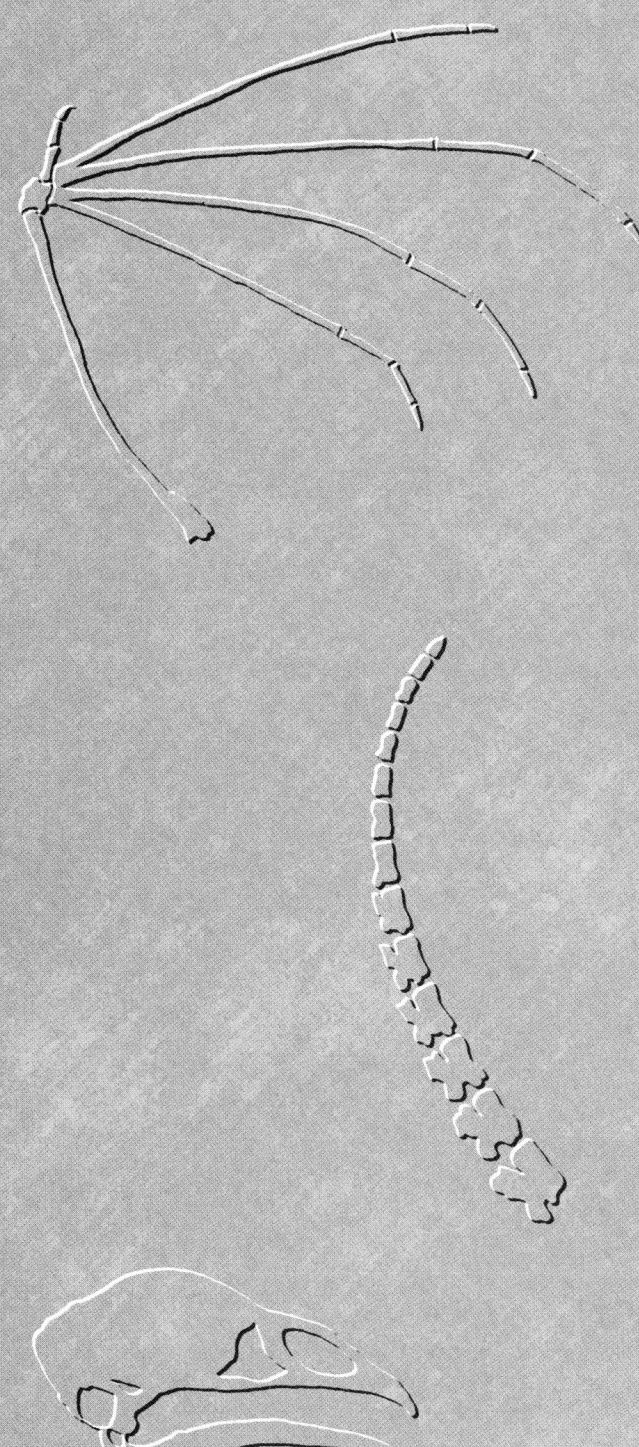

In this chapter, we will focus on skeletal structures that, along with the axial skeleton, participate in locomotion. We will see that the girdles of bony fishes were inherited almost without change by the first amphibians and see how the pelvic girdles of birds and mammals have been modified to facilitate egg laying or birthing. We will find that the skeleton of all fins exhibit a common basic pattern that has diverged in several directions and that the skeletons of all limbs consist of five jointed segments, the distal ones of which have been modified for life in the water, on land, or in the air. We will see how fishes and fishlike mammals differ in their methods of swimming, and we will examine locomotion in terrestrial craniates that have lost their limbs. Along the way, we will examine some hypotheses concerning the origin of fins and a plausible theory of the origin of limbs.

OUTLINE

The pectoral and pelvic girdles and the skeleton of fins and limbs constitute the **appendicular skeleton**. Girdles brace fins and limbs against the counterforces that appendages transmit from the water or from a substrate. (You may recall from a course in physics that every force evokes an equal counterforce.) The girdles, in turn, are braced against one or more components of the axial skeleton, thereby achieving stability. The forces transmitted to the girdles from the appendages are generally greatest in terrestrial amniotes because their limbs elevate the body above the ground.

Agnathans, moray eels, caecilians, snakes, some lizards, and all but one genus of amphisbaenians have no paired appendages. Having only an anterior pair are eels, numerous other teleosts, one family (Sirenidae) of urodeles, a lizard (*Bipes*), cetaceans, and sirenians. There are also lizards and birds with hind limbs only. Craniates that lack one or both pairs of appendages generally have an elongate body and are either aquatic or live in burrows. The evolution of efficient limbless locomotion has been accompanied by adaptations that will be described later.

Tetrapod limbs arise during embryogenesis as paired **limb buds** (see fig. 1.10). Fins arise in the same manner and are called **fin folds**. It is not unusual for limb buds to appear transitorily during embryogenesis in limbless species or to develop incompletely and to persist throughout life as functionless vestiges. Inheritance of genes that have these effects bespeaks an ancestor with functional tetrapod limbs.

PECTORAL GIRDLES

A pectoral girdle is a skeletal complex in the body wall immediately behind the head that articulates with the anterior fins or limbs. Pectoral appendages are present in some extinct ostracoderms although the nature of the girdle associated with these fins is unclear. The pectoral girdles of post-Devonian vertebrates—fishes and tetrapods—are variations of the girdles of Devonian fishes. In the latter, a pectoral girdle consisted of three pairs of replacement bones that were part of the endoskeleton, and at least four pairs of investing bones derived from dermal armor. These bones are named at the bottom of figure 10.1. The pectoral girdle of the basal ray-finned fish *Polypterus* exemplifies this basic pattern (fig. 10.2). The scapula and coracoid receive the forces transmitted to the trunk from the fins, the posttemporal braces the girdle against the caudal angles of the skull, and the clavicle is braced against the opposite clavicle in a midventral symphysis. Thus, the force generated by action of the fins is distributed to the body in such fashion as to diminish the impact.

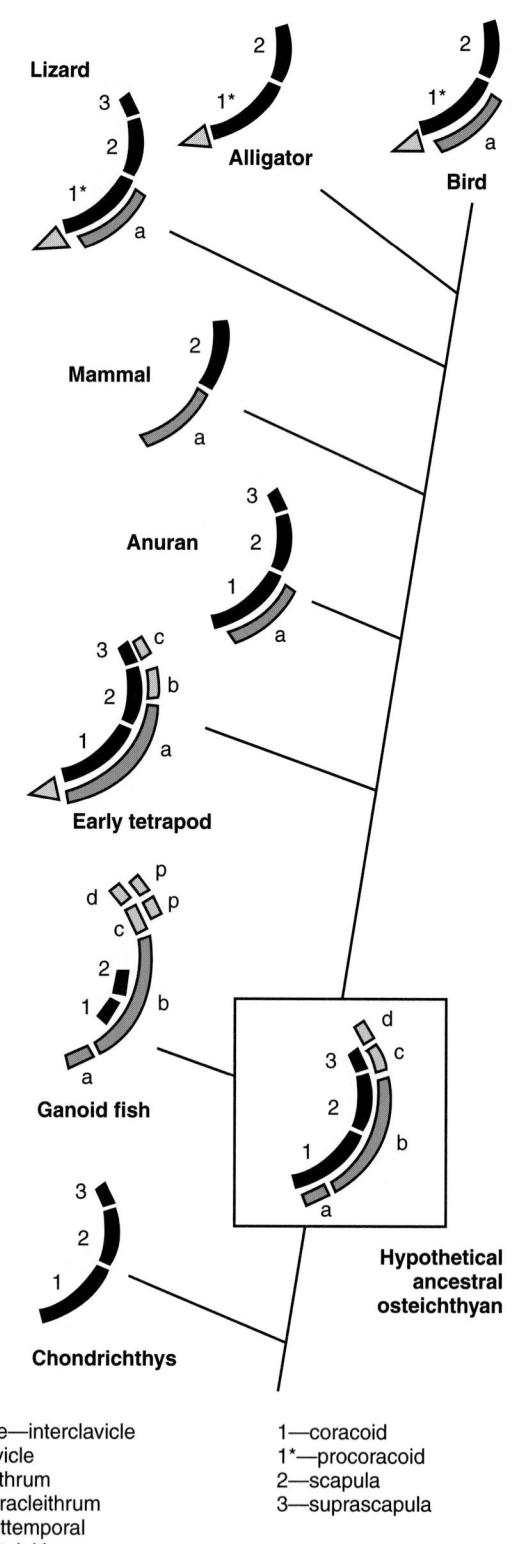

Triangle—interclavicle
a—clavicle
b—cleithrum
c—supracleithrum
d—posttemporal
p—postcleithrum

1—coracoid
1*—procoracoid
2—scapula
3—suprascapula

FIGURE 10.1

Pectoral girdle in selected phylogenetic lines. Dermal bones are *red;* cartilage and replacement bones are *black.* Only one half of each girdle is illustrated, and relationships have been distorted when necessary to emphasize homologies.

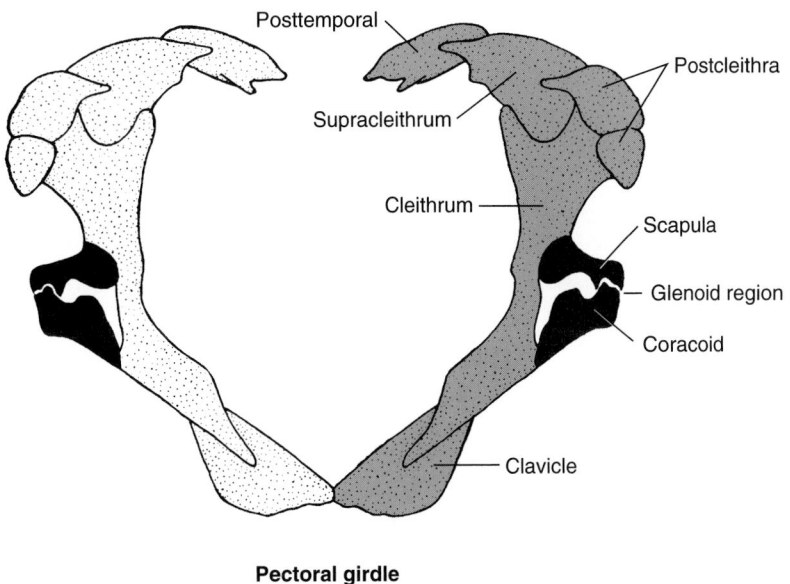

Pectoral girdle
Polypterus

FIGURE 10.2

Pectoral girdle of *Polypterus.* Dermal bones are *red;* replacement bones are *black.* Compare figure 10.1, ganoid fish.

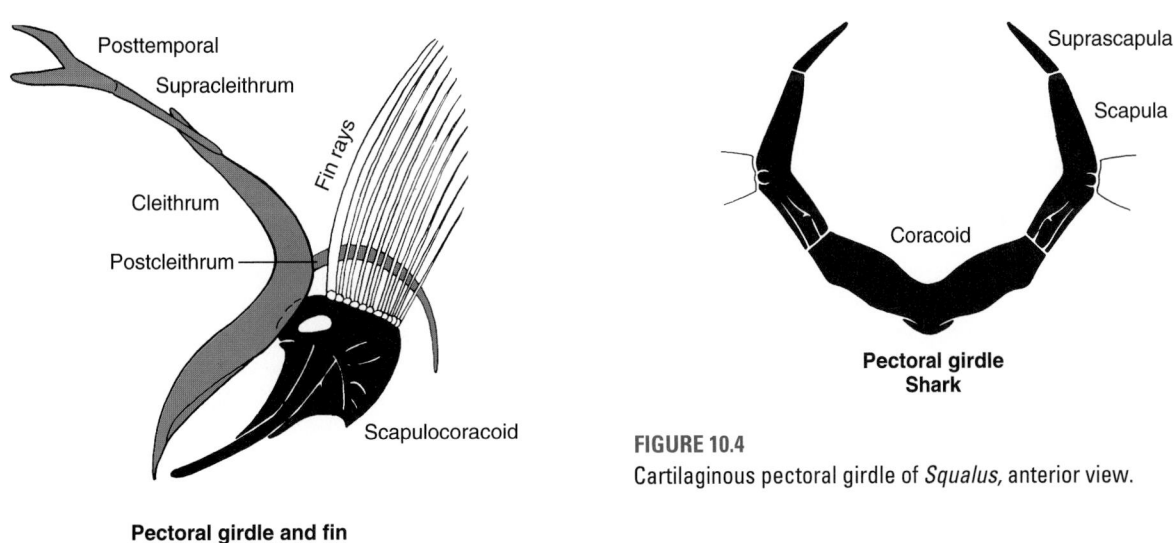

Pectoral girdle and fin
Teleost

FIGURE 10.3

Left half of pectoral girdle and the fin skeleton of a ribbonfish (teleost). Dermal bones are *red;* replacement bones are *black.*

Pectoral girdle
Shark

FIGURE 10.4

Cartilaginous pectoral girdle of *Squalus,* anterior view.

In modern bony fishes—teleosts—the cleithrum has become the major bone of the girdle (fig. 10.3). They have lost the clavicle. Otherwise, modern bony fishes generally have a full complement of primitive dermal bones. The embryonic endoskeletal coracoid unites with the scapula to form a scapulocoracoid. Cartilaginous fishes, on the other hand, have only endoskeletal components and these do not ossify, although they become hardened by calcification of the cartilaginous matrix (fig. 10.4).

The pectoral girdle of early tetrapods were hardly different from those of early bony fishes. If you compare early tetrapods with early bony fishes in figure 10.1, you will see that early tetrapods had acquired an additional membrane bone, the **interclavicle**, and that they had lost the posttemporal, which braced the girdle against the skull in fishes. The girdle of a basal amphibian is seen in figure 10.5. The supracleithrum is missing. The cleithrum, though present in this early amphibian, did not last long once tetrapods became established on land.

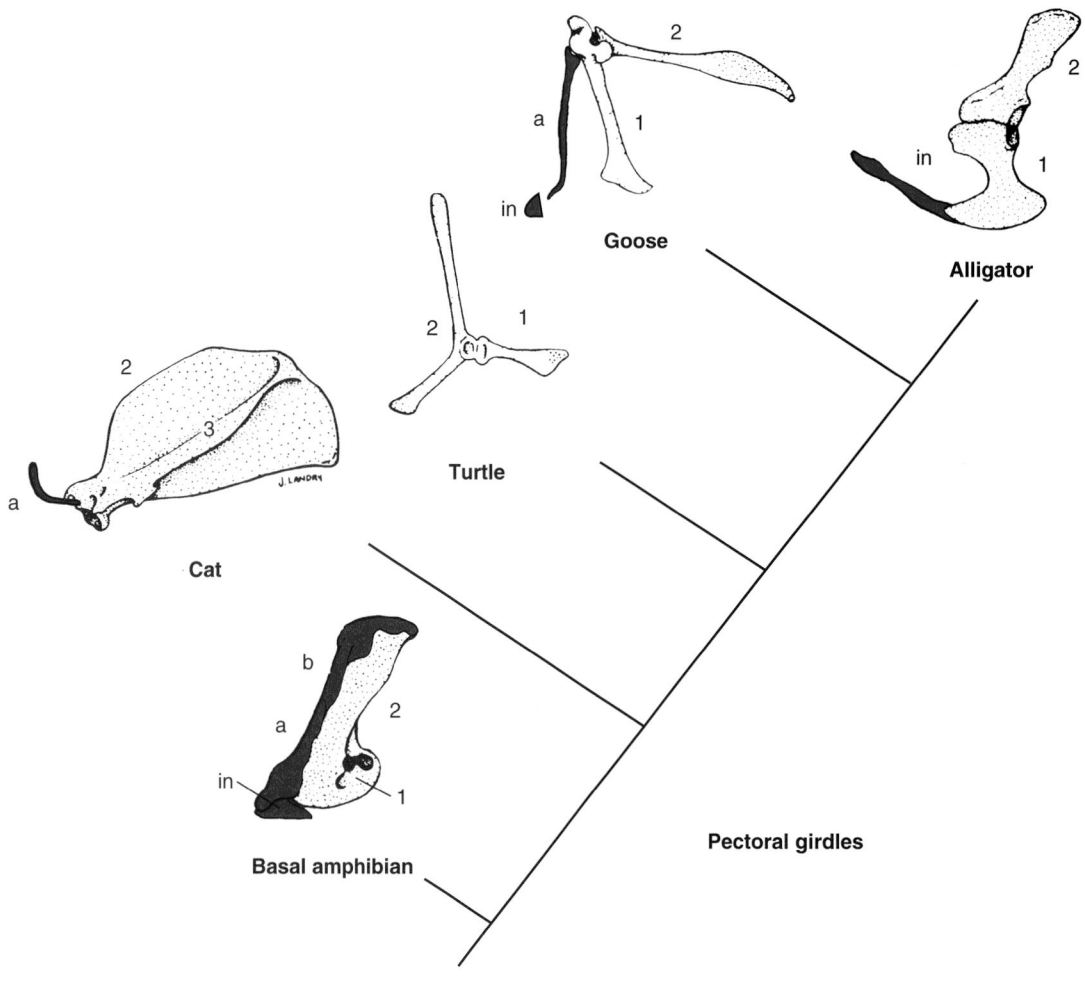

FIGURE 10.5
Left half of the pectoral girdles of selected tetrapods, lateral views on a cladogram showing the hypothesized relationships among represented taxa. Dermal bones are *red.* **1,** coracoid or procoracoid; **2,** scapula; **3,** scapular spine; **a,** clavicle; **b,** cleithrum; **in,** interclavicle. In turtles, the clavicles and interclavicle are fused with the shell. The cat's clavicle is almost vestigial—the coracoid is present only as a medial process on the scapula (not shown).

The interclavicle of stem amphibians appears to have been persistent in lineages leading to amniotes. It is seen as an unpaired bone in an alligator between the base of the two procoracoids (see fig. 8.14). In birds, it is the unpaired cartilage or bone at the tip of the "wishbone." Some paired bones associated ventrally with the clavicles or coracoids in reptiles and monotremes have also been named interclavicles by one authority or another. The interclavicles of an extinct therapsid and of a living monotreme are seen in figure 10.6. Whether or not all bones so labeled from amphibians to mammals are homologues is, at present, uncertain because of insufficient research data. We will now follow the fate of the clavicle, coracoids, and scapula as tetrapod diversity proceeds.

The fate of the **clavicles** is correlated with that of the coracoids. Either a clavicle or a coracoid are likely to brace the scapula against the sternum, and sometimes both do so (figs. 10.6*b* and 10.7). Clavicles are missing in urodeles and apodans and are uncommon in recent nonavian reptiles other than lizards (although vestiges develop transitorily in crocodilian embryos). In birds they are the long bones of the **furculum** (wishbone). They are present in most orders of mammals.

The **coracoids** in tetrapods arise from an embryonic cartilaginous **coracoid plate** in the lateral body wall extending ventrad from the glenoid region of the scapula. Anterior ossification centers in this plate give rise to **procoracoids**; posterior centers give rise to coracoids (see fig. 10.6). Neither procoracoids nor coracoids develop in Eutheria except for a vestige, the **coracoid process of the scapula** overhanging the glenoid fossa (see fig. 11.21*b*). Coracoids and/or procoracoids assist the clavicles or replace them functionally in bracing the scapula against the sternum (see fig. 8.28, iguanid and chicken).

The **scapula** is present in all tetrapods that retain any vestiges of anterior limbs because it bears part or all of

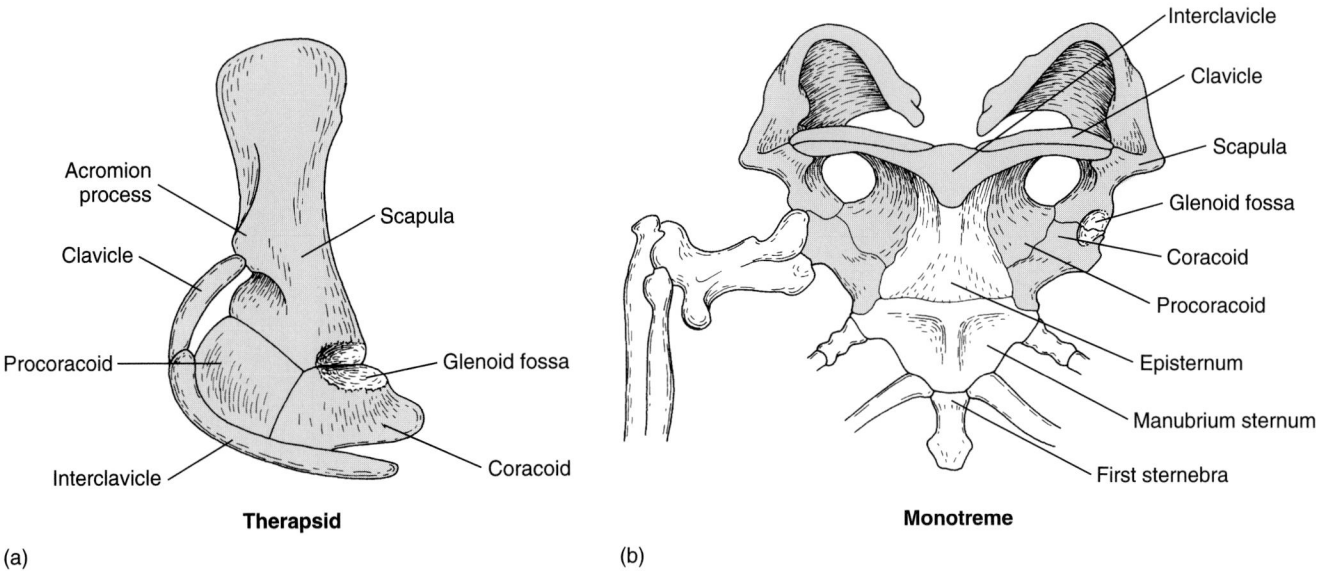

FIGURE 10.6

Pectoral girdles (*red*) of an extinct therapsid and a prototherian mammal, the latter from a ventral view. In (*b*), note the coracoid, procoracoid, clavicle, and interclavicle bracing the girdle against the sternum.

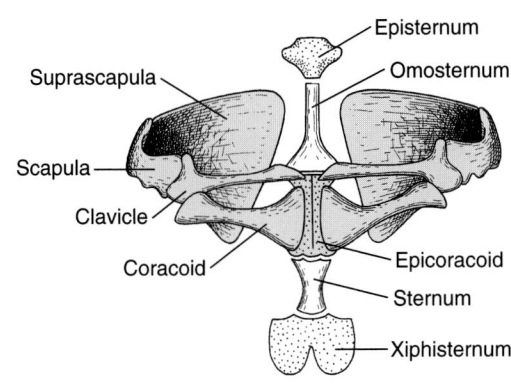

FIGURE 10.7

Pectoral girdle (*red*) and sternum of a frog, ventral view. A procoracoid (not illustrated) lies dorsal to the clavicle. *Stipple* is cartilage.

the glenoid fossa for articulation of the girdle with the head of the humerus. **Suprascapular** ossification centers usually coalesce with scapular centers to become a single bone, but a suprascapula remains independent in urodeles and anurans (see figs. 8.28, salamander, and 10.7).

The mammalian pectoral girdle is the result of evolutionary changes that gave rise to early therapsids and their mammalian descendants. The therapsid girdle consisted of interclavicle, clavicle, procoracoid, coracoid, and scapula (see fig. 10.6*a*). Monotremes have the same bones today (see fig. 10.6*b*). In eutherian mammals, the only bones remaining in the pectoral girdle are a scapula and, generally, a clavicle. As mentioned earlier, a vestige of the coracoid persists as a coracoid process of the scapula. In humans, it remains unattached to the scapula until about 15 years of age.

The lateral aspect of the **mammalian scapula** is divided by a **scapular spine** into **supraspinous** and **infraspinous fossae** (see fig. 10.5, cat). These fossae are the anatomic origins of strong muscles that insert on the humerus. The scapular spine is also an insertion site for some of the appendicular muscles that arise on the vertebral column. An **acromion process** for muscular attachment forms near the glenoid fossa. The process in cats is shown in figure 11.21.

The **mammalian clavicle** is large in the monotremes and in insectivores and primates, the last two being generalized mammals. It also is large in mammals with strong forelimbs that are used in digging, climbing, or flying. In bats, strong clavicles brace the scapula against the sternum. On the other hand, there are mammals whose skeletomuscular system has undergone adaptations that have reduced the clavicle to a mere splinter, as in felines, or eliminated it altogether, as in cetaceans, ungulates, and some other mammals. In cats, the clavicle is a vestigial splinter reaching neither the sternum nor the scapula (see fig. 10.5). This enables cats to withstand the shock of landing upright on their forelimbs after a leap because there are no rigid connections between the scapulae, which receive the initial impact, and any other part of the skeleton. The shock of the impact is dissipated in the musculature of the shoulders and in the muscles and joints of the forelimbs. Absence of a clavicle in ungulates facilitates grazing.

The phylogenetic history of the components of the pectoral girdle inspires this observation: *Dermal bones predominate in the pectoral girdle of bony fishes, whereas replacement bones predominate in tetrapods.* You may recall that dermal bones are derivatives of the bony armor

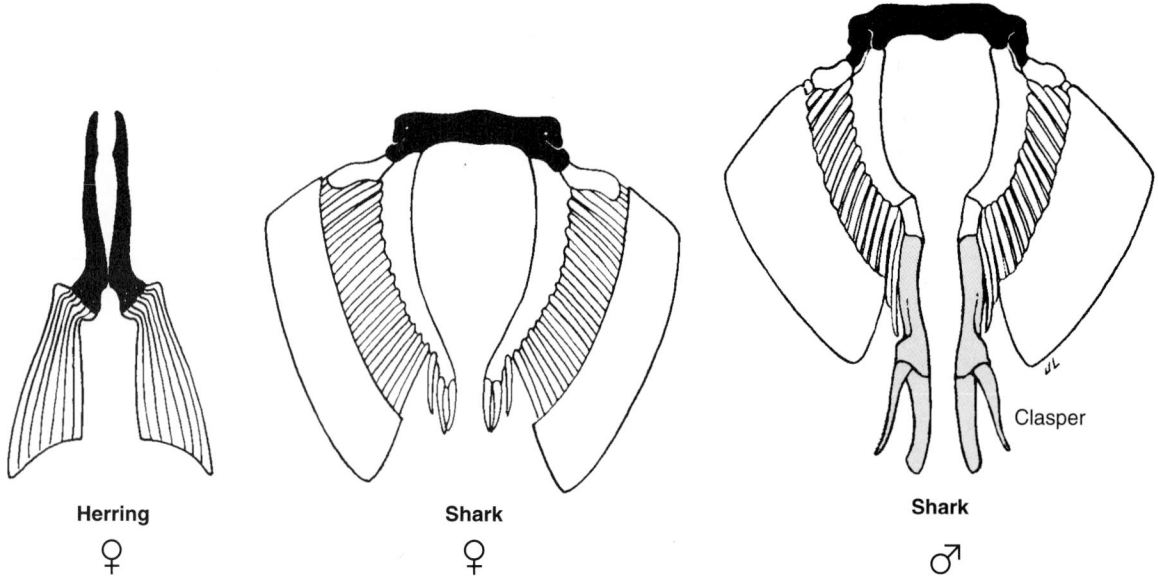

FIGURE 10.8
Pelvic plates (*black*) and fins of a bony fish and a male and female shark. The basal cartilages of the male fin have become claspers.

of early fishes, in which they overlaid, like siding on a house, the replacement bones, the latter being the true (endo) skeleton of the girdle. Little wonder then that the dermal bones of the girdle became reduced or lost during the evolution of tetrapods. Dermal armor elsewhere in the body was undergoing the same fate. The phenomenon is discussed in chapter 6, under "The Bony Dermis of Fishes."

Two loose ends are worth attention. First, urodeles fail to develop any dermal bones in their girdles (or, in fact, elsewhere in their body except in the skull). The girdle of a salamander is seen in figure 8.28. Second, have you noticed that tetrapods almost never brace their pectoral girdles against the skull or vertebral column? It is said to have happened in a few pterosaurs. The story is quite different with respect to the pelvic girdle.

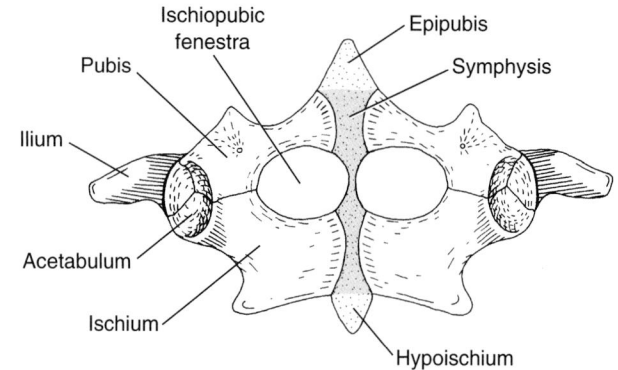

FIGURE 10.9
Pelvic girdle of *Sphenodon,* ventral view, showing ischiopubic symphysis.

PELVIC GIRDLES

Pelvic girdles in most fishes consist of a pair of simple cartilaginous or bony **pelvic (ischiopubic) plates** that meet in a midventral **pelvic symphysis** and provide a brace for the pelvic fins (fig. 10.8, herring). In cartilaginous fishes and lungfishes, the two embryonic cartilages unite to form one adult plate (fig. 10.8, shark). In teleosts that have a short trunk, the pelvic plate lies immediately behind, or sometimes below, the pectoral girdle and is often attached to it. For this reason, pelvic fins may be as far forward as the pectorals, or even appear to be anterior to them (fig. 10.16, perch). There are no dermal bones in the pelvic girdles of either fishes or tetrapods.

Tetrapod embryos also develop cartilaginous pelvic plates. Each plate ossifies at two centers to form a **pubis** (pubic bone) and a more posterior **ischium** (fig. 10.9). In urodeles, however, the pelvic plate may remain cartilaginous except for a small ischial ossification center (see fig. 10.11, *Necturus*). Dorsal to the pelvic plate, an additional blastema gives rise to an ilium (figs. 10.9 through 10.11). At the junction of the pubis, ischium, and ilium, a socket, the **acetabulum,** accommodates the head of the femur.

Dorsally, the tetrapod ilium is braced against the stout transverse processes of the sacral vertebrae, one in amphibians, two in nonavian reptiles, more in birds and mammals. Short sacral ribs intervene (fig. 10.10). Ribs often become ankylosed to the transverse processes, in which case the ribs may become unrecognizable except

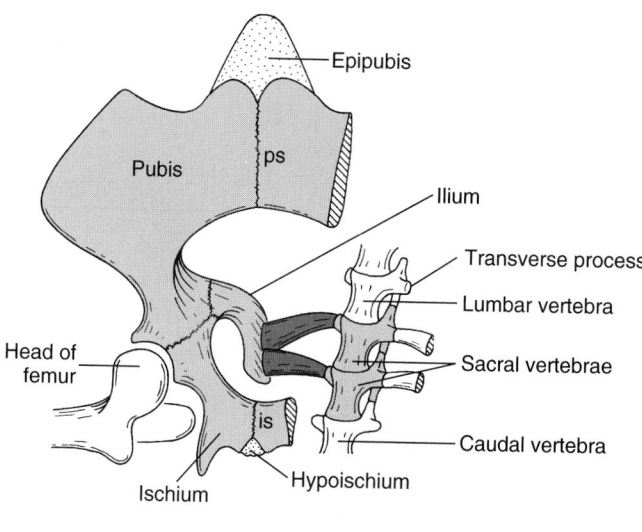

Turtle

FIGURE 10.10

Pelvic girdle, the two sacral vertebrae, and the sacral ribs (*dark red*) of the freshwater turtle *Trionyx*. **is,** ischial symphysis; **ps,** pubic symphysis. The vertebrae and ribs are viewed obliquely upward through the pelvic cavity, from below.

in embryos. Ventrally, except in birds, there is a symphysis between either the two pubic bones (**pubic symphysis**), the two ischia (**ischial symphysis**), or both (see fig. 10.9). The symphysis is in the midventral body wall immediately anterior to the cloaca or its derivatives. The skeletal framework is such that the forces transmitted to the two acetabula as a result of gravity (weight bearing) or locomotion is distributed in two directions: to the vertebral column dorsally via the ilia and to the symphysis ventrally via the ischial and pubic bones. The proportion distributed in each direction depends on the posture of the animal. The joint between the head of the femur and the girdle is stabilized by the joint capsule and by muscles that approach the femur from opposing directions. The sacrum and girdle in amniotes, often rigidly united, form a bony enclosure, the **pelvis,** that encircles the caudal end of the coelom. The resulting **pelvic cavity** contains the urogenital organs and the terminal portion of the large intestine.

Posture and mode of locomotion are correlated with the shape of the ilium, ischium, and pubis, the anatomic relationships of these bones, and their proportional size. A squatting, jumping frog has a different set of vectors affecting the pelvic girdle than does a bird, deer, kangaroo, or marine turtle. In frogs, the ilia (see fig. 8.13) are slender and greatly elongated, and they extend from the sacral vertebra to the end of the urostyle, where they meet the ischia and pubic bones and where the acetabulum is located. The joint between ilium and sacral vertebra is free to move when a frog pushes off at the start of a leap. As the frog lands, the joint, along with others in the leg, helps dissipate the force of the impact, acting as

a shock absorber. In many tetrapods, this **sacroiliac joint** is not freely movable.

Urodeles in general have weak posterior limbs that can scarcely lift the sagging belly off the substrate without the buoying effect of water. Consequently, the pelvic girdle is subject to relatively little stress as long as the animal is resting motionless on a substrate. The pelvic plate, mentioned earlier, differs little from that of fishes other than being braced against a single vertebra by a weak ilium. In urodeles that usually lose their gills and develop lungs, a slim median **prepubic (ypsiloid) cartilage** extends from the girdle forward in the linea alba for two or three somites. It furnishes attachment for certain muscles utilized in respiration.

The structure of the pelvic girdles of reptiles is correlated with their diverse body structure and with their mode of locomotion that varies from lumbering to highly agile. The ilium has become braced against an additional vertebra (fig. 10.10). In most reptiles, the pubis is directed away from the ischium, resulting in a triradiate girdle (fig. 10.11*e, f, i*). However, in ornithischian dinosaurs and independently in birds, the ischium and pubis are parallel and directed caudad (fig. 10.11*g, h*). In some reptiles, a broad *ischiopubic fenestra* develops between the ischium and the pubis on each side (figs. 10.9 and 10.10*d*). It should not be confused with the opening in the wall of the acetabulum of the alligator and carnosaur seen in figure 10.11. Those openings result from incompletion of the wall of the socket.

An **epipubic** and **hypoischial** bone frequently develops in association with the pelvic girdle of reptiles (see fig. 10.9). One or both are present also in monotremes and marsupials. The epipubic bone is also called the **marsupial bone** in marsupials because it supports the marsupial pouch.

The ilia and ischia of modern birds are enormously expanded and united with the synsacrum (see figs. 8.18 and 8.19). Such a girdle provides broad surfaces for attachment of hind limb muscles used in the bipedal stance and for the thrust required by some species for flight takeoff. The pubic bones are reduced to long splinters that are directed caudad parallel to the ischium, and they do not meet ventrally. Therefore, there is no ischial or pubic symphysis. Absence of symphyses provides a wide outlet from the pelvic cavity for laying the large-yoked avian eggs. The pelvic girdles of an ornithischian dinosaur and a goose are illustrated side by side in figure 10.11. Despite the similarity of pelvic girdles, birds are now thought to be descendants of some small saurischian dinosaur. (See chapter 4 for a phylogenetic discussion.) The pelvic girdle of the very early *Archaeopteryx* (see fig. 8.21) was braced against four sacral vertebrae. There was no synsacrum.

In mammals, the ilium, ischium, and pubis ankylose early in postnatal life to form a left and right **innominate**

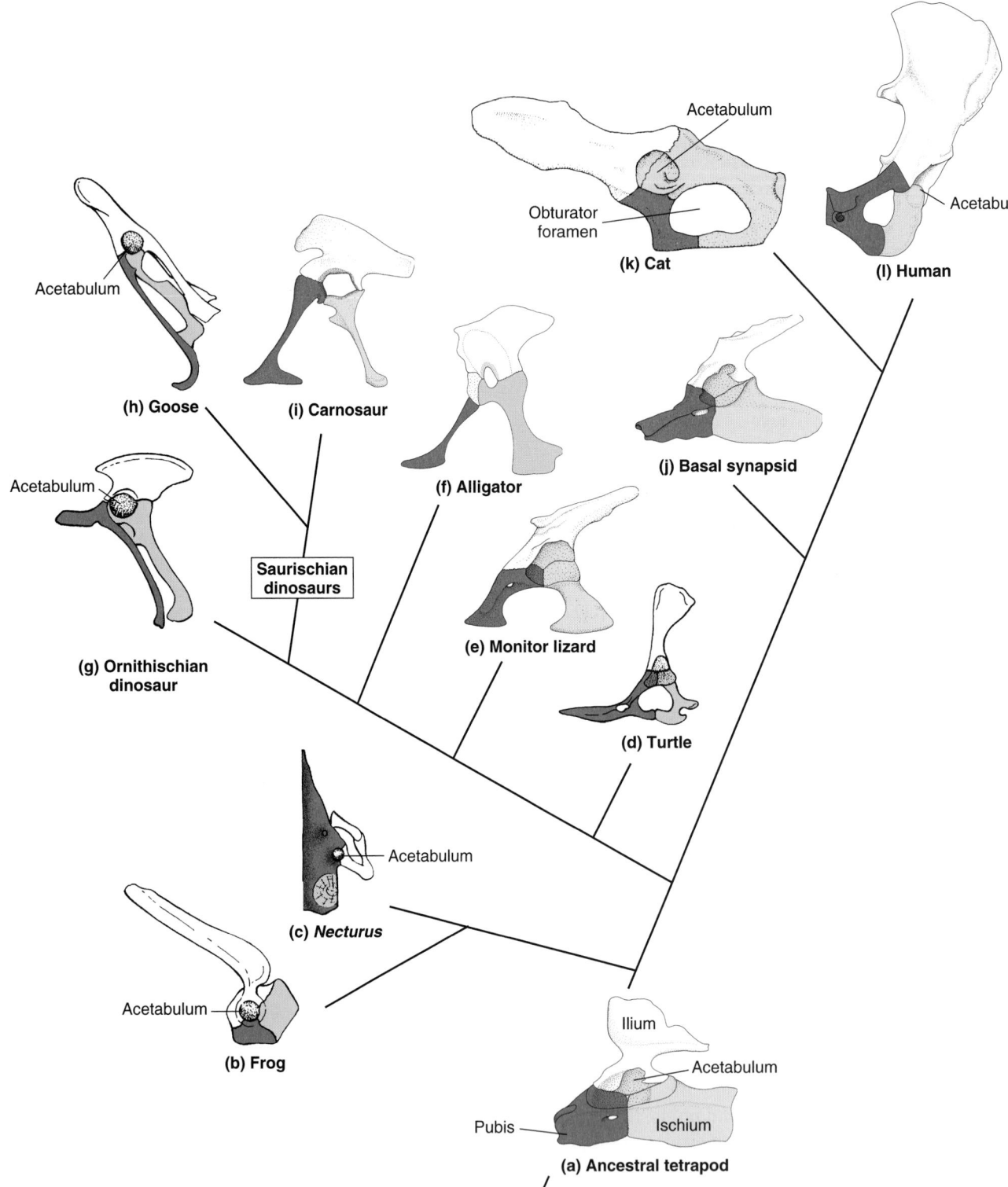

FIGURE 10.11

Left halves of pelvic girdles of selected tetrapods on a cladogram showing the hypothesized relationships among taxa. (*a*)–(*b*) and (*d*)–(*k*) lateral view, head to the left. (*c*) Ventral view. (*l*) Anterior (ventral) view, body upright. The pubis is *dark red,* the ischium *light red,* and the ilium *uncolored. Stipple* in (*c*) and (*f*) is cartilage. In *Necturus* (*c*), the ischium is an ossification center in a cartilaginous ischiopubic plate, and a sacral rib is seen attached to the dorsal end of the ilium. There are no dermal bones in pelvic girdles.

(coxal) bone (fig. 10.11*l*). Dorsally, the ilium on each side ankyloses with the sacrum in an immobile **sacroiliac joint**. Ventroposteriorly, the pubis and, frequently, the ischia meet in a **pubic** or **ischiopubic symphysis** to complete the walls of a pelvic cavity. A pelvic outlet at the base of the tail provides access to the exterior.

Mammalian young are delivered through the pelvic outlet. In late pregnancy, the fibrocartilage separating the bones at the symphysis is softened by **relaxin**, an ovarian hormone, which then permits expansion of the pelvic outlet for delivery. In mice six days pregnant, the gap between the bones in the symphysis was shown on X-ray film to be only 0.25 mm. Thirteen days later, on the day of birth, the gap had widened to 5.6 mm.

FINS

Fins serve as steering devices for changing direction, as stabilizers that prevent the body from rolling or wobbling, and as devices that control the inclination (slope) of the body in swimming away from the horizontal. Paired fins also serve as brakes that slow or halt forward motion, but typically play little role in providing forward thrust. Some use specialized pectoral fins for forward locomotion (e.g., rays). Forward thrust in most fishes is generated by lateral undulation of the posterior part of the trunk, tail, and caudal fin (see fig. 11.8).

Fins, whether paired, median, or caudal, consist mostly of two surfaces of skin back to back, stiffened by flexible fin rays that radiate from a skeletal base (fig. 10.12). The rays are in the dermis and are of two varieties. The rays of bony fishes, **lepidotrichia**, consist of

jointed *bony dermal scales* aligned end to end. **Ceratotrichia**, the rays of cartilaginous fishes, are long *horny rays* composed of a material similar to that comprising the dorsal spines of sharks. Short delicate **actinotrichia** develop distally in both groups. In elasmobranchs and a few other fishes, typical body scales continue onto the fin, growing smaller distally. These further stiffen the fin.

The skeletal base from which fin rays of generalized fishes radiate, whether of paired fins or median ones, consists of a row of cartilaginous or bony **basalia** and one or more rows of **radialia** (figs. 10.13–10.15). The specific anatomy of the base varies with the taxa. The chondrostean *Polypterus* is a generalized ray-finned fish whose pectoral fins (fig. 10.13*a*) have three large basals

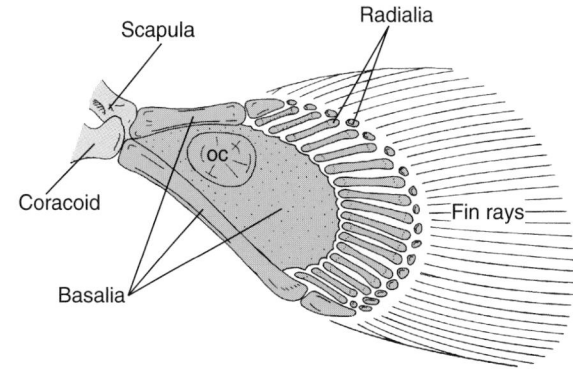

Basal actinopterygian

(a)

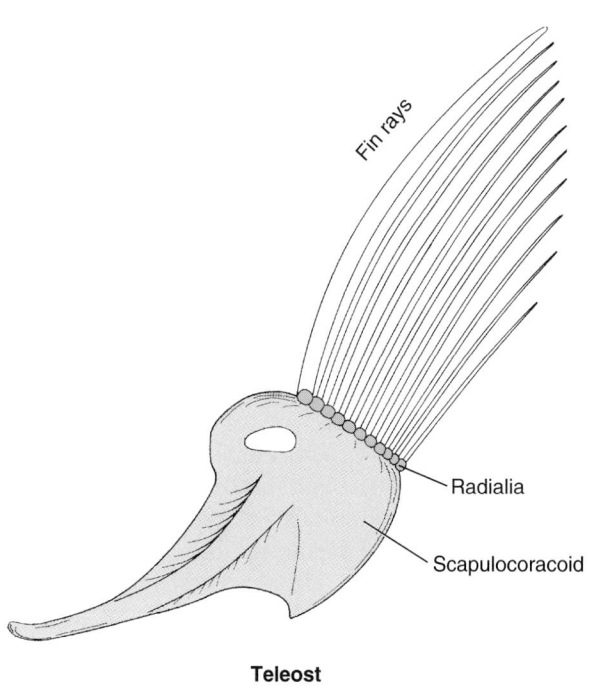

Teleost

(b)

FIGURE 10.13

Pectoral fin skeleton of two ray-finned fishes. (*a*) *Polypterus,* **oc,** ossification center in central basal cartilage. The other basalia are fully ossified. (*b*) Specialized fin of a ribbonfish.

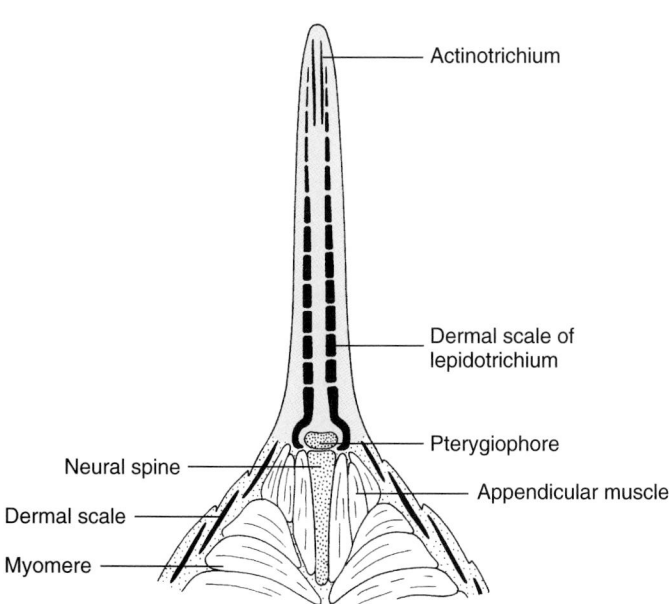

FIGURE 10.12

A pair of fin rays (lepidotrichia) in the dorsal fin of a teleost in diagrammatic cross section. A pterygiophore is a cartilage or bone at the base of rays.

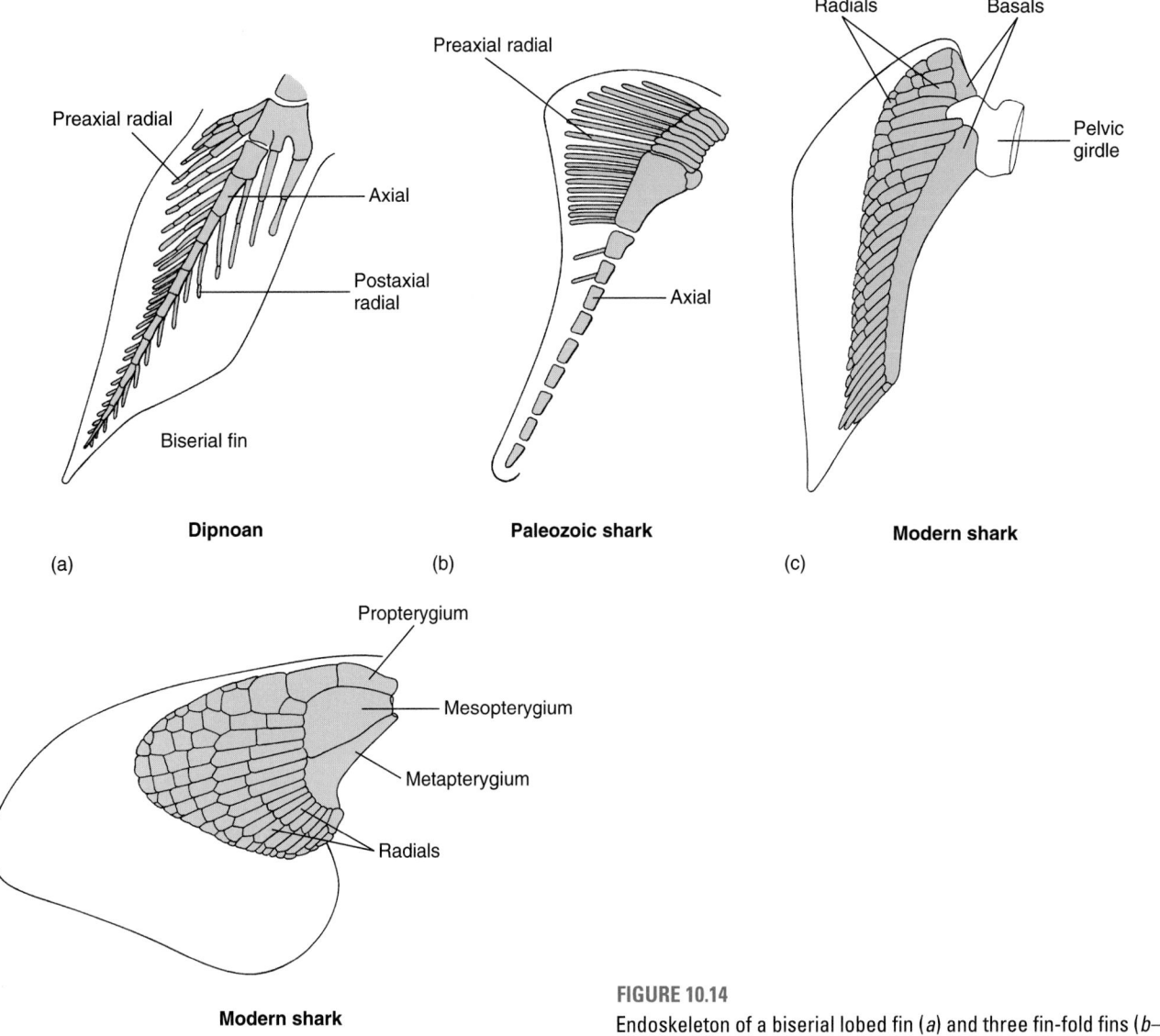

Dipnoan

(a)

Paleozoic shark

(b)

Modern shark

(c)

Modern shark

(d)

FIGURE 10.14

Endoskeleton of a biserial lobed fin (*a*) and three fin-fold fins (*b–d*). (*a*) to (*c*) Pelvic fins. (*d*) Pectoral fin.

and two distal rows of radials. The paired fins of most specialized teleosts, on the other hand, have lost all basalia and retain mere vestiges of distal radials (fig. 10.13*b*). The basal and radial skeleton of median dorsal fins rests on the vertebral column (fig. 10.15). Dipnoans and other rhipidistians exhibit a wide variety of bases in both paired and dorsal fins.

Striated muscle masses from several successive body segments extend into the base of a fin and insert on available skeletal elements. Intensive efforts in the past to identify an ancestral fin type, or **archipterygium**, that might have led to the fins of later cartilaginous and bony fishes have resulted only in speculation. There is too much diversity of morphology and too little evidence.

Paired Fins

Pectoral fins are braced against the glenoid fossa or glenoid region of the pectoral girdle (fig. 10.13*a*). Pelvic fins are braced against a prominence on the lateral or posterior aspect of the pelvic plate (see fig. 10.8). Paired fins are grouped in three categories (fig. 10.16): (1) **lobed fins**, (2) **fin fold fins**, and (3) **ray fins**. These correspond to the three categories of jawed fishes that emerged from the Devonian and have persisted to the present: (1) sarcopterygians (actinistians and rhipidistians), (2) chonrichthyans (elasmobranchs and chimaeras), and (3) actinopterygians (ray-finned fishes). (**Spiny fins**, characteristic of acanthodians, had disappeared by the end of the Paleozoic.)

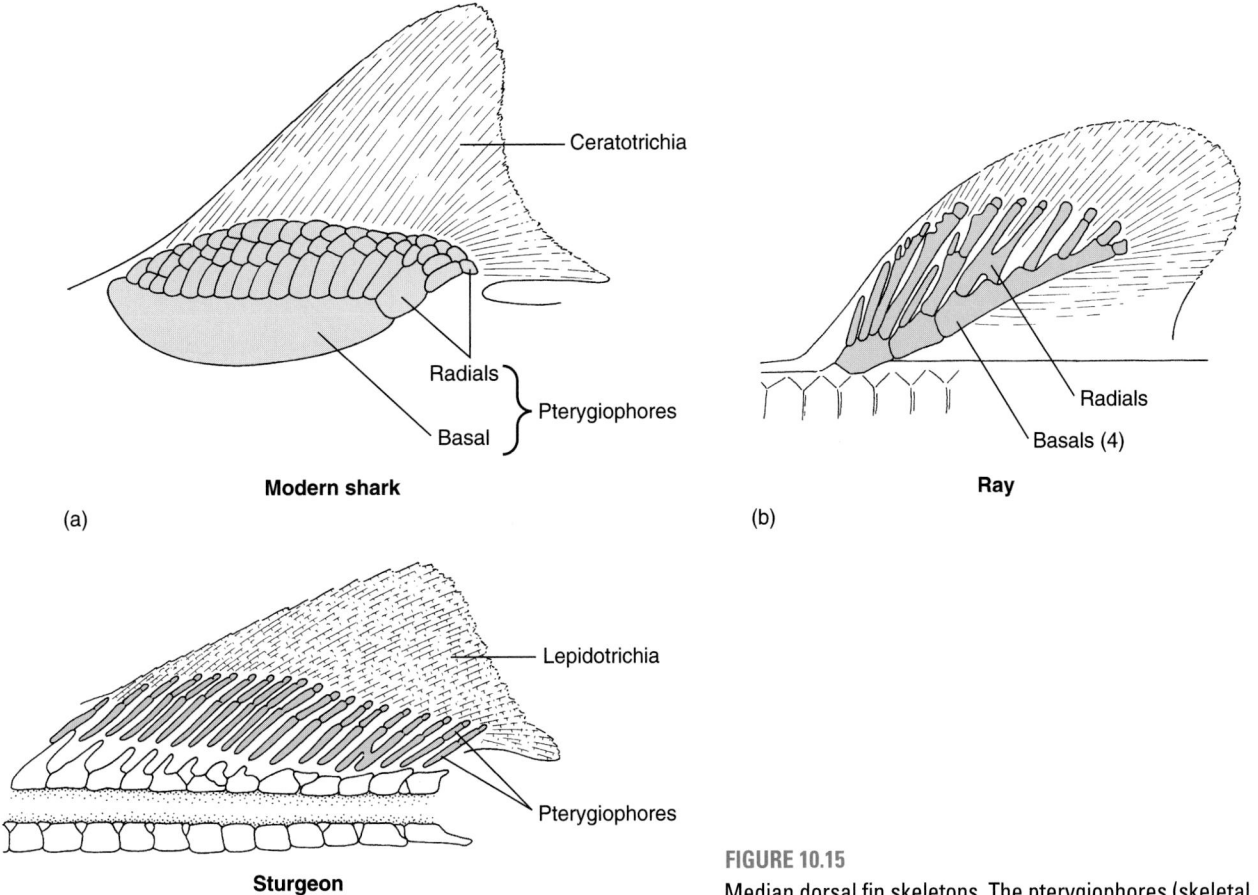

Ceratotrichia

Radials
Basal
Pterygiophores

Modern shark

(a)

Radials
Basals (4)

Ray

(b)

Lepidotrichia

Pterygiophores

Sturgeon

(c)

FIGURE 10.15

Median dorsal fin skeletons. The pterygiophores (skeletal elements at the base of the rays) rest on the vertebral column.

Lobed fins consist of a fleshy proximal lobe containing the fin skeleton and its attached muscles, and a membranous distal portion that is stiffened by fin rays. The fin arises from a narrow base and resembles a paddle. The paired fin skeleton of the dipnoan *Neoceratodus* (see fig. 10.14*a*) consists of a jointed central axis, the bones of which function like basals, and a series of pre- and postaxial radials. Such a fin is said to be **biserial** because of the two series of radials. Other rhipidistians have a wide variant in which the relationship of the axis and radials has been considerably disrupted (see fig. 10.44).

Fin fold fins have a broad base. It is broad in modern sharks and was broader in the Paleozoic. As in other fishes, the three basalia are termed **pro-, meso-,** and **metapterygia** (see fig. 10.14*d*). In male Chondrichthyes, the basalia of the pelvic fin are modified to become an intromittent organ, the **clasper** (see fig. 10.8, male shark).

The ray fins of teleosts have lost components of the basal skeleton in time (compare figs. 10.13*a* and *b*), but the fins have become increasingly flexible. A number of ray-finned fishes have no pelvic fins. Tuna, which are among the largest fishes in the oceans, propel themselves forward by lateral undulation of a highly derived

tail designed for fast continuous swimming. Loss of the pelvic fins may be related to streamlining.

It should be mentioned that a few fishes fly, although for short distances only. Characins, which are voracious teleosts inhabiting fresh tropical waters, get an initial thrust out of the water with the caudal fin and, beating winglike pectorals, fly several yards using appendicular muscles that, for a fish, are exceptionally large. This behavior is said to occur only when the fish are alarmed. If so, it is a fishy way of performing a disappearing act.

Median Fins

Fishes have one, two, or a series of median **dorsal fins,** and many have a midventral **anal fin** (Blieck, 1992, suggests an ostrasoderm origin for this fin) just behind the anus or vent. Dorsal and anal fins are seen in figure 10.16, and over a dozen paired and median fin patterns are illustrated in chapter 4. Median fins act as keels, keeping motionless fishes from rolling to the left and right, and in rare cases they may be used for locomotion. When the fish is swimming, median fins typically minimize deviation from a straight line that inevitably results from side-to-side thrusts of the locomotor tail. Lack of stabilizers of one kind or another would not only increase the cost of locomotion in terms of energy expended but

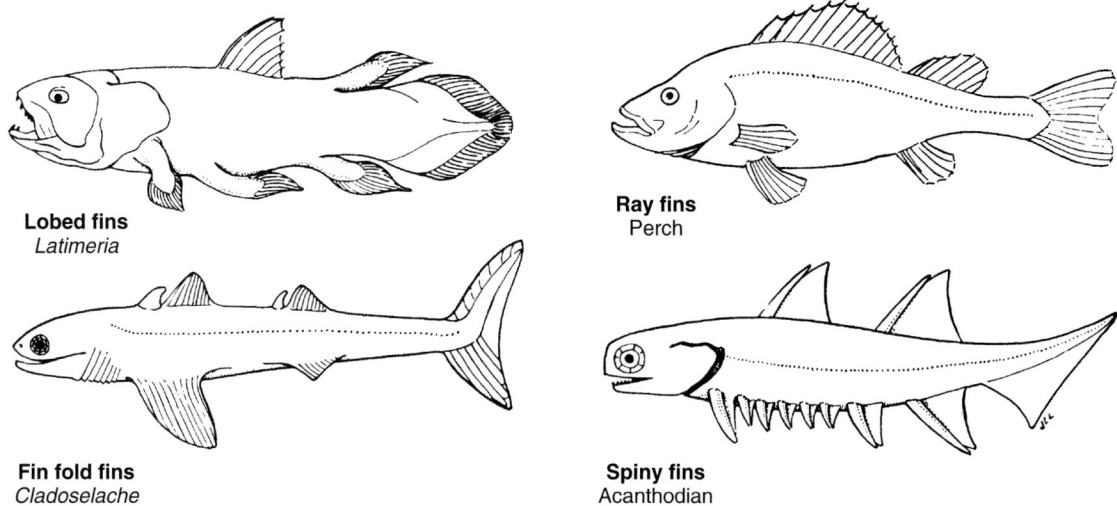

Lobed fins
Latimeria

Ray fins
Perch

Fin fold fins
Cladoselache

Spiny fins
Acanthodian

FIGURE 10.16
The fins of fishes.

might place a species in jeopardy. Bottom-dwellers might be less disadvantaged. Because of the body shape of Raji-formes, one might expect less need for stabilizers; never-theless, rays other than sting rays have two dorsal fins far back on the tail. However, there is no anal fin. Dorsal fins are long in lampreys and bony eels, compensating for the inadequacy or absence of paired fins.

The anal fin of the male of some viviparous teleost species has been elongated posteriorly and modified to transfer sperm to the female. This **gonopodium** is analo-gous to the clasper of cartilaginous fishes.

Caudal Fins

Caudal fins are classified on the basis of their shape and the direction taken by the terminal portion of the notochord and vertebral column (fig. 10.17a to c). A tail that contains dorsal and ventral lobes and in which the notochord turns upward into a large dorsal lobe is said to be **heterocercal**. This was the predominant type in the Paleozoic, having been present in placoderms, Paleozoic sharks, and at least some acanthodians, and it is still seen in modern sharks and two relict chon-drosteans, namely, sturgeons and spoonbills (see fig. 4.14). There is also a rare condition, **hypocercal**, in which the vertebral column turns downward. This was a prominent feature in ichthyosaurs—an example of evolutionary convergence.

Derived from the heterocercal condition were innu-merable variants that make classification of caudal tails less than definitive. Two terms that have been used con-ventionally are diphycercal and homocercal. Both types are *symmetrical externally,* but the notochord and verte-bral column terminate differently in the two. In **diphyc-ercal** tails, the vertebral column ends with very little up-bending, as in dipnoans and *Latimeria* (figs. 4.13 and 10.17b). In **homocercal tails**, the notochord, encased

within a bony sheath, or **urostyle** turns far dorsad (figs. 10.17c and 10.18); as in the similarly named structure in frogs, the urostyle represents an independ-ent fusion of caudal vertebral elements. In the teleost seen in figure 10.18, components (hypurals) of several terminal caudal vertebrae have been modified to be-come the basal skeleton of the fin. Homocercal tails are common in teleosts but not restricted to them. Special-ized teleosts have only two hypurals and no epurals.

During embryonic development, a homocercal tail is initially heterocercal (see fig. 10.17d). Applying von Baer's theorem, we could conclude that modern caudal fins are modifications of a heterocercal condition. That conclusion is reinforced by the observation that hetero-cercal tails predominated during the Paleozoic.

There have been attempts to use the morphologic features of caudal fins in unraveling the phylogenetic re-lationships among early and subsequent fishes. The ap-proach has been thwarted by enormous diversity, an in-complete fossil record, and the realization that similarities may be the result of convergent evolution.

Caudal fins have developed as adaptations to aquatic life in a few amniotes. The fishlike ichthyosaurs devel-oped them, and cetaceans and sirenians acquired hori-zontal flukes rather than vertical fins. The vertebral col-umn provided internal support for the caudal fins of ichthyosaurs, turning ventrad into the ventral lobe of their bilobed tails, but there were no fin rays. The flukes of marine mammals have neither fin rays nor endoskele-tal support. The same is true of the caudal fins of larval amphibians.

Origin of Paired Fins

What was the nature of structures that might have given rise to paired fins? This is one of the more puzzling ques-tions in the study of phylogeny, because to answer it

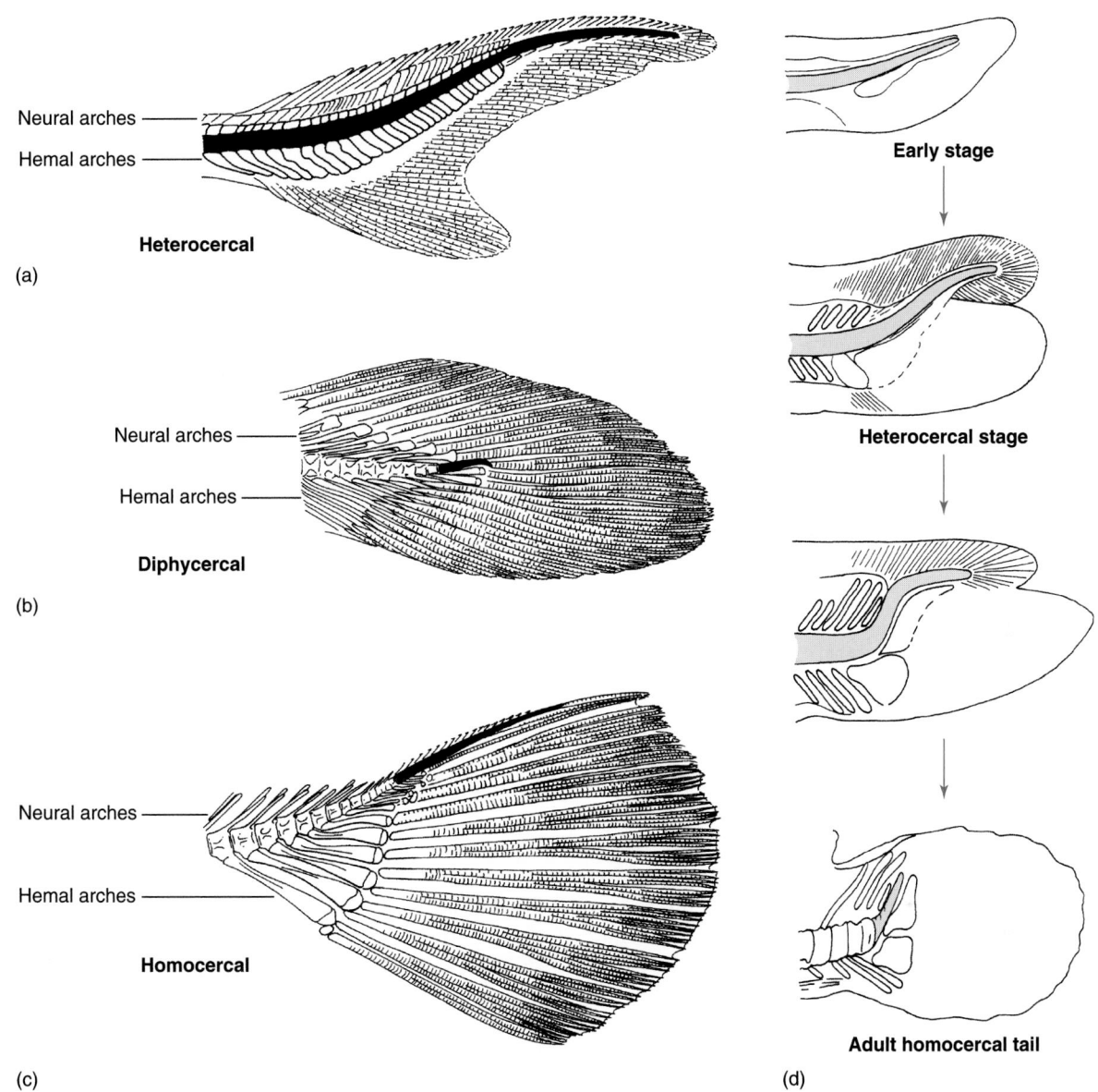

FIGURE 10.17

Caudal fins of fishes. (*a*) to (*c*) Major morphologic varieties, notochord in *black*. The heterocercal fin is probably the most primitive.
(*d*) Successive developmental stages in the flounder *Pleuronectes,* notochord in *red,* showing transition from heterocercal to homocercal.

requires more knowledge than we have of early craniates. This being so, the best we can do to satisfy our curiosity, which is the motivation for basic science, is to look for whatever clues we might find among ostracoderms and early jawed fishes, and then speculate.

The **fin fold hypothesis** states that paired fins are derived from a pair of continuous fleshy folds of lateral body wall analogous to the metapleural folds of an amphioxus. Some ostracoderms had such folds, although they were higher on the body wall. If such a structure were to become interrupted in the middle of the trunk and remaining sections were to be invaded by muscle buds at the base and endowed with an endoskeleton, the result would be paired fins like those of *Cladoselache*

(see fig. 10.16). But there is no evidence this happened, and the hypothesis is of historical interest only.

According to the **gill arch hypothesis** of C. Gegenbaur, pectoral and pelvic girdles are modified gill arches, and the skeleton within the fin is an expansion of the gill rays. It is true that many students of comparative anatomy, seeing a shark skeleton suspended in a museum jar for the first time, think that the pectoral girdle is part of the pharyngeal skeleton because of its location immediately behind the last pharyngeal arch and because of its U shape. It looks as though it might have been a gill arch, but this is quite unlikely.

Another hypothesis is the **fin spine hypothesis.** In early acanthodians (see fig. 10.16, spiny fins), pectoral

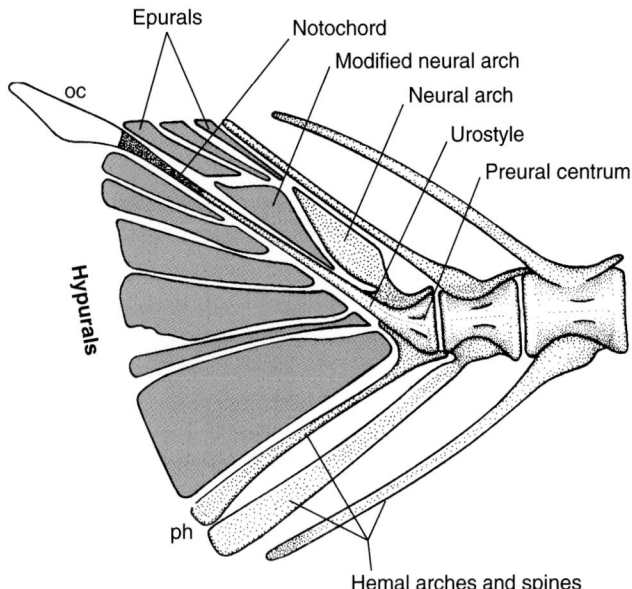

FIGURE 10.18
Skeleton of the homocercal caudal fin of the basal teleost *Clupea.*
The urostyle encloses all but the terminal segment of the upturned
notochord. Fin rays extend from the hypurals and epurals. *Gray*
indicates components of modified caudal vertebrae.
oc, opisthural cartilage; **ph,** parahyural.

and pelvic appendages were the largest of a series of lateral hollow spiny appendages that extended the length
of the trunk and supported a fleshy membrane. Evidently they were practically immobile, and in no way resembled the pectoral and pelvic fins of modern fishes.
In later acanthodians, weak fin rays were present in the
membranes of two pairs, those at the pectoral and pelvic
levels only, and small radial elements supported the
membranes at their base. In time, acanthodians tended
to lose all spines except the two pairs containing fin
rays. This may be a clue to the origin of paired fins, but
it appears that acanthodians were not ancestral to later
fishes. Among placoderms, both pectoral and pelvic fins
were present, supported by a viable number of segmentally arranged basals. In some forms, the pectoral fins
were associated with a spinal plate. The diversity of pectoral fins in placoderms ranges from expanded raylike
fins, sharklike hydrofoils (*Dunkleosteus,* see fig. 4.9) to
armored appendages (*Bothriolepis,* see fig. 4.9). It is
clear that placoderms, like the chondrichthyans, seem
to have been "experimenting" with locomotor designs.

When considering a single hypothesis for the origin
of paired fins, it is important to review the available evidence and the relative time of origin. Pectoral fins first
appeared among the ostracoderms whereas pelvic fins
appear to be a gnathostome feature. The phylogenetic
position of acanthodians suggests that their spinous
condition may represent a derived feature. Given our
current hypothesis of relationships and the paucity of
fossil evidence, it may be that an understanding of fin

development is our best hope of unraveling fin origins.
The fin fold theory can be modified to suggest that there
is a field of potential limb development along the craniate trunk. Activation of this zone by regulatory genes
was initiated in the pectoral region among ostracoderms. A second zone of activation occurred in the
pelvic region among gnathostomes. The origin of accessory fins in acanthodians may represent a more global
activation of the potential fin fold. Given this modified
fin fold theory, there is no need to search for an ancestor
with a complete fin fold.

Alternatively, we can evaluate outgroups to see if they
shed any light on this question. However, no known
protochordate has structures that conceivably could
have given rise to paired craniate fins, even allowing the
imagination free rein, and the earliest known craniates
had none. Reliable clues to the origin of paired fins may
be hidden forever in the obscurity of elapsed time.

TETRAPOD LIMBS

Although tetrapods typically have four limbs, some have
lost one or both pairs, and in others the forelimbs have
been modified as wings or paddles. By employing limbs
with appropriate modifications, tetrapods swim, crawl,
walk, run, hop, jump, dig, climb, glide, or fly to avoid
enemies, seek food and shelter, and find a mate. Each
lifestyle is facilitated by some modification of the appendicular skeleton.

The limbs of early tetrapods were short, the first segment extended nearly horizontally from the trunk, and
the second segment was perpendicular to the first, directed downward (see fig. 10.45*d*). This posture persists
to a considerable degree in urodeles, tortoises, and basal
lizards. In other reptiles and in mammals, there has
been a rotation of the entire appendage toward the body,
so that the long axes of the humerus and femur more
nearly parallel the vertebral column, the elbow being directed caudad and the knee cephalad (see fig. 10.31*a*).
Limbs so oriented are excellent shock absorbers. They
also provide more support for a body suspended above
the ground and greater leverage for locomotion on land.

Terminology describing positions within the limbs is
dependent upon the normal anatomical position for the
organism of interest. In the hypothetical transition from
aquatic to terrestrial locomotion, we see a shift from a
horizontal structure (a fin) to one that is rotated and
flexed at several joints (a limb). Other transitions, as
noted earlier, include the change among tetrapods from
a sprawling stance to an upright one. Finally, there is
the shift from a quadrupedal stance to a bipedal one
that occurs independently in several lineages. Note that
the anatomical position of humans is unusual and is
represented by an erect stance with the arms rotated so
that the palms face forward (thus the preaxial muscles
of the human forearm are postaxial in quadrupeds and

ventral in aquatic ancestors). Keep these differences in mind as you read about or describe anatomical structures. The relationship between proximal and distal is independent of stance and can be applied across taxa.

Tetrapod limb skeletons consist of three segments: **propodium**, **epipodium**, and **autopodium.** In the forelimb, these correspond to the bones of the upper arm, forearm, and **manus**, or hand (fig. 10.19). Table 10.1 lists positionally equivalent components of the fore- and hind limbs. The skeleton within homologous segments in the various tetrapods is remarkably similar despite outward appearances; it is the orientation of the bones, the relative mobility of the joints, and the complexity of the appendicular muscles as much as the skeleton per se that makes possible the variety of locomotor activities of tetrapods. The most striking differences in skeletons are at distal ends of the appendages.

Propodium and Epipodium

The humerus is the bone of the upper arm. The similarity of the humeri of all tetrapods is more striking than any differences (fig. 10.20). Variations in length, diameter, and shape are adaptive modifications. The odd humerus of the mole (fig. 10.21), for example, has expansions for insertion of massive shoulder muscles for digging. The humeri of carinate birds have a slender central cavity containing diverticula from the lungs.

The radius and ulna are bones of the forearm. The radius is a former preaxial bone that has shifted in orientation and articulates proximally with the humerus and distally with wrist bones on the thumb side of the hand (compare figs. 10.45a, e, and f). It bears most of the force being transmitted from wrist to humerus. The ulna is a longer, formerly postaxial bone articulating proximally with the humerus and radius and distally with wrist bones on the side opposite the thumb. The ulna sometimes fuses with the radius, or it may be vestigial, as in frogs and bats (fig. 10.20).

The femur is the bone of the thigh, and the tibia and fibula are bones of the lower leg (shank). The three bones differ relatively little from one tetrapod to another. A sesamoid bone, the **patella**, or kneecap, develops in birds and mammals. It ossifies in the tendon of insertion of the powerful extensor muscle of the thigh where the tendon passes over the complicated knee joint to insert on the tibia. The patella protects the joint from the abrasive action of the tendon. The fibula (fig. 10.22, *red*) may unite partially or completely with the tibia to form a **tibiofibula** as in frogs; it may be reduced to a splinter, as in birds; or it may be lost, as in deer and

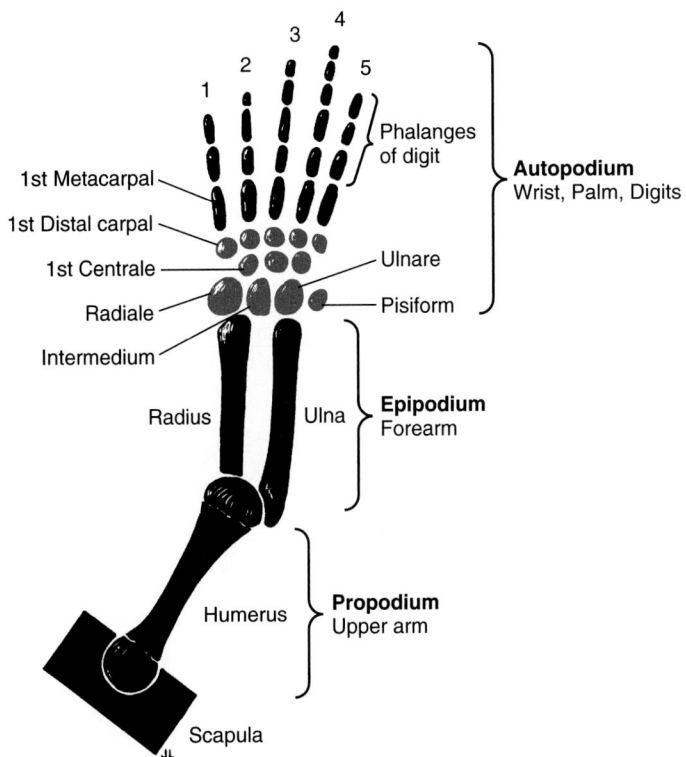

FIGURE 10.19
Skeletal organization of a generalized (primitive) right forelimb, palm down, wrist in *red*. **1–5**, digits.

TABLE 10.1 Positionally Equivalent Components of the Anterior and Posterior Limbs

	Name of Segment		Skeleton
Anterior Limb	1 Upper arm (brachium)		Humerus
	2 Forearm (antebrachium)		Radius and ulna
	3 Wrist (carpus)	Manus	Carpals
	4 Palm (metacarpus)		Metacarpals
	5 Digits		Phalanges
Posterior Limb	1 Thigh (femur)		Femur
	2 Shank (crus)		Tibia and Fibula
	3 Ankle (tarsus)	Pes	Tarsals
	4 Instep (metatarsus)		Metatarsal
	5 Digits		Phalanges

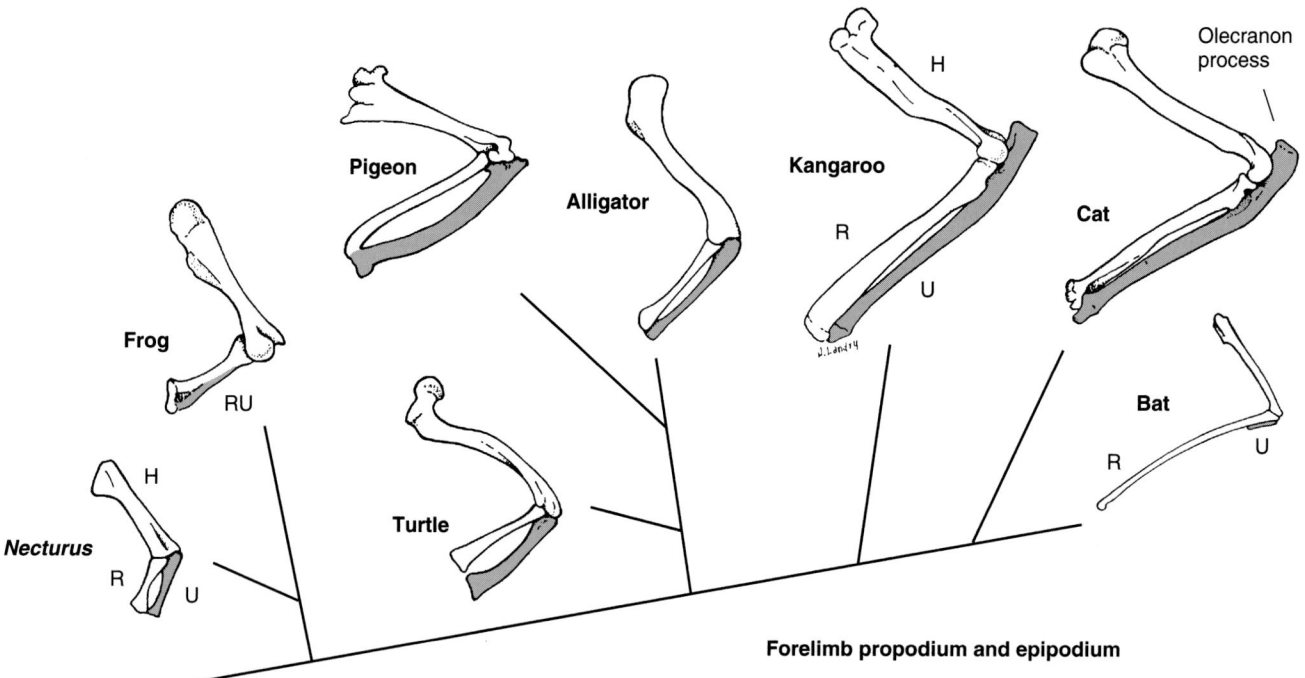

FIGURE 10.20

Humerus, radius, and ulna of the left forelimb (lateral views) on a cladogram showing the hypothesized relationships among represented taxa. **H,** humerus; **R,** radius; **U,** ulna (*red*). In the frog, the radius and ulna have united to form a radioulna, **RU.** In bats, the ulna is vestigial.

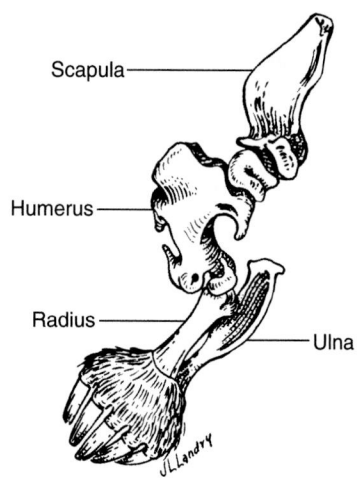

FIGURE 10.21

Right anterior limb of a mole, medial view. The palms turn outward as an adaptation for digging.

other ungulates. In birds, the tibia fuses with the proximal row of tarsals to form a **tibiotarsus** (see fig. 10.39).

Manus: The Hand

The wrist, palm, and digits constitute a functional unit—the hand (manus). Considering the multitude of modifications that have adapted hands (and feet, too) to the challenges of life on land, and eventually, in the air and sea, the skeletons of hands and feet are remarkably similar from frogs to eagles and from moles to seals.

The forelimbs and hind limbs of the three oldest known tetrapods, recovered from the Late Devonian, had from six to eight digits. The complete hind limb of *Ichthyostega*, one of the three, has seven digits, and a complete forelimb belonging to *Acanthostega* has eight (Coates and Clack, 1990). A third Devonian species, *Tulerpeton,* unearthed by a Soviet paleontologist, has six digits on its limbs. None of the three specimens has only five on either limb. The three digits in *Ichthyostega* that occupy the position of the big toe in later tetrapods are small and closely adherent, perhaps fulfilling the same locomotor role as the big toe of a **pentadactyl (five-digit) limb.** These observations suggest that sometime after the Late Devonian, and before the amniote line became established, pentadactyl limbs, anterior and posterior, became dominant. They have persisted until this day in generalized amniotes. The primitive structure of the pentadactyl hand and foot and the adaptations that have modified these during geologic time are the subjects of this and succeeding sections.

The wrist (**carpus**) of a generalized pentadactyl hand consists of three more or less regular rows of carpal bones (see fig. 10.19, autopodium). The proximal row has a radial carpal (**radiale**) at the distal end of the radius, an ulnar carpal (**ulnare**) at the end of the ulna, and an **intermedium** between the two. At the ulnar end

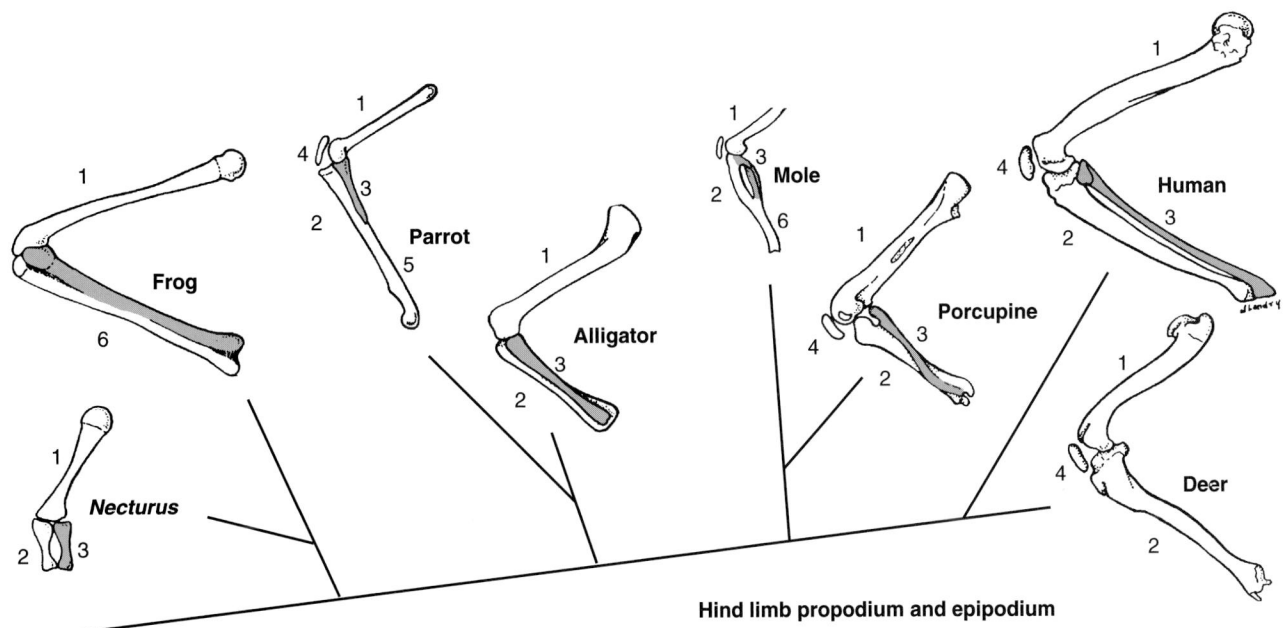

FIGURE 10.22
Left thigh and shank bones of representative tetrapods, lateral views, on a cladogram showing the hypothesized relationships among taxa.
1, femur; **2,** tibia; **3,** fibula; **4,** patella; **5,** tibiotarsus; **6,** tibiofibula, fibular component in *red.*

TABLE 10.2 Synonymy of Carpal Bones

Terms Preferred by Comparative Anatomists	Nomina Anatomica*	Anglicized Names and Synonyms
Radiale	Os scaphoideum	Scaphoid, navicular
Intermedium	Os lunatum	Lunate, lunar, semilunar
Ulnare	Os triquetrum	Triquetral, cuneiform
Pisiform	Os pisiforme	Pisiform, ulnar sesamoid
Centralia (0 to 4)	Os centrale	Central carpal(s)
Distal carpal 1	Os trapezium	Trapezium, greater multangular
Distal carpal 2	Os trapezoideum	Trapezoid, lesser multangular
Distal carpal 3	Os capitatum	Capitate, magnum
Distal carpal 4 ⎫ Distal carpal 5 ⎭	Os hamatum	Hamate, unciform, uncinate

*Terms adopted by the Eighth International Congress of Anatomists at Wiesbaden.

of the proximal row in most reptiles and mammals is a sesamoid bone, the **pisiform.** The middle row of carpals consists of **centralia**—often three or four in early tetrapods, two in early reptiles—one of which is sometimes displaced to the proximal or distal row of carpals. The distal row is composed of five **distal carpals** numbered 1 to 5 commencing on the thumb side. Table 10.2 lists the names of the carpal bones.

Metacarpals are the skeleton of the palm. Primitively, the pentadactyl limb may have had as many distal carpals and metacarpals as there were digits.

Each digit consists of **phalanges.** The early phalangeal formula for a pentadactyl hand, commencing with the thumb, may have been something like 2-3-4-5-3. In late therapsids, it had become 2-3-3-3-3, an almost universal formula for modern mammals with five fingers (see fig. 10.24, human).

Modifications of the manus with few exceptions involve *reduction in the number of bones by evolutionary loss or fusion.* A less common modification is the *disproportionate lengthening or shortening of some of the bones.* Least common is an *increase in the number of phalanges.* Centralia frequently unite with one of the proximal carpals or disappear. As a result, most reptiles and numerous mammals have a single centrale, and it is sometimes found among the proximal row of carpals. Fusion of distal carpals 4 and 5 is common and results in a **hamate bone.** Phalanges or entire digits may be lost. In the latter event, the corresponding metacarpal becomes vestigial or lost.

Most amphibians have five digits on the hind limb and four on the forelimb, although a few have less. Members of the heterochronic urodele genus *Amphiuma* have one to three digits. The metacarpals that correspond to any lost ancestral digits have become vestigial (fig. 10.23, frog), or they have been lost (fig. 10.23, *Necturus*). The number of wrist bones also is smaller in modern than in labyrinthodont amphibians. The embryonic intermedium and ulnare often coalesce during ontogenesis, an embryonic proximal carpal frequently coalesces with an adjacent carpal, and fusion between embryonic centralia and proximal or distal carpals is common. In addition, some ancestral carpals must have been lost completely.

Because the line of descent from the earliest amphibians to urodeles is largely conjectural, there is no present way to ascertain homologies between specific bones of the urodele hand and foot and those of the earliest labyrinthodonts. In fact, we cannot be confident that *any* ancestor of urodeles passed through a pentadactyl limb stage. Indeed, there is a possibility that urodeles were derived independent of other tetrapods directly from a rhipidistian ancestor and that they underwent an evolution that was more or less parallel with that of labyrinthodonts and the latter's descendants (see chapter 4).

Assuming, for a moment, that the skeleton of the manus of *Necturus* is derived from a pentadactyl ancestor, the three distal carpals could be considered to be, commencing on the radial side (fig. 10.23, starting with P), distal carpal 2, distal carpal 3 (two arrows attached), and hamate (fused distal carpals 4 and 5, two arrows attached). This presupposes that the thumb (1, broken lines) is missing. An alternative would be that the fifth finger (V, broken lines), rather than the thumb, is missing. The three distal carpals, starting with P, would then be a **prepollex** (a bone of unknown homology that develops in some species near the thumb, or **pollex**), carpals 1 plus 2, and carpals 3 plus 4. Carpal 5 (broken line) would be missing. This interpretation takes into account that the bone labeled P has no muscle connecting it with a finger (no arrow) and that the muscle from the digits numbered 2 and 3 in figure 10.23 attaches to the second of the three distal carpals (arrows). This carpal also frequently has two separate ossification cen-

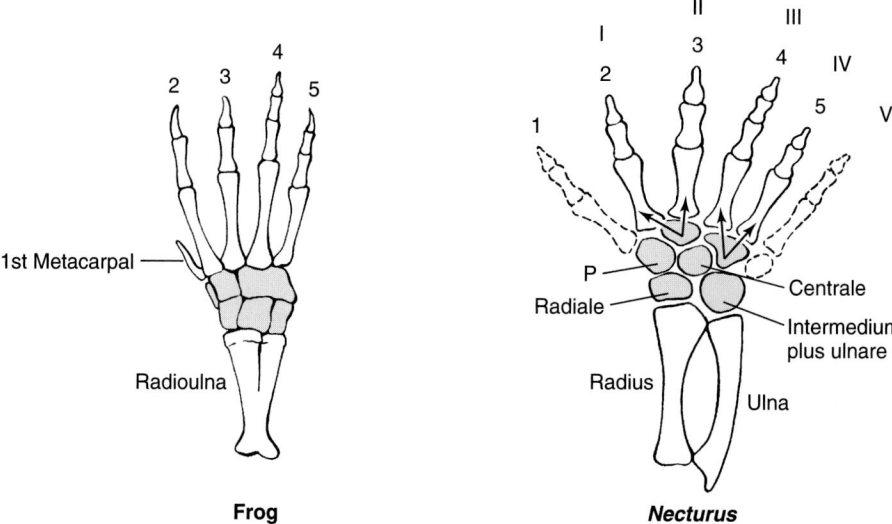

FIGURE 10.23

Hands of *Rana catesbeiana* and *Necturus,* dorsal views. Both have four digits. Which digit is missing in *Necturus*? *Broken lines* represent some of the missing primitive bones. Arabic numerals suggest that in a *Necturus,* the thumb, **1,** is missing, and the little finger, **5,** is present. Roman numerals suggest that the thumb, **I,** is present, and the little finger, **V,** and last distal carpal and metacarpal are missing (*broken lines*). Carpal bones are *red. Arrows* indicate existing muscle attachments, **P,** the prepollex or primitive distal carpal number **2,** depending on interpretation. In the frog illustrated, the history of the bone at the base of the first metacarpal is conjectural.

ters, suggesting that it is a composite. This is one approach used in attempting to determine homologies of the bones of the hand and foot. The hypothesis of Alberch and Gale (1985) provides an alternative mechanism to account for digit reduction. Rather than a truncation of development (a paedomorphic loss of the last digit to form), they argue that a reduction in the number of primordial cells in the developing limb bud causes reorganization within the limb resulting in digit reduction.

The muscles that insert on the hands and feet of urodeles are neither strong nor well differentiated, and the joints between the epipodia and wrists or ankles and between the wrists or ankles and the metacarpals and metatarsals are capable of little mobility. Thus, neither the hands nor the feet of urodeles generate locomotor thrust. They are chiefly platforms, or podia, which, pressing on the substrate, provide friction while muscles higher on the limbs extend the legs. The same is true of the hands, but not the feet, of anurans.

The hands of living nonavian reptiles and mammals such as insectivores and primates tend to remain pentadactyl and to have five metacarpals and a nearly full complement of carpals except central carpals, or centralia (fig. 10.24, turtle and human). In crocodilians, however, the wrist has been reduced to five adult bones (fig. 10.24, alligator), and in birds the entire manus has been reduced (fig. 10.25). When present in mammals, the centrale may lie in the distal row of carpals, as in

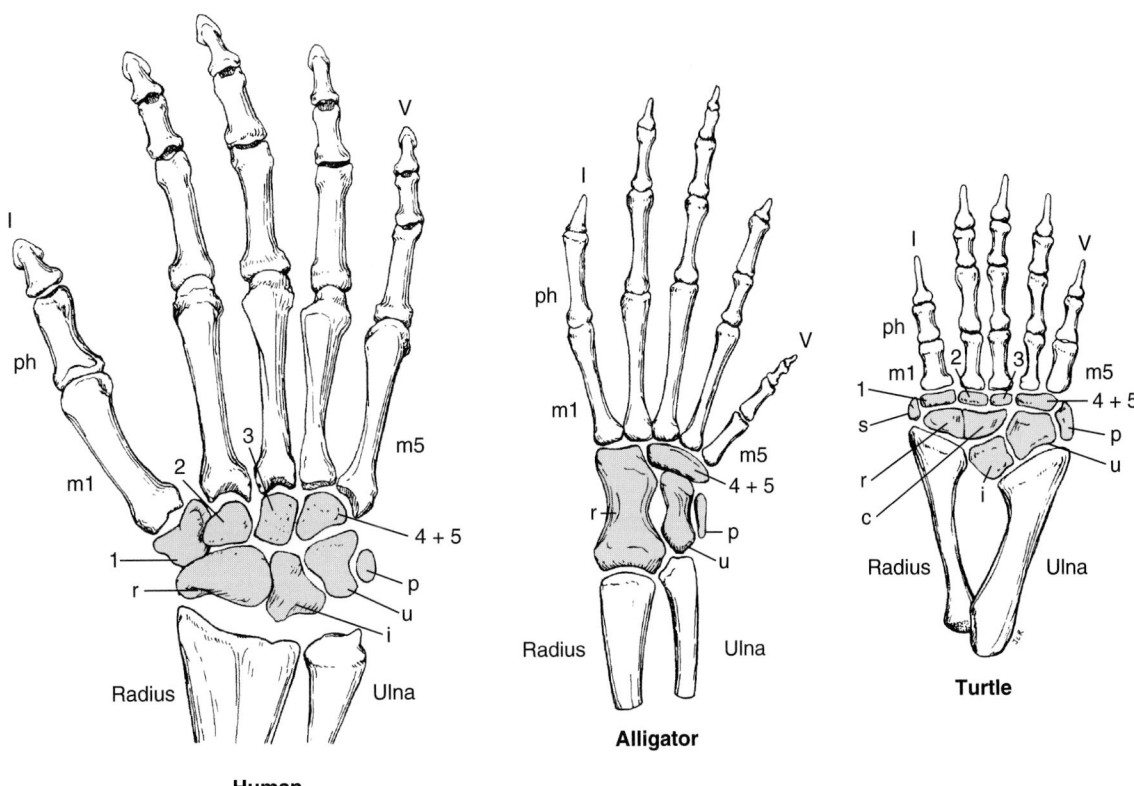

FIGURE 10.24

Right manus of human, alligator, and turtle (dorsal views, carpal bones in *red*). **c,** centrale; **i,** intermedium; **m1** and **m5,** first and fifth metacarpals; **p,** pisiform; **ph,** proximal phalanx; **r,** radiale; **s,** radial sesamoid; **u,** ulnare; **1** to **5,** distal carpals; **I** and **V,** first and fifth digits. The alligator has an additional carpal that cannot be seen from this view.

rabbits (fig. 10.26), or it may unite with the radiale and intermedium to form a scapholunar bone of triple origin, as in cats. Human fetuses have a centrale until the third fetal month, when it fuses with the radiale. Among major modifications of the hand are those for flight, life in the ocean, swift-footedness, and grasping.

Adaptations for Flight

In many birds, the hand has little independent role in propulsion in air, but being at the end of an airfoil, it has an aerodynamic effect. Loss and fusion of bones have reduced the hand to a rigid, tapering structure (see fig. 10.25). Despite this, most of the basic components of tetrapod hands are identifiable in avian embryos. Two carpals (radiale and ulnare) form in the proximal row, and three in the distal row. As development progresses, the three distal carpals unite with the three metacarpals to form a rigid **carpometacarpus**. Three fingers are usually present, and the number of phalanges has been reduced. (Terns often develop four embryonic digits, but only three persist.) The fingers rarely bear claws, and like the rest of the hand, they are covered by feathers. In the juvenile *Opisthocomus*, a

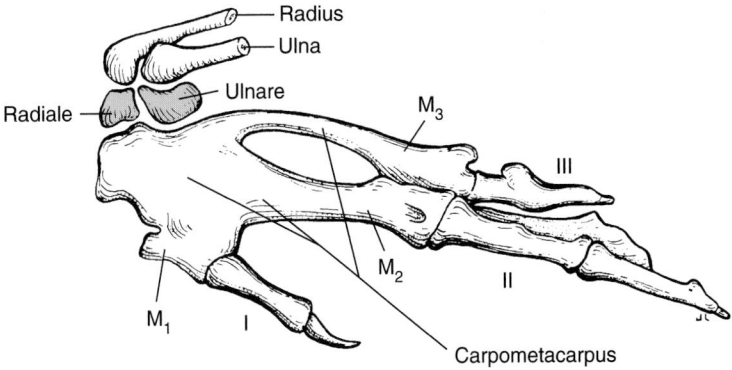

FIGURE 10.25

Left manus of a bird, **I** to **III,** Digits; **M₁** to **M₃,** metacarpals fused with three distal carpals to form a carpometacarpus.

tree-living game bird of South America, claws are present and used in climbing. They are lost in the adults, and despite their similarity to those in *Archaeopteryx*, they most likely represent a secondary feature.

The first finger of birds that maneuver, alight, and take off in limited spaces is elongated, prominent, and independently movable and is called an **alula**. Songbirds have short, broad wings, and the feathered alula serves as an accessory airfoil at the leading edge of the wing as

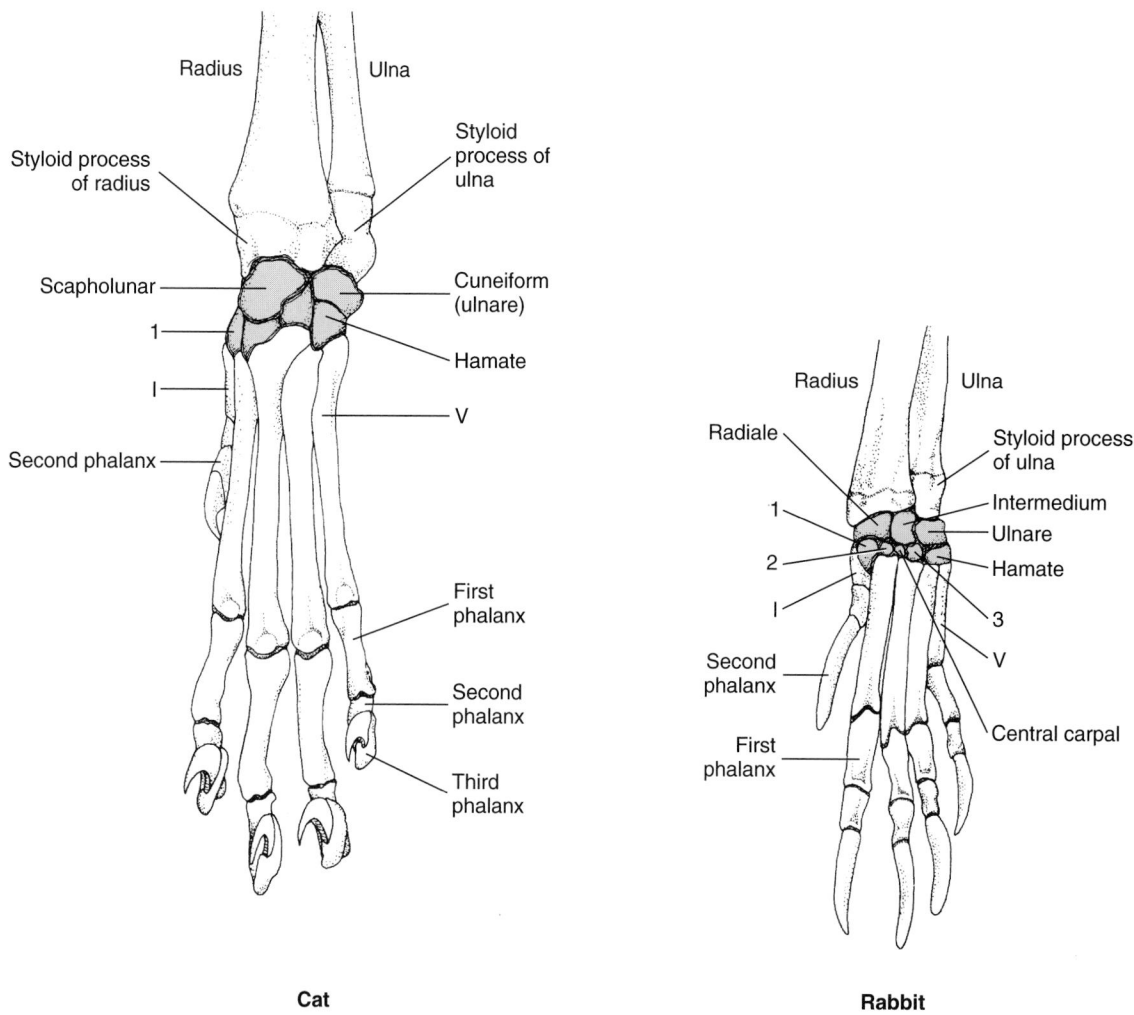

Cat

Rabbit

FIGURE 10.26
Left manus of cat and rabbit, anterior view, carpal bones in *red*. The pisiform bone is on the palmar side and cannot be seen. The scapholunar of cats combines radiale, intermedium, and central carpal, which are separate bones in kittens. The phalangeal formula of both animals is 2-3-3-3-3. **1** to **3,** first through third distal carpals. **I** and **V,** first and fifth metacarpals. Hamate incorporates fourth and fifth distal carpals.

the bird flits among tree branches. Carnivorous birds have moderately long, broad wings adapted for slow-speed flight and quick landings in limited spaces. When these birds are braking, the alula is moved away from the rest of the wing, which creates a slot through which air rushes, making it possible to maintain stable low-speed headway until additional braking is applied with the main wings and tail feathers. There is a joint with considerable flexibility at the elbow, and another between the forearm and wrist, except in birds that hover. When flexed at the wrist, the hands exert a strong braking effect for landing, especially in birds with a large wingspread. Birds that hover have high-speed wings (up to 50 strokes per second in hummingbirds), the hand is as long as the arm or longer, and the entire wing is quite rigid.

Contrary to the condition in birds, the hand in pterosaurs and bats was, or is, the main part of the wing. Pterosaurs (fig. 10.27*b*) had four fingers, three of

which were normal and bore claws. The fourth was embedded in the wing membrane (**patagium**) and consisted of four enormously elongated phalanges that made this finger as long as the entire body. The associated metacarpal was not elongated, but it was much enlarged. Bats have five fingers (fig. 10.27*a*). The thumb is normal and bears a claw. The other four fingers are elongated and are associated with four greatly elongated metacarpals. The metacarpals and the phalanges of the four fingers constitute the skeleton of the patagium. The three proximal carpals are united in a single bone. Movement of the hand is responsible for takeoff and true flight in bats. No one has ever seen a pterosaur take off!

Gliding lemurs have a patagium, but it is less well developed than in bats and pterosaurs, and the fingers, although embedded in it, are not elongated. Gliding lemurs soar but are not capable of true flight. Patagiums

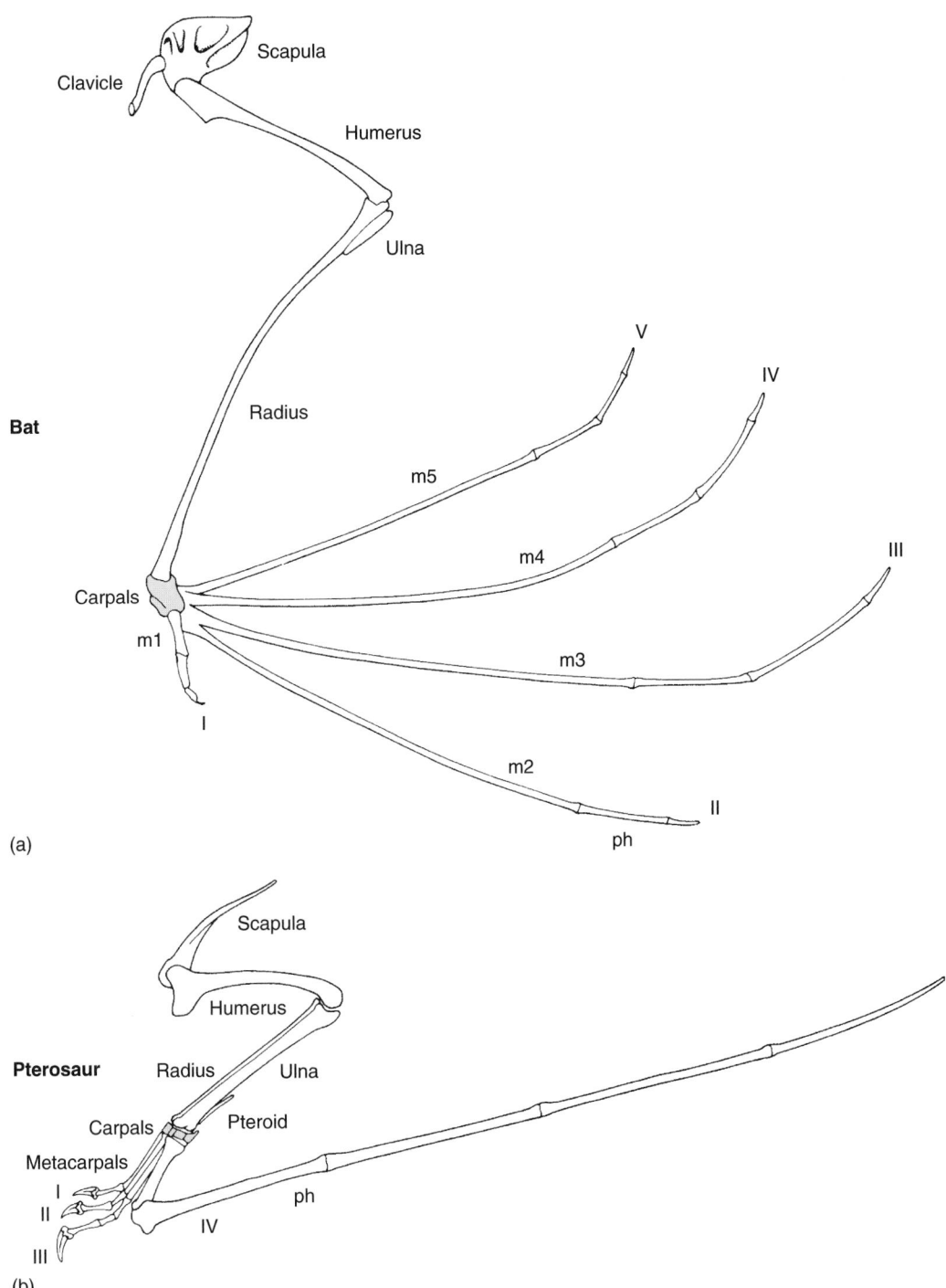

FIGURE 10.27
Pectoral girdle and limb of two flying vertebrates. (*a*) Bat, right wing. (*b*) Jurassic pterosaur, left wing. **m1** to **m5,** first through fifth metacarpals; **ph,** proximal phalanx; **I** to **V,** digits.

in unrelated animals such as pterosaurs, bats, gliding lemurs, and gliding lizards are instances of evolutionary convergence.

Adaptations for Life in the Ocean

The hands have become paddlelike flippers in well-adapted marine amniotes. Among these are sea lions (see fig. 4.45), cetaceans (see fig. 4.50), sirenians (see fig. 4.49), ichthyosaurs and plesiosaurs (fig. 10.28), seals (see fig. 10.43), and penguins. Flippers are generally flattened and stout, and in several taxa the number of phalanges has greatly increased (fig. 10.29). In some ichthyosaurs, there were as many as 26 phalanges per digit and more than 100 in a single hand. Within the

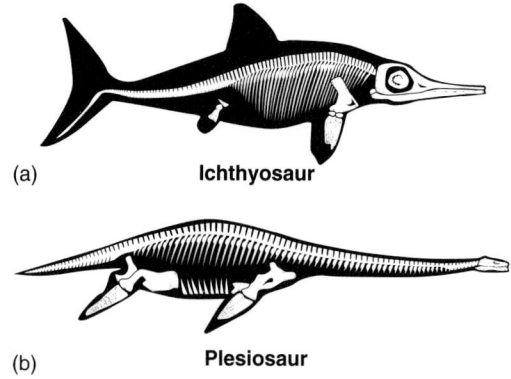

(a) **Ichthyosaur**

(b) **Plesiosaur**

FIGURE 10.28

Marine reptiles of the Mesozoic. (*a*) Ichthyosuar, a porpoiselike reptile with some species attaining a body length of 15 m long. (*b*) Sauropterygian, a plesiosaur, with some species up to 12 m in length.

From Kenneth V. Kardong, *Vertebrates.* Copyright © 1995 Times Mirror Higher Education Group, Inc., Dubuque, Iowa. All Rights Reserved. Reprinted by permission.

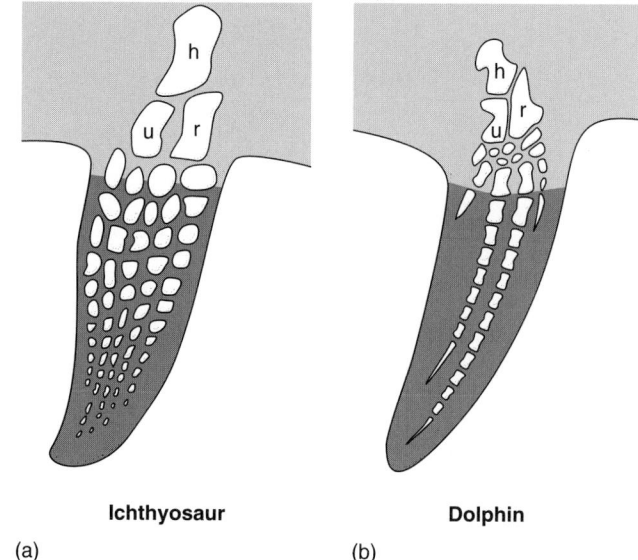

Ichthyosaur   **Dolphin**

(a)   (b)

FIGURE 10.29

Convergent evolution in anterior limbs. (*a*) Extinct, water-dwelling reptile. (*b*) Water-dwelling mammal. **h,** Humerus; **r,** radius; **u,** ulna.

flippers of most other swimmers, however, the skeleton conforms closely to the generalized tetrapod pattern (fig. 10.30). Some aquatic mammals have lost all traces of hind limbs (fig. 10.31*c*).

Penguins obtain thrust for swimming solely from their flipperlike wings, their webbed feet serving as rudders. Sea turtles, fur seals, and sea lions swim by rowing action of their long front flippers. Some marine mammals, however, do not use their foreflippers to produce locomotor thrust when swimming. These include wriggling seals (true seals), walruses, cetaceans, and sirenians. How they swim and how pinnipeds maneuver when on land will be discussed after we have studied the hind limbs.

Adaptations for Swift-Footedness

Mammals with pentadactyl hands and feet are usually **plantigrade**, which means that the wrists, ankles, and digits all rest on the ground. This is a primitive tetrapod stance not associated with fleetness. Generalized mammals—monotremes, marsupials, insectivores, primates—are plantigrade, as are some specialized mammals such as bears and the arboreal raccoons (fig. 10.32, monkey). Mammals in which only the first digit has been reduced or lost tend to be **digitigrade**, which means that they bear their weight on digital arches with wrist and ankle elevated (fig. 10.32, dog). Among digitigrade mammals are rabbits, rodents, and most carnivores. These mammals run faster, walk more silently, and are more agile

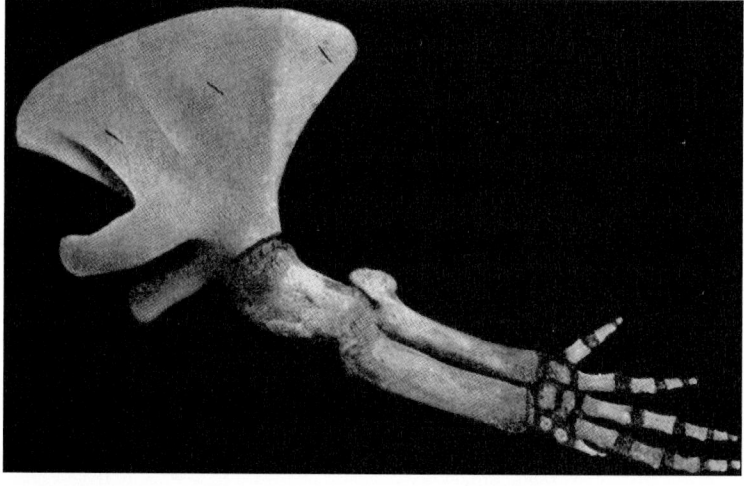

FIGURE 10.30

Right scapula and forelimb of a beaked whale. A remarkable resemblance to basic tetrapod skeleton remains, although the limb has become paddlelike.

Negatives/transparencies #314469, Courtesy Department of Library Services, American Museum of Natural History.

than plantigrade species, and some, such as cats of various kinds, are the fastest runners of all. Cheetahs, for example, can sprint at an estimated 70 miles (112 km) per hour (Hildebrand, 1980). To a carnivore, speed makes eating more likely; to a herbivore such as a hare, it may mean reaching a burrow instead of being eaten.

The extreme modification of reducing the number of digits and walking on the tips of the remaining digits is seen in ungulates. Unguligrade mammals walk on four,

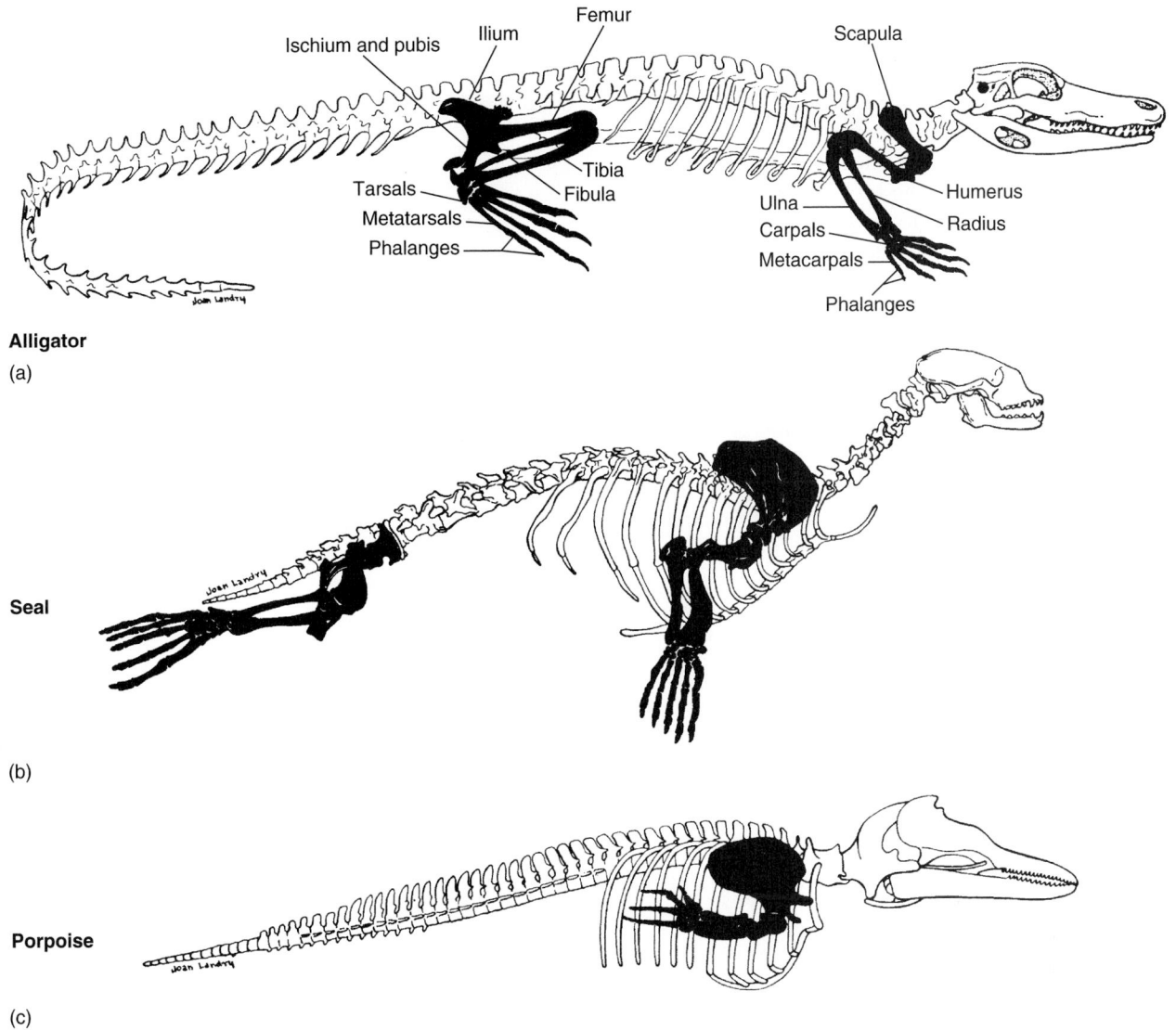

FIGURE 10.31

(*a*) Skeleton of land-dwelling amniote. (*b*) and (*c*) Appendicular skeletal adaptations for life in the water. Appendicular skeleton is shown in *black,* (*b*) is a "wriggling seal" (*Phoca*). In the seal and porpoise, the hand is a paddle in which the phalanges are embedded.

three, two, or even one digit (fig. 10.32, deer), with wrists and ankles elevated well above the ground. The claws have become thick hoofs that bear the body weight and protect the living tissues of the toes from the abrasive action of the substrate (see fig. 6.21, hoof). The metacarpals and metatarsals that correspond to the missing digits are vestigial or lost, and those that remain are much elongated and frequently united (fig. 10.33, horse, deer, camel). This has the effect of providing an additional functional segment to the limb. Although excelled in sprinting speed by some carnivorous digitigrade mammals, ungulates can sustain their speed for much longer. Not only are the feet of ungulates well suited for running, but they also function extremely well in rugged mountainous terrains. However, special-

ization has made their fingers and toes useless for anything else. This is the price of specialization.

Sequential evolutionary steps leading to the most specialized unguligrade stance may be illustrated by placing the fingers and palm flat on a tabletop, with the forearm perpendicular to the surface. This represents roughly the plantigrade position. Raising the palm off the table while keeping the fingers flat on the table illustrates roughly the digitigrade position. Unguligrade conditions may be illustrated by placing only the fingertips on the table and then raising the thumb, the little finger, the second finger, and finally the fourth finger, leaving only the third finger to bear the body weight, as in modern horses. Those fingers that fail to reach the table represent digits that have been successively reduced or lost in ungulates.

The horse underwent these successive changes commencing with the early *Eohippus*, which had four digits on the manus, and culminating in the modern *Equus*, which has a single digit. Despite extreme specialization of the manus of the modern horse, the proximal row of carpals (fig. 10.34) is intact, and the distal row lacks only the first carpal. With the loss of digits I, II, IV, and V, metacarpals 1 and 5 fail to develop and 2 and 4 have been reduced to splinters. Metacarpal 3, associated with digit III, has elongated.

Evolution among the ungulates progressed along two independent lines. In the line leading to **artiodactyls**, the weight of the body tended to be distributed equally between digits III and IV (fig. 10.34, camel). Thus arose the "cloven" hoof (fig. 10.35). Such a foot is said to be **paraxonic** because the body weight is borne on two parallel axes. Artiodactyls of today have an even number of digits. In the evolutionary line leading to **perissodactyls**, the body weight is increasingly tended to be borne on digit III, the middle digit. This is a **mesaxonic** foot (figs. 10.33 and 10.34, horse). Most perissodactyls have an odd number of digits; tapirs, however, have a small fourth digit on the front feet, three on the rear. It is the mesaxonic foot and not the number of digits that defines the perissodactyls.

Adaptations for Grasping

Many mammals are able to flex their hand at the joint between palm and fingers. Rodents, for example, sit on their haunches and nibble on food held between *two hands,* which are flexed in this manner and which *face one another.* A further specialization is the ability to wrap the fingers *around* an object so that it is held se-

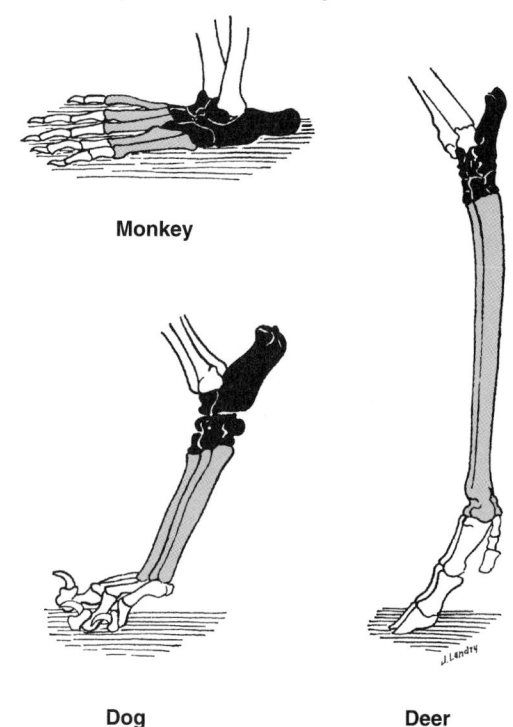

Monkey

Dog   **Deer**

FIGURE 10.32

Plantigrade, digitigrade, and unguligrade feet (from left to right). Ankle bones are *black;* metatarsals are *gray.*

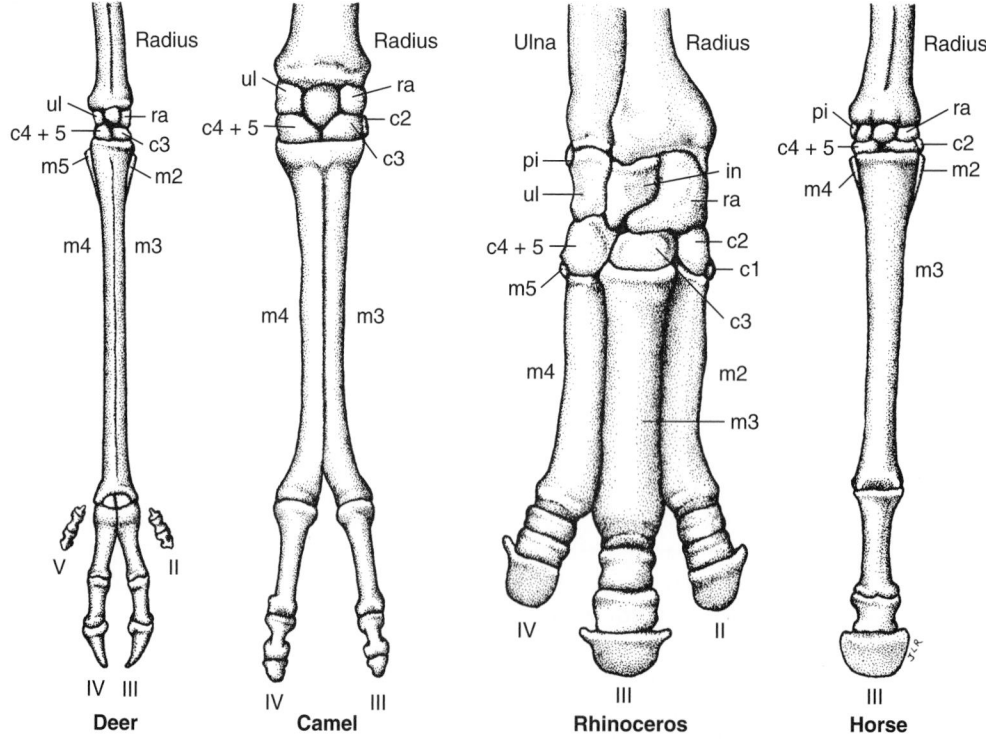

FIGURE 10.33

Right manus of several ungulates as seen from the front. **c2** to **c5,** distal carpals 2 to 5; **in,** intermedium; **m2** to **m5,** metacarpals 2 to 5; **pi,** pisiform; **ra,** radiale; **ul,** ulnare; **II** to **V,** digits. In horses, **m3** is the "cannon bone."

Deer   Camel
Artiodactyls

Rhinoceros   Horse
Perissodactyls

curely in *one hand*. This is accomplished by flexing the fingers at each interphalangeal joint. Primates are among the few mammals that can do this.

Another step in the evolution of the mammalian hand was development of an opposable thumb—one that can be made to touch the tips of each of the other digits. This was accomplished by formation of a saddle joint at the base of the thumb where it meets the palm,

by setting the thumb at increasingly wider and wider angles to the index finger, and by the evolution of strong adductor pollicis (thumb) muscles. True opposability is found in Old World monkeys, but even there the hand does not have the full range of functional capability that has evolved in human beings. Neither New World monkeys nor anthropoid apes have a perfectly opposable thumb. With such a hand, humans were able to fashion increasingly sophisticated instruments, commencing with rocks chipped by design and continuing to the electronic age. It has been said that the implements of early humans were as good (or as bad) as the hands that made them. Of course, evolution of the brain was an essential concomitant. Still, it seems impossible that any species on earth lacking a prehensile hand, even having the brain, could evolve so sophisticated an existence.

Pes: The Hind Foot

A generalized pes is comparable bone for bone with the manus except that there is no consistent equivalent to the pisiform bone (table 10.3). Many early tetrapods had four centralia in the ankle (figs. 10.36 and 10.37, labyrinthodont). The number of centrale bones became smaller as diversity of modes of locomotion expanded. A comparison of the primitive condition (fig. 10.36) with the centralia (dark red) of selected ankles in figures 10.37–10.39 illustrates the phylogenetic trends. The reduction or apparent loss is partly accounted for by the observation that primitive ossification centers for centralia tended to coalesce with those of the proximal or distal row of tarsals, thereby reducing the number of centralia. Some of the primitive centers still appear transitorily in the ontogeny of modern tetrapods.

The pes of three amphibians is illustrated in figure 10.37. The bones labeled **prehallux** in the labyrinthodont and anuran may be vestiges of a tarsal or metatarsal that was associated with an ancestral digit that has been lost. A pentadactyle limb was not the most primitive, so it is unlikely that these specific vestiges are homologues.

The number of ankle bones in anurans is considerably smaller than in urodeles. However, the tibiale and fibulare have elongated and united firmly at each end to articulate distally with the remaining tarsals in an intratarsal joint. This condition, although not unique, is not universal. Such an ankle, coupled with unusually long, webbed metatarsals and phalanges, provides a foot well adapted for pushing off for a leap on land, their only mode of terrestrial locomotion, or swimming in water, their alternate natural habitat. In swimming, the hind limbs are drawn forward against the body alternately, and then quickly thrust backward, toes spread apart, so that the broad fan-shaped foot pushes against the water, providing forward thrust. The forelimbs assist in swimming to some extent but are principally used in maneuvering.

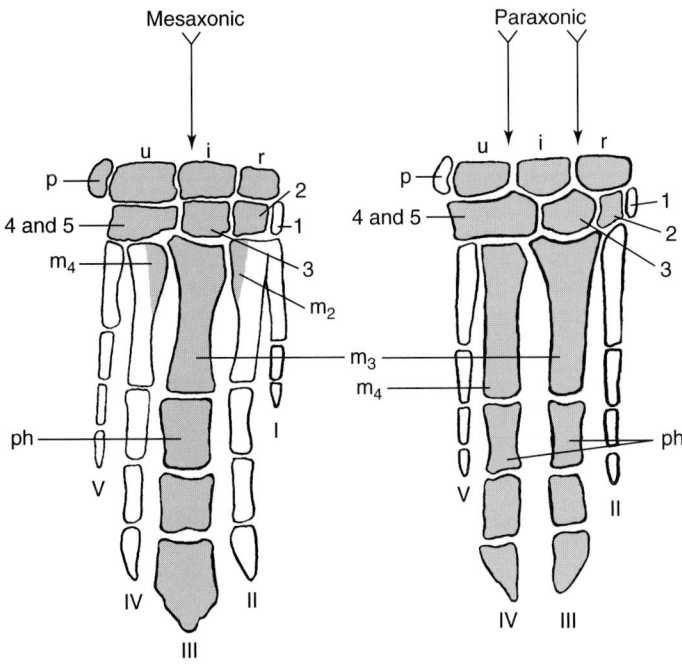

FIGURE 10.34

Mesaxonic and paraxonic manus of representative ungulates, showing bones lost (*white*) and retained (*red*) and indicating distribution of body weight through wrist and digits. The number of bony elements is correct for these animals. **i**, intermedium; **m₂** to **m₄**, second, third, and fourth metacarpals; **p**, pisiform; **ph**, first phalanx; **r**, radiale; **u**, ulnare; **1** to **5**, distal carpals; **I** to **V**, digits.

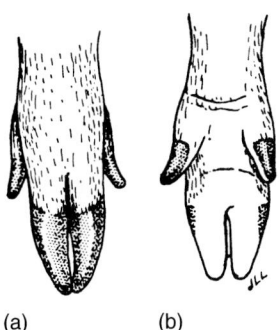

FIGURE 10.35

Hoof of fetal pig. There is a cleft between the weight-bearing digits. Thus, the hoof is "cloven." (*a*) As seen from in front. (*b*) As seen from behind.

TABLE 10.3 Comparable Skeletal Elements of Manus and Pes

Manus*	Pes	Synonyms in Humans
Radiale	Tibiale	Talus or astragalus[†]
Intermedium	Intermedium	
Ulnare	Fibulare	Calcaneus
Pisiform		
Centralis (0 to 4)	Centralia (0 to 4)	Navicular
Distal carpal 1	Distal tarsal 1	Ectocuneiform
Distal carpal 2	Distal tarsal 2	Mesocuneiform
Distal carpal 3	Distal tarsal 3	Entocuneiform
Distal carpal 4 ⎫ Hamate	Distal tarsal 4 ⎫ Cuboid	
Distal carpal 5 ⎭	Distal tarsal 5 ⎭	
Metacarpals (1 to 5)	Metatarsals (1 to 5)	
Digits (I to V)	Digits (I to V)	

*For synonyms, see table 10.2
[†]Incorporates the intermedium and a centrale.

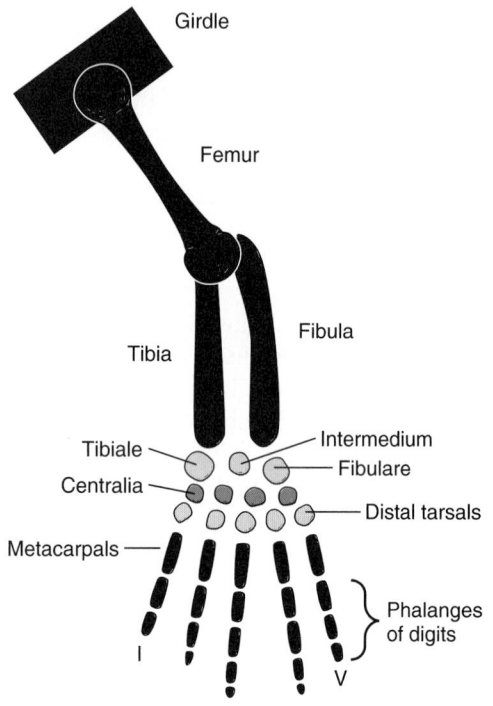

FIGURE 10.36
Skeletal organization of a primitive left hind limb. Tarsal bones (ankle) in *red;* centralia *dark red.* **I** and **V** are first and fifth digits.

Living reptiles display considerable loss and fusion of ankle bones, turtles to a lesser degree than others (see fig. 10.38). The proximal row of tarsals is reduced by coalescence to a single bone in *Sphenodon* and many lizards and has been given the name **astragalocalcaneus.** It may incorporate all proximal tarsal bones and a centrale. A highly flexible intratarsal joint between proximal and distal tarsals helps bipedal lizards to run rapidly in a digitigrade manner, the long tail maintaining balance. Most reptiles have five toes, although alligators and some lizards have four, and some freshwater turtles have three. The phalangeal formula for *Sphenodon,* 2-3-4-5-4, is the generalized formula for reptiles. Alligators have reduced it to 2-3-4-4-0, and turtles to 2-3-3-3-2.

Relative to other living amniotes, a bird's foot is highly modified (fig. 10.39). The proximal tarsals are united with the lower end of the tibia to form a **tibiotarsus,** there are no centralia, and the distal tarsals are united with the upper end of three fused metatarsals to form a long, rigid **tarsometatarsus.** A vestigial first metatarsal remains independent in some species. *There is an intratarsal joint between the tibiotarsus and tarsometatarsus and a joint between the tarsometatarsus and the toes.* The latter permits a digitigrade stance (see fig. 4.28). This stance keeps the bird "ready" for takeoff because extending the leg at the knee and intratarsal joints gives the initial thrust for becoming airborne. The thigh of a bird that is not in flight is not visible because it is quite short and, along with the knee, tucked away behind the wing.

Most birds have four toes, and a few have three; ostriches alone have two. The phalangeal formula is the same as that for the first four toes of *Sphenodon.* Usually three toes are directed radially forward and one comes off the back of the foot, but a few birds, including woodpeckers and parrots, have two toes at the back, the four forming an **X** (**zygodactyly**). This enables woodpeckers to obtain a firm grip on rough bark on a vertical tree trunk while drilling, braced against the trunk by stiff tail feathers. Birds can sleep while on a perch because the long tendons of the flexor muscles of the lower leg pass along the posterior aspect of the ankle to insert on the claw-bearing digits. The weight of the body pulls on the tendons, keeping the claws flexed on the perch (see fig. 11.22).

Mammals, like their therapsid ancestors, lack an intratarsal joint but have a large hinge joint where the tibia and fibula meet the ankle (see fig. 10.32). The tibiale is

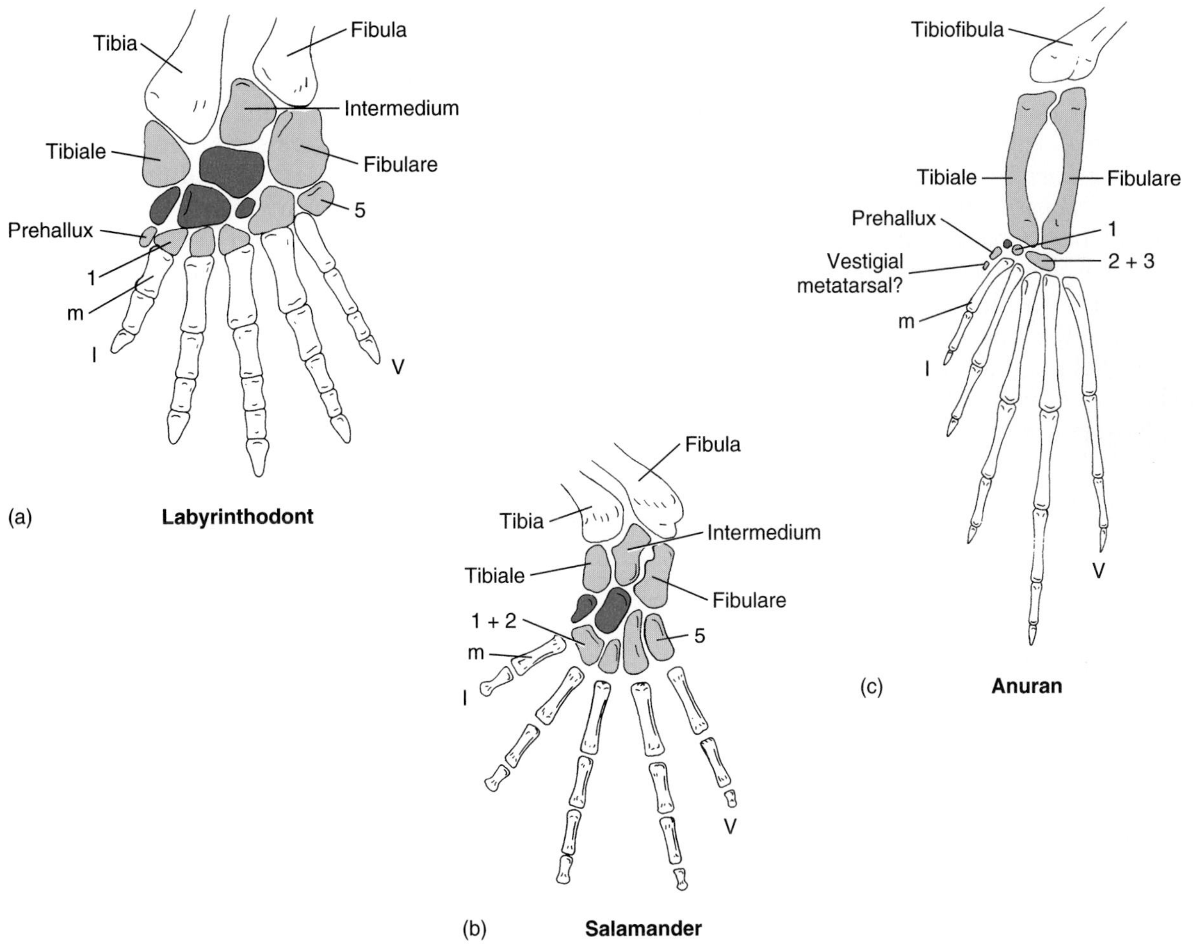

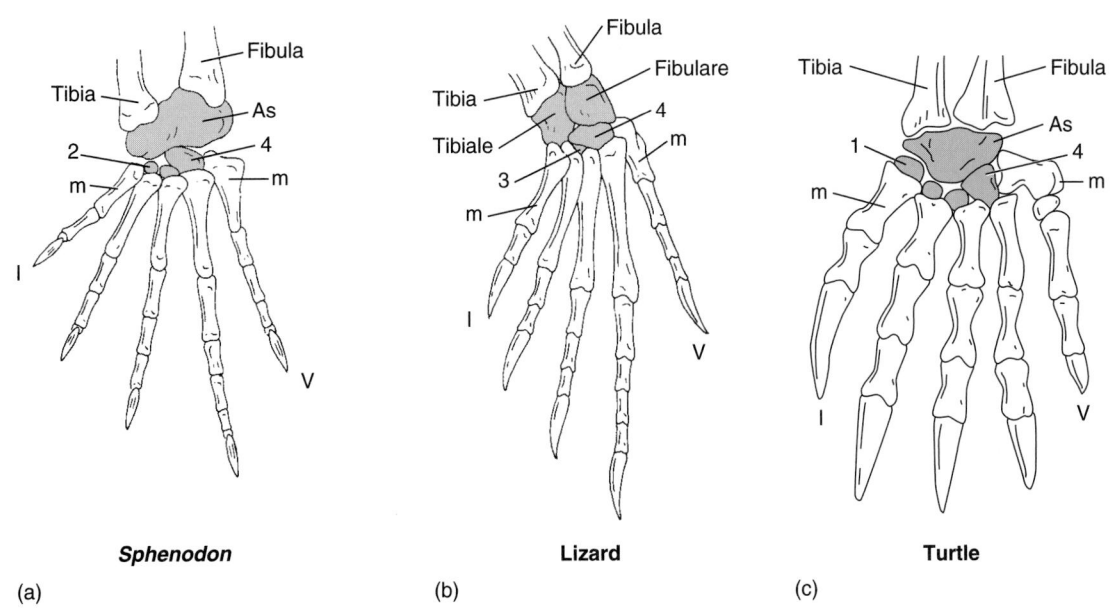

FIGURE 10.37

Left pes of amphibians. (*a*) Labyrinthodont. (*b*) Generalized plethodontid salamander. (*c*) *Rana catesbeiana*. Tarsal bones are *red,* and centralia are darker. **I** and **V,** first and fifth digits; **1** to **5,** distal tarsals; **m,** metatarsal.

FIGURE 10.38

Left pes of reptiles. (*a*) *Sphenodon*. (*b*) The lizard *Uromastix*. (*c*) The snapping turtle *Chelydra*. Tarsal bones are *red*. **I** and **V,** first and fifth digits; **1** to **4,** distal tarsals; **As,** astragalocalcaneous; **m,** metatarsal. There are no independent centralia.

the principal weight-bearing bone of the ankle (fig. 10.40*b*). The other proximal tarsal, the fibulare, is elongated backward in plantigrade mammals, upward in digitigrades and unguligrades (see fig. 10.32). It is the heel bone of plantigrades. Except for a reduction in the number of centralia, the number of mammalian ankle bones has not changed appreciably since the first therapsids. In hominoids, a **metatarsal arch**, or instep, distributes the body weight over four solid bases: the heel and ball of each foot. This has been compared with the four-pedestal architecture of the Eiffel Tower. The arch also absorbs some of the shock generated by bipedal locomotion and provides "spring" for walking and running.

The phalangeal formula for the pes of early therapsids was 2-3-3-3-3, and this is still the typical formula for modern pentadactyl mammals and humans. The great toe, or **hallux**, is opposable in many primates but not in humans (fig. 10.41). The condition, when present, is correlated with brachiation.

Sculling and Galloping in Marine Mammals

Wriggling seals, walruses, cetaceans, and sirenians have unusual modes of swimming for mammals, dictated by the size and shape of the body and the morphology of the limbs they have inherited. In none do anterior flippers provide forward thrust for swimming, these being used chiefly for maneuvering. Cetaceans and sirenians have no pelvic flippers. How then do these marine mammals manage to survive?

Wriggling, or earless, seals swim with their foreflippers adducted against their sides, their necks retracted, and with *lateral undulations* of the posterior portion of their trunks. The tail is short and of no use in locomotion. The posterior flippers, affixed to the side of the trunk by a narrow connecting "neck" enclosing the tibia and directed caudad, are swept alternately to the left and right with each lateral undulation cycle of the trunk. These sculling movements, in conjunction with synchronous limited flexion at the tibiofemoral and ankle joints with each sweep, propel these pinnipeds forward (fig. 10.42). Walruses also swim by sculling, but they are not strong swimmers. Sculling is a unique mode of swimming, but one could say that seals and walruses make the most of what evolution has given them!

Unlike seals, cetaceans derive their forward thrust from a stiff, horizontal bifluked tail, which is operated by alternate *dorsal and ventral undulations* of the caudal portion of the trunk—a sort of legless gallop—rather than by lateral undulations. Both functional modes of swimming—sculling in pinnipeds and galloping in cetaceans—have been shown by kinematic studies to represent highly efficient propulsive mechanisms (Fish et al., 1988).

Sirenians, too, derive their locomotor thrust from dorsoventral undulation of their flukelike tail, but these vegetarian, shallow-water creatures are sluggish swimmers (their food cannot flee!) and can scarcely be said to gallop.

How Pinnipeds Maneuver on Land

Fur seals, sea lions, and walruses come ashore more frequently than wriggling seals. They have an adaptation

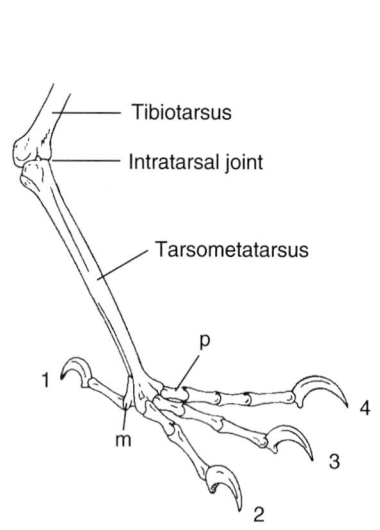

FIGURE 10.39

Left ankle and digits of a passerine bird, medial view, head to the right. **1** to **4,** digits; **m,** first metatarsal; **p,** phalanx. The claws obscure the terminal phalanges.

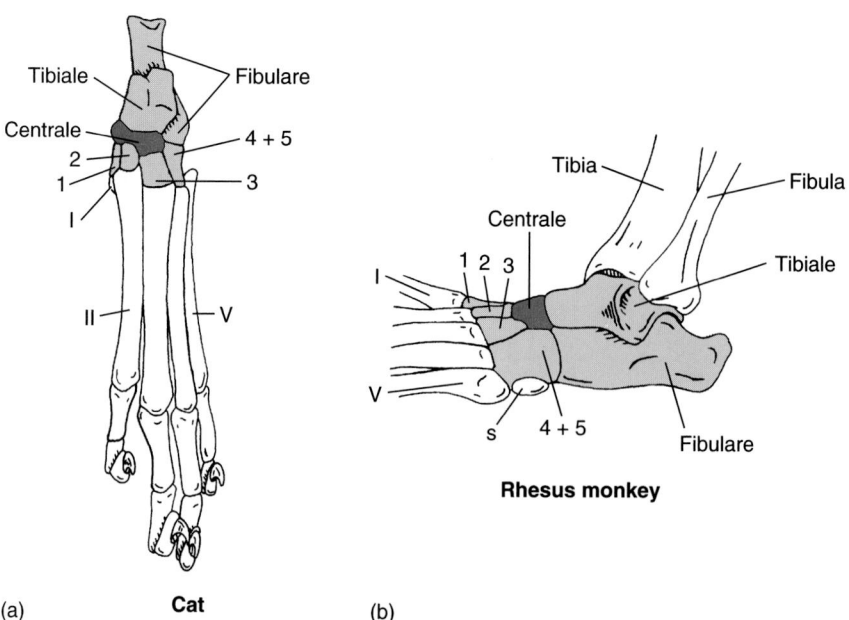

FIGURE 10.40

(*a*) Left pes of cat, anterior view. Ankle bones are *red*. The phalangeal formula is 0-3-3-3-3.
(*b*) Left ankle and associated bones of rhesus monkey, lateral view. **1** to **5,** distal tarsals; **I, II,** and **V,** metatarsals; **s,** sesamoid in the peroneus longus muscle.

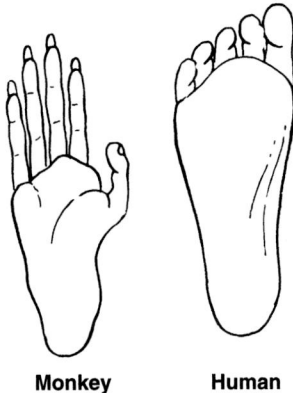

FIGURE 10.41

The partially opposable big toe of an Old World monkey and the nonopposable toe of a human.

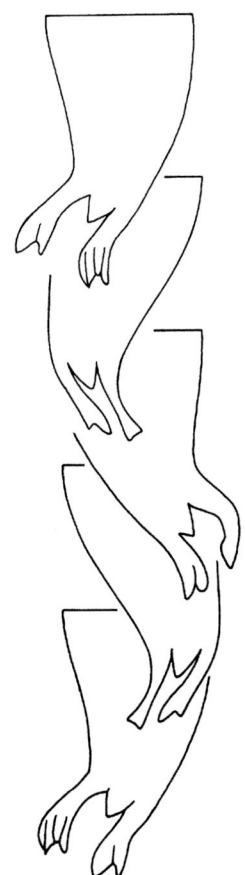

FIGURE 10.42

Sculling in a wriggling seal.
Source: Data from F.E. Fish, et al., "Kinematics and Estimated Thrust Propulsion of Swimming Harp and Ringed Seals" in *Journal of Experimental Biology,* 137:157, 1988.

FIGURE 10.43

Tetrapod stance of an eared seal pup.

that enables the hind flippers to change readily from a caudally directed swimming position to the equivalent of a tetrapod stance. The hind flippers are simply rotated beneath the body from a swimming position to a tetrapod position or the reverse, as the animal leaves or enters the water. (Human beings cannot do this!) A

flexible wrist joint further adapts them for locomotion on land. Wriggling seals are not so gifted. Their hind limbs are permanently bound to the tail; consequently, movements on land consist of pulling themselves along by wriggling, which accounts for their name.

The stance of young fur seals during their first months on land is more tetrapod than pinniped (fig. 10.43). Both pairs of juvenile flippers are disproportionately large compared with those of adults. These oversized flippers, coupled with rotating pelvic appendages, enable pups to run quickly and nimbly over the rookery grounds.

Origin of Limbs

Although the problem of the origin of paired fins may never be solved, the same is not true of the origin of tetrapod limbs. The paired fins of some ancient fish must have been the precursor of tetrapod limbs. The question then arises, did the skeleton of the fin of any known Devonian fish evince the potential for becoming a limb? For an answer we turn to the lobe-finned rhipidistians whose skull and fin skeleton bore a remarkable resemblance to those of the first tetrapods (figs. 9.8 and 10.44).

Two hypotheses on the origin of the tetrapod limb center on either a modification of preexisting structures or the formation of new features. Common to both hypotheses is the origin of the proximal elements of the limb. In the pectoral fins of rhipidistians, a single basal bone, which we will call a humerus, articulates proximally with the scapula and distally with a pair of radials we will call radius and ulna (fig. 10. 44). Loss of fin rays and modifications of the radials distal to the radius and ulna could have produced the skeleton of the first tetrapod limb. In the first hypotheses, the limb axis extends

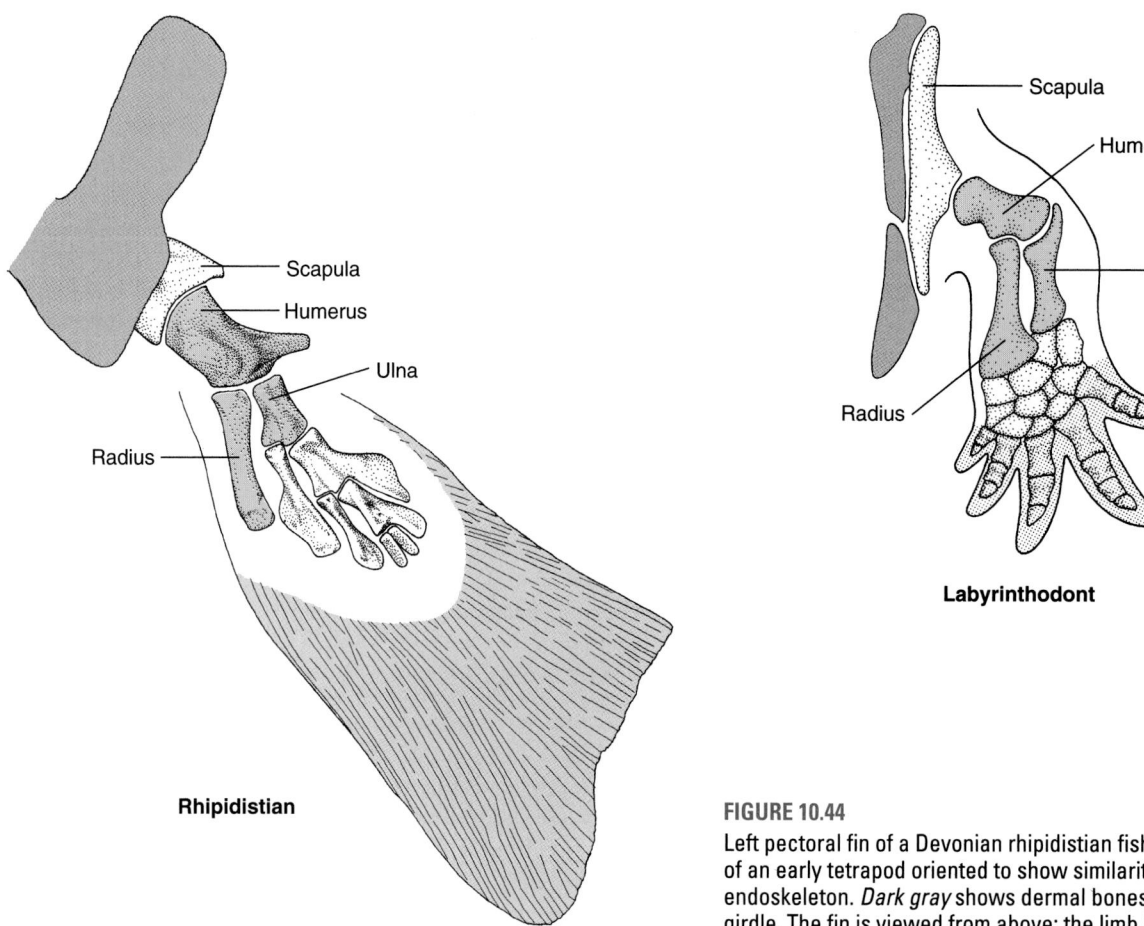

Rhipidistian

Labyrinthodont

FIGURE 10.44
Left pectoral fin of a Devonian rhipidistian fish, and left forelimb of an early tetrapod oriented to show similarities of endoskeleton. *Dark gray* shows dermal bones of pectoral girdle. The fin is viewed from above; the limb is viewed from in front.

through the radials of the fin with pre- and postaxial radials forming the digits. An alternative hypothesis is based on current developmental studies. The development of the tetrapod limb differs from that in ray-finned fishes (the development of living lobe-finned fishes has not been evaluated). The two systems parallel each other initially; however, the tetrapod limb has a second period of cell proliferation at right angles to the limb axis distal to the wrist. It is this secondary proliferation that gives rise to the digits. This pattern suggests that the digits are a novel feature of tetrapods and not a restructing of preexisting features (modification of radials to form the digits). Meanwhile, the girdles of labyrinthodonts remained fishlike.

The changes in external appearance necessitated by the transition of fins into limbs are illustrated in figure 10.45. These involved the following modifications, not necessarily in the sequence listed, not simultaneously, and certainly not exclusively, inasmuch as concomitant adaptive modifications were necessitated in the entire skeletomuscular system subserving each appendage: (1) elongation of the two bones of the epipodium; (2) formation of hinge joints between the propodia and epipodia (now elbow and knee joints), and

between the epipodium and wrist or ankle; (3) rotation of the long axes of the humerus and femur to parallel the vertebral column (see fig. 10.32*a*); and (4) emergence of a definitive manus and pes. The transition appears to have occurred sometime during the Devonian Period.

It is probable that rhipidistian fins were used at times for resting on the water bottom. They would not have borne appreciable weight because of the bouyancy of water. Minor modifications would have permitted "walking" on the muddy bottoms close to shore. Several hundred living species of fishes do this today, including the Australian lungfish. Using pectoral fins, some living fishes move several feet inland nightly. Some climb inclined planes with little fringed fins that have a remarkable resemblance to hands.

The pressures that drove craniates onto the land must, of necessity, be conjectural. It may have been the absence of predators on land or that there was less competition for food, or simply that food was abundant on land. Or perhaps it was simply a manifestation of the tendency of organisms to invade a contiguous environment whenever nothing prevents the invasion. Whatever the explanation, it was inevitable that a limb more suitable for life on land would evolve from the fin of a fish.

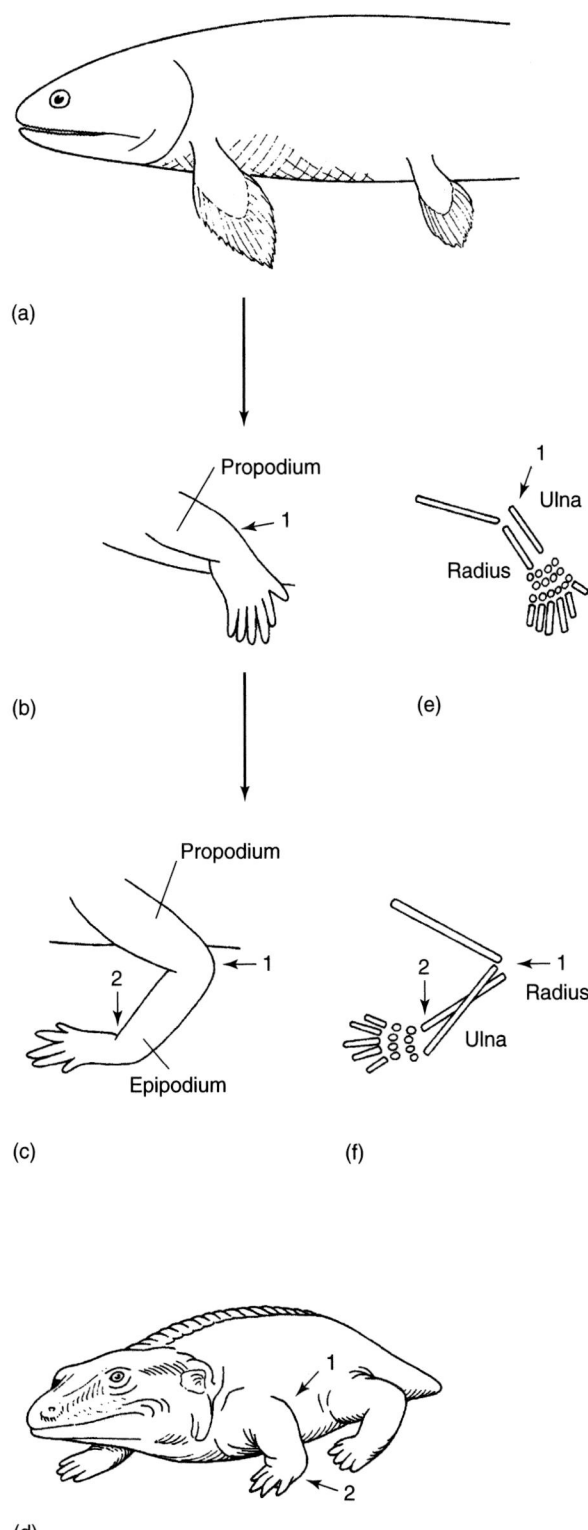

FIGURE 10.45

From fin to limb in the Devonian. (*a*) Rhipidistian fish.
(*b*) Hypothetical transitional stage. Note location of future elbow
joint (arrow **1**). (*c*) and (*d*) Temnospondylous labyrinthodonts. In (*c*),
note elongation of forearm and formation of a wrist joint (arrow **2**).
(*e*) and (*f*) Orientation of skeletal elements of (*b*) and (*c*). In (*f*), note
that some reorientation of the radius and ulna may have been
necessary for the manus to lie flat on the ground.

Locomotion on Land without Limbs

Limbs provide leverage for locomotion on land. Snakes
and other limbless tetrapods cannot propel themselves
in this manner. Nevertheless, snakes are highly successful on land because modifications of the vertebral column, ribs, body wall musculature, and skin have provided them with alternative methods of moving about.

The most common method of locomotion in snakes
and limbless lizards is by forming irregular loops that
become propped against, and push against, any available stationary object on their path, such as a clump of
vegetation, a rock, a root, or simply an irregularity in
the surface (fig. 10.46). While the head is being thrust
forward, the eyes scan the substrate for the next contact
point. When one looms, the head is thrust in that direction. On contact, a wave of muscular contraction commences at the head and travels to the last segment of the
tail, drawing the trunk and tail toward the head. A minimum of three contact sites are necessary for this mode
of locomotion, and no appreciable force is exerted

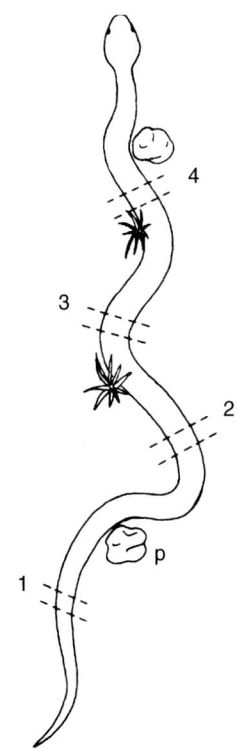

FIGURE 10.46

Serpentine locomotion. Forward progression is achieved by exerting
pressure against contact points, or props, **p,** on the path. The head
moves forward to a new contact point, and the rest of the body
follows the established looping path, the segment at position **1**
moving successively to positions **2, 3,** and **4** while the loops remain
stationary with reference to the contact points. Eventually, each
loop will be at the end of the tail, and new loops will have formed
anteriorly. Any track consists of sinuous waves.

From Carl Gans, *Biomechanics: An Approach to Vertebrate Biology,* 2nd
edition. Copyright © 1980. University of Michigan Press, Ann Arbor, MI.
Reprinted by permission.

downward against the ground. This mode of locomotion is referred to as **serpentine**. A basic necessity is a body that is metameric with respect to the adult skeletomuscular system. The term **lateral undulation** has also been applied to serpentine locomotion, but there are functional aspects of serpentine locomotion that are not operative in fishes.

Some snakes glide—seemingly flow—forward on the substrate while keeping the entire body in a straight line, a spectacle that some viewers find unnerving! The ventral skin acts like a conveyor belt, sections of which stop and go. This has been termed **rectilinear locomotion**, and it depends on generating friction between sections of the ventral skin and the substrate. Unlike snakes that press laterally against contact points for leverage, these snakes press against the ground, using bunched groups of belly scutes as intermittent holdfasts. After a brief interval of holding, during which time the next more caudal group of scutes is catching up, the downward pressure exerted by the first group is released, and the group flows forward, stretched out and elevated from the substrate, until it catches up with the group ahead. Meanwhile, the group immediately behind is holding fast. Thus rhythmic waves of forward flowing and resting scutes sweep along the body from head to tail. The skin, like that of many other snakes, is only loosely attached to the underlying tissues; there is much elastic tissue in the dermis, and the scutes, approximately one body segment long, overlap and are connected to the next by a pleat of cutaneous membrane that unfolds during movement. This combination of factors enables each scute to be released from the ground and commence its forward movement an instant before the next.

Two sets of striated muscles are responsible for rectilinear locomotion. A pair of slender costocutaneous muscles extends downward and backward from high on each rib to the dermis at the edges of each scute. When these contract, they lift the scute off the substrate so that it no longer serves as a holdfast, and they draw it forward while stretching the interscutal membrane. A second, more powerful pair extends less obliquely from a scute to the lower end of a more caudal rib. Contraction of the latter maintains the forward movement of the body mass within the skin envelope. Amphisbaenians use rectilinear progression, but the entire skin moves, not just the ventral skin.

Sidewinding (fig. 10.47) enables rattlesnakes and other snakes to inhabit sandy deserts, where the ground offers too few stationary contact points for serpentine locomotion and is too unstable to provide sustained friction for rectilinear locomotion. Sidewinding is also used by many snakes in other environments when they are temporarily in a situation where, because of the nature of the surface, other methods of locomotion would be clumsy or ineffective. In sand, the body usually occupies two or three tracks at a time. More or less of the an-

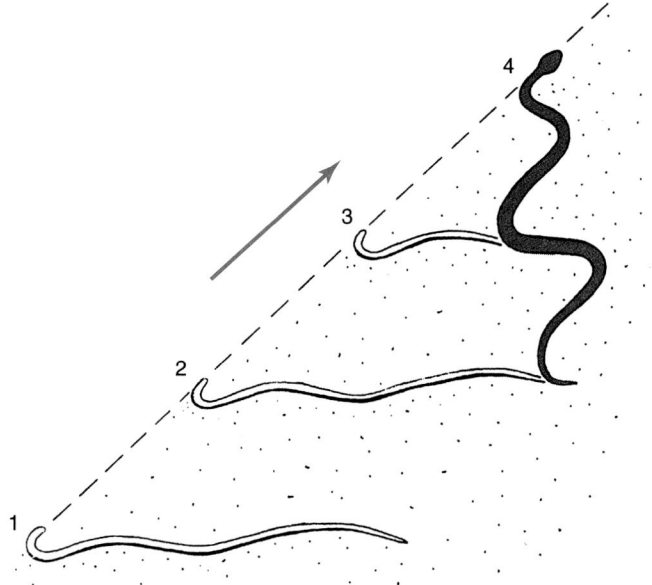

FIGURE 10.47

Locomotion in sidewinder snakes. **1** to **4,** successive tracks in sand left when the head and neck are lifted from the ground, thrust forward, and set down firmly at the next position. The body does not touch the ground between tracks. The rest of the body is "flowing" aboveground to the new position. *Dotted lines* and *arrow* indicate direction of movement.

terior quarter of the snake is thrust forward to start a new one.

Snakes propel themselves in a burrow by modified serpentine movements, bracing S-shaped loops against the burrow wall and exerting horizontal force while thrusting the head and forebody forward. The more caudal sections then advance. The process, reminiscent of the appearance of the bellows of an accordion (concertina) that is being played, has been called **concertina movement**. Snakes often combine several methods of locomotion in appropriate habitats as a common method of going from place to place.

All the foregoing methods of locomotion of snakes on land are made possible by vertebral columns consisting of up to 400 or more highly flexible intervertebral joints; by exceptional ribs that extend from the atlas to the tip of the tail and reach almost to the midventral line; by the unusually large number of muscle bundles that interconnect vertebrae and connect vertebrae with ribs; by wide, smooth, overlapping horny ventral scutes interconnected by pleated membranes; by the exceptional elasticity of the dermis; and by loose skin that allows independent movement of skin and enclosed body mass.

Of course, snakes and limbless lizards are not the only tetrapods that move about on land without limbs. Amphisbaenians use a modified serpentine movement in their burrows. Limbless amphibians on land employ

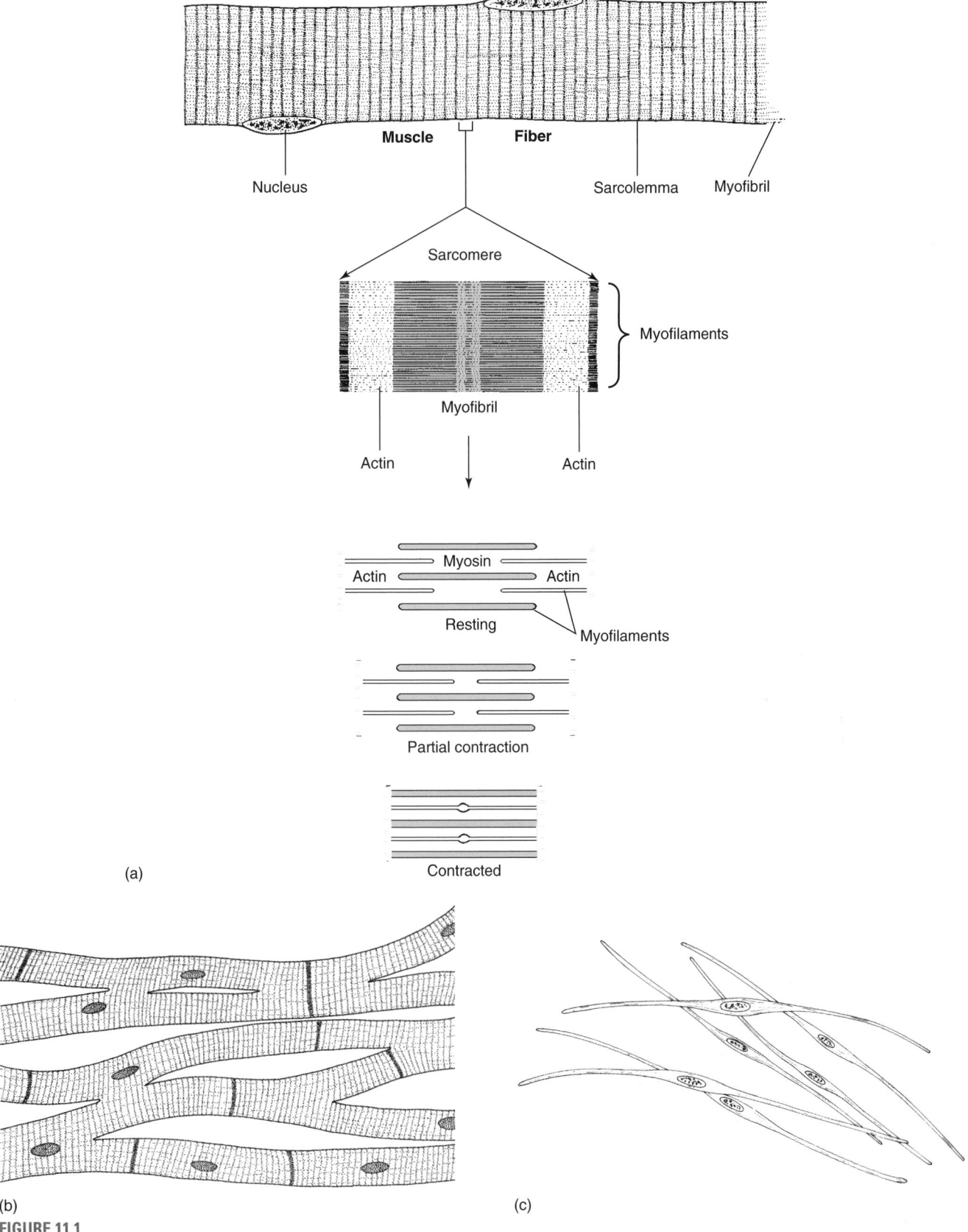

(a)

(b)   (c)

FIGURE 11.1

Muscle tissues. (*a*) Section of a striated muscle fiber, as seen with light microscopy (*top*); a schematic section of one myofibril consisting of myofilaments of myosin (*red*) and actin; and a schematic representation of muscle contraction as a function of sliding myofilaments (*lower.*) The left and right boundaries of the sarcomere are too fine to be seen with light microscopy. (*b*) Cardiac muscle tissue. *Dark bars* are intercalated disks at cell boundaries. (*c*) Isolated smooth muscle cells from intestinal wall.

In general, somatic muscles are said to be **voluntary** if they can be contracted at will. This does not preclude reflex contraction when the body is endangered, such as when the skin unexpectedly comes in contact with a pin, when holding one's breath becomes detrimental to physical welfare, or when a tendon is unavoidably stretched.

Somatic muscles are derivatives, either ontogenetically or phylogenetically, of the myotomes of mesodermal somites (see figs. 5.9, 11.6, and 11.7). For this reason, they are often referred to as **myotomal** or **somitic muscles.**

Visceral Muscles

Visceral muscles in general maintain an appropriate internal milieu. They are the smooth muscles of hollow organs, vessels, tubes, and ducts, the intrinsic musculature of the eyeballs, and the erector muscles of feathers and hair. Visceral muscle also includes cardiac muscle. They are derived chiefly from splanchnic mesoderm

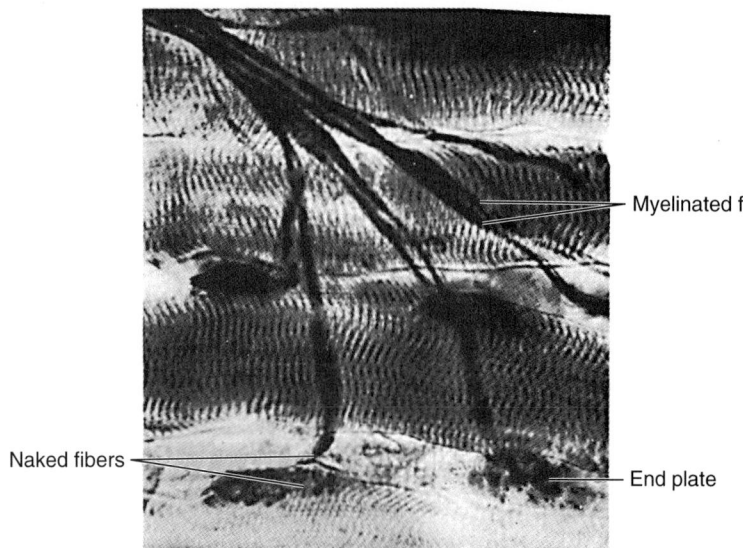

FIGURE 11.2
Innervation of skeletal muscle fibers.
Courtesy Gerrit L. Bevelander and Judith A. Ramaley.

(see figs. 5.11 and 11.7) and are innervated by the autonomic nervous system. Figure 16.35 shows the innervation of a few visceral organs.

Visceral muscles have undergone relatively few evolutionary changes through craniate history because they are less subject to the vagaries of the environment. Therefore, they will receive little attention in this chapter. Some contrasts between somatic and visceral muscles are listed in table 11.1.

Branchiomeric Somatic Muscles

Branchiomeric somatic muscles belong to the pharyngeal arches and their ontogenetic or phylogenetic derivatives from fishes to human beings. They are striated skeletal muscles. Branchiomeric muscles are also myotomal in origin, but they are derived from the most anterior somites and unsegmented paraxial mesoderm in the head. The paraxial mesoderm shows an incomplete segmentation, and the individual subdivisions are called **somitomeres.** Somitomeres, unlike somites of the body, do not fully segmented and lack sclerotome and dermatome components (the derivatives of these missing components are derived from neural crest in the head). They are innervated by cranial nerves. They will be discussed later in the chapter.

The remainder of this chapter focuses primarily on skeletal muscles. These are the ones that have diversified beginning with the earliest tetrapods. Discernment of a basic pattern of disposition of skeletal muscles from fishes to humans, discovery of the commonality of their embryonic origins, and the uncovering of the evolutionary changes that have occurred in muscle systems during geologic time have been achieved by research that began during the Renaissance and that will continue into this twenty-first century aided by increasingly sophisticated techniques of present and future scientists.

TABLE 11.1   Contrasts between Somatic and Visceral Muscles

Somatic Muscles*	Visceral Muscles
Striated, skeletal, voluntary	Smooth, nonskeletal, involuntary[†]
Primitively segmented	Unsegmented
Myotomal	Arise mostly from lateral mesoderm
Mostly in body wall and appendages	Mostly in splanchnopleure[‡]
Primarily for orientation in external environment	Regulate internal environment
Innervated directly by spinal nerves and cranial nerves III, IV, VI, and XII	Innervated by postganglionic fibers of autonomic nervous system

*Branchiomeric muscles, although somatic, are typically considered separately because they do not directly assist in orienting the body in the external environment.
[†]Cardiac muscle is striated.
[‡]Those that erect hairs or feathers or constrict blood vessels are in the skin.

INTRODUCTION TO SKELETAL MUSCLES

Skeletal Muscles as Organs

Skeletal muscles consist of muscular and tendonous portions (fig. 11.3). Surrounding a muscle is a tough, glistening fibrous sheath, the muscle fascia, or **epimysium**. It consists chiefly of collagenous connective tissue and elastic fibers in small amounts that vary with the muscle. Major bundles of muscle fibers (fascicles) within the muscle are surrounded by a **perimysium**, which also penetrates the bundles to encapsulate smaller fascicles. The smallest bundles consist of relatively few muscle fibers and constitute the functional units of the muscle. The muscular, neural, and vascular components of each functional unit are supported by a very fine collagenous reticulum, the **endomysium**, a continuation of the perimysium. It surrounds each individual muscle fiber superficial to the sarcolemma. The epimysium, perimysium, and endomysium comprise a single continuum of high tensile strength that encompasses all contractile units of the organ.

Tendons are continuations of the muscle beyond the site where fascicles end. The collagenous bundles of the perimysium and epimysium continue into and become part of the tendon. Similarly, at the site of attachment of the tendon to the skeleton, the collagenous bundles of the tendon continue into and contribute to the peri-chondrium or periosteum of the bone to which they are attached. Consequently, tension produced by muscle contraction is transmitted throughout the entire organ and from one skeletal attachment to another.

Twitch and Tonic Muscle Fibers

The fibers of striated muscle are of several functional varieties that can be identified based, for example, on their histochemistry, contractile characteristics, innervation pattern, or intracellular characteristics. They have also been identified less precisely as red and white muscle fibers. These terms will be avoided here because the color of the fibers can be due to either extrinsic or intrinsic factors (e.g., increased blood capillaries in the muscle or presence of a pigmented hemoglobinlike molecule, **myoglobin**, respectively). Contractile fiber types provide the best classification across taxa, which includes **twitch** and **tonic** fiber types (table 11.2). These fiber types are found in most taxa and differ in their proportions.

Twitch fibers are the predominant fiber in mammals with tonic fibers restricted to extrinsic eye muscles and ear muscles. Twitch fibers perform a wide range of functions (table 11.3) with the slow twitch fibers paralleling the postural function and low fatigue of the tonic fibers in amphibians and reptiles. Slow twitch fibers are associated with a richer blood supply and large amounts of myoglobin (thus their dark color). Fast glycolytic twitch fibers represent those that are capable of great power in the absence of oxygen during periods of heavy exertion.

Most muscles from fishes to humans are mixtures of fiber types, the proportions of each varying with the role the muscle plays in survival. For example, the muscles that move the head of a rat consist of a very high proportion of fast fibers, whereas in corresponding muscles of domestic cats the frequency of slow fibers is considerably higher. This correlates with the observation that rats (sitting on their haunches) monitor the environment more frequently and with more rapid head movements than domestic cats. On the other hand, the appendicular muscles of both domestic cats and rats are

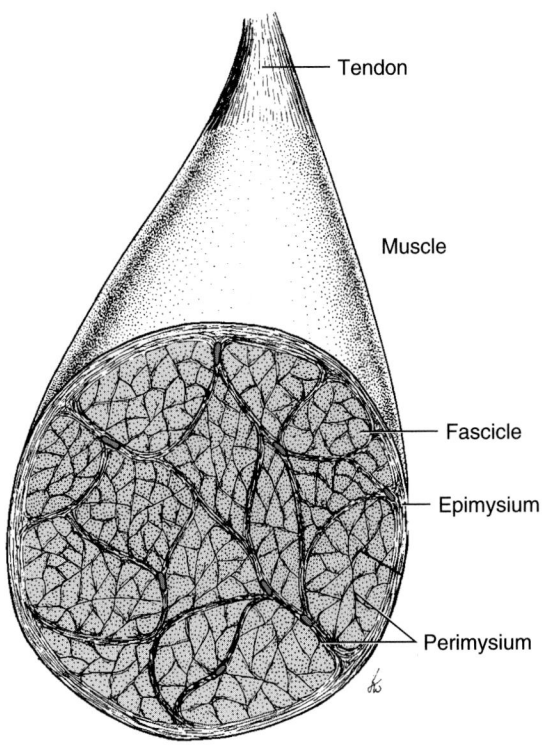

FIGURE 11.3
A mammalian fusiform skeletal muscle in cross section. All visible subdivisions are fascicles. The smallest fascicles and endomysium are too fine to be seen.

TABLE 11.2 Properties of Contractile Fiber Types in Craniates

Twitch	Tonic
Fast to slow contraction	Slow contraction
Slow—mammalian postural muscles	Postural muscles in amphibians and reptiles
Fast—most locomotor muscles	Extraocular and ear muscles of mammals
Innervation—a single axon	Multiple axons
Action potential—all-or-none	A temporal summation with a graded contraction
Variably fatigues	Can maintain tension efficiently

TABLE 11.3   Fiber Type Variations within Twitch Fibers

Slow Twitch (Type I of Mammals)	Fast Oxidative (Type IIA of Mammals)	Fast Glycolytic (Type IIB of Mammals)
Posture or slow repetitive movements	Fast	Powerful and fast
Fatigue slowly	Fatigue slowly	Fatigue quickly
Large number of mitochondria	Large number of mitochondria	Few mitochondria
High oxygen storage proteins (myoglobin), "red muscle"	ATP formed by oxidative phosphorylation	ATP formed by glycolysis—with possible oxygen debt
"Dark" meat of fish and fowl	Bird flight muscles	"White" breast of domestic fowl

rich in fast fibers. Continued use of either fast or slow fibers results in increase in the size of the fiber, hence in the size and strength of the muscle.

During dissection of adult mammals, students frequently notice that skeletal muscles differ in size in the two sexes. Androgens, the predominant gonadal hormones of males, cause amino acids to be linked together into polypeptides and proteins. Inasmuch as muscle is 80 percent protein, androgen produces statistically demonstrably larger muscles in males.

Origins, Insertions, and Muscle Shapes

The **anatomic origin** of a muscle, as opposed to its phylogenetic or ontogenetic origin, is the site of attachment that remains fixed under most functional conditions; that is, the bone on which it originates is not displaced when the muscle contracts. For example, when the biceps muscle of the upper arm contracts (*Bi*, fig. 11.4), the forearm is flexed (that is, drawn toward the upper arm). The origin of the biceps is therefore somewhere above the elbow. The **insertion** of a muscle is the site of attachment that is usually displaced by contraction of the muscle. The biceps inserts on the forearm. A muscle may cause displacement of the bone of origin instead of the bone of insertion if the former is immobilized by other muscles. For example, the geniohyoid muscle, which extends between the hyoid bone and the lower jaw at the chin, either lowers the lower jaw or draws the hyoid forward, depending on which bone is immobilized at the time.

When thinking of muscles, one generally envisions a muscle structured like the biceps brachii of a rabbit, consisting of a belly and two tendons (fig. 11.4). This is a fusiform muscle. (The name *biceps* is a misnomer for this muscle in cats and rabbits because it has only one head, or tendon of origin. In humans it has two.) Some muscles (digastrics) have two bellies; some have three heads or four. Many muscles are straplike as, for example, the geniohyoideus of tetrapods (see fig. 11.15) and the sternomastoid muscles of mammals (see fig. 11.24*c*). Less frequently seen are pinnate muscles, so named because their structure resembles that of a contour feather. Pinnate muscle slips "feather out" from a central or bordering tendon to insert at multiple sites, such as on successive vertebrae. Muscles with still other contours, such

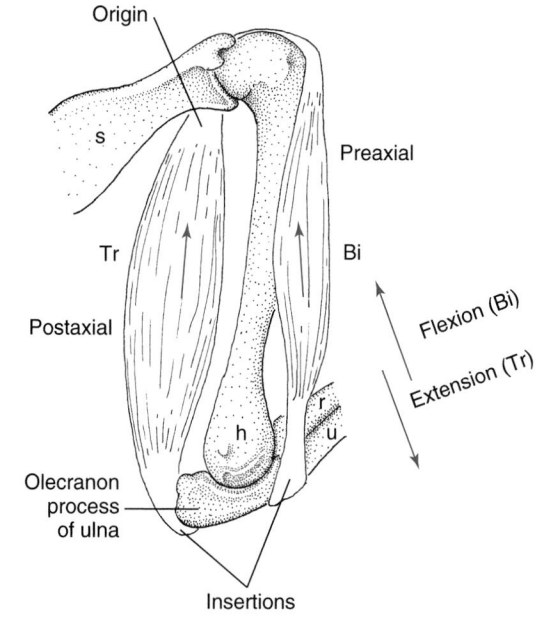

FIGURE 11.4

Origin, insertion, and direction of tensile force exerted by two muscles of the right upper arm of a rabbit, lateral view (i.e., head is to the right). *Arrows* within muscles indicate direction of pull on the forearm; *arrows* at right indicate direction of displacement of forearm when muscles contract. **Bi**, biceps brachii; **h**, humerus at elbow; **r**, radius; **s**, scapula; **Tr**, long head of triceps brachii; **u**, ulna. **Tr** provides the motive power for a first-class lever, and **Bi** for a third-class lever. The two muscles must operate synergistically, monitored by proprioceptive feedback to the central nervous system to effect smooth operation of the forearm against a load.

as the domed diaphragm with a central tendon (see fig. 11.13), are seen in mammalian anatomy laboratories.

Tough, thin, sheetlike expanses of mammalian tendons and ligaments are known as **aponeuroses.** One of these, the **galea aponeurotica,** is the major component of the mammalian scalp, lying in intimate association with the integument and being the common tendon of insertion for a number of thin, broad integumentary muscles of the forehead, temporal, and occipital regions (see fig. 11.25*e* and *f*). If you place your fingers on your own scalp and move it, you will feel the galea aponeurotica move. **Raphes** are long, seamlike tendons, such as the **linea alba** in the ventral midline of the trunk (see fig. 11.14). Muscles inserting on raphes often compress a cavity and the organs within it. Fine nonmetameric myoseptalike **tendonous inscriptions** traverse many straplike muscles or broad muscular sheets, adding tensile strength to the muscle (see fig. 11.15). They are prominent in the rectus abdominis of humans and in the broad oblique and transverse muscle sheets of the abdomen of anurans and many amniotes.

Actions of Skeletal Muscles

Skeletal muscles may be categorized according to their function. **Extensors** tend to straighten two segments of a limb or vertebral column at a joint. **Flexors** tend to draw one segment toward another. **Adductors** draw a part toward the midline. **Abductors** cause displacement away from the midline. **Protractors** cause a part, such as the tongue or hyoid, to be thrust forward or outward; **retractors** pull it back. **Levators** raise a part; **depressors** lower it. **Rotators** cause rotation of a part on its axis. **Supinators** are rotators that turn the palm upward; **pronators** make it prone (turn it downward). **Tensors** make a part such as the eardrum more taut. **Constrictors** compress internal parts. **Sphincters** are constrictors that make an opening smaller; **dilators** have the opposite effect. Most sphincter and dilator muscles are nonskeletal.

Although the action of any muscle is dependent on its origin, insertion, and shape, few muscles act independently. More frequently, they act in functional groups and also synergistically with other functional groups that have an opposing action. While one group is contracting, the opposing group must relax simultaneously and at the same rate; otherwise, a stalemate would be the minimum effect. More likely, a torn muscle or stretched tendon or ligament would result. For both muscle groups to function smoothly, they must be under reflex regulatory control of the cerebellum, which dispatches motor impulses to appropriate muscles on receiving sensory feedback from proprioceptive receptors located in the muscles, in their tendons, and in the bursas and capsules of affected joints (chapter 17).

Names and Homologies of Skeletal Muscles

Skeletal muscles have been named for the direction of their fibers (oblique, rectus), location or position (thoracis, supraspinatus, superficialis), number of subdivisions (quadriceps, digastric), shape (deltoid, teres, serratus), origin or insertion (xiphihumeralis, stapedius), action (levator scapulae, risorius), size (major, longissimus), and for still other features including combinations of these. Insight into the significance of a muscle's name should aid in recall of other information about the muscle.

Names were originally given to muscles in accordance with one of the preceding criteria without regard to homology. Therefore, the fact that two muscles in two different species have the same name is no assurance that they are homologous. The more distantly related the two animals are, the greater is the likelihood that they are not homologous. Similarity of location, origin, and insertion is a good starting point but not a reliable basis for establishing homologies. One reason is, muscles sometimes alter their sites of attachment during evolution as they spread and produce new slips. An example is the genioglossus muscle that inserts on the mammalian tongue. It was inherited, of course, from the common ancestor of amniotes. The probable homologue in some birds inserts on the sublingual seed pouch.

A way to test this hypothesis is through a combination of methods involving, although not exclusively, their embryogenesis and nerve supply. One method of seeking homologues, therefore, is to compare the early developmental stages of two muscles suspected of being homologous. Do the anlagen appear to be identical? If so, the two should be homologues, even though in one species the anlage expands in one direction or another to assume new relationships, whereas the anlage of the other does not do so. If the anlagen are similar, the muscles may be homologous. On this premise, the coracoscapular muscle of reptiles is thought to be homologous with the supraspinatus and infraspinatus muscles of mammals. Neurologic studies often provide substantiating evidence. If the motor cell bodies that innervate two muscles suspected of being homologues are located in homologous nuclei within the cord or brain, this is substantiating evidence. Of course, there must also be some confidence that the two motor nuclei are homologous. The latter method is able to identify muscles of the mammalian head and neck, the homologues of which operated the gills of fishes, because most of the motor cell bodies of branchiomeric muscles are in a different motor column than the motor cell bodies supplying any other group of muscles. Embryologic (including the expression of regulatory genes) and neurologic evidence is the most reliable at present for establishing muscle homologies. However, such data are available for relatively few individual muscles. On the other hand, homologies between *functional groups* of muscles may be deduced with a much greater degree of reliability.

AXIAL MUSCLES

Axial muscles are the skeletal muscles of the trunk and tail. They extend forward beneath the pharynx as hypobranchial muscles and, in amniotes, as tongue muscles. They do not include branchiomeric or appendicular muscles.

The immediately evident feature of axial muscles of fishes and generalized tetrapods is their metamerism (fig. 11.5). This primitive condition along with a

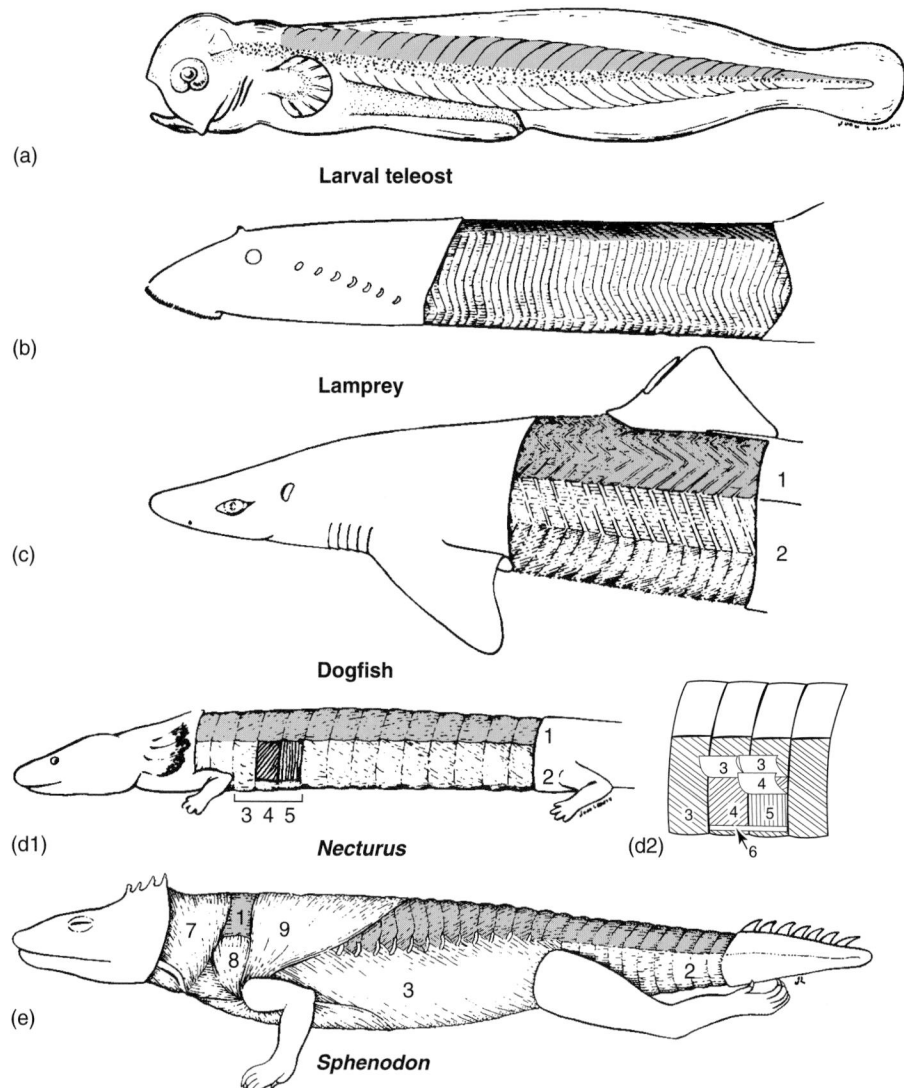

FIGURE 11.5
Trunk muscles of selected vertebrates.
1, epaxials (*red*); **2,** hypaxials;
3, external oblique;
4, internal oblique; **5,** transverse
muscle of abdomen; **6,** rectus
abdominis. In (*e*), the appendicular
muscles are **7,** trapezius; **8,** dorsalis
scapulae; **9,** latissimus dorsi. In (*a*),
stipple indicates position of notochord.
In (*c*) and (*d*), **1** and **2** are separated by
the horizontal skeletogenous septum.
(*d2*) A close-up of the bracketed
myomeres in (*d1*) showing the layered
hypaxial muscles. In (*d*) and (*e*), the
epaxials are the dorsalis trunci
muscles.

flexible metameric vertebral column enables fishes and some aquatic tetrapods to propel themselves in water by lateral undulation (see fig. 11.8). These muscles perform the same function for limbless amphibians living on land. The metamerism of the axial musculature was increasingly obscured as locomotion by lateral undulation was superceded by locomotion by limbs. Nevertheless, even in mammals, traces of metamerism remain in the axial musculature.

The axial muscles are inherently segmental because of their embryonic origin: They arise from segmental mesodermal somites (fig. 11.6). Mesenchyme cells from the myotome of each somite stream into the embryonic lateral body wall and migrate ventrad while undergoing repeated cell division (fig. 11.7). They cease migrating when they reach the midventral line, where the linea alba develops. These myotomal cells give rise to blastemas for body wall muscles. Because the somites are metameric, the blastemas are initially metameric.

Blastemal cells, having become myoblasts, unite to form striated muscle fibers, and the body wall muscles commence to take shape. The metamerism of the somites is expressed as myomeres in those adults in which **myosepta** (**myocommata**) separate the muscle of one body segment from the next.

Myosepta do not form in the abdominal region of anurans and amniotes (see fig. 11.5*c*). As a result, their abdominal musculature consists of broad sheets strengthened by tendonous inscriptions. Nevertheless, these sheets are innervated by as many spinal nerves as there were somites that contributed to them.

Trunk and Tail Muscles of Fishes

The musculature of the body wall and tail of fishes consists of myomeres separated by myosepta to which the longitudinally directed muscle fibers attach (see fig. 11.5*a* and *c*). The first and foremost role of this

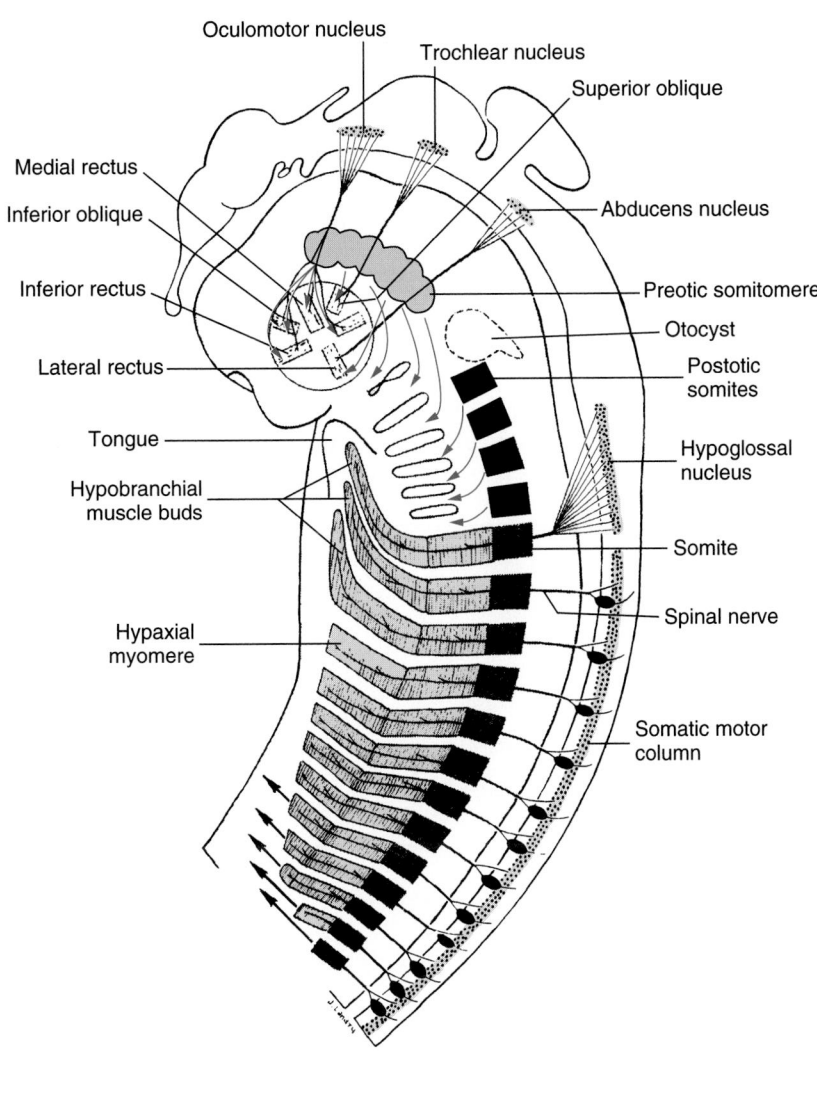

Oculomotor nucleus
Trochlear nucleus
Superior oblique
Medial rectus
Inferior oblique
Abducens nucleus
Inferior rectus
Preotic somitomeres
Otocyst
Lateral rectus
Postotic somites
Tongue
Hypoglossal nucleus
Hypobranchial muscle buds
Somite
Spinal nerve
Hypaxial myomere
Somatic motor column

FIGURE 11.6

Axial muscle origins and innervation in a generalized vertebrate embryo (diagrammatic). The eyeball muscles organize from preotic somitomeres at the level of the oculomotor, trochlear, and abducens nuclei, respectively. Segmental muscles of the trunk arise from trunk somites and are supplied by corresponding segmental nerves. Hypobranchial musculature migrates forward into the floor of the pharynx accompanied by a nerve supply (occipitospinal nerves in nontetrapods and XII in tetrapods). Motor fibers innervating myotomal muscles have cell bodies in the somatic motor column (interrupted in the brain). Postotic somites that contribute to branchiomeric muscles are indicated. The central nervous system has been projected above the embryo. *Red arrows* indicate migration of somitomeres and somites in the head.

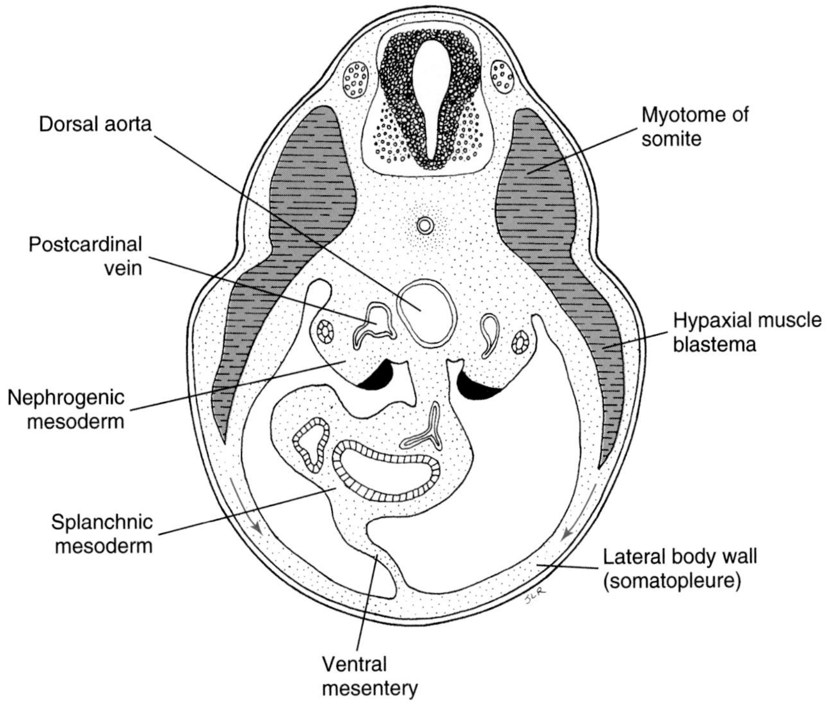

Dorsal aorta
Myotome of somite
Postcardinal vein
Hypaxial muscle blastema
Nephrogenic mesoderm
Splanchnic mesoderm
Lateral body wall (somatopleure)
Ventral mesentery

FIGURE 11.7

Cross section of mammalian embryo showing invasion of myotomal cells (*red*) into lateral body wall to form hypaxial muscles.

musculature is locomotion. Typically, there is a myomere for each vertebra and a spinal nerve for each myomere (see fig. 11.6). Except in agnathans, the myomeres are divided into dorsal and ventral masses, the **epaxial** and **hypaxial muscles**, by a fibrous sheet, the **horizontal skeletogenous septum**. The septum is anchored to dorsal ribs when the latter are present and stretches between the vertebral column and the skin along the entire length of the trunk and tail (see figs. 8.5 and 11.5c). Ventral ribs in the lateral body wall of most fishes develop within the myosepta. Brook lampreys lack not only the horizontal septum but also myosepta.

The myosepta of fishes are seen to zigzag when the skin is removed, but deeper within the wall, the myosepta are elaborately folded, the angles of each zigzag being elongated forward or backward to form muscular cones that fit into the cones of adjacent myomeres like stacked dunce caps. The cones become longer toward the tail, and the apices of caudally directed cones near the end of the trunk are often continued as tendonous extensions that insert on caudal vertebrae. The forces exerted by contractions of the myomeres are thereby distributed over more than one body segment and become most powerful in the tail, where flexibility of the vertebral column is also the greatest. In response to waves of motor impulses that sweep caudad over successive spinal nerves, the wave on one side being in advance of the other, successive myomeres exert a pull on the vertebral column, first on one side then the other, evolving rhythmic lateral undulations of the trunk and tail. These propel the fish forward (fig. 11.8).

A thin sheet of **oblique fibers** lies superficial to the main hypaxial mass ventrolaterally in many fishes, and a thin ribbon of still more superficial fibers parallels the linea alba on each side. The latter resembles the rectus abdominis muscle of tetrapods (see fig. 11.15).

The bluefin tuna, which is among the fastest swimming fishes and weighs an average of 500 lbs (225 kg), is often said to be "warm blooded" by commercial fishers because heat generated by contractions of the voluminous axial musculature warms its body to temperatures far exceeding that of the surrounding water.

The metamerism of the hypaxial muscles of fishes is interrupted where the pectoral and pelvic girdles are built into the body wall, and by the gills. Dorsal to the gills, the epaxial muscles continue to the skull as **epibranchial muscles** (see fig. 11.24a). Beneath the gills, the hypaxials extend to the lower jaw as **hypobranchial muscles**, to be discussed shortly.

Upon emerging from the vertebral column through intervertebral foramina, spinal nerves divide into two major rami (branches) that supply, among other structures, the axial musculature. Dorsal rami supply epaxial myomeres. Ventral rami, which are larger because they innervate a greater muscle mass, innervate hypaxial myomeres. If, during embryonic development, premuscle mesenchyme from a given body segment streams beyond the limits of that segment before differentiating, branches of the spinal nerve from the original metamere will innervate the muscle in its adult location. Later in the chapter, when we study the hypobranchial muscles, we will see an example of this phenomenon. The innervation of the epaxial and hypaxial musculature and their derivatives is no different in tetrapods.

Trunk and Tail Muscles of Tetrapods

Tailed amphibians have retained the primitive metamerism of epaxial and hypaxial muscles (see

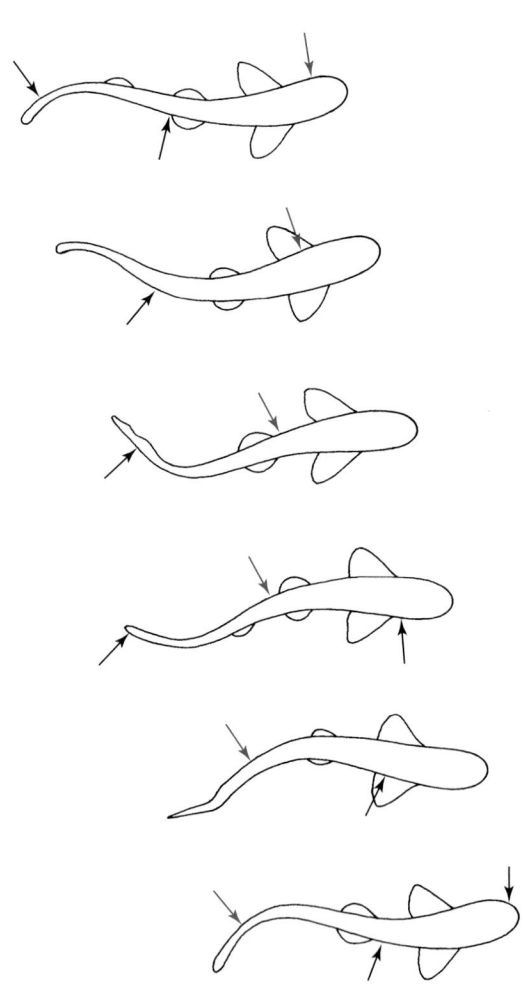

FIGURE 11.8

Propulsion by lateral undulation in a fish. *Red arrows* show progression of counterforces exerted by the surrounding water against one side commencing at the head during passage of a single contractile wave. *Black arrows* show other simultaneous counterforces. Each is at an angle that propels the fish forward. Counterforces are evoked by pressure of the undulating body against the water.

From Carl Gans, *Biomechanics: An Approach to Vertebrate Biology*, 2nd edition. Copyright © 1974. University of Michigan Press, Ann Arbor, MI. Reprinted by permission.

fig. 11.5d). This enables aquatic urodeles to swim by fishlike lateral undulation of the trunk and tail. However, a strictly metameric axial musculature unaccompanied by a well-developed appendicular musculature is ill-suited for locomotion on land. Amniotes gradually lost much of the metamerism of the axial muscles and developed, instead, a complex appendicular musculature well adapted for terrestrial life.

Disappearance of epaxial myosepta in amniotes gave rise to long, straplike or pinnate bundles disposed above the transverse processes, leaving only a vestige of metamerism in the deepest bundles (fig. 11.9). Similarly

modified were the muscle bundles immediately beneath the transverse processes (that is, in the roof of the coelom). This revised architecture of the vertebral musculature, accompanied by modifications of the intervertebral joints, greatly increased the flexibility of the reptilian and mammalian vertebral columns, permitting dorsal arching, especially in mammals, and greater lateral undulation along a larger extend of the trunk, especially in nonavian reptiles other than turtles. Birds forgo these benefits for another: The rigid column of birds requires few epaxial muscles caudal to the neck but makes avian flight possible.

Changes with the advent of limbs were not confined to the vertebral musculature; they occurred also in the lateral body wall. Hypaxial myomeres in that location were gradually replaced by strata of broad muscular sheets, the fibers of which were disposed in a different direction in each layer. This alteration of the lateral body wall (*parietal*) musculature was able to occur without negatively affecting survival because tetrapods had evolved paired limbs for locomotion, freeing the parietal muscles from this role. The new parietal musculature supports the trunk viscera in a muscular sling suspended well above the ground.

Loss of metamerism of the axial musculature, along with adaptations of the vertebral column and appendages, enabled tetrapods to better meet the locomotor needs imposed by competition in the terrestrial habitat. In no tetrapod is the locomotor effectiveness of this revised axial muscle better demonstrated than in the agile anole lizards, in fleet digitigrade carnivorous mammals that must "hump" (hustle) to capture food, or in jackrabbits, who must scamper to avoid being eaten (fig. 11.10).

Intermediate stages in the adaptation of trunk and tail muscles to life on land are seen in *Necturus* and *Sphenodon* (see fig. 11.5d and e). *Necturus* has retained epaxial and hypaxial myomeres, but each hypaxial myomere is split into three layers. *Sphenodon* has retained epaxial myomeres only, and the lateral body wall musculature consists of broad stratified sheets. With loss of metamerism of the axial muscles, the horizontal skeletogenous septa also disappeared.

Figure 11.11 illustrates the components of the axial musculature as they are disposed in the abdominal wall of a mammal.

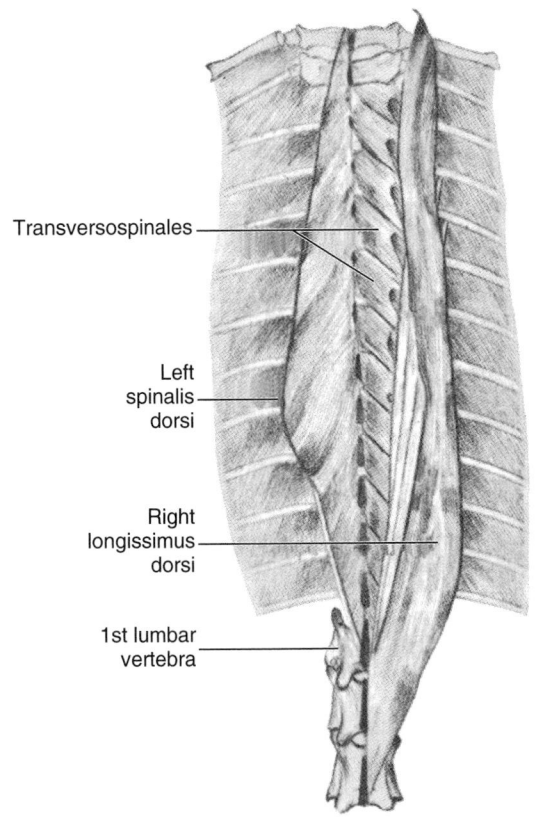

Transversospinales

Left spinalis dorsi

Right longissimus dorsi

1st lumbar vertebra

FIGURE 11.9
Three epaxial muscles of the thorax of a hamster, dorsal view. The right spinalis dorsi has been removed to show the underlying transversospinales. For the relationship of the longissimus to the spinalis, see figure 11.12.

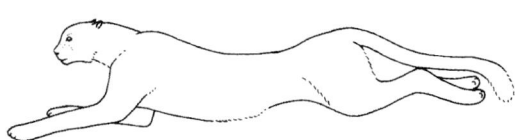
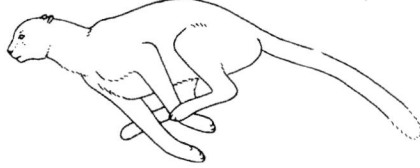

FIGURE 11.10
Extension and flexion of the vertebral column of a fleet digitigrade mammal.

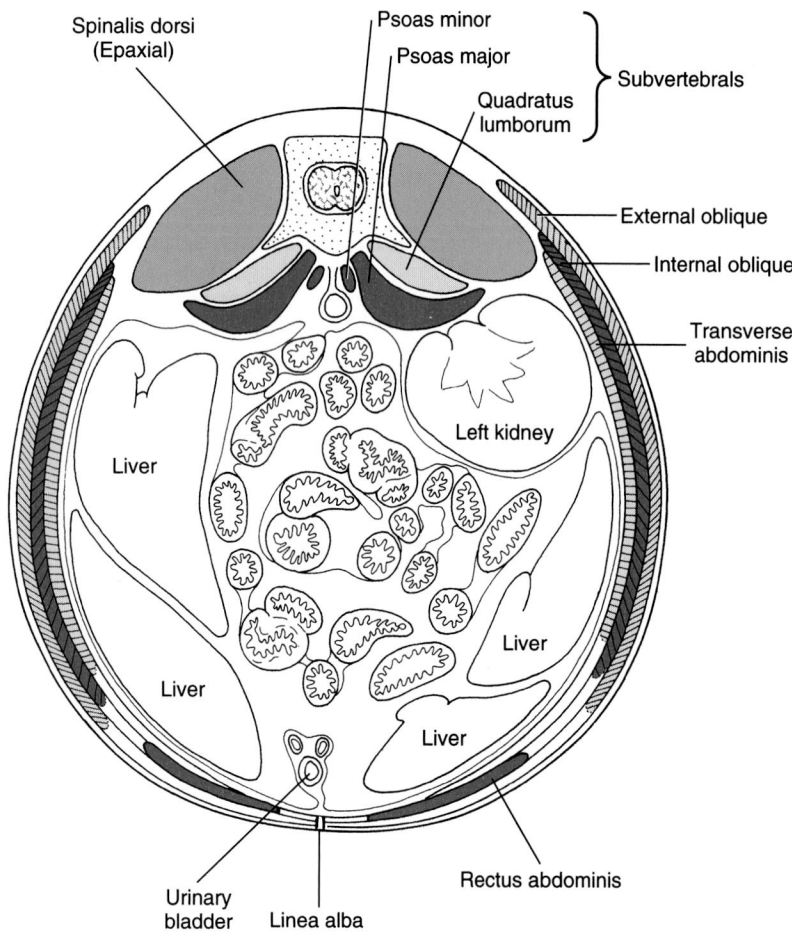

FIGURE 11.11
Disposition of axial muscles in the abdominal region of a mammal, based on cross section of a rabbit, cephalic view. The cutaneous maximus muscle has been removed. The psoas major, inserting on the lesser trochanter of the femur and thus an appendicular muscle, is also shown.

Epaxial Muscles of the Trunk

The epaxial muscles of tetrapods extend from the base of the skull into the tail for varying distances (fig. 11.12a). The most anterior ones attach to the occiput. In amphibians other than anurans, the epaxials retain their primitive metamerism, arising and inserting on myosepta and transverse processes, constituting collectively the **dorsalis trunci**. In amniotes, excepting rhynchocephalians, most of the epaxials are long bundles, some of which extend over many body segments.

Epaxial muscles collectively function in straightening (extending) the vertebral column and in lateral flexion of the body. Keep in mind that muscles do not act independently, but synergistically with others and that this dictates, within limits, their specific effect on the skeleton at any given moment. Note, too, that the action of a given appendicular muscle in a bipedal animal may not be the same as it is in one that walks on all four legs. The epaxial muscles of amniotes are divided arbitrarily into four groups: **intervertebrals**, a **longissimus**, **spinales**, and **iliocostales**.

Intervertebrals are the deepest epaxial muscles and the only ones to retain their primitive metamerism. They extend between two successive transverse processes (**intertransversarii**), two neural spines (**interspinales**), two neural arches (**interarcuales**), or two successive zygapophyses (**interarticulares**), and participate with longer epaxials in maintaining a vertebral posture appropriate to the needs of the moment.

Longissimus and spinales muscles occupy, respectively, lateral and medial positions above the transverse processes, and individual bundles are named according to their location (table 11.4). Capitis bundles insert on the skull and assist in movements of the head. Cervicis bundles are in the neck, and dorsi bundles are in the trunk. The longissimus, so named because it is the longest epaxial mass, consists of bundles that, with their tendons, may extend over many body segments. It is the dominant extensor in mammals, in which lateral undulation is minimal, and plays virtually no role in generalized mammalian locomotion. In the lumbar region of mammals, the longissimus consists of three distinct bundles (fig. 11.12b). The medial bundle continues into the tail.

Spinales include long and medial bundles that connect neural spines or transverse processes with neural spines several or many segments cephalad, and transversospinales that connect transverse processes with the neural spines of the second vertebra forward (see fig. 11.9). Spinales are often grouped with lumbar intervertebrals and named, collectively, **multifidus spinae**. Their chief role in mammals is to assist in maintaining stability (temporary rigidity) of the column in whatever degree of extension or flexion is imposed by other vertebral muscles. In humans, the spinales assist other bundles in maintaining an upright posture while standing.

Iliocostales are lateral to the longissimus. They constitute a thin sheet arising on the ilium and passing forward to insert on the ribs and uncinate processes. They are the dominant epaxials in reptiles, because their lateral location and their insertions along the length of the column provide leverage for lateral undulation, a significant component of reptilian locomotion. Lateral undulation in crocodilians rotates the pectoral girdle and, to a lesser degree, the pelvic girdle on a vertical axis, thereby increasing the length of the stride of both pairs of limbs. The iliocostales continue forward into the neck but not caudad into the tail.

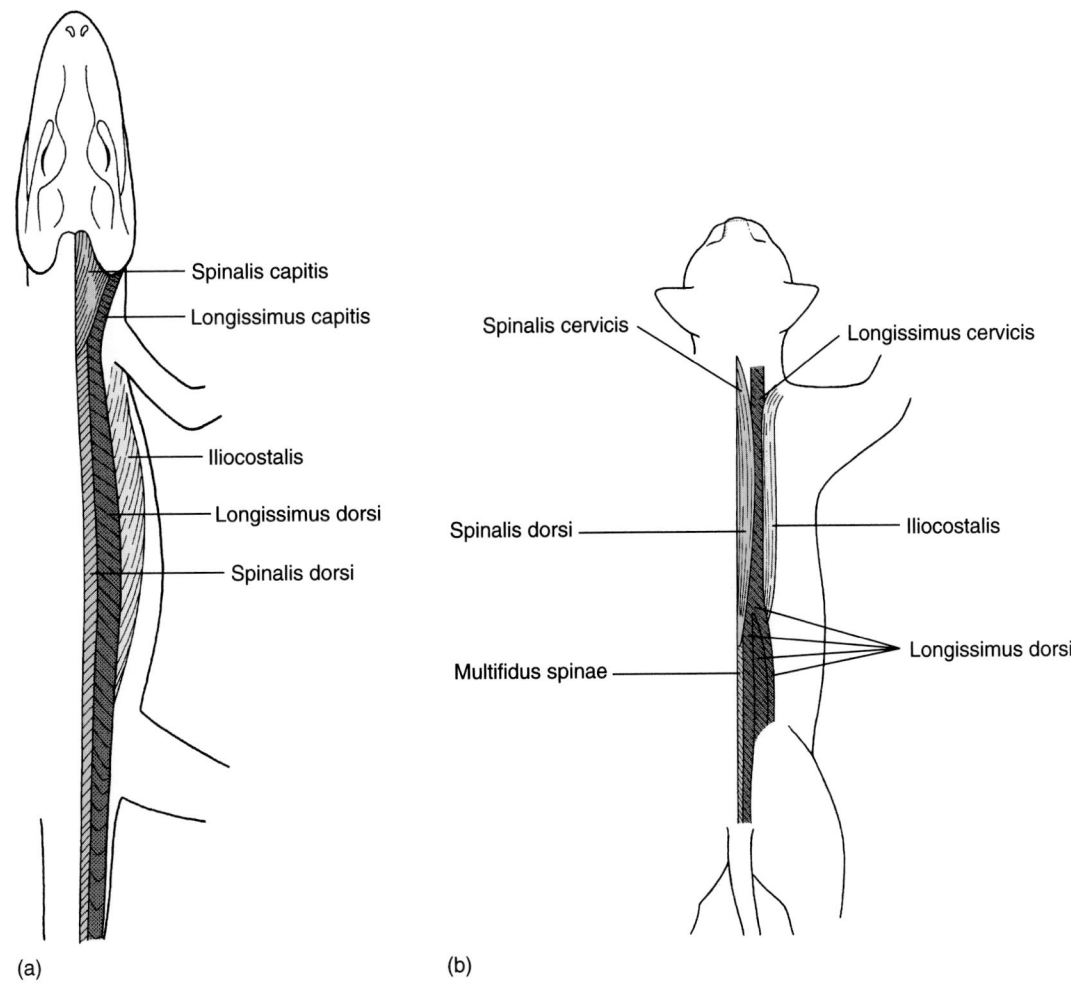

FIGURE 11.12
Disposition of the epaxial muscles. (*a*) Crocodilian. (*b*) Cat. The intervertebrals are not represented except as part of the multifidus spinae in the cat.

In turtles and birds, epaxials are prominent only in the neck because the vertebral column and, in turtles, the ribs are immobilized by fusion with the synsacrum or carapace. The longissimus, for instance, is mostly ligamentous in the trunk. However, well-developed cervical epaxials participate in providing exceptional flexibility to the long necks of birds such as swans. One of the cervical epaxials of birds, the **complexus**, inserts on the interparietal bone and provides power for cracking the eggshell with the beak during hatching. In contrast to turtles and birds, snakes utilize a variety of locomotor strategies, including both dorsoventral flexion and lateral undulation. All epaxial muscles are voluminous and complex.

The epaxial muscles became increasingly hidden by the expansion dorsad of the appendicular muscles and associated lumbodorsal aponeurosis as a derived condition in amniotes (compare *a* and *c* in fig. 11.19).

Hypaxial Muscles of the Trunk

The hypaxials of the trunk of amniotes can be divided arbitrarily into (1) **subvertebrals** (longitudinal bundles beneath the transverse processes in the roof of the coelom), (2) **oblique** sheets, (3) **transverse** sheets in the lateral body wall (parietal muscles), and (4) **rectus abdominis** muscles (longitudinal straplike muscles on either side of the linea alba). These are listed in table 11.4 and represented schematically in figure 11.11.

Subvertebral Muscles Subvertebral muscles form a longitudinal band of fairly powerful flexors of the vertebral column lying *beneath* the transverse processes from the atlas to the pelvis. The portion in the neck of birds and mammals is known as the **longus colli**. Subvertebrals are meager in the thorax but become prominent again in the lumbar region, where they are represented by the **quadratus lumborum** and, in mammals, the **psoas minor**. The quadratus lumborum of mammals originates on the centra of several of the last thoracic vertebrae and the bases of their ribs and on the transverse processes of lumbar vertebrae, and it inserts on the ventral angle of the wing of the ilium, with variations depending on the species. The psoas minor, known commercially as the

TABLE 11.4 Representative Somatic (Myotomal) Muscles of Head, Trunk, and Forelimbs of Mammals

	Eyeball	Tongue (Hypobranchial)	Hypobranchial
Head and neck	Superior oblique Inferior oblique Medial rectus Lateral rectus Superior rectus Inferior rectus	Genioglossus Hyoglossus Styloglossus Lingualis	Geniohyoideus Sternohyoideus Sternothyroideus Thyrohyoideus Omohyoideus
	Branchiomeric Muscles		
Pharynx	Mandibular muscles Hyoid muscles Other branchial muscles		
	Epaxial Muscles		**Hypaxial Muscles**

Trunk	Intervertebrals
	  Intertransversarii
	  Interspinales
	  Interarcuales
	  Interarticulares
	Longissimus
	  L. capitis
	  L. cervicis
	  L. dorsi
	  Extensor caudae lateralis
	Spinales
	  S. dorsi
	  S. cervicis
	  S. capitis
	  Transversospinales
	Iliocostales
			Subvertebrals
			  Longus colli
			  Quadratus lumborum*
			  Psoas minor
			Oblique group (parietals)
			  Internal and external intercostals
			  Internal and external obliques of abdomen
			  Cremaster
			  Supracostals
			    Scalenus[†]
			    Serratus dorsalis[†]
			    Levatores costarum[†]
			    Transversus costarum
			  Diaphragm
			Transverse group (parietals)
			  Transversus thoracis (subcostal)
			  Transversus abdominis
			Rectus muscles
			  Rectus abdominis
			  Pyramidialis

	Extrinsic		**Intrinsic**

Forelimb	Secondary appendicular[‡]
	  Levator scapulae
	  Rhomboideus
	  Serratus ventralis
	Primary appendicular[‡]
	  Latissimus dorsi
	  Pectorales
			See table 11.5

*Secondary appendicular muscle.
[†]May be epaxial by derivation.
[‡]See discussion of these terms under "Appendix Muscles: Tetrapods."

tenderloin, connects the lumbar vertebrae with the pelvic girdle. The action of the subvertebrals is, in general, opposite that of the epaxials.

Oblique and Transverse Muscles The parietal muscles have become stratified into superficial and deep layers, typically **external oblique, internal oblique,** and **transverse muscles of the abdomen;** and in amniotes **external intercostal, internal intercostal,** and **transverse muscles of the thorax.** The fibers of all of these pass more or less obliquely from origin to insertion.

One or another of these sheets may either be split in two or lost. In aquatic urodeles, for example, in which parietal muscles participate in locomotion, the external oblique muscle is split into superficial and deep parts. In crocodilians and some lizards, all three layers consist of two sheets each. In anurans, the internal oblique is sometimes missing, and in birds, all sheets are thin. In turtles, they are less than thin—they are vestigial! But because of the rigid shell, a turtle could not use them anyway. (If you want some turtle soup, it will have to come from the neck, tail, or appendicular muscles!)

Muscle slips from the inferior border of the internal oblique of male mammals, and sometimes from the transverse abdominal muscle, form a **cremaster muscle** (see fig. 15.36) that loops around the spermatic cord commencing at the inguinal ring and inserts on a fibrous sheath in the wall of the scrotum below the testes. It is best developed in mammals with permanently open inguinal canals, such as rabbits, because it retracts the testes into the canals in those species.

The oblique and transverse muscles, particularly intercostals, play a major role in external respiration in most amniotes other than turtles, and an accessory role in mammals. The intercostals are assisted by subdivisions of **supracostal muscles** that differentiate on the surface of the rib cage, chiefly **scalenus, serratus dorsalis, levatores costarum,** and **transversus costarum.** On the basis of their innervation, these muscles are thought to be derivatives of the oblique sheets of primitive pari-

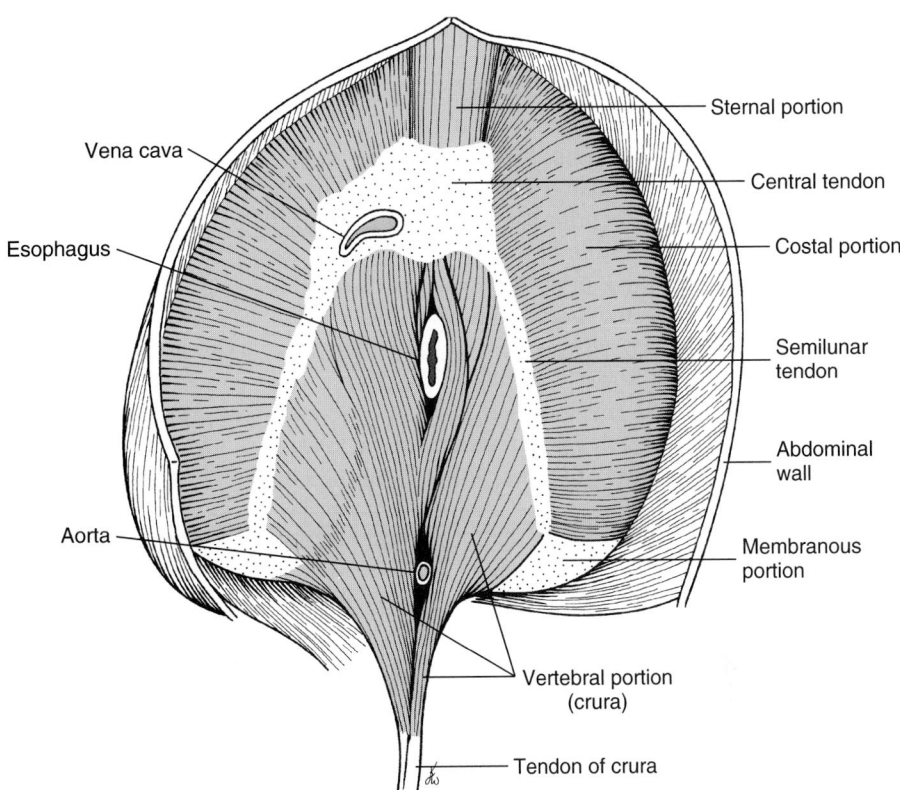

FIGURE 11.13

The domed mammalian diaphragm, viewed from abdominal cavity. The concavity is occupied by the liver. Apertures are present for the postcava (or vena cava in the central tendon), esophagus (*center*), and dorsal aorta (between the crura).

etal muscles. Along with the rectus, the oblique and transverse muscles support the abdominal viscera in a muscular sling, and they compress the viscera for such functions as egg laying, delivery of mammalian young, and emptying of the digestive tract. Although they lack segmental myosepta in amniotes, their metameric history is evidenced by their embryonic origin from successive somites and by their innervation via ventral rami of successive spinal nerves.

Rectus Muscles of the Abdomen　The rectus abdominis extends longitudinally on either side of the linea alba between the pubic symphysis and the sternum and, in amniotes, the base of several sternal ribs. It assists in flexing the trunk and in supporting the abdominal viscera in a muscular sling. In urodeles, it is strictly segmental (see fig. 11.15). In anurans and amniotes, it exhibits irregular transverse tendonous inscriptions.

A pyramidalis muscle in the ventral wall of the marsupial pouch is a slip of the rectus abdominis. Eutherian mammals may have vestiges of the pyramidalis as a species characteristic or as an anomaly.

Mammalian Diaphragm　In mammals, mesenchyme from several embryonic somites at the level of the developing third, fourth, and fifth cervical spinal

nerves, varying with the species, migrates caudad in the somatopleure to invade the embryonic septum transversum, which is thereby converted into a dome-shaped muscular diaphragm separating the thoracic and abdominal cavities (fig. 11.13). When completed, the diaphragm consists of a **central tendon** with a pair of semilunar extensions and a **muscular portion.** The latter converges on the tendonous portion from a circumferential perimeter, arising on the xiphoid process ventrally (**sternal portion**), on the caudalmost ribs or their costal cartilages laterally (**costal portion**), and from several lumbar vertebrae dorsally (**vertebral portion**). The vertebral portion consists of a pair of triangular muscular masses, the **crura,** that are firmly anchored to the lumbar vertebrae by a single short, tough, cylindrical tendon. The mammalian diaphragm has become the major component of the suction pump utilized in mammalian breathing, superceding, but not totally replacing, the ribs and parietal muscles in this role. Innervation of the diaphragm by ventral rami of cervical spinal nerves indicates its hypaxial status.

Muscles of the Tail

The tail musculature of tetrapods is a continuation of the epaxial and hypaxial musculature of the trunk. This is most evident in urodeles and generalized reptiles (see fig. 11.5*e*). Disruption of the continuity, especially of the hypaxial bundles, occurs to one degree or another at the level of the pelvis of amniotes generally and especially of birds and mammals. This is accounted for by the presence of the large pelvis characteristic of amniotes. (The latter is correlated with the increased volume of the extrinsic appendicular muscles of the amniote hind limb.) The volume of the caudal musculature varies, of course. Snakes, for example, employ the tail for locomotion, and epaxial and hypaxial muscles are well developed along the entire extent of the tail.

Intervertebrals and short spinales (collectively, the multifidus spinae) and the longissimus continue into the tail as medial and lateral extensors, respectively, lying on the dorsal aspect of the vertebral column. The tail section of the longissimus, the **extensor caudae lateralis,** arises from sacral and caudal vertebrae and inserts on distal tail vertebrae by many long slender tendons. In general, these epaxial muscles extend the tail and arch it upward.

Long hypaxial bundles similar to the epaxials arise from the medial surface and caudal border of the wing of the ilium, from transverse processes of the last lumbar vertebrae, or from the sacrum and pass into the tail as long bundles paralleling the vertebral column. They are supplemented by long and short flexors that arise within the tail and insert farther along by long or short tendons. Hypaxial muscles bend the tail laterad or downward.

In urodeles and reptiles, a **caudofemoralis** muscle contributes to the fleshy part of the tail. It connects several caudal vertebrae at the base of the tail with the femur, being secondarily an extrinsic appendicular muscle. It exerts a powerful backward pull on the hind limb during locomotion of lizards and crocodilians.

Caudal centra, arches, and processes become increasingly vestigial distally, especially in mammals (see fig. 8.22), and all muscle bundles dwindle until only long tendons of insertion remain. The major fleshy part of the mammalian tail is therefore the proximal portion. Distally, the mammalian tail is mostly skin, stringlike tendons, and cylindrical bones united by interosseus ligaments. The

meatiest oxtail soup comes from the proximal portion of the tail!

Dilator and sphincter muscles of the cloaca and the sphincter muscle of the anus are derived from the hypaxial musculature that anchors the tail to the pelvis. In turtles, for instance, they arise on sacral and proximal caudal vertebrae and insert on the skin encircling the vent. In mammals, they have lost their skeletal attachments.

Hypobranchial and Tongue Muscles

Mesenchyme from several postbranchial somites streams forward in the floor of the pharynx beneath the branchial arches to give rise to hypobranchial muscles and, when present, tongue muscles (see fig. 11.6). These myotomal muscles are anterior extensions of the hypaxial musculature of the trunk. In fishes, hypobranchial muscles extend forward from the coracoid region of the pectoral girdle (via **coracoarcuales**) to insert on Meckel's cartilages (**coracomandibularis**), basihyals (**coracohyoideus**), and the ventralmost segments of the gill cartilages (**coracobranchials,** fig. 11.14). They assist the branchiomeric muscles in respiration and feeding movements by expanding the pharynx and gill pouches, moving parts of the hyoid skeleton, and depressing the

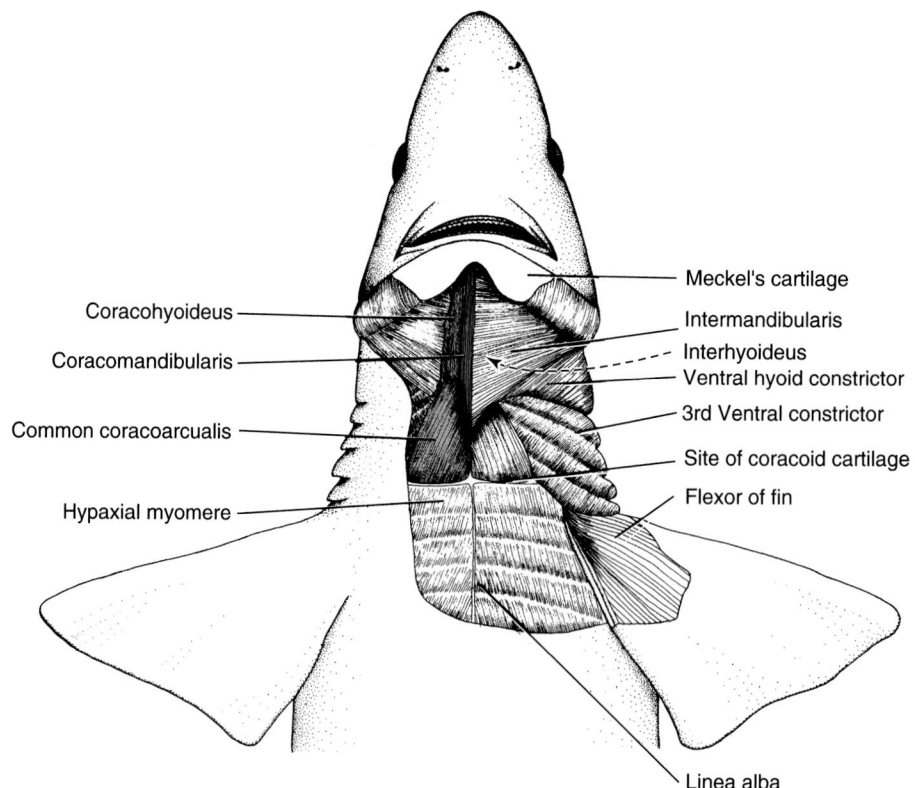

Coracohyoideus —
Coracomandibularis —
Common coracoarcualis —
Hypaxial myomere —

— Meckel's cartilage
— Intermandibularis
— Interhyoideus
— Ventral hyoid constrictor
— 3rd Ventral constrictor
— Site of coracoid cartilage
— Flexor of fin

Linea alba

FIGURE 11.14

Hypobranchial muscles of a shark (*red*) and ventral branchiomeric muscles (at right of midline). The intermandibularis and interhyoideus are superficial to the coracohyoideus and coracomandibularis. They have been reflected on the left. The interhyoideus is continuous with the ventral hyoid constrictor and lies deep to the intermandibularis. The coracobranchials are deep hypobranchial muscles that are not shown.

lower jaw. They perform related functions in anurans and urodeles, depending on the species. The **rectus cervicis** muscle of urodeles is a hypobranchial muscle (fig. 11.15).

With further development of a neck in amniotes, hypobranchial muscles became longer and straplike. They stabilize the hyoid apparatus and larynx and draw these cephalad or caudad, depending on the concurrent actions of other muscles inserting on these same structures. They bear such names as **sternohyoid, sternothyroid, thyrohyoid, omohyoid,** and **geniohyoid.** One of these is seen in a rabbit in figure 9.44.

The tongue of amniotes is essentially a mucosal sac anchored to the hyoid skeleton and stuffed with hypobranchial muscles. Premuscle mesenchyme migrates into the developing tongue from anterior hypobranchial muscle blastemas. This accounts for the observation that, in bats, some tongue muscles extend into the tongue from as far back as the sternum. The chief extrinsic tongue muscles of mammals are the **hyoglossus, styloglossus,** and **genioglossus.** Mammals and some reptiles develop an intrinsic tongue muscle, the **lingualis.** It has no skeletal attachments.

Because hypobranchial and tongue muscles are derived from the anteriormost trunk somites, they are supplied by cervical spinal nerves or, with respect to the tongue muscles, by the last cranial nerve (hypoglossal), a cervical spinal nerve that became "trapped" within the amniote skull. The cell bodies of the motor fibers that innervate these muscles are in the same somatic motor column of the spinal cord and brain that supplies myotomal muscles elsewhere in the body and no others (see figs. 11.6 and 16.34, SE).

APPENDICULAR MUSCLES

Appendicular muscles are those that insert on the girdles, fins, or limbs. Most fishes use axial muscles for locomotion. As a correlate, the appendicular muscles of fishes are uncomplicated, exhibit little variety, have little mass, and perform a restricted function. Those of tetrapods have become increasingly numerous and complex as a consequence of adaptations for life on land.

Fishes

Paired fins arise in embryos as fin folds that protrude from the lateral body wall (fig. 11.16, shark). Thereafter, hollow **muscle buds** sprout from the lower edges of a series of embryonic myomeres near the base of each fin fold (fig. 11.17). The buds split into dorsal and ventral moities, invade the developing fin, and establish blastemas from which dorsal and ventral muscle masses are formed. Dorsal blastemas form extensors (elevators) of the fin; ventral blastemas form the flexors (depressors, fig. 11.18). The resulting musculature establishes attachments to the girdles, basalia, radialia (if any), and the fascia overlying the base of the fin rays. Any muscle tissue that may develop distal to these masses is insignificant except muscles such as those in the claspers of male Chondrichthyes and in the undulating winglike fins of rays and skates.

The muscles of median dorsal fins organize from myotomal mesenchyme that is giving rise to epaxial myomeres. Those of median ventral fins arise from hypaxial mesenchyme. The musculature of median fins is meager, with the exception of anal fins that have become gonopodia.

Tetrapods

The appendicular muscles of tetrapods are far more complex than those of fishes because of the joints in tetrapod limbs. As in fishes, the appendicular muscles of tetrapods are disposed in opposing groups. In fishes, postaxial (posterior; dorsal) muscles extend the appendage, and preaxials (anterior; ventral) flex it. When applied to limbs that have rotated toward the long axis of the body, so that elbows point backward and knees point forward (see fig. 10.45), one or another of these terms becomes meaningless to the layperson. Not only have the segments and the bones within them become reoriented, but

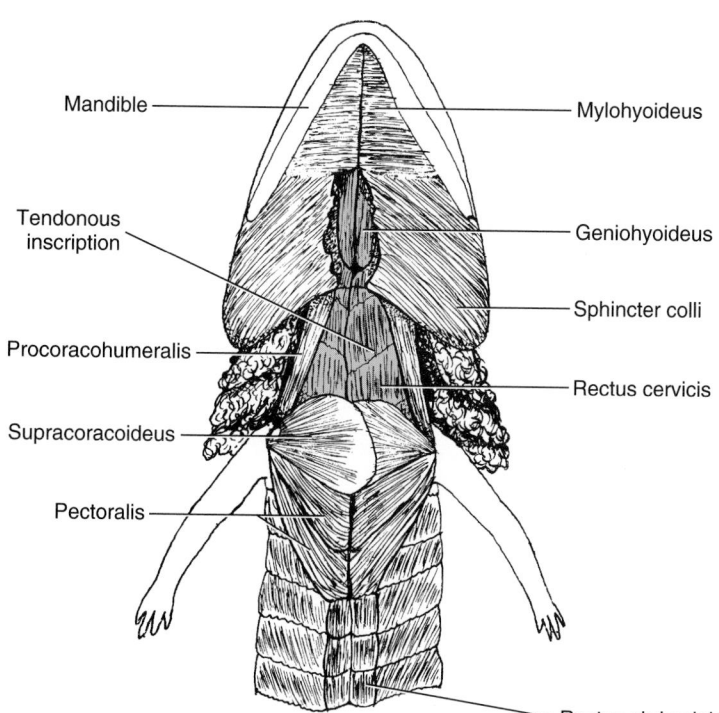

FIGURE 11.15

Anteroventral muscles of *Necturus.* The sphincter colli is a very thin muscular sheet on the surface of the powerful branchiohyoideus (see fig. 11.24*b*). Hypobranchials are shown in *red.*

Mandible — Mylohyoideus

Tendonous inscription — Geniohyoideus

Procoracohumeralis — Sphincter colli

Supracoracoideus — Rectus cervicis

Pectoralis

Rectus abdominis

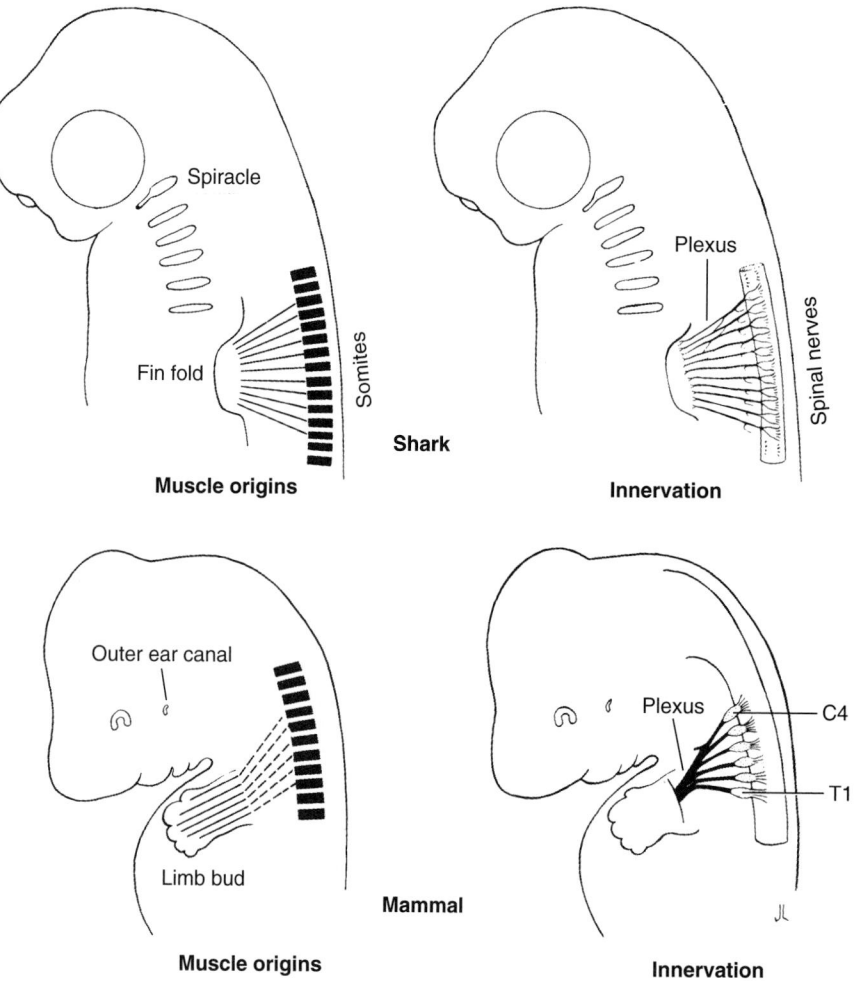

FIGURE 11.16

Top, Origin of shark appendicular muscles from somites and their innervation by corresponding spinal nerves. *Below,* Probable phylogenetic derivation of appendicular muscles of a mammal from six somites based on their innervation. **C4** and **T1,** Dorsal root ganglia of the fourth cervical and first thoracic spinal nerves of the brachial plexus.

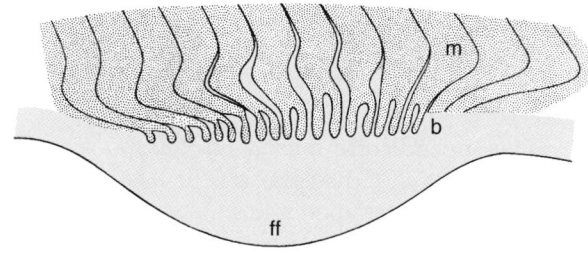

FIGURE 11.17

Budding of fin musculature from embryonic body wall myomeres in a shark. In one electric ray, 26 myomeres participate; **b,** muscle buds; **ff,** fin fold; **m,** myomere.

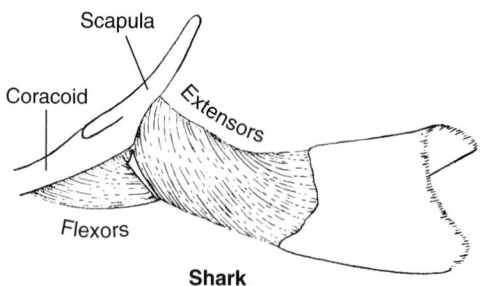

FIGURE 11.18

Muscles of pectoral fin of a shark. Extensors are dorsal (elevators); flexors are ventral (depressors).

so have the muscles, which often spiral around a bone or a joint to insert on an opposite surface of the next segment. For this reason, we will discuss the appendicular muscles of tetrapods with reference to their current anatomic relationships, rather than to their primitive locations.

Anatomists have found it practical for descriptive purposes to divide appendicular muscles into two groups according to their anatomic origins. **Extrinsic appendicular muscles** are those arising on the axial skeleton or fascia of the trunk and inserting on a girdle or limb. **Intrinsic appendicular muscles** are those arising on a girdle or limb and inserting more distally on the limb. When all extrinsic muscles of the pectoral girdle and forelimbs of most mammals have been transected, the clavicle, scapula, and limb of either side can be removed from the body as a unit. The pelvic girdle cannot be similarly removed because the girdle forms an immobile joint with the sacrum.

Most extrinsic muscles of tetrapod limbs begin development from blastemas *within the embryonic body wall.* Buds from these blastemas grow toward and establish insertions on a girdle or proximal bone of the limb. Muscles that develop in this manner are also referred to as *secondary* appendicular muscles because their embryonic origin is not intrinsic to the limb. The levator scapulae, serratus ventralis, and rhomboideus (all of myotomal origin) and the trapezius, sternomastoideus, and cleidomastoideus (all of branchiomeric myotomal origin) develop in this manner. Extrinsic muscles are the most primitive, being extensions of the locomotor musculature of ancestral craniates.

Intrinsic appendicular muscles of tetrapods organize from blastemas *within the developing limb,* rather than as buds from hypaxial myomeres. Muscles that form in this manner are also referred to as *primary* appendicular muscles. In the latter category are a few extrinsic muscles that form when their blastemas spread trunkwise out of the limb and establish an attachment on the axial skeleton. Among these are the latissimus dorsi (fig. 11.19) and the iliopsoas of the posterior limb of generalized mammals.

The mesenchyme that contributes to the intrinsic blastemas originates in somites; therefore, the muscles of the appendages are myotomal by *embryonic origin.* Their innervation by motor fibers from the somatic motor column of the cord via ventral rami of spinal nerves is consistent with their being myotomal by phylogenetic derivation (see fig. 11.16, mammal).

Extrinsic Muscles of the Pectoral Girdle and Forelimbs

Dorsal Group   The most constant dorsal extrinsic appendicular muscle of tetrapods is the **latissimus dorsi**, a primary appendicular muscle that inserts on the humerus. In urodeles (fig. 11.19*a*), it is a delicate trian-

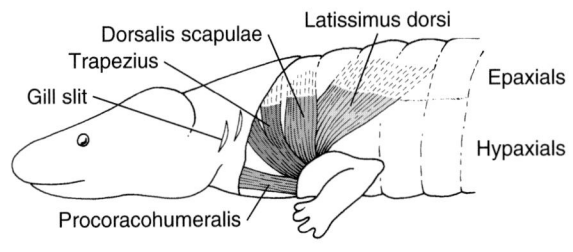

Necturus

(a)

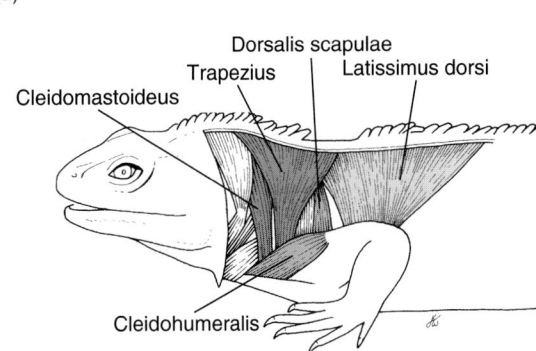

Sphenodon

(b)

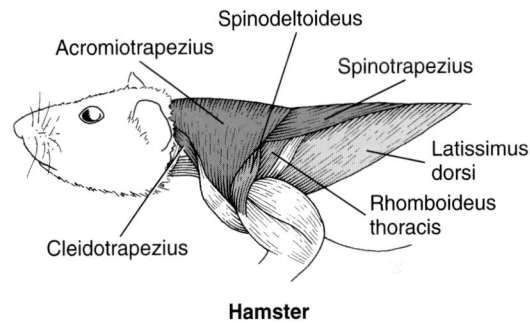

Hamster

(c)

FIGURE 11.19

Superficial shoulder muscles (in color) of a urodele (*a*), generalized reptile (*b*), and rodent (*c*) illustrating the expansion of the shoulder musculature as a derived feature in tetrapods. The dorsalis scapulae and procoracohumeralis are also known as deltoid muscles.

gular muscle arising from the superficial fascia that overlies the epaxial myomeres of the shoulder region. In reptiles, it became stronger. It spread dorsad to acquire a firm attachment to the tough fascia that is anchored to neural spines, and it broadened its axial origins by spreading still farther caudad (fig. 11.19*b*). In mammals, the trend toward a broader dorsal anchorage continued. It now arises from the neural spines of most of the thoracic vertebrae caudal to the first few, and from the tough fibrous lumbodorsal fascia that overlies the lumbar vertebrae and extends all the way to the base of the tail (fig. 11.19*c*). The result of these dorsocaudad expansions was a continuing increment in the force the muscle was able to exert on the forelimb.

Deep to the latissimus dorsi (and to the trapezius, to be discussed next) in most amniotes are three extrinsic muscles that insert on the scapula: two levators of the scapula, a rhomboideus group (found only in crocodilians among living reptiles), and the serratus ventralis (serratus anterior). The mammalian levators of the scapula have their origins on the transverse processes of the atlas or on the basioccipital bone (**levator scapulae ventralis** or omotransversarius) and on the transverse processes of the number of posterior cervical vertebrae (**levator scapulae dorsalis**). The **rhomboideus** group arises from the occiput and neural spines of a series of cervical and anterior thoracic vertebrae. The **serratus ventralis** arises by many separate prominent tendonous slips from a series of ribs near their junction with the costal cartilages (giving the muscles a serrate appearance) and, in some mammals, from a series of posterior cervical vertebrae. All the foregoing insert on the dorsal border of the scapula except the levator scapulae ventralis, which inserts on a process of the scapular spine near the glenoid fossa. (The levator scapulae dorsalis is sometimes considered part of the serratus ventralis.)

The pharyngeal arches make an important contribution to the extrinsic musculature of the pectoral girdle. The **trapezius** (fig. 11.19*b*), a superficial muscle of the shoulder region, is a survivor of the cucullaris muscle of fishes (see fig. 11.24*a*). It acquired attachments to the pectoral girdle and then underwent the same expansion as the latissimus dorsi. It eventually became subdivided into several components, including **cleidotrapezius** (**cleidocervical**), **acromiotrapezius** (**cervical trapezius**), and **spinotrapezius** (**thoracic trapezius**). It receives its motor innervation via branchiomeric nerves. Two other muscles, the **cleidomastoideus** and **cleido-occipitalis** (see fig. 11.24*c*) acquired an attachment to the clavicle, but they do not function as appendicular muscles. When they contract, they move the head.

Ventral Group The ventral extrinsic muscles of the forelimb, subsumed under the general term "pectoral muscles," have undergone expansive changes matching those of the dorsal group (fig. 11.20). These fan-shaped muscles, originating primitively on the coracoid cartilages or bones and the associated midventral raphe, extended their origins to include, in one species or another, the epicoracoid, the entire length of the sternum, some of the costal cartilages, and part of the midventral raphe of the neck. They subdivided into a varying number of superficial and deep muscle masses that converge to insert on the proximal end of the humerus. The insertion sites of the more powerful pectoral muscles of amniotes are marked by prominent stress-induced ridges and tuberosities.

Two pectoral muscle masses, the **pectoralis** and **supracoracoideus**, are seen in their primitive state in a urodele (figs. 11.15 and 11.20*a*). Their different inser-

tion sites on the humerus result in different actions on the forearm, the pectoralis being the chief adductor. In birds, in which the supracoracoideus is deep to the pectoralis (fig. 11.20*c*), these two are the powerful flight muscles that depress (pectoralis) and elevate (supracoracoideus) the wings, providing the downstroke and upstroke, respectively, when the bird is airborne. The supracoracoideus, which in birds underlies the procoracoid bone (see fig. 8.28, chicken), is able to elevate the wing because its tendon of insertion passes to the dorsal surface of the humerus. (Note the caudally directed humerus on a skeleton in fig. 4.28.) The supracoracoid has become an intrinsic muscle in mammals.

Intrinsic Muscles of the Pectoral Girdle and Forelimbs

Dorsal Group Arising from the scapula of modern mammals are five postaxial muscles inherited from ancestral amniotes. They are the **deltoideus** or subdivisions thereof (**spinodeltoideus** and **acromiodeltoideus**), **teres major, teres minor, subscapularis** (on medial surface of the scapula), and the **long head of the triceps brachii** (fig. 11.21). With the exception of the triceps, these insert on the humerus a short distance below its head and align it with the scapula, rotate it about its long axis, or adduct it. The long head of the triceps, which has two additional heads that arise on the humerus, inserts on the olecranon process of the ulna where it exerts a powerful pull that extends the forearm (see fig. 11.4). To test your own triceps, try a few pushups! This exercise extends the arm against resistance. Distal to the triceps in mammals, two **supinators of the manus** connect the humerus with the radius. Finally, an assortment of **extensors of the hand and digits**, with long distal tendons, insert on the skeleton of the wrist and digits, some as far distad as the terminal phalanx (fig. 11.21). The shortest ones are intrinsic to the manus.

The mammalian deltoid muscle is probably a homologue of the dorsalis scapulae or scapular deltoid in other tetrapods, the teres major appears to be a slip of the latissimus dorsi, and the teres minor is probably the muscle that, in reptiles, is called scapulohumeralis anterior. The mammalian subscapular is an expansion of a muscle in the same location in reptiles.

Ventral Group Representing the primitively ventral musculature of the pectoral girdle and forelimb are two strong muscles on the lateral aspect of the scapula, the supraspinatus and infraspinatus (fig. 11.21), and a very short, deep coracobrachialis (not illustrated). The **supraspinatus** and **infraspinatus** are homologues of the **supracoracoid of reptiles**, a broad muscle having a ventral anatomic origin over a broad area of the procoracoid bone near the glenoid fossa. Studies on embryonic

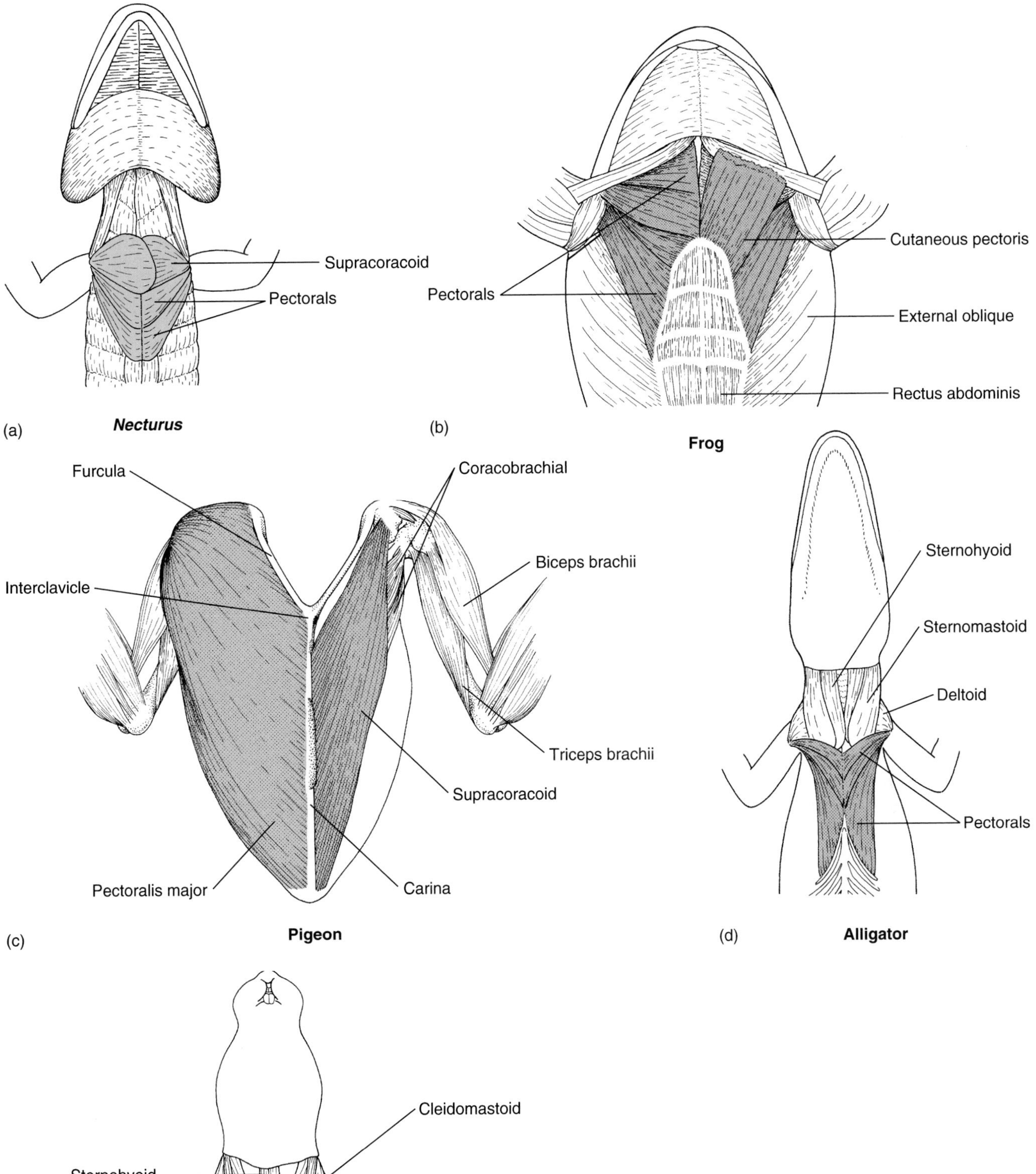

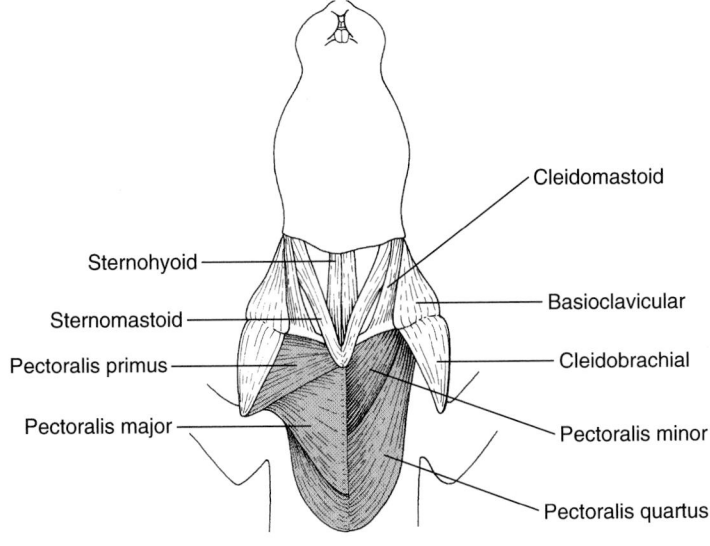

(a) *Necturus*
- Supracoracoid
- Pectorals

(b) **Frog**
- Cutaneous pectoris
- External oblique
- Rectus abdominis
- Pectorals

(c) **Pigeon**
- Furcula
- Interclavicle
- Coracobrachial
- Biceps brachii
- Triceps brachii
- Supracoracoid
- Pectoralis major
- Carina

(d) **Alligator**
- Sternohyoid
- Sternomastoid
- Deltoid
- Pectorals

(e) **Rabbit**
- Cleidomastoid
- Sternohyoid
- Sternomastoid
- Pectoralis primus
- Pectoralis major
- Basioclavicular
- Cleidobrachial
- Pectoralis minor
- Pectoralis quartus

FIGURE 11.20

Pectoral musculature of selected tetrapods (*red*). In (*b*), the cutaneous pectoris is an integumentary slip of the pectoral musculature that has acquired an insertion on the skin between the forelimbs. It is superficial to the other pectoral muscles. In (*c*), the pectoral muscle has been removed on the right to reveal the supracoracoid. In (*e*), the pectoralis primus and pectoralis major have been removed on the right to expose deeper muscles. A pectoscapularis lies unseen deep to the cephalic border of the pectoralis minor.

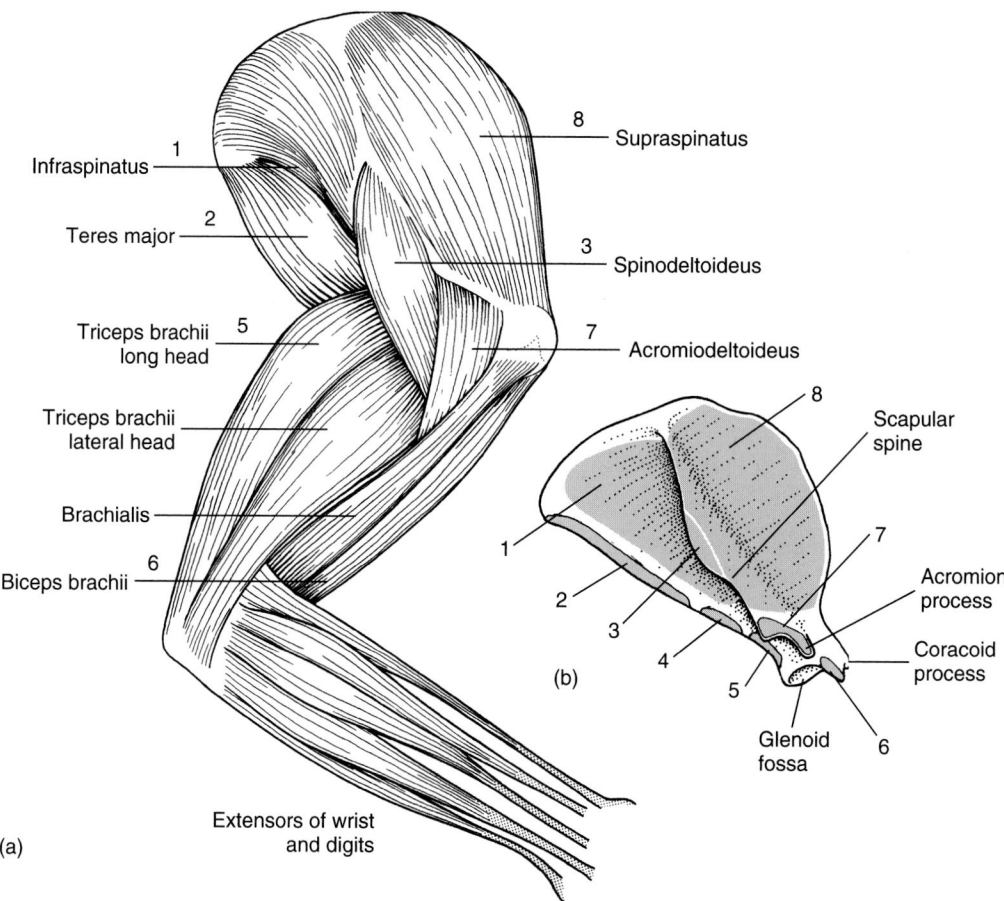

FIGURE 11.21
(a) Superficial muscles of the right scapula and proximal part of the forelimb of a cat, lateral view. (b) Sites of origin of correspondingly numbered muscles in (a); **4,** origin of teres minor, a deep muscle not visible in (a). Also not seen are the subscapularis on medial surface of scapula and the tiny coracobrachialis (origin, coracoid process of scapula).

Labels in figure (a):
Infraspinatus — 1
Teres major — 2
Triceps brachii long head — 5
Triceps brachii lateral head
Brachialis
Biceps brachii — 6
(a)
Extensors of wrist and digits
8 — Supraspinatus
3 — Spinodeltoideus
7 — Acromiodeltoideus

Labels in figure (b):
8 — Scapular spine
7 — Acromion process
— Coracoid process
Glenoid fossa
(b)

mammals have shown that a blastema for a supracoracoid muscle develops in mammals at the reptilian site, but it then spreads dorsad onto the embryonic scapula, pushing its way beneath the developing deltoideus and establishing itself on both sides of the scapular spine to become the supraspinatus and infraspinatus muscles of the adult mammal.

The **coracobrachialis** of mammals originates on the coracoid process of the scapula, the latter being the sole mammalian remnant of the coracoid bones of reptiles. The muscle is present in reptiles, passing from the coracoid bone to the shaft of the humerus. The muscle, like the bone from which it arises, has become very small in mammals.

A **biceps brachii** and a **brachialis** (fig. 11.21) are the major flexors of the forearm of reptiles and mammals. A small deep **anconeus** (not homologous with the anconeus of frogs) extends between the distal end of the humerus and the proximal end of the ulna, and a small transverse **epitrochleoanconeus** partially encircles the elbow joint medially. Distal to these, **pronators of the manus** insert on the radius and rotate this bone, and **flexors of the manus** with origins chiefly on the humerus insert by long tendons on the carpals, metacarpals, and phalanges. Intrinsic to the manus are very short flexors of the digits.

Present in some mammals is a **cleidobrachialis** that extends from the clavicle to the humerus or ulna. Because the clavicles of rabbits and cats are vestigial, the cleidobrachialis has the appearance of being a continuation of the **basioclavicularis** (**cleido-occipital**) of rabbits (fig. 11.20e) and of the cleidotrapezius of cats. However, the raphe in which the vestige is embedded is the boundary between the two. The continuity of some of their fibers is secondary to reduction of the intervening clavicle. The cleidobrachialis is innervated by spinal nerves, whereas the muscle above the raphe is supplied by branchiomeric nerves.

The musculature of birds is essentially reptilian. Much of the weight of a bird is represented by the extrinsic muscles of the forelimb and the intrinsic muscles of the hind limb. The intrinsic muscles of the wing have been reduced, the epaxial musculature of the trunk is vestigial, and the parietal muscles are thin.

Table 11.5 lists the chief intrinsic muscles of the pectoral girdle and forelimbs of mammals and their probable homologues in reptiles.

Muscles of the Pelvic Girdle and Hind Limbs

Pelvic girdles, especially in amniotes, are incapable of independent mobility. They are united with the vertebral column dorsally, and the two halves meet in a sym-

TABLE 11.5 Chief Intrinsic Muscles of the Pectoral Girdle and Forelimbs of Mammals and Their Probable Homologues in Reptiles

	Mammals	Reptiles
Muscles of Girdle		
Girdle to humerus, proximally	Deltoideus	Deltoideus clavicularis / Dorsalis scapulae
	Subscapularis	Subcoracoscapularis
	Teres minor	Scapulohumeralis anterior
	Supraspinatus / Infraspinatus	Supracoracoideus
	Coracobrachialis	Coracobrachialis
	Teres major	Slip of Latissimus dorsi
Muscles of Upper Arm		
Girdle of humerus to proximal end of radius or ulna	Triceps brachii	Triceps brachii
	Biceps brachii	Biceps brachii
	Brachialis	Brachialis
	Epitrochleoanconeus	Epitrocheleoanconeus
	Anconeus	Anconeus
Muscles of Forearm*		
Humerus and proximal end of radius and ulna to hand	Extensors and flexors of carpus and digits	
	Supinators and pronators of hand	
Muscles of Hand†	Extensors, flexors. abductors, adductors of digits	

*Frequently have long tendons of insertion.
†Reduced in species in which digits are reduced.

physis ventrally, except in birds. In birds, the pelvic girdle is united with the immobile synsacrum. Consequently, muscles arising from the axial skeleton and inserting on the girdle of amniotes—the extrinsic muscles of the appendages—play little or no role in locomotion.

A **caudofemoralis muscle** in urodeles and reptiles extends between some of the proximal caudal vertebrae and the femur, but it is not locomotor in urodeles. It exerts its pull on the tail. A part of it has become a locomotor muscle, the pyriformis, in mammals.

The locomotor muscles of the hind limbs are chiefly intrinsic muscles. They extend between the bones of the girdle—ilium, ischium, pubis (one or more of these)—and the femur, tibia, or fibula. In modern reptiles, their names usually indicate their attachments—**puboischiofemoralis, puboischiotibialis, iliofemoralis, iliotibialis, iliofibularis, femorotibialis,** and others. These muscles were derived from a common amniote ancestor, as were also the hind limb muscles of mammals. Consequently, many hind limb muscles of mammals and reptiles are homologues. Unfortunately, the muscles of humans were named first, and the names were often based on something other than origin and insertion. The major muscles of the hind limbs of mammals are the subject of the discussion that follows.

The **iliopsoas,** which in humans and some other mammals appears as separate **iliacus** and **psoas major** muscles, is the **puboischiofemoralis internus** of reptiles. The iliac portion arises on the ilium; the psoas major portion arises from an extensive area of fascia covering the psoas minor and from a series of lumbar vertebrae. The two portions unite and insert by a powerful tendon on the **lesser trochanter,** a protuberance on the femur near its head. The muscle protracts and rotates the femur.

A functional group of three hip muscles, **gluteus** (the reptilian iliofemoralis), **pyriformis** (the reptilian **caudofemoralis brevis**), and **gemelli,** has, collectively, a wide origin on sacral and caudal vertebrae and on the ilium and ischium, and inserts on a large prominence, the **greater trochanter,** near the head of the femur. The group, of which the gluteus is the most powerful, abducts the thigh and, like the iliopsoas, rotates the femur to turn the foot outward. The gluteal and pyriform are primary appendicular muscles that have increased their leverage by spreading from the pelvic girdle onto the axial skeleton. In quadrupeds (for example, the cat), the gluteal complex orginates both anterior and posterior to the axis of the femur. Anterior fibers would rotate the limb medially, whereas those in a posterior position would, like in humans, rotate the limb laterally. Although the functions of homologous muscles among taxa may be similar, a general statement of function should not be extended to all taxa without caution.

A **quadratus femoris,** consisting of four muscles (three **vasti** and a **rectus femoris**), arises on the ilium and the greater and lesser trochanters of the femur and inserts on the ligament in which the patella is embedded. The vasti extend the leg, and the rectus femoris adducts the thigh, rotating it to carry the foot inward.

Other extensors or adductors of the thigh include the **semi-membranosus, adductor femoris, adductor longus, pectineus, sartorius,** and **gracilis.** They arise on the ilium, ischium, or pubis, and insert on the shaft of the femur or on the patellar ligament and tibia. The sartorius, the longest muscle in the human body, is a muscular strap that arises on the ilium and passes diagonally down the medial aspect of the thigh. It is the muscle that may become irritated when an adult human who is not used to doing so sits too long on the floor with the legs crossed in the manner of old-time tailors who, while creating sartorial splendor, sewed by hand.

Two **obturator** muscles flex, rotate, and abduct the thigh. They arise from the lip of the obturator foramen (see fig. 10.11*k*) and from the ischium and pubis, and insert on the proximal end of the femur.

A **biceps femoris** and a **semitendinosus** are primarily flexors of the leg, although the former also abducts the thigh and the latter extends it. They arise on the ischium and insert on the patella or tibia.

The foregoing muscles commencing with the iliopsoas either insert on the femur or parallel it and insert just beyond it. Distal to these are **extensors and flexors of the foot and digits** similar to those of the manus. Among these are the **gastrocnemius,** an extensor of the foot made famous by Achilles. Its long, wide tendon inserts on the calcaneus bone of mammals. In birds (fig. 11.22), it inserts on the tarsometatarsus (which incorporates the homologue of the calcaneus) and also continues behind the heel to insert on the phalanges. The weight of the perching bird stretches the tendon, thereby retracting the claw-bearing toes so that a sleeping bird, muscularly relaxed and with minimal expenditure or energy, maintains its grip on the perch.

As mentioned, the pectoral girdles and forelimbs are joined to the trunk by extrinsic appendicular muscles, vessels, nerves, and skin alone, whereas the pelvic girdles are firmly ankylosed to the axial skeleton. This accounts in part for the fact that cats jump by pushing off with their hind limbs but land on their forelimbs, where much of the shock is absorbed by the extrinsic and intrinsic musculature. Among the reasons frogs cannot do this is lack of a suitable axial skeletomuscular system and of a cerebellum competent to direct precision landings. The result is that frogs land awkwardly—almost abashedly—on whatever part of the body touches down first. The hind limbs of ungulates are good shock absorbers because elongated metatarsals provide what amounts to an extra shock-absorbing joint at the end of the shank.

Branches of several successive spinal nerves innervate the appendicular muscles of fishes and tetrapods (see fig. 11.16). In fishes, these nerves come from the same body segments that contribute muscle buds to the appendage. It

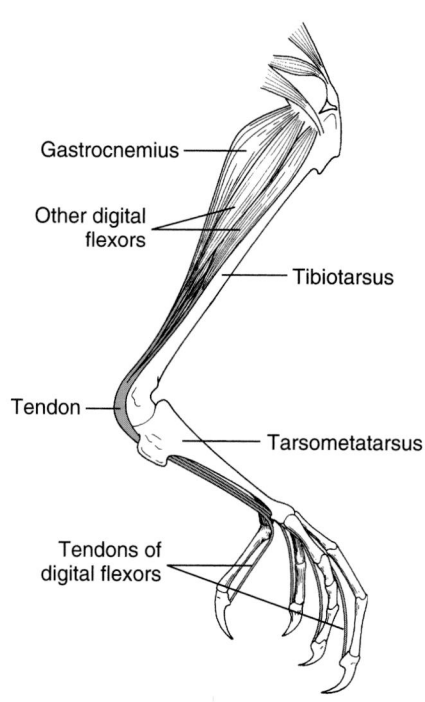

FIGURE 11.22
Perching mechanism of a bird. The weight of the body flexes the digits, causing them to grip the perch.

TABLE 11.6   Innervation of Somatic Muscles

Epaxials of trunk and tail Appendiculars of epaxial derivation	Dorsal rami of spinal nerves
Hypaxials of trunk and tail Appendiculars of hypaxial derivation Hypobranchials Mammalian diaphragm	Ventral rami of spinal nerves
Tongue	Cranial nerve 12

seems probable that the number of spinal nerves entering a tetrapod limb is indicative of the number of somites that contributed mesenchyme to the limb phylogenetically.

The pattern of innervation of somatic muscles is given in table 11.6.

SOMITOMERIC AND SOMITIC MUSCLES OF THE HEAD

Branchiomeric Muscles

In chapter 1, we discussed the basic architectural pattern of the craniate body. One of the discussions was an

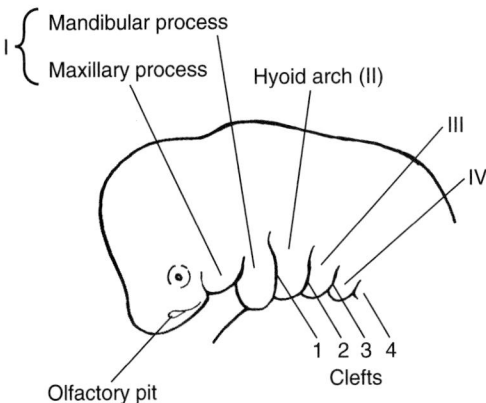

FIGURE 11.23
Embryonic pharyngeal arches of a gnathostome.

introduction to the pharyngeal arches (fig. 11.23). You are urged to read this discussion again because it will have more meaning for you now at this stage of your study.

In chapter 9, we discussed how the pharyngeal skeleton was altered in tetrapods as an adaptation to life on land. Now we will examine the changes that took place in the muscles that operate this tetrapod skeleton.

Operating the skeleton derived from the embryonic pharyngeal arches, whether in fishes or tetrapods, is a series of striated, skeletal, muscles that have been termed *branchiomeric*. In no living craniate is the basic pattern of these muscles, and their motor innervation, better exemplified than in a shark, in which they operate the jaws and gill arches. In tetrapods, they still operate the jaws. However, with loss of gills, the muscles of the former gill arches have acquired new functions for life on land.

Despite the fact that branchiomeric muscles are striated and skeletal, in fishes they perform two primitive visceral functions inherited from their filter-feeding ancestors: food handling within the oropharyngeal cavity, and respiration. They did not in the past, nor do they today, orient the body (soma) in the external environment, which is the role of somatic muscles. Thus, they are often considered separately or as a special case of somatic musculature. The mesenchyme that provides the stem cells for these muscles comes from somitomeres and anterior somites of the head (see fig. 11.6). Like typical somatic muscle, they are myotomal in origin but differ in that their motor neurons lie within a distinct motor column in the central nervous system (nucleus ambiguous in contrast to the somatic motor column of the spinal cord and its extension for the nuclei of cranial nerves XII, VI, IV, and III; see fig. 16.31, BSM and SM, respectively).

Muscles of the Mandibular Arch

In all gnathostomes, the muscles of the first arch are chiefly jaw muscles. In *Squalus*, a **levator palatoquadrati** arises on the otic capsule and inserts on the quadrate end of the upper jaw cartilage (fig. 11.24*a*), a powerful **adductor mandibulae** arises on the quadrate process and inserts on Meckel's cartilage, and an **intermandibularis** extends between Meckel's cartilage and a strong midventral raphe in the pharyngeal floor. A slender **spiracularis** (craniomaxillaris) inserting on the upper jaw completes the first arch muscles of *Squalus*. The levator palatoquadrati raises the upper jaw, which is possible because *Squalus* has hyostylic jaw suspension. The spiracularis assists in this. The adductor mandibulae raises the lower jaw, thereby closing the mouth during the phase of the respiratory cycle when the spiracle is closed and water is being forced over the gills by constriction of the walls of the orobranchial chamber. The adductor mandibulae also enables sharks to hold in a viselike grip any prey unlucky enough to be caught. The intermandibularis is a ventral constrictor that elevates the anterior pharyngeal floor during respiration and feeding. (The lower jaw is lowered indirectly by the coracohyoids, which are hypobranchial muscles.)

The adductor mandibulae is the most powerful muscle of the first arch in all gnathostomes. In tetrapods, it has become split into several muscles—many in snakes, three in mammals. The three in mammals are the **masseter** and **temporalis** (fig. 11.24*c*) and the **pterygoideus**. (The latter lies deep to the other muscles and is not illustrated.) These muscles have spread in three directions to acquire origins on the zygomatic arch (masseter), the temporal bone (temporalis), and the pterygoid fossa (see fig. 9.29). The three muscles on each side constitute a muscular sling for the lower jaw and provide most of the multidirectional tensile forces that produce the side-to-side, up-and-down, forward and back, and rotary chewing movements of such different mammals as herbivores, carnivores, and rodents. The role of temporal fossae in the evolution of the amniote masseter, temporalis, and pterygoideus muscles should be reviewed at this time (chapter 9).

The intermandibular muscle of fishes is the homologue of the **mylohyoideus** of tetrapods (fig. 11.24*b*). One slip probably gave rise to the **digastricus** muscle of tetrapods (anterior belly, when there are two bellies, fig. 11.24*c*). Also, a slip of the first arch muscle that was attached to the articular bone of therapsids remained attached when the bone became the malleus; that muscle, now the **tensor tympani**, tenses the mammalian eardrum. The mandibular arch muscles of all gnathostomes are innervated by motor branches of the fifth cranial nerve.

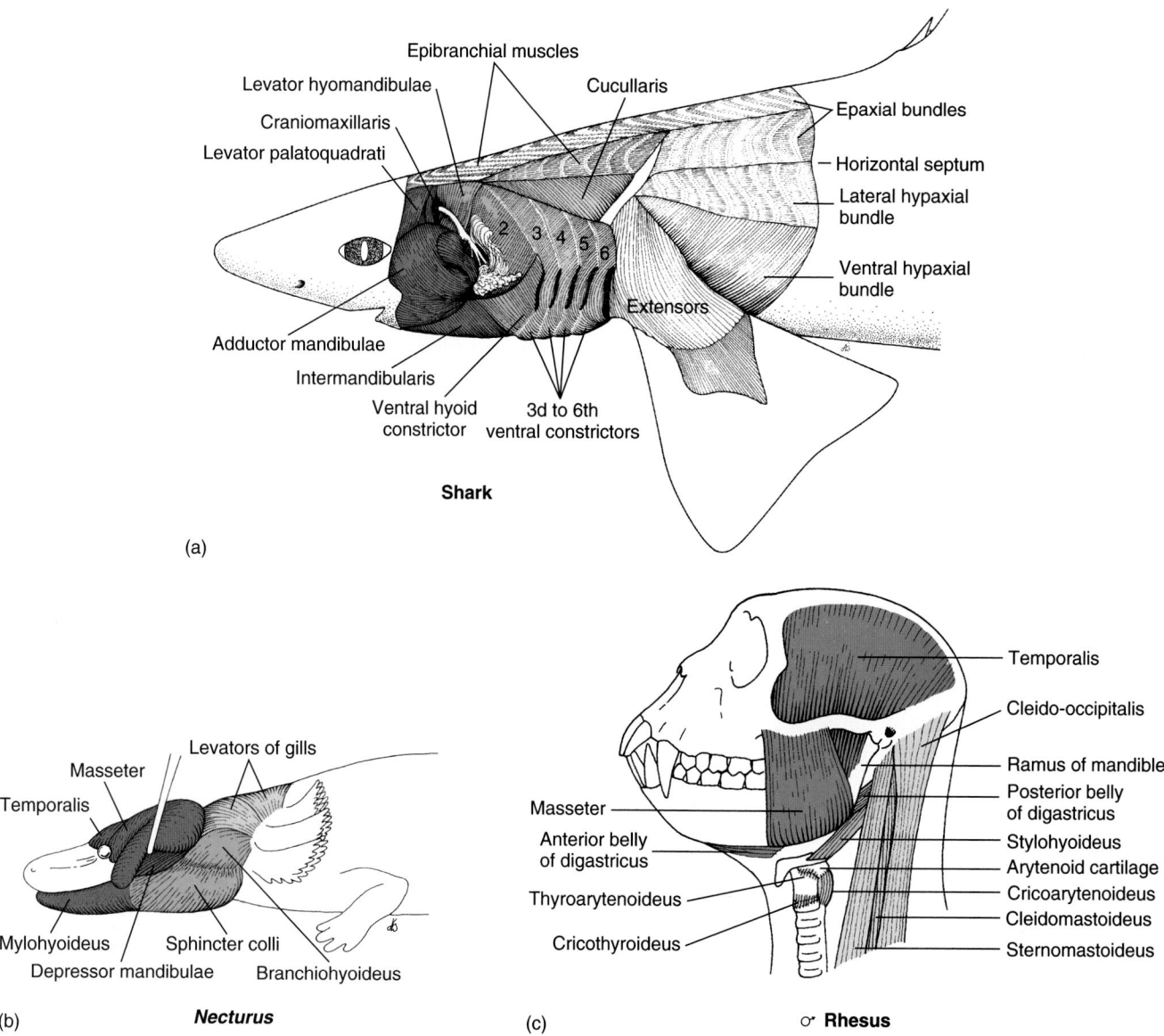

FIGURE 11.24
Selected branchiomeric muscles of three gnathostomes. Muscles of the first and second pharyngeal arches are *dark* and *medium red*, respectively. Muscles of the remaining arches are *light red*. In (*a*), **2** to **6** are the dorsal constrictors of the second and successive arches. (*c*) The temporalis inserts on the coronoid process of the ramus of the mandible. The pterygoideus muscle, not illustrated, can be seen only after the mandibular symphysis has been separated and the dentary bone retracted laterad.

Muscles of the Hyoid Arch

The principal muscles of the hyoid arch of *Squalus* elevate the arch or constrict the pharyngeal cavity, partly as a function of respiration and feeding. A **levator hyomandibulae** and a **dorsal constrictor** arise on the neurocranium and insert on the hyomandibula and ceratohyal cartilage. The ventral constrictor of the hyoid arch is subdivided into a ventral hyoid constrictor (in a position similar to the other branchial ventral constrictors) and an interhyoideus found deep to the intermandibu-

laris (see figs. 11.14 and 11.24*a*). In bony fishes, the dorsal constrictor is subdivided, and part of it operates the operculum. Because of the joint between the upper jaw, lower jaw, and hyomandibula, contraction of hyoid arch muscles results in movements of the jaws.

In *tetrapods*, hyoid arch muscles continue to serve some of the functions seen in fishes, but they have also acquired some quite different roles. In *Necturus*, a **branchiohyoideus muscle** of the hyoid arch arises on the ceratohyal cartilage and inserts on the epibranchial cartilage of the first gill (fig. 11.24b). Along with levators, it waves the gills back and forth in the water for respiration. A **depressor mandibulae** opens the mouth of urodeles and many reptiles, and in some mammals a **posterior belly of the digastricus** participates in chewing movements (fig. 11.24c). A slender **stylohyoideus** (anterior belly when there are two) connects the styloid or jugular process of the skull with the anterior horn or body of the hyoid in mammals (see fig. 9.44).

A thin collarlike **sphincter colli**, overlying the origin of the branchiohyoideus, adheres to the skin of the neck of lower tetrapods (fig. 11.24b). In reptiles, it spreads upward around the rear of the skull to insert on the skin of the head and is called the **platysma**. In mammals, the platysma spreads forward onto the face to become the muscles of facial expression, or **facial (mimetic) muscles** (fig. 11.25c–f). The sphincter colli is thought to be derived from a portion of the interhyoideus of fishes.

A small **stapedius** muscle originates on the posterior wall of the middle ear cavity of mammals and inserts on the stapes, a homologue of the hyomandibular cartilages. It contracts reflexly to impede extra loud airborne sounds that might injure the delicate hair cells of the cochlea.

All these hyoid arch muscles receive their motor innervation via the facial (seventh cranial) nerve, so named because of its distribution to the facial muscles. The innervation of the two muscles of the middle ear by two different branchiomeric nerves, V and VII, is not surprising to those who know the phylogenetic history of the malleus and stapes.

Muscles of the Third and Successive Pharyngeal Arches

The muscles of the arches of fishes caudal to the hyoid are **constrictors** (dorsal and ventral), **levators**, **adductors**, and **dorsal and lateral interarcuals**, which compress or expand the pharyngeal cavity and gill pouches during respiration. Constrictors in sharks lie just under the skin, covered by tough subcutaneous fascia that is not readily removed, and they attach to strong fascia above and below the gill pouches (see fig. 11.24a). They compress the gill pouches, expelling respiratory water. The levators of these arches make up a strong muscle sheet, the **cucullaris**, which raises the pharyngeal wall assisted by the levator hyomandibulae of the second arch. Adductors deep in the gill arches connect epibranchial and ceratobranchial cartilages and cause the lateral pharyngeal walls to bow outward when the muscles contract,

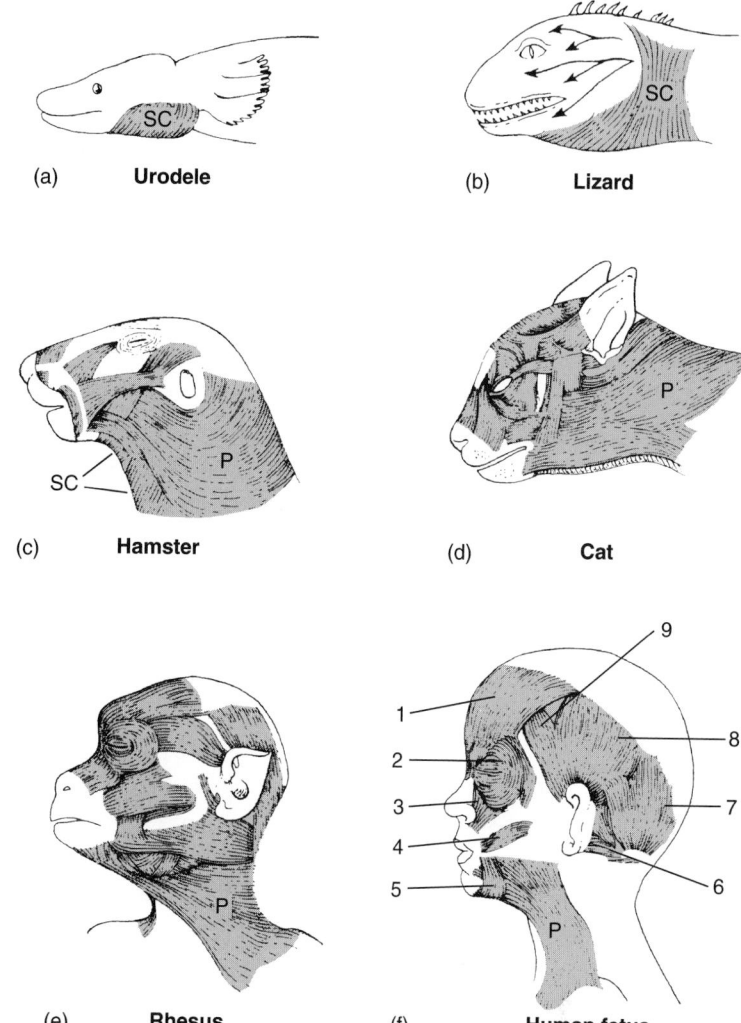

FIGURE 11.25

Evolution of mammalian facial muscles from hyoid arch muscles of tetrapods. The sphincter colli, **SC**, spreads onto the neck in amniotes and becomes the platysma, **P**, in mammals. It then spreads forward onto the head and face, as indicated by *arrows* in (b), to become muscles of facial expression. Note increasing differentiation of facials in the mammals shown.
1, frontalis; **2**, orbicularis oculi; **3**, quadratus labii superioris; **4**, risorius; **5**, triangularis; **6**, posterior auricular; **7**, occipital; **8**, superior auricular; **9**, anterior auricular. In (e) and (f), the white expanse on top of the head is the galea aponeurotica.

thereby expanding the pharyngeal chamber. Two sets of interarcual muscles in the roof of the pharynx just above the mucosa connect successive pharyngobranchial cartilages and individual pharyngo- and epibranchials, drawing these together. Their action assists in further expanding the pharynx. *The coracobranchials in the floor of the pharynx are hypobranchial, not branchiomeric muscles.* In bony fishes, the branchiomeric muscles caudal to the hyoid arch are much reduced as a consequence of the role of the operculum in moving respiratory water across the gills.

TABLE 11.7   Chief Branchiomeric Muscles and Their Innervation in *Squalus* and in Tetrapods

Pharyngeal Arch	Pharyngeal Skeleton in *Squalus*	Chief Branchiomeric Muscles		Cranial Nerve Innervation
		Squalus	Tetrapods*	

I Mandibular arch	Meckel's cartilage	Intermandibularis
		
		Adductor mandibulae
			Intermandibularis
			    Mylohyoideus (anterior part)
			    Digastricus (anterior part)
			Adductor mandibulae
			    Masseter
			    Temporalis
			    Pterygoidei
			    Tensor tympani
				V

	Pterygoquadrate cartilage	Levator palatoquadrati Spiracularis		

II Hyoid arch	Hyomandibula
	Ceratohyal
	Basihyal
		Levator hyomandibulae
		Dorsal constrictor
		Interhyoideus
			Stapedius
			Stylohyoideus (anterior part)
			Depressor mandibulae
			Digastricus (posterior belly)
			Sphincter colli
			    Platysma
			    Mimetics
				VII

III	Gill cartilages	Constrictors Levators Adductors Interarcuals	Stylopharyngeus Stylohyoideus (posterior part)	IX
IV to VI	Gill cartilages	Constrictors Levators Adductors Interarcuals	Thyroarytenoideus Cricoarytenoideus Cricothyroideus	X
		Cucullaris (derived also from dorsal constrictor 3)	Trapezius Sternomastoideus[†] Cleidomastoideus[†] Basioclavicularis[†]	Occipitospinal nerves in shark; spinal roots of XI in amniotes

*Indented muscles in this column may be derivatives of the preceding muscle.
[†]Presumed to be branchiomeric on the basis of their motor innervation.

In tetrapods, branchiomeric muscle has pretty much disappeared from what had been the gill-bearing arches. Remaining from arch III are a **stylopharyngeus** muscle that is used in swallowing and, in some mammals, a **posterior belly of the stylohyoideus** (see fig. 9.44). Remaining from arch IV are the intrinsic muscles of the mammalian larynx: **cricothyroideus, cricoarytenoideus,** and **thyroarytenoideus** (fig. 11.24*c*).

The trapezius muscles of amniotes (see fig. 11.19*b* and *c*), discussed earlier with other extrinsic muscles of the pectoral girdle, are derivatives of the cucullaris muscle of fishes. Whether or not muscles with that name in urodeles are homologues is unknown. The sternocleidomastoid complex (sternomastoideus, cleidomastoideus, and, in lagomorphs and a few other mammals, a basioclavicularis, also known as the cleido-occipitalis) has been thought to be of branchiomeric origin on the basis of its innervation by the spinal accessory nerve. How-

ever, the history of the components of this nerve, discussed in chapter 16, is conjectural. The chief branchiomeric muscles and their innervation are given in table 11.7.

Extrinsic Eyeball Muscles

The extrinsic muscles that move the eyeball are striated, skeletal, voluntary muscles that arise on the wall of the orbit and insert on the fibrous sclerotic coat of the eyeball. In elasmobranchs, their embryonic origin is from preotic somitomeres in the embryonic head (see fig. 11.6). The most cephalic two somitomeres giving rise to eyeball muscles in elasmobranchs are at the level where the third cranial nerve emerges from the midbrain. This nerve innervates four eyeball muscles, the **superior, medial,** and **inferior rectus,** and the **inferior oblique** (table 11.8 and fig. 11.26). The next more caudal somitomere is at the level where the fourth cranial nerve emerges. This nerve innervates the **superior oblique** eyeball muscle. The last preotic somitomere is at the level where the sixth cranial nerve emerges. This nerve innervates the **lateral rectus** eyeball muscle.

The eyeball muscles of other craniates are innervated by the same nerves as in elasmobranchs, and the cell bodies of their motor fibers are in the same motor column as the cell bodies that innervate hypobranchial muscles, tongue muscles, and the axial muscles of the trunk and tail. In view of these observations, it appears that the rectus and oblique muscles of the eyeball of all craniates are derivatives of an ancient segmental axial musculature that, in the head, has been conscripted to operate lateral visual organs. In other words, the eyeball muscles of craniates other than elasmobranchs are myotomal by phylogenetic derivation. As might be expected, craniates with vestigial eyeballs have vestigial eyeball muscles.

In many amniotes, muscles insert on the upper lids and nictitating membrane (**pyramidalis** of reptiles,

quadratus of birds, **levator palpebrae superioris** of reptiles and mammals). The specific source of their myoblasts has not been ascertained, but their innervation by somatic motor fibers with cell bodies in the somatic efferent column indicates that they are somatic muscles (table 11.8). **Protractors and retractors of the eyeballs** of reptiles, and **depressors of the lower lids**, when present, are evidently not of similar origin to the eye muscles because they are innervated by the fifth cranial nerve.

INTEGUMENTARY MUSCLES

In fishes and amphibians, slips of branchiomeric or axial somatic muscles insert on the dermis at one site or another, attaching the skin to the underlying muscle at these locations. **Costocutaneous muscles** of snakes are hypaxial integumentary muscles used in locomotion (chapter 10). But it is only in mammals that integumentary muscles are well differentiated. A **panniculus carnosus (cutaneous maximus)** wraps around the entire trunk of some mammals, enabling armadillos to roll into a ball when endangered, forms a sphincter around the entrance to the abdominal pouch of marsupials, and vigorously shakes flies off horses (fig. 11.27). It probably started as superficial slips of the pectoral musculature that became attached to the undersurface of the skin and then spread, as did the longus colli. The panniculus is poorly developed in monkeys and is absent in humans. A sheet of pectoral musculature, the **cutaneous pectoris**, maintains its original attachment to the chest wall in anurans (see fig. 11.20b); in bats, slips of pectoral muscles insert on the skin of the wing membranes as **patagial muscles**. All these are myotomal muscles innervated by spinal nerves.

The most notable integumentary muscles are the facials mentioned earlier (see fig. 11.25). More than 30 different muscles in humans depress the corners of the mouth in grief, raise them for smiling, wrinkle the forehead, raise the eyebrows quizzically, close the lids tightly, pucker the lips, draw milk when suckling, or dilate the nostrils. Other mammals have fewer of these facial muscles. One of them, the **caninus**, elevates the part of the upper lip that hides the spearlike canine tooth used by carnivores for ripping flesh. When a human uses the muscle, he or she is said to be sneering. **Auricular muscles** in nonhuman mammals direct the pinnas of the ears toward faint sounds. Facial muscles are innervated by cranial nerve VII, a branchiomeric nerve.

TABLE 11.8 Chief Extrinsic Eyeball and Eyelid Muscles and Their Innervation

Cranial Nerve Supply	Extrinsic Eyeball Muscles	Eyelid Muscles
III (oculomotor)	Superior rectus Inferior rectus* Medial (internal) rectus Inferior oblique	Levator palpebrae superioris
IV (trochlear)	Superior oblique	
VI (abducens)	External (lateral) rectus Retractor bulbi	Pyramidalis of eye Quadratus of eye

*In lampreys, the inferior rectus is innervated by cranial nerve VI.

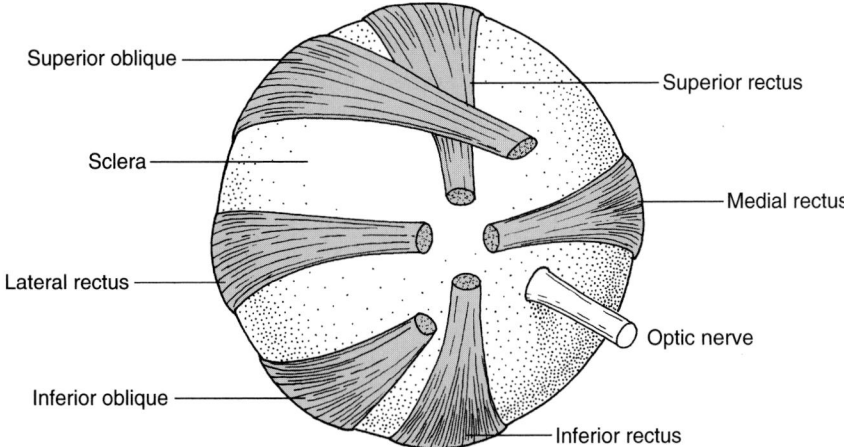

FIGURE 11.26
Extrinsic muscles of the left eyeball of a generalized vertebrate viewed from behind the eye. The rectus muscles of all vertebrates have origins on a single location on the orbital wall near the optic foramen or optic pedicle. The origin of the obliques are typically more anterior.

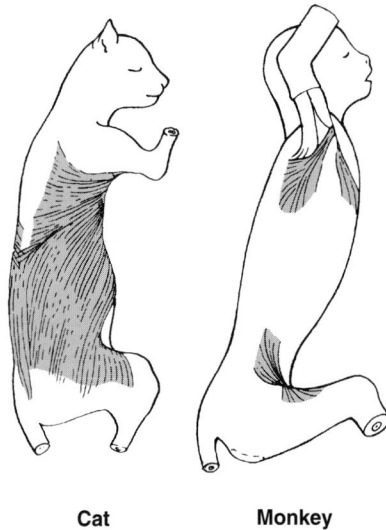

Cat   **Monkey**

FIGURE 11.27
Panniculus carnosus (cutaneous maximus) of a cat and primate. Note difference in the extent of the muscle in the two animals.

All the preceding are *extrinsic* integumentary muscles: Their developmental and anatomical origins are away from the dermis. Their insertions are along the undersurface of the dermis. *Intrinsic* integumentary muscles develop entirely within the skin, in the dermis. They are the **arrectores plumarum** and **arrectores pilorum** that insert on feather or hair follicles and ruffle the feathers or elevate the fur. They are smooth muscles in nearly all species and are innervated by the sympathetic nervous system.

ELECTRIC ORGANS

In several hundred species from at least seven families of cartilaginous and bony fishes, certain muscle masses are modified to produce, store, and discharge electricity. In *Torpedo,* the electric ray, an electric organ lies in each pectoral fin near the gills. It is of branchiomeric origin, being supplied by motor fibers of cranial nerves VII and IX. In *Raja,* a skate, and *Electrophorus,* a South American electric eel, electric organs lie in the tail and are modified hypaxial muscles (fig. 11.28). The potential produced by these organs in eels amounts to 600 volts and is used to paralyze or kill prey or to repel other organisms. Some other fishes have electric organs with a lower electric potential that serve as radarlike mechanisms or for communication. Certain neuromast organs (see chapter 17) are receptors for these signals.

Electric organs consist of a large number of electric discs, up to 200,000 in the tail of one ray, piled in either vertical or horizontal columns. Each disc, or **electroplax,** is a modified multinucleate muscle fiber embedded in a vascular jellylike extracellular matrix sur-

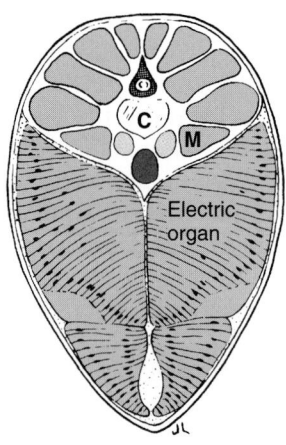

FIGURE 11.28
Electric organs in tail of electric eel. Each nucleated horizontal disc (electroplax) is a single modified hypaxial muscle fiber. **C,** centrum; **M,** epaxial myomere.

rounded by connective tissue. Nerve endings terminating on each disc induce the discharge.

Electric organs seem to have no systematic distribution among fishes and probably evolved independently many times. The electric organ of *Malapterurus,* a freshwater catfish from the Nile, envelops nearly the entire body—head, trunk, and tail—in a jellylike mass located subcutaneously rather than among muscles. It appears to be composed of modified skin glands rather than muscle tissue.

Summary

1. Histologically, muscle tissues are striated, smooth, or cardiac.
2. Functionally, muscles can be classified as somatic, visceral, or branchiomeric.
3. Somatic muscles orient the organism in the environment. Visceral muscles regulate the internal milieu. Branchiomeric muscles operate the pharyngeal arches and their derivatives.
4. Somatic muscles are striated and skeletal. They are myotomal by origin or derivation and they are innervated by spinal nerves.
5. Visceral muscles are mostly smooth or cardiac. Generally, they arise from the splanchnopleure and are innervated by the autonomic nervous system.
6. Branchiomeric muscles are striated and skeletal muscles of the pharyngeal arches. They are also myotomal and are formed from the somitomeres and anterior somites of the head. They receive their motor innervation from branchiomeric nerves.
7. Muscles may flex, extend, abduct, adduct, protract, retract, elevate, depress, rotate (pronate, supinate), constrict, dilate, or otherwise alter the spatial relationships or shape of a body part.

8. Skeletal muscles are named according to the direction of their fibers, location, number of parts, shape, attachments, action, size, a combination of these, and miscellaneous other considerations.

9. The trunk and tail muscles of jawed fishes are locomotor muscles arranged as myomeres divided into epaxial and hypaxial moities by a horizontal skeletogenous septum. Loss of myosepta obscures the primitive metamerism in most tetrapods.

10. The metamerism of the axial muscles of fishes and urodeles is an expression of the metamerism of embryonic somites.

11. Epaxial and hypaxial myomeres are innervated by dorsal and ventral rami, respectively, of spinal nerves.

12. The epaxial muscles as a derived feature in some tetrapods are disposed as longitudinal bundles, some of which extend over many body segments. Hypaxial muscles of the lateral body wall have lost their metamerism except where there are long ribs, and they are disposed in oblique and transverse strata.

13. Hypaxial muscles immediately ventral to transverse processes in amniotes are disposed in long (subvertebral) bundles. Those paralleling the midventral line form long straplike muscles.

14. In all craniates, hypaxial muscles extend forward beneath the pharynx as hypobranchial muscles. In amniotes, they continue forward into the tongue.

15. The mammalian diaphragm and cremaster muscles are hypaxial muscles that have migrated accompanied by their spinal nerves.

16. The muscles of the craniate tail are continuations of the epaxial and hypaxial vertebral musculature of the trunk.

17. Eyeball muscles arise from preotic somitomeres in elasmobranchs and other craniates. They are innervated by cranial nerves III, IV, and VI and are myotomal by phylogenetic and ontogenetic derivation.

18. Extrinsic appendicular muscles have an anatomic origin on the axial skeleton and insert on the girdle or limb. Intrinsic appendicular muscles arise on a girdle or proximal skeletal segment of the limb and insert on more distal segments.

19. The appendicular muscles of fishes are chiefly extrinsic. They differentiate from blastemas within the embryonic trunk and spread into the developing fin fold where they become disposed as ventral (preaxial) flexors and dorsal (postaxial) extensors. Reorientation of the limb in tetrapods generally has obscured these primitive relationships.

20. Appendicular muscles of tetrapods are extrinsic or intrinsic. Blastemas giving rise to extrinsic muscles develop in two different sites: (1) within the embryonic body wall, then spreading onto the girdles and into the developing limb buds, as in fishes (secondary appendicular muscles); and (2) within the limb bud, then spreading trunkward to establish an attachment on the axial skeleton (primary appendicular muscles).

21. Intrinsic appendicular muscles of tetrapods arise from blastemas that develop within the embryonic limb.

22. The muscles of the first pharyngeal arch operate the jaws. In mammals, a derivative inserts on the malleus. They are innervated by the fifth cranial nerve.

23. The muscles of the second arch attach to the hyoid skeleton, lower jaw, and, in fishes, operculum. In mammals, a derivative inserts on the stapes. A sphincter colli of lower tetrapods spreads onto the head of amniotes to become the platysma and facial muscles. The muscles are innervated by the seventh cranial nerve.

24. Muscles of the third and remaining arches operate gills in fishes and perform new functions in tetrapods. Muscles of the third arch are innervated by the ninth cranial nerve, those of the remaining arches by the vagus (tenth cranial nerve).

25. The cucullaris, a branchiomeric muscle of fishes, has given rise to the trapezius and sternocleidomastoid complex of amniotes.

26. Extrinsic integumentary muscles reach peak development as the panniculus carnosus (cutaneous maximus) and the facial muscles of primates. Erectors of feathers and hairs are smooth muscles that are intrinsic to the dermis of birds and mammals.

27. Electric organs are modified axial, appendicular, or branchiomeric muscles, the fibers (electroplaxes) of which produce, store, and discharge electric potential. A few appear to be modified skin glands.

Critical Thinking Questions

1. What are the categories of muscle? Compare these categories.
2. What is the embryonic source of somatic and visceral muscle (be specific)?
3. Why are branchiomeric somatic muscles and somatic muscles grouped separately? Given your knowledge of these muscles, would you classify them separately? Why or why not?
4. Compare the functional types of striated muscle.
5. Epaxial and some hypaxial muscles are situated above and below the axis of the vertebral column, respectively. Describe the consequence of the bilateral contraction of each group separately. What would happen if the right epaxial and hypaxial muscles contracted simultaneously?

6. Can you name an animal that has specialized in the alternate contraction of the epaxial and hypaxial muscles in locomotion?

7. What is the embryonic source for the tongue muscles?

8. What are the six extrinsic eyeball muscles? In what type of organisms might you predict a reduction in these numbers?

9. How do you account for the reduction in hagfish? You should be able to provide two hypotheses.

10. What is the embryonic source for primary and secondary appendicular muscles?

11. In the forelimb and hind limb, name an extensor and flexor muscle for each joint (shoulder–hip, elbow–knee, and wrist–ankle).

12. What are the two embryonic sources of branchiomeric muscles?

13. What derivatives of the hyoid muscles do we find in the human head and neck? (Smile, it is not that hard of a question.)

Selected Readings

Alexander, R. M.: Functional design in fishes, ed. 2. London, 1970, Hutchinson University Library.

Allen, E. R.: Development of vertebrate skeletal muscle, *American Zoologist* **18**:101, 1978.

Armstrong, R. B., and Phelps, R. O.: Muscle fiber type composition of the rat hind limb, *American Journal of Anatomy* **171**:259, 1984.

Bennett, M. V. L.: Electric organs. In Hoar, W. S., and Randall, D. J., editors: Fish physiology, vol. 5. New York, 1971, Academic Press.

Carlson, B. M.: Human embryology and developmental biology. St. Louis, 1994, Mosby.

Cundall, D.: Activity of head muscles during feeding by snakes: A comparative study, *American Zoologist* **23**:383, 1983.

Gans, C: Biomechanics: An approach to vertebrate biology. Philadelphia, 1974, J. B. Lippincott Co. (Copyright now held by University of Michigan Press.)

Gans, C., and DeGueldre, G.: Striated muscle: Physiology and functional morphology. In Feder, M. E., and Burggren, W. W., editors: Environmental physiology of the amphibians, chapter 11. Chicago, 1992, Chicago University Press.

George, J. C., and Berger, A. J.: Avian myology. New York, 1966, Academic Press.

Jayne, B. C.: Muscular mechanisms of snake locomotion: An electromyographic study of lateral undulation of the Florida banded water snake (*Nerodia fasciata*) and the yellow rat snake (*Elaphe obsoleta*), *Journal of Morphology* **197**:159, 1988.

Lauder, G. V.: On the relationship of the myotome to the axial skeleton in vertebrate evolution, *Paleobiology* **6**:51, 1980.

Morgan, D. L., and Proske, U.: Vertebrate slow muscle: Its structure, pattern of innervation, and mechanical properties, *Physiological Reviews* **64**:103, 1984.

Sacks, R. D., and Roy, R. R.: Architecture of the hind limb muscles of cats: Functional significance, *Journal of Morphology* **173**:185, 1982.

Wardle, C. S., and Videler, J. J.: Fish swimming. In Elder, H. Y., and Truman, E. R., editors: Aspects of animal movement. Cambridge, England, 1980, Cambridge University Press.

Links to the Internet

Visit the zoology website at http://www.mhhe.com/zoology to find live Internet links for each of the following references.

1. Master Muscle List Home Page. Muscles of the human body from the Loyola University Medical Education Network (LUMEN).

2. The Muscular System. A site with nice diagrams of the musculature of the entire body, and many links to other sites with information on the muscular system.

3. Complete Muscle Tables of the Human Body. This site contains comprehensive information on human muscles. More than you'd ever want to memorize!

4. Somite Lineages. Development of the musculature of a vertebrate.

5. Histology—The Web Laboratory. Links to units on many systems of the body, including muscle tissue. Extensive coverage of microstructure of all three types of muscles, from Ohio State University.

6. Jay Doc Histo Web. The University of Kansas Histology site. You can click on cartilage and bone to view photomicrographs and electron micrographs of histological sections. Expanded views show much detail.

7. Muscle. Clickable index of histological sections of muscle tissues, and accompanying informative text. The text includes an innovative clickable quiz for interactive learning.

8. Muscular System. Great photos of cat musculature.

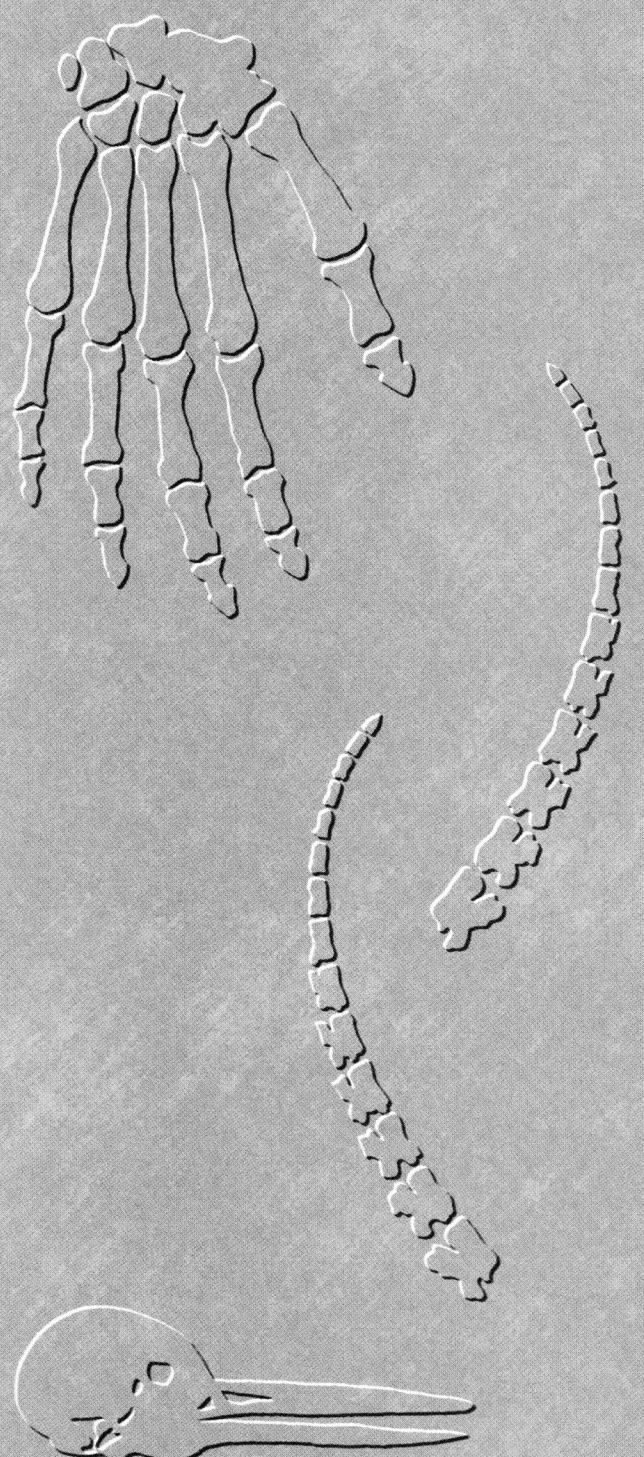

12

Digestive System

In this chapter, we will examine the alimentary canal, the glands and other evaginations that arise from it, and some spectacular modifications associated with varying food habits, such as tongues that are stored under the scalp, tooth plates that crush molluscs, gizzards that macerate, stomachs that return cuds to the oral cavity, and ceca that are the home of cellulose-digesting bacteria. We will start by recalling some of the means employed by craniates for obtaining energy in the form of food from their environment.

PROCURING FOOD

Ancestral chordates and larval lampreys are filter feeders. However, the origin of paired external sense organs and their concentration on the head fits better with a hypothesis of active predation or active acquisition of particulate food in the adults of early fishes. The primitive process of filter feeding, which can be employed only by aquatic organisms, consists of passively filtering organic matter out of the incoming respiratory stream and propelling the food particles to the rear of the pharynx for swallowing. Filter feeding in sea squirts, amphioxus, and larval living agnathans has been described in chapter 3. A more active version of filter feeding (independently derived) is employed by some fishes, such as spoonbills, tiny clupeids, and huge basking sharks. In these, plankton or small fishes are strained out of the respiratory water stream by long filamentous gill rakers that hang into the pharyngeal chamber from the gill arches. The largest mammals, baleen whales, strain tons of small fish, jellyfish, and other invertebrates from the sea each day through sieves of whalebone, or baleen, that hang into their oral cavity. However, whales take water into the oral cavity solely for the food it contains because they do not breathe with gills, and the water spills out of the mouth and back into the sea.

With the advent of fishes with jaws and the elaboration of muscular body walls that can be used for locomotion and pursuit, more aggressive methods of obtaining food became possible. Jaws, at first invested with denticulated bony dermal armor, ultimately were furnished with small, often sharp, single denticles that fit our concept of teeth. Many organisms thereupon became specialized predators, and their method of feeding, as in modern sharks, was a bite-tear-swallow technique that required no tongue or other specializations of the oral cavity.

A less energy-consuming procedure evolved as a result of further adaptations of the skull and hyoid arches. The modifications enabled some teleosts to approach close to organisms that are small enough to be swallowed, extend protrusible jaws, create suction, close the mouth, retract the jaws, and swallow. Anyone who has watched a goldfish feeding on flakes of fish food has observed this technique. Lampreys, being parasitic, have a different feeding technique. They rasp the tissues of the host with their spiny "tongue," a muscularly propelled fleshy and cartilaginous rod armed with horny teeth, and suck the debris into their pharynx.

On land, long sticky tongues are found among amphibians, squamate reptiles, and many birds and mammals. Some snakes impale prey on their upper jaw teeth. Winged tetrapods pick up grubs, seeds, and grains and perform other food-getting acts with appropriately shaped beaks (see fig. 2.4) and shorebirds pierce fish with them, or scoop fish from the sea, pelican fashion. Sanguinivorous bats suck whole blood and use anticoagulants in their saliva to prevent clotting, and baby mammals suckle milk, using muscular cheeks and lips.

Herbivorous ungulates crop grasses, and carnivorous mammals use a snap-bite-tear technique that often includes the piercing effect of a saberlike tooth. Because of appropriate joints in the wrist or digits, primates can grasp food and convey it to the mouth, and rodents can hold food between their hands and nibble. Other methods of food taking will come to mind.

Food taking depends on food finding. This is accomplished by sense organs that monitor the external environment. Chemical receptors such as olfactory organs, mechanical receptors such as inner ears and lateral-line organs of fishes and amphibians, thermal receptors such as the loreal pits of some snakes, capsulated touch receptors such as those on the sensitive snout of pigs, visual receptors, and electroreceptors—one or more of these alert one craniate or another to the presence and location of food. Once the energy-containing food is within the body, the digestive tract can process it, extract needed nutrients, and return unassimilated matter to the environment.

THE DIGESTIVE TRACT: AN OVERVIEW

The digestive tract, or alimentary canal, is a tube, seldom straight and often tortuously coiled, that commences at the mouth and empties either into a vented cloaca or directly to the exterior via an anus (fig. 12.1). It functions in the digestion and absorption of foodstuffs and in elimination of undigested wastes. Food is propelled from the pharynx to the vent or anus by smooth muscle tissue in the walls of the tract. The process is termed **peristalsis.**

Major subdivisions of the tract are an oral cavity and pharynx (oropharyngeal cavity in fishes), esophagus, stomach, and intestine, the latter being divided into small and large intestines in tetrapods. Emptying into the tract are ducts from accessory organs, principally pancreas, liver, and gallbladder. These organs and their ducts arise as evaginations from the embryonic digestive tract. Other evaginations known as **ceca,** having one function or another, are common. The digestive tract and accessory organs constitute the **digestive system.**

The digestive organs of fishes, amphibians, and most reptiles occupy the **pleuroperitoneal cavity,** a major component of the coelom, along with the lungs in tetrapods (see fig. 1.12, fish and amphibian). In mammals, birds, and a few other reptiles, the lungs occupy separate **pleural cavities,** and the digestive organs beyond the esophagus occupy an **abdominal (peritoneal) cavity** (see fig. 1.12). The coelom arises by the delamination, or splitting, of the early lateral-plate mesoderm into two sheets, somatic and splanchnic mesoderm (see

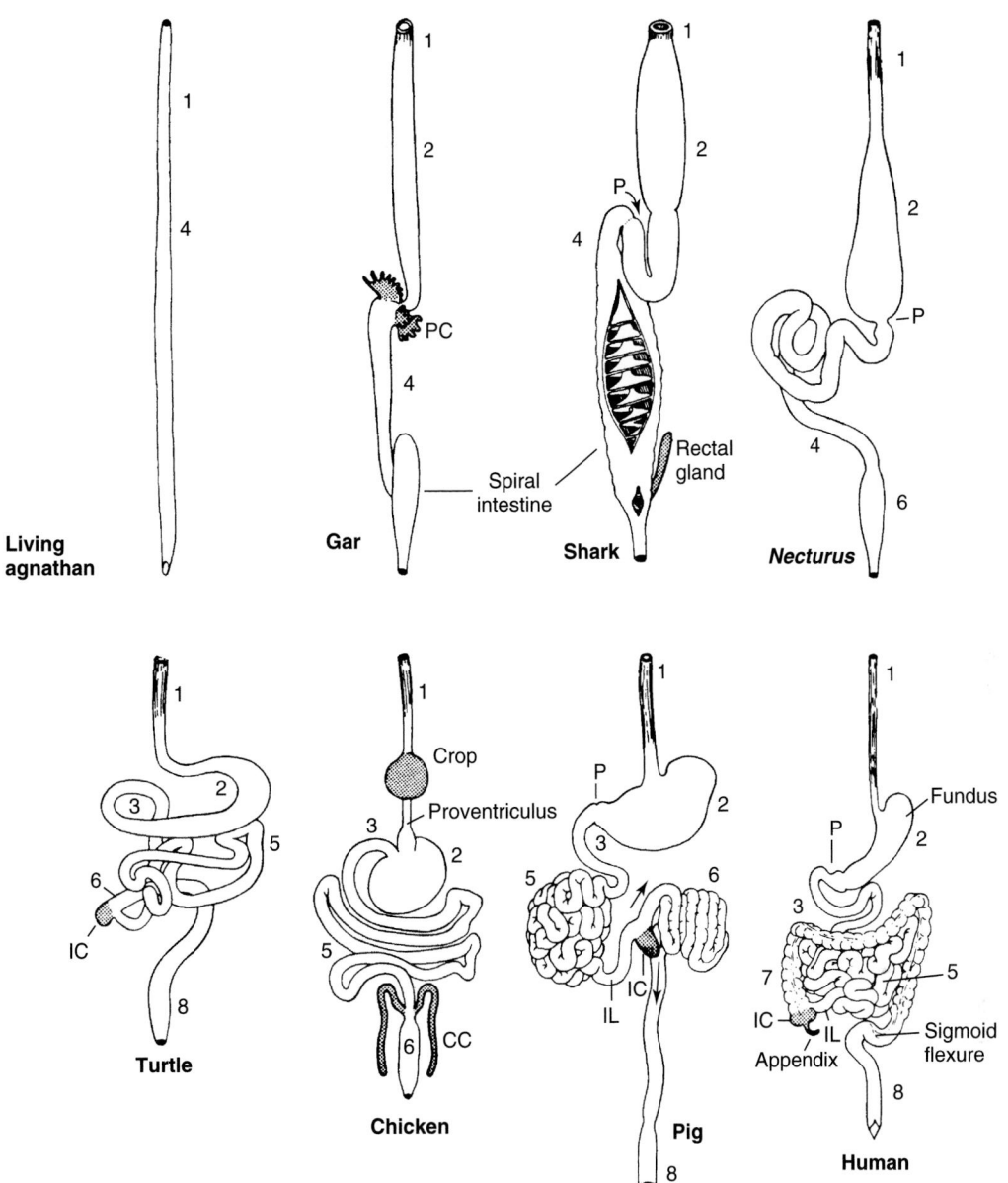

FIGURE 12.1

Digestive tracts of selected craniates. The intestine empties into a cloaca except in the therian mammals. **1,** esophagus; **2,** stomach; **3,** duodenum; **4,** intestine; **5,** small intestine; **6,** large intestine (colon in humans); **7,** colon; **8,** rectum; **CC,** paired ceca of bird; **IC,** ileocolic cecum; **IL,** ileum; **P,** pyloric sphincter; **PC,** pyloric ceca. Ceca are shown in *stipple.*

fig. 5.9*d*). The space between the two sheets becomes the **coelom.**

Most of the embryonic digestive tract other than its endodermal lining arises in cephalochordates and craniates from splanchnic mesoderm (see fig. 5.3*c*). The outer covering of the digestive tract is **visceral peritoneum** (6, in fig. 1.2). It is continuous with the **parietal peritoneum,** which lines the body wall (5, in fig. 1.2). Early in embryonic life, the parietal and visceral peritonea are continuous via dorsal and ventral mesenteries, at which time the coelom is divided into separate right and left cavities. The dorsal mesentery

remains essentially intact throughout life, conducting blood vessels and nerves from the roof of the coelom to the digestive organs. The ventral mesentery eventually disappears except at the level of the liver and urinary bladder. The fate of the ventral mesentery of the liver will be discussed when we study that organ.

The embryonic digestive tract consists of three regions. The part containing the yolk, when present, or to which the yolk sac is attached, is the **midgut** (figs. 5.13*b* and 12.2). Caudal to the midgut is the **hindgut,** and anterior to the midgut is the **foregut.** The foregut elongates to form part of the oral cavity, the pharynx, esophagus,

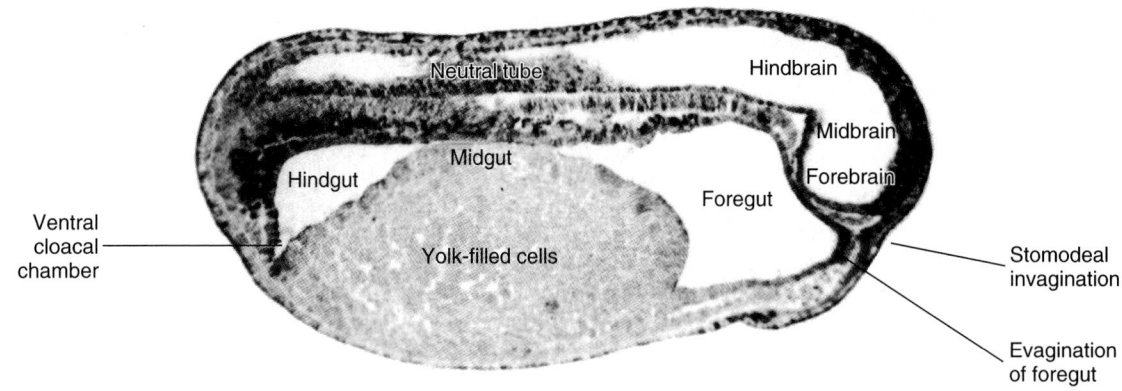

FIGURE 12.2
Sagittal section of 3.5 mm frog tadpole. An oral plate separates the stomodeal invagination from the pharynx.
From Phillips, JB: *Development of Vertebrate Anatomy* 1975 (Fig. 13-1, page 284).

stomach, and much of the small intestine. The hindgut becomes the remainder of the intestine and the cloaca. Little of the midgut remains in adults.

The anterior portion of the oral cavity, or of the oropharynx of fishes, arises as a midventral invagination of the ectoderm of the head, the **stomodeum** (figs. 1.1 and 12.2). A thin membrane, the **oral plate**, temporarily separates the early embryonic foregut from the exterior. It soon ruptures to provide an anterior entrance to the digestive tract. A similar invagination, the **proctodeum**, provides an exit from the hindgut when the **cloacal plate** ruptures (see fig. 1.1). Like all chordates, therefore, craniates, including humans, are deuterostomous.

Differences in the anatomy of digestive tracts caudal to the pharynx are correlated not so much with whether an animal lives on land or in the water as with the nature and abundance of food. Is it readily absorbable when ingested, as in vampire bats, or does it require extensive enzymatic activity or mechanical maceration, as in carnivores? Is the food supply constant, so that whenever an animal is hungry the food is likely to be there, or does it have to be stalked? If the latter, the meal is probably bulky, and there has to be room for it in an appropriately expandable stomach until it can be digested. And what is the shape of the animal's body? If it is long, like that of a lamprey or snake, the tract will probably be straight. If the trunk is short, as in turtles and anurans, the intestine must be coiled in order to provide a sufficient absorptive area. Some fishes have a special membrane that increases the absorptive area.

MOUTH AND ORAL CAVITY

The mouth is the entrance to the digestive tract. In gnathostome fishes, it opens into an **oropharyngeal cavity** that exhibits teeth of one variety or another, and walls perforated by gill slits. The oropharyngeal cavity

terminates at a short esophagus. In tetrapods, the mouth opens into an **oral cavity (buccal cavity)** housing the teeth and tongue. The oral cavity leads to the pharynx.

The roof of the oropharyngeal cavity of fishes and of the oral cavity of amphibians is a primary palate. It is pierced anteriorly in lungfishes and amphibians by internal nares (fig. 12.3*a*). Most reptiles have an incomplete secondary palate. This leaves in the roof of the oral cavity a deep cleft, the **palatal fissure**, that channelizes respiratory air between the choanae (internal nares of amniotes) and the pharynx (fig. 12.4). The secondary palate of crocodilians and mammals provides a cleftless roof for the oral cavity all the way from the mouth to the pharynx (see figs. 9.19 and 12.3*b*, respectively).

In anurans, paired **vocal sacs**, which are reverberating chambers beneath the floor of the pharynx, open into the oral cavity near the angles of the jaws. In all tetrapods, a variety of oral glands, or their ducts, empties into the oral cavity.

In mammals, a trench called the **oral vestibule**, separates the gums, or **alveolar ridges**, from the cheeks and lips. In many rodents, an opening leads from the vestibule on each side into a **cheek pouch** in which hamsters and others transport grain from the fields to the burrow for storage. The pouches of hamsters extend from the first cheek teeth caudad beneath the orbit, beneath the ears, and above the forelimb, to a position lateral to the scapula. The collagenous connective tissue walls are lined by a moderately low keratinized squamous epithelium that protects against abrasion. Inserting on the wall is a slip of the buccinator, a facial muscle that acts as a retractor. The overlying skin is loose, permitting expansion of the pouches that, when full, create the impression that the hamster has a bad case of the mumps! Some seed- and grain-eating birds have a median sublingual **seed pouch** that is employed in a similar fashion. The pouch lies upon the caudal portion

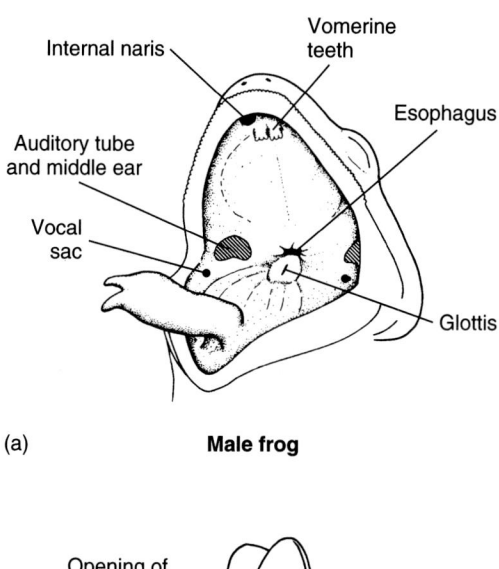

(a)   **Male frog**

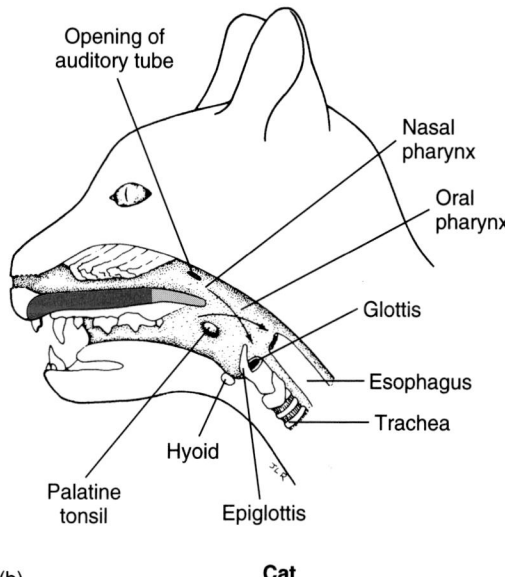

(b)   **Cat**

FIGURE 12.3

Oral cavity of an amphibian and a mammal. The roof of the oral cavity in the frog is a primary palate. In the cat, it is a secondary palate (*red*). The latter consists of a hard (bony) palate (*dark red*) and a soft palate (*light red*). *Arrows* indicate pharyngeal chiasma where food and airstreams cross.

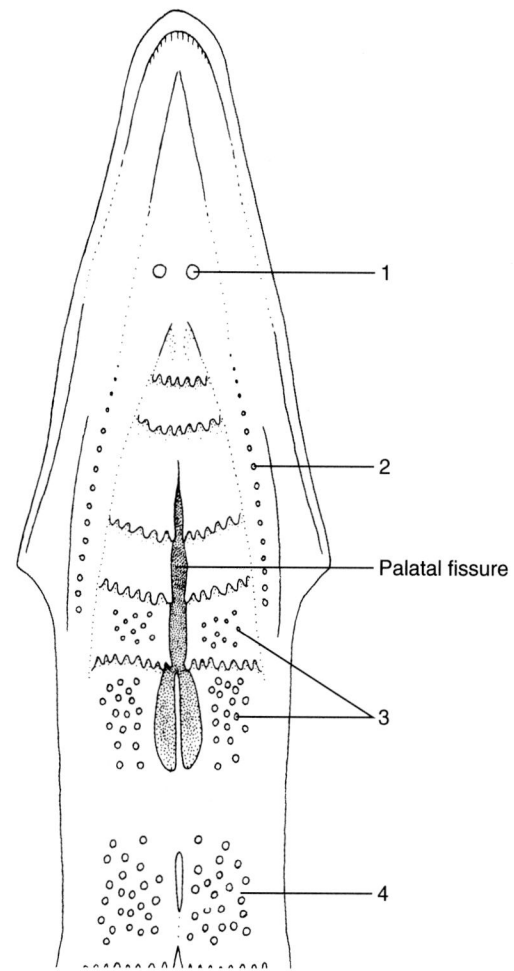

FIGURE 12.4

Palate of domestic hen. **1** to **4**, gland openings as follows: **1**, paired maxillaries; **2**, lateral palatines; **3**, medial palatines; **4**, sphenopterygoids. Internal nares (choanae) are above palatal folds at anterior end of palatal fissure.

of the mylohyoid muscle beneath the oral cavity and is retracted by what may be a homologue of the genioglossus muscle of mammals. When full, it hangs beneath the rear of the oral cavity in a sling composed of the mylohyoid muscle. It is emptied by shaking the head vigorously. The tongue, oral glands, and teeth will be discussed in the sections that follow.

Tongue

The tongue of elasmobranchs, bony fishes, and perennibranchiate amphibians is a mere crescentic or angular elevation of the floor of the oropharyngeal cavity shaped by the underlying basihyal and ceratohyal cartilages that constitute a skeleton. This lean hyoid elevation is a **primary tongue** (fig. 12.5, adult *Necturus*). A primary tongue has no musculature; hence, it cannot be independently manipulated. It may assist the jaws in holding prey within the oropharyngeal cavity until it can be swallowed, especially if the surface of the tongue has teethlike denticles. It is, however, the forerunner of the tongues of tetrapods.

The tongue of terrestrial urodeles and of anurans consists of a primary tongue and of an extension that can be flipped out of the mouth. In both groups, the primary tongue develops from hyoid arch mesenchyme, and the extension develops from an embryonic **glandular field** in the pharyngeal floor anterior to the hyoid arch (fig. 12.5, larval frog). The glandular field is so named because this region of the adult tongue secretes a sticky mucus that entangles insects when the tongue is suddenly thrust out of the mouth. The tip of the tongue of insectivorous amphibians usually terminates in a

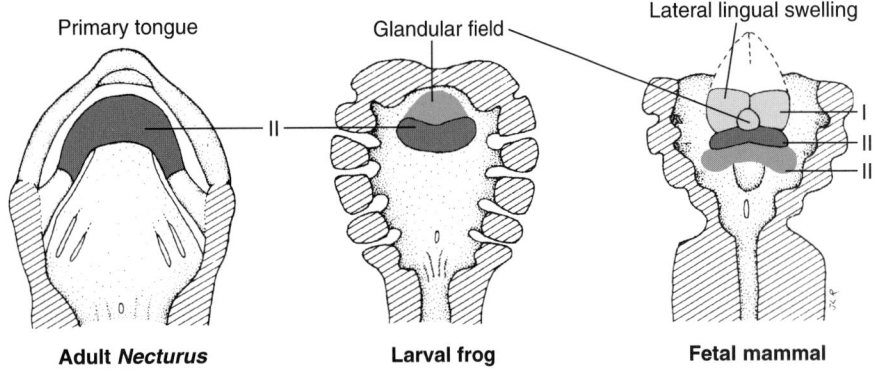

FIGURE 12.5

Floor of the oral cavity depicting stages in evolution of the tongue. **I,II,III,** derivatives of pharyngeal arches 1, 2, and 3, *Dotted* outline in mammal shows final extent of lateral lingual swellings.

broad, sticky, fleshy expansion that increases the probability that the prey will be caught and then delivered into the oral cavity. The root of the tongue of terrestrial urodeles becomes anchored to the basihyal and ceratohyal cartilages in the floor of the pharynx; that of anurans, unlike other tetrapods, becomes anchored to the floor of the oral cavity immediately behind the mandibular symphysis. In the toad family *Pipidae,* no tongue develops.

The tongue of reptiles and mammals has three distinct features: (1) a pair of **lateral lingual swellings** formed by mesenchyme from the mandibular arch, a contribution not found outside of amniotes, (2) a primary component from the hyoid arch that develops a glandular field, and (3) mesenchyme from the third pharyngeal arch that spreads forward over some of the second-arch mesenchyme (fig. 12.5, fetal mammal). Sensory epithelia of the tongue (general sensation and taste) are thus formed from mesenchyme from pharyngeal arches 1, 2, and 3 and thereby innervated by three different cranial nerves—5, 7, and 9, respectively. Hypobranchial musculature invades the entire complex and receives somatic motor innervation from the twelfth cranial nerve. The innervation of the tongue of a mammal is detailed in chapter 16 under "Innervation of the Mammalian Tongue: An Anatomic Legacy." As in basal tetrapods, the tongue of amniotes is anchored to skeletal components of the hyoid arch. However, the tongue of turtles, crocodilians, alligators, and some baleen whales is also affixed to the floor of the oral cavity; hence, it is not protrusible. Nevertheless, the immobilized tongue of baleen whales plays a vital role in food handling. Garter snakes and a few other snakes have no tongue.

In birds, the lateral lingual swellings are suppressed, although less so in birds of prey, and the tongue is almost lacking in intrinsic muscles. The only movement, therefore, comes from muscular operation of the hyoid skeleton to which the tongue is anchored. A unique use

of the posterior horns of the hyoid of woodpeckers for "shooting" the tongue out of the mouth has been described in chapter 9 under "Amniote Hyoid."

The "tongue" of agnathans is not homologous with any component of the tongue of gnathostomes, being simply a rodlike lingual cartilage of unknown homology capped by horny spines and operated by protractor and retractor muscles (see figs. 9.35 and 13.1*b*).

The tongue is widely used for capturing or gathering food. Long, sticky tongues of insectivorous or nectar-feeding tetrapods from salamanders to bats and anteaters dart in and out at lightning speed. The sticky tongue of the 4-foot-long Great Anteater is 2 feet in length! When the mouth of a toad is closed, the tongue lies folded back over itself so that the tip is directed toward the esophagus. Flipping the tongue out of the mouth occurs when the long fibers of the genioglossus medialis muscle, which extend to the tip of the tongue, stiffen to form a complex of intrinsic rods, and the genioglossus basalis muscle, which forms a wedge beneath the anchored anterior end of the tongue, suddenly swells. The combination flips the rigid tongue over the mandibular symphysis. It returns to the mouth by contraction of the hypoglossal muscles (Gans and Gorniak, 1982).

Woodpeckers have a barbed tongue that shoots like an arrow into dark crevices in tree trunks, impaling grubs and carrying them to the mouth (fig. 12.6*a*). The tiny tongue of hummingbirds darts rapidly back and forth between flower and mouth, collecting droplets of nectar at the hollow frayed tip (fig. 12.6*b*). A parrot's tongue is armed with two flexible horny shields in the walls of a seed cup that is used for shucking hardcoated seeds and preparing grain for swallowing (fig. 12.7). Like fingernails, the shields are composed of keratinized epithelial cells that grow forward from a nail-like bed halfway back on the tongue. The anterior edge is constantly being worn away and replaced. Embedded within the tongue of birds and lizards is an entoglossal bone, an anteriorly directed process of the hyoid (see fig. 12.6*a*). In many birds, a pair of long paraglossal bones are attached to the entoglossus near the tip of the tongue and extend caudad embedded in its edges.

The huge immobilized muscular tongue of some baleen whales directs tons of incoming seawater into huge reservoirs under the throat and chest. When these are being emptied by compression of their walls, the water is strained back into the sea through frayed sheets

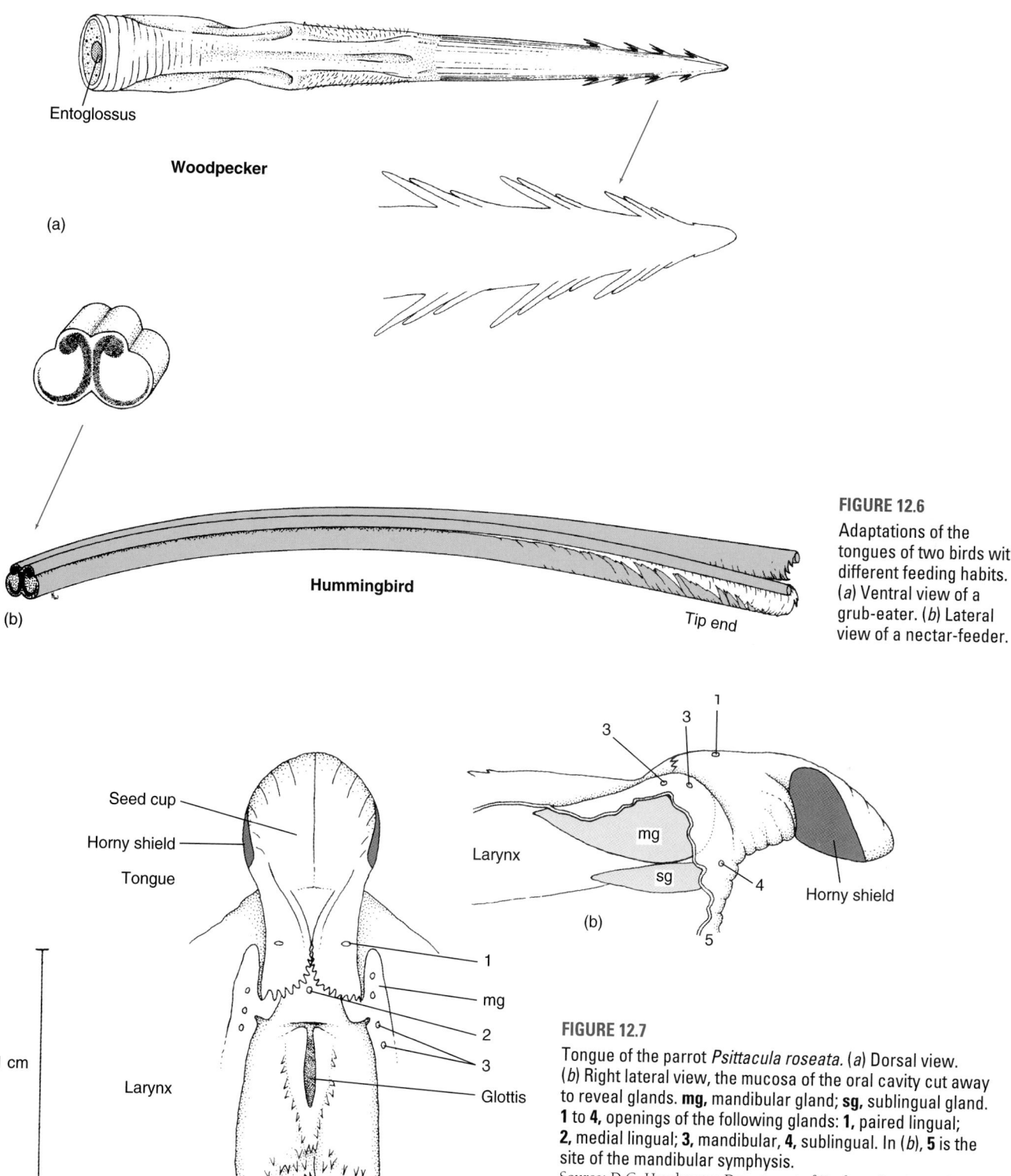

Entoglossus

Woodpecker

(a)

(b)

Hummingbird

Tip end

Source: D.G. Homberger, Department of Zoology, Louisiana State University, Baton Rouge, LA.

FIGURE 12.6

Adaptations of the tongues of two birds with different feeding habits. (*a*) Ventral view of a grub-eater. (*b*) Lateral view of a nectar-feeder.

Seed cup

Horny shield

Tongue

Larynx

1 cm

Larynx

1

mg

2

3

Glottis

(a)

Larynx

mg

sg

Horny shield

4

5

(b)

FIGURE 12.7

Tongue of the parrot *Psittacula roseata*. (*a*) Dorsal view. (*b*) Right lateral view, the mucosa of the oral cavity cut away to reveal glands. **mg**, mandibular gland; **sg**, sublingual gland. **1** to **4**, openings of the following glands: **1**, paired lingual; **2**, medial lingual; **3**, mandibular, **4**, sublingual. In (*b*), **5** is the site of the mandibular symphysis.

of baleen, and organisms strained out of the water accumulate on the deeply furrowed tongue and are then swallowed. Newborn baleen whales have only rudiments of baleen, so there is room in the neonatal oral cavity for a mobile tongue to be manipulated for suckling for about six months before it becomes immobi-

lized. The tongues of most other mammals are protrusible, although tied to the floor of the oral cavity by a ligament, the **frenulum linguae**. In human beings, if the frenulum inhibits movements of the tip of the tongue, making the individual "tongue-tied," it can be snipped at its anterior edge.

The mucosa of the tongue contains receptors not only for taste but for other stimuli, including, in amniotes, **stereognosis** (perception of the shape, weight, and texture of a solid body by feeling, handling, or lifting it). These encapsulated nerve endings on the tip of the tongue enable insectivores to locate food by exploring dark recesses with the tongue. Seed-eating birds use stereognostic information in manipulating a seed that is being husked in the seed cup.

Tongues not used to procure food or water manipulate fluids and solids in the oral cavity, and the tongues of most tetrapods participate in swallowing. Long tongues function in overheated mammals as a site for cooling the blood by evaporation of saliva while panting. Lizards clean their spectacles (transparent eyelids) with them. Spiny papillae on the tongue's surface are used by carnivores for rasping bones and by many mammals for grooming. The latter accounts for hairballs that often form in the stomach of domestic cats. And, of course, human speech as we know it would not be possible without a tongue.

Oral Glands

Terrestrial tetrapods have a variety of multicellular glands that secrete watery or viscous fluids into the oral cavity. The chief ingredient is usually mucus of varying viscosities and chemical composition. It moistens the food as necessary to produce a bolus that can be manipulated by the tongue, when present, and lubricates dry food for passage through the pharynx and down the esophagus. Moisture is also essential for taste buds to function because the stimulant for taste must be in solution to evoke a gustatory response. A taste bud is illustrated in figure 17.23. Other secretions include, in one species or another, serous fluids, toxins, venoms, and an enzyme that digests starch, although the latter is rare outside of mammals. Viscous secretions keep the tongue sticky in species where this is essential, and venom tranquilizes prey or dispatches them promptly.

Oral glands are usually named according to their location. Labial glands open into the oral vestibule at the base of the lips, molar glands lie near the molar tooth, infraorbital glands are in the floor of the orbit, palatal glands open onto the palate, and sublingual and submandibular glands open via common papillae under the tongue. (To observe these papillae in your oral cavity, open your mouth while looking into a mirror, raise your tongue, and note the two papillae just behind the mandibular symphysis at the base of the tongue. A flashlight may help!) Intermaxillary (internasal) glands lie near the premaxillary bone. Frogs have up to 25 small intermaxillary glands, each with its own duct that delivers a sticky secretion onto the palate. The large venom glands of the four families of venomous snakes are palatal glands. Their ducts open at the base of a maxillary tooth, and the venom exudes into a groove or tube

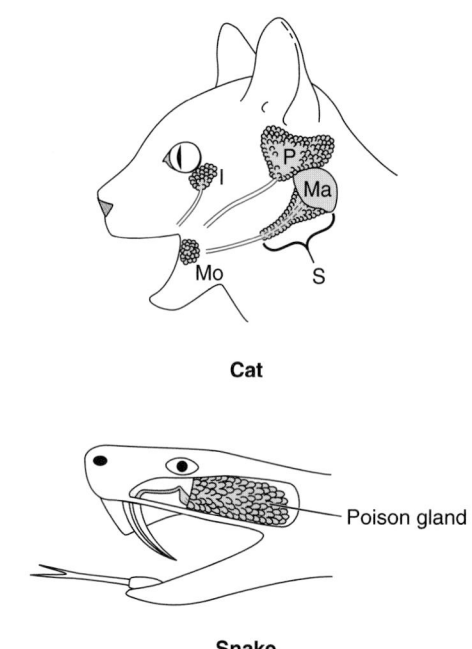

Cat

Snake

FIGURE 12.8

Oral glands in a cat and a snake. **P,** parotid; **Ma,** mandibular; **S,** sublingual; **Mo,** molar; **I,** infraorbital. A proximal portion of the mandibular duct (*dotted*) is surrounded by the sublingual gland. Ducts for molar and sublingual glands are small and cannot be seen. A duct from the poison gland drains into a grooved maxillary tooth.

in the tooth (fang) (fig. 12.8, snake). In *Heloderma*, the only poisonous lizard, sublingual glands secrete the toxin.

Saliva is a mixture of oral secretions, but the term is usually reserved for oral secretions of mammals. The mammalian parotid is the largest tetrapod salivary gland, and ptyalin (amylase), an enzyme that initiates digestion of starch, is always one of its secretions. Mammals have several salivary glands, but not all of them produce ptyalin. The specific mix of mucus, serous secretions, and ptyalin produced by any mammalian salivary gland varies with the species and is correlated with dietary habits. The duct of the parotid gland crosses the masseter muscle to open into the oral vestibule near one of the upper molar teeth (fig. 12.8, cat). The poison glands of reptiles resemble the parotid gland histologically. Birds do not have a copious supply of saliva, but the openings of some of the oral glands of birds are shown in figures 12.4 and 12.7.

Aquatic craniates have adequate moisture in the oral cavity for taste buds to function, but glands that secrete digestive enzymes would be a waste of energy. Any enzymatic secretion would be diluted and washed away. Mucus-producing **goblet cells** are about the only source of oral secretions, and these are common. Mucus lubricates the esophagus. Goblet cells perform a special function in the male population of a few species of catfishes.

Males carry fertilized eggs in temporary folds, or crypts (**brood pouches**), in the mucosa of the palate and retain the hatchlings in the crypts for about two months during which time the male does not feed. The goblet cells in the brood pouches produce a copious nutritious secretion that maintains a suitable environment for development of the eggs and later nourishes the hatchlings. After the hatchlings leave, the brood pouches atrophy in response to altered hormonal ratios. A rare instance of multicellular oral glands in aquatic craniates are the anticoagulant glands of lampreys, which open into the oral cavity beneath the tongue just internal to the buccal funnel. Their secretion inhibits blood clotting of the prey while the lamprey is feeding.

Teeth

Bony teeth are found among jawed fishes, amphibians, reptiles, and mammals with few exceptions, and they were present in the earliest birds. In mammals, they have achieved a peak in regional specialization. Lacking bony teeth are sturgeons, numerous teleosts including sea horses, a few amphibians, all turtles, and modern birds; and among mammals, whalebone whales, South American and scaly anteaters, and the monotreme *Echidna*. But even among toothless species, there are many that develop an embryonic set that simply does not erupt or, after erupting, disappears. Teeth are descendants of the denticles of the dermal armor that covered the head and extended into the oropharyngeal cavity of the early fishes (see fig. 6.29*a*). The relatively small number of gnathostomes that lack teeth have lost the genetic code necessary either to induce their formation or to complete their development.

Bony **dental plates** of dermal origin overlay the endoskeletal components of the jaws of early fishes. Many had pointed, rounded, or jagged surface projections that prevented escape of live food from the oropharyngeal cavity, or that were used for crushing shellfish, biting flesh, or rasping vegetation. The pattern of dermal structures associated with the jaws in placoderms is incompletely known. The jaws of placoderms range from unossified Meckelian cartilages associated with overlying denticles within the skin to fully ossified dental plates overlying or adjacent to Meckel's cartilage. Placoderms did not possess teeth; however, the surface of their dental plates were hardened and shaped to perform toothlike functions. The morphology of the biting surface for these dental plates is correlated with their putative function: flattened surfaces for crushing, sharp edges for shearing, large spikes for impaling prey, numerous individual cusps for holding prey, and loss of specialized biting structures in filter-feeders. Single or paired upper dental plates opposed the lower jaw on each side.

In acanthodians, each denticle, usually broadly based, some conical, some slender and curved, some

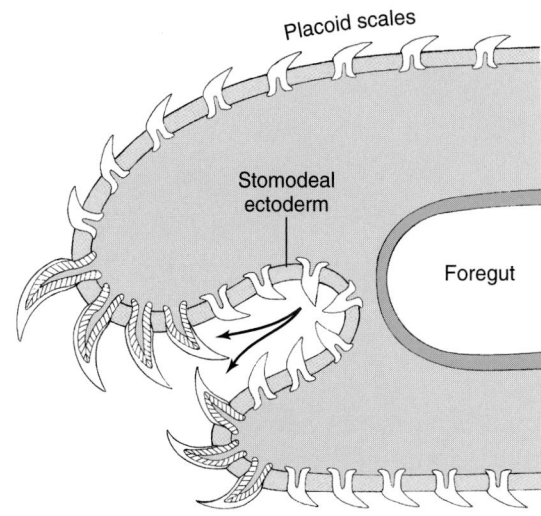

FIGURE 12.9

Representation of the identity of placoid scales, stomodeal denticles, and teeth in sharks. *Arrows* indicate direction of migration of formative to functional teeth at margins of the jaws.

with a dozen spiny cusps arranged in whirls, was attached directly to the endoskeleton of the jaws, as in modern teeth, instead of being borne on dental plates, and the jaws were often flanked by additional denticles. Some acanthodians were toothless or had only lower teeth. These bony denticles of early fishes were the precursors of the teeth of modern gnathostomes.

Until the phylogenetic relationships between placoderms, acanthodians, and later cartilaginous and bony fishes can be ascertained, only the broadest generalization regarding the phylogeny of later vertebrate teeth can be made: *They are derivatives of bony dermal armor.* Evidence comes not only from paleontology but also from the comparative histology of armor and teeth and from the observation that dermal denticles and placoid scales, even today, show a gradual transition to teeth as they approach the cutting edges of the jaws (figs. 12.9 and 12.10).

Teeth, like placoid scales, are composed of **dentin**, which is a variety of bone, surmounted by a crown of **enamel** or **enameloid** (fig. 12.11). The earliest indication of the development of socketed teeth is the ingrowth into the dermis of a longitudinal ridge of ectoderm, the **dental lamina**, that extends more or less the length of the jaws (2, in fig. 12.12*c*). Beneath the lamina (or above it, in the upper jaw), a linear series of dermal papillae, each designating the site of a future tooth (1 in fig. 12.12*c*) form at intervals, indenting the lamina, and organizing blood vessels necessary for further development of a tooth primordium. The cells at the periphery of each papilla become organized into a definitive layer of **odontoblasts** that commence to deposit dentin. As deposition proceeds, the odontoblasts slowly withdraw

FIGURE 12.10

Jaws of the Port Jackson shark, *Heterodontus,* showing regional transition from dermal denticles to placoid scalelike teeth. The rounded teeth are used for crushing molluscs.

Courtesy Ward's Natural Science Establishment, Inc. Rochester, NY.

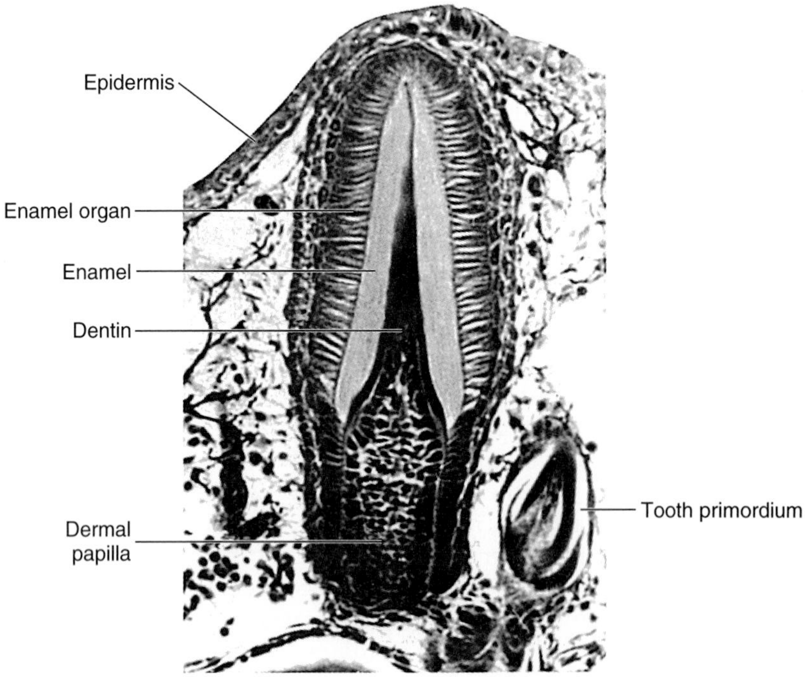

Epidermis

Enamel organ

Enamel

Dentin

Dermal papilla

Tooth primordium

FIGURE 12.11

Unerupted tooth of a gar. Primordium is a developing replacement tooth.

toward the center of the primordium, which is becoming a pulp cavity containing components of the dermal papilla. The withdrawing odontoblasts leave behind evidence of their withdrawal in the form of **dentinal tubules** that contain protoplasmic processes of the odontoblasts (see fig. 7.6). The odontoblasts remain alive throughout the life of the tooth. Meanwhile, the ectoderm of the dental lamina has organized an **enamel organ**, consisting of ameloblasts, that deposits enamel on the surface of the dentin (figs. 12.11 and 12.12*d*). A thin layer of **cementum**, a type of acellular bone (chapter 7), eventually anchors the tooth to the bone of the jaw by means of collagenous fibers. Living remnants of the dermal papilla, including vessels and nerves, remain within whatever is left of the pulp cavity (root canal) throughout the life of the tooth. Vessels and nerves are both essential for maintaining the tooth in a healthy state. The details of tooth development and emergence, the time of initiation of the different stages, and the ultimate fate of erupted teeth vary with the species.

An enamel organ is present but functionless in armadillos and a few other vertebrates, hence, their teeth have no enamel. Enamel in mammals, at least, is deposited by ameloblasts of ectodermal origin. Enameloids of other vertebrates, especially fishes, differ in their physical characteristics, and the ultimate source of the scleroblasts that elaborates them (see fig. 7.1) has been determined to be odontoblasts, which form a compact dentin. Similarities in the development of placoid scales and mammalian teeth are illustrated in figure 12.12.

Teeth vary among gnathostomes in number, distribution within the oral cavity, position with reference to the summit of the jaw, degree of permanence, and shape. They are numerous and widely distributed in the oropharynx of living fishes where they develop on jaws, palatal bones, and even on the pharyngeal skeleton. For example, the blue sucker has 35 to 40 teeth on the last gill arch. In early tetrapods, too, teeth were widely distributed on the palate, and even today most amphibians and many reptiles have teeth on the vomer, palatine, and pterygoid bones and occasionally on the parasphenoid. They are confined to the jaws in crocodilians, fossil toothed birds, and mammals, and they are least numerous among mammals. Teeth, therefore, like dermal armor, have tended toward a more restricted distribution with the passage of time.

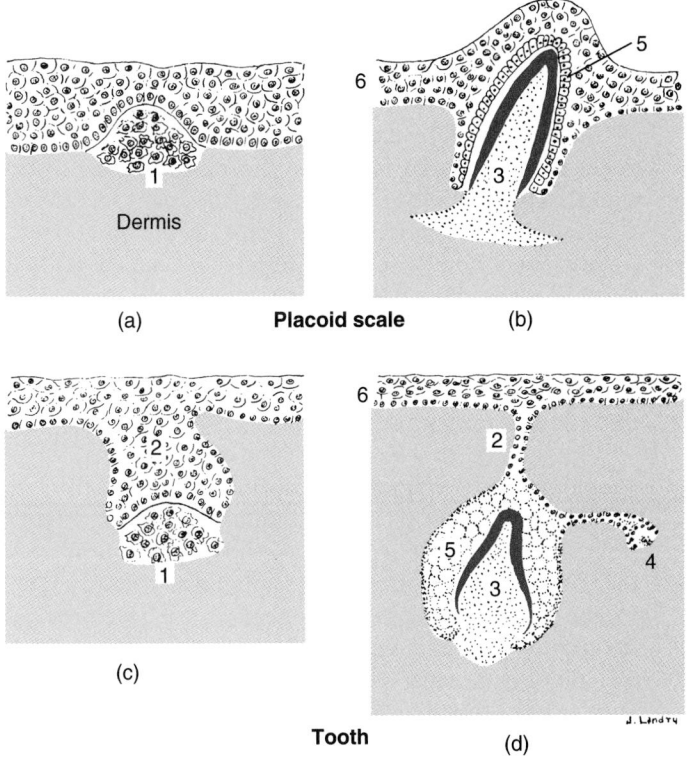

(a)   **Placoid scale**   (b)

(c)

Tooth   (d)

Morphogenesis

FIGURE 12.12
Development of a placoid scale (*a, b*) and mammalian tooth (*c, d*). *Gray* shows dermis; *red* indicates enameloid/enamel. **1,** dermal papilla; **2,** (in *c* and *d*), cross section of dental lamina; **3,** dentin, produced by dermal papilla; **4,** primordium of replacement tooth; **5,** enamel organ; **6,** epidermis.

Jaw teeth may be attached to the outer surface or to the summit of the jawbone, as in many teleosts (**acrodont dentition,** fig. 12.13); they may be attached to the inner side of the jawbone, as in anurans, urodeles, and many lizards (**pleurodont dentition**); or they may occupy bone sockets, or alveoli (**thecodont dentition**). Socketed teeth are found in many fishes, in crocodilians, extinct toothed birds, and mammals. The sockets (alveoli) are deepest in mammals.

Most gnathostomes through amniotes have a succession of teeth, and the number of replacements during a lifetime is indefinite but numerous (**polyphyodont dentition**). It has been estimated that an elderly crocodile may have replaced its front tooth 50 times. They and other nonmammalian gnathostomes that have been studied replace teeth in waves that sweep along the jaws eliminating and replacing every other tooth. Thus in one wave, in tetrapods at least, even-numbered teeth are lost, and in the next wave odd-numbered ones are lost. Meanwhile, tooth germs for the next wave of eruptions are forming. Whether the loss and replacement waves sweep from back to front or reverse is not agreed on at present. There is evidence that they sweep in different directions in different species. The waves ensure a balanced distribution of teeth along the jaws throughout life. In sharks, tooth germs form in the dermis on the oropharyngeal cavity side of the jaws and while growing migrate onto the cutting edge of the jaw (see arrows in fig. 12.9) as the teeth that are being replaced move beyond the edge before falling away. The cause of the migration is not known.

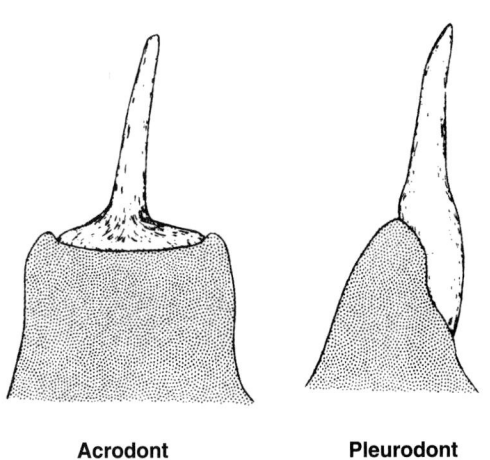

Acrodont   **Pleurodont**

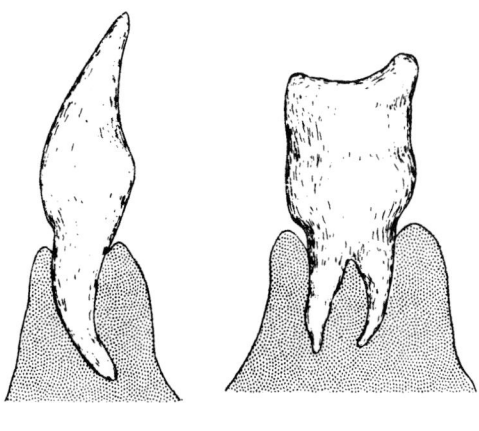

Thecodont

FIGURE 12.13
Variations in the relationship of teeth to jaws. Acrodont teeth are attached either at the outer surface of the jaw or, as shown, at the summit. Pleurodont teeth are attached to the inner lingual surface of the jaw. Thecodont teeth occupy alveoli.

Only in mammals is there a definite number of teeth in a species, with rare exceptions. Most mammals develop two sets, **deciduous**, or **milk teeth**, and permanent teeth (**diphyodont dentition**), and there is a definite sequence in which the teeth erupt. For example, numbering the permanent set in human beings 1 to 8 from front to rear, the sequence of eruption is 6,1,2,4,5,3,7,8. Eruption of number 8, the last molar, is delayed in higher primates, and this "wisdom tooth" is sometimes imperfect, unerupted, or missing. A first set provides the constantly changing infant jaw with small temporary teeth adequate for an infant's diet until the jaws are more stabilized structurally and have elongated sufficiently to accommodate large teeth for macerating coarse foods.

A few mammals develop only a first set (**monophyodont dentition**). In a platypus, the deciduous set is replaced by **horny epidermal teeth**. In toothless whales, the first set, although formed within the jawbone, may not erupt, and if they do they are usually shed. The freshwater manatee from the Amazon River and the Australian Rock Wallaby do not have "sets," teeth being replaced throughout life by the forward migration of new teeth formed at the rear of the jaws. In the manatee, migration is at the rate of 1 or 2 mm a month. Thin, bony sockets separate the roots of successive teeth, and the bony septa are resorbed under pressure from the migrating teeth. The grasses eaten by the manatee contain abrasives that appear to be necessary for the teeth to move forward because babies fed experimentally on milk show no forward migration of teeth. Proboscidians have a slow but constant succession of molar teeth that move forward from the rear.

Morphologic Variants in Fishes

Most sharks are fish eaters and have numerous rows of jaw teeth that are either flat, sharp, notched triangles that are used for cutting, or single or multipointed tusks that curve toward the pharynx and hold a struggling prey until it can be swallowed whole. Each shark tooth has a broad **basal plate** of bone embedded in the dermis. A minority eat shellfish and, although the teeth at the entrance have curved caudally directed spines, the rest form batteries of rounded denticles used for crushing shells (see fig. 12.10). Tiny stomodeal denticles line the pharynx in some sharks. Near the jaws, these have transitional shapes between denticles and teeth.

The dental armor of today's holocephalans and modern lungfishes is like that of early jawed fishes, consisting of a few large plates of enameloid/enamel-covered dentin that bear rows of various-sized rounded moundlike denticles that usually become sharp spines at the entrance to the oropharyngeal cavity. *Chimaera*, a cartilaginous fish that eats molluscs, has on each side of the upper jaw one large anterior and one small posterior dental plate that together cover the entire upper jaw. There is a single large plate on each side below. In modern lungfishes, the plates are restricted to the palate and medial aspects of the lower jaw.

The jaw teeth of actinopterygians, amphibians, and most reptiles are simple pointed cones attached to one or more membrane bones. Small teeth may be interspersed among large ones, and those in front are sometimes larger and curved slightly to the rear. Specialized shapes sometimes appear on one jaw or the other. Gars, for instance, have a few fanglike teeth shaped at their ends like arrows, and the fangs of venomous snakes, borne on the maxillae, are curved or bladelike, grooved on the rear surface, or tubular, for injecting venom. When all teeth are essentially similar, the dentition is described as **homodont**.

Morphologic Variants in Mammals

All but a very few mammals exhibit **heterodont dentition**; that is, the teeth vary morphologically from front to rear, being incisors, canines, premolars, and molars. Premolars and molars are "cheek teeth." Heterodont dentition arose in late synapsids. Most marine mammals of today—cetaceans, sirenians, and some marine carnivores—have reverted to homodont dentition, but their Cretaceous ancestors exhibited heterodonty.

Incisors, located on either side of the mandibular symphysis, have one horizontal cutting edge and a single root. They are best developed in herbivorous mammals, which use them for holding, cropping, or gnawing. The single pair of large chisel-like incisors of rodents and the large front pair of lagomorphs have enamel on the anterior surface only. (Lagomorphs have a small second pair of incisors *behind*, not lateral to, the front pair.) Because dentin is softer than enamel, gnawing wears dentin away faster. This keeps the cutting edges of the incisors sharp. These incisors grow throughout life. Incisors may be lacking on the upper jaw only, as in bovines (fig. 12.14, ox), lacking on the lower jaw only, as in vampire bats, or lacking completely, as in sloths. Elephant and mastodon tusks are modified incisor teeth (fig. 12.14). Incisor tusks are found also in an occasional member of a few other mammalian orders. Tusks, like the incisors of rodents, grow throughout life compensating for wear. Walrus tusks are not incisors but canine teeth.

Canine teeth lie next to the incisors (fig. 12.14, dog). In generalized mammals, incisors and canines scarcely differ in appearance (fig. 12.14, shrew). In carnivores, the canines are spearlike and used for piercing flesh (fig. 12.15*d*). They are the tusks of a walrus (see fig. 12.14). Canine teeth are absent in lagomorphs, so there is a toothless interval, or **diastema**, between the incisors and the first cheek tooth (see fig. 12.14, rabbit). In rodents, premolars are also missing, and the diastema is longer (fig. 12.15, mouse). Canine teeth attained their greatest length on the upper jaw of the now extinct saber-toothed tigers. In these cats, they extended as much as 20 cm below the lower jaw with the

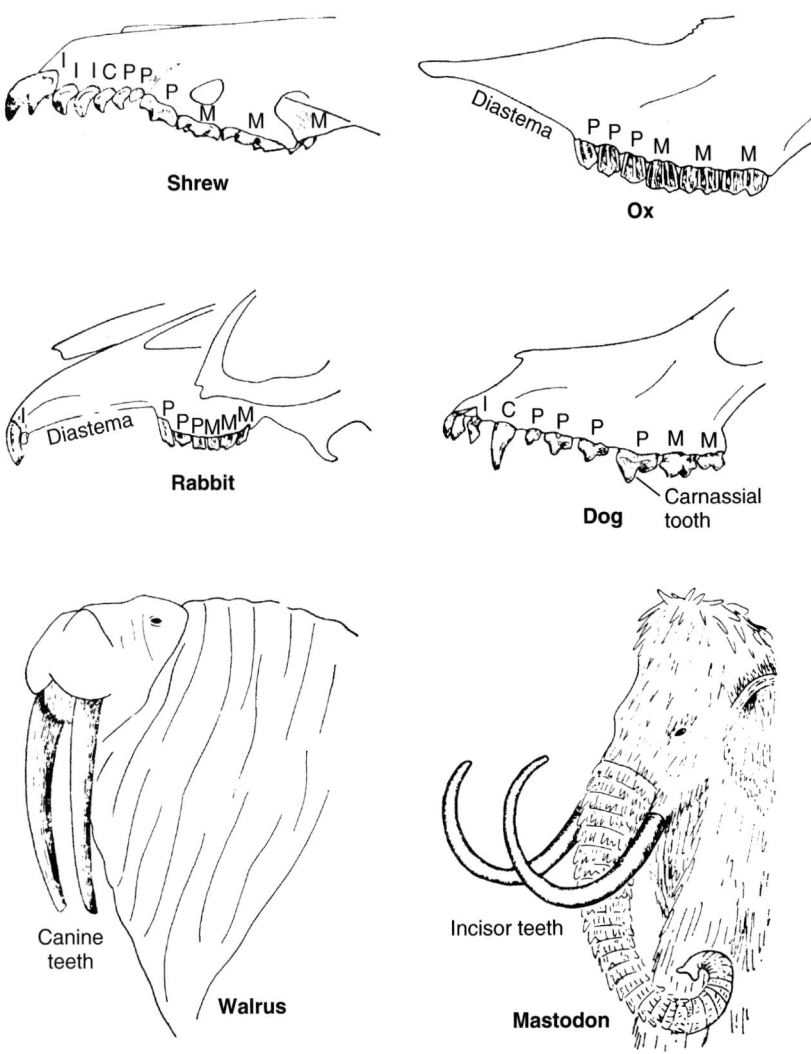

FIGURE 12.14

Mammalian upper teeth, showing generalized pattern (shrew) and specializations in a herbivore (ox), gnawing animal (rabbit), and carnivore (dog). Extreme specialization of canines is seen in the walrus and of incisors in the mastodon. **I**, incisor; **C**, canine; **P**, premolar; **M**, molar.

and molars (cheek teeth) of carnivores and those of herbivorous ungulates show extreme morphological differences. Those of carnivores are adapted for tearing flesh and crushing bone. Those of ungulates and some other herbivores are adapted for macerating vegetation.

The crowns of carnivores' cheek teeth are laterally compressed, and the two or three cusps are interconnected by sharp ridges of enamel. The teeth also have long roots. They are known as **secodont teeth** (fig. 12.15*f*). They are spaced on the jaws in such a fashion that the cusps of the teeth on the upper jaw fit between the cusps of the teeth on the lower jaw. Thus, as the jaws are closing, the sharp enamel ridges of the crowns produce the shearing effect necessary for macerating animal tissues. The last upper premolar and the first lower molar, the **carnassial teeth**, are larger and longer than adjacent cheek teeth and are employed when a particularly tough shearing problem presents itself, as when a dog is chewing on a bone.

The cheek teeth of ungulates are specialized for grinding vegetation. They are wider and longer than those of carnivores, thereby providing broad surfaces for grinding (fig. 12.15*g* and *h*), and the crowns are tall, allowing for plenty of wear. The crowns consist of crescentic columns of dentin, each column surrounded by enamel, and the columns are embedded in additional dentin that is devoid of an enamel overlay. The dentin, being softer than enamel, wears away more quickly, leaving sharp crescentic enamel ridges with a wide variety of configurations. These ridges macerate vegetation during the complex side-to-side and forward-backward chewing movements of opposing jaws. Because the cutting surfaces are crescentic, these teeth are said to be **selenodont**. (For derivation, see the glossary.) The cheek teeth of bovines are selenodont. However, there are no teeth anterior to them on the upper jaw of bovines. The cheek teeth are employed in chewing cud.

The adaptation for grinding is exaggerated in proboscidians, whose teeth are termed **lophodont** (fig. 12.15*h*). The enamel and dentin are intricately interfolded, and the enamel is disposed on ridges (lophs) on enormous plateaus of naked dentin. Lophodont teeth reach a foot or more (30 cm) in length and a third of a foot in width in the largest elephants.

mouth closed. The lower canines were correspondingly reduced.

The **premolars** of most mammals other than ungulates usually have two prominent cusps; hence, they are bicuspids. They also have one or two roots, and the number of roots may differ on the upper and lower jaw and among different individuals of the same population, including humans.

Molars have three or more cusps; hence, they are tricuspids. They usually have three roots, but there are occasional molars with four or five. Molars are not replaced by a second set. In fact, they are late arrivals of the first set.

The **crown** of a tooth is the part above the gum line. It is covered with enamel. The crowns of the premolars

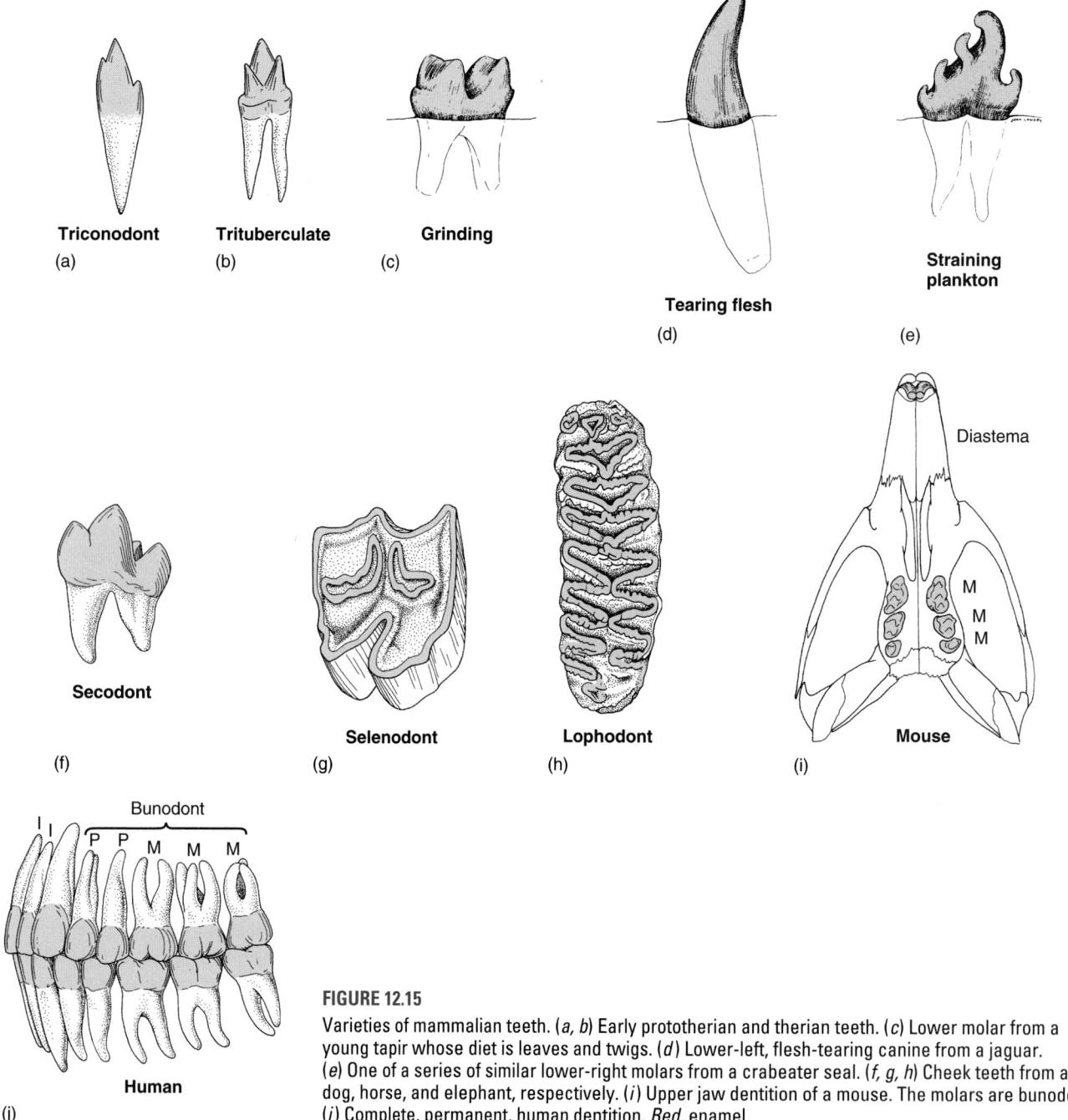

FIGURE 12.15

Varieties of mammalian teeth. (*a, b*) Early prototherian and therian teeth. (*c*) Lower molar from a young tapir whose diet is leaves and twigs. (*d*) Lower-left, flesh-tearing canine from a jaguar. (*e*) One of a series of similar lower-right molars from a crabeater seal. (*f, g, h*) Cheek teeth from a dog, horse, and elephant, respectively. (*i*) Upper jaw dentition of a mouse. The molars are bunodont. (*j*) Complete, permanent, human dentition. *Red,* enamel.

The cheek teeth of the remaining mammals exhibit a diversity of molariform styles. Those of omnivores and some herbivores may lack sharp edges and pointed cusps and have, instead, low rounded cusps, as in higher primates, including humans (fig. 12.15*j*). These are **bunodont teeth.** Among other mammals with bunodont teeth are rhinos, some hogs, some primitive ruminants, and some rodents (white-footed mice, for instance).

Rodents, the largest mammalian order and with the largest variety of diets, exhibit the largest variety of teeth, some low crowned with long roots as in squirrels,

some high crowned with short roots, as in wood rats. Among the most unusual mammalian teeth are those of the crabeater seal (fig. 12.15*e*), whose diet consists chiefly of crustaceans. The teeth are employed to strain small crustaceans and other plankton out of mouthfuls of seawater as it spills back into the sea.

Cheek teeth of early prototherians were **triconodont,** the crown having three conelike prominences arranged in a straight line (fig. 12.15*a*). In early therians (see fig. 4.31), the cones were arranged in a triangle (fig. 12.15*b*). Those teeth are termed **trituberculate.** Whether the latter are derivatives of triconodont teeth is not known because the fossil record of the origins of prototherians and therians is incomplete. It seems certain, however, that trituberculate teeth were the forerunners of today's tricuspids. Formation of enamel crests connecting the cones of trituberculate teeth in various configurations is thought to account for selenodont and lophodont teeth. Most of our knowledge of the evolution of mammals and much of the classification of fossil mammals is based on studies of fossil molariform teeth, and for a good reason: In many cases, these were the

only recovered remnants of the extinct mammal that bore them.

The first eutherians had three incisors, one canine, four premolars, and three molars on each side of the jaw, a total of 44 teeth. This may be expressed by the formula $\frac{3\text{-}1\text{-}4\text{-}3}{3\text{-}1\text{-}4\text{-}3}$. Adults of some of the generalized mammals of today still exhibit this formula. Formulas for a few selected mammals of varied dietary habits are:

Cretaceous eutherian		$\frac{3\text{-}1\text{-}4\text{-}3}{3\text{-}1\text{-}4\text{-}3}$			
Insectivores	*Solenodon*	$\frac{3\text{-}1\text{-}3\text{-}3}{3\text{-}1\text{-}3\text{-}3}$		Mole	$\frac{3\text{-}1\text{-}4\text{-}3}{3\text{-}1\text{-}4\text{-}3}$
Marsupials	American opossum	$\frac{5\text{-}1\text{-}3\text{-}4}{4\text{-}1\text{-}3\text{-}4}$		Numbat	$\frac{4\text{-}1\text{-}3\text{-}5}{3\text{-}1\text{-}3\text{-}6}$
Primates	*Tarsius*	$\frac{2\text{-}1\text{-}3\text{-}3}{1\text{-}1\text{-}3\text{-}3}$		Catarrhines	$\frac{2\text{-}1\text{-}2\text{-}3}{2\text{-}1\text{-}2\text{-}3}$
Carnivores	Canines	$\frac{3\text{-}1\text{-}4\text{-}2}{3\text{-}1\text{-}4\text{-}3}$		Felids	$\frac{3\text{-}1\text{-}3\text{-}1}{3\text{-}1\text{-}2\text{-}1}$
Lagomorphs	Rabbits	$\frac{2\text{-}0\text{-}3\text{-}3}{1\text{-}0\text{-}2\text{-}3}$		Pika	$\frac{2\text{-}0\text{-}3\text{-}2}{1\text{-}0\text{-}2\text{-}3}$
Rodents	Hamster	$\frac{1\text{-}0\text{-}0\text{-}3}{1\text{-}0\text{-}0\text{-}3}$		*Squirrel*	$\frac{1\text{-}0\text{-}2\text{-}3}{1\text{-}0\text{-}1\text{-}3}$
All bovines		$\frac{0\text{-}0\text{-}3\text{-}3}{3\text{-}1\text{-}3\text{-}3}$			

Teeth, along with the tongue and hyoid, constitute a functional triad that procures, manipulates, and, in mammals, masticates foodstuffs at the entrance to the digestive tract, then starts a bolus of food on its way to digestive sites.

Epidermal Teeth

Keratinized (horny) teeth sometimes function like bony ones. Living agnathans have horny teeth in the buccal funnel and on the tongue, which are used for rasping. Anuran tadpoles have several rows on temporary lips perched on poorly developed jaws; they are used for rasping algae and other vegetation, which is the tadpoles' diet. At metamorphosis, the horny teeth of anurans are shed and replaced by bony ones. Before hatching, turtles, crocodilians, *Sphenodon*, birds, and monotremes have a temporary horny **egg tooth** that is used for cracking the shell; after a baby platypus has lost its first set of bony teeth, horny teeth replace them and remain throughout life. The horny beaks of turtles

and modern birds often have serrations that perform some of the functions of teeth.

PHARYNX

The adult tetrapod pharynx is the part of the digestive tract that had pharyngeal pouches in the embryo. As a functional part of the digestive tract, it opens into the esophagus. The pharynx of fishes is a functional part of the respiratory system and will be discussed in the next chapter.

The most constant features of the tetrapod pharynx are the **glottis**, a slit opening into the larynx, the **openings of the paired auditory tubes** that lead to the middle ear cavity, and the **opening into the esophagus** (see fig. 12.3a and b). The pharynx of mammals exhibits additional features. It consists of a **nasal pharynx** above the soft palate and an **oral pharynx** between the oral cavity and the glottis. Nasal passageways empty into the nasal pharynx via choanae (internal nares), and the auditory tubes, derived from the first pair of pharyngeal pouches, open into its lateral walls. A narrow **laryngeal pharynx** is located dorsal to the larynx in those mammals in which the opening to the esophagus is caudal to the glottis (see fig. 13.13*a*).

The oral cavity in mammals leads to the oral pharynx. The transition is a narrow passageway, the **isthmus of the fauces.** The lateral walls of the isthmus each exhibit two **pillars of the fauces.** The pillars are muscular folds that arch upward from the side of the tongue to the soft palate (glossopalatine arch) and from the pharyngeal wall to the soft palate (pharyngopalatine arch). In humans and as a derived condition in some other primates, a fleshy **uvula** hangs from the caudal border of the soft palate into the oral pharynx. You can readily see the arches and the uvula at the end of your own oral cavity by looking in a mirror and saying "Ah!" (Good lighting is necessary.)

In the hollow between the pillars of the fauces on each side is a lymphoid organ, the **palatine tonsil.** These tonsils develop in the walls of the embryonic second pharyngeal pouches. A remnant of a pouch often remains as a pocket-like crypt with a tonsil in its wall. **Pharyngeal tonsils (adenoids)** develop in the mucosa of the nasal pharynx, and **lingual tonsils** develop on the tongue near its attachment to the hyoid bone. This encircling array of lymphoid masses is the body's first line of defense against infective agents that have entered the mouth and external nares.

In mammals, a fibrocartilaginous flap, the **epiglottis,** lies in the floor of the pharynx ventral to the pharyngeal chiasma (see fig. 13.13). It is attached to the hyoid bone. In many mammals, the act of swallowing draws the larynx forward (upward, in humans) against the epiglottis, closing the glottis, and thereby momentarily preventing fluids or particles of food from entering the pathway to the lungs. In other mammals, the epiglottis and part of the larynx can be drawn into the nasopharynx to provide an uninterrupted air pathway to the lungs, while permitting foodstuffs to detour around the larynx and enter the esophagus (see fig. 13.15). The advantages of these arrangements are discussed in the next chapter.

Other tetrapods have in either the air or food pathway, or both, fleshy valves at appropriate locations that are also able to regulate air and food traffic in the pharynx. Among these, for example, are valves that open or close the entrance to the external nares in aquatic tetrapods.

In some teleosts, a pair of elongated muscular tubes, **suprabranchial organs,** evaginate from the roof of the pharynx on each side near the esophagus, extend cephalad above the membranous roof of the pharynx behind the skull, and then turn caudad to terminate as blind sacs. Elongated gill rakers from the last two gill arches form funnel-shaped baskets that extend into the entrances of the tubes, and each tube is surrounded by a cartilaginous capsule to which the striated muscle of the tube is attached. The epithelium at the blind ends has many goblet cells, and the sacs contain quantities of plankton, sometimes compressed into a bolus. It may be that a function of these is to trap plankton from the incoming respiratory water stream and concentrate it into mucus-rich masses that are swallowed. In at least one air-gulping teleost, the cavity is filled with air, and a highly vascular epithelial lining serves as an accessory respiratory membrane.

MORPHOLOGY OF THE GUT WALL

The structure of the gut wall from the beginning of the esophagus to the cloaca or anus is essentially the same throughout its length, consisting of four histological layers: **mucosa, submucosa, muscularis externa,** and **serosa** (fig. 12.16). Differences in the detailed histology of any segment are primarily in the thickness of some of its layers and in the nature of its glands. Variations reflect the specific role of the segment.

The mucosa consists chiefly of (1) a glandular epithelial lining of endodermal origin; (2) an underlying layer of not very dense (areolar) connective tissue supporting the base of cryptlike epithelial glands, lymph nodules of varying sizes, and blood and lymph capillaries that service the glandular epithelium; and (3) a thin coat of smooth muscle fibers (muscularis mucosae, absent in some regions). Mucous glands are ubiquitous, providing, among other benefits, a lubricant that facilitates passage of the contents of the tract during peristalsis.

The submucosa is a thicker layer of connective tissue that supports the base of compound alveolar glands and, of vital importance, a rich plexus of arterioles (arteries of small diameter), venules (small veins), and lymphatics that service the capillary beds of the mucosa, bringing nutrients, oxygen, and raw materials for the synthesis of glandular secretions, and carrying away the waste products of metabolism and the absorbed products of digestion.

The muscularis externa consists of smooth muscle tissue arranged in two clearly defined layers, an **inner circular layer,** in which smooth muscle fibers encircle the gut, constricting the lumen on neural demand, and an **outer longitudinal layer** of muscle fibers that contracts short segments of the gut. The combined action of the two produces the macerating, peristaltic, and, in the colon of mammals, the segmenting actions of the gut. Neural stimuli for contraction are supplied from autonomic (visceral) nerve plexuses located between the longitudinal and circular muscle layers and between the circular layer and the submucosa.

The serosa consists of loose connective tissue (the adventitia) and a covering of visceral peritoneum. The

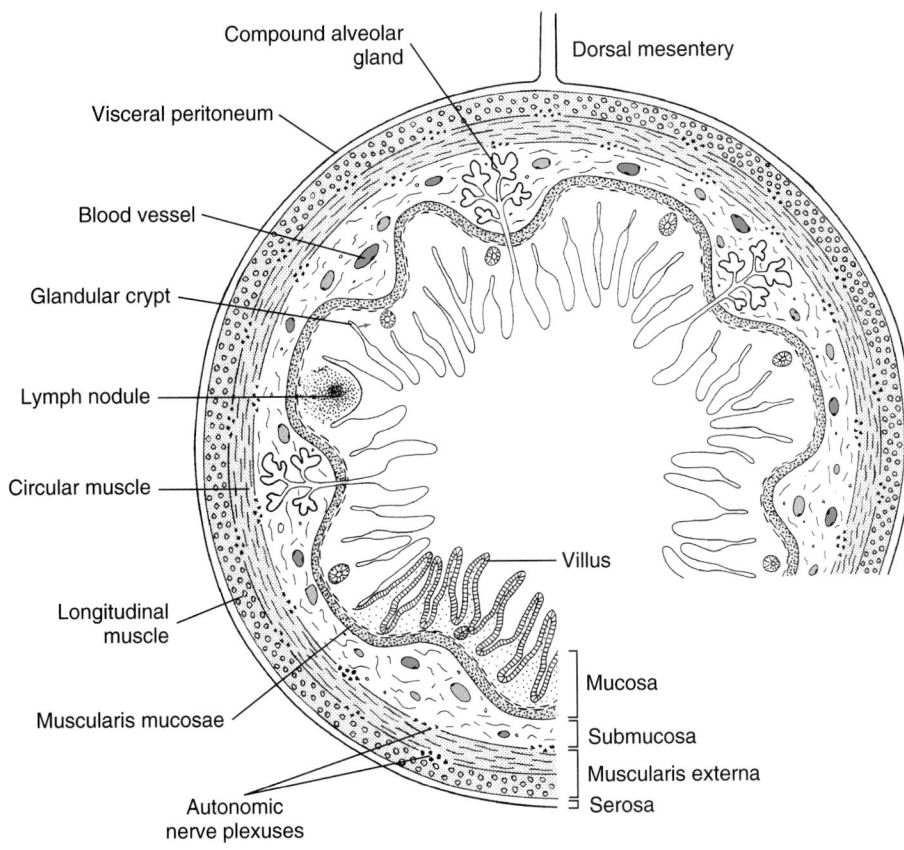

FIGURE 12.16

Cross section of mammalian digestive tract at level of small intestine. The four layers identified at bottom of figure are present from esophagus to cloaca or anus. Villi are confined to small intestine.

serosa exudes small amounts of a serous fluid that lubricates the surface of the viscera, reducing friction generated when the organs rub against one another. Inflammation of the serosa (peritonitis) results in exudation of excessive quantities of fluid.

The esophagus and the caudalmost portion of the intestine generally lie against the body wall; hence, they may be covered by serosa on only one surface, the one that bulges into the coelom.

The entire digestive tract is ciliated in many larval craniates (as in protochordates), and there are cilia in the stomach of many teleosts, in the oral cavity, pharynx, esophagus, and stomach of some adult amphibians, and in other locations in various species. Cilia are present for a time in the stomach of a human fetus. However, peristalsis is chiefly responsible for moving foodstuffs along the alimentary canal of craniates.

ESOPHAGUS

The esophagus is a distensible muscular tube, shortest in fishes and neckless tetrapods, and extending between the pharynx and stomach. In fishes, it serves as a

sphincter that closes the passageway to the stomach during the phase of respiration when water is being forced across the gills. Its only function otherwise in fishes and tetrapods is to conduct foodstuffs to the stomach. The glands in its lining secrete only mucus. Striated muscle at the cephalic end of a long esophagus is gradually replaced farther down by smooth muscle, except in ruminants. In these cud-chewing mammals, striated muscle tissue continues onto the walls of the rumen of the stomach. However, the lining of the rumen is similar to that of the esophagus, and it produces no digestive enzymes. Other features of ruminant anatomy are discussed later in the chapter.

The esophagus is lined by a stratified squamous epithelium, which, in terrestrial turtles, birds, and few mammals, is cornified, enabling the lining to withstand abrasion caused by roughage in the diet. The esophagus of marine turtles is lined by horny papillae that are directed backward, thereby preventing regurgitation while making it easy to swallow slippery seaweed, which is their preferred diet.

One of the few specializations of the esophagus of birds is the **crop** (see fig. 12.20). It is a paired or unpaired membranous diverticulum, or sac, found chiefly in grain eaters, who use it for hoarding seeds and grain until there is room for them in the stomach. Under the stimulus of prolactin from the pituitary, the cells of the glandular area of the lining of the sac undergo fatty degeneration and are shed as a holocrine secretion. This is regurgitated along with partially digested food and fed to nestlings as "pigeon's milk." After the nestlings leave, prolactin levels fall, and the glandular area regresses.

The esophagus of vampire bats has a very narrow lumen through which only fluids can pass. This is correlated with their sanguinivorous diet. The adaptation inspires the speculative question of whether a small esophageal lumen forced these bats to adopt a diet of mammalian blood (the only available fluid containing all the nourishment needed by most mammals), or whether the narrowing of the esophageal lumen resulted from the disuse of the part as a passageway for solid foods (see "Doctrine of Acquired Characteristics," chapter 2). Our present understanding of hereditary mechanisms provides no explanation as to how the use or disuse of a body part could alter the gamete cells necessary for reproduction. This observation makes the doctrine suspect but does not disprove it.

STOMACH

The stomach is a muscular chamber or series of chambers that serves as a receiving site for recently ingested foods, secretes digestive enzymes and lubricatory mucus, and macerates food while mixing it with the gastric juices. The mucus and enzymes partially liquify solid foods before they are injected into the small intestine. The stomach terminates at the **pylorus**, which is the opening from the stomach leading into the duodenum. The opening is surrounded by a ring of smooth muscle, the **pyloric sphincter.**

When there is more than one chamber, examples of which are found in most craniate classes other than amphibians and living agnathans that lack a stomach, the first chamber usually serves simply as a temporary holding site for recently ingested food. Its epithelium is no different from that of the esophagus, having many mucous glands but none of the gastric glands found farther along.

The stomach is straight when it first differentiates in embryos and remains straight throughout life in some basal vertebrates. Most often, flexures develop, producing J-shaped or U-shaped stomachs (figs. 12.17, 12.18, and 12.19). As a result, the stomach may exhibit a concave border, or **lesser curvature**, and a convex border, or **greater curvature**, the latter in mammals having been the embryonic dorsal border connected to the coelomic roof by the dorsal mesentery (**mesogaster**). Not only do most stomachs exhibit a flexure, but those of mammals undergo torsion along with part of their dorsal mesentery, twisting in such manner that stomach and mesentery lie more or less crosswise in the trunk. The mesentery, elongated and still attached to the greater curvature, is draped like a curtain between the ventral body wall and the intestines, where it constitutes the **greater omentum**. In humans, it covers the abdominal viscera like an apron. Within the double-walled omentum is a small part of the coelom, the **lesser peritoneal cavity**, that was trapped when the embryonic stomach and dorsal mesentery were rotating around a dorsoventral axis in the process of assuming their adult orientation in the coelom. The cavity remains continuous with the main abdominal coelom via a small passageway, the **epiploic foramen**.

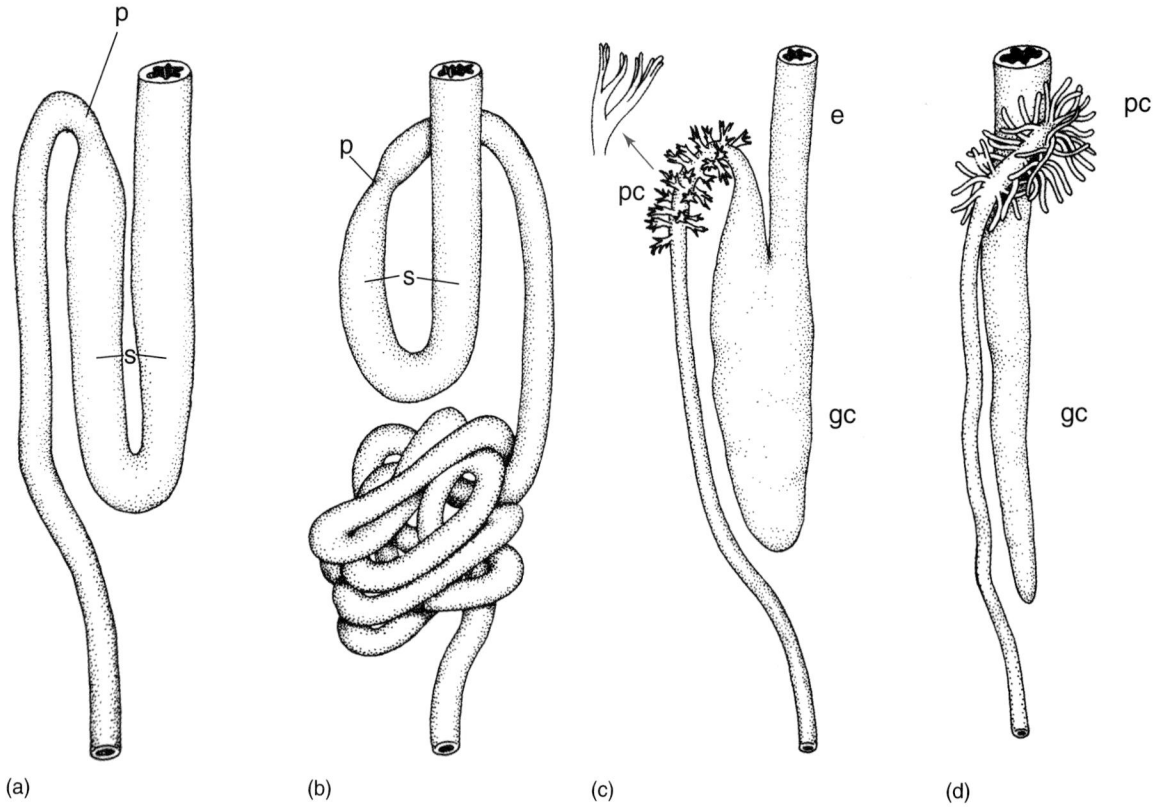

(a)   (b)   (c)   (d)

FIGURE 12.17

Digestive tracts of four teleosts. (*a*) *Fundulus*. (*b*) *Cyprinodon*. (*c*) *Elops*. (*d*) *Trichiurus*. **e,** esophagus; **gc,** cecumlike stomach; **p,** pylorus; **pc,** pyloric ceca; **s,** stomach.

Living agnathans have no definitive stomach. The digestive tract is one long tube from mouth to vent exhibiting no gross differentiation of esophagus, stomach, or intestine (see fig. 12.1). The epithelium of the agnathan tract is a single layer of cells including goblet cells that secrete mucus and flask-shaped cells that synthesize proteolytic (protein-splitting) enzymes. The base of each cell is in contact with the underlying vascularized layer of the mucosa, from which they receive nourishment.

The stomachs of fishes display a wide variety of shapes, and the epithelium is sometimes ciliated. The gar stomach is almost straight; sharks exhibit the more common **J** shape (see fig. 12.1). The entire stomach of some teleosts is one large cecum (see fig. 12.17*c* and *d*). Chimaeras and lungfishes have no definitive stomach, or have one that is poorly differentiated and lacks digestive glands.

The stomach of a frog (see fig. 12.18) is not distinguishable grossly from the esophagus. Both regions are capable of enormous distension. The stomach of a urodele (*Necturus*) is seen in figure 12.1.

The stomach of crocodilians and birds is divided into two parts, **proventriculus** and **gizzard** (figs. 12.19 and 12.20). The proventriculus secrets the digestive enzymes. The gizzard, lined with a horny membrane, is simply a grinding mill that makes a mash of food mixed with gastric secretions. Pebbles that have been swallowed are usually retained in the gizzard where they assist in macerating the food. The proventriculus and gizzard of carnivorous birds are less well differentiated.

Terms derived from human anatomy have been applied to regions of the stomach of other vertebrates, sometimes inappropriately. In humans (fig. 12.21*e*), the region at the base of the esophagus is the **cardiac portion** because it lies close to the heart. Lateral to the cardiac region in humans is the **fundus**. It is characterized by a specific array of gastric glands. The region between the greater and lesser curvatures is the **body** of the stomach, and the region preceding the pylorous is the **pyloric portion**. It houses the **pyloric canal**. X-ray films reveal that the shape and orientation of a human stomach in life are less transverse than in cadavers, and that the shape varies with posture, volume of the contents, and the activity of the intestines on which the stomach rests.

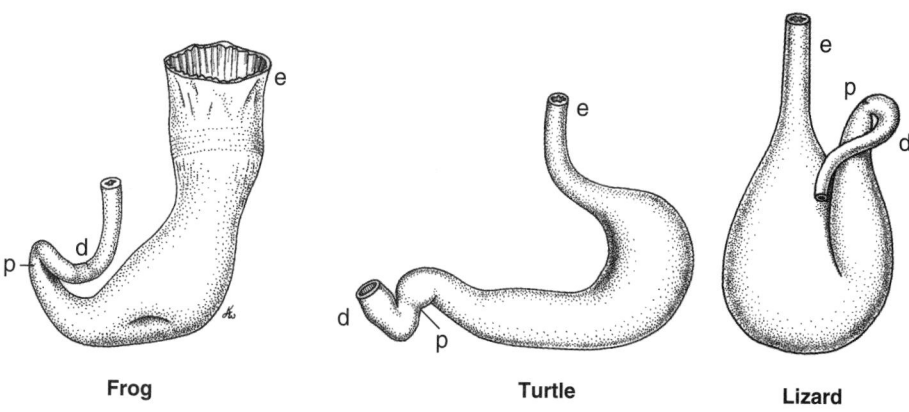

FIGURE 12.18

Stomach of a frog, turtle, and lizard. **d,** duodenum; **e,** esophagus; **p,** pylorus. Shape of stomachs will vary with volume of contents.

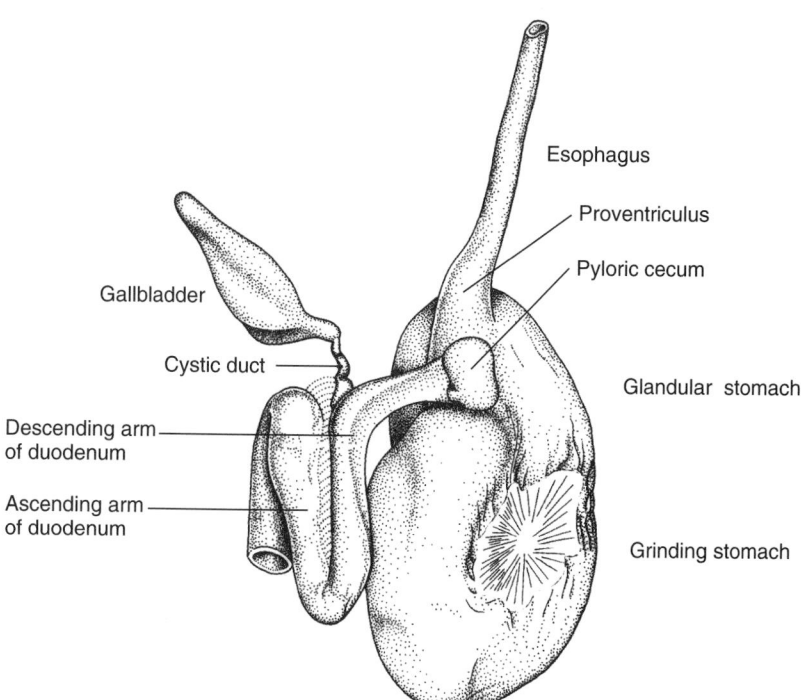

FIGURE 12.19

Stomach and associated structures of a caiman, ventral view. The gallbladder has been lifted cephalad from its normal position between the stomach and descending arm of the duodenum. The grinding stomach has thick muscular walls, except in the midsection where there is a glistening fibrous membrane.

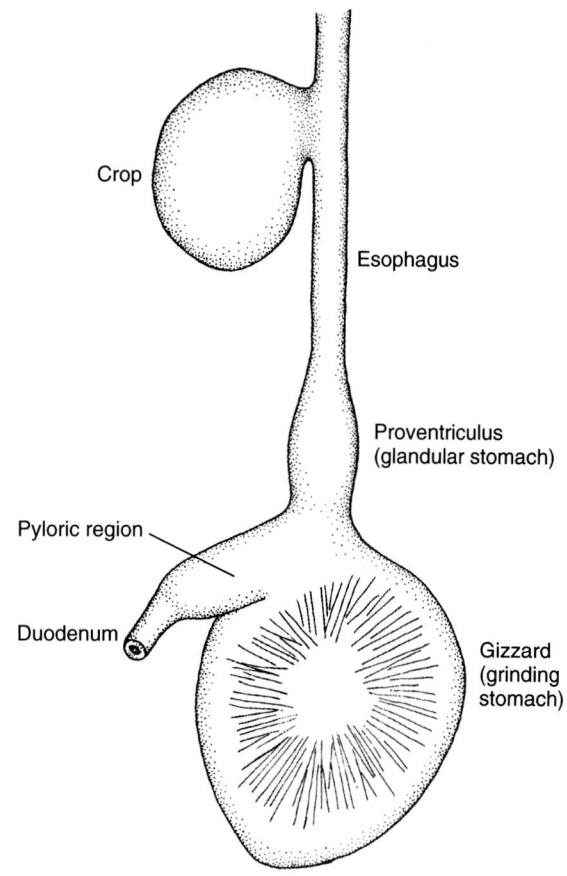

Crop

Esophagus

Proventriculus
(glandular stomach)

Pyloric region

Duodenum

Gizzard
(grinding
stomach)

FIGURE 12.20
Esophagus and stomach of a grain-eating bird.

Fundic glands are elongated *simple tubular* glands, many with two tubules that open into a single terminal duct. There are several types of cells in their epithelium, among them zymogenic cells (**chief cells**) that synthesize and release the preenzyme **pepsinogen**; **parietal cells**, which secrete **hydrochloric acid**; and **goblet cells**, found mostly in the neck of the tubule. When released into the lumen of the stomach, hydrochloric acid splits the active enzyme pepsin (stomach protease) from a larger pepsinogen molecule, and pepsin initiates the process of protein digestion that eventually yields absorbable amino acids. Mucous cells lubricate the lining and provide a vehicle for the accumulating products of digestion.

The cardiac and pyloric regions of the human stomach can be distinguished from other regions on the basis of the histology of their glands. Neither have any zymogenic cells. The mucosa in the cardiac region resembles that of the lower end of the esophagus, the glands being *compound tubular* (see fig. 6.10*d*) with many goblet cells. A few parietal cells are also present. Pyloric glands are *simple branched tubular glands* (see fig. 6.10*c*), and they extend deeper into the mucosa than other types. They have many goblet cells and relatively few parietal cells. Between the fundus and the pyloric region, the

mucosa exhibits typical gastric glands like those of the fundus, with a gradual shift toward a pyloric mucosa in the lower third of the organ. The pattern of distribution of these different varieties of gastric mucosas is not necessarily the same in other mammals.

As mentioned earlier, stomachs are often divided into several distinct chambers in animals whose food requires prolonged processing, as in mammals that consume grasses and grain. The stomach of a ruminant, adapted for processing cellulose, is an extreme example (fig. 12.22). Grasses are taken into the mouth, manipulated, and swallowed without processing. They pass down the esophagus and into the **rumen**. There the food becomes mixed with mucus and with **cellulase**, an enzyme that digests cellulose, a major component of the cell walls of plants. Cellulase is not secreted by the rumen but by a multitude of anaerobic bacteria that live there. The enzyme cleaves cellulose molecules into simpler carbohydrates that can then be further digested by the animal's own digestive juices. The muscular walls of the rumen knead the contents—mucus, enzyme, and vegetation. At intervals, some of the contents pass to the **reticulum**, so named because the lining is honeycombed (reticulated) by ridges and deep pits. Here the process of cellulose fermentation continues, and small boluses, or cuds, of fermenting pulp are regurgitated for further maceration by the teeth. The process of mastication, swallowing, and regurgitation may occur several times. Eventually, thoroughly masticated mash is swallowed, and it passes to the **omasum**, a temporary holding site, before it is moved to the **abomasum**. The latter is the true glandular stomach where gastric enzymes are added to the mash. The lining of the other chambers is much like that of the esophagus.

Other than water, few components of the diet of craniates are absorbed in the stomach. The accumulating product of stomach activity is a soupy mixture, **chyme**. When the pH of chyme in the pyloric canal is appropriately acidic, visceral receptors sense this and trigger a reflex arc that relaxes the pyloric sphincter for the injection of chyme into the duodenum. Figure 12.21 shows the location of gastric glands in the stomachs of mammals that have a variety of diets.

INTESTINE

The intestine begins at the pyloric sphincter and ends at its entrance into the cloaca or, if there is no cloaca, at the anus. The morphology of the intestine from fish to human is as different as the logistics attendant on digesting and absorbing the components of the diet, whether microscopic or gross, plant or animal, and on the frequency of meals and their volume. The intestine of tetrapods consists of small and large segments, each with a distinctive epithelium. That of fishes does not exhibit this differentiation.

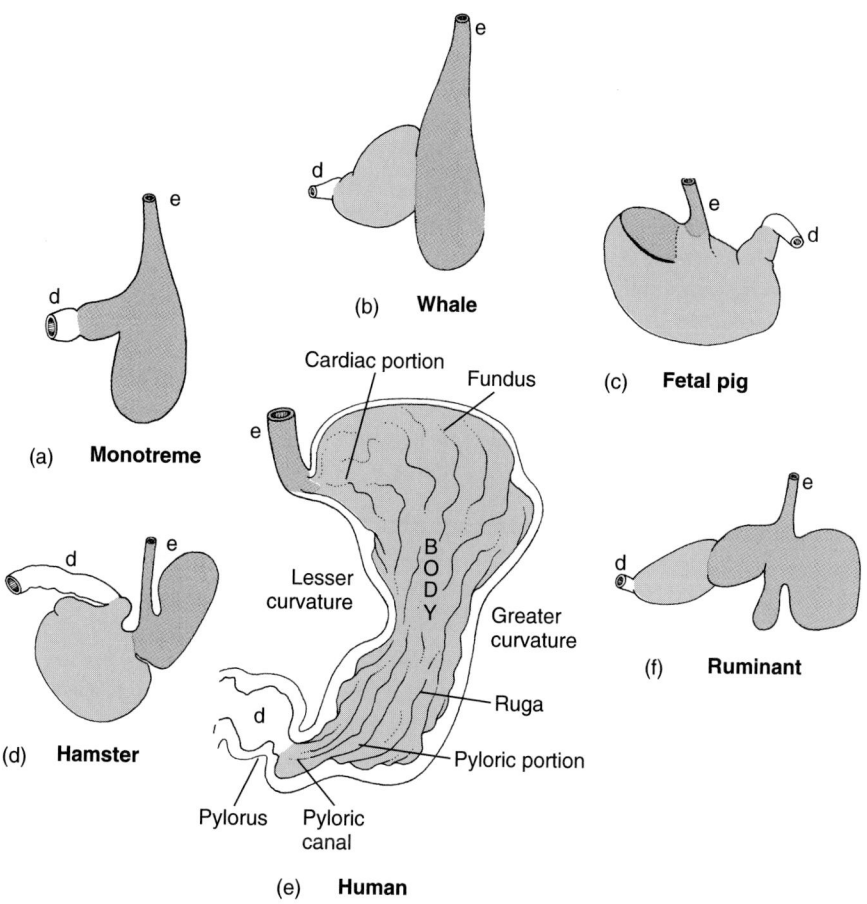

(a) **Monotreme**

(b) **Whale**

(c) **Fetal pig**

(d) **Hamster**

Cardiac portion

Fundus

e

Lesser
curvature

B
O
D
Y

Greater
curvature

Ruga

Pyloric portion

Pylorus   Pyloric
canal

d

(e) **Human**

(f) **Ruminant**

FIGURE 12.21

Distribution of esophageal-like epithelia (*gray*) and glandular epithelia (*red*) in stomach of selected mammals, ventral views. **d**, duodenum; **e**, esophagus. The fetal pig stomach is a dorsal view to show the esophageal diverticulum that evaginates near the base of the esophagus.

Fishes

The intestine of living agnathans, chondrichthyes, and basal bony fishes is quite straight (see fig. 12.1, living agnathans, gar, shark). A few teleosts—killifish, for example—exhibit intestinal coiling, but this is rare (see fig. 12.17b). A **spiral intestine**, so named because a **spiral valve**, or **typhlosole**, is suspended within its lumen, is present in many fishes other than teleosts (see fig. 12.1, shark). It serves the same function as intestinal coils, increasing the epithelial area available for absorption. Beyond the spiral intestine, a short **postvalvular intestine** leads to the cloaca. Living agnathans have a typhlosole, but it is a mere ridge projecting into the intestinal lumen. It makes only a few spirals throughout its length.

Intestinal ceca are the major adaptations for increasing the absorptive area of the teleost intestine. Most common are **pyloric ceca**, which are diverticula near the pylorus (see fig. 12.17c and d). Mackerel have up to 200 pyloric ceca.

The duct of a cecumlike, finger-shaped **rectal gland** empties into the short postvalvular intestine of sharks, but it has no digestive function. It extracts and excretes excess sodium chloride from the blood.

Tetrapods: The Small Intestine

Tetrapods have a small and large intestine. The short curved duodenum is the first segment of the small intestine. The remainder is coiled to one degree or another, except in urodeles and apodans. In lizards, birds, and mammals, it is lined with fingerlike or leaflike **villi**, low in birds, tall in mammals, so densely crowded that they give the lining a velvety appearance (see fig. 12.16). (A hand lens or dissecting microscope will enable you to observe this effect if you are dissecting a mammal.) Villi greatly increase the absorptive surface of the intestine. Digested (hydrolyzed) lipids absorbed through their epithelium enter dead-end lymph vessels, or **lacteals**, within the villi (fig. 12.23). By suddenly shortening, lacteals pump the milky fluid, **chyle**, into larger lymphatics that eventually empty into a major vein near the heart (see thoracic duct of a mammal, fig. 14.41). The lipids are then carried by the bloodstream to the liver, the chief site of synthesis of cholesterol and other sterols and lipids, and to other tissues primarily for storage.

In mammals, the small intestine beyond the duodenum is divided into **jejunum** and **ileum** on the basis of the shape of the villi, the nature of the epithelial lining, and the size of the lymph nodules in the mucosa. Small

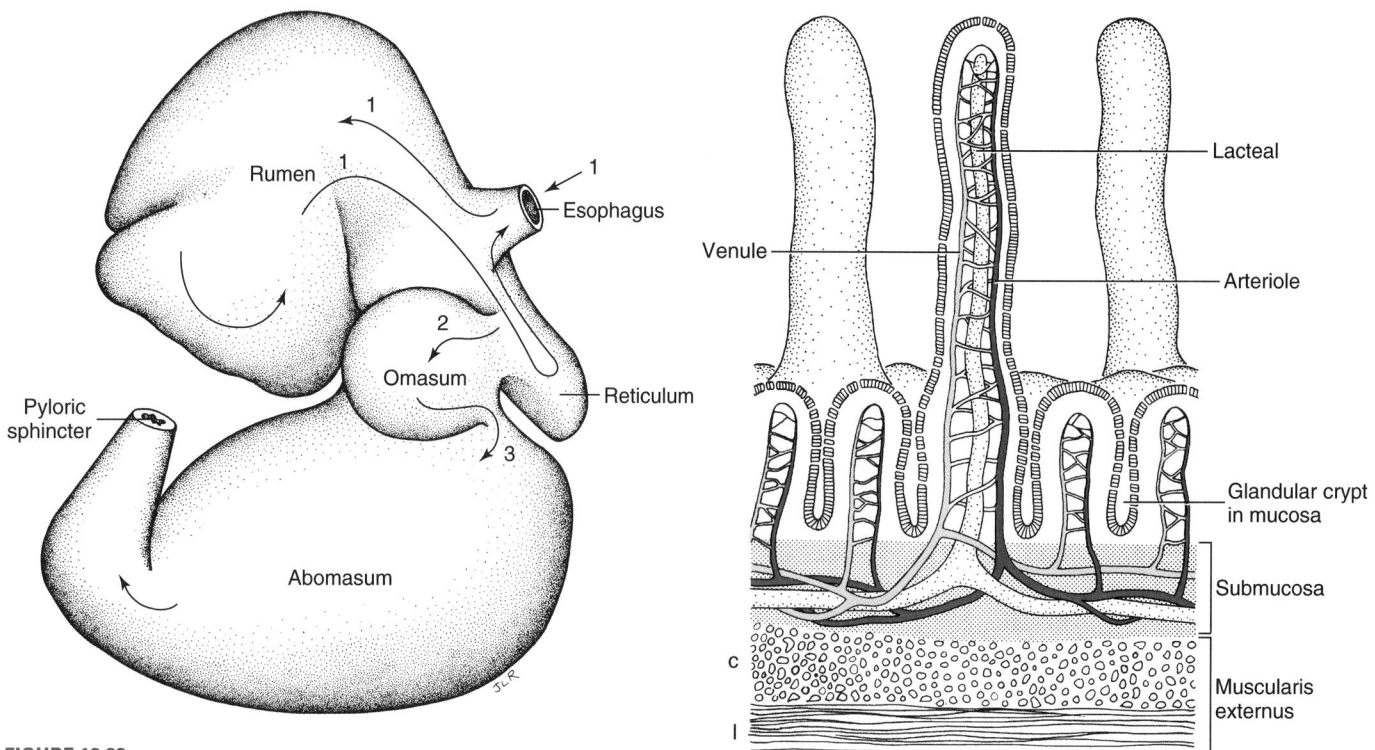

FIGURE 12.22
Stomach of a calf (ruminant). **1,** route of unmacerated food and of regurgitated cud; **2,** route of fermenting mash; **3,** entrance of mash into glandular stomach.

FIGURE 12.23
Blood supply and lymph drainage of a villus. **c,** circular muscle; **l,** longitudinal muscle.

nodules are common throughout the length of the intestine (see fig. 12.16), but in the ileum they aggregate into large masses known as **Peyer's patches.** The small intestine of amniotes terminates at an **ileocolic sphincter** (fig. 12.24), which regulates the ejection of the contents of the ileum into the large intestine.

The small intestine is the chief site of digestion and absorption of nutrients, although, as we have seen, ptyalin, pepsin, and cellulase (the latter from commensal bacteria), split starch, proteins, and cellulose in the stomachs of one species or another. However, the final stage of digestion that results in absorbable nutrients takes place in the small intestine in the presence of intestinal juice and pancreatic enzymes. The enzymes of **intestinal juice** are secreted by glands in the epithelial lining of the crypts and by compound alveolar glands near the pylorus (see fig. 12.16). The enzymes split polypeptides into amino acids, and they split disaccharides into monosaccharides (simple sugars). Amino acids and simple sugars are absorbable molecules. **Pancreatic juices** contribute an **amylase** that acts on carbohydrates, a **lipase** that digests lipids yielding absorbable fatty acids and glycerol, and **proteolytic enzymes** that continue the digestion initiated by pepsin. The digestive process reaches a crescendo in the small intestine. By the time the contents of the ileum arrive at the ileocolic sphincter, the recoverable nutrients that were ingested

earlier have been absorbed into the bloodstream. Except in some herbivores, what is left is chiefly water and undigestible roughage. Most of the water is absorbed in the colon.

The Large Intestine

The large intestine is rarely coiled, but ceca are common. In mammals and in some reptiles including birds, the large intestine is divisible into **colon**, which commences at the ileocolic sphincter, and **rectum**, a straight terminal portion in the pelvic cavity. The colon of some mammals has ascending, transverse, and descending portions with pronounced flexures between them; in humans, the descending colon ends in a **sigmoid flexure** (see fig. 12.1, human). Although ceca are rare beyond the duodenum in fishes and amphibians, ceca just beyond the ileocolic sphincter (**ileocolic ceca**) are common in amniotes. Birds typically have two (see fig. 12.1, chicken). In mammals that feed exclusively on fibrous vegetation, fruits, grains, and seeds, the ileocolic cecum may exceed the large intestine in capacity. An **appendix** with a lumen and histology the same as that of the cecum terminates the cecum in anthropoids, rodents, and many mammals. The cecum and appendix of a rabbit are seen in figure 12.24. As one might expect, the ileocolic cecum of a carnivore is quite abbreviated, as is that of humans.

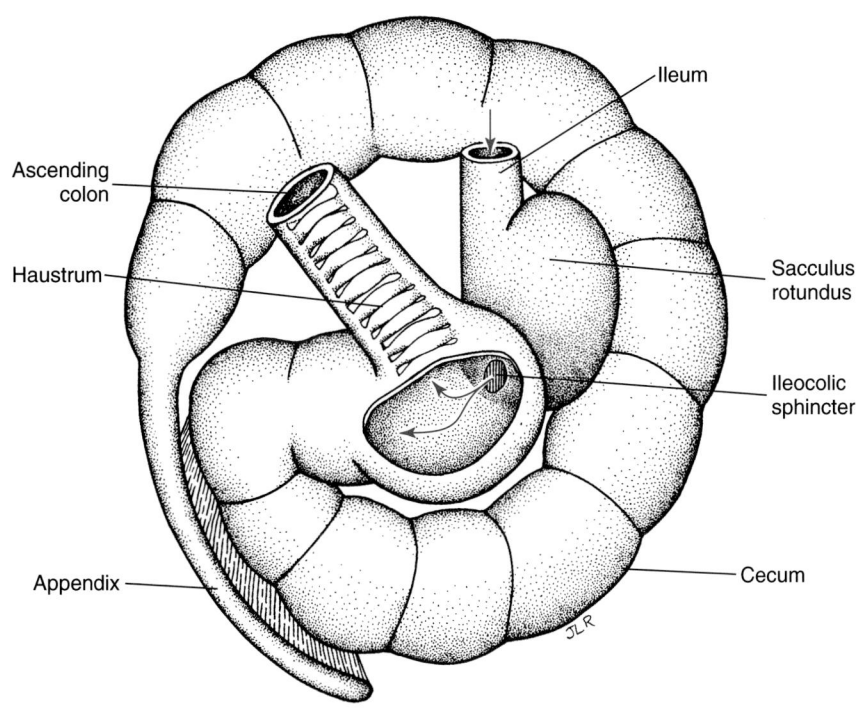

FIGURE 12.24
Ileocolic junction, cecum, and vermiform appendix of rabbit. The haustrum is a pouchlike structure of the large intestine, which is also seen in horses and pigs.

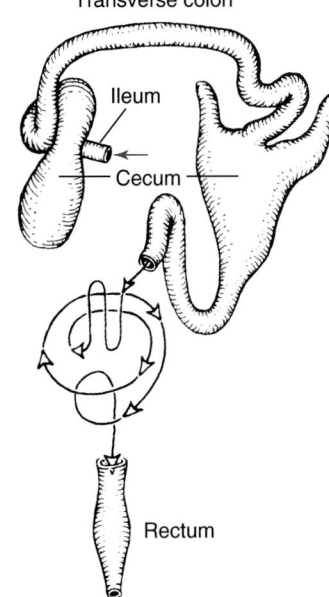

FIGURE 12.25
Large intestine of *Hyrax*. *Arrows* indicate direction taken by missing segment.

The koala, a marsupial 2 to 2½ ft (60 to 75 cm) long that feeds preferentially on eucalyptus leaves, less frequently on eucalyptus bark, buds, and unripened woody fruits, has a cecum that may be up to 6 ft long. This is correlated with the observation that the diet has a low nutritional value, which mandates a large gut capacity. Thoroughly chewed leaves pass to the small intestine where any cytoplasmic nutrients are absorbed, then through the ileocolic sphincter and into the cecum. There, during the course of more than a week of fermentation by enzymes produced by commensal anaerobic bacteria, cellulose (from the cell walls of plants) is converted into absorbable carbohydrates, as in the rumen of ungulates.

A *Hyrax*, whose diet also consists of seeds, fruits, and leaves, has an ileocolic cecum, colic flexures, a bicornuate cecum farther along on the intestine, and then an intestinal coil before reaching the rectum (fig. 12.25).

The colon in all tetrapods recovers water from the residual contents (feces) of the intestine. Water reclamation in terrestrial animals is necessary to replace that "borrowed" earlier from tissues for the digestive process. Failure to recover water would lead to dehydration and possible death.

LIVER AND GALLBLADDER

The liver arises from the midventral aspect of the midgut as a hollow cecumlike diverticulum, the **liver bud** (fig. 12.26*a*). The bud grows cephalad in the ventral mesentery of the stomach. The growing tip of the bud gives rise to numerous sprouts that become the liver and the gallbladder. The cephalic pole of the liver finally becomes anchored to the embryonic transverse septum (see fig. 1.12) by a **coronary ligament**.

The lobes of the adult liver are drained by several **hepatic ducts**, and the gallbladder is drained by a **cystic duct**. These ducts converge very close to the base of the gallbladder at such an angle that bile from the liver can pass via the cystic duct into the gallbladder, where it is stored. The site where the hepatic and cystic ducts converge is the beginning of a **common bile duct** that empties into the duodenum (8 in fig. 12.26). A short terminal segment of the common bile duct is embedded in the wall of the duodenum, where it constitutes an **ampulla of Vater**.

Although most of the embryonic ventral mesentery disappears during subsequent development, the mesentery ventral to the duodenum and stomach that was invaded by the liver bud remains as the **hepatoduodenal ligament** connecting the duodenum and liver, and as the **gastrohepatic ligament** that connects the pyloric

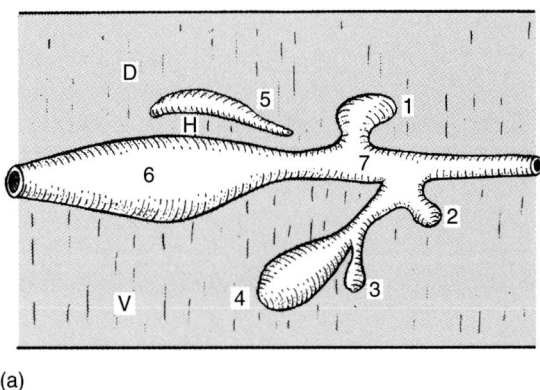
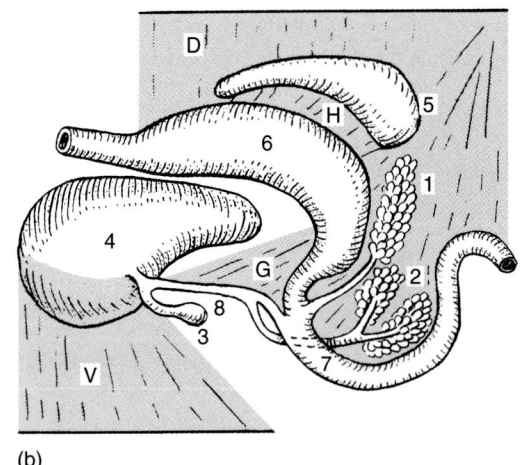

(a)   (b)

FIGURE 12.26

Development of liver, pancreas, spleen, stomach, and associated mesenteries. (*a*) Early stage. (*b*) A later stage. **1,** dorsal pancreatic bud from duodenum; **2,** ventral pancreatic bud; **3,** gallbladder; **4,** liver bud in (*a*) and liver in (*b*); **5,** spleen; **6,** stomach; **7,** duodenum; **8,** common bile duct. **D,** dorsal mesentery; **G,** lesser omentum; **H,** gastrosplenic ligament; **V,** ventral mesentery in *(a),* falciform ligament in (*b*).

stomach and the liver. These two ligamentous mesenteries constitute the **lesser omentum**. It serves as a bridge that conducts the common bile duct to the duodenum and the hepatic artery and hepatic portal vein to the liver. The embryonic mesentery ventral to the liver remains in adults as the **falciform ligament**. (The term describes its shape in humans.) The shape of the liver conforms to the space available in the coelom. In animals with eel-like bodies, the liver is elongate. In those with short trunks, it is short and wide, as in humans.

The liver has many roles. It produces **bile**, an alkaline fluid containing bile salts that emulsify lipids in the small intestine and impart to them an alkalinity necessary for digestion. Without emulsification, only a small proportion of dietary fat would be digested. Lipids are stored in liver cells and in adipose tissue throughout the body.

Some liver cells phagocytose aging red blood cells, splitting the hemoglobin molecule and freeing the iron. The remainder of the molecule is converted into red and green pigments (**bilirubin** and **biliverdin**) that are excreted as part of the bile.* (Red blood cells are phagocytosed also in the spleen and red bone marrow. The pigments from those sites are extracted from the bloodstream in the liver and contribute to the bile.) The fetal liver is a source of red blood cells, but that role is taken over later by other tissues.

The liver removes from the circulation glucose in excess of immediate tissue needs. The glucose is converted to glucose-6-phosphate and then stored in the

liver as glycogen. The latter is converted back to glucose as needed to maintain an appropriate level in the circulation.

The liver removes dietary amino acids from the hepatic portal system and deaminates them. The byproducts of deamination include ammonia, uric acid, and urea. These are eventually excreted in the urine in one form or another, depending on the species (see chapter 15, "Excretion of Nitrogenous Wastes").

Several blood proteins are manufactured in the liver, including **fibrinogen** and **prothrombin**, which are essential for blood clotting. The liver has still other roles that are discussed in physiology textbooks.

A gallbladder develops in most craniates. None develops in lampreys, a few teleosts, many birds, perissodactyls, whales, and some rodents, including rats. Craniates lacking a gallbladder are among those that have little fat in their diet. Human beings can live without a gallbladder, but they must be careful about the amount of lipids they consume. The gastrointestinal hormones associated with the digestive process are discussed in chapter 18.

EXOCRINE PANCREAS

The pancreas consists of two histologically distinct and functionally independent components. An *exocrine* portion produces digestive enzymes in **alveoli (acini)**. The enzymes are transported via **pancreatic ducts** to the duodenum. An *endocrine* portion bears **pancreatic islets (islands of Langerhans)**, lacks ducts, and thus secretes its hormonal products, insulin and glucagon, into the bloodstream (fig. 12.27). These two components are not always part of the same organ (see chapter 18, "Endocrine Pancreas").

*The "black and blue" disfiguration of bruised skin is caused by the pigments released from traumatized red blood cells being phagocytosed at the site of trauma.

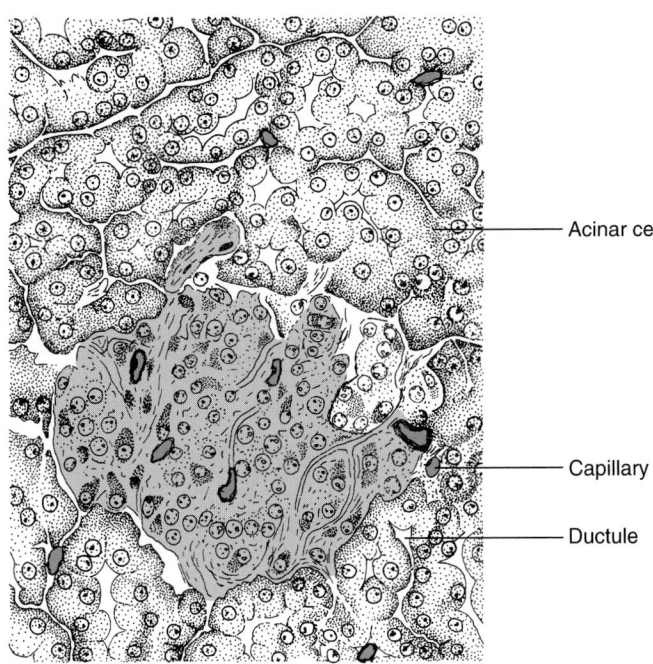

Acinar cell

Capillary

Ductule

FIGURE 12.27
Island of Langerhans (*light red*) and surrounding acini (alveoli) in a mammalian pancreas.

The pancreas varies from diffuse to compact. When diffuse, as in teleosts and many other craniates, pancreatic tissue is distributed along the blood vessels in the ventral mesentery of the stomach and duodenum. When compact, it may consist of several discrete lobes. In living agnathans, there is no definitive pancreas, the exocrine and endocrine components are spatially separated, and many exocrine cells remain in the intestinal epithelium.

The pancreas arises typically as one or two ventral **pancreatic buds** off the liver bud and as a single dorsal pancreatic bud from the foregut immediately caudal to the stomach (see fig. 12.26*a*). The ventral buds invade the ventral mesentery of the duodenum and stomach and generally coalesce to form a ventral lobe (**body**) of the pancreas (see fig. 12.26*b*, 2). The dorsal bud becomes the dorsal lobe (**tail**) of the pancreas. There are variants of this pattern. In sharks, the entire pancreas develops from the dorsal bud, and in most mammals it develops from one ventral and one dorsal bud.

Craniates may have as many pancreatic ducts as there were embryonic pancreatic buds, but more often one or more of the ducts loses its connection with the bile duct or gut from which it evaginated, and the pancreas is drained by the remaining duct or ducts. For example, in sheep the dorsal lobe loses its connection with the gut, and the entire pancreas drains into the common bile duct. In pigs and oxen, the duct of the ventral lobe loses its connection with the common bile duct, and the entire pancreas drains directly into the duodenum. In still

other mammals, both ducts remain. When one duct is larger, as in cats and humans, the other is relegated to the status of **accessory pancreatic duct**. The variations in the drainage of the mammalian pancreas illustrate, once again, that a primitive pattern may be modified in any of several directions during the course of phylogenesis.

CLOACA

In many fishes and in most tetrapods other than therian mammals, the digestive tract terminates in a common chamber, the cloaca (Latin, "sewer"), into which the urinary and genital tracts also empty. The cloaca is then vented to the exterior. In lampreys, chimaeras, living female coelacanths, and ray-finned fishes, the embryonic cloaca becomes increasingly shallow or disappears as development proceeds, and the digestive tract eventually opens independently to the exterior. In therian mammals, the cloaca becomes partitioned during embryonic development into two or three separate passageways, one of which becomes the rectum leading to an anus (see fig. 15.49). Figure 15.48 illustrates the exit from the digestive tract in two anatomically diverse mammals, a monotreme and a female rodent.

Summary

1. Filter feeding in an aquatic environment is the oldest craniate method of acquiring food. It is still employed by larval lampreys and, with modifications, by a few jawed fishes and baleen whales.
2. The early embryonic digestive tract consists of foregut, midgut, and hindgut. Excepting its epithelial lining, the gut forms from splanchnic mesoderm. The epithelium is endodermal in origin.
3. Stomodeal and proctodeal invaginations establish an entrance (mouth) and exit (vent or anus).
4. The gut is suspended within the coelom by a dorsal mesentery that is continuous with the parietal peritoneum lining the body wall (somatopleure) and with the visceral peritoneum on the surface of the organs.
5. An embryonic ventral mesentery disappears except at two locations, one of which is the falciform ligament of the liver.
6. The mouth is the anterior opening to the digestive tract. Major subdivisions of the adult tract are oropharyngeal or oral cavity, pharynx, esophagus, stomach, and intestine. Chief accessory organs are tongue, teeth, oral glands, pancreas, liver, and gallbladder.
7. Nasal passageways open into the oropharyngeal cavity in lobe-finned fishes and into the oral cavity

in tetrapods with primary palates. Multicellular oral glands open onto the roof, walls, and floor of these cavities but are scarce in fishes.

8. Jawed fishes and perennibranchiate amphibians have a primary tongue overlying the ventralmost component of the hyoid skeleton. As a derived condition in amphibians, a glandular field contributes to the tongue, and in amniotes paired lateral lingual swellings contribute. An entoglossal bone stiffens the tongue of lizards and birds.

9. Teeth are vestiges of dermal armor. They consist of dentin formed by odontoblasts in dermal papillae, a hardened outer layer of enamel or enameloid, and cementum. Enameloid, like dentin, is formed by odontoblasts. Enamel, ectodermal in origin, is formed by ameloblasts of enamel organs. In basal vertebrates, teeth are more numerous, more widely distributed in the oral cavity, more frequently replaced, and more alike throughout the oral cavity. A few craniates in every class are toothless. A few have horny teeth.

10. Dentition may be monophyodont, diphyodont, or polyphyodont; acrodont, pleurodont, or thecodont; homodont or heterodont; secodont, selenodont, lophodont, or bunodont. Primitive mammalian cheek teeth were triconodont and trituberculate.

11. The pharynx is the part of the foregut that has pharyngeal pouches in the embryo. The tetrapod pharynx has openings to the middle ear cavity, larynx, and esophagus. Amniotes have nasal and oral pharynges. Some mammals have a laryngeal pharynx.

12. The wall of the digestive tract caudal to the pharynx consists of a mucosa, submucosa, muscularis externa, and serosa (visceral peritoneum overlying a loose connective tissue membrane, the adventitia).

13. The esophagus connects the pharynx with the stomach. In birds, it has a diverticulum, the crop. The esophagus is lined with mucous glands.

14. The stomach is compartmentalized in birds (proventriculus and gizzard), in cud-chewing ungulates the true glandular stomach (abomasum) is united with specializations of the esophagus (rumen, reticulum, and omasum) to form a complex stomach, and is compartmentalized to a lesser degree in other vertebrates. It terminates at the pyloric sphincter.

15. The intestine is the major site of digestion and absorption. Typhlosoles, coils, ceca, and villi increase the absorptive area. Tetrapods have small and large intestines, the latter in amniotes consisting of colon and rectum. In mammals, the small intestine consists of duodenum, jejunum, and ileum.

16. Liver and pancreatic buds arise as evaginations from the foregut. The gallbladder is an evagination from the liver bud.

17. The intestine opens into a cloaca in most craniates. The cloaca opens to the exterior via a vent. In the absence of an adult cloaca, the large intestine opens to the exterior via an anus.

18. Digestive enzymes are synthesized in one species or another by salivary glands, stomach, intestine (small intestine in amniotes), and pancreas. Bile salts from the liver emulsify ingested lipids in preparation for digestion.

19. Craniates produce no known enzyme for digesting the complex carbohydrate cellulose. The enzyme is provided by commensal bacteria in the stomach of ungulates and in the colic ceca in most other herbivores.

Critical Thinking Questions

1. In procuring and processing food, what systems other than the digestion system are employed?
2. The mammalian air and food passages cross over. When breathing through your nose and chewing, how is food delivered to the correct passage when swallowing?
3. What taxa develop a complete secondary palate? What biological roles do the secondary palates serve in these taxa?
4. Compare how aquatic teleosts and terrestrial amphibians acquire particulate food.
5. What can we say about the habits of a particular organism based on the morphology of its teeth? (Give examples.)
6. Describe the cross section of the gut wall. What are the primary embryological sources for the gut?
7. Name one specialization of the esophagus and indicate the organism that possesses it along with its function.
8. How do you distinguish an organism with a stomach from one that lacks a definitive stomach?
9. What modifications of the digestive tract occur in the cow?
10. Outline the mechanical/chemical digestive processes in the digestive tract of a typical mammal.
11. In the rabbit, the site of bacterial digestion occurs distal to the small intestine (site of absorption for nutrients). Given a unidirectional flow through the digestive tract, can you explain this observation?
12. What evolutionary adaptations are there that improve the diffusion of nutrients within the digestive system?

Selected Readings

Bramble, D. M., and Wake, D. B.: Feeding mechanisms in lower tetrapods. In Hildebrand, M., and others, editors:

Functional vertebrate morphology, p. 159. Cambridge, MA, 1985, Harvard University Press.

Butler, P. M., and Joysey, K. A., editors: Development, function, and evolution of teeth. New York, 1978, Academic Press.

Domning, D. P.: Marching teeth of the manatee, *Natural History* 92:5, May 1983.

Gans, C., and Gorniak, G. C.: How does the toad flip its tongue? Test of two hypotheses, *Science* **216**:1335, 1982.

Gorniak, G. C.: Trends in actions of mammalian masticatory muscles, *American Zoologist* **25**:331, 1985.

Lauder, G. V.: Patterns of evolution in the feeding mechanisms of actinopterygian fishes, *American Zoologist* **22**:275, 1982.

Moss, M.: Enamel and bone in shark teeth; with a note on fibrous enamel in fishes, *Acta Anatomica* **77**:161, 1970.

Osborn, J. W.: The evolution of dentitions, *American Scientist* **61**:548, 1973.

Pivorunas, A.: The feeding mechanisms of baleen whales, *American Scientist* **67**:432, 1979.

Pough, F. H.: Feeding mechanisms, body size, and the ecology and evolution of snakes, *American Zoologist* **23**:339, 1983.

Links to the Internet

Visit the zoology website at http://www.mhhe.com/zoology to find live Internet links for each of the following references.

1. Histology—The Web Laboratory. Links to units on many systems of the body, including the digestive system. Extensive coverage of microstructure of all parts of the GI tract, from Ohio State University.

2. Digestive Physiology of Herbivores. Links to other pages on subjects such as digestion in ruminants and non-ruminants.

3. Digestive Physiology of Birds. A description of the unique features of the digestive tract of birds, including a photograph and links.

4. The Digestive System. A series of slides with information on the function of the parts of the GI tract.

Respiratory System

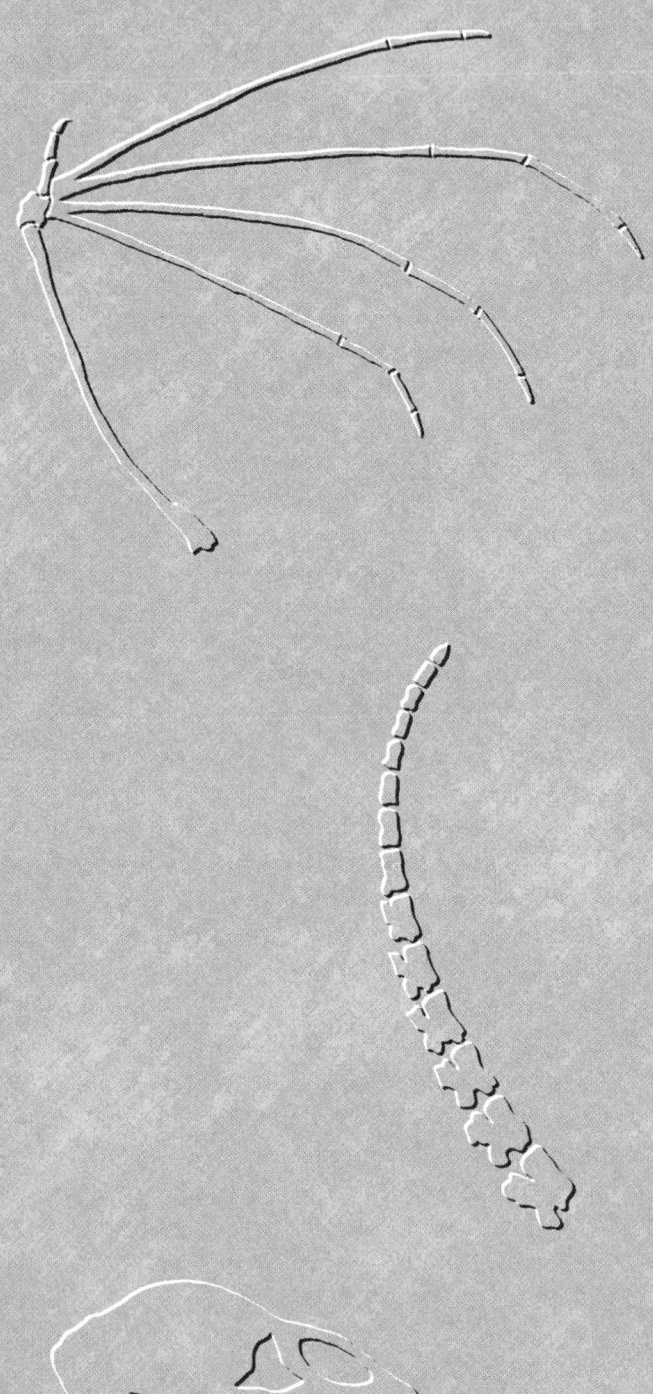

In this chapter, we look at some of the methods and mechanisms that embryonic, larval, and adult craniates have developed for obtaining oxygen and eliminating carbon dioxide in aquatic and terrestrial environments. We will find that many fishes are aerial respirators and that lungs may be older than tetrapod limbs. Along the way, we will see how a crocodilian or a whale can breathe with a mouth full of water, why a nursing kitten does not have to interrupt swallowing during inhalation, and how the position of the human larynx affects human speech.

OUTLINE

he process of obtaining oxygen from the environment and eliminating carbon dioxide is **external respiration.** It is accomplished via respiratory membranes that, except in embryos, are part of some organ. Organs that are essential for external respiration constitute, collectively, the respiratory system.

External respiration precedes **internal respiration,** which is usually defined as the exchange of oxygen and carbon dioxide between capillary blood and tissue fluids. Because carbon dioxide quickly inhibits cellular activity, the continual elimination of this gas from the vicinity of the cell, and then from the organism, is essential. The role of the circulatory system in the total process of respiration is therefore vital.

External respiration is carried on through respiratory membranes. Except in very early embryos, these must be highly vascular, the epithelium must be thin, the surface should be moist to facilitate diffusion of gases through semipermeable membranes, and it must be in contact with the environment, as in external gills; or else the environment must be brought into contact with the respiratory surface, as in lungs.

The chief organs of external respiration in adult craniates are external and internal gills, the oropharyngeal mucosa, air sacs or lungs, and the skin. Less common adult respiratory devices include bushy or filamentous outgrowths of the pectoral fins, as in the male *Lepidosiren,* or of the posterior trunk region and thigh, as in the African hairy frog; the cloacal, rectal, or anal lining; and the lining of the esophagus, stomach, or even of the intestine in one species or another. Embryos employ a variety of respiratory surfaces including extraembryonic membranes.

In both water and air, respiration through the skin is employed extensively by modern amphibians. It is also used by some fishes, especially those that lack scales and therefore have capillary beds close to the surface. Aquatic urodeles, lacking dermal armor and a significant stratum corneum, acquire as much as three-fourths of their oxygen from the water through the skin. Tree frogs acquire only one-fourth by that route, and terrestrial species of *Rana* acquire one-third. Regardless of the proportionate role of skin and lungs in oxygen uptake in amphibians, up to nearly 90 percent of the carbon dioxide is excreted through the skin. Cutaneous respiration is not significant in amniotes because the thick stratum corneum insulates the capillaries from the atmosphere.

PRINCIPLES OF DIFFUSION

The respiratory and circulatory systems are presented separately, but they represent an integrated system for the delivery of oxygen and the elimination of metabolic wastes. Diffusion across the organismal/environmental boundary is influenced by a number of properties of the gas exchange surface. We see in evolution both independent and correlated modifications of these systems that enhance the net diffusion of molecules across the exchanger:

$$diffusion \approx \frac{\text{surface area} \times \text{diffusion gradient}}{\text{thickness} \times \text{material properties}}$$

Surface area is enhanced through such evolutionary novelties as gill lamellae, alveoli, or sacculations. The diffusion gradient is increased by the introduction of external (ventilation) and internal (heart) pumps along with the origin of countercurrent circulation. The thickness of the gas exchange membranes was reduced with the origin of craniates with most membranes consisting of a simple epithelium adjacent to the endothelial lining of a capillary. Finally, the materials that make up the gas exchanger must be easily diffusible to the pertinent molecules. As an example, the aortic arches in amphioxus do not serve a respiratory function and are enclosed within the collagenous skeleton of the pharyngeal arch. In contrast, the aortic arches of craniates extend beyond the pharyngeal arch to form a capillary bed for gas exchange. We have already seen similar concerns with evolutionary novelties of the digestive system (chapter 12) for the absorption of nutrients, and similar concerns will arise again when we consider the elimination of nitrogenous wastes in the urogenital system (chapter 15).

GILLS

The internal gills of fishes arise in the walls of pharyngeal pouches, which are paired evaginations of the embryonic pharynx (see fig. 1.6). These pouches meet ectodermal grooves, which are invaginations of the ectoderm (see fig. 1.7). A thin membrane, the branchial plate, separates pouch and groove, and subsequent rupture of the membrane establishes a passageway for respiratory water from the pharynx to the exterior. The gills of jawed fishes are supported by the pharyngeal skeleton. The development and fate of pharyngeal pouches and arches from fish to humans has been discussed in chapter 1.

Agnathans

Hagfishes in the genus *Myxine* have five or six pairs of well-vascularized gill pouches, rarely seven, associated with the pharynx (fig. 13.1*a*). Various species of *Eptatretus,* another hagfish genus, have from five to 15 pairs. **Afferent branchial ducts** conduct respiratory water from the pharynx to the pouches, and **efferent ducts** lead from the pouches to the exterior. Each efferent duct usually has its own external aperture (see fig. 1.9, *Eptatretus*), but in myxinoids the efferent ducts unite to open via a single common external slit on each side (fig. 13.1*a*).

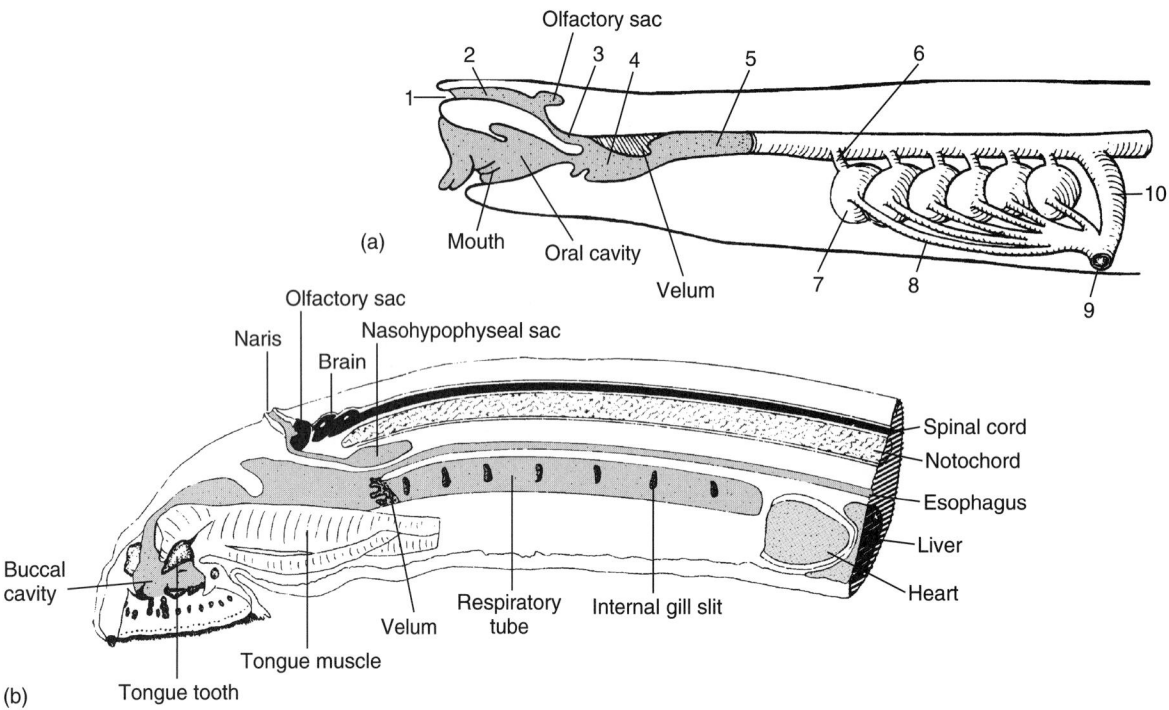

FIGURE 13.1

Head and pharyngeal region of hagfish and lamprey. (*a*) The hagfish *Myxine glutinosa*. **1**, naris; **2,3**, nasopharyngeal duct; **4**, velar chamber; **5**, pharynx; **6**, afferent branchial duct; **7**, gill pouch; **8**, efferent branchial duct; **9**, common external gill aperture (present on both sides); **10**, pharyngocutaneous duct (present on left side only). (*b*) The lamprey *Petromyzon,* sagittal section. Connecting naris and nasohypophyseal sac is the nasohypophyseal duct.

In hagfishes, respiratory water enters the median unpaired naris and passes via a **nasopharyngeal duct** to a **velar chamber** at the anterior end of the pharynx (fig. 13.1*a*). The wall of the chamber is a pulsating muscle, or velum, which pumps water from the velar chamber caudad into the pharynx and, according to some observers, into the gill pouches, thereby creating a vacuum in the velar chamber. The vacuum draws additional water through the naris and into the velar chamber. Muscle tissue in the walls of the gill pouches expels the water to the exterior. The velum is supported by pharyngeal cartilages and pulsates 50 to 100 times a minute in sleeping animals. On the left side only, a **pharyngocutaneous duct** of greater diameter than the afferent branchial ducts connects the pharynx with the last efferent branchial duct (*Myxine*), or, in some hagfishes, with the exterior. Periodically, debris or particles too large to enter the afferent branchial ducts are forcefully ejected through the pharyngocutaneous duct in a manner somewhat analogous to coughing. The pharyngocutaneous duct appears to be a modified last gill pouch.

The mechanism of respiration differs considerably in lampreys in comparison with hagfishes. The buccal cavity of lampreys is pressed tightly into the flesh of a host, and the duct from the unpaired naris (**nasohypophyseal duct**) does not open into the pharynx but ends blindly in a **nasohypophyseal sac** (fig. 13.1*b*). Therefore neither the nasal nor buccal route is available for the passage of respiratory water to the gills. As an adaptation, respiratory water *enters and exits* the pharyngeal pouches via the external gill slits as a result of pulsations of the pharyngeal and gill pouch musculature. The external apertures are guarded by thin flaps of skin that serve as two-way valves. As a further adaptation, the lamprey pharynx becomes subdivided longitudinally at metamorphosis into an esophagus that continues to the stomach and a ventral respiratory tube that ends blindly (fig. 13.1*b*). The entrance to the adult respiratory tube is guarded by a valvelike velum that prevents nutrients rasped from a host from entering the respiratory tube, becoming diluted by respiratory water, and then being lost through the external gill slits.

The seven pairs of voluminous gill pouches are lined with gill lamellae and communicate directly with the respiratory tube and the exterior with no intervening afferent or efferent branchial ducts. The external gill slits of *Petromyzon* are seen in figure 1.9.

The respiratory systems of hagfishes and lampreys have little in common with those of jawed fishes, structurally. The former represent two variant adaptations to parasitism. But adaptations from what preceding condition? The branchial apparatus of jawed fishes is very different. It is a pharyngeal arch mechanism. The phylogenetic history of lampreys and hagfishes and the com-

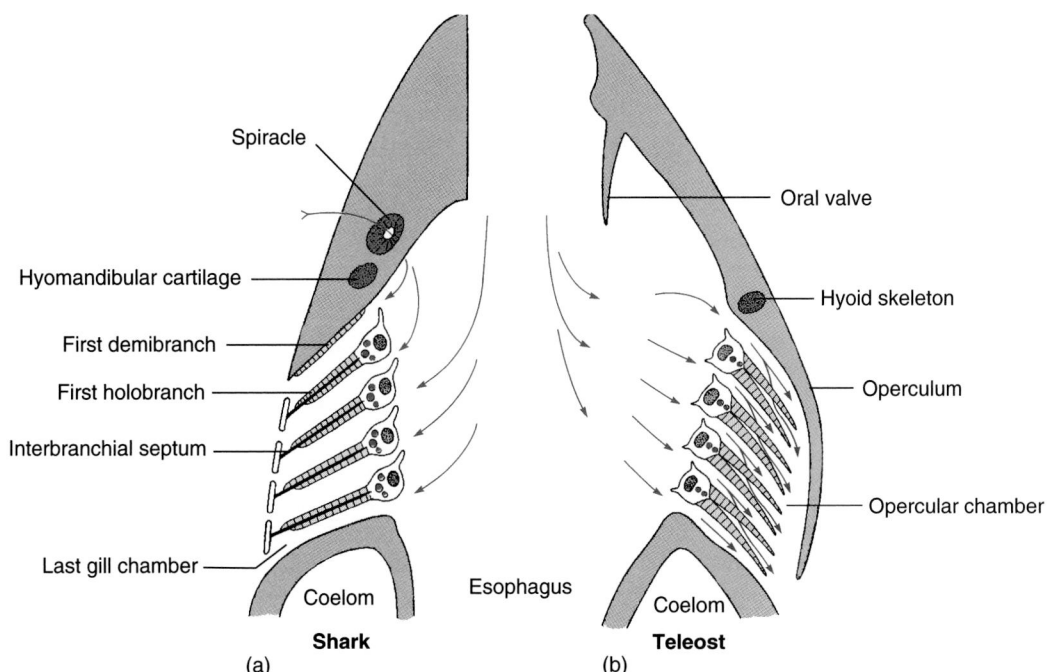

Spiracle

Hyomandibular cartilage

First demibranch

First holobranch

Interbranchial septum

Last gill chamber

Coelom

Esophagus

Shark

(a)

Oral valve

Hyoid skeleton

Operculum

Opercular chamber

Coelom

Teleost

(b)

FIGURE 13.2

Gills of a pentanchid shark and teleost, frontal section, schematic. *Arrows* show routes of respiratory water.

mon ancestor of agnathans and gnathostomous fishes are unknown. However, it is highly unlikely that a system such as that of agnathans provides a glimpse into the ancestral system of jawed fishes.

Cartilaginous Fishes

Most elasmobranchs—sharks, skates, and rays—are pentanchid, having five pairs of gill pouches, each with an internal and external gill slit, and a pair of functional **spiracles** located dorsally immediately in front of the hyomandibular cartilages (fig. 13.2*a*). The shark *Hexanchus* is exceptional, having a spiracle and six gill pouches; *Heptanchus,* another shark, has a spiracle and seven gill pouches, this being the largest number in any gnathostome. No gill surface develops in the posterior wall of the last pouch. The external gill slits in sharks are visible on the side of the head; those in Rajiformes are on the underside (see fig. 13.7). The slits are said to be "naked" because, unlike in other gnathostomes, there is no operculum. In embryos, the spiracle and gill slits are the same size and aligned in a row, but the spiracle does not keep pace in growth with the other slits (see fig. 1.7). A miniature gill-like structure, or **pseudobranch,** consisting of a vascular rete (network) develops in its anterior wall. The rete may regulate blood pressure in the eyeball.

The spiracle has a one-way intake valve and is the exclusive incurrent aperture for respiratory water in Rajiformes and for much of the respiratory water in sharks except some fast-swimming voracious ones in which the spiracle becomes secondarily closed by a membrane of skin. Its dorsal position in rays minimizes the entrance of muddy debris-laden water into the pharynx of these bottom-dwelling species.

If a gill slit is probed, the instrument will enter a **gill chamber** (fig. 13.3). The anterior and posterior walls of the first four chambers each exhibit a gill surface, or **demibranch.** The last chamber lacks a demibranch in its posterior wall. Ceratohyal cartilages support the demibranch in the anterior wall of the first chamber, and epibranchial and ceratobranchial cartilages support the remaining eight demibranchs. The demibranch in the *anterior wall of a gill chamber* is a **pretrematic demibranch.** The one in the posterior wall is a **posttrematic demibranch.** Separating the two demibranchs of a single gill arch is an **interbranchial septum** that is supported by many long, tapering, and in some species branching, cartilaginous **gill rays** that radiate from the gill cartilage into the interbranchial septum along its entire extent. The two demibranchs of a single gill arch, together with the associated interbranchial septum, cartilages, blood vessels, branchiomeric muscles, nerves, and connective tissue, constitute a **holobranch** (white area in upper drawing of fig. 13.3). Stubby **gill rakers** projecting from the pharyngeal border of each gill arch guard the slitlike entrances from the pharyngeal lumen into the gill chambers, protecting the gills from mechanical injury.

The functional surface of each demibranch consists of large numbers of transverse shelflike folds, or lamellae, of gill mucosa, the epithelium of which is very thin. The folds multiply the surface area for gaseous exchange. Underlying the lamellar epithelium is a rich capillary bed supplied by afferent branchial arterioles that receive blood from the ventral aorta via afferent branchial arteries (see fig. 14.17*b*). Because the ventral aorta carries blood low in oxygen, the blood entering the capillary bed is low in oxygen. As oxygen-laden respiratory water flows steadily across the gill lamellae, it

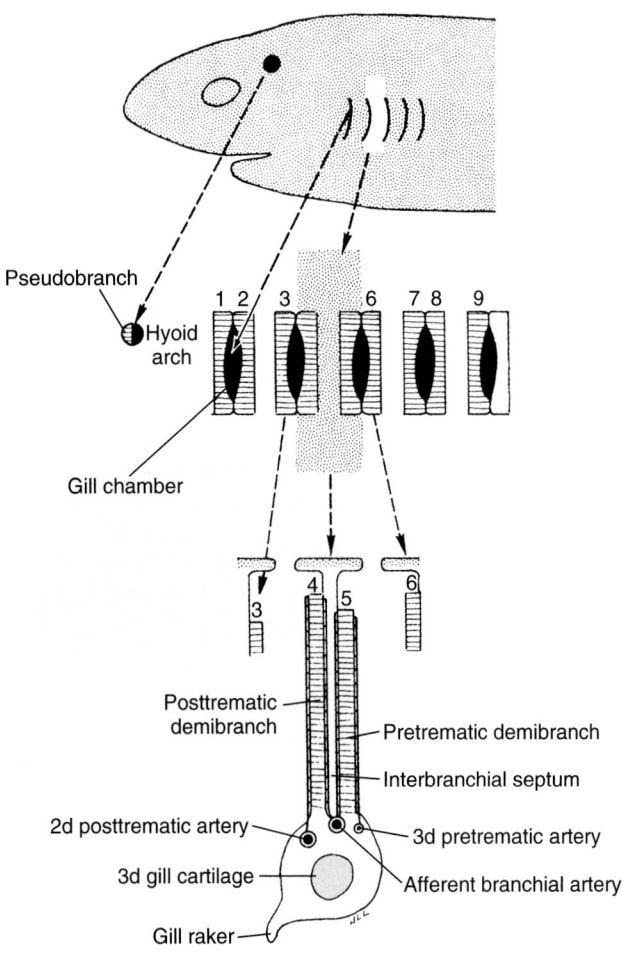

FIGURE 13.3
The gill chambers (*black*) and the nine demibranchs (*lined*) in *Squalus*. A holobranch (location indicated in *white* at top of figure) is represented at bottom in cross section.

does so in a direction opposite that of the oxygen-poor blood flowing in the capillaries. As a result, the partial pressure of oxygen in the water always exceeds that in the blood even though the blood is becoming increasingly oxygenated. This **countercurrent flow** of blood and water maximizes the efficiency of gaseous exchange, enabling the blood to "make the most" of its opportunity to acquire oxygen. Other examples of the countercurrent principle will be found elsewhere in the text. From the capillaries, oxygenated blood passes via efferent branchial arterioles to pretrematic or posttrematic arteries, then into efferent branchial arteries and the dorsal aorta (see fig. 14.17b). From there, it is distributed to all tissues of the body.

It has been mentioned that respiratory water enters the pharynx of most elasmobranchs via both the mouth and spiracle. Most of the water entering via the spiracle in sharks flows into the first two gill pouches, whereas that entering via the mouth enters the last three pouches. Water is sucked through the spiracle and the

open mouth when branchiomeric muscles constrict the external gill slits and expand the pharyngeal chamber, creating a vacuum within the chamber. This inspiration phase slows down and then ceases when the pharynx has become filled with water. The mouth then closes, the gill chambers are expanded by the action of levator and hypobranchial muscles, and the gill chambers fill with water. In the third phase of respiration—expiration—the water is forced out of the gill chambers via the external gill slits by constrictor muscles. In the interval between expiration and the next inspiration, both the mouth and the gill slits are temporarily open. Experiments have shown that pressure is nearly always higher in the pharyngeal chamber than in the gill pouches. This assures a steady, uninterrupted flow of water over the gill lamellae, rather than tidal surges. There is a respiratory rhythm of about 35 per minute when the shark is at rest. Sharks swimming with their mouths open are utilizing the forward motion of the body to accumulate water in the pharynx, thereby reducing the cost of external respiration in terms of energy expended.

The holocephalan *Chimaera* has only four gill pouches (fig. 13.4), the spiracle closes during early ontogeny, interbranchial septa are short and do not reach the skin, and a fleshy flap, the **operculum** extends caudad from the hyoid arch, hiding the gills and deflecting excurrent water caudad. In some of these traits, *Chimaera* resembles teleosts.

Bony Fishes

The gill apparatus of cartilaginous and bony fishes exhibits the same basic pattern. A series of pharyngeal arches support holobranchs, and a stream of respiratory water flows over the demibranchs en route from the pharyngeal cavity to the exterior (see fig. 13.2b). The chief differences lie in the presence of an operculum and an opercular chamber and in the shorter length of the interbranchial septa in bony fishes. The **operculum** is a bony flap that commences at the hyoid arch and extends caudad, covering the gill chambers on each side. Extending from the ventral edge of each operculum is an accordianlike **branchiostegal membrane** supported, in teleosts, by numerous long bony **branchiostegal rays**. Shorter **gular** bones, of more ancient vintage (see fig. 9.10a), support the membrane found in some of the more basal actinopterygians. The two branchiostegal membranes are united midventrally beneath the gills to enclose an **opercular chamber** that respiratory water pours into after passing over the gills and before being expelled via a cleft at the caudal end of the operculum. The cleft varies in size. It is exceptionally small and round in eels. Infrequently in teleosts the chamber empties via a midventral aperture. The interbranchial septa are very short, with the result that the demibranchs of each holobranch are unattached distally, allowing water freer access to their capillaries (fig. 13.5).

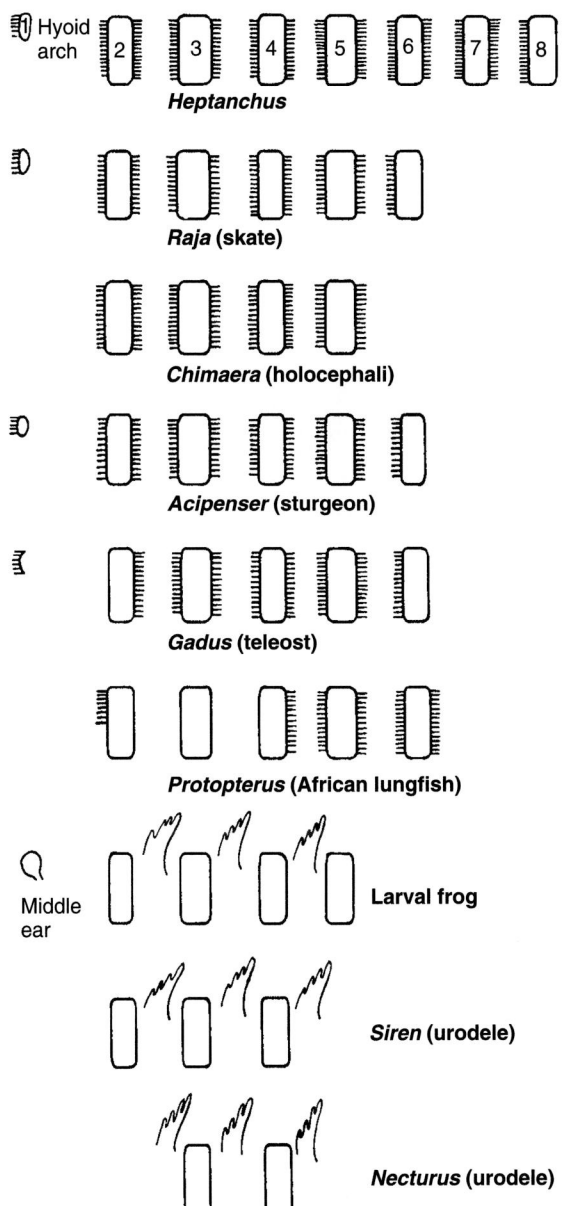

FIGURE 13.4

Open pharyngeal slits in selected aquatic vertebrates and distribution of demibranchs (*lines*) in fishes. *Heptanchus* is a basal modern shark. Note that *Gadus*, a cod, retains a vascular rete in the mandibular arch even though the spiracle is closed.
1 to **8**, pharyngeal slits. The position of external gills is indicated in the three amphibians.

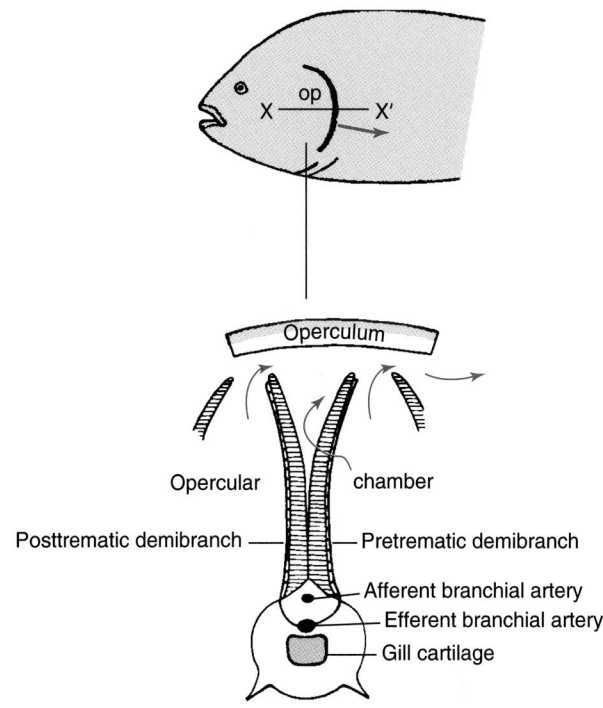

FIGURE 13.5

Operculum (**op**) and holobranch of a teleost. The holobranch, *bottom*, is seen in frontal section (X—X'). *Arrows* indicate direction of efferent water flow.

Water is drawn into the pharynx by lowering the pharyngeal floor with the mouth open and the operculum closed. Simultaneous expansion of the opercular chamber by unfolding of the branchiostegal membrane draws the incoming water across the gills and into the chamber. The mouth is then closed, and the water is forced to the exterior via the opercular cleft. This is accomplished by elevating the pharyngeal floor and compressing the opercular chamber. An **oral valve** at the anterior end of the oropharynx immediately behind the mouth prevents escape of water by mouth. Thus, a suction pump and a pressure pump, operating rhythmically, keep the gills bathed in oxygenated water. A few teleosts, including mackerel and tuna, have a paucity of branchiomeric musculature and are obliged to swim with the mouth open in order to create an adequate flow of water over the gills.

Most bony fishes other than dipnoans, chondrosteans, and gars have four holobranchs and five gill chambers, the demibranch in the anterior wall of the first gill chamber having been lost (see fig. 13.4, *Gadus*). Additional demibranchs have been lost in dipnoans (see fig. 13.4, *Protopterus*). In fact, the gills of *Protopterus* and *Lepidosiren* are so inefficient that these fishes suffocate when forcibly held under water. A spiracle is present in chondrosteans (see fig. 13.4, *Acipenser*), but it closes during embryonic life in other bony fishes.

Larval Gills

Larval gills are of three kinds—**external gills** that are outgrowths from the external surface of one or more gill arches; **filamentous extensions of internal gills** that project through gill slits to the exterior; and the **internal gills** of late anuran tadpoles, which are hidden behind the larval operculum.

External gills usually develop before gill slits open and before any opercular fold has started to develop.

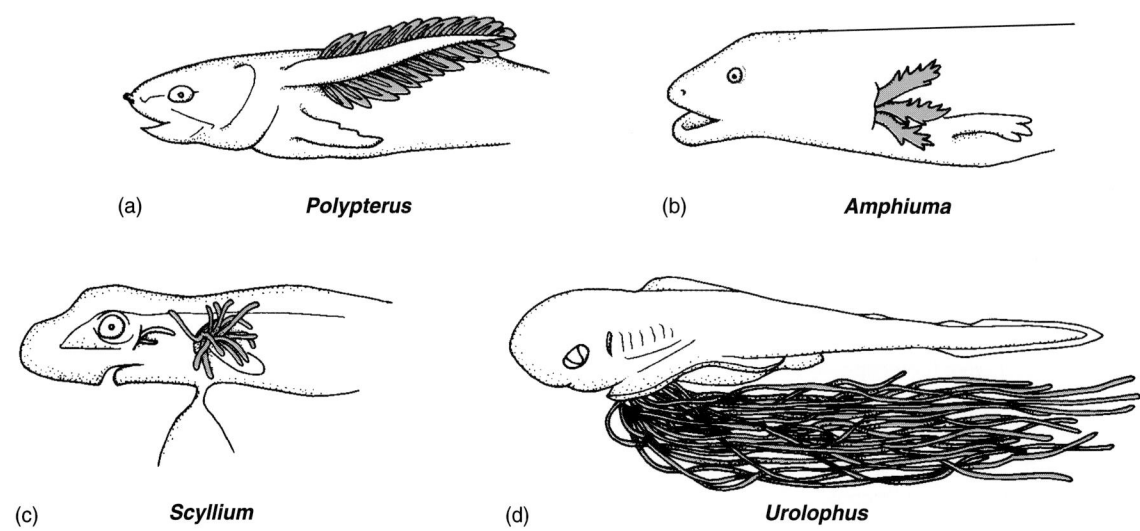

(a) **Polypterus** (b) **Amphiuma**

(c) **Scyllium** (d) **Urolophus**

FIGURE 13.6

Larval external gills. (a–d) A basal actinopterygian, amphibian, shark, and ray, respectively. In *Amphiuma,* the gills are resorbed before the larva emerges from the egg envelope. The spiracle in *Urolophus,* presently behind the eye, will ultimately be located dorsally. Pectoral fins characteristic of Rajiformes have just appeared as fin folds.

They can be waved about by branchial muscles and can often be retracted. They develop in the embryonic or larval stages of most dipnoans, in all amphibians, and in a few ray-finned fishes such as sturgeons and *Polypterus* (fig. 13.6a). Although external gills are larval structures, there are perennibranchiate urodeles and a perennibranchiate lungfish, *Protopterus* (see fig. 14.18b). The latter has four pairs of larval gills and retains three pairs in a reduced state throughout life.

External gills develop in anuran tadpoles on pharyngeal arches III to V. Later, when pharyngeal pouches II to V rupture to the exterior, the pouch linings become folded to form a set of internal gills. Thereupon, a fleshy operculum grows backward from the hyoid arch over the gill region, enclosing external gills and internal gills in an opercular chamber that retains only a single excurrent pore, this being located at the posterior edge of the left operculum. Thereafter, the external gills gradually atrophy, and the internal gills function until metamorphosis. At that time, the internal gills and operculum are absorbed along with the larval tail. All these structures are, of course, very small in a metamorphosing anuran.

Elasmobranchs develop either in a yolk-laden egg case attached to underwater vegetation, or they develop in the maternal uterus. During early development, filamentous outgrowths of the developing internal gills project through the external gill slits and into the fluids within the egg case or the uterine chamber, where they serve temporarily as respiratory organs (fig. 13.6c and d). In viviparous species, they also absorb nutrients from the uterine fluid. A few viviparous chondrostean and teleost larvae utilize filamentous external gills in a similar manner. Absorption of nutrients from uterine or other maternal fluids is known as **histotrophic nutrition.**

Excretory Role of Gills

Although usually thought of as respiratory organs, gills perform vital excretory functions. Marine fishes excrete the common marine salts chiefly via salt-secreting glands located on the lamellae of the gills. (Freshwater species secrete most of their excess chloride via the kidneys.) The gills of lampreys and those of marine fishes that migrate between salt water and freshwater excrete chloride when in salt water, and they absorb chloride in freshwater, thereby assisting in maintaining homeostasis under these stressful conditions. Fishes without exception also excrete their nitrogenous wastes largely through the gills, unlike tetrapods, which excrete them via the kidneys. And, last, fishes that acquire oxygen from air in an air sac release most of their carbon dioxide into the water flowing over the gills. Carbon dioxide dissolves more readily in water than in air; hence, the partial pressure of CO_2 in the water passing over the gills is low enough to assure uninterrupted release of CO_2 from the circulation.

Aerial Respiration in Bony Fishes

Craniates arose in an aqueous environment, and water was the earliest source of oxygen. During the Devonian, however, if not earlier, air became a source of oxygen for many fishes. One compelling factor may have been that the water at that time was warm and swampy and, as such, low in dissolved oxygen. Air, on the contrary, contains 20 times the oxygen that oxygen-saturated water can hold. Little wonder, then, that many Devonian

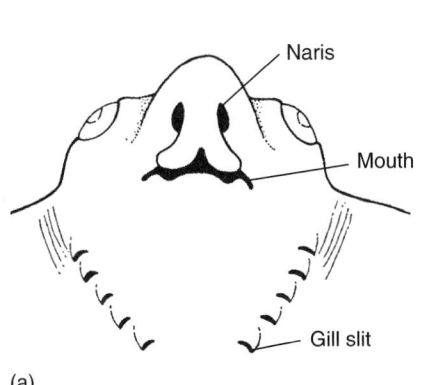

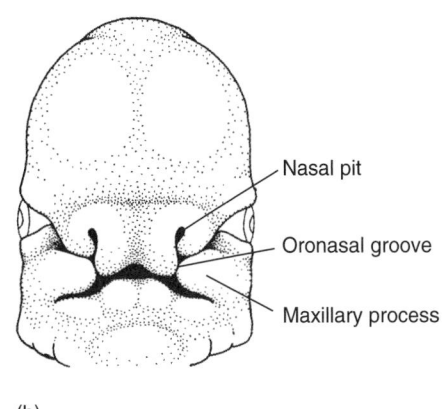

FIGURE 13.7

Oronasal relationships in (*a*), an adult skate, and (*b*), a 6-week (12 mm) human fetus.

fishes obtained some or all of their oxygen from above the surface of the water. Today's air-breathing fishes include dipnoans, basal actinopterygians, gars, *Amia,* and many teleosts. The fishes snatch bubbles of air from just above the water surface, and the air then comes in contact with the oropharyngeal lining or, as we will see shortly, with the lining of swim bladders in those few fishes in which these swim bladders serve as lungs. A few teleosts swallow the bubble and extract oxygen in the stomach or intestine. Although oxygen is acquired from the atmosphere, most of the excess carbon dioxide is eliminated through the gills into the water.

NARES AND NASAL CANALS

The paired nostrils or **external nares** of cartilaginous and ray-finned fishes open directly into blind olfactory sacs that contain the sensory epithelia for smell. Each nostril is divided into a forward-directed **incurrent aperture** through which water is driven as a fish swims forward, and a laterally or ventrally directed **excurrent aperture** through which water exits after having bathed the olfactory epithelium. In lobe-finned fishes, the nostrils are connected with the oropharyngeal cavity by a pair of nasal canals that conduct respiratory water. Their openings into the oropharyngeal cavity are **internal nares.** As we learned earlier, the nasal canals of agnathans are unpaired, and only in myxinoids do they extend to the pharynx. In no living fish, including those that obtain their oxygen from air, are nostrils used for breathing. They are part of a sensory system for monitoring one constituent of the environment—chemicals in solution in the surrounding water.

If the rhipidistian ancestors of labyrinthodonts were not using their nostrils for breathing—and whether they were is not known—labyrinthodonts were opportunists because they commenced to use nostrils for drawing *air* into their oral cavity. With subsequent development of a secondary palate in amniotes, the internal nares opened farther caudad, and, as we have seen, the longer the sec-

ondary palate, the farther caudad the nares are located. In mammals, they open into the nasopharynx above the soft palate (see fig. 12.3, cat).

Nasal canals (now **choanae,** see glossary), arise from paired **nasal pits** and **oronasal grooves,** the dorsolateral walls of which roll together to form a tube (fig. 13.7*b*). Although elasmobranchs lack nasal canals, oronasal grooves form and remain throughout life (fig. 13.7*a*). In mammals, the *olfactory epithelium* is restricted to the upper chambers of the nasal passageways. The lower chambers have a ciliated glandular *nasal epithelium.* Hairs at the entrance to mammalian nasal canals trap insects and coarse particulate matter in the incoming air; venous plexuses in the submucosa overlying the turbinal bones heat cold air; and air sinuses that open into the upper reaches of the nasal passageways of mammals (see fig. 13.13*a*) serve partly as resonating chambers for vocalization.

A fleshy, partly cartilaginous, proboscis (nose) develops in some mammals and carries the nares to positions characteristic of the species. Compare the location of the nostrils of a catarrhine (see fig. 4.40) with that of an elephant. Whales have no nose. The nostrils are far back on the top of the head and are known as **blowholes.** They are equipped with valves that close when the whale dives. (In fetal whales, the nostrils are farther forward.) In some whales, the two nares unite during development to become a median blowhole.

SWIM BLADDERS AND THE ORIGIN OF LUNGS

Nearly every osteichthyan from fish to human develops an unpaired evagination from the foregut that becomes one or a pair of pneumatic sacs (swim bladders or lungs) filled with gases derived directly or indirectly from the atmosphere. The only adult osteichthyans that do *not* have pneumatic sacs—and their ancestors probably lost them—are a few marine teleosts, some bottom dwellers such as flounders, *Latimeria* (the pneumatic sac is replaced by a fat-filled one for buoyancy), and one clade of tailed amphibians representing over one-half of the known species of salamanders (Plethodontidae representing underground, aquatic, terrestrial, and arboreal forms). Some of these develop sacs temporarily as embryos. In one Late Devonian placoderm, imprints of paired sacs with ducts have been described and would represent an independent origin, although the evidence for this is disputed. The term *air (pneumatic) sac* does not imply a function whereas the terms *swim bladder* and *lung* suggest functions for buoyancy and respiration, respectively. This does not preclude an air sac from serving both roles. Among fishes, the presence of an air sac will affect buoyancy, and the term *swim bladder* is typically applied. In contrast, the air sac of tetrapods is referred to as a lung. We will hereafter follow this latter terminology.

After the budlike anlage for a swim bladder evaginates from the foregut, the resulting duct may retain its connection with the foregut, or the duct may close during later development. Fishes are said to be **physostomous** when the duct remains open. Chondrosteans, basal neopterygians, the three living dipnoans, and some teleosts have open ducts. Fishes in which the duct closes, as in many teleosts, are said to be **physoclistous** (fig. 13.8).

Swim bladders may be paired or unpaired. In chondrosteans and dipnoans, the unpaired duct leads from the ventral aspect of the esophagus. It has shifted toward the dorsal side in neopterygians. Infrequently, the duct leads from the pharynx or stomach.

Swim bladders lie close to the kidneys and, like the latter, they are retroperitoneal (see fig. 15.18). As the embryonic anlagen grow longer, they push their way caudad in the roof of the coelom between the embryonic parietal peritoneum and the body wall. However, adult bladders may bulge into the roof of the coelom.

Swim bladders in one species or another serve as hydrostatic organs, respiratory organs, organs that participate in sound detection, or organs of communication. Because their walls contain elastic and smooth muscle tissue, their capacity for gases can be increased or decreased as required by their roles.

Swim bladders in teleosts serve chiefly as hydrostatic organs. To maintain an appropriate depth in a body of water or to hover at a specific location, a fish must achieve, as closely as possible, a body density, or specific gravity, equal to that of the displaced water at the selected depth. This is accomplished by regulating the volume of gas in the swim bladder. Any difference in density will affect its buoyancy, causing the fish to slowly rise or sink. But because disturbances in the water preclude complete stability moment by moment, a hovering fish must also resort to frequent gentle undulations of the paired and unpaired fins to compensate for these disturbances. Some entire populations of fishes

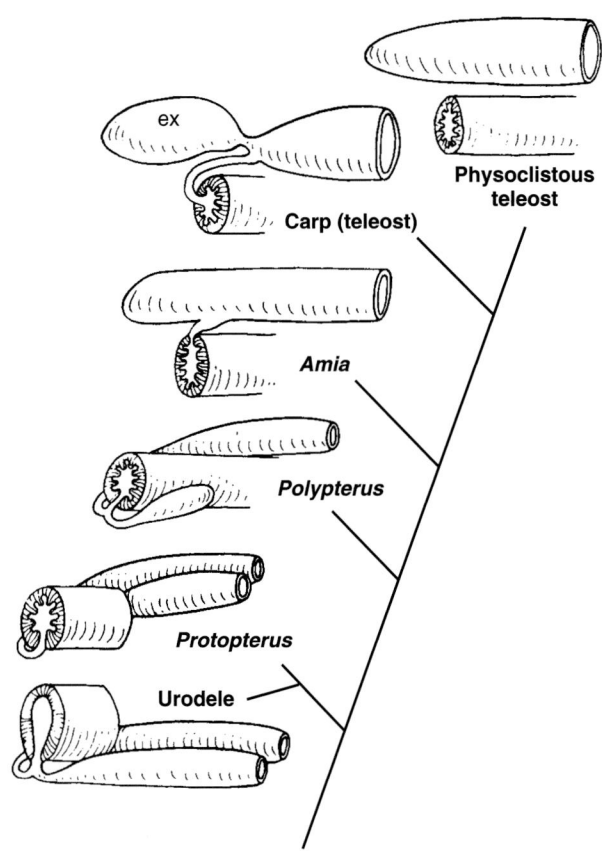

FIGURE 13.8
Swim bladders and urodele lungs. In teleosts, the duct usually closes. Carp are physostomous teleosts with an anterior extension of the bladder, **ex,** connected to weberian ossicles (see fig. 17.7). *Protopterus* is a dipnoan, *Amia* is a basal neopterygian, and *Polypterus* is a basal actinopterygian.

make daily changes in depth, therefore in body density, as a correlate of the time of day.

The gas in the hydrostatic swim bladder usually comes from the blood, being actively transported into the lumen of the bladder from the **red gland,** a localized rete of small arteries in the bladder lining. The blood in the rete inspired the term *red gland.* The gas is eventually resorbed into the bloodstream of the bladder near its caudal end in a pocketlike area of modified epithelium. The pocket can be closed off from the main cavity by a sphincter during passage of gas into the lumen of the bladder, and relaxed when resorption of gas is taking place. In physostomes, the gas may be bubbled to the exterior through the mouth.

The gases in swim bladders differ among fishes. Some bladders contain almost pure (99 percent) nitrogen, some up to 87 percent oxygen, and all contain at least traces of carbon dioxide and argon. In deepwater fishes, nitrogen may be transported from the blood into the lumen of the bladder against a nitrogen pressure as high as 10 atmospheres.

Swim bladders function as lungs to one degree or another in physostomous fishes. Air is gulped at the water's surface, and an oropharyngeal pump forces it from the oropharyngeal cavity into the swim bladder. Air depleted of oxygen is expelled from the bladder into the oropharynx as a result of a vacuum created by lowering the oropharyngeal floor while the mouth and nares are closed. Expulsion is facilitated by the elasticity of the bladder wall, its smooth musculature, and the pressure of the surrounding water on the body wall. Oxygen-depleted air in the oropharynx is then bubbled through the mouth into the water.

Two of the three genera of true lungfishes, *Protopterus* and *Lepidosiren,* require continual gulping of air for survival because their gills (three pairs of external ones in *Protopterus,* see fig. 14.18*b*) are incapable of providing sufficient oxygen for ordinary metabolic needs. These fishes drown if held under water for a sufficient length of time. Their ability to survive tropical summers in burrows when the swamps dry up is attributable in part to the fact that their swim bladders serve as lungs. However, these fish also absorb oxygen through the skin at those times.

The swim bladder of the only other living dipnoan, *Neoceratodus,* those of two genera of ray-finned African freshwater fishes, *Polypterus* and *Calamoichthys* (sometimes inappropriately called "lungfishes"), and the swim bladder of the relict basal neopterygians (*Amia* and gars) function as a lung only when the oxygen content of the water is so low that the gills alone cannot meet the respiratory needs of the organism. The lining of the swim bladders of *Polypterus* and *Calamoichthys* is almost smooth; that of true lungfishes is lined with low septa and may exhibit thousands of tiny air sacs (fig. 13.9, a dipnoan). Unlike those in other fishes, the swim bladders of *Polypterus, Amia,* and dipnoans are supplied by arteries arising from the sixth embryonic aortic arch, a tetrapod condition (see fig. 14.18*b*), and, as in tetrapods, the venous return in dipnoans is directly to the left atrium of the heart. This is not true in any other fish.

In addition to serving as hydrostatic organs, the swim bladders of some teleosts participate in sound detection. In Cypriniformes (goldfish, carp, North American catfish, minnows, others), a series of small bones, **weberian ossicles**, connects the anterior end of the swim bladder with the sinus impar, an extension of the perilymphatic space of the inner ear. In another order of teleosts, Clupeiformes, a thin-walled anterior extension of the swim bladder makes direct contact with the inner ear (fig. 13.10). These fish can hear.

In a few fishes, contraction of striated muscles attached to the swim bladder causes it to emit thumping sounds, or it forces air back and forth between chambers separated by muscular sphincters within the bladder to produce croaking or grunting sounds for communication.

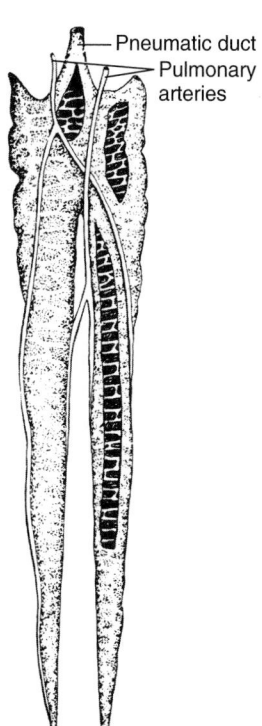

FIGURE 13.9
Swim bladders of the lungfish *Protopterus.*

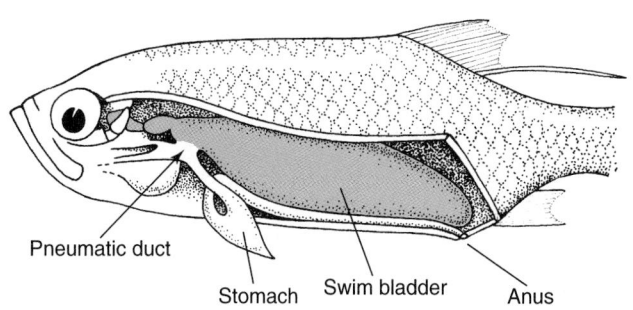

FIGURE 13.10
Unpaired swim bladder of a tarpon, a physostomous teleost. The extension into the skull contacts the inner ear.
Source: Data from J.S. Kingsley, *Outline of Comparative Anatomy of Vertebrates,* 1920. The Blakiston Company, Philadelphia, after de Beaufort.

Swim bladders usually degenerate in fishes that are bottom feeders. Elimination of this hydrostatic organ has optimized their body density, enabling them to hover close to their food supply with a minimal expenditure of energy. Although degeneration of the swim bladder may be an adaptation for bottom feeding, we will never know the nature of the original selective pressures responsible for the advantage. An analogous condition is seen in salamanders who live in swift mountain streams and whose lungs are only a few millimeters long, or absent. Any buoyancy that might be conferred by lungs would be detrimental. Optimal body density and the saving of energy are the benefits. It should be noted, however, that *no* member of the plethodont family has lungs, and not all inhabit swift streams. These urodeles breathe through their skin.

The striking similarities between swim bladders and lungs, especially bladders that are supplied via the sixth aortic arches, suggest that these may be the same organs. In the Devonian, when the freshwaters of the swamps were warm and periodically stagnant, and therefore low in dissolved oxygen, aerial respiration may have made the difference between survival and extinction. A similar argument can be made for marginal marine intertidal zones. This would support a marine origin for tetrapods and is consistent with the known piscine outgroups to tetrapods being found in marine sediments (e.g., *Panderichthys* and *Eusthenopteron*). At that time, most rhipidistians had pneumatic sacs. We can only speculate on which came first, respiratory or hydrostatic bladders, but these two conclusions seem warranted: A pneumatic sac was functioning in aerial respiration long before craniates ventured onto land, and closure of the pneumatic duct in physoclistous fishes is probably derived from a more primitive open-duct condition.

LUNGS AND THEIR DUCTS

Tetrapod lungs arise as an unpaired evagination, the **lung bud**, from the caudal floor of the pharynx just beyond the last pharyngeal pouch (see fig. 1.6). The opening in the pharyngeal floor becomes a longitudinal slit, the **glottis** (see fig. 13.13*a*). The unpaired lung bud elongates only slightly before bifurcating to form bronchi and lungs (see fig. 13.22). The lung primordia push caudad underneath the foregut until they bulge into the coelom lateral to the heart. As they grow into the coelom, they carry along an investment of peritoneum, which becomes the visceral pleura. The part of the lung bud between glottis and lungs develops into larynx, trachea, and bronchi.

Larynx and Vocalization

The larynx is a short air passageway between the glottis and the upper end of the trachea of tetrapods, the walls of which are supported by cartilaginous derivatives of the caudalmost pharyngeal arches (see table 9.4). The glottis and associated cartilages have become an instrument for vocal communication in species in which vocal cords develop. In urodeles, the larynx is a primitive structure consisting of a single pair of lateral cartilages that surround the glottis and support it as their sole role. Most nonmammalian tetrapods have two pairs of laryngeal cartilages, **arytenoids** and **cricoids** (fig. 13. 11), and mammals have a third pair, the **thyroid** cartilages. Additional small cartilages—**cuneiforms, corniculates, procricoid**, and others—develop in some mammals

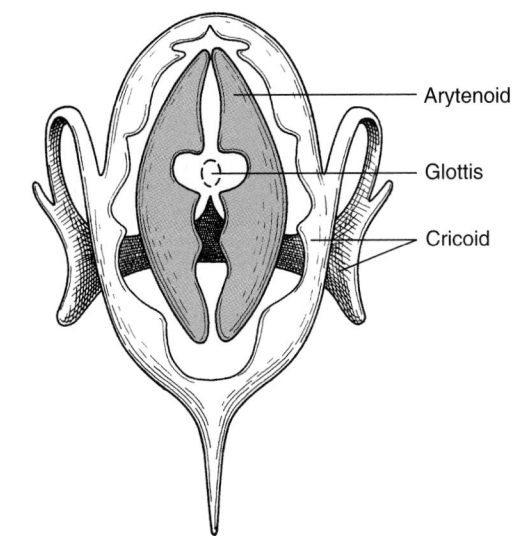

FIGURE 13.11
Laryngeal skeleton of a frog.

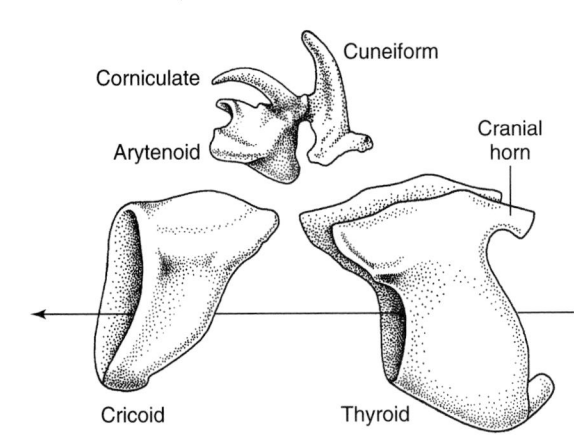

FIGURE 13.12
Laryngeal cartilages of a dog, disarticulated. The arytenoid has been elevated above its articulation sites. Cranial horns of thyroid articulate with caudal horns of hyoid seen in figure 9.43. *Arrow* is in respiratory passageway to trachea.

(fig. 13.12). The paired cricoids and thyroids of mammalian embryos tend to unite across the midventral line during ontogeny (fig. 13.13*b*). The laryngeal cartilages are interconnected by ligaments and by intrinsic laryngeal muscles. Extrinsic laryngeal muscles, particularly the sternothyroid and thyrohyoid of mammals, provide the larynx with the mobility that is necessary during the act of swallowing.

Folded within or stretched across the laryngeal chamber in anurans, some lizards, and most mammals are **vocal folds**, more often referred to as **vocal cords**, which endow these species with the ability to communicate vocally to one degree or another with other members of the species. The cords vary from mere fleshy

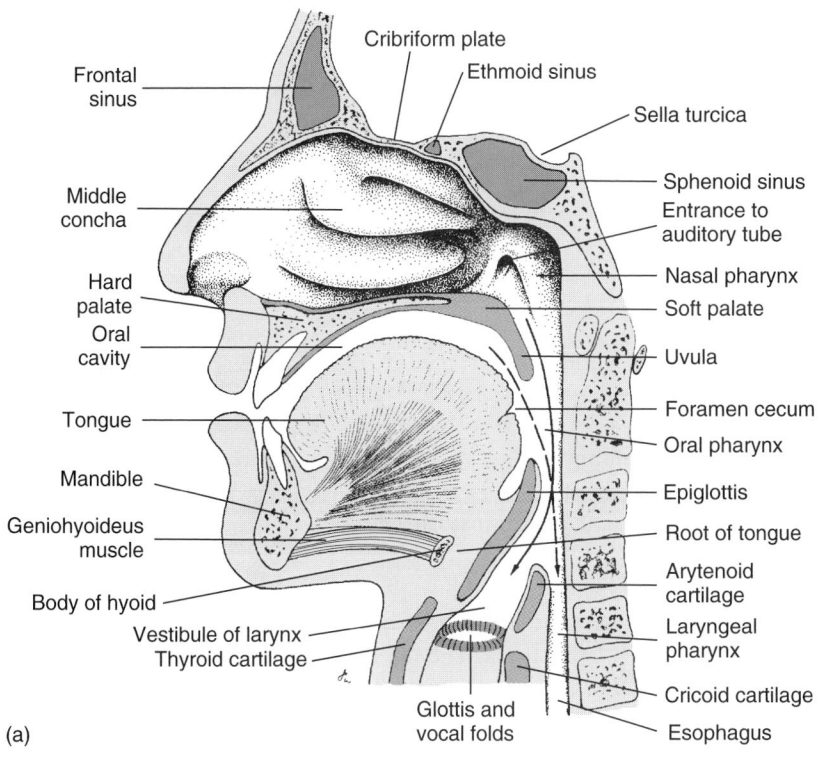

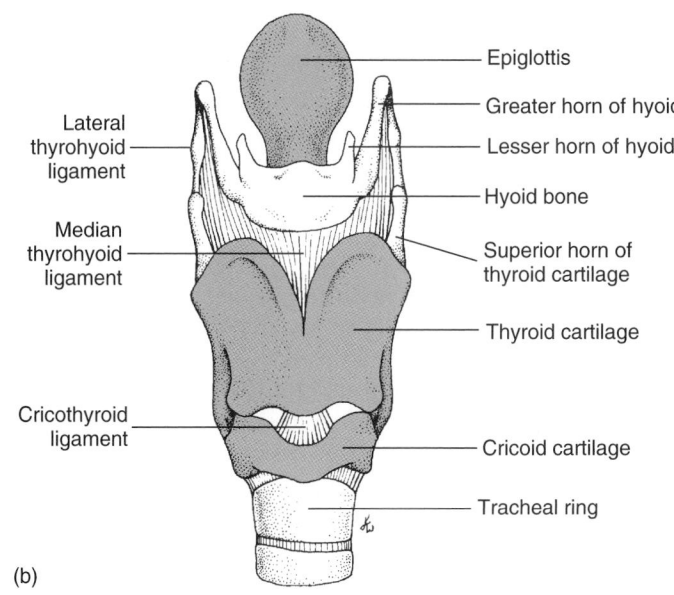

FIGURE 13.13

(*a*) Human upper respiratory tract, sagittal section. (*b*) Hyoid and larynx, frontal view. *Arrows* in (*a*) designate pharyngeal chiasma.

calization, the cords are relaxed, and exhaled air passes between them silently. When under tension, the cords vibrate, giving rise to sounds. During vocalization, *intrinsic* muscles of the larynx alter the position of the thyroid and arytenoid cartilages with respect to one another, thereby regulating the tension on the cords. The pitch (frequency of vibration) of the human voice is a function of the amount of tension within the cords. Among mammals, hippos and a few others lack vocal cords, and one breed of dogs, basenjis, have poorly developed ones and cannot bark. By contrast, the thyroid and associated hyoid bones of howler monkeys are enormous, causing a goiterlike bulge in the neck (fig. 13.14). They surround a voluminous resonating chamber that enables the weird howl of this species to penetrate deep into the jungle.

Except in birds and mammals, vocalization is not a significant means of communication among tetrapods. Apodan and urodele amphibians are mute, as are most reptiles. Male anurans force air out of the lungs, over vocal folds, and into the oropharyngeal cavity with the mouth and nares closed, and from there into **vocal sacs**, which produce a croak each time they refill and deflate (see fig. 12.3*a*). The vociferous croaking of male toads immediately after a rainstorm is a mating call. (The eggs of oviparous females are laid after a good rain has filled the ponds and ditches in which eggs are deposited. Sperm is deposited over the eggs while they are being extruded.) Each species has its own breeding call. There may be other calls also. Birds lack vocal cords; their calls are produced by an avian voice box, the syrinx, at the base of the trachea.

Most mammals have "false" vocal cords, as well as true ones. In some species, these produce unusual sounds such as purring in kittens. They are fleshy folds located at the entrance to the **vestibule of the larynx,** an ante-chamber located just above the true cords. The vestibule in howler monkeys has become the enormous resonating chamber referred to earlier.

Anatomic adaptations prevent fluids from entering the glottis in submerged aquatic tetrapods. Chief among these are valves at the entrance to the external nares.

folds surrounding the glottis and attached to the arytenoid and cricoid cartilages, as in anurans, to a pair of mammalian folds containing strong bandlike elastic ligaments that are stretched between the arytenoid and thyroid cartilage on each side of the sagittal plane. Air expelled from the lungs passes through a narrow slit, the glottis, between the folds. When not in use for vo-

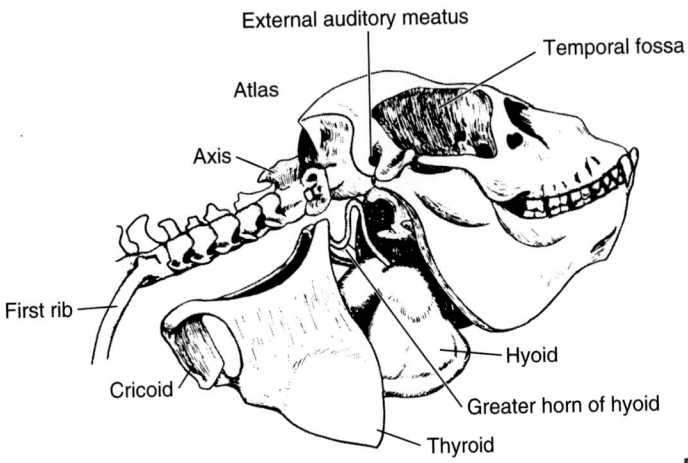

FIGURE 13.14
Hyoid bone and larynx of a howler monkey.

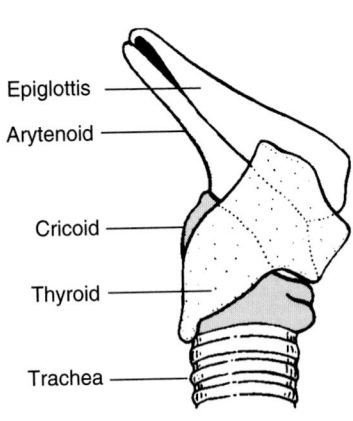

FIGURE 13.15
Laryngeal skeleton of whale. The epiglottis and arytenoid can be inserted into the nasopharynx.

Crocodilians have, in addition, a fleshy valve, the **palatine velum**, which consists of two large transverse folds suspended from the caudal end of the long palate just anterior to the internal nares. When a matching fold at the base of the tongue is drawn up against the velum, the resulting partition prevents the passage of water to the pharyngeal chiasma and provides a continuous passageway for air from the nostril to the lungs. Consequently, a crocodile or alligator can be submerged with only two moundlike nostrils and two elevated eyes protruding above the water, breathing (with a mouth full of water), and eyeing the environment above the water for prey. This is especially useful after an air-breathing victim is caught—a rat, for example. The victim can be carried or dragged into the water and held under by clenched jaws until it drowns. Meanwhile, the crocodile continues to breathe without water getting into the lungs. When the crocodilian is totally submerged, the valves at the entrance to the external nares prevent flooding of the respiratory tract.

In most mammals, terrestrial and aquatic, the larynx is high in the oropharynx where it is able to lock into the nasopharynx during the act of swallowing liquids. This allows adult cats, monkeys, and many other mammals to breathe while simultaneously drinking because a pair of pharyngeal recesses lateral to the larynx serve as bypasses to the esophagus for liquids. It also allows their babies to breathe while suckling. The larynx of human babies is high in the throat until about 18 months of age, at which time it begins to descend in the throat, "altering the way in which a child breathes, swallows, and vocalizes" (Laitman, 1984). It remains high—does not descend—in monkeys or apes. In marsupials, in which the baby's mouth cannot be withdrawn from the nipple, the high larynx permits milk that is being pumped by the mother's mammary glands into the baby's esophagus from being sucked into the lungs

when the baby breathes. In whales that have surfaced to breathe, the beaked cephalic end of the larynx (fig. 13.15) can be extended into the nasopharynx, permitting prolonged inhalation of fresh air through the blowhole while the cavernous mouth may be running over with seawater.

In humans and other mammals with a low larynx, the act of swallowing draws the larynx forward (upward in humans) against the epiglottis, momentarily blocking the entrance to the larynx and thereby preventing food from entering the lower respiratory tract. But each blockage can be only momentary. You cannot eat a hamburger while holding your nose closed!

Of course, a mammal cannot breathe through the mouth if the larynx protrudes into the nasopharynx, nor could air returning from the lungs be used for articulate speech. The latter requires participation of the lips, tongue, and the cavity acquired between the root of the tongue and the glottis when the larynx descends in the course of postnatal development. Thus, the final location of the human larynx facilitates human speech.

The anterior wall of the larynx of some male lizards has a saccular evagination, the **gular pouch** or dewlap, just under the skin of the throat. When the pouch is inflated with air, it causes a ballooning of the throat; when sunlight shines through its well-vascularized walls, the reddish pouch becomes a sex attractant. The pouch deflates noiselessly when the animal is disturbed. A process of the hyoid bone supports the inflated pouch.

Trachea, Syrinx, and Bronchi

The **trachea** is ordinarily as long as the neck. Therefore, it is short in amphibians and long in amniotes. In birds and some turtles, the part of the trachea within the neck is longer than the neck as an accommodation to the stretching, twisting, and looping that these necks are capable of. When a swan's neck is relaxed, or a turtle with-

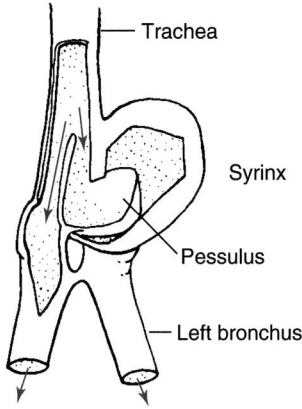

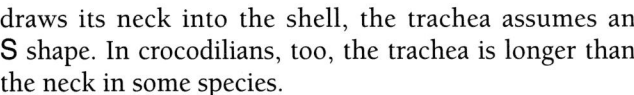

FIGURE 13.16
Asymmetrical bronchotracheal syrinx of a canvasback duck.
Arrows indicate path of inhaled air.

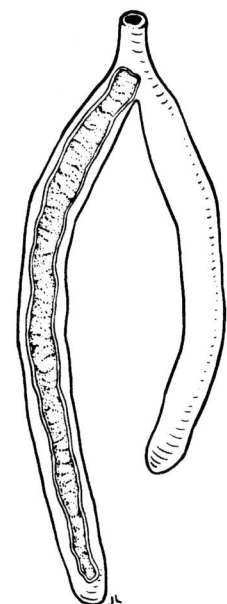

FIGURE 13.17
Lungs of *Necturus.*

draws its neck into the shell, the trachea assumes an S shape. In crocodilians, too, the trachea is longer than the neck in some species.

The walls of the trachea are prevented from collapsing under the negative pressure of inhalation by cartilaginous or bony rings or plates that form within the wall. The rings are usually incomplete dorsally, the ends being united by smooth muscle that can change the diameter of the tube in response to the need for a greater or lesser volume of tidal air. In crocodilians and birds, however, all tracheal rings are complete. Except in urodeles, the trachea bifurcates to become two primary **bronchi** that are similarly strengthened.

Birds have a special voice box, the **syrinx**, at the bifurcation of the trachea (fig. 13.16). A **bronchotracheal syrinx** consists of a resonating chamber with walls strengthened by the last several tracheal rings and the first bronchial half rings. Mucosal folds project into the chamber, and a bony **pessulus** may form in a special semilunar membranous fold. In a **tracheal syrinx**, parts of the last several tracheal rings are missing, enabling the membranous wall of the trachea itself to vibrate. In a **bronchial syrinx**, the membranous wall between two bronchial cartilages folds into the chamber when the cartilages are drawn together, and the fold vibrates with the flow of air.

Amphibian Lungs

The lungs of amphibians are simple sacs, long in urodeles (fig. 13.17) and bulbous in anurans, conforming to the shape of the pleuroperitoneal cavity. The internal lining may be smooth throughout, there may be simple sacculations in the proximal part, or the entire lining may be pocketed. The left lung of caecilians is rudimentary, and the lungs of salamanders that inhabit swift mountain streams may be only a few millimeters long. Plethodontid salamanders have *no* lungs.

The lungs of normally aquatic urodeles function mostly as hydrostatic organs, and respiration in species lacking gills takes place chiefly through the pharyngoesophageal lining and the skin. In water, which is the normal habitat of *Necturus,* only about 2 percent of the oxygen is acquired via lungs. Another perennibranchiate, *Siren,* a gill-breather like *Necturus* when in oxygenated water, frequently expresses air through the gill slits when obligated to breathe air. Urodeles using lungs as either hydrostatic or respiratory organs fill the lungs in the same manner as lungfishes; respiratory air enters the mouth, not the nostrils, and the muscular floor of the oropharynx pumps it to the lungs. Elasticity of the lungs probably plays a major role in emptying them.

The details of respiration in bullfrogs have been described by Gans, DeJongh, and Farber (1969). The mouth remains closed during respiration. In essence, they found that lowering the floor of the oropharyngeal cavity with the glottis closed draws air via the nostrils into the deep inferiocaudal reaches of the oropharynx where it is put on hold. Thereupon the glottis is opened, and the air already in the lungs escapes to the exterior via the nostrils. Emptying of the lungs appears to be the result of the elasticity of the lung and the smooth muscle in its walls. In the third phase, the fresh air stored in the floor of the oropharynx is forced into the lungs by raising the floor of the oropharynx with the nostrils again closed. In the fourth phase, the reservoir of air on hold is replenished. Once this sequence has been completed and the glottis closed, oscillation of the muscular floor of the oropharynx commences. The phenomenon can be observed in any frog by watching the rapid rhythmic pulsations of the throat behind the lower jaw.

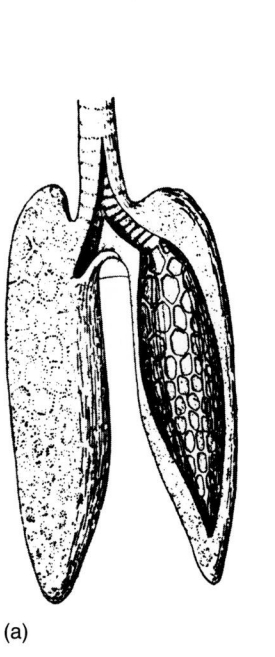

(a)

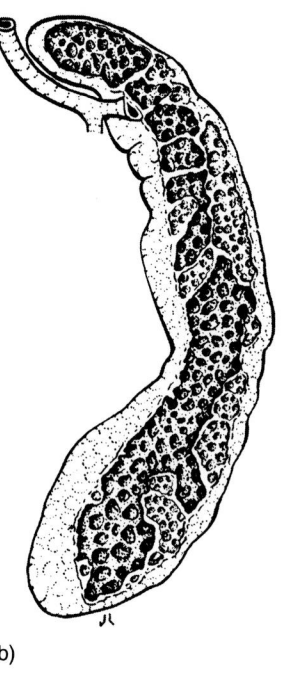

(b)

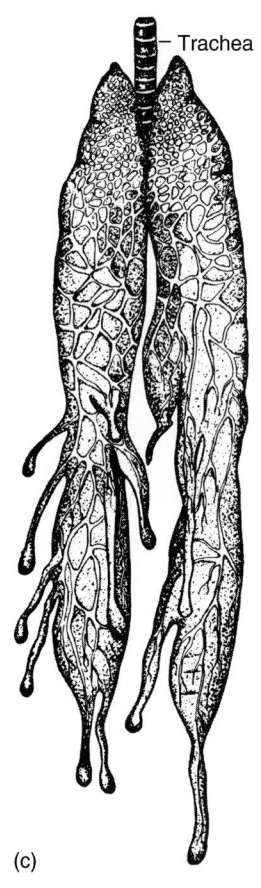

— Trachea

(c)

FIGURE 13.18
Lungs of three reptiles:
(*a*) *Sphenodon;*
(*b*) *Heloderma,* a large
lizard; and
(*c*) a chameleon.

The oscillations maintain a steady flow of fresh air via the nostrils into and out of the oropharyngeal cavity, the lining of which is an accessory respiratory epithelium. The oscillations also probably flush out any residual air remaining from the preceding emptying of the lungs. Air is forced in from the oropharyngeal cavity by a *pressure pump,* and the elasticity of the lungs pushes it out. However, a *vacuum pump* fills the oral cavity with fresh air.

Nonavian Reptilian Lungs

Unlike the positive *pressure pump* of living amphibians, the amniotes utilize negative pressure to draw air into the thoracic cavity (aspiration breathing). The mechanism to achieve aspiration breathing differs among amniote groups (discussed later). The lungs of *Sphenodon* (fig. 13.18*a*) and of snakes are simple sacs, although the caudal third of the lining in snakes is septate and contains residual air. In lizards (fig. 13.18*b*), crocodilians, and turtles, the septa divide the lungs into numerous large chambers. These lungs are somewhat spongy because of the numerous pockets of trapped air. Although asymmetry of the two lungs is usual in tetrapods, it is pronounced in legless reptiles, in which the left lung is much shorter than the other, sometimes atrophied, and occasionally totally absent.

An enormous diverticulum of the left lung of puffing adders extends into the neck. Inflation of this air sac causes the neck to balloon. Similar **air sacs** extend among the abdominal and pelvic viscera in some lizards (fig. 13.18*c*). Dinosaurs and pterosaurs had them, and in these reptiles the sacs also invaded the centra of vertebrae, as they do in today's birds. Air sacs are used by reptiles to inflate the body as a scare tactic or for other defensive purposes; in birds, as will be seen shortly, they have become a vital part of the mechanics of respiration.

Most reptilian lungs, like those of amphibians, occupy the pleuroperitoneal cavity along with the other viscera. In turtles, they lie dorsal to the viscera, flattened against the carapace just caudal to the pectoral girdle on each side, and all but their caudal ends are retroperitoneal. The lungs of crocodilians and a few squamates occupy separate subdivisions of the coelom, the paired **pleural cavities.** These are isolated from the rest of the coelom by a tendonous **oblique septum.** (An oblique septum in a bird is seen in fig. 13.19.) Although the septum plays no active role in respiration, it participates passively when the abdominal viscera that press against it are displaced cephalad or caudad by action of abdominal muscles.

Squamates inhale by employing intercostal muscles to rotate the ribs forward and outward, thereby increasing the volume of the pleuroperitoneal cavity. This action creates a vacuum around the lungs and other viscera, and atmospheric pressure forces environmental air

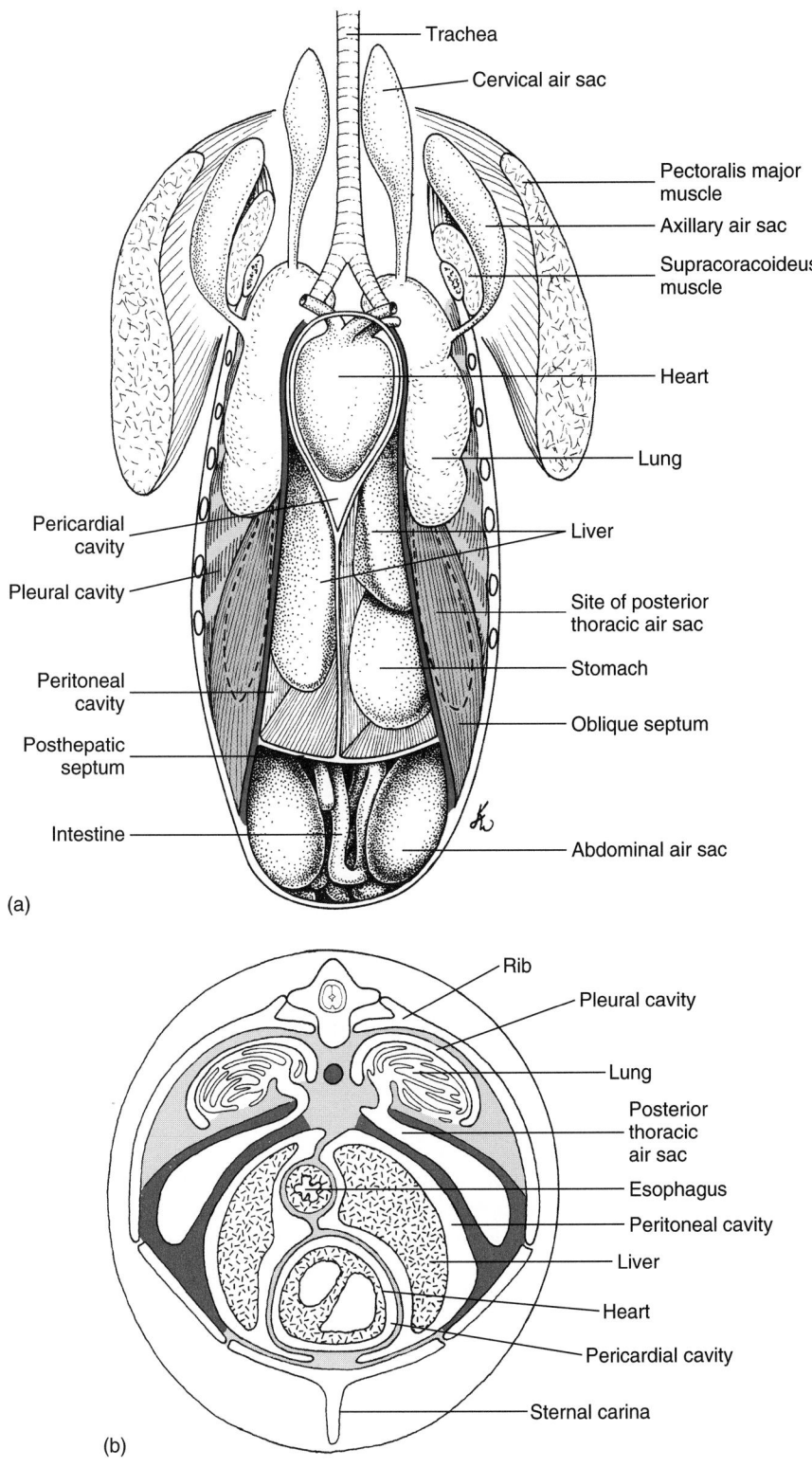

Trachea

Cervical air sac

Pectoralis major muscle

Axillary air sac

Supracoracoideus muscle

Heart

Lung

Pericardial cavity

Pleural cavity

Liver

Site of posterior thoracic air sac

Stomach

Peritoneal cavity

Oblique septum

Posthepatic septum

Intestine

Abdominal air sac

(a)

Rib

Pleural cavity

Lung

Posterior thoracic air sac

Esophagus

Peritoneal cavity

Liver

Heart

Pericardial cavity

Sternal carina

(b)

FIGURE 13.19

Oblique septum (*red*) and subdivisions of the coelom in birds. (*a*) Ventral view, interclavicular and anterior thoracic air sacs omitted to display lungs. The posterior thoracic air sacs (*broken lines*) are embedded within oblique septum. (*b*) Thorax in cross section.
Source: Data from E.S. Goodrich, *Studies on the Structure and Development of Vertebrates,* 1930, Macmillan and Company, Ltd., London.

into the nostrils and throughout the respiratory tract, causing the thin-walled elastic lungs to balloon with oxygenated air. Return of the ribs to their resting position and, in some species, contraction of the abdominal wall musculature compress the lungs and other viscera; this effect, aided by the resilience of the lung walls, results in exhalation.

Crocodilians may employ ribs for respiration to some degree, but they also have a unique method of getting air into the lungs, which, atypically for reptiles, lie cephalad to the large bilobed liver. The liver, in turn, lies tightly against the oblique septum that separates the two organs. A pair of long striated **diaphragmatic** muscles, derived from the musculature of the hypaxial body wall, extend between the pelvic girdle and the fibrous investment of the liver. When these muscles contract (which they presumably do in response to the same reflex stimuli that cause the mammalian diaphragm to contract), the liver and oblique septum are drawn caudad creating a negative pressure within the pleural cavities. Atmospheric pressure then expands the lungs via the respiratory tract, and air rushes into the lungs. Exhalation takes place when the transverse abdominal muscles contract, constricting the coelom and forcibly displacing the viscera. The pressure is transmitted from the coelomic viscera to the lungs, and oxygen-depleted air is forced out. Thus, the body wall musculature plays the major role in respiration in crocodilians but in a unique manner.

The process of respiration in turtles is complicated by the observation that their ribs cannot participate in respiratory movements because they are fused with the carapace. And because the abdominal wall is rigid and its musculature is vestigial, it, too, cannot participate. But all is not lost! Reflex movements of the pectoral girdle play the major role in ventilation of the lungs by altering the volume of the coelom.

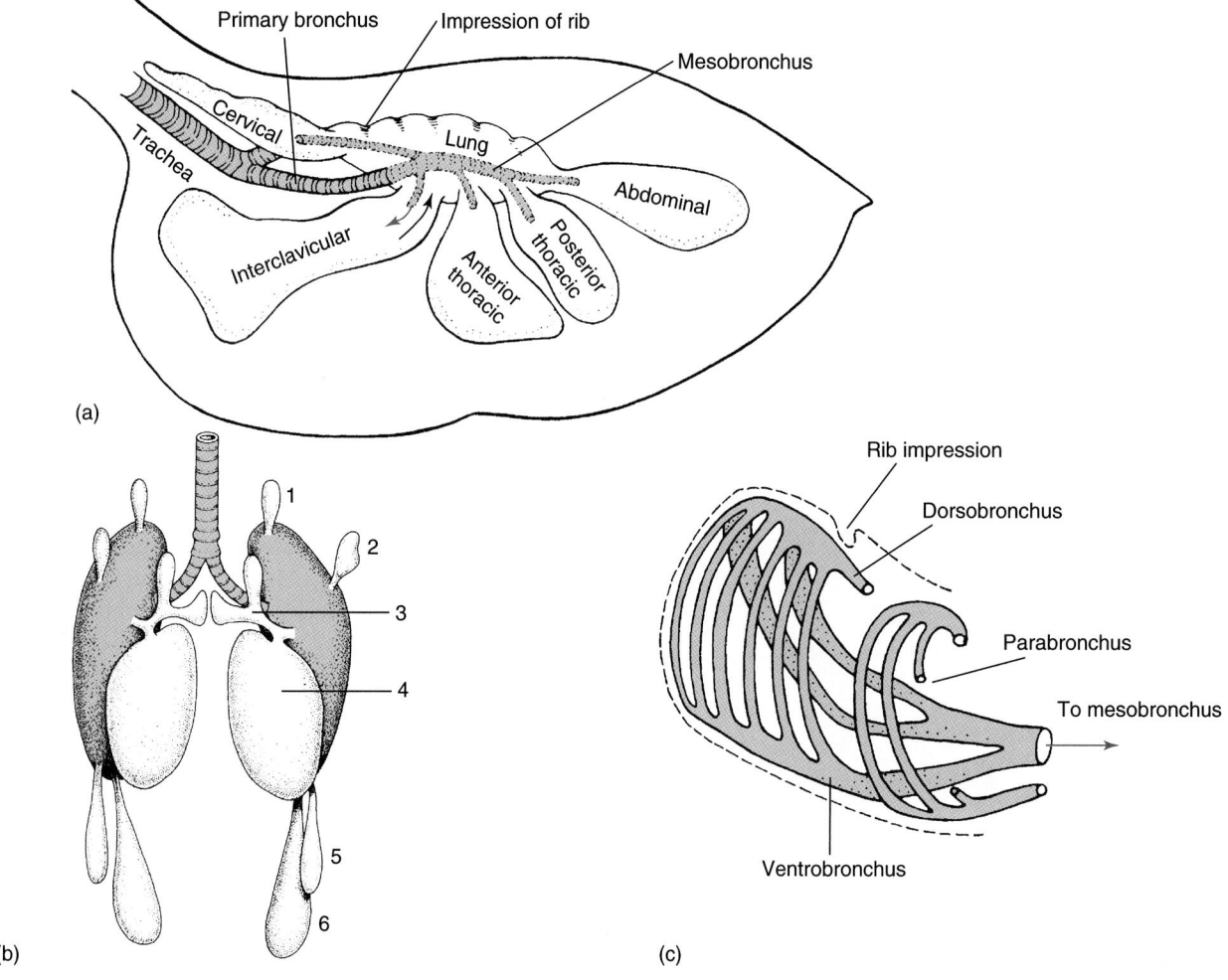

FIGURE 13.20

Lower respiratroy tract of a bird. (*a*) Adult air sacs. Air leaves the sacs via recurrent bronchi (*black arrow*). (*b*) Air sacs (**1–6**) of a 12-day chick embryo. Trachea, primary bronchi, and lungs are in *red*. (*c*) Parabronchi and their connections. In (*b*): **1**, cervical; **2,3,** lateral and medial anlagen of interclavicular; **4** anterior thoracic; **5,** posterior thoracic; **6,** abdominal.
Source: Data from K.F. Locy and O. Larsell, "Embryology of the Bird's Lung" in *American Journal of Anatomy*, 19:447, 1916.

Lungs and Their Ducts in Birds

The lungs of birds occupy paired pleural cavities separated from the rest of the coelom by a tendonous oblique septum (fig. 13.19*a* and *b*). Although the septum plays no active role in altering the pressure within the lungs, it participates passively, as in turtles, when the viscera within the peritoneal cavity are displaced by the action of muscles that move the ribs and sternum.

Avian lungs are unique morphologically. Incoming air enters a lung from the trachea via a **primary bronchus**, continues *nonstop* through the substance of the lung in the **mesobronchus**, and enters spacious air sacs before returning to the lungs (fig. 13.20*a* and *b*).

Air sacs are voluminous, thin-walled, distensible diverticula of the lungs. Birds have five or six pairs (fig. 13.20*a*): (1) cervical sacs at the base of the neck; (2) interclavicular sacs dorsal to the furcula and sometimes united across the midline; (3) anterior thoracic

sacs in their own compartment of the coelom lateral to the heart; (4) posterior thoracic sacs within the double-layered oblique septum; (5) abdominal sacs among the abdominal viscera; and (6) axillary sacs (see fig. 13.19*a*), less common, lying between the major elevator muscle of the wings (supracoracoideus) and the major depressor muscle (pectoralis major, see fig. 11.20*c*). Except in ratites, diverticula from the sacs penetrate most of the bones of the body, including centra, entering via **pneumatic foramina**. For this reason, birds are sometimes said to have hollow bones.

Avian air sacs function in the following manner: When a bird is in flight, contraction of a set of flight muscles on the chest wall expands the coelom. This creates a vacuum around and, hence, within the air sacs, and atmospheric pressure forces fresh air via the mesobronchi into the sacs. Subsequent contraction of an antagonistic set of flight muscles compresses the sacs,

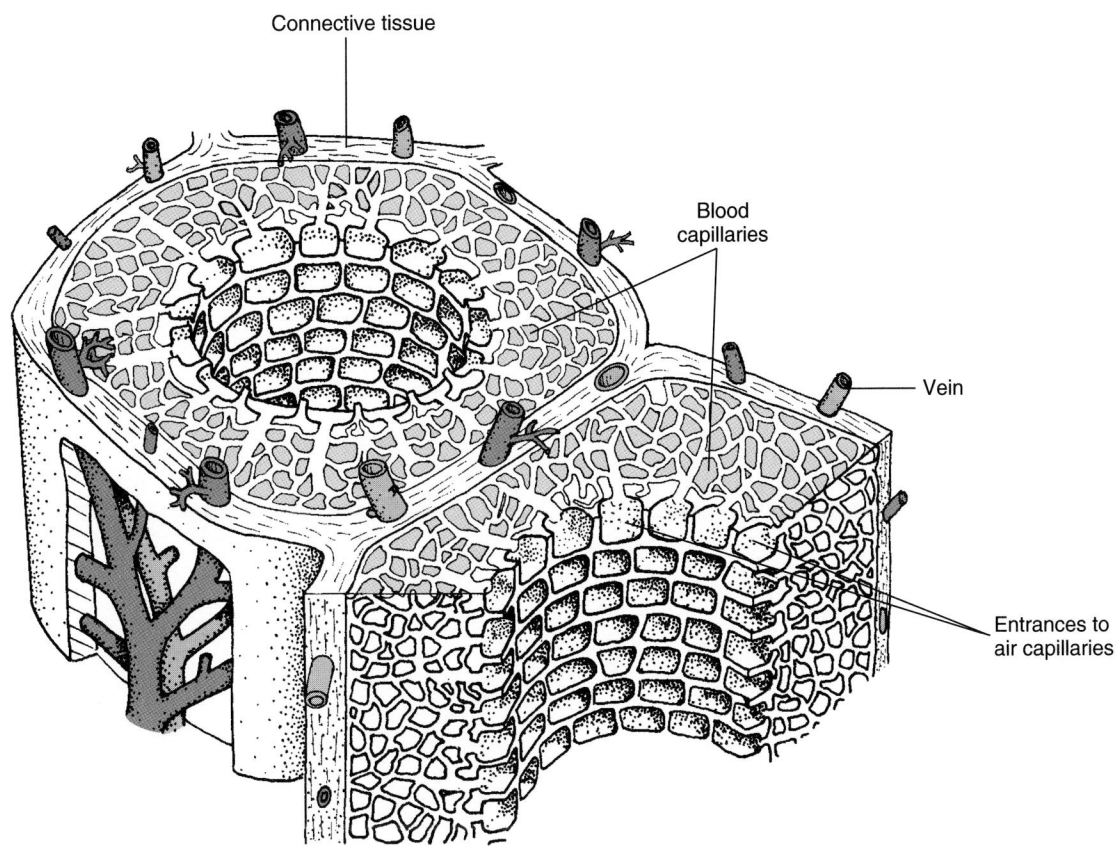

Connective tissue

Blood
capillaries

Vein

Entrances to
air capillaries

FIGURE 13.21
Parabronchi and air capillaries of a bird's lung.

forcing oxygen-rich air out of the sacs via **excurrent bronchi**, into the duct system within the lungs, across the respiratory epithelium, and on to the exterior. This muscular bellows (vacuum-compression-vacuum-compression) maintains an uninterrupted flow of air to and from the lungs as long as the bird is in flight.

When a bird is at rest, the chest wall still functions to create a vacuum, but it is the intercostal and other non-appendicular muscles that expand the coelom. The thoracic wall relaxes as a result of the elasticity of the tissues of the wall, not by muscle contraction. External respiration in resting birds does not require the energy expenditure necessary to maintain a bird in flight.

Dorsobronchi and **ventrobronchi** within the lungs are connected by **parabronchi** (fig. 13.20c). The walls of the latter (fig. 13.21) consist of an anastomosing network of **air capillaries**, each only a few thousandths of a millimeter in diameter, surrounded by **blood capillaries**. It is the lining of the air capillaries that constitutes the respiratory epithelium.

The air capillaries are interconnected in three dimensions. Air flows freely through the network, returns to the parabronchus from which it came and, without interruption, continues to the mesobronchus as an exha-

lation. Further details on the flow of air through the pulmonary duct system are lacking.

Avian air sacs and the open-ended duct system within the lungs result in complete replacement of the air in the lungs with every inflation-compression cycle of the bellows. Consequently, unlike the lungs of other tetrapods, those of birds contain no residual air.

Air sacs are thermoregulatory, dissipating excess heat produced by the muscles during flight. The heat is transferred directly from the tissues to the air sacs rather than via the bloodstream because air sacs have a relatively poor vascular supply. At rest, a bird's chief defense against overheating is a reflexive increase in respiratory rate. In resting doves, excess heat is transferred from an exceptionally vascular esophageal mucosa to the outgoing airstream in the adjacent trachea.

As we have seen, air sacs are not unique in birds. They are found in some lizards and snakes, and they were present in dinosaurs and pterosaurs as revealed by pneumatic foramina in their bones. In *Archaeopteryx*, the pneumatic foramina are restricted to the vertebrae. Foramina in the long bones of birds appear to be a derived feature (apparently absent in Cretaceous birds). The extension of air sacs into bones results in energy conservation during flight by decreasing the density of the bird's body, making it more buoyant in air. The warmer the air in the sacs, the more buoyant is the bird.

Mammalian Lungs

Mammalian lungs differ from those of birds in that incoming air enters a system of passageways that branch and rebranch like the limbs of a tree. At the end of the smallest passageways, **alveoli** lined by a respiratory epithelium are the sites of gaseous exchange. Oxygen-depleted and carbon dioxide–enriched air, along with metabolic water, is exhaled via the same path by which the air entered.

Mammalian lungs usually exhibit several lobes, asymmetrically, with one more on the right (fig. 13.22, 14 mm embryo). The lungs of some aquatic and terrestrial mammals are not lobed (whales, sirenians, elephants, and horses, for example), and some are lobed on the right side only (including monotremes and rats).

Left and right lungs occupy separate pleural cavities separated in the midline by the **mediastinum**, a septum of loose connective tissue enclosing the esophagus, the heart within the pericardial sac, major ascending and descending blood vessels, lymphatics, nerves, and the lower end of the trachea.

The trachea divides into two primary bronchi, each of which penetrates a lung at the **hilus** and divides into one **secondary bronchus** for each lobe, when lobes are present. These give rise to **tertiary bronchi** that branch and rebranch into smaller and smaller **bronchioles**, the last of which opens into several to a dozen thin-walled **alveolar ducts** (fig. 13.22*b*). The walls of the ducts are evaginated to form clusters of alveoli, or respiratory pockets, estimated to number over 300 million in each human lung. It is in these alveoli that gaseous exchange takes place.

The walls of the bronchi and larger bronchioles contain smooth muscle fibers, connective tissue, and irregular cartilaginous plates, and the lining is a **ciliated pseudostratified columnar epithelium** (a single layer of tall ciliated cells with nuclei that lie at different heights, giving the impression that the epithelium is multilayered). As the branches become smaller, the cilia are lost, the epithelium becomes flatter, the cartilage disappears, and then, in some mammals at least, the smooth muscle cells disappear. Alveoli are lined by a **simple squamous epithelium** ("pavement cells") beneath which is a rich plexus of capillaries held together by a close network of reticular connective tissue fibers. Red blood cells in the capillaries take on oxygen to form oxyhemoglobin, and carbon dioxide and some metabolic water pass from the blood plasma into the alveoli to be exhaled.

The parietal epithelium of each pleural cavity lines the chest wall as the **parietal pleura** and covers the cephalic surface of the diaphragm as the **diaphragmatic pleura** (fig. 13.22). At the **root** of each lung, where the primary bronchus and pulmonary vessels enter and

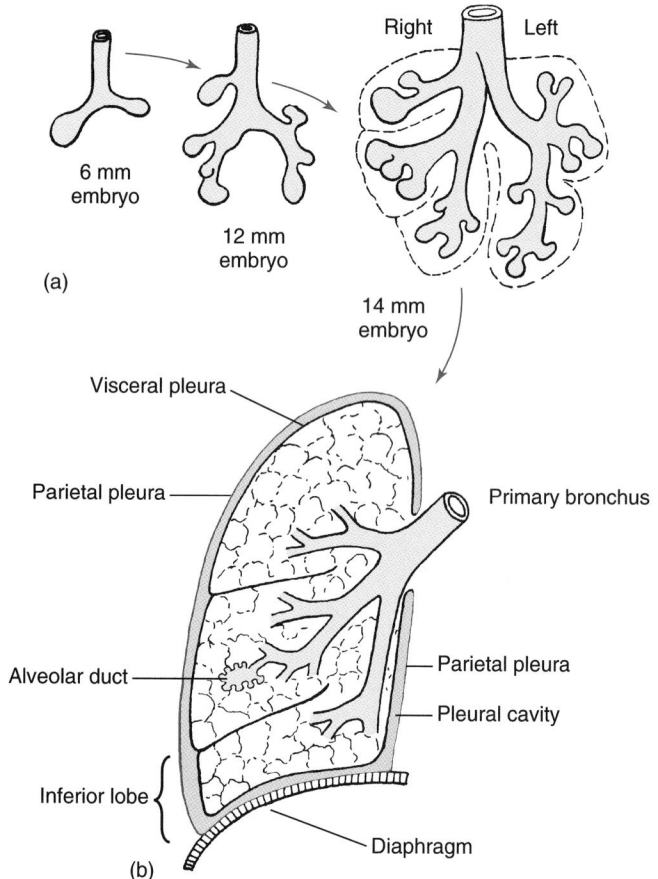

FIGURE 13.22

(*a*) Development of lung of fetal pig. Developmental stages are also applicable approximately to human embryos of the same length. The anlagen of all five adult lobes have been established by the 14 mm stage. (*b*) Relationships of parietal (*red line*) and visceral pleura, pleural cavity, and diaphragm. The width of the pleural cavity is exaggerated.

leave, the parietal pleura is continuous with the visceral pleura that lies on the surface of the lung. The space between the parietal and visceral pleurae is the pleural cavity. It surrounds the lung except at the hilus. A *subatmospheric* pressure exists in the pleural cavity, and the normal atmospheric pressure exerted via the nares keeps the elastic wall of the lung in intimate contact with the thoracic wall at all times, with only a thin layer of serous lubricant intervening. The lubricant is a product of the pleurae. The fluid minimizes friction between lungs and chest wall as the latter rises and falls with each inhalation and exhalation. Inflammation of the pleura, or pleurisy, causes an increase of fluid in the cavity. Perforation of the thoracic wall and the parietal pleura—from a gunshot wound, for example—allows atmospheric air to enter the pleural cavity, resulting in deflation of the lung on that side, a condition known as pneumothorax.

Ventilation of mammalian lungs is accomplished primarily by a dome-shaped muscular diaphragm, unique in mammals, that functions as a suction pump (see fig. 11.13). It is anchored to the xiphoid process of the sternum ventrally, to a half dozen or so of the caudal-most ribs and their costal cartilages laterally, and to several of the more anterior lumbar vertebrae dorsally. When not under tension, the dome bulges cephalad into the thoracic cavity. Contraction of the diaphragmatic muscles flattens the diaphragm. This further decreases the already subatmospheric pressure within the pleural cavity. Consequently, normal atmospheric pressure pushes more air into the lungs to fill the vacuum. This is the phenomenon known as inhalation.

Exhalation in mammals is largely a passive phenomenon attributable to the following: (1) relaxation of the diaphragm, which returns it to the domed position, decreasing the volume of the thorax and increasing the pressure around the lungs; (2) the upward pressure exerted on the diaphragm by the resilient abdominal viscera, which were under compression while the diaphragm was flattened; (3) resilience of the abdominal wall, which bulges when the abdominal viscera are compressed; (4) return of the ribs to a resting position as the costal muscles relax; and (5) elasticity of the lungs, which enables them to conform to these changes. As a result, air is squeezed out of the lungs. During forceful expiration, as in panting, roaring, coughing, or singing, the abdominal wall participates in the expulsion of air.

Marine mammals that feed in the depths of the oceans have exceptionally muscular diaphragms. The waterspout in a whale, for instance—the sign that exhalation is taking place—lasts three to five minutes. Also, deep-sea foraging mammals cannot store oxygen in their lungs in anticipation of a dive because the lungs collapse as the dive takes the animal to increasing depths. Instead, oxygen acquired prior to a dive is stored in retia mirabilia in the trunk (chapter 14 "Rete Mirabilia"). The animal exhales just before commencing a dive, and breathing does not resume until the animal resurfaces, which may be as long as two hours later.

Summary

1. External respiration is the exchange of respiratory gases between organism and environment. It takes place via highly vascular membranes with thin, moist epithelia.

2. Internal respiration is the exchange of gases between capillary blood and the tissues.

3. The chief adult organs of respiration are pharyngeal gills, oropharyngeal mucosa, swim bladders or lungs, and skin.

4. Cutaneous respiration is the chief method of respiration in aquatic urodeles and some scaleless fishes.

5. Internal gills arise in the walls of pharyngeal pouches and are supported by the skeleton of the pharyngeal arches except in agnathans, where a peripherally positioned bronchial basket supports the gill pouches.

6. A typical gill, or holobranch, consists of two demibranchs and an interbranchial septum. The septum is shorter in fishes that have an operculum.

7. Respiratory water usually enters via the mouth, but it enters the spiracle in some elasmobranchs, the unpaired naris in hagfishes, and external gill slits in lampreys.

8. Elasmobranchs have gills that open independently to the exterior with each aperture protected by a flaplike valve. The slits of chimaeras, bony fishes, and larval anurans are covered by an operculum attached to the hyoid arch.

9. A branchiostegal membrane is attached to the ventral edge of each operculum in bony fishes, supported by gular bones or branchiostegal rays. The membranes enclose an opercular chamber, which receives water that has passed over the gills. The water is then expelled to the exterior.

10. The first embryonic pharyngeal slit persists as a spiracle in adult elasmobranchs and chondrosteans. In some species, it houses a pseudobranch on its anterior wall.

11. Functional hyoid arch demibranchs occur in cartilaginous fishes but tend to disappear in teleosts. The number of demibranchs is reduced still further in dipnoans.

12. Larval gills may be external or internal. They are external in dipnoans, a few basal ray-finned fishes, and amphibians. Anuran larvae later develop internal gills. Filamentous internal gills project to the exterior in larval elasmobranchs and some bony fishes.

13. Gills in fishes function also in maintaining salt homeostasis and in excretion of nitrogenous waste and carbon dioxide.

14. External nares lead to blind olfactory sacs in extant jawed fishes other than dipnoans. In dipnoans and tetrapods, they are connected with the oropharyngeal cavity or pharynx via nasal canals. They are used for respiration in tetrapods only. Hagfishes have a nasopharyngeal duct that carries respiratory water. A nasohypophyseal duct in lampreys ends blindly.

15. Internal nares (choanae) open far forward in the oral cavity of dipnoans and amphibians, and farther caudad if there is a secondary palate. Despite the presence of internal nares, dipnoans and aquatic amphibians take in respiratory air through the mouth.

16. Pneumatic sacs arise as median evaginations from the embryonic floor of the foregut. They become swim bladders in fishes and lungs in tetrapods.

17. Physostomous fishes are those in which the embryonic ducts of swim bladders remain open throughout

life. The ducts eventually close in physoclistous fishes, chiefly teleosts.

18. Swim bladders of fishes are primarily hydrostatic organs. They are used for respiration in dipnoans and a few physostomous ray-finned fishes. In some teleosts, they participate in sound detection or sound production.

19. The glottis is protected against the entrance of fluids and solids by fleshy valves at one site or another, and, in mammals, by its occasional location in the nasopharynx, or by the epiglottis.

20. The larynx is supported by a single pair of lateral cartilages in urodeles and by arytenoid and cricoid cartilages in anurans and reptiles. A thyroid cartilage is added in mammals. Other small cartilages develop in mammals.

21. Vocal folds (cords) are chiefly mammalian but are found also in anurans and some lizards.

22. Tracheal walls are reinforced by cartilaginous or bony plates, rings, or half rings. The trachea in most cases bifurcates into two primary bronchi. At the base of the trachea in most birds is an avian voice box, the syrinx.

23. Paired lungs arise as an unpaired evagination from the embryonic pharyngeal floor. They occupy the pleuroperitoneal cavity in amphibians and most reptiles. Each lung occupies its own pleural cavity in crocodilians, birds, and mammals.

24. Fibrous oblique septa separate the pleural cavities from the rest of the coelom in crocodilians and birds. In mammals, a muscular diaphragm does so, and a mediastinum separates one pleural cavity from the other.

25. Lungs are saccular with shallow pocketed linings in amphibians and most snakes, more compartmentalized by septa in other reptiles, and highly spongy because of millions of blind air pockets (alveoli) in mammals. One lung in limbless tetrapods is often rudimentary.

26. Saccular diverticula (air sacs) of the lungs in some lizards and in birds extend among the viscera. In birds, they also extend into hollow bones.

27. In birds, inhaled air flows nonstop through the lungs and into air sacs. It then returns to the lungs via recurrent bronchi, flows uninterruptedly through open-ended air capillaries where gaseous exchange occurs, and is then vented.

28. In addition to maintaining a steady flow of air through the air capillaries of birds, air sacs assist in thermoregulation and lighten the skeleton in flying birds.

29. Inhalation of air in most amniotes is by action of a suction pump (aspiration breathing), chiefly the thoracic wall (ribs and costal muscles). In birds in flight, certain flight muscles constitute a suction pump. In mammals, the thoracic diaphragm partic-

ipates. In anurans, the oropharyngeal floor serves alternately as a suction pump and a pressure pump.

30. Unlabored expiration in amniotes results primarily from the elasticity of lungs and resilience of the thoracic wall and viscera. In birds in flight, some flight muscles act as a pressure pump.

Critical Thinking Questions

1. How are the respiratory and circulatory systems related?
2. Lampreys occlude their oral cavity during feeding. How can they respire given this observation?
3. Compare the holobranchs of a typical teleost and shark and their external openings.
4. How do marine teleosts eliminate excessive salts acquired from their marine environment?
5. What functions are associated with an air sac? (You should be able to name at least three.)
6. In deep-sea and in fast continuous swimming fishes, we observe a reduction or loss of the air sac. Can you functionally explain these observations?
7. What are the functions for the accessory air sacs of birds?
8. Contrast the pattern of airflow between mammals and birds.
9. Among amniotes, how is negative pressure achieved in aspiration breathing?
10. What are the structures of vocalization, and in which groups does vocalization play a significant role?

Selected Readings

Alexander, R. M.: The chordates. Cambridge, England, 1975, Cambridge University Press.

Brainerd, E. L.: The evolution of lung-gill bimodal breathing and the homology of vertebrate respiratory pumps, *American Zoologist* 34:289, 1994.

Brainerd, E. L., Liem, K. F., and Samper, C. T.: Air ventilation by recoil aspiration in polypterid fishes, *Science* 246:1593, 1989.

Britt, B. B., Makovicky, P. J., Gauthier, J., and Bonde, N.: Postcranial pneumatization in *Archaeopteryx, Nature* 395:374, 1998.

Brodal, A., and Fänge, R., editors: The biology of *Myxine*. Oslo, 1963, Universitetsforlaget (Norway).

Feder, M. E., and Burggren, W. W.: Skin breathing in vertebrates, *Scientific American* 253:126, November 1985.

Gans, C., DeJongh, H. J., and Farber, J.: Bullfrog (*Rana catesbeiana*) ventilation, How does the frog breathe? *Science* 163:1223, 1969.

Graham, J. B.: An evolutionary perspective for bimodal respiration: A biological synthesis of fish air breathing, *American Zoologist* 34:229, 1994.

Hughes, G. M., and Morgan, M.: The structure of fish gills in relation to their respiratory function, *Biological Reviews* 48:419, 1973.

Laitman, J. T.: The anatomy of human speech, *Natural History,* p. 20, August 1984.

Laitman, J. T., and Reidenberg, J. S.: Specializations of the human upper respiratory and upper digestive systems as seen through comparative and developmental anatomy, *Dysphagia* 8:318, 1993.

Liem, K. F.: Form and function of lungs: The evolution of air breathing mechanisms, *American Zoologist* 22:739, 1982.

Randall, D. J., and others: The evolution of air breathing in vertebrates. New York, 1981, Cambridge University Press.

Wood, S. C., and Lenfant, C. J. M.: Respiration: Mechanics, control, and gas exchange. In Gans, C., editor: Biology of the reptilia, vol. 5. New York, 1976, Academic Press.

Links to the Internet

Visit the zoology website at http://www.mhhe.com/zoology to find live Internet links for each of the following references.

1. Respiration. A series of slides and nice diagrams describing the human respiratory system.
2. Histology—The Web Laboratory. Links to units on many systems of the body, including the respiratory system. Extensive coverage of microstructure of all parts of the respiratory tract.
3. U. Wisconsin Histology Atlas. A ton of pictures related to the respiratory system. And really neat ways to display them–try them with each of the three options. Not overly intuitive, but very well done. You can also check out their other system sites that are linked.
4. Avian Respiratory Dynamics Shockwave Animation. A very cool animation, a bit of text, and just a few diagrams. But seeing respiration through the avian system animated is worth a thousand words.
5. Fish Respiration. A lengthy description of fish respiration, including great graphics.

14

Circulatory System

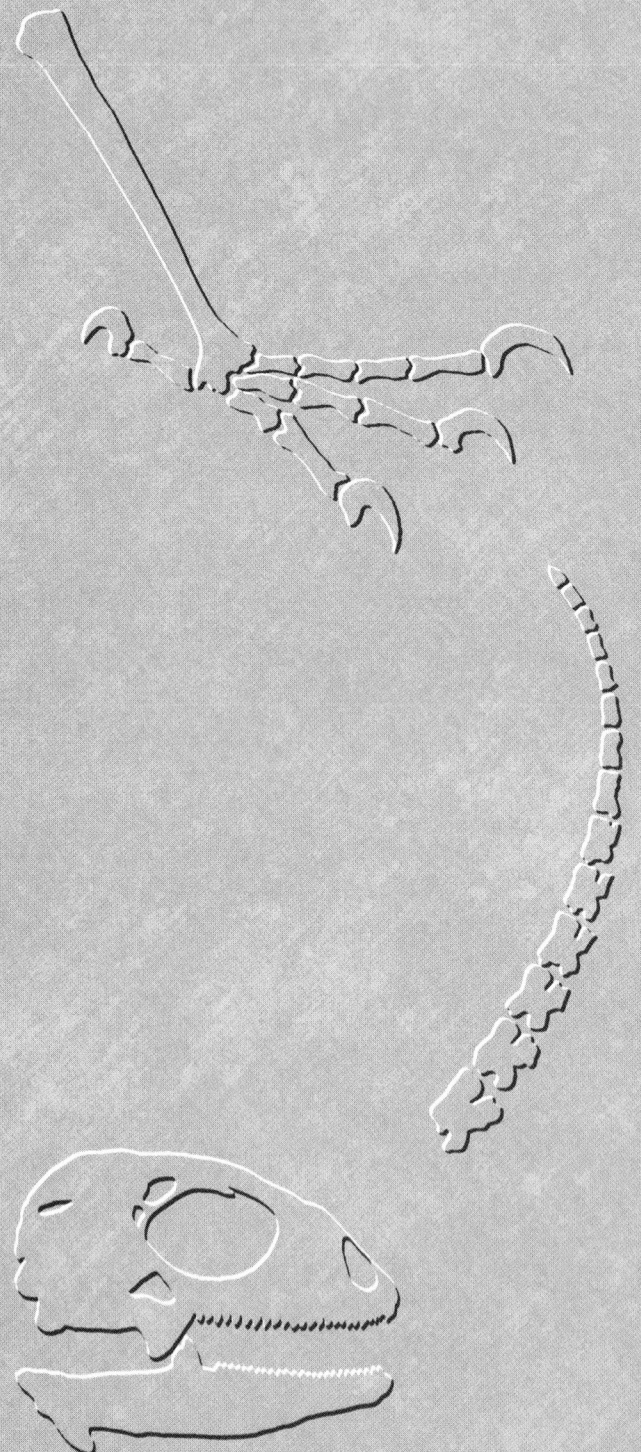

In this chapter, we will look at the heart, blood vessels, and lymph channels of craniates from fishes to human beings. We will find that in embryonic development, structure, and function they conform to a basic pattern, and we will see how the pattern with respect to the blood circulatory system was modified step-by-step during the evolution of modern species. Then we will examine the abrupt changes that occur when a chick or a mammalian fetus shifts from life in amniotic fluid to breathing air. Along the way, we will look at a few more examples of countercurrents, some marvelous vascular retes, and the pacemaker of the craniate heart.

The circulatory system of craniates consists of the heart, arteries, veins or venous sinuses, capillaries or sinusoids, and blood (**blood vascular system**) and of lymph channels and lymph (**lymphatic system**). The blood carries oxygen from respiratory organs; nutrients from extraembryonic membranes, digestive tract, and storage sites; hormones and other substances associated with homeostasis and immunity to disease; and waste products of metabolism to the excretory organs. Blood also conducts heat to and from the skin and other surfaces where heat is exchanged, thereby regulating and equalizing internal temperatures. Lymph channels collect interstitial tissue fluids not taken up by the bloodstream, and they collect lipids from the villi of the small intestine. Lymph vessels terminate in venous channels.

Arteries carry blood away from the heart. They have muscular and elastic walls (figs. 14.1a and 14.2) capable of distention with each intrusion of blood. (Feel your pulse!) The smallest arteries, 0.3 mm or less in diameter, are **arterioles**. They dilate and constrict reflexly and thereby assist in regulating blood pressure. They terminate in blood capillaries. **Veins** commence in capillaries (other than the respiratory capillaries of the gills) and carry blood toward the heart. They have proportionately less muscle and elastic tissue and more fibrous tissue than arteries and are therefore capable of less distention or constriction. The smallest are **venules**, which provide a direct connection to the capillaries. Arteries and veins, like most other organs, have a loose connective tissue covering, the **adventitia**.

Capillaries (fig. 14.3) generally consist of endothelium alone, although certain capillaries are accompanied by a delicate investment of mesenchyme and a scattering of smooth muscle fibers. Their lumen is just large enough to accommodate red blood cells in single file. In fact, the cells must often "squeeze through," becoming temporarily deformed until they "pop out" into the smallest venule. A capillary "bed" or network represents all the individual capillaries served by a single arteriole. At the site where a capillary emerges from an arteriole, a short section of the capillary wall, the **precapillary sphincter,** has smooth muscle fibers that, given an adequate neural or hormonal stimulus, close off the entrance to the capillary bed at that site. Relaxation of the constrictor muscle readmits blood to the capillary. Blanching or blushing of human skin is a result of the constrictor and dilatory effects of these sphincters. The smallest arterioles and venules are connected directly by short vascular **capillary shunts** that assure uninterrupted circulation between the arterial and venous sides of the capillary beds when other capillaries are constricted.

A **portal system** is a system of veins bounded by capillary beds (fig. 14.4). In most craniates blood from the capillaries of the tail passes via a **renal portal system** to capillaries of the kidneys before continuing to the heart. Blood from the digestive tract, pancreas, and spleen passes via a **hepatic portal system** to the capillaries of the liver before continuing to the heart. Blood containing pituitary regulating hormones from the hypothalamus passes via a **hypophyseal portal system** to the adenohypophysis before continuing to the heart.

DEVELOPMENT

The formation of the circulatory system involves a complex interaction of genetic and environmental controls

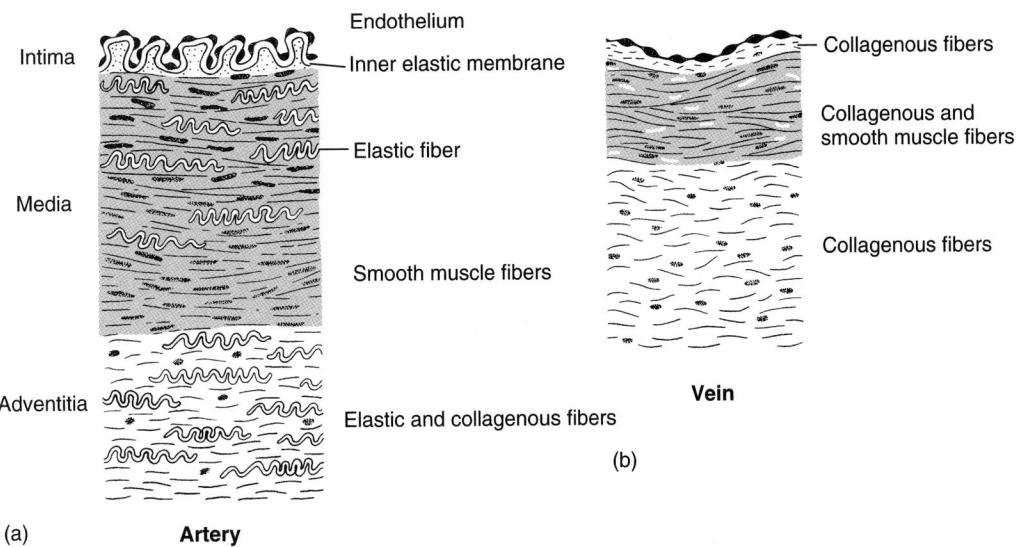

FIGURE 14.1

Structure of a medium-sized artery and its accompanying vein. Note differences in thickness of the muscle layers (*red*) and the absence of elastic fibers in the vein.

on vascularization and angiogenesis, mechanisms of vessel formation. Vascularization is the process of vessel formation from endothelial precursors with the formation of vessels potentially occurring prior to the onset of blood flow. In contrast, angiogenesis represents the remodeling of existing vessels and can occur both prenatally and postnatally, as seen in the vascularization of neoplasms (e.g., tumors).

The pattern of vessel formation is variable. The same vessel in different species may form from a vascular plexus with considerable remodeling (axial vessels in some birds and mammals). On the other hand, some species develop well-formed tubes from the onset of vascularization (seen in the axial vessels of many amphibians and some fish species). There is strong evidence that vascularization may be an inductive interaction with surrounding tissues. The notochord, which has a documented influence on the development of the overlying dorsal hollow nerve cord, also appears to influence the formation of the axial dorsal aorta. Posterior cardinal veins form in close association with trunk endoderm and are influenced by locally produced growth factors.

We will see that the patterns of blood vessels, despite developmental and genetic constraints, demonstrate a degree of variability within species. The vessels show a statistical pattern of occurrence. For example, in humans, most vessels follow a typical pattern, but limited variations are found with decreasing frequency (a major concern for surgeons).

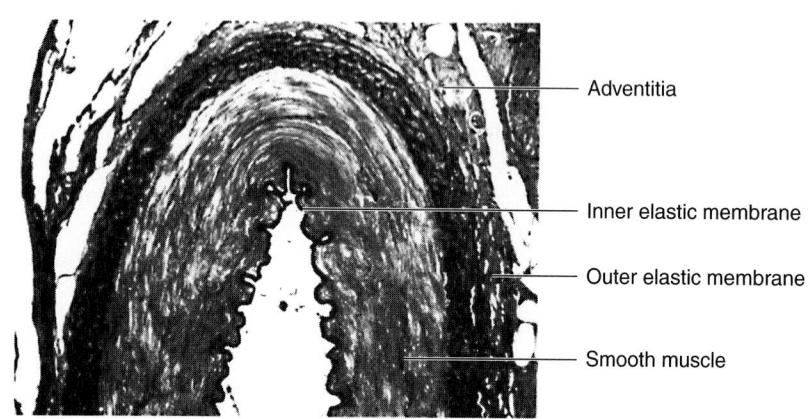

FIGURE 14.2
Microphotograph of a medium-sized artery.
Courtesy Gerrit L. Bevelander and Judith A. Ramaley.

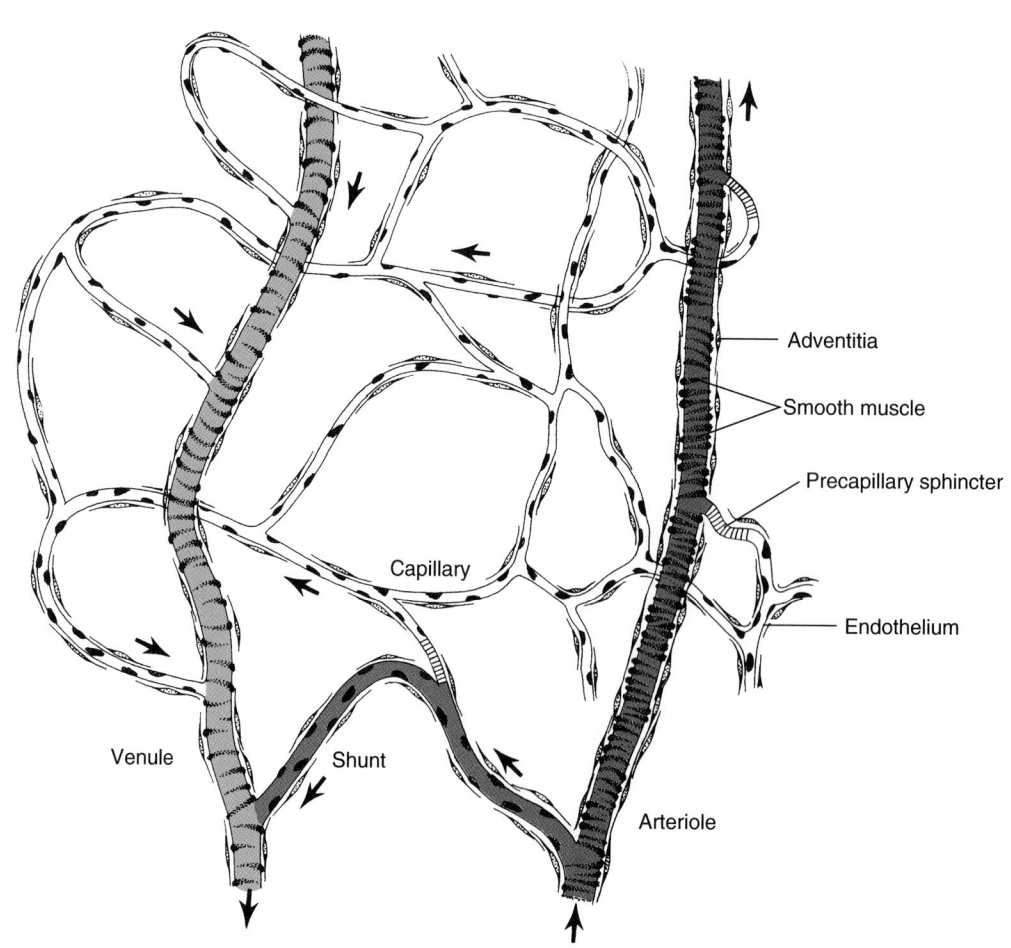

FIGURE 14.3

A capillary bed and its connections. Three crosshatched segments are sites of precapillary sphincters. *Arrows* indicate direction of flow.

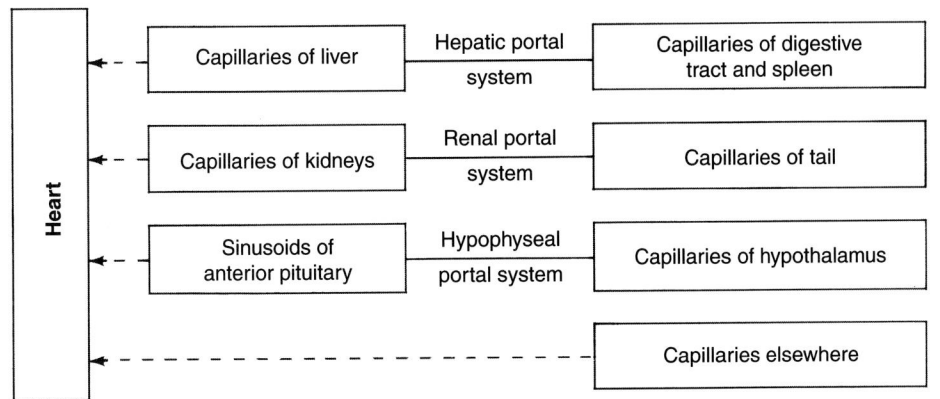

FIGURE 14.4
Chief portal systems of craniates. The renal portal system is lacking in typical mammals, and the hypophyseal portal system has not been demonstrated in most bony fishes. Channels indicated by *broken lines* are not part of a portal system.

Finally, understanding development of the circulatory system has important clinical implications. Current research is looking at angiogenesis as a tool to attack tumors (through induced inhibition of angiogenesis) and to aid the heart patient (through promotion of angiogenesis in the heart).

BLOOD

Blood is one of the tissues comprising the circulatory system. It consists of **plasma** and **formed elements** (fig. 14.5*a*). Plasma, which constitutes 55 percent of human blood, is a viscous fluid that is 90 percent water and 10 percent dissolved solids. The dissolved solids are blood proteins (serum globulin, serum albumin, and fibrinogen), substances required by living cells throughout the body (glucose, fat and fatlike substances, amino acids, ions of necessary salts), essential products synthesized by living cells (enzymes, antibodies, hormones), and some of the waste products of cellular metabolism. **Serum** is essentially plasma from which fibrinogen (clotting substance) has been removed.

Suspended in the plasma, and carried along in it as it flows, are formed elements consisting of red blood corpuscles (**erythrocytes**), white blood corpuscles (**leukocytes**), and platelets (**thrombocytes**). The formed elements can be separated from the plasma by centrifugation. With light microscopy and appropriate staining, the following cell types can be identified (fig. 14.5):

Erythrocytes
Leukocytes
    Granulocytes (polymorphonuclear leukocytes)
        Eosinophils
        Basophils
        Neutrophils
    Agranular leukocytes
        Lymphocytes
        Monocytes
Thrombocytes (platelets)

Hemopoiesis

Hemopoiesis is the formation of blood cells. In nearly every craniate embryo from fishes to humans, the earliest sign of hemopoiesis is seen in the walls of the yolk sac, where large numbers of **blood islands** are producing the first blood cells and vessels (see fig. 5.8). Beginning as concentrations of mesenchyme, blood islands produce **hemocytoblasts**. These are the stem cells from which all later blood cells arise. Blood islands also elaborate an extensive network of endothelial-lined spaces that form around the hemocytoblasts, confining them to vascular channels. Elsewhere in the embryo, other blood channels and a heart begin to form, and these establish connections with the channels in the blood islands. This combined activity quickly provides an early circulatory system adequate to meet the immediate needs of the embryo, which initially are the nutrients stored in the yolk sac and oxygen from whatever extraembryonic respiratory membrane is functioning at the time. Even if the early embryo is not being nourished by yolk (and this applies to only a small minority of craniates), the yolk sac is still the initial source of blood cells.

With continuing development of the embryo, descendants of hemocytoblasts form in tissues within the body itself, eventually including the liver, kidneys, spleen, and bone marrow (but not confined to these) and varying with the class or order. Hemopoietic bone marrow is found in the spongy core of flat bones and in the spongy proximal epiphyses of long bones. Adult red blood cells of all craniates are produced in varying proportions in the liver, kidneys, and spleen; and in amniotes, the red bone marrow is an important source. Granular leukocytes are produced in the spleen of most craniates, in the intestinal submucosa of teleosts, in the bone marrow of amniotes, and elsewhere in lesser numbers. Lymphocytes are produced in the lymphatic tissues discussed later in this chapter.

Deteriorating formed elements in the blood of craniates in general are removed from circulation largely in the sinusoids of the spleen, liver, and bone marrow,

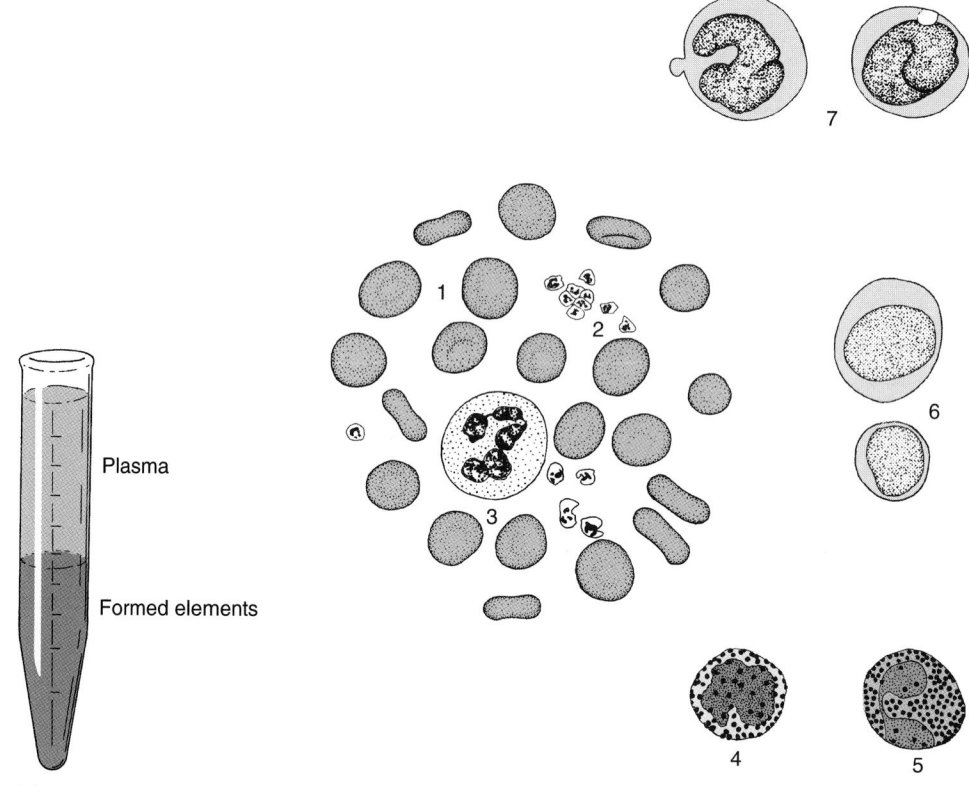

FIGURE 14.5
Cellular constituents of blood.
(*a*) Centrifuged whole blood.
Plasma is a light yellow liquid; the
formed elements constitute a
dark sludge. (*b*) Red and white
blood cells and platelets.
1, erythrocytes; **2,** thrombocytes;
3 to **5,** neutrophilic, basophilic,
and eosinophilic granulocytes;
6, lymphocytes; **7,** monocytes.

Plasma

Formed elements

(a)

(b)

which are lined by agranular **macrophages** that phagocytose (engulf and digest) blood cells in these locations.
Blood cells and other elements are also phagocytosed by
amoeboid white cells in any traumatized tissue of the
body. The breakdown of hemoglobin into bile pigments
was discussed in chapter 12.

Formed Elements

The red corpuscles of most craniates are oval nucleated
cells; those of mammals are biconcave circular discs just
large enough to squeeze through capillaries in a distorted shape, and without nuclei, having lost them before leaving the erythropoietic tissue where they developed. The circulating red cells of camels are an
exception, resembling the oval nucleated cells of more
basal craniates. The functional constituent of red blood
cells is hemoglobin, a compound protein composed of a
simple protein, globin, and an iron-containing pigment,
heme, with which available oxygen immediately binds
loosely to form **oxyhemoglobin** (HbO_2). Iron, therefore,
is indispensable for life. The oxygen is freed from the
hemoglobin in any capillary where the surrounding tissue has a partial pressure of oxygen lower than that in
the capillary. The result is **reduced hemoglobin**. The average life of an erythrocyte in a human being is three to
four months. Because there are about 25 trillion erythrocytes in the human body, it is obvious that the erythropoietic centers are very active.

The white corpuscles are far less numerous than either red cells or platelets. Granulocytes and monocytes
are amoeboid and can pass through capillary walls between adjacent epithelial cells and enter tissue spaces
where, as phagocytes, they serve as scavengers of
broken-down tissues. They have additional functions.
Lymphocytes are abundant in lymph nodes, in the
spleen, and in other lymphoid tissues.

Platelets participate with fibrinogen in the clotting of
blood. They are tiny fragments of stem cells (**megakary-
ocytes**) found in bone marrow. They consist of
membrane-enclosed cytoplasm and lack a nucleus.

Most of the preceding discussion refers to mammalian
blood. Whereas blood comprises 5 percent to 10 percent
of the body weight of amniotes, it represents only about
1.5 percent to 3 percent of the body weight of fishes. The
ratios of the constituents of blood differ among the craniate orders as adaptations to the environmental conditions with which each organism must cope.

THE HEART AND ITS EVOLUTION

The craniate heart is chiefly a muscular pump that occupies the pericardial cavity. The walls consist of an **endocardium**, **myocardium**, and **epicardium**, which correspond, respectively, to the intima, media, and
adventitia of the arteries. However, as described in chapter 11, the myocardium is cardiac muscle, rather than

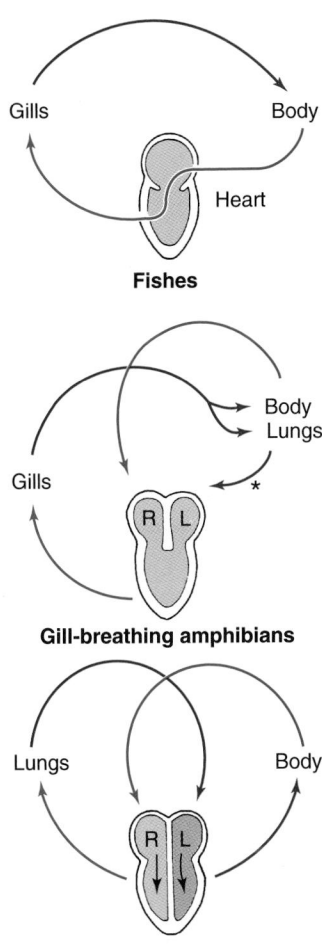

Fishes

Gill-breathing amphibians

Amniotes

FIGURE 14.6
Usual circulatory routes in gill-breathing fishes, perennibranchiate amphibians, and amniotes. *Red* is oxygenated blood. When the gill-breathing amphibian is forced to breathe air, unoxygenated blood bypasses the gills (not shown) and is oxygenated in the lungs, making the oxygen content of the pulmonary veins (*) rise.

smooth muscle as in arteries. It is especially thick in the ventricular wall. Lying on the epicardium is the **visceral pericardium**, the equivalent of the visceral peritoneum or pleuroperitoneum of the main coelom. The parietal and visceral pericardia are continuous with one another, being reflected over the blood vessels that enter and leave the heart. The space between these pericardia is the **pericardial cavity**.

The tissues of the heart are supplied with arterial blood by **coronary arteries** and drained by **coronary veins**. The heart muscle pulsates in response to specific electrolytes in the blood that perfuses it. The rhythmicity of the beat, which will be discussed shortly, is imposed by the autonomic nervous system in vertebrates, although all craniates possess an intrinsic control.

Single- and Double-Circuit Hearts

In fishes, blood passes from the heart to the gills and from there directly to all parts of the body, after which it returns to the heart (fig. 14.6, fishes). Thus, blood makes a *single circuit* during which it is pumped, oxygenated, distributed, and returned to the pump. No blood escapes oxygenation, and none fails to enter a capillary bed where oxygen can be released for use by the tissues.

In craniate classes in which all members have totally abandoned gills, namely, amniotes, a **pulmonary circuit** ordinarily carries oxygen-poor blood from the heart to the lungs and brings back oxygen-rich blood. A **systemic circuit** then carries the oxygenated blood to the other organs of the body and returns oxygen-depleted blood to the heart (fig. 14.6, amniotes). The evolution of *double-circuit* hearts was preceded by a long period of time during which natural selection produced circulatory systems adaptable for either aquatic or aerial respiration. Further adaptive changes culminated in the double-circuit heart. It freed amniotes from the necessity of an aquatic existence but held them in bondage to atmospheric oxygen. We will have a glimpse of these transitional stages in the paragraphs that follow.

Hearts of Gill-Breathing Fishes

The hearts of fishes other than dipnoans have four chambers in a series: a **sinus venosus, atrium, ventricle**, and **conus arteriosus (truncus arteriosus)**. Blood flows through these chambers in that sequence. Such a heart is seen in sharks and hagfishes (figs. 14.7 and 14.8). Designation of fish hearts as "two-chambered" recognizes only atria and ventricles as chambers. The terminology is not used in this text because it is not definitive.

The heart of sharks is typical of jawed fishes in general (fig. 14.7). The sinus venosus has thin walls, little muscle, and much fibrous tissue. Its caudal wall is anchored to the anterior face of the septum transversum. The sinus has some contractility but is chiefly a collecting chamber for venous blood that is returning from all parts of the body. It is filled by suction each time the ventricle contracts and relaxes. Blood from the sinus gushes through the sinoatrial aperture into the atrium between two one-way valves as soon as the atrium begins to relax after emptying.

The atrium is a large thin-walled muscular sac that is a sort of staging area for blood that is about to enter the ventricle to be propelled toward the gills. From the atrium, blood pours into the relaxing ventricle through an atrioventricular aperture that is guarded by a pair of one-way valves. These prevent ventricular blood from being pumped back into the atrium when the ventricle contracts.

The ventricle has very thick muscular walls and is the actual pumping portion of the heart. The anterior end is prolonged as a muscular tube of small diameter, the conus arteriosus, which extends to the extreme cephalic end of the pericardial cavity, at which point it is

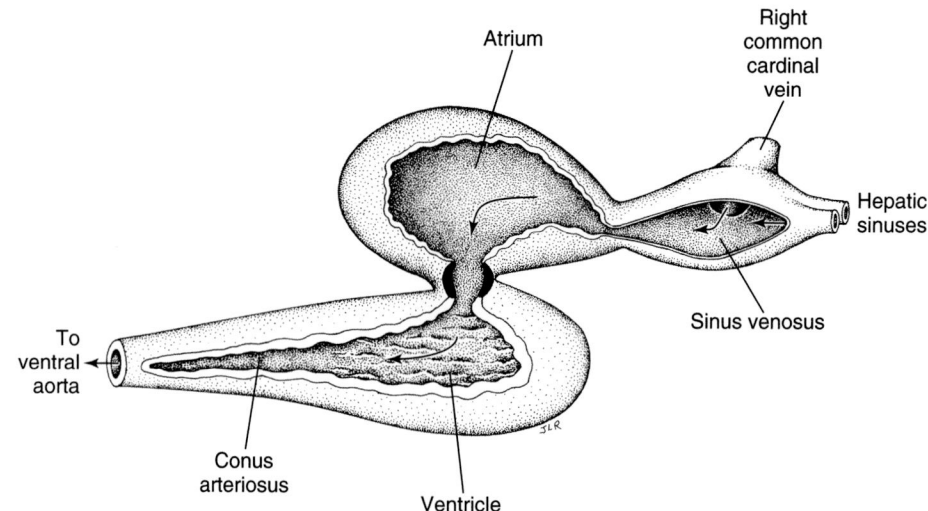

FIGURE 14.7
Heart of *Squalus acanthias* opened to show chambers and blood flow. Left lateral view.

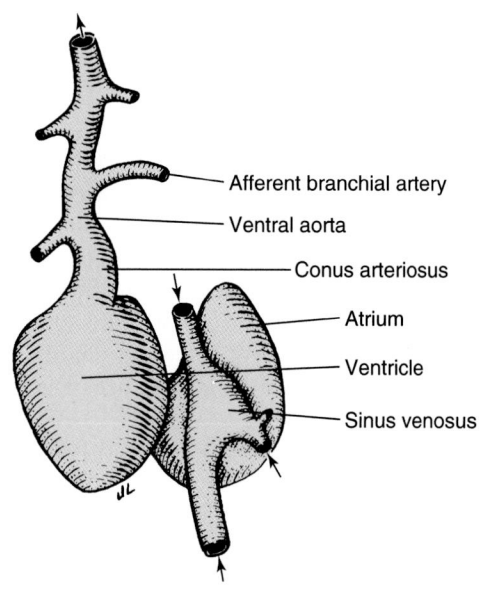

FIGURE 14.8
Heart and associated vessels of an agnathan. *Myxine glutinosa,* ventral view. *Arrows* indicate direction of blood flow.

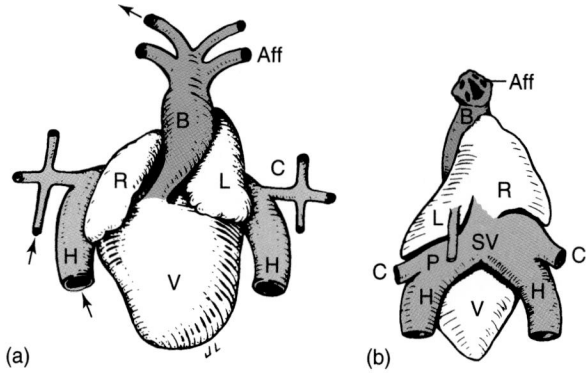

FIGURE 14.9
Heart and associated vessels of *Necturus.* (*a*) Ventral view. (*b*) Dorsal view. **Aff,** common channel leading to fourth and fifth afferent branchial arteries (see fig. 14.18*d*); **B,** bulbus arteriosus of ventral aorta; **C,** common cardinal vein; **H,** hepatic sinus; **L,** left atrium receiving, in (*b*) the pulmonary vein; **P,** pulmonary vein; **R,** right atrium; **SV,** sinus venosus; **V,** ventricle. In (*a*), a short conus arteriosus connects the ventricle with the bulbus. *Arrows* show direction of blood flow. Colors represent arteries (*red*) and veins (*blue*) but not necessarily oxygen content.

continuous with the ventral aorta. The conus is composed chiefly of cardiac muscle and elastic connective tissue. A series of semilunar valves facing forward within the conus prevents backflow of blood into the ventricle. Because of its elasticity, it balloons with each delivery of ventricular blood and then slowly constricts, thereby maintaining a steady arterial pressure in the ventral aorta so that the flow of blood through the gill capillaries, like the flow of countercurrent respiratory water over the gill filaments, is steady despite the rhythmicity of the ventricular beat.

The conus arteriosus of teleosts is shorter than that of sharks, but a muscular swelling at the base of the teleost ventral aorta, the **bulbus arteriosus,** performs the same role. It maintains a steady flow of blood over the gills. A bulbus arteriosus is present also in *Necturus* and some other perennibranchiate amphibians (fig. 14.9*a*).

Hearts of Dipnoans and Amphibians

Modifications in the heart and associated arteries in dipnoans and tetrapods are correlated with aerial respiration. The modifications enable oxygenated blood returning from the swim bladders or lungs to be separated in the heart from deoxygenated blood returning from other organs. Air-breathing ray-finned fishes differ in

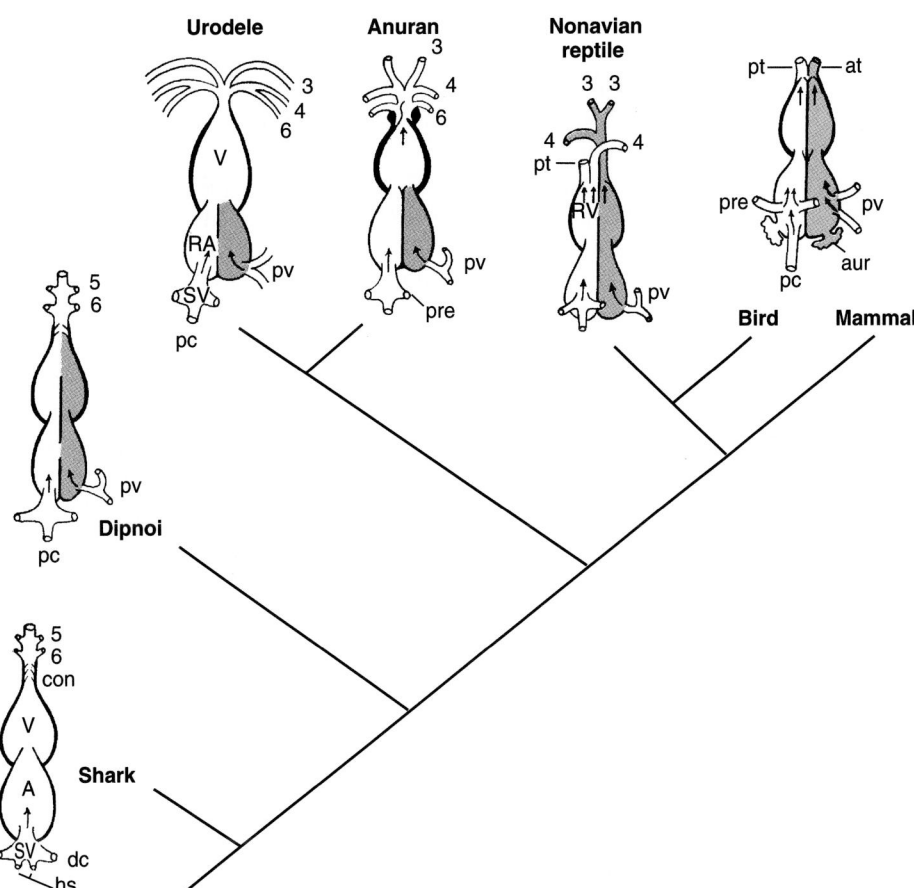

FIGURE 14.10
Heart chambers and oxygenated blood flow (*red*) in some generalized craniates. Distribution of oxygenated blood beyond the ventricle in dipnoan, urodele, and reptile depends on the species and the physiological needs of the organism. The parts of the heart shown are **A**, atrium; **RA**, right atrium; **V**, ventricle; **RV**, right ventricle; **SV**, sinus venosus; **con**, conus arteriosus; **aur**, auricle of mammalian heart. **3** to **6**, third to sixth aortic arches. Other vessels are **at**, aortic trunk; **dc**, common cardinal vein; **hs**, hepatic sinus; **pc**, postcava; **pre**, precava (common cardinal vein); **pt**, pulmonary trunk; **pv**, pulmonary veins.

that oxygenated blood returns to the systemic veins, resulting in mixing prior to arrival at the heart. It should be remembered that aerial respiration in these taxa represents a secondary mechanism that augments gill respiration in times of depleted environmental oxygen. On the other hand, the extant dipnoans (South American and African lungfishes) and tetrapods are obligate air breathers.

One modification in dipnoans and amphibians was the establishment of a partial or complete **interatrial septum** so that there are partial or complete **right and left atrial chambers** (figs. 14.9*a,* and 14.10, Dipnoi, urodele, anuran). The septum is complete in anurans and in some urodeles. Except in *Neoceratodus* (Dipnoi), the veins from the swim bladder or lungs empty directly into the *left* atrium; therefore, the blood in this chamber is oxygen rich. The sinus venosus empties into the *right* atrium (see fig. 14.9*b*); hence, the blood in this chamber is low in oxygen. In lungless urodeles, the atrium remains totally undivided.

A second modification is the formation of a partial **interventricular septum** (chiefly in dipnoans but also in *Siren,* a urodele) or of **ventricular trabeculae** (in amphibians). Trabeculae are shelves or ridges projecting from the ventricular wall into the chamber and running mostly cephalocaudad. Interventricular septa and ventricular trabeculae perform identical functions: They maintain separation of oxygenated and unoxygenated blood that began in the left and right atria.

A third modification is formation of a **spiral valve** in the conus arteriosus in dipnoans and anurans. The valve in dipnoans (fig. 14.11) consists of a pair of longitudinal typhlosolelike folds of the lining of the conus. In anurans, it is a single flap (fig. 14.12). The valves direct oxygen-poor blood into aortic arches that lead to gills or lungs (see figs. 14.11 and 14.18*b*, *blue*; and fig. 14.12), and they channelize oxygenated blood into arches that supply other organs (see figs. 14.11 and 14.18*b*, *red*; and fig. 14.12).

A fourth modification shortened the ventral aorta so that it becomes practically nonexistent as embryonic development progresses (figs. 14.11 and 14.12). As a result, blood moves from the conus arteriosus directly into appropriate vessels. Urodeles, however, retain a prominent ventral aorta (see fig. 14.9).

Many of the adaptations enabling aerial respiration in dipnoans and postlarval amphibians are seen also in amniotes.

Hearts of Amniotes

Amniote hearts have two atria, two ventricles (a unique third ventricular chamber in turtles and squamates), and, except in adult birds and mammals, a

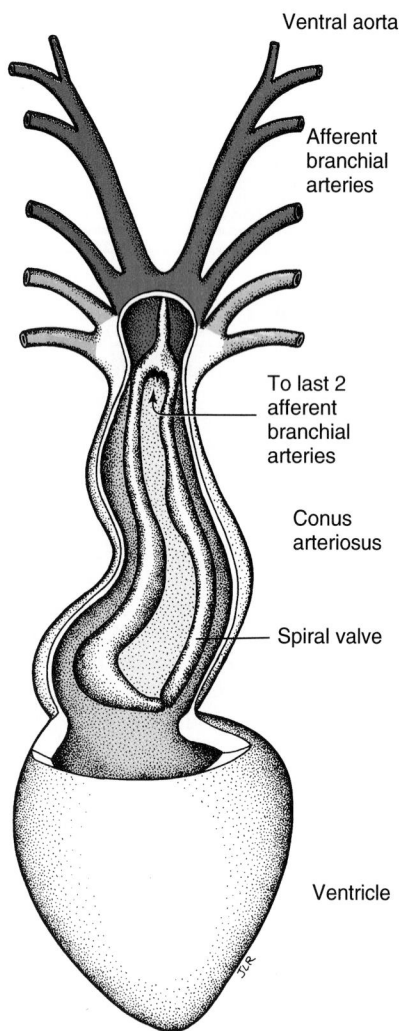

Ventral aorta

Afferent branchial arteries

To last 2 afferent branchial arteries

Conus arteriosus

Spiral valve

Ventricle

FIGURE 14.11

Conus arteriosus and afferent branchial arteries of the lungfish *Protopterus.* The spiral valve distributes oxygen-rich blood (*red*) to the first three afferent branchial arteries and oxygen-poor blood (*blue*) to the last two, which supply the respiratory swim bladder and internal gills, as shown in figure 14.18*b*.

sinus venosus (figs. 14.13*b* and 14.24). The sinus venosus in crocodilians, unlike in other reptiles, is partially incorporated into the wall of the right atrium. Birds and mammals have a sinus venosus during early development, but it fails to keep pace with the growth of the right atrium into which it empties and is finally indistinguishable as a separate chamber. Thereafter, the vessels that emptied into the sinus venosus empty directly into the right atrium. The embryonic location of the sinus venosus is marked in adults by a **sinoatrial (SA) node** of neuromuscular tissue. The node plays a key role in innervation of the heart, to be discussed shortly.

The right and left atria of amniotes are completely separated by an interatrial septum. Nevertheless, they are confluent during embryonic development via an **in-**

teratrial foramen, or **foramen ovale,** which closes near the time of hatching or at birth. The site of the obliterated foramen is marked in adult mammalian hearts by a depression, the **fossa ovalis,** in the medial wall of the right atrium.

The *right* atrium of nonavian reptiles (see fig. 14.24) receives blood from the sinus venosus; that of birds and mammals receives, directly, blood that earlier in phylogeny and during embryonic development emptied first into the sinus venosus (see fig. 14.26). The *left* atrium receives blood from the pulmonary veins. In mammals only, each atrium has an earlike flap, or **auricle,** within which is a blind chamber (see fig. 14.39). The functional advantage of the mammalian auricle has yet to be demonstrated.

The two ventricles are completely separated in crocodilians, birds, and mammals (see fig. 14.26). In turtles and squamates, however, a unique third chamber, the **cavum venosum,** has developed at the upper limit of the interventricular septum (see fig. 14.24). It functions to shunt oxygen-rich and oxygen-poor blood into or away from specific arteries that leave the heart. Its role can best be understood after we have studied those arteries, which we will do shortly.

The muscular walls that line the ventricular chambers of mammals exhibit sturdy interanastomosing muscular ridges and columns, the **trabeculae carneae.** They strengthen the walls of these powerful pumps and increase the force exerted by them.

One-way valves guard the passageways from the atria into the ventricles of craniates. In amniotes, each valve consists of one or more fibrous flaps, or **cusps** (muscular on the right side of the heart in crocodilians and birds), connected, chiefly in mammals, by tendonous cords (**chordae tendineae**) to **papillary muscles** that project from the ventricular walls (see fig. 14.26). During relaxation of the ventricles (**diastole**), blood from the atria flows freely past the cusps and into the ventricles. During ventricular contraction (**systole**), the cusps are forced forward or upward into the atrioventricular passageway, preventing reflux of blood into the atria. Each valve has one or two cusps in reptiles. In most mammals, the left valve has two cusps (**bicuspid,** or **mitral valve**), and the right has three (**tricuspid valve**). **Semilunar valves** at the exits of the ventricles into the pulmonary and aortic trunks prevent backflow into the ventricles as the latter relax (see fig. 14.26).

Innervation of the Heart

The hagfish heart lacks any external innervation. There are modified intrinsic cells that may respond to circulatory signals. The embryonic vertebrate heart begins to pulsate before any nerve fibers have reached it, and adult heart tissue, infused with an appropriate physiological solution, will continue to pulsate after its extrinsic innervation has been severed. Tissue from the heart

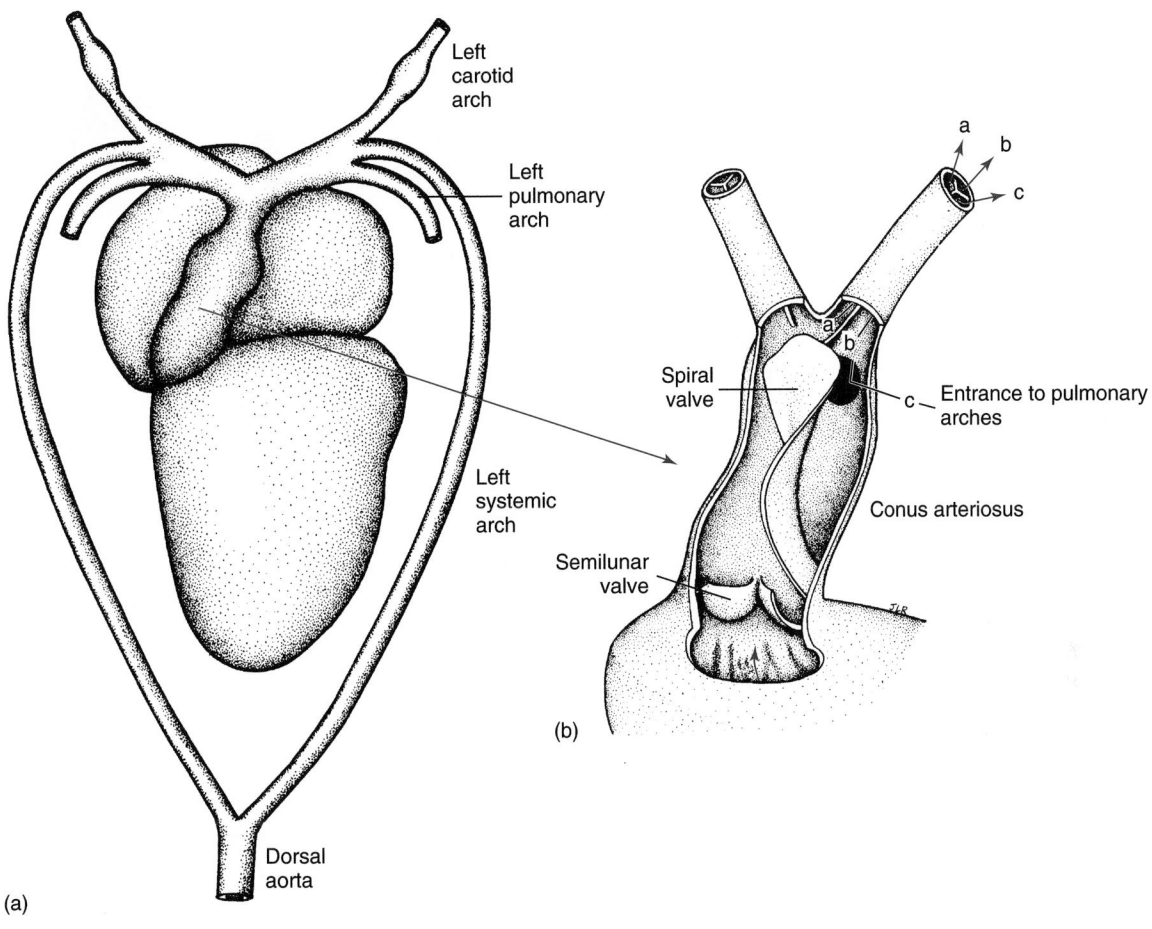

(a)

(b)

FIGURE 14.12

Heart and aortic arches of frog, ventral view. (*b*) Conus arteriosus opened to show passageways to left carotid arch (arrow **a-a**), and left systemic arch (arrow **b-b**). **c,** arrow enters common passageway to left and right pulmonary arches, then turns to enter passageway to left pulmonary arch.

of an embryonic chick, and its descendant cells, was kept alive and pulsating in a French laboratory for many years early in the twentieth century. The contraction of heart muscle is, therefore, autogenic, requiring no extrinsic neural stimulus for its initiation. Pulsation depends on the appropriate concentrations of certain electrolytes, especially Na, K, and Ca ions, in the tissue fluids that bathe cardiac muscle. The rate of autogenic pulsation of a denervated sinus venosus is imposed on the atria and ventricles via an intrinsic conduction system composed of atypical cardiac muscle fibers (**Purkinje fibers**) that constitute a conduction network with high conductile competence.

An extrinsic neural stimulus is necessary, however, to produce a regular beat that can be increased or slowed reflexly by the central nervous system in response to the physiological needs of the animal. Consequently, once nerve fibers have reached the sinus venosus during embryonic development, these fibers impose a rate of contraction on the cardiac muscle of that chamber. Thereafter, in fishes, amphibians, and non-

avian reptiles, this beat is transmitted by the intrinsic conduction system to the atrium (or atria), then to the ventricle (or ventricles), and finally, to the conus arteriosus, if there is one. Thus, the sinus venosus is the *pacemaker* of the heart of these cold-blooded vertebrates. The sinus venosus of fishes is innervated by the vagus nerve (tenth cranial nerve) only. In anurans and amniotes, the heart is innervated also by fibers from the sympathetic trunk, the two sets of fibers having opposite effects on the heart rate (inhibitory fibers via the vagal nerve, accelerator fibers via cardiac nerves from the sympathetic trunk, see fig. 16.35).

In birds and mammals, the embryonic sinus venosus eventually becomes incorporated into the wall of the right atrium, where it remains as a nodular mass of modified cardiac muscle and connective tissue known as the sinoatrial (SA) node. This node continues to receive the incoming fibers of the autonomic nervous system, and it is the SA node that serves as the pacemaker for the heart of birds and mammals. Impulses from the SA node are conducted via Purkinje fibers to the

FIGURE 14.13

Heart and associated vessels of *Sphenodon.* (*a*) Ventral view. The ventricle, **V,** is divided internally into two chambers. (*b*) Dorsal view. **L,** left atrium; **R,** right atrium. *Red* vessels contain oxygenated blood. They are **la,** left aortic trunk; **lcc,** left common carotid artery; **PV,** pulmonary veins entering right atrium; **ra,** right aortic trunk; **rcc,** right common carotid artery. *Blue* vessels carry blood low in oxygen. They are **LP,** left precava; **P,** pulmonary trunk emerging from right ventricle; **PO,** postcava; **RP,** right precava; **SV,** sinus venosus.

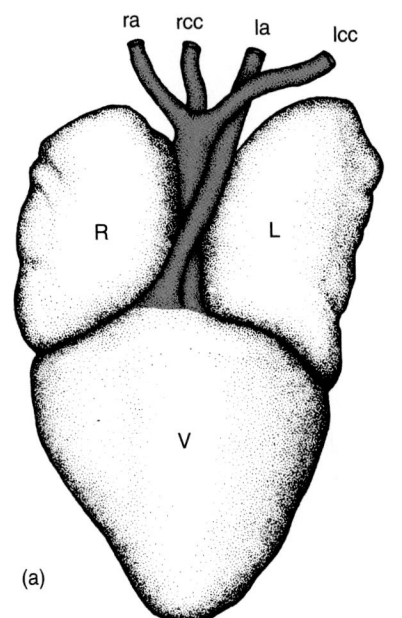

(a)

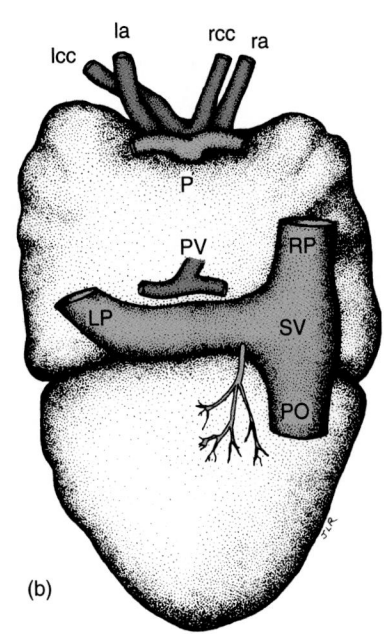

(b)

myocardia of both of the atria, causing them to contract. The stimuli also spread to an **atrioventricular (AV) node** embedded in the heart near the cusps of the right atrioventricular valve. The AV node is unique to birds and mammals. From the AV node, an **atrioventricular bundle** of Purkinje fibers extends into the interventricular septum and splits into two parallel bundles that extend the length of the septum. From these, small bundles of Purkinje fibers ramify throughout the myocardium of both ventricles, providing stimuli that cause the ventricles to contract fractions of a second after the atria have done so. The Purkinje fibers are also innervated directly by autonomic nerve fibers, as is the myocardium in general.

During the short interval between contraction of the atria and that of the ventricles, blood in the atria empties into the dilated ventricles. Contraction of the latter then propels ventricular blood into the outgoing arteries.

Morphogenesis of the Heart

When the heart is first recognizable in craniate embryos, it is an almost straight, pulsating tube in fishes and amphibians (fig. 14.14*a*) or a pair of tubes in amniotes (fig. 14.14*d*). The tubes receive incoming blood at their caudal ends and are continuous with the embryonic ventral aortas at their cephalic ends. The unpaired tube of sharks, basal bony fishes, and amphibians organizes from an unpaired aggregation of mesenchyme beneath the pharynx derived from lateral-plate mesoderm (fig. 14.15*a*). The paired tubes of amniotes organize from the same precursor tissues, differing only in that the tubes organize before the paired lateral plates have met in the midventral line (fig. 14.15, chick and rabbit). The paired tubes of amniotes eventually unite to form one (see fig. 14.14*e*). Mesenchyme cells trapped within the tubes become hemocytoblasts (stem blood cells).

As development progresses, the tube bends to the animal's right and then twists, so that the atrial region, previously at the caudal end, is carried dorsad and cephalad until it reaches its adult location. The location of the atrium in sharks is shown in figure 14.7. The twisting and bending is correlated with the confinement of the rapidly growing heart in a less expansive pericardial cavity.

In amphibians and amniotes, the twisting is carried further than in fishes, so that the atrial region finally lies cephalad to the ventricular region (see fig. 14.14*a* to *c*, frog, and *d* to *i*, chick). Following this, the atrial chamber expands to form bilateral pouches, and a median dorsal fold within the atrial chamber grows ventrad, separating the pouches into left and right atria. As a final major step in amniotes, an interventricular septum commences the separation of the ventricular chamber into right and left sides. The site of the septum is marked externally by a longitudinal **interventricular groove** that creases the surface of the ventricle (fig. 14.14*i*).

As hatching approaches in birds, the sinus venosus becomes almost completely incorporated into the wall of the right atrium. This takes place much earlier in mammals. Because it is imperative that oxygen and nutrients be circulated as early in development as possible, the heart is the first organ to function. It does so even before any nerves have reached it to impose a rhythm.

The initial straight tube that will become the heart of sharks and amphibians organizes from *paired mesenchymal masses* of lateral-plate somatic and splanchnic mesenchyme that aggregate in the midline beneath the pharynx to form a single endothelial tube (fig. 14.15*a*). In amniotes, a *pair of already organized endothelial tubes* comes together beneath the pharynx, fuses, and forms a single tube (fig. 14.15*b* and *c*). In either case, the heart is a bilateral contribution of lateral-plate mesoderm.

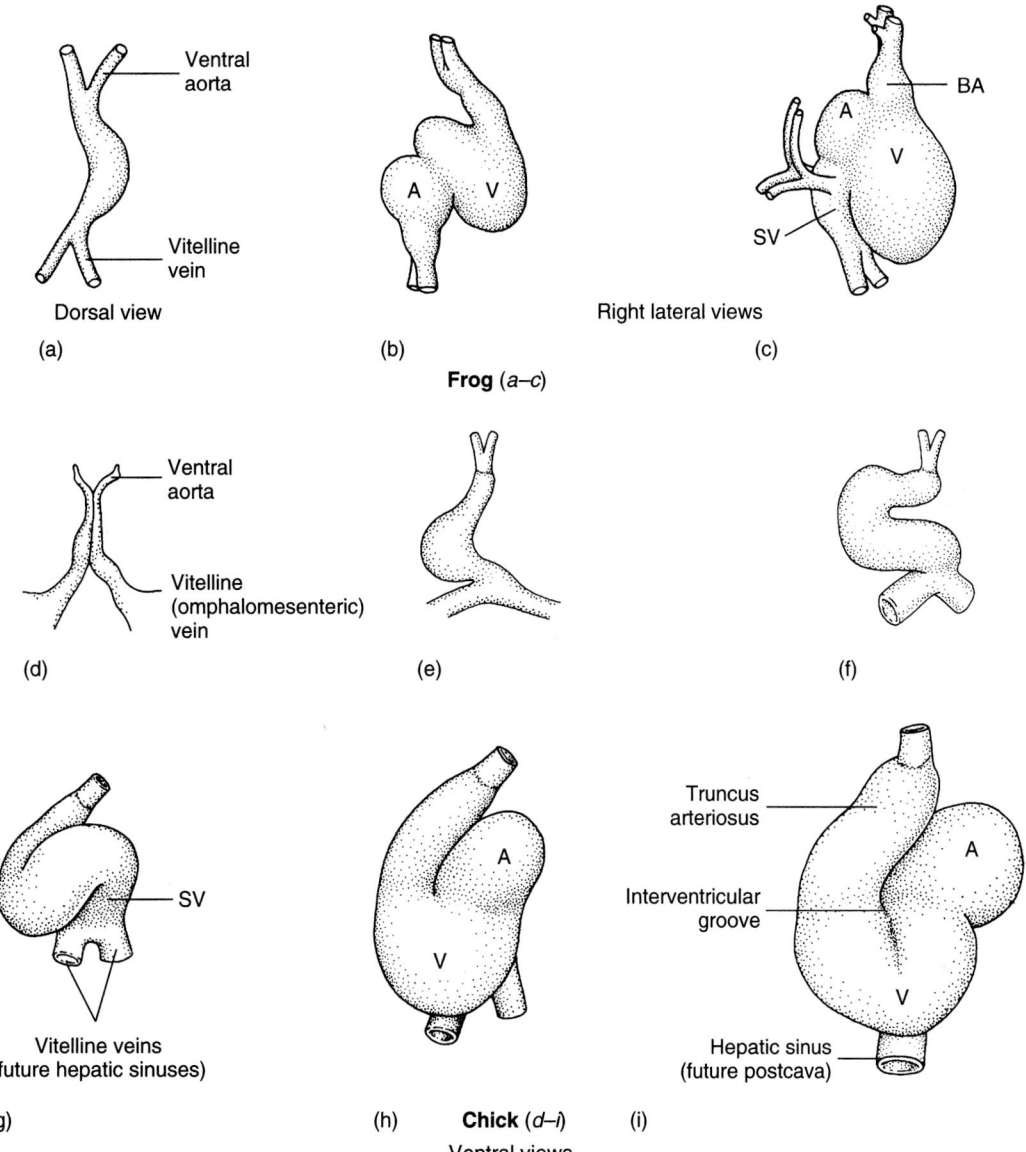

FIGURE 14.14
Development of an anuran heart (*a* to *c*, dorsal and lateral views) and an avian heart (*d* to *i*, ventral views). **A**, atrium; **BA**, bulbus arteriosus; **SV**, sinus venosus; **V**, ventricle. Oriented for optimal viewing of bending and twisting. For a view of the early chick heart in situ, see figure 16.6.

ARTERIAL CHANNELS AND THEIR MODIFICATIONS

Arterial channels supply most organs with oxygenated blood, although they carry deoxygenated blood to respiratory organs. In the primitive pattern of gnathostomes (fig. 14.16), the major arterial channels consist of (1) a **ventral aorta** (paired early in embryogenesis) emerging from the heart and passing forward beneath the pharynx, (2) a **dorsal aorta**, paired above the pharynx only, and extending caudad in the roof of the coelom, and (3) six pairs of **aortic arches** connecting the ventral aorta with the dorsal aorta (the number of arches varies in living agnathans with the number of branchial

arches). Branches of these major channels supply all parts of the body. Modifications of this inherited basic pattern during embryonic development involve, most prominently, the aortic arches, which become modified for respiration by whatever means is employed during the life history of the species.

Aortic Arches of Fishes

Adaptive modifications of the six embryonic aortic arches for adult respiration by gills are illustrated in developing sharks (fig. 14.17). The ventral aorta in *Squalus* extends forward under the pharynx and connects with the developing aortic arches. The aortic arches in the mandibular arch are the first to develop.

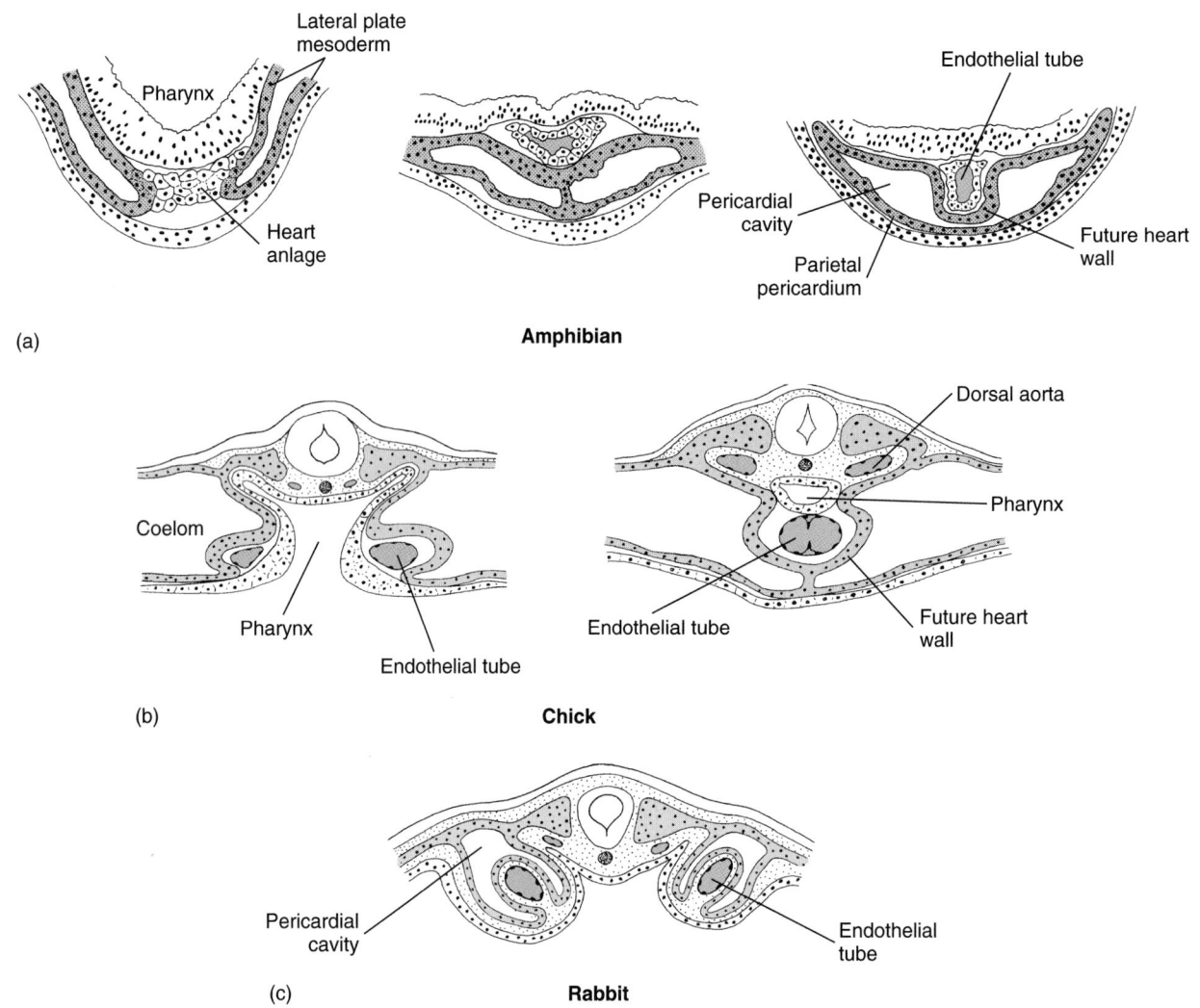

FIGURE 14.15

Origin of heart from lateral-plate anlagen. In amphibians (*a*), the heart forms from a median mass of mesenchyme that has a bilateral origin. In amniotes (*b*) and (*c*), two endothelial tubes of splanchnic mesoderm are brought together to form the primordium of the heart.

Shortly thereafter the other five pairs appear. Before the sixth pair is completed, the ventral segments of the first pair disappear, and the dorsal segments become the efferent spiracular arteries. The second pair sprouts buds that become the first pretrematic arteries. Other buds sprout from the third, fourth, fifth, and sixth aortic arches and give rise to posttrematic arteries. The posttrematic arteries then sprout crosstrunks, which grow caudad in the holobranch and by further budding establish the last four pretrematic arteries. Aortic arches II to VI soon become occluded at one site (broken lines in fig. 14.17*a*). The segments ventral to the occlusions become **afferent branchial arteries.** The dorsal segments (III to VI) become **efferent branchial arteries.** In the meantime, capillary beds are developing within the nine demibranchs. Afferent branchial arterioles connect the afferent branchial arteries with the capillaries. Efferent branchial arterioles return oxygenated blood from the capillaries to the pretrematic

and posttrematic arteries. As a result of these modifications, blood entering an aortic arch from the ventral aorta must pass through gill capillaries before proceeding to the dorsal aorta.

Similar developmental changes convert the embryonic aortic arches of bony fishes into afferent and efferent branchial arteries. The specific number converted determines the number of functional gills. In most teleosts, the first and second aortic arches tend to disappear (fig. 14.18*a*). In *Protopterus* (fig. 14.18*b*), the third and fourth embryonic aortic arches do not become interrupted by gill capillaries, although the fourth pharyngeal arch bears an external gill.

In dipnoans, a pulmonary artery sprouts off the left and right sixth aortic arch and vascularizes a swim bladder. This happens also in two ray-finned fishes, *Amia* and *Polypterus*. This is precisely how tetrapod lungs are vascularized! (In most other actinopterygians, the swim bladders are supplied from the dorsal aorta.)

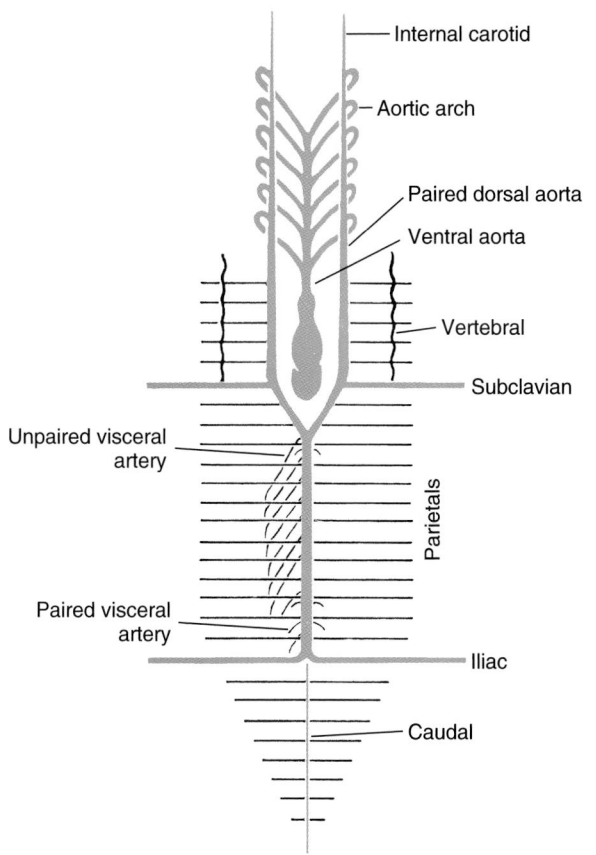

FIGURE 14.16

Basic pattern of the chief arterial channels of gnathostomes. This pattern is also applicable to basal craniates differing primarily in the number of aortic arches.

Aortic Arches of Tetrapods

Embryonic tetrapods, like fishes, construct six pairs of embryonic aortic arches (see fig. 1.6). The first and second pairs are transitory and regress fairly soon (figs. 14.19 and 14.20). After arches I and II disappear, arch III (the carotid arch) and the paired dorsal aortae anterior to arch III constitute the internal carotid arteries (figs. 14.18c, 14.19, and 14.20). Amniotes, with the exception of some limbless squamates, lose the fifth aortic arches during embryonic life (figs. 14.18f–h and 14.19e–h). Frogs (fig. 14.18e and 14.19d) and some salamanders independently lose the fifth aortic arches (compared to the modified, but primitive pattern seen in *Necturus,* figs. 14.18d and 14.21). Pulmonary arteries sprout off the sixth pair to vascularize the lung buds (figs. 1.6 and 14.20d). Other modifications of the aortic arches and associated vessels in tetrapods will be discussed in the sections that follow.

Amphibians

Most terrestrial urodeles retain four pairs of aortic arches (see fig. 14.18c). Perennibranchiate urodeles generally retain three, because the fifth arches either disappear or unite in part, with the fourth during embryogenesis (figs. 14.18d and 14.21). The larval afferent and efferent branchial arterioles of perennibranchiates that carry blood from the aortic arches into the gills and back function throughout life, and a short section of the third, fourth, and fifth (fused to the sixth in *Necturus*) aortic arches becomes a **gill bypass** (see fig. 14.18d). These bypasses are constricted while the animal is using its gills; however, when the dissolved oxygen in the pond becomes low enough to cause the animal to gulp air, the gills shrink, and the bypasses carry more blood. Similarly, when resorption of the gills of *Siren* is brought about by thyroid hormone injections, the animal gulps air, and the bypasses carry all the blood that enters the arches. On the ventral aorta of perennibranchiates, the bulbus arteriosus (see fig. 14.9) maintains a steady nonpulsating arterial pressure in the gills.

Embryonic anurans, like terrestrial urodeles, have four aortic arches (III through VI) at the time they enter the larval state (6 mm stage of development). Aortic arches III, IV, and V supply the larval external gills during the five or six days these gills function; thereafter, they supply internal gills until metamorphosis. Arch VI sprouts a pulmonary artery that vascularizes the developing lung bud.

With loss of gills at metamorphosis, three changes affect the aortic arches and associated vessels of anurans (see fig. 14.18e): (1) Aortic arch V disappears, and (2) the dorsal aorta between aortic arches III and IV (**ductus caroticus**) disappears. As a result of these two changes, blood entering aortic arch III (**carotid arch**) can pass only to the head. Eventually, (3) the segment of aortic arch VI dorsal to the pulmonary artery (**ductus arteriosus**) disappears. Blood entering arch VI (**pulmonary arch**) in anurans can now pass only to the lungs and skin. Aortic arch IV (**systemic arch**) on each side continues to the dorsal aorta to distribute blood to the rest of the body (see figs. 14.12 and 14.19d).

As noted earlier, oxygenated blood from the left atrium and deoxygenated blood from the right are kept remarkably well separated as they pass through the ventricle in amphibians. In anurans, this is accomplished by ventricular trabeculae, by expulsion from the ventricle of right atrial blood first, and by action of the spiral valve in the conus arteriosus. At the start of ventricular systole, the valve is flipped into a position that closes off the entrance to the systemic and carotid arches, thereby diverting deoxygenated blood into the common aperture that leads to the two pulmonary arches (see fig. 14.12b). Then, as back pressure builds up in the pulmonary arteries because of filling of the lung capillaries, the spiral valve flips into an alternate position that directs oxygenated blood into the systemic and carotid arches. (Late in ventricular systole, a little of the left oxygenated blood may enter the pulmonary arches.) In a marine toad studied by injection of radioisotopes

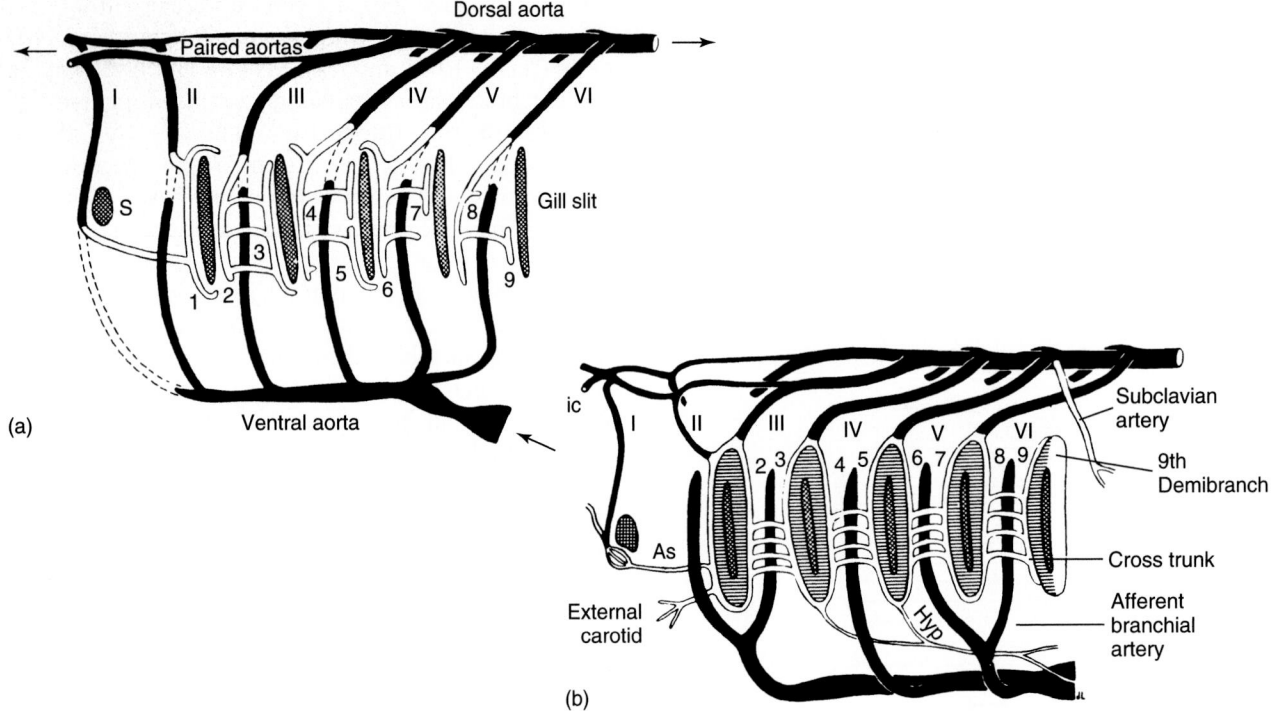

FIGURE 14.17
Developmental changes in embryonic aortic arches I to VI of *Squalus,* lateral view. In (*a*), buds (*white*) off aortic arches II–VI are establishing pretrematic and posttrematic arteries and cross trunks. *Broken lines* indicate sections of the aortic arches that become occluded, forcing blood from the ventral aorta to enter the demibranchs via afferent branchial arterioles (not shown). **1, 3, 5, 7, 9** pretrematic arteries; **2, 4, 6, 8,** posttrematic arteries. In (*b*) **I** has become the efferent spiracular artery, **II** has become the hyoidean efferent artery, and **III** to **VI** have become efferent branchial arteries. **As,** afferent spiracular artery; **Hyp,** hypobranchial; **ic,** internal carotid; **S,** spiracle.

and using angiocardiography, the mixing of oxygenated and deoxygenated blood in the heart and aortic arches was found to be slight.

Adult apodans retain three complete aortic arches (III, IV, and VI) and a ductus caroticus. However, the ductus arteriosus and ductus caroticus are of small diameter and carry little blood.

Nonavian Reptiles

Generalized adult reptiles exhibit three adult aortic arches, III, IV, and the base of VI. The ductus arteriosus and the ductus caroticus are usually obliterated (see fig. 14.19*e*). Exceptions are found among basal lizards and some limbless squamates.

An innovation has been introduced in the ventral aorta of reptiles. Instead of developing a spiral valve to shunt fresh and deoxygenated blood into appropriate arches, reptiles underwent a series of developmental modifications that had the effect of splitting the embryonic ventral aorta into three separate channels: **two aortic trunks (left and right systemic arches)** and a **pulmonary trunk** (figs. 14.18*f*, arv, alv, and prv; 14.19*e*; and 14.22, 2, 3, and 4). One result of these changes can be demonstrated in crocodilians.

In crocodilians, (1) the pulmonary trunk emerges from the right ventricle and leads to the left and right pulmonary arches (see fig. 14.19*e*, vessels in blue). Deoxygenated blood is therefore sent to the lungs. (2) One aortic trunk emerges from the left ventricle and carries oxygenated blood to the right systemic arch and the carotid arches (fig. 14.23). (3) A second aortic trunk emerges from the right ventricle and leads to the left systemic arch. One might expect that blood in the left systemic arch would be low in oxygen because it comes from the right ventricle. However, this is not the case when the animal is breathing. An aperture, the **foramen of Panizza** (fig. 14.23*a*, curved arrow), connects the two aortic trunks at their base. During normal respiration, the valve at the exit from the right ventricle into the aortic trunk remains closed, and blood from the right ventricle can pass only to the lungs. Some oxygenated blood from the left ventricle is shunted through the foramen of Panizza to the left systemic arch (fig. 14.23*a*). This left-right shunt assures delivery of oxygenated blood to all parts of the body.

Crocodilians are aquatic animals that spend a proportion of their time totally submerged. In this state, they cannot breathe, and the oxygen content of their blood

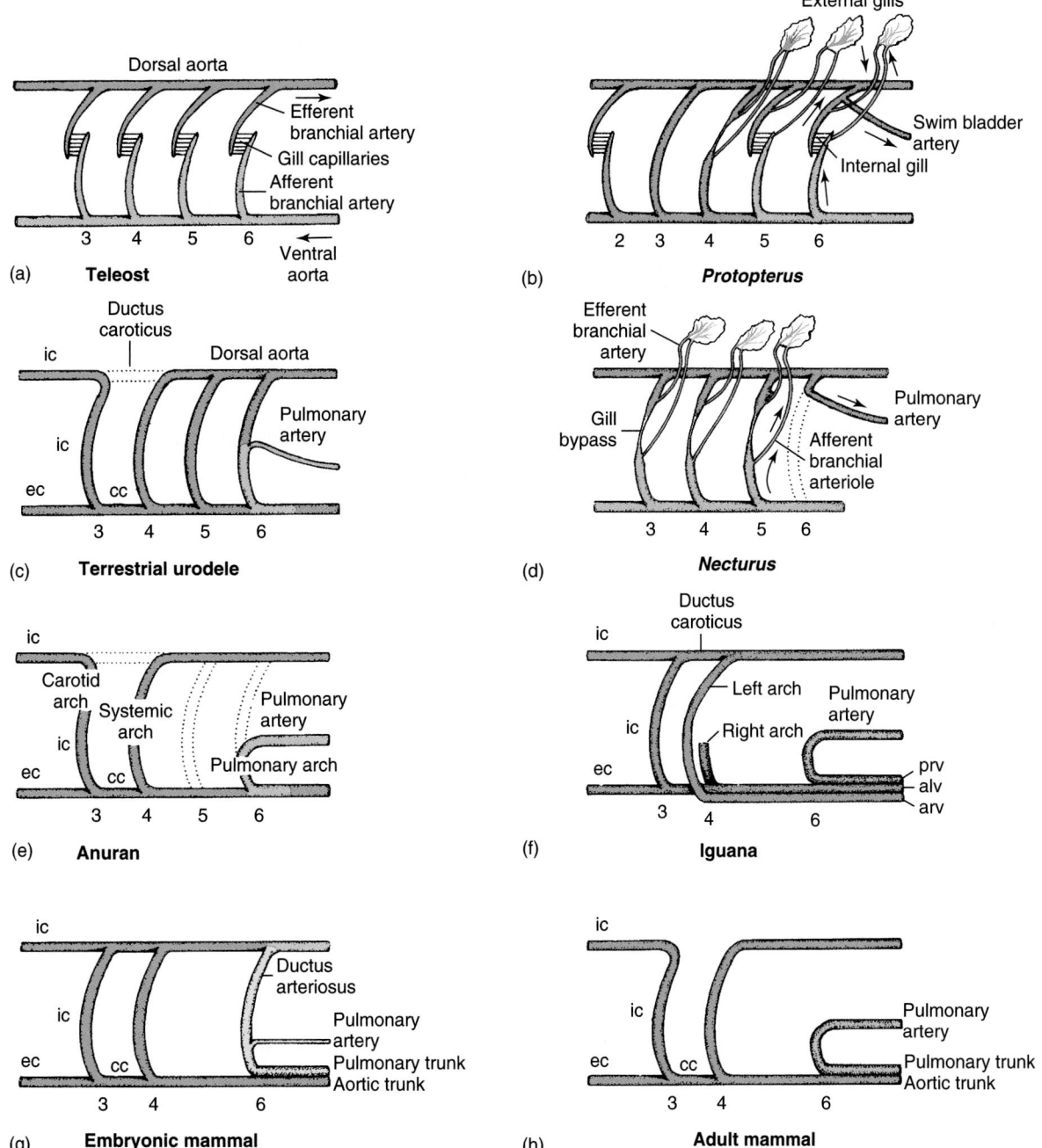

FIGURE 14.18

Persistent left aortic arches in representative craniates. **2** to **6,** second through sixth embryonic aortic arches. *Dotted lines* in (*c*) indicate vessel present in some species; in (*d*), a developmental loss; in (*e*), vessels that are functional in larvae. *Arrows* indicate direction of blood flow. **alv,** aortic trunk from left ventricle; **arv,** aortic trunk from right ventricle; **cc,** common carotid artery, which is the paired segment of the embryonic ventral aorta; **ec,** external carotid; **ic,** internal carotid; **prv,** pulmonary trunk from right ventricle. In (*b*), (*c*), and (*e*), the ventral aorta carries venous blood (*blue*) during one phase of a single ventricular contraction, and arterial blood (*red*) during the next phase. In (*b*), the oxygen content of each vessel depends on the extent to which oxygen is being acquired via the gills as compared to the swim bladder. In (*c*), the sixth arch carries only oxygenated blood after the pulmonary artery is filled with unoxygenated blood. In (*d*), colors indicate conditions while animal is in well-oxygenated water. In (*g*), all blood in arches is mixed, and colors designate predominant condition. Blood in the aortic trunk was oxygenated in the placenta.

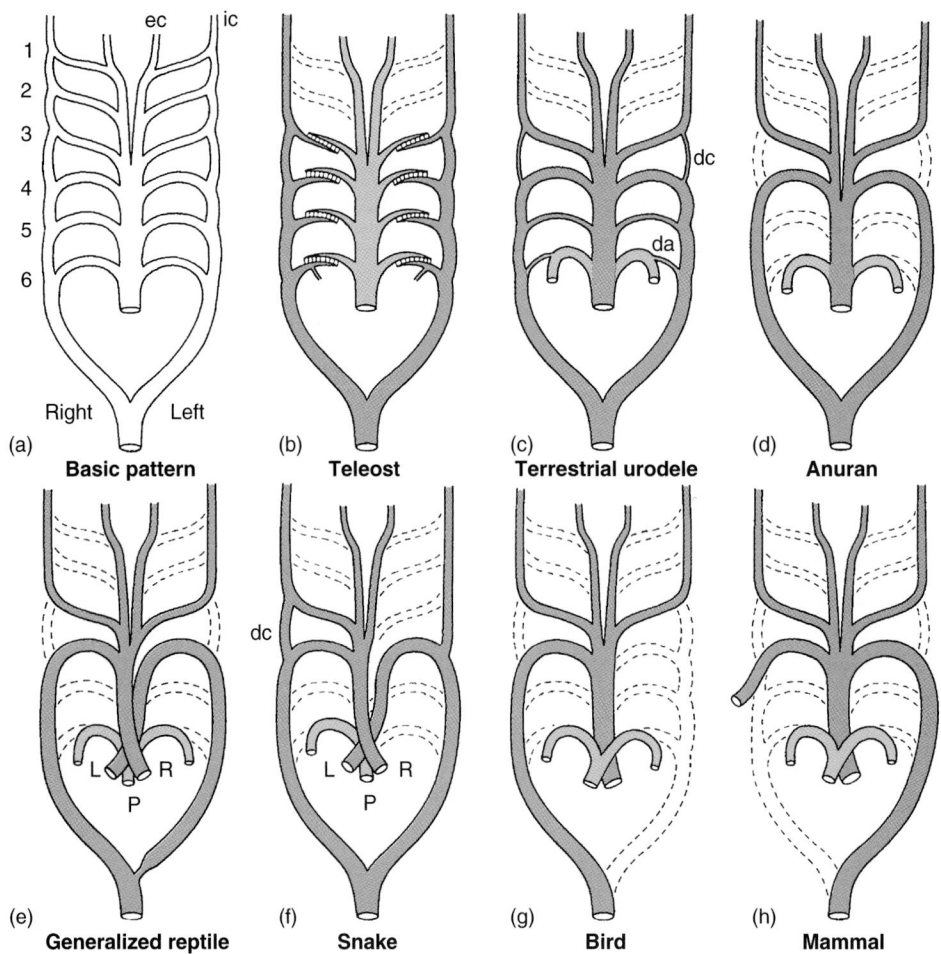

FIGURE 14.19

Adult aortic arches (**1** to **6**) of representative craniates diagrammed from ventral aspect. The snake illustrated is a species with no left lung. **L,** aortic trunk to left systemic arch; **P,** pulmonary trunk; **R,** aortic trunk to right systemic arch. **da,** ductus arteriosus; **dc,** ductus caroticus; **ec,** external carotid artery; **ic,** internal carotid artery.

Source: Data from J. S. Kingsley, *Outline of Comparative Anatomy of Vertebrates,* 1920, The Blakiston Company, Philadelphia.

slowly falls. In response, the pulmonary arteries constrict reflexly, causing a backup of blood in the right ventricle. The valve between the right ventricle and its aortic trunk is thereby forced open (fig. 14.23*b*), and some right-ventricular blood is shunted via the foramen of Panizza to the aortic trunk that emerges from the left ventricle. This right-left shunt diverts considerable blood away from the lungs and into the systemic circulation. Utilization of the shunt is facilitated by a mild reduction in blood pressure within the left ventricle. Eventually, the animal will have to come to the surface to breathe, whereupon the pulmonary circulation will revert to that of an air-breathing amniote.

As described earlier, turtles and squamates have, at the upper limit of the interventricular septum, a cavum venosum that receives deoxygenated blood from the right atrium and is confluent with both the left and right ventricles (fig. 14.24). How do these reptiles avoid the mixing of oxygenated and deoxygenated blood in

the heart? A series of studies using cinefluoroscopy, simultaneous readings of arterial pressure in several chambers and vessels, assay of blood gases, and other cardiovascular techniques have provided the answer. During *ventricular diastole,* blood from the right atrium—venous blood—passes through the cavum venosum and enters the right ventricle, known also as the **cavum pulmonale** in these reptiles. From the cavum pulmonale, the next *ventricular systole* pumps it into the pulmonary trunk and thence to the lungs. Blood from the left atrium during *ventricular diastole* enters the left ventricle (**cavum arteriosum**). By the time the cavum venosum has discharged its unoxygenated blood into the cavum pulmonale, ventricular systole begins. The muscular activity of ventricular systole displaces the septal wall, blocking the passageways between the cavum venosum, the right atrium, and the cavum pulmonale, and opening the passage between the cavum venosum and the cavum arteriosum. Thereupon, blood

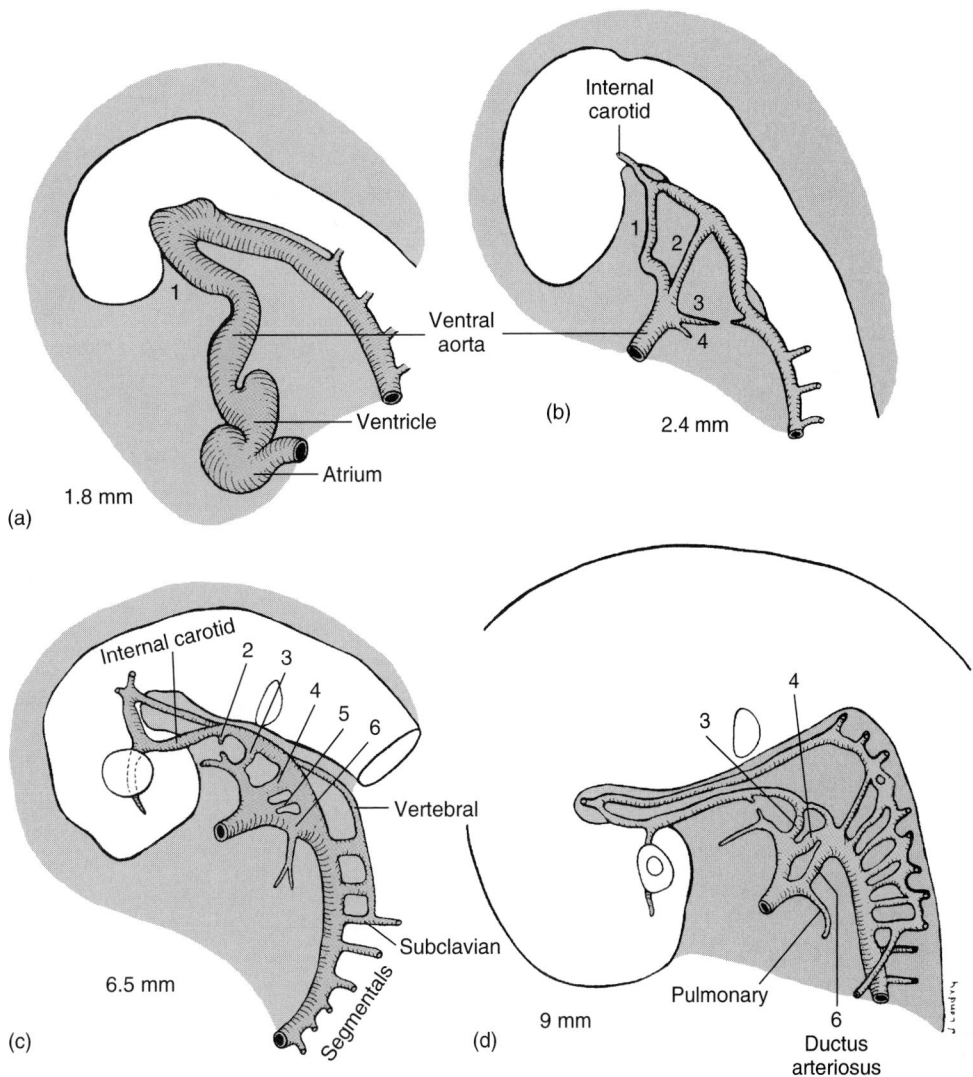

FIGURE 14.20

Embryonic modifications of the left aortic arches (**1** to **6**) of a porcupine. In (*c*) and (*d*), **3** is the base of the internal carotid artery.

From P. H. Struthers, "The Aortic Arches and Their Derivatives in the Embryonic Porcupine" in *Journal of Morphology and Physiology*, 50:361. Copyright © 1930 John Wiley & Sons, Inc. Reprinted by permission of John Wiley & Sons, Inc.

from the cavum arteriosum is pumped into the cavum venosum and from there into the two aortic trunks that lead to the left and right systemic arches (fig. 14.24, 2 and 4). When this circulatory pattern is in effect, as it is when the lungs are ventilating, deoxygenated blood is sent to the lungs, and oxygenated blood is sent to the rest of the body.

When physiological conditions initiate it, as when ventilation in aquatic species is halted by submersion in water, the pulmonary arteries become constricted, and deoxygenated blood in the cavum venosum is shunted away from the cavum pulmonale and into the left aortic arch. A similar response has been elicited when radiant heat was applied to the reptilian body. It has been suggested that the latter response may be thermoregulatory in cold weather, enabling blood warmed by basking in

the sun to be diverted from the lungs, where heat loss would occur during exhalation. When insufficient oxygen is available to these reptiles during submersion, an additional adaptation provides energy for metabolic needs by glycolysis, an anaerobic process.

Basal lizards retain the fifth aortic arch and the ductus caroticus on both sides. Limbless lizards and snakes in which the left lung is atrophied may lose the entire left sixth aortic arch, retaining only a single pulmonary artery; in some snakes, the left third aortic arch also disappears. However, in the latter species, the ductus caroticus persists throughout life, and blood reaches the left internal carotid artery and brain by that route. Of the primitive six aortic arches, some adult snakes (see fig. 14.19*f*) retain only the right third, the left and right fourth, the ductus caroticus, and the ventral segment of

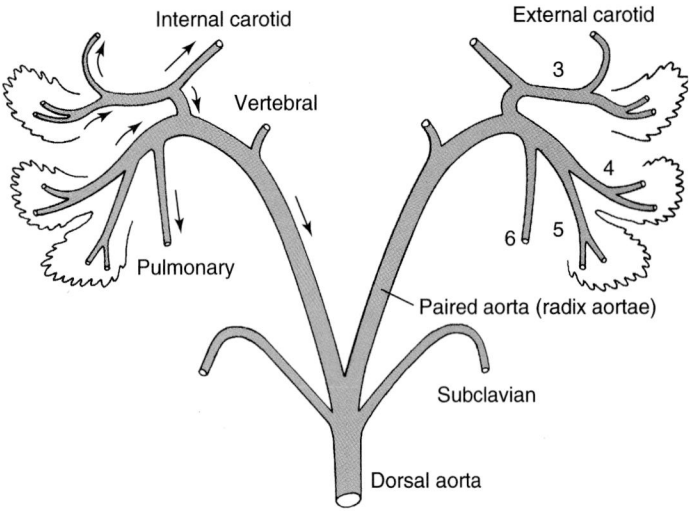

FIGURE 14.21
Efferent branchial arteries (**3, 4, 5, 6**) of *Necturus* as seen from above the gills. Numerals designate dorsal segments of third, fourth, fifth, and sixth embryonic aortic arches. *Arrows* indicate direction of blood flow.

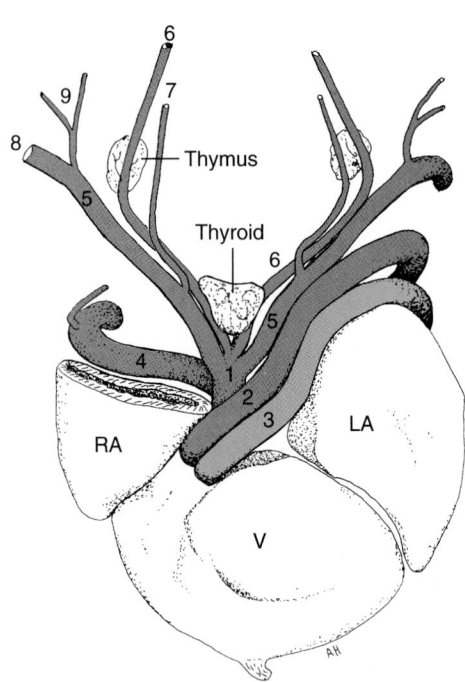

FIGURE 14.22
Turtle heart and emerging aortic arches, ventral view. **LA**, left atrium; **RA**, right atrium; **V**, ventricle. **1**, brachiocephalic artery; **2**, left fourth aortic arch; **3**, pulmonary trunk; **4**, right fourth aortic arch. **1** and **4** arise from a common trunk. **5**, subclavian artery; **6**, common carotid; **7**, ventral cervical; **8**, axillary; **9**, arteries to pectoral and shoulder muscles. *Red*, oxygenated blood; *blue*, deoxygenated blood.

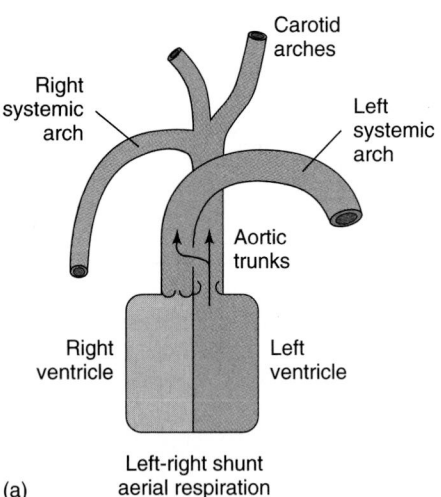

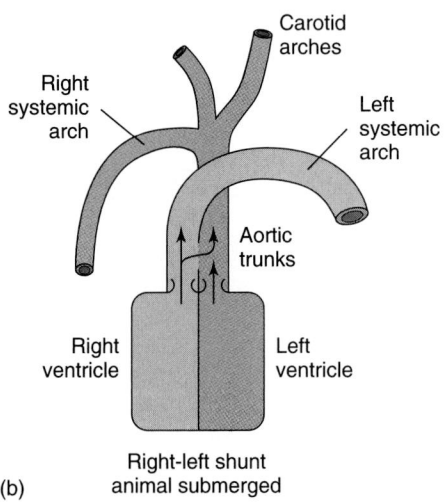

FIGURE 14.23
Crocodilian foramen of Panizza, schematic. Curved *arrows* go through foramen. (*a*) Distribution of oxygenated blood during aerial respiration and (*b*) when lungs are not ventilating.

the right sixth aortic arch to the site where the right pulmonary artery arises.

Left-right and right-left shunts in reptiles, persistent carotid ducts, retention of the fifth aortic arch, and other features of the cardiovascular system that characterize one reptilian species or another enable its members to survive in the specific environment they inhabit.

Birds and Mammals

Birds and mammals became the first tetrapods in which there is no opportunity for the mixing of oxygenated and deoxygenated blood in the heart after birth or hatching (see fig. 14.6, amniotes). This has come about as a result of division of the left and right ventricles by a complete interventricular septum and division of the embryonic ventral aorta into only two trunks, a pulmonary trunk

neath the transverse processes of vertebrae and within the bony vertebral canal that houses the spinal cord. The retia are confluent and are drained by vertebral arteries that are en route to the brain. When the animal dives, the viscera are compressed by the weight of the surrounding water, and blood is forced out of the viscera into the retia, which are protected from compression by their bony surroundings. The retia constitute a reservoir of red blood cells that were oxygenated just before a dive. The oxyhemoglobin supplies the brain with oxygen while the animal is submerged.

Often the tortuous artery is associated with an equally tortuous vein in such a manner that artery and vein lie side by side, with the blood flowing in opposite directions. In some birds that wade in icy waters, these countercurrents in retia above the thigh result in transfer of heat from warm arterial blood entering the leg to cold venous blood returning to the trunk. This conserves body heat, therefore energy, by preventing heat loss to the icy water while also warming the returning blood to body temperature. Polar bears and arctic seals have retia that perform the same function. On the other hand, retia do not occur in arteries en route to the limbs of penguins and some other arctic animals in which swimming generates excessive heat that must be dissipated. Mammals with testes in scrotal sacs have a rete, the **pampiniform plexus**, in each inguinal canal. Heat is transferred from spermatic artery to spermatic vein, assuring that the temperature within the scrotal sacs is lower than body temperature, a necessity for the viability of sperm in those species.

The **red glands** of the swim bladders of fishes are retia. Swim bladders are supplied by branches of the celiac artery and drained partly by the hepatic portal system. Venous blood leaving a rete carries away oxygen that has entered the red cells of the vein because of the high partial pressure of oxygen gas within the lumen of the bladder. This oxygen is reclaimed by arteries that are approaching the gas-secreting area of the rete. The retia therefore provide countercurrents that help to maintain an appropriate level of gaseous oxygen in the bladder.

VENOUS CHANNELS AND THEIR MODIFICATIONS

A generalized venous system consists of the following major streams: **cardinals** (anterior, posterior, and common cardinals), **renal portal, lateral abdominal, hepatic portal**, hepatic sinuses, and coronary veins (fig. 14.29*a*). The hepatic portal system is derived from the embryonic subintestinal vein and the distal portion of the embryonic vitelline veins, and the hepatic sinuses are derived from the vitellines between liver and heart. Two additional streams develop in lungfishes and tetrapods: a **pulmonary stream** from the lungs and a

postcava from the kidneys. These eight channels and their tributaries drain the entire body—head, trunk, tail, and appendages. As development progresses, they are slowly modified by deletion of some vessels and addition of others. Modifications are few in stem craniates, more numerous as derived features within individual clades.

Basic Pattern: Sharks

An adult shark is an ideal living blueprint of the basic venous channels of craniates. Inasmuch as very few changes occur in the embryonic channels of sharks during later development, knowledge of the venous channels of sharks and how they develop is a good introduction to the venous channels of other craniates.

Cardinal Streams

The sinus venosus receives all blood returning to the heart of sharks. Most of this blood, except that from the digestive organs, enters the sinus by a pair of **common cardinal veins** (fig. 14.29*b*). These use the transverse septum as a bridge from the lateral body walls to the heart. They appear early in development and remain essentially unchanged thereafter.

Blood from the head other than the lower jaw is collected by a pair of **anterior cardinal (precardinal) veins** lying dorsal to the gills. The anterior cardinals pass caudad and empty into the common cardinals. The embryonic anterior cardinals, like common cardinals, remain essentially unchanged throughout life.

The earliest *embryonic* **posterior cardinal (postcardinal) veins** are continuous with the caudal vein (fig. 14.29*a*). These embryonic postcardinals pass cephalad lateral to the developing kidneys and receive a series of renal veins from the kidneys en route. They then empty into the common cardinals. Their cephalic ends in adult sharks expand to become **posterior cardinal sinuses**.

While these embryonic postcardinals are functioning, a new pair of postcardinal veins is forming *between* the kidneys from a subcardinal plexus. (A similar plexus in turtles is illustrated in fig. 14.35.) These new veins become confluent with the old postcardinals at the anterior end of the kidneys, and they, too, drain the kidneys. As more and more blood from the kidneys flows into the new veins, the older postcardinals are lost anterior to the kidneys (fig. 14.29*b*). Thereafter, the name *posterior cardinal* is applied to the newer vessels. In adults, they drain chiefly the kidneys, body wall, and gonads.

Renal Portal Stream

At an early stage in development in sharks, some of the blood from the caudal vein continues forward beneath the gut as a subintestinal vein that drains the digestive tract (fig. 14.29*a*). Later, the connection of the caudal vein with the subintestinal is lost (fig. 14.29*b*). When the old posterior cardinals are lost anterior to the kidneys, all

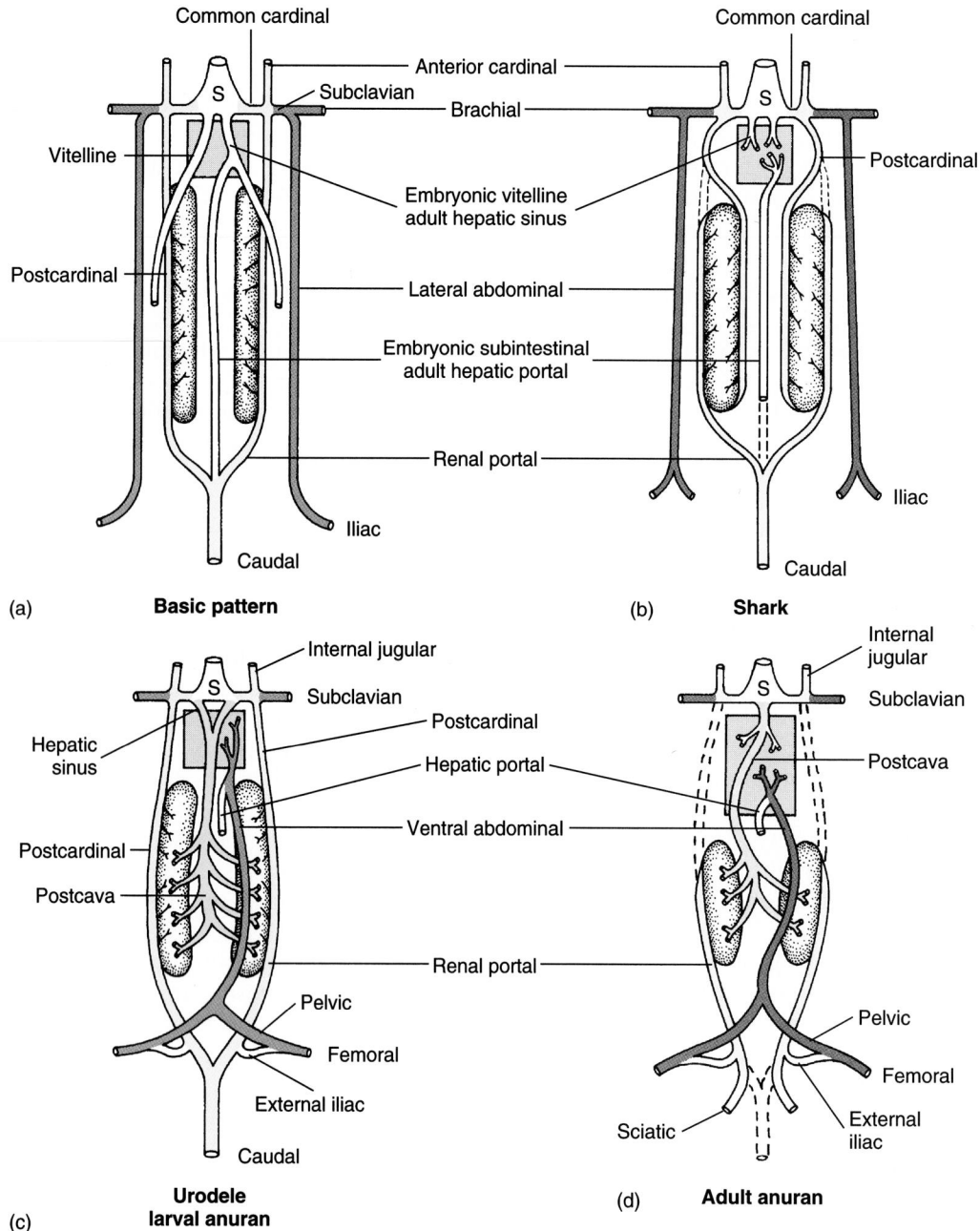

FIGURE 14.29

Modifications of the basic venous pattern in sharks and amphibians. *Broken lines* indicate lost segments. *Light blue,* cardinal and caudal (renal portal) streams; *dark blue,* postcava, receiving renal veins; *red,* abdominal stream. **S,** sinus venosus. The kidneys are *stippled,* and the liver is *gray.* The venous channels medial to the kidneys in (*b*), (*c*), and (*d*) develop from an embryonic subcardinal plexus.

blood from the tail thereafter enters the capillaries sur-rounding the kidney tubules (**peritubular capillaries**). The result is a renal portal system.

Lateral Abdominal Stream

Commencing at the pelvic fin from which it receives an **iliac vein** and passing forward in the lateral body wall on each side is a **lateral abdominal vein** (fig. 14.29*b*). At the level of the pectoral fin, it receives a **brachial**

vein, after which the vessel turns abruptly toward the heart to enter the common cardinal vein. The part of the abdominal stream between brachial and common cardi-nal is the **subclavian vein**. In addition to collecting blood from the paired fins, the abdominal stream also receives a **cloacal vein**, a metameric series of **parietal veins** from the lateral body wall, and minor tributaries. This basic channel remains unmodified during subse-quent development.

Hepatic Portal Stream and Hepatic Sinuses

Among the first vessels to appear in any craniate embryo are paired **vitelline**, or **omphalomesenteric**, **veins** from the yolk sac to the heart, whether the yolk sac is functional or not (see figs. 5.12 and 16.6). One of the vitelline veins is soon joined by the embryonic **subintestinal vein** that drains the digestive tract (fig. 14.29*a*). As the developing liver enlarges, it encompasses the vitelline veins, causing them to be broken into many sinusoidal channels. Caudal to the liver, one vitelline vein disappears, and the other, with the subintestinal vein, becomes the hepatic portal system that drains the digestive organs of the coelom, and the spleen. Between liver and sinus venosus, the two vitelline veins become **hepatic sinuses** (fig. 14.29*b*).

Other Fishes

The venous channels of other fishes are much like those of sharks. Living agnathans have no renal portal system and no left common cardinals, although two common cardinals develop in embryos. Abdominals are lacking in most ray-finned fishes, and the pelvic fins are drained by the postcardinals. In dipnoans, the pelvic fins are drained by an unpaired ventral abdominal vein that ends in the sinus venosus, and the right postcardinal is missing. Blood from the swim bladders of ray-finned fishes empties into either hepatic, hepatic portal, posterior cardinal, or common cardinal veins; in dipnoans, it empties into the left atrium. Coronary veins in all fishes empty into the sinus venosus.

Tetrapods

The early embryonic venous channels of tetrapods are basically the same as those of embryonic sharks. We have just seen how, by adding a vessel here and dropping one there during development, the basic pattern of a shark embryo is converted into the veins of an adult shark. We will now see how the same embryonic pattern is converted into the veins of adult tetrapods.

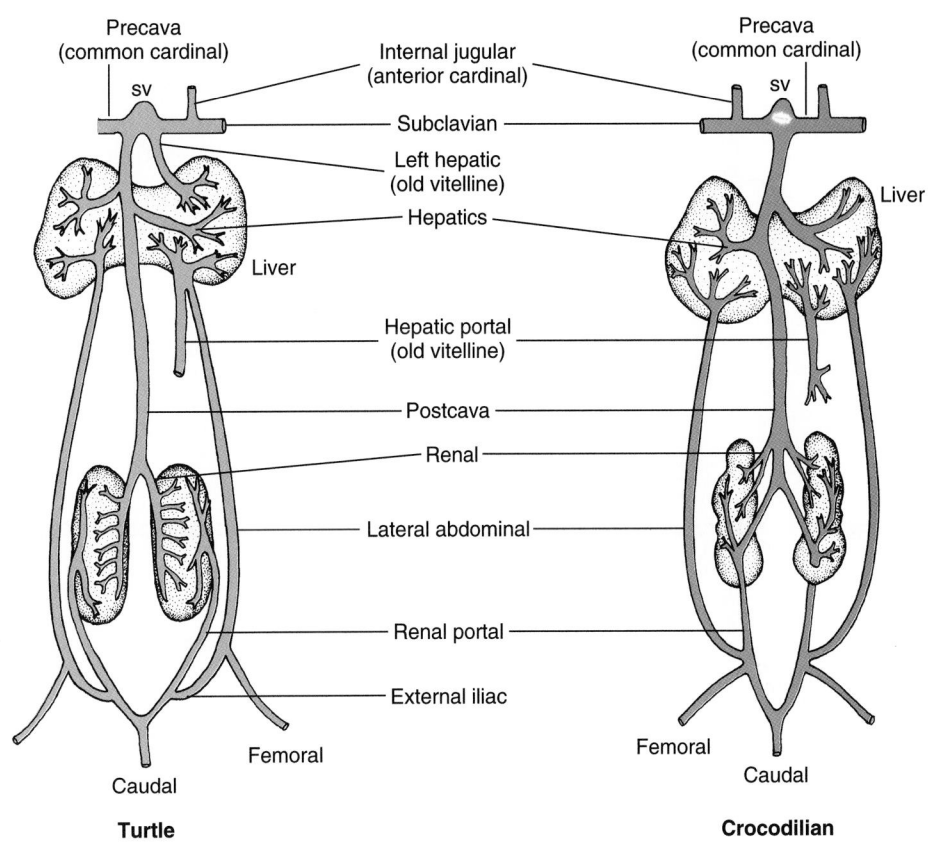

FIGURE 14.30

Systemic veins of two reptiles. Only vessels of the basic pattern illustrated in figure 14.29*a* are shown. A strong branch of the renal portal vein of crocodilians continues through each kidney without ending in capillaries. **sv,** sinus venosus.

Cardinal Veins and the Precavae

Embryonic tetrapods have postcardinals, precardinals, and common cardinals. In urodeles, the embryonic postcardinals persist between the caudal vein and common cardinals throughout life (fig. 14.29*c*). In this respect, a *Necturus* is less modified than a shark. In anurans, the segment of the embryonic postcardinals anterior to the kidneys disappears as development progresses (fig. 14.29), and the connection of the postcardinals with the caudal vein is lost at metamorphosis. In amniotes, postcardinals anterior to the kidneys generally disappear during embryonic development as a newer vessel, the postcava, gradually takes on more and more tributaries (figs. 14.30 and 14.31). In mammals, however, the anterior segment of the right postcardinal persists under the name **azygos**, and part of the left postcardinal persists as the **hemiazygos** (figs. 14.32 and 14.33). These two veins drain intercostal spaces (fig. 14.34). The azygos empties into the old right common cardinal vein (precava). It receives transverse shunts from the hemiazygos.

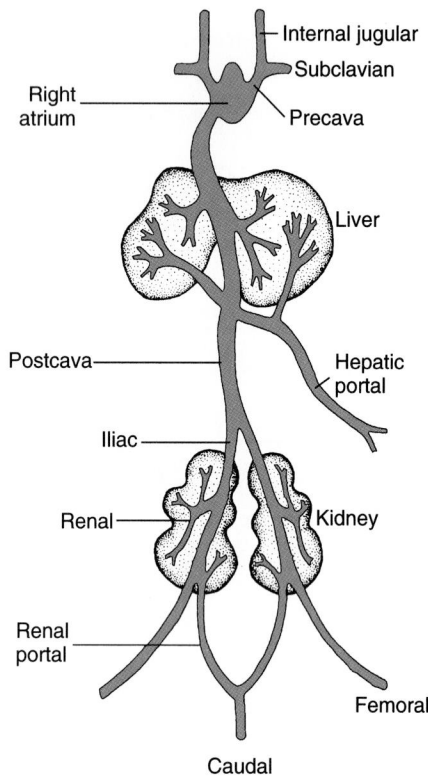

FIGURE 14.31
Major systemic veins of a bird.

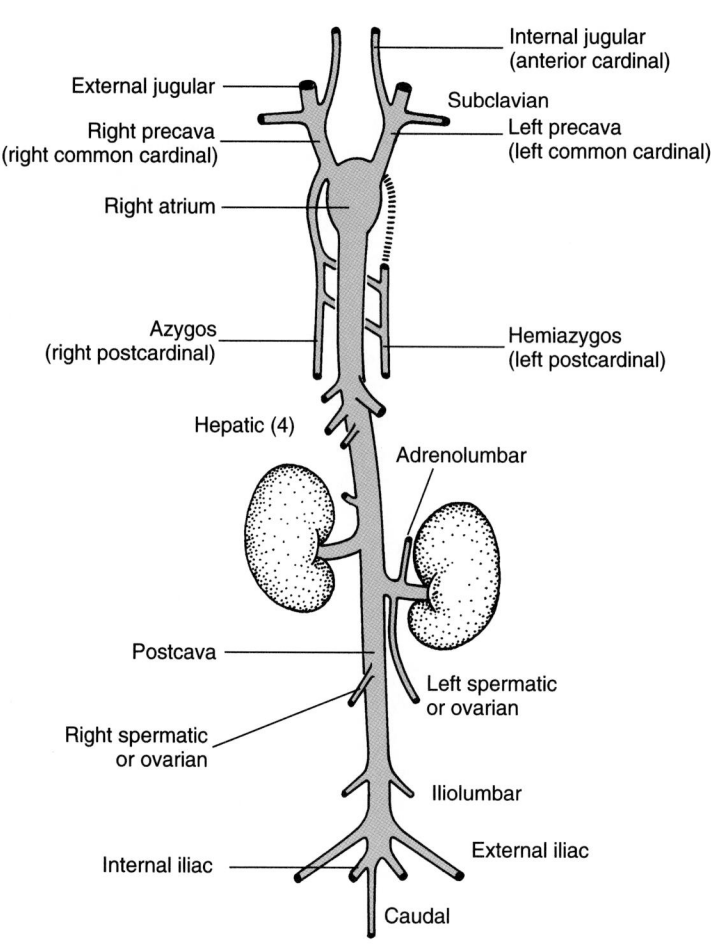

FIGURE 14.33
Major systemic venous channels of a rabbit, ventral view. *Broken line* indicates obliterated segment of left posterior cardinal vein. Entrance of one spermatic or ovarian vein into a renal vein is uncommon in rabbits, common in cats.

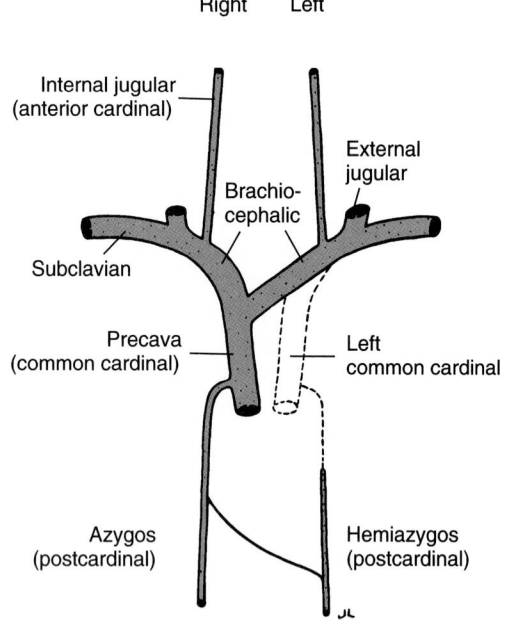

FIGURE 14.32
Basic anterior venous channels of cat and human, ventral view. *Broken lines* indicate vessels obliterated during ontogeny. These sometimes remain as anomalies in adult mammals, including cats and humans. Internal and external jugulars are sometimes confluent.

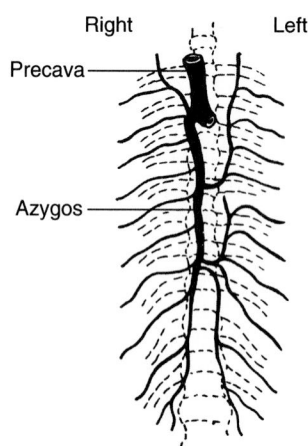

FIGURE 14.34
The azygos (on the animal's right) and the hemiazygos (on the animal's left) in a rhesus monkey, ventral view. This condition is one of many variants in this species. Similar variants are seen in humans.

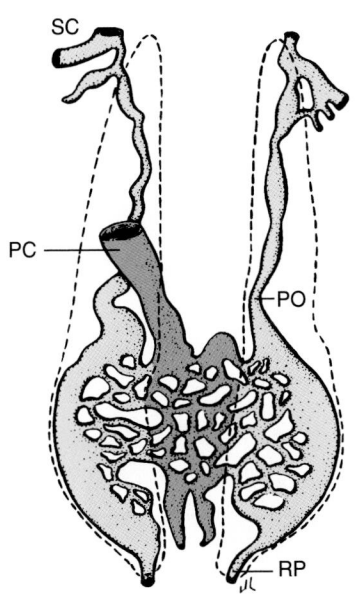

FIGURE 14.35
Embryonic subcardinal venous plexus and origin of the postcava (*dark blue*) in the turtle *Chrysemys,* ventral view. The mesonephros is outlined by broken lines. **PC,** postcava; **PO,** left postcardinal; **RP,** renal portal vein; **SC,** subclavian vein.

Common cardinal veins in amniotes are known as **precavae** and anterior cardinals are known as **internal jugular** veins. Although most mammals retain both the left and right precavae (see fig. 14.33), some, including cats and humans, lose most of the left precava (left common cardinal) during embryonic life (see fig. 14.32). A transverse vessel, the left brachiocephalic vein, then shunts venous blood from the left side of the head and left anterior limb to the right precava (see fig. 14.32). A remnant of the left precava remains as the **coronary sinus,** a venous sinus on the surface of the heart that receives several coronary veins and empties into the right atrium. The persisting right precava in humans is known also as the **superior vena cava** (cranial vena cava of quadrupeds). The precavae empty into the sinus venosus in nonavian reptiles and into the right atrium in birds and mammals.

Postcava

The postcava arises during embryonic life in a subcardinal venous plexus that receives renal veins from the kidneys (fig. 14.35). One subcardinal channel (usually the right) predominates, grows into the mesentery in which the liver is developing, and becomes confluent with the hepatic sinuses. This vessel is the **postcava.** The enlarging liver envelops it but does not break it into capillaries. Thus, the postcava becomes an expressway from kidneys to heart via the hepatic sinuses (see fig. 14.29*a* and *c*). In most tetrapods, the two hepatic sinuses ultimately fuse to form a median vessel that becomes part

of the postcava (see figs. 14.29*d,* 14.30, and 14.31). In mammals, the postcava is also called the **caudal (inferior) vena cava.**

In crocodilians, some of the blood from the hind limbs that enters a renal portal vein bypasses the kidney capillaries and flows directly into the postcava (see fig. 14.30, crocodilian). In birds most, if not all, and in mammals, all blood from the hind limbs flows directly into the postcava (see figs. 14.31 and 14.33).

With establishment of a postcava, blood that previously passed from the kidneys to the heart via postcardinal veins now uses the postcava, and the postcardinals are reduced to draining the intercostal spaces.

Abdominal Stream

In early tetrapod embryos, paired lateral abdominal veins commence in the body wall at the level of the future hind limbs, pass cephalad in the lateral body wall, receive veins from the developing forelimbs, and terminate in the common cardinal veins or sinus venosus (see fig. 14.29*a*). As development progresses in tetrapods, this stream alters its course.

In amphibians, the two embryonic lateral abdominal veins, which lie parallel and close together on either side of the midventral line, unite in the midventral body wall as far forward as the level of the embryonic falciform ligament that connects the ventral body wall with the developing liver (see fig. 14.29*c* and *d*). Blood in this unpaired segment, now known as the **ventral abdominal vein,** establishes connections with venous channels in the falciform ligament, one of these channels enlarges, and soon all the blood in the ventral abdominal vein is crossing the falciform ligament and emptying into the liver (see fig. 14.29*c* and *d*). Thereafter, the abandoned segments of the lateral abdominal veins anterior to the liver disappear. The result is establishment in amphibians of a *portal stream* between the capillaries of the developing hind limbs and those of the liver. This is an addition to the already-established hepatic portal system that drains the digestive organs and spleen in all craniates. Rerouting of the abdominal stream dissociates it from the drainage of the anterior limbs (see fig. 14.29*d*).

In nonavian reptiles, the two lateral abdominals do not unite (see fig. 14.30). Otherwise, they undergo the same modification as in amphibians, using the falciform ligament as a bridge across the coelom to the capillaries of the liver, and losing their connection with the common cardinal veins. They acquire temporary tributaries, the two **allantoic veins,** as they pass along the ventral body wall. (The allantois in fig. 5.13*b* is part of the respiratory membrane of unhatched reptiles.) The allantoic vessels regress when the extraembryonic portion of the allantois is lost prior to hatching. In birds, none of the embryonic abdominal stream remains in adults (see fig. 14.31).

The abdominal stream serves mammals, but during fetal life only. The portion of the **umbilical (allantoic) vein** within the fetal body beginning at the umbilicus, crossing the falciform ligament, and entering the liver is all that is left of the abdominal stream (see fig. 14.38). Like the lateral abdominal of other craniates, it commences early in embryogenesis as a *pair* of vessels in the ventrolateral body wall, but no connection with the drainage of the hind limbs develops. Branches from this very early vessel grow out the umbilical cord and vascularize the placenta. Soon, the paired embryonic vessels unite, forming a single umbilical vein. The abdominal stream in mammals has no function other than to drain the placenta; when, at birth, the umbilical vein is severed along with the umbilical cord, blood no longer flows through the part that remains between the navel and the liver, and this segment becomes the **round ligament of the liver**, visible as a fibrous thickening in the free border of the falciform ligament extending between the umbilicus and the liver. Thus, a history of modifications in ancestral mammals has transformed the very ancient abdominal stream, which at one time drained the posterior appendages, body wall, and anterior appendages. It is now strictly an embryonic vessel that drains the placenta.

The umbilical vein in embryonic mammals carries a heavy load of blood and erodes a broad channel, the **ductus venosus**, directly through the substance of the liver and into the postcava (see fig. 14.38, 3). After birth, the channel becomes a ligament, the **ligamentum venosum**.

Renal Portal System

The renal portal system of amphibians has acquired a tributary, the **external (transverse) iliac vein** *(not homologous with the iliac veins of amniotes)*, which carries some blood from the hind limbs to the renal portal vein (see fig. 14.29c and d). This channel provides an alternate route from the hind limbs to the heart. The connection exists also in reptiles (see fig. 14.30). It may have been one of the factors responsible for the eventual loss of the abdominal stream beyond the allantoic veins in mammals.

Snakes have no hind limbs, so the renal portal system is seen in its primitive relationship (fig. 14.36). In crocodilians, some blood from the hind limbs that enters the kidneys via the renal portal system bypasses the kidney capillaries, going straight through the kidneys and into the postcava (see fig. 14.30, crocodilian). In birds, this has become the principal route from the hind limbs to the heart (see fig. 14.31). In therian mammals, the renal portal system disappears.

From the foregoing descriptions, it can be seen that the posterior appendages of tetrapods have been drained by a succession of vessels (fig. 14.37): the abdominal stream, the renal portal, and finally the postcava. Dis-

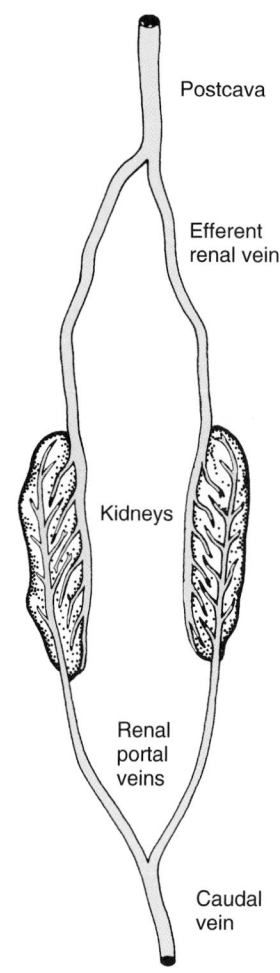

FIGURE 14.36
Renal portal system and venous drainage of the kidneys of a snake.

placement of the caudal end of the nephrogenic mesoderm during embryonic development of the mammalian kidney (chapter 15) may have been a factor in the evolutionary loss of the renal portal system in mammals.

Hepatic Portal System

The hepatic portal system is similar in all craniates. It drains chiefly the stomach, pancreas, intestine, and spleen and terminates in the capillaries of the liver. Its origin in sharks from the embryonic vitelline and subintestinal veins has been described earlier. It arises in a similar manner in tetrapods. The abdominal stream (umbilical in mammals) becomes a tributary of the hepatic portal system, commencing with amphibians. Veins from the swim bladders are often tributaries of the hepatic portal system in bony fishes (seen in teleosts with a physoclistous swim bladder).

Coronary Veins

Many amphibians seem to lack a definitive coronary system. In frogs, one coronary vessel enters the left

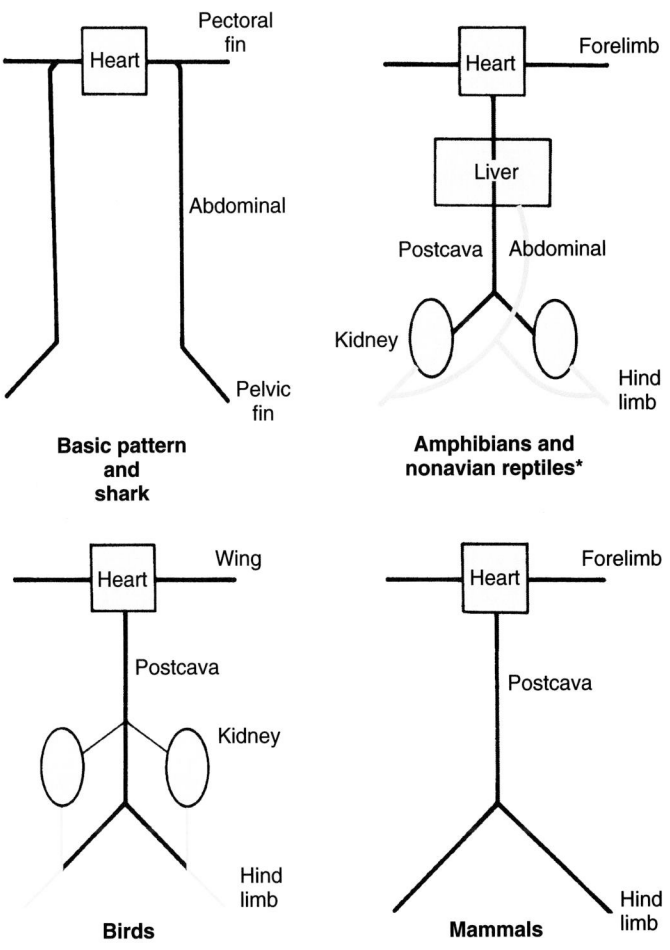

FIGURE 14.37

Venous routes from paired fins and limbs, diagrammatic. Vessels represented by *thin lines* in birds carry relatively less blood. *Blue* in a vessel indicates a portal stream. (*The abdominal stream in reptiles is paired.)

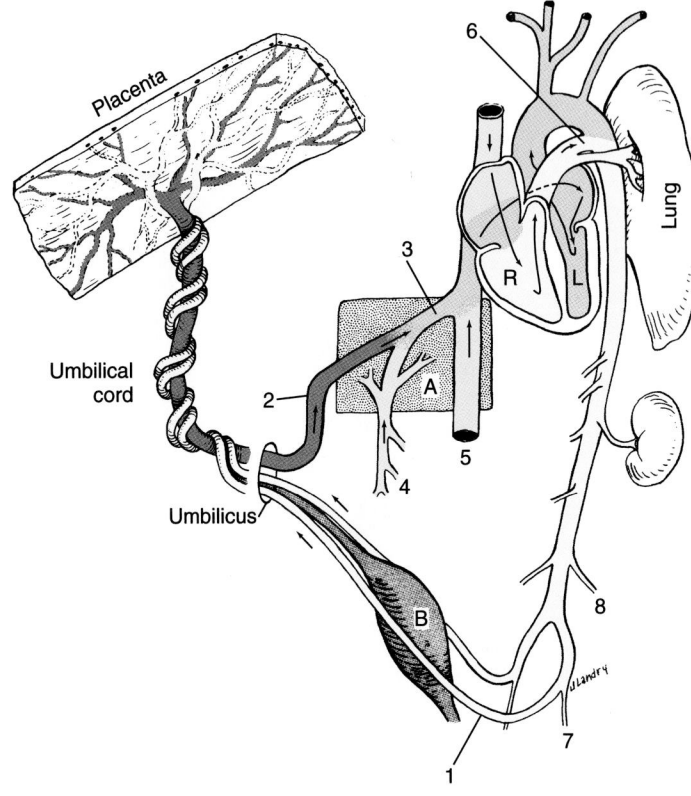

FIGURE 14.38

Circulation in a mammalian fetus. *Dark red* indicates oxygen-rich blood; *dark blue,* blood low in oxygen; *light red,* mixed, considerable oxygen; *light blue,* mixed. **1,** umbilical artery; **2,** umbilical vein in the falciform ligament; **3,** ductus venosus; **4,** hepatic portal vein; **5,** postcava (inferior vena cava); **6,** ductus arteriosus; **7,** internal iliac artery; **8,** external iliac artery growing into limb bud; **9,** umbilicus. *Gray,* allantois. **A,** liver; **B,** base of allantois, which is developing into a urinary bladder; **L,** left ventricle; **R,** right ventricle. After birth, the constricted part of the allantois distal to the bladder becomes the urachus. Much of the blood returning to the right atrium from the placenta passes through a foramen ovale (not illustrated, but indicated by a *dashed flow line*) into the left atrium to be distributed via the left ventricle to the head and anterior limbs.

precava, and another empties into the ventral abdominal vein near the liver. In reptiles and mammals, the several coronary veins empty into the coronary sinus or directly into the right atrium. The coronary sinus lies on the surface of the heart in the **coronary sulcus**, a groove between the left atrium and ventricle.

CIRCULATION IN THE MAMMALIAN FETUS, AND CHANGES AT BIRTH

Blood in the mammalian fetus passes from the caudal end of the dorsal aorta into the umbilical arteries (fig. 14.38). These extend out the umbilical cord to the placenta. From the placenta, oxygenated blood returns to the fetus via an umbilical vein that traverses the falciform ligament to enter the liver. Some of this blood, which contains nutrients as well as oxygen, enters liver capillaries for processing. Most of it continues nonstop via the ductus venosus into the postcava and finally into the right atrium. From the right atrium, it passes through an interatrial foramen (foramen ovale of the heart) into the left atrium, bypassing a trip to the lungs. The foramen is guarded by a one-way, flaplike valve. From the left atrium, it goes to the left ventricle and is expelled at the next systole into the systemic arch. Thus, oxygenated blood is sent to the fetal head, trunk, and limbs (fig. 14.38).

The unoxygenated blood returning to the left atrium from the lungs is a very small quantity compared to that pouring in through the foramen ovale, except near the time for delivery of the fetus; the mixing of the two only slightly dilutes the oxygenated blood.

Most of the blood returning to the right atrium from the major venous channels enters the right ventricle and, at the next systole, is pumped into the pulmonary

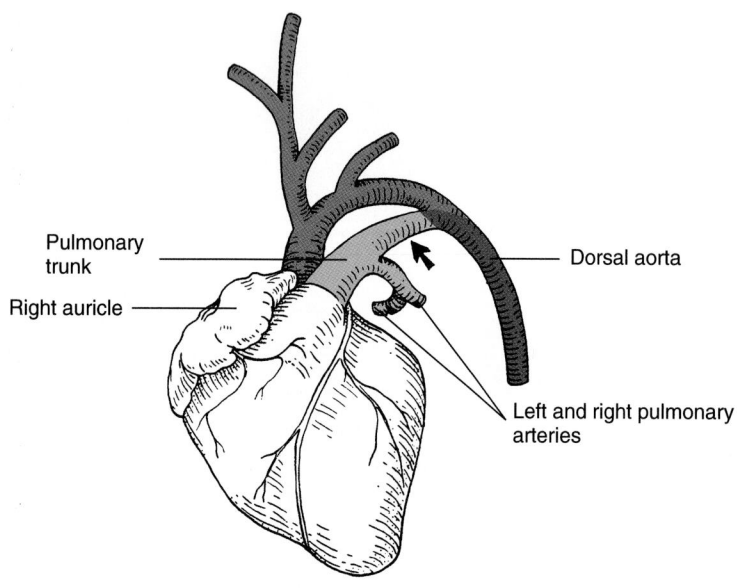

Pulmonary trunk

Right auricle

Dorsal aorta

Left and right pulmonary arteries

FIGURE 14.39
Ductus arteriosus (*arrow*) of fetal pig. The left auricle has been removed to show the pulmonary arteries. Blood in the dorsal aorta is mixed.

trunk. However, the ductus arteriosus is open and functioning, and all but a small amount of this blood is shunted away from the lungs and into the dorsal aorta (fig. 14.39, blue). A quantity sufficient to satisfy the oxygen needs of the fetus reaches the placenta via the umbilical arteries.

From this account, it may be seen that blood in the fetal circulation is either low in oxygen or mixed to one degree or another, except in the umbilical veins, in which it is rich in oxygen; that the umbilical arteries receive blood in need of oxidation; and that the brain and trunk receive the blood that is richest in oxygen. An essentially identical allantoic circulation is seen in unhatched chicks.

It should be pointed out that fetal circulation cannot be exactly as described until the embryo has developed two independent ventricles. In humans, this is not until the beginning of the eighth week of embryonic life.

At birth, four major circulatory changes adapt the organism for pulmonary respiration:

1. The ductus arteriosus closes as a result of nerve impulses passing to its muscular wall. These impulses are initiated reflexly when the lungs are filled with air with the first gasp after delivery. In birds, this is usually the day before hatching, when the imprisoned chick pecks a hole in its extraembryonic membranes and starts breathing the air entrapped between these membranes and the shell. When the chick inside the shell starts to peep, it already has air in its lungs! Shortly thereafter, all blood entering the pulmonary trunk goes to the lungs, and the

ductus arteriosus of birds and mammals becomes converted into an **arterial ligament (ligamentum arteriosum)**.

2. The flaplike interatrial valve is pressed against the interatrial foramen by the sudden increase in pressure in the left atrium that results from the greatly increased volume of blood entering from the lungs. This valve prevents the unoxygenated blood in the right atrium from entering the left atrium, which now contains only oxygenated blood from the lungs. Within a few days, the foramen ovale is permanently sealed, and only a scar, the fossa ovalis, remains.

3. At birth, the umbilical arteries and vein are severed at the umbilicus. Thereafter, no blood passes through the umbilical arteries distal to the urinary bladder, which they continue to supply. Eventually, the umbilical arteries from bladder to navel are converted into **lateral umbilical ligaments** in the free border of the ventral mesentery of the bladder.

4. Blood no longer flows through the umbilical vein because the source has been cut off. Eventually, this vessel becomes converted into the round ligament of the liver, and the ductus venosus becomes the ligamentum venosum. (The latter occurs halfway through gestation in whales, forcing umbilical vein blood through the liver capillaries en route to the right atrium.) As a result of these changes, the fetal mammal (or bird) is changed from an allantoic-respiring or placenta-respiring organism to one capable of breathing air and, indeed, forced to do so.

Failure of the foramen ovale to close or of the ductus arteriosus to fully constrict may result in cyanosis (blueness of the skin, lips, and nail bed in humans) because of the amount of reduced hemoglobin in the red cells in the integumentary capillaries. The condition may be ameliorated if the fluid pressure in the left atrium is sufficient to keep the valve of the foramen ovale pressed against the aperture, thereby blocking the shunt.

SYSTEMATIC SUMMARY OF RESPIRATION AND CIRCULATION

As noted at the beginning of chapter 13, the respiratory and circulatory systems serve a similar role in meeting oxygen demands and elimination of metabolic wastes. Many of the evolutionary innovations represent improved effectiveness for diffusion and involve both systems. Figure 14.40 summarizes the major synapomorphies associated with gas exchange.

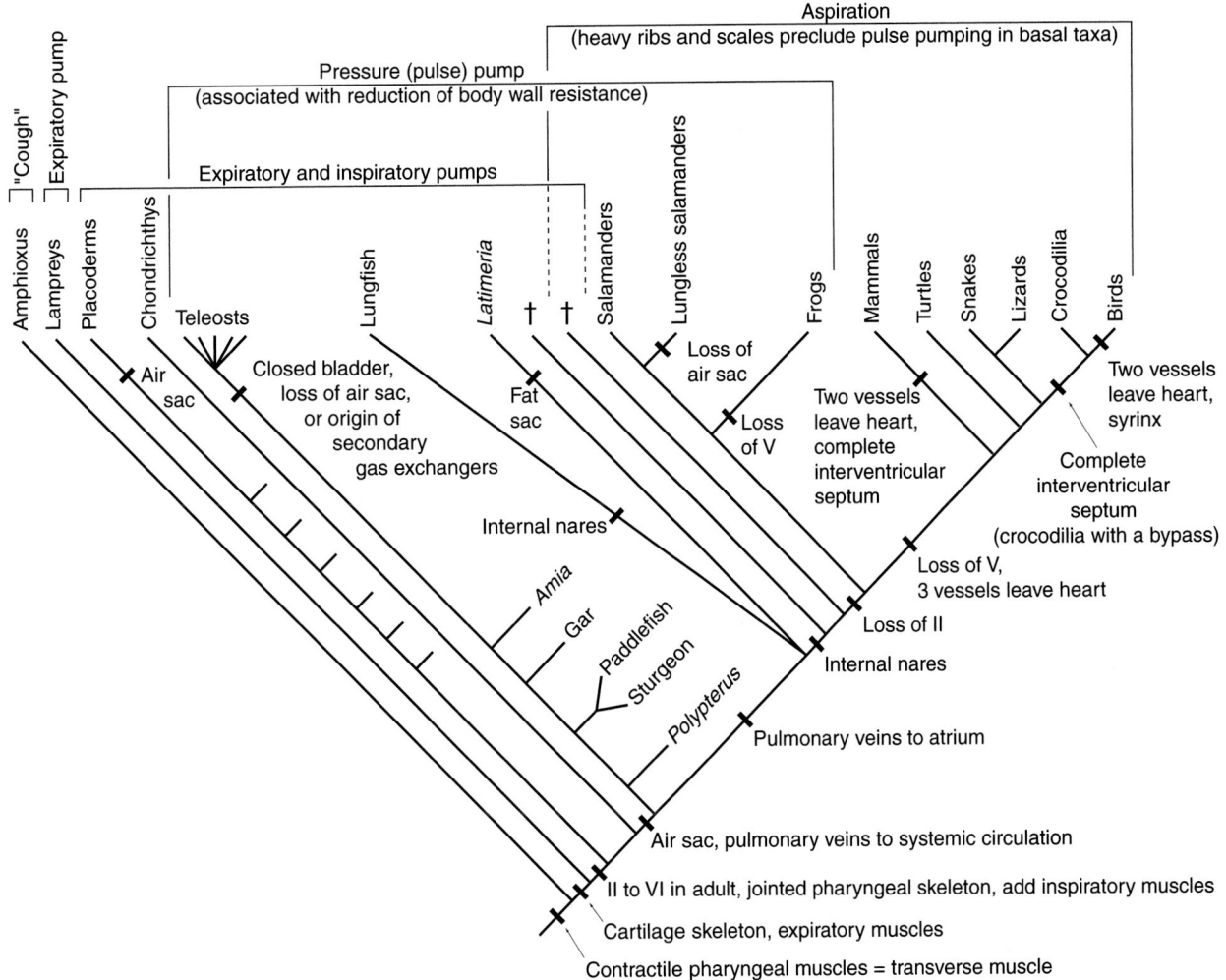

FIGURE 14.40

A systematic review of respiratory and circulatory structures and behaviors. Synapomorphies are indicated by tick marks along the cladogram and respiratory strategies by brackets above. The overlap of strategies represent multiple capabilities, (e.g., lungfish can respire by means of gills or pulse pumping). Other taxa may shift daily, seasonally, or as life history shifts. The "cough" of amphioxus is not respiratory in nature, but it is a reflex to dislodge sand grains during filter feeding. A single occurrence of an air sac is reported in one Late Devonian placoderm. Other, more basal, members of the group (unlabeled clades) lack an air sac. A number of teleosts have independently lost an air sac, developed a physoclistic bladder, or developed secondary gas exchange mechanisms. The origin of internal nares is unclear due to debates concerning the relationships of lungfish (unresolved in this cladogram). The number of vessels leaving the heart refers to the pulmonary artery and one or two systemic aortic arches. **II–VI**, aortic arches II through VI. (†), extinct taxon.

How these structures are used to achieve respiration differs among taxa and within single species. Similar structures may have more than one function. In amphioxus, the transverse muscle of the atrial wall (see fig. 3.7c) provides a "cough" reflex to dislodge grains of sand drawn into the pharynx during filter feeding. As noted before, amphioxus utilizes cutaneous respiration. In living agnathans, water is expelled by muscular contraction of the branchiomeric constrictor muscles of the pharynx. Water is drawn into the pharynx with the elastic expansion of the cartilaginous branchial basket. Thus, these taxa possess a single-phase muscular pump. Gnathostomes, in contrast, possess a complex jointed pharyngeal skeleton requiring both inspiratory and expiratory muscles to pump water over the gills (a double muscular pump). With the origin of an osteichthyan air sac, atmospheric oxygen is available to augment gill respiration during times of low oxygen levels in the aquatic habitat. In these taxa, the same buccopharyngeal muscles used to pump water over the gills are used to pump a pulse of air into an air sac (a positive pressure system). There is a mechanical cost for organisms to compress air sufficiently to overcome the elasticity of the lung and the resistance of surrounding body tissues. Adaptations in pulse pumpers include reduction of these resisting tissues (e.g., frog lungs are dorsally located with the overlying ribs severely reduced).

On the other hand, in aspiration breathing, the pleural cavity is expanded, creating a negative intrathoracic pressure, drawing in atmospheric air. The muscles we saw utilized in gill breathing and pulse pumping are replaced by other muscles that expand the thoracic cavity (e.g., costal muscles of the rib cage and diaphragm in mammals).

The respiratory strategies in figure 14.40 show an overlap, suggesting that some organisms are capable of utilizing more than one behavior. As just noted, the double muscular pump of fishes can be behaviorally modified to pump air into an air sac. Gill-breathing fishes challenged by low-oxygen partial pressures in their habitat can shift to pulse pumping. This pattern is seen in tropical lungfish that may be faced with daily shifts in available dissolved oxygen or seasonally during periods of aestivation when the streams dry up. Alternatively, a species may shift strategies during its life history (e.g., a gill-breathing larva shifting to pulse pumping at metamorphosis). Due to the mechanical cost of pulse pumping, any shift to an aspiration mode would potentially conserve energy. We see aspiration breathing in all amniotes, but its origin predates the tetrapods based on fossil evidence. Many fossil rhipidistian fishes with lungs were incapable of pulse pumping due to their heavy overlapping ribs and covering of interlocking scales. The large body size and ribs of early tetrapods would also preclude pulse pumping. A single South American fish may use gill breathing and pulse pumping that is augmented with thoracic aspiration musculature. If confirmed, it may serve as a model for the evolutionary transition to aspiration breathing.

Given the proposed early origin of aspiration breathing, you may ask, Why do lissamphibians, such as the frog, utilize pulse pumping? It is important to remember that in evolution all that matters is a sufficient design to allow an organism to reproduce and pass on its genes. Optimal designs have to be balanced with the other needs of the organism in its attempts to make a living. Given the known costs of pulse pumping, what might balance these? Vocalization in frogs is very important both in territoriality and mating behaviors. The balance between the costs of pulse pumping and the benefits of vocalization has been an evolutionary success for the frog. Finally, there is the question of whether the respiratory patterns in lissamphibians are primitive or independently derived. This, as of yet, has not been answered. What do you think?

LYMPHATIC SYSTEM

Lymph and lymph channels are found in all craniates (figs. 14.41 and 14.42). A lymphatic system consists of thin-walled **lymph channels**, **lymph** (a fluid in transit), **lymph hearts** as far up the phylogenetic scale as embryonic birds, **lymph nodes** in birds and mammals, and

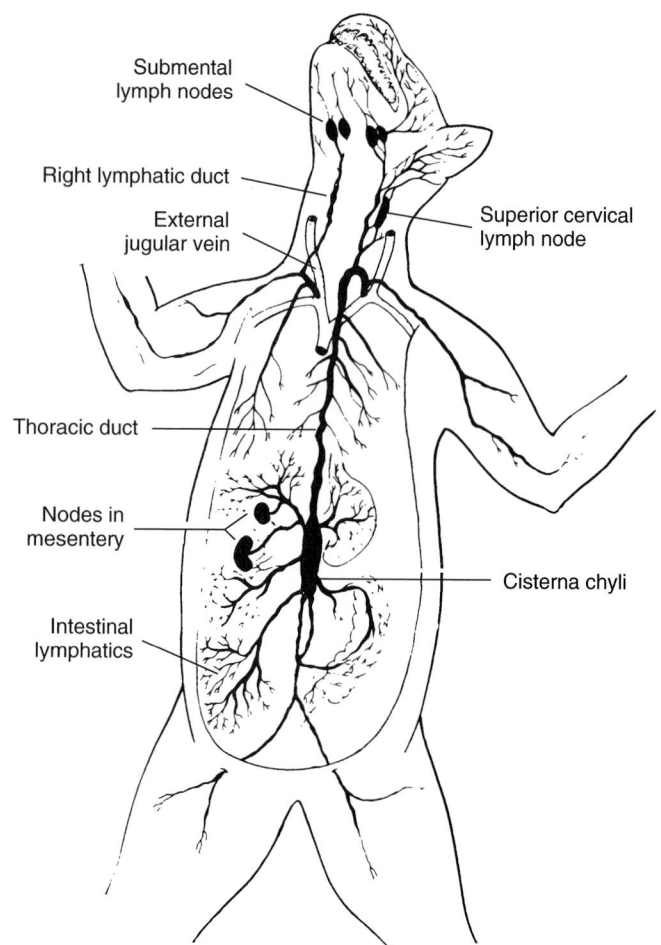

FIGURE 14.41

A few of the superficial and deep lymphatics of a cat.

solitary or aggregated masses of lymph nodules, the largest of which is the spleen. In contrast to blood, lymph moves in one direction only: from the tissues toward the heart. Lymph channels are seldom dissected in the anatomy laboratory because of their delicate and collapsible nature. Lymph nodes are more readily observable (see fig. 14.41).

The system begins in **lymph capillaries,** the walls of which are single-layered endothelial tubes, or in **lymph sinusoids,** which are expansions of capillaries. The capillaries are branching and anastomosing tubes of slightly greater diameter than blood capillaries, and they exhibit constrictions and expansions rather than being of a standard diameter. Capillaries or sinusoids penetrate most of the soft tissues of the body other than the liver and the nervous system. They are not present in skeletal tissues. Capillaries and sinusoids collect interstitial fluids. Once inside the tubes, the fluid is colorless or pale yellow and is called lymph. This fluid passes from one endothelial-lined channel to the next and finally empties into a vein. Valves at these exits prevent the influx of venous blood into the lymph channels. A lymphatic network consisting of long, narrow, discrete tubular vessels with a modicum of smooth muscle in the walls is found only in birds and mammals. A series of valves line these tubes and help counteract the effect of gravity, especially in the limbs.

The collection of fats absorbed from the small intestine after a meal by the lymphatics in intestinal villi is a derived feature within craniates (see fig. 12.23). If the meal was particularly fatty, the lymph in these vessels has a milky appearance. For this reason, these specific lymphatics are called **lacteals.** The lymph within them is **chyle.** Certain lymphatics in living agnathans, cartilaginous fishes, and even humans contain some red blood cells. The fluid in these vessels is **hemolymph.**

Lymph channels that drain the body wall, limbs, and tail of craniates typically empty into nearby veins at the base of the tail, in the trunk, or in the neck. Those draining viscera are often paired in most craniates, but in mammals, a single **thoracic duct** commences in a large abdominal lymph sinus, the **cisterna chyli** (see fig. 14.41), and empties into the brachiocephalic or left subclavian vein, or into the external or internal jugular vein (see fig. 14.41). The mammalian thoracic duct also receives lymphatics from the left side of the head and neck and from the left forelimb. One or more additional major lymphatics drain the right side of the body anteriorly.

The lymphatic system of anurans is notable because of the numerous sinusoids that take the place of capillaries at some locations. They are especially prominent just under the skin over almost the entire body, where they form huge lymph reservoirs separated from each other by connective tissue septa that attach the skin to the underlying muscles at these locations only (fig. 14.43). The skin is very thin, and the sinusoids may buffer the underlying muscles from the drying effect of air when a frog is out of water. Because of these subcutaneous lymph sinuses, the skin of a frog is easily removed from the body. Anyone who has dissected a fluid-preserved frog has seen how the skin balloons when the sacs are distended with fluid.

The flow of lymph results from a number of factors, including lymph hearts situated at advantageous locations along lymph pathways in fishes, amphibians, and reptiles (except postembryonic birds). These are pulsating sinusoidal swellings on lymph pathways with thin walls containing striated muscle fibers of unknown embryonic origin. Frogs have two pairs of lymph hearts, urodeles have as many as 16 pairs, and caecilians as many as 100. The anterior pair of lymph hearts in frogs, located just behind the transverse processes of the third vertebra, pump lymph into the vertebral veins, from which it is carried to the internal jugulars. A posterior lymph heart on each side of the urostyle empties into an external iliac vein. Amphibians, especially aquatic and semiaquatic amphibians, have more tissue fluids to ma-

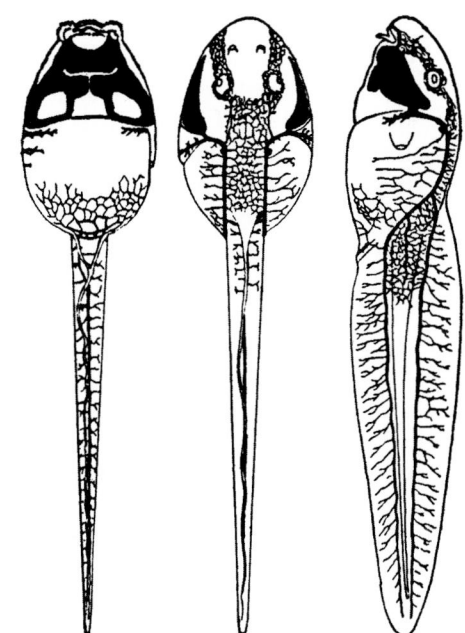

FIGURE 14.42
Superficial lymph channels of a frog tadpole.
Source: Data from J. S. Kingsley, *Outline of Comparative Anatomy of Vertebrates*, 1920, The Blakiston Company, Philadelphia, after Hoyer and Udziela.

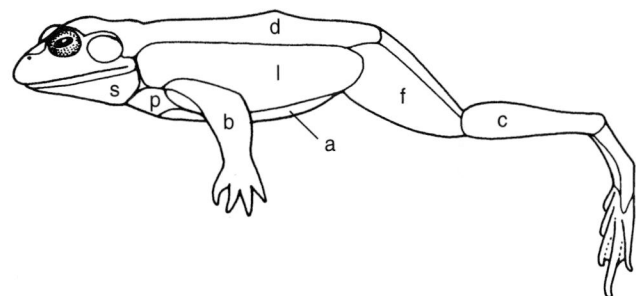

FIGURE 14.43
Subcutaneous lymph sacs of a frog. The outlines of the sacs are the sites of septa that extend between the skin and the musculature.
a, abdominal; **b**, lateral brachial; **c**, crural; **d**, dorsal; **f**, femoral; **l**, lateral; **p**, pectoral; **s**, submaxillary.

nipulate than do other craniates, and as a result their lymph hearts move a proportionately larger volume of fluid hourly than the hearts of most other craniates. Semilunar valves at the exit of the hearts prevent backflow.

Lymph hearts are not present in birds after hatching, although embryonic birds have one on each side of the last sacral vertebra. None has been described in mammals. In these craniates lymph flow is maintained by activity of the skeletal muscles as they contract and relax, by movements of the viscera, and by rhythmical changes in intrathoracic pressure that result from breathing. Some of these factors are also operative in other craniates.

Lymph nodes are masses of hemopoietic tissue interposed along the course of lymph channels of birds and mammals. They may be no larger than a pinhead, or they may be several centimeters in diameter. They are the "swollen glands" that you can feel in the neck, axilla, and groin of humans when there is inflammation in the areas drained. Lymph enters a node via several afferent lymphatics, filters through the node, and leaves via a single large efferent lymphatic. The endothelium of their sinusoidal passageways includes many phagocytes that ingest bacteria and other foreign particles. Enmeshed in their connective tissue stroma are large numbers of small lymphocytes. The nodes are therefore the second line of defense against bacterial infections acquired through the skin, the first line being granulocytes that assemble at the invaded area.

Other **lymphoid masses** develop at one site or another. The largest of these is the **spleen**, which is unique because it is interposed into the blood circulatory stream. Its contribution to that system was discussed earlier. Other masses include the **thymus** (absent in hagfishes), the **bursa of Fabricius** of very young birds, solitary nodules or aggregates of nodules (**Peyer's patches**) in the wall of the small intestine of amniotes, and mammalian **tonsils** and **adenoids**.

Summary

1. The circulatory system consists of a blood vascular system and a lymphatic system.
2. The blood vascular system consists of whole blood, heart, arteries, capillaries, and veins.
3. Blood consists of plasma and formed elements. The formed elements are erythrocytes, granular and agranular leukocytes (the latter includes lymphocytes), and thrombocytes (platelets).
4. Hemocytoblasts are precursors of all blood cells. Their initial source is blood islands of the yolk sac. Eventually they form in liver, kidney, spleen, and bone marrow.
5. The lymphatic system consists of lymph, lymphatic capillaries and veins, lymph hearts, lymph nodes, and miscellaneous lymphoid masses in one major taxon or another. Chyle is lymph collected in intestinal villi. Lymphatics terminate in a systemic vein.
6. Fishes have a single-circuit heart. Venous blood enters a sinus venosus, traverses an atrium, ventricle, and conus arteriosus. The last two are pumps that discharge into a ventral aorta. The latter carries blood to aortic arches that supply gills, where blood is oxygenated. It then passes via arteries to capillaries everywhere in the body, gives up oxy-

gen, takes on carbon dioxide, and returns to the sinus venosus.

7. In craniates that breathe solely by lungs, a pulmonary circuit carries blood to the lungs and back and a systemic circuit carries oxygenated blood elsewhere and returns deoxygenated blood to the heart. This is a double-circuit heart.
8. Double-circuit hearts have two atria and one or two ventricles. The right atrium receives deoxygenated blood from the systemic circuit; the left receives oxygenated blood from the pulmonary circuit.
9. Mixing of oxygenated and deoxygenated blood in the ventricle of a double-circuit heart is avoided by one of several adaptations, including spiral valves when there is a single ventricle, and trabeculae and ventricular septa. A complete interventricular septum separates the blood in crocodilians, birds, and mammals.
10. Right-left shunts of blood (away from the lungs) exist in amniotes to meet metabolic needs imposed by certain behavioral patterns, including remaining underwater for long intervals without breathing.
11. A sinus venosus is present in fishes, amphibians, and reptiles. It becomes partially incorporated into the wall of the right atrium in crocodilians. In birds and mammals, it is not a sinus but a local collection of cells in the right atrium known as the sinoatrial (SA) node.
12. The pulsations of the heart are autogenic, requiring no extrinsic neural stimulus. However, the rate of beat is imposed by the autonomic nervous system (except in hagfish). The stimulus in fishes, amphibians, and nonavian reptiles is spread from the musculature of the sinus venosus. In birds and mammals, it emanates from the SA and an atrioventricular (AV) node.
13. The ventral aorta exhibits a swelling, the bulbus arteriosus, in teleosts and perennibranchiate urodeles.
14. In most reptiles, the ventral aorta is split longitudinally into two aortic trunks and a pulmonary trunk. The two aortic trunks exit from a cavum venosum in the ventricle and supply all functional aortic arches except the pulmonary. The pulmonary trunk emerges from the right ventricle (cavum pulmonale) and supplies the pulmonary circuit. The left ventricle has been termed the cavum arteriosum.
15. The ventral aorta in birds and mammals is split into only two trunks, a systemic trunk that emerges from the left ventricle and supplies all parts of the body except the lungs, and a pulmonary trunk that emerges from the right ventricle and supplies the lungs.
16. Six pairs of aortic arches develop in gnathostome embryos. During ontogeny, the arches are reduced

TABLE 15.2   Environmental Osmotic Challenges and Evolutionary Solutions in Craniates

Environment	Primary Osmotic Problem	Solution	Example Taxa[1]

Freshwater	• Environment is hypoosmotic
	  • Excessive water uptake
	• Must excrete excess water and conserve salts
		• Waste excreted as ammonia
		  • Use excess water as solvent
		• Actively transport solutes (for retention)
			• Freshwater teleosts, aquatic, and semiaquative amphibians

Salt water	• Environment is hyperosmotic
	• Excessive water loss
	• Excessive uptake of salts (in food and drinking water)
	• Must retain water and eliminate salts
		• Isoosmoticity
		• Lose glomerulus (↓ water loss)
		• Become hyperosmotic relative to environment
		  • Retain urea to achieve
		  • Increases water uptake
		  • Retain glomerulus to excrete water
		• Salt excretion glands:
		  • Rectal gland
		  • Salt glands on gills
			• Hagfish
			• Marine teleosts
			• Elasmobranchs and *Latimeria*
			
			• Elasmobranchs
			• Teleosts

Terrestrial	• Environment is dry
	• Water and salts are limited.
	  • Must retain both water and salts
	• A secondary return to water
	  • Reintroduces above aquatic concerns
		• N excreted in all three forms
		  • Balance costs vs. benefits
		  • Variable among taxa
		  • Variable in life histories
		• Reduce glomerulus
		• Solute recovery
		• Loop of Henle or equivalent
		• Salt glands (elimination)
			• Nonavian reptiles and anurans inhabiting arid habitats
			• Birds and mammals
			• Marine taxa[2]

[1]Taxa that have adapted the evolutionary solution presented in the adjacent column.
[2]Terrestrial taxa that return to or make a living from the sea evolved adaptive accessory structures to eliminate excess salts acquired in their diets.

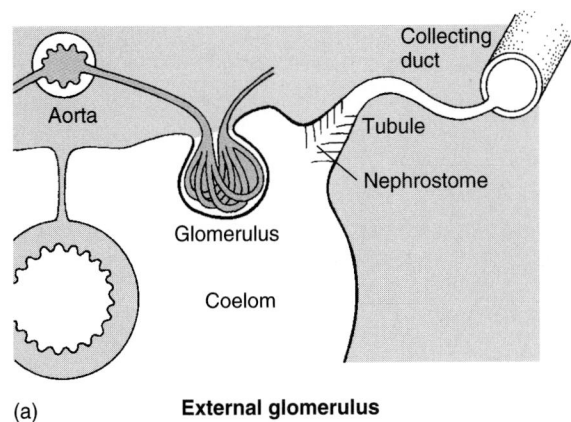

(a)   **External glomerulus**

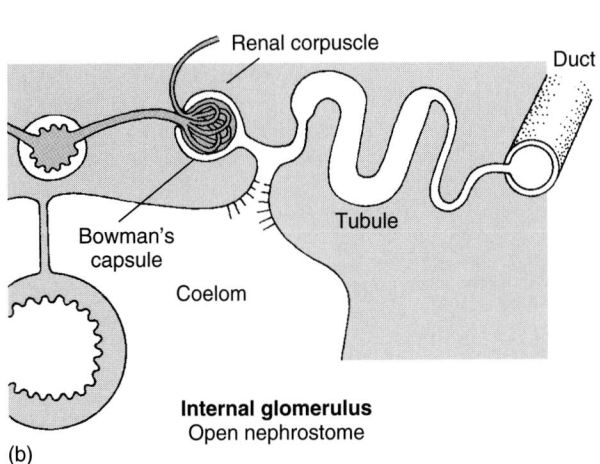

(b)   **Internal glomerulus**
Open nephrostome

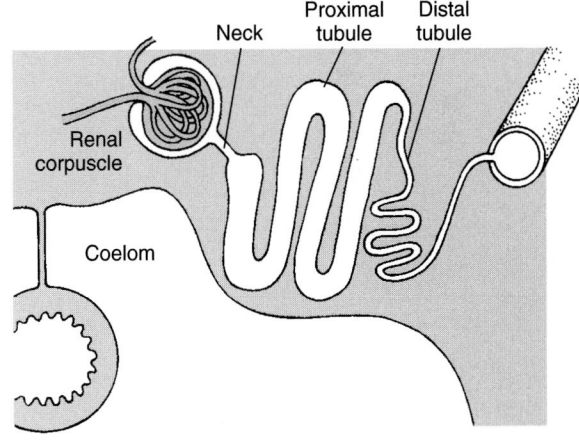

(c)   **Internal glomerulus**
Closed nephrostome

FIGURE 15.1

Basic components of a craniate kidney. (*a*) External glomerulus and tubule in the cephalic end of the segmental kidney of a larval hagfish. (*b*) Tubule with internal glomerulus and open nephrostome, as in an apodan kidney. (*c*) Generalized nonsegmental tubule of freshwater fishes, urodeles, and anurans. The tubules of marine fishes often lack glomeruli.

some fishes, they are large enough to be seen with the naked eye or a hand lens. In others, they are microscopic. The most primitive glomeruli are suspended in the coelom surrounded by peritoneum. They discharge their filtrate into the coelomic fluid, which is then swept into a peritoneal funnel, or **nephrostome**, leading to a tubule (fig. 15.1*a*). These are **external glomeruli**. In today's craniates, external glomeruli are confined to embryos and larvae. Glomeruli in adults are embedded within the dorsal body wall (hence said to be retroperitoneal) and are ensheathed by **Bowman's capsule**, a delicate double-walled outgrowth from a kidney tubule (figs. 15.1*b* and 15.2). Its inner wall adheres to the surfaces of the vascular loops. The capsular cavity collects the glomerular filtrate, which then passes into a renal tubule. These are **internal glomeruli**. A glomerulus and the surrounding capsule constitute a **renal corpuscle** (fig. 15.2). A renal corpuscle, renal tubule, and the associated peritubular capillaries constitute a **nephron**, the functional unit of a gnathostome kidney (fig. 15.3).

Supplying a glomerulus is an **afferent glomerular arteriole**, and emerging from the glomerulus is an **efferent glomerular arteriole** of *lesser diameter*. This has the effect of increasing the blood pressure within the

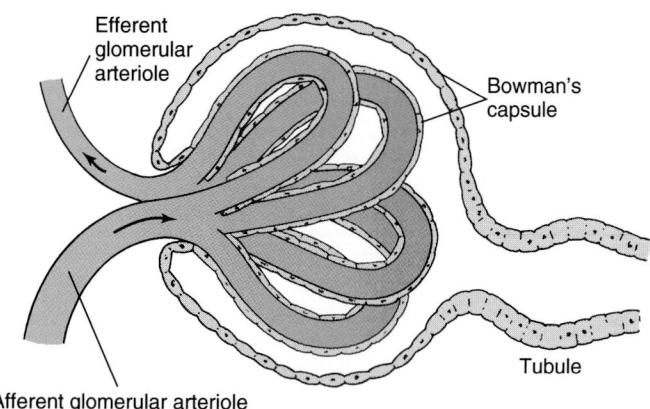

Renal corpuscle

FIGURE 15.2
Renal corpuscle consisting of glomerulus (*red*) and Bowman's capsule (*gray*).

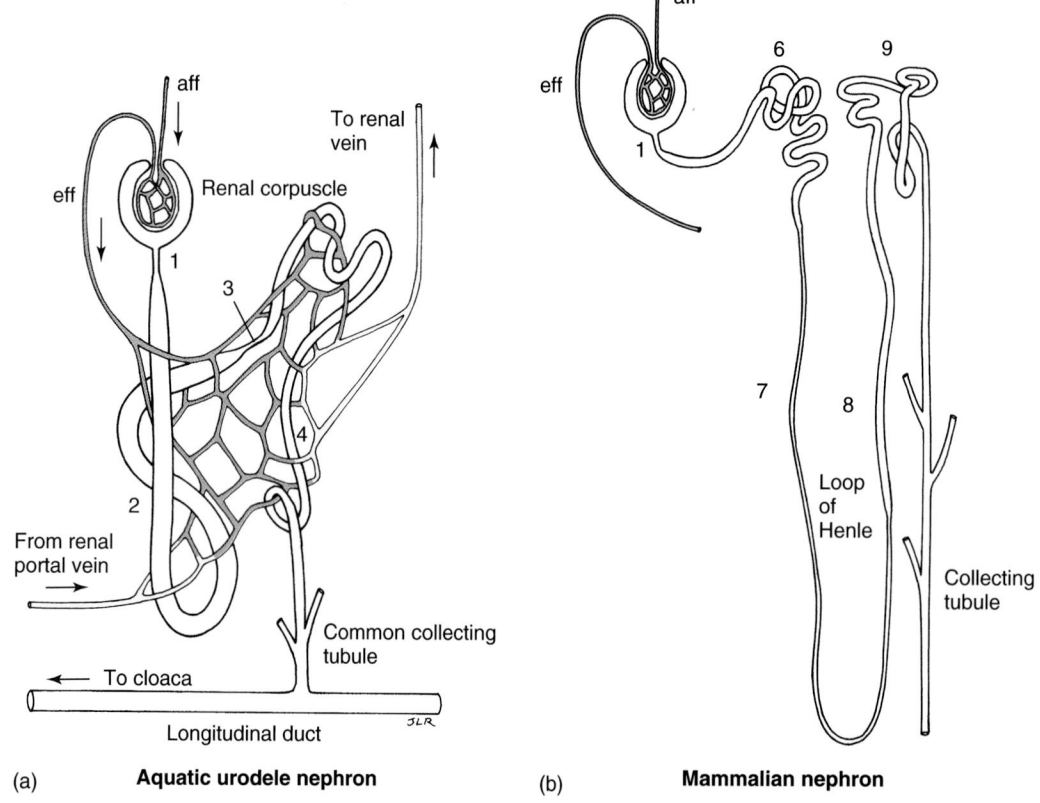

(a)   **Aquatic urodele nephron**

(b)   **Mammalian nephron**

FIGURE 15.3
Anamniote and mammalian nephrons, peritubular capillaries omitted in (*b*). In (*a*): **1,** neck of renal tubule; **2,** proximal segment; **3,** intermediate segment; **4,** distal segment. In (*b*): **1,** neck of tubule; **6,** proximal convoluted tubule; **7,** descending arm of loop of Henle; **8,** ascending arm; **9,** distal convoluted tubule. **aff,** afferent glomerular arteriole; **eff,** efferent glomerular arteriole. The peritubular capillaries of therian mammals receive no portal blood. (Not to scale.)

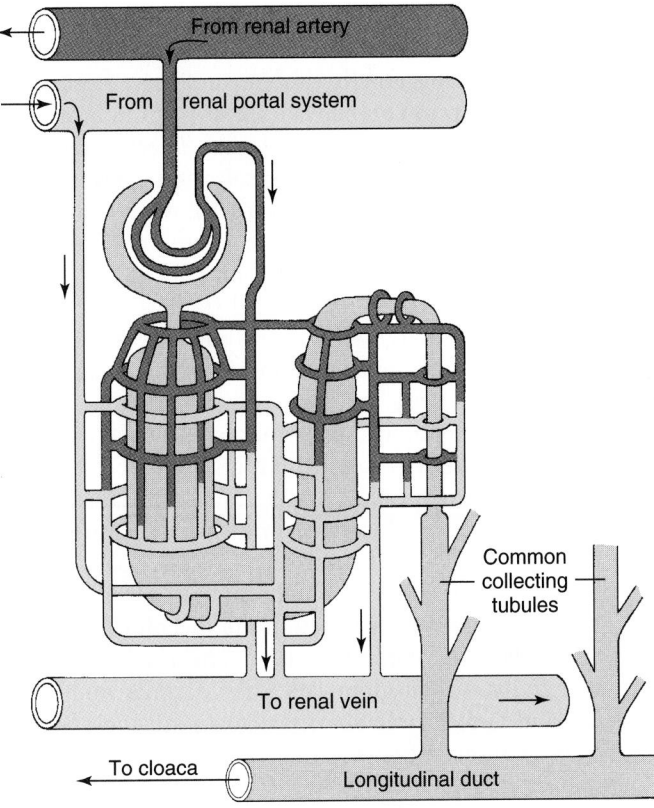

FIGURE 15.4
Schematic representation of blood supply to and from the peritubular capillaries of a generalized nephron. *Red,* arterial blood; *blue,* venous blood.

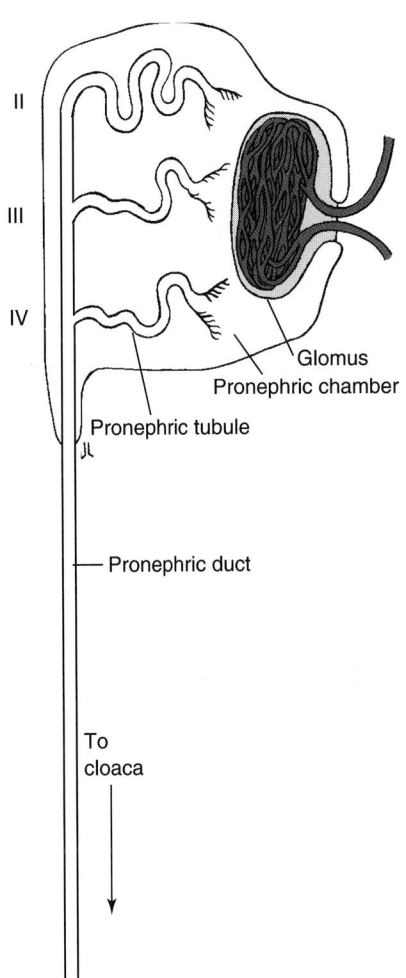

FIGURE 15.5
Pronephric tubules of a 15 mm larval frog. **II, III,** and **IV,** location of the second, third, and fourth somites. The glomus is three fused external glomeruli; the pronephric chamber is a coelomic pocket. The next tubule to form will be a mesonephric tubule at the level of somite VII.

glomerulus. The afferent arterioles of agnathans are supplied directly from the dorsal aorta by segmental arteries. In gnathostomes, they are terminal branches of renal arteries. Efferent glomerular arterioles supply peritubular capillaries along with blood from the renal portal system (figs. 15.3*a* and 15.4). Efferent glomerular arterioles are the sole source of blood in the peritubular capillaries of therian mammals because a renal portal system is lacking. Peritubular capillaries are drained by venules that lead to renal veins. Agnathans alone lack peritubular capillaries, but as compensation, the longitudinal kidney ducts are surrounded by unusually rich capillary beds that take their place functionally.

Renal tubules differentiate from a ribbon of embryonic **nephrogenic (intermediate) mesoderm** that lies lateral to the mesodermal somites and extends the length of the embryonic trunk from immediately behind the head to the cloaca (see figs. 5.9*d* and 15.8). The earliest tubules appear at the anterior end of the ribbon, and additional tubules are added as the embryonic trunk elongates. The paired longitudinal ducts begin development as caudally directed extensions of the first tubules (fig. 15.5). The two ducts grow caudad until they reach the cloaca, into which they establish an

opening. The ducts participate, although not exclusively, in the induction of additional tubules as the body elongates. Adult kidneys remain retroperitoneal throughout life, although in lampreys and *Necturus* (a urodele) they are intraperitoneal and in mammals they bulge into the roof of the coelom.

Except in primitive metameric kidneys, several renal tubules empty into a single **common collecting tubule**, and this, in turn, empties into one of the two longitudinal kidney ducts (fig. 15.6). The longitudinal duct collects fluids from all the common collecting tubules and empties into the cloaca or a derivative thereof.

Renal tubules increase in complexity among the craniate classes, exhibiting increasing degrees of *histologic and functional* specialization along the course of the tubule. They are short and straight in agnathans, longest and with the greatest regional specialization in mammals.

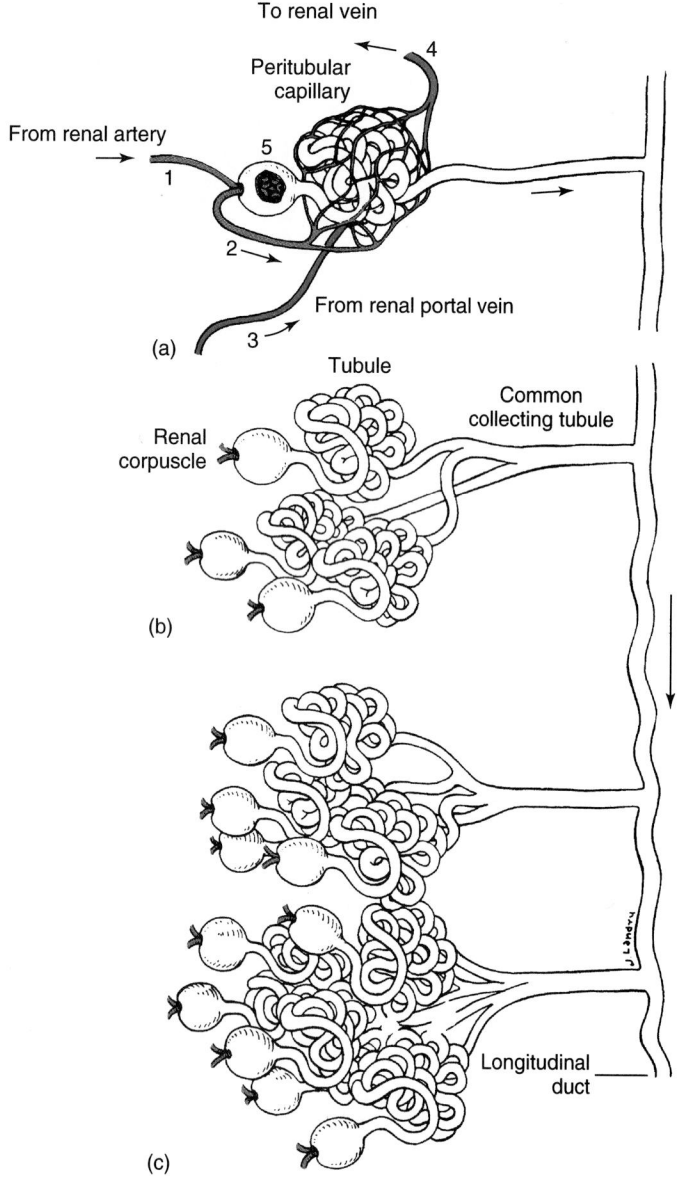

FIGURE 15.6

Effect of multiple nephrons in a single body segment. (*a*) Segmental nephron consisting of renal corpuscle, renal tubule, and peritubular capillary. **1,** afferent glomerular arteriole; **2,** efferent glomerular arteriole; **3,** venule from renal portal system; **4,** efferent venule to renal vein; **5,** renal corpuscle, Bowman's capsule opened to show glomerulus. (*b*) Secondary and tertiary nephrons in a single body metamere. (*c*) Increased number of nephrons in a metamere disrupts the primitive metamerism of the kidney.

Compare the tubules in figure 15.1 with those in figure 15.3.

Many marine teleosts have renal corpuscles that are poorly vascularized, cystic, or vestigial. In some others, the distal segment of their tubules is abbreviated or lost. Loss of glomeruli, the major site of water removal, results in increased water retention. Shortening of the distal segment of the tubules eliminates the major site for reabsorption of salt. Both effects have survival value in

salt water (table 15.2). There are, however, freshwater teleosts with the same specializations. In these, water excretion is chiefly tubular, and these freshwater fishes seem to compensate for any salt deficiency by active uptake of salt via the gills. There is reason to believe that these are former saltwater species that were seasonally migratory into freshwater streams (anadromous, see glossary) and eventually became fully adapted to a freshwater habitat.

Elasmobranchs are exceptional (table 15.2). They have unusually large glomeruli, but they also have tubules that are unusually long for a fish. Elasmobranchs retain urea, and those living permanently in salt water are potentially hyperosmotic relative to their marine environment. This imbalance would provide for an influx of water for use in metabolic processes and the elimination of waste. Feeding in a marine habitat would introduce excessive salts, and to compensate, elasmobranchs have evolved **rectal glands** that secrete large amounts of chlorides into the caudal end of the intestine. The total effect of the activity of their glomeruli, tubules, gills, and rectal gland is to enable sharks to maintain their body fluids in an osmotic state that assures their survival, whether in freshwater or in the sea.

Like marine fishes, terrestrial nonavian reptiles and anurans that inhabit arid environments have very small glomeruli or none at all, a water conservation adaptation. Dehydration is the principal stimulus for release of pituitary hormones that induce water reabsorption from glomerular filtrate in tetrapods. These same hormones induce reabsorption of water from amphibian and reptilian urinary bladders, when necessary.

Several of the most anterior kidney tubules of some adult fishes and many tetrapod embryos and larvae have nephrostomes (see fig. 15.1*a,b*), and vestigial nephrostomes lacking a lumen are often found in avian and mammalian embryos. Nephrostomes may be vestiges of the kidneys of a postulated ancestral protochordate in which there may have been one external glomerulus, one nephrostome, and one unconvoluted tubule in each body segment along the entire length of the coelom. This hypothetical ancestral kidney has been termed an **archinephros** (fig. 15.7). The nearest approach to such a kidney in living craniates is seen in larval hagfishes, in which a transient series of segmental external glomeruli, nephrostomes, and tubules is formed throughout much of the extent of the nephrogenic mesoderm (see fig. 15.1*a*). However, at this stage of body elongation the kidney is still elongating. Segmental tubules with closed nephrostomes and renal corpuscles develop farther caudad. This transient larval hagfish kidney has been termed a **holonephros.** The adult hagfish kidney is a mesonephros (described shortly).

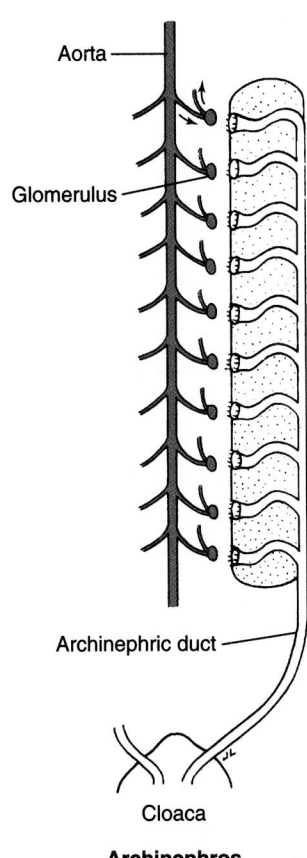

Aorta

Glomerulus

Archinephric duct

Cloaca

Archinephros

FIGURE 15.7
Hypothetical archinephros with one external glomerulus,
nephrostome, and tubule per body segment. See also figure 15.1*a*.

Gathering coelomic fluid for elimination of liquid wastes is not confined to craniates. Nephrostomes in every body segment of marine annelids gather coelomic fluid for subsequent excretion.

Role of Glomeruli and Tubules

Three major processes are at work in urine formation: glomerular filtration, tubular reabsorption, and tubular secretion. Glomeruli function much like a simple filter. Because of an elevated blood pressure within the glomerulus resulting partly from the smaller size of the efferent glomerular arteriole, water, certain salts, glucose, and other solutes in the blood plasma are filtered into the space between the outer and inner walls of Bowman's capsule. Some constituents of the glomerular filtrate may not be expendable, and those are selectively reabsorbed during passage through specific segments of the tubule. As a result, the composition of the filtrate becomes altered in passage. All the glucose, for example, is ordinarily reabsorbed. It is the sole immediately available circulating source of energy, it is not readily acquired, and it is not ordinarily in excess in the bloodstream because, once acquired, it is quickly stored in

the liver. Water and certain salts may or may not be in excess, depending on the environment. When in excess, water is allowed to pass through the length of the tubules and to enter the kidney ducts without having been reabsorbed. Under conditions that tend to lead to dehydration, water is reabsorbed from segments of the tubule. Certain salts are also selectively reabsorbed when appropriate. Finally, tubular secretion removes from the circulation useless or harmful substances that were not removed by filtration. Among these are, for instance, wastes from protein breakdown (nitrogenous wastes) in tetrapods. (Nitrogenous wastes, Na^+ and Cl^- in most fishes, are eliminated by extrarenal means.) In marine fishes, certain salts (Mg^{++}, Ca^{++}, SO^-_4, phosphates) are tubular secretions, glomerular filtration being low. Glomerular filtration, tubular reabsorption, and tubular secretion produce the final excretion, urine, which varies in the amount of water, salts, and other constituents depending on the environment and the species. The constituents of urine may vary in a single organism from hour to hour, if not from minute to minute.

By varying the size and number of glomeruli in species that occupy different habitats, water can be excreted abundantly or sparingly. By varying the length of the kidney tubules, either salt or water, as necessary, can be recovered from the glomerular filtrate, or salt can be excreted abundantly. In general, glomeruli are larger in freshwater fishes and aquatic amphibians, and smaller in marine fishes and tetrapods, especially tetrapods living in arid environments.

Excretion of Nitrogenous Wastes

Nitrogenous wastes (see table 15.1) are excreted chiefly as *ammonia* by freshwater teleosts and aquatic and semiaquatic amphibians, because ammonia is highly soluble in water, and water is plentiful. These are therefore ammonotelic animals. These wastes are also excreted as ammonia in most marine fishes other than elasmobranchs, but the water that serves as a solvent in these species is provided by drinking seawater and rapidly excreting the salt. Nitrogen is excreted as *urea* by elasmobranchs and mammals (ureotelic animals). It is excreted in a semisolid urine as *uric acid* in species in which water is not abundant, as in reptiles (uricotelic animals). The nature of the excreted nitrogenous compound in each instance is dictated in part by the availability of the solvent (water) that serves as a vehicle for elimination.

Pronephros

The earliest embryonic tubules arise from the cephalic end of the nephrogenic mesoderm. Because they are the first tubules to appear and are anteriorly located, they are called **pronephric tubules**, and this region of the nephrogenic mesoderm is the **pronephros** (fig. 15.8,

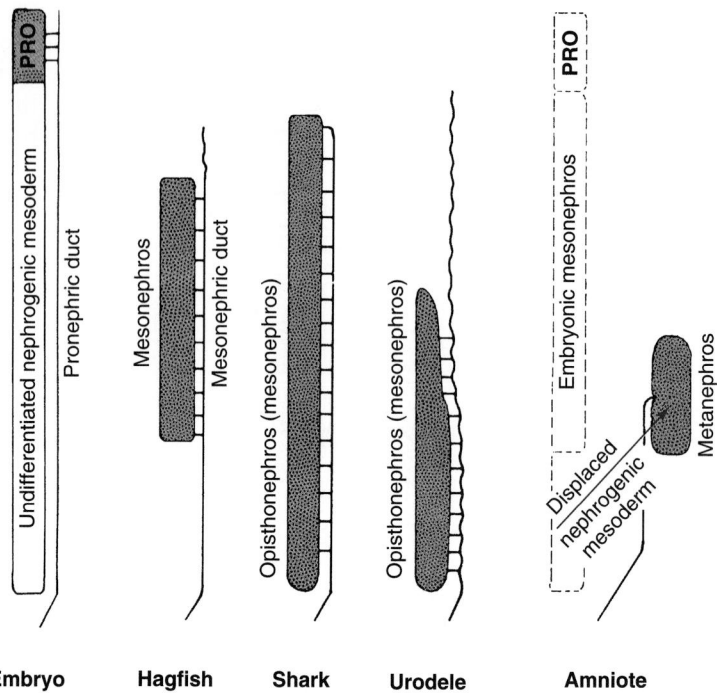

FIGURE 15.8

Fate of the nephrogenic mesoderm (*red*) in representative craniates. **PRO,** pronephros. The pronephric duct persists in adult anamniotes to drain the adult kidney.

embryo). The tubules are segmental, one pair at the level of each of the more-anterior somites. Pronephric tubules arise as solid aggregates of nephrogenic mesodermal cells that organize a lumen and frequently, except in birds and mammals, a nephrostome. Associated with each pronephric tubule in anamniotes may be an external glomerulus. In some dipnoans and in larval urodeles and anurans, the first two or three external glomeruli coalesce to form a single large glomus that lies in a pronephric chamber, an isolated pocket of the coelom (see fig. 15.5).

The number of fairly well-differentiated pronephric tubules is never large—three in larval frogs at the level of somites II, III, and IV; seven in human embryos at the level of somites VII to XIII; and about 12 in chicks commencing at somite V. Additional rudimentary segmental tubules usually appear and regress without reaching full development.

The distal ends of the first several pronephric tubules turn caudad and unite end to end to form a caudally directed duct that eventually establishes an opening into the cloaca. This is the **pronephric (archinephric) duct** (see fig. 15.5).

Pronephric tubules are temporary. As additional, more complicated tubules form farther caudad, the pronephric glomeruli lose their connection with the dorsal aorta and regress. Tubules regress more slowly, and traces may remain in some adult fishes. In teleosts, the site of the embryonic pronephros is converted into a lymphoid organ housing components of the adrenal complex (see fig. 18.9, teleost). *The pronephric duct does*

not regress. It continues to drain the mesonephric tubules that are forming farther back.

Although pronephric tubules develop in all craniates, they function only in embryonic ray-finned fishes and in larval agnathans, dipnoans, and amphibians, and they atrophy during the equivalent of a late larval stage. They were also reported as being functional for a time in turtle and crocodilian embryos; otherwise, in amniote embryos they are vestigial or lost.

Mesonephros

Under the stimulus of the pronephric duct acting as an inductor, additional tubules develop sequentially in the nephrogenic mesoderm behind the pronephric region and establish an entrance into the existing pronephric duct. These are destined to give rise to the **mesonephros.** With regression of the pronephric kidney and development of a mesonephros, the pronephric duct is thereafter called the **mesonephric (archinephric) duct.**

There is usually a gradual transition from tubules characteristic of the embryonic pronephric region to the more complicated nephrons characteristic of a definitive mesonephros. In the transitional zone, the tubules may be segmentally disposed, and in fishes, amphibians, and nonavian reptiles, they have open nephrostomes that usually close later in development. However, nephrostomes remain open in some adult fishes throughout life. Nephrostomes that do not open are common in this transitional region in the embryos of birds and mammals.

Little basis exists for drawing a definite boundary between the pronephros and the embryonic mesonephros, unless it is where one first finds secondary and tertiary tubules forming as buds from the initial segmental tubule in each body segment (see fig. 15.6). As these secondary and tertiary tubules enlarge and encroach on one another, the metamerism of the developing kidney is at first obscured and then lost altogether. The mesonephric tubules remain throughout life to function as the adult kidney (mesonephros) of fishes and amphibians, and it is the functional embryonic kidney of amniotes.

The adult kidney of gnathostome fishes and of amphibians is sometimes termed an **opisthonephros** in recognition of the observation that kidney tubules in anamniotes may develop as far caudad as the cloaca, whereas in amniotes a section of the undifferentiated nephrogenic mesoderm anterior to the cloaca is displaced to give rise to the final (metanephric) kidney of amniotes. (For derivation of the prefix *opistho,* consult the glossary.)

Agnathans

The kidney of the adult hagfish *Myxine* is an expression of the larval holonephros, with modifications. All nephrostomes have closed. The glomerular part occupies a 10-cm segment of the nephrogenic mesoderm commencing some distance behind the regressed pronephros and terminating some distance anterior to the cloaca. This segment consists of 30 to 35 large renal corpuscles up to 1.5 mm in diameter, strictly segmental, and connected to the longitudinal duct by very short tubules. There are no peritubular capillaries and no renal portal system. Between the regressed pronephros and the glomerular segment are a variable number of corpuscles that lack glomeruli or have lost their connection with the longitudinal duct. Caudal to the glomerular segment are additional aglomerular corpuscles. The total number of corpuscles, typical and atypical, is about 70, and they are strictly segmental.

The adult lamprey kidney is a long, thin fold extending about half the length of the coelom and, unlike most other kidneys, hanging into it. Along its free edge is the mesonephric duct. There are large renal corpuscles and more tubules than body segments, but no peritubular capillaries and no nephrostomes.

Gnathostome Fishes and Amphibians

Some of the more anterior mesonephric tubules in male fishes and amphibians become utilized to transport sperm from the testis to the mesonephric duct via the substance of the kidney. This comes about because, instead of associating with a glomerulus, these tubules grow into the immediately adjacent embryonic genital ridge that gives rise to the testis (fig. 15.9). There they

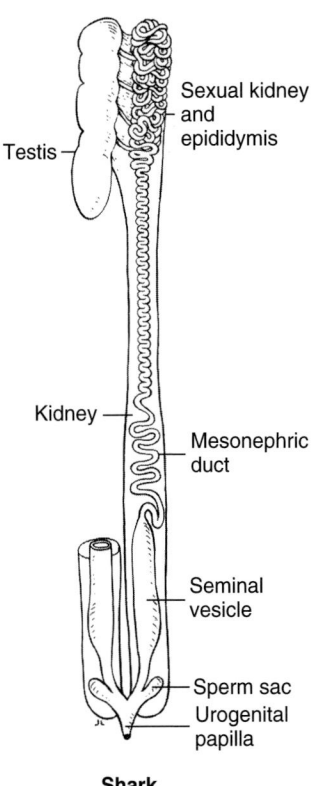

FIGURE 15.9
Urogenital system of a male shark. An accessory urinary duct, not shown, drains the more posterior kidney tubules.

establish continuity with sperm channels within the developing testes, from which they eventually collect sperm. These mesonephric tubules are **vasa efferentia.** This region of the adult male kidney is therefore preempted for sperm transport and is known as the **sexual kidney.** The usually coiled part of the mesonephric duct that drains it is the **epididymis** (figs. 15.9 and 15. 10*a*). The more caudal portion of the male kidney is the major uriniferous portion. There is no sexual kidney in male teleosts because sperm are not conveyed to the kidney (see fig. 15.25*f*).

Accessory urinary ducts frequently supplement, or even replace, the mesonephric duct as a carrier of urine. This is particularly true in elasmobranchs. For an interesting variation in accessory urinary ducts, see figure 15.10*b*, ct.

Amniotes

The mesonephros is the functional kidney of amniotes during part of embryonic or fetal life (fig. 15.11). In embryonic chicks, it reaches a peak on the eleventh day of incubation—halfway through embryonic life. In therian mammals, it peaks earlier (nine weeks of gestation in humans). The first mesonephric tubules in humans appear during the fourth week of embryonic life. A wave

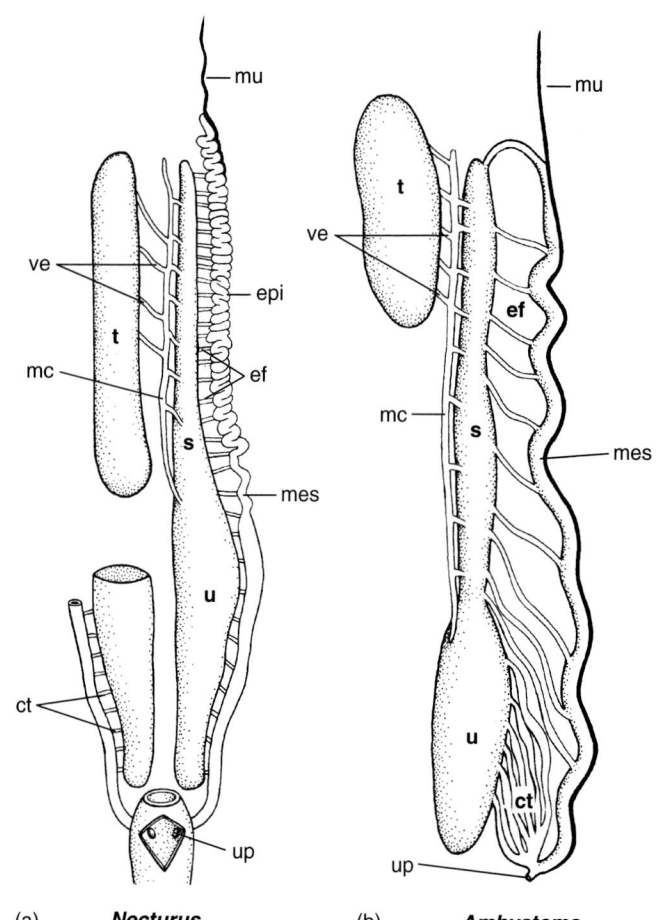

FIGURE 15.10

Left mesonephroi (opisthonephroi) and associated reproductive structures of two male urodeles, ventral view. The testes have been displaced mediad and the mesonephric ducts laterad. The sexual portion of the kidney, **s**, serves solely for sperm transport between testis and mesonephric duct. **ct,** collecting tubules; **ef,** efferent epididymal ducts; **epi,** epididymis (not highly coiled in *Ambystoma*); **mc,** marginal canal, for sperm collection; **mes,** mesonephric duct; **mu,** rudimentary oviduct (muellerian duct); **s,** sexual portion of kidney; **t,** testis; **u,** uriniferous kidney; **up,** urogenital papilla opening into cloaca; **ve,** vasa efferentia.

(a) ***Necturus***   (b) ***Ambystoma***

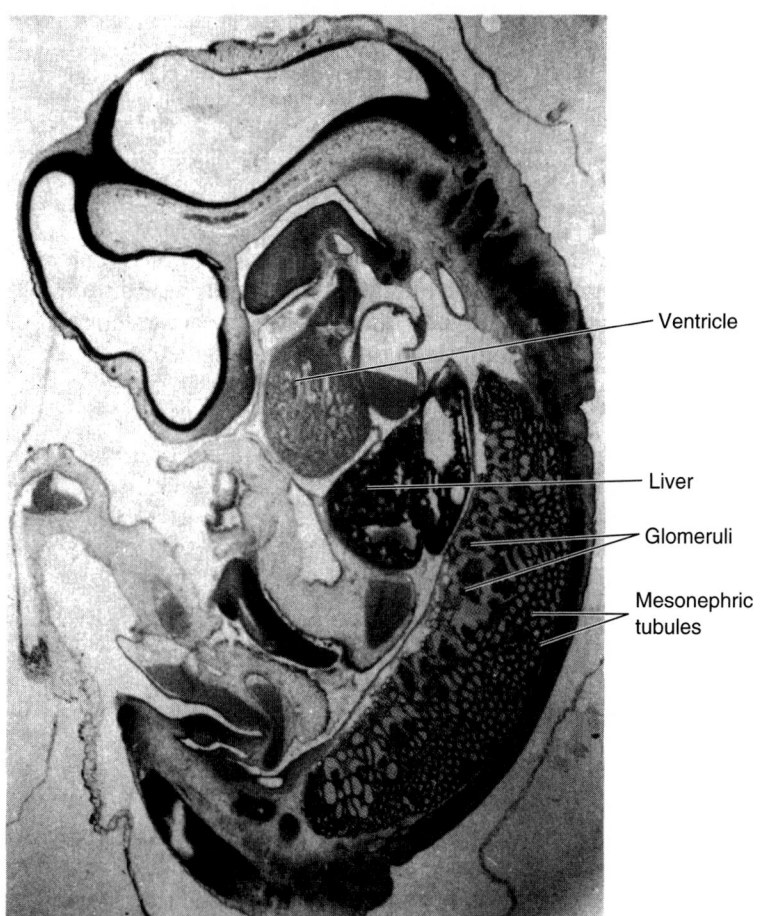

FIGURE 15.11

Mesonephros (*red*) of a 20 mm pig embryo. Parasagittal section.

From Phillips, J. B.: *Development of Vertebrate Anatomy* 1975 (Fig. 18-7, page 440).

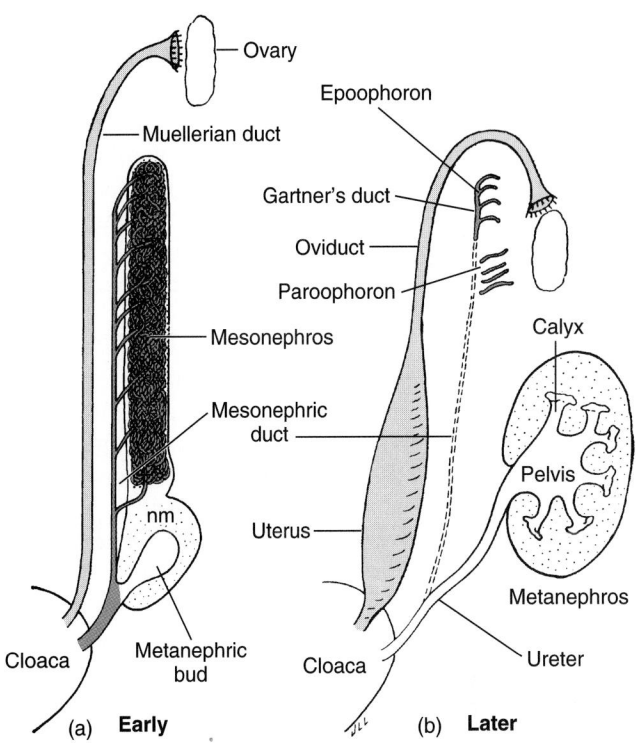

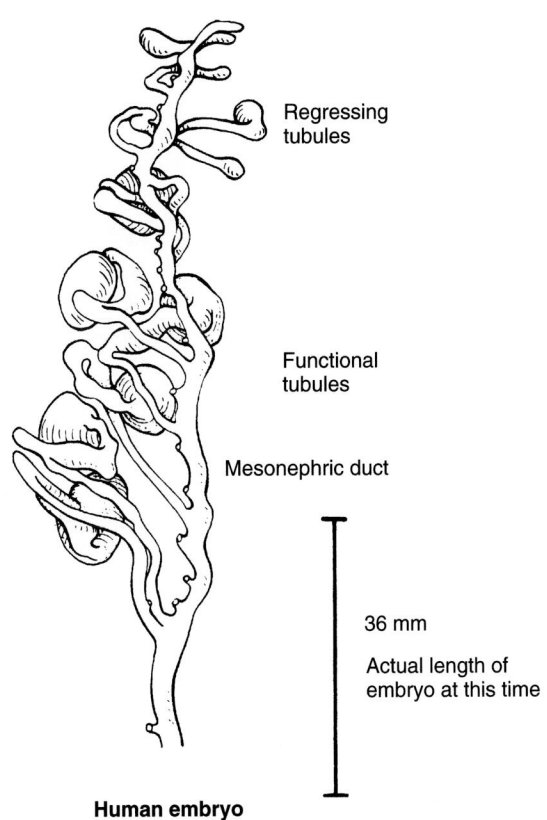

FIGURE 15.12

Right functional mesonephros of a 36 mm human embryo.
From M. D. Altschule, "The Change(s) in the Mesonephric Tubules of
Human Embryos Ten to Twelve Weeks Old" in *Anatomical Record,* 46:81,
1930. Copyright © 1930 American Association of Anatomists. Reprinted by
permission of John Wiley & Sons, Inc.

FIGURE 15.13

Developmental changes in urogenital system of female mammals. In
the early stage **(a)**, the mesonephric kidney and duct are present
(*red*), and the metanephric bud has formed at the caudal end of the
nephrogenic mesoderm **(nm)**. A muellerian duct (oviduct) parallels
the mesonephric duct. In the later stage **(b)**, the mesonephric kidney
and duct have regressed except for remnants (*red*), and the
muellerian duct has differentiated to form a female reproductive
tract.

of differentiation sweeps along the nephrogenic mesoderm, establishing a mesonephros, but before the last mesonephric tubules have formed, the earliest ones have already regressed (fig. 15.12). As a result, the human mesonephros at its peak consists of about 30 functioning renal corpuscles, although as many as 80 have been known to form. During the time the mesonephros is functioning in amniotes, a new kidney, the metanephros, is organizing. When the metanephros is able to function, the mesonephros regresses, except for remnants. However, it functions as late as the first hibernation in some lizards and the first molt in some snakes, and it is still functioning in monotremes and marsupials at birth.

Mesonephric Remnants in Adult Amniotes

Vestiges of the embryonic mesonephros remain in adult mammals as two groups of blind tubules, the **epoophoron** and **paroophoron,** in the dorsal mesentery of the ovary (fig. 15.13), and the **paradidymis** and **appendix of the epididymis,** both near the epididymis in the dorsal mesentery of the testis (see fig. 15.24). The mesonephric ducts remain as sperm ducts in all male amniotes. In females, mesonephric duct remnants in-

clude a short, blind **Gartner's duct** in the mesentery of each mammalian oviduct (fig. 15.13). Blind vestiges of the embryonic mesonephric duct are found in association with the ovaries of some lower amniotes.

Metanephros

The **metanephros,** the adult kidney of amniotes, organizes from the caudal end of the nephrogenic mesoderm, which becomes displaced cephalad and laterad during development (see fig. 15.8, amniote). This is the same mesoderm that gives rise to the caudalmost part of the adult kidney of fishes and amphibians. Differentiation of the metanephric kidney commences when a hollow **metanephric bud** sprouts from the caudal end of the mesonephric duct (fig. 15.13a). Surrounding the bud is nephrogenic mesoderm. The bud grows cephalad, carrying the metanephric blastema along with it. Eventually, the basic components of a metanephric kidney organize in the displaced nephrogenic mesoderm, which continues to enlarge. The hollow stalk connecting the metanephros with the embryonic mesonephric duct becomes the **ureter,** and the end of the stalk surrounded by the developing blastema gives rise to the urinary channels within the kidney up to and including the common collecting tubules. Meanwhile, S-shaped renal tubules are organizing within the blastema. One end of each renal tubule grows toward and encapsulates a glomerulus to form a renal corpuscle; the other end grows toward, and acquires an opening into, a common collecting tubule. Failure to establish this connection may result in a fluid-filled renal cyst. In mammalian kidneys, a **renal pelvis** with funnel-shaped extensions (**calyxes**) collects urine from the common collecting tubules (figs. 15.13*b* and 15.14).

The tubules of mammalian kidneys have a long U-shaped **loop of Henle** of narrow diameter and with very thin epithelia-lined walls. The loop is interposed between proximal and distal convolutions of the tubule (see fig. 15.3*b*). As the loops develop, they grow away from the periphery of the kidney, where the glomeruli are located, and toward the renal pelvis. The kidney therefore consists of a **cortex** housing the glomeruli, convoluted tubules, and the upper ends of the loops of Henle, and of a **medulla** consisting of the loops and collecting tubules (figs. 15.14 and 15.15). Associated with the loops and paralleling each of them are similarly shaped loops of peritubular capillaries, the **vasa recta**. Blood in the vasa recta and filtrate in the loops of Henle flow in opposite directions. The adaptive value of this arrangement is explained in discussions of the counter-

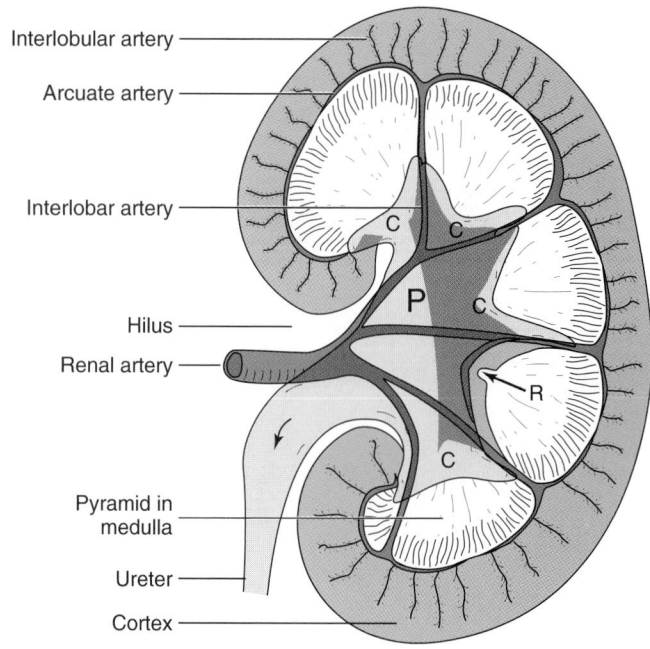

FIGURE 15.14

Mammalian kidney, frontal section, renal vein removed. Glomeruli are confined to the cortex (*red*) and are supplied by interlobular arteries. Loops of Henle, vasa recta, and common collecting tubules constitute the bulk of the pyramids. One calyx has been cut open revealing a Renal papilla (**R**). **P**, pelvis; **c**, calyx.

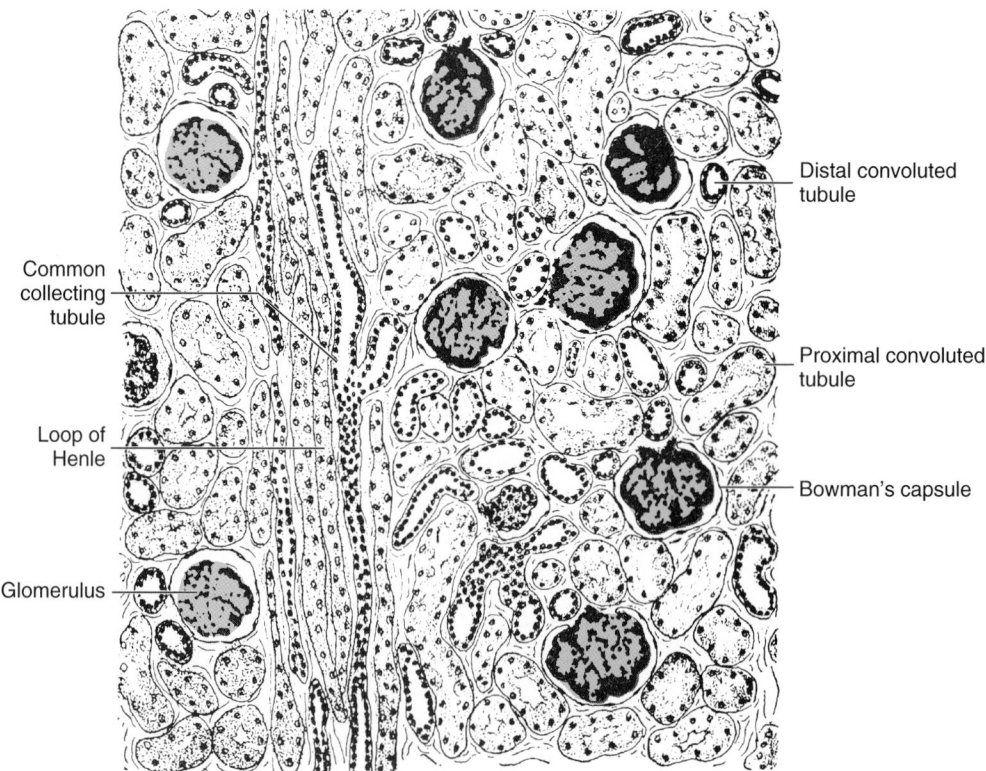

FIGURE 15.15

Cortex of a mammalian kidney, frontal section. The surface of the kidney is near the top.
From G. Bevelander and J. A. Ramaley, *Essentials of Histology*, 8th edition. Copyright © 1979 Mosby-Year Book, Inc. Reprinted by permission of G. Bevelander.

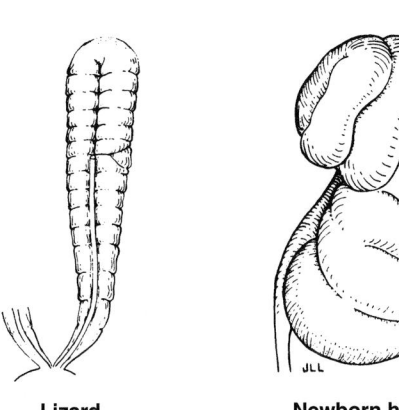

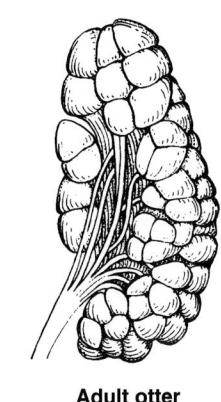

Lizard   **Newborn human**   **Adult otter**

FIGURE 15.16
Lobulated metanephric kidneys.

current multiplier theory in textbooks on physiology. The loops of Henle, the vasa recta, and the collecting tubules are aggregated into **pyramids** to which they give a striated appearance. Each pyramid—as many as 20 in some mammalian kidneys—tapers to a blunt tip, a **renal papilla**, that projects into a calyx along with a few other papillae. The tip of each papilla is perforated by numerous small openings through which urine from the collecting tubules enters the calyx en route to the ureter. Each collecting tubule drains a small number of renal tubules—seven to 10 in humans. Avian kidneys have loops similar to mammalian loops of Henle, but they are much shorter than those of most mammals.

Glomerular filtrate enters the loop of Henle at given concentrations of salt and water. It is the role of the loops of Henle, along with the convolutions and the collecting tubules, to so alter the osmotic pressure of the fluid within the tubule that the urine excreted is either *hyperosmotic* (higher solute concentration) or *hypoosmotic* to the blood, depending on the need to conserve or excrete salts and water. The segments of the tubule from neck to collecting tubule each has its own role. Their collective activity maintains a constant osmotic state of the circulating blood plasma within the narrow limits that will sustain life. Salt moves out of the loop of Henle and into the surrounding interstitial fluid, or it moves from the interstitial fluid into the loop. A difference in osmotic pressure between the two sites and selective tubular permeability results in *passive movement* of salt in one direction or the other. When called upon, hormones cause *active transport* of salt into or out of the loops against an osmotic gradient. Water is reclaimed from the filtrate in the collecting tubules as needed. Excretion of a dilute urine rids the body of excess water.

Reptilian and some mammalian kidneys are lobulated, each lobe consisting of clusters of many tubules (figs. 15.16, 15.28, 15.29). The kidneys of snakes, legless lizards, and apodans are elongated, conforming to the slender body. Those of birds lie snugly against the contours of the sacrum and ilium. In most mammals, the kidneys are smooth and bean shaped, and the arteries, veins, nerves, and ureters enter and leave at a median notch, the **hilus.**

Because of their embryonic origin as buds off the mesonephric ducts, the ureters at first terminate in these ducts, and they do so throughout life in male sphenodons and lizards (see fig. 15.26). In other reptiles and in monotremes, they ultimately open into the cloaca. In therian mammals, the ureters open into the urinary bladder because of differential growth of components of the embryonic cloaca, to be explained later (see figs. 15.30 and 15.31).

Extrarenal Salt Excretion

Craniates that live in an environment laden with salt or that inhabit arid environments and can afford little body water for carrying away accumulated salts have extrarenal structures that supplement the kidneys in this role. Marine teleosts have chloride-secreting glands on the gills, and elasmobranchs and coelacanths have chloride-secreting rectal glands that utilize little water and empty into the caudal end of the intestine. (The rectal glands of bullsharks caught in freshwater are smaller than those caught in salt water and show regressive changes.)

Marine turtles and snakes have salt glands. So do iguanas that live in trees on the shore but feed on seaweed; and so, too, do birds that scoop fish out of seawater. Terrestrial lizards and snakes that inhabit arid environments also have them. These same lizards and snakes have atrophied glomeruli, which also conserves water. The salt glands of reptiles are compact and sometimes lobulated and consist of an abundance of tubular crypts that are drained by a single duct.

The salt-excreting glands of lizards are located below the orbits just external to the olfactory capsule, and their ducts empty into the nasal canals. Whitish incrustations of sodium chloride and potassium can be seen at the nostrils. Those of marine birds are large and are located above each orbit. Their long ducts open at the base of the beak just lateral to the nostrils. A groove on the beak extends from this opening to the tip of the beak. Within 15 minutes after these birds have drunk water containing sodium and potassium salts, minute drops of fluid containing these trickle down the groove and drip or are shaken off the beak. The salt glands of sea turtles open into the orbit, and those of sea snakes open into the oral cavity in the fashion of salivary glands.

The copious mucous coat of hagfishes, which inspired the name slime eels, has been shown to be rich in salt; however, hagfish appear to be isoosmotic relative to their environment. Hagfishes have very short kidney tubules, which deprives them of one route for salt excretion. It may be that integumentary mucous glands play a primary role in osmoregulation in those agnathans.

Sweat glands in mammals eliminate salt, but salt loss by this route is incidental to secretion of water for its evaporative cooling effect. Whereas the excretion of salt by some routes is hormone regulated, salt loss via sweat glands is unregulated. In fact, salt lost by this route, unless in excess in the tissues, must be replaced by dietary salts.

URINARY BLADDERS

Urinary bladders are adaptations to life on land. They serve as reservoirs for water that may be needed later and therefore should not be wasted. Urinary bladders arise during embryonic development as an evagination from the ventral wall of the embryonic cloaca; in amphibians, turtles, basal lizards, and monotremes—the only nontherian tetrapods that have a urinary bladder excepting a few ratite birds—the bladder continues to open into the cloaca throughout life (figs. 15.17 and 15.38, amphibians; 15.26, 15.27b, 15.40, and 15.41, reptiles; and 15.43, monotreme). Because the ureters also open into the cloaca in these tetrapods, urine backs up into the bladder from the cloaca unless the vent is open.

The urinary bladder of therian mammals, like that of all other tetrapods, develops as an evagination from the embryonic cloaca, but the evagination in therian mammals is called an **allantois**. The ureters of therian mammals empty into the urinary bladder. The bladder is drained by the urethra.

The distal tip of the urinary bladder of therian mammals is connected to the umbilicus (navel) by a **middle umbilical ligament**, or **urachus**. This is a fibrous remnant of the part of the embryonic allantois that persists within the coelom distal to the bladder after birth (see fig. 14.38). The umbilicus and the urachus are more prominent in some mammals (primates, for instance) than in others (cats, for example). The urachus occupies the anterior border of the ventral mesentery of the bladder along with the obliterated umbilical arteries. The bladders of amphibians may also exhibit a persistent ventral mesentery. This mesentery, the lesser omentum, and the ventral mesentery of the liver (falciform ligament) are the sole remnants in adult craniates of the ventral mesentery that, in embryos, extended the entire length of the coelom.

Active water reabsorption from the bladder, a water-conservation adaptation, is evoked by **antidiuretic hor-**

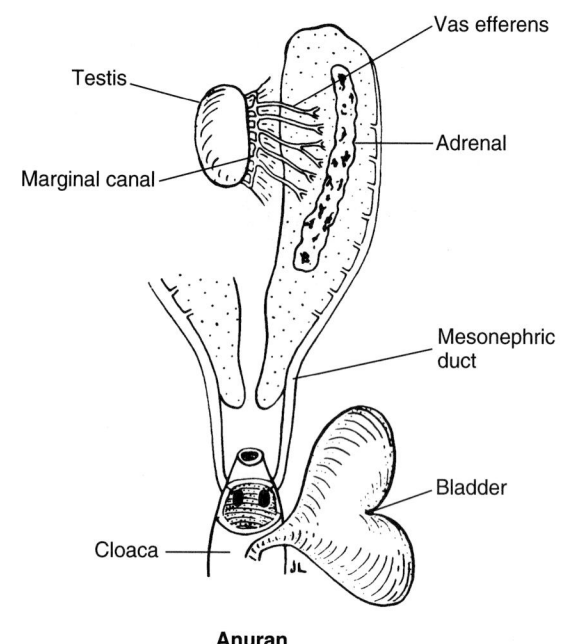

FIGURE 15.17
Urogenital system and adrenal of a male frog, ventral view.

mone. This is a neurosecretory hormone from the hypothalamus that is released from the posterior lobe of the pituitary when dehydration threatens (see fig. 18.2). In one species or another, stored urine is put to an additional use. Some freshwater turtles have voluminous bladders and, sometimes, accessory bladders that are used by females to carry water for softening and moistening the soil when a nest is being prepared for eggs (see figs. 15.27 and 15.41). Urinary bladders in mammals make it possible for pheromones in the urine to be deposited at sites where they can serve as a signaling device to other members of the species, male or female. (Fire hydrants seem popular in cities.) In some basal craniates, essential ions that are scarce in the specific environment are also reclaimed from stored urine.

The smooth muscle layers in the bladder wall are disposed in a reverse pattern to that in the digestive tract, having inner longitudinal and outer circular layers. Much elastic tissue is interspersed among the muscle fibers, and the cells of the epithelial lining slide over one another, thinning out as the bladder fills and heaping as it empties. Consequently, the size of the lumen is at all times just sufficient to accommodate the contained urine, whatever the quantity.

Organs called urinary bladders in fishes, when present, arise from nephrogenic mesoderm rather than from the cloaca. They are not homologues of tetrapod bladders. Some ray-finned fishes have an unpaired, sometimes bicornuate sinuslike enlargement at the caudal junction of the two mesonephric ducts. In some species, it develops into an elongated sac termed a **tubal bladder**

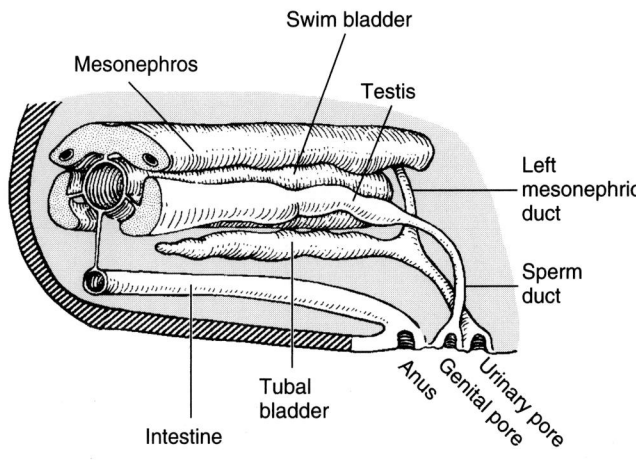

Teleost

FIGURE 15.18

Caudal end of urogenital system of a male teleost (pike), left lateral view. The unpaired urinary bladder arises as a bud off the conjoined caudal ends of the two mesonephric ducts. Note absence of cloaca.

Source: Data from E. S. Goodrich, *Studies on the Structure and Development of Vertebrates,* 1930, Macmillan and Company, Ltd., London.

(fig. 15.18). Lobe-finned fishes have a small diverticulum from the dorsal wall of the cloaca that has been called a urinary bladder. Living agnathans and cartilaginous fishes have no such structures.

GENITAL ORGANS

Gonadal Primordia

Gonads have two major functions: (1) to produce gametes and (2) to synthesize steroidal hormones. The hormones are essential for the differentiation, growth, and maintenance of **accessory sex organs** (reproductive ducts and their glands), **secondary sex characteristics** (sex-specific plumage in birds, for example), and sexual behavior.

Gonads arise as a pair of embryonic **genital ridges**, which are thickenings of the coelomic mesothelium just medial to the mesonephroi (fig. 15.19, *black*). The ridges are longer than the resulting mature gonads, which suggests that at one time gonads may have extended the length of the pleuroperitoneal cavity, as they do in living agnathans. **Primordial germ cells** (progenitors of sperm and eggs) migrate into the genital ridges and establish themselves just internal to the peritoneum (fig. 15.20). This region of a

gonadal primordium thereby becomes the **germinal epithelium.** Although primordial germ cells are first observed in the germinal epithelium, their progenitors are in the endoderm of the yolk sac and elsewhere before genital ridges have developed. From these locations, they migrate to the genital ridges.

During early differentiation, the gonads have the potential to become either ovaries or testes. Under the influence of genes, hormones, and factors not yet fully identified, these bisexual gonads develop into the gonads of a specific sex. As gonads enlarge, they generally acquire a dorsal mesentery, the **mesorchium** in males, the **mesovarium** in females.

In gonadal primordia that are becoming testes, **primary sex cords** invade the deeper region of the genital ridge and become **seminiferous tubules** lined with germinal epithelium (figs. 15.20*a–d,* 15.21, and 15.22). These tubules produce sperm. Primary sex cords form also in females, but they are not the source of ova. Oocytes form from **secondary sex cords** that also arise from the germinal epithelium (see fig. 15.20*b, e,* and *f*). These cords do not grow deep into the medulla. Consequently, with the exception of some teleosts, eggs form just internal to the surface of the ovary and are shed to the exterior when mature (see fig. 15.35). The primary

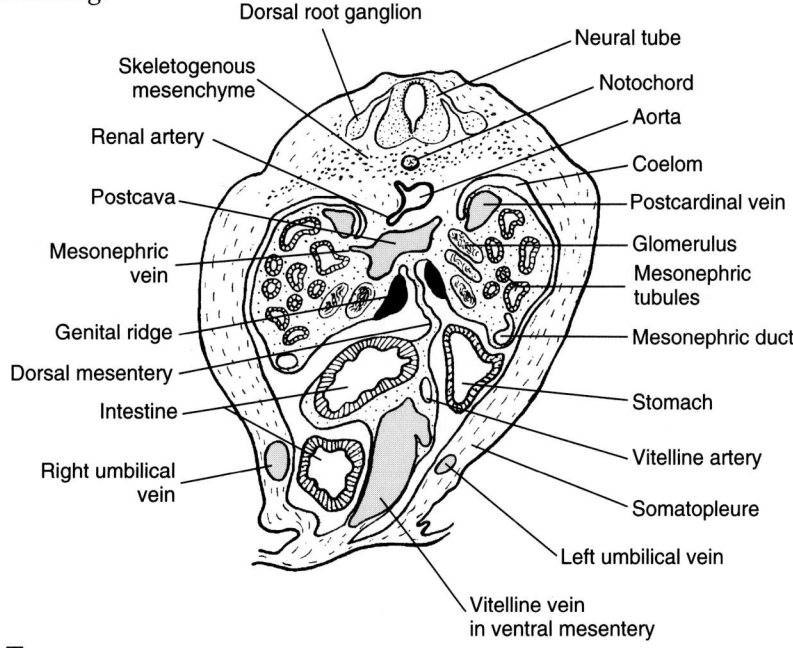

Actual length of above embryo (8 mm)

Actual length of opossum at birth (10 mm)

FIGURE 15.19

Opossum embryo (*Didelphis*) 6½ days after cleavage, cross section. The gonadal primordia are shown in *black.*

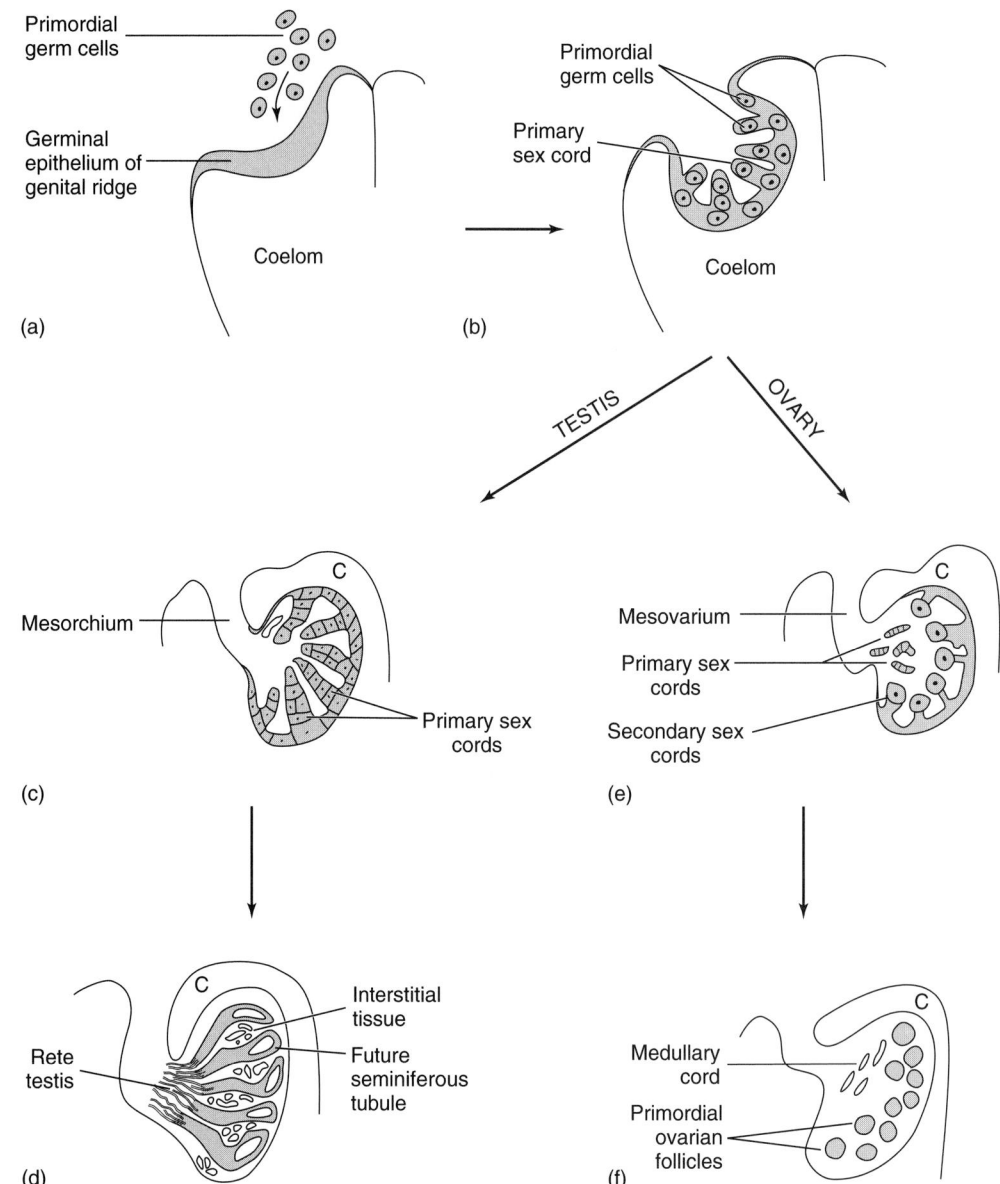

FIGURE 15.20
Early development of mammalian gonads.
(*a*) Germinal epithelium on genital ridge of sexually indifferent primordium.
(*b*) Primary sex cords of bisexual stage.
(*c*)–(*f*) Differentiating testis and ovary. **C,** coelom.

sex cords of females regress, but remnants remain deep in the ovarian medulla, where they are called medullary cords.

A few adult craniates have a single gonad because of fusion of the two ridges across the midline, as in lampreys and a few teleosts, or because one of the juvenile gonads, most commonly an ovary, fails to differentiate. Among craniates with a single functional ovary are hagfishes, some viviparous elasmobranchs, a few nonavian reptiles, and most species of birds. A few female mammals, including the platypus and some bats, have only one ovary.

Instances of natural or experimental sex reversal can be demonstrated in many craniates. The embryonic testis of anurans consists of an anterior segment, **Bidder's organ,** which in frogs usually disappears before sexual maturity, and a more caudal portion that becomes the functional testis. Bidder's organ persists in adult male toads (fig. 15.23) and contains large undifferentiated cells resembling immature ova. If the testes of a toad are removed experimentally and Bidder's organs are allowed to remain with a blood supply, the latter develop into functional ovaries, and the rudimentary female duct system enlarges under the influence of female sex hormones produced by the new ovaries. Hens that have not been subjected to any treatment have been known to cease laying eggs, crow, and develop other male characteristics. This comes about when the left ovary atrophies and the right ovary, which is rudimentary in birds, enlarges and produces male hormones. However, these birds are not known to have produced sperm.

In all craniates, mature *eggs are released into the coelom* or a compartment of it before being swept into an oviduct. *Sperm,* on the contrary, except in living agnathans, *are conducted in a closed system of vessels* that conveys them from the testis to the exterior without

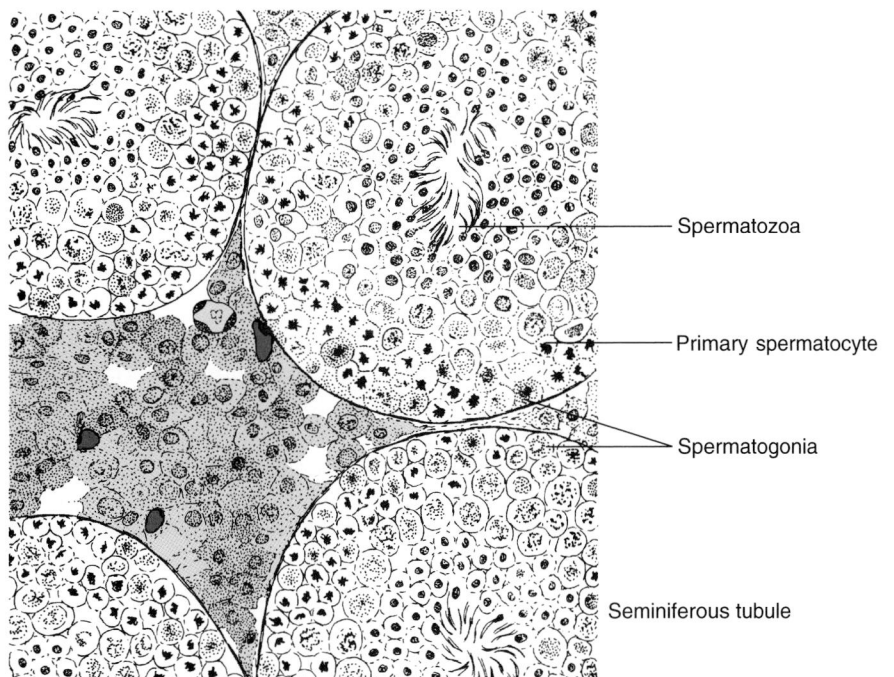

FIGURE 15.21
Section of mammalian testis. Portions of four seminiferous tubules are shown. Interstitial cells (*light red*) are source of male sex hormones (chapter 18).

Spermatozoa

Primary spermatocyte

Spermatogonia

Seminiferous tubule

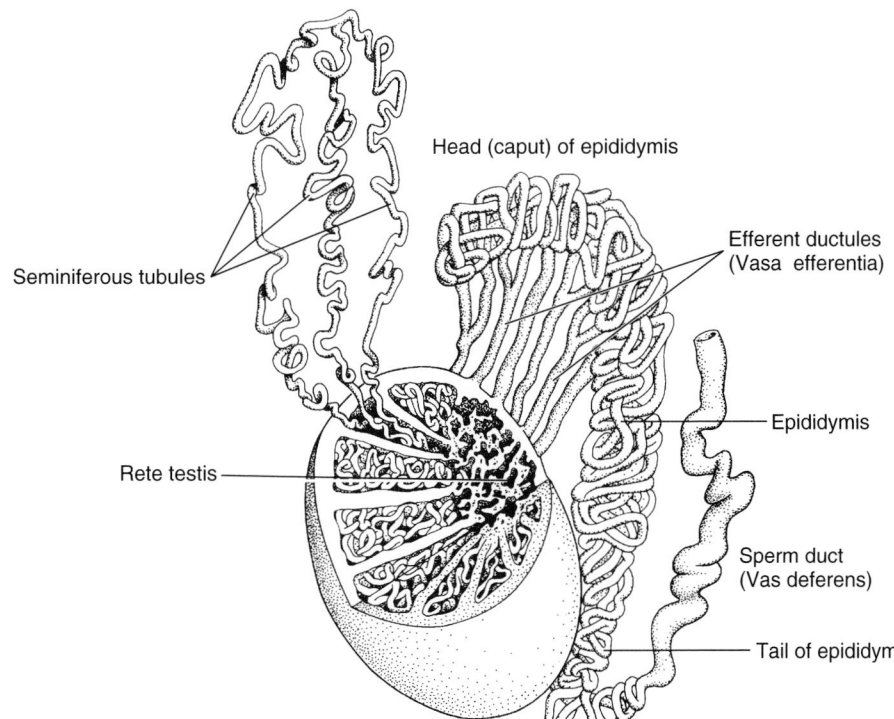

Head (caput) of epididymis

Seminiferous tubules

Efferent ductules (Vasa efferentia)

Epididymis

Rete testis

Sperm duct (Vas deferens)

Tail of epididymis

FIGURE 15.22
Mammalian testis. The efferent ductules (vasa efferentia) are modified mesonephric tubules.

having entered the coelom. All living agnathan gametes are shed into the coelom.

Testes and Male Genital Ducts

Craniate testes are essentially the same with few exceptions. Germinal epithelium lines the seminiferous tubules where sperm are formed. Mature sperm are mi-

croscopic and abundant (3.5 billion/cc in roosters). They separate from the lining of the tubules and, propelled by flagellumlike tails, swim the length of the tubule to reach the vasa efferentia, which leads to the sperm duct. In mammals, sperm are first collected in a network of fine channels, the **rete testis**, before entering the vasa efferentia (see fig. 15.22). Vasa efferentia are mesonephric tubules that invaded the developing testis instead of becoming associated with glomeruli (fig. 15.24). The vasa efferentia (usually called **efferent ductules** in mammals) conduct sperm to the spermatic duct. Efferent ductules number a dozen, more or less, in humans. In most craniates with **mesonephric kidneys**, the mesonephric ducts carry *both urine and sperm* (fig. 15.25a and d).

Hagfishes, lampreys, some jawed fishes, and urodeles lack seminiferous tubules. In these craniates, the germinal epithelium may be on the surface of the testis or deep within, and primordial germ cells migrate from the germinal epithelium into cystlike **seminiferous ampullae**, where they mature before being discharged. At the end of the spawning season, after hundreds of thousands of spermatozoa have been shed from each ampulla, the cysts collapse.

In some fishes and urodeles, there has been a tendency to form a new sperm duct to replace the mesonephric duct as a carrier of sperm (see figs. 15.10, **marginal canal**; and 15.25d and e). In teleosts, this has culminated in a mesonephric duct that carries no sperm whatsoever (fig. 15.25f). *In all other fishes and amphibians, and from reptiles to humans, the embryonic mesonephric duct remains in adult males to carry sperm.* A duct that carries only sperm is termed a **vas deferens (ductus deferens)**.

Spermatic ducts generally empty into the cloaca or a derivative thereof in craniates excluding therian mammals (figs. 15.26, 15.27, 15.28, 15.29). In therian mammals, they empty into the urethra at the prostate gland (fig. 15.30). This development is a result of the complete separation of the embryonic cloaca into a urogenital sinus and rectum (see fig. 15.49e and f, duct in *black*).

As a result of caudal migration of the testes in mammals during late fetal life, the spermatic ducts become "caught" or "hung up" on the ureters, so that thereafter they loop over the latter en route to the urethra (figs. 15.30 and 15.31).

A number of multicellular glands produce secretions that contribute to seminal fluid in amniotes. These are most abundant in mammals, in which one or more open

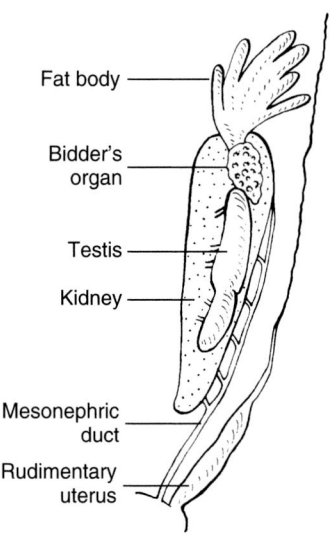

Male toad

FIGURE 15.23
Bidder's organ and the rudimentary female reproductive tract of a young male *Bufo*, ventral view. Only the left organs are illustrated. The mesonephric duct and rudimentary uterus empty into the cloaca.

FIGURE 15.24
Developmental changes in the urogenital system of male mammals. In the early (bisexual) stage, there is a rudimentary female duct (muellerian duct) and a mesonephros. Some of the mesonephric tubules have invaded the genital ridge (testis) to become vasa efferentia. Later (*right*), the muellerian duct has regressed (*broken lines*), the mesonephros has regressed except for remnants (appendix of epididymis, paradidymis), and the mesonephric duct remains to carry sperm. **e,** epididymal portion of mesonephric duct.

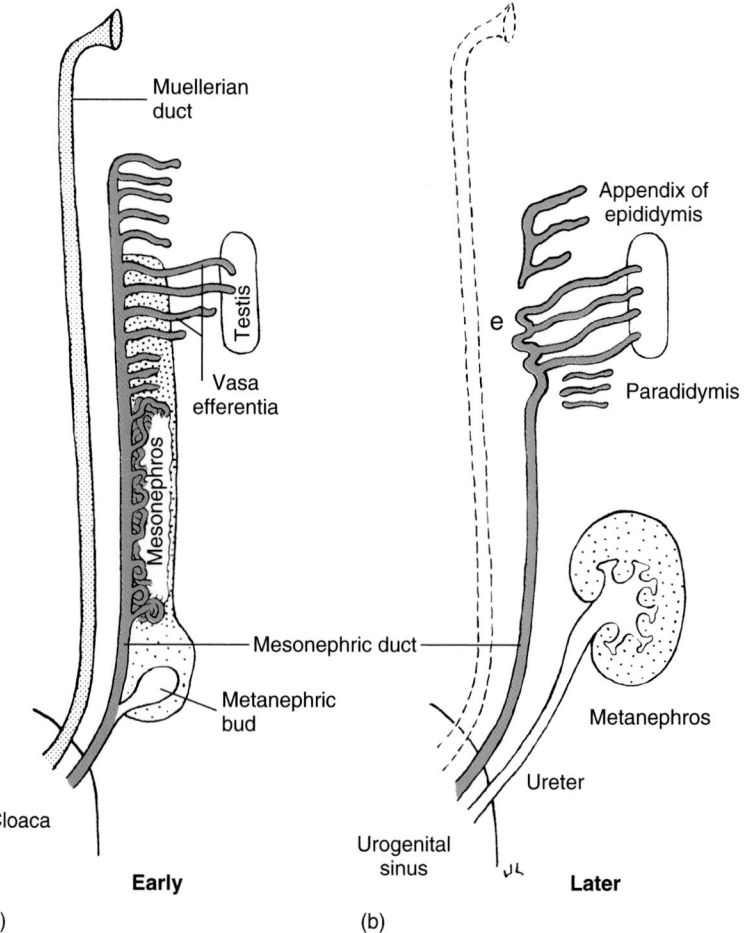

(a)

(b)

into the genital tract at or near the junction of the spermatic ducts and the urethra (fig. 15.31). Among these are **ampullary glands**, several median and paired **prostate glands**, **seminal vesicles**, **coagulating glands**, and **bulbourethral (Cowper's) glands**. Not all are present in every mammalian order. Seminal vesicles have only a glandular role in mammals, although they serve as storage sites for sperm in other craniates. Coagulating glands cause semen to coagulate in the vagina. This results in a **copulation plug** that precludes another mating for several hours or days. Other glands produce mucus, nutritive secretions, or fluids that neutralize the slight acidity of the vaginal contents, thereby providing a more suitable milieu for motile sperm.

The urethra in male mammals is termed the **prostatic urethra** where the prostate glands empty, **membranous urethra** from the prostate glands to the base of the penis, and **spongy urethra** within the penis (see fig. 15.30).

Intromittent Organs

In species in which fertilization is internal, males with few exceptions develop **intromittent (copulatory) organs** for introducing sperm into the female reproductive tract. These are present in fishes with internal fertilization, in the anuran family Ascaphidae (frogs lacking eardrums), in reptiles including a few birds, and in all mammals. In urodeles with internal fertilization and in most birds, the cloaca is eversible, facilitating sperm transfer. Internal and external fertilization are discussed in chapter 5.

The intromittent organs of elasmobranchs are grooved fingerlike **claspers**, which are modified pelvic fins (see fig. 10.8*c*). Embedded in each fin at the base of some claspers is a muscular **siphon sac** that adds copious quantities of an energy-rich mucopolysaccharide to the semen. In many teleosts, a modified anal fin, or **gonopodium**, serves for sperm transfer. In the frog *Ascaphus*, the intromittent organ is a permanent tubular, taillike extension of the cloaca.

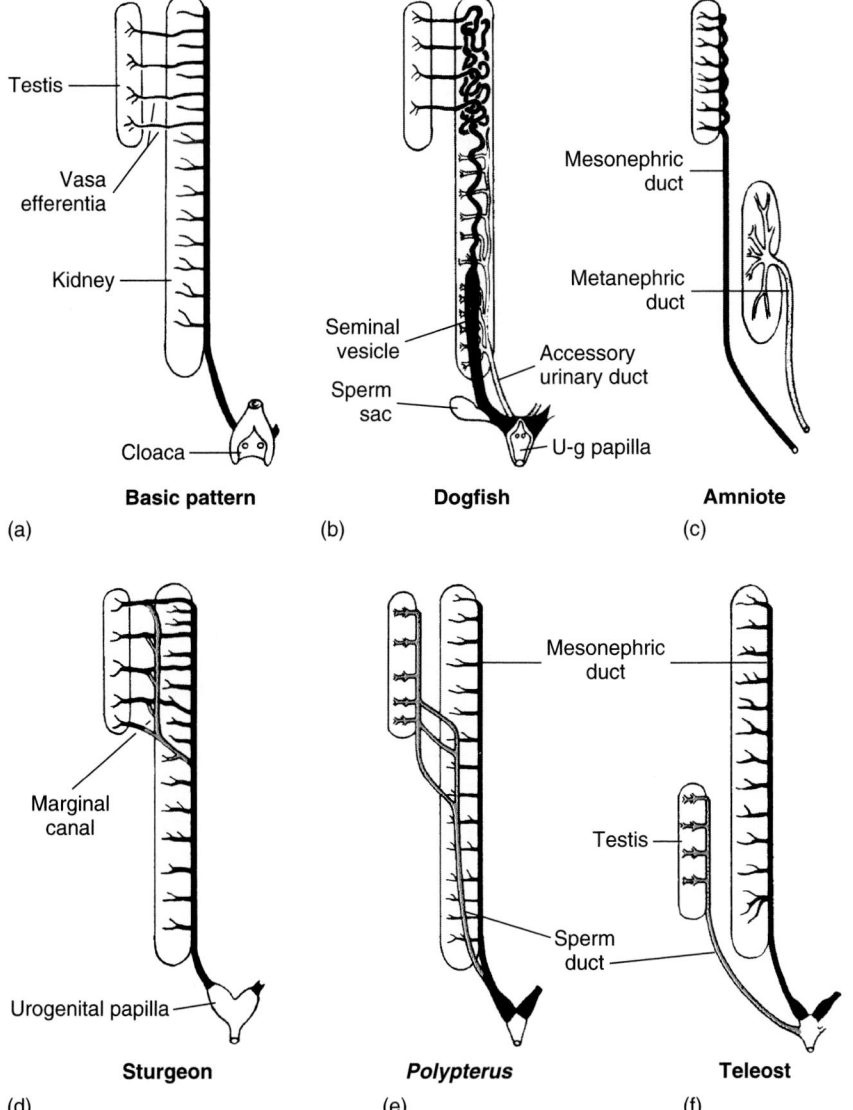

FIGURE 15.25

The mesonephric duct (*black*) as a carrier of sperm and urine. (*a*) Carrying both sperm and urine. (*b*) Carrying urine from the anterior part of the kidney only; chiefly a spermatic duct. (*c*) Carrying sperm only. (*d*) to (*f*) Trend toward a separate sperm duct (*red*) in ray-finned fishes, the mesonephric duct ultimately carrying only urine. Note the reverse in amniotes.

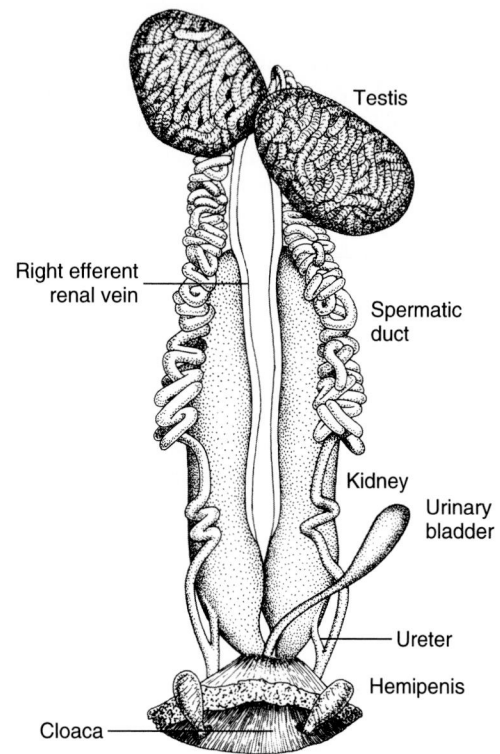

Testis

Right efferent
renal vein

Spermatic
duct

Kidney

Urinary
bladder

Ureter

Hemipenis

Cloaca

FIGURE 15.26

Urogenital organs of the male lizard *Anolis carolinensis,* ventral
view. The sperm duct is the persistent mesonephric duct. The
hemipenes are seen in the everted (erected) position.

Male snakes and lizards have hemipenes, a pair of protrusible saclike diverticula stored in an inverted position
in pockets under the skin at the entrance to the cloaca
(see fig. 15.26). They are held in the pocket by retractor
muscles. Each hemipenis has a deep spiral groove on its
surface. During copulation, the retractor muscles relax,
the sacs turn inside out and protrude on the surface, and
semen flows along the groove and into the cloaca of the
female. Rudimentary hemipenes are present in females.

Male turtles, crocodilians, a few birds (ostriches,
ducks, others) and monotremes exhibit an unpaired
erectile **penis.** In its simplest form, as seen in reptiles,
the penis is a thickening of the floor of the cloaca consisting chiefly of spongy erectile tissue, the **corpus
spongiosum,** which has a **urethral groove** on its surface
and ends in a **glans penis** (see figs. 15.27 and 15.28).
The spongy tissue consists of cavernous blood sinuses
that, when engorged, cause the glans penis to be extruded through the cloacal aperture. The glans is richly
supplied with sensory endings that reflexly stimulate
ejaculation. The urethral groove channels sperm toward
the cloaca. The penis of monotremes is similar to that of
reptiles, being in the cloacal floor, but the groove on the
dorsal surface has become folded into the corpus spongiosum to form a tube, and it carries only sperm, urine
being conveyed to the cloaca by a separate passageway.

The penis in therian mammals develops from an embryonic **genital tubercle** and a pair of **genital folds** that
border the urogenital sinus in the embryos of both sexes
(see fig. 15.48*b*). The folds grow toward one another in
the midline to form a groove and, eventually, a closed
tube within which is the urogenital sinus. This tube,
surrounded by the corpus spongiosum, becomes the
spongy urethra within the penis. It eventually carries
urine and sperm. The glans penis, the expanded distal
end of the corpus spongiosum, is ensheathed by a fold
of skin, the **prepuce,** except during erection. Two additional erectile masses, the **corpora cavernosa,** develop
within the penis of therian mammals (fig. 15.32).

The genital tubercle of females generally develops
into a spongy, penile prominence, the **clitoris,** embedded in the floor of the urogenital sinus (vestibule of the
vulva as a derived condition in some primates). Like the
penis, the clitoris is erectile, sensitive to tactile stimulation, and ensheathed by a prepuce.

The fate of the clitoris of female rodents and hyenas
is slightly different. In rodents, it develops into a closed
tube, as in males, and becomes a urinary papilla (see
fig. 15.48*c*). Like the penis, it encloses the terminal segment of the urethra and conducts urine to the exterior.
In female hyenas, the clitoris also becomes a tube, but
in these mammals it encloses an extension of the urogenital sinus that, in most mammals, is a common terminal passageway for both the urinary and genital systems (see fig. 15.49*c*). Therefore, in hyenas, the clitoris
carries urine, enables copulation, and serves as a birth
canal. Yet it looks precisely like a male penis. Perhaps
this is what hyenas are laughing about.

Ovaries

Although ovaries are usually thought of as being compact organs like those of sharks and therian mammals,
this is not the case in a large number of female craniates. The ovaries of many teleosts have a central cavity
(fig. 15.33). Those of anurans and urodeles are folded
thin-walled sacs, and those of some nonavian reptiles
and of birds and monotremes exhibit numerous internal, irregular, fluid-filled lacunae. On the other hand,
the ovaries of fishes other than teleosts are generally
compact, as are those of turtles, crocodilians, and therian
mammals.

The hollow ovary of teleosts results from entrapment
of a small part of the coelomic cavity within the developing ovary (fig. 15.34). Because of this, the cavity—an
isolated pocket of coelom—is lined by the germinal epithelium; the small eggs or, in viviparous teleosts, the
young, are discharged into that cavity. The latter may be
directly continuous with an oviduct (see fig. 15.33) or
partially surrounded by an oviducal funnel composed of
peritoneal folds that have become so arranged as to
form a tube. The ovary of some other teleosts is also
hollow but for a different reason: It results from a

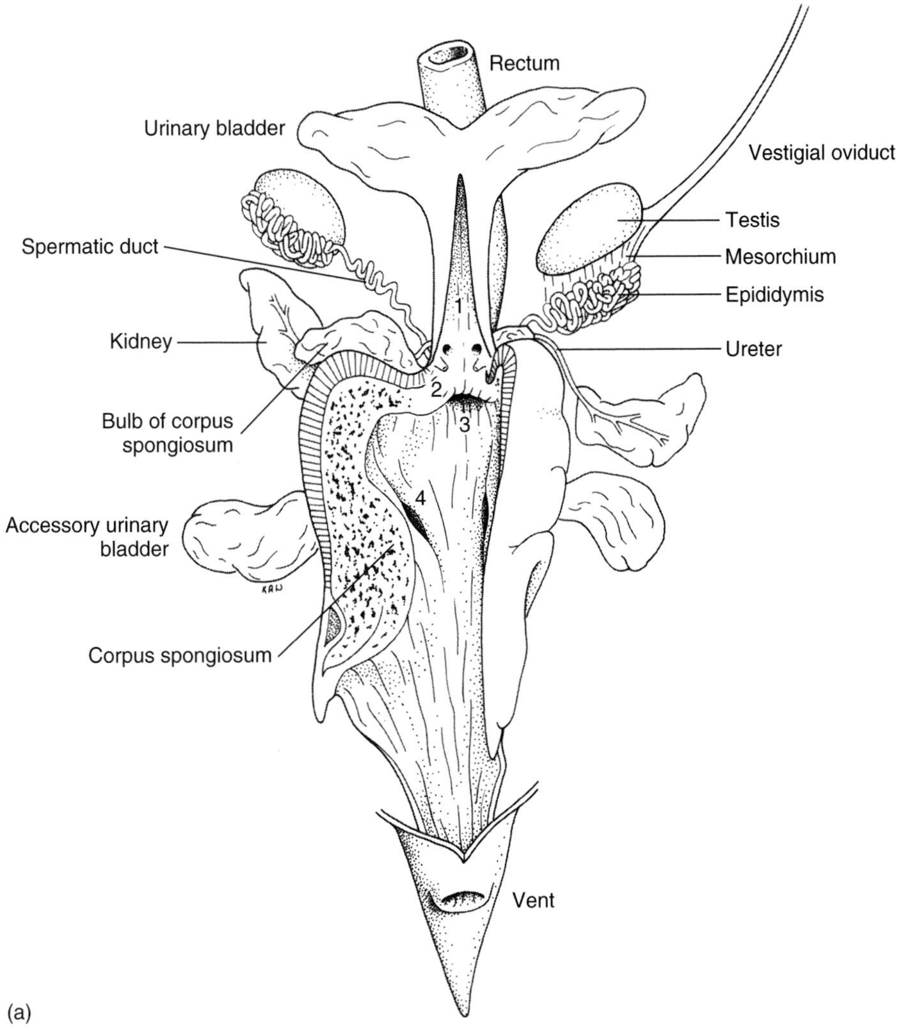

Rectum

Urinary bladder

Vestigial oviduct

Spermatic duct

Testis

Mesorchium

Epididymis

Kidney

Ureter

Bulb of corpus
spongiosum

Accessory urinary
bladder

Corpus spongiosum

1

2

3

4

Vent

(a)

Spermatic duct

Urinary bladder

Bulb of corpus
spongiosum

Ureter

Urethral groove

Accessory urinary
bladder

Corpus spongiosum

Site of vent

Glans penis

(b)

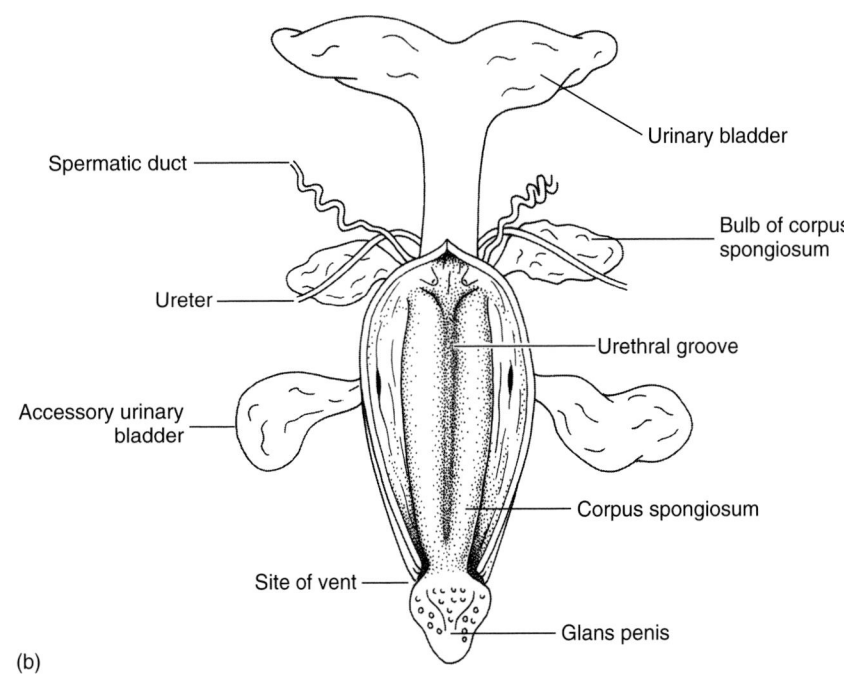

FIGURE 15.27

Urogenital system of a male turtle. (*a*) Ventral
view. The floor of the cloaca has been
sectioned longitudinally and reflected to the
observer's left. The ventral wall of the urinary
bladder has been opened at its junction with
the cloaca. The kidneys have been displaced
caudad from their normal position dorsal to
the testes. **1,** opening from ureter; **2,** genital
papilla from spermatic duct; **3,** entrance of
rectum; **4,** opening from accessory urinary
bladder. (*b*) The cloacal floor viewed from
above, showing the penis extruded. The
rectum has been removed.

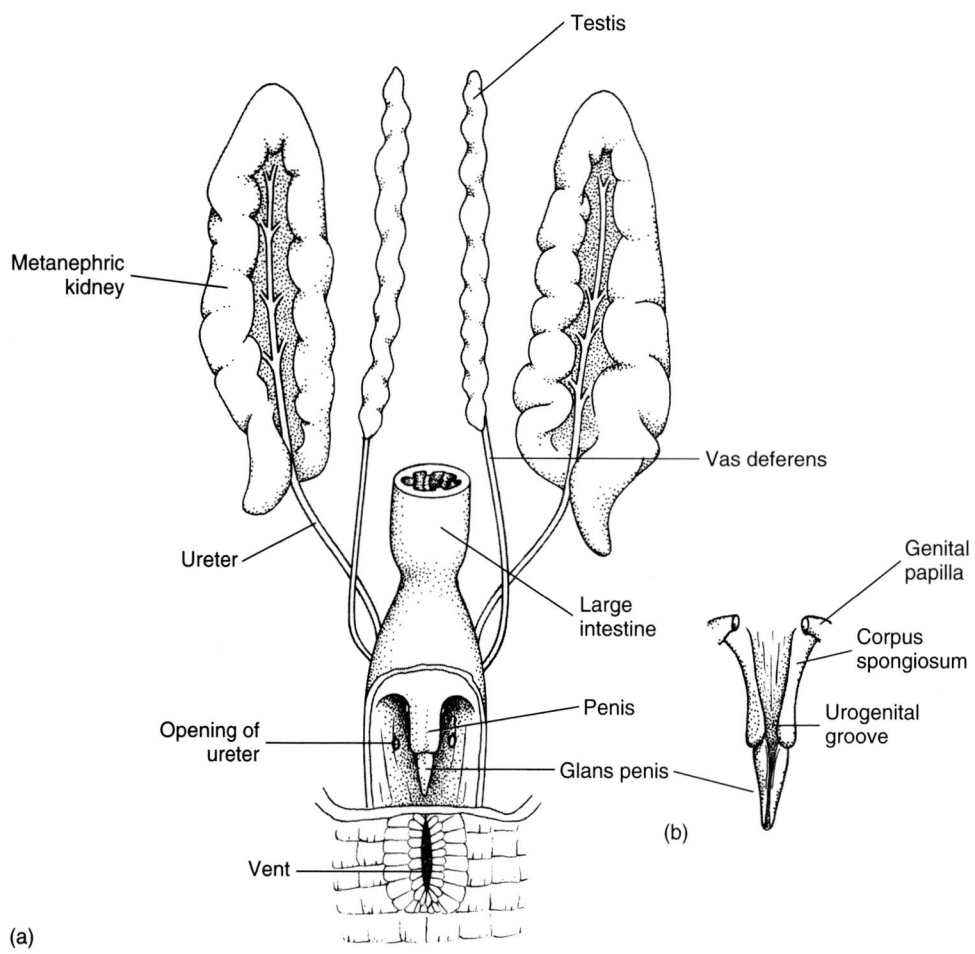

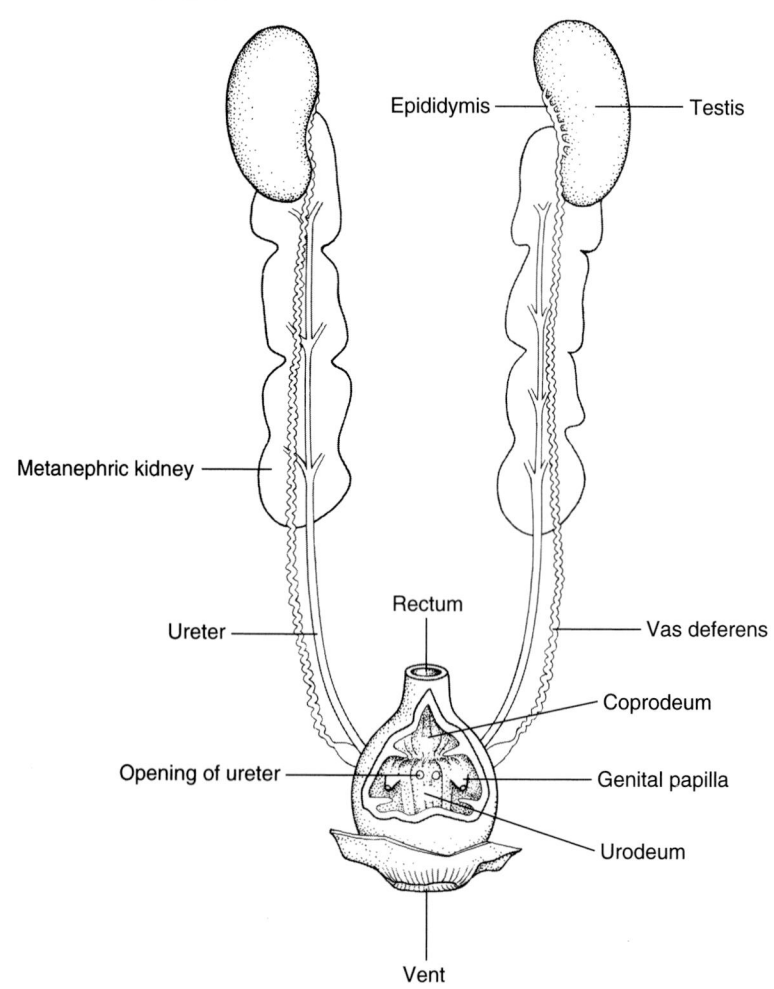

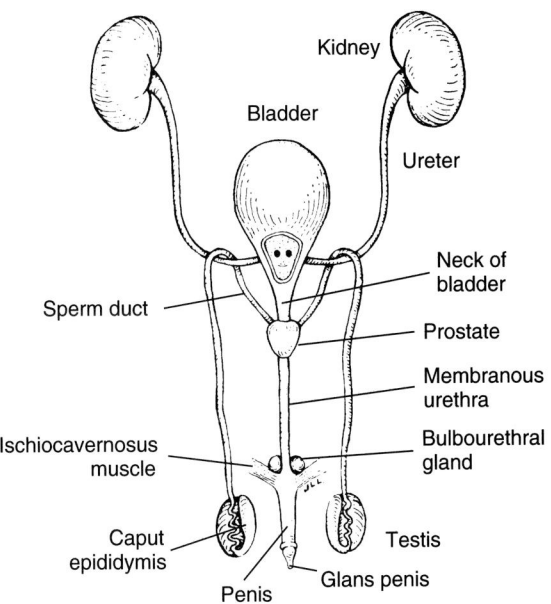

FIGURE 15.30
Urogenital system of a male cat, ventral view.

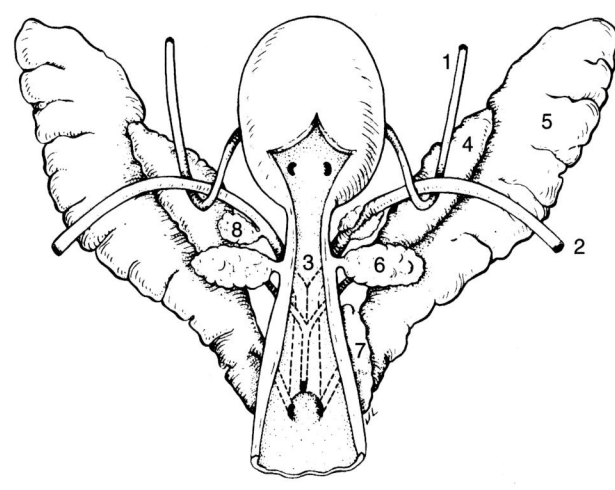

FIGURE 15.31
Secondary sex glands of a male hamster, ventral view. The bladder and urethra have been opened to show entrances of ducts.
1, ureter; **2**, spermatic duct; **3**, prostatic urethra; **4**, coagulating gland; **5**, seminal vesicle; **6**, cranial prostate; **7**, caudal prostate; **8**, ampullary gland. A bulbourethral gland enters the urethra farther caudad.

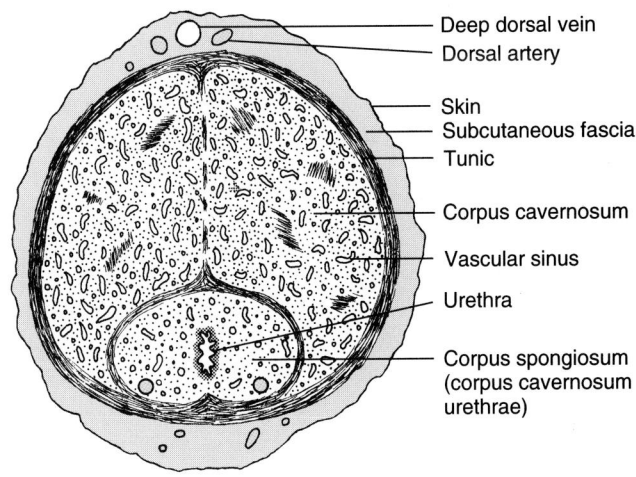

FIGURE 15.32
Proximal cross section of a fetal primate penis.

FIGURE 15.33
Female reproductive system of a teleost. Ova are shed into the ovarian cavity. In some teleosts, the ovary reaches almost to the genital pore.

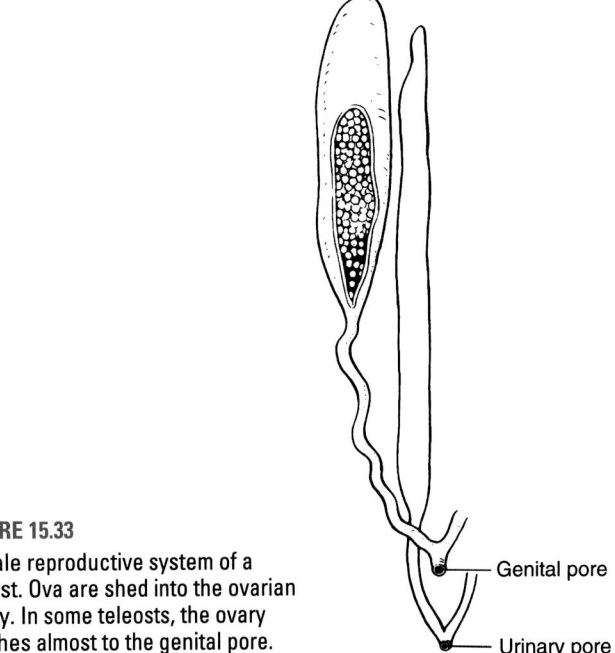

Teleost

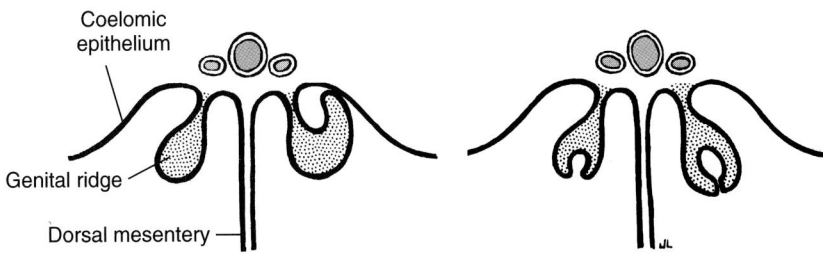

FIGURE 15.34
Two methods of entrapment of coelom to form a permanently hollow ovary in teleosts. The genital ridges are shown in cross section.

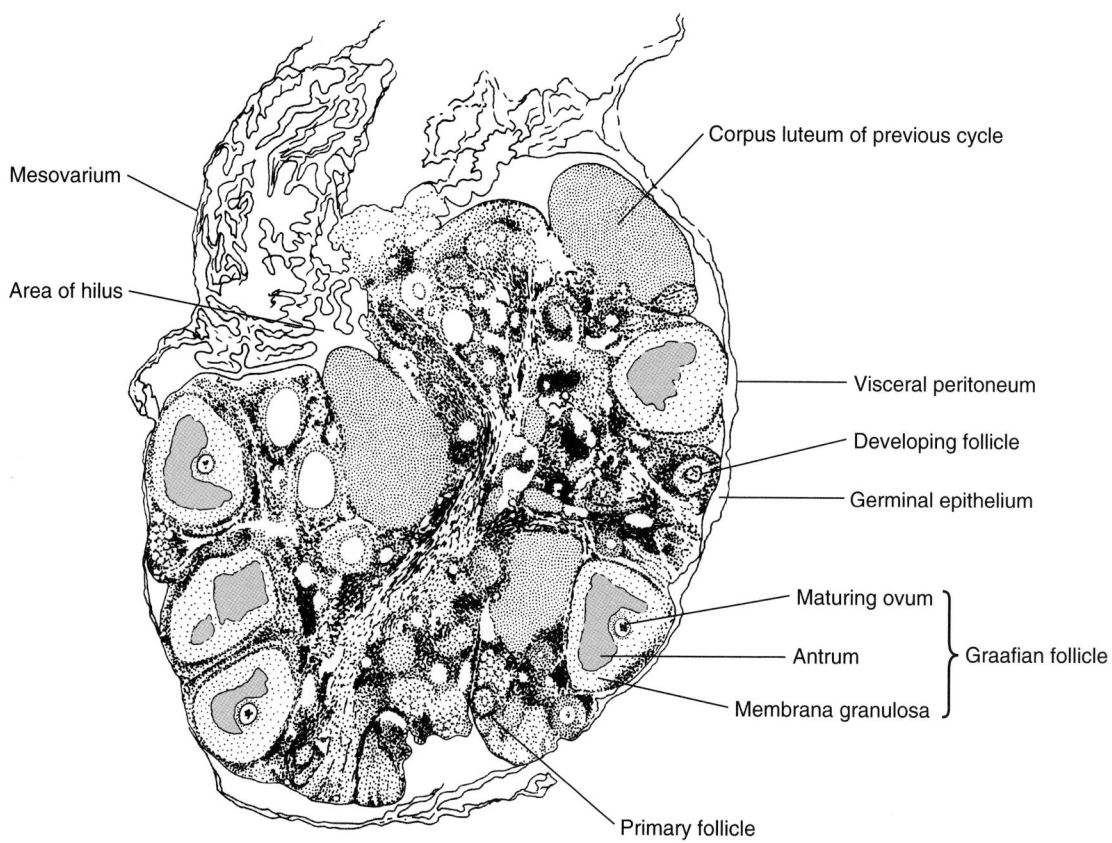

Mesovarium

Area of hilus

Corpus luteum of previous cycle

Visceral peritoneum

Developing follicle

Germinal epithelium

Maturing ovum ⎫
Antrum ⎬ Graafian follicle
Membrana granulosa ⎭

Primary follicle

FIGURE 15.35

Section of hamster ovary at approach of estrus (heat) and ovulation, which occurs every fourth day unless pregnancy intervenes. The follicles are in the germinal epithelium.

secondary hollowing out of the interior of the ovary at each ovulation. However, the eggs are discharged in a similar manner. The germinal epithelium of most other craniate ovaries, whether compact, saccular, or lacunate, is in a cortex immediately internal to the visceral peritoneum, and the eggs are discharged into the coelom through eroded areas in the visceral peritoneum. At the end of each reproductive season, the ovaries of most craniates other than therian mammals regress to a state resembling those of juveniles.

The ovaries of therian mammals (fig. 15.35) are compact, the only cavitation being the fluid-filled **antra** within the egg follicles (**graafian follicles**) as ovulation approaches. The number of ovulated ova is not large in mammals—one to three, generally—although the number in some carnivores and small rodents is larger. The South African elephant shrew is atypical. It ovulates 50 or more eggs, although there is room for only two to implant and develop to term. The average number of mature eggs ovulated in most species is slightly larger than the average number that implants, and the average number that implants exceeds the average number born. Thus, nature compensates for possible prenatal loss just as in teleost fishes, in which millions of eggs may be extruded, but far fewer are fertilized.

The fully mature mammalian graafian follicle bulges on the surface of the ovary. Its wall becomes very thin in the hours immediately preceding ovulation, and it finally ruptures. The ovum, surrounded by a corona of follicular cells, then oozes through a narrow slit in the wall and into the coelomic fluid. The escape into the coelomic fluid is the process of **ovulation.** Following ovulation, the many cells that remain in the ovulated follicle are altered histologically and physiologically under the stimulus of luteinizing hormone from the pituitary, and they organize a **corpus luteum** (fig. 15.35). Thereafter, the corpus luteum is a major source of the progesterone required for the maintenance of pregnancy. Before ovulation, estrogen is the predominant secretion of these cells.

Corpora lutea are not confined to mammals. They have been identified in every craniate class having taxa that retain ovulated eggs or embryos in the female tract for a period of time. The endocrine role of gonads is discussed in chapter 18.

Translocation of Mammalian Gonads

The caudal pole of each embryonic gonad is connected, by a ligament, to the coelomic floor of one of the **genital swellings,** the latter being shallow evaginations of the embryonic coelom (see fig. 15.48 *b*). These swellings evaginate further to become **scrotal sacs** in males and **labia majora** (the major lips of the **vulva**, or external genitalia) in females (fig. 15.36). Partly because elongation of the ligaments does not keep pace with elongation of the fetal trunk and partly as a result of actual shortening of the ligaments during late fetal development, the gonads, anchored to the caudal floor of the coelom, are displaced caudad toward the scrotal sacs in males and toward the labia majora in females. Testes are displaced farther caudad than ovaries, generally entering the scrotal sacs on a temporary or permanent basis. The scrotal sacs are lined by coelomic peritoneum, which, in this location, is the **internal spermatic fascia** (fig. 15.36, male).

In males, the ligament connecting the testis with the floor of a scrotal sac is the **gubernaculum**. In females, it is the **ovarian ligament** between the ovary and the uterus, and the **round ligament of the uterus** between the uterus and the floor of the labium majus (fig. 15.36). The gubernacula appear to play a role in the descent of the testes, although their removal does not always prevent descent.

Testes are housed permanently in the scrotal sacs in many mammals, including ungulates, carnivores, some marsupials, and as a derived state in some primates. In other mammals, the testes can be lowered into the scrotal sacs and retracted at will (rabbits, bats, a few rodents, some basal primates, and others). The passageway between the abdominal cavity and the scrotal cavity is the **inguinal canal**. The entrance to the canal is the **inguinal ring** (fig. 15.37). In species that retract the testes, the canal remains broadly open to the coelom. In species in which the testes are permanently confined to the **scrotum** (the two scrotal sacs), the inguinal ring eventually becomes only wide enough to accommodate the spermatic cord. An abnormally wide ring may become the site of an inguinal hernia.

The **spermatic cord** is a composite structure composed of the spermatic duct and the spermatic arteries, veins, lymphatics, and nerves, all wrapped in the internal spermatic fascia and all dragged into the scrotal cavity along with the testis. Scrotal sacs do not develop in monotremes, elephants, whales, or a few other mammals. In these, the testes remain permanently in the abdomen in a retroperitoneal location (that is, between the parietal peritoneum and the body-wall muscle). Testes that descend do so by sliding

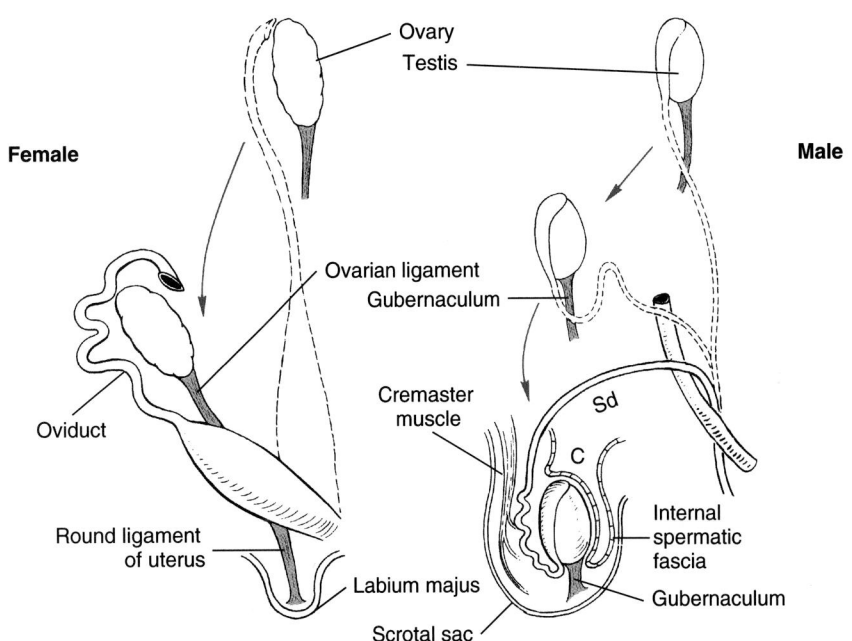

FIGURE 15.36

Caudal displacement of mammalian gonads, ventral view. The ovarian ligament and round ligament of the uterus collectively are homologous with the male gubernaculum. *Arrows* indicate the route of translocation of the right gonads. **C,** scrotal recess of coelom; **Sd,** spermatic duct arching over the ureter. The internal spermatic fascia is part of the coelomic peritoneum.

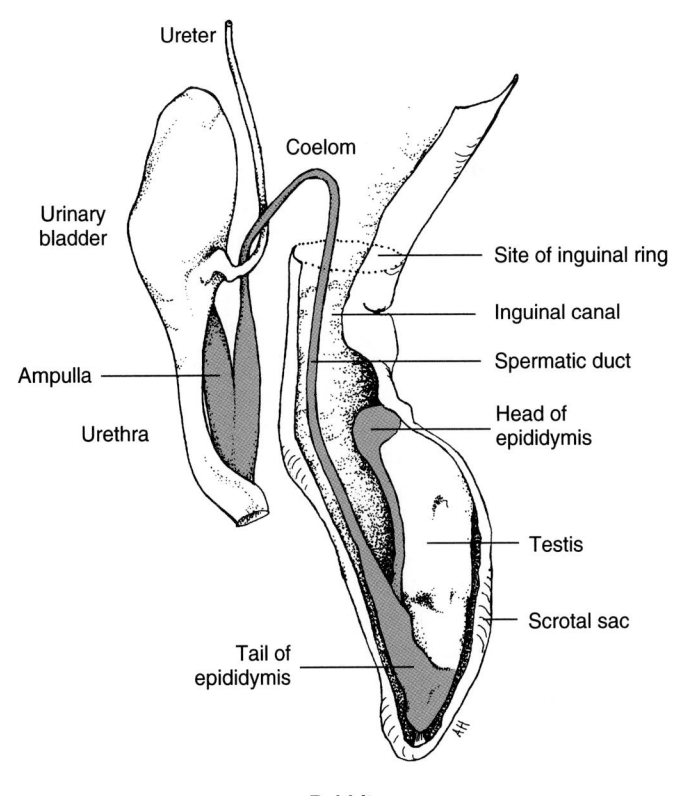

FIGURE 15.37

Rabbit testis in scrotal cavity. The inguinal canals are broadly open to the main coelom at the inguinal ring; therefore, the testes are retractable. Sperm pathway from testis to urethra is shown in *red.*

behind the peritoneum (internal spermatic fascia, see fig. 15.36).

Scrotal sacs are adaptations that maintain mammalian sperm at a temperature cooler than that within the abdominal cavity in which the sperm of most mammals cannot survive. A further adaptation is seen in the relationship between the internal spermatic artery and vein as they approach and leave the testis. These vessels are intimately associated in a rete, the **pampiniform plexus**, where heat is transferred from arterial to venous blood. As a result, blood entering the testis has been cooled in the plexus, and blood returning to the trunk has been warmed. The plexus also conserves body energy in the form of heat.

Female Genital Tracts in Craniates (Excluding Therian Mammals)

With the exception of agnathans, which have no genital ducts, and ray-finned fishes, whose female genital tracts are aberrant, the female duct systems of craniates other than therian mammals all reflect a generalized pattern of structure. It consists of paired muscular and glandular **oviducts** that commence at an **ostium**, an opening to the coelom, leading to the neck of an **oviducal funnel (infundibulum)** (fig. 15.38). Oviducts end in the cloaca or a subdivision of it. Any differences along the length of the tracts are in the degree of regional specializations that support species-specific functions such as collecting eggs from the coelom; coating them with protective or nutrient substances; temporarily housing eggs, embryos, or larvae and maintaining them until the eggs are shed or young are delivered; expelling the eggs or the young; receiving the male intromittent organ, if any; and temporarily housing viable sperm in species in which eggs normally are not mature at the time of mating.

In craniates other than ray-finned fishes, the female duct system differentiates from a pair of embryonic **muellerian ducts** (see fig. 15.13). These develop in both sexes but do not persist in males. The muellerian ducts of elasmobranchs and amphibians arise by longitudinal splitting of the pronephric ducts, and their ostia develop from one or several anterior pronephric nephrostomes. In other craniates, each muellerian duct arises as a longitudinal groove in the coelomic epithelium paralleling the mesonephric duct. The groove subsequently becomes a tube except at the anterior end where the ostium at least partially surrounds the ovary. Eventually, it separates from the mesonephros but remains attached to the body wall by a dorsal mesentery, the **mesotubarium**.

The muellerian ducts of female sharks give rise to an oviduct that develops a **shell gland** and **uterus** (fig. 15.39). The cephalic half of the shell gland secretes albumen. The caudal half secretes a shell that is very thin in viviparous species, leathery-tough in oviparous species, and often with tendrils that attach the egg case to seaweed. Two embryonic ostia unite to form a single

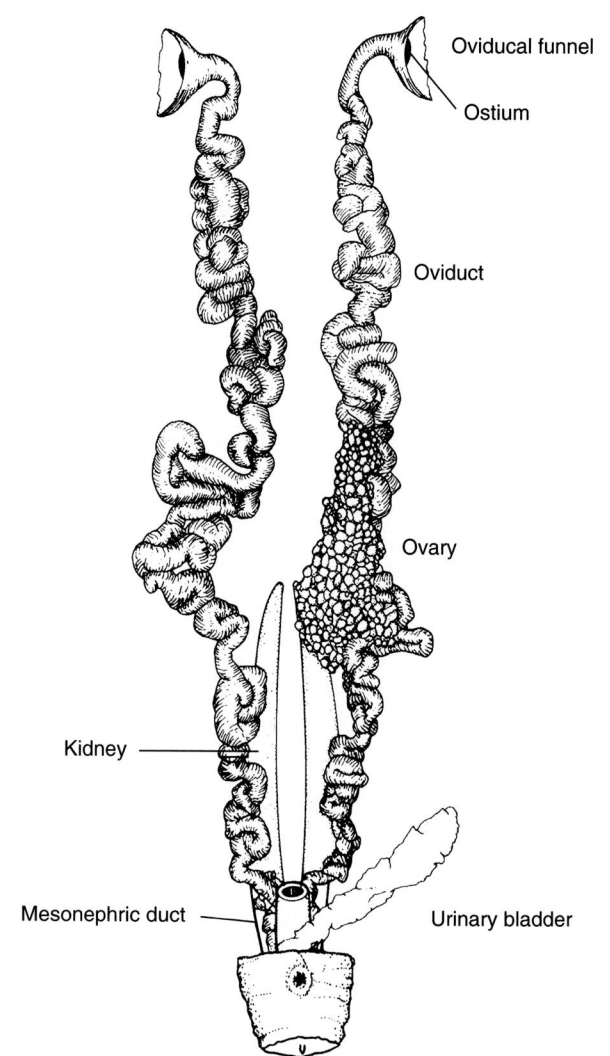

FIGURE 15.38
Generalized female duct system (*Necturus*).

ostium at the neck of an oviducal funnel. The funnel in sharks is supported by the falciform ligament.

After observing the large size of a mature shark egg, laboratory students often ask how such a huge egg can enter the small ostium and traverse the narrow upper region of the oviduct. We know the answer, because we know how this happens in chickens. At the time the egg is released from the ovary, it is a shapeless, almost flowing mass of yolk—no albumin—contained in a nonrigid cell membrane like the yolk of a fresh chicken egg. Under the influence of ovulatory hormones, the fringe (**fimbria**) of the oviducal funnel waves about in a gentle undulating movement. When it comes in contact with the shapeless mass, whether the yolk is still in the ovary or separated from it, the fimbria embraces the yolk, gently at first and then more firmly, drawing the yolk into the funnel until the egg is entirely engulfed (fig. 15.39). Muscular contraction of the funnel then squirts the

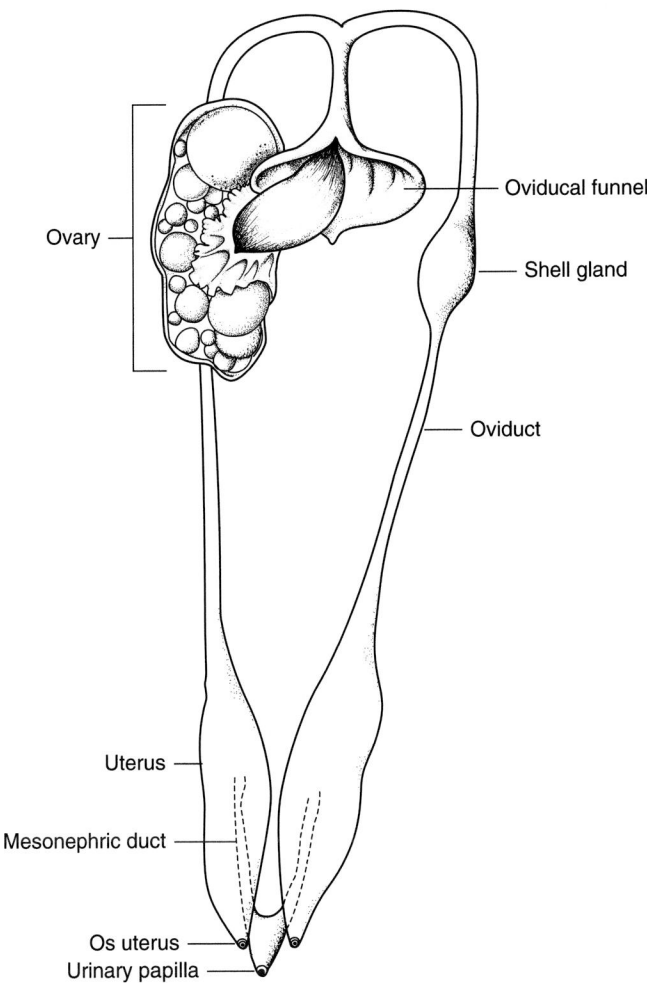

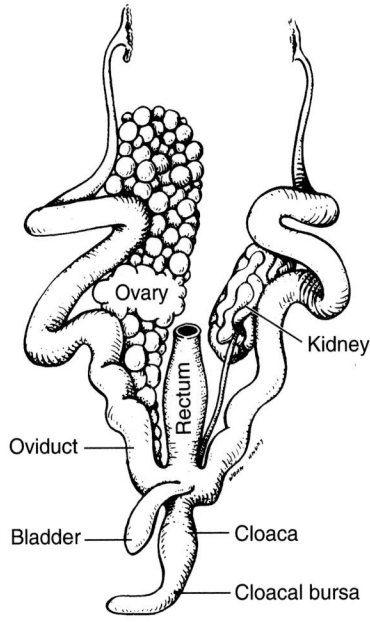

FIGURE 15.40
Urogenital system of female aquatic turtle, *Trionyx euphraticus,* ventral view. The left ovary has been removed.
Source: Courtesy of Mohamad S. Salih, University of Baghdad.

FIGURE 15.39
Reproductive system of female *Squalus,* ventral view. The left ovary has been removed. The right ovary shows an ovulating ovum being drawn into the oviducal funnel. An adult ovary typically exhibits follicles at varying stages of maturation. A urinary sinus precedes the urinary papilla.

shapeless mass through the ostium and into the oviduct. Thereafter, peristalsis of the oviducal wall moves the yolk caudad. Cilia play an insignificant role in retrieval of macrolecithal eggs from the coelom. However, in therian mammals, cilia along with undulations of the oviducal funnel *are* important because the eggs are microlecithal.

The oviducts of teleosts are not homologous with those of other craniates. They arise as caudally directed folds of the coelomic peritoneum created by the genital ridges as the latter entrap the coelom to form a hollow ovary (see fig. 15.34). Consequently, the oviducts are usually continuous with the ovarian cavity throughout life (see fig. 15.33). These oviducts are short funnels that open into paired or unpaired external genital papillae. The unpaired papillae of some teleosts elongate to form tubelike **ovipositors** through which the eggs are extruded. The continuity of the oviducts with the ovar-

ian cavity is lost secondarily in some species, and in these species the eggs are extruded into the coelom.

The linings of amphibian oviducts are richly supplied with glands that secrete several layers of protective jelly envelopes around each egg as it moves down the tube. The caudal portions in anurans become voluminous thin-walled **ovisacs** where ovulated eggs accumulate until they are shed. In terrestrial anurans, this occurs after a heavy rain has provided appropriate pools for egg-laying during amplexus (mating). Similar sacs in ovoviviparous urodeles shelter unborn young. The oviducts of dipnoans are similar to those of oviparous urodeles.

Female tracts of reptiles and monotreme mammals conform closely to the basic pattern (figs. 15.40 to 15.43). However, only one of the two embryonic muellerian ducts may differentiate, as in crocodilians and most birds. The rudimentary right muellerian duct of birds is seen in figure 15.42. Albumen glands line part of the oviduct, and there is usually a prominent shell gland just anterior to the cloaca (figs. 15.42 and 15.43). The shell remains leathery in species whose eggs develop in moist environments; it becomes brittle in eggs that develop in dry air. The difference depends on the nature of the secretions. Ovoviviparous reptiles as well as oviparous ones cover their eggs with a shell.

Although the female tracts of birds are similar to those of other reptiles, some of the anatomic terminology differs. The albumen-secreting area in fowl is termed the **magnum,** the shell gland is a **uterus,** and the short glandular and muscular segment at the

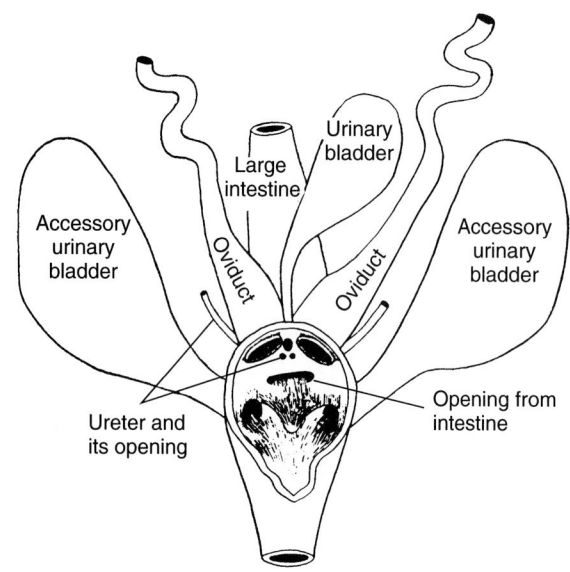

FIGURE 15.41
Cloaca and associated structures of a female terrestrial turtle, ventral view. Clitoris has been removed.

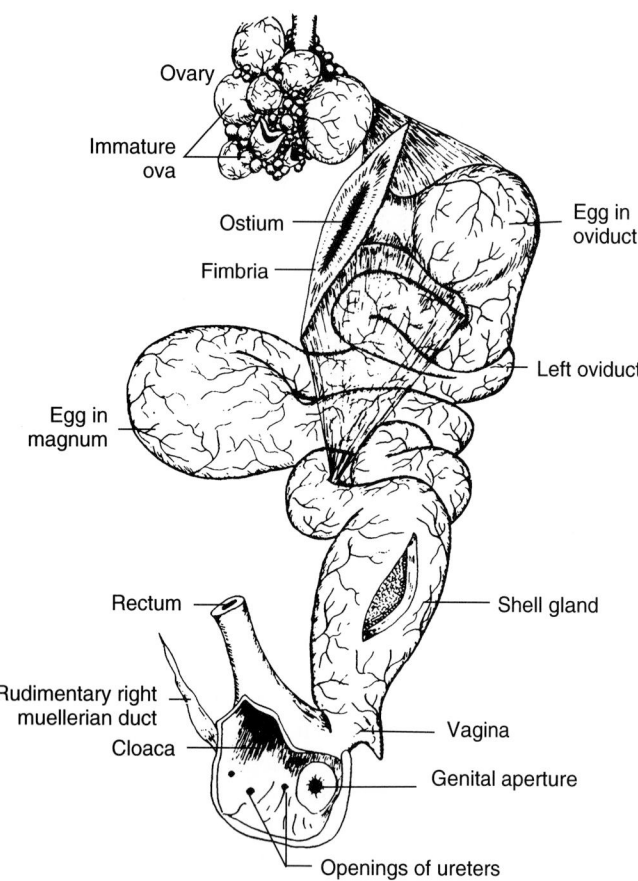

FIGURE 15.42
Genital tract of a hen. The presence of two eggs in the oviduct is unusual.

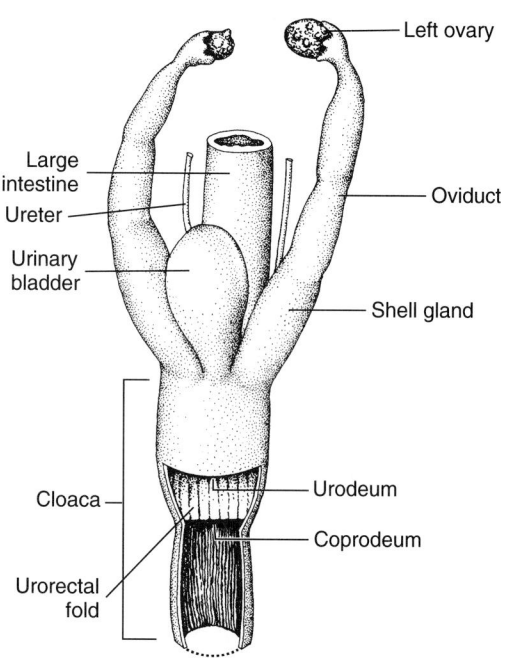

FIGURE 15.43
Genital tract and cloaca of a female monotreme, ventral view. The right ovary is usually the smaller, or it is rudimentary. The cephalic half of the cloaca is partitioned by the urorectal fold into a urodeum and coprodeum. An exterior view of the vent is seen in figure. 15.48*a*.

entrance to the cloaca is a **vagina**. The vagina secretes mucus that seals the pores of the shell, preventing the loss of vital water vapor from within the shell without affecting the supply of oxygen to the chorioallantoic membrane (see fig. 5.13). The vagina then expels the egg to the exterior.

Spermathecae, which are dorsal diverticula of the female cloaca or tubular crypts in the oviducal lining, are present in many urodeles, lizards, snakes, and domestic fowl. They store sperm received during mating. This ensures viable sperm at the time of next ovulation, which, in some species, is weeks or months after mating.

Female Tracts of Therian Mammals

As in other craniates, paired muellerian ducts in mammals give rise to the female genital tract commencing with the ostium of the oviduct. However, the muellerian ducts of eutherian mammals unite at their caudal ends to one degree or another. The extent of the union varies with the mammalian order. The tract of eutherian mammals, therefore, consists of two oviducts, a uterus with or without horns (cornu), and a vagina.

Oviducts

Oviducts, or **fallopian tubes** as they are also called in mammals, are relatively short, small in diameter (accommodating microlecithal eggs), more or less convoluted, and lined with cilia. The oviducal funnel has a fringed border. In many mammals, a membranous fold of peritoneum, the **ovarian bursa**, envelopes the ovary and the oviducal funnel, entrapping them and a small coelomic cul-de-sac. The bursa increases the probability that ovulated eggs will enter the oviduct rather than be lost in coelomic fluid, perhaps to implant somewhere in the coelomic cavity. The bursa may be broadly open to the main coelom, as in cats and rabbits; it may communicate with the coelom by a mere slit, as in most carnivores and rats; or it may be completely closed off from the coelom, as in hamsters. There is no ovarian bursa in humans. Ectopic pregnancies with fetuses implanted in the ventral abdominal wall are occasionally found in cats and rabbits in anatomy laboratories, and they occur in humans. The fetuses, of course, cannot be delivered.

Uteri

The muellerian ducts of marsupials are paired all the way to the urogenital sinus, a subdivision of the embryonic cloaca (fig. 15.44). Marsupials therefore have two completely separate uteri (a **duplex uterus**) and two vaginas. The uteri of eutherian mammals exhibit varying degrees of fusion distally, resulting most often in a **bicornuate uterus**—a uterus with two horns in which the young develop (figs. 15.45 and 15.47b). A uterus with two horns may have two totally separate passageways within the body of the uterus although this is not discernible from external view (figs. 15.46, rabbit, and 15.47a). When this is the case, the uterus is more precisely termed **bipartite**. In some mammals, one uterine horn is much larger than the other, and blastocysts implant in that horn—the right in impalas—even though both ovaries produce viable eggs.

Maximal fusion of the muellerian ducts occurs in armadillos, some bats, and in monkeys, apes, and humans, in which fusion commences at the ends of short oviducts and there are no uterine horns. This is a **simplex** uterus (figs. 15.46, monkey, and 15.47c). Blastocysts implant in the body of the uterus, and there is usually only one fetus per pregnancy. Armadillos are exceptional; they always give birth to identical quadruplets.

The narrow neck, or **cervix**, of the uterus of therian mammals projects into the vagina as the **lips of the cervix**. These surround the opening, or **os uteri**, of the uterus into the vagina (fig. 15.47c). Sperm

deposited in the vagina pass between the lips of the cervix en route to the upper end of the oviduct, where fertilization takes place. The uterine lining (**endometrium**) of therian mammals becomes highly vascular under the stimulus of hormones before implantation

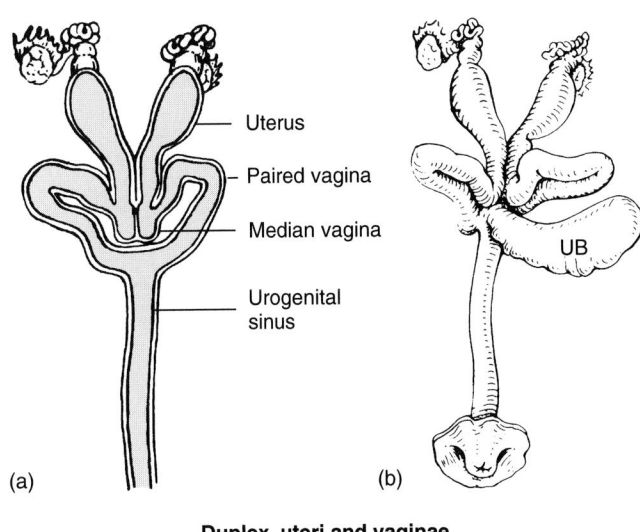

**Duplex uteri and vaginae
Marsupial**

FIGURE 15.44

Reproductive tract of a female opossum. **UB,** urinary bladder. (*a*) Section showing internal structure; urinary bladder is absent. (*b*) Surface view.

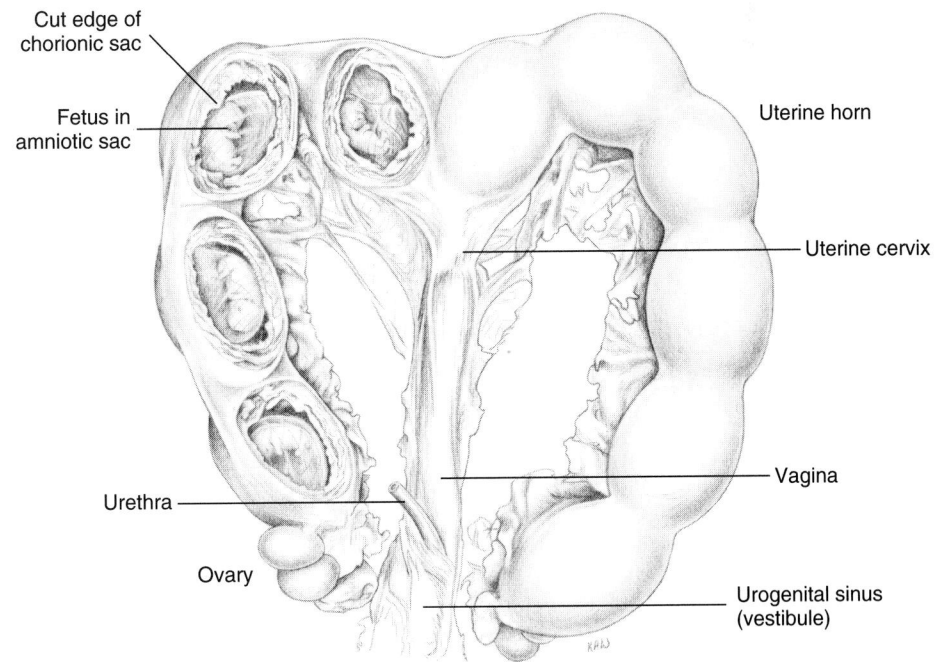

FIGURE 15.45

Gravid bicornuate uterus of a dog. The chorionic sacs have been partially cut away to expose the fetuses within the amniotic sacs. The body of the uterus is insignificant in the pregnant dog.

of blastocysts. The thick muscular layer of the uterine wall (**myometrium**) assists in ejection of the young at birth, provided it, too, has been hormonally prepared. Before delivery, the cervix must dilate to provide a passageway large enough for the young to pass through. Dilation takes place under the influence of the hormone **relaxin** from the ovary. For review of the action of relaxin on the pubic symphysis, see chapter 10, "Pelvic Girdles."

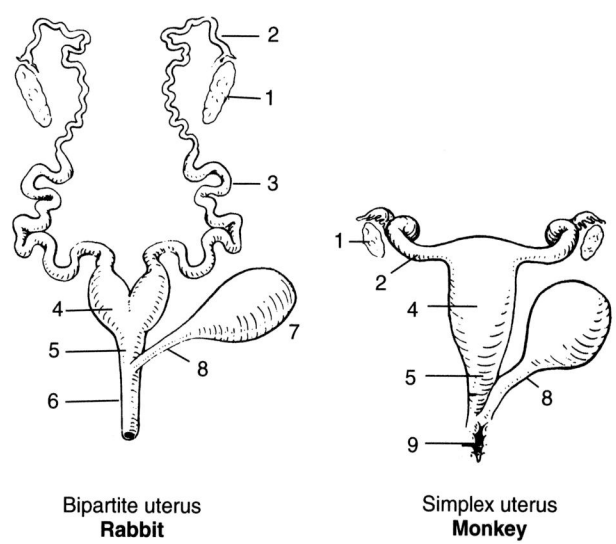

FIGURE 15.46

Bipartite and simplex uteri. **1**, ovary; **2**, oviduct; **3**, uterine horn; **4**, body of uterus; **5**, vagina; **6**, urogenital sinus; **7**, urinary bladder; **8**, urethra; **9**, vestibule of vulva. There are two complete passageways within the rabbit uterus all the way to the vagina.

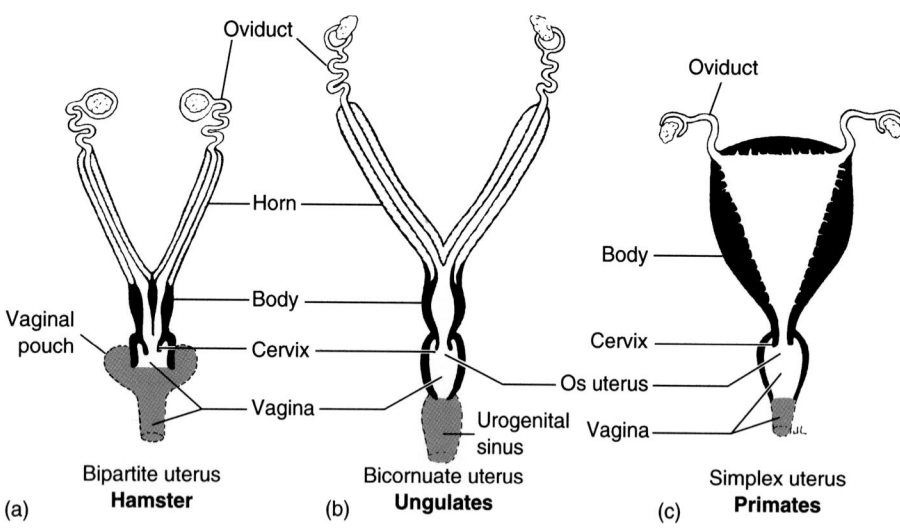

FIGURE 15.47

Varieties of mammalian uteri. *Blackened* regions are the fused caudal ends of the muellerian ducts; *red* is a derivative of the cloaca. Note the two lumens in the body of the bipartite uterus. The urogenital sinus in ungulates, as in most female mammals, receives the vagina and urethra.

Vaginas

The vagina of therian mammals is the unpaired terminal portion of the embryonic muellerian ducts (fig. 15.47b). The cloaca contributes to it in some mammals (fig. 15.47a and c, red). Except in rodents and as a derived character is some primates, the vagina empties into the **urogenital sinus**, a derivative of the embryonic cloaca (figs. 15.44, 15.45, 15.46, rabbit, and 15.47b). In many rodents, the vagina opens directly to the exterior (fig. 15.48c). In apes and humans, it opens into a shallow **vestibule of the vulva**, which also receives the urethra (see fig. 15.46, monkey). The vestibule is bordered by two fleshy folds, the **labia minora**, derived from the genital folds (fig. 15.48b). It is overlaid by two additional folds, the labia majora, derived from the genital swellings. The latter are homologues of the scrotal sacs of males. The vagina is lined by a cornified epithelium for reception of the penis.

The two pouchlike components of the **median vagina** of marsupials (see fig. 15. 44) lie against the urogenital sinus, separated from the latter by thin septa. At birth, the fetus is usually forced through the septa directly into the urogenital sinus, and the passageway thus established may remain throughout life. As an adaptation to paired vaginas, some male marsupials have a bifurcated penis.

Muellerian Duct Remnants in Adult Males

Although muellerian ducts do not mature in males, vestiges are common. In sharks, a pair of rudimentary oviducts commence at an ostium in the falciform ligament and curve around the anterior end of the liver on the surface of the coronary ligament, after which they disappear in connective tissue. Paired sperm sacs of male sharks (see fig. 15.9) are caudal remnants of the muellerian ducts. A more or less complete, but rudimentary, muellerian duct persists in many male amphibians (see fig. 15.23). In anurans, removal of the testes, the source of male sex hormones, has been shown to induce the rudimentary muellerian ducts to develop into functional uteri and oviducts. Vestiges are also common in other male craniates. Remnants in male mammals include an **appendix testis** (not to be confused with appendix epididymis), a small cystlike nodule with an epithelia-lined lumen, a homologue of the oviducal funnel; and a **prostatic sinus (vagina masculina)**, an unpaired saccular remnant of

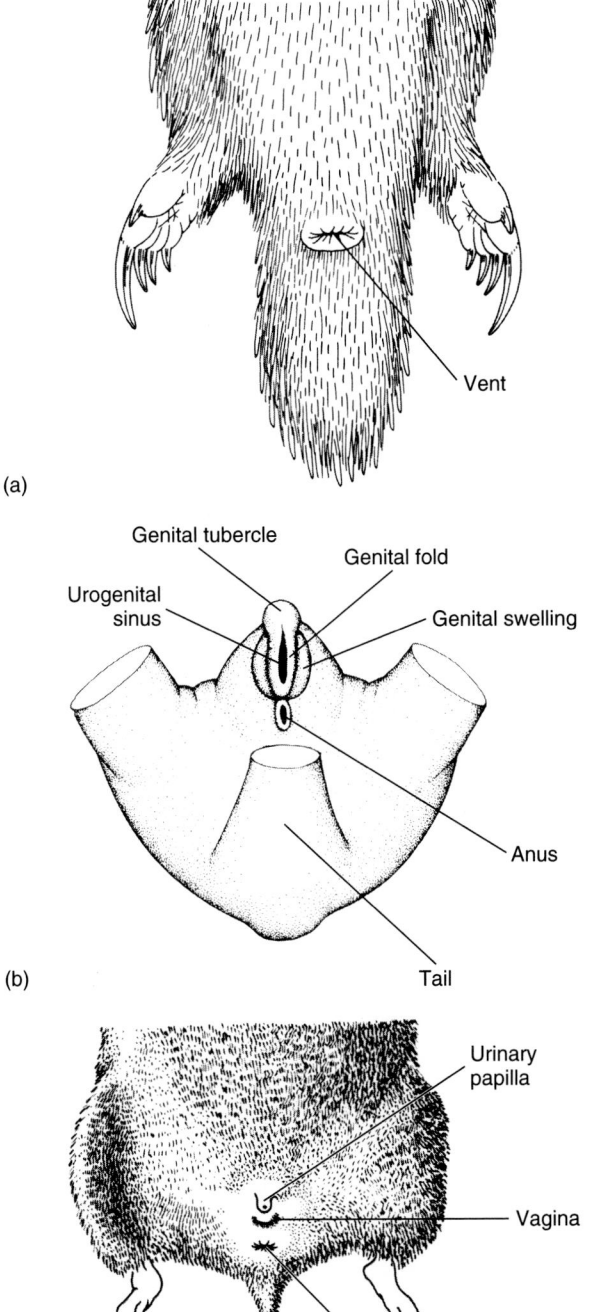

(a)

(b)

Genital tubercle

Genital fold

Urogenital sinus

Genital swelling

Anus

Tail

(c)

Urinary papilla

Vagina

Anus

Vent

FIGURE 15.48

External openings from mammalian cloaca or its derivatives.
(*a*) Cloacal vent of the monotreme *Echidna*. (*b*) External genitalia of bisexual stage of 75 mm human embryo (about 12 weeks of age). The cloaca at this stage has been divided into urogenital sinus and rectum. For adult derivatives of these genitalia, see table 15.3.
(*c*) Perineal region of female hamster. At the tip of the urinary papilla is the urethral aperture.

the fused caudal ends of the muellerian ducts homologous with the vagina and opening into the male urethra near the prostate gland.

CLOACA

Urinary and genital ducts empty into the cloaca in amphibians and in fishes other than those in which an adult cloaca is shallow or absent. In the latter—predominantly teleosts—genital and urinary ducts open directly to the exterior (see figs. 15.18 and 15.33). They also empty into the cloaca in some reptiles. In other reptiles (including birds) and monotremes, a horizontal **urorectal fold** partitions the cephalic end of the cloaca into two channels—a **urodeum** that receives the urogenital ducts and a **coprodeum** that receives the alimentary canal (see figs. 15.29 and 15.43). The terminal (unpartitioned) portion of the cloaca is sometimes called a **proctodeum**, but it is not homologous with the ectodermal proctodeum of embryos.

Therian Mammals

In mammals that deliver live fetuses—that is, marsupials and eutherians—the urorectal fold grows caudad until it reaches the cloacal membrane separating the cloaca from the exterior. This modification completely divides the cloaca into a urogenital sinus and a **rectum** (fig. 15.49*a*, *b*, and *c*; and *a*, *b*, and *e*, male). Rupture of the embryonic cloacal membrane at two sites provides two openings to the exterior, an **anus** and a **urogenital aperture** (see fig. 15.48*b*). The embryonic urogenital sinus of both sexes receives the mesonephric ducts, muellerian ducts, and the future urinary bladder (allantois, fig. 15.49*b*).

As development progresses in males, the muellerian ducts regress, and the urogenital sinus and rectum elongate (compare fig. 15.49*b* and *e*). The urogenital sinus eventually becomes continuous with the spongy urethra that has developed in the penis (fig. 15.49*f*). The terms *urogenital sinus* and *urethra* have now become synonyms when applied to males. By differential growth, the ureters become reoriented to open into the bladder, and the mesonephric ducts (now sperm ducts) continue to empty into the urogenital sinus (figs. 15.30 and 15.49*f*). This is the condition in all males.

As development progresses in females, the mesonephric ducts regress, and the muellerian ducts unite at their caudal ends to form the body of the uterus and the vagina (fig. 15.49*c*, gray). The part of the urogenital sinus between the bladder and the entrance of the vagina is now known as the **urethra**. As a result of these developments, *most adult female mammals have two caudal openings* to the exterior, a urogenital aperture and an anus.

Female apes, monkeys, humans, and some female rodents develop an additional partition in the embryonic

Bisexual stages

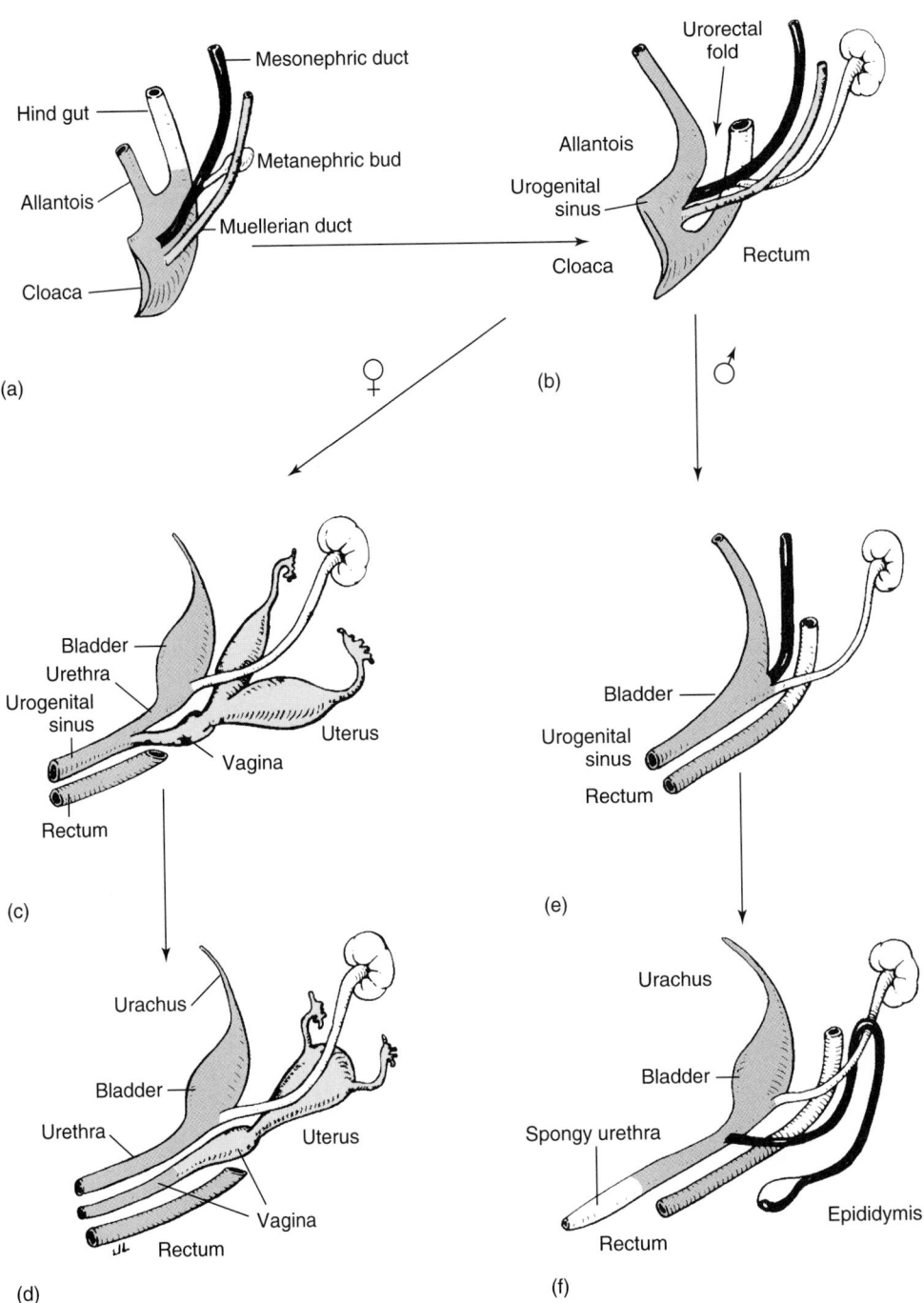

FIGURE 15.49
Fate of the cloaca and allantois (*red*), muellerian ducts (*gray*), and mesonephric duct (*black*) in therian mammals. (*a*) and (*b*) Bisexual stages. Only the left muellerian and mesonephric ducts are shown. In (*b*) the cloaca is becoming subdivided by the urorectal fold into a urogenital sinus ventrally and a rectum dorsally. (*c*) Typical adult female mammal. (*d*) Female primate. In (*c*) and (*d*), the contributions of both the left and right muellerian ducts are shown. (*e*) Developing male, intermediate stage in reorientation of mesonephric and metanephric ducts. (*f*) Adult male.

TABLE 15.3   Some Homologous Urogenital Structures in Male and Female Mammals

Indifferent Structure	Mature Male	Mature Female
Mesonephric duct	Ductus deferens (vas deferens)	Gartner's duct
	Epididymis	
Mesonephric tubules	Vasa efferentia	
	Appendix of epididymis*	Epoophoron*
	Paradidymis*	Paroophoron*
Muellerian duct	Appendix testis*	Oviduct
		Uterus
	Prostatic sinus (vagina masculina*)	Vagina
Gonadal ridge	Testis	Ovary
	Rete testis	Rete ovarii*
Gubernaculum	Gubernaculum	Ovarian ligament
		Round ligament of uterus
Genital swellings	Scrotal sacs	Labia majora
Genital tubercle	Penis	Clitoris
Genital folds	Contribute to penis	Labia minora
Urogenital sinus	Urethra, prostatic, and membranous portions	Urethra
		Urogenital sinus
		Lower vagina in rodents and primates

*Vestigial

cloaca, this one in the urogenital sinus. It divides the urogenital sinus into a urethra and a vagina (fig. 15.49*d*). As a result, the embryonic female cloaca in these mammals becomes subdivided into a *urethra, vagina,* and *rectum,* each with its own exit. In monkeys and rodents *all three exit directly* to the exterior (see fig. 15.48*c*). In apes and humans, the vagina and urethra open separately into the shallow vestibule of the vulva, noted earlier, and the anus opens directly to the exterior. The urogenital systems of these females have therefore evolved further than those of males. The vagina in these females has a dual origin: The cephalic part is derived from the muellerian ducts, and the terminal portion is cloacal (see fig. 15.47*a* and *c*).

From the discussions in this chapter it may have become apparent that all craniate embryos are bisexual; that is, they exhibit embryonic structures competent to become the urogenital organs of either sex. Under the initial influence of genes and, perhaps other unknown factors, and the subsequent stimuli of gonadal hormones, the indifferent embryonic rudiments differentiate to become the adult genital structures of one sex, or they remain rudimentary, become vestigial, or disappear. Abnormal, incomplete development of either or both sets of organs is a result of chromosomal aberrations and/or hormonal imbalances during fetal life.

True **hermaphroditism**,* the production of viable eggs and sperm by the same individual, is common in living agnathans and some species of bony fishes, but it is extremely rare or represents a congenital deformity in other craniates. These species lack sex chromosomes. Sexual differentiation then is determined by nongenetic factors. In hermaphroditic teleosts studied, the social environment, such as the disappearance of a dominant male or female, commonly directs an animal's sexual development.

Parthenogenesis, or asexual cloning, occurs in some lizards. In whiptail lizards, for example, there are no *genetic* males, although individuals engage in mating behavior, alternating between sexually typical male and female behaviors. David Crews (1994) has shown that sexual interactions between parthenogenetic animals yield more eggs laid for each than if they had bred alone.

Table 15.3 summarizes the fate in adult males and females of the chief sexually indifferent urogenital rudiments of mammalian embryos.

Summary

1. The earliest role of kidneys was osmoregulation in an aquatic environment. Elimination of nitrogenous wastes is an added excretory role that has become significant in tetrapods. Kidneys replaced gills in this function.
2. The basic functional components of kidneys are glomeruli, renal tubules surrounded by peritubular capillaries, and a pair of longitudinal excretory

---

*Hermaphroditus, of Greek mythology, was the beautiful mythical son of Hermes and Aphrodite. While bathing in the mythical fountain of Salmacis, he became united in one body with the nymph in the fountain.

ducts that empty into the cloaca or a derivative thereof.

3. Glomeruli are arterial retia that filter water and certain solutes from arterial blood. External glomeruli secrete into coelomic fluid, which is then swept by nephrostomes into the tubules. Internal glomeruli are in the body wall and filter into Bowman's capsules, which are extensions of tubules. Several tubules generally empty into collecting tubules that terminate in the longitudinal duct.

4. A glomerulus and its associated capsule constitute a renal corpuscle. A renal corpuscle and its tubule constitute a nephron.

5. Osmoregulation is affected by the tubules, which add to or subtract from glomerular filtrate in accordance with the metabolic needs of the organism.

6. Kidney tissue arises from a ribbon of intermediate (nephrogenic) mesoderm. Primitive pronephric tubules form at the anterior end of the ribbon, and thereafter a wave of differentiation sweeps along the ribbon, giving rise to tubules of increased complexity. As tubules form farther caudad, many that appear earlier regress.

7. An archinephros is a hypothetical archaic kidney with external glomeruli, nephrostomes, and simple tubules arranged metamerically the length of the trunk. The holonephros of larval hagfishes most nearly resembles the archinephros.

8. The anteriormost renal tubules constitute a temporary pronephric kidney in all craniate embryos and then regress. A mesonephros develops caudal and subsequent to the pronephros. The ducts of the pronephros continue to serve as two longitudinal mesonephric ducts.

9. The mesonephros is the functional kidney of adult fishes and amphibians and a temporary functional kidney in embryonic amniotes. In adult anamniotes, it is also called an opisthonephros.

10. Some of the more anterior mesonephric tubules invade the developing testis and become vasa efferentia that conduct sperm to the mesonephric duct. This duct in generalized fishes and amphibians carries both sperm and urine. It persists in amniotes to serve as the spermatic duct (vas deferens).

11. The caudal end of the nephrogenic ribbon is displaced laterad and cephalad to form the metanephric kidney of adult amniotes. Thereafter, the mesonephros regresses.

12. The metanephros consists of cortex, medulla, and pelvis. Renal corpuscles pack the cortex; long loops of tubules (loops of Henle) and their peritubular capillaries comprise much of the medulla. The tubules empty into the renal pelvis, which is drained by a ureter.

13. Craniates that imbibe excessive quantities of salt (marine craniates, birds whose food comes from the sea, other reptiles that inhabit arid environments) excrete salt extrarenally via gills or rectal, nasal, orbital, oral, or integumentary salt glands.

14. Bladders of most fishes are derived from the conjoined ends of mesonephric ducts. They are absent in living agnathans and elasmobranchs. Those of dipnoans arise as dorsal evaginations of the cloaca.

15. Tetrapod bladders arise from the floor of the cloaca and, except in therian mammals, empty directly into it. Bladders of therian mammals develop from the base of the allantois, a cloacal derivative, and empty into a urogenital sinus or, as a derived feature in some female primates, to the exterior. They are drained by a urethra.

16. The primary role of urinary bladders in tetrapods other than therian mammals is to serve as a reservoir for urine, the water of which is recycled when needed by body tissues.

17. Gonads arise from paired genital ridges adjacent to developing kidneys. Germinal epithelium arises on the surface of the ridges and may remain there or be transposed to a deeper location.

18. Mature eggs are released into the coelom before entering an oviduct. Sperm are transported in a closed system of ducts that begins within the testes.

19. The germinal epithelium of most male craniates lines seminiferous tubules that empty into a rete testis, drained by vasa efferentia. These conduct sperm to the spermatic duct.

20. The testes remain in the abdominal cavity in a few mammals, descend permanently in some, and are retractile in others.

21. Except in some fishes and urodeles the mesonephric duct is the spermatic duct of adult males. Spermatic ducts open into the cloaca, when present, except in therian mammals. In therian mammals, they open into the urethra.

22. Male intromittent organs of fishes are modifications of pelvic fins (claspers) or anal fins (gonopodia). Snakes and lizards have paired hemipenes. Turtles, crocodilians, some birds, and monotremes have an unpaired penis in the cloacal floor containing a corpus spongiosum. Therian mammals have an external penis with a corpus spongiosum and two corpora cavernosa.

23. In females of reptilian species in which there is a male penis and in female mammals, a clitoris develops. It is the female homologue of the penis.

24. The ovaries of teleosts have a central cavity lined by the germinal epithelium. The cavity may be continuous with the cavity of the oviduct. Anuran and urodele ovaries are thin-walled sacs; those of some reptiles including birds and monotremes have irregular fluid-filled lacunae. Other ovaries are generally solid, with the germinal epithelium on the surface. Following ovulation, the ovarian (graafian) follicles

of mammals organize a corpus luteum that becomes a source of the progesterone that assists in the maintenance of pregnancy.

25. Mammalian external genitalia of both sexes differentiate from a median genital tubercle and paired genital swellings and folds. The tubercle gives rise to the penis or clitoris. Genital swellings give rise to scrotal sacs or labia majora. Genital folds contribute to the penis or become labia minora.

26. Ligaments in mammals connect the embryonic testes to the floor of the scrotal sacs and the embryonic ovaries to the labia majora. Shortening of the male ligaments (gubernacula) assists caudal migration of the testes toward or into the scrotum. Shortening of the female ligaments (ovarian ligaments and round ligaments of the uterus) causes lesser displacement of ovaries.

27. Female genital tracts from ostium to cloaca or its derivatives differentiate, except in teleosts, from muellerian ducts present in embryos of both sexes. Muellerian ducts in males regress. Teleost female tracts arise from peritoneal folds rather than from muellerian ducts.

28. A generalized female genital tract consists of an oviduct within a funnel-like ostium and terminates in the cloaca. Segments of the generalized oviduct are specialized for species-specific functions such as secretion of egg envelopes, temporary storage of ova or developing young, and expulsion of eggs or offspring.

29. The female tracts of therian mammals consist of fallopian tubes (oviducts); duplex, bipartite, bicornuate, or simplex uteri; a uterine cervix; and a vagina. The latter empties into a urogenital sinus in most female mammals, into the vestibule of the vulva in monkeys, apes, and humans.

30. The ovaries of some therian mammals occupy an ovarian bursa, a cul-de-sac of the coelom that may be partly or completely closed off from the main coelom. It ensures that ovulated eggs will not become implanted in an ectopic location.

31. An adult cloaca receives the large intestine, urinary bladder when present, and urinary and genital ducts. A cloaca is shallow or absent in many fishes. A single aperture opens to the exterior.

32. In reptiles and monotremes, a urorectal fold divides the anterior portion of the embryonic cloaca into a urodeum that receives the urinary and genital ducts, and a coprodeum that receives the rectum. The caudal portion of the cloaca remains undivided.

33. In female therian mammals except monkeys, apes, and humans, the embryonic cloaca becomes completely divided into a urogenital sinus and rectum, and two apertures open to the exterior.

34. In some female rodents and in female monkeys, apes, and humans, the embryonic cloaca is divided into a urethra that drains the bladder, a vagina, and a rectum, and each opens to the exterior independently.

Critical Thinking Questions

1. Why are structures of osmoregulation and reproduction united within a single system—the urogenital system?

2. What solutes and solvents are maintained in osmoregulation? What structures are employed in this maintenance?

3. Compare the various forms of nitrogenous wastes in craniates.

4. Compare holonephric, mesonephric, opisthonephric, and metanephric kidneys.

5. What are the constituent parts of a gnathostome nephron?

6. Give one specialization of the craniate kidney for life in marine, freshwater, and terrestrial habitats.

7. What are the functions of the pronephric (archinephric) duct? Do these functions differ in different organisms? How?

8. Desert kangaroo rats are able to survive solely on the water they acquire in the food they eat and produce the most concentrated urine known in mammals. How can you explain these observations?

9. What are some of the extrarenal salt excretion organs found in craniates? What taxa possess them?

10. Why do we not find urinary bladders in most fishes? What function do they serve?

11. Compare the passage of gametes in males and females.

12. What are the intromittent organs in sharks, snakes, and therian mammals?

13. A potentially fatal condition in humans is an ectopic pregnancy where the developing embryo implants in the peritoneal cavity. How is this possible?

14. What are the structures that are formed by the caudal fusion of oviducts in therian mammals?

15. In the bisexual stage in eutherian mammals, a single cloaca is formed. How is this structure differently divided in males and females during development?

Selected Readings

Crews, D.: Animal sexuality, *Scientific American* **271**:108–14, 1994.

Dawley, R. M.: An introduction to unisexual vertebrates. In Dawley, R. M., and Bogart, J. P.: Evolution and ecology of unisexual vertebrates. Albany, New York, 1989, Bulletin 466, New York State Museum. An introduction to parthenogenesis and other forms of reproductive unisexuality in all-female fishes, amphibians, and reptiles. Knowledge of genetics recommended.

Eckert, R., Randall, D., and Augustine, G.: Animal physiology, mechanisms and adaptations. New York, 1998, W. H. Freeman and Company.

Fox, H.: The amphibian pronephros, *Quarterly Review of Biology* **38**:1–25, 1963.

Fox, H.: The urinogenital system of reptiles. In Gans, C., and Parsons, T. S., editors: Biology of the reptilia, vol. 6. New York, 1977, Academic Press.

Moffat, D. B.: The mammalian kidney. New York, 1975, Cambridge University Press.

Mossman, H. W., and Duke, K. L.: Comparative morphology of the mammalian ovary. Madison, 1973, The University of Wisconsin Press. Informative, clearly written, richly illustrated.

Pang, P. K. T., Griffith, R. W., and Atz, J. W.: Osmoregulation in elasmobranchs, *American Zoologist* **17**:365, 1977.

Peaker, M., and Linzell, J. L.: Salt glands in birds and reptiles, Monographs of the Physiological Society. New York, 1975, Cambridge University Press.

Prosser, C. L., editor: Comparative animal physiology, ed. 4, vol. 2. Philadelphia, 1990, W. B. Saunders Co.

Rankin, J. C., and Davenport, J: Animal osmoregulation. New York, 1981, John Wiley and Sons.

Rupert, E.E.: Evolutionary origin of the vertebrate nephron, *American Zoologist* **34**:542–53, 1994.

Shapiro, D. Y.: Differentiation and evolution of sex change in fishes, *BioScience* **37**:490–97, 1987.

Stahl, B. J.: Vertebrate history: Problems in evolution. New York, 1974, McGraw-Hill Book Co.

Vincent, A.: A seahorse father makes a good mother, *Natural History*, p. 34, December, 1990. Courtship, mating, and true pregnancy in male seahorses, and a rare photograph of their elaborate filamentous dermal appendages.

Wourms, J. P., and Callard, I. P.: Evolution of viviparity in vertebrates, *American Zoologist* **32**:251, 1992.

Links to the Internet

Visit the zoology website at http://www.mhhe.com/zoology to find live Internet links for each of the following references.

1. Anatomy: The Pelvis. A description of the male and female reproductive systems in humans. Definitions of terms.

2. Anatomy: The Pelvis. Development of the urogenital region.

3. Histology—The Web Laboratory. Links to units on many systems of the human body, including the male and female reproductive systems and the urinary system. Extensive coverage of microstructure of all parts of these systems.

4. The Basics of the Kidney. A good introduction to the basic function of the kidney is presented at this site.

5. Reproduction: A Last Hope for Some Endangered Species. This is a page from the National Zoological Park. It explains the importance of reproductive technologies for some rare animals and the importance of a large gene pool for a population.

16

Nervous System

In this chapter, we will examine the nervous system and its parts. We will see how the components are assembled and what they do to ensure survival. Much of our attention will be focused on neurons, the living cells that perform the actual conductive and secretory activities of the system.

OUTLINE

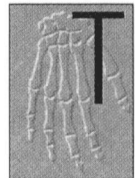he craniate nervous system plays three basic roles. It acquaints the organism with its external environment and stimulates the organism to orient itself favorably in that environment; it participates in regulation of the internal environment; and it serves as a storage site for information. These functions are accomplished by the nerves, spinal cord, and brain in association with **receptors** (sense organs) and **effectors** (chiefly muscles and glands).

The organism must constantly monitor the external environment. Information is supplied by **afferent (sensory) nerves** commencing in sense organs. The response (body movement) is initiated by nerve impulses over **efferent (motor) nerves** that stimulate the skeletal muscles of the body and thus cause the fish to swim or the tetrapod to crawl, run, or fly. Information from the external environment is also employed in the regulation of internal secretions, such as seasonal release of reproductive hormones (see fig. 18.5).

The organism also has an internal environment that must be continually monitored and controlled. Afferent nerves from visceral receptors carry information in the form of nerve impulses to the central nervous system; efferent nerves carry impulses from the center to visceral effectors, chiefly smooth and cardiac muscles and glands.

Memory (information storage) is a function of the nervous system. Without information storage and recall, no animal could modify its behavior in accordance with experience, and every situation would be faced as if it were the first time. As experiences multiply, information accumulates and the penalty of past errors and the rewards of successes modify behavior accordingly.

The nervous system is subdivided for convenience into central and peripheral nervous systems. The **central nervous system (CNS)** consists of the brain and spinal cord. The **peripheral nervous system** consists of cranial, spinal, and autonomic nerves. Cranial nerves emanate from the brain and spinal nerves from the spinal cord.

NEURON

To understand the anatomy of the nervous system, one must be acquainted with the **neuron**—the living nerve cell. The neuron is to the nervous system what a muscle cell is to the muscle system: It performs the specific function of the system. In this instance, it is to transmit nerve impulses.

Neurons exhibit many shapes, but all have a cell body containing **Nissl material** (sites of high protein synthesis) and one or more processes (figs. 16.1 and 16.2). The longest process, distinguished by the absence of Nissl material, is an **axon.** It transmits nerve impulses to other neurons or to an effector. Axons are surrounded by a layer of fatty **myelin** that varies in thick-

ness from heavy to ultramicroscopic, depending on the role of the axon (fig. 16.3). The axon and its myelin constitute a **nerve fiber.** On the surface of nerve fibers outside the CNS is a sheath of living cells, the **neurilemma.** The neurilemma produces and maintains the underlying myelin.

Nerve fibers within the CNS extend for short or long distances up and down the spinal cord or within the brain, aggregated into functional bundles called **fiber tracts** (T in fig. 16.4). Nerve fibers in the peripheral nervous system are in nerves (see fig. 16.3). In fact, a **nerve** is one or more bundles of nerve fibers outside the CNS, wrapped in a fibrous sheath (**epineurium**) and supplied by blood vessels.

The other processes of neurons are **dendrites** (see fig. 16.2). They are short extensions of the cell body that provide additional surface areas for receipt of incoming impulses from other neurons. Their cytoplasm, like that of the cell body, contains Nissl material. To observe how the neuron fits into the peripheral nervous system, we will examine a sensory nerve, a motor nerve, and two mixed nerves.

A typical sensory nerve is diagrammed in figure 16.4*a*. Its fibers commence in a sense organ (in this instance, the membranous labyrinth) and terminate in the brain (this example) or the spinal cord. The cell bodies of sensory nerve fibers, with few exceptions, are found in a **sensory ganglion** on the pathway of the nerve. A **ganglion** is a group of cell bodies outside the central nervous system. A sensory ganglion contains sensory cell bodies. In basal craniates some sensory cell bodies are scattered along the nerve.

A typical motor nerve is diagrammed in figure 16.4*b*. The cell bodies of motor nerve fibers in cranial and spinal nerves are inside the CNS in a **motor nucleus.** Neurologically speaking, a **nucleus** is a group of cell bodies within the brain or cord. Motor nuclei contain the cell bodies of motor nerve fibers. The motor fibers of cranial nerve XII terminate peripherally in striated muscle. There are few purely motor nerves in craniates because most nerves supplying somatic muscles have sensory fibers for proprioception from the muscle (see fig. 17.26).

Mixed nerves contain both sensory and motor fibers (fig. 16.5). Their sensory cell bodies are in sensory ganglia and, excepting some nerves of the autonomic system, their motor cell bodies are in motor nuclei. The different fibers in a mixed nerve are simultaneously carrying impulses in opposite directions and at different speeds. Most craniate nerves are mixed.

The site where a nerve impulse is transferred from one neuron to another is a **synapse.** As an axon approaches a synapse, it sprays into a multitude of fine branches, or **telodendria,** each of which ends in a **synaptic knob (terminal button)** that is in contact with the cell body, dendrite, or axon of the next neuron (see fig. 16.2). Nerve impulses are propagated across the

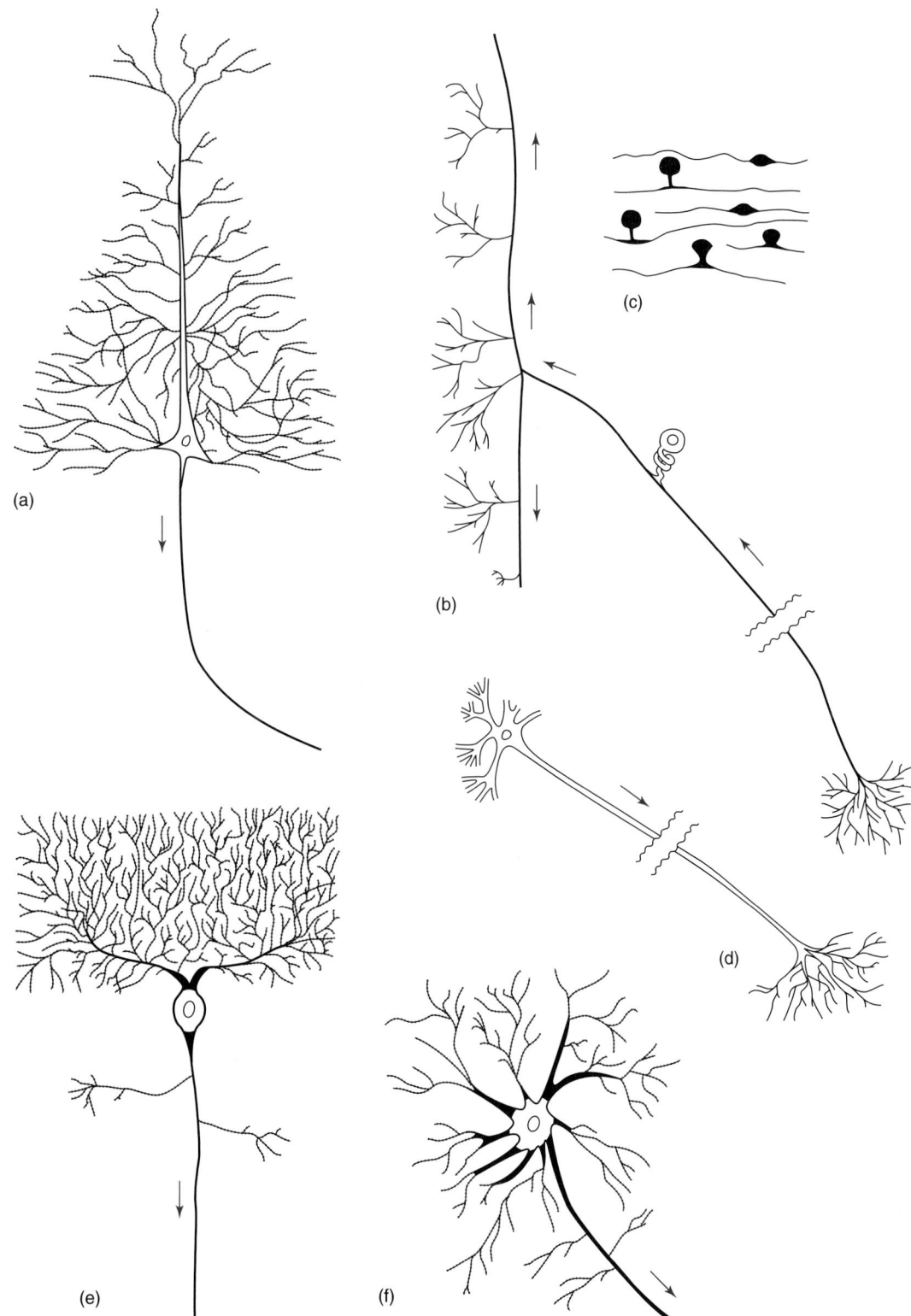

FIGURE 16.1

Several morphological types of neurons. (*a*) Pyramidal cell from motor cortex. (*b*) Dorsal root ganglion cell, unipolar. The single process is an axon. Ascending and descending branches are in fiber tracts in the spinal cord. (*c*) Embryonic dorsal root ganglion cells in transition from bipolar to unipolar. (*d*) Motor cell body from cord or brain stem. The axon ends on a motor end plate (see fig. 11. 2). (*e*) Purkinje cell from cerebellum. (*f*) Motor cell from sympathetic ganglion. *Arrows* indicate direction of impulses.

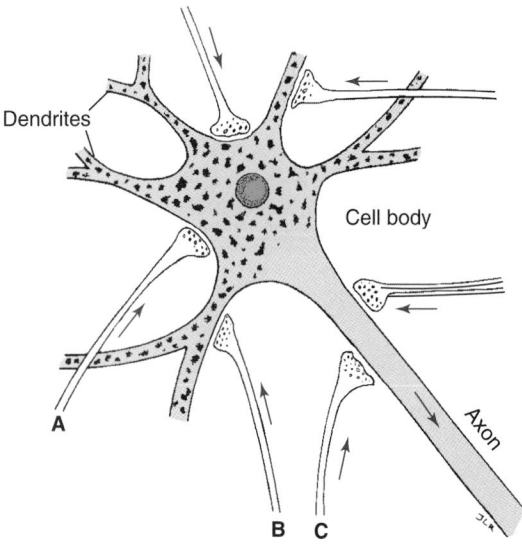

FIGURE 16.2

Synaptic endings on a motor neuron. **A,** synapse between the synaptic knob and a cell body; **B,** between synaptic knob and a dendrite; **C,** between synaptic knob and another axon. Dark splotches in dendrites and cell body are Nissl material. The knobs contain neurotransmitters.

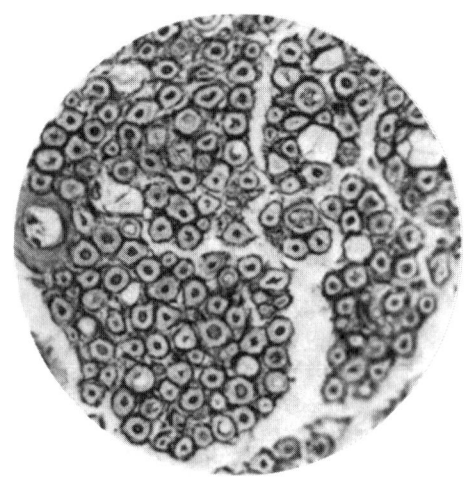

FIGURE 16.3

Cross section of a portion of a nerve. The axons (*black dots*) are surrounded by myelin sheaths (*white*) of varying thickness. The dark ring around the myelin is the neurilemma.

FIGURE 16.4

Typical locations of cell bodies (*black*) of sensory and motor fibers.
(*a*) Sensory nerve with sensory cell bodies in a sensory ganglion. Upon entering the brain, the fibers branch and pass toward synapses in several directions. (*b*) Motor nerve (hypoglossal) with motor cell bodies in a motor nucleus in the brain.
T, descending fiber tract (corticospinal);
X, decussating fibers.
Arrows indicate direction of nerve impulses. Note the exceptional bipolar nature of the cell bodies in the vestibular ganglion. The cell bodies in most sensory ganglia are unipolar as in figure 16.5.

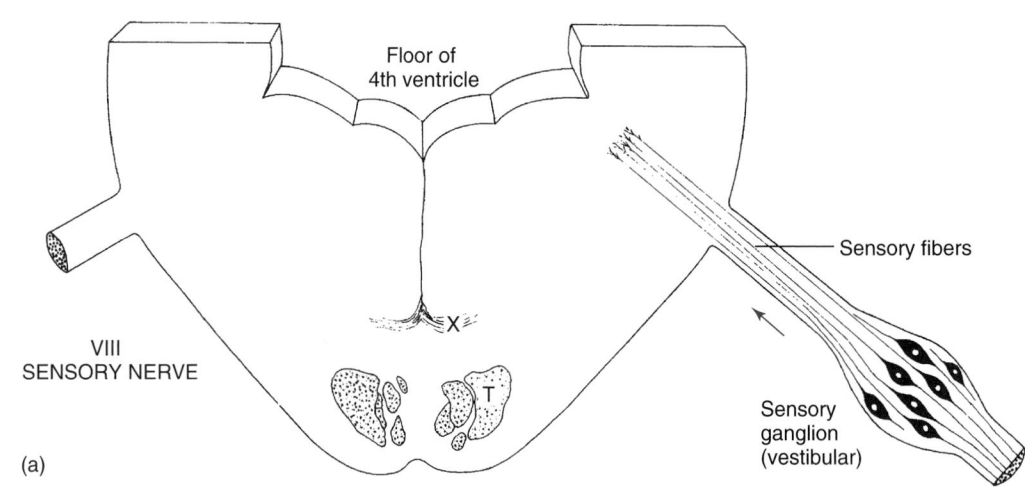

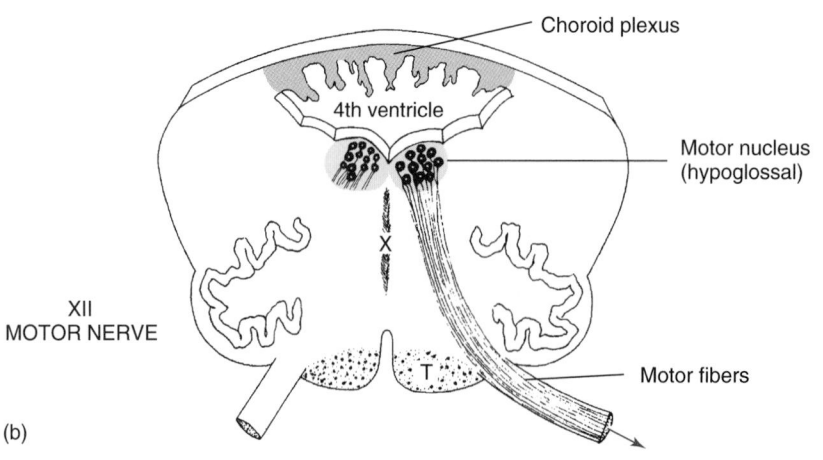

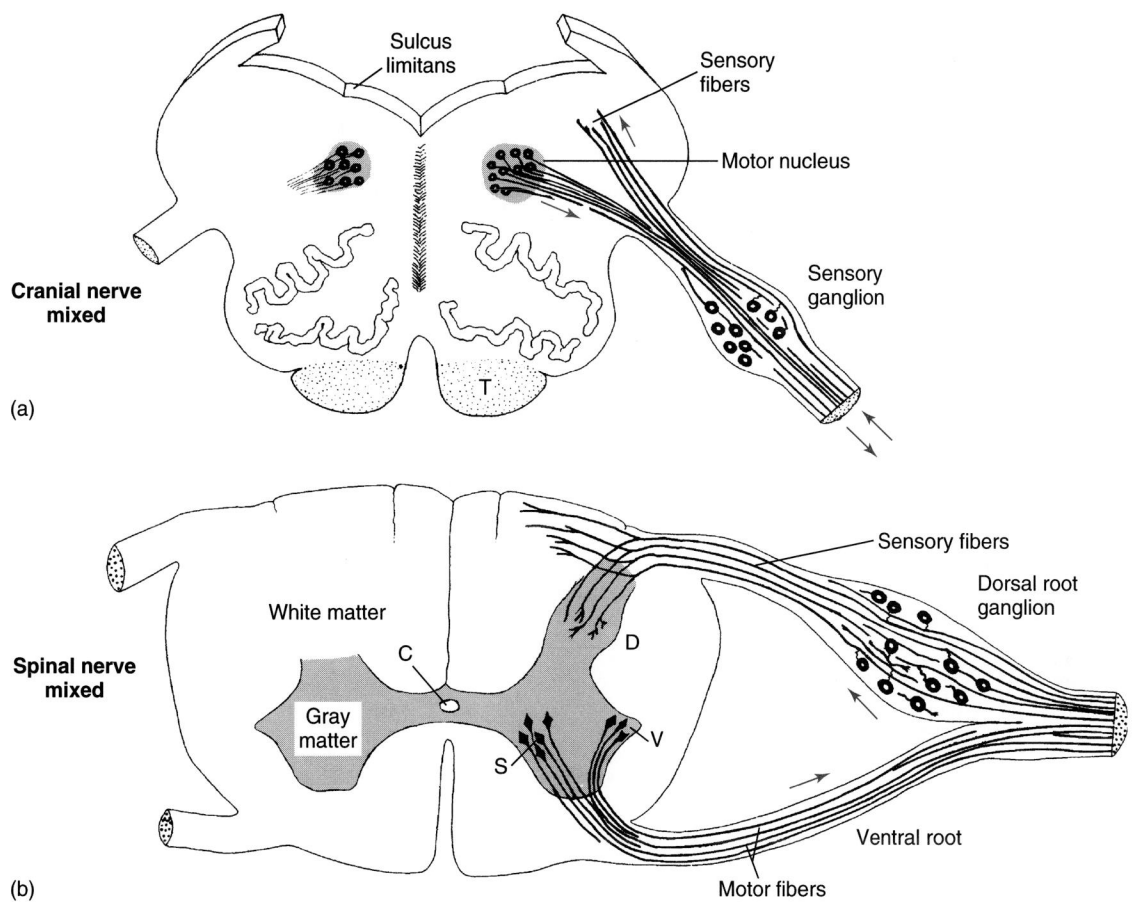

FIGURE 16.5

Locations of cell bodies of mixed nerves. (*a*) Cranial nerve with sensory cell bodies in a ganglion and motor cell bodies in a motor nucleus in the brain. **T,** descending fiber tract. (*b*) Spinal nerve with sensory cell bodies in a ganglion and motor cell bodies in gray matter of ventral horn of cord. **C,** central canal; **D,** dorsal horn of gray matter; **S,** somatic motor nucleus in ventral horn; **V,** visceral motor nucleus in lateral horn. *Arrows* indicate direction of nerve impulses.

synapse by short-lived secretions, chiefly amino acids and amines such as norepinephrine, acetylcholine, serotonin, melatonin, and others, which are released from the synaptic knobs when an electrical nerve impulse arrives. These amino acids and amines are **neurotransmitters.** Neurotransmitters are also released from axon terminals in contact with effectors, causing the effector (muscle, gland, pigment cell) to respond (see fig. 11.2). The short life of neurotransmitters ensures against a protracted response from a single nerve impulse.

Most long axons within the central nervous system give off **collateral branches** along their pathway, thereby distributing a single impulse to sometimes thousands of neurons. Collateral branches are seen on the ascending and descending branches of the sensory neuron illustrated in figure 16.1*b*.

Some neurons (**neurosecretory neurons**) with cell bodies in the central nervous system secrete small polypeptides from their axon terminals. These polypeptides are **neurohormones** (neurosecretions). Instead of terminating in synapses or at effectors, most neurosecre-

tory fibers terminate at sinusoidal vascular channels into which they release their secretions (see fig. 18.1). The antidiuretic hormone, which is released in the posterior lobe of the pituitary gland, is such a secretion (see fig. 18.2, neurosecretory fiber).

GROWTH AND DIFFERENTIATION OF THE NERVOUS SYSTEM

To gain additional insight into the architecture of the nervous system, it is important to understand how the components of the nervous system develop. **Neurulation,** the initial stages in the formation of the nervous system, and the morphogens that regulate its induction, are discussed in chapter 5. (You may be interested in homeotic genes, which are the source of these morphogens.) Neurulation results in formation of a dorsal, hollow, neural tube. The central cavity is the **neurocoel** (see fig. 16.7*a*). The next discussions describe maturation of the neural tube and the embryonic sources of motor and sensory neurons.

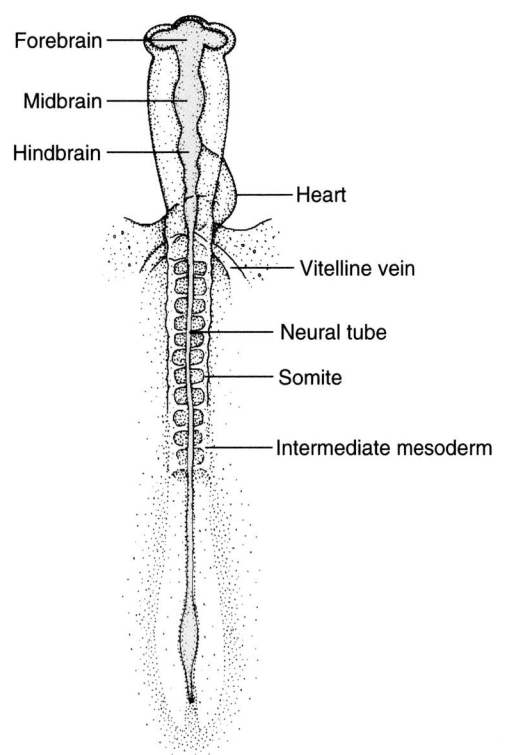

FIGURE 16.6
Neural tube (*red*) of a 33-hour chick embryo viewed through tissue rendered transparent. The optic vesicles are beginning to evaginate from the forebrain.

Neural Tube

An early stage in neurulation is seen in figure 5.9*a–c*. A later stage is illustrated in figure 16.6. The cephalic end of the neural tube becomes the brain. The rest is future spinal cord. A cross section of the embryonic spinal cord at this stage is seen in figure 16.7*a*. Shortly thereafter (fig. 16.7*b*), it displays three zones, a **ventricular zone** (germinal layer) of actively mitotic cells, an **intermediate zone** (mantle layer) of cells proliferated from the germinal layer, and a **marginal zone** that is practically devoid of nuclei. The ventricular zone (site of germinative cells) is the source of cells in the intermediate zone. Germinative stem cells undergo numerous mitotic divisions, which eventually lead to two lineages of cells—neuronal and neuroglial lineages. The former is the source of **neuroblasts**, which sprout dendrites and an axon, and become neurons. The latter lineage forms a variety of **neuroglia**, the supportive cells of the brain and cord. As neuroblasts and neuroglia differentiate, they contribute to the volume of the intermediate zone, and axons from differentiating neuroblasts grow into the marginal zone, contributing to it. Because axons eventually become coated by fatty myelin, the marginal layer appears white, hence, it is **white matter.** The preponderance of cytoplasm in the intermediate zone, which is

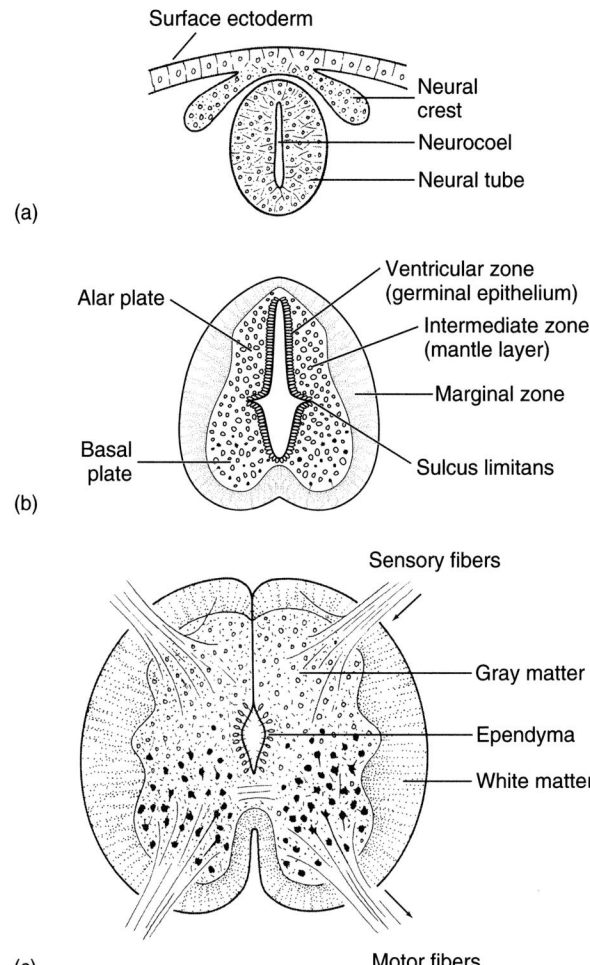

FIGURE 16.7
Differentiation of neural tube at spinal cord level. (*a*) Cross section of neural tube caudal to brain. (*b*) and (*c*) Later developmental stages. In (*c*), sensory fibers from a dorsal root ganglion are seen entering alar plate to synapse with second-order sensory neurons, and motor fibers are emerging from basal plate neuroblasts to innervate striated muscles.

composed chiefly of cell bodies, causes this zone to appear gray—hence the name **gray matter.**

The embryonic cord, hindbrain, and midbrain consist of **alar and basal plates** located, respectively, above and below a **sulcus limitans** (figs. 16.7*b* and 16.31). The cell bodies that develop in the alar plate receive sensory impulses from cranial and spinal nerves and distribute them elsewhere in the cord or brain. Those in the basal plate become the cell bodies of motor neurons (fig. 16.7*c*).

Eventually, the cells of the ventricular zone cease to divide, and thereafter additional neurons are rarely formed. The undifferentiated cells that remain adjacent to the neurocoel then become **ependymal cells.** The **ependyma** is the nonnervous lining of the adult neurocoel (figs. 16.7*c* and 16.8*b*, 5).

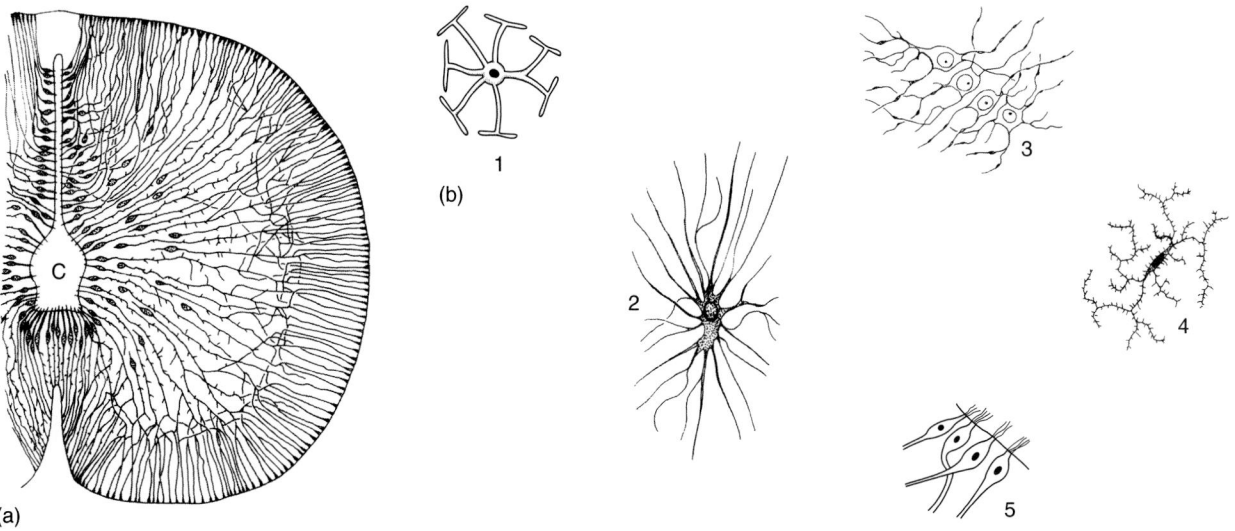

FIGURE 16.8

Neuroglia. (*a*) Neural tube of a 10-week human fetus stained to show neuroglial progenitor cells in the germinal, mantle, and marginal layers. Those that remain in the germinal layer at birth will have become ependymal cells. (*b*) Differentiated neuroglia. **1,** protoplasmic astrocyte; **2,** fibrous astrocyte; **3,** oligodendroglia; **4,** microglia; **5,** ependymal cells. **C,** neurocoel.

(*a*) Source: From Cajal, *Histologie du Systeme Nerveux de L'Homme et des Vertebres,* Vol. 1, 1909, A. Maloine, Paris.

Development of Motor Components of Nerves

Most of the axons that sprout from neuroblasts in the basal plate grow out of, and away from, the neural tube to make contact with striated muscle cells. These axons become motor fibers of the cranial and spinal nerves. Because they sprout from neuroblasts within the central nervous system, the cell bodies of these motor fibers are within the adult brain or cord (see figs. 16.4*b* and 16.5).

Other fibers that sprout from neuroblasts in the basal plate grow out of the neural tube to make contact with neuroblasts in autonomic ganglia (see fig. 16.36, sympathetic ganglion). Neuroblasts in autonomic ganglia sprout **postganglionic fibers** (see fig. 16.36, Post) that grow toward and innervate smooth muscles and glands. Thus, *all motor neurons have their cell bodies in the cord or brain with one exception—postganglionic neurons of the autonomic system have their cell bodies in autonomic ganglia.* Neuroblasts of autonomic ganglia are migrants derived from neural crest (next paragraph).

Development of Sensory Components of Nerves

At the time the neural groove is closing to form a tube, a longitudinal ribbon of ectoderm separates from the developing neural tube dorsolaterally on each side and soon segments to form a metameric series of **neural crest,** one pair in each body segment (see figs. 5.9b and 16.39). Some neural crest cells become neuroblasts that give rise to sensory neurons in spinal and cranial nerves. In doing so, they pass through a bipolar stage (see fig. 16.1*c*) in which one process grows into the alar plate of the cord or brain and the other grows to a sense organ. Thus, there is established a neuronal connection

between sense organ and central nervous system. Ganglionic cells associated with some cranial nerves are derived from ectodermal placodes. Because neural crest (cranial and spinal nerves) and ectodermal placodes (some cranial nerves) give rise to a large number of sensory cell bodies, the result is a swelling, the sensory ganglion, on the nerve close to the cord or brain. Neurons with cell bodies in sensory ganglia are **first-order sensory neurons.** They conduct impulses from a sense organ to the central nervous system. Within the alar plate, they synapse with **second-order sensory neurons (association neurons)** that distribute the impulses elsewhere in the central nervous system. Because the cell bodies of first-order sensory neurons arise from neural crest and ectodermal placodes, we can make the following generalization: *The cell bodies of sensory neurons of cranial and spinal nerves are generally in sensory ganglia on the pathway of the nerves.*

There are three major exceptions to the rule that cell bodies of first-order sensory neurons are in ganglia on the pathway of nerves: (1) Olfactory nerve fibers sprout from neuroblasts (derived from ectodermal placodes) in the embryonic olfactory epithelium, and their long processes grow into the nearest part of the brain, which is the olfactory bulb. Therefore, *the cell bodies of olfactory nerves are in the olfactory epithelia* (see figs. 16.28*a* and 17.21). (2) Neuroblasts that give rise to sensory fibers of the optic nerves differentiate in the nervous layer of the embryonic retinas, and long processes grow brainward along the optic stalk (see fig. 17.6*b*). Therefore, *the cell bodies of optic nerve fibers are in the retina.* The retina is actually part of the brain because it arises as an evagination from the diencephalon and never separates from it (see fig. 17.6b). (3) The cell bodies of **proprioceptive fibers** (sensory fibers that monitor the activity of skeletal muscles) in the cranial nerves (except XI and XII), are not in ganglia. Neuroblasts within the alar plate of the embryonic midbrain sprout long processes that grow out to the muscles and tendons innervated by these nerves, providing them with proprioceptive innervation (see fig. 17.26). Therefore, *the cell bodies of proprioceptive fibers in the cranial nerves (except XI and XII) are in the mesencephalic nucleus of the midbrain.* (The cell bodies of other proprioceptive fibers associated with XI and XII are in sensory ganglia.)

Olfactory neurons and the rods and cones of the retina, all of which serve as first-order sensory neurons, are termed **neurosensory cells** because they receive the stimulus directly instead of via an intervening epithelial cell that usually serves as a transducer of sensory stimuli. An axonic process of the neurosensory cell transmits the impulse to a synapse. Neurosensory cells are the only afferent neurons in many diploblastic invertebrates.

The comparison of this level of diploblastic organization to that in craniates is difficult, if not impossible, without considering triploblastic outgroups to craniates (for example, amphioxus). Cephalochordates possess an epidermal nerve plexus and intraspinal sensory neurons (**Rohon-Beard cells**) but lack sensory structures equivalent to olfaction and paired eyes. In lampreys, we see Rohon-Beard cells (lost in amniotes and hagfish) and sensory cells derived from either neural crest or ectodermal placodes (embryonic cranial ectoderm, chapter 5). Additionally, lampreys possess paired external sensory organs including olfaction and sight. Efforts have been made to homologize the epidermal nerve plexus with neural crest or ectodermal placodes (see Northcutt and Gans, 1983, for a review), but it now appears that neutral crest, ectodermal placodes, and paired external sense organs represent new structures in craniates (Northcutt, 1996).

NEUROGLIA AND NEURILEMMA

Not all the undifferentiated cells of the embryonic neural tube become neuroblasts. Nearly half the bulk of the brain and cord consists of a variety of interstitial cells known as neuroglia, which develop from neuroglial progenitor cells (see fig. 16.8a). They are generally smaller than most neurons, have nonnervous dendritic processes, and fill all the interstices in the brain and cord that are not occupied by neurons or blood vessels.

The oldest neuroglial elements phylogenetically are **ependymal cells.** These line the neurocoel of the cord and brain. Each has cilia that project into the fluid of the neurocoel and a long process that extends radially, sometimes to the circumference of the neural tube (see fig. 16.8a). Ependymal cells are the sole neuroglial elements of amphioxus and agnathans, and they are a stage in the differentiation of neuroglial cells in other craniates. Blood vessels do not enter the spinal cord of an amphioxus; the long processes of the ependymal cells are thought to supply nutrients from the neurocoel to the neurons.

A progressive increase in the variety of glial cells is seen from living agnathans to teleosts and from amphibians to amniotes. Some ependymal cells lose their connection with the periphery of the neural tube, become isolated among nerve cell bodies and their processes, and become functionally specialized. Other glial cells form a **glial membrane** on the surface of the cord and brain. Among the varieties of glial cells are astrocytes, microglia, and oligodendroglia.

Astrocytes, so named because they have the shape of a radiant star, are interposed between blood capillaries and the cell bodies of adjacent neurons. They obtain nutrients from the bloodstream and transfer these to the neurons (see fig. 16.8b). **Microglia** are phagocytic, engulfing bacteria and cellular debris, which they digest. Unlike other glial cells, they are of mesodermal origin (and are not found in the developing brain until after blood vessels have penetrated the substance of the brain). The cytoplasmic processes of **oligodendroglia** wrap around naked axons in the brain and cord and elaborate myelin. When a nerve impulse sweeps along an axon, there is a change in electric potential on the surface of the axon. The fatty sheath insulates axons against action potentials in adjacent fibers and speeds conduction of the nerve impulse within the axon.

In the early stages of formation of nerves, when motor axons are growing out of the basal plate and sensory axons are arriving in the alar plate (see fig. 16.7c), neural crest adjacent to the developing cord and brain migrate outward along the developing nerve root, mature, embrace the naked axons, and deposit myelin on them. These oligodendroglialike cells persist throughout life as **neurilemmal cells** (originally named **Schwann cells** after their discoverer). Heavily myelinated fibers conduct nerve impulses faster than lightly myelinated ones. The thickest sheaths are on fibers from encapsulated nerve endings for touch (chapter 17), on proprioceptive fibers, and on motor fibers that supply skeletal muscles. They are thinnest on visceral fibers. Life in the raw is seldom mild, as the saying goes, and a speedy reaction to environmental threats, as opposed to visceral ones, increases the chances of survival.

SPINAL CORD

The spinal cord occupies the bony vertebral canal formed by the centra of the vertebrae and the successive neural arches. It is flattened in agnathans but tends to be rounded or quadralateral as a derived condition in some tetrapods. Generally, the neurocoel is relatively large in craniates, narrower or even obstructed at some levels in mammals.

Closely applied to the outer glial membrane of the cord and brain in most fishes is a delicate connective tissue membrane, the **meninx primitiva**. A similar meninx appears during embryonic development in nonmammalian craniates, but it eventually differentiates into an inner vascular **leptomeninx** and a dense outer membrane, the **dura mater**. Mammals have three meninges—an inner **pia mater**, a filmy weblike **arachnoid**, and a tough fibrous dura mater. The latter loosely surrounds the other meninges. The pia mater adheres closely to the glial membrane of the cord, and it is attached to the arachnoid by a network of fibrous strands that traverse a subarachnoid space filled with cerebrospinal fluid. A similar subdural space separates the arachnoid from the dura mater, and a peridural space filled with adipose tissue separates the cord and its meninges from the bony vertebral canal. The cerebrospinal fluid and the cushion of fat buffer the mammalian cord against mechanical trauma. (The same meninges surround the brain, but there is no peridural fat, the dura mater having a somewhat different relationship to the periosteal lining of the skull.)

The spinal cord begins at the foramen magnum, but there is no abrupt transition between cord and brain. Instead, there is a gradual rearrangement of the gray matter in the transitional area. In the cord, the gray matter is confined to the area surrounding the neural canal. In the brain stem (brain without the cerebral hemispheres and cerebellum), it is dispersed into many separate gray masses, or nuclei, separated by fiber tracts. The transition from cord to brain in mammals is completed over the length of about one body segment (see fig. 16.10, compare *a* with *b*). In other craniates the transitional region is longer, and the nerves in the transitional region (occipitospinal nerves, see fig. 16.32) are atypical.

The adult cord is coextensive with the vertebral column in craniates with abundant tail musculature, as in some urodeles. In these species, each spinal nerve emerges from the vertebral canal via an **intervertebral foramen** at the same level at which it comes off the cord. Generally, however, the embryonic vertebral column elongates more rapidly than the cord, with the result that the adult cord is shorter than the column. Consequently, the spinal nerves of the lower trunk must pass caudad within the vertebral canal before exiting via their foramen. In doing so, they constitute a bundle, the **cauda equina**, or "horse's tail" (fig. 16.9). The ependyma and meninges of the cord surround the bundle, enclosing it in a connective tissue sheath. After the last nerves have exited, the sheath alone continues for a distance as a threadlike **filum terminale**. It, too, eventually terminates. In humans, the adult cord ends at the upper border of the second lumbar vertebra, in cats and

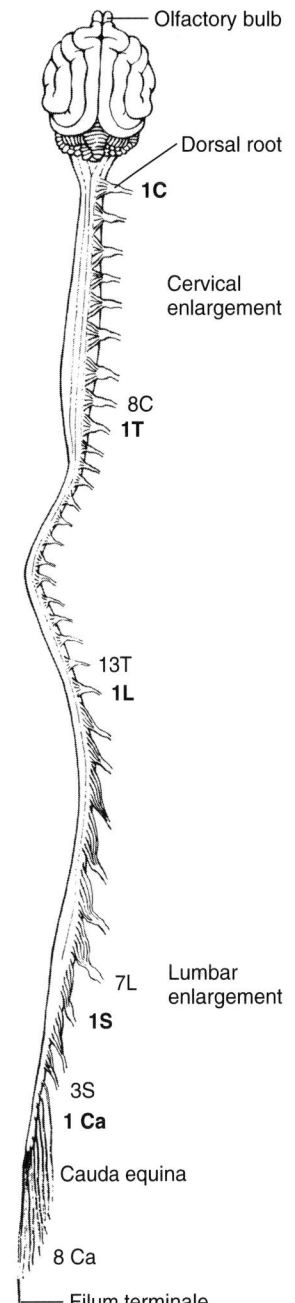

FIGURE 16.9

Brain, cord, and dorsal rootlets of spinal nerves of a cat. The peridural fat and dura mater have been removed. Spinal nerves identified by region are **C**, cervical; **T**, thoracic; **L**, lumbar; **S**, sacral; and **Ca**, caudal. The first and last nerves of each region are numbered.

rabbits at about the middle of the sacrum, and in frogs anterior to the urostyle. In a few bony fishes, the spinal cord is actually shorter than the brain.

The spinal cord exhibits cervical and lumbar enlargements at the levels of the anterior and posterior appendages. The enlargements result from the large number of cell bodies and fibers required to innervate the

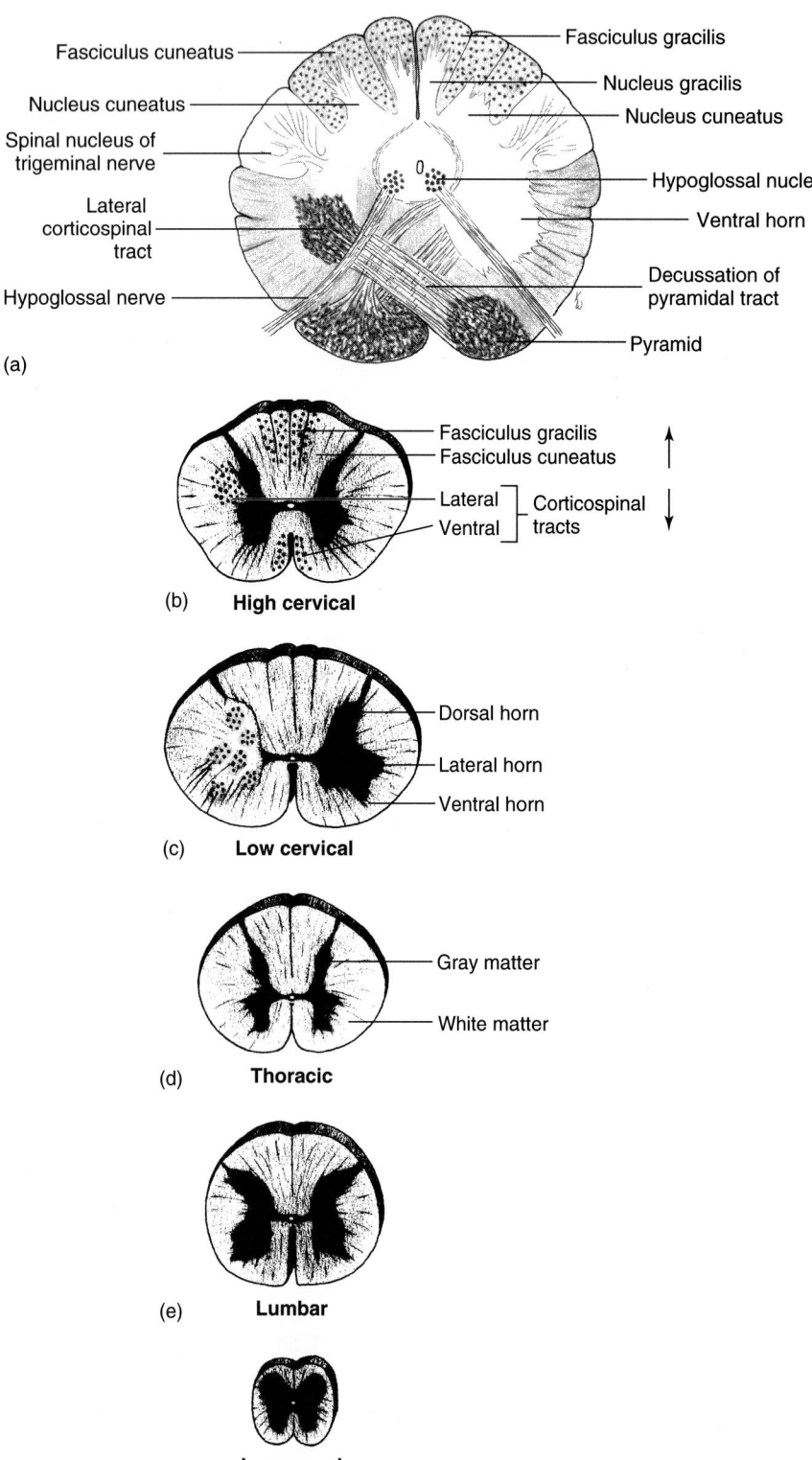

(a)

(b) **High cervical**

(c) **Low cervical**

(d) **Thoracic**

(e) **Lumbar**

(f) **Low sacral**

FIGURE 16.10

Selected cross sections of the human spinal cord. (*a*) Transitional zone between human medulla and cord. Descending fibers in the pyramidal tracts are decussating to continue caudad on the opposite side as the lateral corticospinal tract. The spinal nucleus of the trigeminal nerve is a caudal continuation of the nucleus of that nerve. The ventral horn of the cord is coming into view. In (*b*), two ascending and two descending fiber tracts (*arrows*) are identified. In (*c*), six nuclei of the gray matter are highlighted in *red*. (*a* is a composite drawing of selected features of several successive stained microsections.)

appendage. When one pair of appendages is exceptionally muscular, as were the hind limbs of massive dinosaurs, the corresponding enlargement of the cord is pronounced. In dinosaurs, it was often larger than the brain. Conversely, the spinal cord of turtles is very slender in the trunk because much of the body wall musculature internal to the shell is vestigial. The cord of many fishes exhibits a neuroendocrine swelling, the **urophysis**, near its caudal end (see fig. 18.4).

Cross sections of a cord, appropriately stained, reveal gray matter, white matter, nuclei, fiber tracts, and other features (fig. 16.10). Nuclei are responsible for gray matter. Nerve fibers occupy the periphery of the cord and, with neuroglia, constitute the white matter. Ascending and descending fiber tracts (arrows in 16.10*b*) conduct impulses with a similar function up or down the cord and to or from the brain. The **fasciculus gracilis**, for example, consists of heavily myelinated fibers for touch and proprioception. These fibers synapse in the **nucleus gracilis**. From there, the impulse is carried in other fibers until it arrives in the **somesthetic cortex** (see fig. 16.26), where it evokes a specific sensation. **Corticospinal tracts** conduct motor impulses for voluntary muscle contraction from the voluntary motor cortex. There are many tracts in a mammalian cord. The agnathan cord has relatively few because their sensory and motor systems are relatively simple. The tracts and nuclei in the low cervical region of humans are large, reflecting, in part, the large number of sensory and motor fibers that innervate the primate hand.

SPINAL NERVES

Roots and Ganglia

Spinal nerves emerge from the spinal cord by an uninterrupted series of dorsal and ventral **rootlets** in each body segment (see figs. 16.9 and 16.30). The dorsal rootlets unite while still within the vertebral canal to form a **dorsal root** just proximal to the dorsal root ganglion, and the ventral rootlets unite to form a **ventral root** (see fig. 16.12). Except in lampreys, the roots then unite to form a spinal nerve.

Dorsal roots are predominantly or wholly sensory, and ventral roots are purely motor. However, it is evident that in the earliest craniates (1) dorsal and ventral roots did not unite; (2) dorsal roots contained visceral motor fibers as well as sensory fibers; (3) there were no dorsal root ganglia, bipolar sensory cell bodies being scattered within the nerve along its course; and (4) when sensory cell bodies first aggregated in ganglia, they were still bipolar.

As craniate nervous systems evolved, (1) dorsal and ventral roots united, (2) the dorsal roots became predominantly or totally sensory, (3) sensory cell bodies aggregated in the dorsal root to form ganglia, and (4) dorsal root ganglion cell bodies became increasingly unipolar.

Dorsal and ventral roots alternate in lampreys and do not unite; in hagfishes, they unite in the trunk but not in the tail. Some sensory cell bodies are in ganglia, some are scattered along the nerve, and most are bipolar. The dorsal roots contain visceral motor fibers as well as sensory fibers. Ventral roots are purely motor.

In gnathostomes, dorsal and ventral roots unite. The dorsal roots of many bony fishes contain visceral motor fibers, but in cartilaginous fishes and tetrapods, most or all of the visceral motor fibers have been lost from the dorsal root. The cell bodies of first-order sensory neurons are bipolar in cartilaginous fishes; bipolar, intermediate, and unipolar in bony fishes; chiefly unipolar in amphibians; and almost entirely unipolar in amniotes.

In an amphioxus, spinal nerves have only one typical root, this being the equivalent of the mixed dorsal root of basal craniates. It arises from the cord at the level of each myoseptum and enters the myoseptum. It is then distributed to the integument (sensory) and viscera (visceral motor). The cell bodies for the visceral motor fibers are in the cord, and those for the sensory fibers are either in the cord or scattered along the course of the nerve. There is no dorsal root ganglion. Sensory cell bodies are bipolar.

What grossly resembles a ventral root in amphioxus emerges from the cord between two myosepta and enters a myomere. Electron microscopy reveals that these ventral "roots" contain no nerve fibers. Instead, they consist of filamentous extensions of the striated muscle fibers from the myomere. These myofilaments enter the cord, where they receive their motor innervation directly from neurons within the central nervous system. Thus, the somatic muscles of amphioxus actually "come to the cord" for their stimuli. A similar condition is seen in echinoderms and some other invertebrates.

Occipitospinal Nerves

In many fishes and amphibians, one or more pairs of **occipitospinal nerves** arise between the last cranial nerve and the first pair of typical spinal nerves (see fig. 16.32). They generally lack dorsal roots and supply the hypobranchial musculature, including tongue muscles when present. Embryonic frogs have an occipitospinal nerve between the last cranial nerve and the first spinal nerve, the latter supplying the tongue, but the occipitospinal nerve is suppressed during later development. Cranial nerves XI and XII of amniotes lack dorsal roots and appear to be derived in part from occipitospinal nerves.

Spinal Nerve Metamerism

With the exception of the lampreys and hagfishes, a spinal nerve arises from each segment of the cord except near the end of the tail. The fibers in these nerves, other than fibers to the viscera, are distributed metamerically to the skin and muscles of the body wall and tail. Where fins or limbs develop, nerves at those levels innervate the appendages (fig. 16.11). The segmental distribution of spinal nerves is best observed in fishes because the metamerism of their body wall muscles is so evident. It is this metamerism, inherited with adaptive modifications by tetrapods from fishes to human beings, that enables swimming by lateral undulation (see fig. 11.8). Tadpoles swim like fishes, and they have as many as 40 spinal nerves. However, they lose all but the first 10

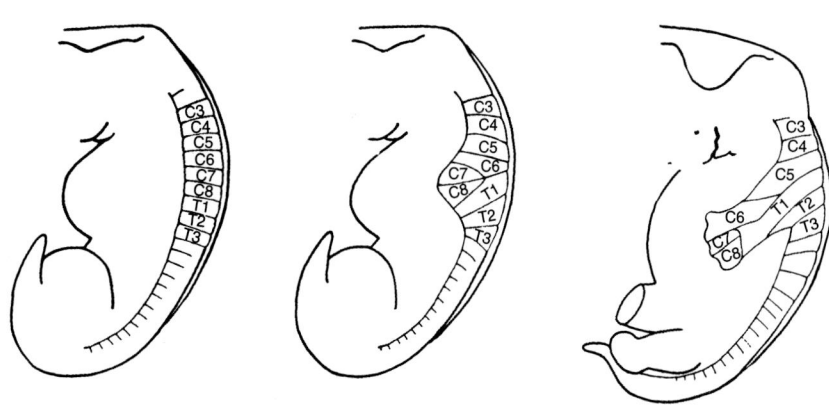

FIGURE 16.11
Metameric innervation of the skin of a mammalian forelimb. **C**, cervical, and **T**, thoracic somites and the skin areas supplied by their associated spinal nerves.

when, at metamorphosis, the tail is resorbed and tetrapod appendages develop.

Rami and Plexuses

Shortly after emerging from the vertebral canal, spinal nerves divide into two or more branches (fig. 16.12). A **dorsal ramus** supplies the epaxial muscles and skin of the back, and a larger **ventral ramus** enters the lateral body wall and supplies the hypaxial muscles and skin to the midventral raphe. In the thoracic and lumbar region of mammals, **white and gray rami communicantes** (sing., ramus communicans) connect the spinal nerves with the ganglia of the sympathetic trunk (figs. 16.12 and 16.35). These rami conduct impulses that have visceral functions.

The ventral rami of successive spinal nerves of craniates with fins or limbs unite to form **spinal nerve plexuses**—sites where two or more nerves or branches of nerves converge and their fibers commingle in one or more common trunks prior to being distributed. These plexuses are relatively simple in fishes but become increasingly complex in tetrapods (compare shark and mammal, see fig. 11.16). The chief spinal nerve plexuses are the **brachial** and **lumbar** (**cervicobrachial**

and **lumbosacral** in tetrapods that have cervical, thoracic, lumbar, and sacral vertebrae). Nerves emerge from spinal nerve plexuses to supply the fins or limbs. Highly complex spinal nerve plexuses with fibers that also supply pelvic viscera are present in the pelvis of mammals, including humans.

Functional Components of Spinal Nerves

The fibers in a typical spinal nerve are either sensory (afferent) or motor (efferent), and they are either somatic or visceral, in accordance with the structures they innervate. They represent, therefore, four functional varieties of nerve fibers: somatic sensory (afferent—SS), visceral sensory (afferent—VS), somatic motor (efferent—SM), and visceral motor (efferent—VM). These fibers are widely distributed throughout the body and are present in some cranial nerves. They are often referred to as *general* fibers to differentiate them from *special* fibers that are found only in cranial nerves (compare the differing terminology in table 16.5). Table 16.1 lists the kinds of structures they innervate.

BRAIN

The cranial end of the early embryonic neural tube of all craniates from fishes to humans exhibits three primary brain vesicles, a **prosencephalon** (future forebrain), **mesencephalon** (future midbrain), and **rhombencephalon** (future hindbrain, fig. 16.13*a*). The rhombencephalon was named for its shape. The primary vesicles soon differentiate to form the anlagen of the five major subdivisions of the adult brain: **telencephalon, diencephalon**, mesencephalon, **metencephalon**, and **myelencephalon** (fig. 16.13*b*).

Differentiation of the primary vesicles to form an adult brain is accomplished by (1) localized thickenings of the lateral walls and floor of some vesicles, and (2) dorsal, lateral, or ventral evaginations, median or paired, in others. The surface structures that develop in all craniates are readily seen to be homologues

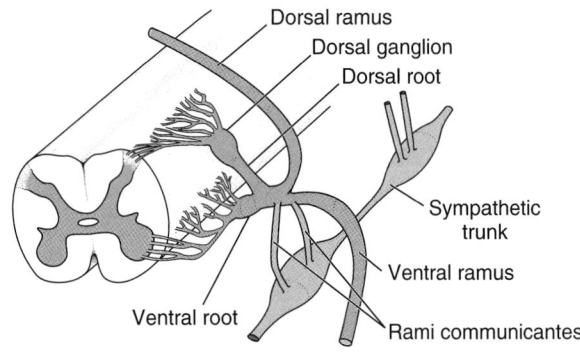

FIGURE 16.12
Primary branches of thoracic and lumbar nerves in mammals.

TABLE 16.1   Fiber Components of Typical Spinal Nerves[1]

Components[2]	Innervation
Sensory	
Somatic sensory (afferent) fibers (SS)	General cutaneous receptors (touch, pain, temperature, and pressure); receptors in striated muscle, tendons, and bursae (proprioceptive)
Visceral sensory (afferent) fibers (VS)	Viscera, including general receptors in endoderm
***Motor*[3]**	
Somatic motor efferent fibers (SM)	Myotomal muscle
Visceral motor efferent fibers (VM)	Smooth and cardiac muscle, and glands (via autonomic ganglia)

[1]These components are found also in some of the cranial nerves.
[2]Visceral (VS and VM) and somatic sensory (SS) of the spinal cord are classified as "general" fibers in some texts to distinguish them from "special" components found in cranial nerves.
[3]Motor fibers are not restricted to innervation of muscles and can include innervation of glands and vessels or modulation of receptors.

(fig. 16.14). Because the huge cerebral hemispheres and the cerebellum of mammals overlie the other structures, it is necessary to remove the hemispheres to reveal the underlying structures from a dorsal view. When they have been removed, what remains is the **brain stem** (fig. 16.15).

The embryonic prosencephalon differs from the rest of the neural tube in not being divided into alar and basal plates. Early in development, except in ray-finned fishes, it exhibits two pairs of evaginations, the **telen-cephalic vesicles**, the walls of which become the cerebral hemispheres, and the **optic vesicles**, which become the retinas of the eyeballs (see figs. 16.13c and 17.6). The optic vesicles become associated with the diencephalon. The mesencephalon does not undergo subdivision. The rhombencephalon becomes the metencephalon and myelencephalon.

The adult brain is surrounded by the meninges described earlier. The major components of the brain are listed in table 16.2. We will begin study with the least specialized region, the hindbrain.

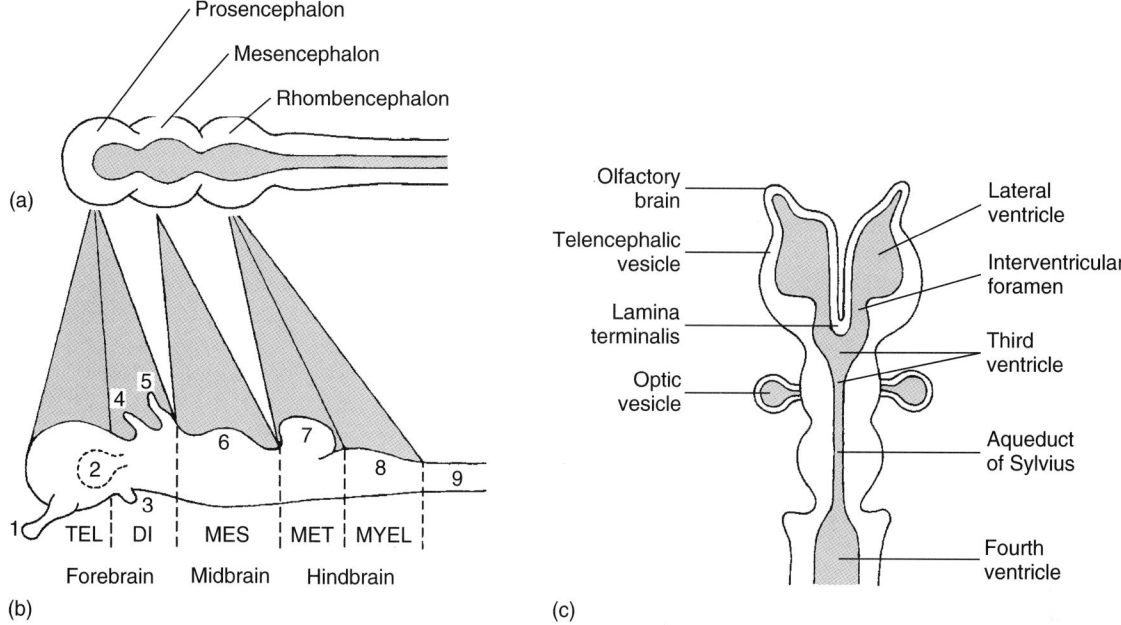

FIGURE 16.13

Primary brain vesicles and initial stages of differentiation, generalized scheme. (*a*) Three-vesicle stage, sagittal view. (*b*) Early differentiation, lateral view. **1,** olfactory bulb; **2,** future site of optic vesicle; **3,** neural lobe of pituitary; **4,** parapineal; **5,** pineal; **6,** optic lobe; **7,** cerebellum; **8,** medulla; **9,** postcranial neural tube. (*c*) Initial formation of cerebral hemispheres, except in actinopterygians, frontal section.

TABLE 16.2   Subdivisions of the Brain and Their Major Components

Subdivisions			Chief Components
Prosencephalon (forebrain)	Telencephalon	{	Cerebral hemispheres Lateral ventricles
	Diencephalon	{	Epithalamus Thalamus Hypothalamus Third ventricle
Mesencephalon (midbrain)		{	Tectum Tegmentum Cerebral aqueduct
Rhombencephalon (hindbrain)	Metencephalon	{	Cerebellum Tegmentum Fourth ventricle
	Myelencephalon	{	Medulla oblongata Fourth ventricle

Note: For a list of all components discussed in this chapter, see chapter summary.

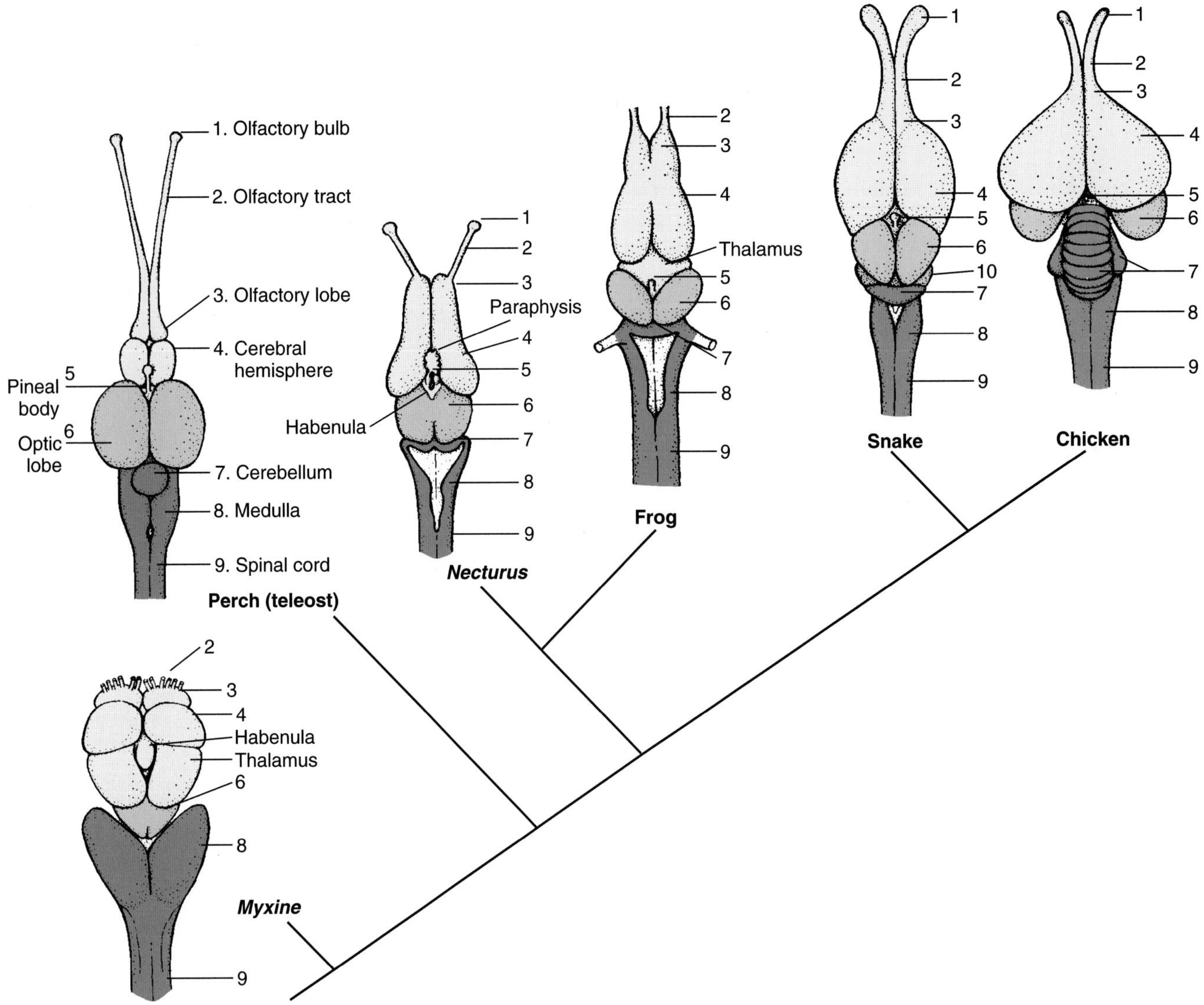

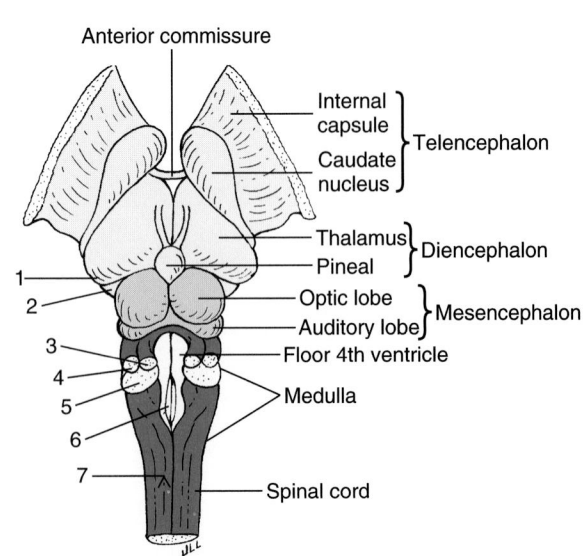

FIGURE 16.14
Craniate brains, dorsal view. The roof of the fourth ventricle has been removed in *Necturus,* frog, and snake. The forebrain, midbrain, and hindbrain are differentially colored. The habenuli, thalami, and pineal bodies are part of the diencephalon. 1 to 9, see perch for key; 10, auditory lobe of amniotes. Mammalian brains are shown in figure 16.23.

FIGURE 16.15
Brain stem of a sheep. The cerebral hemispheres and cerebellum have been cut away to reveal the underlying structures. 1, location of lateral geniculate body of thalamus; 2, medial geniculate body; 3 to 5, anterior, middle, and posterior cerebellar peduncles. The peduncles had to be cut to remove the cerebellum. 6, hypoglossal trigone in floor of the fourth ventricle (location of the hypoglossal nuclei); 7, posterior funiculus (ascending fibers for proprioception). Optic and auditory lobes constitute the corpora quadrigemina. The habenulae lie under cover of the pineal. A paired fiber tract, the stria medullaris, is seen passing toward the habenulae on the surface of the thalamus. Three shades of *red* indicate forebrain, midbrain, and hindbrain.

Metencephalon and Myelencephalon: The Hindbrain

The myelencephalon is represented chiefly by the **medulla oblongata**, which merges imperceptibly with the spinal cord. The area of transition is characterized internally by gradual relocation of the fiber tracts, which are the white matter (see fig. 16.10*a*). As a result, whereas in the cord the gray matter is centrally located, in the brain it is dispersed into small and large nuclear masses of gray matter interspersed among fiber tracts.

A conspicuous feature of the hindbrain is the **cerebellum**, a dorsal evagination of the metencephalon (see figs. 16.14, 16.23, and 16.32). In mammals it is connected to the brain stem by three pairs of **cerebellar peduncles** (stalks) composed of afferent and efferent fiber tracts (fig. 16.15, 3–5). The cerebellum coordinates the response of skeletal muscles to input from the membranous labyrinth, lateral-line canals of fishes, proprioceptors in muscles, joints, and tendons, and from reflex and voluntary motor centers in the brain stem and forebrain. Its size is correlated with the complexity of the activities of the striated musculature. It is larger in fishes than in amphibians because swimming, which involves schooling, vertical movements, adjusting to water currents, and keeping the dorsal part of the body from tipping over, requires more synergistic muscle activity than dragging the belly along the ground or squatting on a lily pad. Lacking a large cerebellum, aquatic urodeles rely to a large extent on spinal cord reflexes and primitive nuclei in the hindbrain for muscle coordination when swimming. The cerebellum is largest in birds and mammals, which require a large computerlike neural center to coordinate the muscles of the head, neck, trunk, and appendages in such diverse activities as flying, running, climbing, balancing, or, in the case of humans, playing a piano. The cell bodies of the cerebellum are on the surface. In living agnathans, the cerebellum is not well developed and does not cause a bulge on the brain.

Other topographical features of the hindbrain include various swellings caused by underlying nuclei, and elevated ridges or transverse bands from superficial fiber tracts. These topographical markings are most prominent in mammals. One sensory nucleus (**nucleus solitarius**) in the alar plate becomes enormous in fishes that have taste buds over the entire surface of the body. The resulting swelling, or **vagal lobe** (fig. 16.16), is the termination of the many incoming sensory fibers for taste. The nucleus contains second-order sensory neurons whose fibers are projected to reflex and relay centers elsewhere in the brain. Among ventral ridges on the mammalian hindbrain are the **pyramids** (figs. 16.10*a* and 16.17), which contain the corticospinal tracts that carry voluntary motor impulses from the cerebral cortex to the cord, and the **ventral decussation of the pons** in those mammals that have relatively large cerebellums (fig. 16.17). The hindbrain also houses a reflex vasomo-

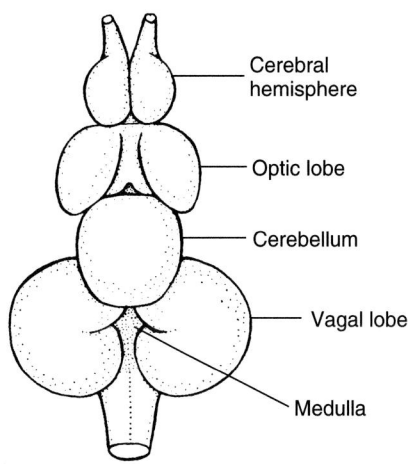

Modification for bottom-feeding

FIGURE 16.16

Brain of the buffalo fish *Carpiodes velifer*. Note unusual bulge (vagal lobe) on the alar plate of the medulla. Here terminate the many incoming taste fibers characteristic of bottom feeders.

tor center that maintains normal blood pressure and a reflex respiratory center that is vital in maintaining pulmonary respiration.

The cavity of the hindbrain is the fourth ventricle. The cerebellum is part of its roof (figs. 16.18 and 16.24). The remainder of the roof anterior and posterior to the cerebellum is a thin membrane that includes the ependyma of the neural tube and, posteriorly, a layer of highly vascular pia mater from which vascular tufts hang into the cavity. The tufts constitute the **choroid plexus of the fourth ventricle**.

Mesencephalon: The Midbrain

The **tectum**, or roof, of the mesencephalon displays a pair of prominent **optic lobes** in all craniates (see figs. 16.14 and 16.15). These bulging gray masses serve partly as reflex and relay centers that receive impulses originating in the retina. They are especially large in birds, which have large eyes and rely on visual stimuli for much information about the environment. A pair of **auditory lobes** lies caudal to the optic lobes in the tectum commencing with amniotes, and the four bodies (optic and auditory lobes) constitute the **corpora quadrigemina** (see fig. 16.15). Fishes have auditory nuclei in this location, but they are not large enough to bulge on the surface. The auditory lobes receive input from the part of the membranous labyrinth that is sensitive to vibratory stimuli, and also from other sources. Phylogenetically, they enlarge as the receptors for sound expand. The **mesencephalic nucleus of the trigeminal nerve** is a prominent gray mass in the alar plate. It contains the cell bodies of the proprioceptive fibers of most, if not all, of the cranial nerves that have them.

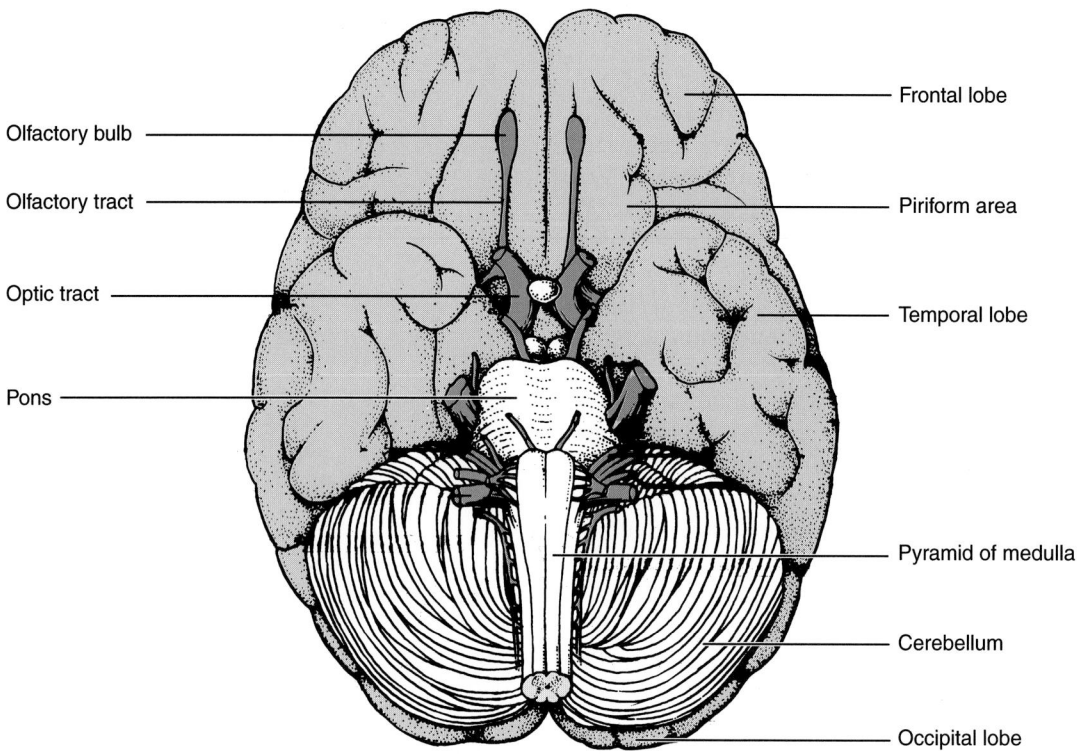

FIGURE 16.17
Human brain, ventral view.
From David Shier, et al., *Hole's Human Anatomy and Physiology,* 7th edition. Copyright © 1996 Times Mirror Higher Education Group, Inc., Dubuque, Iowa. All Rights Reserved. Reprinted by permission.

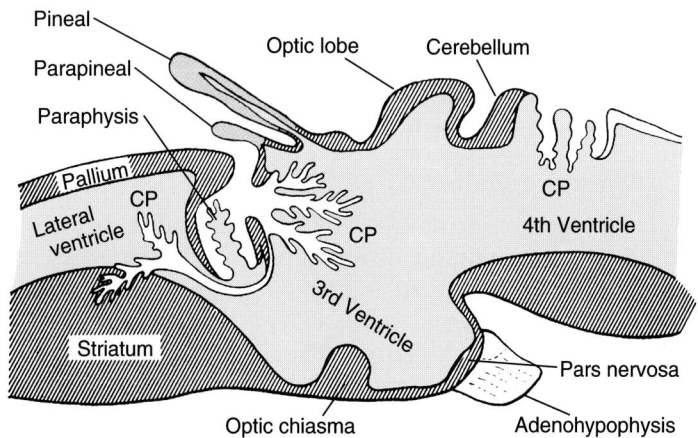

FIGURE 16.18
Diencephalon and adjacent areas of a larval anuran brain, sagittal section, anterior end to the left. **CP,** choroid plexus of lateral, third, and fourth ventricles. The choroid plexus of the lateral ventricle enters the latter via an interventricular foramen. *Red,* brain ventricles.

The basal plate of the mesencephalon, the **tegmentum,** is greatly thickened by nuclei and by aggregates of fiber tracts that connect higher levels of the brain with the hindbrain and spinal cord. In mammals, some of these nuclei and tracts are massive. One of these is the **red nucleus,** a large mass of gray matter that, in fresh brains, has a pink tinge. Most of its afferent fibers come from the cerebellum; some come from the cerebral cortex via a massive pair of fiber tracts, the **cerebral peduncles.** Efferent impulses from the red nucleus influence, directly or via relays, motor functions of the medulla and spinal cord, thus participating in control of the striated musculature.

The ventricle of the midbrain is quite large in fishes and amphibians and extends dorsad into the hollow optic lobes. The derived condition for the optic lobes is that they are not hollow, and the midbrain ventricle in this case is constricted to a narrow canal, the **cerebral aqueduct,** also known as the **aqueduct of Sylvius** and the **iter** (see figs. 16.13c and 16.22, human).

Diencephalon

The diencephalon consists of three major components, **epithalamus, thalamus,** and **hypothalamus** (fig. 16.19). Its neurocoel is the third ventricle.

Epithalamus

The epithalamus is the dorsalmost component of the diencephalon and, as such, constitutes the roof of the third ventricle. It consists of a **pineal** or **parapineal** organ, or both, a choroid plexus like that of the fourth

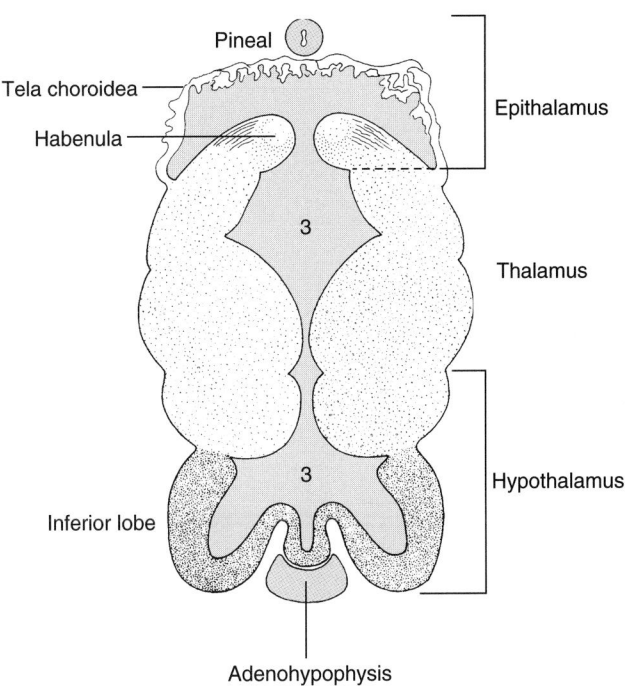

FIGURE 16.19
Major regions of diencephalon of a shark, cross section. Third ventricle, **3**, in *red*.

ventricle, and a pair of elevated thickenings (median in *Myxine*), the **habenulae** (fig. 16.19).

The pineal is a club-shaped or knoblike organ, sometimes threadlike or saccular, projecting above the diencephalon (see fig. 16.18). It is connected to the diencephalon by a stalk that frequently contains an extension of the third ventricle. In lampreys, the pineal is a photoreceptor. In gnathostomes, it functions as an endocrine organ that is stimulated in part by light that first enters the body through the retina. The neural pathway from retina to pineal, and the endocrine role of the pineal are described under "Pineal Organ" in chapter 18. The pineal is vestigial or absent in a few craniates, including hagfishes, crocodilians, and a few permanently aquatic mammals. It is relatively large in primates and sheep (see fig. 16.15).

A **median eye** was a constant feature of bony fishes and placoderms of the Devonian and of ancestral amphibians and amniotes. Living sharks and teleosts possess an endocrine pineal suggesting that the median organ of fossil fishes and early tetrapods also consisted of a pineal (whether it was photosensitive is uncertain). In contrast to lampreys, the parapineal in *Sphenodon* (see fig. 17.19) and lizards serves as a photosensitive organ—a **parietal eye.** When functional as a photoreceptive organ, it lies under a patch of translucent skin. The structure and neural connections of the parietal eye are described under "Median Eyes" in chapter 17.

The habenulae are elevations of a pair of underlying habenular nuclei that perform an ancient role associated with olfaction. They receive fibers from olfactory nuclei, the hypothalamus, and other forebrain nuclei, and discharge impulses to the thalamus and midbrain, eliciting reflex responses to olfactory stimuli. One incoming tract, the **stria medullaris,** is seen in figure 16.15 emerging from the thalamus and passing caudad to disappear beneath the pineal, which overlies the habenulae in sheep. The habenulae are largest in species such as sharks and bloodhounds, which depend heavily on olfaction for locating food. They are inconspicuous in most birds, many of which have a poorly developed olfactory sense. They are poorly developed in permanently aquatic mammals, which do not use their olfactory epithelium in seeking food, and for a good reason—they would drown! The habenula of *Myxine* is a median elevation (see fig. 16.14), but the underlying nuclei are paired.

Thalamus

The thalamus is the largest structure of the diencephalon. It is a paired mass of many nuclei in the lateral walls of the third ventricle, and it bulges dorsally just behind the cerebral hemispheres (see fig. 16.14, *Myxine* and frog). In amniotes, it is hidden by the caudal poles of the cerebral hemispheres and must be observed on a brain stem preparation (see fig. 16.15).

All *sensory pathways* ascending to the telencephalon from the spinal cord, hindbrain, or midbrain, *synapse in a thalamic nucleus* before continuing to the telencephalon. With development of the neocortex in mammals, the thalamic nuclei have become so numerous that the left and right thalami bulge into the third ventricle and meet at one site to form an oval bridge of gray matter, the **massa intermedia,** or false commissure (see fig. 16.24). The term *false commissure* reflects the fact that it does not consist of decussating fiber tracts as do true commissures.

Hypothalamus and Associated Structures

A ventral view of the brain reveals the optic chiasma, inferior lobes or other components of the hypothalamus, a saccus vasculosus in fishes only, and the pituitary (figs. 16.17 and 16.20*a*).

The **optic chiasma** is the cephalic boundary of the diencephalon, ventrally. This is where the optic nerves reach the brain, and where some or all of its fibers cross to the opposite side. Caudal to the optic chiasma are the components of the hypothalamus and, at the caudal boundary of the diencephalon, the pituitary.

The floor of the diencephalon exhibits a midventral evagination within which is an **infundibular recess** of the third ventricle (fig. 16.21, 3). An elongated evagination results in an **infundibular stalk** (fig. 16.21*c*). The diencephalic floor at the end of the infundibular recess or stalk becomes the **posterior lobe** (**neural lobe, pars**

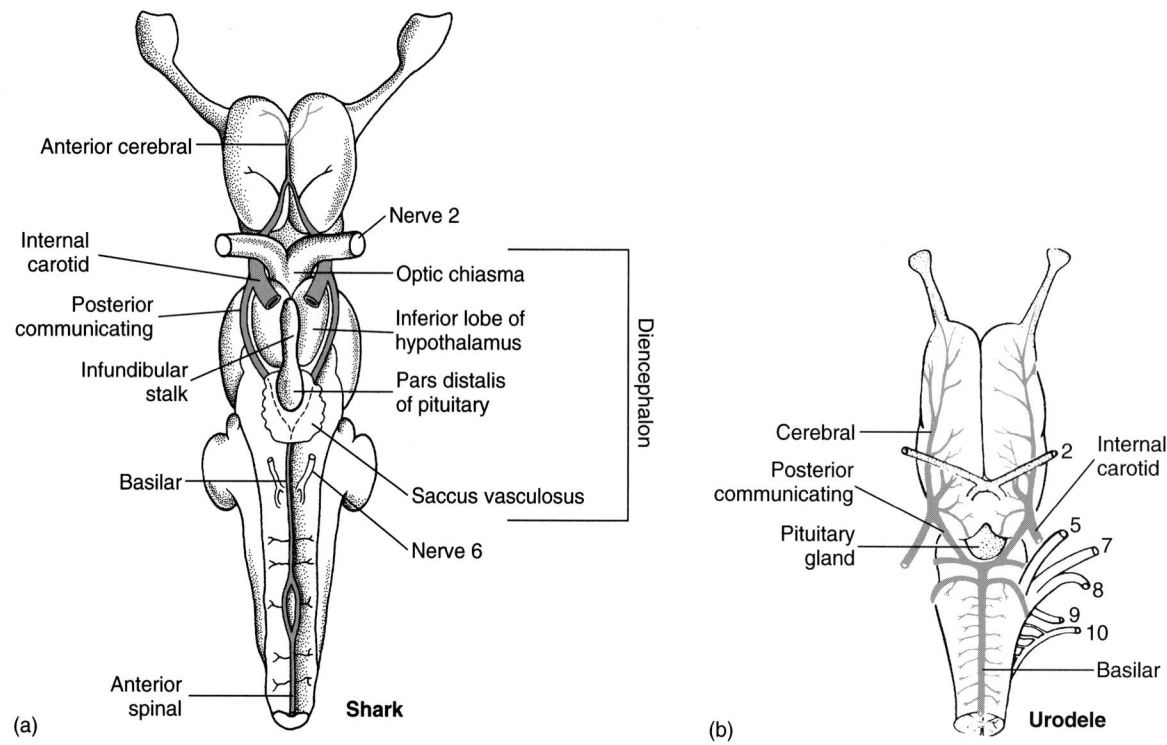

FIGURE 16.20

Brains of *Squalus* (*a*) and *Necturus* (*b*), ventral views. Numerals in (*b*) identify cranial nerve roots.

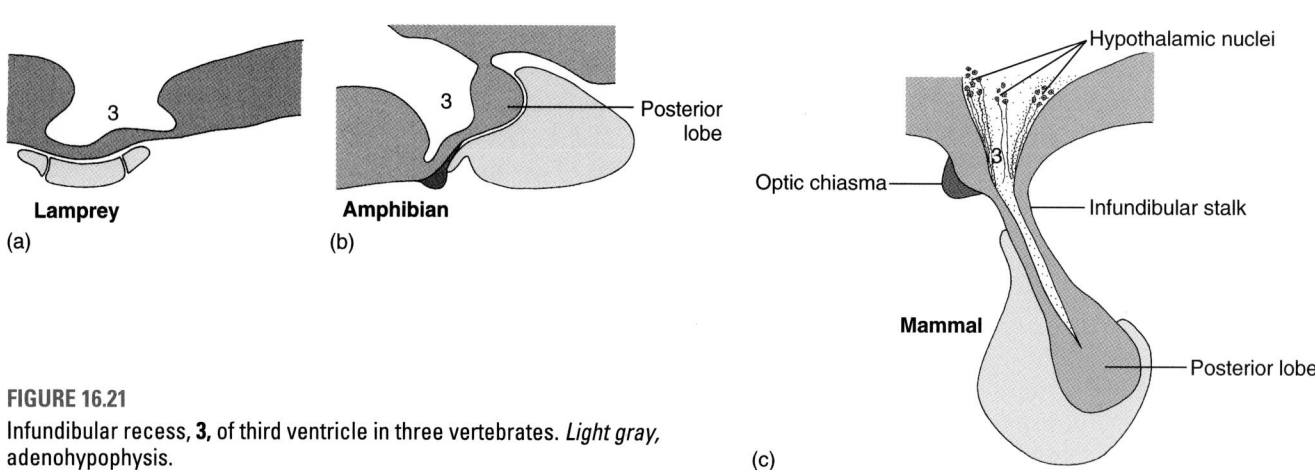

FIGURE 16.21

Infundibular recess, **3**, of third ventricle in three vertebrates. *Light gray,* adenohypophysis.

nervosa) of the pituitary. Intimately associated with it is the adenohypophysis, which arises embryonically from the roof of the stomodeum (see fig. 18.6).

The hypothalamus is the floor and ventrolateral walls of the third ventricle (figs. 16.19 and 16.21*c*, hypothalamic nuclei). It is prominent on the ventral surface of the diencephalon in cartilaginous fishes because of the extraordinary size of the **inferior lobes**, which have connections with the cerebellum (see fig. 16.20*a*). In all craniates, the hypothalamus is an important neural center for homeostasis, a visceral function. Hypothalamic nuclei exert major reflex control over the autonomic

nervous system, produce neurohormones that regulate the pituitary and gonads, and monitor the sodium chloride and glucose content of the blood. There is an appetite regulating center and, in endotherms, a temperature regulating center. In general, the anterior nuclei are associated with parasympathetic functions, the caudal nuclei with sympathetic ones. Hypothalamic nuclei also have connections via fiber tracts with the thalamus, with one or more nuclei of the basal ganglia (see fig. 16.25*f*), and with the **hippocampus**, an ancient olfactory cortex tucked under the temporal lobes of the cerebral hemispheres in mammals. These interconnections

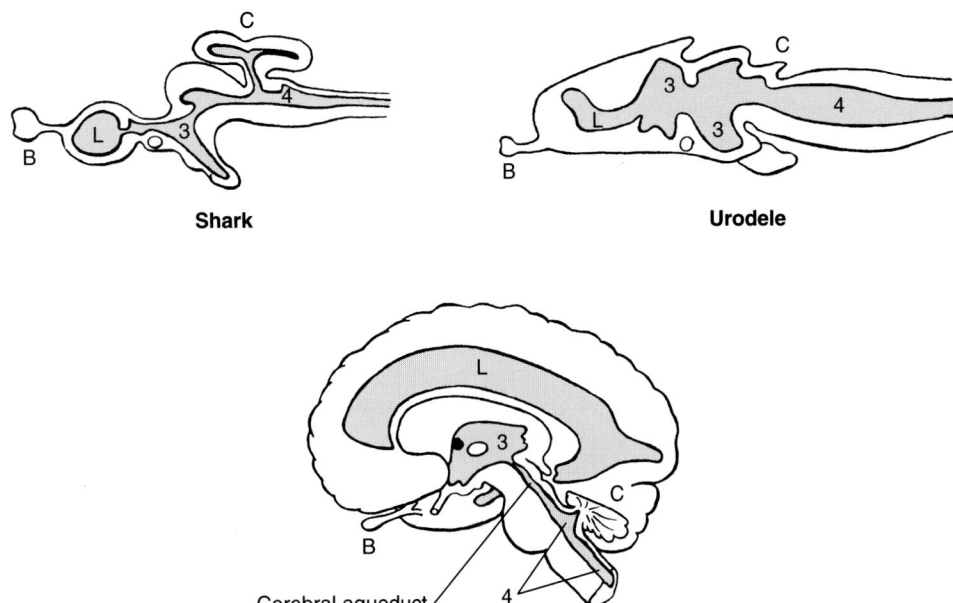

Shark

Urodele

Cerebral aqueduct

Human

FIGURE 16.22

Sagittal brain sections showing the ventricles in *red*. **B**, olfactory bulb; **C**, cerebellum; **L**, **3**, and **4**, lateral, third, and fourth ventricles. The *black* foramen in the third ventricle in the human brain is the interventricular foramen. Immediately behind it (oval) is the massa intermedia.

(the **limbic system**) are involved in evoking emotional responses in primates, an indication of the close relationship between visceral functioning and the emotions.

The **saccus vasculosus** is a highly vascularized, thin-walled ventral evagination of the diencephalic floor of elasmobranch and ray-finned fishes, located just behind the pituitary (see fig. 16.20). Within it is a fluid-filled recess of the third ventricle. The saccus vasculosus, a sense organ, is discussed in the next chapter.

Third Ventricle

The third ventricle is continuous caudad with the cerebral aqueduct of the midbrain and cranially with the ventricle in each cerebral hemisphere via a left and right **interventricular foramen** (figs. 16.13c, and 16.22, human). The third ventricle becomes compressed laterally with expansion of the thalami toward the midline. An optic recess extends toward the optic chiasma, the infundibular recess extends into the infundibular stalk, and a narrow canal extends upward into the stalk of the pineal. A choroid plexus is present in the ventricular roof.

Telencephalon

The telencephalon in all craniates other than ray-finned fishes is the portion of the brain that develops in the walls, floor, and some regions of the roof of the paired embryonic telencephalic vesicles (see fig. 16.13c). The typical craniate telencephalon therefore consists of paired left and right hemispheres, each with a ventricle. These hemispheres, which anatomically are mirror images, are referred to collectively as the **cerebrum**. Extending forward from each hemisphere is an **olfactory**

tract and bulb (see fig. 16.14). In mammals, these are hidden from dorsal view by the enormous overgrowth of the hemispheres, a result of millions of years of hemispheric expansion (figs. 16.23 and 16.24). The olfactory bulbs lie in contact with the ethmoid cartilages or bones and receive axonlike processes from the olfactory cells in the olfactory epithelium. The foramina through which the processes reach the olfactory bulbs in mammals are seen in figures 9.4 and 9.6.

The regions of the cerebrum devoted to olfaction—olfactory bulbs, tracts, lobes, and nuclei that process olfactory information—are prominent in those craniates that depend on olfactory stimuli for survival. Sharks, for example, have olfactory lobes that are larger than the rest of the cerebrum (see fig. 16.32). On the other hand, olfactory components remain rudimentary or are vestigial in many marine mammals, and they are poorly developed in birds other than those that feed on carrion. Olfaction is also a relatively minor source of environmental information in bats and primates, including humans.

At the boundary between the diencephalon and telencephalon, the thin roof of the embryonic neurocoel evaginates dorsad in some members of every craniate class to form a wrinkled thin-walled sac, the **paraphysis** (see fig. 16.18). However, in amniotes other than *Sphenodon*, it is limited to embryos. It is seldom observed during dissection of sharks because it is delicate and readily torn. Little is known of its function. It resembles a choroid plexus but seems to have a different role. It is assignable arbitrarily to either the telencephalon or diencephalon.

Because of its origin from paired embryonic vesicles, the neurocoel of the cerebrum in all craniates other than ray-finned fishes consists of paired lateral ventricles. Each ventricle is continuous with the third ventricle via an interventricular foramen, which marks the site of the initial embryonic evagination of the cerebrum. In mammals, a perpendicular wall, the **lamina terminalis**, separates the two foramina and marks the original cephalic boundary of the embryonic neural tube (see fig. 16.13*c*). Ray-finned fishes employ a slightly different means of producing a cerebrum. The telencephalic neurocoel remains unpaired; hence, there is a single cerebral ventricle and no interventricular septum. As in other craniates, paired nuclei form in the walls of the ventricle, but, in proliferating, the dorsal and lateral walls turn outward (that is, they are "everted").

Fishes

The cerebrum of fishes consists of (1) a primitive sensory and association area (**pallium**, fig. 16.25*a* and *b*) that receives and processes input from the olfactory epithelium and, to a lesser degree, from the thalamus; and (2) a motor area, the subpallial **globus pallidus**, which receives projections from the pallium and from the thalamus and projects efferent fibers into descending tracts that innervate motor nuclei of cranial and spinal nerves. Efferent impulses from the globus pallidus evoke reflex, survival-oriented responses of the muscles of locomotion and feeding. (The globus pallidus is one of a group of striated nuclear masses known as the **striatum** in amniotes. The arrangement of fiber tracts passing through the region is responsible for the striations.) The nuclei of the pallium and the globus pallidus persist in tetrapods, although eventually becoming subsidiary to more recently evolved nuclei.

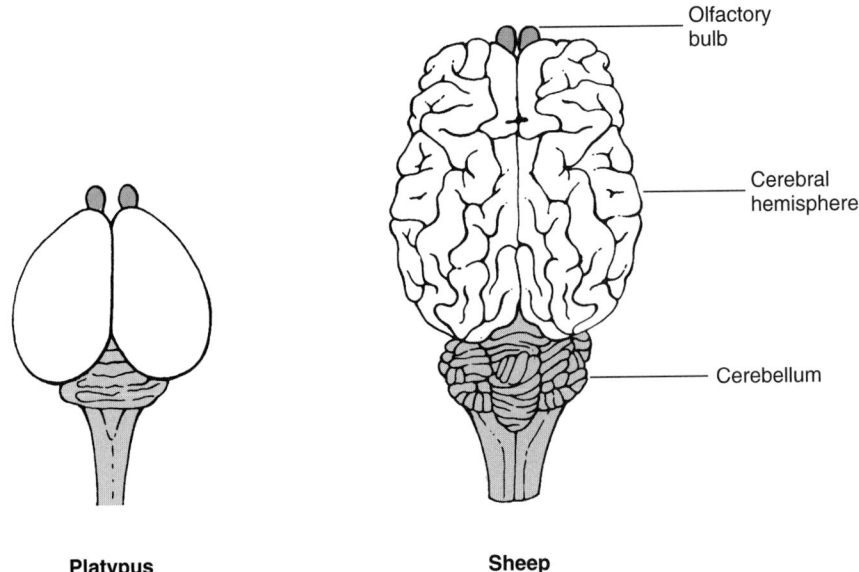

Platypus **Sheep**

FIGURE 16.23
Brain of a monotreme and a sheep.

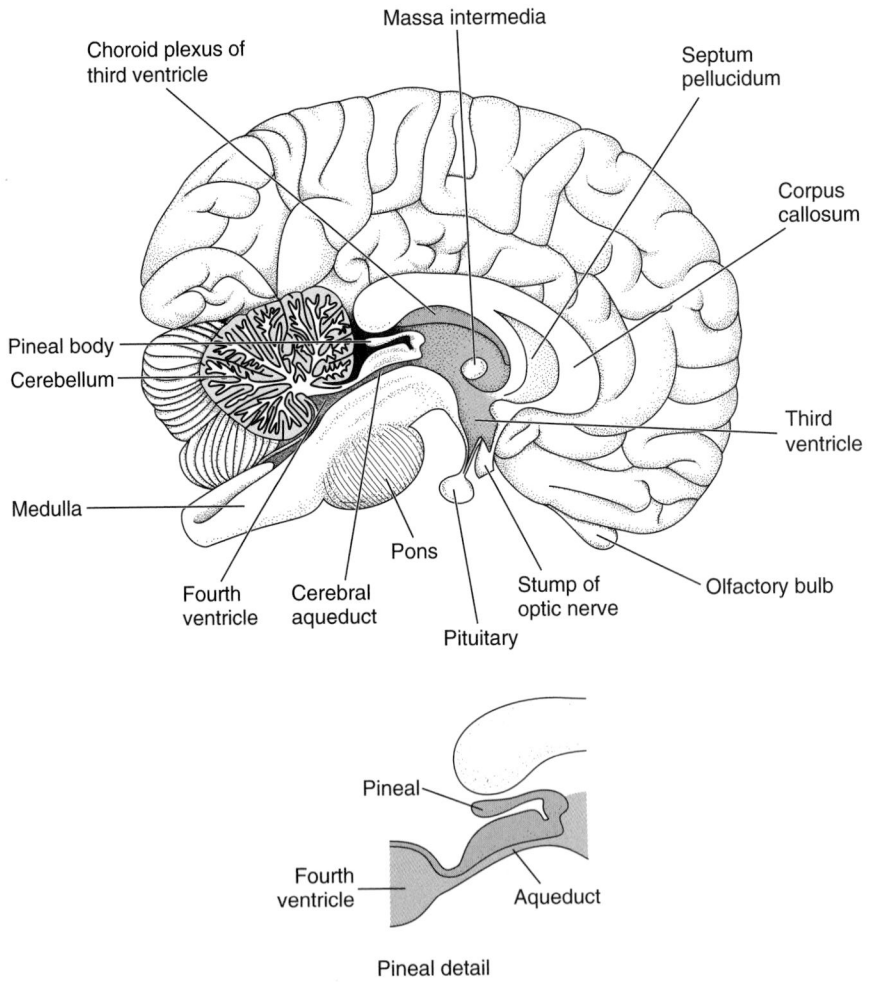

FIGURE 16.24
Human brain, sagittal section.

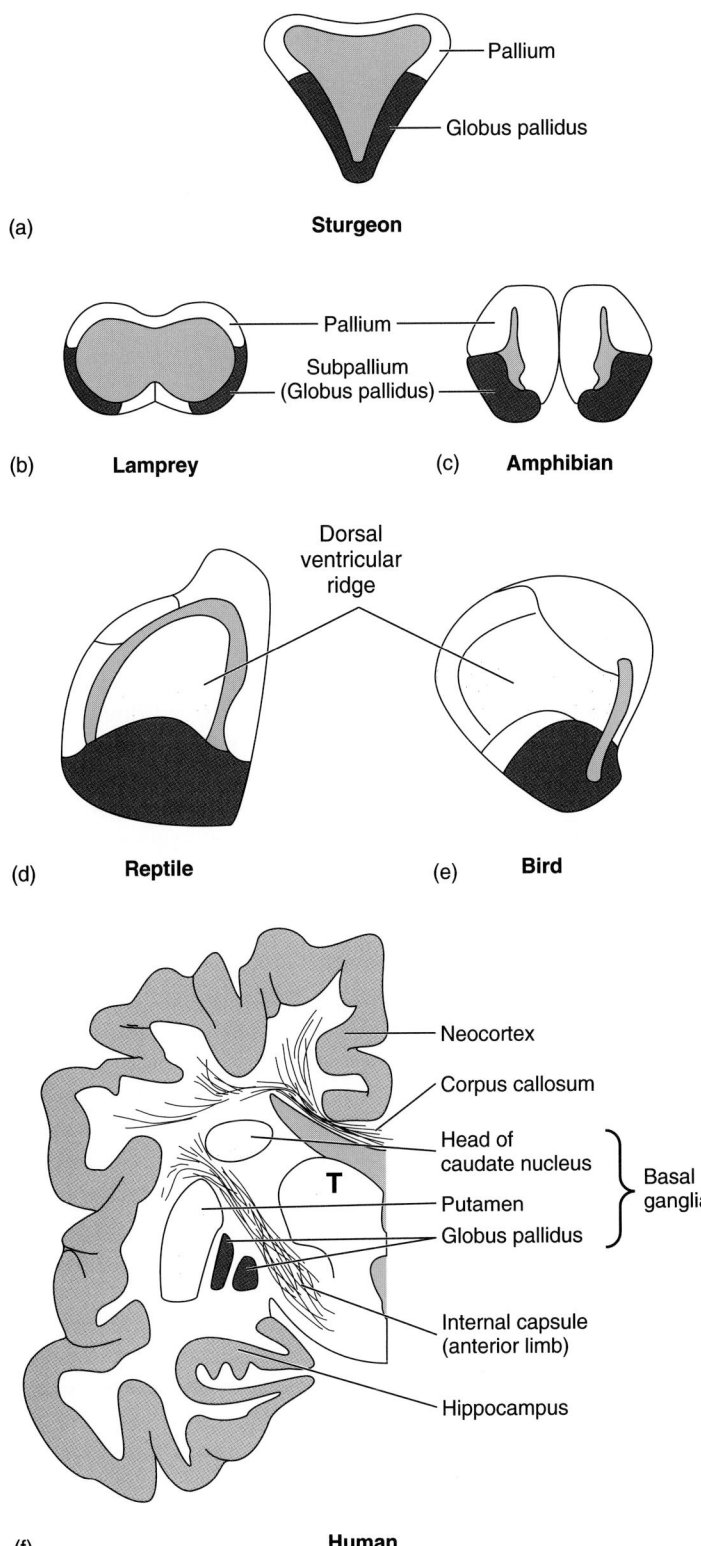

(a) **Sturgeon**

(b) **Lamprey**   (c) **Amphibian**

Pallium

Subpallium
(Globus pallidus)

Dorsal
ventricular
ridge

(d) **Reptile**   (e) **Bird**

Neocortex

Corpus callosum

Head of
caudate nucleus

Putamen   } Basal ganglia

Globus pallidus

Internal capsule
(anterior limb)

Hippocampus

(f) **Human**

FIGURE 16.25
The globus pallidus from fishes to humans. (*d*)–(*f*) Only the left cerebral hemispheres are shown. Ventricles are *red*. **T,** thalamus. Not to scale.

Amphibians

The primitive pallium and globus pallidus that function in fishes are still prominent in amphibians, and they function in a similar relationship (fig. 16.25*c*). Additional nuclei that have evolved in the amphibian subpallium enable the cerebrum to participate more extensively in coordinating incoming sensory information and in directing responses to appropriate somatic musculature, which now includes an appendicular musculature that is more complex than that of fishes.

Nonavian Reptiles

The cerebrum of reptiles has become huge compared to that of amphibians. The hemispheres now bulge laterally, dorsally, and backward over the diencephalon so that the thalamus is no longer visible from above (see fig. 16.14, snake and chicken). One of the factors responsible is a massive new area of association neurons, the **dorsal ventricular ridge**, in the lateral wall of the hemispheres. It increases considerably the bulk of the cerebrum and bulges into the lateral ventricle, converting the latter into a narrow slit (fig. 16.25*d*). The dorsal ridge receives visual, auditory, and somatic sensory stimuli (table 16.1) that have been relayed from the thalamus. The ridge processes this information and projects fibers to the globus pallidus and to other nuclei in the subpallium that constitute, collectively, the striatum. From there, descending fiber tracts conduct efferent impulses to motor nuclei in the brain stem and cord, evoking somatic motor activity appropriate to the exigencies of the environment.

Contributing also to the size of the amniote hemispheres are the additional upper-level neurons necessitated by the musculature of the neck (a tetrapod innovation refined among amniotes), and especially of the limbs. Since the days of amniote origins, amniote limbs have supported the rest of the body above the ground, and, excepting limbless offshoots, many reptiles have been capable of locomotion equal, or superior, to that of many mammals.

Birds

The cerebrum of birds is basically reptilian. The dorsal ridge has been inherited from reptilian ancestors, but another stratum of neurons has been added, capping the ridge (fig. 16.25*e*). The source of those neurons is unknown. The eyes of birds, and especially birds of prey such as hawks and eagles, have unique specializations (chapter 17, "Visual Organs: Lateral Eyes"), and the sensory input via the optic nerves is highly complex. The avian ridge collates and processes this information. Any

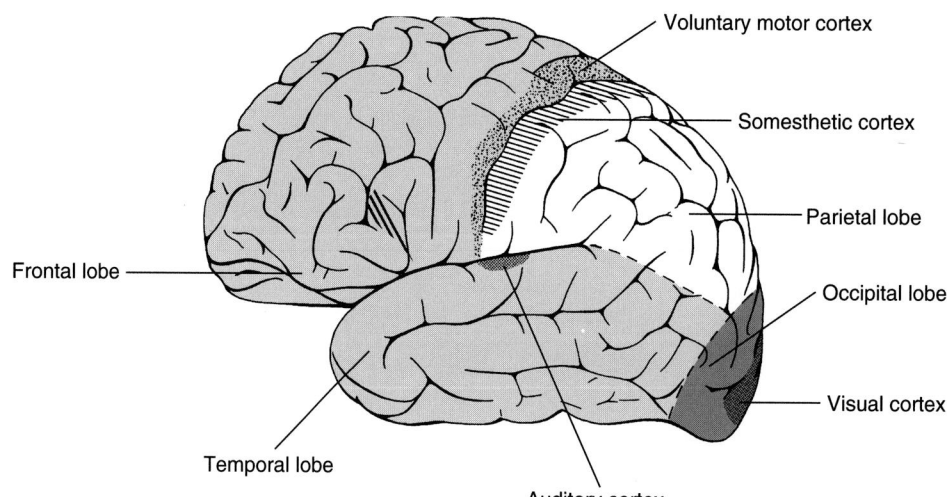

FIGURE 16.26
Topographical lobes of the human cerebral hemisphere and cortical loci for voluntary motor control, hearing, sight, and somesthetic sensibilities. Most of the visual and auditory cortices are on infolded gyri not visible on the surface. Cortical centers for taste and olfaction are less localized.

other competence the avian ridge may have is speculative. It has been suggested that it may have a role in migration, homing, and stereotyped behavior such as nest building. It must be noted, however, that lampreys and many bony fishes exhibit equally remarkable behavior with the very simple brain that they have inherited.

Mammals

The walls and roof of the lateral ventricles of mammals have become expanded in all available directions with the formation of the mammalian **neocortex** consisting, in humans, of the cell bodies of an estimated 13 billion neurons. For convenience in discussing neocortical locations, the neocortex has been divided topographically—and arbitrarily—into **frontal, parietal, temporal,** and **occipital** lobes (fig. 16.26). In most mammals, the cortex is so voluminous that it is folded into ridges (**gyri**) and grooves (**sulci**) that increase the working area of the hemisphere. These folds are lacking in monotremes (see fig. 16.23), a few marsupials, and many rodents. Formation of the neocortex resulted in a massive band of white matter, the **internal capsule,** connecting the cortex with the brain stem (see figs. 16.15 and 16.25). It is the sole pathway for (1) afferent fibers radiating to the cortex of the frontal lobe from the lateral nucleus of the thalamus and (2) the great efferent tracts that descend from the cortex to the red nucleus, to other nuclei in the brain stem, and to motor nuclei in the spinal cord. Among the efferent tracts is the corticospinal, mentioned earlier. These efferent tracts enter the tegmentum of the mesencephalon just anterior to the pons. In the roof of the ventricles, except in monotremes and marsupials, a broad transverse sheet of decussating fibers, the **corpus callosum,** connects the neocortex of the two hemispheres (see figs. 16.24 and 16.25f). Separating the lateral ventricles is a thin, double-walled membrane, the **septum pellucidum.**

An ancient olfactory pallial area that has persisted in mammals, the hippocampus, has become tucked under the temporal lobe of the cerebral hemispheres as a result of overgrowth of the neocortex. It was mentioned earlier as part of the mammalian limbic system. Memory storage has been attributed to the mammalian hippocampus, but ongoing neuroresearch has not confirmed the hypothesis. Modern techniques for identifying fleeting activity within the brain while it is occurring promise to provide uncontroversial evidence for the localization of many cortical functions.

The development of the many additional components of the cerebrum of mammals has displaced the globus pallidus to a position deep within the hemispheres (see fig. 16.25f). It has been joined by adjacent striated nuclei, a **caudate nucleus** (see fig. 16.15) (so called because it has a long tail), a **putamen** (fig. 16.25f), and an **amygdaloid nucleus** (not illustrated). Early homologues of the associated nuclei are present in nonmammalian amniotes. Collectively, these nuclei constitute what has been known as the **basal ganglia.** The globus pallidus, as in fishes, is a source of descending motor fibers that end in synapse with cell bodies of somatic motor fibers of cranial and spinal nerves. This role is shared by the other nuclei, whose total function has not yet been ascertained. However, the motor control exerted by the globus pallidus has been largely preempted by the red nucleus and other recently evolved nuclei in the brain stem, and by motor stimuli from the voluntary motor cortex in the frontal lobe (fig. 16.26). Pathologic changes in the basal ganglia result in certain motor dysfunctions (dyskinesias) such as the rhythmical tremor seen in aged victims of Parkinson's disease (paralysis agitans). Expansion of the hemispheres in mammals has hidden the basal ganglia, thalamus, and midbrain from dorsal view. The large cerebellum overlies almost all the remainder of the brain stem (see fig. 16.23). However, viewing the brain stem from above should convince the most dubious observer that the

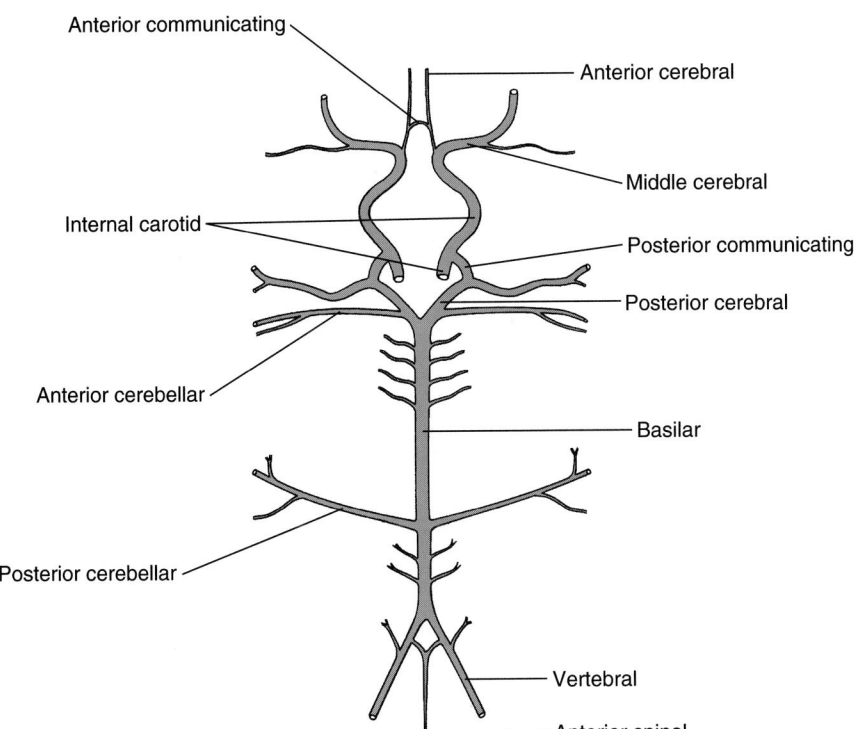

Anterior communicating

Anterior cerebral

Middle cerebral

Internal carotid

Posterior communicating

Posterior cerebral

Anterior cerebellar

Basilar

Posterior cerebellar

Vertebral

Anterior spinal

FIGURE 16.27
Arterial supply to brain of domestic cat. The anterior and posterior communicating arteries link the basilar and internal carotid streams, forming a circle of Willis around the base of the diencephalon.

mammalian brain, including that of humans, is built in accordance with the same architectural pattern as are the brains of other craniates.

The mammalian neocortex has three fundamental roles: (1) It is the center to which sensory impulses from special and general receptors are projected for evoking an awareness of the environment through physical sensation. General receptors evoke sensations of a discriminative (**epicritic**) nature, such as perceiving differences in the temperature, texture, or weight of objects held in the hand. (2) The voluntary motor cortex is the site where voluntary motor activity is initiated. (3) Having received sensory information from the environment, the neocortex collates the information and then shunts it elsewhere in the cerebrum for storage as memory. "Elsewhere" is currently the subject of active investigation. The prefrontal cortex, phylogenetically the newest, retrieves bits of pertinent stored information and employs them as *working memory,* such as recalling a telephone number. The process of decision making is much more complicated, but the manner in which the neocortex is employed in the solution of society's problems will determine the future of civilization insofar as it is under human control.

Blood Supply to Brain

The internal carotids and the anterior (ventral) spinal artery supply the brains of most craniates, the spinal artery continuing onto the ventral surface of the brain as the basilar artery (fig. 16.27). In many reptiles and mammals, the vertebral arteries become a major supply,

emptying into the basilar just anterior to the foramen magnum. Communicating arteries connect the basilar and internal carotid streams in mammals and some other craniates providing an arterial **circle of Willis** around the base of the diencephalon (fig. 16.27). The internal carotids supply principally the cerebral hemispheres, whereas the basilar supplies principally the brain stem. There is considerable variation in the specific configuration of the circle of Willis, even among humans.

Brains are drained of blood by venous sinuses and smaller veins that empty chiefly into the internal jugular vein. The cavernous venous sinuses of mammals, of which the **superior sagittal sinus** is the largest, are located between two layers of the mammalian dura mater. There is no lymphatic drainage of either the brain or the cord.

Choroid Plexuses and Cerebrospinal Fluid

The cavities of the brain, the neural canal of the cord, and the subarachnoid space, when present, are filled with **cerebrospinal fluid.** This is a clear, watery liquid of low specific gravity that is removed from the blood by diffusion and secretion. Most of it is secreted into the ventricles by the choroid plexuses; in mammals, at least, a smaller amount comes from the bloodstream directly. **Choroid plexuses** consist of tufts of a highly vascularized pia mater and ependyma that dangle into the third and fourth ventricles. The plexus in the third ventricle extends forward into the lateral ventricles via the interventricular foramina (see fig. 16.18). Obstruction of these foramina results in accumulation of fluid in the lateral ventricles. When this condition is present in fetal or neonatal mammals, the fontanels of the immature skull allow the head to enlarge, a condition known as hydrocephalus. The plexus in the fourth ventricle is in the roof of the medulla.

Cerebrospinal fluid moves sluggishly caudad from the fourth ventricle into the neural canal of the cord. In mammals, it also enters the subarachnoid spaces via a median and two lateral apertures in the roof of the fourth ventricle under cover of the cerebellum. These spaces surrounding the cord of mammals exhibit several cisterns where fluid normally accumulates. One of these in humans surrounds the cauda equina in the lumbosacral region, from which a *spinal tap* may aspirate some of the fluid for study of its contents or diagnosis of disease.

From the subarachnoid spaces, the cerebrospinal fluid travels along narrow perivascular spaces that accompany blood vessels entering or leaving the substance of the cord and brain; it follows the rootlets of the spinal and cranial nerves where they pierce the subarachnoid spaces and dura mater to emerge from the cord; and it seeps into the cavity surrounding the membranous labyrinth where it contributes to the perilymph.

Cerebrospinal fluid of the brain and cord of nonmammalian craniates is removed by diffusion into adjacent venous channels. In mammals, clusters of **arachnoid granulations**, which are tufts of the pia-arachnoid membrane, project into large subdural venous sinuses (a derived condition in many craniates), and the fluid diffuses out of the granulations into the venous blood of the sinuses.

In addition to cushioning the brain and cord against mechanical trauma, cerebrospinal fluid selectively exchanges metabolites with the tissues it bathes.

CRANIAL NERVES

Traditionally, 12 cranial nerves are presented that are numbered sequentially using Roman numerals. This is the convention we see used for humans (table 16.3); however, it is an incomplete picture that fails to recognize the complex and composite nature of cranial nerves in Craniata. As many as 25 cranial nerves are recognized in craniates, although the exact number is debated (table 16.4). The number present in any given

TABLE 16.3   Human Cranial Nerves[1]

Symbol	Name
0	Terminal
I	Olfactory
II	Optic
III	Oculomotor
IV	Trochlear
V	Trigeminal
VI	Abducens
VII	Facial
VIII	Vestibulocochlear
IX	Glossopharyngeal
X	Vagus
XI	Spinal accessory
XII	Hypoglossal

[1]The traditional numbering of cranial nerves as used in humans. Although not found on gross dissection or evaluated as part of a neurological examination, the terminal nerve (0) is present and represents a thirteenth cranial nerve in humans.

TABLE 16.4   Craniate Cranial Nerves[1]

Symbol	Nomenclature	Innervation
0	Terminal	Nasal epithelium
I	Olfactory	Olfactory epithelium
VN	Vomeronasal	Vomeronasal organ
II	Optic	Retina
E	Epiphyseal	Pineal; parietal eye
III	Oculomotor	Intrinsic and extrinsic eye muscles
P (V$_1$)	Profundus	Skin of snout
IV	Trochlear	Extrinsic eye muscles
V (V$_{2,3}$)	Trigeminal	Jaw musculature; skin of face; snout
VI	Abducens	Extrinsic eye muscles
VII[2]	Facial	Facial (hyoid) musculature; salivary and tear glands; taste buds
ALL (three nerves)	Anterior lateral line	Lateral-line organs
VIII	Vestibulocochlear (octaval)	Vestibular and cochlear organs
PLL (three nerves)	Posterior lateral line	Lateral-line organs
IX[2]	Glossopharyngeal	Pharynx; salivary glands; taste buds
X[2]	Vagus	Visceral organs of thorax and abdomen; larynx; pharynx; taste buds
XI	Spinal accessory	Sternocleidomastoid and trapezius muscle groups
XII	Hypoglossal	Tongue; syrinx

[1]A comprehensive list of cranial nerves found within craniates. Individual species vary in the presence or absence for specific nerves or their components (e.g., terrestrial craniates lack lateral-line nerves).
[2]The taste components of VII, IX, and X are considered as separate cranial nerves by some authors (e.g., Butler and Hodos, 1996).

species has varied phylogenetically due to the gain, loss, or fusion of cranial nerves or their components.

Cranial nerves can be presented numerically, but this neglects the important developmental and functional aspects of these nerves. A full understanding of cranial nerves is achieved only with an appreciation of development. This provides information on the unique origin of some nerves and the composite nature of others. An overview of embryology is provided with the individual cranial nerves discussed in functional groupings. Functional categories, as used here, include sensory, motor, and mixed cranial nerves.

You may ask, "Why abandon a traditional numerical approach to grouping cranial nerves?" for which two examples are provided. First, for the anatomist it provides a clearer picture of the organization of cranial nerves, and it provides the basis for a better understanding of the evolution of the craniate head (see discussion of segmentation in this chapter). Second, for the medical student learning how to perform a neurological examination, it will help to explain the relationship of certain clinical findings to the affected component of a cranial nerve. Through an understanding that the human condition represents a phylogenetic history of loss, fusion, or gain does the student appreciate the variation in function for the individual nerves.

The site where a cranial nerve emerges from the surface of the brain, or enters it, is its **superficial origin** as distinguished from the deep origins of its fibers, which arise (motor fibers) or terminate (sensory fibers) in a number of nuclei located at distances from one another within the brain stem. Cranial nerves III, VI, and XII emerge from the basal plate and resemble spinal nerves that have only ventral roots. Nerve IV is unique in that, although its nucleus is in the basal plate, it emerges from the roof of the brain. Branchiomeric nerves have only lateral roots (see fig. 16.30, nerve X). The purely sensory nerves have superficial origins that will be discussed shortly.

Cranial nerves or their components are derived from three embryological tissues: neurectoderm from the brain, neural crest, and/or ectodermal placodes. Two nerves represent outgrowths of the central nervous system: the retina of the eye and the epiphyseal complex (pineal and parapineal). Ectodermal placodes contribute to the ganglion cells of the terminal, olfactory, vomeronasal, lateral-line, and vestibulocochlear nerves and the taste bud components of cranial nerves VII, IX, and X. Neural crest forms the ganglion cells associated with cranial nerves V, VII, IX, and X. The exact contribution of neural crest and placodes is still under study.

Predominantly Sensory Cranial Nerves

The sensory nerves (0, I, VN, II, E, P, ALL, VIII, PLL) are associated in ontogeny with a series of ectodermal placodes in the developing head. For some of these nerves, neural crest may also contribute to their formation.

Nerve 0 (Terminal)

A **terminal nerve**, distinct from the olfactory nerve, lies close to the olfactory bulb and tract in representatives of most gnathostomes (a probable gnathostome synapomorphy). It arises from the ventral surface of the forebrain and consists of sensory and, at least in some instances, vasomotor fibers that supply a restricted area of the nasal mucosa. These fibers are believed to play a role in reproductive behaviors (possibly as pheromone receptors). Although present in humans, it is seldom seen in human anatomy laboratories.

Nerve I (Olfactory)

The cell bodies of olfactory nerve fibers are located in olfactory epithelium, and the fibers terminate in the olfactory bulb (fig. 16.28*a*). In *Squalus*, the olfactory epithelium lies so close to the bulb that an olfactory nerve cannot be distinguished as an anatomical entity. In *Scoliodon*, another shark, and in some teleosts, the sac containing the olfactory epithelium is sufficiently distant from the olfactory bulb that an olfactory nerve is demonstrable (fig. 16.28*b*). In most craniates, a number of short, discrete bundles of olfactory nerve fibers, the **filia olfactoria**, extend between the olfactory epithelium

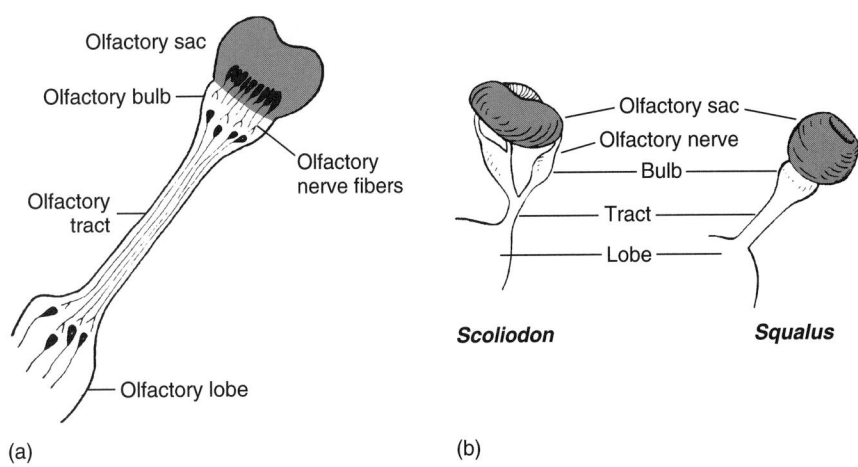

FIGURE 16.28

(*a*) Olfactory sac (*red,* containing the olfactory epithelium), bulb, and tract of *Squalus.* The cell bodies of the olfactory nerve fibers are in the olfactory epithelium. These fibers synapse in the olfactory bulb with second-order sensory neurons whose axons constitute an olfactory tract. The latter terminates in the gray matter of the olfactory lobe. (*b*) Contrasting spatial relationship of the olfactory sacs to the bulbs of two species of sharks. Note the discrete nerve in *Scoliodon* only.

and the bulb, separated only by the olfactory capsule of the skull. These filia collectively constitute the olfactory nerve in these species.

In mammals, the olfactory epithelium is in the upper part of the nasal passage, separated from the olfactory bulb by the cribriform plate of the ethmoid bone. The foramina in the plate (see figs. 9.4 and 9.6) transmit the filia olfactoria. When the brain is lifted from the cranial cavity, the olfactory nerve bundles are torn, and only stumps remain attached to the brain. The number of olfactory fibers is small in birds, which, as explained earlier, have poorly developed olfactory systems. It is also small in the platypus, which spends most of its time in creeks or rivers or in a burrow; hence, it has few natural enemies on land. Tactile organs and electroreceptors on the beaks of the platypus substitute for olfaction as they rout for food in the water bottom. In many marine mammals, the olfactory nerve is vestigial.

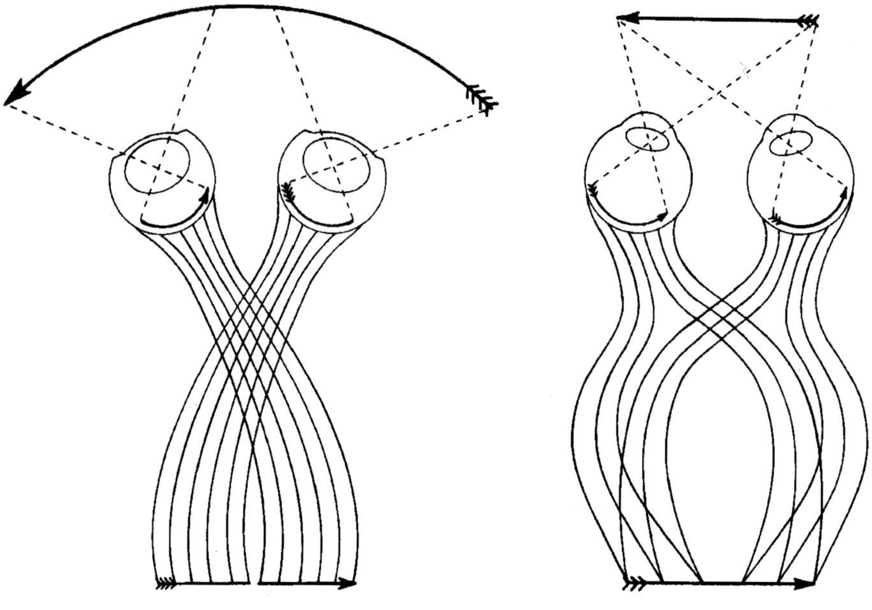

Nerve VN (Vomeronasal)

In tetrapods that have a vomeronasal organ, there is found a **vomeronasal nerve**, whose fibers terminate in an accessory olfactory bulb. The vomeronasal organ, a chemoreceptor, is discussed in chapter 17.

Nerve II (Optic)

The cell bodies of the optic nerve fibers are in the retina. The nerve emerges from the rear of the eyeball and extends to the optic chiasma. Beyond that, it is termed the **optic tract**. Except in mammals, all or nearly all optic nerve fibers decussate in the chiasma and enter the opposite side of the brain (fig. 16.29a). As a derived condition in some primates, the eyes are directed forward, and only the fibers from the nasal side of the retina decussate (fig. 16.29b). These correlated primate features result in overlap of the visual field and concomitant depth perception (binocular vision). Nerve II contains a number of efferent fibers from the brain to the retina. These fibers apparently modulate the receptors.

The term *optic nerve* is, from an embryological viewpoint, a misnomer because the retina arises from paired optic vesicles that never separate from the forebrain (see fig. 16.13). The term *tract* (a bundle of nerve fibers

(a) **Nonmammalian craniate**   (b) **Derived condition**

FIGURE 16.29

Decussation of optic nerve fibers in optic chiasma in non-mammalian craniates contrasted with the derived condition in some primates. In nonmammalians, only half of the *arrow* is projected onto the opposite retina. In some primates, the entire *arrow* is projected onto each retina, and from a slightly different angle of vision, resulting in depth perception.

within the central nervous system) is appropriate for the pathway from the optic chiasma to the tectum of the mesencephalon, where many of the fibers within the tract terminate. An ectoderm placodal component contributes to form the lens of the eye.

Nerve E (Epiphyseal Complex)

Like the optic nerve, the pineal and parapineal represent evaginations of the neurectoderm of the brain. Further paralleling development of the eye, it appears that a lens (see fig. 17.20c), when present, is derived from ectodermal placodes.

Nerve P (Profundus)

The **profundus nerve** has a variable history in terms of its fusion with the trigeminal nerve. It appears that a fused condition is primitive for craniates, and a discrete nerve is a synapomorphy for gnathostomes. Fusion reoccurs in sarcopterygians (lobe fin fishes and their tetrapod descendants; however, *Latimeria* shows a secondarily discrete profundus). The profundus will be discussed further with the trigeminal nerve (mixed nerves).

Nerve ALL/PLL (Anterior and Posterior Lateral Line)

The lateral lines of craniates are mechanoreceptive organs (chapter 17). They are not found in terrestrial organisms and are thus restricted to aquatic anamniotes.

Recognition of the **lateral-line nerves** as distinct nerves has been obscured by their tendency for fibers to accompany other cranial nerves. With the refinement of cellular level techniques, it is now clear that there are up to six lateral line nerves (three preotic and three postotic in distribution). Hagfish possess only two preotic or anterior lateral-line nerves. In contrast to the multicellular neuromast of lampreys and gnathostomes, the hagfish lateral-line mechanoreceptor is unicellular (whether this character is primitive or represents a secondary reduction is not clear).

Nerve VIII (Vestibulocochlear)

The eighth nerve in all craniates comes off the medulla very close to the roots of nerves V and VII and innervates the membranous labyrinth. It consists of two major primary branches in fishes, one innervating the ampullae on the anterior vertical and horizontal semicircular ducts, and the utriculus; the other innervating the ampulla on the posterior vertical duct, and the sacculus and lagena (see fig. 17.4, shark).

In tetrapods, the lagena elongates to eventually become the cochlear duct (see fig. 17.4), and the branch supplying this region thereafter is known as the **cochlear nerve**. The other branch is then the **vestibular nerve**. The nerves have a **cochlear** or **vestibular ganglion** on their pathway. The cell bodies in these ganglia are exceptional for first-order sensory neurons in that they remain in a primitive bipolar state. Because the cochlear ganglion spirals within the cochlea of mammals, it is also called the **spiral ganglion**. As in the retina of the eye, some efferent fibers are present apparently to modulate the receptors.

In some mammals (rat and mouse but not cat or bat), large unipolar sensory cell bodies are distributed within the eighth nerves along their entire length. These are cell bodies of second-order sensory neurons. They are stimulated by collateral (side) branches of incoming fibers, and because their axons end on other second-order neurons in the cochlear nuclei of the medulla, they reinforce (augment) stimuli coming from the receptor.

Motor Nerves

Motor nerves (III, IV, VI, XI, and XII) carry primarily motor (efferent) fibers to muscles of the head and pharynx (or their homologues). Cranial nerves III, IV, and VI innervate the extrinsic muscles of the eye (nerve III also carries autonomic fibers for the intrinsic eye muscles). Cranial nerves XI and XII are clearly defined in tetrapods and innervate muscles of the neck and shoulder, and tongue, respectively.

Nerves III, IV, and VI (Oculomotor, Trochlear, and Abducens)

The third, fourth, and sixth nerves supply the superior oblique (IV), external rectus (VI), the four remaining extrinsic eyeball muscles (III), and certain other myotomal muscles of the eyes (table 11.8). These eyeball muscle nerves resemble spinal nerves that have lost their dorsal roots. In addition to somatic motor fibers, the nerves contain sensory fibers for proprioception from the muscles innervated.

Nerve III arises ventrally from the mesencephalon. Nerve IV is the only nerve arising dorsally from the brain (anterior roof of the fourth ventricle) and one of the few nerves with motor fibers that decussate before emerging. Nerve VI emerges ventrally at the anterior end of the hindbrain. Nerves IV and VI are the smallest of the cranial nerves, having fewest fibers. The cell bodies of all fibers in these nerves, both motor and proprioceptive, are in nuclei within the central nervous system. Therefore, the nerves have no sensory ganglia.

Nerve III also contains visceral motor fibers that end in the ciliary ganglion of the autonomic nervous system (see fig. 16.35). From the ganglion, postganglionic fibers pass to the muscles of the iris diaphragm that constrict the pupil (sphincter pupillae muscles) and to the ciliary muscles of the eye (see fig. 17.18). These govern the position or thickness of the lens for visual accommodation, depending on the species.

Nerve XI (Accessory)

The spinal accessory nerve constitutes an eleventh cranial nerve in tetrapods. It is purely motor. The details of assembly and distribution of its components vary among reptilian and mammalian orders. Its presence in amphibians has been obscured by its close topographic association with the vagus. Modern techniques clearly delineated this nerve as a tetrapod feature (identified, to date, in some salamanders and frogs).

In mammals, rootlets of the accessory nerve emerge from the medulla immediately behind, but not in line with, the roots of the vagus (fig. 16.30). A small number of the fibers in those rootlets innervate small muscles of the pharynx, such as those in the caudal border (velum) of the soft palate. The majority of the fibers in those rootlets join the vagus to be distributed to autonomic ganglia in the coelom. In mammalian terminology, the latter fibers constitute the **internal ramus of the accessory nerve**.

In mammals, additional rootlets of the accessory nerve emerge from the spinal cord along with ventral rootlets of a variable number of cervical spinal nerves. The spinal rootlets of the accessory nerve unite to form a nerve trunk that passes cephalad close to the cord and enters the cranial cavity via the foramen magnum. Within the braincase, the trunk joins the cranial root of XI for a short distance, after which most of the spinal fibers leave the accessory nerve to become its **external ramus**. This ramus exits the cranial cavity via the jugular foramen along with nerves IX and X. It is composed largely of fibers that innervate the trapezius

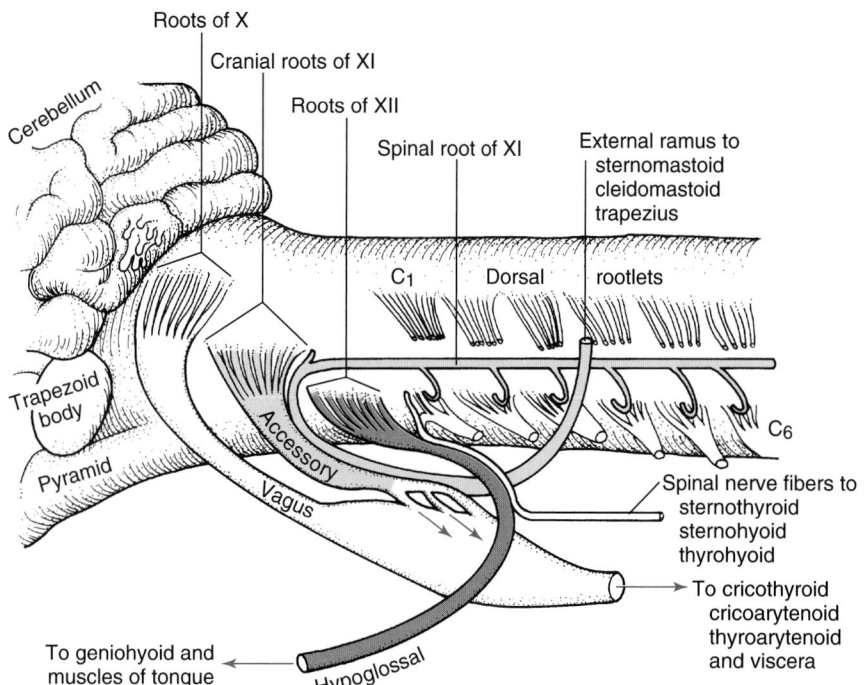

FIGURE 16.30
Superficial origins of the vagus, accessory (*light red*), and hypoglossal nerves (*dark red*) of a generalized mammal. Details vary among mammalian orders. The rootlets of XII are in series with the ventral rootlets of the spinal nerves. **C₁**, dorsal rootlets of the first cervical spinal nerve; **C₆** ventral rootlets of the sixth cervical spinal nerve. Nerve XI contributes an internal ramus (*arrows*) to the vagus.

and sternocleidomastoid complex of muscles. A corresponding nerve in reptiles (including birds) arises from medullary roots, rather than from the spinal cord.

What is the history of the eleventh nerve of amniotes? The components of the nerve are not new; only the way they are assembled is new. If we look at the location of the cell bodies of fibers that arise from the spinal cord in reptiles, we find they are in neither the SM column, which supplies myotomal muscles, nor in the BSM column, which supplies branchiomeric muscles (fig. 16.31). The location varies, but it is very close to the BSM column that supplies branchiomeric fibers to nerve X. In mammals, the BSM column does not extend into the cord, and the cell bodies are close to the SM column, but clearly not part of it. Originally, they could have been part of one or the other of these columns.

As mentioned earlier, there is a series of occipitospinal nerves between the vagus and the first cervical spinal nerve of basal craniates. These nerves are either missing in amniotes, or their fibers are now part of nerve XI. On the basis of available observations, which indeed are slight, it is reasonable to believe that the fibers of the eleventh nerve of amniotes are contributions from the posteriormost branchiomeric nerves of fishes and from one or more ancestral occipitospinal nerves.

Nerve XII (Hypoglossal)

The twelfth nerve of tetrapods is motor except for proprioceptive fibers and is associated with development of a tongue in terrestrial feeding. It arises from the **hy-**

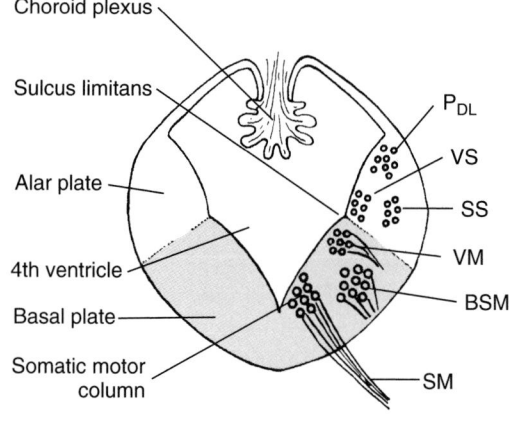

FIGURE 16.31
Cross section of amniote medulla, generalized, showing location of the columns of sensory and motor nuclei of cranial nerves. Sensory nuclei are in the alar plate; motor nuclei, in the basal plate. **SM**, somatic motor fibers with cell bodies in the nucleus in the somatic motor column. **VS**, column for all general and special visceral afferent nuclei except those for smell. Remaining nuclei include: **BSM**, branchiomeric somatic motor; **P_DL**, neurons associated with dorsolateral placodes of the lateral line; **SS**, somatic sensory; and **VM**, visceral motor (refer to tables 16.1 and 16.5).

poglossal nucleus in the medulla by a series of ventral rootlets and passes into the base of the tongue from which location it sends branches to the hyoglossus, styloglossus, genioglossus, and lingualis muscles (see figs. 11.6, 16.4*b* and 16.30). On emerging from the hypoglossal foramen in mammals, the nerve is joined by fibers from one or more anterior cervical spinal nerves, the number depending on the species. Some of these spinal nerve fibers are distributed to the geniohyoid muscle. The rest of the spinal fibers leave nerve XII to join a loose plexus of cervical spinal nerves supplying the hypobranchial muscles of the neck (sternothyroid, sternohyoid, thyrohyoid, or their homologues).

That the hypoglossal is a cranial rather than a spinal nerve is dictated solely by the location of the foramen magnum. *Actually,* it is a spinal nerve that became "locked up" in the braincase. Like spinal nerves, the hypoglossal nerve of some mammals develops an embryonic dorsal root and dorsal root ganglion (**Froriep's ganglion**). The root and ganglion later disappear.

The twelfth nerve is a derivative of the occipitospinal series of fishes, as may be deduced from the following observations: (1) Occipitospinal nerves have been reduced in number above fishes, whereas the number of cranial nerves has increased. (2) The twelfth nerve, like many occipitospinal nerves, lacks a dorsal root. (3) The occipitospinal nerves of nontetrapod craniates supply hypobranchial muscles, and the tongue is hypobranchial muscle.

Mixed Nerves

The mixed cranial nerves (V, VII, IX, X) carry both sensory (afferent) and motor (efferent) fibers. The trigeminal (V) nerve differs from other mixed nerves in that it carries only somatic motor and sensory fibers and lacks fibers for taste. The remaining mixed nerves carry both somatic and visceral components.

Pharyngeal arches were the unique solution arrived at by hemichordates and chordates for getting oxygen into the bloodstream. The arches of craniates required muscles to operate them, and a series of similar nerves—V, VII, IX, and X—perform this function. These are the mixed nerves of craniates that innervate the branchial arches and their muscles.

The primitive pattern of distribution of mixed (branchiomeric) nerves is readily demonstrable in a dissection that reveals cranial nerves IX and X of a modern shark (fig. 16.32). Nerve IX innervates two successive pharyngeal arches via two major branches, a **pretrematic nerve** that enters the arch in front of the first gill slit (Pre, in fig. 16.32) and a **posttrematic nerve** (Post, in fig. 16.32) that enters the arch behind it. The posttrematic branch contains the motor fibers that oper-

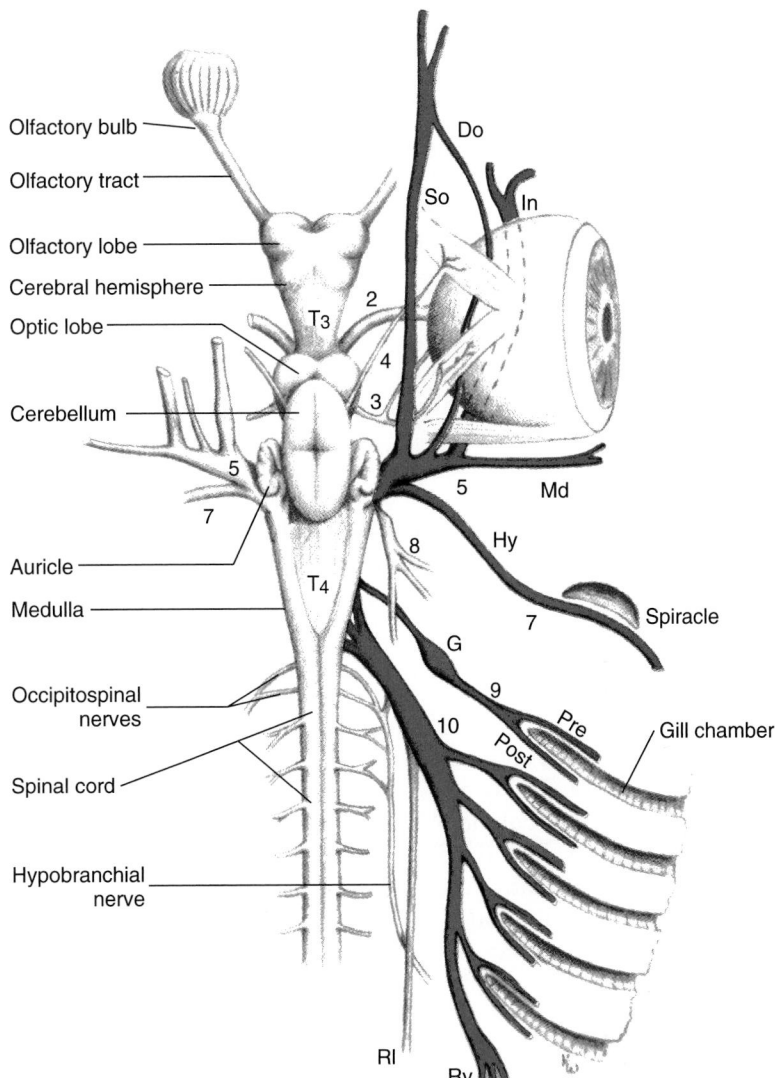

FIGURE 16.32

Brain and cranial nerves 2 to 10 (6 not visible) of *Squalus acanthias,* dorsal view. Branchiomeric nerves are in *red.* **Do,** deep ophthalmic (profundus); **G,** petrosal ganglion; **Hy,** hyomandibular; **In,** infraorbital; **Md,** mandibular; **Pre,** pretrematic; **Post,** posttrematic; **So,** superficial ophthalmic; **Rl,** ramus lateralis of vagus (posterior lateral-line nerve that travels with the vagus proximally); **Rv,** ramus visceralis of vagus; **T₃** and **T₄** tela choroidea of third and fourth ventricles. The pineal body has been removed. The three eyeball muscles shown are, commencing anteriorly, superior oblique, superior rectus, and lateral rectus.

ate the muscles of the arch that it enters. It also has sensory fibers that monitor the condition of the gill filaments in the posterior wall of the first gill chamber. It is the only branch that contains motor fibers to striated muscles. The pretrematic branch monitors the gill filaments in the anterior wall of the chamber. A third branch, the **pharyngeal**, extends as a branch of the pretrematic nerve onto the palate to monitor the pharyngeal mucosa at that body level, including any taste buds that may be present. Nerve X exhibits the identical pattern, four times over. Nerves V and VII, which provide motor innervation to the mandibular and hyoid arches, will be discussed shortly.

In addition to innervating the pharyngeal arches of fishes or their derivatives in tetrapods, branchiomeric nerves have functions that are unrelated to respiration. With the exception of nerve V, they have visceral motor fibers that are components of the autonomic nervous system and sensory fibers that innervate several varieties of sense organs. Each motor branch has proprioceptive fibers. The distribution of branchiomeric nerves in *Squalus* reveals the primitive pattern of distribution to the pharyngeal arches. Alterations in tetrapods result chiefly from elimination of gills and from other adaptations to terrestrial life.

With awareness of the basic pattern of distribution of nerves IX and X, nerves V and VII in sharks become easier to understand. And because tetrapods have inherited all these nerves with modifications, the distribution and roles of the branchiomeric nerves of tetrapods will be more comprehensible.

Nerve V (Trigeminal)

The fifth nerve in all craniates arises from the anterior end of the hindbrain and exhibits two divisions: **maxillary** (V_2; **infraorbital** of fishes) and **mandibular** (V_3). A third branch, the **opthalmic** (V_1) is seen independently in living agnathans, lobe-fined fishes and tetrapods (sarcopterygians), and many ray-finned fishes. The primitive gnathostome condition is where the opthalmic division represents a discrete cranial nerve, the profundus nerve. All branches contain sensory fibers. Only the mandibular, a posttrematic nerve in fishes, contains motor fibers.

Via one or another of the three divisions, cranial nerve V is sensory for general sensation to the skin of the entire head (fig. 16.33), including the conjunctiva of the eye, the lining of the outer ear canal, and the ectodermal surface of the eardrum. It also supplies the jaw teeth, oral mucosa, nasal passageway, and anterior surface of the tongue for general sensation. The cell bodies for all sensory fibers in the nerve, except those for proprioception, are in the trigeminal ganglion unless the ophthalmic division has its own, as in some basal gnathostomes.

The mandibular nerve is motor to the muscles derived from the mandibular arch. These are chiefly the muscles of the jaws. It also innervates the tensor tympani muscle that is attached to the malleus, an ear ossicle. This should not be surprising because the malleus is the displaced posterior tip of the embryonic lower jaw (see fig. 9.40). Table 11.7 lists the muscles supplied by the trigeminal nerve in *Squalus* and tetrapods. The fiber components of the nerve are listed in table 16.5.

Nerve VII (Facial)

The seventh nerve arises from the anterior end of the hindbrain in close association with the fifth nerve. In

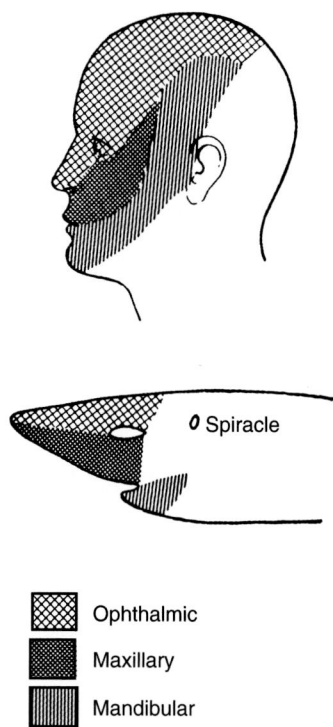

FIGURE 16.33
Cutaneous distribution of the trigeminal nerve in a human and a shark.

Legend:
- Ophthalmic
- Maxillary
- Mandibular

O Spiracle

sharks, the fifth and seventh nerves have a common trigeminofacial root.

Nerve VII is sensory to taste buds in the ectoderm of fishes, which can extend all the way to the tip of the tail. A pharyngeal (**palatine**) branch in fishes supplies taste buds in the mucosa of the pharynx at the level of the second pharyngeal arch. In tetrapods, the taste buds on the anterior portion of the tongue are also derived from the second arch, and they continue to be innervated by the facial nerve. Typically, lateral-line nerve fibers will accompany VII fibers to their final distribution on the head. Nerve VII of fishes also contains sensory fibers for general sensation from the endoderm of the second arch. The cell bodies for sensory fibers in nerve VII, except those for proprioception, are in the facial ganglion. Another name for this ganglion is geniculate, so named because it is at a kneelike bend in the facial canal within the temporal bone of mammals.

The facial nerve is motor to the muscles of the hyoid arch in all craniates. A list of these muscles will be found in table 11.7. They include the facial muscles of mammals and the stapedius muscle, which is attached to the stapes, an ear ossicle derived from the hyomandibular cartilage of the embryonic hyoid arch.

The facial nerve in mammals also contains visceral motor fibers to the submandibular and sphenopalatine (pterygopalatine) ganglia of the autonomic nervous system (figs. 16.34 and 16.35). These ganglia contain the

Preganglionic fibers of the autonomic nervous system are present in nerve IX. They have not been fully explored in all craniates, but in mammals they synapse in the otic ganglion, from which postganglionic fibers supply the parotid salivary gland.

The sensory cell bodies for nerve IX, other than those for proprioception, are found in the **petrosal ganglion** of anamniotes and nonavian reptiles and in the **superior (petrosal)** and **inferior glossopharyngeal ganglia** of birds and mammals. The superior ganglion, chiefly somatic, is derived from neural crest. The inferior ganglion, chiefly visceral, is derived largely from an ectodermal placode.

Nerve X (Vagus)

The vagus arises from the medulla by lateral rootlets. In *Squalus* (see fig. 16.32), it consists of four **branchial trunks** (more in sharks with more gill chambers) that subserve branchiomeric functions and a **ramus visceralis** to the coelomic viscera. As seen in other cranial nerves, lateral-line nerve fibers are intimately associated with the vagus and represent a **ramus lateralis** to the lateral-line canal system.

Each branchiomeric trunk in *Squalus* has a pretrematic, posttrematic, and pharyngeal branch. The posttrematic branches are motor to the fourth and remaining pharyngeal arches and are therefore the chief respiratory nerves of fishes. They are also sensory to the posterior walls of the last four gill chambers. The pretrematics are sensory to the demibranchs in the anterior walls of those chambers. The pharyngeal branches are sensory for general sensation and taste on the palate between the fourth arch and the esophagus.

The ramus visceralis and its branches carry preganglionic fibers of the autonomic nervous system to autonomic ganglia in the coelom. The ganglia innervate coelomic viscera. The same branches contain visceral afferent fibers.

With the rise of land craniates, the vagus lost the sensory fibers from the gill chambers. The remainder of the vagus continues to function as in fishes. The *ramus visceralis* is now the major branch of the vagus, supplying the heart and most of the viscera via autonomic ganglia in the thorax and abdomen (see fig. 16.35). Visceral sensory fibers accompany the motor fibers.

The vagus of tetrapods continues to supply muscles derived from the fourth and successive pharyngeal arches. In amniotes, these are chiefly the cricothyroid, cricoarytenoid, thyroarytenoid, and several small but important muscles of the soft palate and pharyngeal walls that are active in swallowing (palatoglossus, for example, which is seldom seen in routine mammalian dissection).

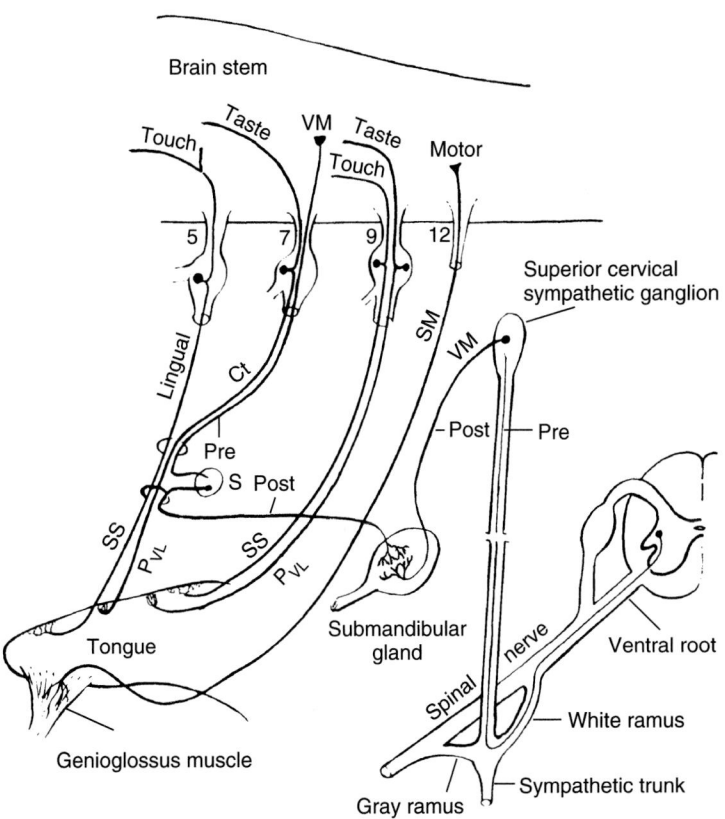

FIGURE 16.34
Innervation of the tongue and submandibular gland of a mammal (based on cat and human). **5, 7, 9,** and **12,** cranial nerves; **Ct,** chorda tympani; **Pre** and **Post,** preganglionic and postganglionic fibers of the autonomic nervous system; **S,** submandibular ganglion of the autonomic system. A key to the fiber components is given in tables 16.1 and 16.5.

cell bodies for postganglionic fibers that innervate the submandibular and sublingual salivary glands, the lacrimal glands, and the mucous membranes of the nose.

Nerve IX (Glossopharyngeal)

The ninth nerve arises from the medulla. In sharks, it has three major branches, **pretrematic, posttrematic,** and **pharyngeal.** The role of these branches was discussed in the introductory paragraphs to mixed nerves. A branch of one lateral-line nerve accompanies IX and represents a small **lateral-line branch** at the junction of the head and trunk. The glossopharyngeal nerve in all craniates is motor to the muscles derived from the third visceral arch (see table 11.7). Of these, only a stylopharyngeus remains in mammals. The muscular branches contain proprioceptive fibers.

With loss of gills in tetrapods, the ninth nerve lost many fibers. It supplies taste buds on the posterior surface of the tongue in mammals, and general receptors (touch, pain, and temperature) on that same mucosa.

Finally, as in fishes, the tetrapod vagus carries fibers for general sensation and any fibers for taste from the pharyngeal mucosa—this being, in tetrapods, the mucosa of the oral pharynx.

The cell bodies of all sensory fibers, except those for proprioception, are in sensory ganglia. In some elasmobranchs, each of the four or more branches to the gills has its own epibranchial ganglion. The vagus of birds and mammals, like nerve IX, has two sensory ganglia on its root, a **superior vagal ganglion** that is chiefly somatic, and an **inferior vagal ganglion** that is chiefly visceral. The superior ganglia are derived from neural crest, the inferior ganglia from epibranchial placodes.

The innervation of the mucosa of the oral cavity and pharynx by nerves V, VII, IX, and X *in that sequence from front to rear, in all craniates from fish to humans,* is evidence of the genetic ties between fishes and tetrapods and of the negligible effect of a terrestrial habitat on the sensory innervation of constantly moist membranes.

The pretrematic, posttrematic, and pharyngeal branches of nerves IX and X of Squalus have their counterparts in nerves V and VII. The ophthalmic (profundus) and mandibular branches of nerve V are functionally equivalent, respectively, to pretrematic and posttrematic branches farther caudad. Nerve V has no pharyngeal branch. The pretrematic branch of VII has joined sensory fibers from nerve V to form an infraorbital trunk; the hyomandibular nerve, behind the spiracle and supplying the second arch, is the posttrematic branch of VII; and the palatine is the pharyngeal branch. Some deviation from the basic pattern is to be expected in the first two arches because they are specialized for procuring and handling food.

Innervation of the Mammalian Tongue: An Anatomic Legacy

The innervation of the mammalian tongue illustrates how a single organ may be served by numerous nerves, depending on the ontogenetic and phylogenetic history of its parts (see fig. 16.34). The mucosa of the anterior part of the tongue is associated with the first arch and is therefore innervated by nerve V for general sensations. The taste buds on this part of the tongue are innervated by nerve VII, which supplies taste buds on the first and second arches in fishes. The mucosa on the posterior part of the tongue is innervated by nerve IX for both general sensation and taste because of the origin of this mucosa from the third pharyngeal arch. The muscles of the tongue are myotomal and are innervated by nerve XII.

Although four cranial nerves innervate the tongue, only three can be traced into it because the branch of nerve VII–conducting taste fibers (chorda tympani, see fig. 16.34) unites with the lingual branch of nerve V just before the latter reaches the tongue.

Functional Components of Cranial Nerves

As noted earlier, fibers in spinal nerves may be classified in four functional categories: SS, VS, SM, and VM. Each supplies general receptors or effectors (see table 16.1). One or more of these components may be found in cranial nerves. In addition to general fibers, cranial nerves contain fibers not found in the spinal cord, of which there are five categories (table 16.5): P_{DL} (hair cell receptors), P_{VL} (taste), placodal (olfactory: 0, I, VN), BSM, and brain outgrowths (photoreceptive: II, E). The functional categories of the spinal cord form columns of gray matter composed of nuclei subserving that specific function, motor columns being in the basal plate, sensory

TABLE 16.5 Functional Categories for Cranial and Spinal Nerves[1]

A[2]	B[2]
Spinal Nerves	
Sensory	
GSA	SS
GVA	VS[3]
Motor	
GSE(SE)	SM[4]
GVE	VM
Cranial Nerves	
Sensory	
SSA (II, VIII, LL)	Brain (II, E)
	P_{DL} (VIII, LL)
SVA (I, taste of VII, IX, X)	Placodal (0, I, VN)
	P_{VL} (taste of VII, IX, X)
GVA (VII, IX, X)	VS[3] (VII, IX, X)
GSA (SS) (Profundus [V$_1$])	SS (Profundus [V$_1$])
Motor	
SVE (EB) (V, VII, IX, X, XI)	BSM[4] (V, VII, IX, X)
GSE (SE) (III, IV, VI)	SM[4] (III, IV, VI, XI, XII)
GVE (III, VII, IX, X)	VM (III, VII, IX, X)

[1]Column (A) represents a traditional nomenclature. Afferent—sensory and efferent—motor are often interchanged. "General" and "special" are used to distinguish spinal and distinct cranial functions, respectively. Column (B) represents the terminology as used in this text. Three ectodermal placode categories are distinguished: P_{DL}—dorsolateral placodes associated with the lateral line; P_{VL}—ventrolateral placodes associated with taste; and placodal (placodes associated with the forebrain). The special visceral efferent (SVE) or efferent branchial (EB) of column (A) is replaced with BSM (branchiomeric somatic motor). The branchiomeric muscles were originally believed to be derived from lateral-plate mesoderm (hypomere, thus visceral in origin), but it is clear now that they are derived from paraxial mesoderm (somitomeres and somites) like all myotomal muscle.
[2]Other abbreviations include: GSA, general somatic afferent; GVA, general visceral afferent; GVE, general visceral efferent; GVE (SE), general somatic efferent; SM, somatic motor; SS, somatic sensory; SSA, special somatic afferent; SVA, special visceral afferent; VM, visceral motor; VS, visceral sensory.
[3]VS (GVA) fibers accompany most VM (GVE) fibers.
[4]Proprioceptive fibers accompany SM (SE) and BSM (SVE) fibers.

columns in the alar plate (fig. 16.31). Purely cranial functional categories may represent continuous, or, more often, discontinuous extensions of spinal columns. Alternatively, their nuclei may be restricted to the brain (e.g., BSM or nuclei associated with paired external sense organs). Table 16.4 summarizes the major distribution of the cranial nerves in craniates in general.

AUTONOMIC NERVOUS SYSTEM

The autonomic nervous system innervates glands and smooth and cardiac muscle. It consists chiefly of autonomic nerves, plexuses, and ganglia (fig. 16.35). It does not, however, constitute an anatomical entity; that is, it cannot be completely dissected away from the rest of the nervous system because its components begin inside the central nervous system and emerge via cranial or spinal nerves. It is entirely a visceral *motor* system. However, sensory fibers from the viscera use autonomic pathways to reach cranial and spinal nerves.

The autonomic nervous system is composed of (1) a **sympathetic system** (**thoracolumbar system** in mammals) that emerges from the cord via most of the spinal nerves of the trunk and (2) a **parasympathetic (craniosacral) system** that emerges from the brain and from the sacral region of the spinal cord. Most visceral effectors except those in the skin are innervated by fibers from both systems (fig. 16.35, iris diaphragm, heart, stomach, urinary bladder). The stimulatory effects of one system modulate the inhibitory effects of the other to bring about an appropriate response. Impulses from the sympathetic and parasympathetic systems that innervate the intestines are delivered to delicate neural plexuses (**Meissner's** and **Auerbach's**) located in the external muscle layer of the gut (see fig. 12.16). This **enteric plexus**, consisting of enteric neurons, is responsible for peristalsis.

Two finely myelinated neurons in series conduct the impulse from the brain or cord to a visceral organ (fig. 16.36). The cell body of the first neuron (preganglionic neuron) is in the VM column of the brain or cord, and its preganglionic fiber terminates in an autonomic ganglion. The cell body of the second neuron (postganglionic neuron) is in an autonomic ganglion. Its fiber extends to the effector. Autonomic ganglia are classified in three categories, **sympathetic (paravertebral)**, **collateral**, and **terminal**.

Sympathetic ganglia are interconnected by autonomic fibers that form a **sympathetic trunk** close to the vertebral column (figs. 16.35 and 16.37). There is generally one ganglion for each spinal nerve except in the cervical region, where there are fewer, and in the sacrum, where there are none. The sympathetic ganglia in the trunk are connected to a spinal nerve by white and gray rami communicantes (see fig. 16.36). White rami contain preganglionic fibers from the cord and afferent visceral fibers with cell bodies in the dorsal root ganglion. Gray rami contain postganglionic fibers to the skin.

In general, fibers with cell bodies in the cervical ganglia of the sympathetic trunk supply visceral organs in the head, neck, and thorax. Fibers with cell bodies in other sympathetic ganglia supply vasomotor, pilomotor, and secretory fibers to the skin via gray rami and spinal nerves. (**Vasomotor fibers** supply cutaneous arterioles, **pilomotor fibers** supply muscles that move hairs, and **secretory fibers** supply skin glands.) In cold-blooded craniates, the cutaneous fibers also supply some types of chromatophores, and in birds they supply **plumomotor fibers** to muscles that move feathers. A considerable majority of preganglionic fibers that enter the sympathetic trunk pass through sympathetic ganglia without synapsing, enter an **autonomic nerve** such as the splanchnic, and terminate in the coeliac and other collateral ganglia in the abdominal cavity.

Postganglionic fibers with cell bodies in collateral ganglia in the head—**ciliary**, **submandibular**, **sphenopalatine (pterygopalatine)** and **otic**—innervate visceral effectors in the head. Their neural connections are given in table 16.6. Postganglionic fibers with cell bodies in collateral ganglia in the coelom, such as the coeliac and inferior mesenteric, innervate organs in the abdomen. The collateral ganglia in the abdomen are part of autonomic plexuses on the major arterial branches of the abdominal aorta (fig. 16.38).

Terminal ganglia are embedded in the walls of visceral organs within the coelom. The cell bodies in these ganglia send very short postganglionic fibers to the organs they innervate.

After passing through the sympathetic trunk, most autonomic fibers join **autonomic plexuses** en route to their destination. These plexuses are entwined around blood vessels, the vertebral column, and other available structures including, in the neck, the esophagus (see fig. 16.37).

The neurotransmitter released at the endings of all preganglionic fibers, both sympathetic and parasympathetic, and the one released at postganglionic endings of the parasympathetic system is chiefly **acetylcholine**. These fibers are therefore said to be **cholinergic**. (Acetylcholine also mediates transfer of impulses from somatic motor fibers to skeletal muscle cells.) The neurotransmitters released at postganglionic endings of the sympathetic system are chiefly **norepinephrine (noradrenaline)** and small amounts of **epinephrine (adrenaline)**. These fibers are said to be **adrenergic**.

The **adrenal medulla** (see fig. 16.35) is functionally equivalent to a sympathetic ganglion. The cells of the medulla, like those of sympathetic ganglia, arise from neural crest, are innervated by preganglionic fibers, and synthesize and release adrenaline and noradrenaline. They differ in that the adrenal medulla does not sprout axons. Consequently, their secretions are carried away by the bloodstream.

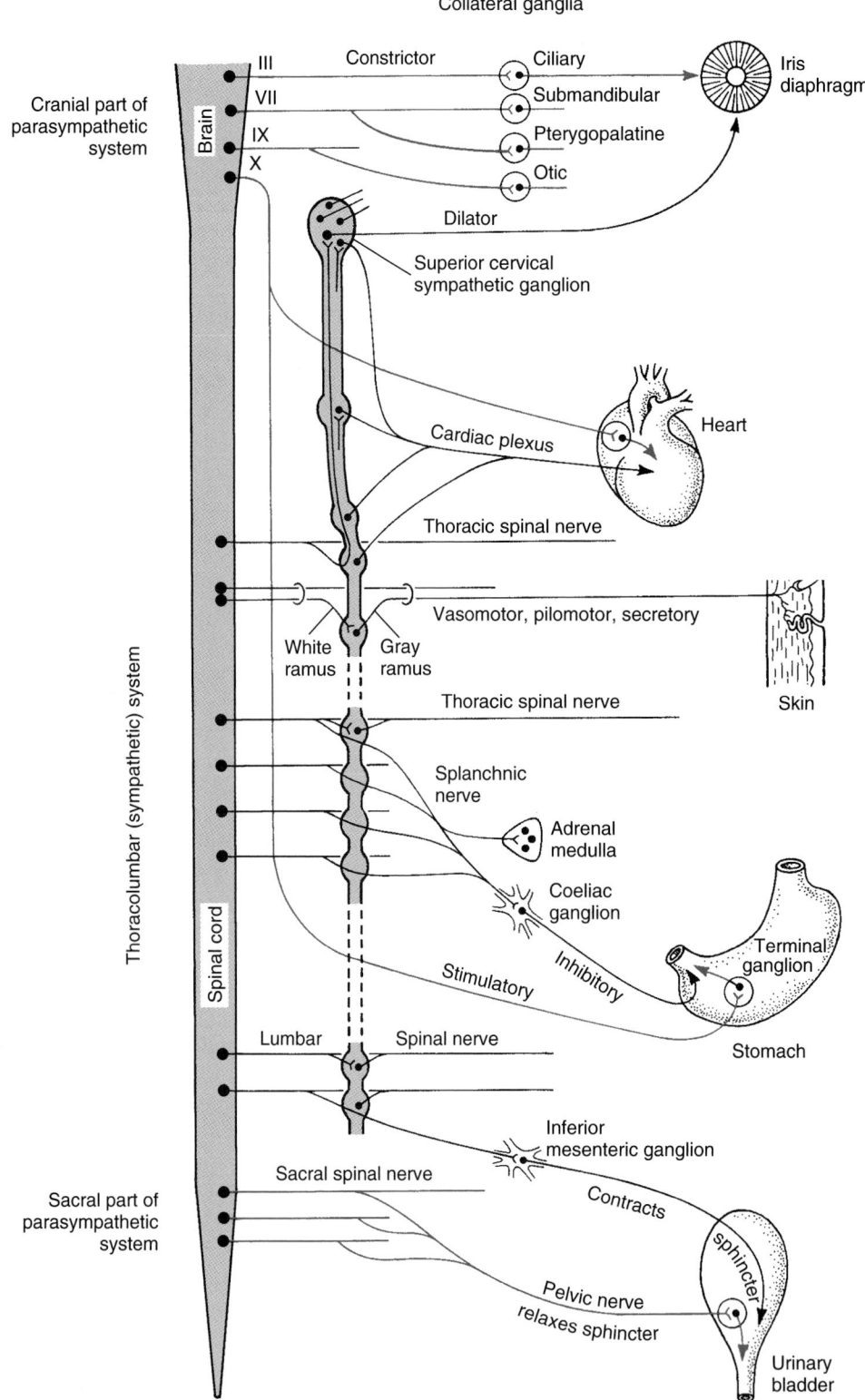

FIGURE 16.35

Representative components of the mammalian autonomic nervous system, sympathetic trunk in *gray*. Autonomic fibers are not confined to the spinal nerves shown. *Arrows* emphasize dual control of visceral effectors by craniosacral and thoracolumbar systems. The cell bodies (*black dots*) of preganglionic fibers are in the brain or cord; those of postganglionic fibers are in a sympathetic, collateral, or terminal ganglion. *Red* indicates parasympathetic innervation (lacking in skin and adrenal).

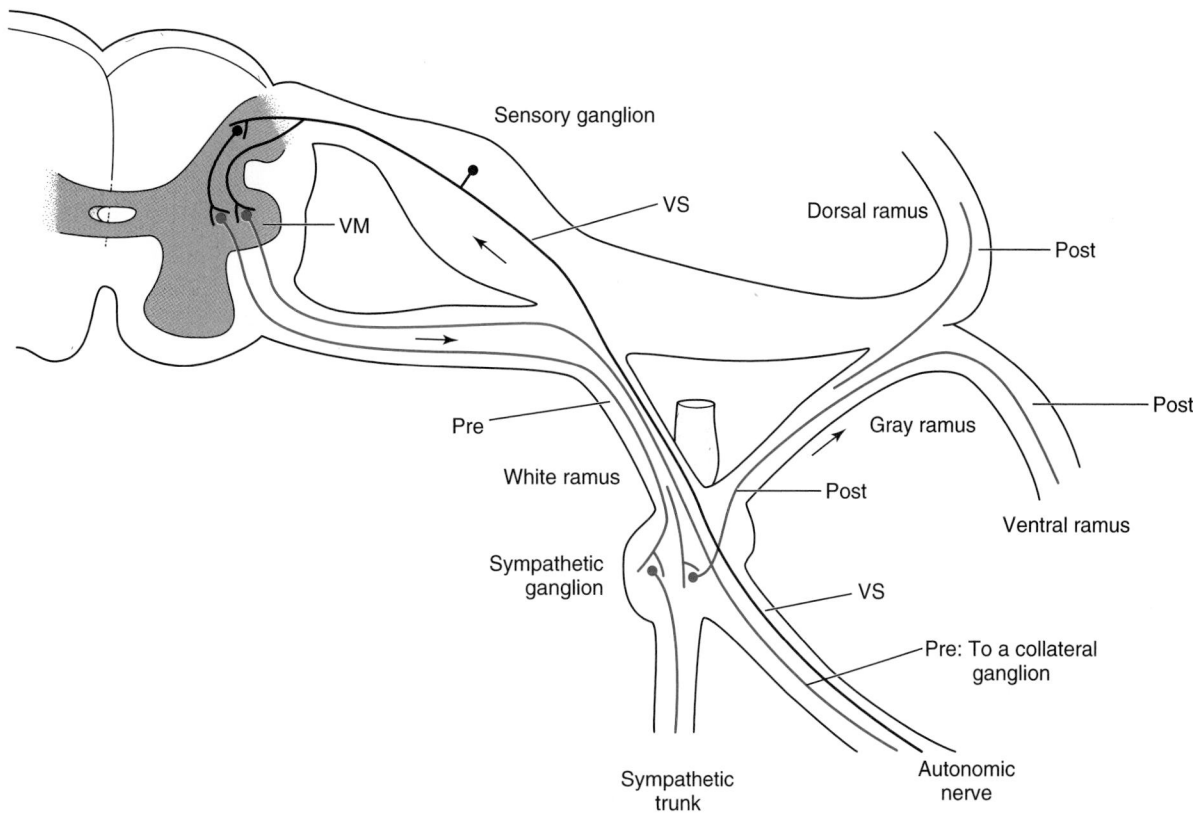

FIGURE 16.36

Motor fibers (*red*) of autonomic system in a thoracic spinal nerve. Fibers in the dorsal and ventral rami are vasomotor, pilomotor, or secretory. The red fiber in the autonomic nerve distal to the ganglion will synapse in a collateral ganglion such as the coeliac (see splanchnic nerve, fig. 16.35). **VS**, visceral sensory fiber; **VM**, visceral motor cell body in lateral horn; **Post, Pre**, postganglionic and preganglionic fiber.

Although the autonomic nervous system is involuntary (you cannot prevent blushing), some voluntary control can be exerted by biofeedback. Autonomic systems are present in all craniates; anurans have the basic components of amniote systems.

SEGMENTATION OF THE CRANIATE HEAD

Now that most of the major tissues and organ systems that constitute the craniate head have been covered, the question of segmentation can be addressed. Craniates retain a plesiomorphic metameric pattern. This is clearly seen in the axial skeleton (vertebrae) and body musculature (myomeres) of most anamniotes. The muscle pattern becomes obscured in many amniotes. Historically, there has been an attempt to reveal the pattern of segmentation in the head. Early hypotheses proposed that the metameric pattern of the body could be extended anteriorly with the skull interpreted as a series of modified vertebrae. The cranial nerves were interpreted as an extension of the segmented spinal nerves. Skeletal and soft tissue structures of the pharynx, also showing a segmented pattern, were superimposed on this anteriorly segmented animal. Although

this story appears logical at a superficial level, it is not so straightforward.

What structures in the head do show segmentation? The hindbrain shows a variable number of visible segments called rhombomeres (fig. 16.39, the numbered segments of the CNS). This pattern has been demonstrated in the mesencephalon and prosencephalon although the pattern is seen primarily in the expression of regulatory genes in these regions. In chapter 11, the paraxial muscles of the head were shown to be segmented as somitomeres anteriorly and somites more posteriorly. The pharynx consisting of a number of visceral arches (mandibular through a variable number of branchial arches) is clearly segmented along with the various structures that compose each visceral arch (skeleton, muscles, and aortic arches, fig. 16.39). As outlined in this chapter, there is an anterior to posterior organization of cranial nerves or their components. So, yes, there is a pattern of segmentation among the constituent parts of the head; however, a number of questions remain in our understanding of this pattern. How do these constituent parts register? What is the developmental relationship among the parts? In other words, how is the pattern determined, and what is the relationship (inductive or otherwise) between the tissues?

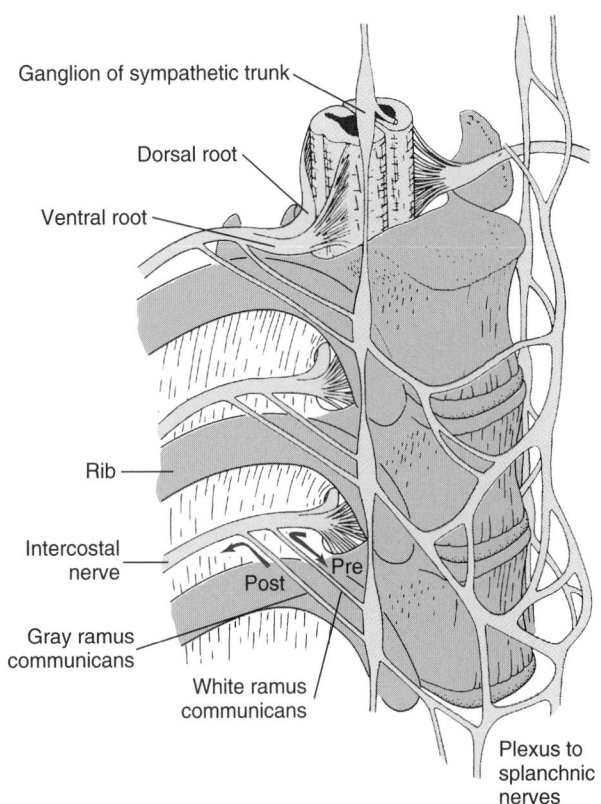

Ganglion of sympathetic trunk

Dorsal root

Ventral root

Rib

Intercostal
nerve

Post

Pre

Gray ramus
communicans

White ramus
communicans

Plexus to
splanchnic
nerves

FIGURE 16.37
Sympathetic (paravertebral) ganglia and their interconnections in
thoracic region of a cat, oblique ventral view. *Arrows* indicate
direction of preganglionic (**Pre**) and postganglionic (**Post**) impulses.

TABLE 16.6   Innervation and Peripheral Distribution of
Autonomic Ganglia of the Head of Mammals

Ganglion	Receives Fibers from	Projects Fibers to
Ciliary	Oculomotor nerve	Ciliary muscle of eye
		Constrictor muscles of iris
Submandibular	Facial nerve	Submandibular gland
		Sublingual gland
Sphenopalatine (pterygopalatine)	Facial nerve	Lacrimal gland
		Glands of nose and pharyngeal mucosa
Otic	Glossopharyngeal nerve	Parotid gland

cestral features (this includes structures involved in the
segmentation of the craniate head). Related to this question are the developmental processes that give rise to
these new structures, including their developmental regulatory genes. As noted above, there is a level of conser-
vatism among regulatory genes. If we look at the homeobox genes expressed in craniates and their outgroups, we
see an increasing level of complexity (fig. 16.40) through
various processes of gene duplication. Is this the source
for genetic regulation of development in the craniate
head? It is argued that through duplication, genes are
"free" to be modified without adversely affecting the phenotype. We are just now coming to the point where current research can address some of these questions.

Skull

Only the postotic occipital region (see fig. 9.26) is serially homologous with the vertebrae. The individual occipital ossifications, as in vertebrae, are derived from
sclerotome. The dorsal (supraoccipital), ventral (basioccipital), and paired lateral (exoccipital) elements surround the spinal cord and parallel the organization of
major elements in the vertebrae. Attempts to extend this
pattern anterior to the occiput fails. The pattern of spatial organization is much more complex and does not
lend itself to a simple one-to-one comparison. What
clearly distinguishes the anterior skull from vertebrae is
its developmental origin from neural crest. In the skull,
we see only a partial serial pattern.

Pharynx

The serial pattern within the pharynx is obvious from
simple examination of the visceral arches and their
components (see figs. 9.1 and 16.39). The pharyngeal
skeleton represents a serially homologous sequence
with mandibular and hyoid arches highly modified. The
five components of a typical branchial arch (e.g., as seen
in *Squalus*) can be homologized with other branchial

These questions go well beyond an introductory text
and are the basis for extensive current research, but it is
important for the student of comparative anatomy to be
aware of the story as it stands at this point.

Genes

Dorsal-ventral and anterior-posterior patterns are determined by regulatory genes. The basic patterns seen in
bilaterally symmetrical organisms are extremely ancient.
Similar genes are responsible for patterning in organisms as diverse as the fruit fly (*Drosophila melanogaster*)
and craniates. These homologous homeotic genes determine an anterior-posterior registry. It is this registry that
establishes developmental domains that provide signals
to other genes. This cascading genetic control
(pleiotropic and polygenetic) establishes the unique nature of each segment.

In the evolutionary transition to craniates, a question
that continually arises is whether the synapomorphies of
craniates represent new structures or modifications of an-

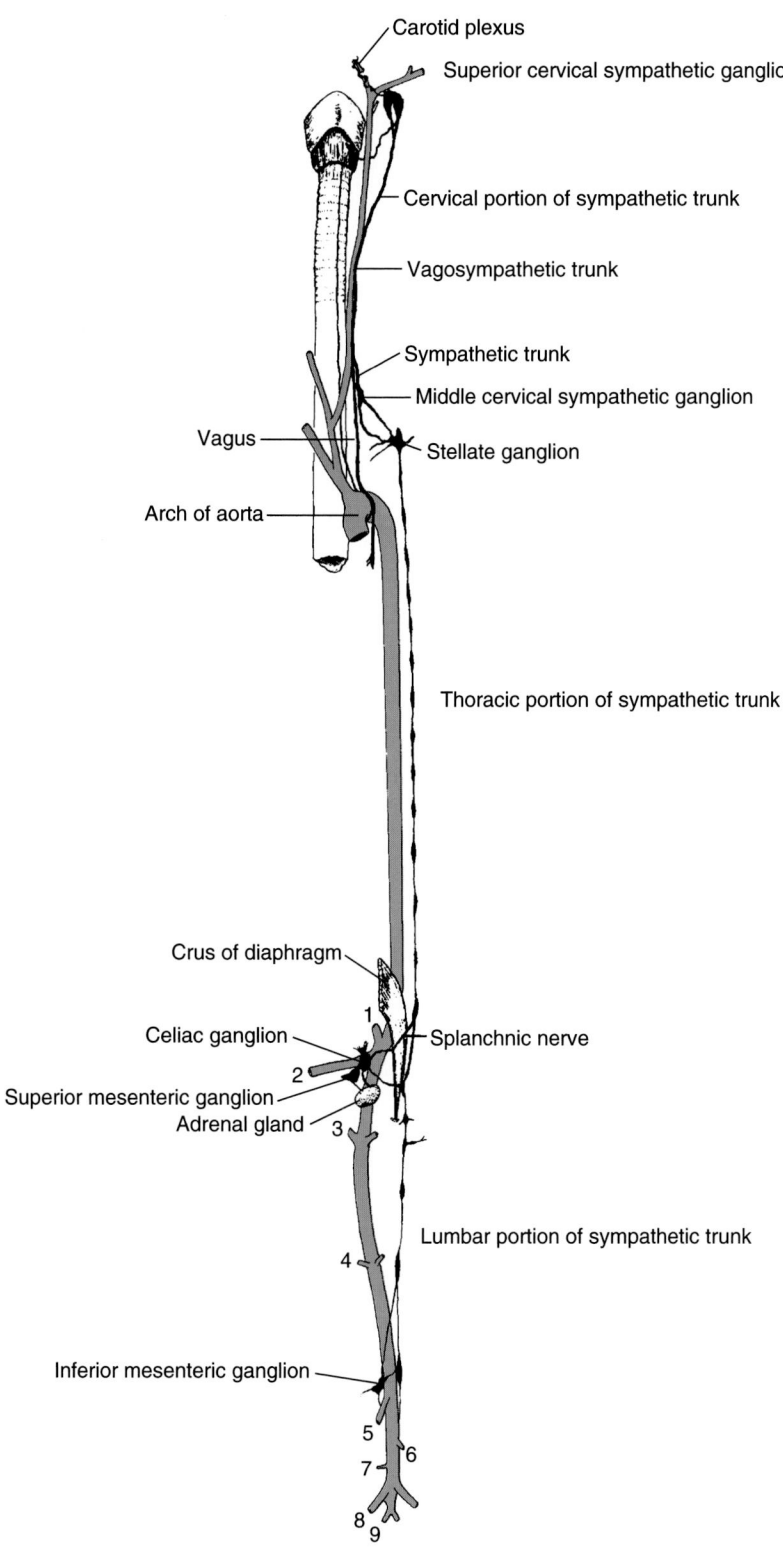

FIGURE 16.38

Left sympathetic trunk and associated structures of a cat. **1** to **9**, major branches of abdominal aorta: **1**, celiac; **2**, superior mesenteric; **3**, renal; **4**, spermatic or ovarian; **5**, inferior mesenteric; **6** and **7**, left and right iliolumbars; **8**, external iliac; **9**, internal iliac. In cats, as in some other mammals, the vagus nerve and sympathetic trunk in the neck are in a common sheath to form a vagosympathetic trunk. Fibers in the carotid plexus have postganglionic cell bodies in the superior cervical sympathetic ganglion.

arch elements and the anterior hyoid and mandibular arches: pharyngobranchial— pharyngobranchial; epibranchial— epibranchial—hyomandibula—palato-quadrate; ceratobranchial—cerato-branchial—ceratohyal—Meckel's cartilage; hypobranchial—hypobranchial; basi-branchial—basibranchial—basihyoid. Forming the pharyngeal skeleton, neural crest segmentally migrates into each arch during early development. Within each arch are associated soft tissues that are also segmented (muscles, aortic arches, nerves; see fig. 16.39).

Cranial Nerves and Brain

The cranial nerves supplying the visceral arches (V, VII, IX, and X) clearly have a segmented appearance. These and other nerves have been hypothesized to represent an anterior continuation of spinal nerves. Numerous attempts have been made to equate individual nerves with the dorsal and ventral roots of spinal nerves.

As already noted, the brain is segmented into a number of neuromeric segments (anterior-to-posterior: prosomeres, mesomeres, and rhombomeres). These segments are related to the various cranial nerves that exit the brain.

Registry of Segments: A Hypothetical Segment?

A large amount of today's research efforts center on studies interpreting the pattern and process of segmentation in individual components within the head. Along with this is an effort to determine the relationships among the components. Figure 16.39 summarizes our current understanding. A hypothetical segment might consist of neural crest, four cranial nerve components and their respective central nervous system (neuromeric) segments, paraxial mesoderm (somitomeres or somites), and other tissues associated with pharyngeal segments (table 16.7). Not all segments will contain each component.

The neural crest forms the connective tissues of each segment (only the splanchnocranium appears to demonstrate a segmental pattern, although the posterior neurocranium, [i.e., the occipitals] represent serial homologues to the vertebrae).

The nerve components of a single segment include a dorsal, ventrolateral, dorsolateral, and ventral nerve component.

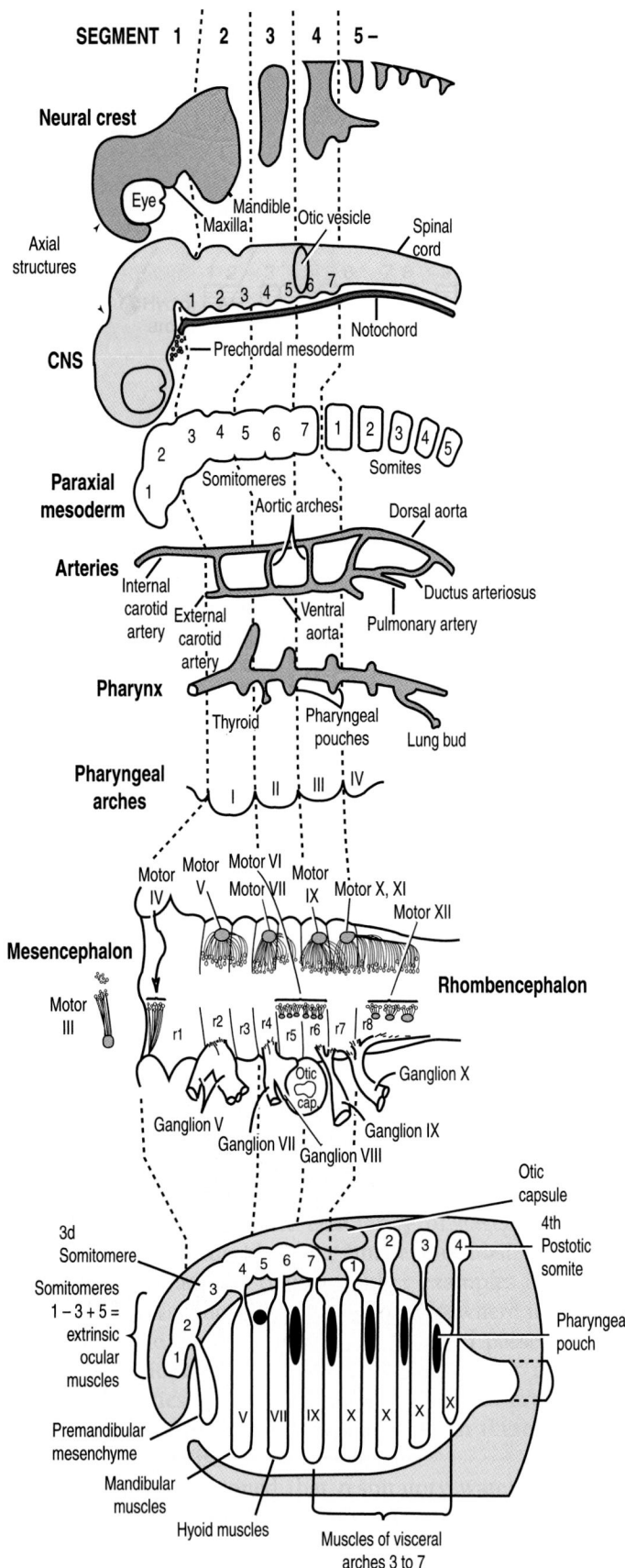

SEGMENT 1   2   3   4   5 –

Pharynx, somitomeres, and somites of the head

TABLE 16.7   Components of Hypothetical Head Segments

Segment	1	2	3	4	5
Nerve[1]					
Dorsal[2]	P	$V_{2,3}$	VII	IX	X
VL[2]	—	—	VII_{VL}	IX_{VL}	X_{VL}
DL	LL	LL	LL	LL	LL
Ventral	III	IV	VI	—	XI[3], XII
Somitomere	1, 2	3, 4	5, 6	7	—
Somite	—	—	—	—	1

[1]Abbreviations: DL, dorsolateral; VL, ventrolateral; VII$_{VL}$, taste component of VII
[2]The dorsal and ventrolateral (VL) components represent a mixed functional cranial nerve.
[3]The spinal accessory nerve, of tetrapods, overlaps the nuclei of XII.

Similar to the spinal cord dorsal root in anamniotes, the dorsal nerve consists of BSM, VM, VS, and SS (note that a spinal nerve does not possess BSM fibers). A ventrolateral nerve is derived from ventrolateral placodes and forms fibers for taste. A dorsolateral nerve, derived from dorsolateral placodes, provides the fibers for lateral-line nerves. Finally, a ventral nerve forms the somatic motor fibers paralleling the ventral root in spinal nerves.

The dorsal and ventrolateral components represent nerves of the mixed function group (VII, IX, X—remember that V lacks taste fibers). Dorsolateral components form the lateral-line cranial nerves (ALL, PLL), and the ventral components form somatic motor cranial nerves (III, IV, VI, XI, and XII). Somatic motor components innervate derivatives of somitomeres (or somites as the case may be) in each segment.

Somitomeres in a hypothetical segment are typically paired. Both somitomeres of the first head segment form extrinsic eye muscles innervated by III. In segments two and three, the anterior somitomere forms eye muscles (innervated by IV and VI, respectively), and the posterior somitomere forms branchiomeric muscles (mandibular and hyoid respectively). In segment four, it appears that the anterior somitomere is lost (no eye muscle associated with this segment), and the single somitomere forms

FIGURE 16.39

Organization of the head based on a 30-day-old embryo. The bottom figure represents a later stage in development with the commencement of brachiomeric somitomere migration. Alignment of segments indicated by *dashed lines.* The head segments are based on those proposed by Butler and Hodos (1996; see Registry of segments: A hypothetical segment? this chapter) and are superimposed on redrawn figures from Carlson (1994) and Northcutt (1990).

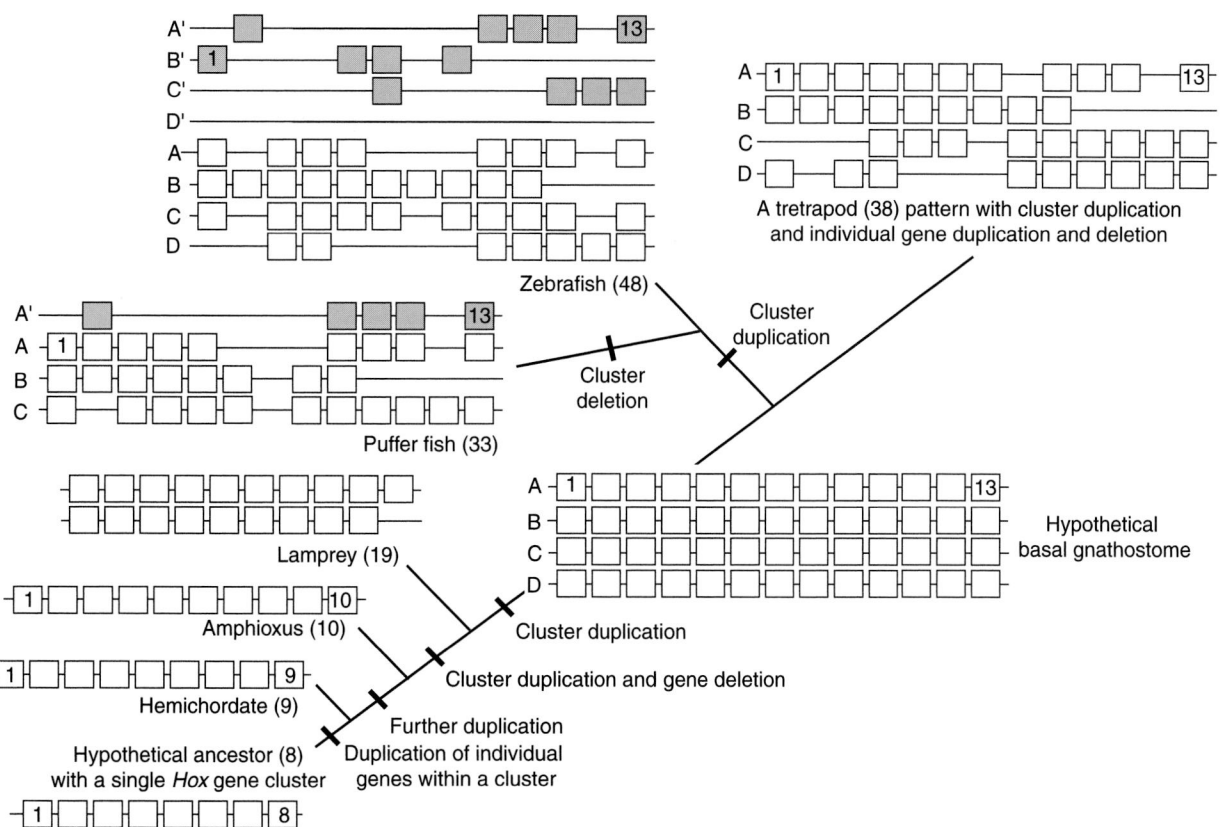

FIGURE 16.40

A hypothesis for the evolution of *Hox* genes in craniates. Retained is the primitive pattern of segmentation, but the duplication of clusters provides the genetic material for introduction and regulation of new tissues and structures (e.g., the possible origin of neural crest and patterns of segmentation in the craniate head). Total *Hox* gene number is indicated in parentheses. The mouse pattern is used to represent tetrapods, which appear to retain the basal gnathostome pattern. Teleosts have apparently duplicated the basal gnathostome pattern (*light red*). This pattern is seen in zebrafish, which possess 7 or possibly 8 clusters. The puffer fish demonstrates a secondary reduction of clusters (retaining 3 of the 4 basal clusters and 1 teleost duplicate cluster). Details of these patterns are subject to change due to the rapid increase in knowledge about *Hox* gene patterns. Teleost patterns after Amores et al. (1998) and Aparicio (2000).

branchiomeric muscles of the first branchial arch (innervated by IX). Subsequent segments appear to be associated with single somites that form the remaining branchiomeric muscles (innervated by X).

Superimposed on the above segmentation pattern are the structures of the pharyngeal arches (aortic arches and pharynx in fig. 16.39). Remember that underlying these patterns is the expression of regulatory genes that appear to determine the developmental domains of individual segments. The mechanism of how this is done is still unknown. Further, the relationships between tissues within a segment are unclear. In the development of the dorsal hollow nerve cord, the notochord (chordamesoderm) provides an inductive signal to the overlying ectoderm. Without this signal, the nerve cord will not form. These tissue-to-tissue relationships remain obscure in the head, but they form the basis of exciting areas of research.

Summary

1. The neuron is the functional unit of the nervous system. It consists of a cell body and one or more processes. The long process is an axon, and short processes are dendrites.
2. A nerve fiber is an axon and its myelin sheath.
3. A nerve is a bundle of nerve fibers outside the central nervous system (CNS). A tract is a bundle of nerve fibers within the CNS.
4. The CNS consists of the brain and spinal cord. The peripheral nervous system includes all neurons and their processes outside the CNS.
5. A nucleus is a group of similarly functioning cell bodies inside the CNS. Nuclei constitute the gray matter of the brain and cord.
6. A ganglion is a group of cell bodies outside the CNS. Ganglia are sensory except those of the autonomic nervous system, which are motor.

7. Neurotransmitters are amines that facilitate or inhibit nerve impulses across synapses. Neurosecretions are polypeptide hormones. They are secreted by neurosecretory neurons and carried in the blood.

8. The cell bodies of motor neurons are inside the cord or brain with one exception. The cell bodies of post-ganglionic fibers of the autonomic nervous system are in autonomic ganglia.

9. The cell bodies of first-order sensory neurons are in sensory ganglia on cranial and spinal nerves with three major exceptions: The cell bodies of olfactory nerve fibers are in the olfactory epithelium, those of optic nerve fibers are in the retina, and those of proprioceptive fibers of cranial nerves are in the mesencephalon.

10. Sensory ganglia arise chiefly from neural crest. A few in the head arise from ectodermal placodes.

11. Olfactory neurons and the rods and cones of the retina are sometimes referred to as neurosensory cells.

12. The embryonic neural tube caudal to the forebrain consists of alar and basal plates that are sensory and motor, respectively. The alar plate gives rise to second-order sensory neurons. The basal plate gives rise to motor neurons other than postganglionic.

13. Neuroglia in the CNS consist of ependyma, oligodendroglia, microglia, and astrocytes.

14. Schwann cells are neural crest cells that surround nerve fibers in the periphery. They form a living membrane, the neurilemma. It secretes the myelin sheath (as do the oligodendroglia of the CNS).

15. The spinal cord consists of nuclei surrounding the central canal and of fiber tracts that ascend, descend, or cross the cord from one side to the other. It is enclosed within one or more meninges.

16. A meninx primitiva surrounds the brain and cord of fishes. A dura mater and leptomeninx are present in basal tetrapods. The latter differentiates into a pia mater and arachnoid membrane in some birds and in mammals.

17. The spinal cord has cervical and lumbar enlargements at the level of the paired appendages. When shorter than the column, the cord terminates in a filum terminale accompanied by a cauda equina. Fishes exhibit an unpaired neuroendocrine swelling of the cord, the urophysis at the base of the tail. It houses neurosecretory cells.

18. Spinal nerves are metameric in origin and distribution. Most spinal nerves exhibit dorsal and ventral rootlets and roots, and dorsal, ventral, and communicating rami. Ventral rami often unite to form simple or complicated plexuses.

19. Most spinal nerves have sensory ganglia on their dorsal roots. The following cranial nerves also have sensory ganglia: V, VII, VIII, IX, and X.

20. Occipitospinal nerves lack sensory roots and supply hypobranchial muscles. They arise between the tenth cranial and the first spinal nerves. They are common in anamniotes. They are represented in amniotes by cranial nerves XI and XII.

21. Spinal nerves contain the following fiber components: SS, SM, VS, and VM. Cranial nerves may contain one or more of the preceding and one or more of the following: Placodal, P_{DL}, P_{VL}, and BSM.

22. Anamniotes have 14–20 cranial nerves; amniotes have 15 (only 13 typically recognized in humans). Nerves 0, I, VN, II, E, P, ALL, PLL, and VIII are purely sensory to the olfactory epithelium, vomeronasal organ, retina, epiphyseal complex, skin of the snout, anterior and posterior lateral lines, and membranous labyrinth. Nerves III, IV, and VI supply myotomal muscles of the eyeball. Nerves V, VII, IX, and X supply the pharyngeal arches with motor and sensory fibers.

23. Cranial nerve V is the chief nerve for cutaneous sensation on the head and the ectodermal part of the oral cavity. Nerves VII, IX, and X, supply also taste buds.

24. Nerve XI has medullar and spinal roots. It is derived from the caudal end of the branchiomeric series and from occipitospinal nerves. The internal ramus contains VM fibers that are distributed with the vagus. The external ramus innervates the trapezius and sternocleidomastoid muscles.

25. Nerve XII represents one or more occipitospinal nerves. It supplies the muscles of the tongue.

26. The autonomic nervous system innervates smooth and cardiac muscles and glands. Preganglionic fibers of the craniosacral (parasympathetic) division emerge from the brain via cranial nerves III, VII, IX, X, and XI, and from the sacral region of the cord via sacral spinal nerves. Preganglionic fibers of the thoracolumbar (sympathetic) division emerge from the cord via thoracic and lumbar spinal nerves.

27. Autonomic ganglia are paravertebral (sympathetic chain), collateral (in head or close to abdominal aorta), and terminal (in trunk adjacent to or within the organ innervated).

28. The autonomic ganglia of the head and their associated nerves are the ciliary (III), sphenopalatine (pterygopalatine) (VII), submandibular (VII), and otic (IX). These are parasympathetic ganglia.

29. Most viscera are supplied by sympathetic and parasympathetic fibers. Visceral effectors in the skin receive only sympathetic innervation.

30. Cerebrospinal fluid is secreted by choroid plexuses in the lateral, third, and fourth ventricles. It fills the brain ventricles and central canal of the cord and escapes to submeningeal spaces via foramina in the roof of the fourth ventricle. It is retrieved directly by venous channels of the brain and cord and by arachnoid villi that return it to major venous sinuses of the brain.

31. The brain has three major subdivisions: prosencephalon (forebrain), mesencephalon (midbrain), and rhombencephalon (hindbrain). The more prominent components of these subdivisions in mammals are listed below.

Prosencephalon (Forebrain)

Telencephalon
- Olfactory bulbs
- Olfactory tracts
- Cerebral hemispheres
  - Neocortex
  - Basal ganglia
  - Corpus callosum
  - Internal capsule
  - Lamina terminalis
- Lateral ventricles

Diencephalon
- Epithalamus
  - Habenulae
  - Pineal
- Thalamus
  - Massa intermedia
- Hypothalamus
- Optic chiasma
- Infundibular stalk and pituitary
- Third ventricle
- Circle of Willis

Rhombencephalon (Hindbrain)

Metencephalon
- Cerebellum
- Pons
- Fourth ventricle

Myelencephalon
- Medulla oblongata
- Vagal lobes
- Pyramidal tracts
- Fourth ventricle

Mesencephalon (Midbrain)

- Optic lobes ⎫
- Auditory lobes ⎭ Tectum
- Tegmentum
  - Cerebral peduncles
  - Red nucleus
- Cerebral aqueduct

Critical Thinking Questions

1. What are the cellular components of the nervous system? Describe them.
2. How do nervous system cells external to the CNS differ from those found within the CNS?
3. It is said that the CNS grows from the inside out. Explain this observation in terms of the layering seen in the neural tube.
4. What are the different types of sensory cells, and how do they differ?
5. How has the pattern of coverings for the brain and spinal cord changed phylogenetically?
6. Draw a typical thoracic spinal nerve in a mammal. Include rootlets, roots, rami, sympathetic trunk, and trace neurons in the four functional groups (VM, VS, SM, and SS).
7. What are the five regions of the brain? Name one primary structure associated with each division.
8. What functions are associated with the cerebellum? What do differences in its size suggest?
9. The tectum is associated with sensory input. What senses are processed in the tectum, and how has it changed phylogenetically?
10. Compare lateral and median eyes.
11. What is the origin for the ventricles? What is one function associated with the ventricles?
12. What functions are associated with the mammalian neocortex?
13. What are the cranial nerves in a typical anamniote? List them by abbreviation, name, and primary function.
14. How do amniote cranial nerves differ?
15. Distinguish the subsystems within the autonomic nervous system. Include location, pre- and postganglionic pattern, neurotransmitters, and function.
16. Is the craniate head segmented? What evidence do we have?

Selected Readings

Ariens Kappers, C. U., Huber, G. C., and Crosby, E. C.: The comparative anatomy of the nervous system of vertebrates, including man, 2 vols. New York, 1936, The Macmillan Co. (Republished by Hafner Publishing Co., 1960, New York.)

Braun, C.B., and Northcutt, R.G.: Cutaneous exteroreceptors and their innervation in hagfishes. In Jørgensen, J.M., Lomholt, J.P., Weber, R.E., and Malte, H., editors: The biology of hagfishes. London, 1998, Chapman & Hall.

Bullock, T. H., and Horridge, G. A.: Structure and function in the nervous systems of invertebrates, 2 vols. San Francisco, 1965, W. H. Freeman and Co., Publishers.

Butler, A.B., and Hodos, W.: Comparative vertebrate neuroanatomy: Evolution and anatomy. New York, 1996, Wiley-Liss.

Finger, T. E.: What's so special about special viscera? *Acta Anatomica* 148:132, 1993.

Finley, B. L., and Darlington, R. B.: Linked regularities in the development and evolution of mammalian brains, *Science* 268:1578, 1995.

Fritzsch, B., and Northcutt, R. G.: Cranial and spinal nerve organization in amphioxus and lampreys: Evidence for an ancestral craniate pattern, *Acta Anatomica* 148:96, 1993.

Hopkins, W. G., and Brown, M. C.: Development of nerve cells and their connections. New York, 1984, Cambridge University Press.

Jacobson, M.: Developmental neurobiology, ed. 2. New York, 1978, Plenum Press.

Morell, P., and Norton, W. T.: Myelin, *Scientific American* 242(5):88, 1980.

Myers, P. Z.: Spinal motorneurons of the larval zebrafish, *Journal of Comparative Neurology* 236:555, 1985.

Nichol, J. A. C.: Autonomic nervous system in lower chordates, *Biological Reviews* 27:1, 1952.

Norris, H. W., and Hughes, S. P.: The cranial, occipital, and anterior spinal nerves of the dogfish, *Squalus acanthias*, *Journal of Comparative Neurology* 31:293, 1920.

Northcutt, R. G.: Ontogeny and phylogeny: A re-evaluation of conceptual relationships and some applications, *Brain Behavior Evolution* 36:116, 1990.

Northcutt, R. G.: The origin of craniates: Neural crest, neurogenic placodes, and homeobox genes. In Gans, C., Kemp, N., and Poss, S., editors: The lancelets (Cephalochordata): A new look at some old beasts (IVth International Congress of Vertebrate Morphology), *Israel Journal of Zoology* 42:273, Supplement, 1996.

Northcutt, R. G., and Bemis, W. E.: Cranial nerves of the coelacanth *Latimeria chalumnae* (Osteichthyes: Sarcopterygii: Actinistia) and comparisons with other craniata, 1993. Reprint of Karger, S.: *Brain, Behavior and Evolution* 42. Basel: Medical and Scientific Publishers.

Northcutt, R. G., and Gans, C.: The genesis of neural crest and epidermal placodes: A reinterpretation of vertebrate origins, *The Quarterly Review of Biology* 58(1):1, 1983.

Pearson, R., and Pearson, L.: The vertebrate brain. New York, 1976, Academic Press.

Rovainen, C. M.: Neurobiology of lampreys, *Physiological Reviews* 59:1007, 1979.

Shuangshoti, S., and Netsky, M. G.: Choroid plexus and paraphysis in lower vertebrates, *Journal of Morphology* 120:157, 1966.

Szekely, G. The morphology of motor neurons and dorsal root fibers in the frog's spinal cord, *Brain Research* 103:275, 1976.

Links to the Internet

Visit the zoology website at http://www.mhhe.com/zoology to find live Internet links for each of the following references.

1. Whole Brain Atlas Top 100 Brain Structures. Actually 106 structures, with photographs and MR images, CT scans etc., of the structures, including pathology.
2. Histology—The Web Laboratory. Links to units on many systems of the human body, including the nervous system. Many interesting histological photomicrographs.
3. The Cranial Nerves. An overview, then specific information on each of the 12 cranial nerves from LUMEN.
4. The Nervous System and Special Senses. The cat nervous system and sheep brains in photographs. Allows an interactive quiz to identify structures.
5. Comparative Mammalian Brain Collection. This site compares the anatomy of many different mammalian brains.

Sense Organs

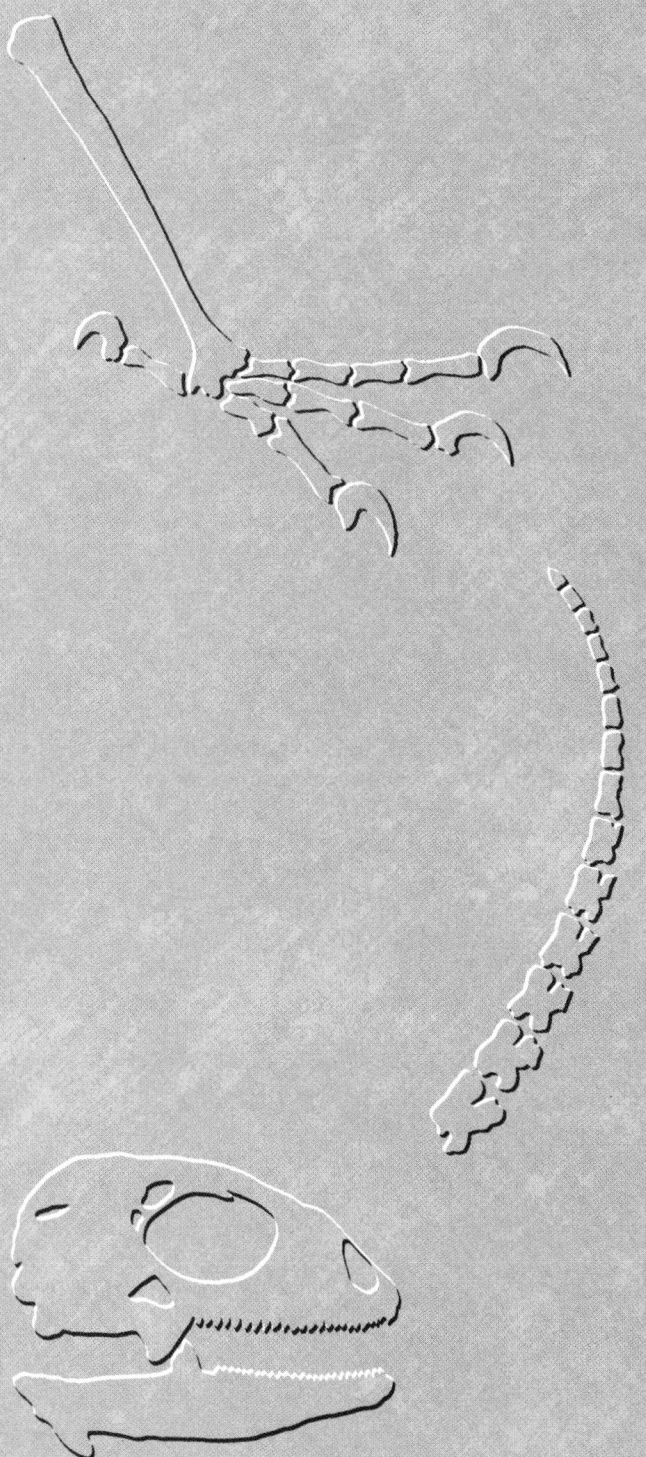

In this chapter, we will study the organs that enable craniates to monitor two environments, the one surrounding them and the one within. Because the former presents more challenges to moment-by-moment well-being, proportionately more time will be devoted to monitors of the ambient environment, the exteroceptors. Along the way, we will look at the sensory feedback mechanism that enables skeletal muscles to respond to any challenge.

ime has resulted in the evolution of a large variety of simple and complex sense organs for monitoring the external and internal environment. Sense organs (receptors) are transducers of specific forms of kinetic energy. They change mechanical, electrical, thermal, chemical, or radiant energy into nerve impulses in sensory neurons.

Craniates arose with an array of essential sense organs and the necessary central nervous system pathways for processing the information: a mechanically stimulated lateral-line canal system, a mechanoreceptive vestibular system that may have had some auditory competence, light receptors (lateral and median eyes) and their central pathways, an olfactory system, chemoreceptors other than olfactory, a general somatic sensory system, and a general visceral sensory system. Electroreceptors arose with the vertebrates. Because many of the receptors are, and have been, protected by skeletal capsules that fossilize readily, or because they leave indelible impressions in fossilized dermal armor, or because of telltale foramina, it can be said with confidence that all these systems were present in the oldest known vertebrates, the ostracoderms. During subsequent vertebrate evolution, receptors have undergone mutation—some more than others. Some, such as the lateral-line system, disappeared among terrestrial vertebrates. Meanwhile, new systems such as thermal receptors in the infrared range have made an appearance and play a role in survival on land.

Somatic receptors provide information about the external environment (**exteroceptors**) and the activity of the skeletomuscular system (**proprioceptors**). Their sensory fibers terminate in nuclei in the *somatic sensory column* of the alar plate of the brain and cord (see fig. 16.31, SS, P$_{DL}$). In reacting to this information, animals make appropriate skeletal muscle responses. For example, stimuli from an approaching potential enemy initiate locomotion or a change in posture, and input from surrounding fishes in a school of fishes reflexly maintains the individual's orientation in the dynamic mass. The category, somatic receptors, does not include chemical receptors such as those for taste and smell.

Visceral receptors provide information about the environment within the animal (**enteroceptors**) and olfactory and gustatory information from the environment. Their sensory fibers, except those for smell, end in nuclei in the *visceral sensory column* of the brain and cord (see fig. 16.31, VS). As a result of information from visceral receptors, an appropriate internal environment is reflexly maintained. Olfactory stimuli evoke both somatic and visceral responses, particularly in amniotes, and they are more or less arbitrarily categorized as visceral.

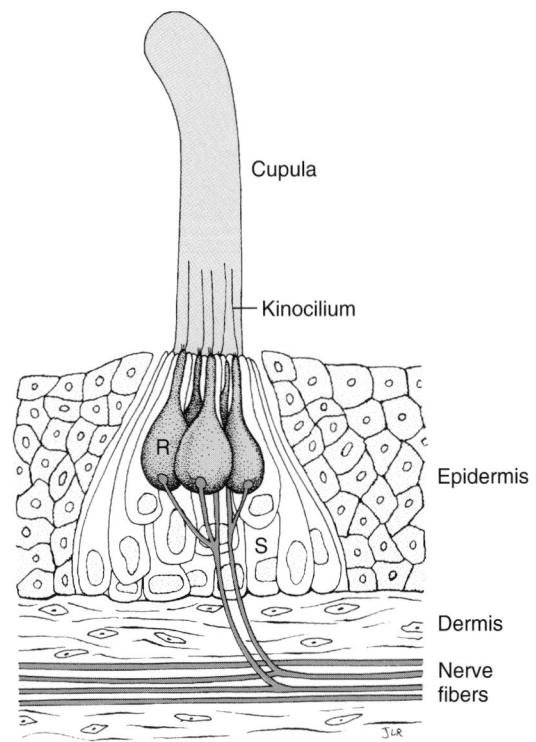

FIGURE 17.1

A neuromast organ in the epidermis of *Necturus*. **R,** hair cell (receptor cell) with a single kinocilium and three or four short delicate stereocilia (at base of kinocilium) extending into the cupula. Contacting the base of the hair cells is a heavily myelinated sensory nerve fiber. **S,** supporting cell.

Although the peripheral endings of many sensory nerve fibers, especially visceral, are stimulated directly, **intermediary nonnervous receptor cells**, or batteries of them, often serve as transducers of the energy actuated by the stimulus, and the receptor cells pass on a stimulus to the sensory nerve fiber. **Hair cells** are one variety of receptor cells (fig. 17.1). They are elongated or bulbous *epithelial* cells with a variable number of short stereocilia and usually a single long kinocilium projecting into the fluid that bathes the sensory epithelium. The cilia are sometimes embedded in a gelatinous mass, the **cupula**. Receptor cells with cytoplasmic apical processes instead of cilia are present in taste buds (see fig. 17.23).

For convenience of discussion, sense organs can be classified as general or special receptors. **General receptors** are widely distributed in the skin and within the body. **Special receptors** are those with a limited distribution. They are confined to the head except in fishes and aquatic amphibians. In the remainder of this chapter, we will examine some of the general and special receptors.

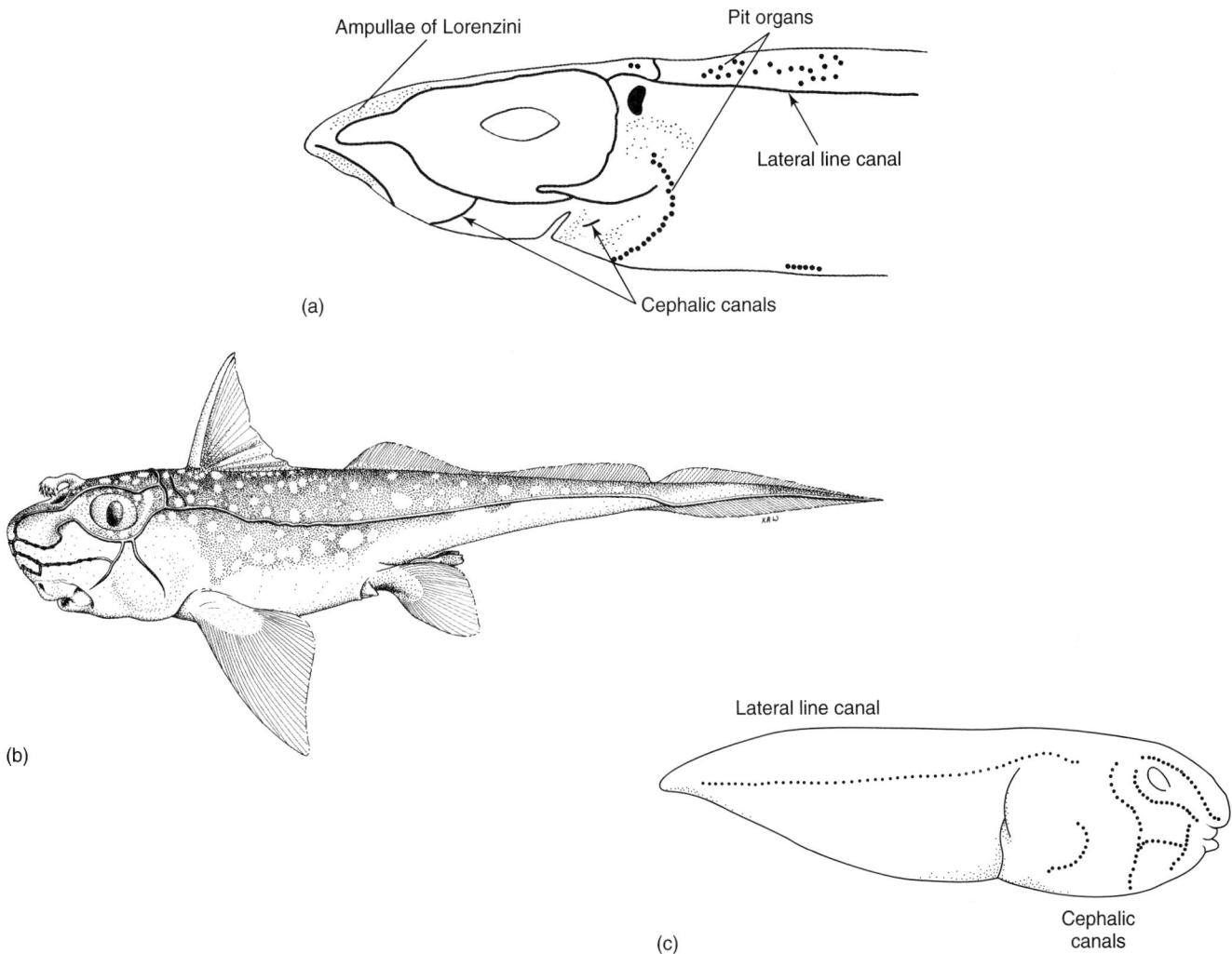

FIGURE 17.2
Distribution of neuromast organs in selected aquatic craniates. (*a*) External openings of several varieties in *Squalus*. (*b*) The elevated canal systems of *Chimaera*. (*c*) An anuran tadpole.

SPECIAL SOMATIC RECEPTORS

Neuromast Organs of Fishes and Aquatic Amphibians

Neuromasts are epithelial receptors that consist of aggregations of hair cells and, generally, a cupula that projects into a fluid (fig. 17.1). Associated with the hair cells are **supportive (sustentacular) cells.** Neuromasts are disposed singly or in groups, depending on their location and the species. In fishes and aquatic amphibians, they monitor principally mechanical stimuli in the pond or seawater, although some serve as electrical or chemoreceptors. The simplest neuromasts occupy shallow pits or grooves in the epidermis, with the cupulae projecting into the pond or seawater. These **external neuromasts** are characteristic of agnathans, larval amphibians, and aquatic urodeles. More specialized are the **internal neuromasts** in the membranous labyrinth (inner ear) of craniates from fishes to humans.

Neuromasts in some jawed fishes occupy fluid-filled pits that lie beneath the epidermis and open to the surface via a pore. These are **pit organs.** Other neuromasts occupy vesicles or ampullae that have a long duct that connects them with the surface. Among these are the **vesicles of Savi** in electric rays and the **ampullae of Lorenzini** in sharks (fig. 17.2*a*). The jellylike fluid in the vesicles and ampullae is secreted by glandular cells that line the organ. *Amia*, a basal neopterygian (a ray-finned fish), has as many as 3,700 ampullae on the head alone. Pit organs, vesicles, and ampullae are found chiefly, but not exclusively, on the head.

The most common systems of neuromast organs are the fluid-filled **lateral-line canal** and the **cephalic canal system** (fig. 17.2). In their simplest manifestation, these are linear series of neuromasts located in numerous shallow canal-like grooves on the surface of the head and in a groove that may extend along the lateral line of the body all the way to the tip of the tail. Identical grooves in the dermal armor of fossil acanthodians and placoderms reveal that, primitively, the system was a network of superficial canals covering the entire body. The network became reduced in later fishes by the loss of many branches and by interruption of the canals at one or more locations. In sharks, the canals have sunk into the skin of the head and trunk but remain as open grooves on the tail. In chimaeras, the canals form elevated ridges on the surface where they are readily visible (fig. 17.2*b*). When neuromasts are in enclosed canals, pores may open to the surface at regular intervals. In modern bony fishes, the canals become even more isolated from the external environment, often being embedded in dermal bone underlying the integument (fig. 17.3). In some teleosts, the canals are restricted to the head.

Free-living amphibian larvae, particularly anuran tadpoles, ordinarily lose the lateral-line and cephalic canal systems at metamorphosis. Among amphibian larvae that do not are those of the urodele *Notophthalmus*. When these larvae metamorphose into red efts and migrate to land, the neuromast organs become buried under the proliferating stratum corneum. Later—several years later in some localities—when the eft returns to water as a sexually mature newt, the stratum corneum is shed, and the canal systems are again exposed.

The major portion of the canal systems of fishes responds to mechanical stimuli such as compression waves of low frequency in the water, water currents, and, perhaps, hydrostatic pressure. Some portions respond to low electric potentials in certain species. Water currents include, for instance, those generated in flowing streams, those caused by locomotor movements of nearby fishes, and waves reflecting from stationary objects. Response to compression waves of low frequency was an early basis for the belief that fishes can hear sounds via the inner ear. (As will be seen shortly, at least some fishes *can* hear.) Input from mechanoreceptors enables fishes to orient appropriately in flowing streams (**rheotaxis**), to participate in schooling, and to avoid potential enemies. Blind carnivorous fishes inhabiting dark caves locate some of their food with this input. The roles of the canal systems vary with the habitat of the species (swiftly flowing mountain streams, for example,

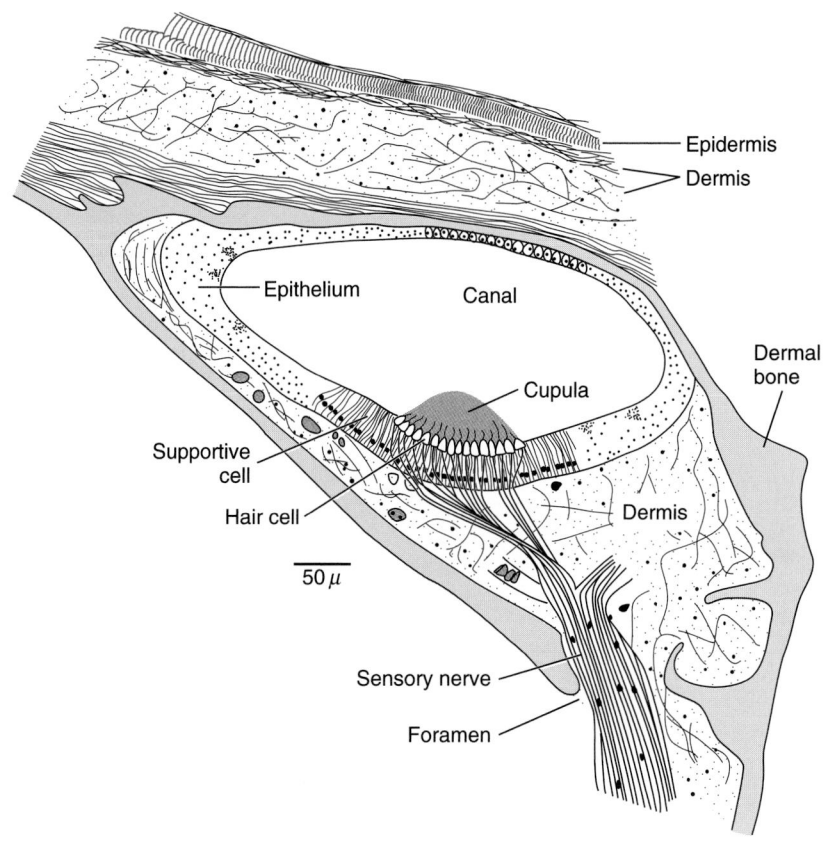

FIGURE 17.3
Neuromast organ in lateral-line canal of a bony fish, cross section. The canal is embedded in bone (*red*). The sensory nerve penetrates the bone via a foramen.

as opposed to ocean depths), with feeding habits of the species (predators as opposed to vegetarians), with schooling proclivities, and with other environmental and behavioral differences.

Electroreception, initially a vertebrate feature, has been demonstrated in ampullary neuromast organs of elasmobranchs and in the cephalic or lateral-line systems of most bony fishes. The mere contraction of respiratory and locomotor muscles in a single fish produces sufficient electric potential to make its presence detectable at short range to other fishes via the electroreceptive neuromasts. Electroreception evidently arose independently several times in fishes. Nerve fibers from electroreceptors terminate in the cord and brain in nuclei that are distinct from those receiving mechanoreceptive input. Also, the nuclei of teleosts do not occupy the same location as those in other fishes. Electroreceptors are present also in larval urodeles, some aquatic apodans, and aquatic monotremes, who use them to detect underwater quarry.

Neuromast organs of the lateral-line system arise from a linear series of embryonic **ectodermal placodes** that extends from the head to the tip of the tail (chapter 5). One pair of these placodes lateral to the hindbrain sinks into the head to become the membranous labyrinth of

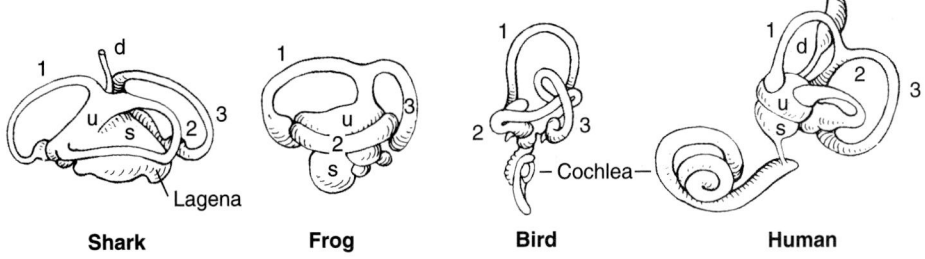

FIGURE 17.4
Left membranous labyrinths of representative craniates. **1,** anterior semicircular duct with ampulla at base; **2,** horizontal semicircular duct; **3,** posterior semicircular duct. **d,** endolymphatic duct in shark, duct and endolymphatic sac in human. In sharks, the endolymphatic duct opens to the exterior. **s,** sacculus; **u,** utriculus.

the inner ear, which is a complex of neuromast organs. The membranous labyrinths and the lateral-line system of fishes constitute, collectively, an **acousticolateralis (octavolateralis) system** of receptors innervated by cranial nerves VIII, ALL, and PLL (the latter two collectively representing up to six lateral-line nerves). In postlarval terrestrial amphibians and in amniotes, the only neuromast organs are those innervated by nerve VIII.

Membranous Labyrinth

All craniates have a pair of fluid-filled membranous labyrinths (**inner ears**) located within a similarly shaped cartilaginous or bony labyrinth that is part of the otic capsule of the neurocranium (fig. 17.4). The membranous labyrinth is a highly specialized complex of neuromast mechanoreceptors. In many fishes, membranous labyrinths function chiefly in equilibration; in some fishes and tetrapods, they are also receptors for sound.

The membranous labyrinth is separated from the walls of the skeletal labyrinth by a **perilymphatic space** filled with **perilymph** (see fig. 17.11, *light red*). This fluid cushions the membranous labyrinths in their cartilaginous or bony canals. Extending across the space between the skeletal and the membranous labyrinth are fine strands of connective tissue that stabilize the membranous labyrinth and anchor it in place. Perilymph consists of cerebrospinal fluid that seeps into the space by one route or another. The fluid *within* the membranous labyrinths is **endolymph**. It resembles interstitial fluids elsewhere in the body. The endolymph and perilymph are not confluent at any point.

Each membranous labyrinth of gnathostomes consists of three **semicircular ducts** and two membranous sacs, the **utriculus** and **sacculus**. In the floor or wall of each sacculus in fishes and amphibians, there is a small outpocketing, the **lagena** (fig. 17.4, shark). The lagena has expanded to become a **cochlea** in mammals and birds. The lumina of the semicircular ducts are continuous with the cavity within the utriculus, and each duct has a dilation, or **ampulla**, near the junction. The lumen of the utriculus is continuous with that of the sacculus

(fig. 17.4). The anterior and posterior vertical semicircular ducts are in vertical planes that are perpendicular to each other. The third, or lateral, duct is in a horizontal plane. Thus, there is a semicircular canal in each of the three planes of space. In lampreys, each labyrinth has only two semicircular ducts, an anterior and posterior vertical. Hagfishes have only the posterior vertical duct with, uniquely, an ampulla at each end. Also, in some agnathan species there is little or no distinction between sacculus and utriculus. One can only speculate whether their labyrinths bespeak a modified craniate condition rather than a primitive one.

Emerging dorsally from approximately the site of confluence of the sacculus and utriculus, except in a few teleosts, is an **endolymphatic duct** that usually terminates in a blind **endolymphatic sac** (figs. 17.4 and 17.11). Uniquely, in elasmobranchs the ducts open to the exterior through two endolymphatic pores on top of the head, rather than ending in blind sacs. In mammals, the endolymphatic sacs lie in the subarachnoid space of the brain beneath the temporal bones. A pair of **perilymphatic ducts** emerge from perilymphatic spaces in most craniates and they, too, usually terminate in a small sac nearby. In sharks, they parallel the endolymphatic ducts and terminate in a small sac in the endolymphatic fossa (a depression on the middorsal aspect of the neurocranium between the two otic capsules). In mammals, the perilymphatic ducts open directly into the subarachnoid space overlying the brain, providing a channel connecting the cerebrospinal fluid and the perilymph. Little is known of the role of the perilymphatic ducts in most craniates.

The membranous labyrinth is a connective tissue complex lined by a simple squamous epithelium. In this epithelium, within each of the components of the labyrinth—sacculus, utriculus, ampullae, and in the lagena or cochlea of birds and mammals—are one or more neuromast sites. Each site contains large numbers of hair cells innervated by sensory fibers of the eighth cranial nerve. A neuromast site within an ampulla is a crista. Those elsewhere are maculae.

Cristae and maculae differ in structural details (fig. 17.5). A **crista** is a papilla of epithelial cells surmounted by hair cells. The kinocilia of the hair cells are embedded in a tall, gelatinous cupula that projects far out into the endolymph. The cilia are mechanically displaced as the inert endolymph responds to rotatory movements of the head. A **macula** in most anamniotes is a mound with a flattened cupula that usually has on

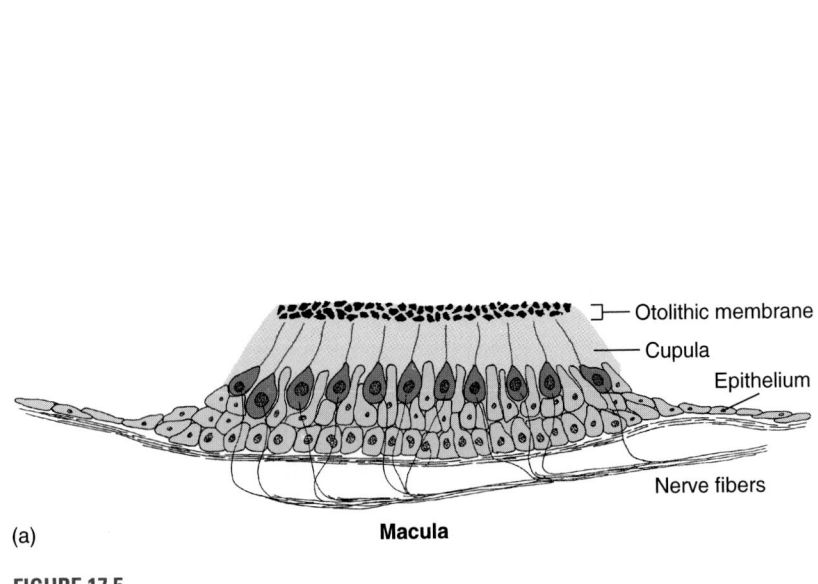

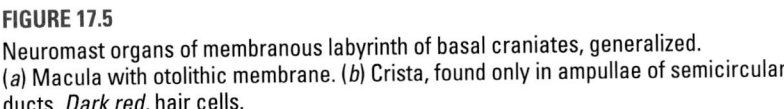

(a)   **Macula**

FIGURE 17.5

Neuromast organs of membranous labyrinth of basal craniates, generalized.
(*a*) Macula with otolithic membrane. (*b*) Crista, found only in ampullae of semicircular ducts. *Dark red,* hair cells.

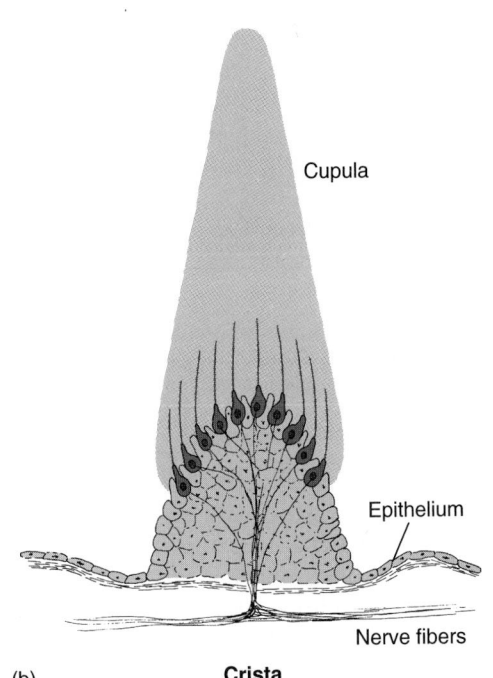

(b)   **Crista**

its surface a layer of otoliths that constitutes an **otolithic membrane**. The maculae of most reptiles (including birds) and mammals lack a cupula, and overlying the hair cells is a tectorial membrane (see fig. 17.10, 2).

Otoliths (otoconia), found only in chondrichthyans, acanthodians, and osteichthyan fishes, are crystals of calcium carbonate combined with a protein. They grow by accretion. In many fishes, a single, large otolith nearly fills the sacculus. In others, otoliths are microscopic, abundant, and constitute an amorphous mass. The crystals spill out of the sacculus of a shark whenever the sacculus is ruptured during dissection. They can be readily seen with a low-power lens, and their crystalline nature can be observed in transmitted light. Species differences in otolith morphology make it possible to assign an otolith to a species without seeing the fish from which it came. Otoliths are generally confined to otolithic membranes in tetrapods.

Membranous labyrinths in all craniates arise during embryogenesis as a pair of ectodermal **otic placodes** on the side of the head lateral to the hindbrain. In fishes, the placodes are in series with those that give rise to the neuromasts of the lateral-line canal system. The otic placodes sink into the embryonic head to become a pair of fluid-filled vesicles, or **otocysts** (fig. 17.6). Eventually, an otic capsule is deposited around the developing labyrinth.

The sacculus, utriculus, and semicircular ducts are primarily organs of equilibration (balance, in a broad sense). However, in at least some fishes, maculae function in hearing to one degree or another.

Equilibratory Role of the Labyrinth

The sacculus, utriculus, and semicircular ducts monitor the position of the head in space, whether at rest (in *static* equilibrium) or in motion (in *dynamic* equilibrium). Information from these organs is employed reflexly by skeletal muscles to maintain a moment-to-moment posture of the body that favors survival. But what is the role of each of these in equilibration?

Answering this question with reference to the sacculus and utriculus in craniates as a group is possible only in the most general terms, largely for lack of sufficient experimental evidence from the several craniate classes. There are facts we are sure of. The maculae of the sacculus and utriculus lie in different planes; not all kinocilia in a single neuromast site are oriented in the same direction; the orientation of the kinocilia is affected by movements of the endolymph and of either an otolithic or tectorial membrane, the latter having no otoliths; and displacement of kinocilia is translated into nerve impulses that are transmitted across synapses in the brain. Thus, different maculae and different hair cells in the same macula are stimulated differentially as the endolymph is displaced by movements of the head. Therefore, the sacculus and utriculus respond to changes in linear velocity of a fish; that is, appropriately oriented hair cells discharge when stimulated as a result of acceleration of the head when the fish is swimming in a straight line. One explanation proposes that linear acceleration tends to carry along the kinocilia but that the otolithic membrane, because of inertia, initially resists acceleration. As a result, certain kinocilia that are in contact with the overlying membrane are bent in a

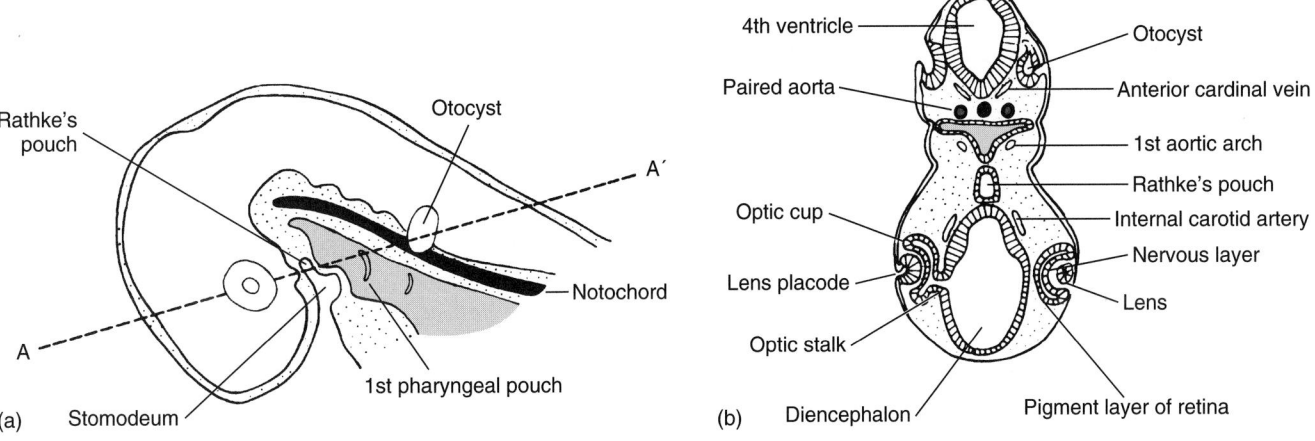

FIGURE 17.6

Anlagen of amniote inner ear (otocyst), retina (optic cup), and lens (lens placode). (*a*) Embryonic head, sagittal plane. (*b*) Cross section of head at level A–A'. In (*b*), the structures are shown as slightly earlier in development on the left. *Red* is the foregut.

direction opposite that of acceleration, and this stimulates the sensory nerve endings at the base of hair cells. It is also thought that a tilted position of the head differentially stimulates one macula as opposed to another.

Stimulation by *rotation* of the head seems to be the province of the semicircular ducts as a derived condition in craniates. The cupula of a crista projects far out into the endolymph, where it is subject to displacement by the shearing forces generated in the endolymph as the head turns in one plane or another. Consequent bending of the enclosed kinocilia is translated into a nerve impulse.

The sensory impulses generated in the semicircular ducts, sacculus, and utriculus are transmitted to nuclei in the brain stem and cerebellum where, as one result, *reflex movements of the eyeballs* are initiated. As a result, the eyes are always looking in a direction compatible with the position of the head. Voluntary effort is required to turn your head swiftly to the side while continuing to gaze straight ahead. (Try it!) The eyeball reflex impulses travel over the vestibular branch of cranial nerve VIII to the anterior end of the medulla, where they synapse in a vestibular nucleus. Second-order sensory fibers from this nucleus enter an ascending tract that distributes them to the motor nuclei of nerves III, IV, and VI. From these nuclei, motor fibers innervate the extrinsic eyeball muscles. This is one of several *vestibular reflexes* initiated by stimulation of the labyrinth.

That the labyrinth controls reflexive eyeball movements can be demonstrated by rapidly spinning someone in a revolving chair a number of times and then stopping the chair abruptly. The individual will feel dizzy, and an observer will note that the eyeballs continue to exhibit rapid, jerky, lateral oscillation (side-to-side movements; *nystagmus*), which stops as the inert endolymph ceases to apply shearing forces on the

cristae of the semicircular ducts. The nystagmus is the cause of the dizzy feeling. Children, in fun, often turn round and round until, for the same reason, they become dizzy.

Input from the equilibratory components of the labyrinth in amniotes also results in *reflex motor activity that alters the position of the head and forelimbs*, keeping these parts oriented with respect to the rest of the body and to gravity. For example, a cat that is turned upside down and dropped from a height will land on its feet. This is the result of a vestibulocollic reflex that terminates in the sternomastoid and other muscles of the neck, and of a reflex pathway that utilizes the vestibulospinal tract to terminate in motor nuclei that supply the brachial plexus. Maintenance of equilibrium is more readily demonstrable in tetrapods because they have a larger number of independently movable members of the body and they are not suspended in a dense medium, but it is equally essential for survival of a fish in water. In its role of maintaining balance, the labyrinth is assisted by proprioceptors, to be discussed shortly.

Auditory Role of the Labyrinth and Evolution of the Cochlea

The labyrinth has an auditory function in some fishes in which maculae respond to longitudinal sinusoidal waves of the amplitude and frequency of sound (low amplitude, high frequency). In Cypriniformes, a group of mostly freshwater teleosts that includes catfish, goldfish, and carp, sound waves in water evoke waves of similar frequency in the gas in the turgid swim bladder, and these are transmitted to the labyrinth by **weberian ossicles** (fig. 17.7). These are a series of modified transverse processes of the first three (occasionally four or five) trunk vertebrae that extend between the swim

bladder and the **sinus impar,** an extension of the perilymphatic space. These fish can hear, although the
mechanism presumably provides no information as to
the direction from which sounds are coming. The specific maculae that detect these sounds are not known. It
is known, however, that the maculae that respond to
sound in *basal tetrapods* are in the sacculus near the lagena. In Clupeiformes, an order of herringlike teleosts,

an anterior extension of the swim bladder comes in direct contact with the labyrinth, but whether this is a
route for transmitting sound waves is not known. Without doubt, there are other teleosts that have maculae
with low enough thresholds to detect sound waves. The
extent of the phenomenon of hearing in fishes and the
receptor sites involved is not yet known.

An account of the evolution of the amniote cochlea
begins with amphibians. In addition to the usual maculae associated with the sacculus and lagena of fishes,
amphibians have two receptor sites that have a **tectorial
membrane** instead of a cupula. (A mammalian tectorial
membrane is illustrated in fig. 17.10.) One amphibian
receptor site is the **amphibian papilla** of the sacculus, so
named because it is not found in any other craniate
(fig. 17.8*a*). The other is the **basilar papilla** in a recess
in the wall of the sacculus. These two papillae are the
receptors for sound in amphibians.

In most reptiles, including birds, the lagena has become a prominent sac suspended from the sacculus
(fig. 17.8*b*). It still houses the lagenar macula, the function of which is unknown. The basilar papilla has become incorporated in its epithelial lining to become, in
crocodilians, birds, and monotremes, the **organ of
Corti,** the sole receptor for sound (fig. 17.8*b*). Elongation of the lagena was accompanied by a corresponding
extension of the surrounding perilymphatic space.

In placental mammals (Metatheria and Eutheria), the
lagena elongated still more to become the **cochlear
duct,** or **scala media** (fig. 17.8*c*). In its epithelial floor is
the organ of Corti. The cochlear duct spirals around a
bony pillar of the petrosal bone, the **modiolus**
(fig. 17.8*c*), accompanied by perilymphatic spaces

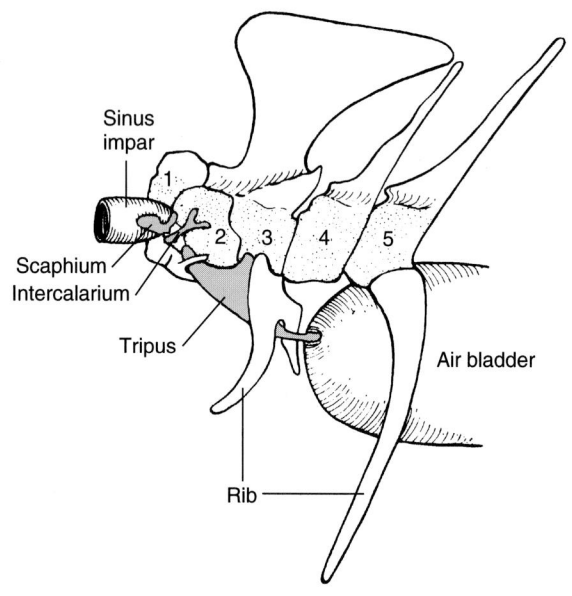

FIGURE 17.7
Weberian ossicles (*red*) of a cypriniform teleost. **1** to **5,** centra of the
first five vertebrae. The sinus impar is a caudal extension of the
perilymphatic space. The ossicles are interconnected by modified
intervertebral ligaments.

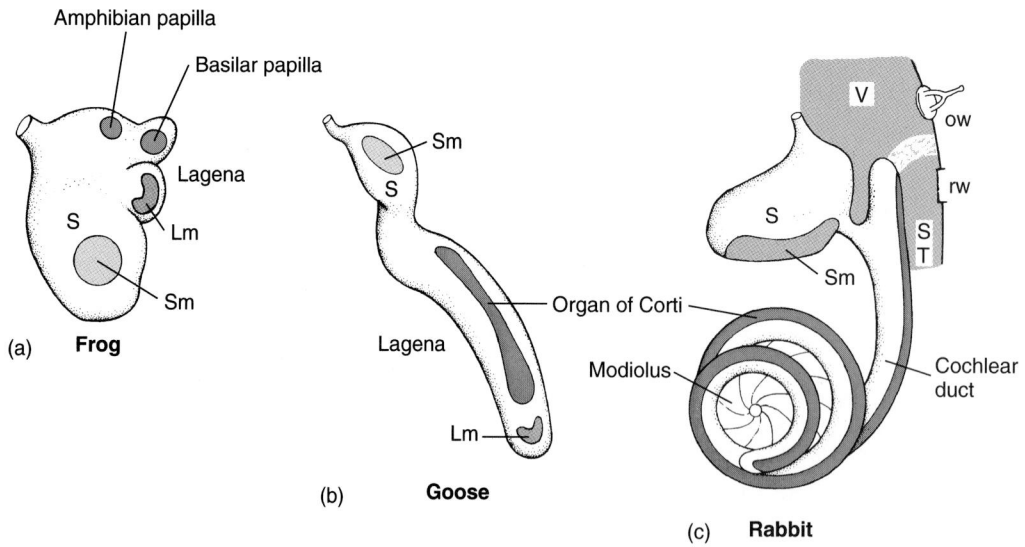

FIGURE 17.8
Receptor sites for sound (*red*) and evolution of cochlear duct in selected tetrapods. **Lm,** lagenar macula; **ow,** oval window; **rw,** round window;
S, sacculus; **Sm,** saccular macula; **ST,** base of scala tympani; **V,** vestibule of cochlea. The organ of Corti of birds and mammals is derived from
the amphibian basilar papilla. For orientation of the cochlear duct within the cochlea, see figures 17.9 and 17.10, **1** (not to scale).

named, in this location, the **scala vestibuli** and **scala tympani** (fig. 17.9). The scala media, scala vestibuli, and scala tympani collectively constitute the mammalian cochlea, so named because it resembles a snail's spiral shell. The spiral exhibits from one to five turns, according to the species. Embedded in the modiolus is the spiral ganglion of the cochlear nerve.

The scala vestibuli ("ladder of the vestibule") begins at the vestibule (fig. 17.9), a chamber of perilymphatic fluid where sound waves from the stapes enter the cochlea. The scala vestibuli spirals *upward* to the apex of the cochlea where it is continuous with the scala tympani via a small aperture, the **helicotrema**. The scala tympani spirals *down* the helix and ends at the membrane stretched across the **round window (fenestra rotunda)** in the lateral wall of the otic capsule.

The mammalian organ of Corti is a long, maculalike neuromast in the epithelium of the floor of the cochlear duct (fig. 17.10). It rests on a **basilar membrane** that is attached to a spiraling, bony shelflike projection of the modiolus, and it projects into the endolymph. A continuation of the basilar membrane separates the scala vestibuli from the scala tympani. Overlying the organ of Corti and suspended in the endolymph is a tectorial membrane, which is attached at one end to the inner lining of the cochlear duct. Each hair cell of the neuromast has several rows of stereocilia of differing lengths that are embedded in the tectorial membrane. The wall of the cochlear duct opposite the organ of Corti is reduced to a thin epithelium, **Reissner's membrane**. It separates the endolymph in the scala media from the perilymph in the scala vestibuli.

Sound waves induced in the perilymph of the vestibule by vibrations of the footplate of the stapes are transmitted up the scala vestibuli (see fig. 17.9), down the scala tympani, and, via Reissner's and the basilar membranes, into and through the endolymph of the scala media and back into the perilymph at all levels of the spiral. Expansion and compression waves induced in the endolymph displace the organ of Corti toward the tectorial membrane at the same frequency, and the hair cells transduce the mechanical stimuli into volleys of nerve impulses in the sensory fibers that innervate the organ. It has been ascertained that hair cells in the first spiral respond maximally to vibrations of higher frequencies, hence higher pitches; those near the apex of the scala media respond only to low frequencies, and those in between exhibit graduated responses. Vibrations in the perilymph are instantaneously discharged back into the middle ear cavity at the round window.

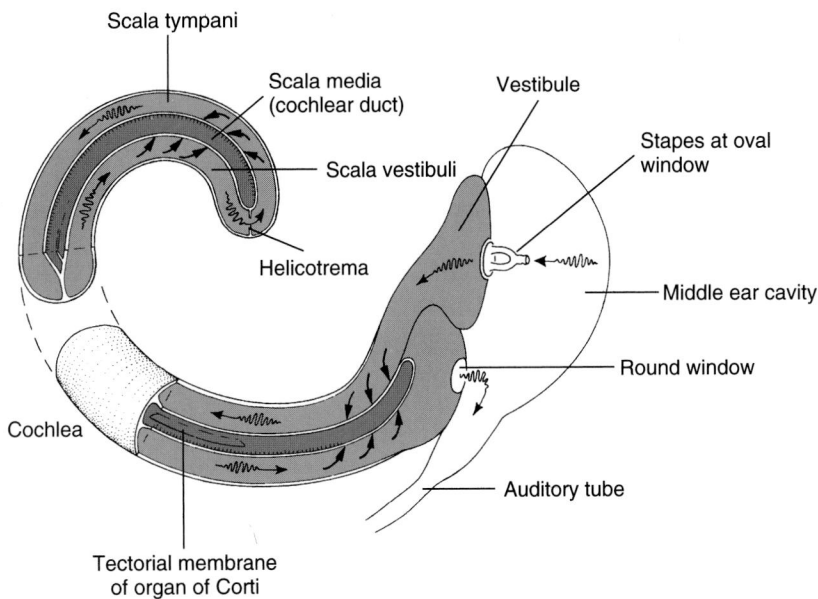

FIGURE 17.9

Pathway of sound waves (*arrows*) from the middle ear cavity, up the scala vestibuli, down the scala tympani, and back to the middle ear cavity in a mammalian cochlea.

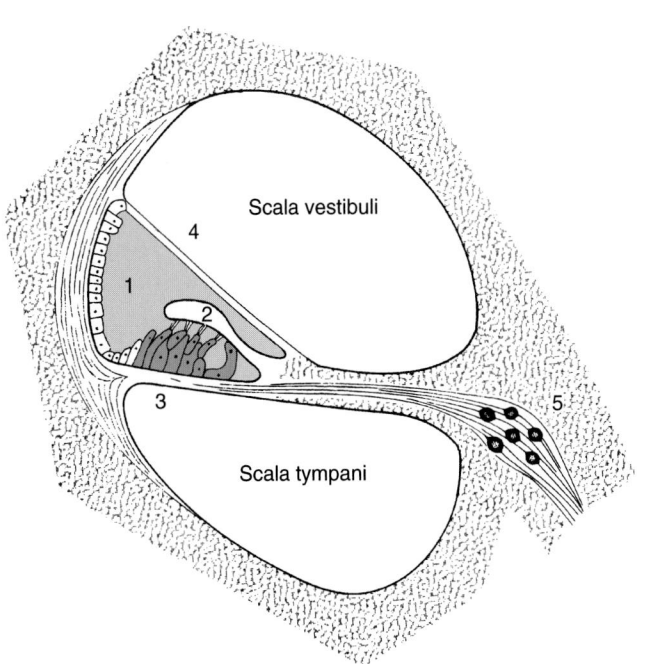

FIGURE 17.10

Cross section of one turn of a mammalian cochlea. **1,** cochlear duct (scala media) containing endolymph (*red*); **2,** tectorial membrane; **3,** basilar membrane (attached at left to the spiral shelf of the modiolus), and organ of Corti (*dark red*) resting on the basilar membrane; **4,** Reissner's membrane; **5,** spiral ganglion and cochlear nerve fibers. The scala tympani and scala vestibuli are perilymphatic spaces. The diameter of the human cochlea is less than 1 mm.

Basal tetrapods, as well as mammals, have oval and round windows in their otic capsules.

With expansion of the basilar papilla to become an organ of Corti, the relay nuclei on auditory pathways in the brain became increasingly larger. Amniotes were the earliest craniates in which the auditory nuclei in the tectum of the mesencephalon bulged above the surface and became auditory lobes (see figs. 16.14, snake, and 16.15). Differentiation of an auditory cortex in the temporal lobes of the cerebral hemispheres (see fig. 16.26) was accompanied by a further increase in complexity of the central pathways that mediate sound.

Middle Ear Cavity

The usual route for conduction of sound waves from the **tympanic membrane** (eardrum) to the inner ear in tetrapods is via one or more ossicles that transmit sound waves across a middle ear cavity (**tympanic cavity; cavum tympanum**) to a **secondary tympanic membrane** in the **oval window (fenestra ovale)** of the otic capsule (fig. 17.11). To perform this function, or to assist in it, all tetrapods have a **columella**, better known as a **stapes** in mammals. The columella in amphibians and reptiles is part of a cartilaginous or bony complex that includes an **extracolumella** attached to the tympanic membrane. The columella, or stapes in mammals, is derived from the dorsalmost segment of the embryonic hyoid arch,

and is, therefore, a homologue of the hyomandibula of fishes.

A stage in the evolution of a columella is found in the urodele *Ranodon* (see fig. 9.12*b*), whose middle ear cavity, like that of other urodeles, is rudimentary and who lacks a tympanic membrane. In this urodele, the columella still occupies its primitive and embryonic location as the dorsalmost segment of the hyoid arch skeleton, abutting against the prootic bone (otic capsule) at the site where the oval window forms in other tetrapods. The columella of anurans, on the other hand, is a strong, rodlike replacement bone that is firmly attached to the large eardrum and traverses a broadly open middle ear cavity to end at the membrane in the oval window of the prootic bone.

Mammals have two additional middle ear ossicles, a **malleus** and **incus**, derived from the articular and quadrate bones, respectively, of ancestral synapsids (fig. 17.12). The history of the three ear ossicles of mammals is related in more detail, along with the evidence, in chapter 9 under "Ear Ossicles from the Hyomandibula and Jaws." Because the surface area of the eardrum of tetrapods approaches 20 times that of the membrane in the oval window, vibratory stimuli impinging on the eardrum have been amplified many times before they are delivered to the vestibule of the cochlea. The leverage achieved by the series of nonlinear articulations between the ear ossicles of mammals intensifies the effect.

Two small muscles are present in the middle ear cavity of mammals. A **tensor tympani**, derived from the first pharyngeal arch and innervated by the fifth cranial nerve, inserts on the malleus (once a lower jaw bone). It regulates the tension in the tympanic membrane. A **stapedial muscle**, derived from the second pharyngeal arch and innervated by the seventh cranial nerve, inserts on the stapes. It restrains the movement of this ossicle. Very loud sounds cause the tensor tympani and stapedial muscles to contract, which dampens, to a degree, the vibrations of the ossicles and protects the hair cells of the organ of Corti. However, it does not protect against repeated loud noises that are not normal in a natural environment.

The middle ear cavity arises ontogenetically as an evagination of the first pharyngeal (spiracular) pouch, which

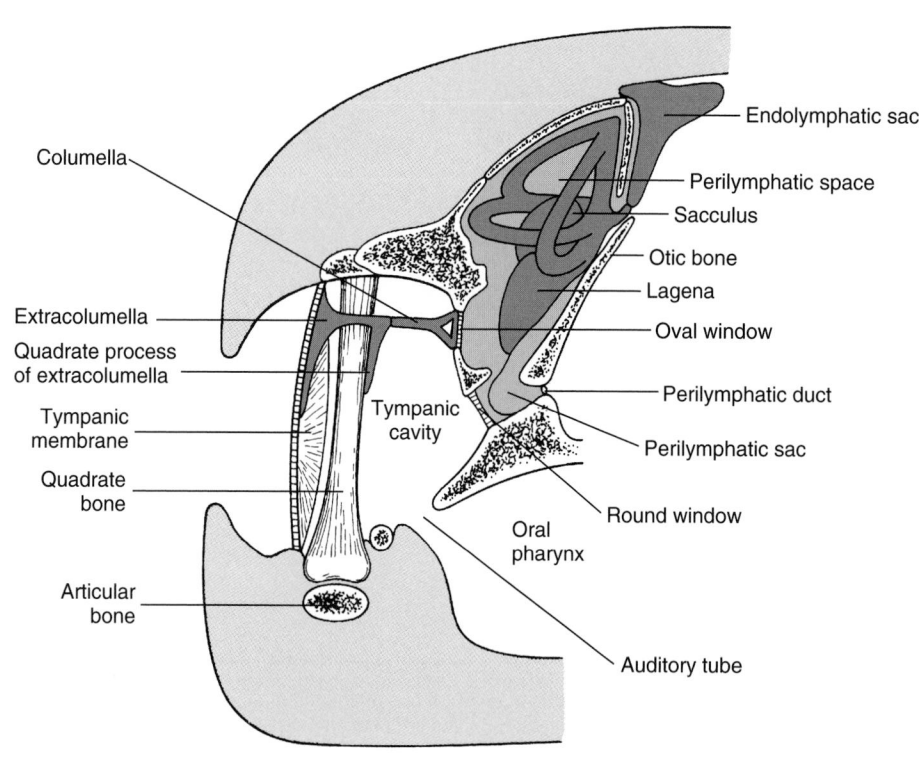

FIGURE 17.11

Columellar complex, perilymphatic space, and membranous labyrinth of a lizard, schematic.
Source: Data from J. S. Kingsley, *Outline of Comparative Anatomy of Vertebrates*, 1920, The Blakiston Company, Philadelphia, after Versluys.

grows toward the developing ear ossicle or ossicles and partially surrounds them, isolating them from other tissues. The cavitation process is completed by erosion of any remaining mesenchyme surrounding these bones. The middle ear cavity remains in communication with the pharynx throughout life via an **auditory tube** (**eustachian tube**), which ensures that the air pressure on both sides of the eardrum will be equal. Anurans and

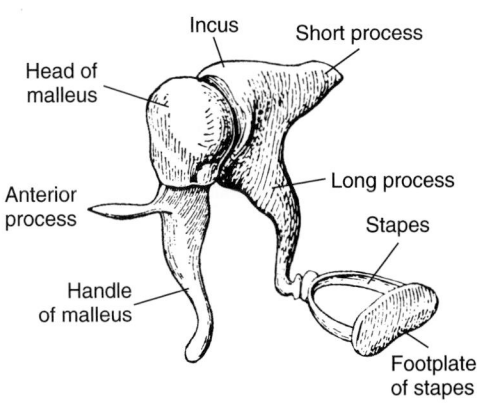

FIGURE 17.12

Human middle ear ossicles. The handle of the malleus is attached by ligaments to the eardrum, the anterior process is anchored by a ligament to the petrous portion of the temporal bone, and the footplate of the stapes rests against the membrane at the oval window.

From B. A. Schottelius and D. D. Schottelius, *Textbook of Physiology*, 17th edition. Copyright © 1973 Mosby-Year Book, Inc. Reprinted by permission of D. Schottelius.

some lizards have a wide, short tube that opens into the buccopharyngeal cavity (see figs. 12.3*a* and 17.11). That of crocodilians, birds, and mammals is longer, of small diameter, and opens in crocodilians and mammals above the secondary palate (figs. 13.13*a* and 17.13). The mammalian auditory tube remains closed at its nasopharyngeal exit except during the act of swallowing. Anyone who has suffered from increased pressure in the middle ear as a result of chronic blockage of the auditory tube will recall the sensation and its effect on hearing. Frequent swallowing relieves the symptoms by opening the entrance to the tube, thereby equalizing the pressure on the two sides of the eardrum.

Outer Ear

The tympanic membrane (eardrum) is the usual portal to the ear ossicles of tetrapods. In anurans, turtles, and primitive lizards, the eardrum is more or less flush with the surface of the head (fig. 17.14). In more specialized lizards, crocodilians, birds, and mammals, the eardrum lies at the end of an outer ear canal (**external auditory meatus**) that may be short or long and has an entrance on the side of the head behind the angle of the jaws (see figs. 6.16*a* and 17.13). In terrestrial mammals, a fibrocartilaginous appendage, the **auricle** (**pinna**) collects sound waves and directs them into the outer ear canal. Cetaceans, pinnipeds, sirenians, and some moles have a very small auricle or, like earless seals, none at all. The outer ear canal of whales is of very small diameter, but whales have excellent hearing, which is used to receive ultrasonic intraspecies communications and for echolocation.

Tetrapod Hearing in the Absence of an Eardrum

Although a tympanic membrane and middle ear cavity are the usual routes to the auditory receptors of tetrapods, they are not always present. Urodeles, apodans, a few anurans (*Ascaphus,* for instance), *Sphenodon,* and most limbless reptiles have no eardrum, the middle ear cavities and auditory tubes are vestigial, and, except in snakes and amphisbaenians, the columella is rudimentary. Yet these tetrapods are sensitive either to airborne or substrate (seismic) sound waves, or both.

Anurans and many urodeles have an aberrant **opercular bone**, a flat plate that is in contact with the membrane in the oval window. An **opercular muscle** extends between the scapula

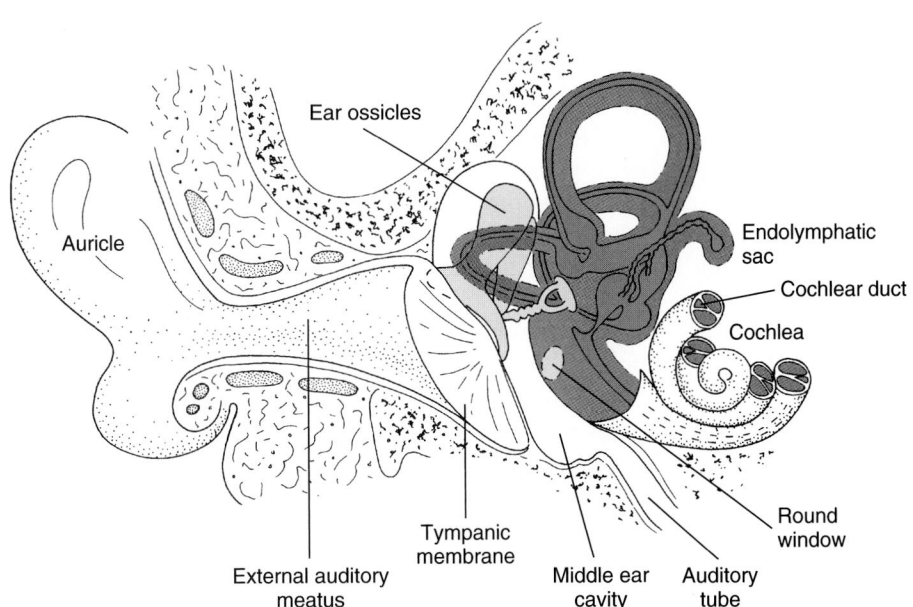

FIGURE 17.13

Mammalian ear complex. *Dark red,* membranous labyrinth; *light red,* middle ear ossicles; *gray,* perilymphatic space within bony labyrinth. The footplate of the stapes rests against the oval window.

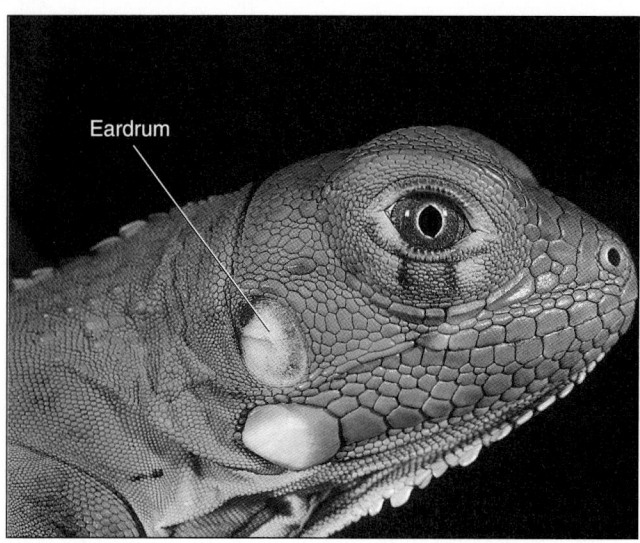

FIGURE 17.14
Head of an iguanid lizard showing slightly depressed eardrum just behind angle of jaws.
© Thomas Gula/Visuals Unlimited.

and the opercular bone. Seismic vibrations detected by the forelimb and scapula are transmitted via the opercular muscle to the opercular bone. This evokes sound waves in the endolymph. The opercular complex of amphibians bears no relationship to the opercular structures of fishes.

In amphisbaenians (squamates that live in burrows in the ground), a very long, slim extracolumella extends forward immediately under the skin of the lower jaw on each side of the head. It vibrates in response to seismic waves in the soil of the burrow and also to airborne sounds. In some species, it ends anteriorly in a broad, flat subcutaneous plate that increases the surface area of exposure to vibrations. The extracolumella articulates with an unusually large columella that ends at an oval window. The hearing complex of amphisbaenians appears to be an adaptation to a subterranean habitat.

Seismic signals are received from the substrate via the lower jaw in many lizards and snakes. In boas, which lack a tympanic membrane and middle ear cavity, the columellar complex connects the quadrate bone and the oval window. The quadrate receives vibrations via the articular bone of the lower jaw (see fig. 9.17c). An articulation of the extracolumella with the quadrate in a lizard is seen in figure 17.11. Many hearing aids for humans employ bone conduction to bypass immobilized middle ear ossicles.

Echolocation

The ability to emit vocal sounds of ultrasonic frequencies in a specific pattern of clicks or structured pulses and to detect the reflection of these from distant objects as small as a flying insect is a sonarlike phenomenon. It is exhibited by mammalian species whose natural environment is such that other sources of sensory input, particularly olfaction and vision, cannot provide sufficient monitoring. Without echolocation, insectivorous bats would starve. A large larynx, grotesque outer ear pinnas, and large and extremely sensitive cochleas combine to provide this capability in bats. Information gleaned at close range includes the size of the insect, its shape, its distance, and escape velocity (Novacek, 1988). Echolocation is less well developed in bats that feed on fruit or nectar, its chief role in these species being obstacle avoidance. Echolocation is also used by shrews and a few other mammals. Whales employ it for navigation, and they also use the sound-emitting competence for intraspecies communication. A few birds also employ echolocation.

Saccus Vasculosus

Suspended from the floor of the diencephalon just caudal to the pituitary in jawed fishes other than dipnoans is a highly vascular, thin-walled ventral evagination of the third ventricle, the saccus vasculosus (see fig. 16.20a). In some species, the sac underlies much of the brain stem. It is lined with hair cells, the cilia of which project into the cerebrospinal fluid. Its sensory nerve fibers pass to the hypothalamus and other brain centers. Its function is unknown. It may be that the organ monitors the pressure of the cerebrospinal fluid, which varies with the depth of the fish, and that the information is used to regulate the volume of gas in the swim bladder. However, elasmobranchs have the organ and no swim bladder. The sac is largest in deep-sea fishes, not well developed in shallow freshwater species, rudimentary in living agnathans, and absent in lungfishes.

Visual Organs: Lateral Eyes

Many cold-blooded craniates have two varieties of photoreceptors, paired (lateral) eyes and unpaired (median) eyes. Lateral eyes focus light from objects in the environment onto a photosensitive epithelium, the retina, where an image is formed. The most primitive response elicited by the information is reorientation of the body with respect to the objects viewed, a necessity for moment-to-moment survival of the individual. Hagfish possess secondarily degenerate eyes, which are also found in some deep-sea and cave fishes. Median eyes form no image; captured light stimulates neuroendocrine reflex arcs that maintain biologic rhythms (see fig. 18.5). Pineal and parapineal median eyes are not found in adult hagfish, with the embryology of these structures, if present, not known.

Retina

The retina is the receptor site of the lateral eye. It is a membrane rich in nervous tissue and is located at the rear of a fluid-filled **vitreous chamber** of the eyeball

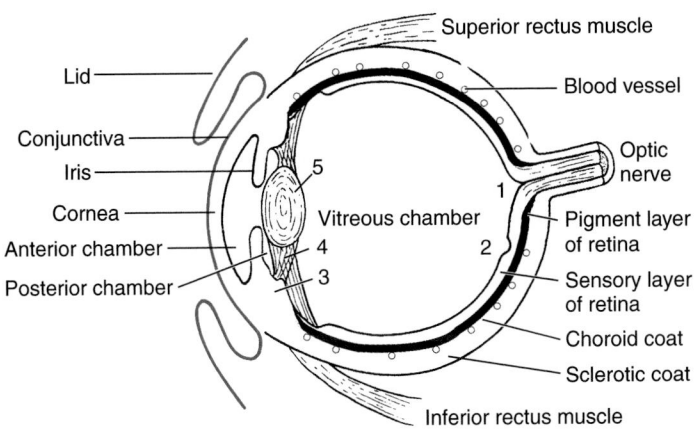

FIGURE 17.15

Generalized spherical eyeball in sagittal section. **1,** blind spot;
2, fovea; **3,** ciliary body; **4,** suspensory ligaments; **5,** lens. *Red* represents
the bulbar conjunctiva (on the cornea) and the palpebral conjunctiva (on
the under surface of the lids).

(fig. 17.15). Retinas arise as lateral evaginations, or
optic vesicles, from the embryonic forebrain (see
fig. 16.13*b*). The vesicles soon become double-
walled **optic cups** (see fig. 17.6*b*). The optic cups
retain an attachment to the brain via **optic stalks.**
The invaginated layer of the optic cup that light
will first strike becomes the nervous or sensory
layer of the retina (figs. 17.15 and 17.16). Some of
the retinal cells become rods and cones, which are
the photoreceptors. Others become bipolar association neurons, and still others become cell bodies of
the optic nerve fibers. These neurons (ganglion
cells) sprout long processes that grow along a
groove (**choroid fissure**) in the optic stalk and establish synapses within the brain. The processes
constitute the optic nerve. From the discussion, it
can be seen that the retina arises from the brain
and never becomes completely detached from it.

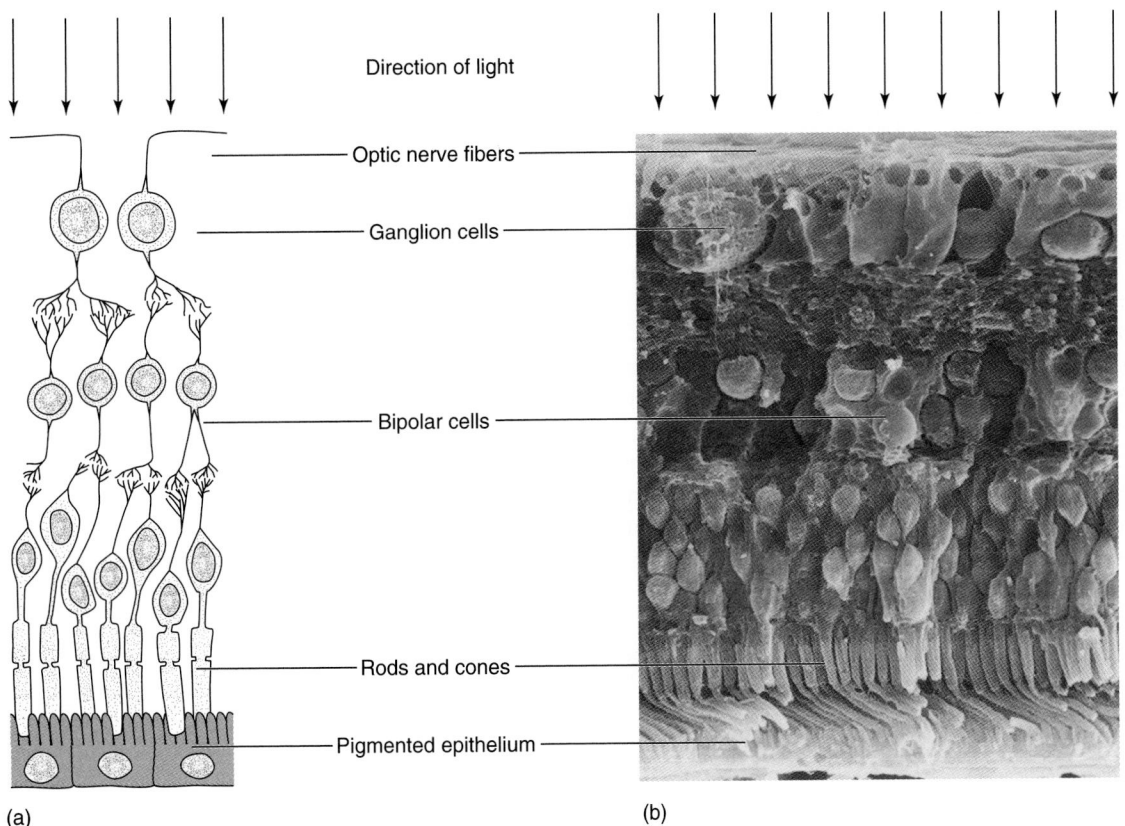

(a)   (b)

FIGURE 17.16

The retina. (*a*) Areas of synapse, diagrammatic. (*b*) A photomicrograph. Light passes through all layers until it reaches the rods and cones, the
cellular photoreceptors. The pigmented epithelium forms a dark choroid coat.
(*b*) SEM From *Tissues and Organs: A Text Atlas of Scanning Electron Microscopy* by Richard G. Kessel and Randy H. Kardon.

Rods and cones are neurosensory cells, so called because their cell bodies lie in an epithelium rather than in a sensory ganglion. Rods are insensitive to color because the pigment within them, **rhodopsin** (visual purple), absorbs light over the entire range of the visual spectrum. They give only black-and-white impressions and are most effective in relatively low intensities of light, in the shade, at dusk, or at night. The pigment is bleached when exposed to light, and in this respect the part of the retina containing rods is somewhat analogous to a photographic film. There are three types of cones, and the pigments in each absorb light in only one part of the visible spectrum—blue, green, or red. The cones function only when there is ample radiation from a source that emits light of these wavelengths.

Not all craniates, nor all mammals, have color vision. Cones are generally few in number or lacking in fishes that inhabit depths where light rays become increasingly weaker. The same is true of many terrestrial craniates that live in dark caves, as do some salamanders, in burrows, like moles and some rodents, or that are nocturnal, like bats and some birds. On the other hand, the retina of many diurnal birds has a preponderance of cones or, in some instances, no rods at all. The environment in which they compete for food is well illuminated by the sun, and rods have little survival value. Other diurnal craniates have a mixture of rods and cones, with the cones sometimes concentrated in a **fovea** (see fig. 17.15).

Raptorial birds (owls, hawks, eagles, other birds of prey) and some lizards have two foveas in each eye. The grand champion of foveas is probably that of hawks, which have a human-sized eyeball with approximately one million cones per square millimeter in each fovea. The few other birds with two foveas in each eye are diurnal birds that catch flying insects.

The greatest concentration of rods is peripheral to the fovea. This can be verified by looking at a very faint star at night. It will not be visible if looked at directly because very little light is falling on the fovea. But by shifting one's gaze to one side of the star, the light will strike peripheral to the fovea, the rods will be stimulated, and the star will come into view. Nocturnal animals have a larger proportion of rods than diurnal ones and may lack a fovea.

Toward the medial quadrant of the retina is a blind spot, the **optic disc**, the epithelium of which lacks both rods and cones. This is where the optic nerve fibers leave the retina and blood vessels enter and leave (see fig. 17.15).

Choroid and Sclerotic Coats and the Pupil

The embryonic mesenchyme surrounding the optic cup gives rise to a highly vascular **choroid coat** and a tough fibrous **sclerotic coat (sclera)** that are external to the retina (see figs. 17.15 and 17.18). The vessels of the choroid coat provide supplementary nutrients and oxygen to the retina, and the pigment assures a pitch-black background for the retinal "screen." The choroid coat is perforated anteriorly by the **pupil,** which appears black because there is no light in the interior of the eyeball. The pigmented ring surrounding the pupil is the **iris diaphragm,** a modified portion of the choroid coat.

The size of the pupil is fixed in most fishes, but in sharks, some teleosts, and tetrapods, the diaphragm has dilator and constrictor muscles that reflexly alter the size of the pupil, chiefly in response to changes in light intensity. These **intrinsic eyeball muscles** are smooth in amphibians and mammals, striated in most reptiles including birds. The effect of light may be demonstrated by shining a flashlight into a dark-adapted human eye and observing the pupil constrict.

Although the pupil is more or less circular, it can be constricted to become *vertically* slitlike in snakes and in mammals such as cats that are active day and night. It is *horizontally rectangular* in a few mammals, including whales. The **dilator fibers** are radially disposed in all types and are innervated by the sympathetic nervous system. The **constrictor** fibers are disposed circumferentially around a circular pupil and parallel to slits that are vertical or horizontal. They are innervated by the oculomotor nerve (see fig. 16.35).

The sclerotic coat, which is the "white" of the eye, is so named because it is often sclerotized more or less by cartilage or bone. In reptiles, the bones take the form of **sclerotic plates** that assist in maintaining the shape of the eyeball (fig. 17.17). In front of the iris diaphragm and pupil, the sclerotic coat is transparent and more convex. This portion of the sclerotic coat is the **cornea.** It is through the cornea that one sees the iris diaphragm and the pupil. Inserting on the outer surface of the sclera are the extrinsic muscles that rotate the eyeball within the orbit (see fig. 11.26). The extrinsic muscles are basically similar from fish to humans except that in birds they are so poorly developed that they are practically useless. As a result, a bird must move its entire head to change the direction of its gaze.

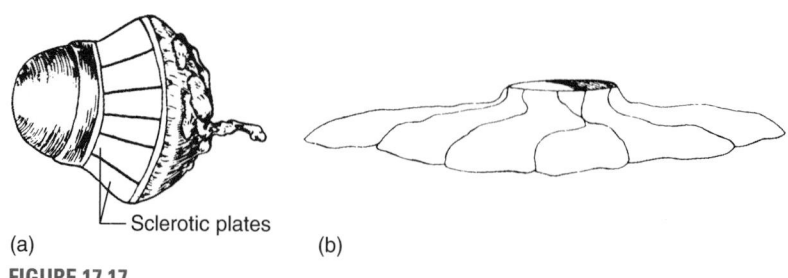

(a)   (b)

FIGURE 17.17

Ossicles of the sclerotic coat of the eye. (*a*) Owl's eye, showing sclerotic plates in place. (*b*) Scleral ring of overlapping ossicles from a lizard's eye.

Lens

The crystalline **lens** consists of many thin strata of epithelial cells layered on one another in a manner similar to the layers of an onion (fig. 17.18). It is mildly elastic and resilient, being capable of regaining its resting shape after having been deformed. It is suspended in the fluid interior of the eye behind the pupil by a system of fine fibers that form the **suspensory ligament**. The ligament is attached to the ciliary body, which completely encircles the periphery of the cornea. The role of the lens is to focus light rays onto the retina.

The lens arises in embryos as an ectodermal **lens placode** that sinks into position in front of the optic cup (see fig. 17.6b). Formation of a lens placode is induced in part by inductors from the developing retina. This has been demonstrated in amphibian embryos by exchanging undifferentiated ectoderm from the region of the future thigh with that at the site where the lens will form. A lens placode is induced in the ectoderm transplanted to the head but not in the potential lens ectoderm that was removed from the influence of the retina. Removal of the optic cup results in no lens placode being formed.

Vitreous and Aqueous Chambers

The large vitreous chamber of the eyeball (fig. 17.18) is filled with a jellylike viscous refracting mass, the **vitreous** (glassy) **humor**. Two smaller chambers are filled with a watery lymphlike **aqueous humor**. These, the **anterior** and **posterior aqueous chambers**, one in front of the iris diaphragm and the other immediately behind it (fig. 17.18), are confluent at the site where the iris diaphragm lies against the lens.

In teleosts, squamates, and birds, a somewhat pigmented and highly vascular fold of the choroid coat protrudes into the vitreous chamber from a site near the blind spot. In fishes, it is a **falciform** (sickle-shaped) **process**. It is a cone-shaped papilla (**papillary cone**) in nonavian reptiles, and a **pecten** (shaped like a comb) in birds. Their intense vascularity enables them to participate in metabolic exchange between the bloodstream and the vitreous humor, providing nutrients and oxygen to the retina and removing metabolites. The processes in reptiles extend nearly to the lens, and it is thought they may alert the animal to nearby moving objects by casting a moving shadow on the retina.

Ciliary Bodies: The Source and Drainage of the Aqueous Humor

Within the eyeball, at the periphery of the iris diaphragm, is the **ciliary body**, a low, wedge-shaped ring of muscle with a richly vascular fimbria (fringe) that projects into the posterior chamber (fig. 17.18, red). Extending between the ciliary body and the lens is the suspensory ligament described earlier. The ciliary muscles and suspensory ligament in amniotes other than snakes play a major role in regulating the curvature of the lens for near and far vision.

The fimbria of the ciliary body consists of delicate **ciliary processes** that project into the posterior chamber and secrete the aqueous humor. The fluid seeps around the border of the iris where it lies against the lens and into the anterior chamber, where it nourishes the cornea. The cornea is devoid of any other source of nutrients and oxygen because it lacks blood vessels. If present, blood vessels would distort or interrupt the visual light rays en route to the retina. The rate of secretion of aqueous humor is controlled reflexly, maintaining an appropriate pressure within the chambers. It drains slowly out of the anterior chamber via tiny canals (of Schlemm), and from these it diffuses into venous

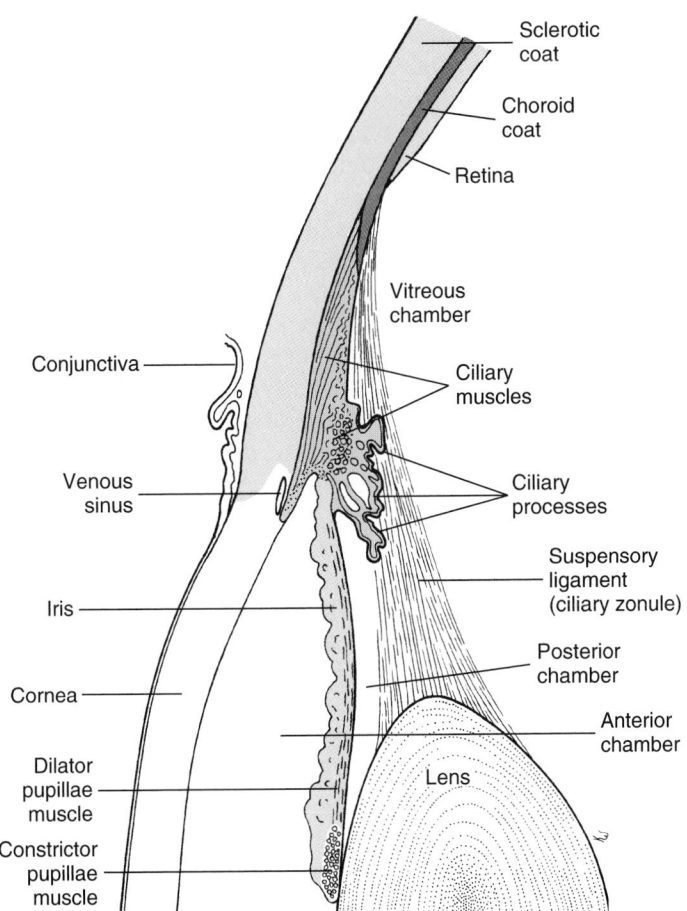

FIGURE 17.18
Ciliary body (*red*) and associated structures of the human eye.

sinuses in the sclerotic coat. Obstruction of the drainage in humans is a major cause of *glaucoma,* a pathologic increase in intraocular pressure.

Shape of the Eyeball

The shape of an eyeball depends on the size, shape, and spatial relations of the parts within, particularly the retina and lens. In general, eyeballs are spherical, like those of mammals and many other craniates, tubular, like those of abyssal fishes and predatory birds (see fig. 17.17*a*); or ovoid, as in most diurnal birds. The shape is a product of natural selection and evinces evolutionary convergence in species with similar habitats and habit. Variability in eyeball shape illustrates adaptations superimposed on an ancestral organ.

Accommodation

Light passes from a medium of a given refractive index (water or air) into a medium of a higher refractive index (the interior of the eyeball). This reduces the speed of the light and deflects the rays (a physical property of light waves summarized in Huygen's Principle and the Law of Retraction). Regardless of whether the light is coming from a distant or near object, it must be redirected to come to a sharp focus on the retina. The combined refractory effects of the lens and vitreous humor, and also of the cornea (except in saltwater fishes), accomplishes this necessary revision of the direction of the rays.

The mechanism of accommodation, which is the *reflex adjustment of the eye for viewing objects at different distances, and in different intensities of light,* varies among the classes and orders. Accommodation for near or distant vision is a natural application of the physical law of refraction of light taught in elementary physics. These are the same principles, of course, that dictate that the lens of a camera must be moved farther from the film for near objects and closer to it for distant objects.

Light from distant objects under water becomes increasingly reduced at increasing depths. However, near vision at any depth usually suffices, especially for fishes whose food, generally algae or smaller fishes, is always in the immediate vicinity. It is not surprising, then, that the lenses of most aquatic craniates are highly concave (spherical), close to the cornea, hence more or less remote from the retina, thereby providing a focus for near objects (nearsighted eyes). The eyes of teleosts at rest are focused in that mode, and a **retractor muscle** *pulls the lens backward* toward the retina for distant vision. The eyes of elasmobranchs and amphibians are focused at infinity (farsighted eyes), and a **protractor muscle** *pulls the lens forward* for near vision. Moving the lens, rather than changing its curvature, is the usual method of accommodation in anamniotes.

Lampreys, which are nearsighted, use the cornea to *push the lens backward.* This is accomplished by con-

tracting an extrinsic **corneal muscle** that arises in the skin and inserts on the cornea just behind the conjunctiva. When this muscle contracts, it pulls on the cornea from the side, causing it to push backward on the lens, with which it is almost in contact at all times. This pushes the lens closer to the retina, causing light from distant objects to be properly focused. When the corneal muscle relaxes, the lens resumes its earlier nearsighted position.

The cornea of marine fishes plays only a minor role in refraction because its refractive index is not very different from that of seawater. As a correlate, many fish lenses are more convex—more spherical—than those of most terrestrial craniates, in which the cornea participates in refraction. If you have examined the lens of a shark in the laboratory, you may have been surprised at how spherical it is.

In snakes, unlike in other amniotes, increased pressure on the vitreous humor generated by muscles near the iris *pushes the lens forward* for near accommodation. As in lampreys, the lens returns passively to the resting position when the force that displaced it is removed.

The process of accommodation is quite different in other amniotes. The resting eyes of reptiles and mammals are focused at infinity, and, except in snakes, the *curvature of the lens* is altered by the action of ciliary muscles. We shall examine accommodation for near objects in mammals, whose resting lens is less convex than that of reptiles.

When the mammalian eye is at rest, there is tension within the suspensory ligament stretched, as it is, between the lens and the ciliary body and choroid coat (fig. 17.18). This causes the lens to be anteroposteriorly flattened and therefore focused at infinity. (Infinity in humans is a distance of 20 feet or more from the retina.) Accommodation for near vision is accomplished by contraction of the ciliary muscles, which draws the choroid coat forward and downward and hence closer to the lens. This reduces the tension within the suspensory ligament, allowing the lens to become more spherical. This increase in convexity increases its refractive power, and near objects are then focused onto the retina.

The ciliary muscles of reptiles are striated. Those of amphibians and mammals are not. For this reason, reptiles can change from infinity to near vision, and reverse, more rapidly than can you and I and other mammals. As in most craniates, accommodation for light intensity is brought about by reflex operation of the constrictor and dilator muscles of the pupil.

Because of the shape of the eyeball, cornea, and lens, the number, ratio, and disposition of rods and cones, the presence of two foveas, the striated nature of the ciliary muscles, and other adaptations, the resolving power of a hawk's eye is said to be at least eight times that of the human eye. Such an eye might be useful in watching a game from the top of a stadium!

Conjunctiva

The surface of the eyeball that you can touch when the lids are open is covered with transparent skin, the **bulbar conjunctiva**. This is continuous with the **palpebral conjunctiva** on the inner surface of the eyelid (see fig. 17.15). In snakes and some lizards, the lids are permanently closed, but they are transparent, constituting a **spectacle**. Each time the snake or lizard molts, the epidermal surface of the spectacle is shed along with the rest of the skin, and this carries away any scratches.

Moisturizing and Lubricating Glands

Epidermal glands in terrestrial tetrapods bathe and lubricate the cornea. They are unnecessary in fishes. The duct of a prominent **lacrimal gland** located behind the upper lid opens into the lateral corner of the eye and secretes a slightly salty fluid (**lacrima**; tears when produced in quantity in humans only). The gland is poorly developed in some reptiles including birds but enormous in marine turtles, in which it functions as an excretory salt gland. In many mammals, the duct of a **Harderian gland** opens into the medial corner of the eye. It secretes a more viscous fluid that lubricates the **nictitating membrane** (third eyelid), which is more prominent in some mammals than in others (reduced to a medial semilunar fold in humans). The gland is absent in primates and some other mammals, especially permanently aquatic ones. Some mammals also have an **infraorbital (zygomatic) gland** that opens into the base of the **palpebral fissure** (the slit between the margins of the eyelids, or **palpebrae**). The fluids secreted into the palpebral fissure generally drain into a **nasolacrimal duct** (tear duct) that leads to the nasal cavity.

Absence of Functional Eyeballs

Species that live in caves or other dark recesses (e.g., some fishes, cave salamanders, caecilians, moles) are frequently blind, and the eyeballs may be vestigial. Often, the lids do not open. In hagfishes, no eyeball whatsoever differentiates. In these species, information necessary for survival is obtained from the environment via other receptors.

Median Eyes

Many nonavian and nonmammalian craniates have a functional third eye on the top of the head (fig. 17.19). Among these are lampreys, ganoid fishes, a few teleosts (especially larvae), anuran larvae, some adult anurans, *Sphenodon*, and some lizards. The median eye is an evagination of the roof of the diencephalon. Primitively, it was one of two evaginations constituting an **epiphyseal complex** (fig. 17.20). The more anterior evagination is the **parapineal**; the posterior one is the **pineal**. It is usually the parapineal that is photosensitive, but in lampreys both are.

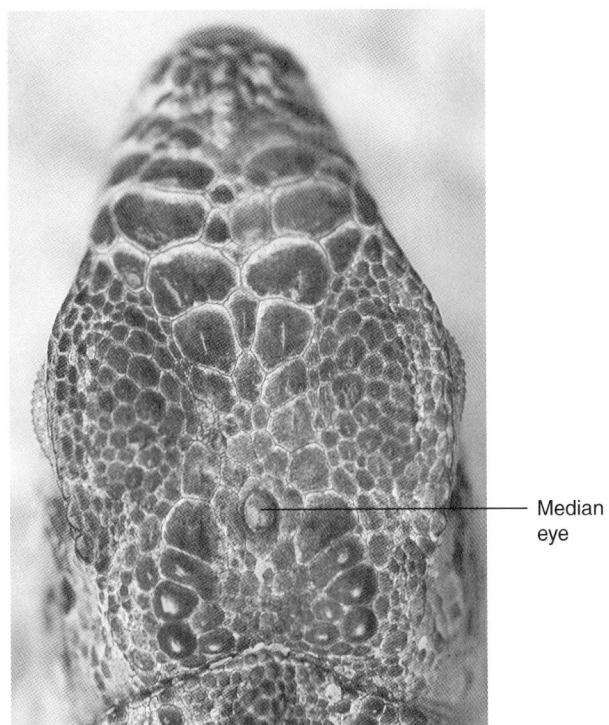

Median eye

FIGURE 17.19
Parapineal eye of an iguana.
© Times Mirror Higher Education Group, Inc./Bob Coyle, photographer.

In lampreys (fig. 17.20*a*), the pineal ends as a hollow knob beneath the cornea, an area of skin devoid of pigment between the lateral eyes. The upper wall of the knob consists of several layers of cells that form a lens. The lower wall contains photosensitive cells and, beneath these, ganglion cells with long processes that pass down the stalk to sensory nuclei in the *right* side of the diencephalon. The parapineal of lampreys is similar, but its fibers terminate on the *left*.

The parapineal of lizards, or **parietal eye**, lies in the parietal foramen immediately under a single, translucent epidermal scale in the midline of the head (fig. 17.20*c*). It consists of a cornea, lens, retina with photoreceptive cells resembling those of the craniate retina, ganglion cells, and a sensory fiber tract that extends down the epiphyseal stalk and enters the roof of the diencephalon.

In larval frogs, the third eye is called a **frontal organ (stirnorgan)**. Authorities are uncertain whether this is the pineal or parapineal part of the epiphyseal complex. At metamorphosis, the photoreceptive part of this organ regresses, leaving only the glandular component that produces the hormone melatonin. In at least one tree frog, however, a functional third eye persists throughout life. The role of melatonin is discussed under "Pineal Organ" in chapter 18.

Median eyes, unlike lateral eyes, form no retinal image. Instead, they monitor the *duration* of environmental

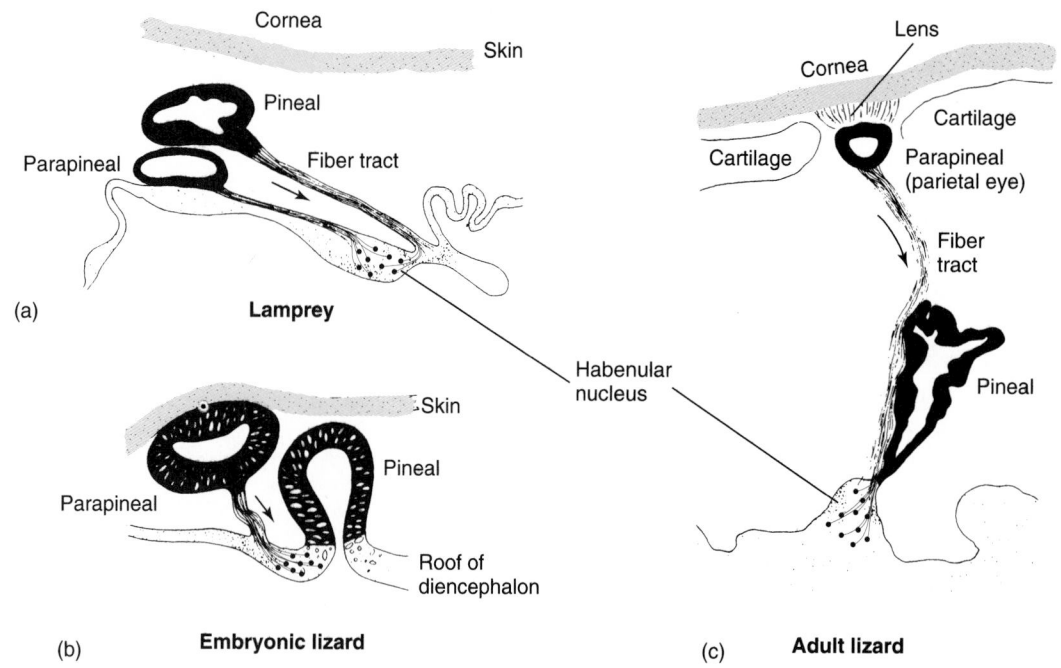

FIGURE 17.20
Epiphyseal complex of lamprey and embryonic and adult lizard. The lizard parapineal occupies the parietal foramen of the skull.
Source: From M. Novikoff, "Untersuchungen uber den Bau, die Entwicklung und die Debeutung des Parietalauges von Saurien" in *Zeitschrift fur eissenschaftliche Zoologie*, 96:118, 1910.

photoperiods, and the input affects internal biologic rhythms such as daily spontaneous motor activity and seasonal changes in the gonads. They probably also monitor the *intensity* of solar radiation. The metabolic rate is correlated with body temperature, which, in turn, depends on how much solar energy has been received. Third eyes were present in all major groups of Devonian fishes and in the earliest amphibians and reptiles before being lost in most later craniates.

Infrared Receptors of Snakes

Pit vipers (venomous snakes in the family Crotalidae, such as rattlesnakes, copperheads, and water moccasins) have an infrared receptor in a deep pit on each side of the head between the eye and the nostril. The pits are lined with an epithelium that houses hair cells that are sensitive to infrared (thermal) radiation. The tissue that underlies the epithelium is highly vascular. Because of their location in the **lore** (the region in front of the eye in reptiles), these are called **loreal pits.** They are readily visible, being several millimeters wide, twice as deep, and directed forward.

Physiologic and behavioral studies have shown that loreal pits can detect temperature changes of as little as 0.001° C at a distance of several feet. This enables pit vipers to detect the presence of warm-blooded craniates such as mice or small birds at night or in daytime when the prey is hidden under the substrate or in a nest or burrow, so long as the environmental temperature is lower than that of the prey. The receptors also enable

the viper to strike accurately in the dark. They can also detect objects that are colder than the surrounding environment.

Pythons (family Pythonidae) and some boas (family Boidae) have a series of similar but smaller and less sensitive **labial pits** with slitlike openings between the scales surrounding the mouth. Some pythons have up to 30 of these infrared receptors. All thermal receptors on the head of snakes are innervated by sensory fibers from the fifth cranial nerve (opthalmic branch). The fibers terminate in a cranial nucleus that is separate from, but closely associated with, the trigeminal nucleus that receives mechanoreceptive input.

Evolution of reptilian pit organs, which lie above the maxillary bones, entailed restructuring of the maxillae to accommodate the pit organs while at the same time continuing to support and move the fangs. Pit organs supplement vision, mechanoreception, and chemoreception in the search for food.

SPECIAL CHEMORECEPTORS

There are two varieties of special chemoreceptors, **olfactory** (for smell) and **gustatory** (for taste). Both are sensitive to certain ions and molecules. Olfactory receptors respond to less specific chemical configurations than taste receptors. Taste and smell are related phenomena, as is evident to anyone who has been unable to savor food because of a head cold.

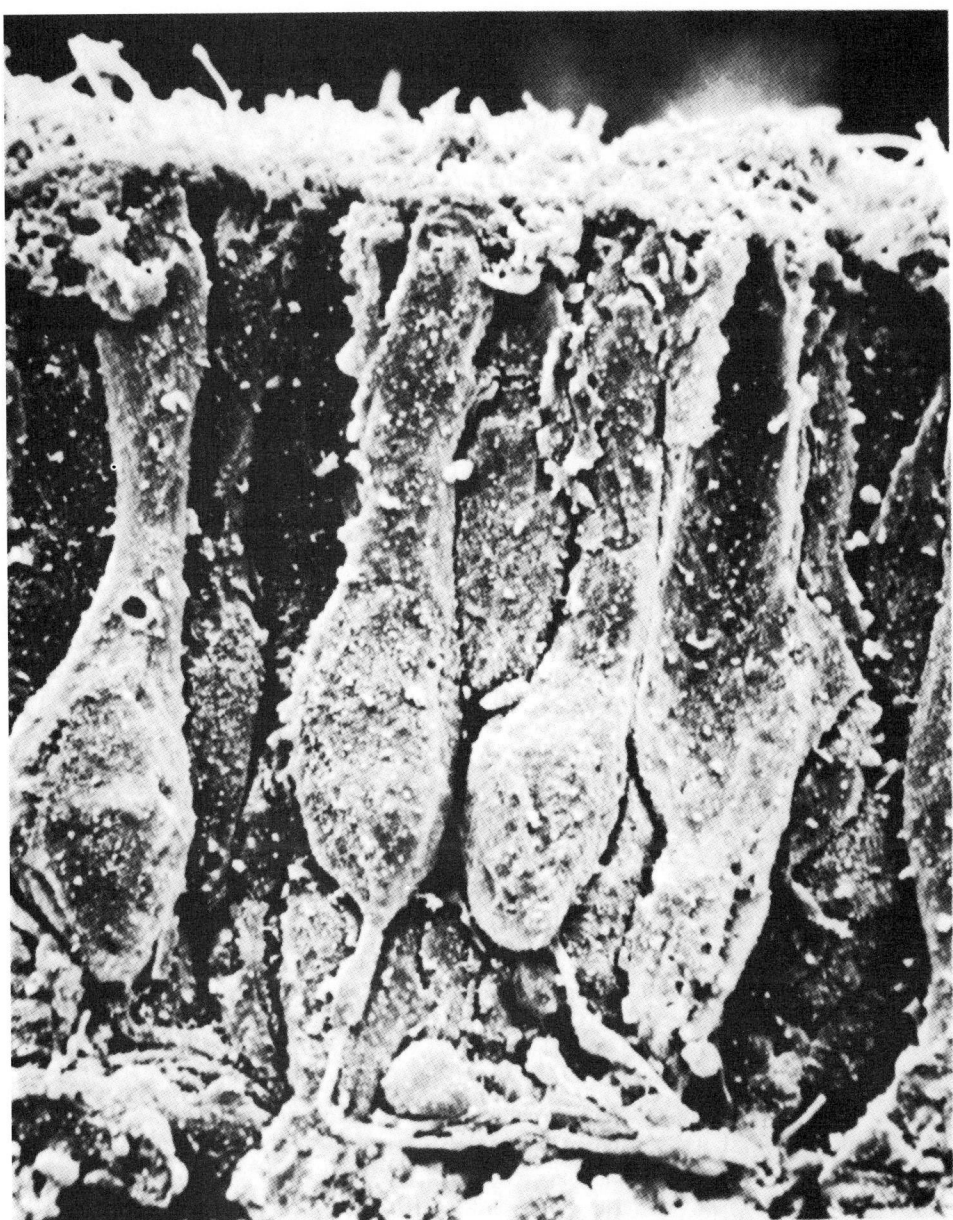

FIGURE 17.21

Section of olfactory epithelium from the channel catfish, *Ictalurus punctatus*. The axons of three olfactory cells are readily identifiable. A cilium at the apex of the olfactory cells projects into the moisture on the surface of the epithelium. Height of the section was approximately 20 microns.

From Caprio, J., and Raderman-Little, R.: *Tissue and Cell,* Vol. 10(1):1, 1978. © Churchill Livingstone.

Special chemoreceptors evoke both somatic and visceral responses. Somatic responses consist of striated muscle activity that reorients the organism in the environment. Unconditioned visceral responses to stimulation of taste buds of teleosts have been demonstrated by applying arginine, an amino acid common in foodstuffs, to the buds, and recording the impulses evoked in branches of the vagus nerve that innervate digestive organs. Unconditioned and conditioned salivation and secretion of digestive juices following olfactory and gustatory stimulation occurs also in mammals.

Olfactory Organs

The olfactory organs of gnathostomes arise as a pair of ectodermal **olfactory placodes** that form just anterior to the embryonic stomodeum. The placodes sink into the head to form a pair of **nasal pits** (see fig. 13.7*b*), the epithelial lining of which differentiates into olfactory, supportive, and mucous cells. Olfactory cells, like rods and cones, are neurosensory cells that differentiate in an epithelium and sprout processes that grow into the brain (fig. 17.21). These processes are the fibers of the olfactory nerve.

There are an estimated 50 million olfactory cells in humans. For a substance to act as an odorant, it must have a chemical configuration compatible with that of a binding substance on the membrane of the olfactory cell, and it must be in solution.

In fishes other than lobe fins, the differentiating olfactory epithelium becomes surrounded by mesenchyme that forms a blind olfactory sac. A current of water into and out of the sac is ensured because each external naris is partitioned into incurrent and excurrent apertures so situated that the forward motion of the fish propels a stream of water into one aperture and out the other. The mucosa containing the olfactory epithelium may exhibit folds that increase the surface area. The olfactory cells monitor the water stream and are stimulated by odorants that may have their source in potential food, mates, or enemies. The most primitive somatic response to olfactory stimuli is reflex contraction of locomotor muscles, which propel the fish closer to, or farther from, the source of the odorant.

In lungfishes and tetrapods, the olfactory pits push deep into the head to acquire an opening into the oral cavity or pharynx. The openings are internal nares. When this occurs, the olfactory epithelium is confined to a portion of the lining of the nasal canal, so that it is appropriate to distinguish the *olfactory* epithelium from the *nasal* epithelium of the rest of the nasal passageway. The olfactory epithelium of tetrapods contains olfactory cell bodies just as in fishes, but in tetrapods, it monitors an airstream instead of a water stream. Compound tubular mucous glands (**Bowman's glands**, see fig. 6.10*d*) keep the epithelium moist. Odorants in the airstream dissolve on the moist olfactory epithelium and stimulate the olfactory cells.

An olfactory epithelium is well developed in fishes but least developed in birds, which therefore have a poor sense of smell. The olfactory nerves of some species of whales disappear during embryonic life. However, a mammal trying to inhale under water would drown, so the evolutionary loss of the olfactory nerves could not have been a disadvantage to whales.

Vomeronasal Organs

In many tetrapods, a ventral section of the olfactory epithelium becomes more or less isolated from the nasal passageway to serve as an accessory olfactory chemoreceptor. This vomeronasal organ, named for its proximity to the vomer bone, receives sensory input via an accessory olfactory bulb and a vomeronasal nerve (VN). The size of the bulb and nerve is proportional to that of the organ.

In urodeles, the organs are a pair of deep grooves in the ventromedial floor of the nasal canals. In anurans and apodans, they are blind sacs that open into the nasal canals. In squamates, they lose their connection with the nasal canals and open into the anterior roof of

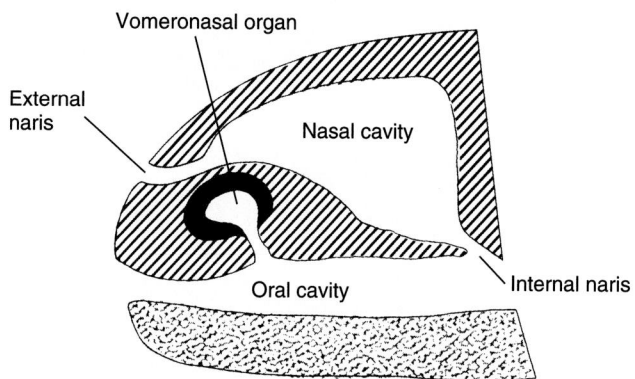

FIGURE 17.22

Location of the vomeronasal organ in a generalized lizard. Only the left organ is shown. *Diagonal lines* designate a parasagittal section.

the oral cavity as two moist pockets (fig. 17.22). These receive the forked tips of the tongue each time it darts in and out of the mouth, monitoring the environment. Because these organs are clearly chemoreceptors, it was long assumed that their role is chemical detection of appropriate foodstuffs. However, removing the forked tip of the tongue of test snakes prevents them from following pheromone trails left by other members of the species. This, then, is one of the roles of the organs in snakes, at least. Garter snakes have no tongue, but autoradiographic studies show that chemicals on the lips, derived from potential foodstuffs, reach the vomeronasal epithelium by diffusion. Vomeronasal organs appear in embryonic crocodilians and birds and then regress—mute evidence of an ancestral function. They have not been found in turtles.

Most mammals have vomeronasal organs just above the hard palate. They open either in the floor of the nasal canals, as in many rodents, or onto the hard palate via nasopalatine ducts that pass through incisive foramina, as in cats. They are especially well developed in monotremes, marsupials, generalized insectivores, and many carnivores. They are absent in some bats, adult non–basal primates, and cetaceans (where ancestral sensory specializations to a terrestrial habitat may have precluded their effectiveness with a return to an aquatic lifestyle). Vomeronasal organs and their nerves develop in human embryos, reach maximum size about the fifth month of gestation, and then regress.

Organs of Taste

Taste buds are barrel-shaped clusters of elongated **taste cells** (nonnervous receptor cells) and **supportive cells** (fig. 17.23). The apices of the receptor cells have prominent apical processes that protrude into a **taste pore** in the moist epithelium. The apical processes are covered with microvilli. Intimately associated with the base of each receptor cell are the sensory nerve end-

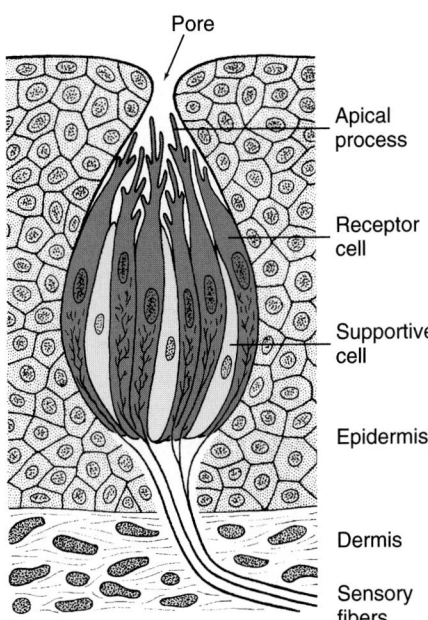

FIGURE 17.23
Receptor cells and their innervation in a taste bud.

ings. The functional life of a taste receptor cell is only about 10 days, at which time it dies from "wear and tear." They are replaced by supportive cells, which are reserve receptor cells. The supply of supportive cells is continually being replenished from the basal layer of the epithelium.

In fishes, taste buds are widely distributed in the roof, side walls, and floor of the pharynx, where they monitor the incoming stream of respiratory water. In bottom feeders or scavengers, such as catfish, carp, and suckers, taste buds are distributed over the entire surface of the head and body to the tip of the tail. They are abundant on the barbels ("whiskers") of catfish. The exaggerated size of the sensory nucleus that receives incoming taste fibers in such fishes is illustrated in figure 16.16.

In most tetrapods, taste buds are restricted to the tongue, posterior palate, and oral pharynx, which are moist. There are fewer on the tongue in reptiles than in mammals, and human embryos have more taste buds than children 7 years old because some that are lost are not replaced.

Taste buds from fish to humans are supplied by the seventh, ninth, and tenth cranial nerves *in that sequence* from mouth to the end of the oral pharynx. Therefore, taste buds on the anterior surface of the human tongue are supplied by nerve VII, those on the posterior surface of the tongue by nerve IX, and those in the vicinity of the glottis by nerve X. Nerve VII also supplies all taste buds in the skin of fishes all the way to the tip of the tail.

GENERAL SOMATIC RECEPTORS

There are two categories of general somatic receptors: (1) **cutaneous receptors** for light touch, pain, temperature, and pressure (touch sufficient to indent the surface) and (2) proprioceptors, found in striated muscles, joints, and tendons.

Cutaneous Receptors

Naked Endings

The skin of all craniates contains free, or naked, endings of sensory nerve fibers that ramify among the epidermal cells and are stimulated by contact (fig. 17.24, naked endings in epidermis). These are the most primitive cutaneous endings in craniates and in agnathans they are the only ones. They give rise to what has been called **protopathic** sensation. This is a crude, poorly localized, phylogenetically ancient sensation. It is purely protective. It is not necessary that a fish know the texture of whatever touches it or whether the object is warm or cool. The mere fact of contact indicates possible danger. The impulses reach the thalamus, where reflex motor activity is initiated to reorient the animal so that it avoids the stimulus. In mammals, naked endings give rise to vague, poorly localized, and sometimes unpleasant sensations including pain, as in toothache or when the skin is pricked. Free nerve endings for touch entwine around the base of each mammalian hair. Displace a single hair on your arm and note the sensation. Localization of the stimulus is a function of the cerebral cortex.

Within the epidermis of some mammals, unique meniscuslike terminals of sensory nerve fibers are applied to the underside of a single concave epidermal cell (fig. 17.24, tactile discs). These **tactile discs** are the simplest receptors in which an epithelial cell combines with the terminal of a nerve fiber to form a functional receptive unit. They are found in clusters in the stratum germinativum of the sensitive snout of moles and pigs, in the epidermal sheaths surrounding the base of the vibrissae of rodents, and in other sensitive sites.

Encapsulated Endings

In addition to naked endings that ramify in the skin, tetrapods have acquired bulbous encapsulated endings that add to the repertoire of receptors that detect localized integumentary stimuli (fig. 17.24). These consist of nerve endings in association with epithelialike cells wrapped in a connective tissue capsule, and, with one exception, located within dermal papillae at the dermal-epidermal interface (see fig. 6.12, touch corpuscle). **Grandry's corpuscles** along the margins of the beaks of many shore birds and freshwater fowl consist of a nerve ending and *two* epithelial cells surrounded by a thin capsule. **Corpuscles of Herbst**, on the beak, tongue, and palate of water birds, have a thick lamellar capsule and a central core of epithelialike cells. **Genital corpuscles** are distributed to the external genitalia of mammals, especially the glans penis and clitoris; and, in humans, on other erogenous areas. **End bulb (corpuscles) of Kraus** and **Ruffini corpuscles** of mammals may be thermal receptors. **Meissner's corpuscles,** found only in primates and chiefly on hairless areas—toes, soles, palms, and especially fingertips—are probably tactile receptors. Still other corpuscles have been described in avian and mammalian skin of one species or another. The sensation of touch as opposed to temperature perception (heat or cold) cannot be ascribed to every known encapsulated ending with certainty. All the foregoing are in dermal papillae.

Pacinian corpuscles, the largest encapsulated receptors, are found deep in the dermis or in subdermal connective tissue (see fig. 6.12) where they are stimulated by compression or, when an object is handled, where they participate in stereognosis (see next paragraph). They are not, however, restricted to the category of cutaneous receptors. They are also present in the bursas of diarthroses where they participate in proprioception, and in coelomic mesenteries and peritonea, including the pericardial peritoneum, where they serve an undefined visceral function. In the latter locations, they are not known to evoke conscious sensations.

Encapsulated exteroceptive endings in mammalian skin are engaged in *epicritic* sensation: those that enable one to differentiate between small differences in warmth or coolness of objects, to distinguish textures, to perceive as separate sites two pinpoints on the surface of the skin that are very close together and are being touched simultaneously, and to perceive the shape and weight of an object being handled, the latter being **stereognosis.** After synapsing in the thalamus, epicritic stimuli are projected to the **somesthetic cortex** of the mammalian brain (see fig. 16.26), the cortical center for epicritic sensibilities.

A few simple exteroceptive corpuscles are found in amphibians and nonavian reptiles, but birds and mammals have the largest number and variety.

Miscellaneous Cutaneous Receptors

Lizards have mechanoreceptors and thermal receptors that are widely distributed over the surface of the body. The most numerous of these occupy **apical pits,** distributed in groups of one to seven at the apices—the posterior free borders—of the epidermal scales. In another variety of cutaneous receptors of lizards, a filamentous hairlike bristle, or **protothrix,** projects to the surface between scales and transmits tactile stimuli to a receptor cell in a pit. Receptors such as these provide sites for input of exteroceptive stimuli in a scale-covered epidermis that is otherwise difficult to penetrate by usual tactile or thermal stimuli.

Some snakes, black snakes for example, have mechanoreceptors located between scales over the en-

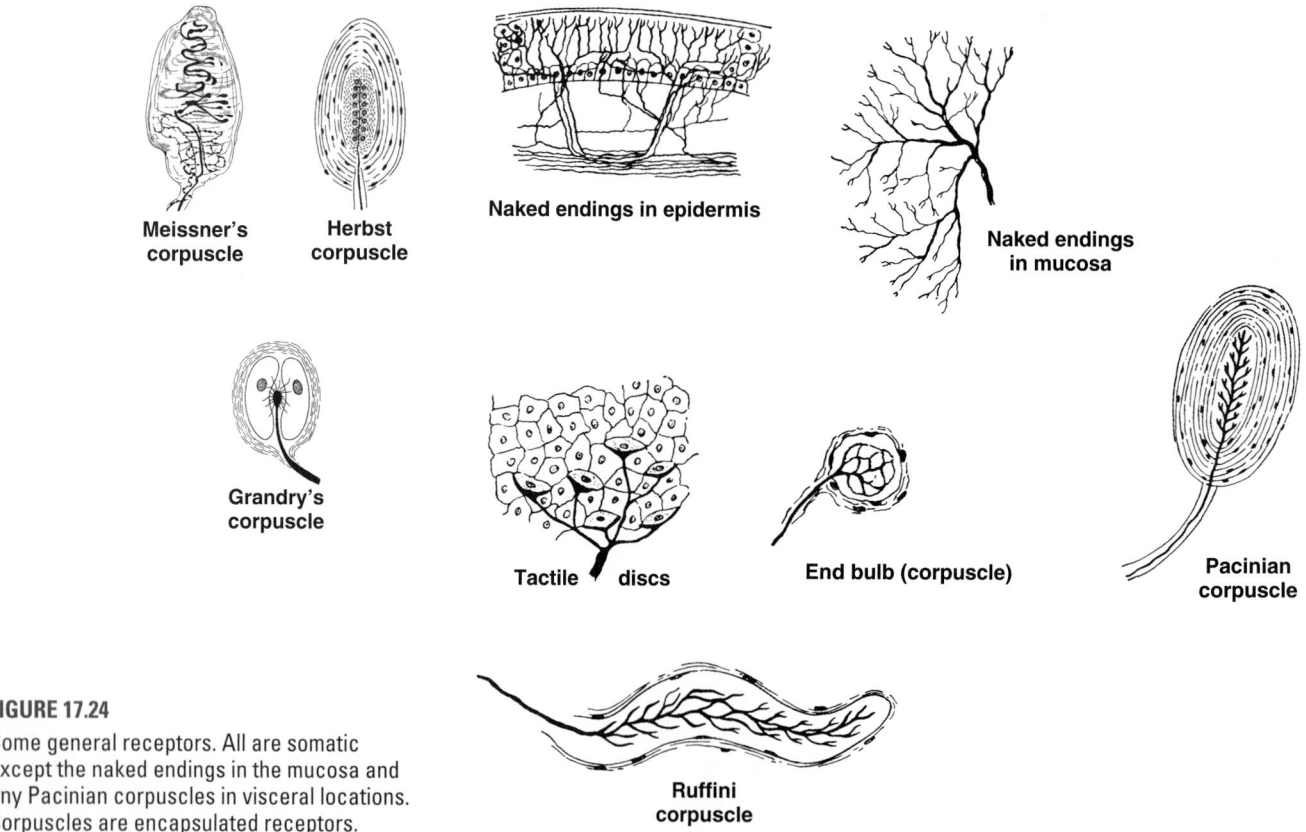

FIGURE 17.24

Some general receptors. All are somatic except the naked endings in the mucosa and any Pacinian corpuscles in visceral locations. Corpuscles are encapsulated receptors.

tire trunk and tail except on the venter. The majority of these receptors have a high threshold and are sensitive to vibrational stimuli in the same frequency range (up to 800 Hz) that induces cochlear potentials in those species.

Proprioceptors

Skeletal muscles, their tendons, and the bursas of joints are supplied with sensory endings that monitor continuously the activity status of muscles and joints. These proprioceptors are distinguished from exteroceptors, which monitor the external environment, and enteroceptors, which monitor visceral effectors. Proprioceptors maintain tonus reflexly in a resting muscle and assure synergy in the activity of functional groups of muscles when the latter are working against a load.

The sensory endings in most mammalian skeletal muscles, especially appendicular, are found in **muscle spindles** (fig. 17.25). In mammals, these are tiny fusiform bundles of three to 20 specialized striated muscle fibers that are wrapped in a connective tissue sheath and interspersed near the extremities of the muscle among and parallel to extrafusal fibers not in the spindle. Each modified muscle fiber (**intrafusal fiber**) is supplied with at least one **annulospiral** proprioceptive ending (fig. 17.26). Intrafusal fibers also exhibit **flower spray** endings, which ramify along the fiber. They have a motor innervation, being supplied by **gamma motor neurons**, a term employed to distinguish them from **alpha motor neurons** that innervate striated muscle fibers that are not in a spindle. Differentiation of mature intrafusal muscle fibers requires both sensory and motor innervation. Spindles are **stretch receptors**, being extremely sensitive to very small changes in the length of a muscle and hence to muscle activity.

Some mammalian muscles have fewer spindles than others. Appendicular muscles have the largest number per unit volume, and the extrinsic eyeball muscles have none. In the latter, annulospiral endings are found on regular striated muscle fibers.

Muscle spindles are not common in amphibians and have not been described in fishes. Proprioceptive endings in the muscles of these craniates are on regular muscle fibers and are relatively simple dendritic sprays of one variety or another.

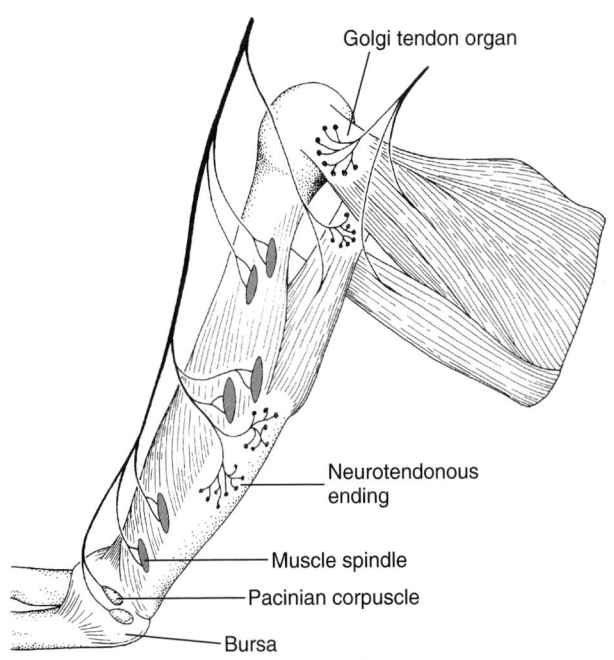

FIGURE 17.25

Representative proprioceptor sites (exaggerated) occurring in muscle, tendon, and bursa.

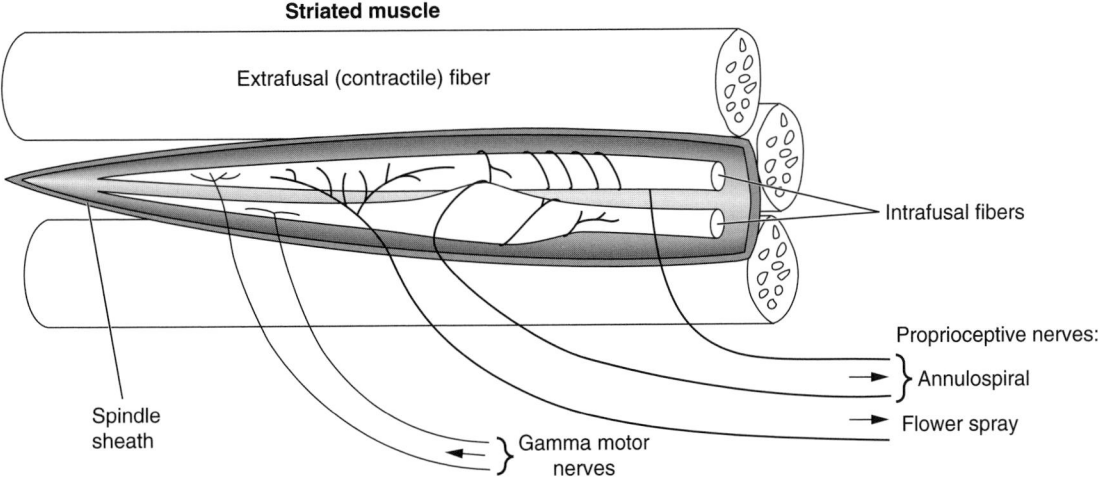

FIGURE 17.26

Muscle spindle (*red*) in section revealing two inner intrafusal fibers, which are modified muscle fibers. Two types of proprioceptive (sensory) nerves are shown, which monitor muscle tension. These nerves synapse in the central nervous system with neurons that stimulate extrafusal muscle fibers.

In maintaining muscle tonus, which is a dynamic state, muscle spindles function as follows. The connective tissue envelope of the spindle is continuous with the endomysium (strands of perimysium, see fig. 11.3) that surrounds extrafusal fibers beyond the spindle so that any change in the length of a muscle affects the spindle. When a muscle is at rest it is elongated, the spindle is slightly stretched, and proprioceptive nerve fibers are firing. Because proprioceptive fibers synapse in the central nervous system with alpha motor neurons, discharge from the spindle triggers contraction of a limited group of extrafusal fibers, and this increases the tonus of the muscle, shortening the muscle slightly. Shortening relieves some of the tension within the spindle, and as a result, firing of the proprioceptive fibers slows. Thereupon, the muscle tends to elongate once again, and this triggers new discharges from the spindle. Thus, feedback from muscles to central nervous system reflexly maintains the muscle in a state of dynamic tonicity while the muscle is at rest.

The role of spindles when muscles are working against a load is more complicated. Alpha and gamma motor neurons in motor nuclei of the spinal cord are being stimulated by motor nerve fibers whose cell bodies are in the brain. Stimulation of large numbers of gamma motor neurons results in contraction of intrafusal fibers in more than one muscle and consequent discharge of many proprioceptive impulses at a high frequency. The resulting steady reflexive firing of alpha motor neurons maintains a steady contraction of large numbers of extrafusal muscle fibers, and this continues, subject to brain control, whether the craniate is locomoting or playing a piano.

Proprioceptive endings on tendons (neurotendonous endings) are flower spray endings that respond to forces generated in the tendon when its muscle contracts. Proprioceptive receptors in mammalian bursas are chiefly Pacinian corpuscles (see fig. 17.25). Naked sensory endings in myosepta, presumably for proprioceptive input, have been described in lampreys.

Most proprioceptive stimuli do not reach centers of consciousness but are shunted into the cerebellum and elsewhere, where they make reflex connections. To demonstrate conscious proprioception, first close your eyes, then extend your arm or leg. Your awareness of the change in position is an example of conscious proprioception, also called *kinesthesia* and *muscle sensibility*. That the act was performed smoothly is a result of reflex responses to volleys of proprioceptive stimuli that did not reach the level of consciousness.

GENERAL VISCERAL RECEPTORS

General visceral receptors are mostly naked endings in the mucosa of the tubes, vessels, and organs of the body, in cardiac muscle, in smooth muscle including that of the blood vessels, and in the capsules, mesenteries, and meninges of the viscera (see fig. 17.24, naked endings in mucosa). Pacinian corpuscles serve a visceral function in mesenteries, in the connective tissue stroma of some viscera, and in the peritonea of the coelom.

General visceral receptors are chiefly stretch and chemoreceptors, but they include a few baroreceptors and osmoreceptors. Among other roles, chemoreceptors monitor the pH of the blood (including oxygen and carbon dioxide content), which affects cardiorespiratory functions, and they monitor the pH of the contents of the stomach and proximal intestine, which affects digestive functions. Baroreceptors monitor blood pressure, and osmoreceptors in the hypothalamus and elsewhere monitor the concentration of certain solutes in the bloodstream. General visceral receptors are stimulated by tactile and thermal stimuli in the pharynx but not beyond. Such input would have no survival value. Most visceral pain from healthy hollow visceral organs results from distension (stretch) and is detected by naked endings.

Three grossly observable vascular monitors are present in amniotes. These are the **carotid bodies**, the **aortic bodies**, and the **carotid sinuses**. The paired carotid bodies lie close to, or are embedded in, the walls of the common carotid or internal carotid arteries; the aortic body, present in mammals only, is on the aortic arch. The carotid and aortic bodies are chemoreceptors. They are richly supplied with sensory endings, the carotid bodies by the ninth nerve and the aortic bodies by the vagus. They are stimulated by a deficiency of oxygen or by an excess of carbon dioxide in the blood. The stimuli evoke discharge of sensory impulses that reflexly increase the depth and rate of respiration. The respiratory center is in the medulla.

The carotid sinuses are slight dilations of the internal carotid arteries at their origin from the common carotid. They are baroreceptors that respond to distension of their walls resulting from an elevation in blood pressure. The sensory nerve impulses travel over the ninth cranial nerve and eventually reach the cardioinhibitory and vasoconstrictor centers in the medulla. From there, autonomic fibers in the vagal nerve (see fig. 16.35) reflexly slow the heart and dilate the peripheral arteries, reducing the blood pressure.

Most of the sensory input from general visceral receptors gives rise to no conscious sensation, although the reflex response may do so. Some of the same kinds of monitoring of the viscera are necessary in all craniates whether in water or on land, so visceral receptors are generally similar from fishes to human beings.

Summary

1. Receptors are transducers of mechanical, electrical, thermal, chemical, or radiant energy. General receptors are widely distributed in the body. Special

receptors have a limited distribution and mostly are paired.

2. Somatic receptors provide information about the external environment (exteroceptors) and about skeletal muscle activity (proprioceptors). Visceral receptors monitor the internal environment (enteroceptors).

3. Neuromast organs consist of hair cells, sensory endings, and supporting cells. They are most abundant in fishes and aquatic amphibians and occupy shallow pits, grooves, ampullae, or canals. They function principally in mechanoreception, less commonly in electroreception and chemoreception. The lateral-line and cephalic canals are neuromast systems.

4. The membranous labyrinth (inner ear) is a neuromast system consisting of semicircular ducts, sacculus, utriculus, endolymphatic and perilymphatic ducts and sacs, and either a lagena or a cochlea. It is filled with endolymph and surrounded by perilymph.

5. The primitive function of the labyrinth was equilibratory. An auditory function employing a cochlea was a later development.

6. Cristae and maculae are the receptor sites in a membranous labyrinth. Otoliths are generally associated with maculae of anamniotes. The organ of Corti of birds and mammals is the most specialized macular complex.

7. Some fishes detect sound waves via air bladders and weberian ossicles. Cephalic and lateral-line canals subserve an analogous function in most fishes and larval amphibians.

8. A middle-ear cavity (cavum tympanum) containing a columellar complex characterizes nonmammalian tetrapods. Mammals have two additional ear ossicles, a malleus and an incus. The columella of mammals is called a stapes. Striated tensor tympani and stapedial muscles attach to the malleus and stapes.

9. Airborne sounds in tetrapods are generally detected by a tympanic membrane (eardrum) and conducted to the membranous labyrinth by the ear ossicles. Seismic waves are normally conducted by the skeleton of the forelimbs or jaws in some amphibians and reptiles.

10. Tympanic membranes are flush with the surface of the head in anurans. They are at the end of a shallow or long outer ear canal (external auditory meatus) in amniotes. Eardrums and middle ear cavities are rudimentary or lacking in urodeles, apodans, and most limbless reptiles.

11. A sound-collecting auricle (pinna) is found on the head of most mammals.

12. A saccus vasculosus of uncertain function evaginates from the floor of the third ventricle of fishes. It is lined by hair cells that project into cerebrospinal fluid.

13. The receptor site for lateral eyes is the retina. It is nourished and protected by a vascular and light-absorbing choroid coat. A fibrous and sometimes bony sclerotic coat (sclera) is the outer covering of the eyeball.

14. Accommodation for near or far vision generally involves altering the distance between the lens and retina in anamniotes and altering the shape of the lens in amniotes. A ciliary body in amniotes participates in accommodation for distance.

15. Accommodation for intensity of light is the role of constrictor and dilator muscles of the iris diaphragm.

16. Median (parietal) eyes are photoreceptors but form no retinal image. They are unpaired and include a pineal and/or parapineal organ. They were present in the earliest fishes and are still present in agnathans, some bony fishes, larval and a few adult anurans, and some lizards.

17. Olfactory and vomeronasal organs are special chemoreceptors that arise from embryonic nasal pits. Neurosensory cells receive the stimuli.

18. Taste buds are special chemoreceptors. In fishes, they may be distributed over much of the surface of the body and also in the entire buccopharyngeal mucosa. They are confined to the buccopharyngeal cavity in tetrapods and to the tongue and oral pharynx of mammals.

19. Infrared receptors (pit organs) are found in pit vipers, pythons, and some boas. They detect thermal stimuli.

20. Cutaneous receptors are naked or encapsulated nerve endings distributed universally in skin. They detect touch, temperature, and pressure on the surface, and they subserve protopathic and epicritic sensibilities. Apical pits are cutaneous mechanoreceptors confined to lizards.

21. Proprioceptors are found in striated muscles, tendons, and bursas of joints. They monitor the activity of skeletal muscles.

22. General visceral receptors are mostly naked endings for chemoreception in the digestive tract and for monitoring smooth muscle activity everywhere. Carotid bodies and aortic bodies house chemoreceptors that monitor the oxygen and carbon dioxide content of blood. Carotid sinuses are mechanoreceptors that monitor blood pressure. Osmoreceptors monitor the concentration of certain solutes in the circulation.

Critical Thinking Questions

1. Sense organs provide information about the external environment. What type of information is necessary for survival?

2. The paired external sense organs are primarily confined to the head, and as we know, the head shows a meristic pattern. Do the sense organs of the head represent serially homologous structures? Why or why not?

3. Compare the use of individual sense organs in aquatic and terrestrial environments.

4. Snakes share the many sense organs of other craniates, but some snakes evolved infrared detectors. What do you know about snakes that might explain this evolutionary novelty?

5. The neuromast appears to be the fundamental unit for several sensory modalities. What is it, and how does it function in the different modalities?

6. Describe two evolutionary novelties for the transduction of sound from the environment to the sensory epithelium for hearing.

7. If an organism has lateral eyes, then what function do medial eyes serve?

8. Why would humans possess a vomeronasal organ and nerve only during development?

9. Given your knowledge of sensory structures, can you describe the deep tendon reflex? (The knee jerk you experience when the doctor taps your knee with a reflex hammer is an example.)

Selected Readings

Adams, W. E.: The comparative morphology of the carotid body and carotid sinus. Springfield, IL, 1958, Charles C. Thomas, Publisher.

Berger P. J.: The reptilian baroreceptor and its role in cardiovascular control, *American Zoologist* 27: 111, 1987.

Bullock, T. H., Bodznick, D. A., and Northcutt, R. G.: The phylogenetic distribution of electroreception: Evidence for convergent evolution of a primitive vertebrate sense modality, *Brain Research Reviews* 6:25, 1983.

Buning, T. d'C.: Thermal sensitivity as a specialization for prey capture and feeding in snakes, *American Zoologist* 23:363, 1983.

Finnell, R. B., editor: Symphony beneath the sea, *Natural History*, p. 36, March 1991. Eleven articles on echolocation and communication sounds of marine mammals, including illustrations of the ear complexes. Fascinating reading.

Gaskell, W. H.: On the structure, distribution, and function of the nerves which innervate the visceral and vascular systems, *Journal of Physiology* (London) 7:1, 1986.

Gilbertson, T. A.: The physiology of vertebrate taste reception. *Current Opinion in Neurobiology* 3:532, 1993.

Gregory, E.: Tuned-in, turned-on platypus, *Natural History*, May 1991.

Jones, D. R., and Milsom, W. K.: Peripheral receptors affecting breathing and cardiovascular function in non-mammalian vertebrates, *Journal of Experimental Biology* 100:59, 1982.

Levine, J. S.: The vertebrate eye. In M. Hildebrand, and others, editors: Functional vertebrate morphology. Cambridge, MA, 1985, Harvard University Press.

McLaughlin, S., and Margolskee, R. F.: The sense of taste, *American Scientist* 82:538, 1994.

Northcutt, R. G.: The brain and sense organs of the earliest vertebrates: Reconstruction of a morphotype. In Foreman, R. E., Gorbman, A., Dodd, J. M., and Olsson, R., editors: Evolutionary biology of primitive fishes. New York, 1985, Plenum Publishing.

Northcutt, R. G.: Distribution and innervation of lateral line organs in the axolotl, *The Journal of Comparative Neurology* 325:95, 1992.

Novacek, M. J.: Navigators of the night, *Natural History*, October 1988. FM listening in bats.

Parker, D. E.: The vestibular apparatus, *Scientific American* 243:98, 1980.

Purves, E., and Pilleri, G. E.: Echolocation in whales and dolphins. New York, 1983, Academic Press.

Tavolga, W. N., Popper, A. N., and Fay, R. R., editors: Hearing and sound communication in fishes. New York, 1981, Springer-Verlag.

Ulinski, P. S.: Design features in vertebrate sensory systems, *American Zoologist* 24:717, 1984.

Weaver, E. G.: The amphibian ear. Princeton, NJ, 1985, Princeton University Press.

Links to the Internet

Visit the zoology website at http://www.mhhe.com/zoology to find live Internet links for each of the following references.

1. Seeing, Hearing and Smelling the World. A full text reprint of a Howard Hughes Medical Institute report on "making sense of our senses."

2. Webvision. The organization of the vertebrate retina.

3. The Physiology of Taste. Much anecdotal information on human taste.

4. Sensation and Perception. Information from the Monell Chemical Senses Center.

5. The Physiology of Taste. Information on taste, and links to other sites on chemoperception.

6. Auditory Physiology. Information and nicely done, labeled graphics on the mammalian ear.

7. Seeing in the Dark and Tuning in Bat Detectors. A lengthy description of bat echolocation with a few interesting links.

18

Endocrine Organs

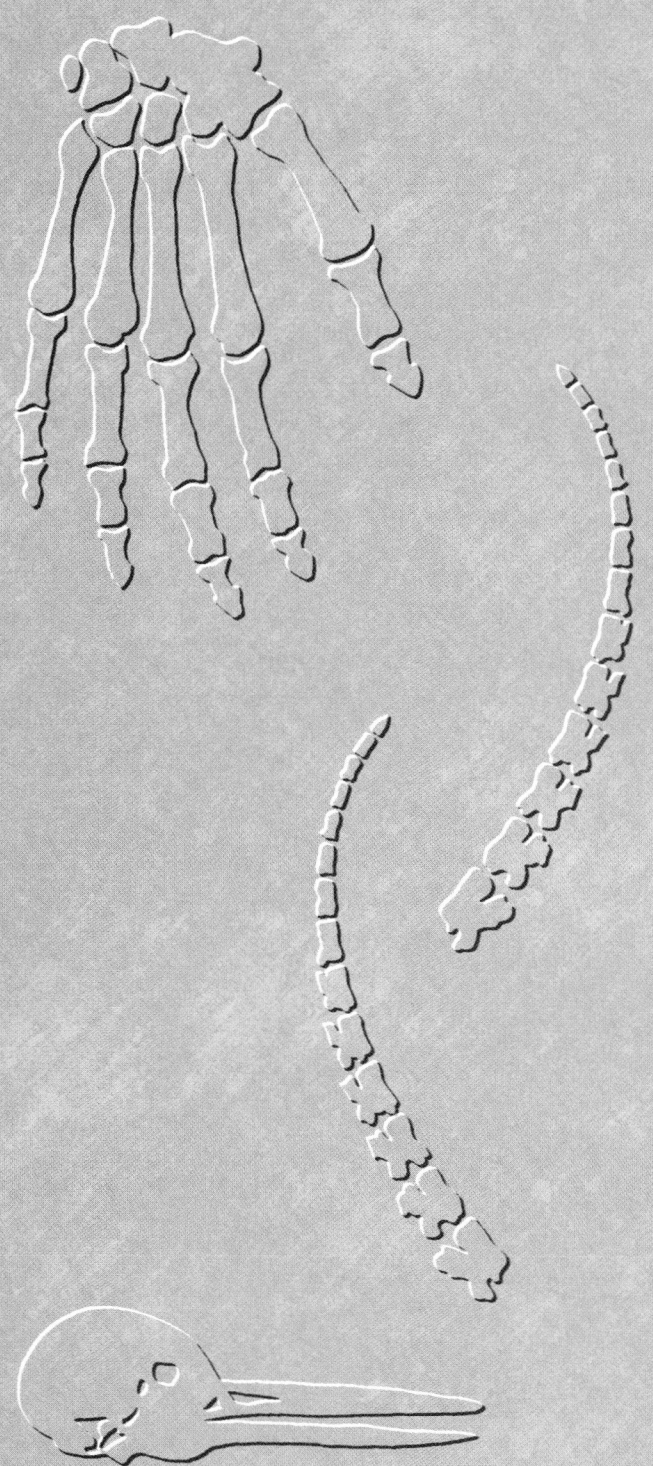

In this chapter, we will look at some of the organs whose hormonal secretions adapt craniates to daily and seasonal fluctuations in the external environment, or which maintain some of the major metabolic activities at levels optimal for survival. We will find that the brain is one of these organs, translating cyclical environmental input into hormonal messages. We will learn that craniates from fishes to humans have the same array of such organs, that they form from the same embryonic precursors, and that they synthesize essentially the same molecules. We will also note some of the anatomical and molecular variations that have occurred during phylogeny. Along the way, we will have glimpses of the endocrine basis for biological rhythms.

OUTLINE

ndocrine organs synthesize hormones. Hormones are organic molecules that have a regulatory effect on other cells, often referred to as "target" cells. Target cells may be right next to the source of the hormone, but usually they are remote, and the hormone arrives via the bloodstream.

The term *target cell* might lead to the mistaken notion that a hormone passes only to the "target." This is not the case. Most hormones are transported in the bloodstream, and the fluid constituents of the stream escape from capillaries everywhere in the body and bathe every living cell. Only those cells having an appropriate molecular "receptor site" on their membranes can be affected by the hormone. For example, thyrotropic hormone from the pituitary gland bathes the cells of the big toe in the same concentration that it bathes the cells of the thyroid gland, which is its target, but only certain cells of the thyroid gland are able to respond. A few hormones, such as insulin and thyroxine, have receptor sites on every cell of the organism. These are **general metabolic hormones.**

The earliest hormones probably had only local action, affecting nearby cells as they diffused from their source, rather than being transported by a closed circulatory system to distant organs. This is a necessity in coelenterates (*Hydra,* for example) because they lack a circulatory system, but hormones that act locally by diffusion are not uncommon in craniates. Certain gonadal hormones, for example, have both local and remote effects.

Most hormones are combinations of amino acids—chiefly polypeptides, proteins, and glycoproteins—or they are steroids, a type of lipid. A few are amines. All these compounds are products of biochemical synthesis not only among craniates but also among many invertebrates and plants. Steroids, for example, are found in yeast, in many green plants, and in invertebrates; thyroxine, a product of the thyroid gland, is found also in seaweed, annelids, molluscs, and other invertebrates.

The synthesis of hormones in craniates is not limited to endocrine organs. There are endocrine cells in the liver, kidney, stomach, small intestine, and many other locations, including the brain. In this chapter, we will discuss only organs that are solely endocrine or that produce hormones as one of their major functions. We will begin with the endocrine roles of the nervous system. Thereafter, we will discuss briefly the endocrine tissues derived from ectoderm, mesoderm, and endoderm, in that order.

ENDOCRINE ROLE OF THE NERVOUS SYSTEM

The brain in all craniates, and the posterior end of the spinal cord in fishes, produces hormones that are synthesized by the cell bodies of **neurosecretory neurons.**

These **neurosecretions (neurohormones)** are small polypeptides and should not be confused with neurotransmitters, which induce other neurons to fire. Neurosecretions are found in the animal kingdom commencing with coelenterates.

Neurosecretory neurons have cell bodies in **neurosecretory nuclei** in the central nervous system (fig. 18.1). The secretions move along the axon (**neurosecretory fiber**), bound chemically to proteins (**neurophysins**), and accumulate in the axon terminals until released reflexly into blood sinusoids. The axon terminals plus the sinusoids constitute a **neurohemal organ.** The major neurohemal organs of craniates are the **posterior lobe of the pituitary gland** (fig. 18.2), the **median eminence** of the diencephalic floor immediately caudal to the optic chiasma (figs. 18.2 and 18.3), and the **urophysis** at the end of the spinal cord in elasmobranchs, basal actinepterygians, and teleosts (fig. 18.4).

Neuron cell bodies with stainable droplets were seen in the caudal region of the cord of fishes early in the twentieth century, but their significance was unknown. Later, similar cell bodies were reported in certain hypothalamic nuclei. It was theorized that these might be secretory cells, but it was not until suitable staining techniques were developed that this was confirmed. Thus, study of neurons in the spinal cord of a fish led to one of the most important discoveries in neurobiology—the endocrine role of the hypothalamus.

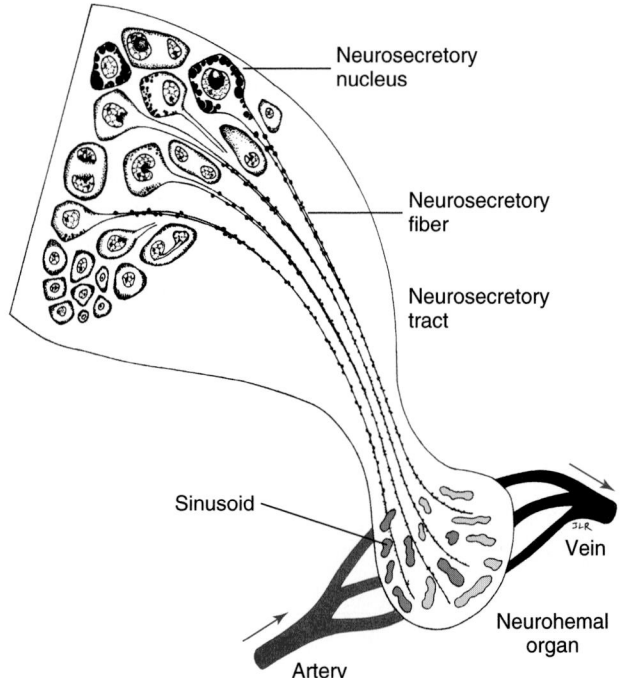

FIGURE 18.1

Neurosecretory neurons in a functional neurosecretory unit. The neurohemal organ consists of vascular sinusoids into which the neurosecretion is released.

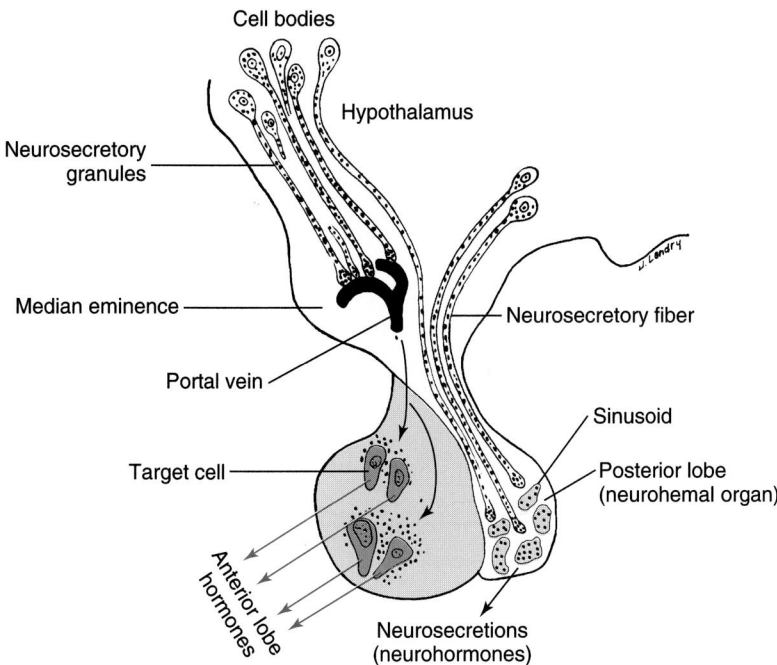

FIGURE 18.2

Hypothalamic neurosecretory neurons. Cell bodies in hypothalamic nuclei synthesize neurosecretions (*black granules*) that flow along the axons (neurosecretory fibers) and are discharged into the hypophyseal portal vein or into sinusoids in the posterior lobe of the pituitary. Neurosecretions released into the portal vein help regulate the hormone-producing cells (*dark red*) of the anterior lobe. *Arrows* show direction of blood flow. Those released in the posterior lobe affect tissue remote from the pituitary.

The majority of hypothalamic neurosecretions are released into blood sinusoids in the median eminence and transported by the **hypophyseal portal system** to the anterior lobe, where the target cells are located (see fig. 18.3). There, a neurosecretion of a specific chemical configuration (thyrotropin-releasing hormone, for example) causes the release of a specific anterior lobe hormone (thyrotropin, in this example). Other neurosecretions in the portal blood (somostatin, for example) inhibit the secretion of a specific anterior lobe hormone (growth-promoting hormone, in this example). Therefore, hypothalamic neurosecretions in the portal stream are either **releasing hormones** (releasing factors) or **inhibiting hormones** (inhibiting factors). Neural stimuli from elsewhere in the brain, delivered to a neurosecretory neuron in the hypothalamus, trigger a nerve impulse that travels down the neurosecretory fiber and releases the neurosecretion into the portal system.

Synthesis and release of these releasing and inhibiting hormones are regulated in part by the cyclic external environment and partly by feedback from the remote endocrine glands affected. The obvious environmental influences include, among other variables, the circadian cycles of light and darkness, and seasonal change in temperature, day length, rainfall, and, in an aquatic environment, salinity. These are monitored by sense organs. Among the results of this input is promotion of gametogenesis and appropriate reproductive behavior (territorial defense, mating behavior, nest building, care of eggs and young) at the precise time of year when environmental conditions are most suitable for survival of offspring (fig. 18.5). Such adaptive **neuroendocrine reflex arcs** (receptor—sensory nerve—association tracts within the brain—hypothalamus—releasing hormone—median eminence—hypophyseal portal system—

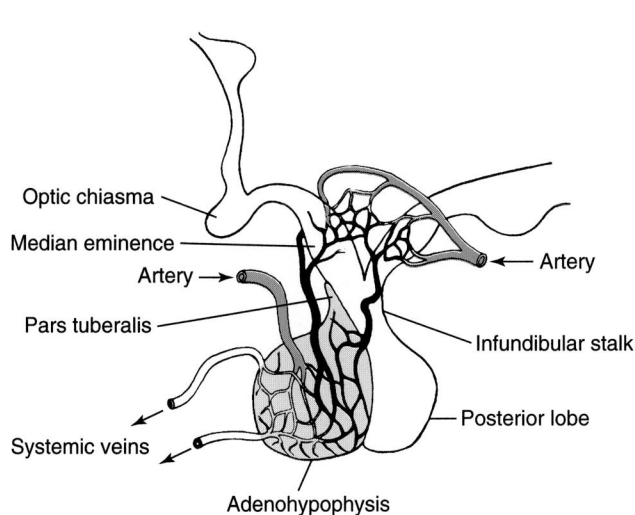

FIGURE 18.3

Schema of the hypophyseal portal system (*black*) of mammals. *Arrows* indicate direction of blood flow.

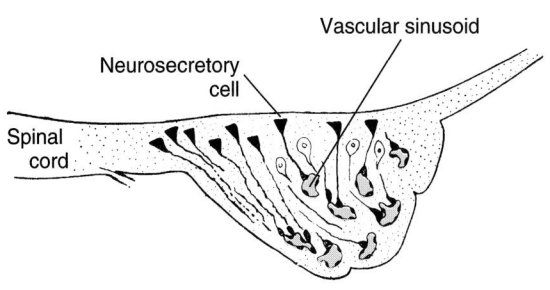

FIGURE 18.4

Urophysis (caudal neurohemal organ) of a carp. In some species, this organ is more pendulous.

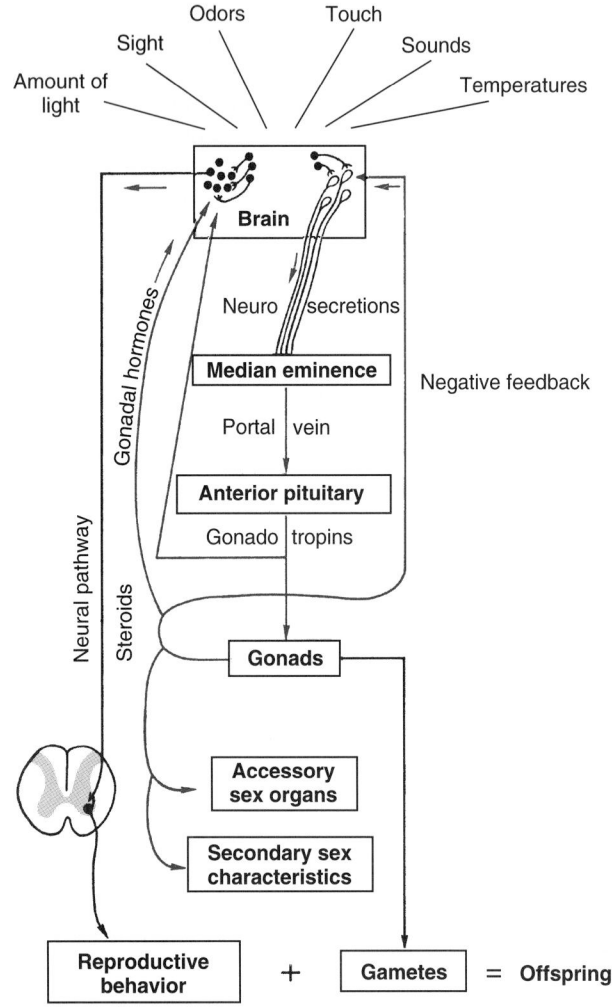

FIGURE 18.5
Regulatory effects of the environment on reproduction. Vascular transport of hormones is indicated by *red* lines.

anterior lobe hormone—effector) are the result of natural selection.

Neurosecretions from other hypothalamic nuclei are released directly into sinusoids in the posterior lobe (see fig. 18.2). These so-called posterior lobe hormones are transported in the systemic bloodstream to their target organs. The role of these neurohormones will be discussed shortly.

Fishes have neurosecretory cells in the spinal cord at the base of the tail (see fig. 18.4). Their axons terminate in the urophysis, a neurohemal organ that varies from an inconspicuous swelling of the spinal cord to a pendulous organ, depending on the species. The neurosecretions of this neurosecretory organ raise the blood pressure of teleosts; hence, they have been termed **urotensins.** The precise target for urotensin is unknown. Experimental evidence suggests that the observed effect of urotensin may be attributable to a role in osmoregulation.

ENDOCRINE ORGANS DERIVED FROM ECTODERM

Pituitary Gland

The pituitary gland, or hypophysis, is suspended beneath the diencephalon, cradled in gnathostomes in the sella turcica. The pituitary consists of two major components with very different embryonic origins, a nonglandular **neurohypophysis** that arises from the floor of the diencephalon, and a glandular **adenohypophysis** that arises from the roof of the embryonic stomodeum (fig. 18.6). The components of these lobes are as follows:

> Neurohypophysis
>   Median eminence
>   Infundibular stalk
>   Pars nervosa (posterior lobe; neural lobe)
> Adenohypophysis
>   Pars intermedia
>   Pars distalis
>   Pars tuberalis

Neurohypophysis

The neurohypophysis forms from the floor of the diencephalon (fig. 18.6, *dark gray*). It contains a recess of the third ventricle that is shallow in anamniotes, deepest in mammals (figs. 18.6 and 18.7). In mammals, the diencephalic floor is drawn out into a long **infundibular stalk** with the posterior lobe (**pars nervosa**), a neurohemal organ, at the end (see figs. 16.21 and 18.7, cat). Just behind the optic chiasma, the neurohypophysis has another neurohemal organ, the **median eminence** (see fig. 18.2).

The posterior lobe *synthesizes* no known hormones. Hypothalamic neurosecretions *released* there are carried to the heart for distribution throughout the body. The most ubiquitous, and probably the oldest, hormone released from the posterior lobe is **arginine vasotocin,** found in living agnathans and all other craniate classes, although in mammals it is secreted only during fetal life. It is the only neurohormone found in living agnathans. Its function in fishes other than lungfishes is unknown, but in many terrestrial craniates, it is an antidiuretic hormone that prevents dehydration by causing the kidney to reclaim water from glomerular filtrate and by inducing absorption of water from the urinary bladder. In estivating lungfishes and in amphibians that seek moist locations during arid conditions, it induces absorption of water from the soil through the skin.

A number of chemical mutants of arginine vasotocin have evolved during craniate phylogeny by the substitution of other amino acids at one or more positions on the peptide molecule (fig. 18.8). **Human antidiuretic hormone** (arginine vasopressin), which regulates water and salt excretion in other mammals as well, is one of

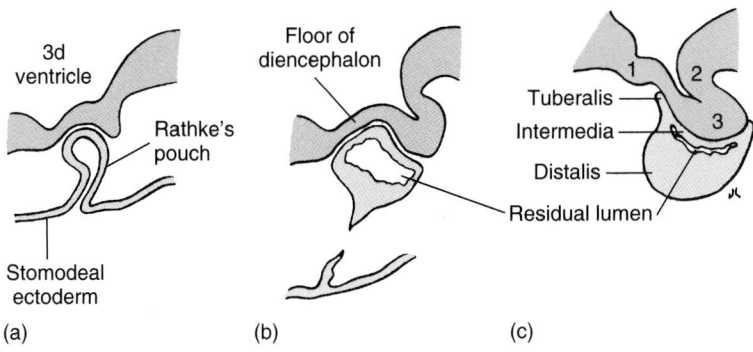

FIGURE 18.6

Embryogenesis of amniote pituitary. (*a*) Rathke's pouch stage. (*b*) Isolation of adenohypophyseal anlage (*light gray*) in contact with the floor of the diencephalon. (*c*) Young pituitary consisting of adenohypophysis (*light gray*) and neurohypophysis (*dark gray*). **1,** median eminence; **2,** infundibular stalk; **3,** pars nervosa (posterior lobe). The subdivisions of the adenohypophysis are labeled.

these variants. **Oxytocin** is another. Oxytocin causes oviducal contractions in turtles, induces strong uterine contractions during birth in humans, and causes letdown of milk into nipples by stimulating the smooth muscle fibers surrounding the milk-secreting alveoli of a lactating mammary gland.

Hypothalamic neurosecretions released into the median eminence enter hypophyseal portal vessels that transport them to the adenohypophysis. There they stimulate or inhibit the release of hormones synthesized in the anterior lobe. The portal vessels in tetrapods and some fishes are discrete portal veins (see fig. 18.3). In teleosts, many capillarylike vessels form a plexus that does not organize into a discrete portal vein, yet performs the same function by emptying into the capillary beds of the anterior lobe. This vascular arrangement is probably correlated with the unusually intimate anatomical relationship between the neurohypophysis and adenohypophysis in teleosts (see fig. 18.7, trout).

Adenohypophysis

The adenohypophysis arises as a bud of ectodermal cells from the roof of the stomodeum. In amniotes, sharks, and some basal bony fishes, the bud is hollow and is known as **Rathke's pouch** (see fig. 18.6*a*). In other fishes and in amphibians, the bud is solid. When the anlage of the adenohypophysis has made intimate contact over a

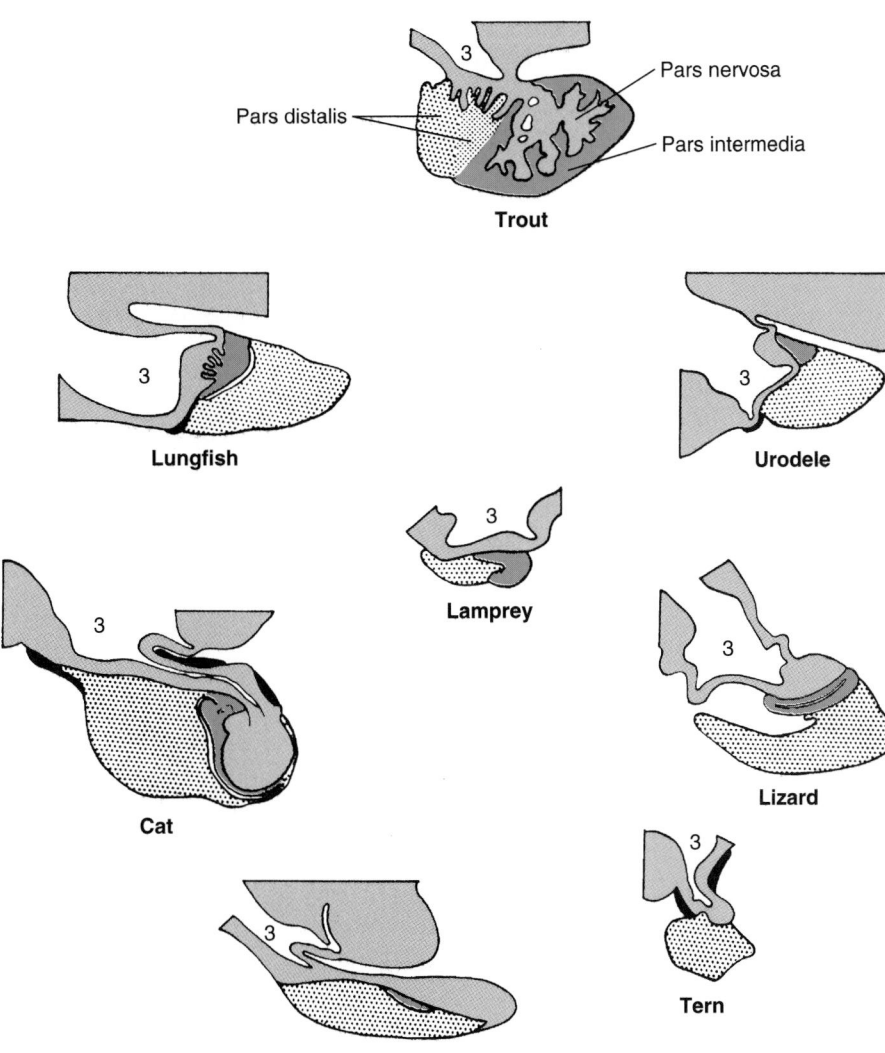

FIGURE 18.7

Pituitary glands of representative craniates, sagittal sections, anterior to the left. **3,** third ventricle. *Gray,* neurohypophysis and contiguous brain stem; *red,* pars intermedia; *stippled* areas, par distalis (anterior lobe); *black,* pars tuberalis. The pars distalis of teleosts, as exemplified by the trout, exhibits only two cytological regions, a rostral part (*coarse stipple*) and a proximal part (*fine stipple*).

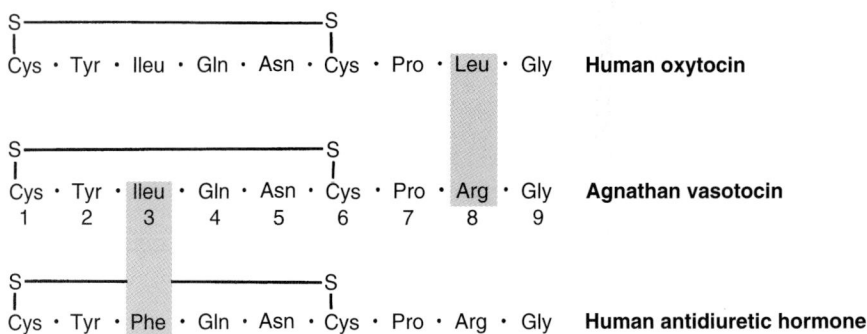

FIGURE 18.8
Two of many evolutionary variants of arginine vasotocin (agnathan vasotocin). *Gray* bars designate the only differences between agnathan and human posterior lobe neurosecretions. The human antidiuretic hormone is arginine vasopressin.

broad area with the floor of the brain, the connection between the stomodeum and the bud usually disappears (see fig. 18.6*b*). However, it remains as an open ciliated duct leading to the oropharyngeal cavity in *Calamoichthys, Polypterus,* and some basal teleosts. Rathke's pouch may remain in the adult gland as a residual lumen (see fig. 18.6*c*).

The origin of the adenohypophysis from the stomodeum suggests that primitively the adenohypophysis may have secreted into the oropharyngeal cavity. It was once thought that the pituitary (*pitua*, phlegm) was the source of phlegm that falls into the throat. Although they were wrong about the source of the phlegm, they may have been correct about the primitive site of secretion of the pituitary!

The adenohypophysis exhibits two chief regions, the **pars intermedia** and **pars distalis** (see fig. 18.7). A pair of thin, narrow extensions, the **partes tuberales** (sing., **pars tuberalis**) develops in a few fishes and in most tetrapods other than squamates. They extend upward on the surface of the infundibular stalk or, in amphibians, forward on the ventral surface of the diencephalon, from which they can be readily removed. Their role, if any, is unknown.

The Pars Intermedia   The pars intermedia is in intimate contact with the neurohypophysis (see fig. 18.7, *red*) It synthesizes and secretes **melanophore stimulating hormone (MSH)**, known also as **intermedin**. The MSH causes pigment granules in some chromatophores of ectotherms to disperse, thus causing the skin to darken. No discrete pars intermedia is present in birds, cetaceans, and some sirenians, but MSH-producing cells have been seen among the cells of the pars distalis. In mammals, MSH is essential for melanogenesis to occur in the hair follicles. It should be mentioned that an entire 13-peptide molecule of alpha-MSH (one of several species of MSH) is incorporated in a molecule of adrenocorticotropic hormone produced in the pars distalis.

The Pars Distalis   The pars distalis is commonly known as the anterior lobe because of its position in mammals. However, the term is not always appropriate (see fig. 18.7). The anterior lobe synthesizes the following hormones: somatotropin (growth hormone) STH (GH); thyrotropin (thyroid stimulating hormone) TSH; adrenocorticotropin (ACTH); gonadotropins—follicle stimulating hormone (FSH), luteinizing hormone (LH) or interstitial cell stimulating hormone (ICSH); and prolactin (luteotropin) PRL (LTH).

Somatotropin is a general metabolic hormone. It promotes growth by stimulating the synthesis of proteins throughout the body. When administered to youths before the epiphyseal plate has ossified (see fig. 7. 5), it increases the height of the recipient. **Thyrotropin** stimulates the thyroid to accumulate iodine, to synthesize thyroid hormone, and to release it into the circulation. **Adrenocorticotropin** regulates the zones of the mammalian adrenal cortex that secrete glucocorticoids. **Follicle stimulating hormone** stimulates the growth of ovarian follicles and their secretion of estrogen. In males, it promotes spermatogenesis. **Luteinizing hormone** induces ovulation of mature ova and conversion of a graafian follicle into a corpus luteum. In males, it induces the interstitial cells of the testes to synthesize androgens. It is usually referred to in males as **interstitial cell stimulating hormone.**

Prolactin has a variety of effects from fishes to human beings, at least some of which are evoked only after the responding tissue has been presynergized by corticosteroids or some other synergizing hormone (see "Hormonal Control of Biological Rhythms" at end of chapter). It is a major freshwater-adapting hormone of euryhaline fishes (fishes that adapt to a wide range of salinity), and it stimulates secretion of maternal skin mucus that nourishes hatchlings of some teleosts. It has marked lipogenic effects, inducing cyclical deposit of fat in many craniates, and it stimulates production of pigeon milk by the avian crop sac and mammalian milk by the mammary gland. The latter effect inspired the name prolactin. Acting on tissue that has been rendered sensitive by synchronizing hormones, it causes some urodeles to migrate to ponds at the approach of sexual maturity; in birds, it induces parental behavior such as nest building, turning and incubation of eggs, and protection of nestlings; and it activates the corpora lutea in rats (luteotropic effect). The many and varied roles of prolactin illustrate how new receptors have arisen for essentially the same hormone. Natural selection has thus far maintained those receptors for prolactin in extant craniates. These varied effects, when first observed, seemed unrelated. However, there is little doubt that ongoing *basic research* will discover a single, unifying principle that accounts for all these observations. Thereafter, *applied research* will design methods to utilize the knowledge for the benefit of humanity.

Only the most obvious roles of the pars distalis hormones have been mentioned above. Each hormone has additional roles—more subtle, but equally important. These may be found in appropriate textbooks of endocrinology.

Pineal Organ

As an outpocketing of the brain, the pineal organ can be considered a cranial nerve in some taxa, as is the optic nerve. The pineal organ, so named because in mammals it is shaped like a pine cone, is an evagination from the roof of the diencephalon that synthesizes **melatonin**. The latter, an amine, is synthesized from serotonin, another brain amine, but only during hours of darkness. Experimental evidence shows that light impedes the synthesis of melatonin by inhibiting the activity of the enzyme *serotonin N-acetyltransferase* that catalyzes the rate-limiting step in melatonin synthesis. The effect of light on the pineal body is direct when the pineal is located under translucent skin, as in numerous craniates. Otherwise, it is mediated by way of the optic nerves, hypothalamic nuclei, descending fiber tracts in the brain stem and cord, thoracic spinal nerves, superior cervical sympathetic ganglion (see fig. 16.35), and the conarial nerves (**nervi conarii**), which pass from the ganglion to the pineal organ.

Mammalian melatonin causes melanin granules in dermal melanophores of living agnathans, larval amphibians, and many fishes to aggregate, thereby blanching the skin. The effect is opposite that of MSH. Amphibian larvae with intact pineals become pale when grown in darkness. Pinealectomy abolishes this response.

Melatonin has a role in preparing the organs of some mammals for reproductive activity, but the effect appears not to be direct. The *duration* of melatonin secretion (rather than total secretion), probably monitored by the hypothalamus, evidently signals the duration of photoperiods, and these signals are transduced into neurosecretory stimuli that influence reproductive cycles. The details of this regulatory mechanism are still to be elucidated. Other aspects of the pineal gland have been discussed in chapters 16 and 17, and its role in certain biological rhythms is discussed later in this chapter.

Aminogenic Tissue and the Adrenal Medulla

The adrenal glands of tetrapods consist of two entirely different components: (1) an **aminogenic tissue** that produces **norepinephrine (noradrenaline)** and **epinephrine (adrenaline)**, which are catecholamines, and (2) a steroidogenic tissue that produces steroid hormones. These two tissues are spatially separated in many fishes, but they aggregate and become wrapped in a single capsule in amniotes. We will therefore discuss the adrenal complex as two separate glands, which is reinforced by the observation that the aminogenic portion is of neural crest origin. We will examine first the aminogenic component, which constitutes the adrenal medulla of mammals. Because of its staining reaction, aminogenic tissue is also called **chromaffin tissue**. Chromaffin tissue is widely distributed elsewhere in the coelom of most craniates, but not all of it is aminogenic. Nodules occur close to the sympathetic ganglia of the trunk and in the gonads, kidneys, heart, and other viscera.

Fishes and Amphibians

In lampreys and elasmobranchs, aminogenic cells lie in clusters along the length of the postcardinal vein (fig. 18.9, ray); in hagfish, chromaffin cells are associated with the arteries, veins, and (at least chromaffinlike cells) in the hagfish heart. In lungfishes, they are scattered along the dorsal aorta of the trunk. These clusters lie close to sympathetic ganglia, whose cell bodies synthesize the same amines. In teleosts, the aminogenic cells are generally at the anterior end of the kidneys within vestiges of the pronephroi, where they are interspersed with the other component of the adrenal complex (fig. 18.9, teleost). In anurans, the two components are interspersed in a diffuse gland on the ventral surface of each kidney. In urodeles, they form small, bright flecks and nodules along the postcava and are difficult to locate without a hand lens.

Amniotes

In amniotes, discrete adrenal glands are generally located at or near the cephalic pole of each kidney, and the two glandular components are usually interspersed within a capsule, as in crocodilians and birds (fig. 18.9, bird). However, in lizards the aminogenic tissue tends to form an almost complete capsule around the steroidogenic tissue (fig. 18.9). In some mammals, including humans, the condition is just the reverse, the steroidogenic tissue forming an adrenal **cortex** and the aminogenic tissue forming a **medulla**. However, there are mammalian species in which the steroidogenic tissue does not form a homogeneous cortex. In sea lions, for example, cortical tissue is scattered also in the medulla, and clumps of medullary cells are scattered throughout the cortex. The adrenal glands are also termed **suprarenal glands** in mammals because of their location in the bipedal human body.

The aminogenic cells of the adrenal and the cell bodies of postganglionic neurons of the sympathetic nervous system are of the same lineage. Both arise from neural crest, both are innervated by preganglionic fibers of the sympathetic system, and both synthesize norepinephrine and epinephrine. In fact, the aminogenic cells

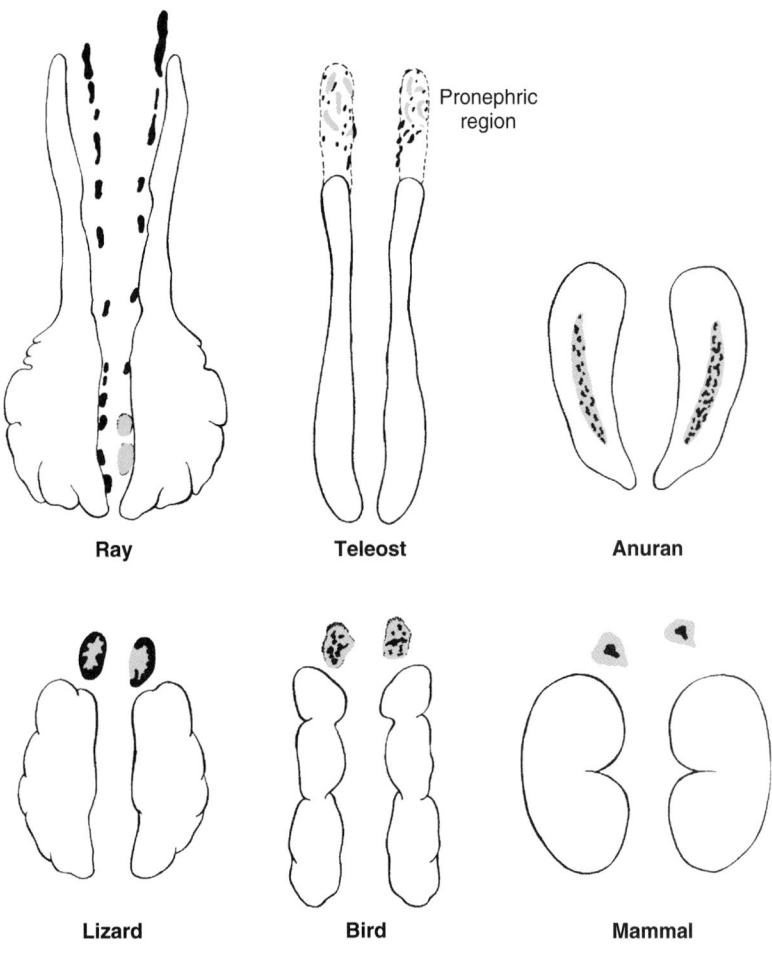

FIGURE 18.9

Adrenal components in selected craniates. Aminogenic tissue (medulla in mammals) is shown in *black,* steroidogenic tissue (cortex in mammals) in *gray.* The kidneys are shown in *outline.* Note the reversed location of the two components in lizards and mammals. The two masses of steroidogenic tissue between the caudal ends of the kidneys in the ray are interrenal bodies.

of the adrenal gland can be thought of as being postganglionic cell bodies of the sympathetic nervous system that fail to sprout processes.

Epinephrine

Norepinephrine is a precursor of epinephrine. It is converted to epinephrine by addition of a methyl group. This conversion generally takes place to a greater degree in the adrenal medulla than in postganglionic cell bodies. Consequently, epinephrine is the predominant amine of the adrenal, whereas norepinephrine is the predominant amine of postganglionic neurons of the sympathetic system. The proportions of each in each tissue are fairly constant in any species.

Among numerous roles, epinephrine released at times of exceptional stress stimulates **glycogenolysis**—the breakdown of glycogen into glucose—in the liver. The effect is the immediate release of glucose (glucose-

6-phosphate) into the circulation. (Glycogen stored in muscles is not directly convertible to glucose. It must first be converted to lactic acid and then transported to the liver, where it is reconverted into glycogen and then to glucose.) Elevated blood glucose levels satisfy the increased metabolic demand of the heart and skeletal muscles for glucose during stress. Epinephrine and norepinephrine have similar actions at some receptor sites, but their relative potency differs with the number of specific receptors at the site. Because norepinephrine is a potent vasoconstrictor, it has a significant role in the day-by-day management of the peripheral circulation. Both epinephrine and norepinephrine relax the smooth musculature of the bronchi and trachea, thereby enhancing the delivery of oxygen to the lungs. The preganglionic fibers that supply the adrenal medulla arrive via a nerve plexus supplied in part by a splanchnic nerve (see fig. 16.35). Autonomic control centers in the hypothalamus initiate these incoming neural stimuli (fig. 18.10).

ENDOCRINE ORGANS DERIVED FROM MESODERM

Steroidogenic Tissue and the Adrenal Cortex

We have just seen that the adrenal complex is composed of two entirely different components, aminogenic and steroidogenic, which may be intimately associated or spatially separated. The aminogenic component becomes the adrenal medulla in mammals, and the steroidogenic component becomes the cortex. Whereas aminogenic tissue receives motor innervation (see fig. 16.35), steroidogenic tissue is regulated by hormones.

The steroidogenic component is derived from mesodermal cells that arise from the epithelium of the genital ridge and from the underlying nephrogenic mesoderm (see fig. 15.19). In sharks and rays, the steroidogenic cells usually form one or more compact **interrenal bodies**, so named because they lie between the caudal ends of the kidneys (see fig. 18.9, ray). The steroidogenic cells of teleosts generally aggregate in the vestigial pronephric region interspersed among the aminogenic cells. All such steroidogenic masses are homologues of the adrenal cortex of mammals. Although the term *cortical tissue* is appropriate for this tissue in mammals only, it is sometimes used to designate tissues in other craniates that produce "corticoids," that is, steroids that resemble those of the

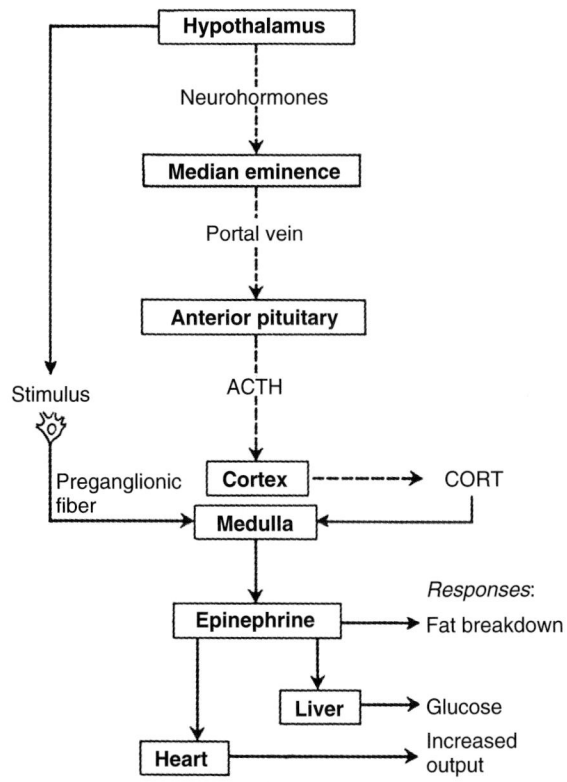

FIGURE 18.10
Some regulatory functions of the adrenal medulla. When presented with a suitable neural stimulus (far left), a preganglionic neuron of the sympathetic nervous system stimulates the medulla to release epinephrine. The latter elicits many responses, three of which are indicated at the right. **CORT**, adrenal corticoids.

mammalian cortex. There are two major categories of corticoids in mammals, **glucocorticoids** and **mineralo-corticoids**. Other species of steroids, some with no known effect, are synthesized in the mammalian adrenal cortex, but in much lower quantities.

Glucocorticoids stimulate the conversion of noncarbohydrates into glucose, a process known as **gluconeo-genesis**. They also have other roles. Among the glucocorticoids of mammals are **cortisol, corticosterone**, and **cortisone**. Cortisol is the chief corticoid produced by the human adrenal cortex, and by apes, monkeys, some other mammals, and fishes. Corticosterone is the chief one in birds, rabbits, and rats. Hamsters produce cortisol and corticosterone in approximately equal amounts. Cortisone is not a significant mammalian corticoid, but it has important effects in some basal craniates. During prolonged stress, one or more of these glucocorticoids stimulates the synthesis of epinephrine in the adrenal medulla. They also have a degree of sodium-regulating competence. The secretion of glucocorticoids is regulated by adrenocorticotropin.

Aldosterone is the most potent mineralocorticoid in tetrapods. Its role is to reclaim sodium from glomerular filtrate in the loop of Henle. The secretion of aldosterone is stimulated by **angiotensin**, a hormonelike compound formed in the bloodstream under the catalyzing effect of the enzyme **renin**, which is produced in the kidneys. (Renin should not be confused with rennin, an enzyme in the stomach of some mammals that curdles milk.)

Only adrenal cortical tissue and gonads produce steroid hormones in craniates. The mammalian adrenal cortex produces about 50 different steroids. Some are intermediate products, some are end products, and some have no known survival value. Included are small amounts of male and female sex hormones. A bearded lady is an example of what may happen when the adrenal cortex of a female produces, as end products, excessive quantities of male hormones (androgens).

Gonads as Endocrine Organs

Gonads arise from the coelomic epithelium as a pair of genital ridges medial to the kidneys (see fig. 15.19). Ovaries and testes of most craniates produce three classes of steroid hormones: **estrogens, androgens**, and **progestogens**.

Estrogens produced by ovarian follicles (see fig. 15.35) and androgens produced by the interstitial cells of the testes (see fig. 15.21) are primarily responsible for differention, growth, and maintenance of **accessory sex organs** (reproductive ducts and their glands) and for **secondary sex characteristics** (characteristics associated with one sex, such as wattles in roosters, mammary glands in female mammals). They are also responsible for characteristic reproductive behavior from fishes to mammals. Most accessory sex organs and secondary sex characteristics atrophy on ablation of the gonads, but atrophy can be prevented by administration of the appropriate androgen or estrogen. Reproductive behavior also ceases in the absence of gonadal steroids.

Androgens are intermediate stages in the biosynthesis of sex steroids, including estrogens, in all craniates from fishes to humans. **Testosterone** is the most potent androgen in mammals, and **17β-estradiol** is the most potent estrogen. As mentioned earlier, androgens are also synthesized in the adrenal cortex, and their overproduction may result in sexual abnormalities.

Progesterone, one of the progestogens, is a precursor in the biosynthesis of androgens, estrogens, and corticosteroids throughout the craniate classes (fig. 18.11). However, its role in nonmammals, other than being a step in the biosynthesis of steroids in general, has not been ascertained. In female mammals, small amounts of progesterone from mature graafian follicles are essential for ovulation and preovulatory preparation of the uterus for implantation. Larger amounts from the corpora lutea (see fig. 15.35) are essential for final preparation of the uterus for implantation. Progesterone from the corpora lutea and eventually the placenta, depending on the

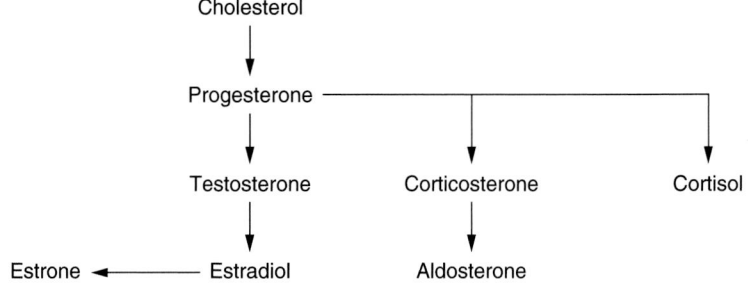

FIGURE 18.11

Synthesis of selected steroid hormones. One or more intermediate molecules occur between most of the hormones shown.

species, maintain the pregnant uterus in a healthy state until term. It is these roles that earned progesterone its name, which means "a steroid-promoting gestation." By negative feedback to the hypothalamus, progesterone inhibits the formation of a new wave of ovarian follicles after ovulation "in anticipation" of pregnancy. During pregnancy, it inhibits the maturation of new ova. Pregnancy is the anticipated outcome of mating in mammals.

The mammalian ovary also produces a peptide hormone, **relaxin.** Relaxin produced during pregnancy softens the ligaments of the pubic symphysis and sacroiliac joints of the mother before birth, thus enlarging the birth canal for delivery of the fetus.

Corpuscles of Stannius

Embedded in the posterior part of the mesonephric kidneys or attached to the mesonephric ducts of ray-finned fishes are spherical epithelioid bodies, the corpuscles of Stannius. They are easily mistaken for interrenal bodies, but their embryonic origin is different because they arise as evaginations of the pronephric duct. In most teleosts there are two, but in *Amia* there are 40 to 50. In large salmon, the corpuscles may reach 0.5 cm in diameter.

Ablation of the corpuscles has been followed by increases in the concentration of calcium in tissue fluids of basal neopterygians and teleosts. The physiological basis for this is not known.

ENDOCRINE ORGANS DERIVED FROM ENDODERM

The endoderm of the embryonic pharynx or pharyngeal pouches of all craniates gives rise to the thyroid, parathyroid, and ultimobranchial glands, and to thymus glands, which may or may not produce a hormone but which have a vital role in establishment of the immune system of the craniate body. The endoderm immediately caudal to the embryonic pharynx gives rise to endocrine components of the pancreas and to cells that produce gastrointestinal hormones. Except in a few basal craniates, the pharyngeal derivatives separate from the pharyngeal lining, sink into the surrounding mesenchyme, and become displaced short or long distances from their origins. We will examine first the endocrine organs that arise from the embryonic pharynx.

Thyroid Gland

Cells in the epithelium of the pharyngeal floor of chordates have the capacity to accumulate iodine and to bind it to molecules of tyrosine, an amino acid, thereby producing an iodinated protein, **thyroxine.** Cells in the hypobranchial groove (endostyle) of protochordates and larval lampreys perform this synthesis. Thyroxine is also found in algae and in various segmented worms, insects, and molluscs. In craniates, the synthesis is stimulated by thyroid stimulating hormone from the adenohypophysis. In peripheral tissues, thyroxine (T_4) is converted to **triiodothyronine** (T_3), the biologically active form of thyroid hormone. Thyroxine is a general metabolic hormone that regulates the rate of cellular oxidation, stimulates thermogenesis in homeotherms ("warm-blooded" animals), induces metamorphosis of larval amphibians, and facilitates the action of other hormones on their target tissues.

The mammalian thyroid synthesizes a second hormone, **calcitonin,** which protects calcium in bones and other tissues from depletion when not needed elsewhere. No hormone is required to cause circulating calcium to be stored, at least in tetrapods. Only the release of calcium from storage is a hormonal function (see Parathyroid Glands). Calcitonin is synthesized in **parafollicular cells** (**C cells,** C for calcium) that lie between the follicles, the C cells having migrated into the mammalian thyroid during embryonic life from developing ultimobranchial glands.

Thyroxine is synthesized in the epithelium of **thyroid follicles** (fig. 18.12). A follicle consists of the epithelium and a colloid-filled cavity that temporarily stores thyroxine as large molecules of **thyroglobulin.** As demanded by metabolic needs, a protease from the epithelium breaks the stored thyroglobulin into smaller molecules that are able to pass through the follicular walls and enter the circulation.

The thyroid gland arises as a median evagination from the pharyngeal floor at the level of the second pharyngeal pouches (fig. 18.13). Typically, the median evagination develops into paired glands in tetrapods

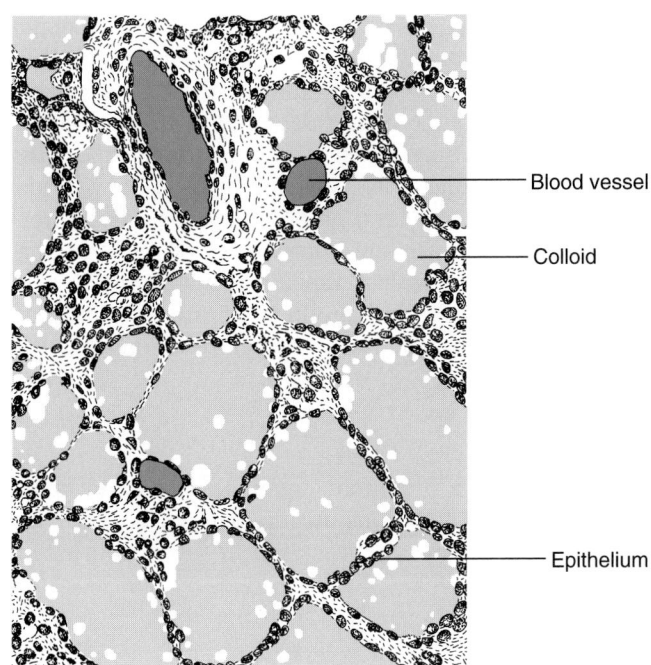

FIGURE 18.12
Actively secreting thyroid follicles. A follicle includes acellular colloid and surrounding epithelial cells.

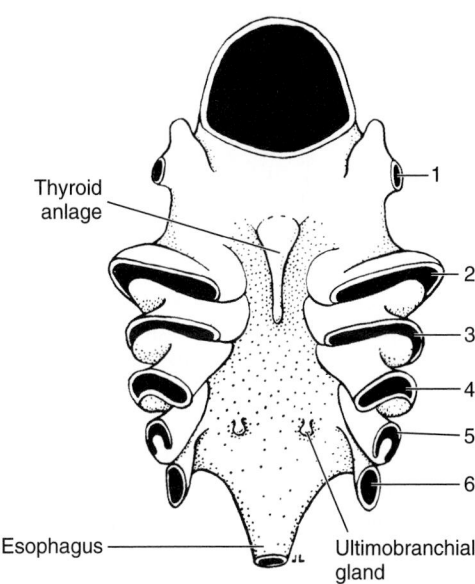

FIGURE 18.13
Pharynx of shark embryo, viewed from below. **1,** spiracle; **2** to **6,** gill slits.

(fig. 18.14, *Necturus,* bird, cat). A median thyroid is characteristic of snakes, turtles, a few birds, and the monotreme *Echidna*. Sharks have a median thyroid just caudal to the mandibular symphysis near the insertion of the coracomandibular muscles.

The larval thyroid of lampreys (endostyle or subpharyngeal gland) opens to the pharyngeal floor until metamorphosis (fig. 18.15). The iodinated protein is secreted into the pharyngeal lumen and absorbed from the digestive tract farther along. After the duct closes at metamorphosis, thyroxine accumulates in follicles and is removed by the bloodstream.

In teleosts, thyroid follicles are usually scattered singly or in small groups along the ventral aorta under the pharyngeal floor. They may also accompany some of the afferent branchial arteries into the gill arches. In a few teleosts, they follow the dorsal aorta caudad and even invade the kidneys. In still others, they form one or two compact masses between the bases of the first gill pouches.

In amphibians, the two glands lie in the floor of the pharynx. In amniotes, they migrate caudad for varying distances, taking a position close to the trachea and to the common carotid arteries from which they receive a rich arterial supply (see fig. 18.14b–d). The gland was named for its proximity to the thyroid ("shield-shaped") cartilage in mammals.

After the evaginating organ has reached its final location, thyroid follicles organize, and the embryonic stalk

connecting the thyroid with the pharyngeal floor usually disappears. However, a thyroid *duct* remains in the shark *Chlamydoselache*. In mammals, a small pit at the root of the tongue, the **foramen cecum**, marks the site where the thyroid evagination took place (foramen cecum, see fig. 13.13a). In humans, short sections of the embryonic stalk may persist as one or more thyroglossal *cysts* that occasionally require surgical removal. The cysts may occur anywhere in the neck between the foramen and the gland.

Parathyroid Glands

Parathyroid glands arise as evaginations from pharyngeal pouches and are so named because they usually lie close to or embedded within the thyroid gland (see fig. 18.14c and d). They are not present in fishes or larval or heterochronic amphibians. A few reptiles have three pairs, which arise as endodermal outgrowths from the second, third, and fourth pharyngeal pouches. However, most tetrapods have only two pairs because second-pouch anlagen usually fail to mature. When there is a single pair, as in a few urodeles, crocodilians, some domestic fowl, and some mammals, the glands develop from either the third or fourth pouch, depending on the species. Occasionally the single gland on each side has contributions from both pouches.

Parathyroid glands produce **parathyroid hormone.** Levels of serum calcium below a normal range, which varies with the species, evoke release of parathyroid hormone. Parathyroid hormone causes calcium to be released from bone and other storage sites, restoring serum calcium levels to their normal range.

Ultimobranchial Glands

Ultimobranchial glands develop from the epithelium of the last pair of pharyngeal pouches. They synthesize calcitonin. They are present as discrete glands in all gnathostomes throughout life, with the exception of adult mammals (figs. 18.13, 18.14*c,* and 18.16). In elasmobranchs, the left gland alone matures. It lies between the caudal end of the pharynx and the parietal pericardium, not far from its origin. In teleosts, the ultimobranchial glands may be median or paired, depending on the species, and they lie in the septum transversum ventral to the esophagus. In reptiles, they lie close to the thyroid.

Adult mammals have no ultimobranchial glands, with the possible exception of scaly anteaters. The fourth pharyngeal pouches of mammals are unusual in that a small extra pouch evaginates from their caudal walls. These extra pouches are the anlagen of mammalian ultimobranchial glands. Potential calcitonin cells migrate from the thickened lining of these pouches and enter the substance of the developing thyroid gland. There they become clusters of calcitonin-producing cells (parafollicular cells) lying among and between the thyroid follicles. Whether or not these caudalmost pouches in mammals are homologues of the fifth pharyngeal pouches of other craniates is not certain. No ultimobranchial glands have been found in living agnathans, and no calcitonin has been demonstrated in their blood serum.

Thymus

The thymus is a lymphoid organ. In birds and mammals, at least, it functions during fetal and juvenile life only, during which time it is vital in the establishment

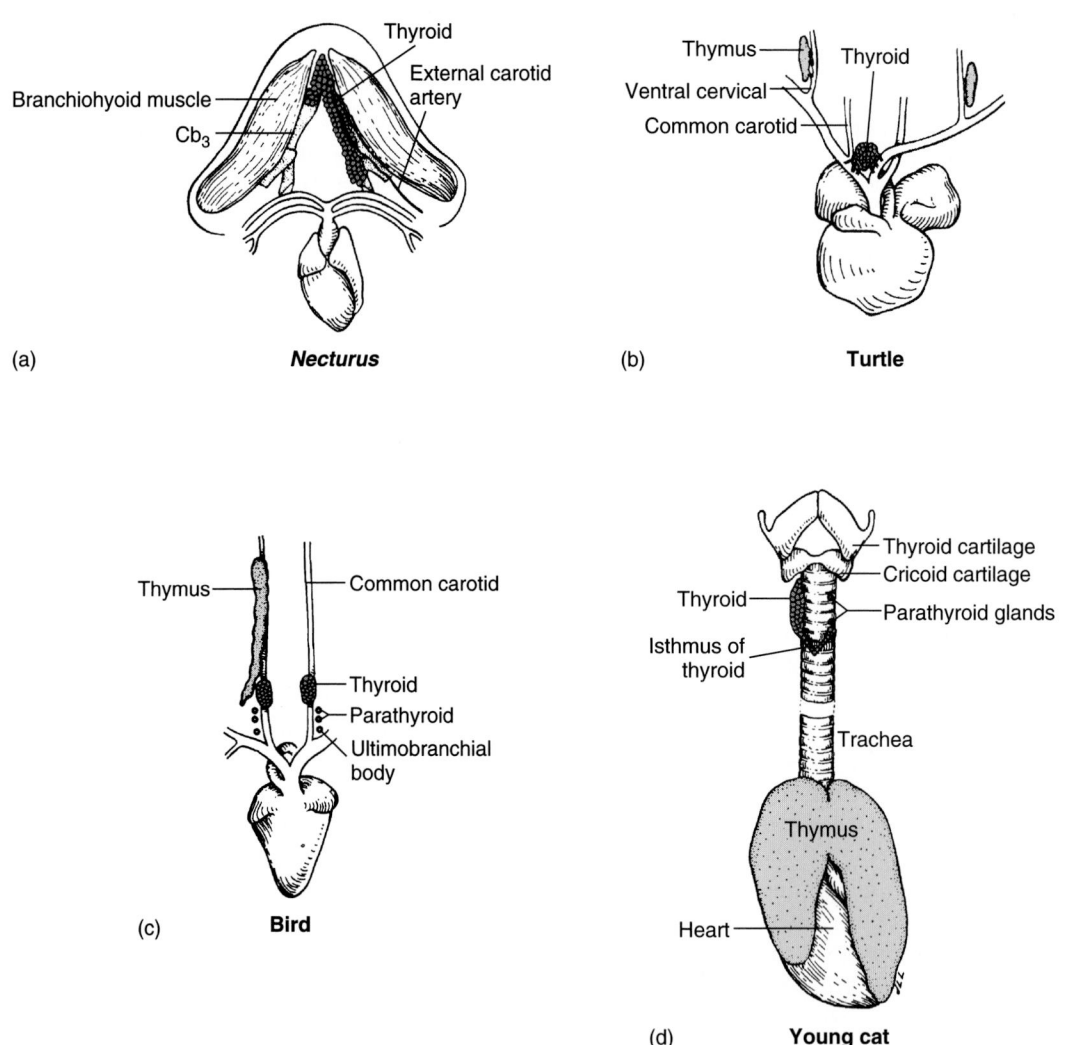

FIGURE 18.14

Thymus (*light red*), thyroid (*dark red*), parathyroids, and ultimobranchial bodies in selected craniates. **Cb₃**, ceratobranchial cartilage of third pharyngeal arch. The parathyroids of turtles are embedded in the thymus. The thymus of birds and thyroid of mammals are paired. Only one is illustrated.

of an immune system that functions for the rest of the individual's life. Without an immune system, the organism cannot survive.

Primitively, thymus tissue arose as a thickening in the epithelial lining of all the pharyngeal pouches. (A thymus gland has not been identified in hagfish, although isolated thymuslike cells may be present.) The role of those thickenings is unknown. As craniate evolution proceeded, fewer and fewer pouches were involved.

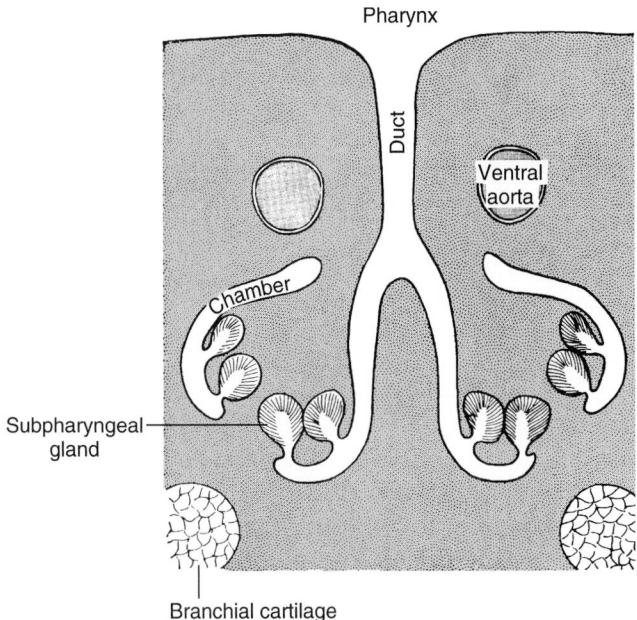

FIGURE 18.15
Cross section of endostyle (subpharyngeal gland) of a larval lamprey. All chambers empty into the duct. At metamorphosis, the duct closes, and some of the glandular cells form thyroid follicles.

Thymus tissue still develops in the walls of all seven pouches in lampreys: from all except the first pouch in most gnathostome fishes (fig. 18.16); from the first six in apodans; from the third, fourth, and fifth in urodeles; and generally from the second pouch only in anurans. In most amniotes, the third and fourth pouches are the sole source (fig. 18.16), and in many mammals, including humans, the thymus arises from the third pouch only.

There is a tendency for successive anlagen to unite during development. In bony fishes, they all more or less coalesce to form an elongated gland above the gill chambers. In the larvae of at least one elasmobranch (*Heptanchus cinereus*), ducts from the first six lobes open into the pharyngeal cavity. The ducts sometimes persist in young adults.

The epithelium of a differentiating mammalian thymus is converted into a reticulum by invading mesenchyme. Immigrant cells from embryonic bone marrow invade the interstices of the reticulum, where they proliferate. These immigrants are *stem cells* whose descendants differentiate into **T lymphocytes (T cells)** of the developing immune system. Mature T cells enter the general circulation and lodge in lymph nodes, the spleen, and other lymphoid tissues throughout the body. Proliferation continues at an intensive pace in the juvenile thymus, during which time it is large or massive (see fig. 18.14d). Once the thymus has completed "seeding" other lymphoid tissues with T cells, it undergoes fatty degeneration and can be dispensed with. Thereafter, T cells are produced in the remaining lymphoid tissues. T lymphocytes are destroyed in acquired immunodeficiency syndrome (AIDS), and immunity to disease is thereby lost. T cells are also responsible for rejection of transplanted organs (foreign proteins) such as

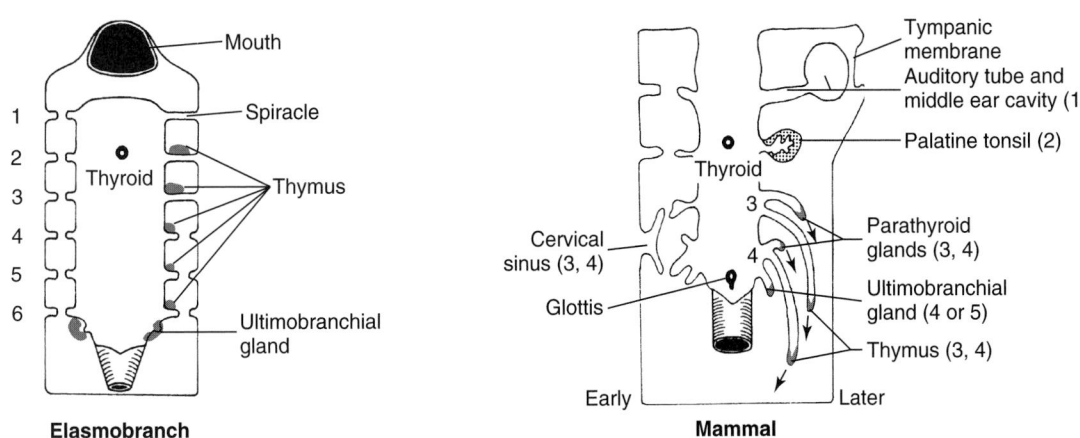

FIGURE 18.16
Pharyngeal derivatives of sharks and mammals (diagrammatic frontal sections looking down onto pharyngeal floor). Left sides are earlier in ontogeny than the right. Numbers identify ectodermal grooves or pharyngeal pouches. *Arrows* indicate caudal growth of anlagen. Whether the ultimobranchial gland of mammals is from pouch 4 or from a vestige of pouch 5 is not certain. Endocrine anlagen, except thyroid, are in *red*.

the heart or liver. Whether or not the thymus produces an endocrine factor is debatable.

Avian Bursa of Fabricius

The role of the thymus in many young birds is supplemented by a **bursa of Fabricius**, a lymphoid organ that arises as a middorsal evagination of the embryonic cloaca and extends into the pelvic cavity, where it is sandwiched between the large intestine and the synsacrum. It resembles the young thymus in structure and regresses completely at sexual maturity. Its role is similar to that of the thymus.

Endocrine Pancreas

The pancreas, like the adrenal, consists of two functionally different components: **acini**, which synthesize digestive enzymes, and **pancreatic islets (islands of Langerhans)**, which synthesize the hormones **insulin** and **glucagon** (see fig. 12.27). These two components are more or less interspersed in tetrapods. In many jawed fishes, however, the exocrine and endocrine tissues are more or less separate from each other, and in living agnathans they are remote to the extent that they have no organ that can be called a pancreas. In lampreys, the cells that secrete insulin are in the submucosa of the intestinal wall near its cephalic end. In hagfishes, they constitute a small lobe, or islet, at the base of the bile duct or encircling it. The digestive enzyme-secreting cells, on the other hand, are scattered diffusely within the submucosa over a considerable length of the intestine and also to some extent in the tissue of the liver.

In elasmobranchs, the endocrine cells lie within the epithelium of the pancreatic ductules that drain the acini. In most teleosts, they are aggregated into several compact macroscopic nodules, often two or three, but in the same mesenteries that support the diffusely disposed exocrine components. The aggregation of insulin-secreting cells in a few nodules that can be readily excised made teleosts good subjects for early insulin research. Although in tetrapods the exocrine and endocrine components are more intimately associated, the pancreas remains rather diffuse in many species, and aggregations of chiefly endocrine tissue or chiefly exocrine tissue are not uncommon.

The seemingly random location of the endocrine cells may be more comprehensible in the light of their ontogeny in craniates as a group. Both the exocrine and endocrine cells arise from the epithelial lining of the foregut. When this lining is disrupted by the formation of pancreatic and liver buds, some epithelial cells from the lining are carried to new locations (see fig. 12.26). Whether no bud forms and the cells remain in the wall of the gut (as in agnathans), whether they become part of a bud but fail to separate from the epithelial lining of the ductules (as in elasmobranchs), or whether they

lose all connections with the epithelium and become isolated islands may be correlated in part with the stage of ontogeny during which their inductor substances are synthesized. In the final analysis, of course, their adult location is an expression of inherited genetic directions.

Insulin stimulates the *uptake* of glucose by all living cells, regulates the *utilization* of glucose in most tissues, promotes *glycogenesis* (the conversion of glucose into glycogen for storage in the liver and skeletal muscles), and it performs other roles associated with carbohydrate metabolism. Also, it is the most potent of the currently known hormones that have a lipogenic (fat-synthesizing) effect.

Glucagon has a role essentially opposite that of insulin. It elevates blood glucose by promoting *glycogenolysis* (the breakdown of glycogen to glucose) in the liver, but not in the striated muscles, where glycogen is protected from the action of glucagon.

Phylogenetically, insulin and glucagon are among the oldest polypeptide hormones. Mammalian and fish insulin are so similar that the former causes a decrease in blood sugar levels when administered to agnathans and jawed fishes. Sheep, beef, or pork pancreases were the major source of insulin for human diabetes mellitus until insulins identical with human insulin were synthesized in pharmaceutical laboratories. Because insulin is destroyed by stomach acids, it cannot be administered orally.

Synergistic Regulation of Blood Glucose

Insulin and glucagon synergistically maintain a stable day-by-day level of blood glucose, causing it to be stored when in excess and to be released when needed to meet normal tissue requirements. In times of stress, epinephrine invokes a rapid release of glucose from the liver (see fig. 18.10). Under conditons of sustained stress, glucocorticoids stimulate gluconeogenesis.

Gastrointestinal Hormones

Several gastrointestinal hormones that affect the activity of the stomach, pancreas, and gallbladder are synthesized in the epithelium of the stomach and small intestine. These include gastrin, secretin, and pancreozymin-cholecystokinin (a single hormone).

Gastrin is synthesized in the pyloric epithelium and released in response to the presence of food in the stomach. It evokes secretion of hydrochloric acid and pepsinogen by the parietal cells and chief cells of the stomach lining (chapter 12). **Secretin** is released by the duodenal epithelium when **chyme**, the macerated, acidified contents of the stomach, enters the duodenum. Secretin stimulates release of water, salts, and bicarbonates into the intestinal lumen, thereby altering the pH of the chyme for optimal functioning of digestive enzymes. **Pancreozymin-cholecystokinin** is secreted by the intestinal epithelium in response to increasing titers of lipids and peptides (partially digested proteins) in the lumen. Carried by the bloodstream, this hormone stimulates the release, and perhaps the production, of digestive enzymes by the pancreas, and it stimulates contraction of the gallbladder, producing a flow of bile.

The autonomic nervous system participates in the release of all these digestive secretions. Most of our knowledge of gastrointestinal hormones comes from mammalian studies. The hormones of other craniates have received much less attention. Secretin, however, has been found in all craniate classes.

HORMONAL CONTROL OF BIOLOGICAL RHYTHMS

All organisms exhibit rhythms of metabolism and behavior. These vary from annual cycles in which gonads may regress and recrudesce and reproductive behavior waxes and wanes, to approximately 24-hour cycles (**circadian rhythms**) in which metabolism exhibits phases of elevated and depressed activity and behavior is rhythmically altered. The daily photoperiod (period of light) is the principal *environmental* operant of circadian rhythms, but other environmental factors such as temperature and more subtle stimuli may also be involved (see fig. 18.5). Environmental photoperiod rhythms are detected by photoreceptors—pineal or parapineal and the retinas of paired eyes. The information is transmitted to hypothalamic nuclei that serve as pacemakers by rhythmically elaborating and releasing neurosecretions that regulate synthesis and release of anterior pituitary hormones (see fig. 18.5). Hormones that regulate circadian rhythms at the tissue level are of two kinds, synchronizers and inducers.

Synchronizing hormones set the *phases* of circadian rhythms. Through their synchronizing effects on responding tissues, they determine the onset, duration, and termination of daily cycles. These effects render specific systems "ready"; that is, the tissues develop a *competence to respond* to the inducer hormones. Adrenal corticoids are among the most important synchronizing hormones. The corticosteroid rhythm is under direct control of the daily light-darkness cycle in craniates exposed to one. A phase of the corticosteroid rhythm determines, for example, the time of day or night that the liver is able to respond to the fat-synthesizing effects of prolactin and insulin. This has been demonstrated in fishes, several species of birds, and mammals. Fat in Syrian hamsters is synthesized by the liver at the onset of light but not throughout the entire day. Seasonal changes in the daily photoperiod also result in seasonal differences in fat synthesis.

Melatonin is also a synchronizing hormone under direct control of the daily photoperiod. The blanching of skin in certain anamniotes discussed previously occurs in the daily dark period as an expression of high melatonin levels during those hours. Hence, melatonin sets the phase for a daily rhythm of skin melanocyte response. In house sparrows, pinealectomy abolishes both the daily body temperature rhythm and a "free-running" locomotor activity rhythm, which occurs in constant lighting conditions. Reproductive cycles in craniates are commonly timed by the photoperiod. In species with a single annual reproductive season, higher melatonin levels occur during the short days and long nights of winter. Melatonin exerts an antireproductive effect by way of inhibiting gonadotropic hormonal secretion and thus contributes to synchronizing the overall reproductive cycle.

Inducing hormones (not to be confused with inductor substances) act on tissues presensitized by synchronizing hormones. Inducing hormones too, are synthesized, stored, and ultimately released rhythmically. For example, pituitary prolactin levels in male and female golden hamsters in Baton Rouge, Louisiana, are higher at 8:00 P.M. than at 8:00 A.M. in January and August. Their release into the circulatory system induces those metabolic processes that have been rendered inducible by the earlier release of synchronizing hormone. *The time interval between release of corticosteroids and the subsequent release of prolactin is critical* and is not necessarily the same for each function of prolactin. Prolactin and the adrenal corticoids "get out of sync" seasonally for certain functions. The result is an annual rhythm superimposed on the circadian rhythm. The phenomenon is adaptive. Reproductive hormones and insulin also act in an inducing capacity. Some of the effects of seasonal environmental changes on reproductive structure, physiology, and behavior were noted earlier in the chapter.

The temporal interaction of synchronizing and inducer hormones determines specific physiological conditions such as growth rates, fat storage, and rhythms of reproductive system and migratory readiness. The greatest net effect occurs when the daily peak of the inducer hormone coincides with the daily peak of greatest tissue responsiveness. Any other temporal relationship between synchronizer and inducer produces a gradation of lesser effects.

The effectiveness of synchronizing and inducing hormones is dependent on general metabolic hormones, among which are thyroid hormone and insulin. A tissue that has an inadequate cellular respiratory rate because of lowered thyroid hormone or that is incapacitated by an insulin deficiency cannot respond fully to the stimuli of inducing hormones. Deficiencies of general metabolic hormones affect not only peripheral target tissues but also the neurosecretory cells of the hypothalamus. General metabolic hormones therefore exert a modifying effect on circadian and annual rhythms.

Biological rhythms can be restored or impeded in humans depending on the schedule of administration of daily medication that affects metabolism. This insight offers an opportunity for new approaches to the medical treatment of metabolic diseases, including diabetes mellitus and cancer.

Summary

1. Hormones are products of specific groups of cells that alter the metabolism of nearby or remote cells of a different nature. They are mostly polypeptides, proteins, glycoproteins, or steroids. A few are amines.

2. Neurosecretions are small polypeptides produced by neurosecretory neurons and released into circulatory channels of neurohemal organs.

3. Known neurohemal organs of craniates are the median eminence, posterior lobe of pituitary, and urophysis of elasmobranchs, basal actinopterygians, and teleosts.

4. The urophysis at the caudal end of the spinal cord of fishes secretes urotensins, which have vasopressor and osmoregulatory effects.

5. The hypothalamus is the major source of craniate neurosecretions. It is regulated by hormonal feedback and, in part, by the external environment. Its secretions are released into the median eminence and posterior lobe of the pituitary.

6. The pituitary consists of a neurohypophysis derived from the floor of the third ventricle and an adenohypophysis derived from the roof of the stomodeum.

7. The neurohypophysis consists of a median eminence, an infundibular stalk, and a posterior lobe (pars nervosa). Hypothalamic neurosecretions released from the posterior lobe prevent dehydration and stimulate smooth muscles in restricted locations. Those released from the median eminence reach the adenohypophysis via hypophyseal portal vessels.

8. The adenohypophysis consists of pars distalis, pars intermedia, and pars tuberalis. The pars distalis secretes STH, TSH, ACTH, FSH, LH (ICSH), and LTH. The pars intermedia produces MSH.

9. The pineal synthesizes melatonin during darkness. It causes dermal melanophore cells of fishes and larval amphibians to aggregate melanin pigment and signals the photoperiod to the hypothalamus.

10. Aminogenic tissue synthesizes amines, including norepinephrine and epinephrine. In fishes, it tends to be isolated from steroidogenic tissue. It is the medulla of most mammalian adrenals.

11. Steroidogenic tissue arises from the epithelium of the genital ridges and produces adrenal and gonadal steroids.

12. Adrenal steroidogenic tissue includes the interrenal bodies of fishes and the adrenal cortex of mammals. Aldosterone regulates sodium excretion, and glucocorticoids elevate blood sugar levels by stimulating gluconeogenesis.

13. The gonads of both sexes synthesize steroids that include androgens (which predominate in males) and estrogens and progesterone (which predominate in females).

14. The ovary of mammals also produces relaxin, a nonsteroid hormone that induces dilation of the uterine cervix in preparation for delivery of a fetus.

15. Corpuscles of Stannius, derivatives of pronephric ducts, may synthesize a calcium-regulating hormone in ray-finned fishes.

16. Thyroid tissue arises from a median evagination of the pharyngeal floor. The adult gland consists of fluid-filled thyroid follicles. These are generally scattered along the ventral aorta in fishes and aggregated beneath the pharynx or in the neck in tetrapods.

17. Thyroxine and triiodothyronine are thyroid hormones that increase metabolic rates. Parafollicular cells in mammalian thyroids produce calcitonin, which impedes withdrawal of calcium from storage sites.

18. Parathyroid glands, found only in tetrapods, are derivatives of several pharyngeal pouches. Parathyroid hormone elevates serum calcium by withdrawal of calcium from storage sites.

19. Ultimobranchial glands develop from the last pharyngeal pouches and produce calcitonin. Adult glands are lacking in living agnathans and mammals.

20. The thymus is a lymphoid organ that arises from one or more pharyngeal pouches. It produces T lymphocyte stem cells whose descendants are carried by the bloodstream to lymph organs where they participate in the immune response.

21. The bursa of Fabricius is a cloacal derivative of birds that supplements the role of the avian thymus.

22. The endocrine pancreas produces insulin and glucagon. Insulin stimulates glucose uptake, glucose utilization, and glycogenesis. Glucagon elevates blood sugar by inducing glycogenolysis. The exocrine and endocrine components of the pancreas are less intimately associated in fishes than in tetrapods.

23. Gastrointestinal hormones are produced in the epithelial lining of the stomach and duodenum. They stimulate release of digestive enzymes in the stomach, small intestine, and pancreas, and the flow of bile from the gallbladder.

24. Circadian and annual rhythms of reproduction, metabolism, and behavior are entrained chiefly by environmental photoperiods that initiate neuroendocrine and subsequent hormonal responses.

25. Predictable daily rhythms of glucocorticoids and melatonin are hormonal expressions of environmental time-givers. These hormones synchronize rhythmic sensitivities of many tissues to neural stimuli or other hormonal factors.

Critical Thinking Questions

1. Comparative anatomists pay particular attention to the size, shape, and relative position of structures; however, the endocrine glands appear to be highly variable and do not lend themselves to any gross structural pattern. Can you explain why many of these glands, which have similar functions, can be so different structurally?
2. The target of individual endocrine glands can be either other endocrine glands or nonglandular tissues of the body. Give an example of each pathway.
3. Compare the "to whom it may concern" control of endocrines with the "direct" control in the nervous system. How can specificity be maintained in the endocrine system?
4. What is the relationship between the primitive craniate pharynx and the various glandular structures of the human neck?
5. What is the dual origin of the pituitary?
6. How do endocrine glands, which respond to environmental conditions, receive information? Give an example.
7. Given that endocrines use a "lock and key" mechanism for action, what would be the impact of introducing a hormone-mimic in industrial pollution?

Selected Readings

Binkley, S.: The pineal: Endocrine and nonendocrine function. Englewood Cliffs, NJ, 1988, Prentice-Hall.

Cincotta, A. H., Wilson, J. M., de Souza, C. J., and Meier, A. H.: Properly timed injections of cortisol and prolactin produce long-term reductions in obesity, hyperinsulinemia, and insulin resistance in the Syrian hamster (*Mesocricetus auratus*), *Journal of Endocrinology* 120:385, 1989.

Gorbman, A., and others: Comparative endocrinology, ed. 2. New York, 1987, John Wiley and Sons.

Hadley, M. E.: Endocrinology, ed. 4. Englewood Cliffs, NJ, 1995, Prentice-Hall.

Marcus, N. H., and others: Photoperiodism in the marine environment, *American Zoologist* 26:386, 1986.

Thorndyke, M. C., and Falkmer, S.: The endocrine system of hagfishes. In Jørgensen, J. M., Lomholt, J. P., Weber, R. E., and Malte, H., editors: The biology of hagfishes. London, 1998, Chapman & Hall.

Links to the Internet

Visit the zoology website at http://www.mhhe.com/zoology to find live Internet links for each of the following references.

1. ECME: Environmental Estrogens. Naturally occurring and synthetic estrogens have an effect on humans and other animals. Look at the EE sources link.
2. Histology—The Web Laboratory. Links to units on many systems of the human body, including the endocrine system. Many interesting histological photomicrographs, from Ohio State University.
3. Endocrines and Reproduction. This site includes valuable information on the endocrine system and hormones vital in reproduction.
4. The Endocrine System. A description of the action of many hormones, from Rutgers.

Synoptic Classification of Chordates

This synopsis has been provided so that readers may readily ascertain the taxonomic assignment of animals they read about in the text. Traditional evolutionary taxonomy is employed for the most part, and as such does not always provide a clear postulate of phylogenetic relationships. In the text, we have tried to consistently limit our discussions to monophyletic groupings. Differences will be noted between the text and the following taxonomy, but it should help to provide a link to other sources using traditional phylogenies. The following taxonomy is based primarily on the Zoological Record.[1] Even among traditional phylogenies, there is variability, which can be seen by comparing Carroll[2] with the Zoological Record. Students interested in alternative and more complete classifications should consult comprehensive works in systematics (refer to "Suggested Readings" for chapter 4). Most generally recognized extant classes are included. *Subtaxa and extinct taxa have been included only if referred to in the text.* Extinct taxa are designated by asterisks.

PHYLUM CHORDATA

    Subphylum Urochordata (Tunicata)
        Class Ascidiacea
        Class Larvacea (Appendicularia)
        Class Thaliacea
    Subphylum Cephalochordata. *Branchiostoma, Asymmetron,* sole genera
    Subphylum Vertebrata (Craniata)
        Superclass Agnatha. Jawless fishes
            Class Myxini. Hagfishes
            *Class Pteraspidomorphi (Diplorhina)
                Order Pteraspidiformes (Heterostraci)
                Order Thelodontiformes (Coelolepida)
                *Order Galeaspidiformes
            Class Cephalaspidomorphi (Monorhina)
                Order Petromyzontiformes. Lampreys
                *Order Cephalaspidiformes (Osteostraci)
                *Order Anaspidiformes
        Superclass Gnathostomata. Jawed vertebrates

---

[1]Refer to Zoological Record in "Links to the Internet," chapter 4.
[2]Carroll, R. L.: Vertebrate Paleontology and Evolution, New York, 1988, W. H. Freeman and Company.

            *Class Acanthodii. Spiny Paleozoic fishes
            *Class Placodermi. Armored Paleozoic fishes. Arthrodires, antiarchs
            Class Chondrichthyes. Cartilaginous fishes
                Subclass Elasmobranchii. Naked gill slits
                    *Order Cladoselachiformes. *Cladoselache*
                    Order Squaliformes. Squaluslike sharks
                    Order Rajiformes (Batoidea). Rays, skates, sawfishes
                    Order Carcharhiniformes. Requiem sharks
                    Order Heterodontiformes. Port Jackson shark
                    Order Hexanchiformes. Sixgill sharks
                    Order Lamniformes. Great white and basking sharks
                    Order Orectolobiformes. Nurse sharks
                    Order Pristiophoriformes. Sawshark
                    Order Squantiniformes. Angle sharks
                Subclass Holocephali. Gill slits covered by operculum. *Chimaera*
            Class Sarcopterygii (Choanichthyes). Lobe-finned fishes
                Superorder Crossopterygii. Coelacanths, rhipidistians. *Latimeria*
                Superorder Dipnoi. Lungfishes. *Lepidosiren, Neoceratodus, Protopterus*
            Class Actinopterygii. Ray-finned fishes
                Subclass Chondrostei. Chiefly Paleozoic and Mesozoic
                    *Order Palaeonisciformes. Paleozoic to Mesozoic ganoids
                    Order Polypteriformes (Cladistia). *Polypterus* and *Calamoichthys.* Sole genera
                    Order Acipenseriformes. Sturgeons and paddlefishes. *Acipenser, Polyodon, Psephurus*
                Subclass Neopterygii
                    Order Lepisosteiformes (Ginglymodi). Gars (*Lepisosteus*)
                    Order Amiiformes. *Amia*
                Division Teleostei. Modern ray-finned fishes
                  Order Clupeiformes. Herring, sardines, others
                  Order Cypriniformes. Minnows, carp, goldfish, others

Order Anguilliformes. Eels
Order Gadiformes. Codfishes, others
Order Perciformes. Perchlike teleosts.
And more than 30 additional extant orders
Class Amphibia. Anamniote tetrapods
   *Subclass Labyrinthodontia. Earliest tetrapods
      Order Ichthyostegalia. *Ichthyostega*
      Order Temnospondyli. *Archegosaurus*
      Order Anthracosauria. *Seymouria*
   Subclass Lissamphibia. Modern amphibians
      *Order Proanura
      Order Anura (Salientia). Frogs, toads, tree toads
      Order Urodela (Caudata). Tailed amphibians
      Order Apoda (Gymnophiona). Caecilians
Class Reptilia
   Subclass Anapsida. No temporal fossae
      *Order Captorhinida. Stem reptiles
      Order Chelonia (Testudinata). Turtles
      Following reptiles with two lateral temporal fossae
   Subclass Lepidosauria
      Order Rhynchocephalia (Sphenodonta). *Sphenodon*
      Order Squamata. Lizards, snakes, amphisbaenians
   *Subclass Euryapsida. Loss of one temporal fossa
      Order Sauropterygia. Plesiosaurs
      Order Ichthyosauria. Aquatic fishlike reptiles
   Subclass Archosauria
      *Order Thecodontia. Stem archosaurs
      *Order Pterosauria. Winged reptiles. Pterodactyls
      *Order Saurischia. Dinosaurs with reptilelike pelvis
      *Order Ornithischia. Dinosaurs with birdlike pelvis
      Order Crocodilia. Crocodiles, alligators, caimans, gavials
   Subclass Synapsida. Independently evolved one temporal fossa
      *Order Pelycosauria. Early synapsids
      *Order Therapsida. Precursors to mammals
Class Aves. Feathered vertebrates
   *Subclass Archaeornithes. Earliest known birds. *Protoavis, Archaeopteryx,* and the newly discovered Chinese fossil birds
   Subclass Neornithes. All other birds
   *Superorder Odontognathae. Toothed Cretaceous marine birds. *Hesperornis* and *Ichthyornis*
   Superorder Palaeognathae. Ratites. Ostriches, emus, rheas, cassowaries
   Superorder Neognathae. Carinates

Order Columbiformes. Pigeons and other doves
Order Pelecaniformes. Pelicans, cormorants
Order Anseriformes. Ducks, geese, other waterfowl
Order Galliformes. Quail, peacocks, domestic fowl
Order Falconiformes. Hawks, eagles, vultures
Order Psittaciformes. Parrots, paroquets
Order Passeriformes. Perching birds, up to 64 families, including song birds
And about 20 other modern orders
Class Mammalia
   Subclass Prototheria. Oviparous mammals
      Order Monotremata. Platypuses and *Echidna*
   Subclass Theria
   Infraclass Metatheria. Yolk sac placentals
      Order Marsupialia. Opossum, kangaroo, others
   Infraclass Eutheria. True (chorioallantoic) placentals
      Order Insectivora. Shrews, moles, hedgehogs
      Order Xenarthra. Armadillos, sloths, South American anteaters
      Order Tubulidentata. Aardvark, an African anteater
      Order Pholidota. Pangolins (scaly anteaters)
      Order Chiroptera. Bats
      Order Primates
         Suborder Prosimii. Lemurs, lorises, tarsiers, tree shrews
         Suborder Anthropoidea
            Infraorder Platyrrhini. Nostrils open to the side. New World monkeys and marmosets.
            Infraorder Catarrhini. Nostrils open downward
               Superfamily Cercopithecoidea. Old World monkeys
               Superfamily Hominoidea
                  Family Hylobatidae. Gibbons
                  Family Pongidae. Apes: orangutans, chimps, gorillas
                  Family Hominidae. Extinct and modern man
                     Ardipithicus
                     Australopithecus. Several species
                     Homo erectus
                     Homo neanderthalensis
                     Homo sapiens

Order Lagomorpha. Pika, rabbits, hares

Order Rodentia

    Suborder Sciurognathi

        Infraorder Sciuromorpha. Squirrels, marmots (woodchucks), prairie dogs

        Infraorder Castorimorpha. Beavers

        Infraorder Myomorpha. Pocket gophers, mouselike rodents, rats

    Suborder Hystricognathi

        Infraorder Caviomorpha. Porcupines, cavies, nutria

Order Carnivora. Terrestrial carnivores. Canines, hyenas, bears, others

Order Pinnipedia. Aquatic carnivores. Seals, sea lions, walruses

    Family Phocidae. Earless (wriggling) seals

    Family Otariidae. Eared (fur) seals and sea lions

    Family Odobenidae. Walruses

Order Perissodactyla. Ungulates with mesaxonic foot. Usually odd-toed. Horses, tapirs, rhinos

Order Artiodactyla. Ungulates with paraxonic foot. Usually even-toed

    Suborder Suiformes. Pigs, hippos, peccaries. Relatively primitive

    Suborder Tylopoda. Camels, llamas

    Suborder Ruminantia. Cud chewers with complex stomachs

        Family Cervidae. Deer, caribou, reindeer

        Family Giraffidae. Giraffes

        Family Antilocapridae. Pronghorn antelope

        Family Bovidae. Oxen, sheep, goats, true antelopes, others

        Family Tragulidae. Chevrotains (mouse deer)

Order Hyracoidea. Hyraxes

Order Proboscidea. Elephants, mastodons

Order Sirenia. Manatees and dugongs (sea cows)

Order Cetacea. Whales, dolphins, porpoises

GLOSSARY

Following are definitions for boldface terms and some classical components of terms used in the text, with examples. Familiarity with the entries should enable a reader to deduce, within useful limits, the meanings of many additional words, such as *hemangioepithelioblastoma,* which otherwise may be a meaningless jumble of letters. The list is no substitute for a general unabridged or standard medical dictionary, but it is sufficiently long and varied to motivate a reader toward habitual use of those standard works. Such a practice will extend the reader's intellectual horizons far beyond the boundaries of comparative anatomy. Pragmatically, it will result in better recognition and recall and also in more accurate spelling of technical terms.

Meanings are those relevant to the subject matter of the text. Stems (e.g., acanth-) are usually the smallest combinations of letters that are common to the derivatives. The symbol > means *hence* and separates a classical from a derived meaning. Abbreviations and their meaning are as follows: adj.—adjective, n.—nerve, nn.—nerves, pl.—plural, *q.v.*—indicates a term is defined elsewhere in the glossary, and var.—variation of.

Muscle definitions include three parts: (1) muscle group assignment, (2) attachments, and (3) innervation. The dog is used as a model with data for each muscle based on H. E. Evans, *Miller's Anatomy of the Dog* (3rd edition), W. B. Saunders Company, Philadelphia, 1993. The level of detail used here is such that most of the dog muscles can be extended to other typical lab animals (e.g., the cat or rabbit).

A

a- lacking, without; as in *acelous, Agnatha.*

aardvark African mammal of the Order Tubulidentata.

ab- from, away from.

abdominal (peritoneal) cavity joins right and left coelomic cavities containing the viscera.

abducens n. VI; nerve that abducts (*q.v.*) the eyeball.

abduct to move a part away from the longitudinal axis, as in raising the arm laterally.

abomasum glandular stomach of ruminant mammals.

acanth- spine, spiny.

acanthodian member of extinct sister group to Osteichthyes within Teleostomi.

accessory pancreatic duct smaller of two pancreatic ducts seen in cats and humans.

accessory sex organ reproductive duct or gland.

accessory urinary duct supplements or replaces archinephric duct in the transport of urine in some anamniotes.

acellular bone (aspidin) bone lacking lacunae for the osteocytes, deposited appositionally.

acelous *a + coel* (*q.v.*); vertebrae lacking concavities at ends.

acetabulum cup for holding vinegar; the socket in the innominate bone.

acetylcholine neurotransmitter seen in autonomic preganglionic fibers and postganglionic parasympathetic fibers.

acinar resembling a cluster of grapes. See *acinus.*

acinus (pl., acini) a grape; a glandular alveolus (*q.v.*).

acousticolateralis (octavolateralis) system membranous labyrinth and lateral-line systems.

acr- at the top, highest.

acrodont dentition tooth on the summit of the jaw.

acromiodeltoideus dorsal muscle group of the shoulder; acromion to deltoid crest; n. axillaris.

acromion process projection at proximal extremity of the shoulder (*-omo*).

acromiotrapezius (cervical trapezius) branchiomeric muscle group; middorsal raphe of neck to spine of scapula; dorsal branch of n. accessorius.

actin- ray.

Actinistia coelacanth fishes from the Middle Devonian to Recent; includes *Latimeria.*

Actinopterygii ray-finned fish (see *pter-*); sister group to Sarcopterygii within Osteichthyes.

actinotrichia distal delicate fin rays of bony and cartilaginous fishes.

ad- to, toward, upon.

adaptation evolutionary change increasing the probability of survival; change within an individual due to environmental stress.

adduct (adj., adductor) to draw toward.

aden- gland.

adenohypophysis glandular part of the pituitary that arises from the roof of the embryonic stomodeum. Contrast *neurohypophysis.*

adenoid resembling a gland; a nasal tonsil.

adhesive papillae site of attachment in metamorphosing sea squirts.

adrenal gland on the kidney.

adrenal medulla aminogenic portion of the mammalian adrenal gland.

adrenergic nerve fibers with norepinephrine or epinephrine neurotransmitters.

adrenocorticotropin a pituitary hormone affecting the mammalian adrenal cortex.

advanced modification in the direction of further adaptation; a problematic term that inappropriately implies progress.

adventitia loose connective tissue covering of the arteries and veins.

-ae nominative plural ending, as in *chordae tendineae* (tendinous chords); genitive singular ending as in *radix aortae* (root of the aorta).

aestivate to spend summer in a state of lowered metabolism.

af- same as *ad-* (to, toward); the *ad-* is changed to *af-* when combined with *-ferent.*

afferent carrying something to or toward something else. See *-ferent.*

afferent branchial artery vessel carrying blood toward the gills.

afferent branchial duct carries water from the pharynx to the branchial pouch in hagfishes.

afferent glomerular arteriole vessel supplying blood to a glomerulus in the kidney.

afferent (sensory) nerve collection of axons carrying sensory signals to the central nervous system.

afterfeather a secondary feather found at the superior umbilicus of contour feathers.

Agnatha (adj., agnathan) *a- + gnath-* (*q.v.*); a paraphyletic group of basal craniates.

air capillary passageway in avian lung adjacent to blood capillaries; site of gas diffusion.

air sac   distensible diverticulum of the avian lung.

ala-   wing. See *ali*.

alar   winglike.

alar plate   embryonic dorsal portion of the spinal cord, hindbrain, and midbrain located above the sulcus limitans.

alba   white.

aldosterone   most potent mineralocorticoid hormone; serves to reclaim sodium from urine.

alecithal   lacking yolk.

ali-   wing.

alisphenoid   wing of the sphenoid.

alisphenoid bone   mammalian skull bone homologous, in part, to the palatoquadrate.

allantoic (umbilical) artery   embryonic terminal branch of the dorsal aorta in amniotes carrying blood to the allantois.

allantoic vein   temporary embryonic vein in reptiles that drains the allantois emptying into lateral abdominal veins; in therian mammals, the vein supplying maternal oxygen via the umbilical cord to the postcava.

allantois   extraembryonic membrane; synapomorphy of Amniota.

alpha motor neuron   innervates contractile striated muscle fibers. Contrast *gamma motor neuron*.

alula   elongated first finger in some birds.

alveolar duct   most distal tubular branches of the mammalian lung, directly connected to alveoli.

alveolar ridge   gum line in mammals.

alveolus (pl., alveoli; adj., alveolar)   small chamber or sac.

amel-   enamel.

ameloblast   cell that produces enamel.

aminogenic tissue   the adrenal gland component producing norepinephrine and epinephrine. See *chromaffin tissue*.

amnion   extraembryonic membrane; synapomorphy of Amniota.

amniotic fluid   liquid within the amniotic sac or space surrounding the embryo.

amphi-   on both sides; in two ways.

amphiarthrosis   joint with severely limited movement.

amphibian papilla   sound receptor site in the sacculus unique to amphibians.

amphicelous   vertebrae having concavities at both ends.

amphioxus   animal with both ends pointed (*-oxy*); common name for *Branchiostoma*.

amphisbaenian   group of limbless burrowing reptiles within Squamata.

amphistyly   jaw suspension with palatoquadrate and hyoid support.

amplexus   precopulatory pairing by the male grasping the female.

ampulla   flask > small dilation.

ampulla of Lorenzini   neuromast organ in sharks; electroreceptor.

ampulla (papilla) of Vater   terminal segment of the bile duct in the duodenum.

ampullary gland   multicellular gland that contributes to the seminal fluid.

amygdaloid nucleus   collection of neurons (nucleus) in the cerebrum of the brain.

amylase   enzyme of saliva, pancreatic juice, and intestinal juices that catalyzes the breakdown of carbohydrates.

an-   without.

ana-   upward; again.

anadromous   *ana-* + *-dromos* (running) > able to migrate up from the sea into freshwater streams.

anal fin   medial fin posterior to anal vent.

analogy   two structures that have the same function.

anamniote   animal lacking an amnion.

anapsid   *an-* + *apsid* (*q.v.*) > lacking an arch; a skull without a temporal fenestra; the plesiomorphic condition in vertebrate skulls.

Anapsida   group of reptiles possessing anapsid skulls.

anastomose   to unite end to end.

anatomic origin   site of muscle attachment (either

proximal, stationary, or broad based).

anatomy   *ana-* + *-tome* (*q.v.*).

anconeus   brachial muscle; caudal distal humerus to lateral proximal ulna; n. radialis.

andr-   male.

androgen   class of steroid hormones including testosterone (*q.v.*).

angi-   vessel.

angiotensin   hormonelike compound that stimulates the secretion of aldosterone (*q.v.*).

angular   dermal bone of the lower jaw; homologue to tympanic part of temporal bone in mammals.

animal pole   relatively yolk-free end of developing embryo during cleavage and blastulation. See *vegetal pole*.

ankyl-   a growing together of parts.

ankylose   to fuse in an immovable articulation.

ankylosis   *ankyl-* + *-osis* (*q.v.*); a condition of fusion.

anlage (pl., anlagen)   embryonic rudiment or precursor of a developing structure.

annulospiral   proprioceptive axon ending named after nature of contact with intrafusal muscle fiber (forms a spiral ring around fiber).

annulus   ring.

annulus tympanicus   See *tympanic*.

ante-   before.

antebrachium   forearm; the part before the brachium.

anterior aqueous chamber   fluid-filled cavity between the iris and the cornea in the eye.

anterior cardinal vein   systemic vein draining the head and dorsal pharynx in fishes.

anterior (superior) mesenteric (artery)   cranial tributary of the dorsal aorta; (vein) hepatic portal system vessel.

anterior neuropore   cranial opening in the developing dorsal hollow nerve cord during neurulation.

Anthracosauria   extinct group of amphibians closely related to the Amniota.

anthrop-   refers to human beings.

Anthropoidea   primate group that includes Old World and New World monkeys.

anti-   against, opposite.

antiarch   placoderm characterized by dorsal eyes and pectoral fins covered by dermal bone.

antidiuretic   inhibiting loss of water (diuresis) via kidneys.

antidiuretic hormone   hypothalamus hormone released by the pituitary that limits water loss.

antrum (pl., antra)   cavernous space; fluid-filled space in eutherian mammal egg follicles.

Anura (Salientia)   *an-* + *uro* (*q.v.*); tailless amphibians, a group within Lissamphibia.

anus   terminal opening of the digestive tract or rectum.

aortic arch   vessel connecting the ventral and dorsal aortae; five arches present in adult basal gnathostome (three in humans).

aortic body   amniote blood monitor (oxygen and carbon dioxide), a chemoreceptor on the aorta.

aphiarthrosis   condition of a joint with limited movement (e.g., mammalian centra).

apical   at the apex.

apical pad   epidermal thickening (toris, *q.v.*) found at the ends of digits.

apical pit   site of miscellaneous cutaneous receptors on the posterior free edge of epidermal scales.

apo-   from, away from.

apocrine gland   gland whose secretion is formed in the cell and "pinched" off.

Apoda (Gymnophiona)   *a-* + *pod-* (*q.v.*) without legs; a group within Lissamphibia.

aponeurosis   *apo-* + *neuron* (a tendon); a broad, flat, tendinous sheet.

apophysis   outgrowth or process.

appendage   extension or addition to the body or an organ (e.g., fin, limb).

appendicular skeleton   pectoral and pelvic girdles and

the skeleton of associated fins or limbs.

appendix appendage; appendage of the cecum in many mammals.

appendix of the epididymis vestiges of the embryonic mesonephros (archinephric duct) found in the dorsal mesentery of the testis.

appendix testis remnant of the female muellerian duct in male mammals; forms a small cystlike structure on the testis.

-apsid refers to an arch.

aqueduct of Sylvius cerebral aqueduct (*q.v.*).

aqueous humor watery lymphlike solution filling the aqueous chambers of the eye.

arachnoid the middle of three mammalian meninges resembling a weblike structure.

arachnoid granulation tuft of pia-arachnoid meninges that projects into adjacent venous sinuses; site of diffusion for cerebrospinal fluid.

arch- first, primary, ancient.

Archaeornithes paraphyletic group of extinct birds.

archenteron primitive gut.

archetype early model.

archinephros hypothetical primitive kidney.

archipallium first roof of the telencephalon.

archipterygium hypothetical ancestral fin type.

Archosauria all descendants from the common ancestor for crocodiles and birds.

arcuate arched.

area opaca extraembryonic region of developing blastula in contact with the underlying yolk giving it an opaque appearance; site of blood islands.

arginine vasotocin primitive pituitary hormone in craniates; antidiuretic hormone in some tetrapods.

armadillo insectivorous mammal of the Order Xenarthra; unique among living mammals in having a dermal armor.

arrector pili (pl., **arrectores pilorum**) muscle that erects a hair.

arrector plumari (pl., **arrectores plumarum**) muscle that erects a feather.

arterial ligament (ligament arteriosum) closed portion of the sixth aortic arch connecting the pulmonary artery with the dorsal aorta.

arteriole intermediate-sized vessel carrying blood away from the heart.

arteriovertebral canal vascular passageway through the successive transverse foramina of the cervical vertebrae.

artery relatively large vessel carrying blood away from the heart.

arthro- joint, articulation.

arthrodire placoderm with joints, because of dermal plates, in the neck (*-dire*).

arthrosis a joint.

articular bone cartilage replacement bone formed in Meckel's cartilage; forms articulation with quadrate bone.

artio- even number.

artiodactyl having an even number of digits; subgroup of ungulate mammals.

arytenoid resembling a ladle.

arytenoid cartilage paired cartilages of the tetrapod larynx.

Ascaphus genus of frog lacking an eardrum (scapha).

astragalocalcaneus variable fusion of proximal wrist bones seen in *Sphenodon* and many lizards.

astrocyte star-shaped cell of the nervous system that provides nutrients to neighboring neurons.

ataxia *a-* (lacking) + *taxia* (order); disorder of the neuromuscular system.

-ate having the property of; as in *septate* (meaning "with septa").

atlas vertebra that supports the head (like the mythical Atlas holds up the earth).

atriopore opening to the atrium for the expulsion of water in urochordates and cephalochordates.

atrioventricular bundle collection of Purkinje fibers in the interventricular septum; carries signals from the AV node.

atrioventricular (AV) node modified muscle cells of the heart ventricle in birds and mammals that provide a stimulus to the ventricles.

atrium courtyard of a Roman home > cavity that has entrances and exits (e.g., the atrium of the heart).

auditory lobe prominence on the tectum that acts to process auditory sensory input; caudal pair of lobes within the corpora quadrigemina.

auditory tube (eustachian tube) connection of middle ear to nasopharynx.

Auerbach's plexus one of two autonomic plexuses in the external layers of the gut. See *Meissner's plexus.*

auricle (pinna) ear or earlike flap.

auto- self.

autonomic nerve peripheral nerve carrying only autonomic fibers.

autonomic plexus network of autonomic fibers usually entwined with blood vessels.

autopodium distal part of the tetrapod limb; the hand (manus) and foot (pes).

autostyly condition in which the upper jaw braces (*-styly*) itself against the skull.

autotomy cutting one's self, as when a lizard breaks off the end of its tail.

Aves subgroup of saurischian dinosaurs; birds.

axial in the longitudinal axis.

axial skeleton longitudinal component of the skeleton consisting of the skull and vertebral column.

axilla (adj., **axillary**) armpit.

axis modified second cervical vertebra of amniotes.

axon elongate process of a neuron.

azygos *a-* + *zyg-* (*q.v.*) > on one side only.

B

baculum os penis; heterotopic bone in the septum between spongy bodies of the penis.

barb (of feather) appendages from the shaft (rachis) when aligned in parallel form the vane of contour feather; base for flanges and barbules.

barbule (of feather) appendage from a barb; possesses hooklets that interlock with flanges of the adjacent feather.

baro- pressure.

baroreceptor sense organ that monitors pressure.

basal taxa with lineages that diverge at a point close to the common ancestor within a clade; positionally toward the base.

basal ganglion collection of neurons (nucleus) in the cerebrum of the brain.

basal plate (of the spinal cord) embryonic ventral portion of the spinal cord, hindbrain, and midbrain located below the sulcus limitans.

basal plate (root) bony base of a placoid scale.

basalia proximal elements of the fins in fishes.

basi- most ventral; pertaining to a basal location.

basibranchial cartilage cartilaginous (or bony) midventral element in the pharyngeal skeleton.

basidorsal cartilage one of two vertebral elements in fishes forming the vertebral canal (e.g., sturgeon).

basihyal ventral midline element of hyoid skeleton.

basilar membrane site of attachment for the mammalian sensory cells for hearing.

basilar papilla lissamphibian sound receptor found in the wall of the sacculus.

basioccipital bone ventral ossification within the occipital region of the neurocranium.

basioclavicularis (cleidooccipital) trapezius complex of the branchiomeric muscle group; dorsal part of neck to clavicle; n. XI.

basisphenoid bone ventral ossification within the sphenoid region of the neurocranium.

basiventral cartilage one of two ventral vertebral elements in fishes, (e.g., sturgeon).

beak pointed biting structure with a keratinized covering in turtles and birds.

Belon, Pierre (1517–1564) French comparative anatomist known for his comparisons of numerous vertebrates and invertebrates.

bi- double, twice.

biceps brachii ventral appendicular muscle group; dorsal edge for glenoid fossa to proximal ulna and radius; n. musculocutaneous.

bicipital having two heads, as in most tetrapod ribs.

bicornuate having two horns (cornua).

bicornuate uterus uterus with two horns.

bicuspid having two cusps.

Bidder's organ embryonic anterior testis not functional in the adult; capable under experimental manipulation to convert to ovaries.

bilateral symmetry midsagittal plane divides the organism into right and left mirror images; characteristic of Bilateria.

bile alkaline fluid containing bile salts produced by the liver; emulsifies lipids.

bili- bile.

bilirubin red pigment produced in the breakdown of red blood cells.

biliverdin green pigment produced in the breakdown of red blood cells.

bio- life.

bipartite having two parts.

biserial fin pattern where two series of radials extend from a central axis, as in *Neoceratodus*.

-blast- embryonic precursor; a germ of something.

blastema embryonic concentration of mesenchyme.

blastocoel cavity of the blastula.

blastocyst mammalian blastula.

blastoderm multilayered developing embryo seen in massively yolked eggs (e.g., chicken).

blastodisk multilayered developing embryo in therian mammals.

blastomere individual cell within a developing blastula.

blastopore opening to the archenteron in the developing gastrula.

blastula a little (*-ula*) embryo.

blood capillary smallest blood vessel connecting arterial and venous systems; site of diffusion of nutrients and wastes.

blood island site of first blood cell and vessel formation in the embryo, located in the walls of the yolk sac.

blood vascular system blood component of the circulatory system.

blowhole dorsal nasal opening in cetaceans.

body main part of an organism or organ.

body wall outer part of body or trunk surrounding the coelom.

bovine pertaining to oxen or cows.

bovine horn keratinized epidermal covering with a bony core.

bowfin common name for *Amia*.

Bowman's capsule delicate double-walled outgrowth from a kidney tubule that surrounds a glomerulus (*q.v.*).

Bowman's gland compound tubular mucous gland of the tetrapod olfactory epithelium.

brachi- (adj., **brachial**) arm.

brachialis ventral appendicular muscle group; proximal humerus to proximal ulna; n. musculocutaneous.

brachial vein proximal vein draining the arm (forelimb).

brachiocephalic associated with the arm and head.

brain stem that part of the brain excluding the cerebral and cerebellar hemispheres.

branchi- gill.

branchial plate membrane separating the developing pharyngeal pouches from adjacent ectodermal grooves.

branchial trunk branch of the vagus nerve (n. X) innervating individual pharyngeal arch.

branchiomeric referring to parts of the branchial arches.

branchiostegal membrane membrane extending from the ventral edge of the operculum in bony fishes to beneath the jaws.

branchiostegal ray parallel plates or rods within the branchiostegal membrane.

bregmatic bone independent dermal ossification in the anterior fontanel (between ossifying frontal and parietal bones) in humans.

bristle type of feather resembling filoplumes (hair-like feathers) but lacking terminal barbs.

bronchial syrinx avian voice box where the dividing wall between bronchi forms the sound-producing fold.

bronchiole distal branch of a bronchus.

bronchotracheal syrinx avian voice box with the resonating chamber formed with both tracheal and bronchial rings.

bronchus (pl., **bronchi**) distal division of the trachea.

brood pouch temporary fold or crypt in the palatal mucosa of some catfishes for carrying fertilized eggs or hatchlings.

buccal cirri fingerlike projections from the oral hood margin in amphioxus; serve to strain water and possess chemoreceptors.

bucco- cheek.

Bufonidae family of toads uniquely characterized by the presence of Bidder's glands (*q.v.*).

bulb expanded portion of a structure or organ (e.g., hair bulb, olfactory bulb).

bulbar conjunctiva epithelial covering of the eyeball that can be touched.

bulbourethral (Cowper's) gland amniote multicellular gland that contributes to the seminal fluid.

bulbus bulb.

bulbus arteriosus muscular swelling on ventral aorta.

bulbus cordis term sometimes applied to conus arteriosus in lungfishes, amphibians, and mammalian embryos; part of the heart.

bulla bubble > bubblelike part, such as the tympanic bulla.

buno- hill or mound.

bunodont (tooth) tooth with low cusps.

bursa a sac or pouch.

bursa of Fabricius cloacal evaginations of developing birds that supplement the thymus in function.

C

C cell calcitonin-producing parafollicular cell in the thyroid gland.

caecum (pl., **caeca**) blind pouch.

calamus stem or reed; stem of a feather.

calcaneo- pertaining to the heel bone.

calcified cartilage cartilage with calcium salts deposited within its substance.

calcitonin thyroid or ultimobranchial gland hormone important in calcium metabolism.

callus temporary thickening of the stratum corneum at sites of friction.

calyx funnel-shaped extension of the medulla into the renal pelvis in the mammalian kidney.

canaliculus (pl., **canaliculi**) little canal; canal connecting osteocytes in bone.

cancellous bone spongy bone consisting of trabeculae and bone marrow.

canine dog; heterodont tooth form; single cuspid tooth that lies behind the incisor.

capillary smallest of blood vessels; site of gas exchange.

capillary shunt capillary bypass between an arteriole and venule.

capitulum little head.

caput head.

cardi- heart.

cardiac pertaining to the heart.

cardiac portion region of the stomach near the heart.

cardinal of basic importance; chief.

cardinal vein main drainage of embryos and basal tetrapods.

carina keel.

carn- flesh.

carnassial teeth modified molars characteristic of carnivores.

carnivore flesh-eating animal.

carotid from a word meaning "heavy sleep"; compression of the carotid artery cuts off blood to the brain.

carotid arch aortic arch III.

carotid body amniote blood monitor (oxygen and carbon dioxide), chemoreceptor on the carotid artery.

carotid sinus amniote blood baroreceptor; dilation of the common carotid.

carpo- wrist.

carpometacarpus three fused carpals and metacarpals in a bird's hand.

carpus wrist.

cata- down, complete.

catarrhine having a nose (*-rhin*) with nares directed downward; group including Old World monkeys, apes, and humans.

cauda tail.

caudad toward the tail.

cauda equina terminal spinal nerves that resemble a horse's tail.

caudal (vertebrae) regional specialization of vertebrae in the tail.

caudal artery continuation of the dorsal aorta in the tail.

caudal vein venous drainage of the tail.

caudate nucleus collection of neurons (nucleus) in the cerebrum of the brain.

caudofemoralis (muscle) ventral pelvic appendicular muscle of urodeles and reptiles; proximal caudal vertebrae to femur; ventral rami of spinal nerves.

cavum arteriosum left ventricle in the heart of reptiles possessing a cavum venosum.

cavum pulmonale right ventricle in the heart of reptiles possessing a cavum venosum.

cavum tympanum tympanic or middle ear cavity.

cavum venosum unique third ventricular chamber in the hearts of turtles and squamates; shunt between arteries leaving the heart.

cecum (pl., ceca) See *caecum.*

cel- See *-coel-.*

celiac (artery) unpaired visceral branch of the dorsal aorta.

cellulase enzyme that digests cellulose.

cementing substance (in bone) binds hydroxyapatite crystals to a collagenous matrix.

cementum form of acellular bone that anchors a tooth to the jaw.

cen- new.

cenogenic of recent origin.

centralia middle row of carpals (0 to 4); os centrale.

central nervous system (CNS) brain and spinal cord.

central tendon centrally positioned tendon of the diaphragm.

centrum part of vertebra that occupies the position occupied early in ontogeny by the notochord.

cephal- head.

cephalaspidomorpha having a head shield.

Cephalaspidomorphi paraphyletic group of extinct agnathan vertebrates.

cephalic canal system linear series of neuromasts in a fluid-filled canal on the head.

cephalization origin and development of a head in phylogeny.

Cephalochordata sister group to Craniata that includes amphioxus and *Pikaia.*

cera wax.

cerat- horn.

ceratobranchial paired ventral element in the branchial skeleton between the hypo- and epibranchial elements.

ceratohyal paired ventral element in the hyoid arch skeleton; serially homologous to ceratobranchials; horn of the mammalian hyoid.

ceratotrich (pl., ceratotrichia) horny, hairlike fin support.

-cercal tail.

Cercopithecoidea Old World monkeys.

cerebellar peduncle fiber tract connecting cerebellum to brain stem.

cerebellum dorsal evagination of the metencephalon; site of coordination of skeletal muscles.

cerebral aqueduct (aqueduct of Sylvius) central cavity of the brain that connects the third and fourth ventricles.

cerebral peduncle ventral fiber tract of the cerebrum traversing the mesencephalon.

cerebrospinal fluid watery liquid filling the cavities of the brain, neural canal, and subarachnoid space.

cerebrum evagination of the telencephalon consisting of two cerebral hemispheres.

cerumen secretion of mammalian external ear glands. See *cera.*

ceruminous gland sebaceous gland of the outer ear canal that secretes cerumen.

cervical pertaining to the cervix (*q.v.*).

cervical trapezius trapezius complex of the branchiomeric group (extrinsic forelimb muscle); midline raphe of neck to spine of scapula; n. XI.

cervical vertebra vertebra in the neck.

cervicobrachial (plexus) spinal nerve network of the arm and neck.

cervix neck > cervix of the uterus.

chalazion cyst of the Meibomian gland of the eye due to blockage of the duct.

cheek pouch evagination of the oral vestibule used for storage of grain in rodents.

Chelys tortoise.

chemoreceptor sense organ sensitive to chemical stimuli.

chiasma shaped like the Greek letter chi (χ) > a crossing.

chief cell zygogenic (enzyme-producing) cell of the stomach that produces pepsinogen.

chir-, cheir- hand.

Chiroptera mammals in which the hand is modified as a wing; bats.

choana (pl., choanae) funnel-shaped opening > internal nares (posterior choana) (*q.v.*).

chole- bile.

cholinergic (fiber) nerve endings that produce acetylcholine.

chondr- cartilage.

chondrocranium neurocranium composed of cartilage.

chondrocyte cell that deposits cartilage.

Chondrostei group of basal ray-finned fishes (e.g., sturgeon).

chordae tendineae tendinous cords that connect the atrioventricular valves to the muscular walls of the ventricle.

chordal cartilage chondrified notochord forming the centrum in many fishes.

chordin developmental morphogen involved in formation of the dorsal body axis.

chordomesoderm developmental tissue giving rise to the notochord.

chorioallantoic membrane fusion of the chorion and allantois; respiratory structure in oviparous reptiles.

chorioallantoic placenta membrane (chorion and allantois) in direct contact with maternal uterus in therian mammals; embryonic site of nutrient and metabolic waste transfer.

chorion extraembryonic membrane; synapomorphy of Amniota.

chorionic villi fingerlike outgrowths of the chorionic sac extending into the uterine lining.

choriovitelline placenta membrane (chorion and yolk sac) in direct contact with maternal uterus in metatherian mammals; embryonic site of nutrient and metabolic waste transfer.

choroid, chorioid resembling a membrane. See *chorion.*

choroid coat highly vascularized layer of the eye.

choroid fissure groove in the optic stalk; pathway of axons from eye to brain.

choroid plexus (of the fourth ventricle) reduced layers (pia mater and ependyma) covering the cavity (third or fourth ventricle) of the brain; source of cerebrospinal fluid.

chrom-, chromato- color.

chromaffin tissue characterized by its staining properties; includes aminogenic tissue.

chromatophore pigment cell.

chyle lymphatic fluid milky in color due to lipid content.

chyme soupy product of the stomach that is passed on to the small intestine.

ciliary (ganglion) collection of autonomic postganglionic cells bodies in the head.

ciliary body muscle and vascular tissue at the periphery of the eye diaphragm.

ciliary process extension of the ciliary body that produces aqueous humor.

ciliated pseudostratified columnar epithelium lining of the bronchi and bronchioles in mammalian lungs.

circa, circum around, about.

circadian about a day in length.

circadian rhythm biological daily cycle.

circle of Willis right, left, and transverse communicating arteries that encircle the base of the diencephalon.

cisterna chyli abdominal lymph sinus of mammals.

clado- branch.

cladogram a branching diagram of phylogenetic relationships.

Cladoselachii extinct genus of cartilaginous fishes.

clasper male chondrichthyan intromittent organ.

clava club.

clavicle dermal element of the tetrapod pectoral girdle.

clavo-, cleido- clavicle.

cleavage segmentation of the zygote leading to the blastula stage.

cleidobrachialis dorsal forelimb muscle group; clavicle to distal humerus; ramus of brachial plexus (n. brachiocephalic).

cleidocervical See *cleidotrapezius*.

cleidoic closed, locked up > reptilian or monotreme egg with much yolk and a shell.

cleidomastoideus sternocleidomastoid complex of branchiomeric muscle group (extrinsic forelimb muscle); mastoid part of temporal bone to clavicle; n. XI.

cleidooccipitalis supernumerary muscle of the stern-ocleidomastoideus in some taxa; occipital region to clavicle; n. XI.

cleidotrapezius (cleidocervical) trapezius complex of the branchiomeric muscle group (extrinsic forelimb muscle); back of neck to clavicle; n. XI.

clitoris genital tubercle of females.

cloaca sewer > common terminus for digestive and urinary tracts.

cloacal plate developmental separation of proctodeum and hindgut.

cloacal vein part of the abdominal stream draining the cloaca.

coagulating gland multicellular gland that contributes to the seminal fluid.

coccyx fused caudal vertebrae of humans.

cochlea a snail with a spiral shell > spiral labyrinth of the inner ear.

cochlear duct (scala media) extension of lagena in therian mammals; site of mechanoreceptors for sound.

cochlear ganglion site of sensory neuron cell bodies for hearing.

cochlear nerve portion of cranial nerve VIII innervating cochlea (sound detection).

-coel- hollow, a cavity.

coelom body cavity.

coelomic (mesodermal) pouch developmental outpocketing of the early gut in amphioxus that initially forms a segmented body cavity.

collagen gelatinous, glue-like material; a proteinaceous fibril.

collagen bundle woven compact network of connective tissue (collagen fibers).

collagen fiber aggregation of proteinaceous fibrils.

collar nerve cord nerve cord of acorn worms located in the collar.

collateral (ganglion) classification of autonomic ganglia.

collateral branch branch of a long axon.

colli- of the neck.

colliculus little hill.

colon proximal portion of large intestine.

columella little (-*ella*) pillar or column.

com-, con-, cor- with, together.

comb skin crest on the head of a fowl, especially males.

common bile duct junction of hepatic and cystic ducts carrying liver products to duodenum.

common cardinal vein common site of drainage for several vessels, emptying into the sinus venosus.

common collecting tubule site of drainage of renal tubules.

compact bone layered bone.

complexus cervical epaxial muscle of birds used by hatchlings to crack egg shell.

conari of the pineal. (See *pineal*).

conceptus products of conception; embryo plus extraembryonic membranes.

concertina movement accordionlike movement seen in snakes.

conch- shell.

concha pinna of the ear, shaped like a clamshell; scroll-like turbinal bone in the walls of the nasal passageways.

conjunctiva epithelial covering of the eyeball (*bulbar-*) and inner surface of the eyelids (*palpebral-*).

constrictor muscle that compresses internal parts or narrows a cylindrical structure.

constrictor (fiber) intrinsic muscle fibers of the iris to close the pupil; n. III.

contact (nondeciduous) placenta one where extraembryonic membranes are in simple contact with the maternal lining and the lining is not shed. Contrast *deciduous placenta*.

contour feather give the general shape to the bird.

contra- opposite.

contralateral on the opposite side.

conus cone.

conus arteriosus chamber of heart after the ventricle.

copr- feces.

coprodeum fecal passage derived from cloaca.

copulation plug coagulated semen in the vagina; precludes another mating.

cor- See *com-*.

coracoarcuales hypobranchial muscle group in sharks; pectoral girdle to other muscles in group; n. hypobranchial (spino-occipital nerves).

coracobrachialis ventral forelimb muscle group; coracoid process of scapula to proximal humerus; n. musculocutaneous.

coracohyoideus hypobranchial muscle group in sharks; coracoarcuales to basihyoid; n. hypobranchial (spino-occipital nerves).

coracoid shaped like a crow's beak; ventral replacement bone in the tetrapod pectoral girdle.

coracoid plate embryonic cartilaginous plate in the lateral body wall giving rise to adult coracoid.

coracoid process of the scapula vestigial coracoid in eutherian mammals fused to the scapula.

coracomandibularis hypobranchial muscle group in sharks; coracoarcuales to lower jaw; n. hypobranchial (spino-occipital nerves).

corn- horn > temporary "horny" thickening of the stratum corneum at sites of friction.

cornea transparent part of sclerotic coat of the eye.

corneal muscle extrinsic muscle of the cornea in lampreys; adjacent skin to the cornea; used in accommodation by pushing the lens backward.

corniculate small cartilage in the larynx of some mammals.

cornified changed to horn by keratinization.

cornu (pl., cornua) horn of the hyoid.

corona wreath or crown.

coronary artery supplies oxygenated blood to the heart.

coronary ligament attaches cephalic pole of liver to embryonic transverse septum.

coronary sinus forms a "wreath" around the heart; site of drainage for coronary veins and left cranial vena cava (when present).

coronary sulcus external groove overlying the coronary sinus.

coronary vein venous drainage of heart.

coronoid dermal bone of the lower jaw.

corpora cavernosa paired erectile masses in the penis of therian mammals.

corpora quadrigemina two pairs of twins (-*gemini*) of the roof of the mesencephalon of amniotes.

corpus (pl., corpora) body.

corpus callosum broad sheet of transverse fibers connecting cerebral hemispheres in mammals.

corpus luteum yellow body of the ovary; ovulated follicle with endocrine function.

corpuscle of Herbst encapsulated cutaneous receptor on the beak, tongue, and palate of water fowl.

corpuscle of Stannius endocrine of mesodermal origin in ray-finned fishes; related to calcium levels.

corpus spongiosum spongy body.

cortex outermost layer; cortex of mammalian kidney, steroidogenic component of mammal adrenal gland, or cerebral cortex.

corticospinal tract carries voluntary motor impulses from the cerebral cortex to the spinal cord.

corticosterone adrenal steroid.

cortisol adrenal steroid.

cortisone adrenal steroid.

costa- rib.

costal cartilage cartilage at the ventral end of a rib.

costal portion (of diaphragm) part of muscular portion attaching to the ribs.

costal rib portion of rib attaching to vertebrae.

cotyledonary cup-shaped.

cotyledonary placenta chorionic villi distributed as isolated patches.

countercurrent flow parallel flow of two fluids in opposite directions; maximizes concentration gradient for more effective diffusion.

coxa hip.

cranial kinesis movement of a functional component of a skull independent of another.

cranial nerve peripheral nerve extending from the brain (some cranial nerves are evaginations of the brain, [e.g., optic nerve]).

cranial sinus air-filled space in the bony cranium.

cremaster muscle oblique or transverse hypaxial muscle fibers that wrap around the spermatic cord; ventral rami of spinal nerves.

cribriform sievelike.

cribriform plate sievelike part of ethmoid; site of foramina for olfactory nerve.

cricoid resembling a ring.

cricoid cartilage laryngeal cartilage.

crin- shed.

crista ridge or crest; papilla of epithelial cells surmounted by hair cells and a cupula in the membranous labyrinth. Compare *macula*.

crocodilian member of Crocodilia, subgroup of reptiles (e.g., alligator).

crop outpocketing of the esophagus in birds for the temporary storage of seed.

cross section partition in a transverse plane.

crotalum rattle > *Crotalus*, a rattlesnake.

crown part of a tooth above the gum line.

crown group group that includes all living members of a clade and their closest common ancestor.

crus (pl., crura) leg; portion of vertebral part of diaphragm muscle.

cten- comb.

ctenoid resembling a comb.

ctenoid scale dermal scale with a comblike free edge.

cucullaris from a word meaning "a hood"; the two cucullaris (trapezius) muscles collectively resemble a hood or shawl; branchiomeric muscle group; branchial levators in fishes; nn. IX and X in anamniotes, n. XI to its homologue in amniotes.

cuneiform wedge-shaped > small cartilage in the larynx of some mammals.

cuneus calluslike cornified pad at the base of a hoof in ungulates.

cupula structural peak.

cusp peak or point.

cutaneous referring to skin.

cutaneous receptor sensory organ of the skin.

cuticle thin transparent covering of a mammalian hair.

Cuvier, George (1769–1832) founder of comparative anatomy.

cyclo- circular.

cycloid scale circular-appearing scale.

cyclostome agnathan with round mouthlike funnel.

cyno- dog.

cynodont dog-toothed.

cyrtopodocyte excretory cell found in amphioxus.

cystern site of milk accumulation at the base of the nipple (mammary gland).

cystic duct drains the gallbladder.

cysto- sac, bladder.

cyt-, cyto-, -cyte cell.

D

dactyl- finger, toe.

Darwin, Charles (1809–1882) author of *The Theory of Natural Selection*.

de- away from.

decidu- fall off or be shed.

decidua invaded portion of maternal uterine lining that is shed.

deciduous (tooth) milk tooth; first generation tooth in mammals with two sets.

deciduous placenta one in which the uterine lining is partly shed at parturition. Contrast *contact placenta*.

deferens *de-* + *-ferent* (q.v.).

degenerate value-judgment term applied to taxa that have phylogenetically lost a number of characters (e.g., lampreys).

deltoid resembling the Greek letter delta (Δ).

deltoideus dorsal appendicular muscle group; spine and acromion process of the scapula to the deltoid crest of the humerus; n. axillaris.

demi- half.

demibranch gill on one face of a gill arch.

dendrite neuronal arborization.

dendro- tree.

dent- tooth.

dental lamina longitudinal ingrowth of ectoderm along the jaw; site of tooth formation.

dental plate dermal plate covering the jaw analogous to teeth.

dentary primary tooth-bearing dermal bone of the lower jaw.

denticle elevation on the surface of primitive bone composed of dentin.

dentin (dentine) bone like that in teeth.

dentinal tubule cellular canal within dentin (analogous to canaliculi).

depressor muscle that lowers a structure.

derived (modified) any state of change from an ancestral condition.

derm- skin.

dermal bone formed within the dermis of the skin.

dermal papilla fingerlike outgrowth of dermis; seen in development of hair and feathers.

dermato- referring to skin.

dermatocranium skull bones phylogenetically derived from skin.

dermatome layer of a somite giving rise to skin.

dermis mesodermal deeper layer of skin.

Dermoptera subgroup of mammals in which the skin forms a wing membrane. See *-pter*.

-deum, -daeum passageway.

deuter- two.

deuterostome animal that uses the blastopore as an anus and forms a second mouth.

di- twice, double.

dia- through, apart, completely, between.

diaphragm separation (*-phragma*) between two parts; muscle of mammals; hypaxial muscle group; ribs to central tendon; n. phrenic.

diaphragmatic (muscle) aspiratory muscle of crocodilians; pelvis to liver; moves the liver like a piston.

diaphragmatic pleura epithelial covering of the pleural surface of the diaphragm.

diaphysis   ossifying shaft of a long bone.

diapophysis   one of two lateral (transverse) processes on a vertebra.

diapsid   having two arches.

Diapsida   subgroup of reptiles.

diarthrosis   freely movable joint between two bones or cartilages.

diastema   intervening space.

diastole   heart cycle where filling occurs. Contrast *systole*.

-didym-   twins > testes.

diencephalon   posterior region of the forebrain.

diffuse placenta   chorionic villi distributed diffusely over the entire surface.

digestive system   organ system for the acquisition, processing, temporary storage, digestion, and absorption of food and for elimination of unabsorbed residue.

digiti-   fingers or toes.

digitigrade   walking (*-grade*) on the digits.

dilator   muscle that enlarges an opening.

dilator fiber   intrinsic muscle fiber of the iris to open the pupil; n. III.

dino-   fearful, terrible.

dinosaur   fear-inspiring reptile; subgroup of Archosauria.

dioecious   condition where ovaries and testes do not develop in the same individual.

diphy-   double.

diphycercal   caudal fin in which the vertebral column ends with very little or no upbending.

diphyodont (dentition)   having two successive sets of teeth.

diplo-   double, two.

diplospondyly   two vertebrae in each body segment.

Dipnoi   fish with two paired breathing apertures (external and internal); lungfish.

dis-   separation, apart from.

discoidal placenta   chorionic villi distributed as a single large discoidal area.

dissect   to disassemble.

distal carpal   member of the outer (distal) row of wrist bones.

diverticulum   outpocketing.

Doctrine of Acquired Characteristics   Lamarck's principle of characters states that those that are acquired through use or disuse are inheritable.

dorsad   toward the back.

dorsal (vertebrae)   regional specialization of vertebrae; trunk vertebrae.

dorsal aorta   primary mid-dorsal artery supplying the body.

dorsal fin   median fin on the back (dorsum).

dorsal hollow central nervous system   brain and spinal cord of chordates.

dorsal intercalary plate   cartilaginous plate located between lateral neural arches in sharks.

dorsalis trunci   epaxial muscles of salamanders retaining their primitive metamerism; dorsal rami of spinal nerves.

dorsal lip   in development, the upper edge of the blastopore; site of involution of surface cells during gastrulation.

dorsal mesentery   dorsal remnants of the midline coelomic linings.

dorsal mesoderm (epimere)   mesoderm located adjacent to the chordamesoderm in the developing embryo.

dorsal plate   See *neural arch.*

dorsal ramus   dorsal branch of a spinal nerve.

dorsal rib   tetrapod rib or fish rib extending into the horizontal skeletogenous septum. Contrast *ventral rib.*

dorsal root   dorsal component of the spinal nerve formed by the fusion of rootlets that leave the cord; characterized by the presence of a dorsal root ganglion.

dorsal ventricular ridge   area of association neurons extending into the lateral ventricles in reptiles.

dorsobronchus   dorsal bronchus in a bird.

dorsum   the back.

down feather   small fluffy feather with a crown of barbs lacking hooklets.

ductus arteriosus   patent portion of sixth aortic arch connecting the pulmonary artery with the dorsal aorta; in development, it serves as a bypass for the lungs.

ductus caroticus   portion of the dorsal aorta connecting the third and fourth aortic arches.

ductus venosus   vessel through the liver carrying blood from the umbilical vein to the postcava and the heart.

duodenum   twelve, in Latin; the length of the human duodenum is about the breadth of 12 fingers.

duplex uterus   two completely separate uteri, as seen in marsupials.

dura   tough, hard.

dura mater   outer tough meninx in mammals.

dys-   faulty, painful, difficult.

E

e-   without.

ecdysis   periodic shedding of the epidermis in squamates.

ect-   outer.

ectethmoid   ethmoid ossification in lateral walls of the nasal passageway of *Sphenodon.*

ectodermal groove   evagination of ectoderm over developing pharyngeal pouches; future site of pharyngeal slits.

ectodermal placode   thickening of the ectoderm that gives rise to neuroblasts and to sensory epithelia of certain sense organs.

-ectomy   *ex-* + *-tome* (q.v.), as in *appendectomy.*

ectopic   *ex-* (out of) + *topo-* (place).

ectopic pregnancy   fetus is implanted elsewhere than in the uterus.

ectopterygoid   outer pterygoid bone; dermal bone of the palate.

ectotherm   animal whose temperature varies with the environment.

Edentata   subgroup of mammals lacking teeth.

ef-   variant of *ex-.*

effector   muscle, gland, or organ acted upon by the nervous system.

efferent   that which carries away from.

efferent branchial artery   vessel carrying blood away from the gills.

efferent (branchial) duct   carries water from branchial pouch to the external opening in hagfishes.

efferent ductules   name applied to vasa efferentia (*q.v.*) in mammals.

efferent glomerular arteriole   vessel carrying blood away from the glomerulus.

efferent (motor) nerve   impulses travel to muscles or glands.

eft   temporary terrestrial stage in the life history of the salamander *Notophthalmus.*

egg tooth   temporary epidermal tooth of some reptilian hatchlings used for cracking the egg.

elasmo-   plate.

elasmobranch   with gills, composed of flat plates.

Elasmobranchii   sharks.

elasmoid   dermal scale consisting of thin lamellar bone that may be associated with a fibrous plate.

elastic cartilage   possesses a network of elastic fibers.

electroreceptor   sensory organ sensitive to electrical fields.

-ella   a diminutive, as in *columella.*

embryonic disk   See *blastodisk.*

en-   in, into.

enamel   mineralized covering of teeth or dermal denticles; ectodermal tissue deposited by ameloblasts.

enameloid   mineralized covering of teeth or dermal denticles; compact outer layer of dentine.

enamel organ   ectoderm of the dental lamina that deposits enamel on the developing tooth.

encephalon   brain, a structure in the head.

end bulb (corpuscles) of Kraus   encapsulated cutaneous receptor of mammals; possible thermoreceptor.

endo-   within, inner. See also *ento-.*

endocardium   epithelial lining of the cavities of the heart.

endochondral within cartilage.

endodermal lining covering over the surface of the yolk.

endolymph fluid within the membranous labyrinth.

endolymphatic duct passage connecting the membranous labyrinth to the surface in sharks.

endolymphatic fossa depression in the chondrocranium into which endolymphatic and perilymphatic ducts empty in sharks.

endolymphatic sac duct that terminates as a blind sac.

endometrium uterine lining.

endomysium *endo-* + *mys-* (*q.v.*); collagenous covering of a muscle functional unit.

endosteum thin connective tissue lining of bone marrow cavities.

endostyle midventral mucus-producing groove in the pharynx of protochordates; homologue of craniate thyroid.

endotherm warm-blooded; an animal that maintains a relatively constant body temperature regardless of environmental fluctuations.

endothermy ability to be an endotherm.

enteric plexus neural network in the external muscle layer of the gut, responsible for peristalsis.

enteroceptor sense organ providing information about the environment within an animal.

enteron gut.

Enteropneusta subgroup of hemichordates that includes acorn worms.

ento- within, inner. See also *endo-*.

entoglossal within the tongue.

entoglossus elongated bony process of the hyoid extending into the tongue in lizards and birds.

entotympanic cartilage replacement ossification in the tympanic bulla; new structure in some mammals.

ep-, epi- upon, above, over.

epaxial above an axis.

epaxial muscle axial muscle of the trunk above the horizontal skeletogenous septum; dorsal rami of spinal nerves.

ependyma outer garment > membrane covering the part of the central nervous system exposed to the neurocoel.

ependymal cell undifferentiated cell with the ependyma.

epiblast presumptive tissues of the developing embryo in the blastoderm of a massively yolked egg.

epiboly embryonic cell division resulting in migration over a surface.

epibranchial paired dorsal element in the branchial skeleton between the cerato- and pharyngobranchial elements.

epibranchial muscle axial muscle extending above the pharynx.

epibranchial placode ectodermal thickening (placode) located above the pharynx.

epicardium epithelial covering of the heart.

epicritic perception of differences within a single sense.

epidermal gland gland of epidermal origin.

epidermal scale keratinized scale derived from the stratum corneum.

epidermis ectodermal outer layer of the integument.

epididymis (pl., epididymides) structure lying on the testis (*didym-*).

epiglottis fibrocartilaginous flap in mammals that covers the glottis in swallowing.

epihyal teleost hyoid arch skeletal element; an ossification center in the ceratohyal.

epimere dorsal mesoderm.

epimysium *epi-* + *mys-* (*q.v.*); fibrous covering of a muscle; muscle fascia.

epinephrine (adrenaline) secretion of the adrenal gland; one neurotransmitter of postganglionic neurons in the sympathetic system.

epineurium covering of the peripheral nerve fiber.

epiotic *epi-* + *-oto* (*q.v.*); replacement ossification of the otic region of the neurocranium.

epipharyngeal groove midline groove in the roof of the pharynx in amphioxus.

epiphyseal complex pineal and parapineal evaginations of the diencephalon.

epiphyseal plate site of growth between long bone shaft and each of two ends.

epiphysis (pl., epiphyses) *epi-* + *-physis* (*q.v.*); pineal complex; ossification center in the ends of long bones.

epiploic relating to greater omentum (the epiploon).

epiploic foramen opening between lesser and greater peritoneal cavities.

epipodium limb bones immediately above the hand or foot.

epipubic (bone) accessory ossification in the pelvis of reptiles and nontherian mammals.

epithalamus dorsalmost component of the diencephalon.

epithelium surface layer of cells.

epitrichium *epi* + *trich-* (*q.v*); temporary layer of epidermal cells overlying the hair of fetal mammals; mammalian periderm.

epitrochleoanconeus brachial muscle; median condyle to olecranon process; n. ulnaris.

epoophoron vestiges of the embryonic mesonephros (archinephric duct) found in the dorsal mesentery of the ovary.

erythro- red.

erythrocyte red blood corpuscle.

erythrophore pigment cell (chromatophore) containing red granules.

eso- carrier.

esophagus carrier of substances that have been eaten (*phag-*).

estr- female.

estrogen steroid hormone of the gonads.

ethmoid sievelike; anterior region of neurocranium.

ethmoid plate the anterior floor of the neurocranium between the prechordal cartilages in development.

eu- true.

eury- wide.

euryapsid variation of temporal region with a single temporal fenestra; condition found in plesiosaurs and ichthyosaurs.

euryhaline able to live in waters with a wide range of salinity (*halo-*).

Eutheria chorioallantoic placental mammals (e.g., humans).

euviviparity viviparous condition where the developing embryo must receive maternal nourishment.

evolutionary convergence similarity due to independent evolution.

ex-, exo- out, out of, away from, outer.

excretion elimination of waste products from the body.

excurrent aperture opening for outward flow.

excurrent bronchus passageway to carry air from an air sac into the ducts of the lung in birds.

excurrent pore opening for the exit of a current of water.

excurrent siphon opening in a tunicate (urochordate) for the exit of water.

exoccipital bone paired lateral replacement ossifications of the occipital region of the neurocranium.

exoskeleton skeleton in the skin.

extant now living.

extensor muscle that tends to straighten two segments of a limb.

extensor caudea lateralis epaxial muscle group, caudal extension of longissimus; sacral and proximal caudal vertebrae to distal caudal vertebrae; dorsal rami of spinal nerves.

extensor of the hand or foot and digits muscle that moves the hand or foot and digits anterodorsally in a quadruped.

external auditory meatus passageway or opening of the ear to the exterior.

external gill lamellae extending external to the body.

external glomerulus capillary network extended into the coelomic cavity. Contrast *internal glomerulus*.

external (transverse) iliac vein vessel from the

femoral vein to the common iliac vein.

external intercostal hypaxial muscle group; outer layer of muscle between adjacent ribs; ventral rami of spinal nerves.

external naris opening of nasal passageway to the exterior.

external neuromast hair cell exposed to the external environment.

external oblique hypaxial muscle group of the abdominal wall; thoracolumbar fascia and caudal ribs to linea alba; ventral rami of spinal nerves.

external ramus (of the accessory nerve) branch of n. XI supplying sternocleidomastoid and trapezius muscle complexes.

external respiration obtaining oxygen from the environment.

exteroceptive role to function as an exteroreceptor.

exteroceptor sensory organ sensitive to signals from the external environment.

extra- beyond, outside of.

extracolumella accessory skeletal structure attached to the columella in amphibians and reptiles.

extraembryonic outside of the embryo.

extrinsic appendicular muscle extends from the limb to the axial skeleton or fascia of the trunk.

F

falciform shaped like a sickle.

falciform ligament vestige of the ventral mesentery between the ventral midline and liver.

falciform process vascular fold of the choroid coat near the blind spot in fishes; site of metabolic exchange between the bloodstream and the vitreous humor.

fallopian tube oviduct of mammals.

fascicle bundle.

fasciculus gracilis nerve tract in the spinal cord for touch and proprioception.

fauces throat.

feather follicle pit lined with epidermis for the developing feather.

feather primordium pimplelike elevation of the dermis and overlying epidermis early in the development of a feather.

feather sheath epidermal covering of the developing feather.

femoral (artery) vessel of the thigh.

femoral gland granular epidermal gland on hind limb in lizards; assists male during copulation.

femorotibialis hind limb muscle group of nonmammalians; femur to tibia.

fenestra window > aperture.

-ferent, -ferous carrying, as in *afferent* (q.v.).

fiber tract collection of axons within the central nervous system.

fibril proteinaceous unit of collagen.

fibrinogen product of liver essential for blood clotting.

fibrocartilage cartilage with a dense network of collagenous fibers.

fil- thread.

filamentous extension of internal gill projection of the internal gill through the gill slit to the exterior.

filia olfactoria (pl.) discrete olfactory nerve fibers that extend between the olfactory epithelium and the brain.

filoplume hairlike feather with a shaft and a few barbs with barbules.

filter feeder organism that uses the oropharynx to strain food particles from the water.

filum terminale threadlike sheath of the spinal cord that extends beyond the last spinal nerves.

fimbria fringe.

fin fold in development, the foldlike outgrowth of the primordial fin.

fin fold fin broad-based fin typical of modern sharks.

fin fold hypothesis suggests that paired fins evolved from a continuous ventrolateral fin fold.

fin spine hypothesis suggests that fins evolved from a condition similar to that seen in the spiny-finned acanthodians.

first-order sensory neuron sensory neuron with cell body located in a ganglion.

flange branch of feather barb that interlocks with barbules of adjacent feather.

flat bone characterized by a core of spongy bone sandwiched between two layers of compact bone.

flexor muscle that tends to draw one segment toward another in a limb.

flexor of the manus or foot and digits muscle that moves the hand or foot and digits posteroventrally in a quadruped.

flower spray proprioceptive axon ending named after nature of contact with intrafusal muscle fiber (forms a splayed pattern on fiber).

follicle stimulating hormone (FSH) hormone stimulating growth of ovarian follicles and their secretion of estrogen.

fontanel soft spot of child's head; gap in the dermal bone covering of the skull.

foramen small opening; usually transmits something.

foramen cecum pit at base of the tongue in mammals marking the site of embryonic thyroid evagination.

foramen magnum large foramen in the occipital region.

foramen of Panizza aperture connecting the two aortic trunks in crocodilians; functional shunt used while diving.

foramen ovale aperture connecting right and left atria of the heart in amniotes; functional shunt in the embryo to bypass the lungs.

fore- before, in front.

forearm part of the arm before the upper arm.

foregut anterior portion of developing gut.

formative (morphogenetic) movement cell migrations during gastrulation establishing the three germ layers and bilateral symmetry.

formed element corpuscles or platelets suspended in the blood plasma.

fossa pit, cavity, depression, vacuity.

fossa ovalis depression in the interatrial wall at site of embryonic foramen ovale.

fovea small pit; site of most acute vision in retina.

frenulum little bridle.

frenulum linguae membrane that bridles (ties) the tongue to the floor of the oral cavity.

frontal anatomical plane formed by the right-left and longitudinal axes; dermal bone of the skull.

frontal organ (stirnorgan) median eye of frogs.

frontal section partition of an organism along the frontal plane.

Froriep's ganglion embryonic dorsal root ganglion of n. XII in some mammals.

frug- fruit.

frugivore fruit eater.

fundus bottom of a cavity; part of the mammal stomach to the left of the cardiac region and cranial to the opening of the esophagus.

furculum paired clavicles (wishbone) of a bird.

fusiform spindle-shaped.

G

galea aponeurotica sheetlike tendon (aponeurosis) underlying the mammalian scalp; common tendon of insertion for a number of muscles of the head and neck.

Galen (circa A.D. 165–200) Greek philospher-physician whose writings served as the basis for anatomy for the next millennium.

gamma motor neuron neuron that innervates a modified striated muscle fiber found within a muscle spindle. Contrast *alpha motor neuron*.

gan- bright.

ganglion swelling > group of cell bodies outside the central nervous system.

ganoid shining bony scale.

gar basal neopterygian fish.

Gartner's duct remnant of the mesonephric (archinephric) duct in a female amniote.

gastr- belly, stomach; the digastric muscle has two bellies.

gastralia ventral abdominal ribs.

gastrin gastrointestinal hormone that evokes gastric secretions.

gastrohepatic ligament remnant of the ventral mesentery connecting the liver and pyloric region of the stomach; part of the lesser omentum.

gastrula little stomach > stomachlike embryo.

gen- origin.

generalized connotes a state of potential adaptability.

general metabolic hormone possesses receptor sites on every cell of the organism (e.g., insulin).

general receptor sensory receptor with wide distribution in the skin and within the body. Contrast *special receptor*.

genetic drift random changes in gene frequency.

geniculate ganglion kneelike bend of facial nerve. See *genu*.

genio- chin.

genioglossus prehyoid hypobranchial muscle group of tetrapods; symphysis of jaw to tongue; spino-occipital nerves of anamniotes, n. XII of amniotes.

geniohyoid prehyoid hypobranchial muscle group of tetrapods; symphysis of jaw to hyoid; spino-occipital nerves of anamniotes, n. XII of amniotes.

genital corpuscle encapsulated cutaneous sensory receptor in the external genitalia of mammals.

genital fold embryonic fold in therian mammals that borders the urogenital sinus; forms labia minora in the female and, in part, the male penis.

genital ridge embryonic thickening of the coelomic mesothelium; site of gonad formation.

genital swelling shallow evagination of the embryonic coelomic floor adjacent to genital folds; site of future scrotal sac in males and labia majora in females.

genital tubercle embryonic outgrowth forming the female clitoris or, in part, the male penis.

genotype specific assortment of genes in an individual.

genu knee.

geo- earth.

germinal (layer) mitotic layer of the epidermis.

germinal epithelium region of the genital ridge serving as the gonadal primordium.

gill respiratory-exchange surface in aquatic organisms.

gill arch structures between successive gill slits.

gill arch hypothesis suggests the origin of paired fins from modified gill arches.

gill bypass portion of the aortic arch in amphibians that serves as a bypass to the external gills.

gill chamber space adjacent to the internal gills.

gill raker projection from the pharyngeal border of a gill arch.

gill ray skeletal structure extending from the gill skeleton to support the interbranchial septum.

gizzard grinding portion of archosaur stomach. Contrast *proventriculus*.

glandular field embryonic source of developing tongue in the pharyngeal floor of urodeles and frogs.

glans acorn.

glans penis tip of the penis.

glenoid resembling a socket.

glia glue.

glial membrane glial cell covering of the brain and spinal cord.

globus pallidus motor subpallial region of cerebrum.

glomerulus (pl., glomeruli) little glomus (ball or skein) > tiny plexus of blood vessels.

gloss- tongue.

glottis opening into the larynx.

glucagon pancreatic hormone promoting the breakdown of glycogen to glucose.

glucocorticoid mammalian adrenal steroid; promotes conversion of noncarbohydrates into glucose.

gluconeogenesis conversion of noncarbohydrates into glucose.

glycogenolysis breakdown of glycogen into glucose.

gnath- jaw.

Gnathostomata group characterized by the presence of jaws.

goblet cell goblet-shaped epidermal cell that secretes mucus; type of fundic gland in the stomach.

gon- seed > generative, as in *glucagon* (giving rise to glucose).

gonad source of gametes.

gonopodium modified anal fin of some teleosts to serve in transfer of sperm.

graafian follicle egg follicle of therian mammals.

gracilis slender.

Grandry's corpuscle encapsulated cutaneous sensory receptor along the beak of many shore birds and freshwater fowl.

granular cell (of fish) cell that secretes mucus and additional ingredients.

granular gland epidermal gland secreting irritating or toxic alkaloids and many pheromones.

gray matter area of brain consisting mostly of cell bodies.

gray rami communicantes communi-cations between the sympathetic trunk and the spinal cord carrying postganglionic fibers.

greater curvature (of stomach) outer or longer edge of a stomach that has been folded.

greater horn longer process (cornua) of the amniote hyoid.

greater omentum portion of dorsal mesentery attached to greater curvature of the stomach.

gubernaculum rudder > governor, as the ligament that (partly) governs the position of the testis.

gula throat.

gular bone found in the opercular membrane of bony fishes.

gular fold fold at the throat of some tetrapods.

gular pouch (dewlap) saclike evagination in the throat of some male lizards

that can be expanded with air.

gustatory related to gustation (taste).

gymn- naked.

gyrus (pl., gyri) ridge between grooves.

H

habenula elevated thickening of the epithalamus, associated with olfactory nuclei.

haem-, hem- blood.

hair cell receptor cell with cilia (e.g., neuromast organ).

hair horn rhinoceros horn composed of hairlike epidermal fibers.

hallux great toe.

hamate bone fusion of distal carpals four and five in the manus.

hamulus (adj., hamate) little hook.

Harderian gland lubricating gland of the eye; salt-secreting gland of marine turtles.

haversian canal neurovascular channel through compact bone.

haversian system (osteon) haversian canal and its surrounding concentric lamellae.

head in craniates, anterior portion of the body containing the brain or the proximal part of a limb bone.

helico- helix, spiral.

helicotrema opening connecting the scala tympani and scala vestibuli of the membranous labyrinth.

hemal arch vertebral enclosure of the caudal artery and vein.

hemi- half; same as *demi-* (*q.v.*).

hemiazygos remnant of the embryonic left postcardinal vein in tetrapods.

Hemichordata group that includes acorn worms.

hemocytoblast stem cell from which all later blood cells arise.

hemolymph lymph containing blood cells.

hemopoiesis formation of blood.

Hensen's node thickened nodule of closely packed blastoderm cells that defines the caudal end of the future embryo; analogous to the dorsal lip of the blastopore.

hepat- liver.

hepatic duct drains bile from liver.

hepatic portal (system) venous system bounded by capillary beds carrying blood from visceral organs to the liver.

hepatic sinus vein draining liver to sinus venosus.

hepatoduodenal ligament remnant of the ventral mesentery connecting the liver and duodenum; part of the lesser omentum.

hept- seven.

herbivore animal that devours grasses (herbs).

hermaphroditism production of viable eggs and sperm by the same individual.

hetero- other, different; opposite to *homo-* (*q.v.*).

heterocelous (vertebra) avian vertebra with a caudal saddle-shaped articulation and cranial end of next vertebra shaped to accommodate this configuration.

heterocercal (caudal fin) tail fin where the notochord turns upward into a larger dorsal lobe.

heterochrony a time difference.

heterodont (dentition) teeth that differ morphologically from front to back.

heterotopic bone endochondral or intramembranous ossification at site of continuous stress in amniotes.

hex- six.

higher (taxon) misleading term to denote relative position of major taxa.

hilum See *hilus.*

hilus (var. hilum) small notch or recess at entrance or exit of an organ (e.g., lung or kidney).

hindgut posterior part of embryonic digestive tract.

hipp- horse.

hippocampus ancient olfactory cortex tucked under the temporal lobes of the cerebral hemispheres in mammals.

hist- tissue.

histogenesis formation of a tissue.

histotrophic (embryotrophic) nutrition absorption of nutrients from uterine or other maternal fluids.

holo- entire, whole.

holobranch two demibranchs of a single gill arch and the associated interbranchial structures.

Holocephali (chimaeras) subgroup of chondrichthyans.

holocrine gland one in which the cells themselves constitute the secretion.

holonephros kidney extending the length of the coelom.

hom-, homeo-, homo-, homoio- like, similar.

homeostasis control of the internal environment of the organism.

homeotherm animal that maintains a steady body temperature despite ambient (surrounding) temperature; an endotherm.

homeotic gene anterior to posterior segmental gene.

Hominidae group (family) uniting humans and extinct hominids.

homocercal tail symmetrical tail fin with an upturned notochord.

homodont teeth all alike.

homology similarity due to common ancestry.

homoplasy similarity due to any cause other than common ancestry.

hooklet hook on a feather barbule that interlocks with a flange on the next barb.

horizontal skeletogenous septum fibrous sheet extending from vertebral column to periphery that separates epaxial and hypaxial muscle groups.

horn (cornu) extension of the amniote hyoid.

horn of pronghorn antelope keratinized true horn of antelope.

horny epidermal teeth (of the platypus) keratinized replacement teeth for the deciduous set in a platypus (Prototheria).

Hox gene cluster chromosomally related group of homeotic genes.

human antidiuretic hormone neurohypophysis hormone that regulates water and salt excretion in mammals.

hyaline clear, glassy.

hyaline cartilage form of cartilage with little collagenous matrix; precursor cartilage of replacement bone.

hydrochloric acid secretion of the fundic glands of the stomach.

hydroxyapatite crystal mineralized component of bone.

Hylidae group (family) of tree frogs.

hyoglossus hypobranchial muscle group of mammals; hyoid to tongue; n. XII.

hyoid shaped like the Greek letter upsilon (γ).

hyoid arch second pharyngeal arch.

hyomandibular cartilage epibranchial-equivalent structure in the hyoid arch.

hyostylic jaw suspension (hyostyly) condition where the hyoid serves to support the jaws.

hyostyly See *-styly.*

hyp-, hypo- under, below, less than ordinary.

hypapophysis midventral vertebral projection (process) from the centrum in snakes and some other amniotes.

hypaxial below a given axis.

hypaxial muscle axial muscle group located below the horizontal skeletogenous septum; innervated by ventral rami of spinal nerves.

hyper- above, beyond the ordinary.

hypoblast lower sheet of cells in the embryonic blastoderm.

hypobranchial paired ventral element in the branchial skeleton between the cerato- and basibranchial elements.

hypobranchial muscle axial muscle that migrated in development to a position below the pharynx; innervated by spino-occipital nerves in anamniotes, n. XII in amniotes.

hypobranchial plate common midventral branchial element for the paired branchial cartilages in larval frogs.

hypocentrum vertebral centrum element of basal tetrapods; homologue of living amniote centrum.

hypocercal form of caudal fin where the notochord turns downward.

hypoglossal nucleus collection of cell bodies in the medulla oblongata for n. XII.

hypoischial (bone) accessory ossification in the pelvis of reptiles and nontherian mammals.

hypophyseal fenestra opening between elements in the developing floor of the chondrocranium for the pituitary.

hypophyseal portal system venous system bounded by capillary beds from the hypothalamus to the anterior lobe of the pituitary.

hypophysis growth under the brain; the pituitary body.

hypothalamus floor and ventral walls of the third ventricle.

hypothetical ancestor in a cladogram, the putative ancestor represented by the node between sister lineages.

hypso- height.

hypural bone below the urostyle of fishes.

I

ichthy- fish.

ichthyosaur extinct group of euryapsid aquatic reptiles.

Ichthyostega extinct group (genus) of basal tetrapods.

-iform having the shape of.

ileo- pertaining to the ileum of the intestine.

ileocolic ceca blind outpocketings just distal to the junction of the ileum and colon.

ileocolic sphincter constrictor muscle at the junction of the ileum and colon; regulates the ejection of the contents of the ileum into the large intestine.

ileum distal portion of the small intestine in mammals.

iliac (artery) segmental artery of the dorsal aorta supplying the hip and leg.

iliacus pelvic appendicular muscle group; ilium to lesser trochanter of femur; ventral rami of lumbar spinal nerves.

iliac vein vessel draining the pelvis and hind leg.

ilio- pertaining to the ilium of the pelvis.

iliocostales epaxial muscle group; ileum to ribs, ribs to seventh cervical vertebra; dorsal rami of spinal nerves.

iliofemoralis hind limb muscle group of nonmammalians; ilium to femur; ventral rami of spinal nerves.

iliofibularis hind limb muscle group of nonmammalians; ilium to fibula; ventral rami of spinal nerves.

iliopsoas fusion of the iliacus and psoas major muscles of the pelvic appendicular muscle group; combined attachments of psoas major and iliacus to lesser trochanter of femur; ventral rami of lumbar spinal nerves.

iliotibialis hind limb muscle group of nonmammalians; ilium to tibia; ventral rami of spinal nerves.

impar unpaired.

in- not.

Inca bone unfused postparietal bone seen in some human populations.

incisor heterodont tooth form; teeth found on each side of symphysis typically with a cutting edge; tusk of elephants.

incurrent aperture opening for inward flow.

incurrent siphon opening in a tunicate (urochordate) for the entrance of water.

incus anvil middle ear ossicle of mammals; homologue to the quadrate bone of nonmammalian vertebrates.

inducing hormone acts on tissues presensitized by synchronizing hormones.

inferior glossopharyngeal ganglion collection of sensory cell bodies for n. IX in birds and mammals.

inferior lobe enlargement of the hypothalamus on the ventral surface of the chondrichthyan brain.

inferior umbilicus opening at the base of a feather shaft.

inferior vagal ganglion collection of sensory cell bodies for n. X in birds and mammals.

inferior vena cava postcava of mammals.

infra- beneath, under.

infraorbital *infra- + orb- (q.v.)* > beneath the eye. See *jugal.*

infraorbital (zygomatic) gland accessory lubricating gland for the eye in some mammals.

infraspinatus ventral appendicular muscle group; infraspinatus fossa of the scapula to proximal humerus; n. suprascapularis.

infraspinous fossa lateral depression on the mammalian scapula ventral to the scapular spine.

infratemporal (zygomatic) arch the arch of bones that underlies the infratemporal fossa.

infundibular recess evagination of the third ventricle in the midventral floor.

infundibular stalk outgrowth of the brain connecting to the pituitary.

infundibulum little funnel.

ingroup members of a taxonomic group of interest.

inguen groin.

inguinal in the region of the groin.

inguinal canal passageway between the abdominal cavity and the scrotal cavity.

inguinal ring abdominal entrance to the inguinal canal.

inhibiting hormone inhibits the activity of an affected organ.

inner cell mass blastoderm of mammals.

inner circular layer circularly arranged muscle fibers of the gut wall.

inner ear fluid-filled membranous labyrinth.

innominate not named.

innominate (coxal) bone fused ilium, ischium, and pubis of mammals.

insertion site of muscle attachment (either distal, mobile, or narrow).

insulin pancreatic general metabolic hormone; stimulates cellular uptake of glucose.

inter- between.

interarcuale intervertebral subgroup of epaxial muscles extending between two successive neural arches; dorsal rami of spinal nerves.

interarticulare intervertebral subgroup of epaxial muscles extending between two successive zygapophyses; dorsal rami of spinal nerves.

interatrial foramen (foramen ovale of the heart) See *foramen ovale.*

interatrial septum wall separating the right and left atria of the heart.

interbranchial septum part of a gill arch between the gill lamellae.

intercalary plate part of neural arch between neural plates in fishes.

intercalated disk unique intercellular boundary within cardiac muscle.

interclavicle midventral dermal bone of the pectoral girdle.

intercostal *inter- + costa- (q.v.)*; (adj.) to describe structures found between the ribs (e.g., artery, nerve, vein, or muscle).

interdorsal cartilage one of two vertebral elements in fishes forming the vertebral canal (e.g., sturgeon).

interhyal ossification center within the hyomandibular cartilage of bony fishes.

intermediary nonnervous-receptor cell sensory transducer that passes on a stimulus to a sensory nerve fiber.

intermediate (nephrogenic) mesoderm (mesomere) unsegmented mesoderm just lateral to the somites giving rise to structures of the urogenital system.

intermediate zone middle layer of developing neural tube.

intermedin See *melanophore stimulating hormone (MSH).*

intermedium proximal wrist bone between the radiale and ulnare.

internal capsule massive band of white matter connecting the neocortex with the brain stem.

internal gill lamellae located within the body.

internal glomerulus capillary network within the kidney. Contrast *external glomerulus.*

internal iliac vessel supplying or draining the organs of the pelvic cavity.

internal intercostal hypaxial muscle group; inner layer of muscle between adjacent ribs; ventral rami of spinal nerves.

internal jugular anterior cardinal vein of amniotes.

internal naris (posterior choana) opening of the nasal canal into the oropharyngeal cavity.

internal neuromast hair cell organ of the membranous labyrinth (inner ear).

internal oblique hypaxial muscle group of the abdominal wall; thoracolumbar fascia and caudal ribs to linea alba; ventral rami of spinal nerves.

internal ramus of the accessory nerve visceral motor branch of n. XI that joins the vagus in mammals.

internal respiration exchange of oxygen and carbon dioxide between capillary blood and tissue fluids.

internal spermatic fascia coelomic peritoneum lining the scrotal sacs.

interopercular dermal bone in the opercular series of bony fishes.

interrenal (body) steroidogenic tissue in sharks and rays found between the kidneys (equivalent to the adrenal cortex of mammals).

interspinale intervertebral subgroup of epaxial muscles extending between two successive neural spines; dorsal rami of spinal nerves.

interstitial cell stimulating hormone adenohypophysis hormone in males stimulating the interstitial cells of the testes to produce androgens.

intertemporal temporal dermal bone in basal tetrapods.

intertransversarii intervertebral subgroup of epaxial muscles extending between two successive transverse processes; dorsal rami of spinal nerves.

interventral cartilage   one of two ventral vertebral elements in fishes (e.g., sturgeon).

interventricular foramen   connection between the third ventricle of the brain and the lateral ventricle of the cerebral hemisphere.

interventricular groove   superficial crease in the heart signifying the position of the interventricular septum.

interventricular septum   wall separating the right and left ventricles of the heart.

intervertebral   subgroup of epaxial muscles extending between two successive vertebrae; dorsal rami of spinal nerves.

intervertebral disc   fibrocartilaginous structure between successive vertebrae in mammals.

intervertebral foramen   vertebral foramen for the exit of the spinal nerve.

intestinal cecum   evagination of the gut in amphioxus.

intestinal juice   enzymatic secretion of the intestinal epithelium; splits polypeptides and disaccharides.

intra-   within.

intrafusal fiber   modified muscle fiber in a muscle stretch receptor.

intramembranous   within a membrane.

intrasegmental   within a segment.

intrinsic appendicular muscle   a muscle that arises and inserts entirely within the limb.

intrinsic eyeball muscle   internal striated muscle of the eye.

intromittent (copulatory) organ   male structure for the introduction of sperm into the female reproductive tract.

involution   rolling-in of cells during gastrulation.

ipsi-   same.

ipsilateral   on the same side.

irid- **(pl., irides)**   rainbow.

iridophore   pigment cell containing refractory bodies that result in iridescence.

iris   See *irid-*.

iris diaphragm   pigmented ring of the eye surrounding the pupil; modified portion of the choroid coat.

ischi-   hip, pelvis.

ischial callosity   cornified epidermis overlying the ischium in monkeys and apes for sitting.

ischial symphysis   midventral connection between right and left ischial bones in the pelvis.

ischiopubic symphysis   combined ischial and pubic midventral connection in the pelvis.

ischium   posterior ossification in the tetrapod pelvis.

iso-   equal, alike.

isolecithal (egg)   egg having even distribution of yolk.

-issimus   superlative ending; the longissimus dorsi is the longest muscle of the back.

isthmus of the fauces   narrow passageway between oral and pharyngeal cavities in mammals.

iter   Roman road > passageway.

-itis   inflammation of.

J

jejunum   empty; part of intestine that is often empty at death; portion of mammalian small intestine between duodenum and ileum.

juga-   yoke > something that joins.

jugal (infraorbital)   dermal skull bone located beneath the orbit.

jugal bone   in mammals, a yoke uniting maxilla and temporal bones.

jugular   pertaining to the neck.

juxta-   next to, near.

juxtaglomerular   near glomeruli.

K

kat-   down; same as *cat-*.

keratin   from a Greek word meaning "horn"; a relatively insoluble substance in cornified cells.

keratinize   to become cornified with keratin.

kinetic   capable of moving.

knee pad   horny epidermal thickening on a camel's knee.

L

labial cartilage   found at the angle of the mouth supporting the lips in sharks.

labial pit   slitlike opening along the lips of some snakes for infrared sensory reception.

labia majora   major lips of the external genitalia in females; homologues of the male scrotal sacs.

labia minora   fleshy folds bordering the vestibule of the vulva in females; homologue, in part, to the male penis.

labium   lip.

labyrinth   maze.

labyrinthodont   early tetrapod with greatly folded dentin in the teeth.

lac-, lact-   milk.

lacrima   tears.

lacrimal   pertaining to tears; from *lachryma* (a teardrop).

lacrimal (bone)   circumorbital dermal bone; in amniotes, associated with the nasolacrimal duct.

lacrimal gland   tear gland to lubricate the eye.

lacteal   dead-end lymph vessels draining lipid laden lymph from the intestine.

lacuna   lake; fluid-filled pool containing an osteocyte in bone.

lag-   hare.

lagena   flask > flask-shaped part of inner ear.

Lamarck, Jean Baptiste de   (1744–1829) author of the Doctrine of Acquired Characteristics.

lambdoidal   having the shape of the Greek letter lambda (λ).

lamella   thin plate.

lamina   thin sheet, plate, or layer.

lamina terminalis   median wall in mammals separating the two interventricular foramina of the brain.

lancelet   common name for amphioxus, noting its small spearlike shape.

laryng-   larynx.

laryngeal pharynx   portion of pharynx adjacent to the larynx in those mammals where the opening to the esophagus is caudal to the glottis.

lateral abdominal (stream)   venous drainage from the paired appendages, metameric tributaries, and the cloaca to the sinus venosus.

lateral abdominal vein   main vessel of the lateral abdominal stream.

lateral-line branch   of n. IX; posterior lateral-line fibers innervating a segment of the lateral line at the junction of the head and trunk.

lateral-line canal   linear series of neuromasts in a fluid-filled canal along the body (may be applied to the cephalic lines of the head).

lateral-line nerve   cranial nerve supplying the neuromast canal system.

lateral lingual swelling   paired component of the tongue in amniotes derived from mandibular mesenchyme.

lateral neural cartilage   agnathan vertebral element.

lateral-plate mesoderm (hypomere)   unsegmented ventral extension of mesoderm in the embryo.

lateral temporal fossa   single temporal opening of synapsids.

lateral umbilical ligament   remnant of the umbilical arteries in the free border of the ventral mesentery of the bladder in mammals.

lateral undulation   side-to-side movement resembling a sine wave; eel-like locomotion, serpentine locomotion.

laterosphenoid bone   lateral ossification of the sphenoid region of the chondrocranium in archosaurs.

latissimus   broadest. See *-issimus*.

latissimus dorsi   dorsal muscle group; spinal processes of lumbar and posterior thoracic vertebrae to proximal humerus; n. thoracodorsalis.

lecith-   yolk.

left atrial chamber   thin-walled cavity within the heart holding oxygenated blood before entering the ventricle.

lemmo-   sheath or envelope.

lemur   from a word meaning "a nocturnal being or ghost"; subgroup of primates.

lens transparent light–refracting structure of lateral and median eyes.

lens placode embryonic ectodermal thickening giving rise to the lens.

lepid-, lepis-, lepo- scale, husk.

Lepidosauria subgroup of diapsid reptiles that includes tuatara and squamates.

lepidotrichia fin rays of bony fishes.

Lepisosteus genus name for a gar.

lepto- weak, thin, delicate.

leptomeninx inner vascular layer covering the brain in nonmammalian tetrapods.

lesser curvature (of stomach) inner or shorter edge of a stomach that has been folded.

lesser horn shorter process (cornua) of the amniote hyoid.

lesser omentum portion of ventral mesentery attached to the lesser curvature of the stomach and duodenum; hepatoduodenal and gastrohepatic ligaments.

lesser peritoneal cavity portion of the peritoneal cavity enclosed within the greater omentum.

leuco-, leuko- white, colorless.

leukocyte white blood corpuscle.

levator raises a part; levator muscle (of the branchial arches). See *cucularis*.

levator scapulae dorsalis (serratus ventralis cervicis or levator scapulae of humans) hypaxial muscle group; medial aspect of scapula (facies serrata) to transverse processes of last five cervical vertebrae; ventral rami of cervical spinal nerves.

levator scapulae ventralis (omotransversarius) hypaxial muscle group; spine of scapula to cervical process of the atlas; n. XI (although listed in references as a hypaxial muscle, its n. XI innervation suggests homology with the human omocervicalis, a trapezius variant).

levatores costarum supracostal hypaxial muscle group; transverse processes of first–twelfth thoracic vertebrae to second–thirteenth ribs; branches of nn. intercostales.

lien- spleen.

ligament connects bone to bone.

ligamentum arteriosum See *arterial ligament*.

ligamentum venosum ductus venosus becomes a ligament after birth.

limb bud embryonic outgrowth giving rise to the limbs.

limbic system interconnections within the brain stem associated with emotional responses in primates; an ancient system associated, in part, with olfaction.

linea alba ventral midline seamlike tendon (raphe) of the trunk.

linear series of placodes embryonic ectodermal thickenings organized linearly.

lingua tongue.

lingual cartilage supportive structure for the lamprey tongue.

lingualis prehyoid hypobranchial group of mammals and some reptiles; intrinsic muscle of tongue without skeletal attachments; n. XII.

lingual tonsil lymphoid tissue located on the tongue near its attachment to the hyoid.

lip- fat.

lipase pancreatic enzyme that breaks down lipids.

lips of the cervix projection of the narrow neck of the uterus (cervix) into the vagina in mammals.

liss- smooth.

Lissamphibia subgroup of Amphibia that includes living amphibians.

lith stone.

liver bud embryonic evagination of the midventral gut that gives rise to the liver.

lobed fin contains muscular and skeletal extensions from the body wall. Contrast *ray fin*.

locomotion movement from place to place.

long head of the triceps brachii appendicular muscle group; caudal edge of scapula to olecranon process of ulna; n. radialis.

longissimus longest; subgroup of epaxial muscles; ilium to lumbar, thoracic, and cervical vertebrae and ribs; dorsal rami of spinal nerves. See *-issimus*.

longitudinal excretory duct main collecting tube of the craniate kidney.

longus colli hypaxial muscle group; ventral surface of vertebral bodies of first six thoracic vertebrae to sixth and seventh cervical transverse processes, and sixth–third cervical transverse processes to preceding midventral vertebral body; ventral rami of cervical spinal nerves.

loop of Henle U-shaped portion of the kidney tubule in mammals.

lopho- ridge, crest.

lophodont specialized grinding tooth type seen in proboscidians.

lore region in front of the eye in reptiles.

loreal pit deep pit on each side of the head between the eye and nostril in pit viper snakes for infrared sensory reception.

lorises subgroup of prosimian primates including bush babies and pottos.

lower (taxon) misleading term to denote relative position of major taxa.

lumbar pertains to the loins; the region of the vertebral column just cephalad to the sacrum > lumbar vertebra; lumbar (artery) segmental artery in the lumbar region.

lumbosacral pertains to the combined loin and sacral regions > lumbosacral nerve plexus.

lumen light > an opening that light can pass through, cavity in a tube.

lunar, lunate moon-shaped.

lung bud embryonic unpaired evagination of caudal pharyngeal floor giving rise to the air sac of bony fish and their descendants.

luteinizing hormone adenohypophyseal hormone that induces ovulation of mature ova and conversion of a graafian follicle into a corpus luteum in mammals.

luteo- yellow.

lymph body fluid in transit within the lymph system.

lymphatic system collection and transport tubules of lymph and their accessory hearts, nodes, and other lymphoid masses.

lymph capillary smallest vessel of lymph system consisting of single-layered endothelial tube.

lymph channel vessel in the lymph system.

lymph heart accessory pump for movement of lymph.

lymph node mass of hemopoietic tissue along course of lymph channels.

lymphoid mass isolated collection of lymphoid tissue (e.g., spleen and thymus.)

lymph sinusoid expanded lymph capillaries.

M

macrolecithal egg with massive amount of yolk.

macrophage cell capable of engulfing and digesting large particles > agranular macrophages of the liver for processing deteriorated blood cells.

macula spot; mound of epithelial cells with a flattened cupula in the membranous labyrinth. Compare *crista*.

magnum albumen-secreting area in an avian oviduct.

magnus, -a, -um large.

maintenance of homeostasis support of a controlled internal environment.

malleus hammer; middle ear ossicle of mammals; homologue to the dermal articular bone of nonmammalian vertebrates.

mammalian clavicle dermal element of the tetrapod pectoral girdle as seen in mammals.

mammalian scapula replacement bone in the tetrapod pectoral girdle as seen in mammals.

mammary gland compound epidermal alveolar glands of both sexes in mammals; synapomorphy of therian mammals.

mandibula (adj., mandibular) jaw.

mandibular arch first pharyngeal arch associated with the jaws.

manubrium   handle.

manus   hand.

marginal canal   new sperm duct of some fishes and urodeles replacing the mesonephric duct for that purpose.

marginal zone   embryonic outer layer of the neural tube particularly devoid of nuclei.

marrow   collection of connective tissue fibers, blood vessels, nerve fibers, and adipose tissue (yellow marrow) or hemopoietic tissue (red marrow) in the cavities of bones.

marsupial bone   supports the marsupial pouch of metatherian mammals.

marsupium   pouch.

massa intermedia   medial swelling of the gray matter in the thalamus that approaches or meets its opposite in the third ventricle.

mastoid   like a breast (*mast-*).

mastoid portion   part of temporal bone new to mammals.

mastoid process   projection from the mastoid portion of the temporal bone in mammals.

mater   mother.

maxilla   dermal bone covering the palatoquadrate; tooth-bearing dermal component of the upper jaw.

maxillary   (**infraorbital** of fishes) division of n. V; sensory branch of n. V.

maximus, -a, -um   largest.

meatus   canal or passageway.

mechanoreceptor   sensory receptor sensitive to mechanical stimuli.

Meckel's cartilage   lower jaw of chondrichthyans; embryonic mandibular cartilage; serial homologue to ceratobranchial elements.

median eminence   neurohemal organ of the ventral diencephalon.

median eye   photoreceptive pineal or parapineal (or both).

median vagina   site where paired vaginas of marsupials are adjacent at the uterine openings; path of fetus through intervening septum to the urogenital sinus.

mediastinum   septum or space separating two portions of an organ > the pleural linings and potential space separating the right and left pleural cavities.

medulla   bone marrow; inner or central part of an organ > renal medulla, adrenal medulla.

medulla oblongata   enlarged part of spinal cord within the skull.

medulla spinalis   marrow of the backbone > spinal cord.

meg-   great, very large.

megakaryocyte   blood stem cell.

meibomian gland   sebaceous gland of the eyelid that moistens the conjunctiva.

Meissner's corpuscle   encapsulated cutaneous tactile receptor on the soles and hands of primates.

Meissner's plexus   one of two autonomic plexuses in the external layers of the gut. See *Auerbach's plexus*.

melan-   dark, black.

melanocyte   pigmented cell in the light-sensitive ocellus of amphioxus; deep pigment cells containing only dispersed pigment granules.

melanophore   bearing (*-phore*) dark pigment > pigment cell containing melanin responsible, in part, for skin coloration.

melanophore stimulating hormone (MSH) (or intermedin)   hormone of the adenohypophysis affecting coloration (pigment dispersal).

melanosome   melanophore cellular organelle containing melanin pigment granules.

melatonin   pineal synchronizing hormone.

membrane bone   deposited directly within a membranous blastema.

membranous urethra   portion of urethra in male mammals between the prostate and base of the penis.

meninx (pl., meninges)   membrane.

meninx primitiva   delicate connective tissue covering of the brain in most fishes.

meniscus   crescent.

mental foramen   foramen on mandible near chin.

mento-   chin.

-mer-   segment, part, one of a series.

merocrine gland   secretes its products via the cell membrane.

mes-   middle, midway, intermediate.

mesaxonic foot   one in which the weight-bearing axis passes through the middle toe.

mesectoderm   mesenchyme of ectodermal origin.

mesencephalic nucleus of the trigeminal nerve   collection of neuron cell bodies in the alar portion of the diencephalon.

mesencephalon   midbrain.

mesenchyme   tissue (*-enchyme*) that is not yet differentiated.

mesentery   associated with the midline of the enteron (gut).

mesethmoid   amniote ossification center of the ethmoid region contributing to the nasal septum, turbinals, and cribriform plate.

mesobronchus   passage of incoming air in birds between the primary bronchus and air sacs; nonstop tube through the substance of the lung.

mesodermal somite   segmented epimere.

mesogaster   dorsal mesentery.

mesolecithal   egg with moderate amount of yolk.

mesonephric duct   archinephric duct draining a mesonephric kidney.

mesonephric kidney   intermediate kidney; functional kidney of the nephrogenic mesoderm caudal to the pronephric region; adult kidney of fishes and amphibians, embryonic kidney of amniotes. See *opisthonephros*.

mesopterygia   middle basal element in the paired fins of fishes.

mesorchium   remnant of the dorsal mesentery supporting the testis.

mesotubarium   remnant of the dorsal mesentery supporting the oviduct in non-mammalians.

mesovarium   remnant of the dorsal mesentery supporting the ovary.

met-   after, succession, change, behind.

metacarpal   bone distal to a carpal.

metamere   one of a series of segments.

metamerism   serial repetition of structures in the longitudinal axis of the body.

metamorphosis   change in form.

metanephric bud   embryonic evagination of the caudal mesonephric duct; in association with nephrogenic mesoderm, gives rise to the adult amniote metanephric kidney.

metanephros   hindmost kidney; amniote kidney.

metapterygia   posterior basal element in the paired fins of fishes.

metatarsal arch   instep of hominoid foot.

Metatheria   subgroup of therian mammals consisting of the marsupials.

metencephalon   anterior division of the hindbrain.

microglia   nonneuron interstitial phagocytic cell (neuroglia) of the central nervous system.

microlecithal   egg with little or no yolk.

Microsauria   extinct subgroup of Amphibia; possible ancestral group to cecaelians. See *Temnospondyli*.

middle umbilical ligament   remnant of the embryonic allantois connecting the urinary bladder with the umbilicus (naval) in therian mammals.

midgut   embryonic part of the gut containing yolk or connected to the yolk sac.

midgut ring   midgut constriction in amphioxus.

milk line   elevated ridge of ectoderm from the axilla to groin; site of adult mammary glands.

milk teeth   deciduous teeth.

mimetic   capable of mimicking; having the characteristics of a mime.

mineralocorticoid   major category of corticoid steroids in mammals.

mitral refers to a bishop's miter or headdress.

mitral valve bicuspid valve of the mammalian heart.

modern lizard extant lizard.

modified diapsid skull loss of one or both arches beneath the temporal fossae.

modiolus bony pillar of the petrosal bone around which spirals the cochlear duct in placental mammals.

molar heterodont tooth form; posterior teeth.

mono- one.

monophyletic group (clade) group of organisms sharing a common ancestor and containing all descendants.

monophyodont dentition single set of teeth in some mammals.

Monotremata subgroup of Prototheria consisting of platypuses and *Echidna*.

monotreme mammal with one caudal opening (cloaca).

morph- shape, structure, form.

morphogen protein that provides an inductive signal in development.

morphogenesis development of form.

morphogenetic movement See *formative movement*.

morphologic color change relative long-term synthesis of pigment granules to affect color change. Contrast *physiologic color change*.

morphology study of form; anatomy.

morula mulberry.

motor end plate neuromuscular junction in striated muscle.

motor nucleus collection of motor neuron cell bodies within the central nervous system.

motor unit muscle fibers innervated by many branches of a single motor neuron.

mucosa mucous membrane; membrane lining of the gut wall.

mucus glycoprotein covering membranes; nonliving covering of the epidermis in aquatic craniates.

muellerian duct female embryonic reproductive duct

in craniates other than ray-finned fishes.

multifidus spinae spinales epaxial muscle group; mammillary, transverse, or articular processes to spinal processes of more cranial vertebrae (typically two vertebrae are passed over); dorsal rami of spinal nerves.

muscle bud embryonic evagination of myomeres into the developing fin of fishes.

muscle fiber long, cylindrical, multinucleate, contractile unit of striated muscle.

muscle spindle specialized muscle fibers forming stretch receptors in mammalian muscle.

muscularis externa smooth muscle of the gut consisting of inner circular and outer longitudinal layers.

muscular portion muscular portion of diaphragm.

myel- marrow.

myelencephalon marrow inside the skull > medulla; posterior division of the hindbrain.

myelin fatty material; covering of axons.

mylo- from a word meaning "a millstone."

mylohyoid muscle attached near the grinding teeth; hypobranchial muscle group; dentary to dentary; n. V.

myo-, mys- muscle.

myocardium muscle of heart.

myofibril longitudinal threadlike subunit of a muscle fiber. See *sarcomere*.

myofilament myosin and actin threadlike subunits of a sarcomere; contractile proteins of muscle.

myoglobin pigmented hemoglobinlike molecule found in muscle.

myomere one of a series of muscle segments.

myometrium thick muscular layer of the uterine wall.

myoseptum (myocomma) (pl., myosepta, myocommata) connective tissue separating successive myomeres.

myotomal muscle striated muscle derived from the myotome of an embryonic somite or somitomere.

myotome part of epimere giving rise to muscle.

Myxini subgroup of agnathans consisting of the hagfishes.

N

nasal paired dermal roofing bone of the skull.

nasal (olfactory) placode embryonic ectodermal thickening giving rise to the olfactory organ.

nasal epithelium glandular epithelial lining of the nasal canals. Contrast *olfactory epithelium*.

nasal pharynx part of the pharynx above the soft palate.

nasal pit embryonic invagination of the olfactory placode.

nasal process of the palatine projection of the palatine bone to form the lateral wall of the nasopharynx separating it from the orbit.

nasohypophyseal duct blind passage in lampreys connecting the unpaired naris with the nasohypophyseal sac.

nasohypophyseal sac terminal enlargement of the nasohypophyseal duct in lampreys.

nasolacrimal duct (tear duct) drains tears from the eye to the nasal cavity in amniotes; duct in lacrimal bone.

nasopharyngeal duct passage in hagfish connecting the unpaired naris with the velar chamber at the anterior end of the pharynx.

neck the cervical region; the region of the body between the head and shoulders.

neo- new, recent.

neocortex expanded walls and roof of the telencephalon in mammals.

Neognathae subgroup of birds characterized by the presence of a well-developed carina.

neomorph evolutionary novelty lacking an ancestral precursor.

Neopterygii subgroup of Actinopterygii excluding the basal chondrosteans and *Polypterus*.

Neornithes subgroup of birds excluding

Archaeopteryx (and possibly other basal Aves).

nephr- kidney.

nephridial tubule collecting tubule transferring metabolic wastes from the cyrtopodocyte of amphioxus to the atrium.

nephrogenic (intermediate or mesomeric) mesoderm embryonic mesoderm giving rise to the kidney.

nephron functional kidney unit.

nephrostome opening connecting the coelom with the urinary collecting duct.

nerve collection of nerve fibers external to the central nervous system.

nerve fiber axon and its myelin cover.

nervi conarii conarial nerves passing from the superior cervical sympathetic ganglion to the pineal.

neural arch portion of a vertebra covering the spinal cord.

neural canal channel or canal for the spinal cord formed by the neural arches.

neural crest embryonic ectodermal tissue found adjacent to the neurectoderm; synapomorphy of craniates; gives rise to numerous adult structures (e.g., pigment cells, bone, muscle, and neurons).

neural groove embryonic longitudinal depression in the neurectoderm giving rise to the spinal cord.

neural keel wedge-shaped neurectoderm that forms the nerve cord in living agnathans and ray-finned fishes.

neural plate (floor plate) band of thickened neurectoderm on the dorsal surface of the developing gastrula.

neural tube dorsal hollow nerve cord of chordates.

neurectoderm presumptive neural ectoderm in the embryo.

neurilemma living membrane that ensheaths all nerve fibers.

neurilemmal cell (Schwann or oligodendroglial cells) individual cell forming the axonal covering.

neuroblast precursor neuron cell.

neurocoel lumen or internal cavity of the central nervous system (brain and spinal cord).

neurocranium (endocranium or chondrocranium) skeletal structure enclosing the brain and sensory organs of the head; arises in cartilage.

neuroendocrine reflex arc feedback loop involving sensory receptor to endocrine to hormonal release to effector.

neuroglia supportive cells of the brain and spinal cord.

neurohemal organ complex of neurosecretory neurons and vascular sinusoids.

neurohormone hormonal neurosecretion that acts via the circulatory system.

neurohypophysis nonglandular portion of the pituitary that arises from the floor of the diencephalon. Contrast *adenohypophysis*.

neuron nerve cell.

neurophysin binds neurosecretions for transport within the axon.

neurosecretion (neurohormone) hormone produced by neurons.

neurosecretory fiber axon of a neurosecretory neuron.

neurosecretory neuron nervous system cell (neuron) that produces hormones.

neurosecretory nucleus collection of neurosecretory neuron cell bodies within the central nervous system.

neurosensory cell sensory cell that directly receives a stimulus instead of via an intervening epithelial cell (e.g., olfactory neurons and the rods and cones of the retina).

neurotransmitter short-lived secretion of a neuron that propagates a nerve impulse across a synapse.

neurula embryo at stage of development involving the formation of the dorsal hollow nerve cord.

neurulation developmental process in which the nerve cord forms.

newt terminal aquatic stage in the life history of the salamander *Notophthalmus*.

nictitating membrane third eyelid of some mammals; medial semilunar fold in humans.

Nissl material site of high protein synthesis characteristic of neurons.

nomen- (pl., **nomina**) name.

nondeciduous placenta See *contact placenta*.

nonstriated (smooth) muscle primarily found in the visceral organs and vessels; fusiform, uninucleate muscle cells that lack striations.

norepinephrine (noradrenaline) secretion of the adrenal gland; one neurotransmitter of postganglionic neurons in the sympathetic nervous system.

noto- back.

notochord cordlike skeleton of the back.

notochordal process embryonic movement of chordamesoderm forward from Hensen's node.

notochord sheath fibrous coat covering the notochord.

nourishment food, nutriment.

nuchal refers to the nape of the neck.

nucleus collection of neuron cell bodies within the central nervous system.

nucleus gracilis collection of neuron cell bodies within the central nervous system that synapses with a nerve tract in the spinal cord for touch and proprioception (fasciculus gracilis).

nucleus solitarius collection of neuron cell bodies within the central nervous system that synapse with incoming taste fibers. See *vagal lobe*.

O

oblique (muscle sheets of trunk) subgroup of hypaxial trunk muscles.

oblique fibers thin sheet of muscle fibers lying superficial to the main hypaxial muscles in many fishes.

oblique septum tendonous sheet separating the pleural cavity from the remainder of the coelom in living

archosaurs and some squamates.

occipital pertaining to the occiput or back part of the head.

occipital bone replacement bone forming in the posterior chondrocranium.

occipital condyle vertebral articulation located on the occipital bone.

occipital lobe posterior part of the cerebral hemisphere.

occipitospinal nerve paired segmental nerve exiting the central nervous system between the caudal cranial nerve and first pair of typical spinal nerves.

occiput part of the head surrounding the foramen magnum; in mammals, the back of the head.

ocellus pigmented light-sensitive organ in amphioxus.

octo- eight.

ocul- eye.

odon-, *odont-* tooth.

odontoblast scleroblast cell depositing dentin or enameloid.

odontognath bird with teeth on the jaws.

Odontognathae extinct subgroup of birds; toothed marine birds.

odontoid resembling a tooth.

odontoid process anteriorly projecting process (dens) of the axis (second vertebra).

-oid like, having a resemblance to, as in hominoid (humanlike).

-ole small, as in arteriole.

olecranon head (*cranium*) of the ulna (*olene*) at the elbow joint.

olfactory chemoreceptor sense.

olfactory (nasal) capsule neurocranial structure partly surrounding the olfactory epithelium.

olfactory bulb outgrowth of the brain that receives olfactory sensory nerve fibers.

olfactory epithelium sensory epithelial lining of the nasal canals. Contrast *nasal epithelium*.

olfactory foramen (pl., foramina) opening for transmission of olfactory nerve fiber.

olfactory placode See *nasal placode*.

olfactory sac cavity containing the olfactory epithelia.

olfactory tract collection of nerve fibers within the brain connecting the olfactory bulb to the brain stem.

oligo- few, small.

oligodendroglia glial cell within the central nervous system that wraps around axons in the formation of myelin.

-oma swelling.

omasum temporary holding chamber between the reticulum and true glandular stomach (abomasum) in ruminant mammals.

omentum free fold of peritoneum.

omni- all.

omnivore animal that eats plants and animals.

omo- shoulder.

omohyoid hypobranchial muscle group (mammals, absent in carnivores); scapula to hyoid; n. XII.

omphalo- navel.

omphalomesenteric vein See *vitelline vein*.

ontogeny (ontogenesis) *onto-* (individual) + *-genesis* (origin); the development of an individual.

oö- (pronounced "oh-oh") egg.

oöcyte egg cell.

oöphoron *oö-* + *-phore-* (q.v.); ovary.

opening into the esophagus opening between the pharynx and esophagus.

opening of the paired auditory tubes paired openings from the auditory (eustachian) tubes to the pharynx.

opercular bone dermal bone element within the operculum of fishes; independent bone in frogs and many salamanders used in the transmission of sound for hearing.

opercular chamber cavity between the gills and operculum.

opercular muscle connects the scapula and operculum in frogs and many salamanders; transmits seismic